NOVEL
SUPERCONDUCTIVITY

NOVEL SUPERCONDUCTIVITY

Edited by

Stuart A. Wolf
Naval Research Laboratory
Washington, D.C.

and

Vladimir Z. Kresin
Lawrence Berkeley Laboratory
University of California, Berkeley
Berkeley, California

PLENUM PRESS • NEW YORK AND LONDON

Library of Congress Cataloging in Publication Data

International Workshop on Novel Mechanisms of Superconductivity (1987: Berkeley, Calif.)
Novel superconductivity.

"Proceedings of the International Workshop on Novel Mechanisms of Superconductivity, held June 22–26, 1987, in Berkeley, California"—T.p. verso.
Includes bibliographies and index.
1. Superconductivity—Congresses. I. Wolf, Stuart, A. II. Kresin, Vladimir, Z. III. Title.
QC612.S8I57   1987   537.6′23   87-25731
ISBN 0-306-42691-9

Proceedings of the International Workshop on Novel Mechanisms of Superconductivity,
held June 22–26, 1987, in Berkeley, California

International Workshop on Novel Mechanisms of Superconductivity

Organizing Committee

R. Brandt, *Office of Naval Research*
E. Edelsack, *Office of Naval Research*
V. Kresin, Chairman, *Lawrence Berkeley Laboratory*
N. E. Phillips, *Lawrence Berkeley Laboratory and University of California, Berkeley*
S. A. Wolf, *Naval Research Laboratory*

Local Committee

A. Stacy, *University of California, Berkeley*
P. Yu, *University of California, Berkeley*
A. Zettl, *University of California, Berkeley*

Program Committee

M. L. Cohen, *Lawrence Berkeley Laboratory and University of California, Berkeley*
T. Geballe, *Stanford University*
D. Gubser, *Naval Research Laboratory*
W. A. Little, *Stanford University*

International Advisory Board

J. Bardeen (USA)
G. Deutcher (Israel)
D. Finnemore (USA)
H. Fukuyama (Japan)
V. Ginzburg (USSR)
D. Jerome (France)

Proceedings

V. Kresin
S. A. Wolf

Sponsor: Office of Naval Research

PREFACE

The Novel Mechanisms of Superconductivity Conference was initially conceived in the early part of 1986 as a small, 2-1/2 day workshop of 40-70 scientists, both theorists and experimentalists interested in exploring the possible evidence for exotic, non phononic superconductivity. Of course, the historic discoveries of high temperature oxide superconductors by Bednorz and Müller and the subsequent enhancements by the Houston/Alabama groups made such a small conference impractical.

The conference necessarily had to expand, 2-1/2 days became 4-1/2 days and superconductivity in the high T_c oxides became the largest single topic in the workshop. In fact, this conference became the first major conference on this topic and thus, these proceedings are also the first major publication. However, heavy fermion, organic and low carrier concentration superconductors remained a very important part of this workshop and articles by the leaders in these fields are included in these proceedings.

Ultimately the workshop hosted nearly 400 scientists, students and media including representatives from the major research groups in the U.S., Europe, Japan and the Soviet Union.

Although the potential applications of the high T_c have captured the attention of the media, the discoveries were made by the scientists doing basic research and it is the basic science that was covered by this workshop. In fact, this meeting became a kind of celebration of the history of superconductivity and the quest for high T_c. Many of the scientists involved in prior major events were present at this workshop as well as the major scientists of the present breakthroughs as can be seen from the program. In this spirit, the first article in this proceedings traces the rather rocky road to our present state.

There are many people and organizations responsible for the success of a conference and we thank the Office of Naval Research for their foresight in supporting us even before the monumentous discoveries. We thank the Lawrence Berkeley Lab and Materials and Chemical Sciences Division for hosting this meeting. We would also like to thank Michael Suhr, Harry Lam, Yougtae Kim, Renata Wentzcovitch and Steve Fahy, students of U.C. Berkeley, and Kathie Shaughnessy of the Naval Research Laboratory, for their help in the preparation of this manuscript. We are grateful to the Berkeley Marina Marriot for their help, and cooperation during our growing pains. Finally, we thank Cris Meyer and Kathy Pepe, without whom this conference would not have happened.

Stuart A. Wolf and Vladimir Z. Kresin

CONTENTS

I. CONVENTIONAL SYSTEMS

IV. HIGH T_c OXIDES: GENERAL PROPERTIES

V. THEORIES OF HIGH T_c

VI. RESEARCH ON HIGH T_c SUPERCONDUCTIVITY

SUMMARY

THE ROCKY ROAD TO HIGH TEMPERATURE SUPERCONDUCTIVITY

E.A. Edelsack,[*] D.U. Gubser and S.A. Wolf

Naval Research Laboratory
Washington, DC   20375-5000

The story of high temperature superconductivity has its genesis in the stars, particularly in one star, our sun. In the 1860's, unusual spectral lines were observed from the emitted light of the hot incandescent gases in the chromosphere of the sun. These spectral lines appeared unrelated to any then known substance on earth. The gas was given the name "helium" deriving from the Greek word "helios" - sun. At the turn of the century, helium was discovered on earth, and in 1908, the Dutch scientist, Kamerlingh Onnes, succeeded in liquifying helium gas at a temperature a few degrees above absolute zero. This set the stage for the discovery of superconductivity.

In 1911, Onnes observed that below a temperature of 4K, the electrical resistance of mercury completely vanished.[1] He called this a new state of matter and called it superconductivity. Immediately the race was on to discover new materials with higher superconducting transition temperatures, T_c. Initially this meant surveying the elements and simple alloys to determine their superconducting properties. Intermetallic compounds presented more of a materials challenge, but work on these materials also began in the 1920's and 1930's. This work produced a major milestone in 1941 when Aschermann, Friederich, Justi and Kramer[2] reported superconductivity in NbN with a T_c near 16K. In 1937, F. London became the first to speculate that supercurrents might exist in non-metal systems, namely -- aromatic organic molecules.[3] The first experimental high T_c report was made by R. Ogg Jr. in 1946 when he claimed that dilute alkali metal-ammonia solutions became superconducting near 185K[4] if the solution was rapidly cooled. This result was neither reproducible, nor widely accepted by the scientific community.

During the 1950's, efforts in superconductivity revolved around two main themes: 1) development of a microscopic theory, and 2) development of empirical rules to guide the search for new superconducting materials. The first theme included the discovery of the exponential specific heat dependence (energy gap in the electronic spectrum)[5] , and the discovery of the isotope effect[6] (importance of lattice vibrations), eventually leading to the Bardeen, Cooper, Schrieffer[7] theory of superconductivity and its subsequent refinements. The second theme included development of such empirical rules as the electron per/atom, e/a ratio[8], inverse correlations

[*]Georgetown Cyrogenics Information Center, 3530 W. Place, Washington, DC

with Debye temperatures[9], direct correlations with the specific heat[10], and symmetry preferences (cubic symmetry favored over lower symmetry structures)[11]. The search for exotic materials and reports of very high T_c materials were subliminal during this decade. A major materials advance in the 1950's was the discovery of superconductivity in the cubic A15 structure type materials by Hardy and Hulm.[12]

The decade of the 1960's saw rapid advances in superconductivity on four fronts: 1) applied superconductivity was born with the advent of the discoveries of the Josephson effect[13] and high field, high current materials[14]; 2) materials research needed to augment these growing technologies increased significantly; 3) the search for higher T_c materials continued, led primarily by the empirical rules established in the 1950's; and finally 4) a discernable amount of, to paraphrase the words of Alexander Graham Bell, "off the beaten path" theory and experiment began to emerge which would ultimately lead to the discovery of truly high T_c superconductivity in 1986.[15] This paper will briefly discuss the latter two items concerning the search for high T_c materials.

Although the field of superconductivity was growing rapidly in applied areas and in materials processing (films, wires, coatings, etc.), the search for new superconducting materials did not increase concomitantly. In fact, as the push for rapid technology transfer became stronger and stronger, funding of scientific studies to search for new superconducting materials began to decline. Materials research became a secondary goal in many programs. The remainder of this article is dedicated to those scientists who did not let it die!

Although many scientists contributed, one in particular deserves special credit and recognition for keeping the field of superconducting materials research vibrant and healthy for more than 3 decades. His name is Professor BerndMatthias. His contributions during the 1950's, 1960's, 1970's were immense. He is missed today by all.

<u>Materials Search</u>

The majority of researchers searching for high T_c superconductors in the 1960's used empirical rules and stayed within the standard classes of metallic alloys and compounds. Niobium became the favored element and the cubic A15 structure type became the favored structure. Empirical rules such as the e/a ratio[8], atomic volume[16] and atomic mass correlations[17] etc., identified Nb_3Ge and Nb_3Si as candidates to raise T_c above 20K. Neither of these compounds has a stoichiometric equilibrium A15 phase, which was thought to be necessary to obtain the high T_c; thus, researchers began to develop fabrication methods to make metastable phases of the desired compounds.

Rapid cooling techniques leading ultimately to film preparation techniques (sputtering, thermal evaporation, E-beam evaporation, etc.) were developed leading to the discovery in 1971 of a record high T_c of 23K in Nb_3Ge.[18] Researchers next turned to preparation of Nb_3Si which was expected to have a T_c near 30K. Whereas Nb_3Ge had an off stoichiometric equilibrium A15 phase, Nb_3Si had none. Therefore, the preparation of stoichiometric A15 Nb_3Si was expected to be more difficult than Nb_3Ge, but also more rewarding. In addition to film growth techniques, high pressure synthesis techniques were used in attempts to produce this material. A15 Nb_3Si structures have subsequently been prepared by both techniques in the 1980's, but T_c has been disappointingly low (20K).[19]

Not all research was on A15 structure materials. A significant advance on the road to high T_c materials occurred in 1972 when superconductivity was discovered in $PbMo_6S_8$--a ternary superconductor![20] The significance of this discovery was that it broke the hold of binary superconductors as being the only high T_c materials. Most of the empirical rules developed for the binaries were invalid for the ternaries and synthesis became much more sophisticated. Chemists and material scientists became heavily involved with the search for new materials.

In the late 1970's and early 1980's, superconductivity was discovered in the "heavy Fermion" systems[21] and in nearly magnetic systems.[22] Such research became fashionable even though these systems did not necessarily have high T_c values. New pairing interactions were sought with the hope of eventually using the new interaction for high T_c superconductors.

With the advent of high speed computing, more exact and more precise calculations of T_c in superconducting materials become possible. Theorists began predicting T_c. MoN in the cubic B1 structure was predicted to be a superconductor at 30K.[23] For several years in the mid 1980's a significant amount of experimental research went into attempts to produce this compound. To date, this research has been unsuccessful.

"Off the Beaten Path - Organics"

There were those who decided to forge revolutionary paths in the quest for high T_c. Among the more revolutionary paths was that of looking for superconductivity in organic materials. In 1964, W. Little generated tremendous interest in organic and one dimensional superconductivity when he discussed specific molecular arrangements which would produce superconductivity at room temperature.[24,25] This work generated much excitement, but not much immediate success. Although chemists worked hard to produce structures of the type suggested, and biological and organic molecules of all sorts and types were examined, all of the early reports of superconductivity in these materials proved to be false (or at least nonreproducible and unconfirmed.)

There were many reports of high temperature superconductors in organic materials in the early 1970's. In 1969, Ladik predicted (based on Little's theory) room temperature superconductivity in DNA molecules.[26] In 1972, Wolf claimed superconductivity in bile cholates at temperatures near 140K.[27] Evidence for superconductivity was in magnetic susceptibility anomalies which were not seen in resistance measurements. In 1973, Heeger reported superconducting fluctuations in TTF-TCNQ molecules at temperatures as high as 77K.[28] It was claimed that an electronically driven structural transformation (Peierl's instability) occurred at a temperature slightly higher than the superconducting T_c; hence, bulk superconductivity was not observed. Also in 1971, Cope reported on superconductivity in certain biological systems at temperatures as high as 30C.[29] Evidence for this was the exponential nature of the nerve-muscle response as well as the exponential growth statistics for E. coli. These reponses were claimed to arise from the exponential rise of Josephson currents which followed an energy gap dependence. In 1973, dilute alkali solutions of ammonia were resurrected with a Russian report of superconductivity at 180K.[30] This system is, in certain situations, a very good conductor and at times appeared to exhibit superconductivity. It was unstable, and although several groups tried to reproduce superconductivity, no confirmation was forthcoming.

As reports of very high T_c in organic materials began to fade, significant advances began to occur. In 1975, superconductivity was discovered in a polymeric material SN_x.[31] Although T_c was low (<1K), the discovery did show that superconductivity need not be limited to the conventional alloy systems.

Organic superconductivity was finally discovered in TMTSF-PF$_6$ by Jerome in 1980.[32] As with SN$_x$, the T$_c$ was low, but unlike SN$_x$ new organic superconductors were rapidly discovered and T$_c$ began to rise. Research in this field involves significant efforts in organic synthetic chemistry coupled to careful physical measurements. Many new phenomena have been seen in these organic materials in addition to superconductivity. At present the maximum T$_c$ (under pressure) is 8K in β-(BEDT-TTF)$_2$I$_3$.[33] There is some evidence that the mechanism for superconductivity in the organic materials is not the electron-phonon interaction and there are also speculations that "p" wave pairing interactions are occurring.

"Off The Beaten Path - Layered Compounds"

Later in 1964 (the year that Little revived interest in organic materials), V.L. Ginzburg discussed a new mechanism and a new structure for producing high temperature superconductivity[34] - namely, the excitonic mechanism in layered, or two dimensional structures. Several refinements and variations including the possibility of excitations between two overlapping bands, a three dimensional mechanism proposed by Geilikman,[35] occurred during the next nine years, culminating in the Allender, Bray, Bardeen[36] theory of excitonic superconductivity in 1973. No definitive confirmation of the mechanism has been reported and until recently, there were no high T$_c$ reports in layered structures. Other theories were proposed in this period including various plasmon coupling mechanisms.[37] No experimental demonstration of these models has yet been confirmed.

In 1980, superconductivity was reported in the eutectic Ir-Y at 3K.[38] Since neither Y or Ir have T$_c$'s above 1K, and only these elementary phases were seen in the eutectic, superconductivity was suggested to be a result of the layered nature of the eutectic. This result spurred interest in multilayered metallic systems with the hope of reproducing and improving on what nature had supplied in the eutectics materials. To date no similar enhancements of superconductivity have been reported.

In 1983, Japanese scientists reported superconductivity at temperatures as high as 200K in a Nb layer grown on Si.[39] This result has not gained wide acceptance by the scientific community.

"Off the Beaten Path - Oxides and Hole Carriers"

The last "off the beaten path" that we wish to trace is the path which ultimately led to the recent breakthrough in high temperature superconductivity - namely, the oxides and low carrier density materials. This story begins in 1964 with the publication by M. Cohen predicting superconductivity in semiconducting type materials.[40] The experimental search for superconductivity culminated in 1964 when R. Hein reported superconductivity is p-type GeTe.[41] Shortly thereafter, superconductivity was discovered in SrTiO$_3$ - the first oxide superconductor and the first perovskite superconducting material.[42] Although T$_c$ of these materials were below 1K, history must regard these reports as major milestones in the road to high temperature superconductivity since they started the interest in these types of materials, which persisted on a limited basis until the recent discoveries of superconductivity as in LaBaCuO[15] and YBaCuO.[43]

The next major milestone in this direct route to high T$_c$ occurred in 1973 when Johnston discovered superconductivity in LiTiO$_3$ at temperatures as high as 13K,[44] thus removing the belief that superconductivity in the oxide materials was limited to very low temperatures. In 1975, superconductivity was discovered in PbBiBaO$_3$ at 14K to represent another member to the growing class of higher temperature oxides.[45] PbBiBaO$_3$ had interesting properties which made it potentially useful as sensors of electromagnetic radiation; hence, research on this material persisted over the next eleven years even though T_c was not raised. It was scientists working on these materials who first recognized the significance of the high temperature oxide discoveries and who so rapidly assumed leadership in the early discoveries. But we got ahead of ourselves, for the road to success was not so direct or easy. First came a few false, or at least unconfirmed and nonreproducible results.

In 1975, hints of superconductivity was found in CuCl under high pressures.[46] In 1978, the superconductivity world was rocked by a Russian report of superconductivity in CuCl at temperatures near 140K.[47] This report in fact marked the beginning of the New York Times becoming the premiere journal for reporting high temperature superconductivity discoveries. In the May 9th, 1978 issue of the New York Times, we find reports by B. Matthias "this is a completely false result and probably deliberately meant to deceive", while in the same article we find C.W. Chu suggesting that "there may indeed be some truth to such high temperature superconductivity reports" and that it deserved a further look. Thus, twelve years before the discoveries in LaBaCuO some of the main actors in the eventual explosive discovery were already searching "off the beaten path".

Work on CuCl persisted for years, polarizing many scientists into believers, and nonbelievers. In fact, there is still interest in CuCl even though the results are nonreproducible, thermal history dependent, and subject to a variety of interpretations.

By 1980, interest in the low carrier materials gained new impetus with the report of superconductivity in pressure quenched CdS at temperatures as high as 150K.[48] Like the work on CuCl, the experimental data were highly nonreproducible, and subject to different interpretation.

In 1980, $TiBe_{1.6}$ was reported to be superconducting at 22C.[49] This result received little scientific interest as, once again, it was nonreproducible and subject to reinterpretation in terms of nonsuperconducting phenomena.

These early reports of high temperature superconductivity are an interesting part of the story of high temperature superconductivity. They set the stage for the discovery which was to come in 1986. The early results and their lack of confirmation had led many scientists to treat such reports as coming from scientists who were more interested in gaining fame than in doing creditable scientific research. It is noteworthy that Bednorz and Mueller[15] waited many months to publish their original discovery, due primarily to this climate of distrust for such reports. It is a credit to them and to those who immediately recognized the importance of their announcement that we now have advanced so far in our understanding of the materials.

Where will the search for high temperature superconductivity lead us next; new materials, new technology, new scientific insights? Yes, probably all of these. But perhaps more important is the realization that the scientific world needs to renew its committment and provide proper recognition and support for those scientists who are not afraid to occasionally stray from the beaten path and delve into the forest. The woods are full of delicious fruit. Let us not be afraid to look for them.

<u>Acknowledgement</u>

We acknowledge W.W. Fuller and R.A. Hein for their critical reading of this manuscript and ONR, SDIO/IST and DARPA for their partial support of our research.

References

1. H.K. Onnes. Leiden comm. 124C (1911).

2. G. Ascherman, E. Friederich, E. Justi and J. Kramer, Phys. Zeir., $\underline{42}$ 349 (1941).

3. F.J. London, "Superconductivity in Aromatic Molecules", J. of Chemistry and Physics, $\underline{5}$ (1937).

4. R.A. Ogg, Jr., Phys. Rev. $\underline{69}$, 243 (1946).

5. W.S. Corak, B.B. Goodman, C.B. Satterthwaite and A. Wexter, Phys. Rev. $\underline{96}$, 1442 (1954).

6. E. Maxwell, Phys. Rev. $\underline{78}$, 477 (1950); $\underline{79}$, 173 (1950), Reynolds, Sevin, Wright and Nesbitt, Phys. Rev. $\underline{78}$, 487 (1950).

7. J. Bardeen, L.N. Cooper and J.R. Schrieffer, Phys. Rev. $\underline{106}$, 162 (1957); Phys. Rev. $\underline{108}$, 1175 (1957).

8. B.T. Matthias, Phys., Rev. $\underline{97}$ 74 (1955).

9. J. DeLauney and R. Dolecek, Phys., Rev. $\underline{72}$, 141 (1947).

10. H.W. Lewis, Phys. Rev. $\underline{101}$, 939 (1956).

11. B.T. Matthias, T.H. Geballe and V.B. Compton, Reviews of Modern Phys. $\underline{35}$, 1 (1963).

12. G.F. Hardy and J.K. Hulm, Phys. Rev. $\underline{93}$, 1004 (1954).

13. B.D. Josephson, Phys. Lett. $\underline{1}$, 251 (1962).

14. J.E. Kunzler, E. Benkler, F.S.L. Hsu and J.H. Wernick, Phys. Rev. Lett. $\underline{6}$, 89 (1961).

15. J.G. Bednorz and K.A. Mueller, Z. Phys. $\underline{B64}$, 189 (1986).

16. L.R. Testardi, J.E. Kunzler, H.G. Levinstein, J.P. Maita and J.E. Wernick, Phys. Rev. $\underline{B3}$, 107 (1971).

17. L. Gold. Phys. Stat. Sol. $\underline{4}$, 261 (1964); D. Dew Hughes and V.G. Rivlin, Nature $\underline{50}$, 723 (1974).

18. J.R. Gavaler, App. Phys. Lett. $\underline{23}$, 480 (1973).

19. J.D. Dew Hughes and V.D. Linse, J. Appl. Phys. $\underline{50}$, 3500 (1979); R.E. Somekh and J.E. Evetts, IEEE Trans. Magn. $\underline{MAG-15}$, 494 (1979).

20. B.T. Matthias, M. Marezio, E. Corenzwit, A.S. Cooper and H.E. Barz, Science, $\underline{175}$, 1465 (1972).

21. F. Steglich, J. Aarts, C.D. Bredl, W. Lieke, D. Meschede, W. Franz and H. Schafer, Phys. Rev. Lett. $\underline{43}$, 1892 (1979).

22. H.R. Ott, H., Rudiger, Z. Fisk and J.L. Smith, Phys. Rev. Lett. $\underline{50}$, 1595 (1983).

23. W.E. Pickett, B.M. Klein and D.A. Papaconstantopoulos, Physica B&C, 265 (1981).

24. W.A. Little, Phys. Rev. A134, 1416 (1964).

25. W.A. Little, Report on the International Symposium on the Physical and Chemical Problems of Possible Organic Superconductors, 31, (1969).

26. J. Ladik and A. Bierman, Phys. Lett. 29A, 636 (1969).

27. E.H. Halpern and A.A. Wolf, Advanced Cryog. Eng., 17, 109 (1972).

28. L.B. Colemen, M.J. Cohen, D.J. Sandman, F.G. Yamagishi, A.F. Garito and A.J. Heeger, Solid State Commun. 12, 1125 (1973).

29. F.W. Cope, Physiol. Chem. and Physics, 3, 403 (1971).

30. I.M. Dmitrenko and I.S. Shchetkin, Zhetf. Pis. Red. 18, 497 (1973).

31. R.L. Greene, P.M. Grant and G.B. Street, Phys. Rev. Lett. 34, 89 (1975).

32. D. Jerome, A. Mazaud, M. Ribault, and K. Bechgaard, J. Phys. (Paris) Lett. 41, L95 (1980).

33. V.N. Tauklin, E.E. Khostyuchenko, Yu. V. Sushko, I.F. Schnegolve, and E.B. Yagubskii, Pis'ma Zh. Eksp. Teor. Fiz. 41, 68 (1985) (JETP Lett. 41, 81 (1985]. K. Murata, M. Tokumoto, H. Anzai, H. Bando, G. Saito, K. Kajimura, and T. Ishiguro, J. Phys. Soc. Jpn. 54, 1236 (1985).

34. V.L. Ginzburg, Phys. Letters 13, 101 (1964); V.L. Ginzburg and D.A. Kirzuits, Zh. Etisperim, i Theor. Fiz. 46, 397 (1964) [Trans.0:ov. Phys. JETP 19, 269 (1964).]

35. B.T. Geilikman, Usp. Fiz. Nauk. 88, 327 (1966) [Trans.: Sov. Phys. Usp. 9, 142 (1966).

36. D. Allender, J. Bray, and J. Bardeen, Phys. Rev. 7B, 1020 (1973).

37. L.M. Kahn and J.R. Ruvalds, Phys. Rev. B19, 5652 (1979).

38. B.T. Matthias, G.R. Stewart, A.L. Giorgi, J.L. Smith, Z. Fisk and H. Barz, Science, 208, 401 (1980).

39. T. Ogushi, K. Obara, T. Anayama, Jap. J. Appl. Phys. 22, L523 (1983).

40. M.L. Cohen, Phys. Rev. 134, A511 (1964).

41. R.A. Hein, J.W. Gibson, R. Mazelsky, R.C. Miller and J.K. Hulm, Phys. Rev. Lett. 12, 320 (1964).

42. J.F. Schooley, W.R. Hoster, E. Ambler, J.H. Becker, M.L. Cohen and C.S. Koonce, Phys. Rev. Lett. 14, 305 (1965).

43. M.K. Wu, J.R. Ashburn, C.J. Torng, P.H. Hor, R.L.Meng, L. Gao, Z.J. Huang, Y.Q. Wang, and C.W. Chu, Phys. Rev. Lett. 58, 908 (1987).

44. D.C. Johnston, H. Prakash, W.H. Zachariasen and R. Viswanathan, Mat. Res. Bull. 8, 777 (1973).

45. A.W. Sleight, J.L. Gillson and P,.E. Bierstedt, Sol. State Comm.
     17, 27 (1975).

46. A.P. Rusakov, V.N. Laukhin and Yu A. Lisovskii, Phys. Stat.
     Sol (b) 71, K191 (1975); C.W. Chu, S. Early, T.H. Geballe,
     A.P. Rusakov and R.E. Schwell, J. Phys. C8, L241 (1975).

47. N.B. Brandt, S.V. Kuoshiunikov, A.P. Rusakov, and V.M. Semenor,
     Pisma Zh. Ehsp. Theor. F12, 27, 37 (1978) [Trans.: JEPT Lett.
     27, 33 (1987)]; C.W. Chu, A.P. Rusakov, S. Huang, S. Early,
     T.H. Geballe, and C.Y Huang, Phys. Rev. B18, 2116 (1978).

48. E. Brown, C.G. Homan and R.K. MacCrone, Phys. Rev. Lett. 45, 478
     (1980).

49. F.W. Vahldiek, C & EN, Sept. 8th (1980) pg. 36.

ELECTRIC-FIELD MODULATION OF LOW ELECTRON DENSITY

THIN-FILM SUPERCONDUCTORS

A. F. Hebard and A. T. Fiory

AT&T Bell Laboratories
Murray Hill
New Jersey, 07974

ABSTRACT

Past investigations of electric-field modulation of the normal-state and super-conducting properties of thin films are reviewed and compared with recent work on amorphous-composite In/InO_x thin films with electron carrier density as low as 10^{20} cm^{-3}. Electric charge induced in the In/InO_x films by a capacitatively-coupled gate electrode has a particularly strong influence when the disorder (resistivity) is increased towards a critical value where superconductivity rapidly disappears. The experimentally-measured field-effect mobility, the unperturbed electron density, and the transition temperature are strongly dependent on oxide content, substrate material, and dielectric capping overlayers. The field-effect measurements confirm the importance of electronic states at the metal-dielectric interfaces and enable a description of "interface dominated superconductivity" in this and possibly other low electron density superconductors. Implications for the high T_c oxides are also presented.

I. INTRODUCTION

In 1960 Glover and Sherrill[1] reported that the transition temperature T_c of thin-film In and Sn superconductors could be reversibly changed by application of an electrostatic field normal to the plane of the film. The shift in T_c was observed to be on the order of 10^{-4} K. The generally accepted explanation of this very important result is that the normal electric field, usually produced by capacitive coupling, gives rise to a surface electron density Σ which acts as a perturbation on the ambient electrons and thus affects both T_c and the normal state conductivity σ_N. For a film with thickness d and volume electron density n the fractional change in both T_c and σ_N would be expected to scale in proportion to the fractional change in the total number of electrons, Σ/nd. To achieve a reasonable value for this number one might use a thin-film capacitor with an oxide dielectric such as SiO_2 or Al_2O_3 having a typical charge storage capability[2] of $2\,\mu C cm^{-2}$, equivalent to a surface electron density $\Sigma = 1.25 \times 10^{13}$ cm^{-2} on each electrode. Using, for example, a 100 Å thick pure In film with free electron density 1.15×10^{23}

as one plate of such a capacitor, a value for Σ/nd of 1.1×10^{-4} can be calculated. Typically the changes $\Delta\sigma_N/\sigma_N$ and $\Delta T_c/T_c$ are the same order of magnitude as Σ/nd.[3]

In spite of the discouragingly small size of this estimate there are a number of ways to improve its magnitude. One approach is to use ferroelectric charging to obtain a higher N, an approach taken by Stadler[4] in work on 160 Å-thick Sn films deposited on triglycine sulfate substrates. The change in T_c, $\Delta T_c = 1.3\,\mathrm{mK}$, although small, was an increase by an order of magnitude over the results reported in Ref. 1. A second approach is to use low-electron density superconductors with n significantly reduced compared to the values for typical metals. Candidate systems are amorphous composite In/InO_x films[5] with $n \cong 10^{20}\,\mathrm{cm}^{-3}$, La-S films[6] with n varying from 7×10^{19} to $10^{22}\,\mathrm{cm}^{-3}$ and doped $SrTiO_3$ surface layers[7]. The electron density n is inferred[8] to be as low as $4 \times 10^{15}\,\mathrm{cm}^{-3}$ in a bulk polycrystalline superconducting sample with composition $SrTi_{0.97}Zr_{0.03}O_3$. Finally, electric-field effects can become particularly pronounced as the disorder (resistivity) is increased towards a critical value where superconductivity disappears. This increased sensitivity to electric field charging near critical disorder has been demonstrated in our own work on amorphous composite In/InO_x films[5] and will be more fully elaborated in the following sections.

Having a material with low electron density is of little use in electric field effect investigations unless the material can also be made thin, usually with $d < 100\,\text{Å}$. Accordingly, it is important that the film microstructure occurs on a scale fine enough to ensure connectivity and uniformity for such thicknesses. With decreasing d the properties of the interfaces at the surface of the superconductor become increasingly important in determining the response of the superconductor to electric-field charging. This aspect of interface-dominated superconductivity is particularly amenable to electric-field effect investigations.

Our intention in this paper is to present a detailed discussion of the above issues as they relate to electric-field modulation of superconductivity in thin films. For the sake of brevity we will not include past investigations of superconductivity on semiconducting surfaces[9] or electric-field modulation of proximity-coupled superconductors.[10,11] Section II will begin by briefly reviewing past work on the electric-field effect on thin-film superconductors and includes a classification of common aspects to the patterns of behavior observed in different metal systems. The important role of interfaces and how superconductivity is affected by oxidation, dielectric overlays, and inert gas coverage will also be discussed. Sections III and IV will then treat in some detail our latest results on the electric-field effect on In/InO_x films which advantageously are very thin and homogeneous, have low electron density, and can be fabricated with a resistivity close to a critical value where superconductivity is rapidly suppressed by disorder occurring on microscopic scale lengths.[12] Particular emphasis will be placed on how a Boltzmann-equation interpretation of field-effect data (Section III) can be used to characterize the dependence (Section IV) of the electron mobility, the electronic mean free path, the electron density, the normal state resistivity and the transition temperature on annealing schedules, use of different substrates, or passivation with dielectric overlays. The field-effect induced shifts in T_c will also be described and compared with theoretical estimates. Finally, in Section V we discuss implications for future work on high-T_c oxide superconducting films.

TABLE I. Effect of increasing the electron number (charging) on the normal state conductance σ_n and transition temperature T_c. The effect of oxidation on T_c is shown in the third column.

	$\delta\sigma_n$ (charging)	δT_c (charging)	δT_c (oxidation)
In	↑ [a]	↓ [a]	↑ [b]
Sn	↓ [a]	↑ [a]	↓ [b]
Tl	↑ [c]	↓ [c]	↑ [b]
Al	↓ [c]	↑ [d]	↑ [b]
Bi	↑ [c]	↑ [c]	↓ [c]
Ga	↓ [c]	↓ [c]	↓ [c]
Pb	↓ [c]	↓ [c]	↓ [b]
In/InO$_x$	↑ [e]	↑ [e]	NA

[a] Ref. 1   [c] Ref. 3   [e] Ref. 5
[b] Ref. 13   [d] Ref. 19

II. ELECTRIC-FIELD EFFECT ON THIN FILM SUPERCONDUCTORS

Past investigations of the electric-field effect on thin-film superconductors are summarized in Table I. The first two columns list the variations observed in σ_N and T_c when electrons are added to thin films of the elements listed at the left. The rightmost column complements these data by identifying the changes in T_c when a freshly prepared film is allowed to oxidize. These oxidation experiments were originally performed by Ruhl[13] and interpreted in terms of a diffusion-limited oxidation process which results in an electric field across the oxide with a sign such as to remove electrons from the film.[14] Microscopically, this field drives metal ions towards the surface and arises because of the tunneling of electrons from the metal to oxygen acceptor levels near the oxide surface.

Although there is no single pattern in Table I which universally describes the response of these superconductors to electric-field charging there are significant correlations which should be noted. Firstly, for the weakly coupled superconductors, In, Sn, Tl, and Al, the charge-induced changes in σ_N and T_c are opposite in sign, whereas for the strongly coupled superconductors, Bi, Ga, and Pb, the normal and superconducting shifts are of the same sign. Interestingly, for In and Tl, negative charging increases σ_N and decreases T_c whereas the reverse behavior is observed for Sn and Al. Also, with the aforementioned assumption that oxidation removes electrons from the film, we note that for In, Sn, and Tl, oxidation charging has the same effect on T_c as direct field-effect charging. A second important correlation in the data of Table I is that oxidation causes the T_c's of the group III metals In, Tl, and Al to shift to higher temperatures whereas the T_c's of the group IV metals Sn and Pb plus the low-temperature modifications of Bi and Ga are shifted to lower temperatures.[15] Finally, we note that *only* for low-temperature amorphous Bi and In/InO$_x$ do σ_N and T_c increase simultaneously with negative charging. Such behavior is consistent with the use of a free electron model for σ_N together with a BCS equation description in which an increase in the density of states, proportional to electron density, would concomitantly give rise to an increase in T_c.

A common element to all of the entries in Table I is that the charging effects are all odd in the applied field and that the shifts in T_c are approximately proportional to the change in the number of electrons per unit volume, both for the charging experiments[1,3] and the oxidation experiments[15]. Glover and Sherrill[1] have shown that their T_c shifts cannot be due to strains induced by a piezoelectric substrate. Furthermore, Maxwell stresses at the charged metal surfaces are quadratic, hence even, in the applied field[16] and thus cannot describe the results. Bardasis[17] has constructed a theoretical model to explain the results for the weak coupled superconductors Al and Sn in an analysis of a Friedel-type sum rule in which the imposition of charge neutrality is relaxed. In effect the electrons occupy a volume slightly larger than the geometrical one occupied by the positive background charge and the effect on the T_c equation gives reasonable agreement with the sign and magnitude of the results for Al and Sn. The theory also correctly predicts smaller shifts in T_c for Bi, Ga, and Pb but does not explain the like signs of the normal and superconducting response to charging in these materials.

Charging phenomena similar to those discussed above can also be obtained, for example, by applying dielectric overcoats of Ge on thin films of Sn and Tl[18] to produce changes in T_c which have the same sign shown in Table I when these elemental films undergo oxidation. Charge transfer across the interface is not the only explanation however. This point is clarified in the work of Naugle $et\ al.$[19] who find that noble gas overlayers (Ar and Ne) decrease σ_N and T_c for the weak coupled superconductors Sn, Tl, and Al, and increase σ_N and decrease T_c for strong coupled amorphous Bi. Similar results are reported by Felsch and Glover[20] with additional data on Ga and Pb: in all cases T_c is suppressed by noble gas overlayers. Although the exact mechanism for this reduction is not clear, the most likely explanation is that the phonon spectrum in the thin film is modified by the noble gas overlayer.[19] As in the case of electrostatic charging the shifts in T_c are inversely proportional to the film thickness. Alternation of superconducting layers (Al, In, Pb, Sn, and Zn) with a variety of dielectric barriers has also been observed to cause reproducible and sometimes significant enhancements (more than a factor of two for Al) of T_c.[21] Interestingly, a capping layer of varying thickness can also cause an oscillation in the T_c of the film. This has been observed by Sixl[22] using SiO layers of varying thickness on top of Al films. Mechanisms involving the quantum size effect or Friedel oscillations[22] are presented as plausible explanations. An additional factor to be considered is the hybridization between conduction electrons in the metal and localized electrons in the adjacent dielectric[23] which can account for pair weakening, polarization effects, and leakage currents.

III. A BOLTZMANN DESCRIPTION OF DISORDERED In/InO$_x$

From the above discussion there does not appear to be any obvious means of ascertaining by experiment the exact mechanism by which an interface, either with the substrate or with a capping dielectric overlay, affects the superconductivity of a thin film. A simplification arises if σ_N can be described by the free-electron Boltzmann conductivity σ_B, i.e.

$$\sigma_N = \sigma_B = ne^2\tau/m = e^2 k_F^2 \ell/3\pi^2\hbar \ , \tag{1}$$

where n is the volume electron density, e the electron charge, m the electron mass, τ the electron scattering time, k_F the Fermi wave vector, and ℓ the electron mean free path. The appropriateness of using this approach to describe σ_N of In/InO_x films has been demonstrated in previous work[5] where field-effect measurements have established a linear dependence of the ratio $\sigma(0)/\sigma_B$ and T_c on the reciprocal square of the disorder parameter $k_F\ell$.

The assumptions underlying these measurements are straightforward[24,25] and briefly reviewed here. An electric field applied normal to a film terminates with a surface charge density distributed over a charge screening length Λ into the film. This distance is on the order of a few Å for typical metals. For a film with mobility μ and thickness $d > \Lambda$ the sheet conductance $G = d\sigma_N = dne\mu$ can be broken up into a series combination of two conductances, the first with value $(d-\Lambda)\sigma_N$ is unperturbed by the applied electric field and the second with value $\Lambda\sigma_N$ is perturbed by the applied electric field. The total change in conductance increases as the film becomes thinner and the contribution of the unperturbed shunting conductance is reduced. We now make the assumption that because the degenerate electrons have wave functions which spread out over the entire film, the effective mobility $\mu = e\tau/m$, which reflects scattering processes both in the bulk and at the interfaces, has the same value in the charge-perturbed region as it does in the bulk. With the additional approximation that ℓ is independent of n, it is straightforward to show that the field-effect mobility can be calculated as[24,25]

$$\mu = \frac{3}{2}\frac{\partial G}{\partial Ne} \ , \tag{2}$$

where $N = nd$ is the areal electron density. We note that μ is an experimentally determined quantity proportional to the ratio of the change in sheet conductance induced by the known change in areal charge density caused by the capacitatively-coupled gate electrode. The quantities $N = nd = G/e\mu$, $k_F = (3\pi^2 n)^{1/3}$, and $\ell = \hbar\mu k_F/e$ can now be directly calculated.

Amorphous-composite In/InO_x, which is fabricated by the technique of reactive ion beam sputter deposition,[26] is in many ways ideally suited for field-effect studies. For the work reported here, it is not granular[26] and hence considerations believed to be relevant to field-effect charging in granular films are not appropriate[27]. The films contain a significant amount of oxygen, almost 60 at. %,[5] which localizes most of the valence electrons and yet allows metallic conduction with an electron density which can be as low as $10^{20}\mathrm{cm}^{-3}$. The films are relatively stable in air and because the microstructure is predominantly amorphous the films can be made advantageously very thin (~ 50Å) and continuous. Most importantly, however, the films can be fabricated with a resistivity close to a critical value where superconductivity is rapidly suppressed with increasing disorder.[5,12] This aspect is illustrated in resistive transitions for films with the same thickness made with slightly different resistivities. The salient feature of such curves is the extreme sensitivity of T_c to small changes in the normal-state sheet resistance $R_N = 1/\sigma_N d$. Accordingly, if a film is near critical disorder, any small field-induced variation in σ_N will have a large effect on T_c.[24,25]

To place these statements on a more quantitative footing we draw on previously published work[28] in which it was found that T_c for films with varying thickness scales with film resistivity rather than sheet resistance. This dependence on bulk properties enables a comparison[28] with microscopic theory[29] which includes localization and interaction effects and in which the T_c has an explicit dependence on the disorder parameter $k_F\ell$. We have found empirically that by using the BCS equation

$$T_c = 1.13\Theta_D\exp(-1/g') \; , \tag{3}$$

together with an expansion of the coupling constant g' of the form

$$g' = g[1 - A(k_F\ell)^{-2}+...] \; , \tag{4}$$

a very adequate description of the disorder-induced suppression of T_c is obtained. The dependence of T_c on $k_F\ell$ predicted by these equations not only agrees qualitatively with theoretical dependences but gives excellent agreement with measurements of a 600 Å-thick film annealed in stages to give a change in T_c by more than a factor of four.[28] The constants used to get this agreement were the Debye temperature $\Theta_D = 112\,\mathrm{K}$, $A = 1.15$ and $g = 0.282$.

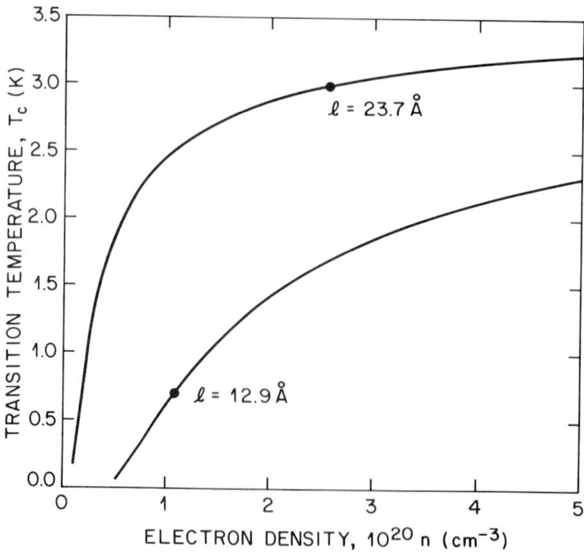

Fig. 1   Dependences of the transition temperature on electron density computed for the same film at two different stages of anneal indicated by the solid points. The mean free path is kept constant at the indicated value for each curve.

In the foregoing analysis based on the free-electron model of the Boltzmann conductivity we have assumed that room-temperature measurements of σ_N are good estimates of σ_B and hence $k_F\ell$. The observed scaling of $\sigma(0)/\sigma_B$ with $(k_F\ell)^{-2}$ is in agreement with localization theory and tends to confirm this approach.[5] In like manner the values of $k_F\ell$ determined from measurements of σ_N at room temperature have been found to determine (Eqs. 3 and 4) the dependence of T_c on disorder. To translate this behavior into an explicit dependence on n we plot in Fig. 1, using Eqs. 3 and 4, the dependence of T_c on n for two cases: the first for a film with initial $T_c = 0.74\,\mathrm{K}$ and $\ell = 12.9\,\text{Å}$ (lower point) and the second for the same film annealed[5] to a higher conductance with $T_c = 2.97\,\mathrm{K}$ and $\ell = 23.7\,\text{Å}$ (upper point). The solid lines indicate the expected dependence of T_c on n with ℓ constant on each curve.

IV. OPTIMIZATION OF THE FIELD EFFECT IN $\mathrm{In}/\mathrm{InO_x}$ THIN FILMS

From the foregoing discussion we have seen that to maximize the field effect it is not only necessary to have thin films with low electron density but it is also advantageous to be near critical disorder where the the superconducting properties

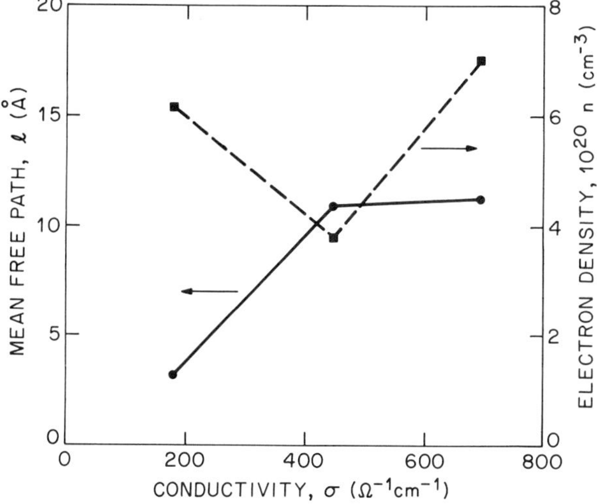

Fig. 2   Plot of the electron mean free path (left hand axis) and the electron density (right hand axis) as a function of the normal state conductivity after annealing and then capping with a magnesium oxide dielectric film.

are very sensitive to small perturbations in the normal state sheet conductance $G_N = d\sigma_N = Ne\mu$. Since G_N has a minimum value of approximately $10^{-4}\,\Omega^{-1}$ consistent with the occurrence of superconductivity,[24] then it is clear that a small N and large μ are desirable. This section addresses the role of interfaces in determining the magnitude of these parameters and concludes with data which indicates that a significant field-effect modulation of T_c on suitably prepared In/InO_x thin films can be obtained.

Fig. 2 illustrates typical behavior of ℓ, n, and σ_N when a $50\,\text{\AA}$-thick nonsuperconducting In/InO_x film is annealed and then capped with a dielectric. Annealing this initially nonsuperconducting film at 170°C for 8 hours increased σ_N (initially at $180\,\Omega^{-1}\text{cm}^{-1}$) by a factor of 2.5, decreased n by a factor of 1.62, and increased ℓ by a factor of 3.43. Following this anneal the sample was capped with a $100\,\text{\AA}$-thick layer of magnesium oxide which increased σ_N further by a factor of 1.55, increased n by a factor of 1.85, and left ℓ relatively unchanged. The film in this final state (rightmost point of Fig. 2) had a T_c of $1.32\,$K. We have found the behavior exemplified in Fig. 2 to be quite typical for many films: annealing affects primarily ℓ and capping with MgO_x affects primarily n. Having a low electron density film then is not necessarily advantageous if the gate dielectric acts like a voltage source to increase n. Similar phenomena apply to the type of substrate used: In/InO_x films deposited on glass and oxidized silicon usually have lower σ_N than when deposited on $LiNbO_3$ or $SrTiO_3$.

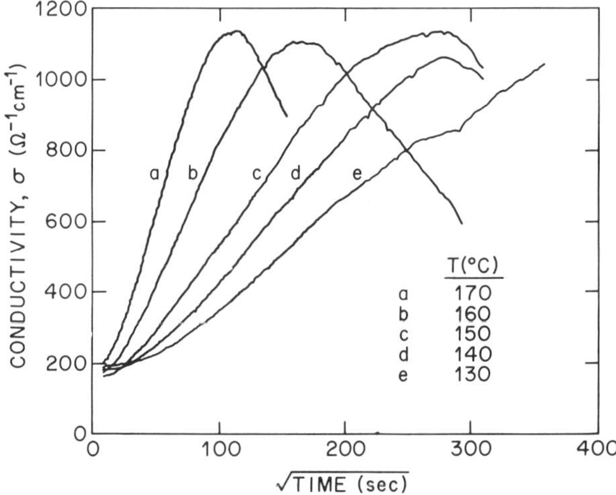

Fig. 3   Dependence of the conductivity on the square root of time for five pieces of the same $100\,\text{\AA}$-thick film annealed at the indicated temperatures.

A confirmation of the effect which annealing has on ℓ is shown in the isothermal annealing traces of Fig. 3 where the conductivities of five pieces of the same film are measured at the indicated temperatures and plotted as a function of the square root of time. Since $\sigma \propto \ell$ for fixed n (cf Eq. 1), the straight line portion of these plots indicates that with isothermal annealing there is a diffusive growth of ℓ, that is $\ell \propto (\text{time})^{1/2}$. The samples remain amorphous in TEM observation[30] until the conductivity peaks and begins to decrease, at which point the films become transparent, with a microstructure consisting of a mixed phase of In_2O_3 crystallites, metallic In precipitates, and a small amount of amorphous component. The activation energy for this process, most likely oxygen diffusion, is approximately 1 eV.

Accordingly, modification of bulk properties by annealing, which increases μ, and modification of interfacial properties by the presence of gate or substrate dielectrics are important aspects in determining the correct materials combinations and processing procedures to optimize the response of a thin-film superconductor to electric charging effects. Typical results are illustrated in the logarithmic plot of the resistive transitions of Fig. 4 for a 63 Å-thick In/InO_x film separated from an Al gate electrode by a reactive ion beam sputter deposited Al_2O_3 dielectric.

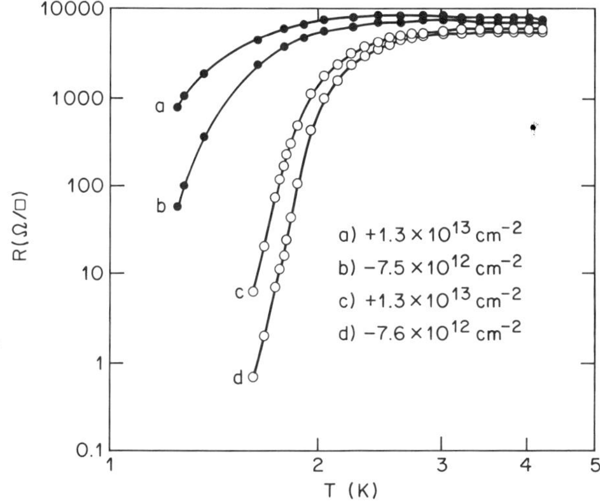

Fig. 4 Resistive transitions on logarithmic axes showing the effect of electric field charging on a 63 Å-thick In/InO_x film before (solid circles) and after (open circles) annealing at 90°C for 15 minutes. The sign and magnitude of the capacitatively-coupled induced charge are indicated for each of the curves in the inset.

Curves (a) and (b) represent the resistive transitions for positive
$(+1.3\times10^{13}$ cm$^{-2})$ and negative $(-7.5\times10^{12}$ cm$^{-2})$ charging respectively. These
numbers represent a greater than 2% perturbation of the ambient charge density,
measured to be 4.27×10^{14} cm^{-2} for this film.

Curves (c) and (d) of Fig. 4 represent the resistive transitions for this same
film after annealing at 90°C for 15 minutes. Note that the normal-state sheet resistance decreases 11.2% from 3676 to 3264 $\Omega/\square$ whereas T_c, measured using a 40%
of normal state criterion,[31] increases 34.3%, from 1.631 to 2.190 K. Not only does
the sensitivity of T_c to R_N increase as the film becomes more disordered but also
the absolute change in resistance for a given amount of charging is similarly
affected. This is illustrated in Fig. 5 which shows respectively the differences in

sheet resistance for positive and negative charging as a function of temperature for
the two films in Fig. 4. The maximum change in resistance of 2250 $\Omega/\square$ for the
more disordered film (upper curve) is almost a factor of three greater than the
780 $\Omega/\square$ maximum of the annealed film (lower curve).

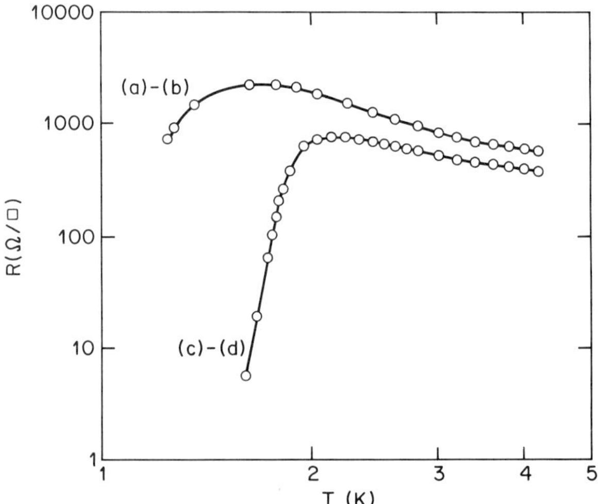

Fig. 5   Logarithmic plot showing the temperature dependence of the maximum
field-induced change in sheet resistance corresponding to the differences (a)-(b) and (c)-(d) of the resistive transitions shown in Fig. 4.

The theoretical dependence of T_c on n embodied in Eqs. (3) and (4) can be compared with the magnitudes of the field-induced T_c shifts shown in Fig. 4. This is done by using the relation $k_F = (3\pi^2 n)^{1/3}$ together with Eqs. (3) and (4) to calculate the derivative

$$\frac{\partial T_c}{\partial n} = \frac{2gA}{3g'^2 (k_F \ell)^2} \frac{T_c}{n} \, .\tag{5}$$

Using the previously established values[28] for g and A together with the field-effect determined values of $k_F \ell$ both before ($k_F \ell = 1.94$) and after ($k_F \ell = 2.13$) the anneal, we calculate theoretical T_c-shifts of $0.114\,$K between curves (a) and (b) and $0.104\,$K between curves (c) and (d). The respective experimental shifts of $0.194\,$K and $0.087\,$K derived from the data of Fig. 5 are in good qualitative agreement, especially considering that we simply assume that the values of A and g determined for a thicker $600\,$Å film[28] are the same as for the $63\,$Å-thick film considered here.

V. In / InO$_x$ AND THE HIGH-T$_c$ OXIDES

Although we have modeled In/InO_x as a BCS superconductor, we also recognize commonality between this material and the high-T_c oxide superconductors, a family with characteristicly low carrier densities. We have prepared samples of

Table II.   Parameters for indium, two examples of In/InO_x, and the oxide superconductors: critical temperature, free-carrier density, Sommerfeld constant, free-carrier-model Fermi wavevector, critical-disorder resistivity.

Superconductor	T_c (K)	n (cm^{-3})	γ (Jm^{-3}K^{-2})	k_F (Å^{-1})	ρ_{crit} (mΩcm)
In	3.4	1.2×10^{23}	115	1.54	0.46
In / InO$_x$ [a]	3.2	5×10^{21}	120	0.53	1.3
In / InO$_x$ [b]	2.5	2×10^{20}	50	0.18	3.9
BaPb$_{.75}$Bi$_{.25}$O$_3$ [c]	12	3×10^{21}	7.9	0.45	1.6
La$_{1.85}$Sr$_{.15}$CuO$_4$ [d,e]	40	3×10^{21}	54	0.39	1.9
Ba$_2$YCu$_3$O$_7$ [f,g]	95	6×10^{21}	86	0.56	1.3

[a] Ref.32   [b] Ref.12   [c] Ref.33   [d] Ref.34   [e] Ref.35   [f] Ref.36   [g] Ref.37

In / InO$_x$ with varying oxide content, hence producing superconductors with free-electron volume densities spanning the range 2×10^{20} to 5×10^{21} cm^{-3}. As shown in Table II the density of states parameter γ — the Sommerfeld constant — as computed from the upper critical field slope using the dirty-limit expression $\gamma = 2.3 \times 10^{-4}$ [MKS] $\sigma_N \left(-dH_{c2}/dT \mid_{T=T_c} \right)$, covers a range with an upper bound equal to that of pure bulk indium. After correcting for depression by disorder, the T_c of the In / InO$_x$ is also the same as for In.[28] Hence, the material acts like In, albeit with substantially-reduced electron density.

Strong localization and interaction effects give a temperature coefficient of resistance which is negative in as-prepared samples; after annealing, whereupon T_c approaches 3.4 K, it changes sign, becoming positive at room temperature. Analogous behavior has been observed in bulk samples of the high-T_c oxides. To make a comparison with the other oxides, we also show in Table II parameters taken from variously reported very recent work [33–37], where our preference here is to express γ in volume density rather than in mole units. The other quantities are best estimates of carrier densities, which for the two high-T_c oxides are holes, and the Fermi wavevector, which we computed from a free-electron model, ignoring for the present the actual band structure.

At this point we conjecture commonality among low-carrier density systems for the effect of disorder on superconductivity, with justification also based on the observation that volume densities of states are no higher than for an ordinary superconductor like indium, so that a model of T_c depression by localization and interaction effects should be similarly applicable. Specifically, if the point of critical disorder were precisely the same, expressed as $k_F \ell = 3^{\frac{1}{2}}$, then from Eq. (1) our imputed correspondence implies a critical Boltzmann resistivity given by

$$\rho_{\text{crit}} = 3^{\frac{1}{2}} \pi^2 \hbar / e^2 k_F . \tag{6}$$

For YBa$_2$Cu$_3$O$_7$ our value 1.3 mΩcm is quite reasonable, since samples with resistivities above about 2 mΩcm usually show reversed, i.e., semiconducting-like, temperature coefficients of resistivity concomitant with broadened and depressed transitions. The implication is that it may be straightforward to prepare a layer of material appropriately near critical disorder, either through Y-Ba framework disorder, Cu-O bond disorder, or O-vacancy disorder, and thus be able to modulate T_c with conductivity in the manner demonstrated above for In / InO$_x$ films.

ACKNOWLEDGEMENTS

The authors acknowledge useful discussions with M. Gurvitch, S. Nakahara, and M. Paalanen. The very capable technical assistance of R. H. Eick is also greatly appreciated.

REFERENCES

[1]   R. E Glover and M. D. Sherrill, Phys. Rev. Lett. **5**, 248(1960).

[2]   R. Mach and G. O. Müller, Phys. Stat. Sol. **69a**, 11(1982).

[3]   W. Felsch and R. E. Glover, J. Vac. Sci. Technol. **9**, 337(1971).

[4]   H. L. Stadler, Phys. Rev. Lett. **14**, 979(1965).

[5]   A. T. Fiory and A. F. Hebard, Phys. Rev. Lett. **52**, 2057(1984).

[6]   A. D. Kent, A. Kapitulnik, and T. H. Geballe, Bull. Amer. Phys. Soc. **31**, No. 3, 316(1986).

[7]   M. Gurvitch, H. L. Stormer, R. C. Dynes, and J. M. Graybeal, Bull. Amer. Phys. Soc. **31**, No. 3, 438(1986).

[8]   D. M. Eagles, Sol. St. Commun. **60**, 521(1986).

[9]   For a review see, W. J. Gallagher, IEEE Trans. Mag., Mag-**21**, 709(1985).

[10]   T. Nishino, M. Miyake, Y. Harada, and U. Kawabe, IEEE Electron Devices Lett. **EDL-6**, 297(1985).

[11]   H Takayanagi and T. Kawakami, Phys. Rev. Lett. **54**, 2449(1985).

[12]   A. F. Hebard and M. Paalanen, Phys. Rev. B**30**, 4063(1984).

[13]   W. Rühl, Z. Physik **186**, 190(1965).

[14]   N. Cabrera and N. F. Mott, Rept. Prog. Phys. **12**, 163(1949).

[15]   W. Rühl, Proceedings of the IX International Conference on Low Temperature Physics, ed by J. G. Daunt, D. O. Edwards, F. J. Milford, and M. Yaqub, (Plenum Press, New York, 1965), p. 475.

[16]   D. J. Lischner and H. J. Juretschke, J. Appl. Phys. **51**, 474(1980).

[17]   A. Bardasis, Phys. Rev. **26**, 1477(1982).

[18]   D. G. Naugle, Phys. Lett. **25A**, 688(1967).

[19]   D. G. Naugle, J. W. Baker, and R. E. Allen, Phys. Rev. **B7**, 3028(1973).

[20]   W. Felsch and R. E. Glover, Solid State Commun. **10**, 1033(1972).

[21]   M. Strongin, O. F. Kammerer, D. H. Douglass, and M. H. Cohen, Phys. Rev. Lett. **19**, 121(1967).

[22]   H. Sixl, Phys. Lett. **53A**, 333(1975).

[23]   J. Halbritter, Solid State Commun. **34**, 675(1980).

[24]   A. T. Fiory and A. F. Hebard, Physica **135B**, 124(1985).

[25]   A. F. Hebard, A. T. Fiory, and R. H. Eick, Proceedings of the 1986 Applied Superconductivity Conference, Baltimore, MD, to be published in IEEE Trans. Mag.

[26]   A. F. Hebard and S. Nakahara, Appl. Phys. Lett. **41**, 1130(1980).

[27]   A. J. McGeown and C. J. Adkins, J. Phys. C **19**, 1753(1986).

[28] A. F. Hebard and A. T. Fiory, Phys. Rev. Lett. **58**, 1131(1987).

[29] H. Fukuyama, H. Ebisawa, and S. Maekawa, J. Phys. Soc. Jpn. **53**, 3560(1984).

[30] S. Nakahara, private communication.

[31] M. A. Paalanen and A. F. Hebard, Appl. Phys. Lett. **45**, 794(1984).

[32] P. L. Gammel, private communication.

[33] B. Batlogg, Physica **126B**, 275 (1984).

[34] B. Batlogg, A. P.Ramirez, R. J. Cava, R. B. van Dover, and E. A. Reitman, Phys. Rev. B35, 5340 (1987).

[35] G. Aeppli, R. J. Cava, E. J. Ansaldo, J. H. Brewer, S. R. Kreitzman, G. M. Luke, D. R. Noakes, and R. F. Kiefl, Phys. Rev. B35, 7129(1987).

[36] R. J. Cava, B. Batlogg, R. B. vanDover, D. W. Murphy, S. Sunshine, T. Siegrist, J. P. Remeika, E. A. Reitman, S. Zuhaurak, and G. P. Espinosa, Phys. Rev. Lett. **58**, 1676 (1987).

[37] D. R. Harshman, G. Aeppli, B. Batlogg, J. H. Brewer, J. F. Carolan, R. J. Cava, M. Celio, A. C. D. Chaklader, W. N. Hardy S. R. Kreitzman, G. M. Luke, D. R. Noakes, and M. Senba, preprint.

SUPERCONDUCTIVITY AT CONTACT OF ULTRATHIN GOLD FILMS

WITH AMORPHOUS GERMANIUM

B. Dwir and G. Deutscher

School of Physics and Astronomy, Tel-Aviv University
Ramat-Aviv, Israel

Abstract

Ultrathin films of Indium, Aluminum and Gold, in intimate contact with Germanium, show interesting electrical properties. Due to good wetting properties, continuity is reached at very small thickness. Au films (8-24Å) have been found to be superconducting ($T_c \simeq 1K$) with relatively high critical current density and high critical fields, and to have a reduced number of carriers with a high effective mass. The interplay between disorder and superconductivity is observed as a function of the Au film thickness.

Introduction

It has been known for some time that intimate contact with metals such as In, Pb and Al can considerably reduce the crystallization temperature of amorphous Ge films[1,2]. Intimate contact can be achieved either by vaccuum co-deposition of the metal and Ge[2] or by deposition of the metal onto a predeposited Ge amorphous film[1,3]. In the case of the co-deposited films, a random percolating structure is observed in the crystalline state[2], pointing out to good wetting properties of the constituents. A similar conclusion is reached when In films are deposited on amorphous Ge[3]. In that case, two related remarkable observations are reported. First, electrical continuity is achieved at thickness of the order of 20Å (see fig.1), as compared with about 1000Å when In is deposited on glass. Second, In deposition causes the crystallization of the Ge underlayer already at room temperature (fig.2). This can only be explained by the existence of a strong interface interaction. It should be emphasized that in the solid state In and Ge show only very limited mutual solubility (less than 1%).

It was conjectured that 30Å In films on Ge actually consist of a much thinner (~5 to 10Å) layer spread over most of the Ge film's surface, topped by a more islandic structure[3]. Magnetoresistance (MR) measurements showed a behavior typical of weak localization with strong spin-orbit interaction. However, the MR data could only be fitted with theory if it was assumed that the thin In layer had a percolative structure. In that case, it has been shown that the expression for the MR must be corrected by a reducing prefactor p that measures the ratio between the local sheet resistance and the measured macroscopic one[4]. This additional fitting parameter, as well as some assumptions made in the estimation of the coefficient of diffusion, introduced some uncertainty in the calculation of the inelastic time from the MR data. Also, superconductivity observed in such In layers around 0.5-1K could be ascribed to the reduced In T_c rather than to a property of the In-Ge interface

In this paper, we concentrate on new results obtained on Au-Ge contacts. Thin (8-24Å) Au films on Ge, or intercalated between two Ge films, are found to be homogeneous in thickness; their behavior points out to the existence of an interface superconductivity mechanism.

<u>Experimental</u>

The samples were prepared by electron-beam vacuum ($\sim 10^{-6}$ torr) evaporation of germanium (40 Ω-cm) and gold (99.999%) onto room-temperature glass-substrates. First, 300-1000Å of Ge was evaporated, then 8-24Å of Au, and in some of the samples, a second 300-1000Å of Ge was deposited, to provide environmental protection, as well as doubling the interface. The film thickness was measured by quartz-crystal microbalances, which also enabled control of the deposition rate. The geometry of the samples was either simple stripes, 10x1 mm (LxW), or a special geometry of a strip with voltage-sensing terminals and middle contacts for Hall-effect measurements. Both geometries were defined during the evaporation, by mechanical masks. Usually, thick (300Å) Au contact pads were evaporated at the ends of the samples, to give better contact for the voltage and current leads.

Microscopic structure of the samples was determined by using a TEM to get pictures and electron-diffraction patterns of the same films that were measured electrically, by removing some of the evaporated film onto microscope grids.

The electrical properties of the samples were measured in the 4-terminal DC method, using a 1μA current-source and a sensitive (0.1 μV) voltmeter. The samples were mounted during the measurements in a cryostat, either pumped ^{4}He (for temperatures down to 1.6K and fields up to 1.3T) or pumped ^{3}He (down to 0.6K and up to 3T).

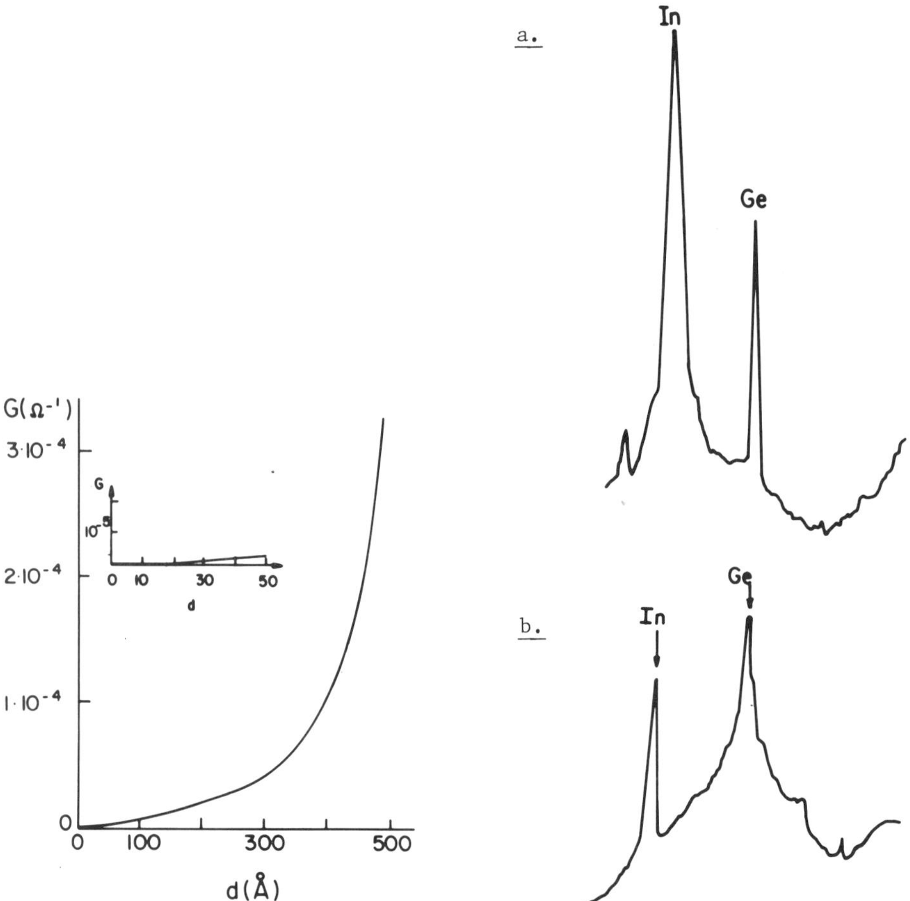

Fig.1 In-situ measured conductance G vs.
thickness d, of In grown on 300Å Ge.
Inset: enlargement of d=0-50Å part.

Fig.2 Microdensitographs of the first two
diffraction rings of: a.200Å In on 50Å Ge.
b.50Å In on 250Å Ge.

24

The structure of a relatively thick (36Å Au on 300Å Ge) sample, as obtained from the TEM, is shown in fig.3. The very small (50-100Å) grains of crystalline Au can be seen, as well as the featureless amorphous Ge underlayer. In the thinnest Au films (like 8Å) the grains can not be resolved. Samples prepared on a hot (we tried the range 70-150°C) substrate showed an increasing Au grain-size with temperature, as well as crystallization of the Ge layer (from ~100°C). However, the electrical conductance (percolation) threshold increased with temperature, especially on crystalline Ge, from <8Å at room-temperature to ~30Å at 150°C. We therefore used only room-temperature evaporations for our superconductivity measurements.

The dependence of the film's sheet resistance R/■ as a function of thickness is shown in fig.4, where it is plotted vs. $1/d^2$, where d is the Au film's thickness. In the case of diffuse reflections at the surfaces in <u>continuous</u> films, this plot gives roughly a straight line as a first approximation to Fuchs' law. Although there is considerable scatter in the data, particularly for the thinner films, it is clear that no systematic deviation from Fuchs' law is observed. In particular, no sharp increase of R/■ is observed close to the thinnest film studied (d=8Å).

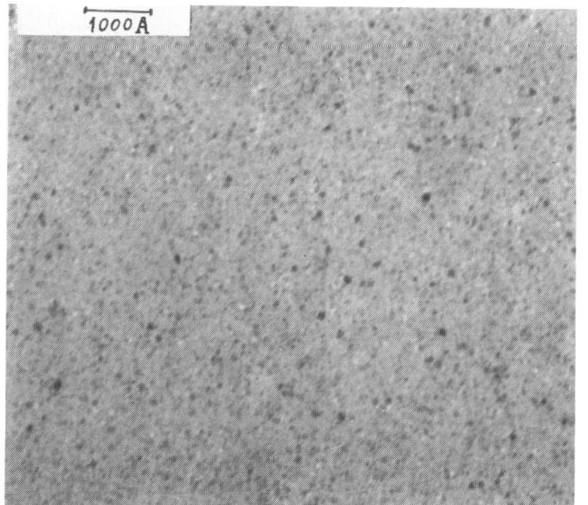
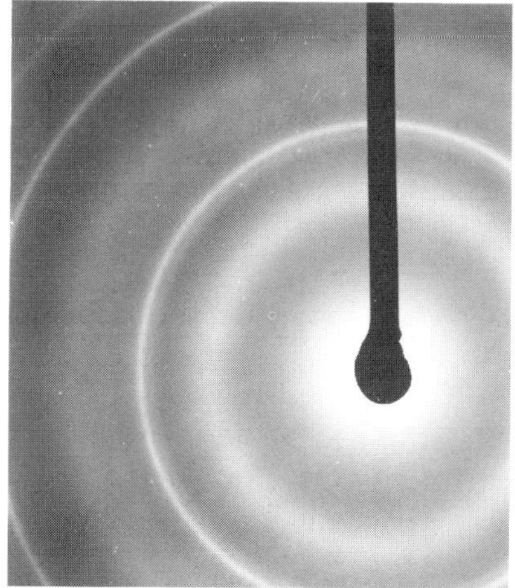

Fig.3 TEM micrograph of 36Å Au film on 300Å amorphous Ge.
a.Bright-field picture.   b.Diffraction pattern.

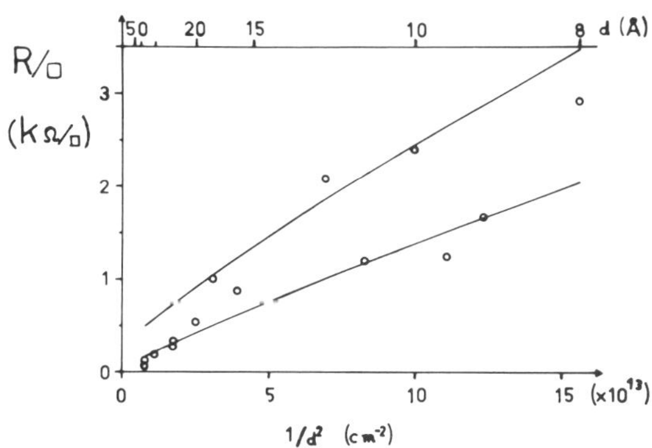

Fig.4 Resistance of several Au films vs. $1/d^2$. Lines are fitted
to Fuchs' formula for ℓ_b=90Å (top), ℓ_b=400Å (bottom).

The nature of the electrical conduction in the Au films was further investigated through Hall effect measurements. In a film with d=11Å, this measurement indcated that conduction is through electrons, but with a much reduced density of carriers ($n=6.1\cdot10^{21}$ cm^{-3}) compared to that of bulk gold ($5.9\cdot10^{22}$ cm^{-3}). We estimate that the contribution of the Ge underlayer is negligible in these measurements (done at T=4.2K) because of its very high resistivity. Also, Au is essentially insoluble in Ge ($<10^{16}$ cm^{-3} at T=850°C, extrapolated to $<10^{11}$ cm^{-3} at room temp.) and thus cannot increase its conductivity in any significant way.

The resistance temperature dependence of the Au films deposited on Ge, without Ge overcoating, all showed weak localization without any superconductive fluctuations, down to T=1.6K (fig.5a) . Samples with Ge overcoat showed mostly either superconductive fluctuations or a complete transition, depending on sample thickness and our measurement capabilities (down to 1.6K or 0.6K - see fig.5b).

The weak localization was also analyzed by measuring the magnetoconductance of samples at various temperatures (above Tc), as can be seen in fig.6. In one sample, we also measured the breakdown of superconductivity due to parallel and perpendicular magnetic fields, giving $dH_{c\perp}/dT \simeq 10000$ Gauss/K (near T_c) with an anisotropy factor of about 10 at a reduced temperature t=0.13. The critical current near Tc (fig.7) was also measured, showing relatively high critical current densities, pointing out to a continuous rather than percolative film.

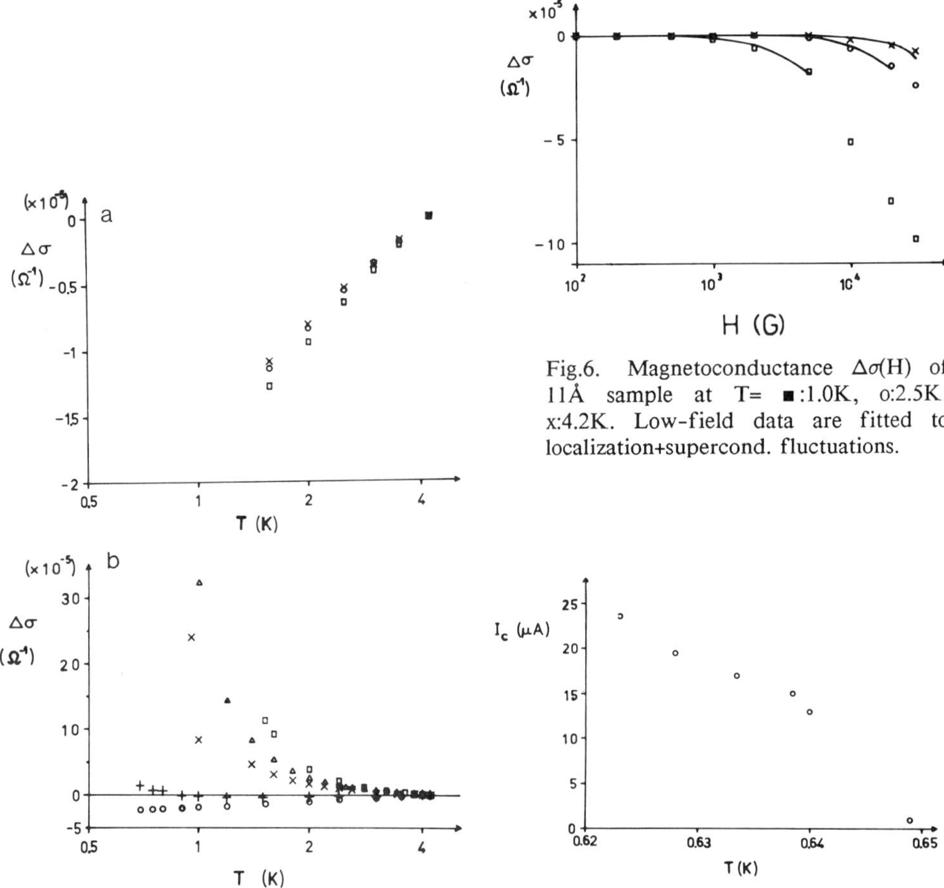

Fig.6. Magnetoconductance $\Delta\sigma$(H) of 11Å sample at T= ■:1.0K, o:2.5K, x:4.2K. Low-field data are fitted to localization+supercond. fluctuations.

Fig.5 Temp. dependence (relative to 4.2K, log scale) of $\Delta\sigma$(T) for several Au films:
a.(top) on 300Å Ge.
b.(bottom) between 300Å Ge layers.

Fig.7 Critical current of 11Å sample near T_c.

Discussion

The most striking features of the experiments are : 1. The metallic character of the Au films down to the smallest thickness studied (apart from weak localization effects) 2. The appearance of superconductivity when the Au film is sandwiched between two Ge layers. Superconductivity is however confined to a finite range of thicknesses ($10\text{\AA} \leq d \leq 30\text{\AA}$). superconductivity disappears, presumably because the net interaction averaged over the film's thickness is then too small.

We interpret this behavior as reflecting the interplay between superconductivity and disorder (localization effects). The origin of superconductivity must reside at the Au/Ge interface, since neither Au nor Ge are superconducting in bulk form (although we note that metastable AuGe amorphous alloys prepared by codeposition on liquid-He-cooled substrates are superconducting[5]). A strong modification of the electronic structure at the interface is clearly indicated by the Hall effect results. The effective interaction parameters $(NV)_{eff}$ averaged over the film thickness is twice as large in Ge/Au/Ge sandwiches as compared to Au/Ge for a given Au thickness. This appears sufficient to induce superconductivity around 1K for d~15Å At larger thicknesses, superconductivity vanishes because $(NV)_{eff} \alpha (d_0/d)$, where d_0 is some thickness of the order of an atomic distance; at smaller thicknesses superconductivity disappears because disorder (i.e. $R/\blacksquare$) increases and localization wins out.

In order to gain a more detailed understanding of the properties of the Au/Ge contacts, we have analyzed the MR data in some detail. The main results are as follows.

1. The MR curves can be fitted at all temperatures using Larkin's parameter[6] $\beta(T/T_c)$, calculated from the measured T_c, and assuming a large spin-orbit interaction (fig.7). Close to T_c, we have observed the saturation of the MR predicted by dos Santos and Abrahams[7]. From the behavior of the MR at high fields, we have been able to calculate the coefficient of diffusion D. We emphasize that the above fit has been obtained without introducing the fitting parameter p, which provides an additional indication for the continuity of the Au layers.

2. The value of D obtained from the MR fit was found to be in reasonable agreement with that obtained from the value of $dH\perp/dT$. For a layer with d=11Å, we obtain D_{MR}=2 cm²/s and $D_{H\perp} \simeq 1$ cm²/s . Using D=2 cm²/s together with the measured carrier density and $R/\blacksquare$, it is then possible to compute in the quasi-free electron model the Fermi energy and the electron effective mass. We obtain E_F=0.23 eV and m_e^*=5.4m_e , again pointing out to very strong interface effects.

3. Using the above parameters for the same sample, and Fuchs' relation[8] :

$$\rho_f = \frac{4\rho_b \cdot \ell_b}{3(\ln(\ell_b/d)+0.4228)}$$, one can then calculate the "bulk" value of the mean free path ℓ_b. We find $l_b \simeq 400$ Å, again pointing out to the fairly "clean" nature of the layers. For the various samples shown in fig.4, we obtain for ℓ_b values varying from 400Å to about 100Å

4. The value of the inelastic time obtained from the MR analysis and D=2 cm²/s is in fair agreement with the theoretical electron-electron time predicted by theory. In particular, $1/\tau_{in}$ shows the expected increase when T_c is approached.

Conclusions

The data that we have presented point out to the existence of strong local interactions at Au/Ge contacts. From the metallurgical point of view, they lead to very good wetting properties that allow the production of extremely thin, continuous, metallic films. However, the contact interactions also strongly modify the electronic structure of the metal. The carrier density and the Fermi energy are strongly reduced, while the effective mass is increased. Low carrier density superconductivity is observed. It seems that it is disorder effects that prevent the occurence of superconductivity at signifiantly higher temperatures. If disorder effects could be reduced by the preparation of films with smoother surfaces, thus allowing a reduction of the thickness while keeping $R/\blacksquare$ down, much higher critical temperatures should be observed.

Acknowledgements

This research was partly supported by the US-Israel BiNational Science Foundation and the Oren Family Chair for Experimental Solid State Physics.

References

1. F.Oki, Y.Ogawa,Y.Fujiki, Japan J. of Appl. Phys. $\underline{8}$, 1056 (1969).
2. See for instance G.Deutsher, in "Percolation, Structures and Processes", eds. G.Deutscher, R.Zallen and J.Adler, p.207 (vol.5 Annals of Israel Physical Society, 1983).
3. B.Dwir and G.Deutscher, Bulletin of Israel Physical Society, pp.20,66 (1985), and to be published.
4. A.Palevski and G.Deutscher, Phys. Rev. B34, 431 (1986).
5. B.Stritzker and H.Wühl, Z. Physik $\underline{243}$, 361 (1971).
6. A.I.Larkin, JETP Lett. $\underline{31}$, 219 (1980).
7. J.M.B.Lopes dos Santos and E.Abrahams, Phys. Rev. B31, 172 (1985).
8. K.L.Chopra, Thin Film Phenomena, p.349, McGraw-Hill, 1969.

SUPERCONDUCTIVITY IN THIN FILMS OF Au/Si BY ION IMPLANTATION

N. Jisrawi*, W.L. McLean*, and N.G. Stoffel**

*Rutgers University
Serin Physics Laboratory
Piscataway, NJ 08855

**Bell Communications Research
Red Bank, NJ 07701

Gold films on silicon substrates have been implanted with
ions of Si, Ar, and Xe at tens of keV for the purpose of studying
metastable phases inaccessible by other means of preparation. The
recent prediction and discovery of superconductivity in the high-
pressure hexagonal phases of silicon has called into question
previous possible explanations of superconductivity in the Au/Si
system.

INTRODUCTION

The amorphous gold-silicon system Au_xSi_{1-x} prepared by electron beam
evaporation from Au/Si ingots was found in the extensive studies by Nishida
et al.[1] to be superconducting for x in the range $0.14 < x < 0.42$ with a
maximum T_c of 0.8K at x = 0.42. It was suggested that the superconductivity arose from metallic electrons in an impurity band formed by the gold
atoms. Transition temperatures between 1 and 2.3K were found in quench-
condensed Au/Si films produced by sputtering by Möckel and Baumann.[2]
Superconductivity was attributed to a liquid-like amorphous phase.

The occurrence of superconductivity in the Au/Si system is of interest
for a number of reasons in addition to the fact that neither of the
constituents by itself under ordinary conditions is a superconductor and
that no stable phases occur in the standard phase diagrams[3] of the Au/Si
system. One of these reasons is that the decrease of transition
temperature from its maximum value with increasing silicon concentration
has some aspects in common with the behaviour of superconductor-insulator
mixtures and composites.[4] Fig. 1 shows the variation of T_c with room-
temperature resistivity of a variety of systems, three of them random
mixtures of superconductor and insulator,[4] while the others are granular
aluminum,[5] molybdenum-germanium,[6] and gold-silicon.[1] While the results in
the random mixtures of superconductor and insulator have been explained in
terms of the interplay of percolation and localization,[4] there is obviously
an additional aspect to contend with in gold-silicon where the system is
not superconducting when x=1 (or x=0). Since the system is amorphous, the
disorder is on a fine scale and so percolation is not likely to be
important.[4] However, the decrease of T_c from its maximum value is expected
to be partly due to localization effects.[1]

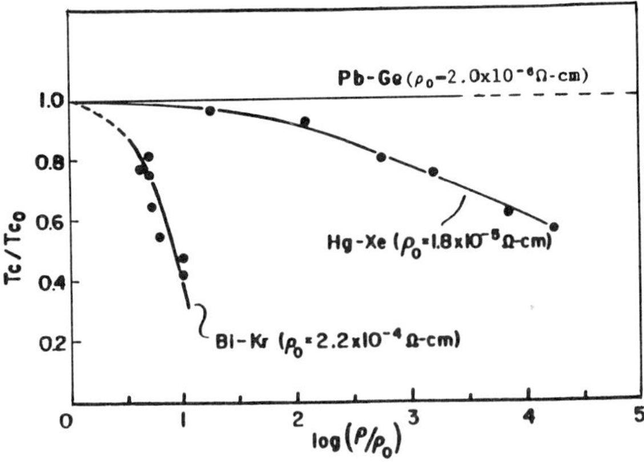

Fig. 1a. Transition temperature dependence on room temperature resistivity for random mixtures of superconductor and insulator. (From Ref. 4)

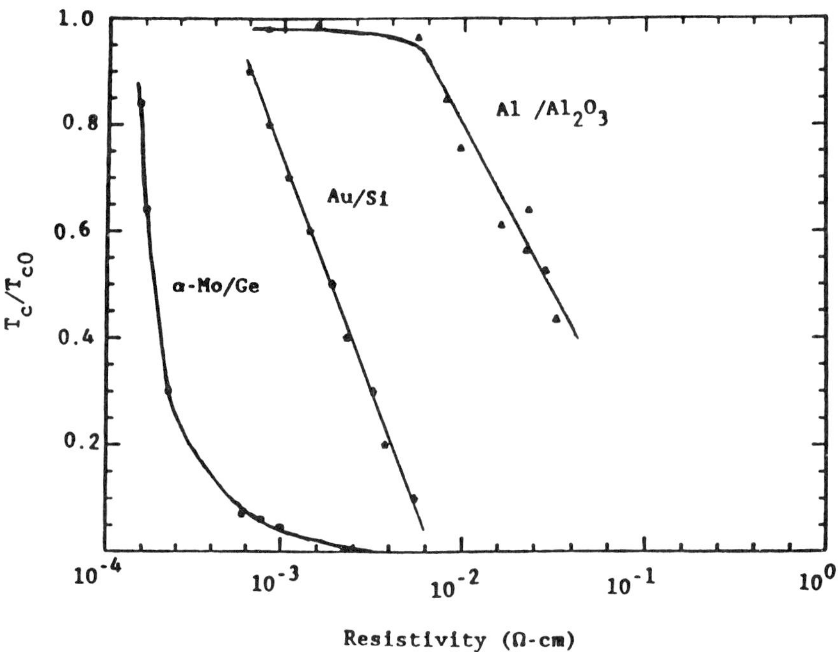

Fig. 1b. Transition temperature dependence on resistivity in normal state just above transition. (Au/Si from Ref.1, $T_{c0} = 1K$; α-Mo/Ge from Ref.6, $T_{c0} = 8.5K$; granular Al/Al$_2$O$_3$ from Ref.5, $T_{c0} = 2.4K$)

A second reason for interest in superconducting Au/Si is the recent prediction and subsequent experimental confirmation[7] of superconductivity in hexagonal phases of silicon formed under high pressure, raising the question of whether or not the addition of gold produces some similar effect.

In their pioneering work on quench-condensed germanium-noble metal films, Stritzker and Wühl[8] found superconducting transition temperatures as high as 3.6K in Au/Ge. They concluded that a liquid-like structure of Ge had been formed. This was later confirmed by Haug et al.[9] from in situ electron diffraction measurements. It was concluded that the role of the noble metal atoms was to stabilize a higher coordination-number structure of the germanium. Germanium by itself, like silicon, also becomes super-conducting under pressure with a maximum T_c of about 5.3K.[10]

The present work on Au/Si was begun to explore whether metastable phases not accessible by other means could be formed by implantation of silicon ions into films of gold.

ION IMPLANTATION

Gold films of thicknesses between 400 and 1000Å were thermally evaporated using 99.999% gold on to single crystal silicon substrates in a vacuum chamber with pressure less than 1×10^{-6} torr. The thicknesses of the gold films were such that at low doses most of the bombarding ions came to rest within the film. Sputtering of the Au film precluded the formation of high Si concentration by direct implantation. The sputtering yield (number of atoms sputtered from target per incident projectile ion) for Si on gold at 40 keV is 10. A saturation concentration of approximately 10% Si in Au is implied by this sputter yield.

We do not find evidence for superconductivity due to this low concentration Au/Si alloy formed by direct implantation. However, we do find superconductivity for doses sufficient to sputter a large fraction of the gold film off the Si substrate.

In addition to substantial sputtering, these doses of about 0.004 Coul/cm^2 (2.5×10^{16}/cm^2) Si at 40 keV induce diffusion and intermixing at the Au-Si interface. The superconductivity we observe is probably due to a metastable compound formed by ion-beam mixing at the interface, not by direct ion implantation of Si into Au. In support of this conclusion, we found that Si implanted Au films on sapphire did not produce super-conductive films, but Ar and Xe implants into Au films on Si substrates did result in superconductive films as discussed below.

EXPERIMENTAL RESULTS

The variation of resistance with temperature for gold films subjected to different doses of Si ions is shown in Fig. 2. There was a dependence of the critical temperature on the dose and also on the energy of the bombarding ions--e.g. for a dose of 6.25×10^{16} ions/cm^2 the mid-point of the transition was increased from 0.75K to 0.82K when the incident energy was raised from 40 to 50 keV.

The variation of T_c with ion dose of Si at 40 keV is shown in Fig. 3. Here T_c is defined as the temperature at which the resistance is half its normal state value just above the transition.

In order to study the extent to which the results depend on the
interface mixing of the gold film with the silicon substrate a series of
gold films evaporated on to silicon have been bombarded with Ar^+ and Xe^{++}
instead of with Si^+ ions. The Ar and Xe doses were selected to guarantee a
complete mixing of the films. The energies (200 keV for Ar and 300 keV for
Xe) were chosen so that the projected range in gold was roughly equal to
the thickness of the virgin gold layer.

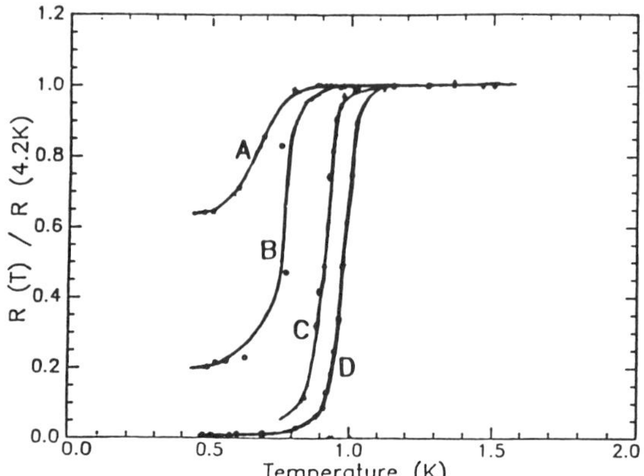

Fig. 2. Superconducting resistive transitions in gold films on
silicon substrates with different doses of Si ions at 40 keV.
A: 1.88×10^{17} ions/cm^2; B: 6.25×10^{16} ions/cm^2; C: 1.56×10^{17} ion/cm^2;
D: 1.25×10^{17} ions/cm^2.

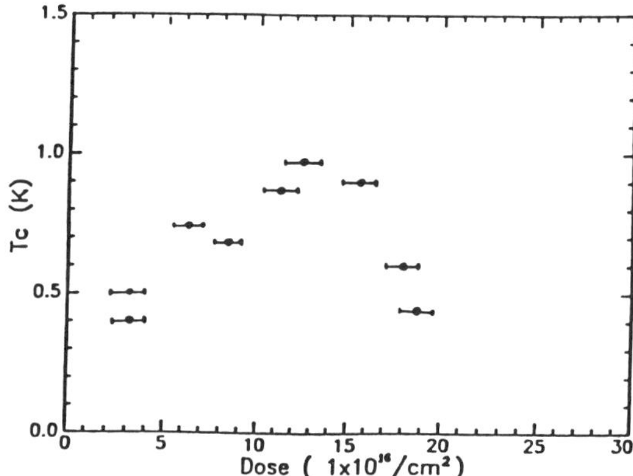

Fig. 3. Variation of superconducting transition temperature
with dose for gold films on silicon substrates implanted by
Si ions at 40 keV.

Our interpretation of this data is that the interface-mixing increases as the sputtering brings the interface nearer to the front surface. This leads to higher T_c's until one reaches doses of about 1.5×10^{17} ions/cm^2. At higher doses, sputtering removes the superconducting layer, leading to an abrupt drop in T_c and eventually to a loss of superconductivity.

The highest superconducting onset temperature achieved so far is just over 1K. The transitions for the set of films that were given similar doses of Si at different energies and among which the maximum T_c was found, are shown in Fig. 4. A surprising feature in the normal state is the linearity of the variation of the resistivity with temperature over a wide range of temperature, as shown in Fig. 5. The results also imply a lower "resistivity" Debye temperature (144K) than that of bulk gold (165K).

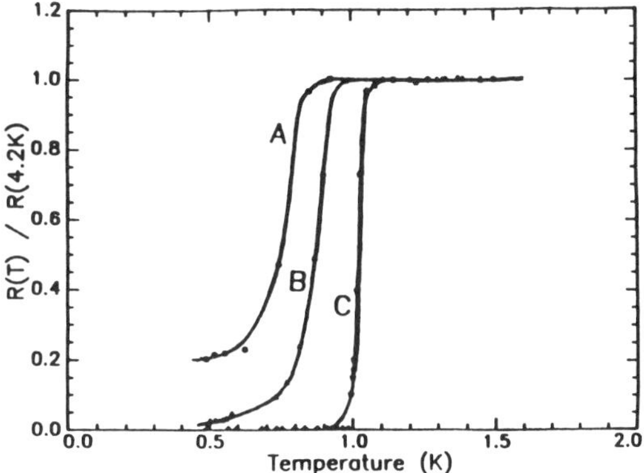

Fig. 4. Resistive transitions for samples implanted with same dose of 6.25×10^{16} ions/cm^2 at different beam energies. A: 40 keV; B: 50 keV; C: 60 keV.

The composition profile of the films before and after implantation has been studied using the Rutherford back-scattering spectrometry (RBS)[13] with alpha particles of initial incident energy of 2 MeV. A typical RBS spectrum is shown in Fig. 6. Also shown is a simulation of a $Au_{0.56}Si_{0.44}$ layer on a silicon substrate with a nominal thickness of 800 Å.

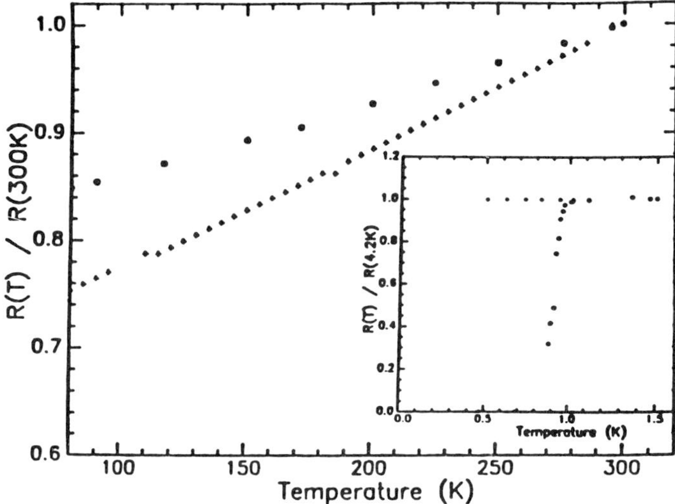

Fig. 5. Variation of resistance at temperatures well above T_c for implanted film (dots) (R(4.2K)/R(300K) = 0.79) and virgin film (crosses) (R(4.2K)/R(300K) = 0.64). Inset shows enlarged transition region.

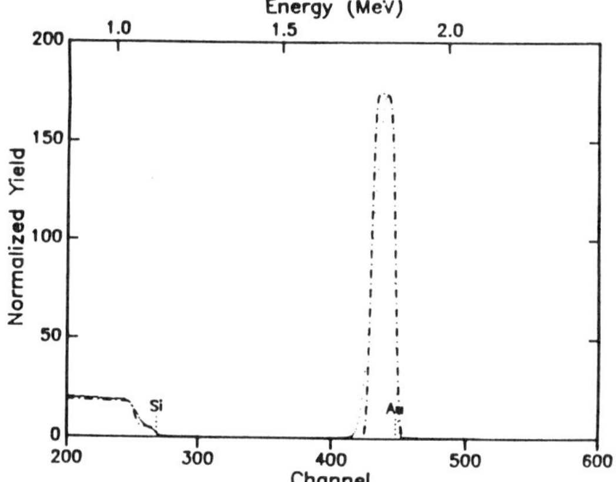

Fig. 6. (·····): RBS spectrum of Au film on silicon substrate implanted with dose of 6.25×10^{16} ions/cm^2 of Si. (—·——·——·): simulation (see text).

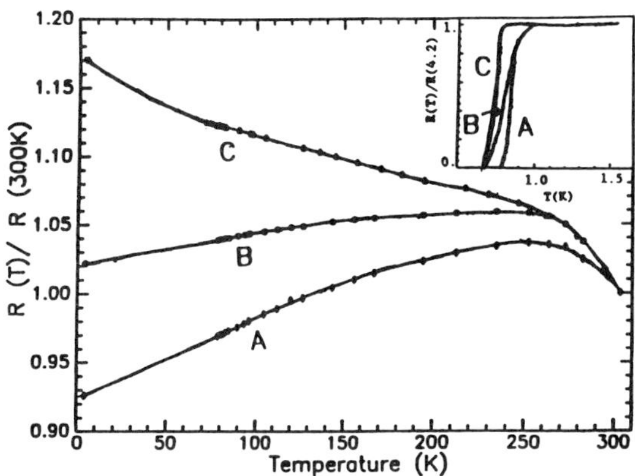

Fig. 7. Temperature dependence of resistance for gold films on silicon substrates implanted with Ar ions at 200 keV. Inset shows superconducting transitions. Doses were A: 1.7 x , B: 1.0x, C: 5x 10^{16}ions/cm^2. The resistance ratios R(4.2)/R(300) are respectively: 0.925, 1.02, 1.17.

Films bombarded in this way were found to be superconducting with transition temperatures almost as high as those implanted with Si, as can be seen from the inset in Fig. 7. However, as shown in the main part of this figure, the dependence of resistance on temperature is not linear as it was in the case of the Si implants. The results of RBS measurements are given in Fig. 8 and indicate that more than 50% of the original Au film has been sputtered from the sample.

DISCUSSION

The linearity of resistance with temperature above the superconducting transitions in the Si implanted films suggests in this case that the films remain metallic. The results of interface mixing by use of the Ar or Xe beams instead of Si indicates that the superconductivity is probably occurring in all three cases because of interface mixing. The marginal increase of T$_c$ in the Si implanted films may be associated with a less disordered metallic structure than was obtained with Ar or Xe. It appears as though the Si implanted films are the closest to some ideal metallic phase with low Debye temperature. Just why there should be a significant softening of the phonon spectrum remains to be clarified. The excitonic mechanism has been suggested[14] as the cause of the transition-temperature enhancement of aluminum implanted with Ge but at the present time it is not obviously necessary to abandon the electron-phonon interaction as a possible source of superconductivity in Au/Si.

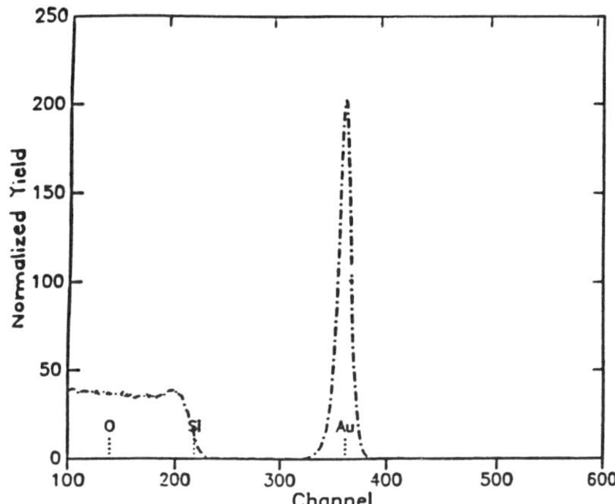

Fig. 8. Rutherford backscattering spectra for one of the
films in Fig. 7.

FUTURE DIRECTIONS

We have recently become aware of an ion-mixing experiment by Tsaur and
Mayer[14] in which gold films on silicon substrates were bombarded with Ar or
Xe. <u>After thermal annealing</u> a homogeneous layer of a metastable phase with
a well-defined hexagonal crystal structure and stoichiometry Au_5Si_2 was
formed. Unfortunately the system was not cooled below 4.2K to check for
superconductivity. Our current work includes reproducing this structure
and cooling it to lower temperatures to look for superconductivity.
Attempts are also being made to look for evidence of hexagonal Au_5Si_2 in
the Si-implanted gold films on silicon substrates.

ACKNOWLEDGMENTS

We are very grateful to B. Wilkins for his help with the RBS
measurements and to D. Hart for assisting with the ion implantations. This
work was partially supported by NSF-DMR-85-11982.

REFERENCES

1. N. Nishida, T. Furubayasha, M. Yamagushi, K. Morigaki, Y. Miura, Y.
    Takano, H. Ishimoto and S. Ogawa, Proc. LT-17 (North Holland
    Physics Publishing, Amsterdam, the Netherlands, 1984) p.729 .
2. D. Möckel and F.Baumann, phys. stat. sol.(a) <u>54</u>, 585 (1980).
3. H.Okamoto and T.Massalaki, Bulletin of Alloy Phase Diagrams, <u>4</u>, 190
    (1983).

4. G.Deutscher, A.M. Goldman and H. Micklitz, Phys. Rev. B $\underline{31}$, 1679
       (1985).
5. Y.Z. Zhang, M. Kunchur, T. Tsuboi, P. Lindenfeld and W.L. McLean,
       Proc. LT-18, to be published .
6. S.Yoshizumi, Ph.D. Thesis , E.L. Ginzton Laboratory, Stanford
       University, 1986 .
7. K.J. Chang, M.M. Dacorogna, M.L. Cohen, J.M. Mignot, G. Chouteau and
       G. Martinez, Phys. Rev. Lett. $\underline{4}$, 2375 (1985); D. Erskine, P.Y. Yu,
       K.J. Chang and M.L. Cohen, Phys. Rev. Lett. $\underline{57}$, 2741 (1986).
8. B. Stritzker and H. Wühl, Z. Physik $\underline{243}$, 361 (1971).
9. E.Haug, N. Hedgecock, and W. Buckel, Z. Physik B$\underline{22}$, 237 (1975) .
10. W. Buckel and J. Wittig, Phys. Letters $\underline{17}$, 187 (1965).
11. P.Sigmund , Phys. Rev. $\underline{184}$, 383 (1969).
12. B.Y. Tsaur, S.S. Liau, and J.W. Mayer, Appl. Phys. Lett. $\underline{34}$, 168
       (1979).
13. W.K. Chu, J.W. Mayer, and M.A. Nicolet, <u>Backscattering Spectrometry</u>
       (Academic Press, New York, 1978).
14. H. Bernas and P. Nedellec, Nucl. Instr. and Methods, $\underline{182}/\underline{183}$, 845
       (1981).
15. B.Y.Tsaur and J.W.Mayer ,Phil.Mag.A $\underline{43}$,345(1981) .

TEST OF T_c-PREDICTIONS USING THE RIGID

BAND MODEL FOR REFRACTORY COMPOUNDS

Ernst L. Haase and Jiri Ruzicka*

Kernforschungszentrum Karlsruhe
Institut für Nukleare Festkörperphysik
P.O.B. 3640, D-7500 Karlsruhe, FRG

ABSTRACT

B1-compounds are particularly suitable to test T_c-predictions, as
they form continuous solid solutions with intermediate properties reflecting the changing number of valence electrons (VE). Thin films of the
$Nb_{1-x}Mo_xC$, $Nb_{1-x}Mo_xC_{1-y}N_y$ and MoC_xN_y systems were prepared by sputtering.
There are serious discrepancies between theory and the experimental results. In particular, the predicted minimum at 10.45 VE was not observed.
It was possible to prepare $Nb_{0.3}Mo_{0.7}C$ with 9.7 VE stoichiometrically.
Instead of the expected T_c of 17 K only 13 K was measured. In the
$NbC_{1-y}N_y$, $Ti_{1-x}Nb_xN$ and $Ti_{1-x}Nb_xC_{1-y}N_y$ systems for around 9.7 VE a T_c of
up to 17.3 K is observed in agreement with theory. However, where theoretically a narrow maximum is predicted, high T_c values are observed for a
wide range of VE.

INTRODUCTION

In the past several years numerous calculations [1-7] of the electronic
density of states and T_c of refractory materials have been published. Experimental tests of the T_c-predictions have been carried out for MoC [8],
MoN [9-11] and for $Nb_{1-x}Mo_x(C_{1-y}N_y)_z$ [6], where however most of the samples
were understoichiometric. It is the aim of the present investigation to
provide more extensive tests for the $Nb_{1-x}Mo_xC_{1-y}N_y$ system and the ternary $Ti_{1-x}Nb_xN$, $NbC_{1-y}N_y$ and $Nb_{1-x}Mo_xC$ systems. These systems provide good
testing grounds, as the number of valence electrons varies continuously
between 9 and 11, and they form continuous solid solutions in the B1-
phase. The situation is less favourable for binary alloys of transition
elements, where there is little structure and for A15 materials, where in
most cases due to the sharp peaks in the electronic density of states $N(E)$,
T_c drops rapidly as ternary systems are formed.

By arc-melting together somewhat substoichiometric carbides and nitrides of transition element solid solutions were obtained [14]. Although these
extensive data give many trends, they will not be considered in the following because of the substoichiometry.

---

*On leave of absence from the Physical Institute, Czechoslovak Academy
 of Sciences, Prague, CSSR

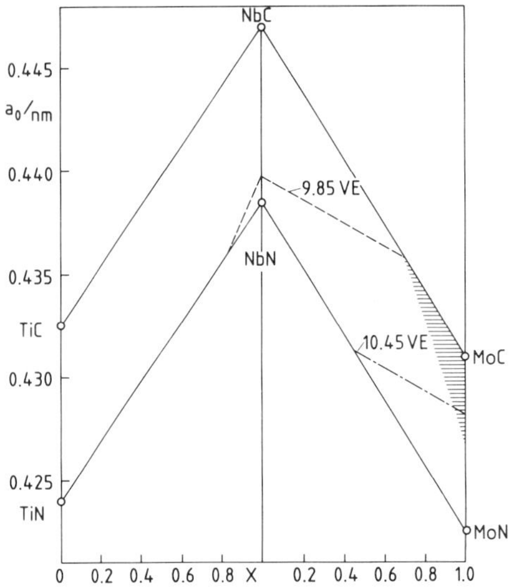

Fig. 1. The lattice parameter a_0 vs. the atomic fraction x. Along the dashed line a T_C maximum is predicted, while along the dot-dashed line there should be a minimum.

THE THEORETICAL SITUATION

If one compares the shape of the N(E) calculated curves for HfC (8 valence electrons VE)[1], NbC(9VE)[1], NbN(10VE)[3] and MoN(11VE)[3] one realizes that they are practically identical, confirming the rigid band model. As the number of VE increases, the Fermi level moves to the right, i.e., in the direction of increasing number of integral VE. With growing number of VE the Fermi level shifts from near a minimum for HfC(0.0 K) via a growing N(E) for NbC (11 K) to near a first maximum for NbN (17 K) and after crossing a minimum to a second maximum for MoN (29 K predicted)[3,4]. MoC, has 10 VE as NbN. Both N(E) structures and the positions of the Fermi level are very similar and MoC should have a T_C comparable to that of NbN[4].

In Fig. 1 are plotted the lattice parameter a_0 vs. the at% fraction, assuming Vegards law to hold. In the six corners we have the a_0-values for TiC, TiN, NbC, NbN, MoC and MoN. Theoretical predictions exist for some of the corners and the $NbC_{1-y}N_y$[5] and $Nb_{1-x}Mo_xN$[3,4] ternary systems. For the $NbC_{1-y}N_y$ system a maximum in T_C is predicted for 9.85 VE, whereas the experimental maximum is found to lie near 9.7 VE[14]. The curve is shown in Fig. 3. For the $Nb_{1-x}Mo_xN$ system Ref. 6 predicts a minimum of 7 K at 10.27 VE, while the later Ref. 7 predicts a minimum of 10 K at 10.45 VE. This curve is shown in Figs. 8 and 9. Because of the similarities of the N(E) curves, and assuming the rigid band model to hold, these extrema should also occur along the dashed and dot-dashed lines in the   quaternary systems in Fig. 1. In particular, the maximum of about 17 K should also occur for $Nb_{0.3}Mo_{0.7}C$ and the minimum should also occur for $MoC_{0.45}N_{0.55}$. The hatched region is inaccessible to experiment, as this material forms only with a metalloid deficiency[8]. While this work establishes the deficiency limit along the $Nb_{1-x}Mo_xC$ line, the limit along the $MoC_{1-y}N_y$ line is not clear. MoN can be formed stoichiometrically[11].

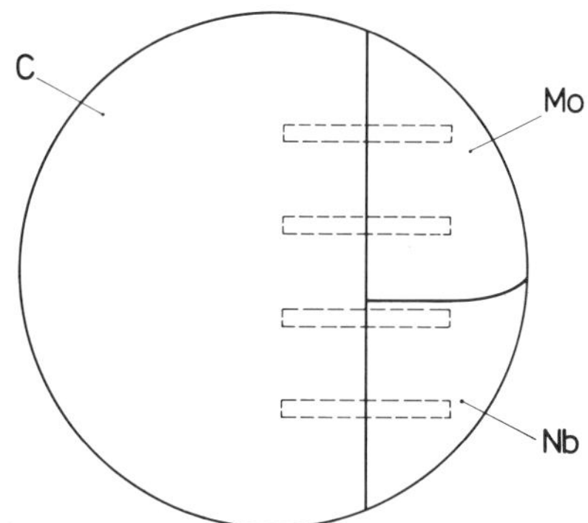

Fig. 2. A typical geometric arrangement for the sputter cathode. The substrates are mounted opposite at the dashed positions.

EXPERIMENTAL DETAILS

About 500 samples were prepared by sputtering onto single crystal sapphire substrates 50x5x1 mm. A typical geometrical composition of the cathode is shown in Fig. 2, there are three areas of C, Nb and Mo. Usually four substrates were clamped to a resistively heated Ta-strip about 35 mm below the cathode at the positions indicated by dashed lines. Along the substrates, there is a C-gradient and the samples were usually measured at positions each 10 mm apart. From the bottom to the top, there is an increase in the Mo-fraction. In addition to the Ar-sputter gas, N_2 was admitted when wanted. The substrate temperature varied between 100-900°C but for almost all of the data presented was held at 750 - 800°C.

The base pressure before sputtering was about 10^{-8} Torr. The Ar-pressure was varied between 1×10^{-3} to 2×10^{-2} Torr, but for all of the data presented was kept fixed at 2×10^{-2} Torr. When present, the N_2 pressure was varied between 2×10^{-4} and 3.3×10^{-2} Torr. With increasing N_2-pressure, the substrates were moved to the C-deficient side. Both gases were continuously pumped to avoid contamination. With an RF-power of 500 W the sputter rate was 0.2 nm/sec and the layer thickness was either 0.3 or usually 1 µm. The elemental analysis was carried out using an electron microprobe, Rutherford backscattering, or indirectly via a_o. Very important is the T_C-optimization. To obtain stoichiometric samples, the substrate temperature of the geometrical position and/or the nitrogen pressure was varied, until a T_c maximum was obtained near the middle of the substrates. The T_c-onset (99%) is given. ΔT_c was usually around 0.5 K when optimized. T_c was measured resistively by the usual four-point method. Occassionally it was checked inductively.

EXPERIMENTAL RESULTS AND DISCUSSION

In the course of time a number of binary refractory compounds were prepared with the RF sputtering apparatus stoichiometrically without any difficulties by optimizing T_c. Only near the MoC_x corner it was impossible to obtain stoichiometric samples. Except for the MoC_x corner, whenever they were checked, the samples were found to be stoichiometric. Fig. 3 shows a_o and T_c versus y for the $NbC_{1-y}N_y$ system. For the incircled crosses y was determined via electron microbeam analysis. Then a smooth a_o-line was drawn through the encircled crosses and the endpoints. The line shows a substantial positive departure from Vegards law (straight line). A similar departure was found for slightly substoichiometric $NbC_{1-y}N_y$ [12]. For the crosses

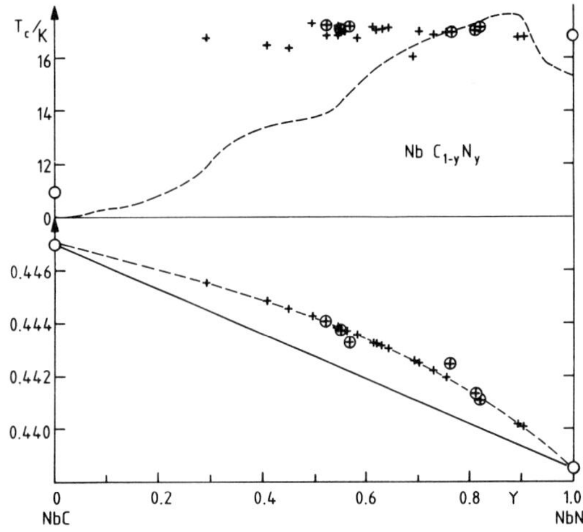

Fig. 3. T_c and a_o for the $NbC_{1-y}N_y$ system. The predicted T_c curve is shown dashed.

the y-value was established by plotting the a_o-value on the smooth line. The corresponding T_c-values are plotted in the upper half of the picture. As is evident the T_c values show a very broad maximum from y=0.3 to 0.9, peaking around y=0.55. The theoretical prediction by Papaconstantopoulos[5] is shown as dashed line, peaking at y=0.85 and having a much sharper peak. Especially the T_c-values for y=0.3 to 0.6 are much higher than the theoretical curve. Altogether one can say that the experimental maximum is much broader and it is shifted from y=0.85 to about 0.55.

It is now interesting to see what happens when one replaces Nb by Ti rather than N by C. Fig. 4 shows the T_c versus x plot for the $Ti_{1-x}Nb_xN$ system. Again we have a broad maximum of about 17 K extending from x=0.3 to 0.95. For small x values T_c falls to 6.1 K for TiN[16]. The corresponding theoretical curve is not drawn in, but should be very similar to the curve in Fig. 3. Experimentally the $Ti_{1-x}Nb_xN$ and the $NbC_{1-y}N_y$ data are very similar, as one expects if the rigid band model holds and the phonon spec-

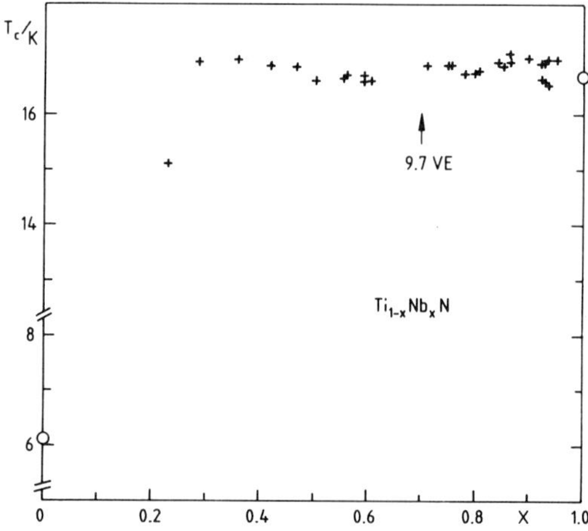

Fig. 4. T_c vs. atomic fraction x for the $Ti_{1-x}Nb_xN$ system.

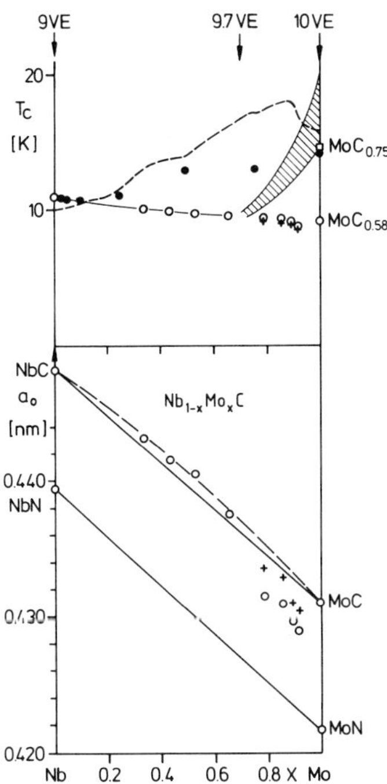

Fig. 5. T_c and a_0 for the $Nb_{1-x}Mo_xC$ system. The dashed line is the infered theoretical expectation for T_c.

tra do not change significantly. No a_0-data are given, as in many instances the X-ray lines are split or are shifted by as much as half a degree, pointing to a distorted unit cell. Numerous samples were prepared for the $Ti_{1-x}Nb_xC_{1-y}N_y$ systems with widely varying values of x and y. It was not possible to obtain precise values for x and y. The T_c values range from 15.8 to 17.3 K and thus also show a broad maximum.

It is also of interest to investigate the situation for the number of VE larger than 10, that is when Nb is replaced by Mo, which matches most closely in atomic radius. Fig. 5 shows the data for the $Nb_{1-x}Mo_xC$ system. The region for x > 0.7 falls into the hatched region of Fig. 1 and samples can be prepared only substoichiometrically as evidenced by the a_0-values falling below the line. Ref. 8 yields T_c=20 K and a_0=0.431 nm by extrapolation. The T_c-value is expected (top of hatched region) and agrees with theory[4]. The a_0-value seems reasonable in view of the positive departure from Vegards law for x<0.7 shown in the bottom half of Fig. 5. The open circles and crosses show our data. The crosses stem from the same samples, but measured 10 mm closer to the C-rich side. The point at x=0.53 has 51, the one at x=0.66 has 49 at% C. The points for x>0.7 also have about 50 at% C, so there must be C-precipitates. The solid circles stem from splat-cooled samples of Ref. 15. The square point for $MoC_{0.75}$ stems from Ref. 13. The $MoC_{0.58}$ point stems from Ref.8. It is evident that the T_c-values of about 13 K for x=0.50 and 0.75 fall below the 17 K for the $NbC_{1-y}N_y$ and $Ti_{1-x}Nb_xN$ systems. The dashed line is the theoretical prediction of Papaconstantopoulos[5] assuming the rigid band model to hold. For $Nb_{0.3}Mo_{0.7}C$ with 9.7 VE one would expect 17 K rather than the observed 13 K. As Mo is substituted for Nb along the 9.7 VE line T_c obviously drops.

If the N(E) calculations are correct, one of the other physical properties influencing T_c must be responsible for the decline. This decline makes it improbable that MoN has the predicted 29 K [3,4].

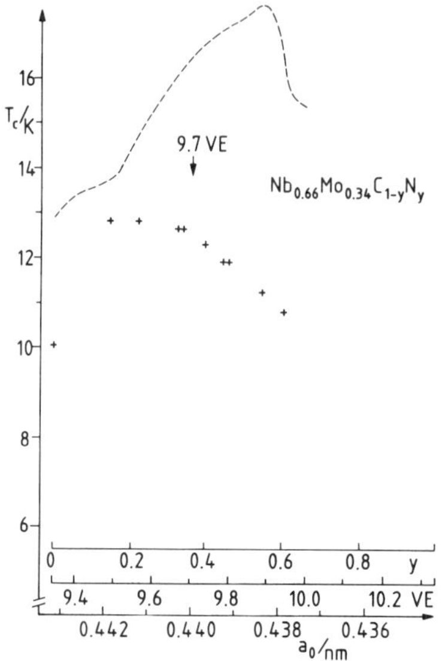

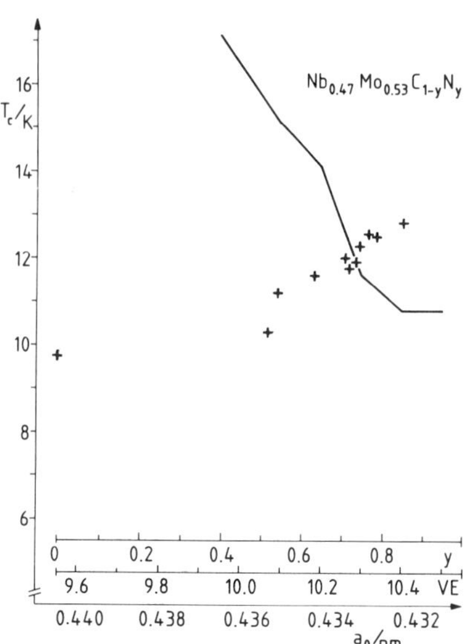

Fig. 6. T_c vs. a_0 for the $Nb_{0.66}Mo_{0.34}C_{1-y}N_y$ system.

Fig. 7. T_c vs. a_0 for the $Nb_{0.47}Mo_{0.53}C_{1-y}N_y$ system.

In the $Nb_{1-x}Mo_xC_{1-y}N_y$ field we have prepared samples with the four values in x of 0.34, 0.53, 0.86 and 0.91. These samples were not analyzed for x, but were made at the same geometrical positions as for $Nb_{1-x}Mo_xC$ shown in Fig. 5.

Fig. 6 shows the data for the $Nb_{0.66}Mo_{0.34}C_{1-y}N_y$ system. Plotted are T_c versus a_0. The corresponding y and VE scales are also plotted, assuming a reasonable departure from Vegards law in Fig. 1. The data show a fairly narrow maximum near 9.5 VE. The theoretical curve inferred from the $NbC_{1-y}N_y$ system is shown dashed. The agreement is rather poor. In particular, the data fail to reach the expected 17 K for around 9.7 VE.

Fig. 7 shows the situation for the $Nb_{0.47}Mo_{0.53}C_{1-y}N_y$ system. With increasing VE the data have a rising tendency while theory inferred from the $Nb_{1-x}Mo_xN$ prediction has a falling trend. The highest experimental points lie at the position of the predicted minimum. Again the experimental points are below the theoretical 17 K.

Fig. 8 and 9 give the situation near the Mo-edge. For low y-values the samples are almost certainly substoichiometric, as they lie in the hatched region of Fig. 1. The data do not show any trends, there is no evidence for a minimum and no rise to the predicted 29 K. Rather they are in agreement with the T_c-values around 12 K reported for MoN[9,10]. As we have not yet fully optimized the substrate temperature, a T_c increase of 1-2 K is still expected.

The excellent reproducability of the data indicates that an optimum has been nearly reached. It has been said[7] that disorder decreases T_c of MoN. Experimentally it has been observed for MoC_{1-x}[8] and $Mo(C_{1-y}N_y)_{1-z}$ that disorder-induced by low substrate temperature enhances T_c, perhaps due to phonon softening. It is now time to compare theory and experiment for the more extensive data compiled in Ref. 14.

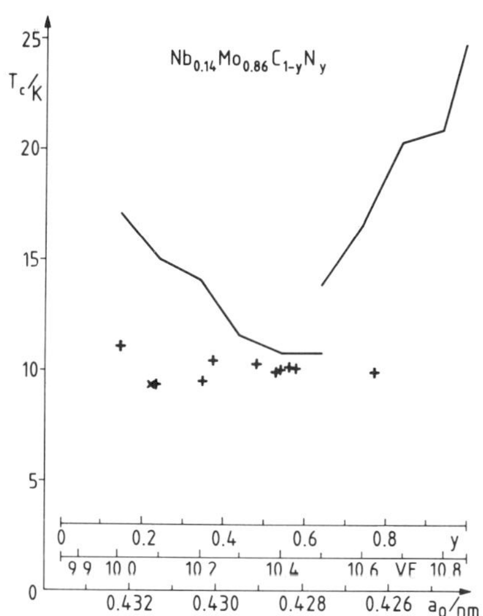

Fig. 8. T_c vs. a_O for the $Nb_{0.14}Mo_{0.86}C_{1-y}N_y$ system

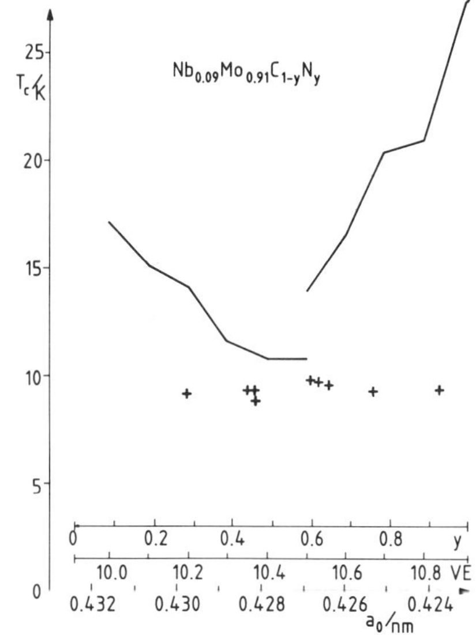

Fig. 9. T_c vs. a_O for the $Nb_{0.14}Mo_{0.86}C_{1-y}N_y$ system

CONCLUSIONS

Instead of the expected constant T_c of about 17 K for about 9.7 valence electrons VE, experimentally T_c moves from 17 K for $Ti_{1-x}Nb_xN$ to 17.3 K for $NbC_{0.3}N_{0.7}$ via 13 K for $Nb_{0.66}Mo_{0.34}C_{0.8}N_{0.2}$ to a value of 13 K for $Nb_{0.3}Mo_{0.7}C$. With rising Mo-content the maximum T_c moves to a lower value of VE. The predicted minimum for 10.45 VE has nowhere been observed, instead of it there are high points near $Nb_{0.47}Mo_{0.53}C_{0.1}N_{0.9}$. Towards the MoN corner no T_c rise is observed. Rather, T_c values around 10 K are observed, tending towards the experimentally observed T_c values for MoN.

Altogether there are serious discrepancies between the experimentally observed T_c-values and those predicted assuming the validity of the rigid band model and gradual variations in the physical properties influencing T_c. It is very unlikely that the observed discrepancies can be accounted for by rapid variations in the phonon spectra. Theoretically a phonon softening is expected with an increasing number of VE[4].

If gradual variations in the properties other than the electronic density of states are responsible for the observed lowering of T_c, the minimum should also be lowered, in contrast with the results. In view of the T_c results of Fig. 5 a slightly enhanced T_c might be expected by preparing MoN by splat cooling. To avoid reliance on the rigid band model, it would be desirable to carry out a T_c calculation for the $Nb_{0.47}Mo_{0.53}C_{1-y}N_y$ system.

ACKNOWLEDGEMENTS

The authors want to thank R. Smithey for preparing some of the samples and help with the T_c measurements. Thanks are also due to H. Kleykamp and H. Späte for the electron microprobe measurements. The critical reading of the manuscript by O. Meyer is gratefully acknowledged.

REFERENCES

1 B.M. Klein, D.A. Papaconstantopoulos, and L.L. Boyer, Phys. Rev. $\underline{B22}$, 1946 (1980)
2 K. Schwarz and H. Ripplinger, Z. Phys. $\underline{B48}$, 79 (1982)
3 D.A. Papaconstantopoulos, W.E. Pickett, B.M. Klein and L.L. Boyer, Phys. Rev. $\underline{B31}$, 752 (1985)
4 W.E. Pickett, B.M. Klein and D.A. Papaconstantopoulos, Physica $\underline{107B}$, 667 (1981)
5 D.A. Papaconstantopoulos, in Physics of Transition Metals 1980, ed. P. Rhodes (IOP, London, 1981) p. 563
6 S.B. Qadri et al., J. Vac. Sci. Technol. $\underline{A3}$, 664 (1985)
7 D.A. Papaconstantopoulos and W.E. Pickett, Phys. Rev. $\underline{B31}$, 7093 (1985)
8 J. Ruzicka, E.L. Haase and O. Meyer, Proc. LT17, U. Eckern, A. Schmid, W. Weber and H. Wühl, eds., North Holland, Amsterdam (1984) p. 113
9 H. Terada, M. Naoe and Y. Hoshi, Int. Conf. Cryogenic Materials, Boston, Aug. 1985, DP4
10 H. Yamamoto, T. Miki and M. Tanaka, Int. Conf. Cryogenic Materials, Boston, Aug. 1985, DP5
11 G. Linker, R. Smithey and O. Meyer, J. Phys. $\underline{F14}$, L115 (1984)
12 P. Duwez and F. Odell, J. Electrochem. Soc. $\underline{97}$, 299 (1950)
13 W. Krauss, Ph.D. Thesis, Karlsruhe University (1986)
14 N. Pessall, R.E. Gold and H.A. Johansen, J. Phys. Chem. Sol. $\underline{29}$, 19 (1968)
15 R.H. Willens, E. Buehler and B.T. Matthias, Phys. Rev. $\underline{159}$, 327 (1967)
16 T. Wolf, Ph.D. Thesis, Karlsruhe University, (1982) p. $\underline{97}$.

SUPERCONDUCTIVITY AND METAL CLUSTERS

W. D. Knight

Physics Department
University of California
Berkeley, CA 94720

INTRODUCTION

It has been generally accepted[1] that there must be some lower particle
size limit below which superconductivity does not occur. However, the idea
has not been adequately tested in the size range which is now possible for
the production and study of metal microclusters containing from two to two
hundred atoms per cluster[2]. It is the purpose of this paper to point out
that further exploration of the question of the superconductivity of metal
clusters deserves consideration, and to suggest some experimental approaches.

SMALL PARTICLE LIMITS ON SUPERCONDUCTIVITY

Several criteria have been suggested based on physical ideas of electron
scattering[3], quantum size effects[4], or fluctuations[5]. Whether the considerations are based on statistical[6] methods or impurities and boundary scattering[7]
they may be expressed in terms of the Anderson criterion[3]

$$\delta \overset{\sim}{>} \Delta \tag{1}$$

where δ is the level spacing and Δ is the superconducting energy gap parameter. The level spacing increases for smaller particles, and if it exceeds
the superconducting energy gap parameter, excitations associated with superconductivity would be forbidden. Eq. 1 has been used to estimate the minimum
particle size for which superconductivity can occur.

Since the estimate of level spacing depends on the degeneracies of the
model used[6], and estimates of gap parameter depend on possible enhancements
of T_c (ref. 4) in small particles, the criterion for quenching of superconductivity in small particles is somewhat imprecise. It also has become
clear that experimental observations of particles embedded in support materials are subject to particle-matrix interactions[8]. It is uncertain whether
results pertain purely to the particle properties or whether the matrix acts
as a kind of impurity.

At the present time microclusters can be produced, size selected by mass
spectrometry, and studied as isolated systems[2], so that impurities are
irrelevant. The simple metals are characterized in terms of quantum level
structures[9] which are based on delocalized electrons in potential wells of

spherical or ellipsoidal symmetry and can be described in a first approximation by a jellium shell model which reflects the symmetries. Properties of the alkali metal clusters stem from degeneracies in the level structure with major shell closings for the set of spherical[2] clusters, and minor ellipsoidal[10] closings for the remaining clusters. Corresponding spherical shell closings are observed in cluster ions of the noble metals[11], and of some divalent and trivalent metals[12].

Because of the high degeneracies in the spherical clusters the average level spacing can be orders of magnitude larger than the values according to the Kubo[13] model used for particles of size 10 nm. The possibility for enhanced T_c (ref. 4) in small particles also makes the application of Eq. 1 uncertain. Since no data on superconductivity exist in the size range for microclusters it would seem desirable to perform appropriate experiments. In fact some experiments[14] were performed for particles close to the predicted size limits, but quenching of superconductivity was not observed. Since the experiments involved embedded particles one might conclude[1] that particle-matrix interactions may have influenced the results.

The energy gaps following spherical closings[9] are of the order of 0.1 e.V. for the alkali micro clusters. These should be typical of other metal clusters and are large compared to kT_c for Type 1 bulk superconductors. According to Eq. 1 it might seem that cluster sizes are far below the superconductivity limit. However, it is necessary to consider other properties of free clusters in molecular beams such as vibrations and rotations. While in bulk superconductors electron and phonon states are considered separately, vibronic states become important in clusters, and the meaning of electron phonon interactions must be carefully defined. When these are included in the total manifold of states, it is clear that the ultimate energy gaps are small and the density of states at the Fermi level is high. Some estimates of phonon effects have been made for jellium models[15,16] but it isn't clear yet how the situation in clusters may compare with that in the bulk materials.

POSSIBLE EXPERIMENTS

Magnetic Deflection

In principle one may investigate both paramagnetic and diamagnetic effects by magnetic deflection of the cluster beam. The former have been studied for small odd-electron alkali clusters in experiments[17,18] analogous to the Stern-Gerlach experiment on atoms, and it has been verified that even-electron clusters show no deflections. Because of spin-rotation interactions the paramagnetism of the larger paramagnetic clusters has so far not been observed.

On the other hand observation of diamagnetism associated with superconductivity is a good possibility and is technically feasible. These experiments should shed some light on a question which has not received extensive theoretical treatment.

Electric deflection

The static dipole polarizabilities have been measured for the alkali clusters[19]. The results show a per atom polarizability decreasing from the atomic polarizability toward the bulk value with features at the spherical closings for 8 and 20 atoms. Contrary to early expectations and in agreement with more recent calculations[20] the measured polarizabilities indicate complete screening of external fields. Calculated effects of superconductivity should be interesting, and the appropriate experiments are straightforward.

Ionization and spectra

Ionization potentials have been measured[21] for clusters of potassium
containing up to N = 100 atoms, with a trend toward the asymptotic work
function for the bulk metal. Features in the curve of ionization potential
vs N are consistent with known spherical and ellipsoidal shell closings,
and include odd-even alternations suggesting higher stability for even-
electron systems. The pair stabilities are quite apparent for N < 20,
and the photoionization efficiency profiles show alternations which are
reminiscent of nuclear spectra which are used to illustrate a superconducting model of the nucleus[22]. Thus analogous pairing interactions appear
in small systems, and it would seem useful to pursue the idea for clusters.
The experiments on ionization potential represent a beginning of a spectroscopy of clusters which ultimately will reveal details of the level
structures, and may be employed to search for superconducting energy gaps
if they exist.

Surface Plasma Resonance

The photoabsorption cross sections for sodium clusters have recently
been measured[23,24] for clusters up to N = 40 atoms per cluster and at
several wavelengths. The cross sections are large, consistent with a high
photon-plasmon cross section, and the marked frequency dependence of the
cross section is consistent with calculated resonance peaks. The resonance
peaks are predicted from an ellipsoidal harmonic oscillator model, and
include single peaks for spherical clusters, double peaks for ellipsoids
of revolution, and triple peaks for triaxial shapes. The experimental
results agree extraordinarily well with the predictions of the simple
theory, and the magnitudes and frequency dependences of the cross sections
are basically understood. The cluster shapes may be determined directly.

The observations are based on a series of processes, starting with
absorption of a single photon, and followed by dissipation of energy into
internal cluster modes, resulting in a temperature rise. Finally an atom
evaporates with high probability and the cluster is removed from the
collimated beam because of the transverse momentum generated by the evaporation. The dissipation of photon energy into internal energy should
be related to electron-phonon interactions. Plasma resonance spectroscopy
could contribute to the detection of transitions in superconducting
clusters.

The opportunity to study the fundamental interactions in possibly
superconducting clusters seems to be significant, especially in view of
the fact that impurities and surface interactions are absent in such a
clean experiment. Typical clusters cool to temperatures of the order of
100 K, and modifications of the conditions of the supersonic expansion
are expected to reduce this figure by an order of magnitude. In any case
the study of the frequencies and damping factors for the plasma resonance,
now in progress, will reveal significant fundamental physics of the
clusters, and there is the possibility that something about superconductivity may be learned as well.

ACKNOWLEDGEMENTS

This work has been partly supported by the U. S. National Science
Foundation under DMR 84-17823 and DMR 86-15246. The author acknowledges
stimulating conversations about the work with Prof. Marvin L. Cohen and
and Dr. Walt A. de Heer.

REFERENCES

1. J. A. A. Perenboom, P. Wyder, and F. Meier, Electronic Properties
    of Small Metallic Particles, Physics Reports 78:173 (1981).
2. Walt A. de Heer, W. D. Knight, M. Y. Chou, and Marvin L. Cohen,
    Solid State Physics, Vol. 40, to appear in September (1987).
3. P. W. Anderson, J. Phys. Chem. Solids, 11:26 (1959).
4. R. H. Parmenter, Phys. Rev. 166:392 (1968).
5. B. Mühlschlegel, D. J. Scalapino, and R. Denton, Phys. Rev. B6:1767
    (1972).
6. R. Denton, B. Mülschlegel, and D. J. Scalapino, Phys. Rev. B7:3589
    (1973).
7. A. Kawabata, J. Phys. Soc. Jap. 29:902 (1970).
8. R. L. Filler, P. Lindenfeld, T. Worthington, and G. Deutscher,
    Phys. Rev. B21:5031 (1980).
9. W. D. Knight, Keith Clemenger, Walt A. de Heer, and Winston A.
    Saunders, Phys. Rev. Lett. 52:2141 (1984).
10. Keith Clemenger, Phys. Rev. B32:1359 (1985).
11. I. Katakuse, I. Ichihara, Y. Fujita, T. Matsuo, T. Sakurai, and
    H. Matsuda, Int. J. Mass Spect. Ion Proc. 67:229 (1985).
12. I. Katakuse, T. Ichihara, Y. Fujita, T. Matsuo, T. Sakurai, and
    H. Matsuda, ibid. 69:109 (1986).
13. R. Kubo, J. Phys. Soc. Jap., 17:975 (1962).
14. I. Giaever and H. R. Zeller, Phys. Rev. Lett. 20:1504 (1968).
15. P. Sheng, M. Y. Chou, and M. L. Cohen, Phys. Rev. B34:732 (1986).
16. J. Bardeen and D. Pines, Phys. Rev. 99:1140 (1953).
17. Walt A. de Heer, Ph.D. Thesis, University of California, Berkeley,
    1985.
18. W. D. Knight, Helv. Phys. Acta 56:521 (1983).
19. W. D. Knight, Keith Clemenger, Walt A. de Heer, and Winston A.
    Saunders, Phys. Rev. B31:2539 (1985).
20. W. Ekardt, Surf. Sci. 152:180 (1985).
21. Winston A. Saunders, Keith Clemenger, Walt A. de Heer, and W. D.
    Knight, Phys. Rev. B32:1366 (1985).
22. A. de Shalit and H. Feshbach, Theoretical Nuclear Physics, Vol. I,
    Ch. VII, Wiley, New York (1974).
23. Walt A. de Heer, Katherine Selby, Vitaly Kresin, Jun Masui, and
    A. Chatelain, Bull. Am. Phys. Soc., 32:484 (1987).
24. Walt A. de Heer, et al., to be published.

PROXIMITY EFFECT AND ANDERSON LOCALIZATION

Hidetoshi Fukuyama

Institute for Solid State Physics
University of Tokyo
7-22-1 Roppongi, Minato-ku, Tokyo 106

ABSTRACT

Proximity effect of interacting electrons in dirty two-dimensional metals is theoretically examined in view of the recent achievement of MOS type three-terminal Josephson junction devices. It is shown that both the amplitude and the coherence length of the Cooper pair propagator are affected by the combined effects of mutual Coulomb interaction and randomness.

INTRODUCTION

Proximity effect has been a subject of extensive investigations both theoretically and experimentally [1-10]. There exists renewed interest in the effect due to the realization of MOS type three-terminal Josephson junction devices [11-14], whose structures are shown in Figs.1(a) and (b). Such a possibility of MOS type Josephson field effect transistor had been proposed by Clark et al. in 1980 [7]. The basic principle of the switching is simple[7]; the gate voltage controls the electron density, which determines the degree of the proximity effect, and hence the coupling between two superconductors. Especially in the experiment of NTT group the normal electrons are confined to the surface due to the MOS structure [15] and hence two-dimensional. Since the electron density of normal metals can be varied in a wide range and may not always be in very metallic regime in MOS , the effect of Anderson localization will be important and interesting, which is the subject of this paper. Such effects of Anderson localization have been examined extensively so far [16-18], especially in two-dimensional systems where these effects have dramatic consequences. Hence we confine ourselves to the effects in two-dimensional systems in the following.

PROXIMITY EFFECT IN NON-INTERACTING DIRTY METALS

In G-L region, to which we confine ourselves in this paper, the proximity effect is characterized by the correlation function of the Cooper pair wave function, $b(r)=\psi_\uparrow(r)\psi_\downarrow(r)$, in the region of normal metals [1]

$$F(r) = <b^+(r)b(0)> \quad ,$$

$$= \sum_Q e^{iQr} F(Q) = T^2 \sum_{\varepsilon_n,\varepsilon_{n'}} \sum_Q e^{iQR} F(Q,2\varepsilon_n,2\varepsilon_{n'}) \quad , \tag{1}$$

where $\varepsilon_n=(2n+1)\pi T$ and $F(Q,2\varepsilon_n,2\varepsilon_{n'})$ is diagramatically given by Fig.2 In the case of non-interacting electrons with normal impurity scattering this $F(Q)$, defined as $F_0(Q)$, is given by [1]

$$F_0(Q) = 2\pi N(0)T \sum_{\varepsilon_n} [DQ^2+2|\varepsilon_n|]^{-1} \quad , \tag{2}$$

$$= N(0) \left[\ln 1.13\omega_D / T + \psi(\tfrac{1}{2}) - \psi(\tfrac{1}{2} + \tfrac{DQ^2}{4\pi T}) \right] \quad , \tag{3}$$

$$\equiv N(0)f_0(Q,T) \quad , \tag{4}$$

where $N(0)$, D and ω_D are the density of states, the diffusion constant and the Debye cut-off energy, respectively, and $\psi(z)$ is the di-gamma function. The Fourier transform of $F_0(Q)$ leads to the spatial variation, $F(x)$,

$$F(x) = 2\pi TN(0) / D \sum_{\varepsilon_n} \xi_n e^{-x/\xi_n} \equiv N(0)f_0(x) = F_0(x) \quad , \tag{5}$$

where $\xi_n=\sqrt{D/2|\varepsilon_n|}$ and we assumed here that the boundary between superconducting and normal metals is perpendicular to the x-axis with the origin therein. For sufficiently large x, $F_0(x)$ is characterized by the longest ξ_n, which corresponds to ε_n with n=0 and -1, i.e.

$$F_0^{(0)}(x) = N(0) \sqrt{2\pi T/D} \; e^{-x/\xi(T)} \quad , \tag{6}$$

where $\xi(T)=\sqrt{D/2\pi T}$ is usually called a coherence length and characterizes the coupling between two superconductors; if junctions are apart by L, the effective coupling is proportional to $\exp[-L/\xi(T)]$ as far as $L>\xi(T)$.

However for $x<\xi(T)$, $F_0(x)$, eq.(5), and $F_0^{(0)}(x)$, eq.(6), differ appreciably as seen in Fig.3, where $\omega_D/T=30$ is assumed.

EFFECTS OF INTERACTIONS WITHIN GORKOV THEORY

In normal metals there exist quite generally interactions, g, between electrons carrying proximity effect. By taking g into account we obtain

$$F(Q) = N(0) f_0(Q,T) / [1 + g f_0(Q,T)] \equiv F_I(Q) \quad . \tag{7}$$

In Fourier transforming eq.(7) we note that there is an essential difference between g>0 and g<0, since $f_0(Q,T)$ at Q=0 is logarithmically divergent as $T\to0$. In the case of g<0, the "one-frequency approximation" proposed by de Gennes [1] will be valid, because $1+gf_0(0,T)$ can be very small and then Q dependence of $1+gf_0(Q,T)$ should be strong and will properly be taken into account by this approximation. On the hand, if $g>0$, which is of our interest here, $1+gf_0(0,T)$ is appreciable and the Q-dependence of $1+gf_0(Q,T)$ can be considered as weak. In this case we

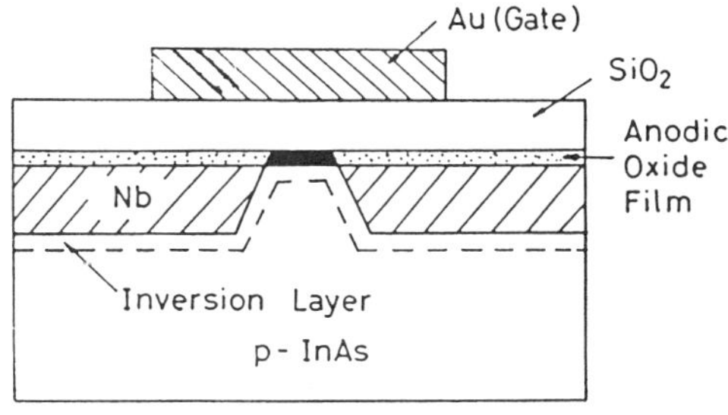

Fig.1(a) NTT structure for three-terminal Josephson junction [after ref.11].

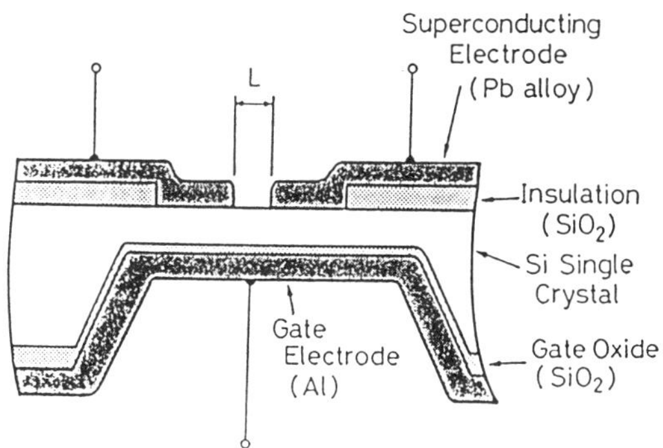

Fig.1(b) Hitachi structure for three-terminal Josephson junction [after ref.14].

proceed as follows: in eq.(7), where we have $f_0(Q,T)^{\ell}$ ($\ell \geq 1$) in general, we note that in the region of the spatial distance of our interest, i.e. for x larger than but of the order of $\xi(T)$, we may introduce the following approximation

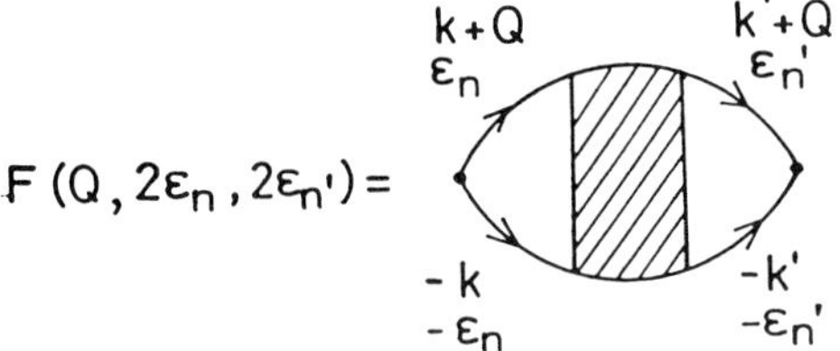

$$F(Q, 2\varepsilon_n, 2\varepsilon_{n'}) =$$

Fig.2 Cooper pair propagator.

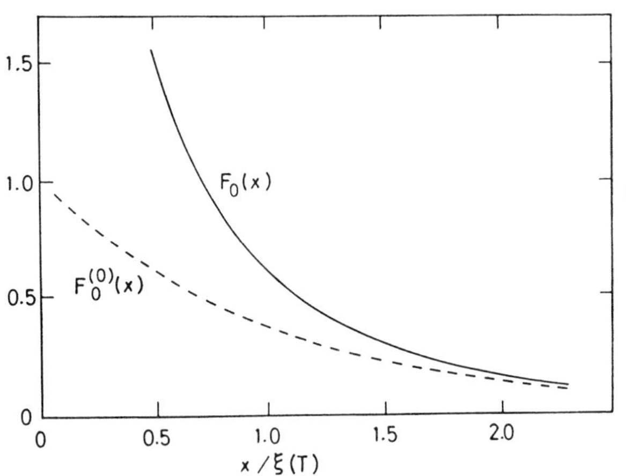

Fig.3 Cooper pair correlation function $F_0(x)$, eq.(5), and $F_0^{(0)}(x)$, eq.(6), in unit of $N(0)\sqrt{2\pi T/D}$.

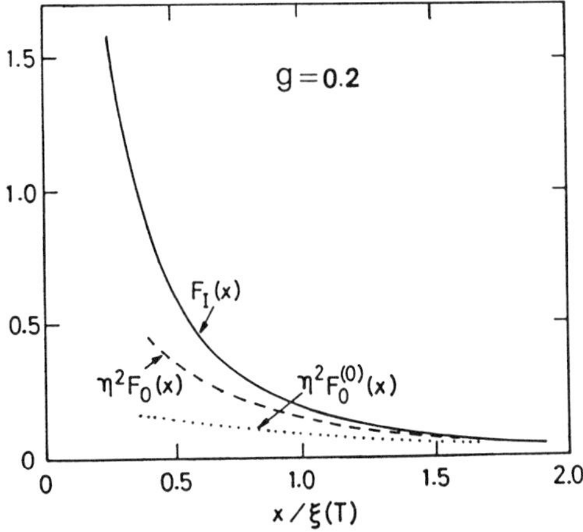

Fig.4 Cooper pair correlation function in the presence of interaction, $F_I(x)$, compared with the approximations $\eta^2 F_0(x)$ and $\eta^2 F_0^{(0)}(x)$. Unit is the same as in Fig.3.

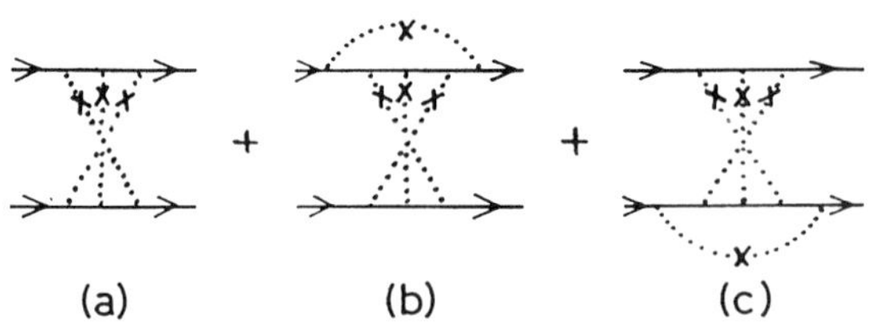

Fig.5 Quantum corrections to the diffusion constant due to the interference effect.

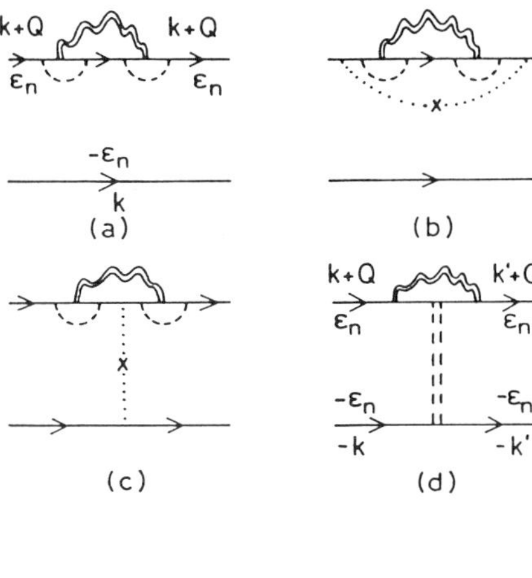

Fig.6 The self-energy type quantum corrections due to the dynamically screened Coulomb inteaction.

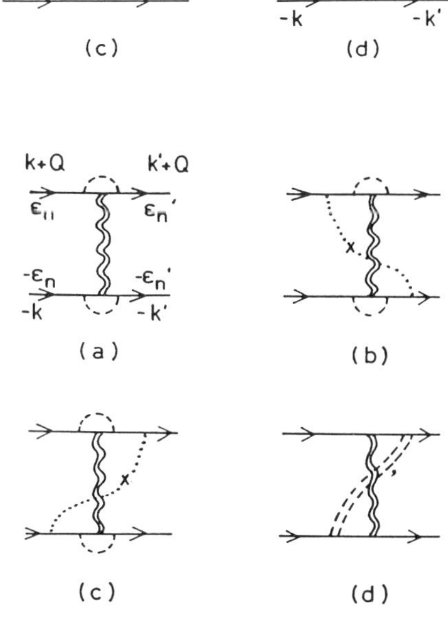

Fig.7 The vertex correction type quantum corrections due to the dynamically screened Coulomb interaction.

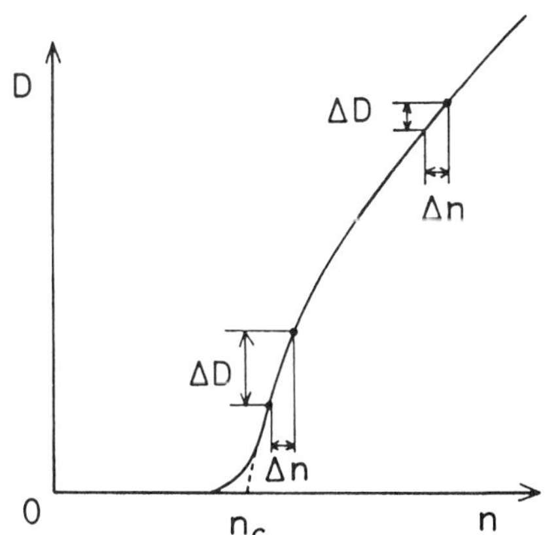

Fig.8 The schematic figure of the carrier number dependence of the diffusion constant.

$$\sum f_0(Q,T)^{\ell} \ e^{iQr} \simeq \ell f_0(x) \ f_0(0,T)^{\ell-1} \quad ,$$

where $f_0(x)$ is defined in eq.(5). In this approximation we obtain

$$F_I(x) = \sum e^{iQr} F_I(Q) \simeq N(0)f_0(x)\eta(T)^2 \equiv \eta^2 F_0(x) \quad . \tag{8}$$

Here the reduction factor $\eta(T)$ is defined by

$$\eta(T) = [1 + gf_0(0,T)]^{-1} \quad . \tag{9}$$

The comparison of $\eta(T)^2 F_0(x)$ with $F_I(x)$ is shown in Fig.4 for a choice of g=0.2 with the same cut-off parameter as in Fig.3. We see that, for $x>\xi(T)$, $F_I(x)$ is reasonably approximated by $\eta(T)^2 F_0(x)$. In this figure is also shown $\eta(T)^2 F_0^{(0)}(x)$, $F_0^{(0)}(x)$ being defined by eq.(6).

This result indicates that unless g (>0) is small the proximity effect can not survive at low temperatures, since $f(0,T) \simeq \ell n \omega_D/T$.

QUANTUM CORRECTIONS ASSOCIATED WITH INTERACTION EFFECTS

As has been clarified in recent years, the interaction processes in random systems are modified in an important way leading to the enhancement of their effects [19]. Especially the modifications of the long range part of dynamically screened Coulomb interaction is overwhelming the contributions from the attractive part of the interactions mediated by phonons, and then the critical temperature, T_c, of superconductivity is reduced as the randomness is increased [20,21]. These have been examined in detail by treating the effect of randomness perturbatively with respect to the randomness parameter, $\lambda = (2\pi\varepsilon_F\tau)^{-1}$, where ε_F and τ are the Fermi energy and the life time, respectively. Since λ is proportional to the Planck constant, corrections associated with λ are called quantum corrections. Similar kinds of effects are also expected in the proximity effect of our present interest. The construction of a general scheme to treat these effects is straightforward; by classifying all possible processes in terms of the irreducible part with respect to g, which is defined as f(Q,T), we obtain instead of eq.(7)

$$F(Q) = N(0)f(Q,T) \ / \ [1 + gf(Q,T)] \quad . \tag{10}$$

In this case the Fourier transfer of F(Q) will also effectively be written as for $F_I(r)$, eq.(8),

$$F(r) = \sum e^{iQr} F(Q) \simeq N(0)\eta(T)^2 f(x) \quad , \tag{11}$$

where f(x) and $\eta(T)$ are defined by

$$f(r) = \sum_Q e^{iQr} f(Q,T) \quad , \qquad \eta(T) = [1 + gf(0,T)]^{-1} \quad . \tag{12}$$

The leading contribution to f(Q,T) is given by $f_0(Q,T)$ defined by eq.(4) and one can perform systematic perturbative treatment of f(Q,T) with respect to λ,

$$f(Q,T) = f_0(Q,T) + \delta f(Q,T) \quad . \tag{13}$$

The quantity of interest here, $\delta f(Q,T)$, has already been examined in another context: in the case of actual superconductors, where g<0, the critical temperature, T_c, is determined by the divergence of F(Q=0), i.e.

$1+gf(0,T_c)=0$ or

$$\ln(T_c^0/T_c) = f(0,T_c) - f_0(0,T_c) \quad ,$$

$$= \delta f(0,T_c) \quad , \tag{14}$$

where $T_c^0=1.13\omega_D\exp(-1/|g|)$. The results obtained there can be applied here for $\eta(T)$. It is to be noted that the quantum corrections to $\eta(T)$ are determined by the sum over ε_n and ε_n' in the process given by Fig.2. The spatial dependence, $f(x)$, however, is governed by its low energy part. Hence we will be concerned with the modifications on $[DQ^2+2|\varepsilon_n|]^{-1}$ in eq.(2) at small and fixed values of $\varepsilon_n\simeq\pi T$. (DQ^2 of interest is also of comparable order of magnitude.)

First corrections by impurity scattering alone of the type as shown in Fig.5, where dotted lines with crosses represent impurity scattering, lead to the energy dependent diffusion constant in eq.(2) [22],

$$\bar{D}_n = D(1 + \lambda\ln2|\varepsilon_n|\tau) \quad . \tag{15}$$

Next we consider the quantum corrections associated with long range Coulomb interaction $v(q,\omega_\ell)$, which are dynamically screened,

$$v(q,\omega_\ell) = v(q) / [1 + v(q)\pi(q,\omega_\ell)] \quad ,$$

$$= 2\pi e^2 [Dq^2 + |\omega_\ell|]/ q[D\kappa q + |\omega_\ell|] \quad , \tag{16}$$

where $v(q)$ and $\pi(q,\omega_\ell)$ are bare Coulomb interaction and the polarization function, respectively, and $\kappa=4\pi e^2 N(0)$. These corrections are given by Figs.6 and 7, where double wavy lines, broken lines and double broken lines are $v(q,\omega_\ell)$, diffuson and Cooperon, respectively. The processes of Figs.6 yield contributions diagonal in energy variables, i.e. $\varepsilon_n'=\varepsilon_n$, while those of Figs.7 have contributions $\varepsilon_n'=\varepsilon_n$ as well as $\varepsilon_n'\neq\varepsilon_n$. Since only processes with $\varepsilon_n'=\varepsilon_n$ are important for small ε_n, we ignore in the following those $\varepsilon_n'\neq\varepsilon_n$. Contributions from Figs.6 and 7 defined as $-2\pi N(0)\tau^2 L(Q,\varepsilon_n)$ can easily be extracted from our previous calculations of $\delta f(0,T_c)$, eq.(14), and they are given by [23]

$$L(Q,\varepsilon_n) = 2(DQ^2 + 2|\varepsilon_n|) \; T \; {\sum_{\varepsilon_n'}}' \sum_q \frac{v(q,\varepsilon_n-\varepsilon_n')}{(Dq^2+|\varepsilon_n-\varepsilon_n'|)^2}$$

$$+ 2\, T \; {\sum_{\varepsilon_n'}}' \sum_q \left(\frac{v(q,\varepsilon_n-\varepsilon_n')}{Dq^2+|\varepsilon_n-\varepsilon_n'|} - \frac{v(q,\varepsilon_n+\varepsilon_n')}{D(q-Q)^2+|\varepsilon_n-\varepsilon_n'|}\right)$$

$$+ 2\, T \sum_q \frac{v(q,0)}{D(q-Q)^2+2|\varepsilon_n|} \quad . \tag{17}$$

Here the last term in eq.(17) is due to those in Fig.7 with $\varepsilon_n=\varepsilon_n'$, and the summations over ε_n' in the first and the second terms are for the region of $\varepsilon_n\cdot\varepsilon_n'<0$. For small values of DQ^2, $\varepsilon_n \simeq 2\pi T$, eq.(17) yields

$$L(Q,\varepsilon_n) = 2(DQ^2 + 2|\varepsilon_n|)I_1 -DQ^2 I_2 + \phi(\varepsilon_n) \quad , \tag{18}$$

where we employed the similar notation for I_1 and I_2 as those by Castellani et al. [26], who examined the combined effects of the interaction and the randomness for the diffusion propagator,

$$I_1 = 2T \sum_{\omega_\ell>0} \sum_q \frac{v(q,\omega_\ell)}{[Dq^2+\omega_\ell]^2} \quad , \tag{19}$$

$$I_2 = 4T \sum_{\omega_\ell>0} \sum_q \frac{Dq^2 v(q,\omega_\ell)}{[Dq^2+\omega_\ell]^3} \quad . \tag{20}$$

By use of eq.(16), I_1 and I_2 are evaluated as follows to the leading logarithmic orders

$$I_1 = -\frac{\lambda}{4} \ln\tau T \ln(D\kappa^2)^2\tau / T \quad , \tag{21}$$

$$I_2 = -\lambda \ln\tau T \quad . \tag{22}$$

In eq.(18), $\phi(\varepsilon_n)$ is the contribution of the term including $v(q,\varepsilon_n+\varepsilon_{n'})$ in eq.(17) from the region of $|\varepsilon_n|>|\varepsilon_{n'}|>0$ and is given explicitly in ref.23. It is seen that $\phi(\varepsilon_n)=0$ for $n=0$. (For $n\neq0$ $\phi(\varepsilon_n)$ has a logarithmic temperature dependence $\ln D\kappa^2/T$ as well as $\ln T\tau$.)

By use of eqs.(18) we see that $[DQ^2+2|\varepsilon_n|]^{-1}$ is modified as

$$[\bar{D}_n Q^2 + 2|\varepsilon_n| + L(Q,\varepsilon_n)]^{-1} = [D_n Q^2 + 2\varepsilon_n z]^{-1} \equiv D(Q,\varepsilon_n) \quad , \tag{23}$$

where D_n and z are given by

$$D_n = D[1 + \lambda\ln2|\varepsilon_n|\tau + 2I_1 - I_2] \quad . \tag{24}$$

$$z = 1 + 2I_1 \quad , \tag{25}$$

in the lowest order in λ. Following Castellani et al. [26] we may be able to write eq.(23) in the following suggestive form,

$$D(Q,\varepsilon_n) = \zeta^2 / [D_n' Q^2 + 2|\varepsilon_n|] \quad . \tag{26}$$

where $\zeta=1-I_1$, $D_n'=D_n(1-I_2)$. Hence ζ represents the renormalization of overall amplitude and D_n' is a frequency-dependent renormalized diffusion constant. The coherence length $\xi(T)$ determined by eq.(26) is given as

$$\xi(T) = [D(1 + \lambda\ln\tau T)(1 - I_2) / 2\pi T]^{1/2} \equiv \sqrt{D(T)/2\pi T} \quad . \tag{27}$$

The foregoing explicit calculations have been carried out based on the perturbational treatment with respect to randomness parameter, λ, and then the results are valid only if $\lambda|\ln\tau T|< 1$ is satisfied. Hence even though the degree of randomness is small, i.e. $\lambda<<1$, the theory breaks down in the limit of low temperatures. In such low temperatures the scaling theory [24-27] will lead to power law dependences $D(T)\simeq T^\gamma$, $\zeta(T)\simeq T^\delta$, though the correct values of γ and δ are hard to estimate.

CRITICAL CURRENT

The critical current, j_c, in dirty metals has usually been given by [4,10]

$$j_c = A\Delta_n(T)^2 \sqrt{D/T} \exp[-L/\xi(T)] \quad , \tag{28}$$

where A, $\Delta_n(T)$ and L are the constant independent of temperature and the mean free path, ℓ, the pair potential in the normal metal at the interface and the separation between two superconductors, respectively, and $\xi(T)=\sqrt{D/2\pi T}$. Equation (28) is derived under the assumption that $L>\xi(T)$ and $\ell<\xi(T)$, and the normal electrons are free from mutual interaction. Based on the results obtained in the previous sections we see that in the presence of mutual interactions, eq.(28), will be modified as

$$j_c = A\Delta_n(T)^2 \eta(T)^2 \sqrt{D/T} \exp[-L/\xi(T)] \quad , \qquad (29)$$

where $\eta(T)$ is defined by eq.(9) together with eq.(4). The effects of Anderson localization are represented by the modifications of D, which is now dependent on temperature and $\xi(T)$ together with overall renormalization constant, $\zeta(T)$,

$$j_c = A\Delta_n(T)^2 \eta(T)^2 \zeta(T)^2 \sqrt{D(T)/T} \exp[-L/\xi(T)] \quad , \qquad (30)$$

where $\zeta(T)$, $D(T)$ and $\xi(T)$ are given by eqs.(26) and (27) in the region of weak localization. This expression, however, will hold even in the strongly localized regime as far as $\zeta(T)$, $D(T)$ and $\xi(T)$ arc properly understood. Hence the temperature dependence of j_c will result from various factors, whereas the length dependence is governed by $D(T)$ only.

DISCUSSIONS

We have investigated the effect of Anderson localization on the proximity effect by focusing on the Cooper pair correlation function in two-dimensional systems in view of the recent realization of MOS type three-terminal Josephson devices. We also examined how the critical current is affected by the Anderson localization. One of the motivations of the present research is that verry little has been known theoretically about the Josephson tunneling across very dirty metals near the (effective) mobility edge on one hand and that the switching will be more effective in dirty metals on the other hand since the change of the diffusion constant by the gate voltage is larger. This will be understood by the schematic representation of the carrier number dependence of the diffusion constant shown in Fig.8. The same effect can now be pursued at high temperatures if oxides are used and in this case even thinner junction is of interest.

ACKNOWLEDGEMENT

The author thanks S. Maekawa and H. Ebisawa for useful discussions in early stage of the study. He also thanks H. Takayanagi, T. Kawakami and K. Inoue for informative discussions.

REFERENCES

1. P.G. de Gennes, Rev . Mod. Phys. 36: 225 (1964)
2. G. Deutscher and P.G. de Gennes, "Superconductivity" ed. R.D. Parks (Marcel Dekker, 1969) p.1005.
3. J. Clarke, Proc. Roy. Soc. London Ser. A., 308: 447 (1969).
4. J. Seto and T. Van Duzer, Proc. 13 Intern. Conf. on Low Temperature Physics, ed. W. O'Sullivan, K. Timmerhaus and E. Hammal (Plenum, 1972) vol.3 p.328.
5. H.J. Fink, Phys. Rev. B14: 1028 (1976).
6. K.K. Likharev, Phys. Rev. Phys. 51: 101 (1979).
7. T.D. Clark, R.J. France and A.D.C. Grassie, J. Appl. Phys. 51: 2736 (1980).

8. R.B. van Dover, A. de Lozanne and M.R. Beasley, J. Appl. Phys. 52: 7327 (1981).
9. H.C. Yang and D.K. Finnemore, Phys. Rev. B30: 1260 (1984).
10. V. Kresin, "Josephson Effect" : Achivements and Treds edited by A. Barone (World Scientific, 1986); Phys. Rev. B34: 7587 (1986).
11. H. Takayanagi and T. Kawakami, Phys. Rev. Lett. 54: 2449 (1985).
12. T. Kawakami and H. Takayanagi, Appl. Phys. Lett. 46: 92 (1985).
13. H. Takayanagi, Ph.D. Thesis (Univ. of Tokyo, 1987).
14. T. Nishino, M. Miyake, Y. Harada and U. Kanbe, IEEE Elect. Device Lett. 6: 297 (1985); Phys. Rev. B33: 2042 (1986).
15. T. Ando, A Fowler and F. Stern, Rev. Mod. Phys. 54: 437 (1982).
16. "Anderson Localization" (Springer, 1982) edited by Y. Nagaoka and H. Fukuyama.
17. "Localization, Interaction, and Transport Phenomena" (Springer, 1984) edited by B. Kramer, G. Bergmann and Y. Bruynseraede.
18. P.A. Lee and T.V. Ramakrishnan, Rev. Mod. Phys. 57: 287 (1985).
19. A.L. Altshuler and A.G. Aronov, "Electron-Electron Interaction in Disordered Systems" (North Holland, 1985) edited by L. Efros and M. Pollak p.1, H. Fukuyama, ibid p.155.
20. S. Maekawa and H. Fukuyama, Physica 107B: 123 (1981); J. Phys. Soc. Jpn. 51: 1370 (1982).
21. H. Fukuyama, Physica 126B+C: 306, (1984); ibid 135B: 458 (1985).
22. S. Maekawa, H. Ebisawa and H. Fukuyama, J. Phys. Soc. Jpn. 52: 1352 (1983).
23. H. Fukuyama and S. Maekawa, J. Phys. Soc. Jpn. 55: 1814 (1986).
24. A.M. Finkelstein, Sov. Phys. -JETP 57: 97 (1983); ibid 59: 212 (1984); Z. Phys. B56: 189 (1984).
25. C. Castellani and C. Di Castro, Varenna Summer School oh Highlights of Condensed Matter (1983) ed. by Bassain, F. Fumi and M. Tosi (North Holland, 1985).
26. C. Castellani, C. Di Castro, P.A. Lee and M. Ma, Phys. Rev. B30: 527 (1984).
27. C. Castellani, C. Di Castro, M. Ma, S. Scorella and E. Tabet, Phys. Rev. 33: 6169 (1986).

SUPERCONDUCTIVITY AND DISORDER IN LOW CARRIER DENSITY La-S

A. Kapitulnik, A.D. Kent, T.H. Geballe and J.H. Kaufman*

Dept. of Applied Physics, Stanford Univ., Stanford CA 84305
*IBM Almaden Research Center, San Jose CA 95120

Abstract

La-S films near the metal-insulator transition were prepared and characterized. Localization and percolation effects were observed in the films and the interplay between them was studied.

Introduction

Low carrier density superconductors provide a unique opportunity to study disorder effects on superconductivity. The reason is that in most of these superconductors it is possible to vary the carrier density as well as the disorder in a controlled way. Thus, in the language of localization phenomenna[1], we can either vary the disorder, i.e. the mobility edge in order to come close to the metal-insulator transition, or vary the Fermi energy for a fixed amount of disorder. This is important in particular because it is not clear that varying the disorder by known methods (e.g. alloying with an insulator or irradiation damaging the material) does not change the microscopic parameters of the material, hence bringing us near the mobility edge in an unknown trajectory.

La_3S_4 is a low electron density superconductor[2-5] in which the electron density can be varied by small changes in the composition while maintaining the same structure and lattice constant[2]. Studies of La-S and analogous lanthanum chalcogenides has already shed much light on the nature of superconductivity in low carrier density materials. These properties which make them ideal for the study of superconductivity near the metal-insulator transition are also useful in thin film applications. Here, the ability to tune the electronic properties and preserve the lattice constant and crystal structure might make possible novel synthetic multilayered structures and devices.

In the Th_3P_4 cubic structure (γ phase) of La-S there is a continuous series of solid solutions between La_3S_4 (a superconductor with $T_c \sim 8K$) and La_2S_3 (a semiconductor with $E_\mu \sim 3eV$). In fact, the entire range of solid solutions corresponds to only a 3 at.% change in the elemental composition. It was first pointed out by Holtzberg et al.[?] that the constancy of structure with changes in electron concentration indicates that the metallic binding in these compounds is negligible suggesting a very simple electronic structure. Thus the material is primarily an ionic compound with only little covalence binding, and we can simply balance the La (3+) and S (2-) to obtain a 1 electron per formula unit imbalance for La_3S_4 which constitutes the conduction band. Varying the composition from the superconducting La_3S_4 compound corresponds to introducing vacancies at a fraction of the La sites . To include the presence of vacancies explicitly the formula unit is usually written as:

$$La_{3-x} [\ ]_x S_4$$

where the $[\ ]_x$ corresponds to a vacancy concentration x. In the semiconducting composition La_2S_3 there is on the average 1/3 vacancy per formula unit (i.e. x=1/3) , or in the ionic picture, the La(3+) together with the negatively charged vacancies , balance the S (2-).   The electron density in the conduction band can be written in terms of x as:

$$n = 6 \times 10^{21} x \, (\, 1 - 3x \,) \text{ electrons/cm}^3$$

The vacancies are believed to be distributed at random (no superlattice peak was observed in the x-ray diffraction pattern to indicate vacancies ordering). Fig.1. illustrates this effect in the optical absorbtion spectrum of thin films of La_2S_3 and La_3S_4 showing no major differences between the insulating and the superconducting films, and in particular the 3eV absorbtion edge is apparent in both of them. The transition into a metallic state is accompanied by a big increase in the oscillator strength in the gap as expected. The plasma frequency is easily identified as ~1eVwhich is the expected value from the free electron model using an effective mass of four times the electron mass[2].

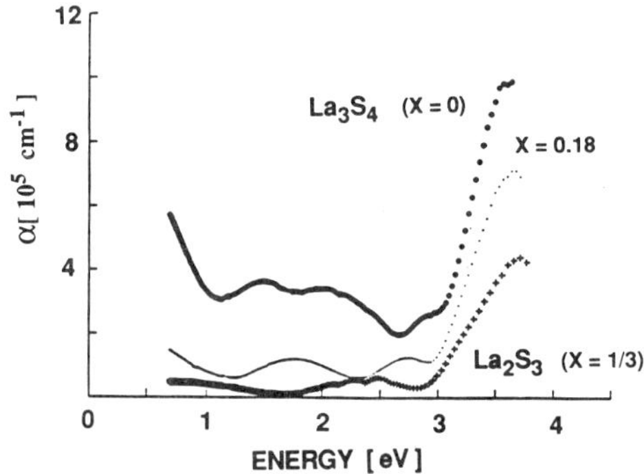

Fig. 1:   Absorbtion spectra of $La_{3-x}S_4$ at room temperature.

Near the metal-insulator transition , small changes in composition δx will have little effect on the number of vacancies (i.e. $6 \times 10^{21}(1/3 - \delta x)$) but a comparitively large effect on the number of electrons in the La d-band $6 \times 10^{21} \delta x$ .The vacancies introduce a strongly fluctuating random potential due to their negative charges, hence the electron density can be varied virtually independently of the disorder, realizing the situation for localization of electron states due to disorder[6], first introduced by Anderson[7]. Adding superconductivity to this picture brings another effect that should coexist with this random potential. Evidently the highest T_c superconducting composition corresponds to x=0.

In this paper we discuss the superconducting properties of thin polycrystalline films of La-S. We show that indeed the properties of the films are similar to those of the single crystal bulk previously grown and reported. The metal-insulator transition ocurs at a finite x, larger than 1/3. The disappearance of superconductivity is associated with this metal-insulator transition and in fact suggests a semiconductor to superconductor transition. We further show that the columnar way in which these samples grew, and the high resistance grain boundaries introduce percolation effects into the films, resulting in a very rich superconducting behaviour near the metal-insulator transition.

Films Preparation and structure

Films of La-S have been prepared by reactive d.c. planar magnetron sputtering of La in a H_2S and Ar atmosphere[8]. The Ar is used to maintain the plasma above the La source while the H_2S is inserted near the substrates to promote the compound formation at the substrate surface. This effectively decouples the reaction and sputtering plasma regions which allows higher sputtering rates and better control of the film stoichiometry. A schematic of the sputtering aparatus is shown in Fig.2. The reactive species flows through the gas ring, producing an even flow of gas across the substrate surface. This is necessary to produce homogeneous single phase films. The stoichiometry is controlled by setting the ratio of the La rate to H_2S partial pressure.

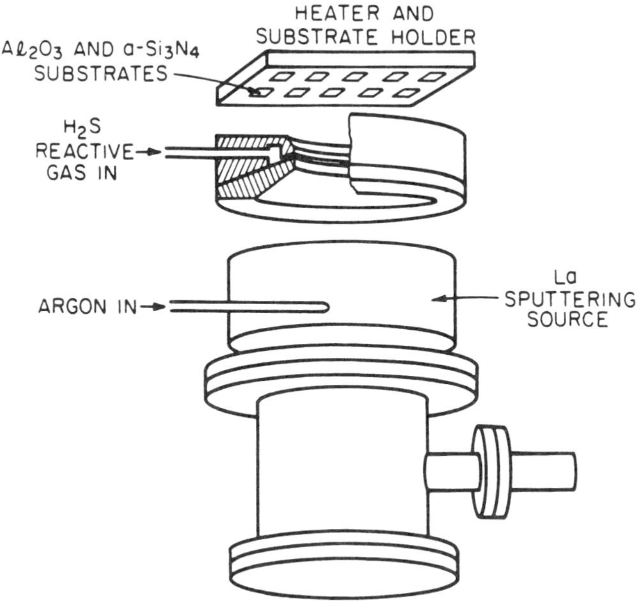

Fig.2:   Schematics of the sputtering system.

The La-S films were grown both on Al_2O_3 (1102) orientation and amorphous Si_3N_4 substrates. The substrate temperature during deposition was 300°C and played a crucial role in the formation of the Th_3P_4 phase. At this temperature we were able to obtain the highest mobility samples and x-ray analysis showed only the presence of the Th_3P_4 phase. Complementary TEM analysis were performed in order to determine the structure and grain size. Close to the La_3S_4 composition the grain size was 250Å, while near the La_2S_3 stoichiometry the grain size was reduced to 50Å. This happened because in order to increase the Sulfur concentration we had to increase the H_2S partial pressure hence reducing the La surface mobility. The electron diffraction pattern indicated the presence of randomly oriented crystallites of γ phase La-S with no additional second phase material. Cross sectional SEM pictures of the thicker films indicated that the grains are predominately columnar. On the scale we could measure there was no change in the lattice constant (a = 8.75Å) as we varied the film composition. Also the films grown on different substrates showed no observable effect of the substrate on the grain size or sample orientation[8].

Localization and Superconductivity

As discussed above, the La-S system is excellent for studying the effects of strong disorder on superconductivity. In particular, since the electronic description was shown to be very simple, and since it was previously shown by Holtzberg et al.[2] that the pure material in almost all composition range,behaves like a BCS superconductor, it is our hope that a single electron description will suffice to explain the experimental results. The theory of Dirty superconductors and Anderson localization was worked out by Kapitulnik and Kotliar[9], Ma and Lee[10], and Bulaevskii and Sadovskii[11]. The main prediction that come out of these theories is that in principle, superconductivity might persist even below the metal-insulator transition, if there are enough states in the gap to build a pair wave function. On the other hand it was shown by Kapitulnik and Kotliar [9] that close to the mobility edge there is a strong renormalization of the zero temperature coherence length and the breakdown of the Ginzburg criterion. Thus, strong fluctuations wil cause the phase transition to change its nature and T_c will degrade substantially.

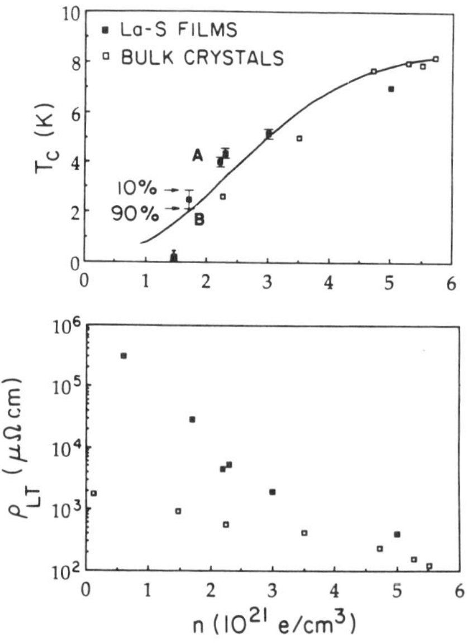

Fig.3: Supercontucting transition temperature (upper curve) and normal state resistivity (lower curve) vs. the carrier concentration. The bulk data are taken from Ref. 4.,solid line is a fit to the theory of Seiden (Ref. 12)

Superconductivity in La-S was studied before and was shown to dissapear very close to the x=0.26 . The equivalent carrier density is ~$1.32x10^{21}$ electrons/cm^3. In the case of our films, near the metal-insulator transition, the grain size is ~50Å, only six times the unit cell size. The carrier density that corresponds to the disappearance of superconductivity in our polycrystalline films was ~$1.45x10^{21}$. This slight difference is attributed to additional disorder introduced by the polycrystallinity of the sample. Since the carrier density in Ref. 4 was calculated from the composition and not measured , some discrepancies, especially near the metal-insulator transition might appear. Clearly more work should be done to clarify this important point. Fig.3 depicts the variation of Tc as the number of carriers was changed. The electron density was inferred from a low temperature (~ 20K) Hall measurements. This varied, depending on the composition from 0.6x10^{21} to 5x10^{21} . All the transitions were fairly sharp as indicated by the 10% and 90% points on the graph.

Of particular interest is the last point showing the transition temperature is ~0.2K. We could not measure temperatures below 0.34K. Thus, we inferred Tc from measuredments of the magnetoresistance of this sample[13,14]. Fig. 4 shows the magnetoresistance of the sample at 4.2K and 0.35K. One sees very large effect at the lowest temperature, indicative of magnetic field reduced superconducting fluctuations. It was pointed out by Kapitulnik et al.[13] that at a field H*, the ghost critical field, all the fluctuations will be quenched, hence the magnetoresistance will go down for higher fields. Knowing the normal state resistivity of the sample, the effective mass and the electron density, we could calculate Tc to be - 0.2±0.1 .The result is of course only an approximation, but it gives a good idea on the rate that the superconductivity dissapears at this range of carrier density. In fact it is easily seen that for higher temperatures this sample behaves like a semiconductor with an activation energy of ~2.6K . This gives a localization length of ~130Å. It is interesting to note that this length is of the order of the BCS coherence length for this material. If indeed the metal-insulator transition in this material is of the Anderson like, It was argued that one should be able to observe superconductivity in the insulating side as long as the Ginzburg-Landau coherence length, (which is shorter than the BCS coherence length), is smaller than the width of the wavefunctions, i.e. the localization length. In fact, It can be argued that in general, if the relevant metal-insulator transition is of the second order type, and superconductivity exists in the region where the Cooper pair size is smaller than the localization length in the metallic side (the length over which the amplitude of the wave function fluctuates), it should persist also in the equivalent length scales region in the insulating side[9]. The incorporation of interactions into the theory of the metal-insulator transition should not change this result as long as the later is characterized by a diverging localization length.

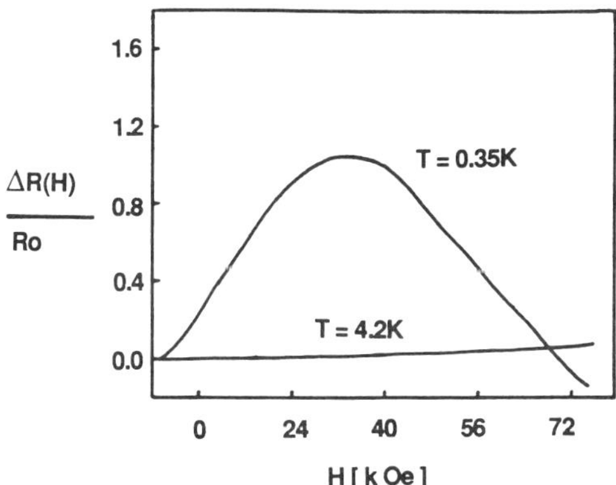

Fig.4:   Magnetoresistance of the measured sample with the lowest carrier density that showed superconductivity (see text).

Fig.5 shows a series of three samples near the metal-insulator transition the bottom curve shows superconductivity at ~2K. The intermediate curve is slightly higher resistance and it shows semiconducting behaviour above 4.2K. The upper curve is deeper into the semiconducting region and the localization length inferred from it is ~50Å. The fact that this behaviour of resistance versus temperature is continuous from the samples that exhibit superconductivity through the samples that show semiconducting behaviour, covinced us that indeed the sample shown in Fig.4 is below the mobility edge.

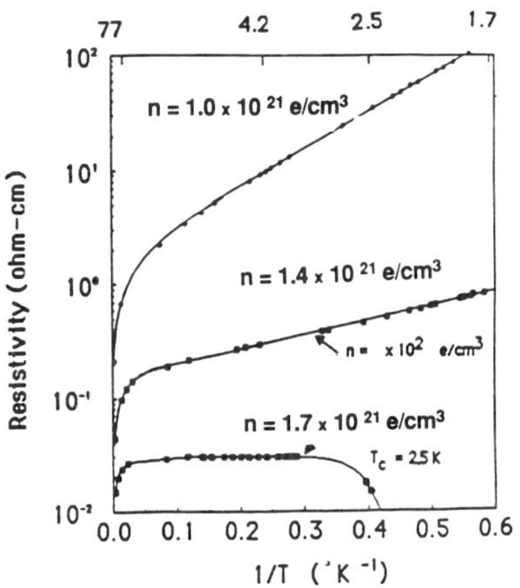

Fig.5:   Resistive measurements near the metal-insulator transition

The validity of our description idepends on if the samples are homogeneous on the relevant length scales. Especially we are concerned with the validity of the Anderson transition descriptin in the presence of percolation effects[15-17] as discussed in the next section. In this system small compositional changes can cause big differences in metallic and superconducting behaviour. Nevertheless standard structural and compositional analysis might not be sufficient. The superconducting properties are a very sensitive probe to the film homogeneity. All the superconducting transitions were very sharp also in the presence of a magnetic field up to 10T. The very dirty samples, near the metal-insulator transition (~x=0.26) are on the other hand broader but the magnetic field does not broaden them anomalously. This means that all the samples do not have weak links that connect below T_c. In fact, a more stringent criteria for film homogeneity obtained from specific heat measurements. The specific heat for x=0.18, 2μm thick film is shown in Fig. 6. below[8].

66

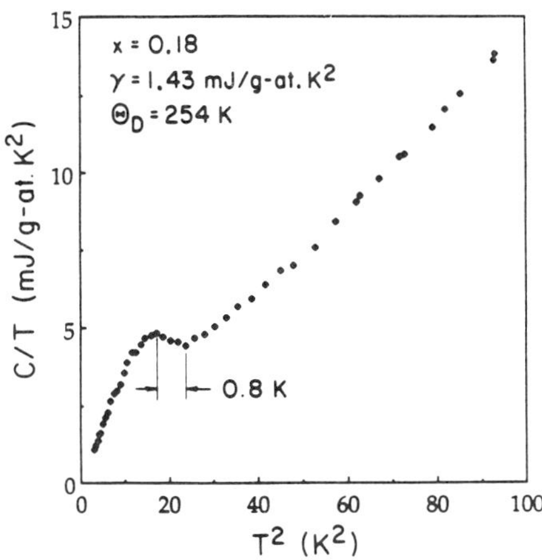

Fig. 6:   Specific heat of x=0.18 sample plotted vs. T^2.

Analysis of this data indicates bulk superconductivity with electronic coefficient γ=1.43 mjoule/g-atK^2, and Θ_D=254K, both are very close to the bulk values. The relative sharpness of the transition (0.8K) indicates that the sample is homogeneous on the scale of the coherence length. The film had a normal state resistivity of 1600$\mu\Omega$-cm which is very high. Statistical concentration-variations-induced-changes in the local Tc, would predict a narrower width of the heat capacity transition (~0.25K) .

Percolation effects

As discussed above, the sputtered films were not single crystals and in fact showed columnar growth. Thus, inhomogeneities , especially very close to the metal-insulator transition were unavoidable. The inhomogeneities in these films we believe are associated with the grain boundaries. This may be due to chemical effects which favor the formation of a greater concentration of La vacancies at the grain boundaries[8]. The precise nature of these inhomogeneities is unknown. The transport properties however provide important clues to the effect of these inhomogeneities. We already argued that the normal state resistivity showed evidence of localization phenomena, by which we mean negative temperature coefficients of resistance are observed in all cases where $\rho >$ 800$\mu\Omega$-cm .The superconducting transition is on the other hand very sharp with no evidence of a double transition due to a superconducting percolation process via Josephson junctions[18]. The resistive transition broadens only slightly in an applied field as shown in the inset of Fig.7. Nevertheless the width remains of constant breadth as the applied field is increased. Note the rise in resistance at low temperatures in high field when the superconducting transition is suppressed.

This suggest that superconductivity is established in regions which are connected and that Josephson interaction play a small role except for posibly in zero field. This analysis substantiate our claim in the former section that the localization effects are incipient. To detect those inhomogeneities effects we have employed critical field measurements. The upper critical field of a high resistivity film is shown in Fig.7, for the applied field oriented both parallel and perpendicular to the film surface. The transition temperature was taken to be the midpoint of the resistive transition.

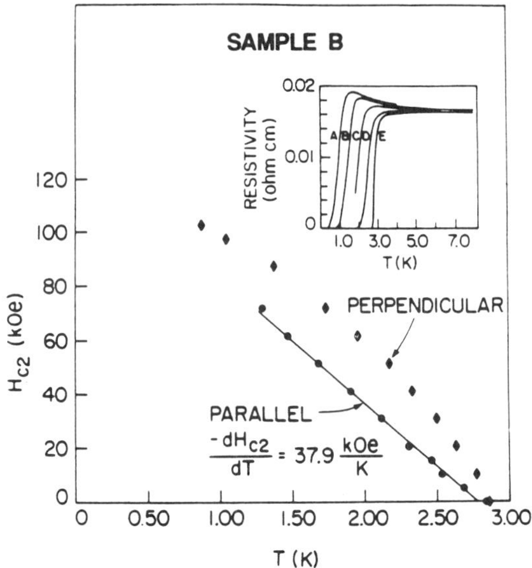

$$\frac{-dH_{c2}}{dT} = 37.9 \ \frac{kOe}{K}$$

Fig.7: H_{c2} data for sample B (see Fig. 2). Inset shows the resistive transitions in applied fields: A Through E are: 10.3T, 8.2T, 6.2T, 3.1T and zero field respectively.

The first thing to note is that these results are the reverse of the usual behaviour observed in a thin film. Here the parallel critical field curve is linear and the perpendicular critical field curve is larger and has a curvature such that it looks as if it comes to Tc with infinite slope. It is interesting to note that this effect is very common to disordered and to low carrier density superconductors when local fluctuation in composition or carrier density can result in substantial local degradation of Tc. In fact, it is interesting to note that it was argued by Sadovskii and Bulaevskii[19], following the analysis of Kapitulnik and Kotliar[9] that incipient inhomogeneities in the superconducting order parameter will emerge in Anderson localization -type superconductors as a direct consequence of the strong superconducting fluctuations near the metal-insulator transition. The angular dependence of H_{c2} has a smooth dependence on angle near the maximum when the film is perpendicular. The angular dependence is adequatly fitted by the usual expression for three dimensional anisotropic superconductors. The rounded angular dependence excludes the possibility of surface superconductivity.

The anisotropy can be easily understood in terms of a percolation process[18] In general, $H_{c2} = \Phi_0/(2\pi\xi_s^2(T))$. where Φ_0 is a flux quantum and $\xi_s(T)$ is the superconducting coherence length. For the parallel field, the curve implies that the superconducting regions are strongly coupled and the usual linear dependence on temperature is recovered. Fig.8 shows the slope analysis of H_{c2} vs. temperature as we approach Tc for two samples that are fairly close to the metal-insulator transition, denoted by A and B in Fig.2. One clearly sees a crossover in the slope towards a slope smaller than unity[20], as we go slightly above Tc for sample A and in sample B it looks as if H_{c2} approaches Tc all the way with an exponent of 0.73.

If transport in the material is indeed governed by percolation, the diffusion on length scales smaller than the percolation correlation length should depend on the length scale[21] as $D(L)\sim L^{-\theta}$. (in two dimensions which is therelevant dimensionality to our problem because of the columnar growth, $\theta\cong0.8$). Using the time dependent Ginzburg-Landau formalism, we can rewrite the renormalized coherence length in the region where the percolation correlation length,ξ_p, is larger than ξ_s, as - $\xi_s \cong [d^\theta \xi_s^2]^{1/(2+\theta) \ 22,23}$. Solving this for H_{c2} we get that $H_{c2} \propto [T - Tc]^{2/(2+\theta) \ 23}$. In 2D the exponent becomes ~0.7., very close to what is observed experimentally. Of course, close enough to Tc the material should look homogeneous again as $\xi_s > \xi_p$. Thus the linear dependence on temperature is recovered. Since the percolation process is only in the film's plane, the parallel field should not be different from a non columnar films. Samples that are more percolative and presumably are closer to the metal-insulator transition, will exhibit this self-similar behaviour closer to Tc[18,24].

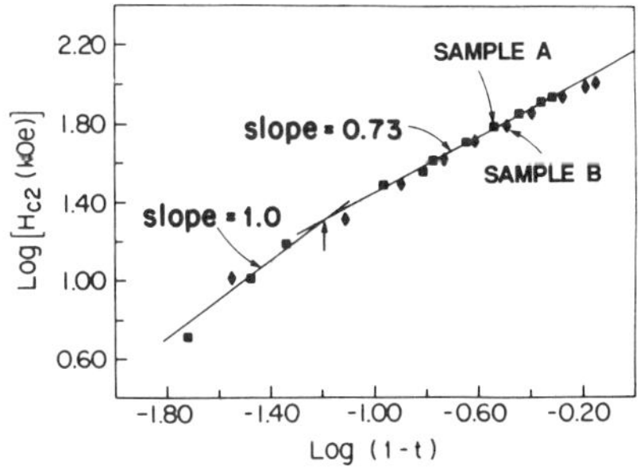

Fig.8:   Log-log plot of H_{c2} vs. reduced temperature for samples A and B (see Fig.2).

The crossover point will roughly give an estimation of the percolation correlation length up to a numerical factor. For these experiments to make sense, it seems that this prefactor should be greater than one (of the order of 10)[20]. In the case of sample B the lower bound for the crossover point was found to be larger than ~5.5 grains times that numerical factor.

Discussion

We have tried to show above some experimental results that might shed some light on the behaviour of superconductors near the metal-insulator transition. Although the La-S system seems to be a good model system to study superconductivity near the Anderson transition, the apparent inhomogeneities make any definitive conclusion somewhat premature. We have shown that over wide range of carrier densities Tc of the films follow the Tc of the bulk samples measured elsewhere although the normal state resistivity is as high as two orders of magnitude of that of the equivalent bulk samples. This give an excellent manifestation of the Anderson theorem[25]. Close enough to the metal insulator transition, Tc of the films degrades very fast and it is not clear whether superconductivity disappears above or below the metal insulator transition. In fact, resistive measurements above the superconducting transition of the samples with the highest resistivity show signs for activation type of transport. A reasonable localization length can be extracted that agrees with the proximity to the metal insulator transition.

The apparent percolation effects do not seem to affect the localization -superconductivity interplay. In fact the grains are still large enough such that localization effects exist over relatively small number of grains. There as shown by our heat capacity measurements, the material is homogeneous. Certainly , further measurements are needed, especially in the region near the disappearance of superconductivity to decide definetly whether we have in this material a semiconductor to superconductor transition.

In particular lower temperatures measurements are needed.

To conclude, it is worth noting that preliminary tunneling measurements indicate that this material is a weakly coupled superconductor with $2\Delta/kTc \cong 3.55$. Fig.9 is a typical tunneling data showing a ahrp S-I-S junction with Pb as a counter electrode. The barrier was made by flushing the film with H_2S before taking it out of the chamber hence producing La_2S_3 phase at the surface, which seems to be an excellent natural barrier .

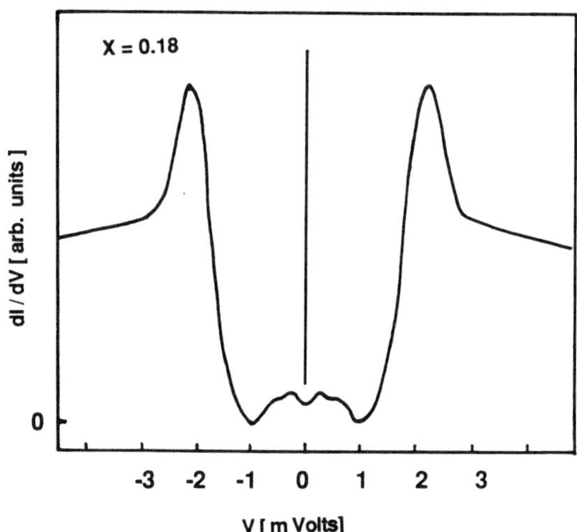

Fig.9:   Tunneling with Pb counter electrodefor sample with 3.5×10^{21} carriers/cm^3. Note the gaps difference inside the gap.

More studies to decide on the strength of the coupling in the various ranges of carrier density are underway.

Acknowledgements

We wish to thank B. Oh for the specific heat measurements and A. Marshall for the TEM study. This work was supported by a grant from NSF/DMR-85-19753 and by the Air Force Office of Scientific research. A.K wishes also to acknowledge the Alfred P. Sloan Fellowship. Materials were prepared and characterized at the center for materials research at Stanford, supported in part by the NSF through the MRL program.

References

1.   For a review, see e.g. P.A. Lee and T.V. Ramakrishnan, Rev. Mod. Phys. **57**, 287 (1985).
2.   F. Holtzberg, P.E. Seiden, and S. von Molnar, Phys. Rev. **168**, 408 (1968).
3.   R. M. Bozoroth, F. Holtzberg, and S. Methfessel, Phys. Rev. Lett. **14**, 952 (1965).
4.   K. Ikeda, K.A. Geschneider, B.J. Beaudry, and U. Atzmony, Phys. Rev. B **25**, 4604 (1982).
5.   K. Westerholt, F. Timmer, and H. Bach, Phys. Rev. B **32**, 2985 (1985).
6.   M. Cutler and N.F. Mott, Phys. Rev. **181**, 1336 (1969).
7.   P.W. Anderson, Phys. Rev. **109**, 1992 (1958).
8.   A.D. Kent, B. Oh, T.H. Geballe, and A.F. Marshall, Materials Lett. **5**, 57 (1987).

9. A. Kapitulnik and G. Kotliar, Phys. Rev. Lett. 54, 473 (1985). and Phys. Rev. B 33, 3146 (1986).
10. M. Ma and P.A. Lee, Phys. Rev. B **32**, 5658 (1985).
11. L.N. Bulaevskii and M.V. Sadovskii, Pis'ma Zh. Eksp. Teor. Fiz. **39**, 524 (1984) [JETP Lett. **39**, 640 (1984)]; J. Low Temp. Phys. **59**, 89 (1985) .
12. P.E. Seiden, Phys. Rev. **168**, 403 (1968).
13. A. Kapitulnik, A. Palevski, and G. Deutscher, J. Phys. C **36**, 1305 (1985).
14. W. L. Mclean and T. Tsuzuki, Phys. Rev. B **29**, 503 (1984).
15. G. Deutscher, A.M. Goldman, and H. Micklitz, Phys. Rev. B 31, 1679 (1985).
16. Y. Imry and M. Strongin, Phys. Rev. B 24, 6353 (1981).
17. Y. Shapira and G. Deutscher, Phys. Rev. B 27, 4463 (1983).
18. O. Entin-Wohlman, A.Kapitulnik, and Y.Shapira, Phys. Rev. B 24, 6464 (1981).
19. L.N. Bulaevskii and M.V. Sadovskii, Pis'ma Zh. Eksp. Teor. Fiz. 43, 76(1986) [JETP Lett. 43, 99 (1986)].
20. G. Deutscher, I. Grave, and S. Alexander, Phys. Rev. Lett. 48, 1497 (1981).
21. Y. Gefen, A. Aharony, and S. Alexander, Phys. Rev. Lett. 50, 77 (1983).
22. S. Alexander, Phys. Rev. B 27, 1541 (1983).
23. O. Entin-Wohlman, A.Kapitulnik, S. Alexander, and G. Deutscher, Phys. Rev. B 30, 2617 (1984).
24. A.D. Kent, A. Kapitulnik, and T.H. Geballe, preprint (1987).
25. P.W. Anderson, J. Phys. Chem. Solid 11, 26 (1959).

BOUNDS ON SUPERCONDUCTING PROPERTIES

IN ELIASHBERG THEORY

J.P. Carbotte

Physics Department
McMaster University
Hamilton, Ontario   L8S 4M1

ABSTRACT

We establish bounds on some of the properties of an Eliashberg superconductor. It is shown that the normalized specific heat jump at T_c and the thermodynamic critical field deviation function, are bounded above by a value not much greater than is observed in some known superconductors. Also, a lower limit exists for the dimensionless ratio $\gamma(0)T_c^2/H_c^2(0)$. Here $\gamma(0)$ is the Sommerfeld constant, T_c is the critical temperature and $H_c(0)$ the zero temperature thermodynamic critical field. On the other hand, the ratio of twice the gap edge to T_c can rise to 11.5 and the reduced upper critical magnetic field in the clean limit to 1.5. Finally, some results are presented for the very strong coupling limit by which we mean that T_c is taken to be comparable in size to a typical phonon energy.

Introduction

In contrast to what is the case for conventional superconductors[1-4], the mechanism responsible for superconductivity in the high T_c oxides is, as yet, not firmly established[5-9]. The observation that there is no isotope effect[10] in some members of this family and the very large scale for the size of T_c as compared with ordinary systems for which $T_c < 23.2(Nb_3Ge)$, is important evidence against the electron-phonon interaction. On the other hand, there are many papers in the literature, both experimental[11-13] and theoretical[8,14], which would indicate that the electron-phonon interaction does play some role in, for example, $La_{1.85}Sr_{0.15}CuO_4$ although it is always possible that an additional mechanism is responsible for the very large observed values of T_c.

In view of the above remarks, it is clearly of interest to understand any limit on superconducting properties that might exist as a result of the mathematical structure of the strong coupling Eliashberg equations themselves[4] which would be quite independent of the assumed strength of the electron-phonon coupling. Eliashberg results in the very strong coupling limit which obtains when we assume T_c to be of the same order of magnitude as a typical phonon energy, are also of importance. In this paper we review some recent work which deals with both these topics.

<u>Thermodynamics</u>

The finite temperature Eliashberg equations, from which the thermodynamics follows, are two coupled non-linear equations for the Matsubara pairing energy $\tilde{\Delta}(i\omega_n)$ and renormalized frequency $\tilde{\omega}(i\omega_n)$,[3,4] which depend only on the electron-phonon spectral density $\alpha^2F(\omega)$ and on the coulomb pseudopotential μ^*. In this work we are not directly interested in the specific form of the electron-phonon spectral density for any particular material. Rather we would like to treat it as an arbitrary function to be chosen at will. We think of it as characterized by some shape and by an overall strength. As a single measure of strength it will be convenient, in what follows, to use the

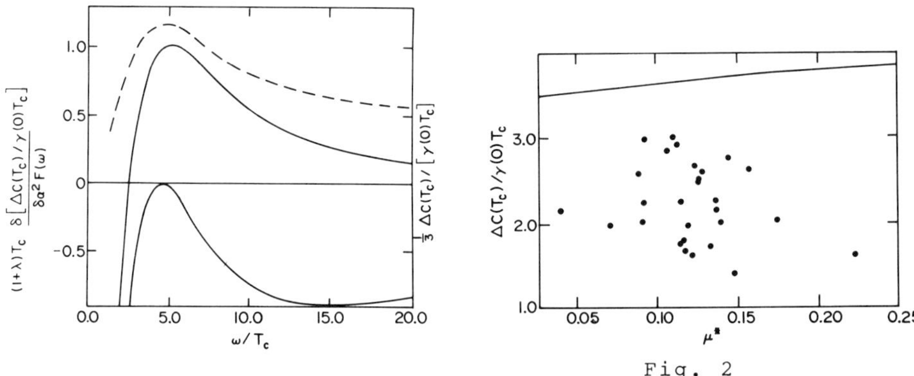

Fig. 1

Fig. 1 The functional derivative of the specific heat jump
        at T_c normalized to $\gamma(0)T_c$ where $\gamma(0)$ is the Sommer-
        feld constant. The upper solid curve is for Pb. The
        value of $\frac{1}{3}\Delta C(T_c)/(\gamma(0)T_c)$ (right hand label) for a Ein-
        stein spectrum as a function of Ω_E/T_c (dashed curve).
        The functional derivative (lower solid curve) for the
        case of a delta function spectrum with Ω_E^* the
        frequency of the maximum in the dashed curve.
Fig. 2 The maximum possible value for $\Delta C(T_c)/(\gamma(0)T_c)$ as a
        function of μ^*. The solid dots represent theoretical
        values for the following materials in order of
        decreasing value of $\Delta C(T_c)/(\gamma(0)T_c)$: $Pb_{.7}Bi_{.3}$, $Pb_{.65}$
        $Bi_{.35}$, $Pb_{.8}Bi_{.2}$, $Pb_{.9}Bi_{.1}$, Pb, $Pb_{.8}Tl_{.2}$, Nb_3Sn, Nb_3Al,
        Nb_3Ge, $Pb_{.6}Tl_{.4}$, Hg, $Pb_{.75}Bi_{.25}$, $Pb_{.4}Tl_{.6}$, V_3Ga, $Pb_{.5}$
        $Bi_{.5}$, La, Ga(am), Bi(am), V_3Si, Mo(am), Nb, In, $Tl_{.9}$
        $Bi_{.1}$, Tl, Sn, Ta, V, Al(BCS).

area under $\alpha^2F(\omega)$ which we denote by A. The question we address and attempt to partially answer is the following. Given an arbitrary choice for $\alpha^2F(\omega)$, are there bounds on predicted superconducting properties that exist because of the mathematical form of the Eliashberg equations themselves as well as the corresponding mathematical prescriptions for a given property in terms of the gap solutions? It is clear that, in such an approach, we need not concern ourselves directly with additional bounds that might exist in real materials due to, for example, lattice instability.

To calculate the specific heat jump $\Delta C(T)$ at temperature T we need the free energy difference $\Delta F(T)$ between normal and superconducting state which is given by a standard formula [3,4] that depends only on $\tilde{\Delta}(i\omega_n)$, $\tilde{\omega}(i\omega_n)$ and the electronic density of states at the Fermi energy. This last factor drops out of normalized jump $\Delta C(T_c)/\gamma(0)T_c$ with $\gamma(0)$ the Sommerfeld constant. Since in this work we fix μ^* to a few typical values, the only remaining parameter is the function $\alpha^2F(\omega)$ which we keep arbitrary.

Our approach to a search for limits on $\Delta C(T_c)/[\gamma(0)T_c]$, independent of the shape and strength of the spectral density, is suggested from a consideration of functional derivatives[15-16]. The functional derivative of the normalized specific heat jump with respect to the electron-phonon spectral density gives the normalized change of the jump when the spectral density is augmented at some specific frequency by an infinitesimal delta function. The larger the functional derivative the more effective are these phonon modes in the specific heat jump. Numerical results, based on work by Marsiglio et al[16] are given in Fig. 1 for the case of a Pb base spectrum (upper solid curve). We see that $\delta(\Delta C(T_c)/\gamma(0)T_c/\delta\alpha^2F(\omega)$ is positive but small at high frequencies, increasing gradually as ω is lowered until a maximum is reached for $\omega/T_c \cong 5.0$. After the maximum, the functional derivative drops sharply through zero and tends towards minus infinity in the limit $\omega\to0$. This singularity can be traced directly to the normalization $\gamma(0)$ which contains a factor $(1 + \lambda)$. Here λ the electron mass enhancement and its functional derivative is proportional to $1/\omega$. Marsiglio et.al.[16] have calculated additional curves based on several other materials besides Pb and have found that the shape of the functional derivative is universal. This observation suggests that for realistic spectra with fixed A, the specific heat jump can be increased through the transfer of some weight from a general frequency to the frequency at the maximum in the functional derivative. This suggests, but does not prove, that for a given A we can maximize the jump if we use a delta function spectral density with Einstein frequency chosen to coincide with the optimum frequency in its own functional derivative.

Results for $\Delta C(T_c)/\gamma(0)T_c$ in the case of a delta function base spectral density of the form $\alpha^2F(\omega) = A\delta(\omega - \Omega_E)$ are shown as the dashed curve in Fig. 1 for which the right hand scale applies. As we expected on the basis of our previous functional derivative arguments, this curve displays a maximum at a frequency $\Omega_E^*/T_c \cong 4.55$. It is important at this point to give a few more technical details about our calculations. First, the calculations of Fig. 1 are for $\mu^* = 0.052$ although we have considered other values. More importantly the results are quite independent of A, the strength of the model electron-phonon spectral density. It can be shown[17] that for a delta function the specific heat jump is independent of A and is only a function of Ω_E/T_c and μ^* i.e. $\Delta C(T_c)/\gamma(0)T_c = K(\Omega_E/T_c, \mu^*)$. Thus for a delta function spectrum and fixed μ^*, there is a single curve for the normalized specific heat jump which depends only on the dimensionless variable Ω_E/T_c. Since this unique curve displays a maximum at a definite $\Omega_E^* = 4.55\ T_c$ we are lead to the idea of an optimum base spectrum which maximizes the normalized specific heat jump.

We can prove further that the optimum delta function $A\delta(\omega-\Omega_E^*)$ for arbitrary A corresponds to a local maximum in $\Delta C(T_c)/\gamma(0)T_c$. The necessary evidence is provided by the lower solid curve of Fig. 1 which is the functional derivative calculated with the optimum spectrum as base. We see that $\delta[\Delta C(T_c)/\gamma(0)T_c]/\delta\alpha^2F(\omega)$ in this case is negative definite and, in addition, is exactly zero at the base frequency Ω_E^*. This means that if we should take spectral weight from the base delta function spectrum and place it at any other frequency, $\Delta C(T_c)/\gamma(0)T_c$ would drop. Thus, we have a local maximum for this quantity.

We are not able to give a mathematical proof that an absolute maximum has been reached but we can give evidence that this is the case for realistic physically observed spectral densities. Consider our Fig.2 where the solid curve is the maximum value of $\Delta C(T_c)/\gamma(0)T_c$ obtained in our analysis for various values of μ^*. Also placed on the same figure are results of theoretical Eliashberg calculations for the same quantity in the case of many well studied superconductors identified in the figure caption. It is to be noted that all solid points fall below our calculated maximum, and that several are not so much smaller than our limit. In a recent measurement on $La_{1.85}Sr_{0.15}CuO_4$, the normalized specific heat ratio was estimated[18] to fall between 2 and 10. It is clear that the upper range of these values cannot be described by an Eliashberg superconductor. Even if excitons were included, as in the work of Allender et.al.[7] by adding a high energy part to the spectral density but leaving the Eliashberg equations unchanged, our upper limit would still apply. This holds more generaly for any boson exchange mechanism that can be described in terms of a spectral density in our basic equations.

We have made similar calculations for other thermodynamic coefficients. Because of lack of space here we can only summarize these. A minimum is found for $\gamma(0)T_c^2/H_c^2(0)$ with $H_c(0)$ the zero temperature thermodynamic critical field. For the thermodynamic critical magnetic field deviation function $D(t)$ we find a maximum at an intermediate temperature value between 0 and 1. These results are displayed in table 1 for three values of μ^* namely 0.052, 0.15 and 0.25. We note that, the maximum in the normalized specific heat jump increases slightly with μ^* as does the maximum in $D(t)$ while the minimum in $\gamma(0)T_c^2/H_c^2(0)$ decreases. The changes are in all cases small.

<u>Energy Gap to Critical Temperature Ratio</u>

Solutions of the Eliashberg equations can be used to calculate the zero temperature gap edge provided a method of analytic continuation to the real frequency axis is introduced[19]. The upper solid curve in Fig. 3 shows results for the functional derivative of the ratio $2\Delta_0/k_BT_c$ in the case of a delta function base spectral density with $\Omega_E/T_c = 4.7$. It is positive definite and goes to zero both for $\omega \to 0$ and $\omega \to \infty$. Many other cases have been treated by Mitrovic et.al.[19]. For real materials all functional derivatives have the same shape as our curve for $\Omega_E/T_c = 4.7$ with a pronounced maximum around $\omega/T_c = 4/3$. Thus the shape is universal for realistic spectra. The same arguments as we have used in the case of the normalized specific heat jump apply here and they lead to the introduction of delta function spectra with the result that, $2\Delta_0/k_BT_c$ is independent of A and is only a function of Ω_E/T_c and of μ^*. The actual functional relation is shown in Fig. 4 for the case $\mu^* = 0$. In contrast to the equivalent results for

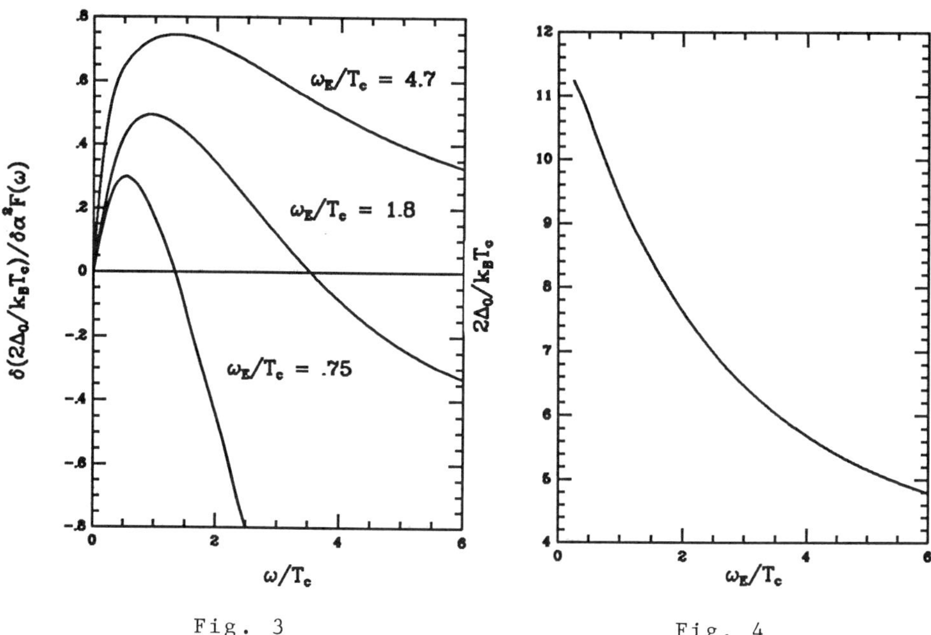

Fig. 3   Fig. 4

Fig. 3   The functional derivative of the gap Δ_0 to critical temperature ratio $2\Delta_0/k_BT_c$ for three delta function based spectrum labled by normalized Einstein frequency ω_E/T_c.

Fig. 4   The ratio $2\Delta_0/k_BT_c$ of the gap Δ_0 to the critical temperature T_c for a delta function based spectrum with coulomb pseudopotential $\mu^*=0$.

Table 1

$(\Delta C/\gamma T_c)_{max}$	$(\gamma T_c^2/H_c(0))_{min}$	$(D(t))_{max}$	μ^*
3.54	.113	.05	0.052
3.74	.112	.054	0.15
3.85	.110	.056	0.25

$\Delta C(T_c)/\gamma(0)T_c$, we now find that $2\Delta_0/k_B T_c$ keeps rising as Ω_E/T_c is lowered with a maximum value of 11.5 reached only at $\Omega_E = 0$.

That a base delta function at $\Omega_E^* = 0$ gives a local maximum for $2\Delta_0/k_B T_c$ can be verified by working out a series of functional derivatives of $2\Delta_0/k_B T_c$ for base delta functions with ever decreasing values of Einstein frequency Ω_E. These are shown in Fig. 3 where it is seen that, as Ω_E/T_c is lowered towards zero, the corresponding functional derivative curves still show a positive peak, but this peak gets progressively reduced in amplitude and shifts towards lower and lower energies while, at the same time, the rest of the curve becomes negative for much of the frequency range. The trend is towards a negative definite curve in the limit $\Omega_E \to 0$ with maximum value zero right at $\omega/T_c=0$. This implies that the point $\Omega_E/T_c=0$ in Fig. 4 represents a local maximum. Recent tunneling measurements[20-23] for the gap in $La_{1.85}Sr_{0.15}CuO_4$ do not agree with each other. In most experiments a range of gaps is observed with Kirtley et. al.[20] giving a largest value $2\Delta_0/k_B T_c \cong 4.5$ while Hawley et.al.[22] quote 5.2 to 9 and still higher values are found by Naito et.al.[23] who quote 8 to 18. Even assuming some anisotropy, these last values are inconsistent with our theoretical work. If confirmed as representative of the bulk superconducting state they rule out any mechanism which can be described, in a first approximation, by Eliashberg like equations with arbitrary spectral density due to phonon, plasmons, excitons or some other yet to be discussed boson exchange mechanism.

Upper Critical Magnetic Field

The equations determining the upper critical magnetic field $H_{c2}(T)$ for a strong coupling superconductor have been derived, for arbitrary impurity content, by Schossmann and Schachinger[24]. Here we ignore Pauli limiting, and consider the reduced quantity $h_{c2}(t)$ normalized to T_c and the slope of $H_{c2}(T)$ at T_c. The functional derivative of $h_{c2}(t)$ has been worked out by Marsiglio et.al.[25] and suggest a maximum in $h_{c2}(0)$. Schossmann et.al.[26] have worked out the details of the maximization of $h_{c2}(0)$. It follows the same general ideas as given previously and we get the results shown in Fig. 5. For the clean limit the lower solid curve applies. It is seen to have a maximum only as $T_c/\Omega_E \to \infty$ with the value approximately 1.5. For the dirty limit, given by the upper solid curve, no maximum has yet been reached in the range accessible to our computer programs.

Very Strong Coupling Regime

Another problem which is related to the one just discussed, is the question of thermodynamics and other properties of superconductors in the limit when T_c is of the order of a typical phonon energy. Here we will taken as a measure of this energy the parameter $\omega_{\ell n}$ first introduced by Allen and Dynes[27] within a discussion of analytic T_c equations for phonon superconductors. Other choices of characteristic phonon frequency are certainly possible but $\omega_{\ell n}$ is preferred here because it has played a major role in describing semi-quantitatively, the thermodynamic indices of a large number of conventional superconductors[10] for which $T_c/\omega_{\ell n}$ is $< .25$.

Weber[8] has calculated a model electron-phonon spectral density for $La_{1.85}Sr_{0.15}CuO_4$, which gives a value $\omega_{\ell n} \cong 14.0$ meV. If we take this value as characteristic of the oxides and if we consider a superconductor with a T_c of 96K, we obtain a value of $T_c/\omega_{\ell n} \cong 0.6$ which is well beyond the coventional range for strong coupling superconductors and into what we will call the very strong coupling regime characterized by a T_c value of the order of $\omega_{\ell n}$. Marsiglio

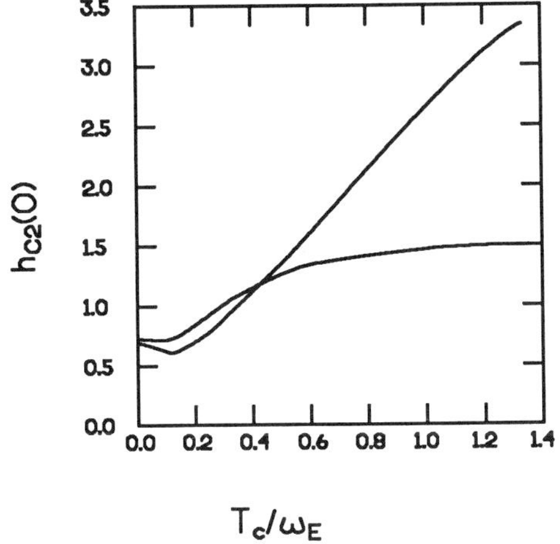

Fig. 5

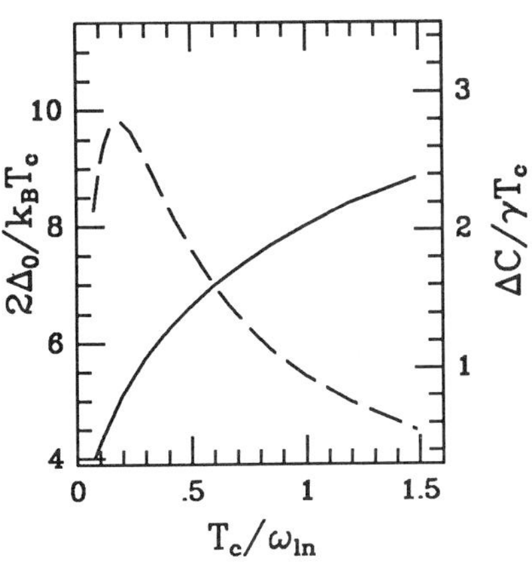

Fig. 6

Fig. 5 The variation of the zero temperature reduced upper critical magnetic field $h_{c2}(0)$, for a delta function spectral density, as a function of T_c/Ω_E with Ω_E the position of the Einstein frequency. The lower curve which shows saturation as T_c/Ω_E becomes large, applies to the clean limit while the upper curve is for the dirty limit.

Fig. 6 The normalized specific heat jump $\Delta C(T_c)/\gamma(0)T_c$ (left hand scale) as a function of $T_c/\omega_{\ell n}$. Also shown is the ratio $2\Delta_0/k_BT_c$ for which the scale on the right hand side applies.

et.al.[28] has worked out the thermodynamic properties of superconductors in this region. They use as a base for $\alpha^2 F(\omega)$, the spectral density computed by Weber[8] which they then scale by a constant factor on the vertical as well as on the horizontal axis to get a T_c value of 96 and various values of $T_c/\omega_{\ell n}$. In Fig. 6 we show results obtained for $2\Delta_0/k_B T_c$ (right hand scale) and for $\Delta C(T_c)/\gamma(0)T_c$ (left hand scale) as a function of $T_c/\omega_{\ell n}$ which is varied continuously from zero up to 1.2. While the gap ratio continues to rise indefinitely with increasing value of $T_c/\omega_{\ell n}$ the specific heat does not. Suprisingly, after the usual strong coupling region in which the specific heat is seen to be increased from its BCS value of 1.43 at $T_c/\omega_{\ell n} = 0$, $\Delta C(T_c)/\gamma(0)T_c$ and reaches a maximum around $T_c/\omega_{\ell n} = 0.2$, it then drops to values which are considerably smaller than 1.43. The case $T_c/\omega_{\ell n} \cong .6$ is well within this new regime for which the usual ideas, based on the case of conventional strong coupling, fail completely.

<u>Summary and Conclusions</u>

In this paper we have not tried to address directly the question: what is the mechanism responsible for the high T_c observed in the oxides? Instead, we have considered an Eliashberg superconductor with arbitrary spectral density and have derived certain limits on the possible values for such properties as the normalized specific heat jump at T_c. Other quantities considered were i) $\gamma(0)T_c^2/H_c^2(0)$ with $H_c(0)$ the zero temperature thermodynamic critical field and $\gamma(0)$ the Sommerfeld constant ii) the critical field deviation function $D(t)$ iii) the ratio of the gap edge to the critical temperature and iv) the zero temperature reduced upper critical magnetic field $h_{c2}(0)$. As examples of the results obtained for $\mu^* = 0.052$ $[\Delta C(T_c)/\gamma(0)T_c] \leqslant 3.54$ $[\gamma(0)T_c^2/H_c^2(0)] \geqslant .113$ and $D(t)_{max} \leqslant .05$. For $\mu^* = 0$ $2\Delta_0/k_B T_c \leqslant 11.5$ and for $\mu^* = 0.1$, in the clean limit, $h_{c2}(0) \leqslant 1.5$.

We have also considered the thermodynamic properties that characterize a superconductor in the very strong coupling regime when T_c is of the same order as a typical phonon energy $\omega_{\ell n}$. In the limit $T_c/\omega_{\ell n} \rightarrow 0$ we find the BCS value 1.43 for $\Delta C(T_c)/\gamma(0)T_c$. As $T_c/\omega_{\ell n}$ is increased the jump increases until a maximum is reached around $T_c/\omega_{\ell n} \cong 0.2$, which is the limit of the conventional range. As we increase $T_c/\omega_{\ell n}$ further $\Delta C(T_c)/\gamma(0)T_c$ begins to drop gradually and can fall well below 1.43 when $T_c/\omega_{\ell n} \cong 1.0$. This unexpected and unusual behaviour could never have been guessed at from an extrapolation to higher values of $T_c/\omega_{\ell n}$ of the results that apply in the conventional strong coupling regime. On the other hand the gap ratio $2\Delta_0/k_B T_c$ is found to increase monotonically with increasing coupling and to saturate only in the limit $T_c/\omega_{\ell n} \rightarrow \infty$.

Acknowledgement

It is a pleasure to thank my collaborators in research, E. Schachinger, M.Schossmann, F. Marsiglio and J. Blezius, with whom much of the work reviewed here was carried out. Partial support from the Natural Sciences and Engineering Research Council of Canada (NSERC) is gratefully acknowleged.

References

1. J.P. Carbotte, Science Progress, Oxf. 71, 329 (1987).
2. W.L. McMillan and J.M. Rowell, in Superconductivity, (edited by R.D. Parks, Marcel Dekker Inc., New York, 1969) Vol. pp. 561.
3. J.M Daams and J.P. Carbotte, Jour. Low Temp. Phys. 43, 263 (1981).
4. F. Marsiglio and J.P. Carbotte, Phys. Rev. B33, 6141 (1986).
5. P.W. Anderson, Science, 235, 1196 (1987).
6. A. Alexandrov and J. Ranninger, Phys. Rev. B23. 1796 (1981).
7. D. Allender, J. Bray and J. Bardeen, Phys. Rev B7, 1020 (1973).
8. W. Weber, Phys. Rev.Lett. 58, 1371 (1987).
9. C.M. Varma, S. Schmitt-Rink and E. Abrahams (preprint).
10. B. Batlogg, R.J. Cava, A.Jayaraman, R.B. van Dover, G.A. Kourouklis, S.Sunshine, D.W. Murphy, L.W. Rupp, H.S. Chen, A. White, A.M. Mujsce and E.A. Rietman Phys. Rev.Lett.
11. R.J. Cava, R.B. van Dover, B. Batlogg and E.A. Rietman, Phys. Rev. Lett. 58, 408 (1987).
12. W.K. Kwok, G.W. Crabtree, D.G. Hinks, D.W. Capone, J.D. Jorgensen and K. Zhang, Phys. Rev. B35, 5343 (1987).
13. M.E. Hawley, K.E. Gray, D.W. Capone and D.G. Hinks, Phys. Rev B35, 7224 (1987).
14. W.E. Pickett, H.Krakauer, D.A. Papaconstantopoulos and L.L. Boyer, Phys. Rev. B35, 7252 (1987).
15. F. Marsiglio and J.P. Carbotte, Phys. Rev. B31, 4192 (1985).
16. F. Marsiglio, E. Schachinger and J.P. Carbotte, Jour. Low Temp. Phys. 65, 305 (1986).
17. J.P. Carbotte, F. Marsiglio and B. Mitrovic, Phys. Rev. B33, 6135 (1986).
18. B.D. Dunlap, M.V. Nevitt, M.Slaski, T.E. Klippert, Z. Sungaila, A.G. McKile, D.W. Capone, R.B. Poeppel and B.K. Flandermeyer, Phys. Rev B35, 7210 (1987).
19. B. Mitrovic, C.R. Leavens and J.P. Carbotte, Phys. Rev. B21, 5048 (1980).
20. J.R. Kirtley, C.C. Tsuei, S.I. Park, C.C. Chui, J. Rozen and M.W. Shafer, Phys. Rev. B35, 7216 (1987).
21. S. Pan, K.W. Ng, A.L. de Lozanne, J.M. Tarascon and L.H Greene, Phys. Rev B35, 7220 (1987).
22. M.E. Hawley, K.E. Gray, D.W. Capone and D.G. Hinks, Phys. Rev. B35, 7224, (1987).
23. N. Naito, D.T.E.Smith, M.D. Kirk, B. Oh, M.R. Hahn, K. Chai, D.B. Mitzi, J.Z Sun, D.J. Webb, M.R. Beasley, O. Fischer, T.H. Geballe, R.H. Hammond, A. Kapitutnik and C.F. Quate, Phys. Rev B35, 7228 (1987).
24. M.Schossmann and E. Schachinger, Phys. Reb. B33, 6122 (1986).
25. F. Marsiglio, M. Schossmann, E. Schachinger and J.P. Carbotte, Phys. Rev. B35, 3226 (1987).
26. M.Schossmann, J. P. Carbotte and E. Schachinger (to be published).
27. P.B. Allen and R.C. Dynes, Phys. Rev. B12, 905 (1975).
28. F. Marsiglio, R. Akis and J.P. Carbotte, Phys. Rev. (sub.).

DISORDER-INDUCED PAIR BREAKING IN SUPERCONDUCTORS

Thomas R. Lemberger and Soon-Gul Lee

Department of Physics
Ohio State University
Columbus, OH 43210

By enhancing thermal supercurrent fluctuations, disorder can cause
significant pair-breaking effects on the transition temperature, density
of states, and order parameter in superconductors. We have observed this
pair-breaking mechanism in two-dimensional Aℓ films by using a novel
tunneling technique.[1] A crude model explains the results rather well. An
extension to 3 dimensions of that model is used below to estimate pair-
breaking rates in Heavy-Fermion and High-T_c compounds. We find that the
disorder-induced pair-breaking rates are likely to be significant.

We have seen evidence of disorder-induced pair-breaking in weakly
disordered Aℓ films by using a dc tunneling technique that measures the
pair-breaking time directly. From measurements of the low-voltage
resistance of low-resistance Superconductor-Insulator-Normal metal (SIN)
tunnel junctions, we can obtain the charge-imbalance relaxation time, and
thence the pair-breaking time.[2,3] A key to the measurement is the
modulation of the junction resistance with an applied pair-breaker such as
a supercurrent or a magnetic field parallel to the junction.

A rather crude model fits the observed effect. The idea is that all
of the superconducting electrons within a coherence length of each other
undergo Brownian motion as a single particle, with an average kinetic
energy given by the equipartition theorem. Extended here beyond two
dimensions, the model predicts a pair-breaking rate:

$$1/\tau_s \approx \frac{2D}{\hbar^2} \frac{d\, k_B T\, m}{n_s \xi^d(T) L^{3-d}}, \tag{1}$$

where D is the electron diffusion constant, d is the dimensionality of the
superconductor, $n_s(T)$ is the density of superconducting electrons, m is
the electron mass, $\xi(T)$ is the Ginzburg-Landau coherence length. L is the
film thickness in two dimensions, and in one dimension L^2 is the
crossectional area of a wire.

This simple model neglects the complicated time and space dependence
of the fluctuations. Nevertheless, it is interesting to look at its
predictions. By using free-electron relations and dirty-limit expressions
for $n_s(T)$ and $\xi(T)$ near T_c, we find

$$1/\tau_s \approx d \times 10^7 \text{ s}^{-1} (T_c/K\text{-}\Omega)(\rho/L)(L/\xi)^{d-2} \qquad \text{(near } T_c\text{)} \qquad (2)$$

$$\approx 1 \times 10^7 \text{ s}^{-1} (T_c/K\text{-}\Omega)(\rho\xi/L^2) \qquad (d=1) \qquad (3)$$

$$\approx 2 \times 10^7 \text{ s}^{-1} (T_c/K\text{-}\Omega)\ (\rho/L) \qquad (d=2) \qquad (4)$$

$$\approx 3 \times 10^7 \text{ s}^{-1} (T_c/K\text{-}\Omega)\ (\rho/\xi). \qquad (d=3) \qquad (5)$$

Note that the rate increases with increasing disorder for all cases since $\rho \propto 1/\ell$ and $\xi \propto \sqrt{\ell}$.

How large might this pair breaker be in interesting three-dimensional superconductors? In the High-T_c compounds, with $T_c \approx 100K$, $\rho \approx 1000\mu\Omega$-cm, and $\xi \approx 3nm$, we find $1/\tau_s \approx 10^{13} \text{ s}^{-1}$ (75K). A pair-breaking rate of this size would have large effects. For values appropriate to[4] UBe$_{13}$, $\rho \approx 200\mu\Omega$-cm, $\xi \approx 3nm$, and $T_c \approx 1K$, we find $1/\tau_s \approx 2\times10^{10} \text{ s}^{-1}$ (0.15K). This is[13] also a large pair breaker.

Disorder can cause pair breaking in other ways, besides enhancing supercurrent fluctuations. In the Heavy-Fermion compounds, the order parameter is thought to be highly anisotropic[5] so that elastic scattering from nonmagnetic impurities is pair breaking. Dramatic effects on the density of states and other properties are expected except in the purest samples.

In the High-T_c compounds, there are other important pair breaking mechanisms in addition to supercurrent fluctuations. Coffey and Cox[6] have shown that disorder in the Resonant Valence Bond[7] model causes pair breaking. In addition, in these materials the transition temperature is so high that inelastic electron-phonon scattering may be important. For comparison, the electron-phonon scattering rate at 100K in pure Cu, in which electron-phonon coupling is rather weak, is roughly[8] $1.3\times10^{13} \text{ s}^{-1}$ (100K). A rate this large in the High-T_c compounds would cause a depression in T_c of about 50K. That is, at low temperature where electron-phonon scattering and supercurrent fluctuations are small, the material should behave as if its transition temperature were about 50K higher than the measured T_c.

Acknowledgements This material is based upon work supported by the National Science Foundation under grant DMR-85-15370. One of us (TRL) gratefully acknowledges support from the Alfred P. Sloan Foundation. One of us (SGL) gratefully acknowledges a Presidential Fellowship from Ohio State University.

REFERENCES

1. S.-G. Lee and T.R. Lemberger, preprint.
2. T.R. Lemberger, Y. Yen, and S.-G. Lee, Phys. Rev. B **35**, 6670 (1987).
3. Y. Yen and T.R. Lemberger, preprint.
4. G.R. Stewart, Rev. Mod. Phys. **56**, 755 (1984).
5. K. Ueda and T.M. Rice, 1985, in: "Advances in Solid State Physics,"
    P. Grosse, ed., Vieweg, Braunschweig.
6. L. Coffey and D.L. Cox, this conference.
7. G. Baskaran, Z. Zou, and P.W. Anderson, preprint.
8. C. Kittel, "Introduction to Solid State Physics," Wiley, NY (1986).

FLUX LATTICE MELTING IN AMORPHOUS COMPOSITE IN/INO$_x$ THIN FILM SUPERCONDUCTORS

P.L. Gammel, A.F. Hebard and D.J. Bishop

AT&T Bell Laboratories
600 Mountain Ave.
Murray Hill, NJ 07974

ABSTRACT

The mechanical response, comprising longitudinal sound velocity and damping, of amorphous In/InO$_x$ films is studied as a function of normal state sheet resistance. The films are sputtered directly onto a high Q oscillator fabricated from single crystal silicon. With a resonant frequency near 3 KHz, the oscillators have Q's of 10^5 and a frequency stability of 10^{-8}. This allows for a unique measurement of the dynamics of magnetic vortices, induced by a field perpendicular to the plane of the sample. The melting of the vortex lattice is signaled by a peak in the attenuation and a frequency shift of the silicon oscillator. The zero field melting temperature and the magnitude of the frequency shift are described by a model of Kosterlitz-Thouless melting due to D.S. Fisher. The temperature dependence of the melting does not agree with this calculation.

INTRODUCTION

A type-II superconductor in a magnetic field is permeated by an array of vortex lines, each containing one quantum of flux. At low temperatures and fields, these flux lines are arranged into a triangular lattice. This vortex lattice behaves as a solid with a well defined shear modulus[1]. As the temperature is raised, this solid may undergo a melting transition. It has been predicted theoretically that this melting should be a Kosterlitz-Thouless (K-T) type continuous symmetry melting for superconducting films which are effectively two-dimensional[2].

In this experiment the superconductor, a 30 nm film of amorphous In/InO$_x$, is deposited onto a high Q ($>10^5$) mechanical oscillator. High Q oscillators have previously been used to study the K-T transition in two-dimensional helium films both with[3] and without[4] a uniform vortex background. The K-T transition in In/InO$_x$ films has been studied in zero field using current-voltage and ac inductance measurements[5,6].

The geometry of this experiment is shown in fig. 1. By monitoring the resonant frequency and Q of the oscillator, the longitudinal sound velocity and dissipation in the vortex lattice at the oscillator frequency (3 KHz) could be measured. As the magnetic field, and hence the vortex lattice density, is changed, a melting transition is signified by a frequency shift and dissipation peak. Here, the oscillator technique and film preparation will first be discussed, followed by a summary of the results of the experiment and its analysis in terms of a K-T melting transition.

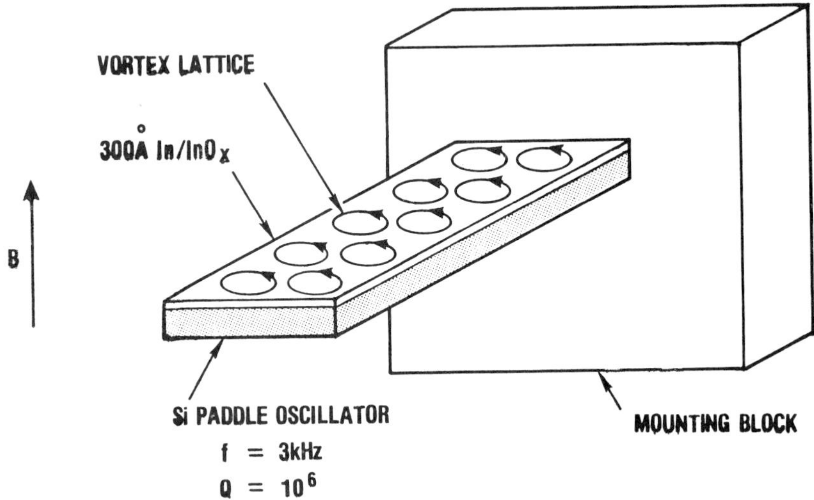

Fig. 1. A magnetic field applied perpendicular to the plane of the sample induces an array of quantized vortices. By monitoring the resonant frequency and Q of the substrate oscillator as a function of magnetic field, changes in shear modulus and dissipation in the vortex array are observed.

I. HIGH Q SILICON OSCILLATORS

High Q silicon oscillators have been discussed by Kleiman et al[7]. The oscillators for this experiment were fabricated in the same fashion from P-type (110) silicon wafers .010" thick. The room temperature resistivity of the silicon was only .03Ω−cm. An oxide layer 100 nm thick was left on the wafer, which allowed for resistance measurements to be made on the films at high temperatures. Earlier work with In/InO$_x$ films has been performed mostly on glass substrates. The oxide layer was intended to provide a similar amorphous substrate potential. The physics of the vortex processes in these films is controlled strictly by the sheet resistance, and is only affected by the substrate insofar as it modifies R$_\square$.

The oscillator was patterned using photolithography and anisotropic etching. The silicon oscillator was epoxied into a copper mounting block. One side was coated with 10 nm Cr followed by 100 nm Au to form part of a capacitor. The superconducting film was deposited on the other side of the oscillator. The resulting oscillator has a Q of 2x10^5 at 4K, a frequency of 3 KHz and a frequency stability of 10^{-8}. Oscillators with resonant frequencies between 500 Hz and 20 KHz can be made with this technique. A variety of normal modes are present in these oscillators. While the torsional mode generally has the highest Q, we use the simplest bending mode of the beam shown in fig. 1. In this mode, the distortions of the vortex lattice are well defined.

The motion of the oscillator was driven and detected capacitively using a phase locked loop as discussed by Kleiman[7]. Since the geometry of the capacitors was well known, the actual motion of the oscillator could be determined. To avoid nonlinear effects, an amplitude at the end of the bar of 100 nm was used.

Even in the absence of the superconducting film, the oscillator has a temperature and field dependent response. This was determined for our case using an identical oscillator with a gold film on both sides. The temperature dependence of the dissipation and frequency were similar to that reported by Kleiman et al[8]. In the range 4K-1K, the frequency increased by 5 ppm/K and the Q was approximately independent of temperature. The magnetic field response was temperature independent and can be seen in fig. 2. Both the resonant frequency and dissipation have a component proportional to H^2, which is attributed to the motion of the gold film in a magnetic field. When a superconducting film was present, the magnetic field dependence above T_c was subtracted off as background.

II. FILM MORPHOLOGY AND PREPARATION.

The In/InO_x films were deposited directly onto the oscillator using reactive ion beam sputtering in an oxygen environment. Films produced by this technique have been shown to be reproducible and uniform[9]. In addition to the oscillator, samples for measuring resistance were deposited onto glass and a piece of silicon taken from the same wafer as the oscillator.

Depending on oxygen concentration, the morphology of the films varies[9]. At low oxygen concentrations, the films are granular, with a grain size roughly equal to the film thickness. As the concentration is increased, the films become smooth with a predominantly amorphous microstructure as seen in TEM. All experiments reported here used these amorphous composite films. While TEM samples were not prepared for the films used, the morphology is known to be reproducible. Unfortunately, it was not possible to make amorphous films with sheet resistivities below about $300\Omega/\square$.

The resistive transition in zero field was quite sharp, 0.1K, indicating unform films. T_c was estimated from a 40% criterion, rather than a detailed fit to the Adamazov-Larkin theory[10]. In a finite field, the transitions broadened. The width increases to 0.8K in 40kG, but saturates at this level up to our highest field, 100kG.

The magnetoresistance of the films was used to determine the zero field Kosterlitz-Thouless transition temperature. Hebard et al[11] have observed that the K-T transition occurs a the point where the magnetoresistance is approximately linear. For the values of $R_\square$ used here, T_{K-T} was less than 0.2K from T_c.

III. EXPERIMENTAL DETAILS

The schematic of the experiment is shown in fig. 1. With a magnetic field applied perpendicular to the sample, an array of quantized flux lines penetrates the sample. For typical fields, 10 kG, the average spacing is 100nm. Because the two-dimensional penetration depth is several mm, the vortices are strongly overlapping. The sample oscillates in the simplest bending mode. For small oscillations, to first order the average vortex density is unchanged. The contribution of the vortex array to the oscillator response can then be separated into two pieces. First, the oscillator frequency will shift proportional to the longitudinal sound velocity of the vortex array. Second, the motion of the vortices in the external field will induce a dissipation.

If the sample is rotated so the external field is parallel to the plane of the sample, no vortices should penetrate the sample, and both effects should vanish. There will an effect due to flux exclusion by the sample, which has been studied by Brandt[12]. The experiment has not yet been performed in this orientation.

The dynamic response of the vortex array is examined at fixed temperature by sweeping the magnetic field, proportional to the average vortex density. The basic result is shown in fig. 2. At low densities and temperatures, the vortex array forms a solid. In this regime, there is a frequency increase and additional dissipation. As the solid melts with increasing density, both effects vanish. The maxima in the frequency and dissipation occur at the same density. We associate this maximum with the melting density, H_M.

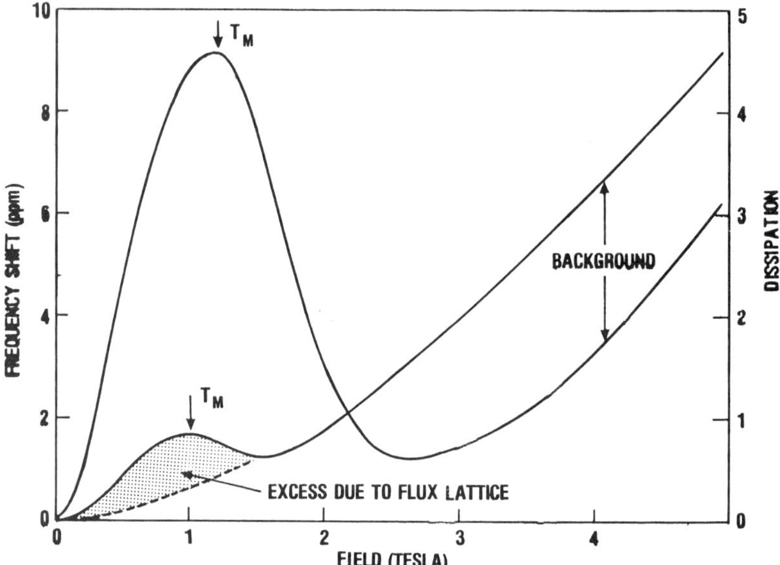

Fig. 2.   When the vortex lines form as solid, there is an additional stiffness due to the shear modulus. In addition, the motion of the solid in the external field causes an additional damping. The melting temperature, T_M is identified as the maximum in both these effects.

The data in fig. 2 are for a temperature about 1K below the superconducting transition. As the temperature is lowered further, there are three effects. First, the maximum in the frequency and dissipation move to higher magnetic field. Second, the magnitude of both the frequency shift and dissipation feature increases. Third, the width of the transition increases. This data can be combined with H_{c2}, determined from magnetoresistance, to form a phase diagram, shown in fig. 3.

For samples with different normal state sheet resistances, the basic picture presented in fig. 3 remains. The melting temperature, however, is reduced relative to T_c, although the slope of the melting line is unchanged. In addition, the critical field slope increases. For sheet resistances greater than $1K\Omega/\square$ there was no evidence of a melting transition in our experiment.

The width of the transition may be evidence for the hexatic phase, predicted theoretically for this system. It may also be due to sample inhomogeneities. We are presently examining these data in more detail.

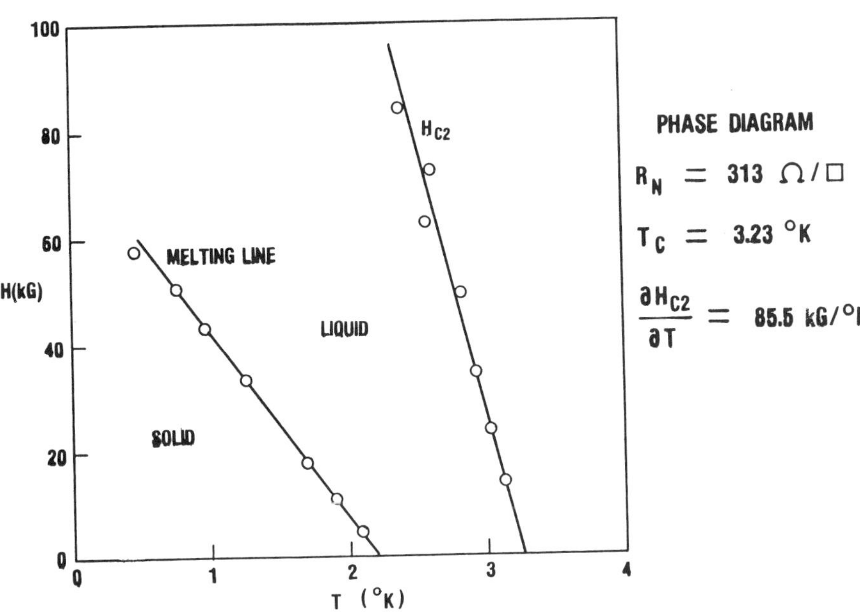

Fig. 3. H_{c2} is obtained from magnetoresistance. Together with T_M from the oscillator, this is used to generate a phase diagram. For these films, the zero-field Kosterlitz-Thouless transition is 0.1K below T_c.

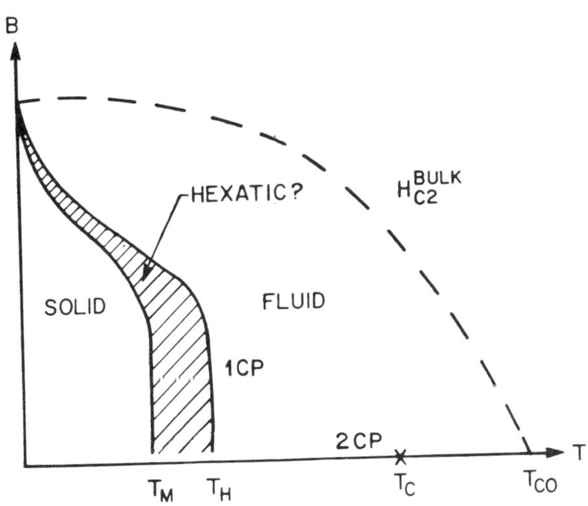

Fig. 4.   The theoretical phase diagram has been calculated in the limit $T_M \ll T_c$. The general features agree with our result. However, T_M is expected to be independent of density except near H_{c2}.

IV. ANALYSIS

The phase diagram has been calculated for $T_M \ll T_c$ using a renormalization group approach. The phase diagram is shown in fig. 4. While this calculation does not include the effects of pinning, it is expected that only for $H \sim H_{c2}$ will T_M be affected. In the intermediate field regime, T_M is found to be independent of magnetic field. This does not agree with our phase diagram. The final upturn of the melting line to equal H_{c2} at $T=0$ cannot be examined in our experiment, as $H_{c2}(T=0) \sim 300\text{kG}$. However, the data to 0.4K show no departure from $H_M \sim (T_M - T)$.

In the limit of low normal state sheet resistance, $R_\square / R_c \ll 1$, the melting temperature is:

$$\frac{T_M}{T_c} = \left(\frac{3.8}{A_1} \frac{R_\square}{R_c} \right)^{-1}$$

$R_c = h/e^2 = 4.12\text{k}\Omega/\square$ is a characteristic resistance near which superconductivity vanishes. This result can be compared to our melting temperature extrapolated to zero field. For a variety of normal state sheet resistances, these data are shown in fig. 5. The value of A_1 which fits our data is 0.53. For a K-T type transition, $0.4 < A_1 < 0.75$. If the value of melting temperature is taken at a finite field, the estimate for A_1 increases. If the sheet resistance is increased above $1\text{k}\Omega/\square$, there was no melting transition observed, so a more detailed fit to the high resistance calculation was not possible.

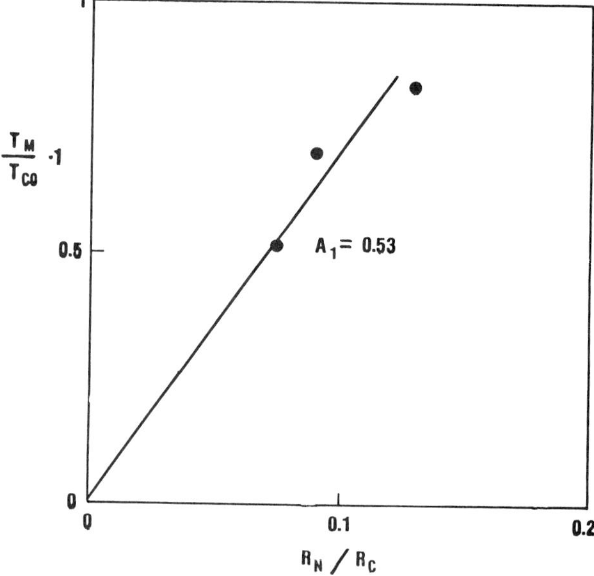

Fig. 5. The extrapolation of the observed melting line to zero field can be compared to the theoretical melting temperature in the low $R_\square / R_c$ limit. The value of the renormalization parameter A_1 inferred from this is within the range expected for a K-T melting transition.

The frequency shift is proportional to the change in longitudinal sound velocity in the vortex lattice. In two dimensions, one has $c/\rho \sim B + \mu$, where B is the bulk modulus and μ is the shear modulus. On melting, there should be no change in the bulk modulus, but the shear modulus should vanish. The magnitude of this change may be compared to theory. The simplest estimate for the shear modulus may be obtained from the total interaction energy of the vortex lattice:[13]

$$\int n(r)\ln(r-r')n(r')$$

The integral is prevented from diverging either by the finite size of the system (1 cm) or by the two dimensional penetration depth (1 cm near T_c). With a vortex spacing of 100nm, the effective shear modulus estimated from this is 100 kbar, on the same order as bulk silicon.

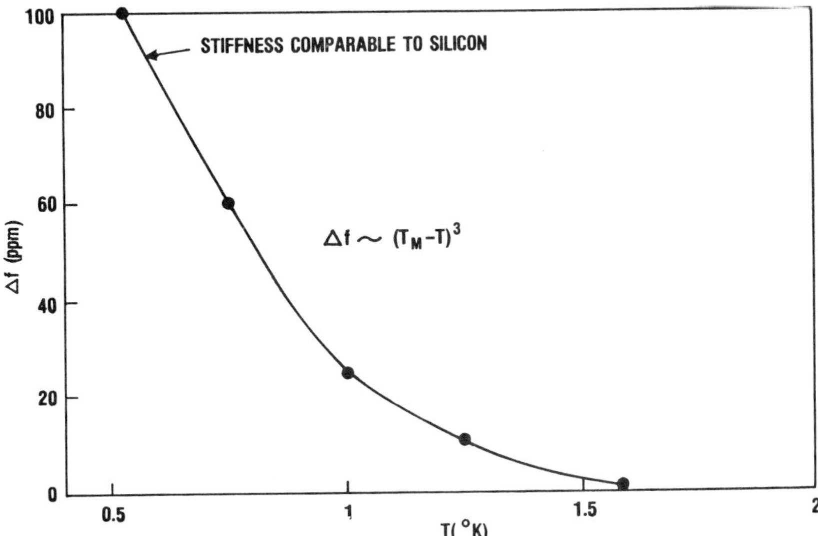

Fig. 6.   The frequency shift on melting, defined as the deviation from background at T_M is found to scale as $(T-T_M)^3$. The magnitude of this frequency shift can be estimated from the interaction of the vortices cut off at the two dimensional penetration depth $(\Lambda(T) \sim 1cm)$. At low temperatures, the measured shear modulus is similar to that of crystalline silicon.

The frequency shift observed on melting, taken as the departure from background at T_M is shown as a function of temperature in fig. 6. At low temperatures, the frequency shift is of order 10^{-4}. Since the ratio of the film thickness to the oscillator thickness is 10^{-4}, this implies that the shear modulus of the solid is equal to the shear modulus of silicon at this temperature.

As the temperature is varied, the frequency shift is found to scale as $(T_M-T)^3$. Because the density at which melting occurs varies as (T_M-T), the change in shear modulus is proportional to $(T_M-T)^2$. A detailed theoretical estimate with which to compare this dependence is not available.

The previous analysis is into the low oscillation amplitude limit. As the amplitude is increased, the frequency shift associated with the melting transition decreases. The width of the transition remains the same. We believe this is evidence for shear induced melting of the lattice. It would be useful to compare this critical amplitude with flux-flow data. However, as can be seen in fig. 7, the critical power levels are approximately 10^{-16}W. As flux flow data a generally presented in the 10^{-12}W regime, where the melting is completely suppressed, this is not possible.

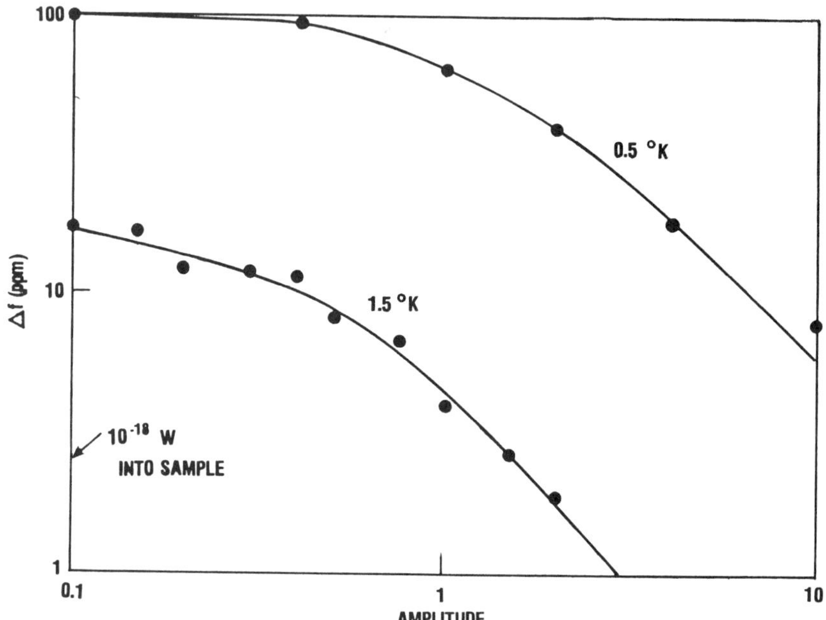

Fig. 7. As the oscillation amplitude is increased, the melting transition is suppressed. The critical velocity for this increases as the temperature is lowered. Note that melting is strongly suppressed when the power dissipated in the experiment exceeds 10^{-16}W, making comparison to flux-flow measurements difficult.

V. SUMMARY

In conclusion, we have used a novel high Q mechanical oscillator technique to study flux lattice melting in amorphous In/InO_x thin film superconductors. The zero field extrapolation of the melting temperature as a function of normal state sheet resistance agrees with a K-T type melting. The shear modulus of the vortex lattice approaches that of bulk silicon at 10 kG at 0.4K due to the long range interactions. The details of the phase diagram do not agree with the theory.

We would like to thank D.S. Fisher and A. Millis for helpful discussions and R.H. Eick for technical assistance.

REFERENCES

[1]   A.L. Fetter and P.C. Hohenberg, Phys. Rev. 159, 330 (1967)

[2]   D.S. Fisher, Phys. Rev. B22, 1190 (1980)

[3]   M. Kim and W.I. Glaberson, Phys. Rev. Lett. 52, 53 (1984)

[4]   D.J. Bishop and J.D. Reppy, Phys. Rev. Lett. 40, 1727 (1980)

[5]   A.F. Hebard and A.T. Fiory, Phys. Rev. Lett. 50, 1603 (1983)

[6]   A.T. Fiory and A.F. Hebard, Phys. Rev. B28, 5075 (1983)

[7]   R.N. Kleiman, G.K. Kaminsky, J.D. Reppy, R. Pindak and D.J. Bishop, Rev. Sci. Instrum. 56, 2088 (1985)

[8]   R.N. Kleiman, G. Agnolet and D.J. Bishop, submitted to Phys. Rev. Lett.

[9]   A.F. Hebard and S. Nakahara, Appl. Phys. Lett. 41, 1130 (1982)

[10]   M.A. Paalanen and A.F. Hebard, Appl. Phys. Lett. 45, 794 (1984)

[11]   A.F. Hebard and M.A. Paalanen, Phys. Rev. Lett. 54, 2155 (1985)

[12]   E.H. Brandt, Phys. Lett. 113A, 51 (1985)

[13]   D.S. Fisher, private communication

EVIDENCE FOR NONPHONONIC SUPERCONDUCTIVITY IN Nb_3Ge

K.E. Kihlstrom (a), P.D. Hovda (a),
Vladimir Z. Kresin (b), and S.A. Wolf (c)

(a) Department of Physics, Westmont College
    Santa Barbara, CA 93108
(b) Materials and Chemical Sciences Division,
    Lawrence Berkeley Laboratory, University of
    California, Berkeley, CA 94720
(c) Naval Research Laboratory
    Washington, DC 20375-5000

We present the results of a quantitative test to separate phononic from nonphononic contributions to the superconductivity in A15 Nb_3Ge. The method proposed by V.Z. Kresin has been used. This method is based on analysis of heat capacity and tunneling $\alpha^2F(\omega)$ data and neutron scattering data. We obtained evidence that the superconductivity in Nb_3Ge is substantially caused by a nonphononic mechanism. In contrast, our results suggest the superconducting state in Pb and V_3Si is due to phonons.

Introduction: For many years there has been considerable theoretical work done on the possibility of nonphononic mechanisms that could give rise to superconductivity.[1-7] This interest has greatly intensified with the discovery of the new class of high-T_c superconductors[8] which give T_c's that are much higher than was thought possible by conventional phonons.[9] However there has been no direct experimental evidence for the existence of a nonphononic mechanism that gives rise to superconductivity. In part the problem is that even if there were a nonphononic mechanism present in a superconductor, how would that be experimentally verified? Recently one of us proposed a method for separation of phonon and nonphonon mechanisms.[10] In this paper we describe the analysis of several systems (Pb, V_3Si, and Nb_3Ge). Our main result is that Nb_3Ge is a nonphonon superconductor. We think this is a first experimental observation of nonphononic superconductivity.

Theory: The method requires measuring tunneling $\alpha^2F(\omega)$, heat capacity and neutron scattering on the same material. The basic idea was that from the tunneling $\alpha^2F(\omega)$ data the electronic component of the heat capacity can be calculated (see ref. 10):

$$C_e(T) = \gamma(T)T \ ,$$

$$\frac{\gamma(T)}{\gamma(0)} = 1 + \rho \left[\frac{\kappa(T)}{\kappa(0)} - 1 \right]$$

where $\gamma(0) = m^*(0)p_F/3$, $m^*(0)$ is the renormalized value of the effective mass, $\rho = \lambda/(1 + \lambda)$,

$$\lambda = 2 \int d\omega \alpha^2 F(\omega)\omega^{-1} \ , \quad \kappa(T) = 2 \int d\omega \alpha^2 F(\omega)\omega^{-1} Z(T/\omega)$$

(Note $\kappa(0) = \lambda$)

where $Z(x)$ is a universal function given in ref. 10. Another quantity affected by the electron-phonon interaction (EPI) is the effective mass which becomes temperature dependent. This dependence is described by the expression[11,12]:

$$\frac{m^*(T)}{m^*(0)} = 1 + \rho \left[\frac{\phi(T)}{\phi(0)} - 1 \right] \ , \quad \phi(T) = 2 \int \frac{d\omega}{\omega} \alpha^2 F(\omega)g(T/\omega)$$

where $g(x)$ is a universal function given in ref. 11. This procedure can be applied when the electronic heat capacity cannot be accurately calculated (as is the case for lead).

The Eliashberg equations, which generate the electron-phonon spectral function $\alpha^2 F(\omega)$ through an inversion procedure[13], assume only phonon mechanisms are contributing to the superconducting state. If there are nonphononic contributions as well (we consider the case that the nonphonon modes are located outside the region corresponding to phonon energies) the resulting phonon spectrum would be distorted. This is because while the nonphononic interaction would affect the gap function the inversion program would be trying to solve the equations using only phonon contributions. The phonon peak positions would be correct but their amplitudes and shapes would not. Thus when the electronic heat capacity is calculated, this distortion would introduce an error which could be either positive or negative (because $Z(x)$ has both positive and negative values).

The electronic heat capacity can also be determined by first measuring the total heat capacity then subtracting off the lattice contribution. The lattice contribution can be calculated[14,15,16] from the phonon spectrum $F(\omega)$:

$$C_{ph}(T) = 3R \int \frac{(\omega/T)^2 e^{\omega/T}}{\left(e^{\omega/T} - 1\right)^2} F(\omega)d\omega$$

The nonphononic interaction (NPI) also contributes to
$C_e(T)$. However, this contribution is small if $T \ll \Delta\varepsilon_e$
($\gamma'_{NPI} (T/\Delta\varepsilon_e')^2$; $\Delta\varepsilon_e$ is the characteristic energy of the
virtual electron transitions, and note that $\Delta\varepsilon_e \gg \omega_D$). One
should distinguish the contributions of NPI to $C_e(T)$ and to
Cooper pairing. The second can be very noticable. The
situation is analogous to the effect of EPI: In the region
$T \to 0$ the effect of EPI on $C_e(T)$ is small [$(T^2/\Omega^2_D)\ln(\Omega_D/T)$];
however, EPI plays a key role in the pairing. Virtual
transitions in the phonon subsystem providing the pairing and
the contribution of EPI to the thermodynamic properties are
described by different regularities. An analogous situation
appears for NPI, and we would like to emphasize that NPI,
connected with interaction of different electron groups, makes
a much more noticeable contribution to the pairing than it
does to $C_e(T)$. Thus any nonphononic contribution should
result in a descrepancy between the calculated and
experimental $C_e(T)$. If this descrepancy is greater than the
experimental uncertainty of the measurements, the material is
exhibiting nonphononic superconductivity. The proposed method
intended for determination of the contribution of NPI is based
on the substantial progress in tunneling spectroscopy. Use of
artificial barriers and the development of proximity electron
tunneling spectroscopy allow the determination of the function
$\alpha^2 F(\omega)$ for complicated materials.

<u>Pb and V$_3$Si</u>: Let us first describe our results for Pb and
V$_3$Si. Analysis based on the methods described above indicate
that superconductivity for both materials is caused by the
phonon mechanism (although V$_3$Si might have a small nonphononic
contribution).

For Pb we used the calculation of the temperature
dependence of effective mass[17] which gives excellent agreement
with experimental data from cyclotron resonance data[18] as can
be seen in figure 1 below. This provides strong evidence that
the superconductivity in Pb is due to the conventional phonon
mechanism.

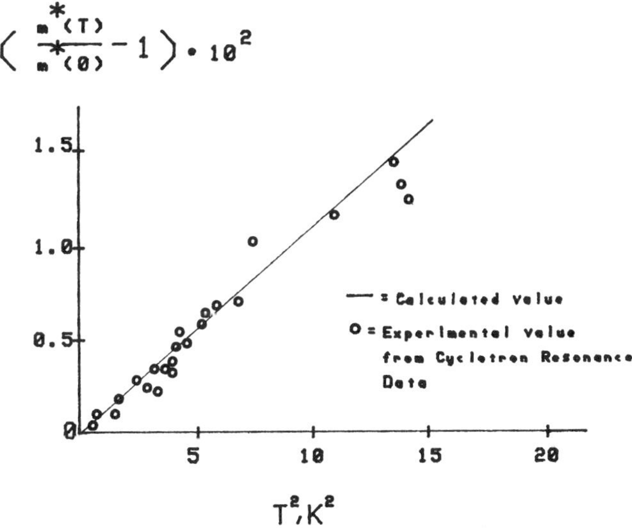

Figure 1.

We also analyzed data for lead to determine the electronic heat capacity where we obtained the tunneling $\alpha^2F(\omega)$ from McMillan and Rowell,[13] the neutron scattering data from Stedman et al.[19], and the heat capacity data from Meads et al.[20] The results were inconclusive because the heat capacity is dominated by the lattice contribution even at relatively low temperatures (lead has a Debye temperature of about 100 K). Thus calculating the electronic heat capacity requires subtracting two large numbers resulting in a very samll one. This gives an uncertainty that is greater than the number itself. The results of this analysis for lead are given in table 1:

Table 1. Experimental and Calculated Heat Capacity Values
for Lead

T(K)	C_{total}	C_p (from $F(\omega)$)	C_e (from $\alpha^2F(\omega)$)	$C_e + C_p$
15	7233	7651	30	7681
20	11018	11767	29	11796
25	14065	15075	29	15104
30	16505	17582	32	17614

The heat capacity values are all in mJ/mole-K.
Note the last column is $C_e + C_p$ to compare with C_{total}

While not relevant to the question of nonphononic superconductivity, there is good agreement (within 10%) between the measured total heat capacity and the sum of the electronic and the lattice heat capacity demonstrating consistency between the heat capacity and neutron scattering measurements.

We chose V_3Si to compare to Nb_3Ge (see below) since V_3Si is an A15 compound with a lower T_C (17 K versus 23 K) but with a substantially higher density of states. Thus V_3Si seems less likely to have a nonphononic contribution to the superconductivity. The heat capacity data (T_C = 16.8 - 17.3K) is from Viswanathan and Caton,[21] the neutron scattering data (T_C = 16.7 K) is from Schweiss et al.[22] and the tunneling $\alpha^2F(\omega)$ data is our own.[23] Table 2 gives the relative values for C_e at 30 K and 25 K as calculated from tunneling $\alpha^2F(\omega)$ and determined using heat capacity and neutron scattering measurements for V_3Si.

Table 2. Relative values of the Electronic Heat Capacity
between 25 K and 30 K for V_3Si.

Compound	Calculated C_e(30 K)/C_e(25 K)	Experimental C_e(30 K)/C_e(25 K)	Per Cent Difference
V_3Si	1.19	1.40	18 %

For V_3Si there is some descrepancy between the calculated
and experimental electronic heat capacities but it potentially
can be explained by experimental error (roughly 20%--this will
be discussed more fully in the Nb_3Ge section). It is
important to note that we focused on the temperature
dependence of the heat capacity, in particular the ratio
C_e(30 K)/C_e(25 K). This allows us to exclude the unknown $\kappa(0)$
factor. The choice of temperatures is dictated by the need to
be above T_C (we are looking at the normal state heat capacity)
and the availability of heat capacity data.

<u>Nonphonon contribution to superconductivity in Nb_3Ge</u>: The
choice of Nb_3Ge was based on several factors. Until recently
Nb_3Ge had the highest transition temperature known (a record
it held for 13 years). Klein et al.[24] made band structure
calculations on many of the A15 compounds and in all cases but
Nb_3Ge found their calculations consistent with the observed
transition temperatures. Nb_3Ge was anamalous in having a high
T_C but a low value (relative to other high-T_C A15 compounds[25])
for N(0), the density of states at the Fermi level. Ruvalds[26]
has proposed Nb_3Ge as a possible candidate for accoustic
plasmons. Tunneling $\alpha^2F(\omega)$ measurements on Nb_3Ge[27,28] have
indicated mode softening as a possible explanation for the
high T_C but the above arguments make Nb_3Ge an good candidate
to test for nonphononic superconductivity.

Tunneling $\alpha^2F(\omega)$ and heat capacity data were available
for the same sample[28] (which had a T_C of 20.3 K). Neutron
scattering experiments were done on Nb_3Ge with a T_C of 20 K by
Muller et al.[29] We believed this was as close to doing the
test on one sample as was experimentally possible for a
compound at present. The error between $F(\omega)$ and $G(\omega)$ (which
arises in a compound due to the different ratios of nuclear
cross section to mass for the component elements) for Nb_3Ge
from neutron scattering was estimated by Muller et al[31]. to be
10%. They also calculated the lattice contribution to the
heat capacity which agreed with our results.

Table 3 gives the relative values for C_e at 30 K and 25 K
as calculated from tunneling $\alpha^2F(\omega)$ and determined using heat
capacity and neutron scattering measurements for Nb_3Ge.

Table 3. Relative values of the Electronic Heat Capacity
between 25 K and 30 K for Nb_3Ge.

Compound	Calculated C_e(30 K)/C_e(25 K)	Experimental C_e(30 K)/C_e(25 K)	Per Cent Difference
Nb_3Ge	1.07	1.63	52 %

The sources of error include uncertainties in the
original experimental data which in all three measurements was
about 10%. This could account for roughly a 20% uncertainty
in the comparison of the electronic heat capacities. In the
case of V_3Si the descrepancy is potentially explainable from
experimental error (although a nonphononic contribution is not
ruled out) but for Nb_3Ge the difference is beyond reasonable

experimental error. While this method does not specify what
mechanism may be at work (i.e. virtual one particles in the
presence of overlapping bonds, accoustic plasmons, etc.), it
is suggestive that there is a substantial nonphononic
contribution to the superconductivity in Nb_3Ge.

<u>Conclusions</u>: The results of the analysis suggest there is a
substantial nonphononic contribution to the super-conductivity
in Nb_3Ge. Of course the analysis assumes the reliability of
the experimental data on which the analysis was based. We can
also conclude, based on the same analysis, that the
superconducting state in Pb and V_3Si is due to the phonon
mechanism. This described method can be used for other
materials. As the experimental data becomes available it
would be interesting to apply this analysis to look for the
possible contribution of high energy modes to the
superconductivity of the oxide superconductors.

ACKNOWLEDGEMENTS

We would like to express our thanks to N. Nucker and D.
Mael for providing important data. We wish to express our
appreciation to Douglas Collins, Briant McKellips, Ned
Divelbiss, David Marten, and H. Michael Sommermann at Westmont
College for valuable assistance. The work was supported by
the Office of Naval Research Contract Nos. N0014-85-K-0345 and
N3014-86-F-0015.

REFERENCES

1. W. A. Little, Phys. Rev. A <u>134</u>, 1416 (1964).
2. B. Geilikman, Usp. Fiz. Nauk <u>88</u>, 327 (1966); <u>109</u>, 65
       (1973) [Sov. Phys. Usp. <u>8</u>, 2032 (1966); <u>16</u>, 17
       (1973)].
3. B. Geilikman and V. Z. Kresin, Fiz. Tekh. Poluprovodn. <u>2</u>,
       764 (1968) [Sov. Phys. Semicond. <u>2</u>, 639 (1968)].
4. D. Allender, J. Bray, and J. Bardeen, Phys. Rev. B. <u>7</u>,
       1020 (1973).
5. H. Gutfreund and W. A. Little, in <u>Highly Conducting One-
       Dimensional Solids</u>, edited by J. Devreese, R. Evrard,
       and V. van Doren (Plenum, New York, 1979), p. 305.
6. J. Ihm, M. L. Cohen, and S. Tuan, Phys. Rev. B <u>23</u>, 3258
       (1981).
7. <u>High Temperature Superconductivity</u>, edited by V. Ginzburg
       and D. Kirzhnits (Plenum, New York, 1982).
8. J. G. Bednorz and K. A. Muller, Z. Phys. B. <u>64</u>, 189
       (1986).
9 . C. M. Varma, <u>Superconductivity in d- and f-Band Metals
       1982</u>, edited by W. Buckel and W. Weber,
       (Kernforschungszentrum, Karlsruhe, 1982), p. 603.
10. Vladimir Z. Kresin, Phys. Rev. B <u>30</u>, 450 (1984).
11. G. Grimvall, J. Phys. Chem. Solids <u>29</u>, 1221 (1968); Phys.
       Condens. Mater. <u>9</u>, 283 (1969); Solid State Commun. <u>7</u>,
       213 (1969).
12. P. B. Allen and M. L. Cohen, Phys. Rev. B <u>1</u>, 1329 (1970).
13. W. L. McMillan and J. M. Rowell, <u>Superconductivity</u>, edited
       by R. D. Parks (Decker, New York, 1969), Vol. 1 p. 561.

14. A. P. Miller and B. N. Brockhouse, Can. J. Phys. $\underline{49}$, 704 (1971).

15. G. Knapp and B. Brockhouse, Phys. Rev. B $\underline{6}$, 1761 (1972).

16. G. Kh. Panova, B. I. Savel'ev, M. N. Khlopkin, N. A. Chenoplekov, and A. A. Shikov, Zh. Eksp. Teor. Fiz. $\underline{85}$, 1308 (1983) [Sov. Phys. JETP $\underline{58}$, 759 (1983). (note: the equation given on p. 761 should have the denominator squared).

17. V. Z. Kresin and G. O. Zaitsev, Zh. Eksp. Teor. Fiz. $\underline{74}$, 1886 (1978). [Sov. Phys. JETP $\underline{47}$, 983 (1978)].

18. Ya. Kransnopolin and M. S. Khaikin, Zh. Eksp. Teor. Fiz. $\underline{64}$, 1750 (1973) [Sov. Phys. JETP $\underline{37}$, 883 (1973)].

19. R. Stedman, L. Almqvist and G. Nilsson, Phys. Rev. $\underline{162}$, 549 (1967).

20. P. F. Meads, W. R. Forsythe, and W. F. Giauque, J. Am. Chem. Soc. $\underline{63}$, 1902 (1941).

21. R. Viswanathan and R. Caton, Phys. Rev. B. $\underline{32}$, 2891 (1985).

22. B. P. Schweiss, B. Renker, E. Schneider, and W. Reichardt _Superconductivity in d- and f-Band Metals_, edited by D. H. Douglass (Plenum Press, New York, 1976) p. 189.

23. K. E. Kihlstrom, Phys. Rev. B. $\underline{32}$, 2891 (1985).

24. B. M. Klein, L. L. Boyer, and D. A. Papaconstantopoulos, and L. F. Mattheiss, Phys. Rev. B $\underline{18}$, 6411 (1978).

25. G. R. Stewart, _Superconductivity in d- and f-Band Metals 1982_, edited by W. Buckel and W. Weber (Kernforschungszentrum, Karlsruhe, (1982) p. 19.

26. J. Ruvalds, Adv. in Phys. $\underline{30}$, 677 (1981).

27. K. E. Kihlstrom and T. H. Geballe, Phys. Rev. B $\underline{24}$, 4101 (1981).

28. K. E. Kihlstrom, D. Mael, and T. H. Geballe, Phys. Rev. B $\underline{29}$, 150 (1984).

29. P. Muller, N. Nucker, W. Reichardt, and A. Muller, _Superconductivity in d- and f-Band Metals 1982_, edited by W. Buckel and W. Weber (Kernforschungszentrum, Karlsruhe, (1982) p. 19.

ONE-DIMENSIONAL CORRELATIONS IN ORGANIC SUPERCONDUCTORS :

MAGNETISM AND SUPERCONDUCTIVITY

D. Jérome and F. Creuzet

Laboratoire de Physique des Solides (associé au C.N.R.S.)
Université Paris-Sud
91405 Orsay, France

ABSTRACT

The examination of far infrared optical conductivity and nuclear spin-lattice relaxation experiments performed in the conducting phase of $(TMTSF)_2X$ organic superconductors reveals strong deviations at low temperature to the Drude and Korringa behaviours of usual metallic conductors.

These experimental data suggest the existence of divergent one dimensional $2k_F$ correlations in a large temperature domain extending down to about 8K. They have also contributed to the development of a model for organic superconductivity which relies on the possibility of a non-phonon mediated Cooper pairing, boosted by an interchain exchange of spin fluctuations.

INTRODUCTION

The search for superconductivity in an entirely new class of conductors, the organic materials, has been stimulated by the suggestion of Little (1964) that superconduction in such materials might proceed via a non-phonon mediated pairing mechanism through the so-called exciton interaction.

Superconductivity has been found in these compounds after 15 years of an intense and oriented research effort in the field of low dimensional organic materials.

During this period the quest for organic superconductivity has been marked by several milestones : one of them is particularly important since it has revealed in 1979 the interest of the TMTSF molecule which is responsible for the stabilization of an highly conducting phase at low temperature for the charge transfer compound TMTSF-DMTCNQ.

The existence of a conductivity larger than 10^5 $(\Omega cm)^{-1}$ at helium temperature in TMTSF-DMTCNQ (Andrieux et al. 1979a) has triggered the suggestion that such large values could only be achieved if fluctuations towards superconductivity (with no long range ordering) exist below ≈ 50 K (Andrieux et al. 1979b). The TMTSF molecule became less than a year later the major ingredient for the first series of organic superconductors.

The suggestion for fluctuation effects was reiterated for materials of the $(TMTSF)_2X$ family (Bechgaard et al. 1980) showing a true superconducting state in the 1K range (Jérome et al. 1980). The possibility of well developed superconducting precursor effects related to one-dimensional (1D) effects raised much controversy in the research community.

The group suggesting the fluctuation picture emphasized the role of one dimensionality (1D) on the electronic properties of $(TMTSF)_2X$ conductors below 50 K (Schulz et al. 1981). Furthermore the same group suggested that superconducting correlations could be rather large in this class of new superconductors but that the modest temperature for long range ordering was a consequence of the weakness of the interchain kinetic coupling. On the other hand, other people (Greene et al. 1983) remained more dubious about the experimental evidence for the existence of superconducting fluctuations in $(TMTSF)_2X$ materials. They also proposed that the properties of the superconducting state could be understood in terms of a conventional although anisotropic BCS pairing mechanism. The latter outlook did not prevent the continuation of the research activity in a number of places all over the world and at Orsay in particular.

The purpose of the present review is to summarize some of the salient experimental results and to show that they can be understood only within the framework of a theory strongly rooted on low dimensional arguments.

MATERIALS PRESENTATION

Organic superconductivity has been observed so far in essentially two series of materials : the $(TMTSF)_2X$ (Parkin et al. 1981, Jérome and Schulz 1982) and the $(BEDT-TTF)_2X$ series (Williams et al. 1985). These single-chain cation radical salts constitute families of organic conductors different in several respects from the TTF-TCNQ like conductors. All organic superconductors exhibit the same triclinic zig-zag packing of radical-cation molecules, (figure 1).

Figure 1 indicates that the lattice periodicity along the stacking

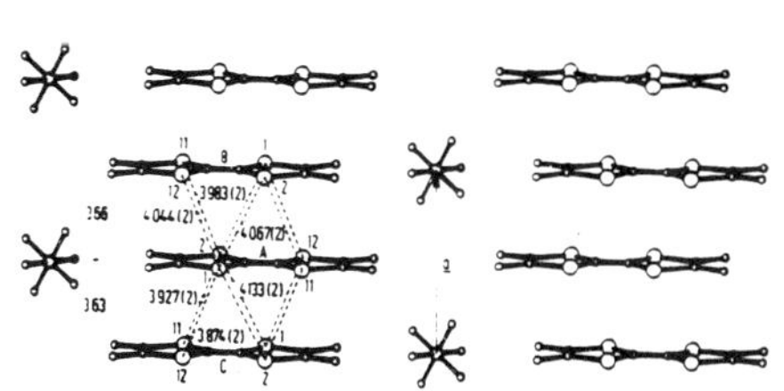

Figure 1 : Some of the molecules used in organic conductors (top), Side view of $(TMTSF)_2PF_6$ (bottom)

direction is determined by the anion spacing. Therefore, the 3D unit-cell contains two organic molecules carrying half a hole per molecule according to the stoechiometry. The energy band formed from the overlapping molecular wave functions of the organic molecules is thus half-filled.

The existence of a structural dimerization can be derived from the examination of the interchain bond lengths. It can even be visible on the intrachain intermolecular distance in case of a pronounced dimerization (case of the sulfur analog $(TMTTF)_2X$ series for example, Galigné et al. (1979)).

The actual structure of $(TMTTF)_2X$ systems allows electron-electron scattering with a wave vector $4k_F$. Such a scattering channel (g_3-terms) has been recognized by Emery et al. (1982) to be important for the determination of the ground state in these systems.

The electronic structure of these conductors is characterized by some basic parameters which are the various overlap interactions, called t'_s and the amplitude of the Coulomb repulsion. The latter interaction is not expected to be small for these organic molecules.

A most significant contribution towards the understanding of the band parameters and the estimate of the correlation strengths has been performed by Jacobsen (1986) with an extensive study of a large variety of these materials by infrared spectroscopy. Polarized-light reflectance experiments reveal a typical metallic character for polarization of the light along the stacking axis of the molecule. Jacobsen (1986) has shown that much information can be extracted from a Drude fit of the observed sharp plasma edge. In the Drude model the frequency dependent dielectric function and the conductivity read

$$\varepsilon(\omega) = \varepsilon_\infty - \omega_p^2/(\omega^2 + \gamma^2) \qquad\qquad\qquad 1.a$$
$$\sigma(\omega) = \omega_p^2\gamma/\varepsilon_o(\omega^2 + \gamma^2) \qquad\qquad\qquad 1.b$$

where ε_∞ is the polarizability arising from all high frequency transitions, γ is the optical electron relaxation rate and ω_p is the plasma frequency.

With a tight-binding model for the energy dispersion the plasma frequency is related to the band parameter by the following relation :

$$\omega_p^2 = \frac{4t\ a^2\ e^2\ \sin(\pi\ \rho/2)}{\pi\ \varepsilon_o\ \hbar\ V_m}$$

where a is the molecular spacing, V_m the molecular volume and ρ the charge transfer ($\rho = 0.5$ for $(TMTSF)_2X$).

For a transverse polarization the light is not absorbed in the visible and near infrared domains. This feature is a signature of the 1-D character of the band structure. However for organic superconductors with appreciable interchain interactions transverse plasmon frequencies can be observed provided the temperature is low enough. The ratio of plasma frequencies is thus proportional to the ratio of transfer integrals (in the open Fermi surface limit). Band parameters derived from optical data for some materials are given in Table I.

Moreover, additional informations about the strength of Coulomb repulsions can also be derived from the optical data in the infrared range. As discussed by Jacobsen (1986) short range Coulomb interactions tend to localize the electrons and consequently to reduce the infrared oscillator strength which is related to the degree of electron delocalization.

In a strongly correlated electron gas, the intra-band oscillator strength

$$I_\sigma(\omega) = \frac{2}{\varepsilon_o \pi} \int_o^\omega \sigma(\omega') \, d\omega'$$

saturates when ω reaches about ω_p and its reduction with respect to ω_p^2 gives a measure of the Coulomb coupling. The reduction of oscillator strength due to Coulomb on-site interaction is characterized by the factor $\beta + (\omega_p^2 - I_\sigma^{sat})/(\omega_p^2 - \omega_p^2 \frac{\sqrt{2}}{2})$, see Table 2.

Table 1 : Data of overlap integrals along the stacking axis and along the axis of intermediate coupling b').

	$t_{\parallel}$ (meV)	$t_{\perp}^{b'}$ (meV)
$(TMTTF)_2PF_6$	200	
$(TMTSF)_2ClO_4$	250	24 (T = 30 K)
$(TMTSF)_2PF_6$	250	20 (T = 25 K)
$(BEDT-TTF)_2I_2$	190	80 (T = 40 K)
$(BEDT-TTF)_2AuI_2$	240	

Table 2 : Optical data, spin susceptibility enhancement and strength of Coulomb coupling for some organic conductors.

	$\omega_p^2 (10^7 \, cm^{-2})$	β	χ_s/χ_{Pauli}	$U/4t_{\parallel}$	U
$(TMTTF)_2PF_6$	7.9	0.91	3.3	1.7	1.4
$(TMTSF)_2ClO_4$	9.8	0.52	2	1.0	1.1
$(BEDT-TTF)_2I_3$	8.3	1.29	3	2.5	1.9

In this model, the larger the value of β the stronger the effect of correlations. It is interesting to notice that the two compounds $(TMTSF)_2ClO_4$ and $(BEDT-TTF)_2I_3$ which give rise to superconductivity at 1.2 K (Bechgaard et al. 1981) and 8.1 K (Laukhin et al. 1985) respectively are also associated to significantly different effects of correlations. In particular Coulomb correlations are fairly large in the system providing the highest T_c so far in organic matter. The optical determination of β is in fair agreement with the enhancement of the susceptibility. The last column in Table II gives the values obtained for U/4t and U with a Hubbard model for two electrons on a four molecule chain assuming U = 3V (V is the nearest neighbour Coulomb repulsion). Table II shows that strong correlations are also found for 2:1 sulfur compounds displaying either spin-Peierls or antiferromagnetic ground states. The role of correlations in organic conductors has already been emphasized by other techniques in the early days of the research activity in this field by Soda et al., (1977) and Torrance (1977).

Before closing this section we recall that the band structure anisotropy can also be derived from conductivity data at room temperature and the anisotropy of the H_{c_2} critical fields in the superconducting state. All these measurements lead to a ratio $t_{\parallel}/t_{\perp} \approx 10$ and 3 in TMTSF and BEDT-TTF series respectively. These values indicate that the Fermi surface is open for all members of the TMTSF family. BETD-TTF compounds lie at the border between open and closed Fermi surface.

Concluding this section we may say that there are definite evidences for strong repulsive Coulomb interactions in organic superconductors. In the conventional BCS picture of superconductivity this Coulomb repulsion has to be overcome by a strong electron-phonon interaction to produce the attractive pairing. Hence, if the electron-phonon coupling is that large in organic conductors why is it that the Peierls channel is not diverging in $(TMTSF)_2X$ materials ?

THE SUPERCONDUCTING PHASE

When studying materials giving rise to organic superconductivity two condensed states are usually encountered : the superconducting state or an antiferromagnetic semiconducting (semimetallic) state. In the present article we shall not discuss the semiconducting ground state resulting from an ordering of non-centrosymmetric anions leading to a folding of the Brillouin zone with the concomitant onset of a gap at the Fermi level (Moret et al. 1986).

The superconducting state can be observed under various conditions :
(i) At $T_c \approx 1.2$ K in $(TMTSF)_2PF_6$ and $(TMTSF)_2AsF_6$ under a pressure exceding 8 kbar (figure 2) or in $(TMTSF)_2ClO_4$ at ambient pressure,

provided the sample is cooled slowly enough below 30 K to allow a nearly perfect ordering of the ClO_4^- anions leaving unaffected the lattice periodicity along the stacking direction (Tomic et al. 1984).

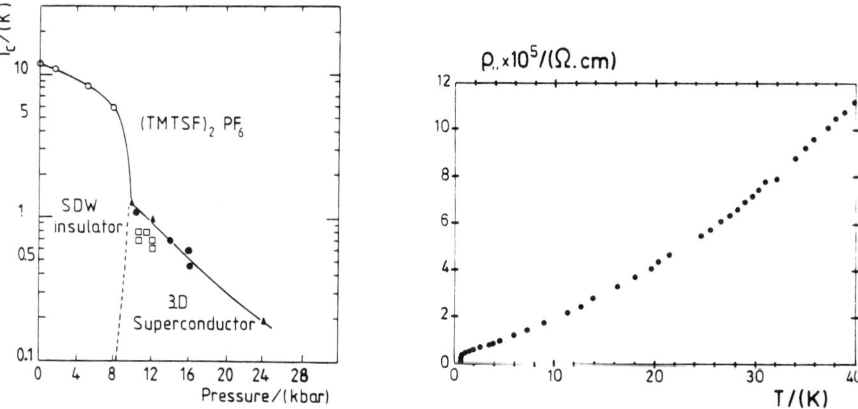

Figure 2 : P-T phase diagram of $(TMTSF)_2PF_6$ or $(TMTSF)_2AsF_6$ (left). Temperature dependence of the resistivity of $(TMTSF)_2PF_6$ at P=11 kbar $T_c \approx 1$ K (right)

(ii) At $T_c \approx 8.1$ K in $(BEDT-TTF)_2I_3$ and at ambient pressure after a special pressurization-cooling process is applied avoiding the onset of an incommensurate lattice distortion at 175 K and P = 1 bar (Creuzet et al. 1985, Ginodman et al. 1985). A thermodynamic study has clearly shown that one of the two phases of $(BEDT-TTF)_2I_3$ can be stabilized at low temperature at will depending on the cooling process (Kang et al. 1987). The T_c's of the β-L and β-H phases are 1.4 K and 8.1 K respectively.

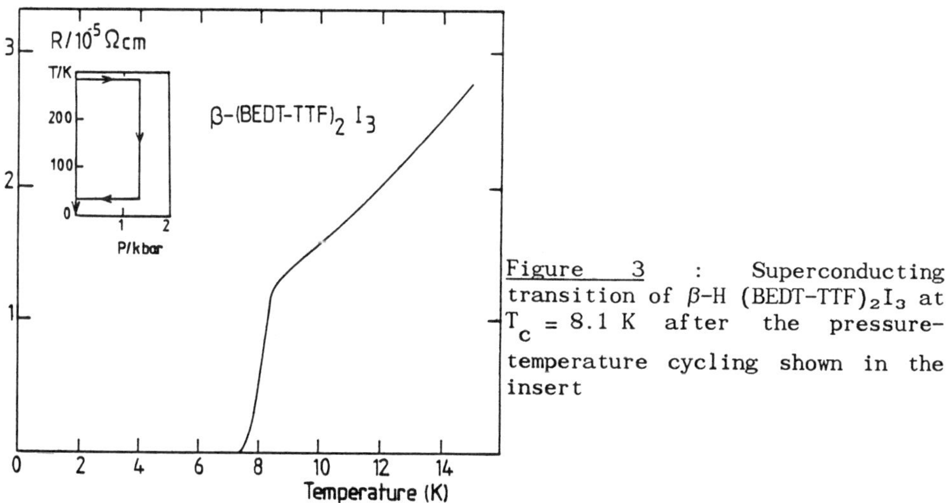

Figure 3 : Superconducting transition of β-H $(BEDT-TTF)_2I_3$ at $T_c = 8.1$ K after the pressure-temperature cycling shown in the insert

Figure 3 displays the transition towards superconductivity which is observed in the β-H phase of $(BEDT\text{-}TTF)_2I_3$ at ambient pressure after cooling down the sample to 100 K under a minimum pressure of about 0.4 kbar. Since NMR (Creuzet et al. 1986) and recent EPR data (Creuzet et al. 1987) indicate no significant differences between electronic spin susceptibilities of β-L and β-H phases the low T_c of the β-L phase can probably be attributed to the existence of the incommensurate lattice distortion (Emge et al. 1984).

So far, studies of the isotope effect performed on deuterated samples of both $(TMTSF)_2X$ (Schwenk et al. 1983), and $(BEDT\text{-}TTF)_2I_3$ (Heidman et al., 1986) have failed to show the depression of T_c expected in the usual BCS mechanism. Instead, a slight increase of T_c has even been observed upon deuteration in both cases. This may be attributed either to a better crystallization of deuterated samples or to the possible existence a non phonon mediated pairing.

The anisotropy of the upper critical field reflects fairly well the band structure ansiotropy. However, the origin of the critical field along the stacking is still unclear. Two possible limitations have been claimed for $H_{c_2}^a$: Pauli (Greene et al. 1982) or orbital pair breaking (Gor'kov and Jérome 1985). A Pauli limited critical field would imply spin singlet pairing. We think that the experimental determination of H_{c_2} needs some improvement especially for the $(TMTSF)_2X$ series before drawing any definite conclusion.

An other signature of the superconducting transition is given by specific heat data. The jump of specific heat amounts to $\frac{\Delta C}{C} = 1.67$ in $(TMTSF)_2ClO_4$ (Garoche et al. 1982). Combining zero and high magnetic field specific heat data of $(BEDT\text{-}TTF)_2AuI_2$ (Andres et al. 1986) and the electronic specific heat of $(BEDT\text{-}TTF)_2I_3$ ($\gamma = 24$ mJ/K^2 mole) (Stewart et al. 1986) the estimated jump for $(BEDT\text{-}TTF)_2AuI_2$ is about $\frac{\Delta C}{C} = 1.33$.

Several attempts have been made to determine the value of the superconducting gap by tunneling techniques in both TMTSF and BEDT-TTF superconductors. Schottky barrier techniques or SIN junctions have been used for $(TMTSF)_2PF_6$ and $(TMTSF)_2ClO_4$ respectively (Fournel et al. 1983, 1985) whereas $(BEDT\text{-}TTF)_2AuI_2$ has been investigated by point contact tunneling (Hawley et al. 1986). For all cases the quality of the experimental data does not allow to draw unambiguous conclusions. However, all experiments have given evidences of a structure in the density of states related to a gap which is always much larger than the value predicted from a BCS theory in the weak coupling limit. The measured gaps amount to $2\Delta \approx 3.6$-3.8 meV and ≈ 5-6 meV in TMTSF and BEDT-TTF respectively. These values are admittedly more than four times larger than the ratio $2\Delta/kT_c = 3.5$ of the BCS theory implying either an extremely strong coupling for organic superconductors or that the large gap 2Δ is not directly related to the onset of 3D superconductivity. Instead, it could be related to the establishment of 1D correlation effects above T_c as we shall discuss at the end of this article. Indeed, an anomaly in the far infrared reflectance spectrum has been found for $(TMTSF)_2ClO_4$ (Ng et al. 1983) at ≈ 30 cm^{-1} (3.8 meV) persisting well above T_c, up to about 30 K. We shall see in the following section that the structure (pseudogap in the FIR conductivity around 30 cm^{-1}) is observed even at high temperatures for all investigated samples of the $(TMTSF)_2X$ series, irrelevant of the nature of the ground state (superconductivity or antiferromagnetism).

ANTIFERROMAGNETIC PHASES

The finding of an antiferromagnetic phase in the system $(TMTSF)_2PF_6$ at 1 bar where superconductivity is also observed under pressure has been one of the highlights of the research activity. The itinerant antiferromagnetic state of $(TMTSF)_2AsF_6$ and $(TMTSF)_2PF_6$, $T_N \simeq 12$ K, has been fairly well studied by the anisotropy of the susceptibility in single crystals (Mortensen et al. 1982) and antiferromagnetic resonance techniques (Torrance et al. 1982). A detailed ^{1}H NMR line shape analysis in $(TMTSF)_2PF_6$ has enabled the determination of the factors characterizing the local magnetic field modulation (spin density wave,

SDW), leading to the wave vector $\underset{\sim}{Q} \approx (\frac{a^*}{2}, \frac{b^*}{4}, 0\ c^*)$ and a SDW amplitude of about 8 % μ_B/molecule. (Delrieu et al. 1986, Takahashi et al. 1986). The SDW state is known to be significantly affected by a high pressure as shown by figure 2. The pressure dependence of the SDW state is indeed particularly large above 6kbar when the SDW state vanishes quite suddenly around 8 kbar. However in both cases, $(TMTSF)_2PF_6$ under pressure ($P > P_c$) or $(TMTSF)_2ClO_4$ at ambient pressure, the SDW phase can be restored by the application of a large enough magnetic field (figure 4). This is a fascinating phenomenon which has attracted much attention in the past few years. The existence of a phase transition occuring under high magnetic field has been first mentionned by ^{1}H NMR T_1 measurements at 94 kG and $T = 1.9$ K in $(TMTSF)_2PF_6$ under pressure (Azevedo et al. 1981). Then, the magnetic character of the high field state of $(TMTSF)_2ClO_4$ has been established by the ^{77}Se NMR line broadening and the critical divergence of the relaxation rate associated with the phase transiton (Takahashi et al. 1982). Magnetotransport data have revealed that the onset of a

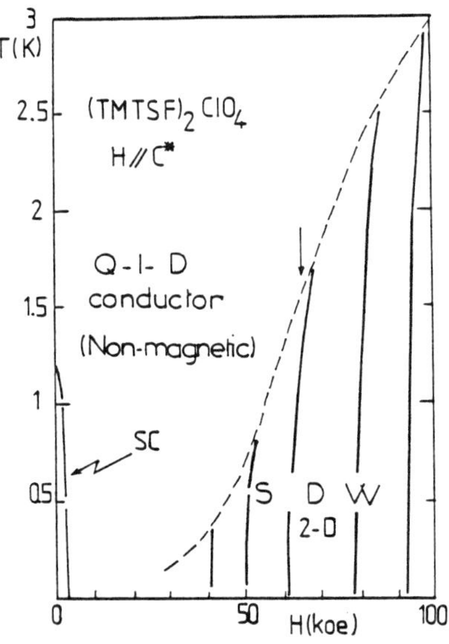

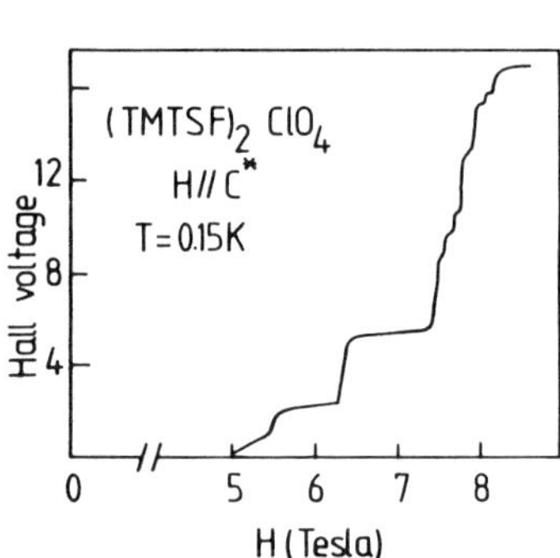

<u>Figure 4</u> a) : H-T phase diagram of (TMTSF)$_2$ClO$_4$ the vertical arrow indicates the magnetic field for the NMR data in fig. 14.
     b) : Hall voltage of (TMTSF)$_2$ClO$_4$ in the FISDW states showing the step and plateau behaviour (after Ribault et al. 1983).

magnetic state at high fields is more subtle than it was first believed (Kajimura et al. 1983). The resistivity increases when entering the magnetic state and oscillations approximately periodic in 1/H are observed in the magnetoresistance. Hall effect (Ribault et al. 1983, Chaikin et al. 1983) and calorimetric data have proved that above a threshold field a SDW semimetallic ground state is stabilized. Furthermore, the magnetic field induces a succession of phase transitions between different SDW states (Pesty et al. 1985) (figure 4). The clue for the understanding of the field induced SDW states in (TMTSF)$_2$X compounds has been given by the nearly field independent Hall voltage in each SDW subphases (figure 4), (Ribault et al. 1983, Chaikin et al. 1983). Gor'kov and Lebed (1984) have explained the onset of magnetism at high fields by the one-dimensionalisation of quasi 1D electrons in a strong magnetic field leading to a transition from a quasi 1D (open Fermi surface) non-magnetic conductor to a 2D magnetic semimetal at a threshold field. Héritier et al. (1984) have interpreted the transition between different SDW subphases by a quantized nesting vector argument. Each sub-phase is a semimetal, i.e. some carriers remain unpaired since the nesting of the Fermi surface is not perfect. Consequently, the density of unpaired carriers (and thus the nesting vector) adapts slightly to the field when it varies in order to maintain completely filled the occupied Landau levels below the Fermi level. However, this situation cannot be maintained in too wide a range of magnetic fields. At certain values of the magnetic field, the number of Landau levels below the Fermi level jumps discontinuously by one unit leading to a first order transition between near neighbour SDW phases (fig. 5).

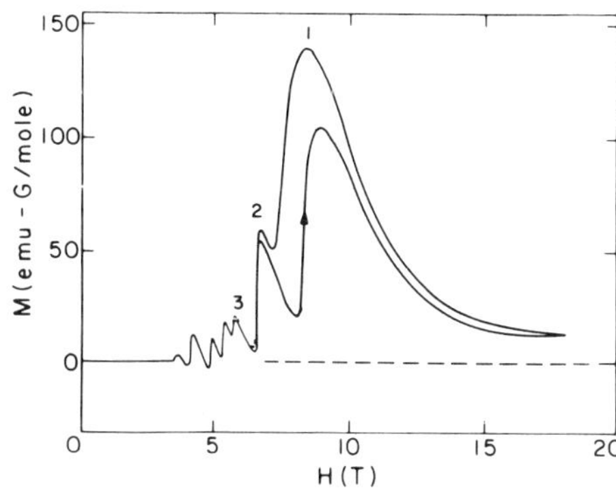

Figure 5 : Magnetization jumps in the field induced SDW phases of $(TMTSF)_2ClO_4$ (after Naughton et al. 1985).

This model gives a possible explanation for the observation of steps and plateaus of the Hall voltage. Recent calorimetric experiments (Piveteau et al. 1987) and a magnetization study (Naughton et al. 1985) have shown that the phase transitions between SDW subphases are indeed first order showing some hysteretic behaviour and a latent heat of transformation.

$(TMTSF)_2PF_6$ under pressure does exhibit a rather similar behaviour at high fields, although giant Hall voltage oscillations are observed instead of the steps and plateaus sequence (Piveteau et al. 1986).

The differences between the high field SDW states of PF_6 and ClO_4 salts of TMTSF may be ascribed to the anion ordering in $(TMTSF)_2ClO_4$ at 24K which opens a small gap on the Fermi surface at $k_y=\pm\pi/2b$ in this salt.

The orbital origin for the stabilization of SDW states at high field is also corroborated by the orientation dependence of the onset field. The angular dependence ($\sin\theta$ law) of the threshold field when H is rotated in the $b^* - c^*$ plane agrees with the 2D nature of the orbits in the field-induced states.

THE CONDUCTING STATE : DIMENSIONALITY CROSS-OVER

We turn now to the discussion of the electronic properties of organic superconductors in the temperature domain between 100 K and the ordering temperature. By ordering temperature we mean the temperature where the conducting state undergoes a transition towards either a SDW state ($(TMTSF)_2PF_6$ at low pressure, or field induced SDW states) or superconductivity. We want to present in this section some experimental observations which are relevant in the discussion about the role of 1D effects in organic superconductors.

Emery (1983) has raised the point that the mixture of e-e and e-h correlations and their simultaneous divergence, which is the essence of 1D physics (Bysckov et al. 1966) can only occur in a quasi 1-D conductor with a finite value of $t_\perp$ as long as T is larger than some cross-over temperature T_x between 1D and 2D (3D) regimes. Within a tight-binding description of the quasi 1D band, T_x is about $t_\perp/\pi$ (Emery, 1986). As $t_\perp$ is of the order of 20-24 meV in the $(TMTSF)_2X$ series according to the IR data, the cross-over should occur in the region of $T_x = 80$ K. Therefore, one may be tempted to argue that since phase transitions, (in particular superconductivity) always occur at temperatures much lower than 80 K the use of 3D physics and mean-field treatments is perfectly justified.

A striking feature of the low temperature DC conductivity of organic superconductors is the huge value $[\sigma \sim 10^6 \ (\Omega \ cm)^{-1}]$ reached at liquid helium temperature with no sign of residual resistance down to the lowest temperature (see figure 2). Moreover, an anomalously large magnetoresistance is observed up to 30-40 K in spite of the existence of an open Fermi surface which is clearly supported by low field Hall effect data (Ribault et al. 1983) giving a density of carriers, equal to 1/2-hole per molecule which is expected from stoechiometry considerations. Furthermore, the strong angular dependence of the magnetoresistance is similar to the law followed by either the H_{c_2} anisotropy or the threshold field for the stabilization of the SDW states at high fields. This provides an indication that the magnetoresistance is related to an orbital and not to a spin effect of the magnetic field.

FAR INFRARED OPTICAL CONDUCTIVITY

We have already seen at the beginning of this review that infrared spectroscopy constitutes a very important tool in attempts to measure the band parameters of organic superconductors. Moreover, the infrared oscillator strength gives an estimate of the role of short-range electron-electron interactions which correlates fairly well with independent evaluation from spin susceptibility data. As infrared spectroscopy has played a very important role in superconductivity for a direct measurement of the energy gap it was tempting to apply similar techniques to organic superconductors in order to check for the existence of superconducting or SDW gaps and also for the validity of a Drude interpretation of the optical conductivity. According to eq. 1b the Drude conductivity should be constant at low frequencies and should fall off as $1/\omega^2$ at a frequency of the order of $\omega = 1/\gamma$. Thus, the study of the FIR conductivity could discriminate between a regular (Drude-like) and an unusual metallic behaviour. Jacobsen et al (1983 a), Ng et al. (1985) and Kornelsen et al. (1987) have noticed that the FIR conductivity of $(TMTSF)_2PF_6$ at low temperature for the light polarized along the stacking axis is indeed much lower than the DC conductivity. For example at $T = 30$ K and $\omega = 60$ cm^{-1} the conductivity is about 500 $(\Omega cm)^{-1}$ whereas the DC conductivity approaches 20 000 $(\Omega cm)^{-1}$ (figure 6 a, b). Kornelsen et al. (1987) have been able to measure a gap of ≈ 30 cm^{-1} in the SDW state of $(TMTSF)_2AsF_6$. In the SDW state there exists a fair agreement between the value of the conductivity at DC and low frequency.

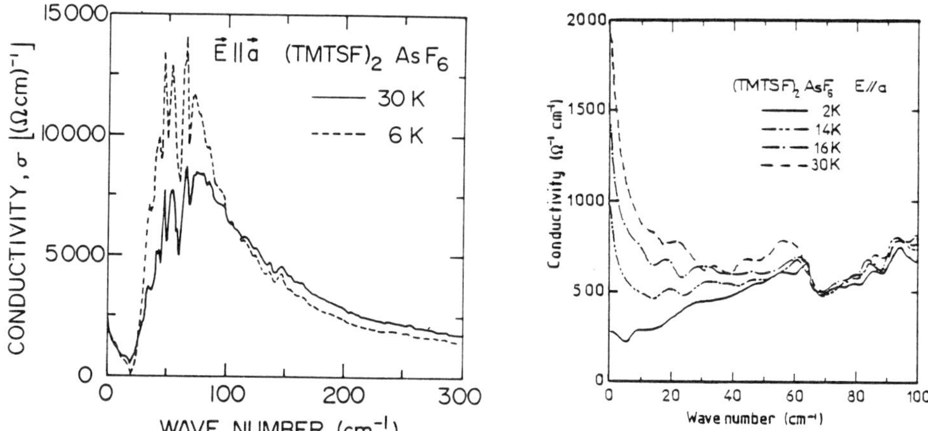

Figure 6 : Far infrared conductivity of $(TMTSF)_2AsF_6$. The FIR gap is clearly seen in the data of Kornselsen et al. (1987) (left). The collective mode at zero frequency exists in the conducting phase above T_N (Ng et al. 1985) (right)

This is consistent with the idea that the material undergoes a metal-semiconductor transition, figure 6 a. However, a gap persists in the FIR spectrum above the SDW transition.

Above 12 K, the discrepancy between DC and low frequency conductivity implies the existence of a conduction peak centered at zero frequency. This discrepancy exists in all studied organic superconductors. It is a characteristic behaviour of the conducting state at low temperature independent of the nature of the ground state. Similar conclusions have been reached for $(TMTSF)_2ClO_4$ where the conduction in the 40 cm^{-1} range at 2 K is about 100 times smaller than the observed DC value (Ng et al. 1983). The proposal by Marianer et al. (1982) that the reflectance measured at the lowest frequencies can be depressed from the bulk value by damaged surface effects has been ruled out by Kornelsen et al. (1987) since the penetration depth is large at these low frequencies.

We are left therefore with a zero frequency collective mode whose width can be estimated from the real part of the frequency dependent dielectric constant following a method proposed by Jacobsen et al., (1983 b) fitting the zero frequency mode with a damped oscillator at zero frequency, (figure 7). The dielectric constant related to this mode reads

$$\varepsilon_1(\omega) = \varepsilon_H - (\Omega_p/\omega)^2$$

in the frequency range ($\gtrsim 6$ cm^{-1}) where it crosses zero. Fitting the data of figure 7 leads to the determination of Ω_p (the collective-mode plasma frequency) and of the collective mode frequency width $1/\tau$ through the relation $\sigma_{dc} = \Omega_p^2\tau/4\pi$. Hence we get at T = 30 K the values $\Omega_p = 600$ cm^{-1}

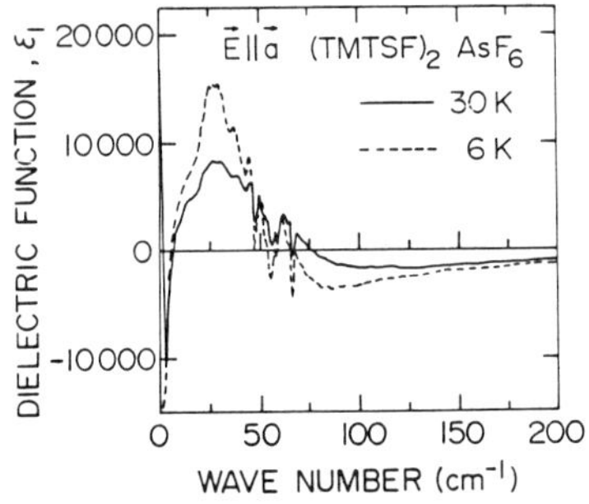

Figure 7 : The real part of the dielectric function for E ∥ a (according to Kornelsen et al. 1987). The zero crossing which is relevant for the collective mode occurs near 6 cm^{-1}.

and $1/\tau = 1$ cm^{-1} (30 GHz).

The existence of such a narrow collective mode is consistent with the measurement of the microwave conductivity in the 20-30 K temperature domain. Data in $(TMTSF)_2PF_6$ (Javadi et al., 1985) show that the conductivity at 35 GHz is lower than the dc value by a factor about 2. This is also in agreement with an earlier study performed on similar systems (Bechgaard et al. 1980).

A straightforward application of the single particle model for the lifetime of the mode, $1/\tau = \omega_p^2/4\pi\sigma$ leads to $1/\tau = 55$ cm^{-1} with $\sigma_{dc} = 2\ 10^4$ $(\Omega cm)^{-1}$ and $\omega_p = 10.000$ cm^{-1} determined by IR optical data. A peak about 50 cm^{-1} wide would dominate the far-infrared spectrum giving a conductivity of $\approx 10^4$ $(\Omega\ cm)^{-1}$ at 50 cm^{-1} and a reflectivity of the order of 98 % up to about 40 cm^{-1} (fig. 8). Both features are not observed in the low frequency spectra of organic superconductors. Notice that the pseudo-gap seen by tunneling techniques is of the same order of magnitude (see figure 8 for $(TMTSF)_2PF_6$ data). From the inspection of FIR data taken in slowly cooled $(TMTSF)_2ClO_4$ at 2 K (Ng et al. 1983) it was also concluded that the width of the zero frequency mode must be less than 0.5 cm^{-1}. Assuming a temperature independent oscillator strength of the zero frequency mode we get from the 20-30 K data the estimate of $1/\tau \approx 0.1$ cm^{-1} at 2 K which is consistent with the FIR results. A careful and quantitative study of the $(TMTSF)_2ClO_4$ microwave conductivity would

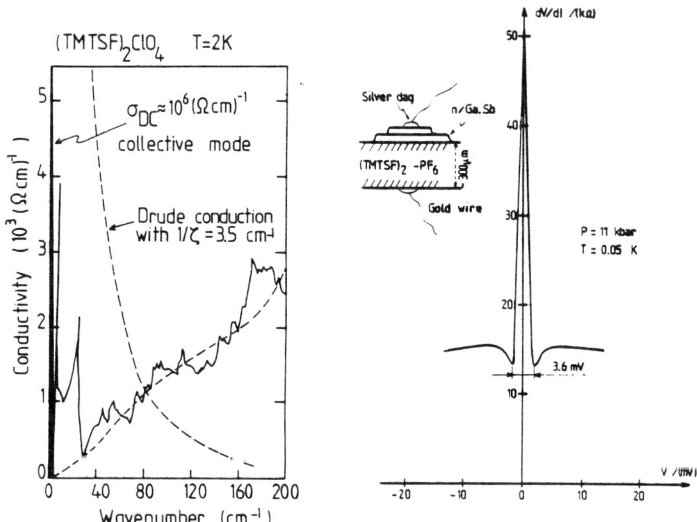

<u>Figure 8</u> : (Left) FIR conductivity of $(TMTSF)_2ClO_4$ with E ∥ a at T = 2 K according to the data of Ng. et a;. (1983). The Drude behaviour with $1/\tau \sim 3.5$ cm^{-1} and $\omega_p = 10.000$ cm^{-1} does not fit the experiment in the FIR regime. (Right) Pseudogap in the density of states of $(TMTSF)_2PF_6$ under pressure (2Δ ~ 3.6 meV) observed by Schottky tunneling techniques in the superconducting states and also above T_c (from Fournel et al. 1983).

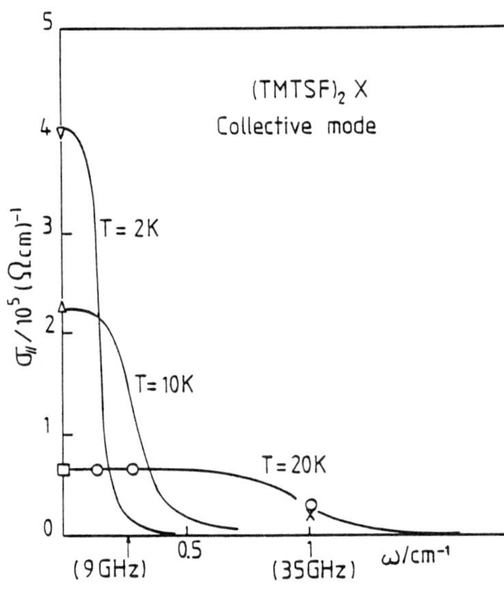

Figure 9 : Schematic picture showing the development of the collective mode at low temperature. At T = 20 K, 4.5, 9 and 35 GHz points (o) are taken from Javadi et al. (1985), the DC and 35 GHz points (, x) are taken from Bechgaard et al. (1981). At T = 10 and 2K only the DC data are known. $\sigma(\omega)$ is drawn assuming a temperature independent oscillator strength in the collective mode.

thus be a very valuable experiment to check the prediction of figure 9 regarding the existence of a fairly narrow collective mode at 2K.

Other studies have indicated that the laws which are valid for a single-particle electron model are not followed at low temperature. This is for instance the absence of a Matthiessen's rule for the resistivity. The increment of resistivity due to a certain concentration of impurities is , larger at low temperature than at high temperature (T > 50 K) (Bouffard et al., 1981). Similarly, the Wiedeman-Franz relation between the thermal conductivity and the electrical conductivity is followed neither when temperature is varied nor under large magnetic field at fixed temperature (Djurek et al. 1985).

Summarizing this section we wish to emphasize that the non-single particle character of the electron-gas below 50 K relies on several experimental evidences, in particular the non-Drude behaviour of the FIR optical properties which has been advocated in various studies of the $(TMTSF)_2X$ family.

MAGNETISM IN THE CONDUCTING STATE

An other experimental technique has proved to be rewarding for understanding the behaviour of the conducting state : this is the NMR of ^1H, ^{77}Se and ^{13}C nuclear spins belonging to the organic molecules. All spins are I = 1/2 i.e. their spin lattice relaxation rate will be mediated by the electron hyperfine interaction. As far as ^1H spins are concerned, classical rotation and quantum tunneling of the CH_3 groups in

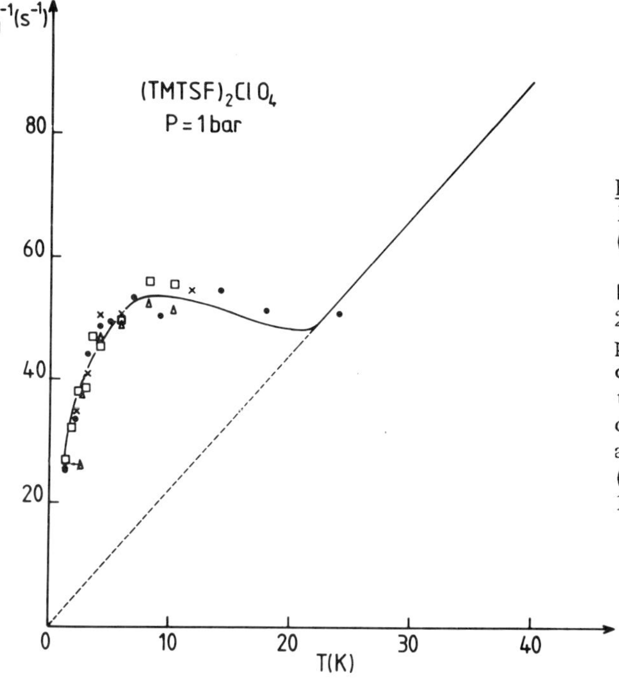

Figure 10 : ^{77}Se spin lattice relaxation in $(TMTSF)_2ClO_4$ at 31.9 kG ‖ c* (□), ⊥ G' (●), 21 kG ‖ b' (x) GkG powder (△). The continuous line shows the Korringa behaviour observed at T > 30 K and H = 64 kG ‖ b', (Bourbonnais et al. 1984).

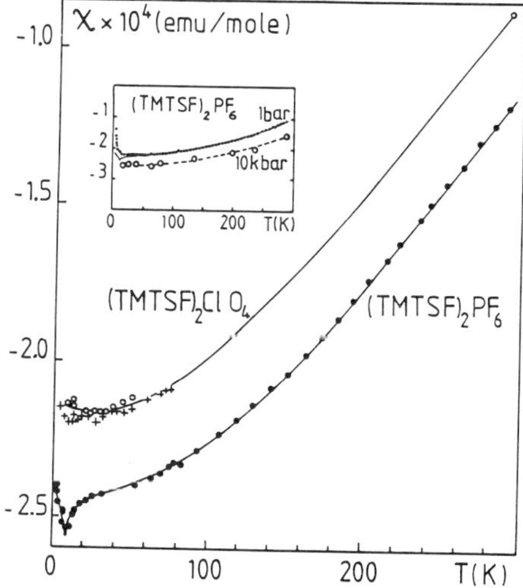

Figure 11 : Temperature dependence of the spin susceptibility for $(TMTSF)_2PF_6$ and $(TMTSF)ClO_4$ according to Miljak et al. 1983). The insert shows the pressure dependence.

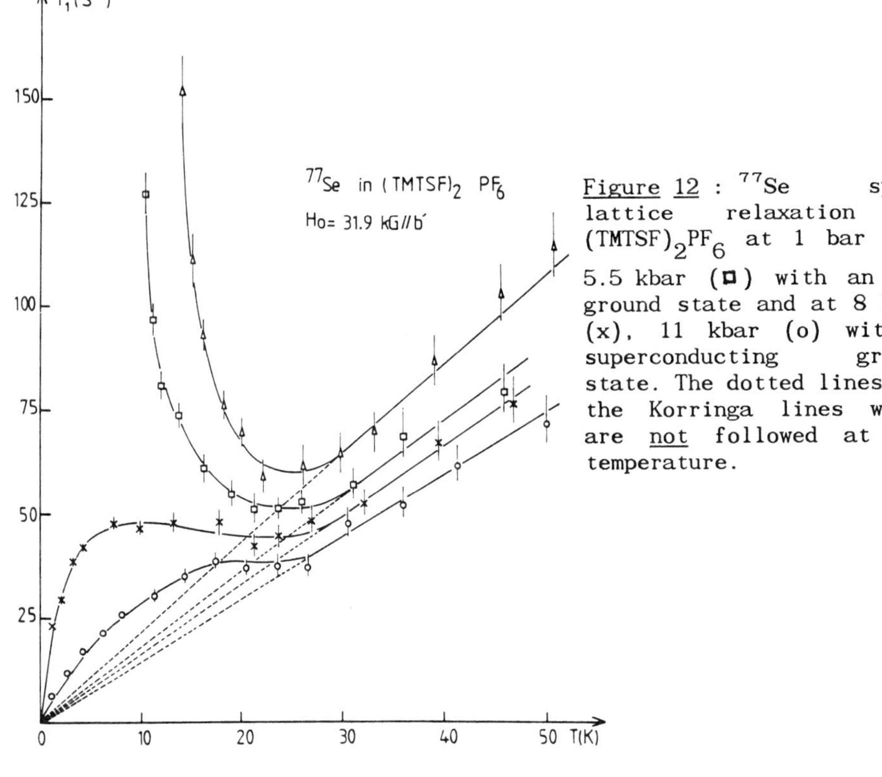

Figure 12 : ^{77}Se spin-lattice relaxation of (TMTSF)$_2$PF$_6$ at 1 bar ($\triangle$) 5.5 kbar ($\square$) with an SDW ground state and at 8 kbar (x), 11 kbar (o) with a superconducting ground state. The dotted lines are the Korringa lines which are not followed at low temperature.

the three-well potential can be the source of an additional relaxation channel. However, experiments have shown that this extrinsic source of relaxation is negligible either at room temperature or at low temperature below 20 K (Takahashi et al. 1984). Measurements of ^{1}H and ^{77}Se relaxation rates in (TMTSF)$_2$ClO$_4$ have revealed a striking feature at low temperature, namely a large enchancement of T_1^{-1} which results in strong deviations from a Fermi liquid type of relaxation ($T_1 T$ = const), (Abragam 1961) occuring below 25 K (Bourbonnais et al., 1984), (figure 10) whereas in the same temperature domain the spin susceptibility remains temperature independent (Miljak et al. 1983), (figure 11). This domain of temperatures lies very near the anion ordering temperature at 24 K (Moret et al. 1983) and not too far from the ^{1}H peak of relaxation due to quantum tunneling (Takahashi et al. 1984) so that the purely electronic origin of T_1^{-1} enhancement could not be established unambiguously. In order to be sure of the electronic origin of the T_1^{-1} enhancement, relaxation studies have also been performed in (TMTSF)$_2$PF$_6$ both below and above the critical pressure for the occurence of superconductivity. Figure 12 displays the results of T_1^{-1} at various pressures (1 bar, 5.5, 8 and 11 kbar) in the T-domain 1.2 < T < 50 K. At 1 bar and 5.5 kbar, the strong enhancement of T_1^{-1} together with the drop of ^{77}Se signal intensity are signatures of transitions towards the SDW state at 12.2 and 8.7 K respectively. On the contrary at 8 and 11 kbar, the Curie law which is followed by the NMR intensity implies that all nuclear spins contribute to an NMR line of constant width. In the same temperature domain, with a magnetic field of 31.9 kG parallel to the b'-axis no SDW state can be

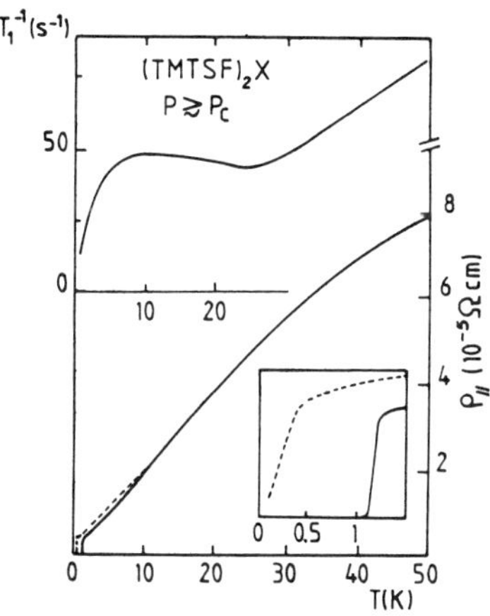

Figure 13 : Comparison between the temperature dependence of ^{77}Se-T_1^{-1} and $\rho_{\parallel}$ in $(TMTSF)_2PF_6$ (or AsF_6) at $P > P_c$ ($\approx$ 9 kbar) in a field of 31.9 kG $\parallel$ b' (dotted line). The insert shows the behaviour of the resistivity at very low temperature in zero (dotted line) and 31.9 kG (continuous line) magnetic field (after Brusetti et al. 1982).

stabilized above 1.2 K and the conductivity remains practically unaffected by the magnetic field ($\frac{\Delta\rho}{\rho_o} \approx$ 10 % at T = 1.2 K in a field of 40kG, Brusetti et al. 1982), (see figure 13). Therefore, the striking deviation from a Korringa type relationship $T_1^{-1} \propto$ T occurring below 30 K cannot be attributed to the onset of any static SDW state. These experiments provide an additional confirmation for the violation of the Korringa behaviour mentionned previously for the ^{1}H and ^{77}Se data in $(TMTSF)_2ClO_4$. Notice that in a field of 4kG//b' $(TMTSF)_2ClO_4$ undergoes a transition towards superconductivity at $T_c \approx$ 0.8 K (Mailly et al. 1983). Thus, we may infer that the violation of the Korringa law is related to electronic phenomena which are precursor to the establishment of superconductivity. Since the uniform spin susceptibility (q = 0) follows a regular behaviour at low temperature, figure 11, we feel justified to attribute the violation of the Korringa law to a divergence of the staggered contribution (q = $2k_F$) to $\chi(q)$. As a matter of facts, a study of the magnetic field dependence of T_1^{-1} in organic conductors (Soda et al. 1977) has shown that the component $\chi(q = 2k_F)$ contributes predominantly to the spin-lattice relaxation at low temperature.

The analysis of the relaxation data in $(TMTSF)_2PF_6$ under pressure, figure 12, as well as those in $(TMTSF)_2ClO_4$ at ambient pressure shows that the relaxation rate follows a power law temperature dependence between 30 and 8 K, namely $(T_1T)^{-1} \propto T^{\gamma}$ where γ is an exponent slightly larger than unity ($\gamma \approx 1.1$).

At very low temperature ($T < 5$ K) an other Korringa behaviour is recovered ($T_1T = $ const) though the Korringa constant is significantly enhanced above its value at higher temperature ($T > 30$ K).

DISCUSSION

We turn now to the discussion of the behaviour of the spin-lattice relaxation which appears rather universal among all superconducting members of the $(TMTSF)_2X$ family.

The temperature dependence of T_1^{-1} can be understood in terms of the Moriya's formulation (Moriya 1963) adapted to the specific problem of 1D conductors by Bourbonnais et al. (1984).

The basic expression for T_1^{-1} is given by the relation

$$(T_1T)^{-1} = 2 \gamma_N^2 |A|^2 \int dq \, \frac{\chi''(q, \omega)}{\omega}$$

where the hyperfine contact matrix element A is supposed to be q-independent and ω is the nuclear Larmor frequency. The dynamic scaling hypothesis for anisotropic conductors (Bourbonnais 1987) implies for the $2k_F$ contribution to T_1^{-1}

$$(T_1T)^{-1}_{2k_F} = \left[\frac{T}{E_o} \right]^{-\gamma_{1D}} \left[E_o \, T_1(E_o) \right]^{-1}$$

where γ_{1D} is the exponent governing the power law divergence of $\chi_{SDW}(2k_F)$ in the 1D regime and E_o ($E_o \ll E_F$) is a certain energy cut-off. Notice that the existence of a power-law divergence for the staggered part of the spin correlation function implies a 1D regime, i.e. $T > T_{x_1}$ where T_{x_1} is the single-particle 1D to 3D(2D) crossover temperature. The exponent γ_{1D} is non universal and depends on the strength of the 1D couplings g_1, g_2, g_3 (Solyom, 1979). Below T_{x_1}, the divergence of χ_{2k_F} is frozen in and a Korringa-like behaviour is recovered but the Korringa constant is enhanced by the factor $\left[\dfrac{T_{x_1}}{E_0}\right]^{-\delta_{1D}}$, namely

$$(T_1 T)^{-1} = \left[\frac{T_{x_1}}{E_0}\right]^{-\delta_{1D}} \left[E_0\, T_1(E_0)\right]^{-1}$$

Our main concern now is the estimate of the actual crossover temperature T_{x_1} since band parameters of the $(TMTSF)_2X$ series would suggest a crossover in the vicinity of 80 K. With such a value of T_{x_1} in mind the unusual behaviour of T_1 would entirely occur in the 3D regime. T_{x_1} should thus be larger than 30 K or so. However, the first non-critical corrections in the 3D domain below T_{x_1}

$$(T_1 T)^{-1}_{2k_F} \sim \left[\frac{T_{x_1}}{E_0}\right]^{-\delta_{1D}} \left[1 + c\ \ell n\ \frac{T_{x_1}}{T} + \ldots\right]$$

are logarithmic (therefore small) and cannot explain the large enhancement of relaxation which is observed between 30 and 8 K. On the other hand, the behaviour of T_1^{-1} in a 1D regime $(T > T_{x_1})$

$$(T_1 T)^{-1}_{2k_F} \sim \left[\frac{T}{E_0}\right]^{-\delta_{1D}} \left[1 + c(T) + \ldots\right] \tag{2}$$

with $c(T) < 1$ and $\delta_{1D} \simeq 1$ could account for the observed temperature dependence between 30 and 8 K. Therefore, we are practically forced to accept that the dimensionality crossover occurs around 8K and not at the value of 80 K calculated for a non-interacting quasi 1D electron gas. Summarizing, the comparison between the theory, eq. (2) and the experimental temperature dependence of T_1, leads to $\delta_{1D} \approx 1.1$ and $T_{x_1} \sim 8$ K. The relatively large value of δ_{1D} indicates that a regime of strong 1D spin correlations is reached in the vicinity of 8 K. The finding of T_{x_1} in the helium temperature range is not too surprising within the framework of the model proposed by Bourbonnais et al. (1984). 1D correlations tend to localize electrons on a given chain, suppress the efficiency of the interchain tunneling thereby decreasing $t_\perp$. In the presence of correlations, the effective T_{x_1} is renormalized (Bourbonnais et al. 1984)

$$T_{x_1} = T^o_{x_1} \left[\frac{t_\perp}{E_F} \right]^{\theta/1-\theta} \tag{3}$$

where $T^o_{x_1}$ in eq. (3) is the free-electron crossover temperature (~ 80 K) and θ an exponent related to the exponent of the 1D power law divergence. Given $T_{x_1} \approx 8$ K, $t_\perp/E_F \sim 0.1$-0.3 we get a θ value in the range of 0.25-0.5 which is compatible with the strength of 1D antiferromagnetic correlations.

Below the critical pressure for the suppression of the SDW state ($P_c \lesssim 8$ kbar) an antiferromagnetic state is stabilized (at $T_N = 8.7$ K under 5.5 kbar). Thus, the 3D antiferromagnetic ordering under 5.5 kbar occurs at a temperature higher than T_{x_1} (5.5 kbar) since T_{x_1} is believed to increase under pressure and is effectively observed (Creuzet et al. 1985). The relation $T_N > T_{x_1}$ means that the 3D character of the band structure is not pertinent for the transverse coupling which is needed for the establishment of 3D antiferromagnetism. Bourbonnais and Caron (1986) and Caron and Bourbonnais (1986) have argued that a finite antiferromagnetic exchange interaction called IEX interaction can still exist between 1D $2k_F$ spin correlations (1-D electron-hole pairs) even in the absence of strong Umklapp g_3 interactions to localize particules on molecular dimers.

The interchain coupling which is responsible for the stabilization of the SDW phases under high magnetic field is however of a quite different nature. According to Gor'kov and Lebed (1984) the SDW stabilization occurs when the 3D character (nesting) of the band structure is relevant, that is below the single-particle dimensionality crossover T_{x_1}. The value of T_N for a nested antiferromagnet (NAF) below T_{x_1} is given by a RPA equation for a 3D Fermi liquid

$$1 - \lambda \, \chi_o(\underset{\sim}{Q}_o, T_N) = 0$$

where $\chi_o(q, T)$ is logarithmically singular in temperature at the nesting vector $\underset{\sim}{Q}_o$.

NMR experiments on $(TMTSF)_2ClO_4$, (figure 14) provide again a nice illustration for the NAF mechanism. The sharp square root increase of T_1^{-1} is related to the existence of 3D magnetic fluctuations in the close vicinity of the NAF transition at 1.55 K ($H = 64$ kG $\parallel c^*$) which remains smaller than $T_{x_1} \sim 8$ K.

It is also interesting to notice in figure 14 that for $H \parallel b'$ the ground state is conducting in a field of 65 kG and an other Korringa regime is recovered below 4K. However, the behaviour of T_1^{-1} at higher temperature is not much influenced by the orientation of the magnetic field. Once more the enhancement of relaxation which is observed at high temperature (T ~ 10-15 K) for both orientations of the magnetic field can be attributed to the effect of 1D correlations. These NMR data are important as they allow a clear distinction between AF correlations leading to 3D critical effects in T_1^{-1} near T_N and those taking place near 8 K and above which are associated to 1D correlations.

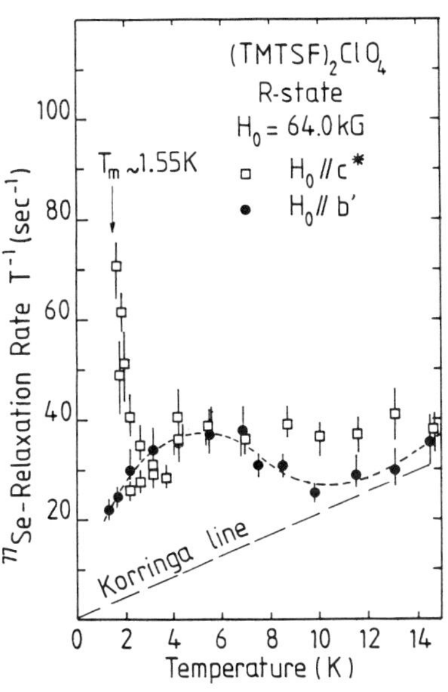

Figure 14 : ^{77}Se spin-lattice relaxation in (TMTSF)$_2$ClO$_4$ under large magnetic field (after Takahashi et al. (unpublished). For H ‖ c* a SDW state is stabilized below 1.55 K (see figure 4). The divergence to T_1^{-1} in the vicinity of 1.55 K is attributed to 3D critical magnetic fluctuations. The enhancement of T_1^{-1} observed at higher temperatures is understood in terms of 1D correlations effects above $T_{x_1} \approx$ 4-6K. This enhancement does not depend much on the orientation of the magnetic field as indicated by the data with H ‖ b'. The Korringa line is taken from the data at high temperature (T > 30 K).

In contrast to $T_1^{-1}(2k_F)$, the uniform part T_1^{-1} (q = 0) is non singular and is expected to be dominant at high temperature. This is supported by experimental results on ^{1}H relaxation (Stein et al, 1985) which give T_1^{-1} (q = 0)/T_1^{-1} (q = 2k$_F$) $\approx$ 3 at ambient temperature, making highly probable a cross-over between the uniform and the staggered contribution at lower temperature. This cross-over probably occurs around 30 K. Thus, above 30 K the linear regime of relaxation, (figure 12), might be attributed to the uniform contribution

$$T_1^{-1} (q = 0) \propto T \chi_s^2 (q= 0)$$

The uniform spin susceptibility χ_s $(q = 0)$ has been studied under pressure (Forro et al. 1986) and is approximately constant in temperature below 50 K, (figure 11), thus leading to a linear behaviour for T_1^{-1} versus T (in a small T-domain at least).

CONCLUSION

The FIR and NMR data of the $(TMTSF)_2X$ series which have been discussed in this article strongly support the existence of 1D correlations exhibiting an enhanced conductivity and magnetic fluctuations simultaneously.

The 1D correlations extend down to fairly low temperatures as shown by the behaviour of $(TMTSF)_2PF_6$. In this compound no sign of saturation of the 1D power law divergence is observed before the phase transition towards the SDW state at $T_N = 12$ K under ambient pressure.

The transverse interaction which stabilizes 3D magnetic order could proceed via an exchange of spin fluctuations between neighbouring chains (the IEX model). This is the same interaction which gives an adequate interpretation of relaxation rate measurements in the sulfur series $(TMTTF)_2PF_6$ (Creuzet et al. 1987) with a plateau of relaxation due to $2k_F$ spin correlations extending in a wide range of temperature (between 75 and 25 K) and also a well defined cross-over to a uniform spin correlations type of relaxation at high temperature which is very similar to the one observed in the $(TMTSF)_2X$ series.

Above the critical pressure of 8 kbar in $(TMTSF)_2PF_6$ or at ambient pressure in $(TMTSF)_2ClO_4$ the 1D divergence levels off at $T_{x_1} \approx 8\text{-}10$ K which is the dimensionality cross-over temperature. Below T_{x_1} the concepts of 1D physics no longer apply and both channels of correlations can be treated separately. Superconducting pairing can thus take over when the nesting properties of the Fermi surface are not good enough to allow the stabilization of the SDW state (this is the situation which prevails for $(TMTSF)_2PF_6$ at $P > P_c$). However, the nesting of the Fermi surface can be recovered under high magnetic fields enabling the restoration of an antiferromagnetic ground state via the NAF mechanism below T_{x_1}.

Bad nesting conditions appear to be a necessary requirement for the occurence of long range superconducting order. However the existence of a common border between SDW and superconducting states in $(TMTSF)_2PF_6$ has led Schulz et al (1981) to infer that magnetism could play a dominant role for the attractive pairing in organic superconductors. Recently, Caron and Bourbonnais (1986) have suggested that 1D AF correlations that are produced above T_{x_1} by the bare <u>repulsive intrachain</u> Coulomb interaction can be the origin of an <u>attractive interchain</u> pairing between electrons with opposite spins via the IEX mechanism, at low temperature.

The interchain spin singlet superconducting correlations grow as T_{x_1} is approached from above and therefore coexist with the intrachain antiferromagnetic fluctuations. Figure 15 gives a schematic illustration for the mechansim which provides the attractive interchain

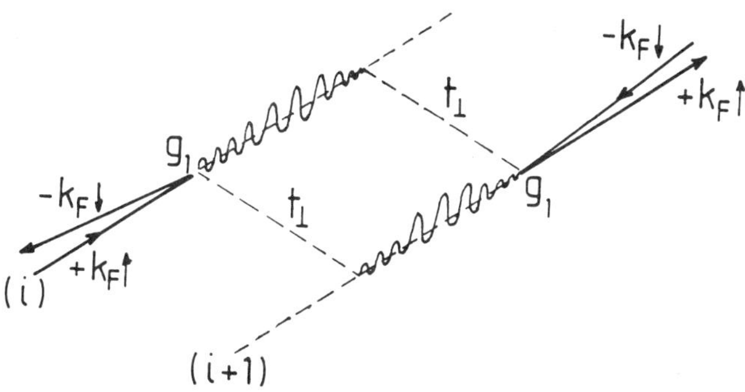

<u>Figure 15</u> : Sketch of the e-e pairing mechanism with an interchain exchange of a spin fluctuation leading to an attractive interaction between electrons of opposite spins.

pairing. The two in (out) going particles on different chains have opposite spins. On chain (i) a $+k_F\uparrow$ electron is backscattered by the g_1 scattering (forward scattering can also be considered) with the emission of a $2k_F$ spin raising $\Delta m_s = +1$ boson. This boson (e-h pair) propagates to the (i+1) chain and is absorbed by an incoming electron in the $-k_F\downarrow$ state. The amplitude of the attractive interchain pairing is of the order of $(g_2 + g_3)^2 \, t_\perp^2 \, \pi v_F/T_{x_1}^2$ (Bourbonnais, 1987). If the compound lies in a region of the P-T diagram not too far from the SDW state the IEX-mediated attractive pairing can overcome the repulsive contribution coming from the intrachain Coulomb interaction and lead to a superconducting transition at a finite temperature.

In the superconducting phase the gap should display a k vector dependence on account of the non-local character of the pairing, namely $\Delta(\underset{\sim}{k}) \sim \Delta_o \cos k_\perp d_\perp$. Hence lines of zeros for the gap on the Fermi surface are expected at $k_\perp d_\perp = \frac{\pi}{2}$.

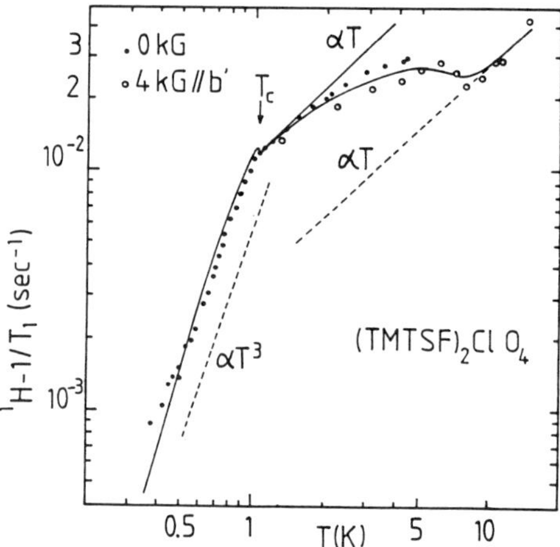

Figure 16 : Composite figure showing the behaviour of the proton relaxation of (TMTSF)$_2$ClO$_4$ at low temperature using Takigawa et al. 1987 data below 5 K and Bourbonnais et al. 1984 data above 1.2 K). The enhancement of T_1^{-1} is clearly observed below 10 K and T_1^{-1} follows nearly a T^3 temperature dependence in the superconducting state.

Emery (1986) has suggested an attractive interchain pairing produced by 2D or 3D-antiferromagnetic correlations below T_{x_1} when the nesting is no longer perfect (see also Béal-Monod et al. 1986).

With this anisotropic coupling in mind Hasegawa and Fukuyama (1987) have calculated the NMR relaxation rate in the superconducting phase. Instead of the exponential temperature dependence which is derived in case of BCS pairing $1/T_1$ follows a power-law temperature dependence

$(1/T_1 \propto T^3)$. Furthermore, the order parameter vanishing along lines on the Fermi surface no strong enhancement of $1/T_1$ would be observed just below T_c. Similar arguments have been proposed to explain the temperature dependence of the ultrasonic attenuation and of the spin-lattice relaxation rates in heavy fermions systems by Bishop et al. (1984) and D. Mc. Laughlin et al. (1984).

The recent finding of a T^3 temperature dependence for $^1H - 1/T_1$ in $(TMTSF)_2ClO_4$ at $T < T_c$ (Takigawa et al. 1987) strongly supports the existence of an anisotropic gap parameter.

On figure 16 we have displayed the data of proton relaxation in $(TMTSF)_2ClO_4$ between 15 and 0.4 K using the 4kG II b' result above T_c (≈ 0.8 K) and the data of Takigawa et al. (1987) in the superconducting state. This figure summarizes the unusual magnetic features of $(TMTSF)_2X$ at low temperature. A strong enhancement of relaxation due to antiferromagnetic fluctuations is observed in the vicinity of $T_{x_1} \approx 8$ K. At lower temperature the Fermi liquid behaviour is recovered down to T_c with an enhanced Korringa constant. Finally in the superconducting state the $1/T_1 \propto T^3$ law discussed above is fairly well obeyed.

We have seen that the zero frequency collective mode and the low FIR conductivity are also manifestations of well developed 1D correlations (magnetic and superconducting). This collective mode contributes to the longitudinal conduction and also to the transverse one. The drop of conductivity anisotropy which is observed below 40 K (Schulz et al. 1981) may be related to the onset of 1D correlations since the interchain pairing is likely to enchance the transverse conductivity faster than the longitudinal one. In this model the occurence of a fairly broad gap (≈ 30 cm^{-1}) in the density of states at the Fermi level which is detected by FIR and tunneling techniques provides the necessary oscillator strength needed for the growth of the zero frequency collective mode.

Summarizing, we feel that our understanding of organic superconductors has been considerably improved in the past few years by the accomplishment of detailed NMR and FIR experimental studies. We have suggested that the violation of two basic laws of non-interacting electrons (Drude and Korringa) can be understood if 1D correlations effects are important in these conductors even at fairly low temperatures. We have emphasized the role of Coulomb interactions which we believe are important in both $(TMTSF)_2X$ and $(BEDT-TTF)_2X$ series for the stabilization of superconductivity via a paramagnon mediated interchain pairing.

Numerous experimental features of 2:1 conducting compounds can be discussed in the framework of the 1D theory. This is the case for example for the properties of the sulfur series $(TMTTF)_2X$ exhibiting spin-Peierls or antiferromagnetic orderings at low temperature. Indeed, the study of the sulfur series has been particularly helpful for the understanding of the mechanisms giving rise to magnetism discussed in some details in the thesis work of F. Creuzet (1987).

So far, no magic recipe can yet be given for the improvement of organic superconductors. However, we feel that large values of $t_{||}$, reduced band structure anisotropy $(t_{||}/t_\perp \approx 2\text{-}3)$ and strong on site Coulomb repulsions could help for the increase of T_c in future materials.

These requirements are partly fulfilled in $\beta\text{-H}$ $(BEDT\text{-}TTF)_2I_3$ the T_c of which is significantly higher than the less Coulomb correlated and more one dimensional $(TMTSF)_2X$ series.

The main conclusion of this short review is that a new kind of superconductivity is probably observed in organic superconductors when spin-singlet Cooper pairing occurs via the interchain exchange of spin fluctuations. Organic superconductivity has shown that the pairing can be of the spin singlet kind and yet unusual. The continuation of the experimental and theoretical effort on these fascinating compounds so rich in providing new phenomena is needed to confirm the proposed picture.

Finally, the non-phonon mediated pairing of organic superconductors could also be relevant for the interpretation of the pairing in the recently discovered high T_c superconductors (Bednorz and Müller, 1986). But so far, a major interest of organic superconductors (in terms of understanding basic mechanisms) is that the existence AF fluctuations has been clearly established by NMR experiments.

Acknowledgements   We thank C. Bourbonnais and L. Caron for their continuous and fruitful interaction with us during the development of the NMR work at Orsay.

REFERENCES

Recent reviews can be found in

Jérome D. and Schulz H. J., 1982, Adv. in Physics, 31, 299 for the $(TMTSF-TMTTF)_2X$ series.

Williams J. M. and Carneiro K., 1985, Adv. in Inorganic Chemistry and Radiochemistry 29, 249 for the $(BEDT-TTF)_2X$ family

Gor'kov L. P., 1985, Sov. Phys. Usp. 27, 809 for the theory

Jérome D. and Caron L. G., 1987, Proceedings of a NATO Advanced Study Institute on Low-Dimensional Conductors and Superconductors, Magog, 1986, Plenum Press, New-York, for an up to date overwiew in theory and experiment.

Abragam A., 1961, The principles of nuclear magnetism, Oxford Univ. Press

Andres K., Schwenk H. and Veith H., 1986, Physica 143 B, 335

Andrieux A., Duroure C., Jérome D. and Bechgaard K., 1979 a, J. Phys. Lett. 40, L-381

Andrieux A., Chaikin P. M., Duroure C., Jérome D., Weyl C., Bechgaard K. and Andersen J. R., 1979 b, J. Physique 40, 1199

Azevedo L. J., Shirber J. E., Greene R. L. and Engler E. M., 1981, Physica B 108, 1183

Béal-Monod M. T., Bourbonnais C. and Emery V. J., 1986, Phys. Rev. B 34, 7716

Bechgaard K., Jacobsen C. S., Mortensen K., Pedersen H. J. and Thorup N., 1980, Solid State Comm. 33, 1119

Bechgaard K., Carneiro K., Olsen, Rasmussen F. B. and Jacobsen C. S., 1981, Phys. Rev. Lett. 46, 852

Bednorz J. G. and Müller K. A., 1986, Z. Phys. B 64, 189

Bishop A., Varma C. M., Batlogg B., Bucher E., Fisk Z. and Smith J. L., 1984, Phys. Rev. Lett. 53, 1009

Bouffard S., 1981, Thesis Université d'Orsay (unpublished)

Bourbonnais C., Creuzet F., Jérome D. and Moradpour A., 1984, J. Phys. Lett. 45, L-755

Bourbonnais C. and Caron L. G., 1986, Physica 143 B, 450

Bourbonnais C., 1987, Synth. Metals 19, 57

Bourbonnais C., 1987 in Low-Dimensional Conductors and Superconductors, editors D. Jérome and L. G. Caron, Plenum, New York

Brusetti R., Ribault M., Jérome D., Bechgaard K., 1982, J. Physique 43, 801

Bysckov Y. A., Gor'kov L. P. and Dzyaloshinskii I. E., 1966, Sov. Phys. JETP, 23, 489

Caron L. G. and Bourbonnais C., 1986, Physica 143 B, 453

Chaikin P. M., Choi M. Y., Kwak J. F., Brooks J. S., Martin K. P., Naughton M. J., Engler E. M. and Greene R. L. 1983, Phys. Rev. Lett. 51, 2333

Creuzet F., Creuzet G., Jérome D., Schweitzer D. and Keller H. J., 1985, J. Phys. Lett. 46, L-1079

Creuzet F., Jérome D., Bourbonnais C., Moradpour A., 1985, J. Phys. C Solid State 18, 821

Creuzet F., Jérome D., Bourbonnais C., Schweitzer D. and Keller H. J., 1986, Europhysics Lett., 1, 467

Creuzet F., Hurdequint H., Monod P. and Jérome D., 1987, to be published

Creuzet F., Bourbonnais C., Caron L. G., Jérome D., Bechgaard K., 1987, Synth. Metals, 19, 289

Creuzet F., 1987, Thesis Université d'Orsay (unpublished)

Delrieu J. M., Roger M., Toffano Z., Wope Mbongue E., Fauvel P. Saint-James R., Bechgaard K., 1986, Physica 143 B, 412

Djurek D., Knezovic S., Bechgaard K., 1985, Mol. Cryst. Liq. Cryst. 119, 161

Emery V. J., Bruinsma R. and Barisic S., 1982, Phys. Rev. Lett. 48, 1039

Emery V. J., 1983, J. Physique C-3, 44, 977

Emery V. J., 1986, Synth. Metals, 13, 21

Emge T. J., Leung P. C. W., Bero M. A., Schultz A. J., Wang H. H., Sowa L. M. and Williams J. M., 1984, Phys. Rev. B 30, 6780

Forro L., Cooper J. R., Rothaemel B., Schilling J. S., Weger M., Bechgaard K., 1986, Solid State Comm. 60, 11

Fournel A., More C., Roger G., Sorbier J. and Blanc C., 1983, J. Physique C 3, 44, 879

Fournel A., Oujia B., Sorbier J. P., 1985, Mol. Cryst. Liq. Cryst. 119, 37

Galigné J. L., Liautard B., Peytavin S., Brun G., Maurin M., Fabre J. M., Torreilles E., Giral L., 1979, Acta Cryst. B 35, 2609

Garoche P., Brusetti R. and Jérome D., 1982, J. Phys. Lett. 43, L-147

Ginodman V. B., Gudenko A. V., Zherikina L. N., 1985, Sov. Phys. JETP Lett. 41, 49

Gor'kov L. P. and Lebed A. G., 1984, J. Phys. Lett. 45, 433

Gor'kov L. P. and Jérome D., 1985, J. Phys. Lett. 46, L-643

Greene R. L., Haen P., Huang Z, S. Z., Zngler E. M., Choi M. Y. and Chaikin P. M., 1982, Mol. Cryst. Liq. Cryst. 79, 183

Greene R. L., Gutfreund H. and Weger M., 1983, Advances in Superconductivity, p. 225, editor B. Deaver and J. Ruvalds, B100, Plenum New-York

Hasegawa Y., Fukuyama H., 1987, J. Phys. Soc. Japan, 56, 877

Hawley M. E., Grey K. E., Terris B. D., Wang H. H., Carlson K. D. and Williams J. M., 1986, Phys. Rev. Lett. 57, 629

Heidman C. P., Andres K., Schweitzer D., 1986, Physica 143 B, 357

Héritier M. Montambaux G., Lederer P., 1984, J. Physique 45, 945

Jacobsen C. S., Tanner D. B., and Bechgaard K., 1983 a, J. Physique C-3, 44, 859

Jacobsen C. S., Tanner D. B., Bechgaard K., 1983 b, Phys. Rev. B 28, 7019

Jacobsen C. S., 1986, J. Phys. C Solid State 19, 5643

Javadi H. H. S., Sridhar S., Grüner G., Long Chiang and Wudl F., 1985, Phys. Rev. Lett. 55, 1216

Jérome D., Mazaud A., Ribault M. and Bechgaard K., 1980, J. Phys. Lett. 41, L-95

Jérome D. and Schulz H. J., 1982, Adv. in Physics, 31, 299

Kajimura K., Tokumoto H., Tokumoto M., Murata K., Ukachi T., Anzai H., Ishiguro T. and Saito G., 1983, J. Physique C-3, 44, 1059

Kang W., Creuzet G., Jérome D. and Lenoir C., 1987, J. Physique 48,

Kornelsen K., Eldridge J. E. and Bates G. S., 1987, Phys. Rev. B, in press

Laukhin V. N. et al. 1985, Sov. Phys. JETP Lett. 41, 81

Little W. A., 1964, Phys. Rev. 134, A, 1416

Mc Laughlin D., Cheng Tien, Clark W. G., Lan M. D., Fisk Z., Smith J. L. and Ott. H. R., 1984, Phys. Rev. Lett. 53, 1833

Mailly D., 1982, Thesis Université d'Orsay (unpublished)

Mailly D., Ribault M. and Bechgaard K., 1983, J. Physique C-3, 44, 1037

Marianer S., Kaveh M. and Weger M., Phys. Rev. B 25, 5197

Miljak M., Cooper J. R. and Bechgaard K., 1983, J. Physique, C 3, 44, 893

Moret R., Pouget J. P., Comès R., Bechgaard K., 1983, J. Physique, C-3, 44, 957

Moret R. and Pouget J. P., 1986, in Crystal Chemistry and Properties of Materials with Quasi One Dimensional Structures, p. 87, editor J. Rouxel, D. Reidel Publisher, Dordrecht

Moriya T. J., 1963, J. Phys. Soc. Japan, 18, 516

Mortensen K., Tankiewicz Y. and Bechgaard K., 1982, Phys. Rev. B 25, 3319

Naughton M. J., Brooks J. S., Chiang L. Y., Chamberlin R. V. and Chaikin P. M., 1985, Phys. Rev. Lett. 55, 969

Ng H. K., Timusk T. and Bechgaard K., 1983, J. Physique C-3, 44, 867

Ng H. K., Timusk T., Jérome D. and Bechgaard K., 1985, Phys. Rev. B, 32, 8041

Parkin S. S. P., Ribault, Jérome D., Bechgaard K., 1981, J. Phys. C 14, 5305

Pesty F., Garoche P. and Bechgaard K., 1985, Phys. Rev. Lett. 55, 2495

Piveteau B., Brossard L., Creuzet F., Jérome D., Lacoe R. C., Moradpour A., Ribault M., 1986, J. Phys. C 19, 4483

Piveteau B., Cooper J. R. and Jérome D., 1987, Solid State Comm. 62, 313

Ribault M., Jérome D., Tuchlendler J., Weyl C. and Bechgaard K., 1983, J. Phys. Lett. 44, L-953

Schulz H. J., Jérome D., Mazaud A., Ribault M., Bechgaard K., 1981, J. Physique 12, 991

Schwenk H., Andres K., Wudl F., Aharon-Shalom E., 1983, J. Physique C 3, 44, 1041

Soda G., Jérome D., Weger M., Alizon J., Gallice J., Robert H., Fabre J. M. and Giral L., 1977, J. Physique 38, 931

Solyom J., 1979, Adv. in Physics 28, 201

Stein P. C., Moradpour A. and Jérome D., 1985, J. Physique Lett. 46, L-241

Stewart G. R., O'Rourke J., Crabtree G. W., Carlson K. D., Wang H. H., Williams J. M., Gross F. and Andres K., 1986, Phys. Rev. B 33, 2046

Takahashi T., Jérome D. and Bechgaard K., 1982, J. Phys. Lett. 43, L-565

Takahashi T., Jerome D. and Bechgaard K., 1984, J. Physique, 45, 945

Takahashi T., Maniwa Y., Kawamura H., Saito G., 1986, Physica 143 B, 417

Takigawa M., Yasuoka H. and Saito G., 1987, J. Phys. Soc. Japan 56, 873

Tomic S., Jérome D., Bechgaard K., 1984, J. Phys. C 17, L-11

Torrance J. B., 1977, Chemistry and Physics of One-Dimensional Metals, Plenum B 25, H. J. Keller editor, p. 137

Torrance J. B., Pedersen H. J. and Bechgaard K., 1982, Phys. Rev. Lett. 49, 881

ANISOTROPIC SUPERCONDUCTIVITY AND NMR RELAXATION RATE IN ORGANIC

SUPERCONDUCTORS

Yasumasa Hasegawa and Hidetoshi Fukuyama

Institute for Solid State Physics
University of Tokyo
7-22-1, Roppongi, Minato-ku
Tokyo 106, Japan

ABSTRACT

Possible types of superconductivity in both quasi-one- and quasi-two-dimensional systems are examined theoretically. The NMR relaxation rate, T_1^{-1}, is calculated for each state. The comparison with the experiment in $(TMTSF)_2ClO_4$, which is the quasi-one-dimensional organic superconductor, indicates that the superconductivity of this family has lines of zeros of the gap on the Fermi surface. It is suggested that β-$(BEDT-TTF)_2X$ can be in the triplet state.

INTRODUCTION

The organic conductors, $(TMTSF)_2X$, $X=PF_6$, AsF_6, ClO_4, have unique features [1]. In the cases of $X=PF_6$ and AsF_6 there exists the spin-density-wave (SDW) transition at $T_{SDW}\simeq10K$ under ambient pressure. The transition temperature is sharply reduced by pressure about 6 kbar. Under higher pressure superconductivity of $T_c\simeq1K$ appears and T_c is decreased by further pressure. The existence of SDW can be understood in terms of the on-site Coulomb repulsion and the nesting of the Fermi surface. Due to the strong anisotropy of the transfer integrals, $t_a:t_b:t_c\simeq100:10:1$, the Fermi surface consists of two warped planes which are nested sufficiently to be unstable against SDW. The mean field approximation may be applicable [2,3] as long as the dependence of T_{SDW} on the degree of nesting is considered, since $t_b\simeq300K$ is much larger than T_{SDW} and the characteristic energy of imperfect nesting, t_b^2/t_a, is of the order T_{SDW}. The fluctuations due to quasi-one dimensionality 4 will be treated as a renormalization of parameters. The effect of pressure is to increase t_b, resulting in the decrease of the degree of nesting. In $(TMTSF)_2ClO_4$ the superconductivity appears even under ambient pressure when the sample is cooled slowly and the SDW state is stabilized when the sample is rapidly quenched. This feature will be due to the difference of the value of t_b according to the cooling rate; in the slowly cooled $(TMTSF)_2ClO_4$ t_b will be as large as that in the $(TMTSF)_2PF_6$ under pressure, while in the quenched sample it will be small and the nesting of the Fermi surface is sufficient to stabilized the SDW state.

In another family of organic superconductors, β-$(BEDT-TTF)_2X$, $X=I_3$,

AuI_2, IBr_2, the Fermi surface is cylindrical. The absence of SDW in this family should be due to the absence of nesting of the Fermi surface. The on-site Coulomb interaction, however, will be as large as that in $(TMTSF)_2X$. The interesting feature in β-$(BEDT$-$TTF)_2I_3$ is that there exist two superconducting states, low T_c and high T_c states [5,6]. The low T_c states appears only in the presence of incommensurate super-structure [7].

The superconductivity realized in the presence of large on-site Coulomb repulsion will be anisotropic as has been discussed in the context of heavy electron systems [8-14]. The possibility of the aniso-tropic superconductivity in organic superconductor has also been discussed by Emery [15] and the present authors [3,16,17]

We classify the possible types of superconductivity in both quasi-one- and quasi-two-dimensional systems and calculate the NMR relaxation rate, 16,17 T_1^{-1}.

QUASI-ONE-DIMENSIONAL SYSTEMS

The Fermi surface is assumed to consist of two planes, which are warped of the order of $t_b/t_a \ll 1$ in the k_y direction. The warping in the k_z direction is neglected for the sake of simplicity. The attractive interactions are assumed between the nearest sites along the most conductive a axis ($V_{/\!/}$) and b axis ($V_\perp$) besides the on-site interaction V_0^*, which is the sum of the attractive interaction and the renormalized Coulomb repulsion. In this model we find that the superconductivity can be classified into four types, whose order parameters on the Fermi surface are given as [3]

$$
\begin{aligned}
&s1: \ \Delta(k) = \Delta_{s1}(1 + C_{s1}\cos bk_y)\\
&s2: \ \Delta(k) = \Delta_{s2}(\cos bk_y + C_{s2})\\
&t1: \ \Delta(k) = \Delta_{t1} \ \mathrm{sgn}(k_x)\left(1 + \frac{\cos ak_F t_b}{\sin^2 ak_F t_a}\cos bk_y\right)\\
&t2: \ \Delta(k) = \Delta_{t2} \sin bk_y \ ,
\end{aligned}
\tag{1}
$$

where b is a lattice spacing in the b direction and C_{s1}, $C_{s2} \ll 1$. The former two states are spin singlet and the latter two are spin triplet. The temperature dependences of the order parameters, Δ_i, can be calculated numerically due to the simplicity of the model.

The density of states for quasi particles, $N_s(E)$, which can be observed in the tunneling experiment, is given by

$$
\frac{N_s(E)}{N(0)} = \int_{-\pi}^{\pi} \frac{d(bk_y)}{2\pi} \ \mathrm{Re} \ \frac{E}{\sqrt{E^2 - \Delta(k)^2}} \ ,
\tag{2}
$$

where $N(0)$ is the density of states in the normal state. In the case of $t_b/t_a = 0$, eq.(2) for the s2 and t2 states leads to

$$
\frac{N_s(E)}{N(0)} = \begin{cases} \dfrac{2}{\pi} \dfrac{E}{\Delta} K(\dfrac{E}{\Delta}) \ , & (E \leq \Delta) \ ,\\[2ex] \dfrac{2}{\pi} K(\dfrac{\Delta}{E}) \ , & (E \geq \Delta) \ , \end{cases}
\tag{3}
$$

where $K(z)$ is the complete elliptic integral of the first kind, while for the s1 and t1 states $N_S(E)/N(0)$ is same as that in the ordinary BCS superconductivity. As shown by the solid line in Fig.1, $N_S(E)$ is linear in E for $E \ll \Delta$, which is due to the lines of zeros of the gap. At $E=\Delta$ it diverges as $N_S(E)/N(0) \sim (1/\pi)\ln|\Delta/(E-\Delta)|$, which is due to the fact that the energy gap has its maximum along the lines on the Fermi surface. When $t_b/t_a \neq 0$, $C_{S2} \neq 0$ and $N_S(E)/N(0)$ diverges at $E=\Delta \pm C_{S2}$ for the s2 states as shown by the broken line in Fig.1, while that for the t2 state is given by the solid line. These divergences disappear when the transfer integral along the c direction is taken into account. Although several tunneling experiments have been reported in $(TMTSF)_2X$ [18-20], the unambiguous identification of the types of superconductivity has not be realized yet because of the difficulty of experiment.

The NMR relaxation rate for the anisotropic superconductivity, T_{1S}^{-1}, is calculated as

$$\frac{T_{1N}}{T_{1S}} = \int_0^\infty dE \; \frac{1}{2T\cosh^2(E/2T)} \left[\left(\int_0^{2\pi} \frac{d(b_k y)}{2\pi} \; \mathrm{Im} \; \frac{E}{\sqrt{\Delta(k)^2 - E^2}} \right)^2 \right.$$

$$\left. + \left(\int_0^{2\pi} \frac{d(bk_y)}{2\pi} \; \mathrm{Im} \; \frac{\Delta(k)}{\sqrt{\Delta(k)^2 - E^2}} \right)^2 \right] \quad , \tag{4}$$

where T_{1N}^{-1} is the relaxation rate in normal state. The result is shown in Fig.2. In the s1 state T_{1S}^{-1} is enhanced just below T_C and reduced exponentially at low temperatures. In the t1 state the enhancement of T_1^{-1} just below T_C, though still large, is smaller than that in the s1 state, since the second term is zero, i.e., the so called coherence factor is 1. On the other hand T_1^{-1} for the s2 and the t2 states, which are essentially same in the case of $t_b/t_a \ll 1$, is enhanced just below T_C very slightly. Its low temperature behavior is $T_{1S}^{-1} \propto T^3$, which is the consequence of the fact that $N_S(E) \propto E$ for $E \ll \Delta$. . The observed T_1^{-1} in $(TMTSF)_2X$[21] agrees fairly well with the theoretical curve for s2 and t2 states.

QUASI-TWO-DIMENSIONAL SYSTEMS

For quasi-two-dimensional systems we employ the model where the Fermi surface is cylindrical and the effective interaction is assumed to be attractive between the neighboring sites in the plane (V_a) and out of plane (V_c) on the tetragonal lattice, but the on-site interaction (V_0) is either repulsive or attractive. Then the following types of superconductivity are possible within the limitation of unitary states for spin triplet states:

$$
\begin{aligned}
s1' &: \Delta(k) = \Delta_{S1'}(\cos ak_x + \cos ak_y + C) \\
s2' &: \Delta(k) = \Delta_{S2'}(\cos ak_x - \cos ak_y) \\
s3' &: \Delta(k) = \Delta_{S3'} \cos ck_z \\
t1' &: \Delta(k) = \Delta_{t1'} \; \vec{d} \; \sin ak_x \\
t2' &: \Delta(k) = \Delta_{t2'} \; \vec{d} \; (\sin ak_x + i \sin ak_y) \\
t3' &: \Delta(k) = \Delta_{t3'} \; (\vec{d}_1 \sin ak_x + \vec{d}_2 \sin ak_y) \\
t4' &: \Delta(k) = \Delta_{t4'} \; \vec{d} \; (\sin ak_x + \sin ak_y) \\
t5' &: \Delta(k) = \Delta_{t5'} \; \vec{d} \; \sin ck_z \quad ,
\end{aligned}
$$

where a and c are the lattice spacings in and out of plane, respectively, and $\vec{d}$, $\vec{d}_1$, and $\vec{d}_2$ ($\vec{d}_1 \perp \vec{d}_2$) are the unit vectors in spin space. As in

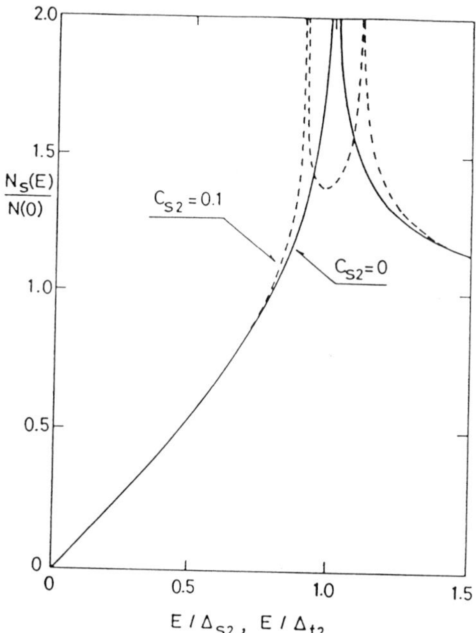

Fig. 1 Quasi particle density of states for the s2 state (solid line for $C_{S2}=0$, broken line for $C_{S2}=0.1$) and the t2 state (solid line) as a function of energy.

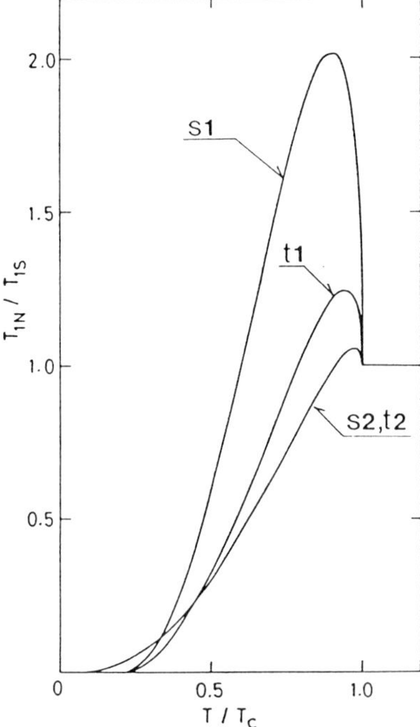

Fig. 2 NMR relaxation rate in quasi-one-dimensional system as a function of temperature.

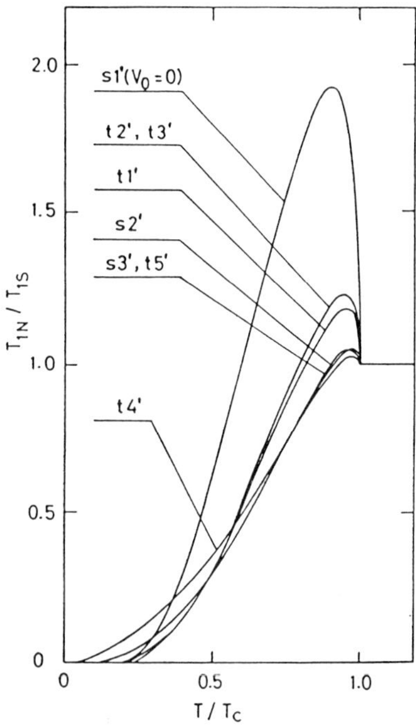

Fig. 3 NMR relaxation rate in quasi-two-dimensional system as a fucntion of temperature.

the quasi-one-dimensional systems, T_{1s}^{-1} is calculated for these states as shown in Fig.3. In the numerical calculation the quarter-filling of the band is assumed.

DISCUSSIONS

We have calculated T_{1s}^{-1} for various possible types of superconductivity in both quasi-one- and quasi-two-dimensional systems, where the attractive interactions between the neighboring sites are assumed. The comparison with the experiment in $(TMTSF)_2ClO_4$ indicates that the superconductivity in these quasi-one-dimensional systems has lines of zeros of the gap. Then there are two possible types of superconductivity in our classification, spin singlet s2 and spin triplet t2 states, which give the same temperature dependences of T_1^{-1}. By the following reason we believe that the s2 state is realized. As mentioned before the superconductivity appears as soon as SDW is suppressed, and then the antiferromagnetic spin fluctuation is expected to be large. The possibility of the attractive interaction between the neighboring sites for spin singlet state in the presence of such fluctuations has been discussed by Emery [15]. The same conclusion has been obtained by Hirsh [22] based on the Monte Carlo simulations. Miyake et al. [23] and Scalapino et al. [24] have shown by means of RPA that antiferromagnetic spin fluctuations cause the attraction between the neighboring sites for spin siglet state on one hand and the repulsion for spin triplet state on the other hand. The latter result has also been obtained by Beal-Monod et al [25].

In terms of this mechanism for the attractive force the suppression of T_c by pressure will be understood; the attractive interaction is reduced when the antiferromagnetic spin fluctuation is reduced by the breakage of nesting condition. The existence of lines of zeros of the gap will also be observed in the specific heat measurement. Experimentally the specific heat in $(TMTSF)_2ClO_4$ has been fitted by the exponential temperature dependence for temperatures above 0.4K [26]. Although this feature appears to be inconsistent with the s2 state for which T^2 dependence is expected, the difference between exponential and power low dependences can only be seen at the very low temperatures, as in the case of T^3 dependence in the heavy electron system, UBe_{13} [27].

In β-$(BEDT\text{-}TTF)_2X$, however, the experimental results [28,29] of T_1^{-1} are not close to any of the theoretical results. In the quasi-two-dimensional systems with cylindrical Fermi surface the antiferromagnetic spin fluctuation cannot be large, so the anisotropic singlet state will not be favored. In these systems spin fluctuation near q=0 will be important and triplet state will be favored. Since the triplet state has the degeneracy in spin space, the superconductivity glass state, which has been discussed in the context of heavy electrons by Volovik and Khmelnitskii [30] will occur below T_c. Then T_1^{-1} will be enhanced at the temperature where the transition from superconducting glass to uniform superconductivity takes place. Further experiments is needed to verify this proposal. The existence of two superconducting states in β-$(BEDT\text{-}TTF)_2I_3$ can be understood as due to the effect of randomness on the anisotropic superconductivity in general independent of singlet or triplet, i.e., the ethylene group is disordered in the presence of the incommensurate superstsructure and then T_c is suppressed. The effect of irradiation on the transition temperature [31] may also be understood in the same context. In $-(BEDT\text{-}TTF)_2AuI_2$ energy gap has been observed to be more than four times larger than the BCS value by the tunneling experiment [32], which has been discussed in the context of strong coupling. When superconductivity is anisotropic, the maximum of the

energy gap is observed in the tunneling experiment. However the
tunneling spectra at lower temperatures is necessary to determine the
type of superconductivity.

REFERENCES

1. for a review see e.g. D. Jerome and H.J. Schulz, Adv. Phys. 31:
    299 (1982).
2. K. Yamaji, J. Phys. Soc. Jpn. 51: 2787 (1982).
3. Y. Hasegawa and H. Fukuyama, J. Phys. Soc. Jpn. 55: 3978 (1986).
4. D. Jerome, F. Creuzet and C. Bourbonnais, Physica 143B: 329 (1986).
5. K. Murata, M. Tokumoto, H. Anzai, H. Bando, G. Saito, K. Kajimura and
    T. ishiguro, J. Phys. Soc. Jpn. 54: 1326 (1985); ibid. 2084.
6. V.N. Laukhin, E.E. Kostynchenko, Yu.V. Sashko, I.F. Shchegolev and
    E.B. Yagubskii, JETP Lett 41: 81 (1985).
7. A.J. Schultz, M.A. Beno, H.H. Wang and J.M. Williams : Phys. Rev.
    B33: 7823 (1986).
8. P.W. Anderson, Phys. Rev. B30: 4000 (1984).
9. G.E. Volovik and L.P. Gor'kov, Sov. Phys.-JETP 61: 843 (1985).
10. K. Ueda and T.M. Rice, Phys. Rev. B31: 7114 (1985).
11. E.I. Blount, Phys. Rv. B32: 2935 (1985).
12. R.A. Klemm and k. Scharnberg, Physica 135B: 53 (1985).
13. K. Miyake, T. Matsuura and H. Jichu, Prog. Theor. Phys. 72: 652
    (1984); K. Miyake, T. Matsuura, H. Jichu and Y. Nagaoka, Prog.
    Theor. Phys. 72: 1063 (1984).
14. F.J. Ohkawa and H. Fukuyama, J. Phys. Soc. Jpn. 53: 4344 (1984).
15. V.J. Emery, Synthetic Metals 13: 21 (1986).
16. Y. Hasegawa and H. Fukuyama, J. Phys. Soc. Jpn. 56: 877 (1987).
17. Y. Hasegawa and H. Fukuyama, submitted to J. Phys. Soc. Jpn.
18. Y. Maruyama, R. Hirose, G. Saito and H. Inokuchi, Solid State
    Commun. 47: 273 (1983).
19. H. Bando, K. Kajimura, H. Anzai, T. Ishiguro and G. Saito, Mol.
    Cryst. Liq. Cryst. 119: 41 (1985).
20. A. Fournel, B. Oujia and J.P. Sorbier, J. Phys. Lett. (Paris) 46:
    L417 (1985).
21. M. Takigawa, H. Yasuoka and G. Saito, J. Phys. Soc. Jpn. 56: 873
    (1987).
22. J.E. Hirsh, Phys. Rev. Lett. 54: 1317 (1985).
23. K. Miyake, S. Schmitt-Rink and C.M. Varma, Phys. Rev. B34: 6554
    (1986).
24. D.J. Scalapino, E. Loh and J.E. Hirsch, Phys. Rev. B34: 8190 (1986).
25. M.T. Beal-Monod, C. Bourbonnais and V.J. Emery, Phys. Rev. B34:
    7716 (1986).
26. P. Garoche, R. Brusetti, D. Jerome and K. Bechgaad, J. Phys. Lett.
    (Paris) 43: L147 (1982).
27. H.R. Ott, H. Rudigier, T.M. Rice, K. Ueda, Z. Fisk and J.L. Smith,
    Phys. Rev. Lett. 52: 1915 (1984).
28. F. Creuzet, C. Bourbonnais, D. Jerome, D. Schweitzer and H.J. Keller,
    Europhy. Lett. 1: 467 (1986).
29. M. Takigawa and Y. Miyake, private communications.
30. G.E. Volovik and D.E. Khmelnitskii, JETP Lett 40: 1299 (1984).
31. M. Tokumoto, I. Nashiyama, K. Murata, H. Anzai, T. Ishiguro and G.
    Saito, Physica 143B: 372 (1986).
32. M.E. Hawley, K.E. Gray, B.D. Terris, H.H. Wang, K.D. Carlson and
    J.M. Williams: Phys. Rev. Lett. 57: 629 (1986).

NUCLEAR MAGNETIC RELAXATION IN THE ORGANIC SUPERCONDUCTOR $(TMTSF)_2ClO_4$

M. Takigawa[1,2], H. Yasuoka[1], G. Saito[1], Y. Maniwa[3] and T. Takahashi[3]

[1]Institute for Solid State Physics, University of Tokyo, Roppongi 7-22-1
Minato-ku, Tokyo 106
[2]Physics Division Los Alamos National Laboratory, Los Alamos, New
Mexico, 87545
[3]Department of physics, Gakushuin University, Mejiro 1-5-1, Toshima-ku
Tokyo 171

INTRODUCTION

Superconductivity in certain family of charge transfer organic salts has
been a subject of intense study since it was first discovered in $(TMTSF)_2X$
series[1]. Two families of organic solids have been widely investigated so far,
namely, the $(TMTSF)_2X$ ($X=PF_6$, AsF_6, ClO_4 etc.) or the Bechgaard salts and the
$(BEDT-TTF)_2X$ ($X=I_3, IBr_2, AuI_2$ etc.) families, although these two series of
materials show very different properties. $(TMTSF)_2X$ is characterized by the
fairly one-dimensional band structure and the open fermi surface, that gives
rise to the strong nesting feature leading to the spin density wave (SDW)
instability. (The magnitudes of the transfer integrals along the three
crystallographic axes have been estimated as ta~260meV, tb~25meV, tc~0.1tb[2].)
Most of this family ($X=PF_6$, AsF_6) undergoes a transition into the insulating
SDW state at ambient pressure. The SDW state, however, is suppressed by
moderate pressure of several kbar. Then, the transition from normal to
superconducting state takes place[1]. On the other hand, the band structure of
the $(BEDT-TTF)_2X$ series is quite two dimensional with little anisotropy in the
conducting plane and the geometry of the fermi surface is considered to be
like a cylinder. None of this family shows magnetic instability and most of
them go into superconducting state at ambient pressure[3].

$(TMTSF)_2ClO_4$ has a unique property that different low temperature states
are obtained at depending on the cooling speed of a sample around the
structural phase transition temperature (24K) at which the orientational order
of tetrahedral ClO_4 anions takes place, doubling the unit cell along the b
direction[4]. The superconducting state is obtained below 1K only under slow
cooling[5] whereas rapid cooling results in the SDW state below 5K[6]. Many of
the superconducting characteristics such as the specific heat jump[7] and the
energy gap deduced from tunnelling spectroscopy[8] are consistent with the
simple BCS theory. However, a few outstanding features of this system such as
the close proximity with the magnatic instability or rather large suppression
of Tc by small amount of disorder produced either by alloying[9,10] or
irradiation[11,12] stimulated the various discussions on the exotic nature of
the superconductivity including the possibility of the triplet paring[13]. So
far, no microscopic experiment has been done that is directly concerned with
the

elementary excitation in the superconducting state such as ultrasonic
attenuation or nuclear relaxation rate. In this paper we present the results
of proton NMR relatation rate $(1/T_1)$ in $(TMTSF)_2ClO_4$ with the emphasis on its
temperature dependence in the superconducting state. A brief report has been
published.[14]

EXPERIMENTAL

A single crystal of $(TMTSF)_2ClO_4$ grown by electrochemical technique was
used for NMR experiment. All the measurements were made in the slowly cooled
state from 50K to 4.2K with cooling rate less than 0.1K/min.. A ^{3}He
evaporation cryostat was used to obtain temperatures down to 0.37K. The
measurements in high field (8.6KOe) were performed by detecting the recovery
of free induction decay (FID) signal after the saturating $\pi/2$ pulse. For
measurements at zero field, the field cycling method was employed which has
been widely used for NMR in superconductors[15]. Initially the nuclear spin
system is kept in thermal equilibrium in the polarizing high field (H=8.6kOe).
Then the magnetic field is adiabatically turned off within 2~10 sec., the
entropy of the nuclear spin system being kept at the high field value. $1/T_1$
at zero field is determined from the FID intensity when field is raised again
after a variable time (τ) spent at zero field, by the relation,
$M(\tau)=M_0(h+(1-h)exp(-\tau/T_1))$, where M_0 is the equilibrium magnetization at the
polarizing field, $h=H_L/H$ the ratio of the local field due to nuclear spin-spin
interactions to the polarizing field. In both the high field and zero field
measurement, time evolution of the nuclear magnetization was always
exponential with a single time constant over two decades. Superconducting
transition temperature (Tc) was determined to be 1.06K from the onset of
change in the resonance frequency of the LC circuit containing the NMR coil.
The midpoint of resistive transition whose width was 0.06K has given Tc=1.03K
under similar cooling process as employed in the NMR measurement.

EXPERIMENTAL RESULTS

The temperature dependence of $1/T_1$ in high field (8.6kOe) applied roughly
along the b-direction is shown in Fig. 1. It was confirmed that the sample
remained normal in this field down to 0.4K from the field dependence of the
high-frequency susceptibility. (Hc$_2$~6kOe at 0.4K) The Korringa relation,
T_1T=const, is satisfied below 1K as is expected for normal metals, although
deviation from this relation is clearly observed above 1K. Unfortunately,
$1/T_1$ of protons is dominated by methyl rotation and the contribution from the
conduction electrons is masked above 10K[16]. Therefore, we show in Fig. 2. the
temperature dependence of $1/T_1T$ of ^{77}Se up to 40K, which is entirely due to
the hyperfine interaction with the conduction electrons. $1/T_1T$ is temperature
independent above 20K, however, it is greatly enhanced below 20K and $1/T_1T$ at
the low-temperature limit is more than 5 times larger than that above 20K.
This non-Korringa behavior has been already reported both for protons and ^{77}Se
nuclei[16,17].

Generally the nuclear spin relaxation rate due to spin flip scattering
with conduction electrons is expressesd in terms of the dynamical
susceptibility $\chi(q, \omega)$ as

$$1/T_1 \propto k_B T \Sigma_q A^2(q) \frac{Im\ \chi(q,\ \omega_0)}{\omega_0}$$

where $A(q)$ is the wave number dependent hyperfine coupling constant and ω_0 is
the nuclear Larmor frequency. Because the system is close to the SDW
instability, the amplitude of the spin fluctuation or $Im\chi(q,\ \omega_0)/\omega_0$ will grow
rapidly with decreasing temperature if q is close to the nesting wave vector

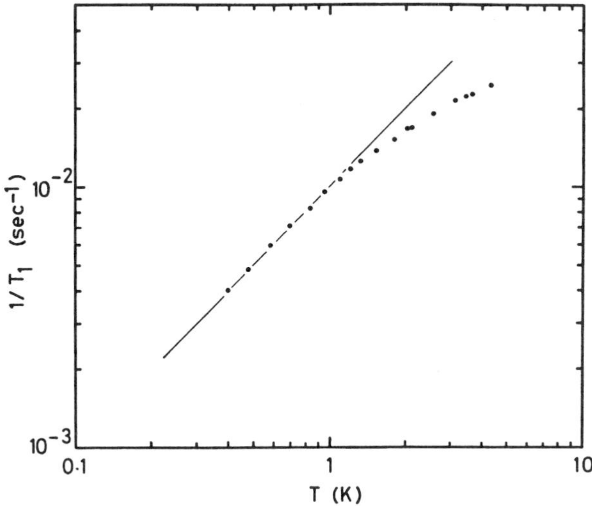

Fig. 1. Temperature dependence of $1/T_1$ of protons in $(TMTSF)_2ClO_4$ at 36.5Mhz (8.6kOe) with magnetic field applied roughly along b. The straight line shows the Korringa relation ($T_1T=1.0X10^2$ sec.K) observed below 1K.

q_0. On the other hand, it will be small and almost temperature independent for small q, as indicated by the temperature independence of the uniform susceptibility below 60K[18]. Thus, the rapid increase of $1/T_1T$ at low temperatures should be a consequence of the development of the antiferromagnetic spin correlations. In three dimensional itinerant electron antiferromagnetic systems, $1/T_1T$ is known to exhibit a temperature dependence as $1/T_1T$ $\sqrt{\chi_{q_0}}$, where χ_{q_0} is the staggered susceptibility[19], although it is not clear whether this can be applied to the present quasi one dimensional system.

It is well established that $1/T_1$ in normal metals shows magnetic field dependence around $H=H_L$[15]. The magnitude of H_L can be estimated either by the field dependence of $1/T_1$ or from the ratio $M(\tau\to\infty)/M(\tau=0)$. The field

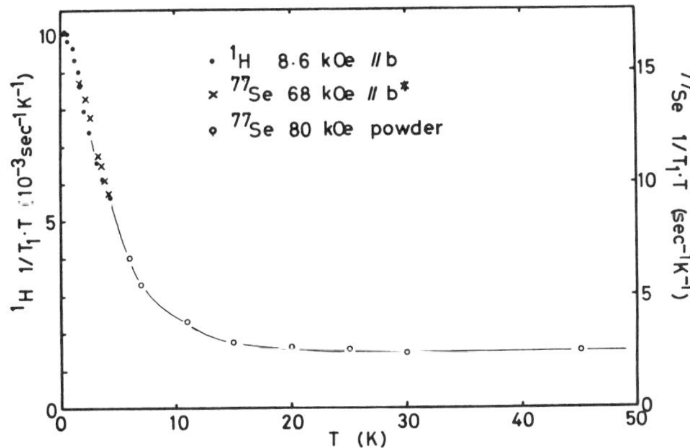

Fig. 2. Temperature dependence of $1/T_1T$ of protons (same as in Fig. 1) and ^{77}Se nuclei.

dependence of $1/T_1$ of protons in $(TMTSF)_2ClO_4$ was measured at 2.1 K as shown in Fig. 3. The field dependence shows a few steps and is not monotonic. Field independent value of $1/T_1$ is obtained only above 4 kOe. From the ratio $M(\tau \rightarrow \infty)/M(\tau=0)$, H_L was estimated to be about 2.5 kOe, which is an exceptionall large value but consistent with the fact that $1/T_1$ is field dependent up to 4kOe. Although the origin of such a large local field is not yet understood, it may be related to the tunnelling rotation of methyl groups that leads to a strong interproton scalar coupling[20]. This large local field must be related to the interaction among the nuclei and we consider it has nothing to do with the electronic properties.

The temperature dependence of $1/T_1$ at zero field is shown in Fig. 4. Th most striking result is that $1/T_1$ decreases rapidly just below Tc in contrast to the typical superconductors where $1/T_1$ increases below Tc reaching maximum at T~0.9Tc. This enhancement of $1/T_1$ is associated with the divergence of th density of the quasiparticle states at the edge of the energy gap and the coherence factor characteristic to the transition probability due to perturbations breaking the time reversal symmetry[15]. Below Tc, $1/T_1$ varies a T^3 fairly well between 0.5K and 0.9K. At lower temperatures, $1/T_1$ takes slightly larger values than the T^3 dependence extrapolated from the higher temperature region.

It is noticed that these features are strongly reminiscent of the heavy electron superconductors with f electrons such as $CeCu_2Si_2$[21] and UBe_{13}[22]. In the heavy electron systems, the above features have been considered as evidences for the anisotropic order parameter vanishing along certain lines o the Fermi surface. A possible example is the polar state with triplet pairing. With such anisotropic order parameter, the density of state has no strong singularity and varies linearly with the quasiparticle energy in low energy region, which are responsible for the disappearance of the enhancement of $1/T_1$ just below Tc and the T^3 dependence in low temperature limit. These arguments are also applicable to the present case. We would like to point ou the following related to the present system.
1) Both the electrical resistivity and the susceptibility measurement have shown a rather broad transition. However, sudden decrease of $1/T_1$ below Tc cannot be reproduced by any choice of distribution of Tc.
2) The absence of the enhancement of $1/T_1$ below Tc might be explained by pair breaking effect due to magnetic impurities. Then according to the theory of Griffin and Ambegaokar[23], the pair breaking parameter must be as large as $\alpha \sim 0.7\alpha_{cr}$, where α_{cr} is the critical value required to suppress the superconductivity. However, such large value of α is not compatible with th

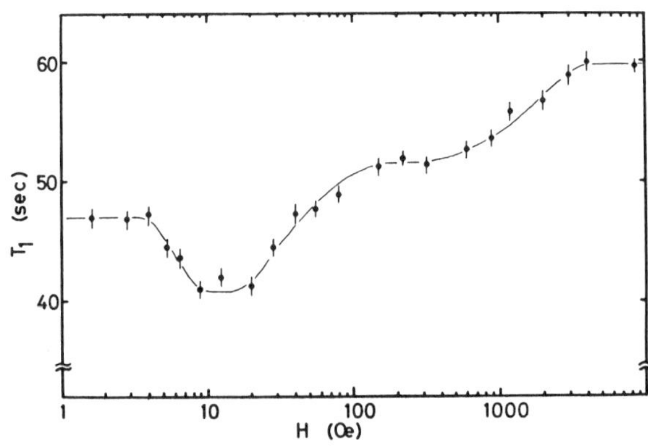

Fig. 3. Magnetic field dependence of T_1 of protons at 2.1K.

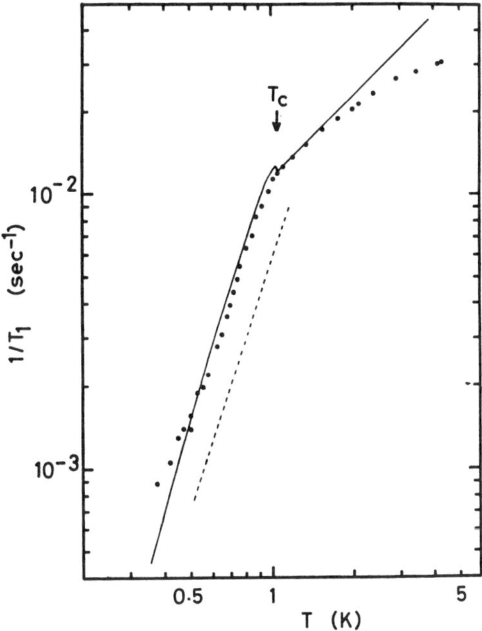

Fig. 4. Temperature dependence of $1/T_1$ of protons at zero
field. The solid curve shows the calculation for
the s_2 or t_2 states in Ref. 29 normalized to the
data at Tc (1.06K). The dashed line indicates the
T^3 dependence.

raped T^3 like decrease of $1/T_1$ down to 0.5K and also with the large specific
heat jump[7].
3) It may be possible in the field cycling experiment that trapped vortices
maintained at zero field suppress the enhancement of $1/T_1$. In the present
experiment, however, it is confirmed that the density of trapped vortices does
not exceed what would be produced by magnetic field of a few Oe from the
diamagnetic response of the high-frequency susceptibility immediatily after
the field is turned off.

DISCUSSIONS

The possible types of the Cooper pair with anisotropic order parameter
have been extensively discussed in relation with the heavy electron
superconductors, where the existence of such anisotropic order parameter is
strongly suggested by various kinds of experiments. The remarkable feature of
the heavy electron superconductors may be the strong on-site Coulomb repulsion
between f electrons, which favors the pair wave function having zero amplitude
at the origin even if the exchange of phonons is the main cause of the
attractive interaction. Recently, several authors have dicussed the non
phonon mechanism for attractive interaction. Hirsch demonstrated by the Monte
Carlo simulation of the Hubbard model that the attractive interaction between
the nearest neighbor electrons with antiparallel spins can be caused by the
on-site Coulomb repulsion[24]. This intcraction leads to an anisotropic singlet
pairing, where the order parameter is given by, $\Delta(k)=\Delta(\cos k_x+\cos k_y+\cos k_z)$ in
the cubic crystals, but suppresses both the tiplet and isotropic singlet
pairng. Beal-Monod et al. investigated the pairing interaction mediated by
antiferromagnetic spin fluctuations and found both the triplet and the
isotropic singlet paring are depressed in contrast to the nearly ferromagnetic

case where the triplet pairing interactin is enhanced[25]. Miyake et al. discussed, however, that the anisotropic singlet pairing is assisted by the antiferromagnetic spin fluctuations which have been observed in some of the heavy electron superconductors[26]. In the $(TMTSF)_2X$ family, the importance of the antiferromagmetic spin fluctuation is indicated by the close proximity between the SDW and the superconducting states in the phase diagram and the deviation of the nuclear relaxation rate from the Korringa relation above 1K in the normal state of ClO_4 salts. The possibility of the anisotropic order parameter in the Bechgaard salts has been pointed out by Emery[27]. He showed that the interaction constant mediated by spin fluctuations are always positive for the value of q close to the nesting wave vector. However, since it oscillates in configuration space, it is possible to confine the pair wave function to regions where the interaction is attractive.

Recently, Hasegawa and Fukuyama have investigated possible types of superconductivity for systems with quasi one-dimensional band structure and open fermi surface by treating the attractive interactions working on the same site (V_0), and between neighboring sites along the most conducting a-axis (V) and the intermediate b-axis ($V\perp$) as variable parameters[28,29]. Owing to the simple band structure and the one-dimensional fermi surface, the number of possible types of the superconductivity is limitted to four, labelled as s_1, s_2, t_1 and t_2 states. The s_1 and s_2 states are associated with spin singlet pairing, while triplet pairing is realized in t_1 and t_2 states. The wave vector depedence of the order parameters are given as, $\Delta_{s1}(k)=\Delta_{s1}$, $\Delta_{s2}(k)=\Delta_{s2}\cos bk_y$, $\Delta_{t1}(k)=\Delta_{t1}\mathrm{sgn}(k_x)$, $\Delta_{t2}(k)=\Delta_{t2}\sin bk_y$, where k_x and k_y are the component of wave vectors along a and b, respectively. Small correction of the order of $t_b/t_a \sim 0.1$ are neglected in these expressions. Δi ($i=s_1$, s_2, t_1 and t_2) are to be determined as a function of temperature by solving the gap equations. The temperature dependence of $1/T_1$ for each state has been also calculated by the same authors[29]. Since s_1 state (isotropic singlet state) and t1 states have finite gap on the entire fermi surface, these states are not compatible with the experimental results. On the other hand, the order parameter is reduced to zero along lines on the fermi surface in s_2 (anisotropic singlet state) and t_2 (triplet state) states.

In Fig. 4 is shown the calculated results for s_2 and t_2 states. $1/T_1$ for these two states are identical as long as the correction of the order of t_b/t is neglected. Below Tc, $1/T_1$ reaches maximum around $T \sim 0.99$Tc and has 5% larger value than that at Tc. Between 0.9Tc and 0.5Tc, $1/T_1$ varies approximately as T^3 although it deviates to lower values at lower temperature and the true T^3 dependence in low temperature limit is realized below 0.1Tc. (See Fig. 2 in ref.29.) The calculation reproduces the experimental results fairly well, as shown in Fig. 4, strongly suggesting either s_2 or t_2 states are realized in $(TMTSF)_2ClO_4$. Although we can not discriminate between s_2 and t_2 states from the data of $1/T_1$, s_2 state seems to be more likely because the antiferromagnetic spin fluctuations favor the s_2 state and there is some indication that Hc_2 along a axis seems to be limitted by the spin paramagnetism (Pauli limit)[30].

It is noticed that we can see minor disagreement that the small enhancement of $1/T_1$ just below Tc appeared in the calculation is not observed experimentally and the calculated values lie slightly above the experimental data below Tc. This may be attributed to pair breaking effect since it can been shown that non-magnetic impurities cause pair breaking when the order parameter is highly anisotropic[31]. It should also be noted that the calculation is based on the assumption that the hyperfine interaction is of short range. Since the contact and the dipolar interaction are shown to be of the same order for metyl protons in this material[32], the hyperfine coupling constant has appreciable wave vector dependence that may slightly change the calculated results when the gap is anisotropic.

It may be of great interest to compare this result to that for
$(BEDT-TTF)_2I_3$. It has been well established that $(BEDT-TTF)_2I_3$ shows two
different superconducting state depending on the pressure when the sample is
cooled[33]. The low Tc state (Tc=1~1.5K) is obtained at ambient pressure while
the high Tc state is obtained by applying small pressure. (Tc=7K at 1kbar.)
Creuzet et al reported that the proton spin relaxation rate in this material
in the high Tc states measured in the stationary field shows a huge peak at Tc
which is more enhanced at lower field[34]. This quite surprising result was
attributed to the effect of critical superconducting fluctuation. We have
measured $1/T_1$ of protons in $(BEDT-TTF)_2I_3$ both in the low Tc (ambient
pressure) and high Tc (2kbar) state in zero and finite field, the detail of
which will be published elsewhere. It was found that the relaxation of the
nuclear magnetization was always non exponential not only in the
superconducting state but in the normal states. This non exponential
relaxation was seen both in the low Tc and high Tc states irrespective of the
field value. The value of $1/T_1$ is distributed over a wide range and , in some
cases, the largest component of $1/T_1$ was more than a decade larger than the
smallest component. In the high Tc states, the smallest component of $1/T_1$
shows a peak at about 0.5Tc, which is obviously different form the peak seen
in typical superconductors. In the low Tc states, the value of $1/T_1$ and its
width of distribution was found to depend on samples. At moment, we cannot
make quantitative analysis or comparison with model calculations.

In conclusion, the temperature dependence of $1/T_1$ in $(TMTSF)_2ClO_4$
indicates the existence of anisotropic order parameter having lines of zero on
the Fermi surface. On the other hand, large distribution of $1/T_1$ is observed
in $(BEDT-TTF)_2I_3$, which suggests some intrinsic inhomogeneity of this system.

REFERENCES

1.   D. Jerome and H. J. Schulz, Adv. Phys. 31:299 (1982).
2.   C. S. Jacobsen, D. B. Tannen and K. Bechgaard, Phys. Rev. B28:7019
     (1983).
3.   See Proceedings of the Yamada Conference XV on Physics and Chemistry of
     Quasi One Dimensional Conductors, Physica 143B (1986).
4.   J. P. Pouget, G. Shirane, K. Bechgaard and J. M. Farbe, Phys. Rev.
     B27:5203 (1983).
5.   T. Takahashi, D. Jerome and K. Bechgaard, J. Phys. (Paris) 43:L-565
     (1982).
6.   H. Schwenk, K. Andres and F. Wudl, Phys. Rev. B29:500 (1984).
7.   P. Garoche, R. Brusetti, D. Jerome and K. Bechgaard, J. Phys. (Paris)
     43:L-147 (1982).
8.   H. Bando, K. Kajimura, H. Anzai, T. Ishiguro and G. Saito, Mol. Cryst.
     &Liq. Cryst. 119:41 (1985).
9.   C. Coulon, P Delhaes, J. Amiell, J. P. Monceau, J. M. Farbe and L. Giral,
     J. Phys. (Paris) 43:1721 (1982).
10. S. Tomic, D. Jerome, D. Mailly, M. Ribailt and K. Bechgaard, J. Phys.
     (Paris) 44:C3-1075 (1983).
11. S. Bouffard, M. Ribault, R. Brusetti, D Jerome and K. Bechgaard, J.
     Phys. C15:2951 (1982).
12. M. Y. Choi, P. M. Chaikin, S. Z. Huang, P. Haen, E. M. Engler and R. L.
     Green, Phys. Rev. B25:6208 (1982).
13. L. P. Gor'kov and D. Jerome, J. Phys. (Paris) 46:643 (1985).
14. M. Takigawa, H. Yasuoka and G, Saito, J. Phys. Soc. Jpn. 56:873 (1987).
15. L. C. Hebel and C. P. Slichter, Phys. Rev. 113:1504 (1959).
16. T. Takahashi, D. Jerome and K. Bechgaard, J. Phys. (Paris) 45:945 (1984).
17. C. Bourbonnais, F. Creuzet, D. Jerome, K. Bechgaard and A. Moradpour, J.
     Phys. (Paris) 45:L-755 (1984).

18. M. Milijak and J. R. Cooper, Mol. Cryst. &Liq. Cryst. 119:141 (1985).

19. K. Ueda and T. Moriya, J. Phys. Soc. Jpn. 38:32 (1975).

20. F. Apaydin and S. Clogh, J. Phys. C1:932 (1968).

21. Y. Kitaoka, K. Ueda, T. Kohara, K. Asayama, Y. Onuki and T. Komatsubara, J. Mag. Mag. Mat. 52:341 (1985); Y. Kitaoaka, K. Ueda, T. Kohara, Y. Kohori and K. Asayama, to be published in the 5th Int. Conf. on Valence Fluctuation., Bangalore, 1987.

22. D. E. Maclaughlin, Cheng Tien, W. G. Clark, M. D. Lan, Z. Fisk, L. J. Smith and H. R. Ott, Phys. Rev. Lett. 53:1833 (1984).

23. A. Griffin and V. Ambegaokar, Low Temperature Pysics (LT9), (Plenum Press, NY 1965) Part A, p. 524.

24. J. E. Hirsch, Phys. Rev. Lett. 54:1317 (1985).

25. M. T. Beal-Monot, C. Bourbonnais and V. J. Emery, Phys. Rev. B34:7716 (1986).

26. K. Miyake, S. Schmitt-Rink. and C. M. Varma, Phys. Rev. B34:6554 (1986).

27. V. J. Emery, Synthetic Metals, 13:21 (1986).

28. Y. Hasegawa.and H. Fukuyama, J. Phys. Soc. Jpn. 55:3978 (1986).

29. Y. Hasegawa and H. Fukuyama, J. Phys. Soc. Jpn. 56:877 (1987).

30. K. Murata, Private Communication, to be published.

31. K. Ueda and T. M. Rice, "Theory of Heavy Fermion and Valence Fluctuation", K. Kasuya and T. Saso ed. Springer, NY (1985).

32. J. M. Delrieu, M Roger, Z. Toffano, A. Moradpour and K. Bechgaard, J. Phys. (Paris).47:839 (1986).

33. K. Murata, M. Tokumoto, H. Anzai, H. Bando, G. Saito, K. Kajimura and T. Ishiguro, J. Phys. Soc. Jpn. 54:1236, 2048 (1985), V. N. Laukhin, E. E Kostyuchenko, Yu. V. Sushko, I. F. Shchegolev and E. B. Yagubskii, JETP. Lett. 41:81 (1985).

34. F. Creuzet, C. Bourbonnais, J. Jerome, D. Schweitzer and H. J. Keller, Europhysics Lett. 1:467 (1986).

ELECTRONIC BAND STRUCTURE AND POINT-CONTACT SPECTROSCOPY OF THE ORGANIC
SUPERCONDUCTOR β-(BEDT-TTF)$_2$I$_3$

M. Weger*, J. Kubler°, and D. Schweitzer#

*Racah Institute of Physics, Hebrew University, Jerusalem, Israel
°Technical University, Darmstadt, W. Germany
#Max Planck Institute, Heidelberg, W. Germany

ELECTRONIC BAND STRUCTURE OF β-(BEDT-TTF)$_2$I$_3$

Several band structure calculations of organic metals have been
carried out, employing the extended Huckel (EH) method [1,2,3]. The
molecular states are calculated using some quantum-chemistry method like
CNDO, MINDO, etc. The overlap between states of neighboring molecules is
then calculated in some approximation, and transfer integrals are evaluated
from the overlap integrals using the Huckel approximation. From these
transfer integrals the bandwidth and anisotropy are determined. This is
illustrated in Fig. 1 for a case of one molecule per unit cell (a) and two
molecules per unit cell (b). The EH method has been applied to β-BEDT-TTF$_2$I$_3$
(ET) by Mori et al (5) and Whangbo et al (3). They obtain an (essentially)
2-dimensional band structure (Fig. 1c), with a closed, nearly cylindrical
Fermi surface of holes around point Γ ($\vec{k}$ = (0,0,0)).

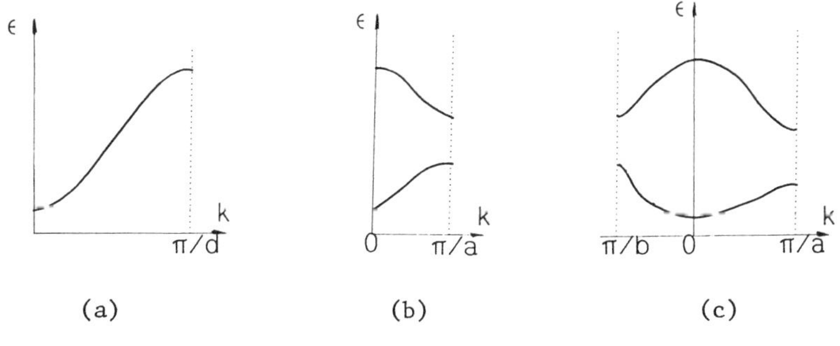

Figure 1

Electronic Band structure in the Extended Huckel Approximation,
for one (a) and two (b) molecules per unit cell. (c) illustrates
the band structure of β-(BEDT-TTF)$_2$I$_3$ calculated by Mori et al [5].

The EH method, as well as the LCAO scheme on which it is based, assumes that the bandwidth W is small compared with the separation between the bands ΔE, so that admixture of states from neighboring bands can be ignored. The justification for this was given by a self-consistent calculation by Salahub et al [6] for TTF dimers, which showed that the splitting between dimer bonding and antibonding states is small compared with the separation $\Delta E \cong 2$ eV between the molecular levels. A bandwidth of $W \cong 0.2-0.5$ eV for TTF was estimated, and a similar value for TTF-TCNQ was estimated by Ladik [4]. Therefore, the assumption $W \ll \Delta E$ seemed to be justified. However, in one case [7] it was found necessary to consider admixtures from a neighboring band to account for the metallic conductivity of HMTSF-TCNQ at low temperatures.

We [8] carried out an ab-initio self-consistent electronic band structure calculation for ET, using the ASW method of Williams [9]. In this method it doesn't matter whether W is smaller or bigger than ΔE. This is a standard method that applies very well to metals, insulators, and crystalline solids in general. It is also well suited for loose structures, where the standard muffin-tin approximation for the potential (i.e. the assumption that the potential is constant outside the atomic spheres) is not good. The space between the atoms must be filled with spheres; the spheres overlap somewhat and their total volume equals the volume of the unit cell. The potential is calculated at each iteration from Poisson's equation, together with the local exchange-correlation term. A linearized basis set is employed, and this may introduce a slight error. Also, the correction terms for the overlaps between the spheres are not absolutely precise. However, the method is relatively fast and therefore suitable for large unit cells with a low symmetry, where the use of a larger basis set with present computers is not practical.

In the present case, there are 55 atoms in a unit cell with $P\bar{1}$ symmetry i.e. 28 atoms are inequivalent. In addition, 32 "empty spheres" were used to fill-up the space between the atoms (60% of the total volume). Each iteration takes about 4 hours "user time" on a very large computer (IBM 3090, $\cong 40$ Mbytes effective CPU), and approximately 40 iterations are required to attain convergence. All "chemically" equivalent carbons, sulphurs, and hydrogens are assumed to possess the same charge; (thus there are 3 inequivalent carbons, 2 inequivalent sulphurs, 2 inequivalent iodines, and all hydrogens are assumed to be equivalent). The E vs. k curves are shown in Fig. 2. It is seen that some bands are approximately 0.5 to 1 eV wide, while the separation between bands is about 0.1 to 0.15 eV. Thus the condition for the validity of the extended Huckel approximation does not apply at all. The band structure looks more similar to that of a metal with a large unit cell, than that of a molecular crystal. The large density of bands is caused not only by the large size of the BEDT-TTF molecule, and the presence of 2 molecules per unit cell, but also there are additional states that are *not* molecular orbitals (LCAO's) but originate from the continuum (Fig. 3). These states are concentrated mainly in the "empty spheres" between the molecules. The location of these "empty spheres" for states with energies close to E_F is illustrated in Fig. 4. We were surprised to find that these "empty sphere" states sometimes form extremely narrow bands; in Fig. 5 we show a simplified partial density of states diagram, calculated assuming that all carbons, and all sulphurs, are equivalent. The narrow virtual state just above E_F corresponds to the empty sphere illustrated as a circle in Fig. 4. The value of the partial DOS of the empty spheres at E_F is very sensitive to details of the calculation; the value is smaller when we relax the condition that different carbons (and sulphurs) are inequivalent. Another surprising result of this calculation is the large partial DOS on the iodines at the FS. This suggests an appreciable participation of the iodines in the electronic properties of this compound.

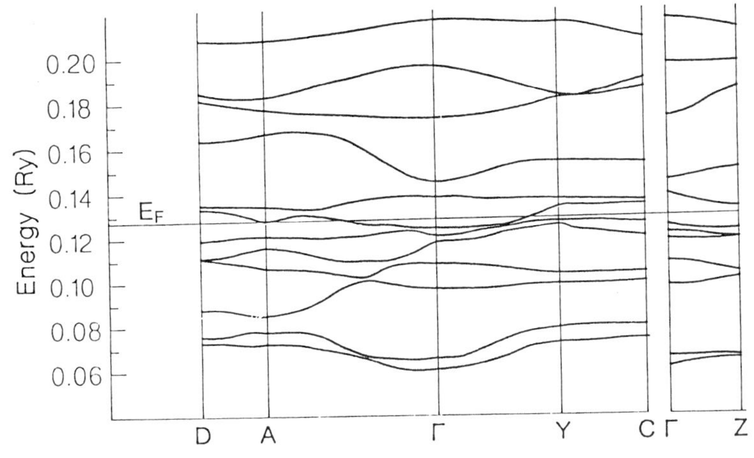

Figure 2

Energy vs. $\vec{k}$ curves for $\beta\text{-(BEDT-TTF)}_2\text{I}_3$

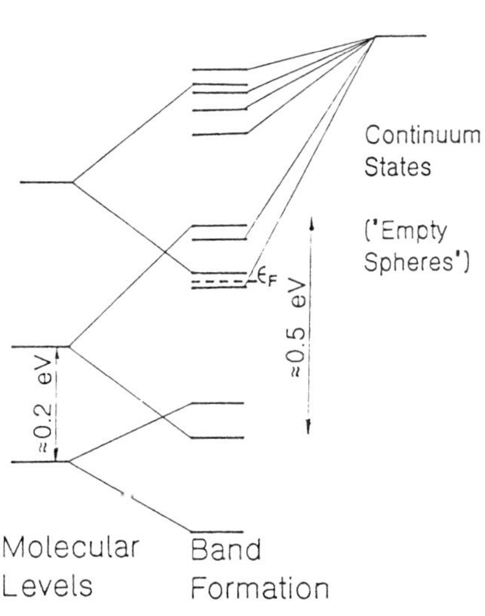

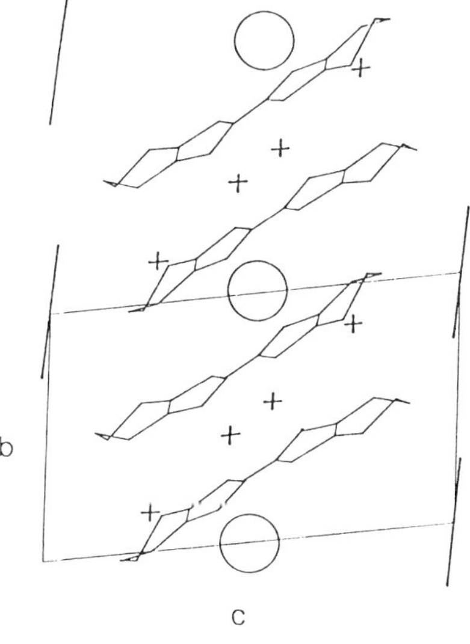

Figure 3

Illustration of the band formation
from molecular and "empty sphere"
states. The bandwidth greatly
exceeds the distance between levels.

Figure 4

Illustration of the position of
the empty spheres (Fig.5) E1
(circle), E2 and E3 (crosses)
between the molecules.

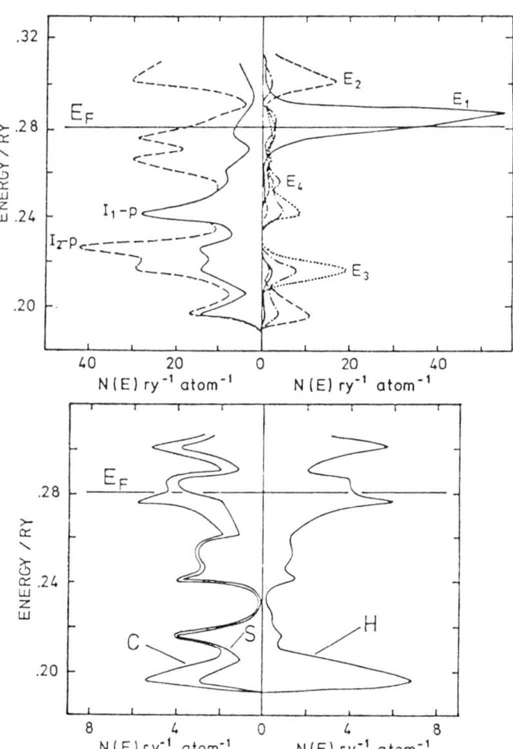

Figure 5

Partial densities of states
for the various atoms and
empty spheres.
The empty sphere states are
narrow and close to the
Fermi energy.

The Fermi surface (Fig.6) is in form of "warped planes" approximately
perpendicular to the [110] direction. This is to be expected from the
stacking of the molecules (Fig.7); however, the warping is very large and
as a result the hole states form cylinders around points A=$(.5,0,k_c)$ and
Y=$(0,.5,k_c)$. This FS is very different from that derived by the EH calculation; as a matter of fact, the states around Γ are electron states, rather
than hole states. The hole cylinders around A and Y are approximately
2-dimensional, as predicted by the EH calculation, and also in agreement
with the approximate 2-d nature of the conductivity and of H_{c2}.[10]

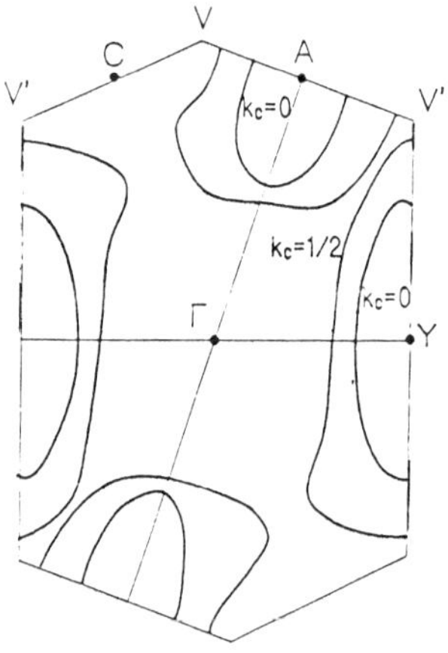

Figure 6

Cuts of the Fermi surface in the
k_a - k_b plane.

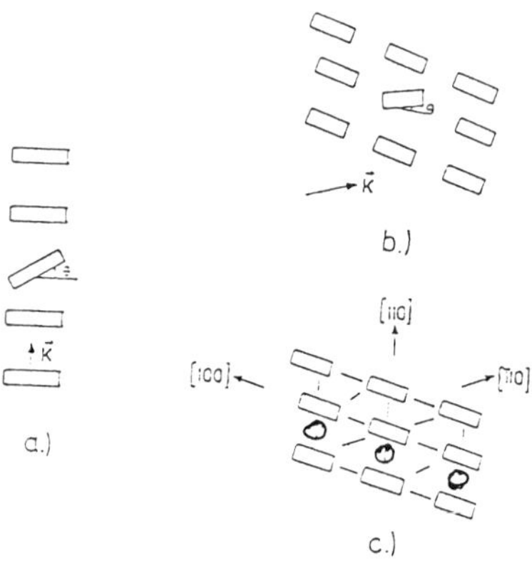

Figure 7

The stacking of the molecules (c),
and illustration why the 2-d nature
makes possible a *linear* electron-
libron coupling (b), in contrast
with 1-d stacks (a).

Superconductivity in organic metals was predicted on basis of the electron-phonon coupling theory, and the Eliashberg equations [11]. After the observation of superconductivity in TMTSF$_2$X, the question arose whether other mechanisms may be responsible for superconductivity [12], and if the electron-phonon mechanism is responsible, which phonons play the most important role: The high-frequency stretching modes, as suggested originally, the medium-frequency internal modes (bond twisting and bending) [13], or the low-frequency external modes (rigid translations and librations [14]). This question is also relevant nowadays, because of the new super-conductors of the YBa$_2$Cu$_3$O$_7$ and La$_{2-x}$Sr$_x$CuO$_4$ families, where the Cu – O stretching mode may or may not play a role in the superconductivity mechanism.

This question can be investigated experimentally using the techniques of tunneling and point-contact spectroscopy to determine the Eliashberg function $\alpha^2(\omega)F(\omega)$ [15]. We used point-contact spectroscopy to determine the Eliashberg function, as well as tunneling to determine the superconducting gap. Preliminary results on α^2F were reported a year ago [16], and on the superconducting gap by the Argonne group [17]. They reported a value of $2\Delta/k_BT_c$ about 5 times larger than the BCS weak-coupling value of 3.52, and suggested that this is due to soft phonons.

We repeated the point-contact measurements using a contact between two crystals of ET in the a-b plane [18]. The d^2V/dI^2 curve is shown in Fig. 8, and the Eliashberg function $\alpha^2(\omega)F(\omega)$ derived from it is shown in Fig. 9.

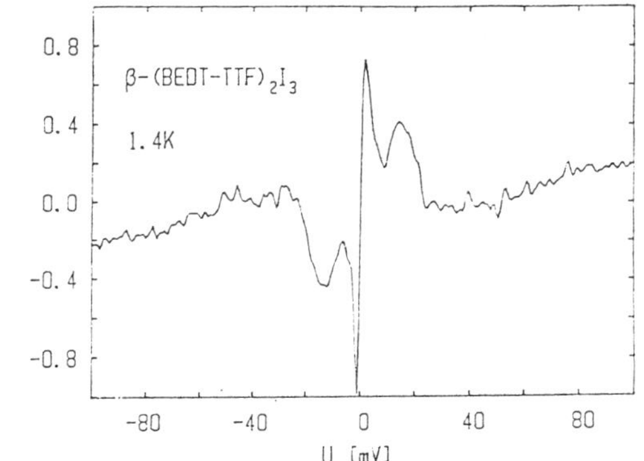

Figure 8

Second derivative of the point-contact spectroscopy I-V curve for ET.

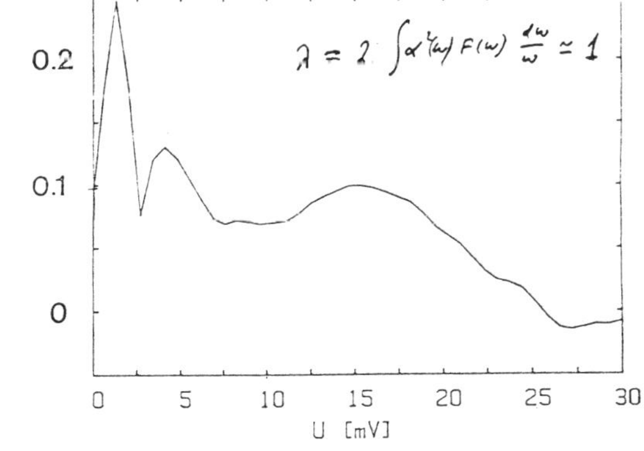

Figure 9

The Eliashberg fuction α^2F derived from the I-V curves (Fig. 8) with no adjustable parameters.

It is seen that phonons with energies of $\cong 1$ meV and 4 meV are strongly coupled with the electrons, as well as a broad band of phonons around 15 meV, up to 25 meV, above which frequency $\alpha^2 F$ vanishes. The 1 meV phonon is exceedingly soft, and such phonons have not been seen before in organic metals (but predicted by Hawley et al[17]). The phonon at 4 meV fits well with a Raman line observed at 35 cm^{-1} [19]. The band around 15 meV is probably due mostly to external modes of the BEDT-TTF molecule, though a stretching mode of the I_3 molecule is observed around 120 cm^{-1} by Raman spectroscopy [20].

Integration of the observed $\alpha^2(\omega)F(\omega)$ curve using the Eliashberg equations, and the Bergmann-Rainer algorithm [21] yields a value of T_c in the range of the observed value of T_c=1.35 K; the absolute precision of such a measurement is not very high, because of uncertainties concerning the surfaces of the crystals that actually touch [15]. Nevertheless, the value of $\lambda = 2 \int \alpha^2(\omega)F(\omega) \, d\omega/\omega$ (we find $\lambda \cong 1$) is probably reliable to better than a factor of 2. This leads strong support to the electron-phonon theory of the superconductivity of organic metals.

The value of $2\Delta/k_B T_c$ determined by the point-contact method is somewhat uncertain, since the shape of the d^2V/dI^2 curve is "gaussian" (Fig.10a) and does not agree precisely with the theoretical curve [22]. Measurements were carried out by U. Poppe using the vacuum-tunneling technique, in which the surfaces of the crystals do not touch, and therefore there is no stress on them [18]. This yields curves closer to the theoretical one (Fig.10b). For this curve, the value of $2\Delta/k_B T_c \cong 4.0\pm0.2$ is close to the BCS value, or more precisely, close to the value of strong-coupling superconductors like Pb, (and agrees with the value we expect for the value $\lambda \cong 1$ obtained from the integration of $\alpha^2 F$). Also, a theoretical determination, from the Eliashberg equations, using the Bergmann-Rainer algorithm, and the $\alpha^2 F$ function of Fig.9 with the soft phonon at 1 meV strongly coupled to the electrons, yields this value, and not the very large value reported by Hawley et al [17]. Various other curves, and values of $2\Delta/k_B T_c$, are given in ref [18], and large values of $2\Delta/k_B T_c$ are indeed observed under certain experimental conditions.

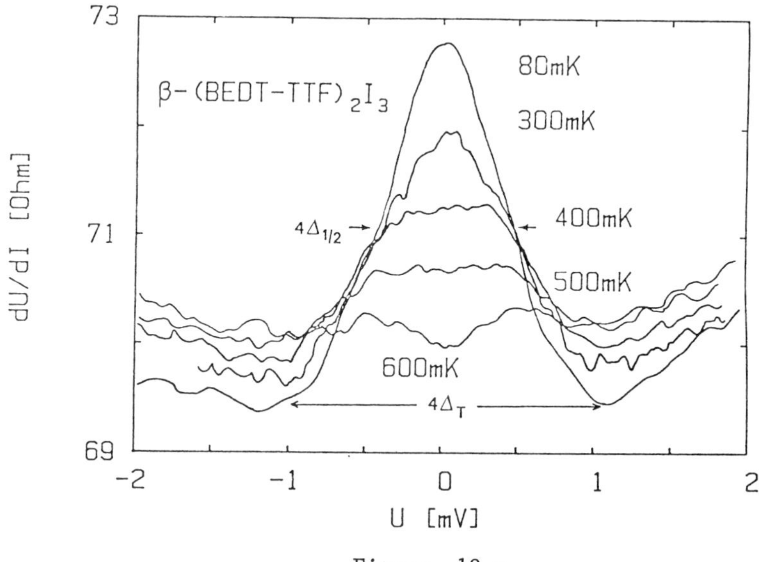

Figure 10 a

Tunneling data at various temperatures for two touching β-(BEDT-TTF)$_2$I$_3$ crystals. $\Delta_{1/2}$ is close to the BCS value for T_c=1.35 K, while Δ_T is considerably larger.

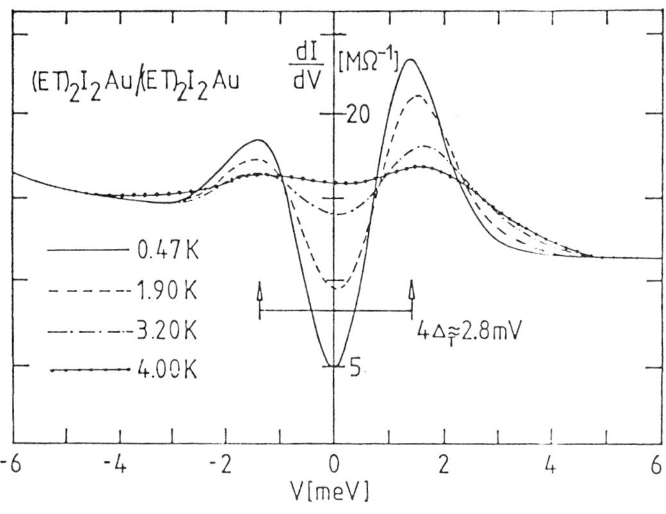

Figure 10 b

Vacuum tunneling data for $(BEDT-TTF)_2IAuI$ $(T_c=4.1$ K).
Here $2\Delta_T/k_BT_c \cong 4$.

THE ELECTRON-PHONON COUPLING

In the 1-d organic metals TTF-TCNQ, TMTSF$_2$X, etc. the resistivity
follows a T^2 law, indicating quadratic electron-phonon coupling [23].
A T^2 law was recently observed also in polyacetylene [24]. In ET the
resistivity does not seem to follow a T^2 law; at low temperatures,
$\rho = \rho_0 + aT + bT^2$ with a significantly large value of a, indicating an
electron-phonon coupling constant $\lambda \cong 1$ [25]. Thus, the strong *linear*
electron-phonon coupling differs from the other organic metals.

The cause for the large linear coupling may be the 2-d electronic
band structure, which makes it possible for librons to couple linearly
with the electrons (Fig. 7; see also ref. 16).

Another possibility is the virtual state close to E_F (Fig.5). A linear
matrix element $\langle\psi_M|\partial V/\partial\theta|\psi_L\rangle \neq 0$ is allowed by symmetry; here ψ_M
indicates a molecular state, ψ_L a localized state("empty sphere" state),
and θ a libration angle. Thus, if ψ_L has a contribution at E_F, a linear
electron-phonon coupling is allowed by symmetry (Fig.11a). We may say that
the degeneracy of ψ_M and ψ_L makes possible a band Jahn-Teller effect .
If the localized state does not have a contribution at E_F , but its energy
E_L is still close to E_F, a second-order term

$$\langle \psi_M |\partial V/\partial\theta| \psi_L \rangle^2/(E_F - E_L)$$

gives rise to a strong quadratic electron-phonon coupling (Fig. 11b). This
may be the case in TTF-TCNQ, TMTSF$_2$X, etc. where this quadratic coupling is
exceedingly large [23,26]. A direct evidence for this is provided by X-ray
determinations of the charge density, that shows maxima in the region
between the molecules [27].

155

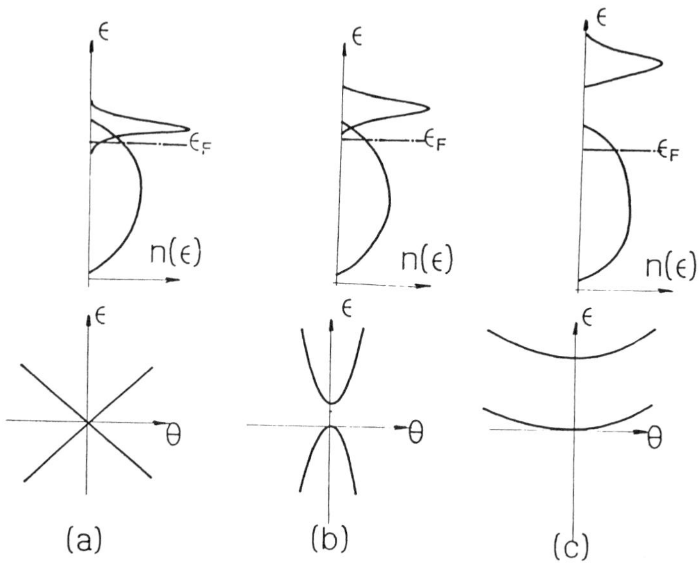

Figure   11

Illustration how a narrow "empty sphere" state at E_F (a) can give
rise to linear electron-libron coupling, and near E_F (b) to a
large quadratic electron-libron coupling.

REFERENCES

1. A.J. Berlinsky, J.F. Carolan and L. Weiler, Solid State Commun. 15,
    795, (1974).
2. P.M. Grant and I.P. Batra, J. Phys. (Paris) Coll. 44, C4-437, (1983).
3. M.H. Whangbo, J.M. Williams, A.J. Schultz, T.J. Emge and M.A. Beno
    JACS 109, 90, (1987).
4. J. Ladik, A. Karpfen, G. Stollhoff and P. Fulde, Chem. Phys. 7, 267,
    (1975).
5. T. Mori, A. Kobayashi, Y. Sasaki, H. Kobayashi, G. Saito, H. Inokuchi,
    Chem. Lett. 1984, 957, (1984).
6. D.R. Salahub, R.P. Messmer and F. Herman, Phys. Rev. B 13,4252,(1976).
7. M. Weger, Solid State Commun. 19, 1149, (1976).
8. J. Kubler, M. Weger and C.B. Sommers, Solid State Commun.1987 (in press)
9. A.R. Williams, J. Kubler and C.D. Gelatt, Phys. Rev. B19, 6094,(1979).
10. K. Bender, I. Hennig, D. Schweitzer, K. Dietz, H. Enders, H.J. Keller,
    Mol. Cryst. Liq. Cryst. 108, 35, (1984),
11. B. Horovitz, H. Gutfreund and M. Weger, J. Phys. C 7, 383, (1974); Mol.
    Cryst. Liq. Cryst. 79, 235, (1982); B. Horovitz, Phys. Rev. B 16, 3943,
    (1977).
12. D. Jerome and H.J.Schultz, Adv. Phys. 31, 299, (1982).
13. E.M. Conwell, Phys. Rev. Lett. 39, 777, (1977).
14. H. Gutfreund and M. Weger, Phys. Rev. B 16,1753,(1977); O. Entin-
    Wohlman, M. Kaveh, H. Gutfreund, M. Weger and N.F. Mott, Phil. Mag.
    B 50, 251, (1984).
15. A.G. Jansen, A.P. Van Gelder and P. Wyder, J. Phys. C 13, 6073,(1980).
16. A. Novack, M. Weger, D. Schweitzer and H. Keller, Solid State Commun.
    60, 199, (1986).
17. M.E. Hawley, K.E. Gray, B.D. Terris, H.H. Wang, K.D. Carlson and
    J.M. Williams, Phys. Rev. Lett. 57, 629, (1986).

18.   A. Novack, U. Poppe, M. Weger, D. Schweitzer and H. Schwenk,
      Z. Physik B (Condensed Matter) 1987 (in print).
19.   S. Sugai and G. Saito, Solid State Commun. $\underline{58}$, 759, (1986).
20.   H. Enders, H.J. Keller, R. Swietlik, D. Schweitzer, K. Angermund and
      C. Kruger, Z. Naturforsch. $\underline{41}$a, 1319, (1986); R. Swietlik, D.
      Schweitzer and H.J. Keller, Submitted to Phys. Rev.
21.   G. Bergmann and D. Rainer, Z. Physik $\underline{263}$, 59, (1973).
22.   G.E. Blonder, M. Tinkham and T.M. Klopwijk, Phys. Rev. B $\underline{25}$,4515(1982).
23.   K. Bechgaard, C.S. Jacobsen, K. Mortensen, H.J. Pedersen and N. Thorup
      Solid State Commun. $\underline{33}$, 1119, (1980); M. Weger, M. Kaveh and H.
      Gutfreund, Solid State Commun. $\underline{37}$, 421, (1981).
24.   K. Kume, S. Masubuchi, K. Mizoguchi and K. Mizuno, Synthetic Metals
      $\underline{17}$, 533, (1987).
25.   K. Bender, Thesis, Heidelberg University 1987.
26.   H. Gutfreund, C. Hartzstein and M. Weger, Solid State Commun. $\underline{36}$, 647
      (1980); Chemica Scripta $\underline{17}$, 51, (1981).
27.   F. Wudl, D. Nalewajek, J.M. Troup and M.W. Extine, Science $\underline{222}$, 415,
      (1983).

THE ORIGIN OF PAIRING INTERACTION IN ORGANIC SUPERCONDUCTORS

Claude Bourbonnais

Centre de Recherche en Physique du Solide
Université de Sherbrooke
Sherbrooke, Québec, Canada J1K-2R1

I-INTRODUCTION

A problem of great importance in the study of organic conductors is
the characterization of the microscopic conditions giving rise to
superconductivity in these compounds. Looking at the phase diagram of the
Bechgaard salts series $(TMTSF)_2X$ (X= PF_6, AsF_6, ClO_4, ...),[1,2] this problem
seems to be intrinsically linked to the puzzling proximity between
superconducting and antiferromagnetic long range order. The existence of
antiferromagnetism and the absence of a Peierls transition is known to be a
strong indication in favor of dominant repulsive interactions among
electrons. This has been used to challenge the traditional role played by
phonons in the superconductive pairing interaction and to infer a
participation of electron spin correlations in the pairing mechanism.[3-5] In
this paper we will be mainly concerned with clarifying the microscopic
origin of this proximity, especially to specify the mechanism by which
antiferromagnetism is itself stabilyzed and to what extend it can exert an
influence on the metallic state where the pairing interaction is formed.
More specifically, we overview the implications of recent calculations for
the quasi-1D electron gas model. The existence and the importance of the
interchain antiferromagnetic exchange coupling for the stabilyzation of
antiferromagnetism are demonstrated.[4] We also discuss how this mechanism
can lead to high-temperature superconducting correlations. In contrast to
the predictions of recent models for high-T_c superconducting oxides, the
true superconducting transition temperature in quasi-1D conductors like the
Bechgaard salts. is strongly reduced by anisotropy.

It is very well established that the origin of antiferromagnetic (AF) correlations in these compounds comes from the short range repulsive interactions along the chains[6]. The mechanisms by which 3D AF long range ordering is stabilyzed however are triggered by interchain kinetic coupling, that is, the interchain single electron transfer $t_\perp$. Owing to the strong anisotropy of the electronic spectrum of these compounds ($t_\perp \ll t_{\|}$) the Fermi surface is open and consists of two warped planes centered at $\mp k_F$, the Fermi wavevector of the isolated chains[1]. The energy scale below which the curvature of the Fermi surface becomes relevant and the electronic motion is considered as 3D, is given by the single particle dimensionality crossover temperature[7,8] T_{x^1} which is itself much smaller than the single chain Fermi energy $E_F \sim .5ev$[1]. This implies that all the AF ordering mechanisms that use the 3D properties of the Fermi surface as a starting point are only effective for electronic energies or temperature below T_{x^1}[9]. This is particularly true for the 3D Overhauser mechanism for which AF long range ordering relies on perfect 3D nesting properties of the Fermi surface. In the following, it will be referred to as the 3D NAF mechanism. It does not mean however that AF long range order is restricted to the temperature domain $T < T_{x^1}$ but the mechanism for its stabilyzation at $T > T_{x^1}$ is different. Indeed, when the transverse single electron band motion is disrupted by thermal effects, the only possibility for electrons to hop coherently on neighbouring chains is via a quantum virtual transfer[4]. In the presence of 1D AF correlations , this inevitably leads to an interchain AF kinetic exchange coupling which in turn favors the transverse propagation of AF order and its long range character[4]. In the following, this will be called the IEX mechanism.

The IEX mechanism is active for all $T > T_{x^1}$ irrespective of the weak- (metallic) or strong (insulating) coupling state of the single chains. Actually, the existence of an AF exchange coupling for *weak* electron-electron coupling constants in the Bechgaard salts is somewhat specific to the rather large anisotropy of theirs electronic spectrums ($t_{\|}/t_\perp \gtrsim 10$). Consequently, there must be a large 1D temperature domain and T_{x^1} must be considered as a relevant low energy scale[7,8]. Besides the weak interaction and for all $E_F > T > T_{x^1}$, there is another small paramater, $\bar{t}_\perp/\pi T$, allowing for the interchain virtual transfer of single electrons[4]. Here, it is the thermal effects that localize electrons on single chains. This contrasts with "strong-U" correlation effects that are well known to act as the source of electronic localization in isotropic systems. Note that here, we have used the renormalized amplitude $\bar{t}_\perp$, instead of the larger bare value $t_\perp$, which take into account the effect of electronic correlations along the chains (see section III).[8,9] When $\bar{t}_\perp/\pi T \simeq 1$, the transverse band character is relevant and the single particle crossover is achieved namely, $T_{x^1} \simeq \bar{t}_\perp/\pi$.

As the AF IEX coupling is equivalent to a coherent interchain tunneling of electron-hole pairs, it is the effective local Coulomb interaction g that assures the coherence of the particles on each chain. From these qualitative arguments, it immediately follows that the IEX amplitude in weak coupling ($g < \pi v_F$) should be proportional to $(g/\pi v_F)^2 (\bar{t}_\perp/\pi T)^2$, which has a finite amplitude as long as the Coulomb interaction is non-zero. We will see in the next section that the renomalization group calculations[4] confirm the presence of such a term and demonstrate that it can lead to AF long range ordering above T_{x^1}. When this occurs, there must be some kind of a dimensionality crossover for correlations just above the transition temperature T_N but it is not single particle like as for T_{x^1} but is rather two-patrticle like since it is achieved via the transverse propagation of electon-hole pairs. An important consequence of the presence of such a coupling is that it assures a continuity between the single chain weak and strong coupling regimes.[4] In the latter for example, $g \simeq \pi v_F$ and a 1D correlation (insulating) gap $\Delta_\rho \simeq \pi T_\rho$ becomes relevant below the characteristic temperature T_ρ at which the occurence of T_{x^1} is no longer possible. The IEX amplitude for $T < T_\rho$ becomes proportional to $(\bar{t}_\perp/\Delta_\rho)^2$, a well-known perturbation theory result for the IEX amplitude in terms of the small paramater $\bar{t}_\perp/\Delta_\rho$.[9,10] It is now quite well understood that it is precisely in this limit that the IEX mechanism stabilyze the AF order in the *sulphur* series of organic conductors $(TMTTF)_2X$ at low pressure.[4,6,10,11] These compounds have a well-defined correlation gap Δ_ρ that produces a metal-insulator "transition" below $T_\rho \simeq 100...200K$.[12] The corresponding Neel temperature T_N range is around $7...20K$.[12] The effect of hydrostatic pressure is quite interesting since Δ_ρ is found to decrease rather rapidly under pressure and is no longer observable when $T_N \simeq T_\rho$.[13] At the weak coupling boundary, T_N is observed to have a rather smooth variation[13] whereas at higher pressure, the weak coupling characteristics of the sulphur series AF transition become quite similar to those found in the Bechgaard salts. This pressure-induced continuity between both series is experimentally and theoretically well documented.[4,6,13] The IEX mechanism appears therefore as a necessary ingredient for its characterization. Otherwise, a change of mechanism (IEX → NAF) would imply that the value of T_{x^1} accidentally coincides with T_N and T_ρ at the boundary. Consequently, the irrelevance of T_{x^1} leads to conclude that the 3D NAF mechanism should not play any role for the AF phases in the Bechgaard (sulphur) salts at low (high) pressure and low magnetic field. Furthermore, near some critical pressure P_c, T_N drops very rapidly and for $P \gtrsim P_c$, superconductivity is stabilyzed below 1K or so. The analysis of the nuclear relaxation rate data performed by Creuzet et al.[8], for several members of the Bechgaard salts at $P \gtrsim P_c$ have shown that the relaxation keeps its 1D strongly AF enhanced character down to 8K or so (see Fig. 1.a), which was in turn identified as the manifestation of the effective low value of $T_{x}1$ in this pressure domain. Furthermore, the 3D NAF mechanism is likely to be frustrated below $T_{x}1$ by sizeable deviations from 2D or 3D perfect nesting conditions. Therefore, in the present approach, it is the emergence of $T_{x}1$ that suppresses AF long range order (see Fig.1.b). The absence of good nesting conditions at low temperature has been remarkably supported by the existence of high field-induced AF phase transitions at $P > P_c$. It follows that in zero field, the 3D NAF contribution to spin fluctuations is probably quite small.

Many authors have suggested that for organic conductors, the 3D NAF-induced spin fluctuations near an AF phase transition favor anisotropic superconductive pairing[3,5] From the above discussion however, this contribution will be strongly reduced. As we will see in sec.III and IV, there is another contribution to interchain pairing that originated from the IEX mechanism[4] Its amplitude is much more important since it takes place in the very large temperature domain $T > T_{x}1$ where precisely AF correlations are seen in NMR experiments (see Fig. 1.a). It can lead to *high temperature (T $\gg$ $T_{x}1$) coexistence of antiferromagnetic and superconducting fluctuations.* The true long range superconducting order is however restricted to the region below $T_{x}1$.

III-RENORMALIZATION GROUP APPROACH TO THE INTERCHAIN KINETIC COUPLING

The delicate role played by the interchain single electron transfer in the Bechgaard salts will be analyzed in the context of a "g-ology" model for a quasi-1D conductor. We are taking advantage of the fact that the partition function can be formally written as a functional integral form $Z = \mathrm{Tr}\, \exp(-\beta H) = \int\!\!\int \delta\psi{*}\delta\psi\, \exp(S[\psi{*},\psi])$, over the anticommuting fields ψ.[4,7] The action S corresponds to:

$$S[\psi{*},\psi] = \sum_{\tilde{k},p,\sigma} G^{-1}_{p\sigma}(\tilde{k})\, \psi^{*}_{p\sigma}(\tilde{k})\, \psi_{p\sigma}(\tilde{k}) \tag{1}$$

$$+\ 1/2LN_{\perp} \sum_{\langle\tilde{k},p,\sigma\rangle} g^{p_1 p_2 p_3 p_4}_{\sigma_1\sigma_2\sigma_3\sigma_4}\, \psi^{*}_{p_1\sigma_1}(\tilde{k}_1)\, \psi^{*}_{p_1\sigma_2}(\tilde{k}_2)\, \psi_{p_3\sigma_3}(\tilde{k}_3)\, \psi_{p_4\sigma_4}(\tilde{k}_4)$$

Here, $\tilde{k}=[k,k_{\perp},\omega_n=(2n+1)\pi T]$, $G_{p\sigma}(\tilde{k})=[G^{-1}_{1D,p\sigma}(k,\omega_n) + 2t_{\perp}\cos(k_{\perp}d_{\perp})]^{-1}$,

$G_{1D,p,\sigma}(k,\omega_n) = [i\omega_n - v_F(pk - k_F)]^{-1}$, $p = \mp$ refers to left $(-)$ and right $(+)$ going 1D electrons and σ is the spin. The g's are the standard decomposition of the Coulomb interaction into backward (g_1), forward (g_2), and Umklapp (g_3) scatterings. The 1D theory is naturally regularized by the bandwidth cutoff energy $E_o = 4t_{\parallel} \gg 4t_{\perp}$. Note that here we have considered for simplicity a 2D electronic system. The generalization to a 3D system is straightforward and does not change the conclusions.

A Kadanoff-Wilson renormalization group procedure can be applied to Z.[4,7] For $T > T_x 1$, it consists of integrating out iteratively for each chain all fermion degrees of freedom in the outer energy shell of thickness $\frac{1}{2} E_o(\ell)d\ell$ on each side of the Fermi level and for all the Matsubara frequencies ω_n. $E_o(\ell) = E_o e^{-\ell}$ is the effective (scaled) bandwidth and $d\ell$ is the infinitesimal generator of the RG. This procedure leads for S to a renomalization of the single particle propagator (G) and the two-particle vertices Γ_i associated to the g_i.

<u>Results for the Interchain Pair Tunneling processes</u>

At $T > T_x 1$, the influence of the interchain hopping on 1D correlated pair of particles is illustrated by the first order vertex corrections of the Figure 2.a. The zero order term in $t_\perp$ is associated with the renormalization of the purely 1D part of the Γ_i.[4,15] On the other hand, the contribution of the second term generates interchain tunneling of all type of correlated pairs of particles in the electron-hole and electron-electron channels.[4] The evaluation of the diagram at each outer shell RG integration gives the infinitesimal contribution $f_\mu^m(\ell)d\ell = (-1)^{|m|+1} 2\pi v_F \, [\bar{t}_\perp/E_o(\ell)]^2 [g_\mu^m(g_1',g_2',g_3')]^2 \, d\ell$. It follows that a new set of interchain pair tunneling interactions has to be added to S. These new couplings will have the form $\sum_{\langle i,j \rangle} \sum_{\mu,m} V_\mu^m(\ell) \, O_{\mu,i}^{m*} \, O_{\mu,j}^m$. From higher RG corrections[4], the V_μ^m will obey to theirs own RG equations:

$$dV_\mu^m(q_\perp) = f_\mu^m(\ell) \cos(q_\perp d_\perp) \, d\ell + V_\mu^m(q_\perp) \, d\ln \chi_\mu^m \, -\tfrac{1}{2} \, [V_\mu^m(q_\perp)]^2 \, d\ell \qquad (2)$$

The composite fields O_μ^m refer to scalar singlet or charge $(\mu=0)$, vector triplet or magnetic $(\mu=1,2,3)$ pairs of particles in the electron hole (density wave) site $(m = +1)$, bond $(m = -1)$ and Cooper pairing $(m = 0)$ channels of correlations. The respective combinaison of renormalized couplings for the generating interchain pair amplitude f_μ^m terms are $g_0^{\pm 1} = 2g_1'-g_2'\pm g_3'$, $g_{1,2,3}^{\pm 1} = -g_2'\mp g_3'$, $g_0^0 = g_1'+g_2'$ and $g_{1,2,3}^0 = g_2'-g_1'$. The second term of (2) is the pair vertex corrections expressed in terms of the 1D auxiliary pair correlation function[15] $\bar{\chi}_\mu^m \equiv -\pi v_F \partial \chi_\mu^m/\partial\ell$. The last term is the RPA interchain ladder contribution which can produces a pole or long range order at $T > T_x 1$. In the repulsive sector, it is the IEX coupling

($V^{+1}_{\mu=1,2,3}$) that gives the highest critical temperature. It is interesting
to note that the integration of (2), far from the critical point, is
dominated by the generating term and leads to:

$$V^{+1}_{\mu=1,2,3}(q_\perp,\ell) \;\alpha\; 2\pi v_F[\,\bar{t}_\perp/E_o(\ell)\,]^2[\,(g_2'+g_3')/\pi v_F\,]^2 \cos(q_\perp d_\perp)$$

Identifying formally $E_o(\ell)$ with temperature ($E_o(\ell) \simeq \pi T$), this form fully
agrees with the one obtained using the qualitative arguments of the section
II. Whenever the electronic system scales to a repulsive strong coupling
regime where $g_2' + g_3' \simeq \pi v_F$ and a correlation gap emerges at $E_o(\ell) \simeq \pi T_\rho \simeq$
Δ_ρ, the coupling smoothly joins to the familiar strong coupling result:[9,10]
$$V^{+1}_{1,2,3}(q_\perp,T\simeq T_\rho) \;\alpha\; 2\pi v_F(\bar{t}_\perp/\Delta_\rho)^2 \cos(q_\perp d_\perp).$$ The IEX mechanism is therefore
continuous at the weak-strong coupling boundary.[4]

The Single Particle Dimensionality Crossover

In second order of the RG there are two major contributions to the
renormalization of the single particle propagator G(see Figure 2.b).
At $T > T_X1$, these corrections have a purely 1D character since they affect
G_{1D} as $G_{1D} \rightarrow Z_1^{-1}(\ell)G_{1D}$. For the total propagator, this is equivalent[7,8,9]
to write $t_\perp \rightarrow \bar{t}_\perp = Z_1^{-1}t_\perp$ namely that we have an effective renormalization of
$t_\perp$. The multiplicative factor Z_1 is the product of two contributions:

$$Z_1^{-1} \simeq \left[E_o(\ell)/E_o\right]^\theta \left[1 + 11\theta/1-\theta \; (t_\perp/E_o)^2 \left[E_o(\ell)/E_o\right]^{2\theta-2}\right]^{-\frac{1}{2}} \qquad (3)$$

The first diagram of the Fig.2.b leads to the term in front of the square
brackets and corresponds to the renormalization of the purely 1D ($t_\perp=0$)
density of states with $\theta > 0$ as the corresponding power law exponent.
Therefore, the correlation cloud that surrounds each electron along the
chain will reduce (screen) the efficiency of the interchain single electron
transfer. According to a recent work of Caron and the author[7] however, if
interchain quantum virtual transfer is included for electrons that form the
cloud (see the second diagram of the Fig.2.b), the "screening" of $t_\perp$ is
further enhanced by the quantum transverse extension of the correlation
cloud. Such a contribution is given by the term in square brackets.
Therefore, each term will contribute to the reduction of T_X1, which reads
for $t_{,,} \gg t_\perp$:[7]

$$T_X1 \simeq T_X^O1 \; (t_\perp/t_{,,})^{\theta/1-\theta}\left[1-11\theta/1-\theta\right]^{1/2-2\theta} \qquad (4)$$

where $T_X^O1 \simeq t_\perp/\pi$ is the free electron gas crossover temperature. From (4),
if the correlations become sufficiently strong, namely $\theta \gtrsim 1/11$, the effect
of lateral extension of the polarization cloud given by the term in square

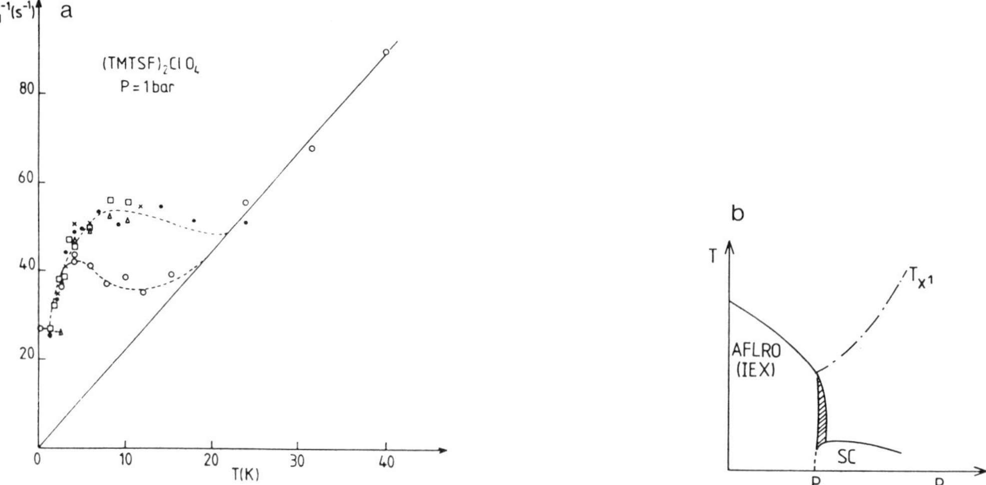

Fig. 1. (a) Nuclear relaxation rate T_1^{-1} vs T of (TMTSF)2ClO4 at P=1bar > Pc. Note the strong AF enhancement at below 25K and the change of regime at $Tx_1 \approx$ 8K. After Creuzet et al., ref. 8.; (b) interplay between the IEX mechanism for AF order, the crossover Tx_1 and superconductiviy in the model proposed for the phase diagram of (TMTSF)2X (see text).

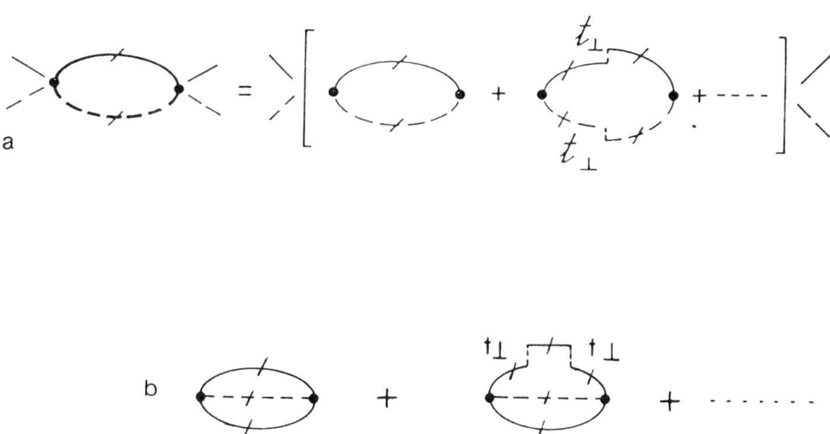

Fig. 2. (a) First order two-particle vertex corrections for the RG described in the text. The thick [thin] dashed (full) lines represent the 2D [1D] propagators for left (right) going electrons in the outer energy shell. The dots are linear combinaisons of the g's. The expansion in $t_\perp$ leads to interchain pair tunneling couplings; (b) 1D self-energy RG corrections. The second diagram and the equivalent permutations gives the quantum transverse extension of the correlation cloud that surrounds the electron. After refs. 4 and 7.

brackets leads to the strongest reduction and forces $t_\perp$ to become an
irrelevant variable in the RG sense. This means that there is no
possibility to observe T_Xᵡ and long range ordering can only occur via
interchain two-particle like processes (eq.(2)). It is interesting to note
that for the reduction factor in parenthesis coming from the isolated
chains, the irrelevance of $t_\perp$ is only obtained for $\theta > 1$. From the first
diagram of the Figure 2.b, we get $\theta \sim \frac{1}{4} (g/\pi v_F)^2 + ..,$. This would imply
that in relatively weak coupling systems like the Bechgaard salts, strong
reductions of T_Xᵡ are likely to occur.

IV- THE NATURE OF THE SUPERCONDUCTING CORRELATIONS

An important consequence of the generation of the interchain
electron-hole pair tunneling processes at $T > T_X$ᵡ is that the formation of
transverse density wave correlations can also be seen as additonal
interactions that introduce new type of correlations but this time, in the
superconducting channel. This is especially true for the IEX coupling
$V^{+1}_{1,2,3}$ which can be viewed as a exchange of 1D $2k_F$ antiferromagnetic
paramagnons between two electrons on neighbouring chains. This boson
exchange leads to an effective interchain singlet pairing interaction
between the two in- and out-going electrons that participate to the
process[4]. Including the effect of all other most relevant but weaker and
non-singular interchain interactions, like the effective charge density
wave ones $V^{\pm 1}_0$, the bond IEX $V^{-1}_{1,2,3}$ and the direct Coulomb repusilve
interactions $g^\perp_1$ and $g^\perp_2$, the effective interchain interaction between
electrons of opposite spins is likely to change continuously from repulsive
to attractive as T decreaes. In the case of a net attraction at $T > T_X$ᵡ
interchain pairing fluctuations can be favorable. These superconducting
fluctuations will therefore coexist with the AF ones. This interesting
possibility can be illustrated by looking at the interchain singlet pairing
correlation function which can be defined as:

$$\chi^S_{i,i+1}(x,\tau) = - \langle T_\tau\, O_{i,i+1}(x,\tau) O^+_{i,i+1}(0,0)\rangle$$

while the interchain singlet pairing operator is written as:

$$O^+_{i,i+1}(x,\tau) = \sum_{\sigma,p=\pm} \sigma\, \psi^+_{p\sigma,i}(x,\tau)\, \psi^+_{-p-\sigma,i+1}(x,\tau)$$

where i is the chain indice. At $T > T_X$ᵡ, the first order RG equation of
the auxiliary quantity $\bar{\chi}^S_{i,i+1}(T) = -\pi v_F\, \partial \chi^S_{i,i+1}/\partial \ell$ with $\ell = \ln E_F/T$, gives:

$$d\ln \bar{\chi}^S_{i,i+1}/d\ell = -1/\pi v_F \left[(g^\perp_1 + g^\perp_2) + \tfrac{1}{2} (|V^{-1}_0| + |V^{+1}_0|) \right.$$
$$\left. -3/2 (|V^{+1}_{1,2,3}| + |V^{-1}_{1,2,3}|) \right] \tag{5}$$

Here, we have supposed that we are sufficiently far from the AF critical
point in order to use the approximation $V_\mu^m(q_\perp,\ell) \approx |V_\mu^m(\ell)|\,\cos(q_\perp d_\perp)$. From
(5), we first note that interchain couplings that couple the charge degrees
of freedom ($g_{1,2}^\perp$ and $V_0^{\pm 1}$) compete with the paramagnon exchange and tend to
suppress interchain pairing. However, these repulsive contributions are
expected to be quite small in organic materials like the Bechgaard salts.
This is confirmed by the absence of any charge density and bond
(spin-Peirls) wave transitions in these materials.[1,2] Furthermore, the 1D
$2k_F$ lattice precursors observed in X-ray experiments[16] are found to be
vanishingly small. On the other hand, AF fluctuations are known to become
sizeable as we drop the temperature. As already mentionned, nuclear
relaxation rate experiments[8] on several members of the Bechgaard salts at P
> P_c have revealed the presence of non-critical AF fluctuations above T_x1
(Figure 1.a). This leads to believe that the repulsive contribution in (5)
should be completly screened by the three site IEX components $V_{1,2,3}^{+1}$ and to
a lesser degree by the three bond IEX contributions $V_{1,2,3}^{-1}$. In that case,
$\bar{\tau}_{i,i+1}^{-s}$ and therefore $\tau_{i,i+1}^{s}$ will be significantly enhanced by
superconducting correlations in the 1D temperature domain. These
correlations are likely to influence the physical properties and this is
especially true for the conductivity. In this respect, the giant zero-
frequency collective mode commonly observed in far-infrared conductivity
experiments[17] for the Bechgaard salts at low temperature could result from
such correlations. One must note that this collective effect on the
conductivity is even present well above the AF critical domain at P < P_c.

The only way the system has to stabilyze a true long range
superconducting ordering stemming from this non-phonon pairing mechanism,
is to "frustrate" the natural tendency toward AF ordering. As we have
already mentionned in section II, the occurence of the single particle
dimensionality crossover can contribute to this frustration.[4] Actually, it
is well known that deviations from perfect 2D or 3D nesting conditions are
able to freeze the NAF mechanism for AF ordering below T_x1. These
deviations depend strongly on the shape of the band structure in the
transverse direction . For example, second nearest-neighbour interchain
hopping terms and interaction of electrons with the anions (X) are possible
sources among many of deviations in the Bechgaard salts.[14] These can be
amplified by the application of external pressure. The absence of good
nesting conditions at P > P_c is supported by the existence of high
field-induced AF phase transitions.[2] Indeed, according to the theory of
Gorkov and Lebed good nesting conditions are restored under the application
of a sufficiently high transverse magnetic field so that the NAF mechanism

becomes no longer frustrated and AF ordering can occur below T_X^1. The theory together with his subsequent refinements are known to give a rather good description of the high field data.[18]

In contrast to the electron-hole channel, the intrinsic instability of the superconducting channel relies on time inversion symmetry and it is not affected by nesting deviations. Therefore, if the interchain attractive pairing contribution coming from the IEX coupling is sufficiently important at T_X^1, superconducting order can be stabilyzed below this temperature. This can be illustrated by the mean field gap equation for *singlet* superconductivity :[4]

$$\Delta(k_\perp) = \tfrac{1}{2} \sum_{\vec{k}'} \Delta(k'_\perp)/\pi v_F \Big[\; [g_0^0 + g_1^\perp + g_2^\perp + \tfrac{1}{2} \sum_{m=\pm 1} V_0^m(k_\perp+k'_\perp)]$$
$$- [\; |V_0^0| + \tfrac{1}{2} \sum_{\mu\neq 0, m=\pm 1} V_\mu^m(k_\perp+k'_\perp)] \; \Big] \; \tanh[\varepsilon(\vec{k}')/2T_C]/\varepsilon(\vec{k}') \qquad (6)$$

$\varepsilon(\vec{k}')$ is the electron energy. All the couplings that appears in r.h.s of (6) take theirs renormalized values at T_X^1 and the effective cut off on the energy summation is T_X^1. Compared to (5), there is an additional contribution to the repulsive part appearing on the first line of (6) which corresponds to the *local intrachain* repulsion $g_0^0 = g_1' + g_2'$. On the other hand, another local contribution comes from the interchain Josephson coupling term V_0^0 0 which is always attractive despite the presence of repulsive g's. This term never becomes singular and at T_X^1, $V_0^0 \sim -2\pi v_F [(g_1' + g_2')/\pi v_F]^2$ which is small in weak coupling. Here again the most important part of pairing will come from from the IEX coupling and if T_X^1 is not too far from an AF critical domain, a net attraction will be possible. The interchain character of the pairing is reflected by the $k_\perp$ dependence of $\Delta(k_\perp)$ which in turn vanishes on lines for a 3D Fermi surface. Taking the approximate form $V_\mu^m(q_\perp=k_\perp+k'_\perp) \approx V_\mu^m \cos[(k_\perp + k'_\perp)d_\perp]$, this leads to $\Delta(k_\perp) = |\Delta| \cos k_\perp d_\perp$. Starting the full second order 1D RG procedure with reasonable weak coupling values for the g's and realistic band paramaters $t_\parallel$ and $t_\perp$ for the Bechgaard salts, Caron and the author[4] have shown that T_X^1 is significantly reduced while the AF correlations are enhanced near this temperature. Preliminary results for the numerical solution of (6) below T_X^1 have shown a finite superconducting $T_C \sim .03 T_X^1$ which has the rigth order of magnitude if one takes $T_X^1 \sim 10K$. T_C has been showed to be strongly depressed by pressure in agreement with experiments.[1,2]

IV-CONCLUDING REMARKS

In the light of the results presented in this paper, it is clear that the existence of a large 1D domain of temperature for correlations in a quasi-1D conductor introduces a richness of interchain mechanisms that play

a important role in the stabilization of long range ordering. The
inevitable interplay between these various mechanisms is certainly not
irrelevant to the succesion of different ground states observed as a
function of pressure for the combined phase diagram of $(TMTTF)_2X$ and
$(TMTSF)_2X$ series of compounds[4]. Although the important role played by the
IEX mechanism for the antiferromagnetism of both series can be assessed by
many experiments, one needs more exprimental works for its relevance for
superconductivity. However, the recent observation by Takigawa et al.,[19] of
a power law temperature profile of the nuclear relaxation rate in the
superconducting state of $(TMTSF)_2ClO_4$ supported the variation of the
superconducting gap on the Fermi surface predicted by the IEX mechanism.
The presence of strong 1D AF enhancement of the NMR relaxation down to 8K
together with the observation of a zero frequency collective mode in
conductivity at relatively high temperature also militate in favor of the
applicability of the IEX induced paring mechanism in organic conductrors.

ACKNOWLEDGMENTS

I would to thank Prof. L. G. Caron, Dr. F. Creuzet and Dr. D. Jerome
for theirs collaboration and encouragment in the development of the ideas
presented in this work.

References

1. D. Jerome and H. J. Schulz, Adv. Phys. $\underline{31}$, 299(1982).
2. D. Jerome, this volume.
3. V. J. Emery, Synthetic Metals $\underline{13}$, 21(1986); M. T. Beal-Monod, C.
   Bourbonnais and V. J. Emery, Phys. Rev. B$\underline{34}$, 7716(1986).
4. C. Bourbonnais and L. G. Caron, Physica $\underline{143B}$, 451(1986); L. G. Caron
   and C. Bourbonnais, Ibid., 453; C. Bourbonnais, Proceedings of the
   NATO Advanced Study Institute on Low-Dimensional Conductors and
   Superconductors, Magog, Quebec, August 1986, to be published by
   Plenum; L. G. Caron, Ibid.; C. bourbonnais and L. G. Caron,
   preprint.
5. D. J. Scalapino, E. Loh. Jr. and J. E. Hirsch, Phys. Rev. B$\underline{34,}$
   8190(1986).
6. S. Barisic and S. Brazovskii, in: "Recent developments in condensed
   matter physics", J. T. Devreese ed., Vol. 1, Plenum N.Y. 1981.; V.
   J. Emery, R. Bruisma and S. Barisic, Phys. Rev. Lett. $\underline{48}$,
   1039(1982).
7. C. Bourbonnais, Mol. Cryst. Liq. Cryst. $\underline{119}$, 11(1985) and Ph.D thesis,
   Université de sherbrooke III-374 (1985), unpublished.; L. G. Caron
   and C. Bourbonnais preprint.
8. C. Bourbonnais, F. Creuzet, D. Jérome, K. Bechgaard and A. Moradpour,
   J. Physique Lett. $\underline{45}$, L-755(1984); F. Creuzet, D. Jerome, C.
   Bourbonnais, and A. Moradpour, J. Phys. C: Solid State Phys. $\underline{18}$,
   L82(1985); F. Creuzet, C. Bourbonnais, L. G. Caron, D. Jerome and A.
   Moradpour, Synthetic Metals, $\underline{19}$, 277(1987).
9. Y. A. Firsov, V. N. Prigodin and Chr. Seidel, Phys. Rep. $\underline{126}$, 245(1985)
10. S. Brazovskii and V. Yakovenko, J. Physique lett. $\underline{46,}$ L-111(1985).
11. C. Bourbonnais, F. Creuzet and L. G. Caron, J.M.M.M. $\underline{54-57}$, 1249(1985);
    F. Creuzet, C. Bourbonnais, L. G. Caron, D. Jerome, and K. Bechgaard,
    Synthetic Metals, $\underline{19}$, 289(1987).

12. C. Coulon, P. Delhaes, S. Flandrois, R. Lagnier, E. Bonjour and J. M.
    Fabre, J. Physique $\underline{43}$, 1059(1982).
13. F. Creuzet, S. S. P. Parkin, D. Jerome and J. Fabre, J.
    Physique(Colloque, $\underline{44}$, C3-1099(1983).
14. L. P. Gorkov and A. G. Lebed, J.Physique Lett. $\underline{45}$, L433(1984); M.
    Héritier, G. Montambaux and P. Ledderer, Ibid., L943.
15. J. Solyom, Adv. Phys. $\underline{28}$, 201(1979); M. Kimura, Prog. Theor. Phys. $\underline{53}$,
    955(1975).
16. J. P. Pouget, R. Moret, R. Comes, K. Bechgaard, J. M. Fabre and L.
    Giral, Mol. Cryat. Liq. Cryst. $\underline{79}$, 129(1982).
17. H. K. NG, T. Timusk and K. Bechgaard, Mol. Cryst. Liq. Cryst. $\underline{119}$,
    191(1985); J. E. Eldridge et al., preprint.
18. J. F. Kwak, Phys. Rev. B$\underline{28}$, 3277(1983); M. Ribault, D. erome, T.
    Tuchendler, C. Weyl and K. Bechgaard, J. Physique lett. $\underline{44}$,
    L953(1983); P. M. Chaikin, Mu-Yong Choi, J. F. Kwak, J. Engler,
    Phys. Rev. Lett. $\underline{51}$, 2333(1984).
19. M. Takigawa, H. Yasuoka and G. Saito, I.S.S.P. preprint; Y. Hasegawa
    and H. Fukuyama, I.S.S.P. preprint.

SYNTHESIS, STRUCTURE AND PROPERTIES OF BEDT-TTF DERIVATIVES

Paul J. Nigrey, Bruno Morosin and James F. Kwak

Sandia National Laboratories, P.O. Box 5800
Albuquerque, N.M. 87185-5800

INTRODUCTION

Because of the unique transport properties of charge-transfer (C-T) complexes based on the π-electron donor molecules, 4,5,4',5'-Bis(alkyldithio)-tetrathiafulvalene[1], the properties of partially selenium substituted donors bis(alkyldiseleno)-tetrathiafulvalene (BADSe-TTF) were investigated. These molecules are of particular interest to us since much of the current research in this area[2] has focused on increasing the dimesionality of the transport[3,4] in C-T salts derived from these donors.

One of the simplest ways of increasing dimensionality is to incorporate either selenium or tellurium into the donor molecules[5,6]. The increased overlap expected from the larger orbitals of these chalcogens should increase both inter- and intramolecular overlaps, thereby imparting greater two-dimensional character to their C-T salts. In addition, the increasingly greater steric requirements of the methylene-, ethylene-, and propylene-diseleno groups in BADSe-TTF should lead to an expansion of the unit cell and thereby favorably affect electronic properties. In this paper, we report the synthesis of one member of these partially selenium substituted donor molecules, BMDSe-TTF, some structural data for the organometallic compound, $(Bu_4N)_2[Ni(dsit)_2]$, which is the precursor to the BADT-TTF donors, and the pressure dependence of conductivity in the C-T salt, $(BMDT-TTF)_2Au(CN)_2$.

EXPERIMENTAL

Bis(tetrabutylammonium)-bis(2-thione-1,3-dithiole-4,5-diselenolato) nickelate (II), $(Bu_4N)_2[Ni(dsit)_2]$, (1)

A fresh solution of lithium diisopropylamide was prepared by the dropwise addition of n-butyllithium in hexane (2.5M, 32 mL, 0.08 moles) to a cooled (-77°C) solution of tetrahydrofuran (70 mL) which contained diisopropylamine (7.6 g, 0.075 moles). The addition rate was controlled so that the reaction temperature did not rise above -65°C. After stirring under an argon atmosphere for 50 minutes at -76°C, a solution of vinylene trithiocarbonate (5 g, 0.037 moles) in THF was added over 15 minutes. This solution was allowed to react at -72°C for 3 hours. To the resulting yellow-brown solution was added powdered selenium (5.9 g, 0.075 moles) in one portion and then allowed to warm up to ambient overnight (17 hrs.).

This solution was then concentrated by blowing argon across it until an oil was obtained. The viscous residue was dissolved in absolute methanol (80 mL). A solution of hydrated nickel chloride (4.43 g, 0.0325 moles) dissolved in a mixture of methanol (30 mL) and conc. ammonia (30 mL) was then added. After stirring the mixture at ambient for 10 min., a solution of tetrabutylammonium bromide (TBABr, 12 g, 0.0375 moles) was added dropwise. The resulting solid was collected by filtration and washed with methanol and ether. This material was vacuum dried to give 18.4 g of an olive-green colored solid. Recrystallization was carried out by dissolving the crude solid in acetone (600 mL) and concentrating to ca. half the original volume. This recrystallization gave 15.1 g (82%) of $\underline{1}$ as dark-purple needles, mp 196 - 198°C: IR (KBr) 2960m (C-H), 1438m (C=C), 1046s, 1036s (C=S), 425 (Ni-Se) cm^{-1}; UV-Vis (acetone) 420 nm (ϵ 23,700), 608 nm (ϵ 7310); <u>Analysis</u>. Calcd: C, 40.53; H, 6.21; N, 2.49; S, 17.16; Se, 28.35; Ni, 5.20%; Found: C, 40.82; H, 6.46; N, 2.48; S; 17.39; Se, 28.30%.

<u>4,5-Methylenediseleno-1,3-dithiole-2-thione (2)</u>

Lithium bromide monohydrate (1.8g, 20 mmoles) was added to a warmed mixture of $\underline{1}$ (4.0 g, 3.6 mmoles) in acetonitrile (200 mL) and the resulting mixture refluxed for 1 hour. Methylene bromide (74.3 g, 0.42 moles) was added and refluxing continued overnight (17 hrs.). The mixture was filtered while hot and the residue was washed with methylene chloride until the washings were pale yellow in color. These combined filtrates were concentrated to dryness on a rotary evaporator. The solid was washed with hot water (200 mL) and then was vacuum dried to give 1.4 g (58%) of (2)as a brown crystalline solid, mp 178-182°C: IR (KBr) 1050s, 1030s cm^{-1} (C=S); UV-Vis (methylene chloride) 423nm (ϵ 10,900); <u>Analysis</u>. Calcd. C, 15.79,; H, 0.66; S, 31.62; Se, 51.92%; Found: C, 15.51; H, 0.68; S, 31.37; Se, 52.13%.

<u>4,5-Methylenediseleno-1,3-dithiole-2-one (3)</u>

$\underline{2}$ (0.914 g, 3 mmoles) was dissolved in a refluxing mixture of chloroform (360 mL), glacial acetic acid (130 mL) and water (10 mL). Mercuric acetate (1.02 g, 3 mmoles) was added to the refluxing mixture in one portion and the suspension was stirred at reflux for 19 hours. This mixture was filtered while still warm through a fine fritted funnel and the filtrate concentrated to dryness. Crude $\underline{3}$ (0.77 g, 89%) was obtained as light-yellow solid. Gradient sublimation[7] at 90°C and 10^{-5} Torr yielded $\underline{3}$ as golden crystals (0.61 g), mp 135-137°C: IR (KBr) 1640s cm^{-1} (C=O); UV-Vis (methylene chloride) 307 nm (ϵ 3730); <u>Analysis</u>. Calcd. C, 16.67; H, 0.70; S, 22.26; Se, 54.81, O, 5.55%; Found: C, 16.81; H, 0.71; S, 22.56; Se, 54.55; O, 6.08%

<u>Bis(methylenediseleno)-tetrathiafulvalene (BMBSe-TTF, 4)</u>

A mixture of $\underline{3}$ (0.4 g, 1.4 mmoles) and triethylphosphite (5 mL) was brought to reflux under an argon atmosphere. After 45 minutes of refluxing, methanol (10 mL) was added and the solution was then placed in an ice bath. The orange colored solid was collected by filtration, washed with methanol, followed by ether, and then vacuum dried to give BMDSe-TTF (0.261 g, 69%) as a orange solid. Recrystallization from dichlorobenzene (35 mL) gave 0.22 g of BMDSe-TTF as red needles, mp 257°C (dec.): IR (KBr) 3020w, 1640w, 1375 , 1135 , 1040 , 970 , 900 , 800 , 765 cm^{-1}; UV (dichlorobenzene) 326 nm (ϵ 12,200); <u>Analysis</u>. Calcd. C, 17.65; H, 0.74; S, 23.57; Se, 58.04%; Found: C, 17.93, H, 0.76; S, 23.37; Se, 57.80%.

RESULTS AND DISCUSSION

The incorporation of selenium into the vinylene trithiocarbonate molecule (1,3-dithiole-2-thione) was achieved by lithiation of the vinylene moiety using _in-situ_ prepared lithium diisopropylamide followed by reaction with elemental selenium. Since most other synthetic methods rely on either the reduction of carbon disulfide[8] or the chemistry of 1,3,5,7-tetrathiapentalene-2,5-dione[9], both of which techniques result in the formation of a symmetrically substituted ethylene functionality, i.e tetrathioethylenes, our approach appears to be the only method for functionalizing the 1,3-dithiole ring system with other chalcogens. The air and moisture sensitivity of these 4,5-dichalcogeno-1,3-dithiole-2-thione dianions prevents their use as starting reagents. It is not surprising, therefore, that most preparative routes immediately alkylate these dianions using either alkyl halides or alkyl dihalides in order to obtain the various 4,5-bis(alkyldithio)-1,3-dithiole-2-thiones. However, as has been shown by Steimecke, et al[10], complexation of the above dianions with nickel chloride followed by the addition of tetrabutylammonium bromide (TBABr) results in the formation of a stable organo-metallic complex which can subsequently be used as an "off-the-shelf-reagent" in the synthesis of 4,5-alkylchalcogeno-1,3-dithiole-2-thiones. Similarly, we have used such a technique to obtain (<u>1</u>) in good yields as shown in Scheme 1.

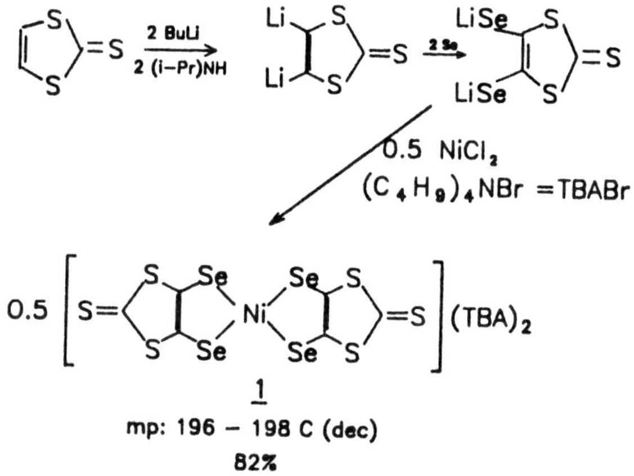

Scheme 1.

Refluxing an acetonitrile solution of (TBA)$_2$[Ni(dsit)$_2$] (<u>1</u>) with four equivalents of lithium bromide for 1 hour is believed to result in the _in-situ_ formation of 4,5-bis(lithioseleno)-1,3-dithiole-2-thione. Addition of excess alkylating agent (1,3-dibromomethane) to the reaction mixture and refluxing at 78°C over a 17 hour period resulted in the formation of (<u>2</u>) in 58% yields. When ethylenedibromide or propylenedibromide are used as the alkylating agents, 4,5-(ethylene-1,2-diseleno)-1,3-dithiole and 4,5-(propylene-1,3-diseleno)-1,3-dithiole-2-thione[11], respectively, were obtained. It should be noted that relatively minor amounts of uncharacterized dark solids are always obtained from the filtered reaction mixture. Based on their solubility characteristics, these solids appear to be polymeric.

In addition to being a synthetic intermediate for the mixed sulfur/selenium heterocycles, the Ni(dsit)$_2$ dianion may also serve as a novel source for a new class of organometallic superconductors. Since Brossard, et al[12] have suggested some three dimensional character in the electronic

interactions of the TTF salt of the all-sulfur nickel complex, TTF
[Ni(dmit)$_2$], it may be possible, similarly, to probe the effects of
selenium incorporation on the electronic interactions in C-T salts of the
Ni(dsit)$_2$ dianions.

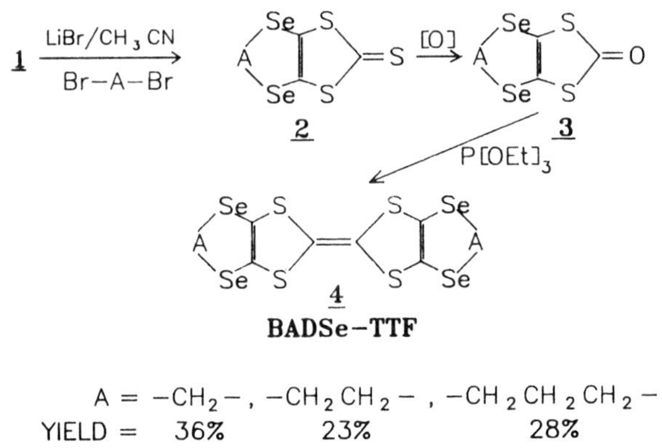

$$1 \xrightarrow[\text{Br—A—Br}]{\text{LiBr/CH}_3\text{CN}} \mathbf{2} \xrightarrow{\text{[O]}} \mathbf{3} \xrightarrow{\text{P[OEt]}_3} \mathbf{4}$$

BADSe—TTF

A = $-CH_2-$, $-CH_2CH_2-$, $-CH_2CH_2CH_2-$

YIELD = 36%   23%   28%

Scheme 2.

The conventional coupling procedures[13], consisting of refluxing (2)
in either neat triethylphosphite or triethylphosphite in benzene solutions, failed to give significant amounts of BMDSe-TTF. These results are
similar to those found in the synthesis of the all-sulfur analog of (4),
BMDT-TTF[1]. However, coupling could easily be achieved by oxidation of 4,5-
(methylenediseleno)-1,3-dithiole-2-thione (2) to 4,5-(methylenediseleno)-
1,3-dithiole-2-one (3) followed by reaction of the latter material with
triethylphosphite as shown in Scheme 2. The oxidation of (2) was performed in a refluxing mixture of chloroform, aqueous acetic acid and mercuric
acetate[14]. This reaction gave crude (3) in virtually quantitative yields.
Purification of this product was achieved by gradient sublimation[7] at 90°C
and 10^{-5} Torr. These combined procedures resulted in good yields of analytically pure (3). When (3) was refluxed in neat triethylphosphite, a
orange solid formed shortly after the solution began to reflux. This
solid was subsequently identified as the coupled product, BMDSe-TTF which
was isolated in 69% yields (overall yield of BMDSe-TTF from (1) was 36%).

The coupling reactions using 4,5-(ethylenediseleno)-1,3-dithiole-2-
one and 4,5-(propylenediseleno)-1,3-dithiole-2-one, results in the isolation of BEDSe-TTF (overall yield 23%) and BPDSe-TTF[11] (28%), respectively.
For comparison, some selected properties of the BADSe-TTF donors are
given in Table 1.

From the cyclic voltammetry data, it can be seen that BMDSe-TTF has
the lowest oxidation potential for removal of one electron to form a
radical cation at +0.10 V vs. Ag/Ag$^+$. BEDSe-TTF and BPDSe-TTF, respectively, undergo this oxidation reaction at somewhat higher potential. It
is interesting to note that, in these donors, oxidation to the cation
radical species occurs at virtually the same potential. This result
suggests that the effects of the ethylenediseleno and propylenediseleno
functionalities have little influence on the redox state of these donors.
One feature, however, which was found common in all these donors was that
oxidation of the cation radical to the dication at the second half-wave
potential ($E^2_{1/2}$) was irreversible in this solvent system. This result is
very much in contrast to the all sulfur substituted donors, BMDT-TTF,

Table 1.   Properties of BADSe-TTF

Donor	M.P./°C	U.V.(log ϵ)[a] λ/nm	Redox Potentials[b] $E^1_{1/2}$ /V	$E^2_{1/2}$ /V
BMDSe-TTF	257 (dec)	326 (4.09)	+0.10	+0.37[d]
BEDSe-TTF	233 (dec)	345 (4.20)[c]	+0.14	+0.47[d]
BPDSe-TTF	272-277	340 (4.20)	+0.15	+0.47[d]

[a]In dichlorobenzene. [b]Versus Ag/Ag$^+$ at a platinum bead working electrode, 0.1M Bu$_4$NAsF$_6$/benzonitrile, scan rate 100 mV/sec. [c]Chlorobenzene. [d]Irreversible wave.

BEDT-TTF, and BPDT-TTF, where no such behavior was observed. Since similar irreversible behavior has been observed in other solvents such as chlorobenzene and methylene chloride, it would appear that special precautions are required during electrochemical crystal growth when attempting to form C-T salts from these donors.

<u>The Crystal Structure of Bis(tetrabutylammonium) Bis(2-thione-1,3-dithiole-4,5-diselenolato)nickelate (II)</u>

The crystal structure of $(Bu_4N)_2[Ni(dsit)_2]$ was determined in order to give some indication of what effects selenium incorporation has on planarity and crystal packing in the Ni(II) complex. This information could provide some guidance for the preparation of C-T salts derived from TTF donors.

The Ni(II) complex crystallizes in the space group $P2_1/c$ with unit cell parameters a = 8.535, b = 14.906, c = 19.494 Å, β = 96.295° , Z = 2 and V = 2465 Å^3. The structure was solved from the Fourier Difference maps using the SHELXTL[15] Crystallographic package and least-squares refinement. Because of the large thermal motion of the tetrabutyl ammonium groups, the hydrogen atoms were omitted in the final refinement. Final coordinates of the non-hydrogen atomic coordinates are given in Table 2. The final agreement factors R(F) are 0.043.

The structure contains nearly planar $[Ni(dsit)_2]$ dianions and is isomorphic with the corresponding all sulfur analog[16]. In both these materials, the tetrabutylammonium cation shows considerable thermal motion. The most significant differences between these two materials is the much shorter C=C bond (1.344 Å vs. 1.39 Å for the S analog). The bond distance for the Ni-Se bond was 2.302(1) Å. Other bond distances and the atomic numbering scheme are given in Fig. 1. No short intermolecular contacts between the atoms of the $[Ni(dsit)_2]$ anions were observed since the dianions are separated from each other by the tetrabutylammonium anions. The shortest Ni-Ni distance was 8.532 Å which was for the nickel atoms located on the <u>a</u> axis. These intermolecular contacts are best observed without the butylammonium counterions as shown in the unit cell perspective of Fig. 2.

Table 2. Atomic Parameters of $(Bu_4N)[Ni(dsit)_2]$

	x	y	z	$U_{eq}*10^4$ [a]
Ni	0.0000( )	0.0000( )	0.0000( )	417(4)
Se1	0.0360(1)	0.1060(1)	-0.0827(1)	526(3)
Se2	-0.1875(1)	-0.0768(1)	-0.0709(1)	644(4)
S 1	-0.1046(2)	0.1082(1)	-0.2410(1)	527(7)
S 2	-0.2994(3)	-0.0450(1)	-0.2285(1)	543(8)
S 3	-0.3140(3)	0.0463(2)	-0.3658(1)	633(8)
C 1	-0.0926(8)	0.0592(5)	-0.1594(4)	431(25)
C 2	-0.1842(8)	-0.0141(5)	-0.1544(4)	420(24)
C 3	-0.2417(9)	0.0377(5)	-0.2817(4)	468(27)
N	0.4652(6)	0.7427(4)	0.0656(3)	403(21)
C11	0.5446(9)	0.6673(5)	0.0297(4)	460(27)
C12	0.6960(9)	0.6971(5)	0.0002(4)	507(28)
C13	0.7698(9)	0.6125(6)	-0.0295(4)	557(29)
C14	0.9295(10)	0.6418(8)	-0.0559(5)	791(40)
C21	0.3019(8)	0.7076(5)	0.0833(4)	505(29)
C22	0.3179(9)	0.6352(6)	0.1407(4)	632(32)
C23	0.1559(10)	0.5868(6)	0.1412(5)	689(35)
C24	0.0278(10)	0.6530(7)	0.1634(5)	719(38)
C31	0.5736(9)	0.7680(6)	0.1303(4)	554(30)
C32	0.5050(11)	0.8442(7)	0.1724(5)	755(38)
C33	0.6305(11)	0.8642(8)	0.2375(5)	908(45)
C34	0.6359(24)	0.7894(10)	0.2848(7)	1797(97)
C41	0.4340(9)	0.8244(5)	0.0196(4)	491(27)
C42	0.3342(10)	0.8069(6)	-0.0492(4)	619(32)
C43	0.3335(11)	0.8936(7)	-0.0919(4)	708(36)
C44	0.2304(13)	0.8788(8)	-0.1627(5)	969(49)

[a]The complete temperature factor is $\exp[-2\pi^2 U_{eq}(\sin^2\theta)/\lambda^2]$

where $U_{eq} = 1/3\Sigma_i\Sigma_j U_{ij} a_i^* a_j^* a_i a_j$ in units of $Å^2$.

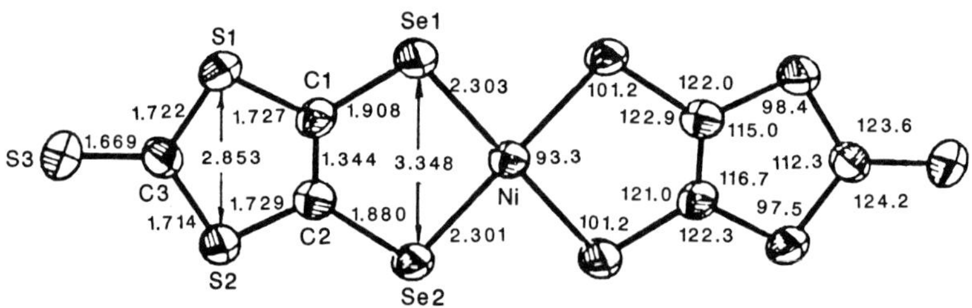

Fig. 1 ORTEP drawing and numbering scheme for the $[Ni(dsit)_2]$ dianion, showing 50% probability ellipsoids for the atoms. Bond lengths and bond angles are shown with the tetrabutylammonium cation omitted for clarity.

176

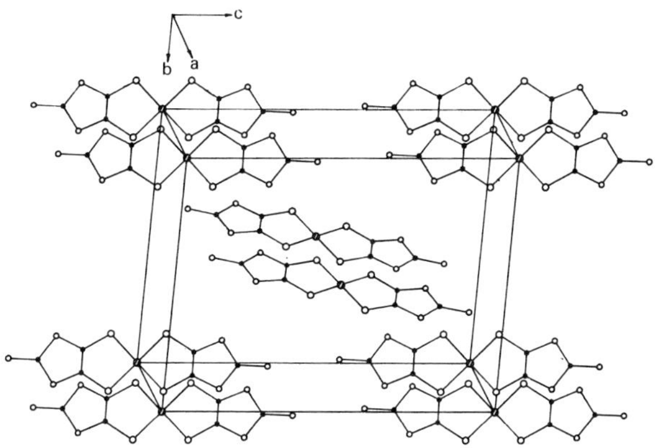

Fig. 2 A projection of the unit cell for the [Ni(dsit)$_2$] dianion.

<u>Pressure Dependence of Resistivity in (BMDT-TTF)$_2$Au(CN)$_2$</u>

The conductivity studies were carried out on single crystal of (BMDT-TTF)$_2$Au(CN)$_2$ prepared by electrochemical oxidation in 0.02 M Bu$_4$NAu(CN)$_2$/

chlorobenzene[1]. These crystals had a plate-like morphology with the plane of the plate corresponding to the crystallographic ab plane. Contacts were applied by evaporating gold pads onto the ab plane of the crystals and attaching gold wires with conductive paint. Depending on the sample size, the measurements were made using a linear four probe or van der Paaw arrangement of contacts along the ab plane of the crystals. Hydrostatic pressures were applied to the sample in a Cu-Be vessel using helium gas as the pressure medium.   The conductivity is given as "in the ab plane" because we did not know the specific directions of the crystal axes in the samples measured.

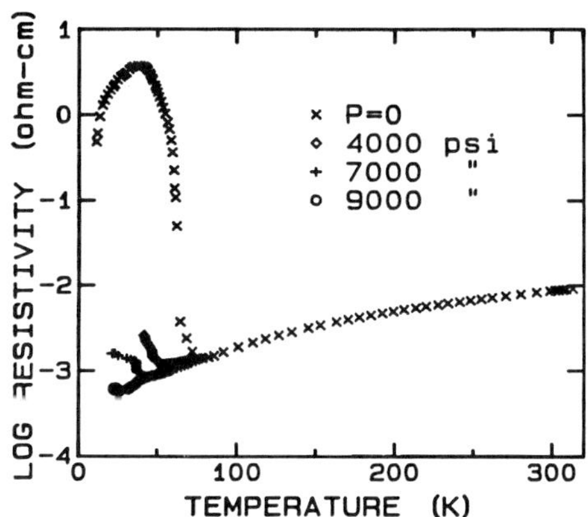

Fig. 3.   Pressure dependence of the ab plane resistivity in
(BMDT-TTF)$_2$ Au(CN)$_2$.

The pressure dependence of the ab plane resistivity is plotted versus temperature   in   Fig. 3. At ambient pressure and temperature, the material has a conductivity of nearly 300 (ohm-cm)$^{-1}$. The conductivity increases by a factor of ten down to 76 K, at which point the material undergoes a transition to a semimetallic state. While the conductivity falls by three

orders of magnitude below 76 K, the low temperature state is not semiconducting since the conductivity actually recovers slightly below 40 K.
Application of a mild pressure of 4000 psi results in a shift of the
transition temperature from 76 K to nearly 50 K. With the application of
progressively larger pressures, the transition is further depressed in
temperature while the conductivity remains high. When 9000 psi of helium
pressure was applied, the sample remained metallic down to 20 K. These
results point out the extreme pressure sensitivity of this material.   In
fact, the rate (-80 K/kbar) at which the transition was supressed is
unprecedently high for organic C-T conductors. The rate of supression of
the transition is even more remarkable in view of the two-dimensional
dimerized cation network observed in the structure of this material[1].
Since we presently believe that the origin of this transition is crystallographic in nature, the use of the partially selenium substituted donor,
BMDSe-TTF, in place of BMDT-TTF may result in the complete suppression of
this transition and possibly induce superconductivity in this class of
materials.

CONCLUSIONS

    In summary, we have shown that the organometallic reagent ($\underline{1}$) can be
used as a synthetic reagent in the preparation of the partially selenium
substituted BADSe-TTF donors. This generalized procedure was exemplified
by the synthesis of BMDSe-TTF which was obtained in an overall yield of
36%. Since the structure of ($\underline{1}$) was found to be isostructural with that of
the all sulfur nickel complex, in whose TTF C-T salt three dimensional
superconductivity was thought to occur, it may be possible to utilize ($\underline{1}$)
for the synthesis of similar materials. We have also presented data on the
pressure dependence of conductivity in $(BMDT\text{-}TTF)_2Au(CN)_2$ for which we

find that modest pressures cause the retention of the metallic state to
low temperatures. Through the appropriate choice of donor/anion combinations, it may be possible to stabilize the metallic state in this material
and induce superconductivity in this class of materials.

ACKNOWLEDGEMENT

    This research was supported by the U.S. Department of Energy (DOE),
Office of Basic Energy Sciences, Division of Materials Sciences, under
Contract No. DE-AC04-76DP00789.

REFERENCES

1a. P. J. Nigrey, B. Morosin, J. F. Kwak, E. L. Venturini, and R. J.
      Baughman, Organic Metals: Synthesis, Structure and Properties
      of $(BMDT\text{-}TTF)_2Au(CN)_2$, Synthetic Metals 16:1 (1986).

1b. P. J. Nigrey, B. Morosin, E. L. Venturini, L. J. Azevedo, J. S.
      Schirber, S. E. Perschk and J. M. Williams, Synthesis, Structure,
      and Properties of $(BPDT\text{-}TTF)_2IBr_2$, Physica B 143B:290 (1986).

2a. F. Wudl, From Organic Metals to Superconductors: Managing Electrons
      in Organic Solids, Acc. Chem. Res. 17:227 (1984).
2b. J. M. Williams, M. A. Beno, H. H. Wang, P. C. W. Leung, T. J. Emge,
      U. Geiser and K. D. Carlson, Organic Superconductors: Structural
      Aspects and design of New Materials, Acc. Chem. Res. 18:261 (1985)
2c. Proceedings of the International Conference on Science and Techno-
      logy of Synthetic Metals (ICSM'86), Kyoto, Japan, 1986, Synthetic
      Metals 19 (1987).
3.   F. Wudl and E. J. Aharon-Shalom, Heaxamethylenetetratellurafulvalene,

J. Amer. Chem. Soc. 106:8303 (1984).

4. K. Bechgaard, D. O. Cowan and A. N. Bloch, Synthesis of the Organic Conductor Tetramethyltetraselenofulvalenium 7,7,8,8-Tetracyan-p-quinodimethanide (TMTSF-TCNQ) [4,4',5,5'-Tetramethyl-bis-1,3-diselenolium 3,6-Bis-(dicyanomethylene)cyclohexadienide], Chem. Commun. 937 (1974).

5. E. Ahron-Shalom, J. Y. Becker, J. Bernstein, S. Bittner and S. Shaik, A New Electro Donor: Synthesis of 2,3,6,7-Tetra(ethyltellurio)tetrathiafulvalene, Tetrahedron Lett., 26:2183 (1985).

6. V. Y. Lee; Chemistry of Organic Metals: Synthetic Approach to Mixed Chalcogenide BEDT-TTF, Synthetic Metals 19:980 (1987).

7. A. R. McGhie, A. F. Garito and A. J. Heeger, A Gradient Sublimer for Purification and Crystal Growth of Organic Donor and Acceptor Molecules, J. Cryst. Growth 22:295 (1974).

8. M. F. Hurley and J. Q. Chambers, Electrochemical Reduction of Carbon Disulfide. Synthesis of Carbon Sulfide Heterocycles, J. Org. Chem. 46:775 (1981).

9. R. R. Schumaker and E. M. Engler, Thiapen Chemistry. 2. Synthesis of 1,3,4,6-Tetrathiapentalene-2,5-dione, J. Amer. Chem. Soc. 99:5521 (1977).

10. G. Steimecke, H.-J. Seiler, R. Kirmse and E. Hoyer, 1,3-Dithiol-2-thion-4,5-dithiolat aus Schwefelstoff und Alkalimetall, Phosporus and Sulfur 7:49 (1979).

11. P. J. Nigrey, Synthesis of 4,5,4',5'-Bis(propylene-1,3-diseleno)-tetrathiafulvalene (BPDSe-TTF), J. Org. Chem., submitted.

12. L. Brossard, M. Ribault, L. Valade and P. Cassoux, The First 3D Molecular Superconductor under Pressure?: TTF[Ni(dmit)$_2$], Physica B 143B:378 (1986).

13. M. Mizuno, A. F. Garito and M. P. Cava, "Organic Metals": Alkylthio Substitution Effects in Tetrathiafulvalene-Tetracyanoquinodimethane Charge-transfer Complexes, J.C.S. Chem. Comm. 18 (1978).

14. I. D. Rae, Synthesis of 1,3-Dithiole-2-ones, Phosphorus and Sulfur 8:273 (1973).

15. G. M. Sheldrick, Nicolet SHELXTL Operations Manual, Nicolet XRD Corporation, Madison, Wisconsin (1983).

16. O. Lindqvist, L. Sjolin, J. Sieler, G. Steimecke and E. Hoyer, Nickel Chelates of Trithione- and Isotrithionedithiolate- A New Class of 1,2-Dithiolates. Part I. The Crystal Structure of Tetrabutylammonium Bis(isitrithionedithiolato)nickelate(II), Acta Chem. Scand. A 33:445 (1979).

ON THE POSSIBILITY OF HIGH-TEMPERATURE SUPERCONDUCTIVITY IN ORGANIC

MATERIALS

János J. Ladik

Institute for Theoretical Chemistry of the Friedrich-
Alexander-University Erlangen-Nürnberg, Egerlandstr. 3
D-8520 Erlangen, F.R.G.

Thomas C. Collins

Physics Department, University of Tennessee at Knoxville
404 Andy Holt Tower, TN 37996

ABSTRACT

First the different models of excitonic superconductivity in 2D and
quasi 1D systems are shortly reviewed. Afterwards the possibility of superconductivity in the TCNQ molecular crystal, which has a narrow valence
band and a broad conduction band is discussed. As next step the possibility
of making the layers in graphite superconducting through doping will be
mentioned. Subsequently a theoretical model (based first of all on the
localization of the wave functions on the highly polarizable Y, La etc.
atoms) for high T_c superconductivity in $Ba_2YCu_3O_7$ systems will be
presented. Finally, the necessary quantum mechanical calculations to
prove this theory will be outlined.

1. INTRODUCTION

Fritz London has postulated the existence of superconductivity at
higher T_c-s already in 1950 /1/. He has not given, however, as an example for
any concrete system (even any kind of type of system) in which this
phenomenon could be realized. In 1970 Ginzburg /2/ (in a more con-
cretized excitonic model of high-temperature superconductivity) has postulated that if one has alternating conductor and semiconductor (insulator)
layers, the mobile electrons are tunnelling over to the layer of the localized electrons, polarize them, in this way they can induce an effective
attraction between two mobile electrons (which form a Cooper pair) and this
can lead to superconductivity. Allender, Bray and Bardeen /3/ investigated
this model in detail (assuming the presence of one conducting and one
insulating layer) with the result that such a phenomenon most probably can
happen.
In 1964 Little /4/ has proposed a quasi 1D model of high T_c superconductivity with a spine of mobile electrons (π electrons) and localized
side chain electrons whose polarization would cause again an effective
attraction between two electrons forming a Cooper pair. This model was severely

criticized on the basis that in a quasi 1D system thermodynamic fluctuations of the electronic distribution would at $T \neq 0$ temperature destroy
the superconductive state. In a second paper Little has shown that though
this might be true for a very long chain, superconductive type enhanced
conductivity still can exist in longer segments of a chain /5/.

Ladik et al. has pointed out as first already in 1965, that to have delocalized mobile electrons and localized ones, one does not need a <u>spatial</u>
separation of the two electronic systems, but the two electronic systems
have to be separated in a <u>quantum mechanical</u> sense (for instance the mobile
π electrons and the localized electrons) in a stack of ethylene molecules /6/
or in a nucleotide base stack /7/.

Subsequent detailed quantum mechanical calculations of a cytosine stack
both at the semiempirical π electron Pariser-Parr-Pople /8/ and at the <u>ab
initio</u> /9/ (though minimal basis) HF crystal orbital level have shown,
that for certain scattering processes around the Fermi level the Coulomb
repulsion and for other ones the effective attraction due to the virtual
σ-excitons are larger by a few eV-s (in absolute value) /7,10/. In this way
it was impossible to design purely theoretically which polymer chain can
become at higher T_c-s superconducting.

In a more recent paper Collins et al. /11/ interpretating the diamagnetic anomalies and increase of the specific conductivity by several
orders of magnitude in CuCl and CdS /12/, have called the attention to O^{2-}
impurities in CuCl and Li and Cl impurities in CdS. Applying a generalized
Eliashberg equation /13/ in a certain approximation /11/ they have been
able to interpret the observed superconductor-type behavior, if they have
taken into account that the dominantly d-type valence band in both systems
is rather narrow and the (due to impurities partially populated) conduction
band is rather broad. Namely, in this case the polarization of the rather
localized electrons in the narrow valence band can introduce an effective
attraction between the mobile electrons in the broad conduction band which
may lead to superconductivity. In their model they have taken into account a
linear combination of Cooper pairs (which antiparallel spins) and electron
pairs with parallel spins (like in superfluide He^3). One should point out
that also in this case the mobile and localized electrons are not separated
in space, but only in a quantum mechanical sense.

Most recently Bednarz and Müller have reported the experimental finding
of superconductivity in the Ba-La-Cu-O system with $T_c \approx 30^\circ K$ /14/. In the
meantime a large number of papers have appeared which show in Ba-Y-Cu-O
systems superconductivity up until 90-100°K (stand of Phys. Rev. Lett.
of June 1) (see for instance /15/). One should mention that according to
different rumors in different countries (USA, Japan, China, Soviet Union,
etc.) they have reached already $T_c \approx 240^\circ K$ (see in point 4, the remarks
on this system).

It should be further mentioned that it is very difficult to reach even
90°K transition temperature through electron-phonon coupling. There are,
however, publications in the latest literature (Phys. Rev. Lett. June 1,
1987) according to which the substitution of ^{16}O by ^{18}O had no effect on
the measured T_c of superconductivity /16/. The authors of both articles
conclude that this makes a mechanism based on phonon coupling rather
improbable.

Finally, they have substituted also partly magnetic ions in the La/Y
plane of these superconductors without any mesurable effect on T_c/17/. This
rules out rather probably any explanation based on magnetic properties of
the high T_c of these ceramic materials. This is the reason that we have
taken granted above excitonic mechanisms for the interpretation of the
new results on high-temperature superconductors.

2. THE TCNQ MOLECULAR CRYSTAL

Ab initio HF band structure calculations /9/ performed for a TCNQ
stack in the TCNQ-TTF mixed crystal (but before charge transfer that is
with neutral TCNQ units) have provided a conduction band width of 1.174 eV
(at the experimental stacking distance of 3.17 Å) and a narrow valence
band of 0.1 eV) /18/. The calculations have been performed only with an
STO-3G basis set /19/, their qualitative features most probably would be
the same also with a better basis set. MINDO/3 all-valence electron
/20/ calculations have resulted in $\delta\epsilon$ = 0.56 eV and $\delta\epsilon$ = 0.10 eV /21/
where $\delta\epsilon_c$ and $\delta\epsilon_v$ are the conduction and valence band widths, respectively.

Though in a pure TCNQ crystal the relative positions of the TCNQ
molecules is different /22/ than those in the TCNQ-TTF mixed crystal, one
can still assume that the qualitative picture (broad conduction and narrow
valence band) would stay the same. Therefore this system seems to be a good
candidate for high T_c-superconductivity, if 1) one would dope it by electron
donor to get free electrons in its conduction band, 2) one would apply pressure
to possibly bring nearer the lower edge of the conduction band to the
impurity (donor levels) and 3) one starts to cool the sample to find out T_c.

3. GRAPHITE LAYERS AS POSSIBLE SUPERCONDUCTORS

It is well known that undoped graphite has a specific conductivity (σ)
of $23 \cdot 10^3 \Omega^{-1} cm^{-1}$ at 300°K /23/. At the same time if it is doped with
different F-containing electron-acceptors σ_{310} goes up until $2.70 \cdot 10^5 \Omega^{-1} cm^{-1}$
(in the case of the dopant $C_{16} SO_3F$ /23/) which is about the half of σ_{310}
of Cu ($5.88 \; 10^5 \Omega^{-1} cm^{-1}$). (In the case of $C_{16}AsF_5$ its σ_{310} is even
$3.00 \; 10^5 \Omega^{-1} cm^{-1}$ /24/.)

The negatively charged electron acceptors possibly may cause an effective attraction between two positive holes in the valence band of a
graphite layer. On the other hand, probably one has a better option, if
one starts to work with electron donor atoms bound to a molecule which can
intercalate between the graphite layers and contains also highly polarizable
atoms bound to the same molecule. Such experimental work is in progress
at the Institute of Inorganic Chemistry (Chair I) of the University
Erlangen-Nürnberg.

4. AN ATTEMPT FOR THE THEORETICAL INTERPRETATION OF HIGH-TEMPERATURE SUPER-
CONDUCTIVITY IN $Ba_2YCu_3O_7$-TYPE CRYSTALS

According to the X-ray investigations on single crystals this system
/25/ has a layer structure six layers forming the unit cell. In one layer
one Cu atom is surrounded by 4 O atoms and vice versa, while in the layer
between two such Cu-O layers one finds only highly polarizable atoms like
Y or La (their first singlet excitation energy is 0.13 or 0.06 eV, respectively /26/). In the other sides of the two Cu-O layers one finds Ba-O
layers and finally the sixth layer is again a Cu-O layer with half of the
O atoms missing (vacancies).

To interpret superconductivity with high T_c in this system we should
like to propose the following mechanism: 1) The $3d^{10}4s$ electronic configuration of a <u>free</u> Cu-atom certainly becomes changed to a $3d^9 4s^2$ configuration if the Cu atoms enter into chemical binding. In this way a narrow
filled essentially Cu 3d band is formed and the dominantly 4s band of Cu
gets below it (probably overlapping with it). 2) Due to charge transfer
from the Y (La etc.) plane to the Cu plane the originally empty $2B_z$ band
of O becomes partially filled and the charge transport takes place in this
band of the Cu plane. In this case the filled narrow d band can play the
role of a polarizable medium and the electrons getting into the conduction
band are the mobile ones. 3) Since the distance between the Y atoms is
~ 3.9 Å in their plane and the nearest Y-O distance in the next plane is
~ 2.4 Å, they will behave as nearly free atoms. Even if the Y atoms lose

one or two electrons to the O atoms [though their first ionization potential, (6.22 eV, /26/) is higher than that of the Ba atoms (5.21 eV) /26/]
rather probably they still remain well polarizable. In this way their
dynamic polarization could create enough effective attraction between a
pair of electrons to stabilize a superconductive state even at higher
T_c-s.

To prove the validity of this theory and to determine the weights of
mechanisms 2) and 3) we have started the following calculations:
1) Cluster calculations on the double unit, $(Ba_2YCu_3O_7)_2$ (using core potentials fitted to experiment for the inner shell electrons of O, Cu, Y and
Ba /27/) to find out the amount of transferred charge from the Y or Ba
layers to the Cu layer. 2) Two dimensional ab initio HF calculations for
the 2d Cu layer with a double ζ /28/ basis set and with the above mentioned core potential to determine its band structure and 3) A slab calculation for all the six layers described above using again a core potential
for the inner shell electrons.

After performing all these calculations one can think to calculate
the Coulomb repulsion (with appropriate screening),the effective attraction
terms (due to virtual excitations on the Y-atoms) and to determine in this
way the necessary condition for superconductivity. To fulfill also the
sufficient condition for it of course one has to satisfy further conditions
if the superconductor gap is not zero /30/.

One is fully aware that this is the hard way to attack the problem of
high T_c in these ceramic systems but we do not see any easier serious
ways to do this. At the same time with the help of the rapidly developping
computer technology such calculations are nowadays possible and the extreme
importance of the problem justifies the necessary great amount of computer
time.

ACKNOWLEDGMENT

We should like to express our gratitude to Professors K. Brodersen,
B. Kunz, P. Otto and M. Seel for very fruitful discussions on different
aspects of the problem. We are very much indebted to Dr. C.-M. Liegener
collecting many important data from the literature. The financial support
of the "Fond der Chemischen Industrie" is gratefully acknowledged.

REFERENCES

/1/ F. London, in Superfluids, Vol. I, John Wiley and Sons, New York
     (1950).
/2/ V.L. Ginzburg, Sov. Phys. Usp. 13, 335 (1975); JETP Letters 14, 396
     (1971).
/3/ D. Allender, J. Bray and J. Bardeen, Phys. Rev. B7, 1020 (1973);
     ibid B8, 4433 (1973).
/4/ W.A. Little, Phys. Rev. 134A, 1416 (1964).
/5/ W.A. Little, Phys. Rev. 156, 396 (1967).
/6/ J. Ladik, G. Biczó and A. Zawadowski, Phys. Lett. 18, 257 (1965).
/7/ J. Ladik, G. Biczó and J. Rédly, Phys. Rev. 188, 710 (1969).
/8/ R. Pariser and G. Parr, J. Chem. Phys. 21, 466, 707 (1953); J.A. Pople,
     Trans. Far. Soc. 49, 1375 (1953).
/9/ G. Del Re, J. Ladik and G. Biczó, Phys. Rev. 155, 997 (1967);

J.-M. André, L. Gouverneur and G. Leroy, Int. J. Quant. Chem. $\underline{1}$, 427, 451 (1967).

/10/ J. Ladik, R.D. Singh and S. Suhai, Phys. Lett. $\underline{81A}$, 488 (1981).

/11/ T.C. Collins, M. Seel, J. Ladik, M. Chandrasekhar and H.R. Chandrasekhar, Phys. Rev. $\underline{B27}$, 140 (1983); T.C. Collins, M. Chandrasekhar and M. Seel, Int. J. Quant. Chem. $\underline{25}$, 831 (1984).

/12/ See for instance: N.M. Brandt, S.W. Kushimikov, A.P. Rusakov and W.M. Smernov, JETP Lett. $\underline{27}$, 33 (1978); C.W. Chu, A.P. Rusakov, S. Huang, S. Early, T.H. Geballe and C.Y. Huang, Phys. Rev. $\underline{B18}$, 2116 (1968).

/13/ G. Eliashberg, Sov. Phys. – JETP $\underline{11}$, 696 (1960); see also J.R. Schrieffer, Theory of Superconductivity, Benjamin Press, Reading, 1964.

/14/ J.G. Bednarz and K.A. Müller, Z. f. Physik $\underline{13}$, 190 (1986).

/15/ L. Bourne, F. Crommie, A. Zettl, H.-C. zur Loye, S.W. Keller, K.L. Leary, A.M. Stacy, K.J. Chang, H.L. Cohen and D.E. Davis, Phys. Rev. Lett. $\underline{58}$, 2337 (1987); B. Batlogg, R.J. Cara, A. Jayraman, R.B. van Dover, G.A. Kourouklis, S. Sunshine, D.W. Murphy, L.W. Rupp, H.S. Chen, K.T. Short, A.M. Mujsce and E.A. Rieman, Phys. Rev. Lett. $\underline{22}$, 2333 (1987).

/16/ S.Y.Hwu, S.N. Song, J. Thiel, K.R. Poeppelmeier, J.B. Ketterson and J.J. Freeman, Phys. Rev. B $\underline{35}$, 7119 (1987).

/17/ T.C. Collins (personal communication).

/18/ S. Suhai and J. Ladik, Phys. Lett. $\underline{77A}$, 25 (1980).

/19/ W. Hehre, D.F. Stewart and J.A. Pople, J. Chem. Phys. $\underline{18}$, 932 (1967).

/20/ R.C. Bingham, M.J.S. Dewar and D.H. Lo, J. Am. Chem. Soc. $\underline{97}$, 1285 (1975).

/21/ R.D. Singh and J. Ladik, Phys. Lett. $\underline{65A}$, 264 (1978).

/22/ J. Commandeur (personal communication).

/23/ S. Karunamithy and F. Aubke, Synth. Metals $\underline{16}$, 41 (1986).

/24/ N. Bartlett, E.M. Carron, B.W. McQuillan and T.C. Thompson, Synth. Metals $\underline{1}$, 22 (1979).

/25/ See for instance: T. Siegrist, S. Sunshine, D.W. Murphy, R.J. Cava and S.M. Zahurak, Phys. Rev. B $\underline{35}$, 7137 (1987).

/26/ A.A. Rodzig and B.M. Smirnov "Reference Data of Atoms, Molecules and Ions", Springer Berlin-Heidelberg-New York-Tokyo (1985) p. 147.

/27/ M. Dolg, U. Wedig, H. Stoll and H. Preuß, J. Chem. Phys. $\underline{86}$, 866 (1987).

/28/ R. Ditchfield, J.W. Hehre and J.A. Pople, J. Chem. Phys. $\underline{54}$, 726 (1971).

/29/ M. Seel, T.C. Collins, F. Martino, D.K. Rai and J. Ladik, Phys. Rev. B $\underline{18}$, 6460 (1978).

/30/ J.R. Schrieffer, "Theory of Superconductivity", W.A. Benjamin Inc., Reading, Massachusetts (1964).

/31/ J.R. Schrieffer, "Theory of Superconductivity", W.A. Banjamin Inc. Reading, Massachusetts (1964).

HEAVY-ELECTRON SUPERCONDUCTIVITY

H.R. Ott

Laboratorium für Festkörperphysik
ETH Hönggerberg
CH-8093 Zürich, Switzerland

ABSTRACT

A brief review of properties of heavy-electron superconductors is given. Especially emphasized are indications for unconventional superconductivity in these materials.

INTRODUCTION

Superconductivity of heavy electrons was certainly a highlight in condensed-matter physics before the tide of superconductivity of oxides at elevated temperatures buried almost everything else. But it is clear that this field is still of great interest, at least in terms of basic science, because of the possibility of unconventional superconductivity involving heavy electrons in metals. In this sense, any new insight in this type of superconductivity may also be of importance with respect to the superconducting state of the copper oxides, because certain experimental facts indicate an exotic type of superconductivity also in these ceramic materials. Below we review the principal facts that give evidence for unconventional superconductivity in heavy-electron materials and for some examples we try to evaluate their significance by comparing them with results of theoretical models and calculations. In the next section we briefly demonstrate that heavy-electron superconductivity occurs under seemingly unfavourable conditions. The following section is then devoted to display various experimental facts that indicate unconventional superconductivity by simply comparing the behaviour of some physical properties in the superconducting state with the respective standard BCS predictions.

OCCURRENCE OF HEAVY-ELECTRON SUPERCONDUCTIVITY

The first conjectures about the unconventional nature of this superconductivity were based on more or less heuristic arguments considering the circumstances under which superconductivity was observed. After the discovery of the first heavy-electron superconductor $CeCu_2Si_1$ [1] it took quite some time to establish this observation of an unusual phenomenon. With the subsequent discovery of UBe_{13} [2] and UPt_3 [3], however, it was no longer considered as a singularity of nature and it was recognized that the same electrons, those with predominant f symmetry, which in many other compounds are responsible for magnetic-order phenomena, were now involved in the formation of a superconducting state. This led first

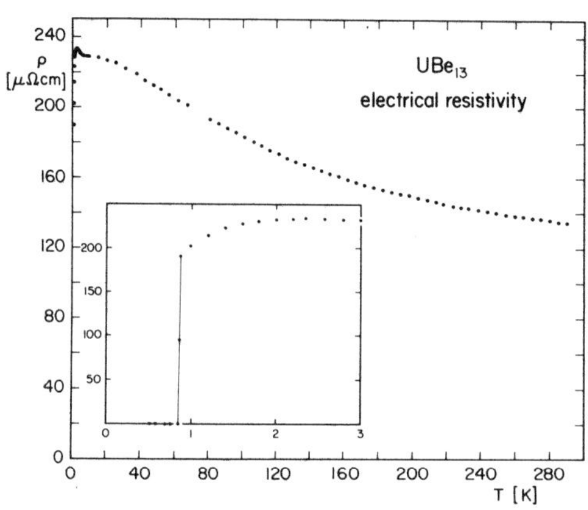

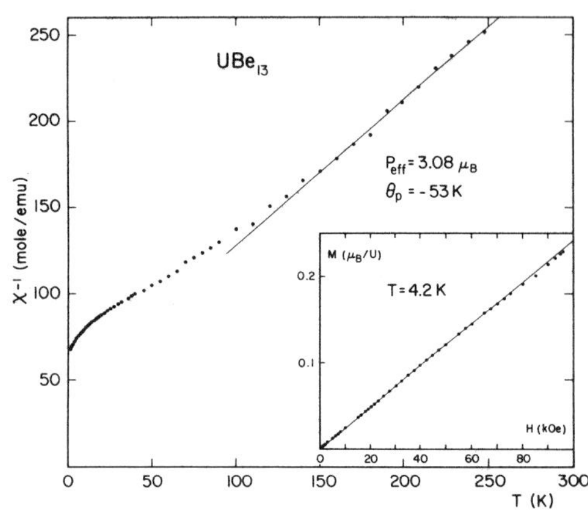

Fig. 1. Temperature dependence of the electrical resistivity of UBe_{13} between 1 and 300 K. The inset shows $\rho(T)$ between 0.5 and 3 K on an expanded temperature scale.

Fig. 2 . $\chi^{-1}(T)$ of UBe_{13} between 1.5 and 250 K. The solid line is compatible with the indicated values for ρ_{eff} and Θ_p. The inset illustrates M(H) up to 100 kOe at low temperatures.

Varma[4] and also Anderson[5] to claim that this kind of superconductivity was not caused primarily by the usual electron–phonon mechanism but rather by some magnetic interactions.

These conjectures were mainly triggered by some experimental observations of properties of the normal state of these materials. As examples we show the temperature dependence of the electrical resistivity $\rho(T)$ and of the inverse magnetic susceptibility $\chi^{-1}(T)$ of UBe_{13} in figures 1 and 2. $\rho(T)$ is dominated by a distinct increase of the resistivity with decreasing temperature and, after passing over a narrow maximum, ρ is still large at the transition although $\partial\rho/\partial T$ is now positive. At T_c, the magnetic susceptibility is very large and $\chi^{-1}(T)$ indicates the presence of fairly well defined ionic moments due to the unfilled 5f–electron shell of U atoms. The actual proof that the superconducting state is formed by electrons with very large effective masses was obtained from observations of the temperature dependence of the specific heat $c_p(T)$ at low temperatures. This is shown in fig. 3. The increase of the c_p/T ratio with decreasing temperature below 7 K is obviously due to a contribution of heavy-mass electrons since the c_p anomaly at the transition is compatible with the large c_p/T ratio in the normal state at T_c. The solid

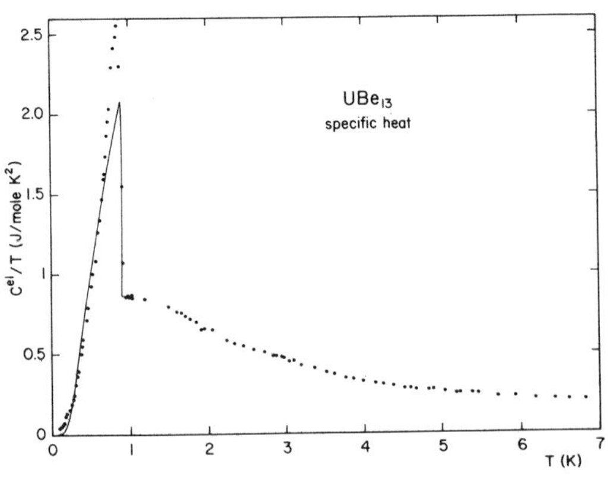

Fig. 3. The electronic specific heat of UBe_{13} plotted as c_p/T versus T between 0.15 and 7 K. The solid line is the prediction of the standard BCS theory for the superconducting state.

line in fig. 3 is the expected $c_p(T)$ curve from BCS theory[6], assuming
that the specific heat at T_c is entirely of electronic nature. The mass
enhancement of the electrons that is compatible with the observed specific heat is, of course, very unlikely due to electron-phonon interaction,
which is often the reason for a small mass enhancement in ordinary
metals. One exception of these general trends in the low-temperature properties of heavy-electron superconductors is the low resistivity at T_c
of UPt_3 but also in this case $\rho(T)$ is still varying considerably with
temperature following a well defined T^2 dependence[7].

If indeed magnetic interactions were responsible for the superconducting state, this would automatically make it very likely that the
simplest BCS state with an overall non-zero gap in the electronic excitation spectrum is no longer a good description of the superconducting
state of heavy electrons. States with more complicated symmetries would
have to be considered and electronic-energy distributions with gap zeroes
on certain parts of the Fermi surface would result. This in turn would
mean that many physical properties in the superconducting state of these
materials which are governed by electronic excitations would no longer
show the familiar BCS exponential temperature dependence. A discussion of
such investigations follows in the next section.

INTRINSICALLY ANISOTROPIC SUPERCONDUCTIVITY?

This classification should distinguish superconductors whose anisotropic properties are entirely due to non-cubic crystal structure or anisotropies of the Fermi surface, from those where the order parameter displays anisotropies in the same way as it does in the A phase of superfluid ^{3}He [8] and which I would label as intrisically anisotropic superconductors.

In the case of superconductivity in metals, such anisotropies would
manifest themselves by points or lines of zeroes of the superconducting
energy gap which in turn would distinctly influence the thermal- and
transport properties of the investigated material below T_c. Instead of
the usual exponential temperature dependence of these properties as T approaches 0 K, one would observe other temperature dependences in experimental measurements of, e.g., the specific heat, the ultrasound attenuation, the thermal conductivity or NMR relaxations rates, to mention a few
examples.

Anderson[9] was the first to warn from simple analogies with the case
of ^{3}He, mainly because the symmetry of a solid crystal is not the same as
that of a liquid and because strong spin-orbit interactions which are to
be expected in these particular compounds are absent in (non-rotating)
superfluid ^{3}He. Rice and co-workers later discussed similarities and differences to ^{3}He in more detail[10]. Nevertheless, the mere possibility of
these anisotropies in metallic superconductors triggered theoretical work
using group-theoretical methods in order to establish the numerous possibilities for the arrangement of the order parameter, depending on the
symmetry of the crystal structure of the particular materials. The most
extensive discussions in this respect are due to Blount[11], Volovik and
Gor'kov[12], and also Ueda and Rice[13].

In fig. 4 we show the result of a search for the above-mentioned
non-exponential temperature dependence, in this case by measuring the
specific heat c_p in the superconducting state of UBe_{13} [14]. The renormalized specific heat $C_s/C_n(T_c)$ is plotted versus T_c/T and one recognizes
immediately the distinct deviations from the BCS curve, representing an
exponential T dependence for large values of T_c/T. The deviations close
to T_c indicate the additional complication of strong coupling effects.

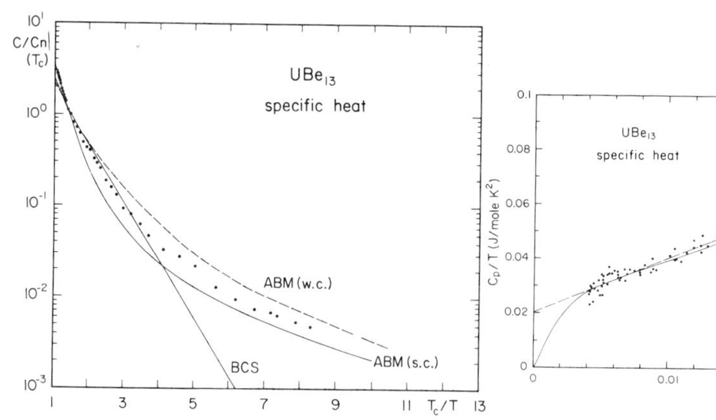
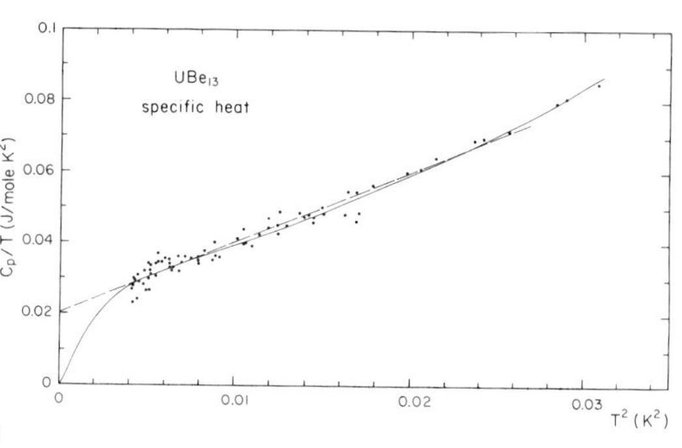

Fig. 4. $C_s/C_n(T_c)$ for superconducting UBe_{13}.
Dashed line: weak-cou-
pling ABM state; solid
lines: BCS and strong-
coupling ABM state, re-
spectively.

Fig. 5. c_p/T versus T^2 for superconducting UBe_{13} between 0.065 and 0.18 K.
The broken line is a con-
ventional fit, the solid
line is a calculation as
mentioned in the text.

The curves labelled ABM represent the temperature dependence of the
specific heat in a state that corresponds to the anisotropic A phase of
superfluid ^{3}He (Anderson-Brinkman-Morel state) and which is characterized
by points of zeroes of the energy gap, often denoted as axial state. Calculations for the weak-coupling (w.c.) and the strong-coupling (s.c.)
case are presented in ref. 14. Recent measurements to even lower temperatures confirm the non-exponential behaviour of $c_p(T)$ of UBe_{13} to values
of T_c/T of about 13 [15]. This is shown in fig. 5 where c_p/T is plotted
versus T^2 between 0.065 and 0.18 K. The solid line in fig. 5 is a calculation of $c_p(T)$ in an axial state of superconductivity, considering resonant impurity scattering of electrons. The effects of impurities on
anisotropic superconducting states beyond the Born approximation were
studied by Pethick and Pines[16] and more detailed calculations can be
found in the work of Schmitt-Rink and co-workers[17] and of Hirschfeld et
al.[18]. Details of the calculations resulting in the solid line of fig. 5
are found in ref. 15.

Non-exponential behaviour of $c_p(T)$ was also found in the superconducting state of UPt_3. There, even large residual terms varying linearly
with T were observed as T → 0 K [19,20], which could, to some extent, be
reproduced by calculations assuming the influence of impurity scattering[18]. Very recent $c_p(T)$ data obtained down to 70 mK using a sample of
UPt_3 that was cut from a crystal where the adjacent part was successfully
used for de Haas-van Alphen studies give, by conventional extrapolation,
a negligible linear-in-T contribution to c_p at T = 0 K [21]. The observed
T^2 dependence of c_p is, however, compatible with lines of gap nodes and
they are characteristic for the so-called polar state. This observation
supports the idea that impurity scattering may drastically alter the
$c_p(T)$ behaviour in these superconducting states and also, of course,
that one deals indeed with unconventional superconductivity.

Analogous non-exponential temperature dependences were found in
other experiments. In fig. 6 we show the example of the attenuation of
ultrasound in superconducting UPt_3 where, instead of the usual exponential decrease, a T^2 dependence is observed over an extended range of temperature[22]. Subsequent measurements[23,24] probing the anisotropy of ultra
sound attenuation on UPt_3 showed that its temperature dependence is dif-

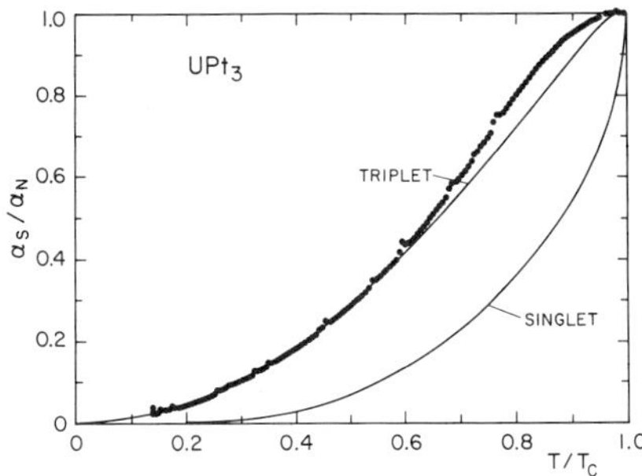

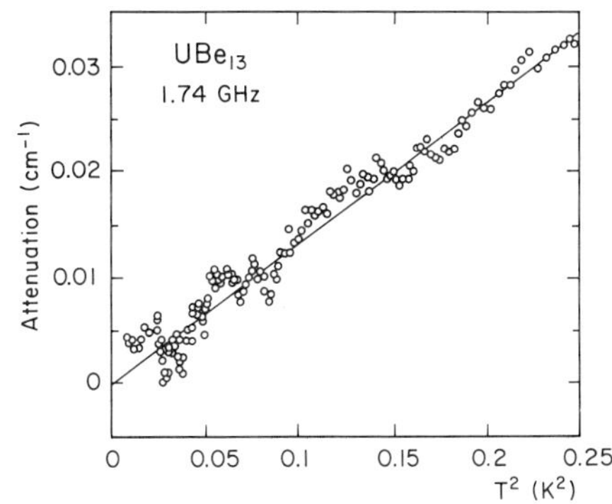

Fig. 6. Normalized ultrasound attenuation in superconducting UPt$_{13}$. The curve labelled "triplet" is compatible with a T^2 variation at very low temperatures (see ref. 22).

Fig. 7. Temperature dependence of the attenuation of ultrasound in superconducting UBe$_{13}$ between 0.1 and 0.5 K.

ferent for different crystallographic directions and polarizations of the ultrasound waves. A non-exponential decay of the absorption coefficient at very low temperatures was also observed for superconducting UBe$_{13}$, as shown in fig. 7 [25]. Unfortunately it is much more difficult to make reliable theoretical predictions in this case and therefore it is not clear yet whether these results are also compatible with the superconducting states that are inferred from the $c_p(T)$ measurements. Similar non-exponential T dependences were found for both UBe$_{13}$ and UPt$_3$ for the thermal conductivity[19,21,26,27] and for the nuclear-spin-lattice relaxation rates of UBe$_{13}$ [28] and CeCu$_2$Si$_2$ [29]. Examples of this behaviour are shown in figures 8 and 9.

Another unusual feature is the observation of a peak in the ultrasound attenuation just below T$_C$ of UBe$_{13}$, as shown in fig. 10. It has been argued that it might be due to collective modes of the order parameter in an anisotropic superconductor[25] but with an alternative explanation it was interpreted as a Landau-Khalatnikov-type damping of the order-parameter amplitude which is, in principle, also possible in the

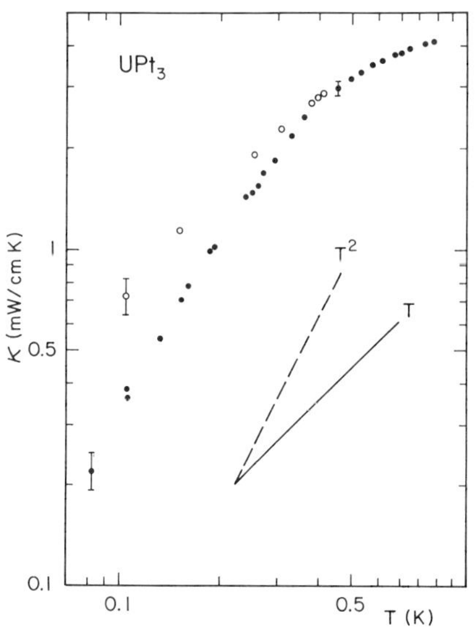

Fig. 8. Temperature dependence of the thermal conductivity of UPt$_3$. The open circles are data for the magnetic-field induced normal state (see ref. 19).

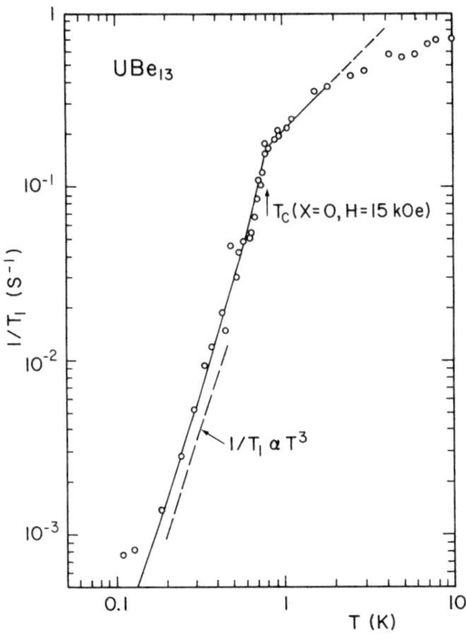

Fig. 9. $T_1^{-1}(T)$ for superconducting UBe$_{13}$.

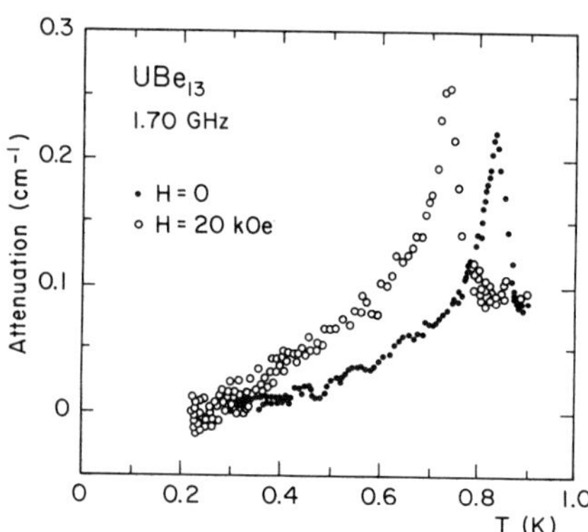

Fig. 10. Accoustic attenuation
peak in superconducting
UBe$_{13}$ below T_c. The
peak shifts to lower
temperatures with in-
creasing magnetic field.

case of isotropic superconductivity[30]. Further studies investigating possible anisotropy effects on this attenuation peak are still being made[31]
and it appears that the last word has not been said in this case. The
shift of the anomaly in an external magnetic field demonstrates that it
is related with the superconducting transition. Again unexpected is the
increase of the anomaly with increasing field. Peak-like but weaker anomalies close to T_c are also observed in the ultrasound attenuation of
UPt$_3$ [23].

It is, of course, also of interest to probe the temperature dependence of quantities which are observed only because superconductivity is
present. As examples we mention measurements of the upper critical field
$H_{c2}(T)$ and of the London penetration depth $\lambda_L(T)$. Fig. 11 shows the

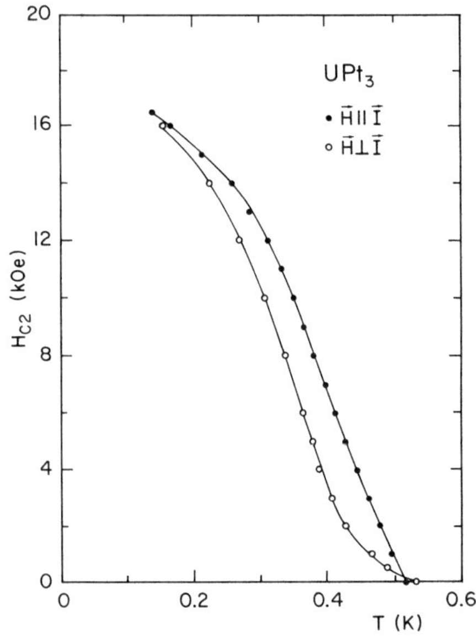

Fig. 11. Temperature dependence
of the upper critical
field of UPt$_3$. The current I was applied pa-
rallel to the hexagonal
axis (see ref. 32).

192

temperature dependences of the upper critical field along the principal
directions of the crystal structure of UPt_3 [32]. The most striking feature
here is the anisotropy close to T_c and its temperature dependence. It
is still not clear whether this observation is a true sign for the anisotropic character of the superconducting state.

The penetration depth of a magnetic field into a superconductor, the
so-called London penetration depth λ_L, may also reveal information on
the character of the superconducting state. Data of this sort have been
published by Einzel and co-workers[33] for UBe_{13} and we show these results
in fig. 12. The technique employed in these measurements did not allow to
determine the absolute values and therefore, only the temperature dependence is shown. But it is instructive to compare it with $\lambda_L(T)$ of tin
that was measured with the same experimental technique in the same region
of reduced temperature T/T_c. λ_L of tin is known to follow the temperature dependence given by the standard BCS theory rather well and it is
obvious from fig. 12 that this is not the case for λ_L of UBe_{13}. Elaborate calculations and comparison with the experimental data lead to the
conclusion[33,34] that λ_L of UBe_{13} is most compatible with expectations
assuming an axial superconducting state. Without going into details, we
only mention that this same assumption fixes the temperature dependence
of λ_L and therefore, an estimate of the absolute value of λ_L can be
made using the data shown in fig. 12. As expected, $\lambda_L(0)$ turns out to
be a few thousand Å, compatible with the large effective mass m* which
enters the classical expression for this quantity. Based on mass-renor-
malization arguments, Varma and co-workers[35] argue that the data can also
be explained by assuming a polar-like superconducting state for UBe_{13}.

INFLUENCE OF IMPURITIES

It is an old wisdom that superconductivity with $\ell \neq 0$ pairing is extremely susceptible to any kind of impurities and, considering this, it
is often argued that an anisotropic superconducting state cannot develop
in any real metal. Since, however, many other experimental facts, some of
which we mentioned above, suggest the possibility of such a state, it is
definitely of interest to investigate the influence of deliberately introduced impurities in these materials. Here we confine ourselves to some
interesting observations concerning UBe_{13} and UPt_3.

In the usual superconductors it is predominantly magnetic impurities
that lead to an effective suppression of the superconducting state because of the pair-breaking effect of the magnetic moments[36]. In anisotro-
picsuperconductors, however, impurities of any kind are predicted to be

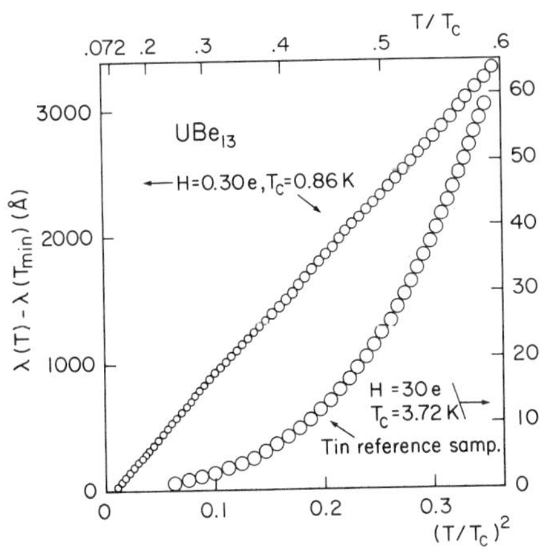

Fig. 12. Incremental magnetic-field
penetration depth of
UBe_{13}. Also shown, for
comparison, are the data
for a tin reference
sample.

pair-breaking. Since our materials are potential magnets anyway, it is obvious that the effect of non-magnetic impurities is of particular interest[37]. In fig. 13 we show that various non-magnetic impurities replacing U are indeed harmful for the superconductivity of UBe_{13}. Amounts of less than 2% of Lu, Y or Zr reduce T_c and the corresponding specific-heat anomaly very effectively. We note that Th impurities appear to be a special case and, indeed, they are. We shall discuss their influence in more detail below. For 1.5% Zr, T_c is still 0.7 K but, as may be seen from fig. 13, $c_p(T)$ is quite different from that of pure UBe_{13}. For Lu and Y impurities of similar concentration, T_c is shifted to below 0.3 K and the corresponding c_p anomalies have virtually vanished. This implies that these impurities very rapidly lead to a gapless state in the sense that the superconducting gap is zero on most parts of the Fermi surface. That this may happen with non-magnetic impurities is suggestive for an extension of the intrinsic gap-zeroes of an anisotropic superconducting state by normal impurity scattering as was discussed by Ueda and Rice[38]. If this interpretation is correct, it would give additional support to the view that we are dealing with some kind of unconventional superconductivity.

Quite unusual is the influence of Th atoms replacing U atoms on the superconducting state of UBe_{13}. In fig. 14 we show the concentration dependence of T_c of $U_{1-x}Th_xBe_{13}$ for small values of x. First, T_c is reduced surprisingly strongly, considering that Th is a non-magnetic impurity. When x exceeds about 0.017, T_c starts to rise again and passes over a broad maximum centered around $x \simeq 0.03$. For still higher values of x, T_c decreases again. The really unexpected feature, however, is the appearance of a second phase transition in the superconducting state of $U_{1-x}Th_xBe_{13}$ for $x \geqslant 0.02$, as first observed by specific-heat measurements[39]. The temperatures T_{c2} at which these second transitions occur, are indicated by solid triangles in fig. 14. It appears that they do not depend very much on x. An example of two consecutive specific-heat anomalies in such a case is shown in fig. 15. Other manifestations of the second transition are observed in measurements of the thermal expansion[40] and the ultrasound attenuation[41].

There is still some uncertainty with respect to the interpretation of the transition at T_{c2}. It definitely does not destroy the superconducting state. Microscopic measurements employing nuclear-magnetic resonance (NMR)[42] and muon-spin-rotation (μSR)[43] techniques suggest that no

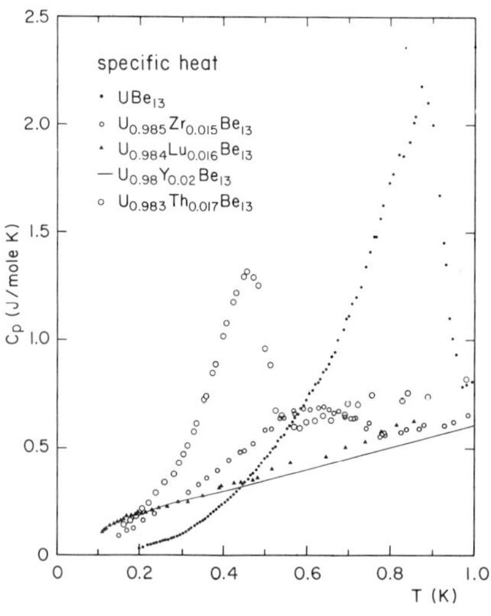

Fig. 13. Influence of various non-magnetic impurities on the specific heat of UBe_{13} below 1 K.

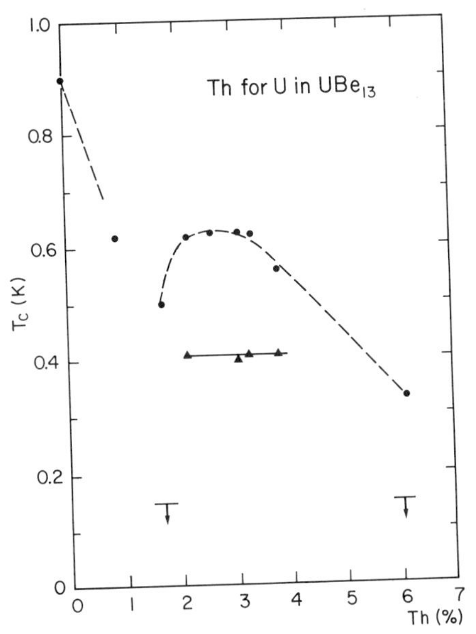

Fig. 14. Critical temperatures for $U_{1-x}Th_xBe_{13}$ from c_p measurements. Dots indicate the superconducting transition, triangles the second transition at T_{c2}. At 1.7 and 6% Th, no second transition was observed above 0.15 K.

ordered moments of size bigger than 0.01 μ_B/U exist below T_{c2} although especially the μSR-results suggest the presence of a static magnetic field at the muon site[44]. A structural transition cannot be ruled out completely but it is very unlikely because in an external magnetic field the lower transition is shifted to lower temperatures very much in parallel with the superconducting transition at T_{c1}. This is shown in fig. 16, where the transition temperatures T_c of UBe_{13} as well as T_{c1} and T_{c2} of $U_{0.9669}Th_{0.0331}Be_{13}$ from c_p measurements in external magnetic fields are plotted[40].

Based on these experimental facts it was suggested that this lower transition leads from one anisotropic superconducting state to another[45], a feature that is again only possible in unconventional superconductors. In ref. 45 it was also concluded that the superconducting states at T_c for UBe_{13} and at T_{c1} for $U_{0.9669}Th_{0.0331}Be_{13}$ should be of different symmetry. Experimental support for different superconducting states at different values of x was recently obtained from T_c measurements under external pressure[46]. The shifts of the superconducting transition temperature with increasing pressure for x = 0 and x = 0.033 have the same

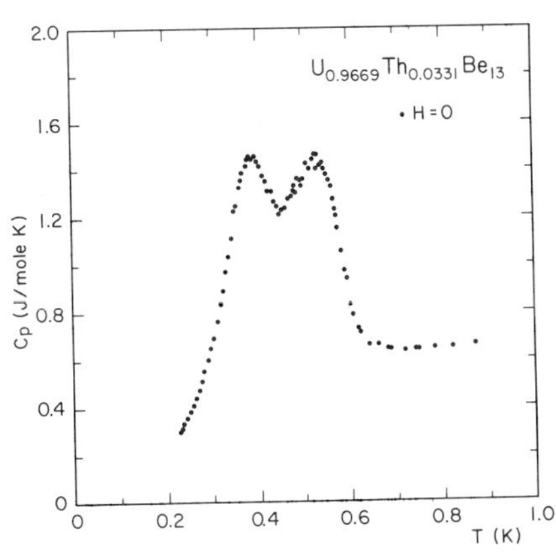

Fig. 15. Specific-heat anomalies for Th-doped UBe_{13} below 1 K.

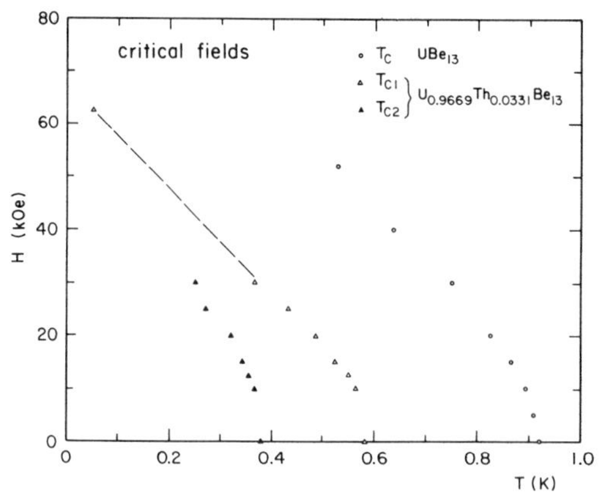

Fig. 16. Critical temperatures for the phase transitions of UBe_{13} and $U_{0.9669}Th_{0.0331}Be_{13}$ below 1 K as a function of external magnetic field.

negative sign but they differ significantly in magnitude. Detailed measurements for various values of x and at pressures up to 12 kbar indicate that the minimum that appears in the plot of fig. 14 for $x \simeq 0.017$ can be squeezed to T = 0 K for values of x between 0.022 and 0.038. This in turn leads to some kind of reentrant superconductivity in $U_{1-x}Th_xBe_{13}$ as a function of x if the external pressure exceeds 9 kbar. The authors of ref. 46 conclude that in this system, two different types of superconducting state occur and that one of them, namely the one that occurs at T_{c1} for $x \simeq 0.03$, can easily be suppressed by moderate external pressure.

Based on recent measurements of the lower critical field H_{c1} of $U_{0.97}Th_{0.03}Be_{13}$[47], Rauchschwalbe and co-workers gave an alternative interpretation for the two phase transitions in this system[48]. They found that at T_{c2}, the temperature dependence of H_{c1} changes abruptly in the sense that the rate of increase with decreasing temperature is considerably larger below T_c than above it. Their result is shown in fig. 17. This means that the tendency to superconductivity increases below this temperature. The authors of ref. 47 then interpret this observation as a sign for inhomogeneous superconductivity in k-space. They propose that at T_{c1} the superconducting transition affects about half of the Fermi surface and that only at T_{c2} the remaining part of the Fermi surface becomes unstable with respect to a superconducting transition. Although this alternative interpretation seems to lead to some consistency in the data analysis, it has already found its critics, mainly because it appears difficult to justify, on theoretical grounds, an only partial instability of any Fermi surface with respect to superconductivity[49].

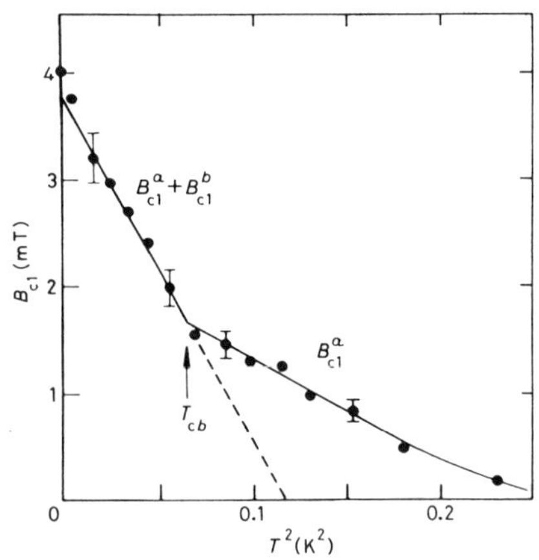

Fig. 17. Temperature dependence of the lower critical field of $U_{0.97}Th_{0.03}Be_{13}$ as a B_{c1} versus T^2 plot. Details of the nomenclature and the lines may be obtained from ref. 47.

It should also be noted that the T_c vs x phase diagram which is subsequently proposed in ref. 48 has some serious differences to that shown in fig. 14 and is not compatible with the data that led to fig. 14.

Impurities are definitely very effective in changing the low-temperature properties of UPt_3. Already in the experiments which led to the discovery of superconductivity in this compound[3], it was realized that probably even disorder in nominally pure UPt_3 could prevent the material from becoming superconducting above 0.1 K. It is therefore clear that very tiny amounts of impurities in well ordered UPt_3 are sufficient to suppress superconductivity. An interesting observation was subsequently made by de Visser and co-workers[50], who investigated the low-temperature properties of $UPt_{3-x}Pd_x$ compounds. We show the results of their c_p measurements in fig. 18. It may be seen, that an increasing Pd content leads to an enhancement of the specific heat at the lowest temperatures and, for x = 0.15, develops even into a distinct anomaly which signals a cooperative phase transition. For still higher values of x, this anomaly is no longer observed. A similar behaviour is also observed when small amounts of Pt are replaced by Au or when Th atoms are substituted for U [51,52]. Intriguing is the fact that for all these types of impurities, the same approximate amount of concentration leads to a cooperative phase transition at about the same temperature of 5 to 6 K. In the case of Th impurites it was demonstrated by neutron-diffraction experiments[53] that the phase transition reflects the onset of magnetic order. It is thus quite obvious, that UPt_3 is indeed very close to a magnetic instability and this fact was discussed in some detail by Batlogg and co-workers[54]. More recent neutron-diffraction experiments seem to indicate that even pure UPt_3, which is superconducting below 0.5 K, may undergo a magnetic phase transition around 5 K, but the ordered moments appear to be very small[55]. All these observations strongly suggest some unconventional superconductivity for UPt_3 as well.

Very similar observations as those mentioned in the penultimate sentence were made in URu_2Si_2, where magnetic order at about 17 K is followed by a superconducting transition around 1 K [56-58]. However, in this latter case, the transition at 17 K is accompanied by a very distinct specific-heat anomaly whereas that at 5 K in UPt_3 is apparently not.

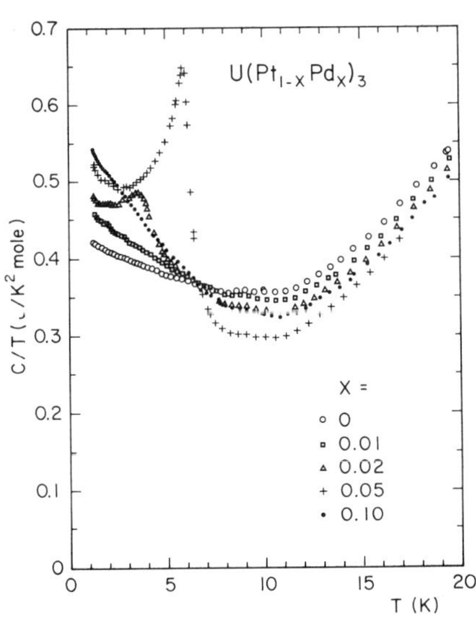

Fig. 18. Influence of Pd substitution on Pt sites on the low-temperature specific heat of UPt_3 (see ref. 50)

CONCLUSIONS AND OUTLOOK

At this point, nobody can claim that the occurrence of unconventional superconductivity in heavy-electron materials has been proven beyond any doubt. However, there is still growing experimental evidence that we are indeed dealing with some other mechanism than the usual electron-phonon interaction and therefore also observe unusual properties in the superonducting state. One very nice experiment that appears to support this claim, but was not mentioned above, is certainly the Josephson-tunneling measurement of Han and co-workers[59]. Here I believe, however, that a lot more theoretical insight is necessary to judge what the outcome of the experiment, the observation of a negative even-parity proximity effect in superconducting UBe_{13}, really means. The same situation holds for many other cases, in particular for the interpretation of transport-property measurements. In this sense it should be stressed that still many experimental observations await theoretical models and calculations to unravel the truth.

Nobody probably argues against the statement that the same is true in the case of the oxide superconductors. Since also here, some signs indicate an unconventional type of superconductivity, I express my hope that the two fields may have some roots in common and that solving one problem might help to solve the other.

ACKNOWLEDGEMENTS

I should like to thank many friends and colleagues for their ever-lasting enthusiasm to collaborate in experimental and theoretical efforts to understand the phenomenon of heavy-electron superconductivity and among them I should like to mention Z. Fisk, H. Rudigier, E. Felder, J.L. Smith, B. Batlogg, and T.M. Rice.

REFERENCES

1. F. Steglich, J. Aarts, C. D. Bredl, W. Lieke, D. Meschede, W. Franz, and H. Schäfer, Phys. Rev. Lett. 43:1892 (1979).
2. H. R. Ott, H. Rudigier, Z. Fisk, and J.L. Smith, Phys. Rev. Lett. 50:1595 (1983).
3. G. R. Stewart, Z. Fisk, J. O. Willis, and J. L. Smith, Phys. Rev. Lett. 52:679 (1984).
4. C. M. Varma, Moment Formation in Solids, ed. W. J. L. Buyers (Plenum, New York, 1984) p. 83.
5. P. W. Anderson, Phys. Rev. B 30:1549 (1984).
6. J. Bardeen, L. N. Cooper, and J. R. Schrieffer, Phys. Rev. 108:1175 (1957).
7. A. de Visser, J. J. M. Franse, and A. Menovsky, J. Magn. Magn. Mat. 43:43 (1984).
8. A. J. Leggett, Rev. Mod. Phys. 47:331 (1975).
9. P. W. Anderson, Phys. Rev. B 30:4000 (1984).
10. T. M. Rice, K. Ueda, and H. R. Ott, J. Magn. Magn. Mat. 54-57:317 (1986).
11. E. I. Blount, Phys. Rev. B 32:2935 (1985).
12. G. E. Volovik and L. P. Gor'kov, Zh. Eksperim. Teor. Fiz. 88:1412 (1985) [Sov. Phys. JETP 61:843 (1985)].
13. K. Ueda and T. M. Rice, Phys. Rev. B 31:7114 (1985).
14. H. R. Ott, H. Rudigier, T. M. Rice, K. Ueda, Z. Fisk, and J. L. Smith, Phys. Rev. Lett. 52:1915 (1984).
15. H. R. Ott, E. Felder, C. Bruder, and T. M. Rice, Europhys. Lett. 3:1123 (1987).
16. C. J. Pethick and D. Pines, Phys. Rev. Lett. 57:118 (1986).

17. S. Schmitt-Rink, K. Miyake, and C. M. Varma, Phys. Rev. Lett.
     57:2575 (1986).
18. P. J. Hirschfeld, D. Vollhardt, and P. Wölfle, Solid State Commun.
     59:111 (1986).
19. F. Steglich, U. Rauchschwalbe, U. Gottwick, H. M. Mayer, G. Sparn,
     N. Grewe, U. Poppe, and J. J. M. Franse, J. Appl. Phys. 57:3054
     (1985).
20. A. Sulpice, P. Gandit, J. Chaussy, J. Flouquet, D. Jaccard,
     P. Lejay, and J. L. Tholence, J. Low Temp. Phys. 62:39 (1986).
21. H. R. Ott, E. Felder, A. Bernasconi, Z. Fisk, J. L. Smith,
     L. Taillefer, and G. G. Lonzarich, submitted to LT-18.
22. D. J. Bishop, C. M. Varma, B. Batlogg, E. Bucher, Z. Fisk, and
     J. L. Smith, Phys. Rev. Lett. 53:1009 (1984).
23. V. Müller, D. Maurer, E. W. Scheidt, Ch. Roth, K. Lüders, E. Bucher,
     and H. E. Bömmel, Solid State Commun. 57:319 (1986).
24. B. S. Shivaram, Y. H. Jeong, T. F. Rosenbaum, and D. G. Hinks,
     Phys. Rev. Lett. 56:1078 (1986).
25. B. Golding, D. J. Bishop, B. Batlogg, W. H. Haemmerle, Z. Fisk,
     J. L. Smith, and H. R. Ott, Phys. Rev. Lett. 55:2479 (1985).
26. D. Jaccard, J. Flouquet, P. Lejay, and J. L. Tholence, J. Appl.
     Phys. 57:3082 (1985).
27. D. Jaccard, J. Flouquet, Z. Fisk, J. L. Smith, and H. R. Ott, J.
     Physique Lett. 46:L811 (1985).
28. D. E. MacLaughlin, C. Tien, W. G. Clark, M. D. Lan, Z. Fisk, J. L.
     Smith, and H. R. Ott, Phys. Rev. Lett. 53:1833 (1984).
29. Y. Kitaoka, K. Ueda, T. Kohara, and K. Asayama, Solid State Commun.
     51:461 (1984).
30. K. Miyake and C. M. Varma, Phys. Rev. Lett. 57:1627 (1986).
31. B. Golding, private communication.
32. J. W. Chen, S. E. Lambert, M. B. Maple, Z. Fisk, J. L. Smith, G. R.
     Stewart, and J. O. Willis, Phys. Rev. B 30:1583 (1984).
33. D. Einzel, P. J. Hirschfeld, F. Gross, B. S. Chandrasekhar, K.
     Andres, H. R. Ott, J. Beuers, Z. Fisk, and J. L. Smith, Phys. Rev.
     Lett. 56:2513 (1986).
34. F. Gross, B. S. Chandrasekhar, D. Einzel, K. Andres, P. J.
     Hirschfeld, H. R. Ott, Z. Fisk, J. L. Smith, and J. Beuers, Z.
     Phys. B 64:175 (1986).
35. C. M. Varma, K. Miyake, and S. Schmitt-Rink, Phys. Rev. Lett. 57:626
     (1986).
36. A. A. Abrikosov and L. P. Gor'kov, Sov. Phys. JETP 12:1243 (1961).
37. J. L. Smith, Z. Fisk, J. O. Willis, A. L. Giorgi, R. B. Roof, H. R.
     Ott, H. Rudigier, and E. Felder, Physica 135B:3 (1985).
38. K. Ueda and T. M. Rice, Phys. Rev. B 31:7114 (1985).
39. H. R. Ott, H. Rudigier, Z. Fisk, and J. L. Smith, Phys. Rev. B
     31:1651 (1985).
40. H. R. Ott, H. Rudigier, E. Felder, Z. Fisk, and J. L. Smith, Phys.
     Rev. B 33:126 (1986).
41. B. Batlogg, D. Bishop, B. Golding, C. M. Varma, Z. Fisk, J. L.
     Smith, and H. R. Ott, Phys. Rev. Lett. 55:1319 (1985).
42. C. Tien, D. E. MacLaughlin, M. D. Lan, W. G. Clark, Z. Fisk, J. L.
     Smith, and H. R. Ott, Physica 135B:14 (1985).
43. R. H. Heffner, D. W. Cooke, Z. Fisk, R. L. Hutson, M. E. Schillaci,
     J. L. Smith, J. O. Willis, D. E. MacLaughlin, C. Boekema, R. L.
     Lichti, A. B. Denison, and J. Oostens, Phys. Rev. Lett. 57:1255
     (1986).
44. R. H. Heffner, private communication.
45. R. Joynt, T. M. Rice, and K. Ueda, Phys. Rev. Lett. 56:1412 (1986).
46. S. E. Lambert, Y. Dalichaouch, M. B. Maple, J. L. Smith, and Z.
     Fisk, Phys. Rev. Lett. 57:1619 (1986).
47. U. Rauchschwalbe, F. Steglich, G. R. Stewart, A. L. Giorgi, P.
     Fulde, and K. Maki, Europhys. Lett. 3:751 (1987).

48. U. Rauchschwalbe, C. D. Bredl, F. Steglich, K. Maki, and P. Fulde, Europhys. lett. 3:757 (1987).
49. T.M. Rice; D. Pines, private communications.
50. A. de Visser, J. C. P. Klaasse, M. van Sprang, J. J. M. Franse, A. Menovsky, and T. T. M. Palstra, J. Magn. Magn. Mat. 54-57:375 (1986).
51. A. P. Ramirez, B. Batlogg, A. S. Cooper, and E. Bucher, Phys. Rev. Lett. 57:1072 (1986).
52. G. R. Stewart, A. L. Giorgi, J. O. Willis, and J. O'Rourke, Phys. Rev. B 34:4629 (1986).
53. A. I. Goldman, G. Shirane, G. Aeppli, B. Batlogg, and E. Bucher, Phys. Rev. B 34:6564 (1986).
54. B. Batlogg, D. J. Bishop, E. Bucher, B. Golding, Jr., A. P. Ramirez, Z. Fisk, J. L. Smith, and H. R. Ott, J. Magn. Magn. Mat. 63&64:441 (1987).
55. C. Broholm, J. K. Kjems, and G. Aeppli, private communication.
56. W. Schlabitz, J. Baumann, B. Politt, U. Rauchschwalbe, H. M. Mayer, U. Ahlheim, and C. D. Bredl, Z. Phys. B 62:171 (1986).
57. T. T. Palstra, A. A. Menovsky, J. van den Berg, A. J. Dirkmaat, P. H. Kes, G. J. Nieuwenhuys, and J. A. Mydosh, Phys. Rev. Lett. 55:2727 (1986).
58. M. B. Maple, J. W. Chen, Y. Dalichaouch, T. Kohara, C. Rossel, M. S. Torikachvili, M. W. Elfresh, and J. P. Thompson, Phys. Rev. Lett. 56:185 (1986).
59. S. Han, K. W. Ng, E. L. Wolf, A. Millis, J. L. Smith, and Z. Fisk, Phys. Rev. Lett. 57:238 (1986).

HEAVY ELECTRON SUPERCONDUCTIVITY: FROM 1K TO 90K TO ?

C. J. Pethick[*] and David Pines[**]

Department of Physics, University of Illinois at Urbana-Champaign
1110 West Green Street
Urbana, IL 61801

INTRODUCTION

Heavy electron systems are intermetallic compounds containing elements
with unfilled f-electron shells, such as U or Ce, which at room temperature
and above behave like a weakly interacting collection of f-electron moments
and conduction electrons with ordinary masses, while at low temperatures
the conduction electron specific heat becomes typically some hundred times
larger than that found in most metals.[1] These highly correlated low temperature states display remarkable behavior whether the system remains
normal down to the lowest temperature measured, becomes antiferromagnetic,
or becomes superconducting. While in ordinary metallic superconductors a
dilute concentration of magnetic impurities destroys superconductivity, in
heavy electron systems superconductivity and antiferromagnetism can
coexist; a transition to either ordered state may be followed by a second
transition to a phase containing both states. Thus in both UPt_3 and
URu_2Si_2 one finds on lowering the temperature that an antiferromagnetic
transition is followed by a transition to the superconducting state, while
in $U_{0.97} Th_{0.03} Be_{13}$ the order of the transitions is reversed.

In this talk we shall review the experimental results and physical
arguments which led us to conclude that in heavy electron systems the physical mechanism responsible for superconductivity is an attractive interaction between the heavy electrons which results from the virtual exchange
of antiferromagnetic f-electron moment fluctuations.[2] In these systems,
then, the superconductivity is of purely electronic origin; the phonon-
induced interaction between electrons which leads to superconductivity in
ordinary metals plays little or no role.

From the perspective of scientists searching for high temperature
superconducting materials heavy electron systems thus provide both good
news and bad news. The good news is a purely electronic mechanism for

---

[*]Also at Nordita, Blegdamsvej 17, DK-2100 Copenhagen O, Denmark

[**]Also at Center for Materials Science, Los Alamos National Laboratory,
Los Alamos, NM 87545

superconductivity has been discovered; hence the "phonon-barrier," the
existence of a maximum superconducting transition temperature of ~30K for
metals in which electron-phonon interactions are responsible for superconductivity,[3] is broken; the bad news is that although the superconductivity is of non-phononic origin, the superconducting transition typically
is found only at $T \lesssim 1K$.

Does this always have to be the case? In the next part of this talk
we consider the possibility that the high temperature superconductors are
indeed part of the heavy electron family, albeit a collateral branch in
which magnetic excitations are present but the carrier density is so low
that the screening of magnetic interactions by itinerant electrons or holes
is negligible. Under these circumstances the itinerant carrier magnetic
susceptibility will be appreciably enhanced by exchange effects, while the
coupling of charge carriers to spin fluctuations could easily give rise to
substantial carrier effective masses; the characteristic temperature for
the spin-fluctuation induced superconductivity would be far closer to 100K
than 1K.

Self-consistent fits to the available experimental data on the
specific heat jump, carrier density, critical field slope, and the London
penetration depth yield a coherence length ~14A for $YBa_2Cu_3O_4$;[4] this
small coherence length in turn makes possible _intrinsic_ pinning of the
magnetic vortices in the 90K superconductors by the barium or rare earth
atoms which lie outside the copper-oxide planes, a possibility we consider
briefly in the latter part of our presentation.

PHYSICAL PICTURE

At high temperatures heavy electron systems behave like a collection
of weakly interacting f-electron moments and conduction electrons, while at
very low temperatures, so far as thermal and transport processes are concerned, they behave like a system of strongly interacting itinerant electrons which scatter against impurities, against low frequency f-electron
spin fluctuations, and against one another.

A physical picture of the transition between these two regimes is that
as the temperature is lowered the local moments and conduction electrons
become more and more strongly coupled. The magnetic behaviour is quenched
while the effective mass of the itinerant electrons becomes substantially
enhanced. As a consequence of this interaction, the f-electrons are no
longer confined to the magnetic sites, but can hop into the conduction
band, as in the Anderson model. The itinerant heavy electron states at low
temperatures are therefore superpositions of localized f-electrons and conduction electrons. Their quite strong interaction reflects not so much
their direct Coulomb interaction, as it does an interaction induced by
their coupling to spin fluctuations on the magnetic sites, and it provides
a natural explanation for the large finite temperature corrections to the
low-temperature form of the specific heat, the strong temperature dependence of the electrical resistivity and other transport coefficients, and
the appearance of superconductivity.

In the very low temperature limit the thermal and transport properties
of heavy fermion systems in the normal state should be those expected for
heavy electron Fermi liquids. However, in most cases experiments have not
yet been carried out in the _Landau limit_, that is at temperatures sufficiently low that one can neglect, in first approximation, the frequency
dependence of the quasiparticle energies and quasiparticle scattering amp-

litudes associated with the coupling of the conduction electrons to the localized f-electrons. If we define θ_{coh} as the temperature below which the electronic specific heat is linear in T, and the electrical and thermal resistivities fall off sharply with decreasing temperature, then it is only at temperatures $T \ll \theta_{coh}$, that one expects to observe the Landau temperature dependence of the electrical resistivity, ρ, the thermal resistivity, W, and the ultrasonic attenuation coefficient α, in which the finite temperature corrections to the low temperature limiting behaviour are proportional to T^2. Such Landau limiting behaviour is observed for UPt_3 at temperatures below ~1.5K, while UBe_{13} at zero pressure becomes a superconductor well before it reaches a temperature at which Landau theory would apply.[5]

The strong coupling between the f-electrons and the conduction electrons which is responsible for the heavy itinerant quasiparticles gives rise to a compensating electron cloud which alters the magnetic response of the local moments. If magnetization were a conserved quantity then local moments and their corresponding electron clouds would not contribute to the long wavelength magnetic susceptibility $\chi(T)$ at low temperatures; that quantity would be entirely determined by the heavy electron quasiparticle contribution χ_{qp}. Because magnetization is not conserved there can be a significant non-quasiparticle contribution χ_{loc}, to $\chi(T)$, which arises from the polarization of the local moments and their compensating clouds, that is from virtual excitations at finite frequencies.

To see how this comes about, consider the exact expression for the magnetic susceptibility at zero temperature,

$$\chi^M_{\sim}(q,\omega) = \sum_n \frac{|\,(M^+_q)_{no}\,|^2\, 2\omega_{no}}{\omega^2_{no} - (\omega+i\eta)^2} \tag{1}$$

where 0 denotes the ground state, n an excited state, ω_{no} the excitation energy of the state n with respect to the ground state, and $M_{\underset{\sim}{q}}$ is the magnetic moment operator. At long wavelengths the excited states may be divided into two classes: i) States obtained by destroying a quasiparticle below the Fermi surface and creating a quasiparticle just above the Fermi surface in the same band. These quasiparticle-quasihole pair states have an energy of order $v_F q$, where v_F is the Fermi velocity. ii) All other states, such as one containing a quasihole in one band and a quasiparticle in another band (an interband transition), a state containing two or more quasiparticle-quasihole pairs, or, in the case of Kondo and similar systems, a state obtained by polarizing a localized spin and its compensating electron cloud. χ^M therefore takes the form

$$\chi^M = \chi^M_{Landau} + \chi^M_{loc}, \tag{2}$$

where the first term comes from the single pair states (i), and the second from states (ii). By performing neutron scattering experiments at small $\underset{\sim}{q}$, one can in principle distinguish between these two contributions, since the frequencies associated with the Landau contribution all vanish for small $\underset{\sim}{q}$.

If magnetization is conserved, χ_{loc} vanishes at long wavelengths. This may be seen from the fact that the total magnetization commutes with the Hamiltonian,

$$\left[H, M_{\underset{\sim}{q}=0}\right]_{no} = \omega_{no}\left[M_{q=0}\right]_{no} = 0, \tag{3}$$

and therefore, assuming $\omega_{no}(M_{\underset{\sim}{q}})_{no}$ for $\underset{\sim}{q} \to 0$ tends to its value at $\underset{\sim}{q} = 0$, it

is easy to see that χ_{loc}^{M} must vanish. This shows that for such systems the Landau contribution to the susceptibility at long wavelengths must be the total susceptibility. This is true for liquid ^{3}He: in this case the magnetization is proportional to the spin, which is conserved if one neglects the nuclear dipole-dipole interaction. Conservation of magnetization also leads to the conclusion that the magnetic moment associated with a quasiparticle is equal to the bare moment.

In heavy fermion systems, magnetization is not conserved, due to the existence of both spin and orbital contributions to it, and to spin-orbit coupling. Consequently χ_{loc}^{M} is finite in the limit $\underset{\sim}{q} \to 0$. The absence of magnetization conservation also means that the effective magnetic moment of a quasiparticle is not related in a simple way to the bare moments of either an f-electron or a conduction electron.

Inelastic neutron scattering experiments give no evidence for a component of Im χ^{M}, the magnetic structure factor, whose frequency tends to zero as $\underset{\sim}{q} \to 0$. This region is difficult to investigate directly, but the fact that the contribution to χ^{M} from the frequencies which are accessible experimentally can account for all of the measured long-wavelength susceptibility to within experimental accuracy suggests that χ_{Landau}^{M} cannot contribute more than 10-20% of the total.[6]

A PHENOMENOLOGICAL DESCRIPTION OF HEAVY FERMION BEHAVIOUR

Recently it has been shown that neutron scattering results for UPt_3, $CeCu_3$, and U_2Zn_{17} may be fit by a model for the spin-spin correlation function in which fluctuations of the magnetic moment at the f-atom site are coupled to those at other sites by an effective exchange interaction.[7] If one assumes that all the magnetic moment is associated with electrons in f-orbitals, this leads to an expression for the wavenumber- and frequency-dependent spin-spin correlation function of the form

$$\chi(\underset{\sim}{q},\omega) = \frac{\chi_\mu(\omega,T)}{1-J(\underset{\sim}{q},\omega,T)\,\chi_\mu(\omega,T)} \tag{4}$$

where $\chi_\mu(\omega,T)$ describes the correlations of the spin at a single f-site, including the effects of interaction with the compensating electron cloud, and $J(\underset{\sim}{q},\omega,T)$ is an effective exchange interaction which describes the

coupling between spins at different sites. (For simplicity we restrict
ourselves to the case where the sites of all magnetic ions are equivalent.)

In the fits to the data, J was taken to be a temperature-dependent
nearest neighbour interaction, and χ_μ was taken to be of the form

$$\chi_\mu = \frac{\chi_o \Gamma}{\Gamma - i\omega} ,\tag{5}$$

which is known to give a good description of the properties of a single
Kondo impurity. Here χ_o is the susceptibility of a single ion and Γ is a
measure of the typical excitation energies for a single ion and its screening cloud, energies which lie between 50K and 250K for the systems thus far
studied; for an isolated impurity, Γ would be of order the Kondo temperature T_K.

We have proposed that an expression of the form (4) provides a useful
starting point for the examination of all aspects of heavy fermion behavior.[2] From a microscopic point of view the induced spin-spin interaction is given by an expression of the form

$$J(\underset{\sim}{q},\omega,T) = - \sum_{\underset{\sim}{K}_n} |V_{\underset{\sim}{q}+\underset{\sim}{K}_n}|^2 \chi_c(\underset{\sim}{q}+\underset{\sim}{K}_n,\omega,T)\tag{6}$$

where V describes the coupling of a conduction electron-hole pair to the
local spin fluctuations described by $\chi_\mu(\omega,T)$, K_n is a reciprocal lattice
vector, and χ_c is the conduction electron-hole spin-spin response function.
As a result of the coupling, J, the characteristic energies which enter
into the low frequency limit of χ, (Eq. 4), become wavevector- and temperature-dependent, being given by

$$\theta_{loc}(\underset{\sim}{q},T) \cong \Gamma[1-J(\underset{\sim}{q},0,T) \, \chi_\mu(0,T)] .\tag{7}$$

We argued that the presence of a second energy scale, lower than T_K, is a
characteristic feature of all heavy fermion systems, and may be a necessary
condition for observing heavy fermion behavior. Put another way, if
$J\chi_\mu \ll 1$, one is likely in a weak coupling limit, and no heavy fermion behavior results. One the other hand if, as in U_2Zn_{17}, for some wavevector
$\underset{\sim}{q}$, and temperature T, $J\chi_\mu = 1$, then an antiferromagnetic phase transition
occurs. (Indeed, Broholm et al.[7] have shown that this transition is
driven by a temperature dependent coupling, J, which below 18K increases
with decreasing temperature until it drives the antiferromagnetic transition at 9.7K.) Normal and superconducting heavy fermion compounds would
seem to lie in the strong coupling regime, $J\chi_\mu \sim 1$.

The coupling between the heavy quasiparticle pairs and the local spin
fluctuations gives rise to an induced wavevector-, frequency- and temperature-dependent, heavy electron interaction,

$$U_{ind}(\underset{\sim}{q},\omega,T) = - V^2_{eff} \, \chi(\underset{\sim}{q},\omega,T) = \frac{V^2_{eff} \, \chi_\mu(\omega,T)}{1-J(\underset{\sim}{q},\omega,T) \, \chi_\mu(\omega,T)} .\tag{8}$$

The matrix element, V_{eff}, includes vertex corrections to the electron-local moment coupling; to the extent that V_{eff} depends only weakly on q, the momentum dependence of $U(q,\omega,T)$ would arise from that of $J(q,\omega,T)$. For frequencies low compared to the characteristic frequencies which enter into χ, that interaction will be attractive between like spins and repulsive between unlike spins; to the extent that χ exhibits antiferromagnetic correlations (and neutron scattering experiments suggest that this might quite generally be the case), $U_{int}(q,\omega)$ will behave in similar fashion. This induced interaction is the physical origin of both the $T^3 \ell n\, T$ corrections to the specific heat (where these are observed) and of the pairing instability which gives rise to superconductivity. The proposed approach is quite reminiscent of the electron-phonon interaction problem, with the local moment spin fluctuation frequency-dependent susceptibility playing the role of a phonon propagator. However, there is no reason to expect that a Migdal theorem exists for the heavy electron local moment fluctuation interaction. Indeed, in the present theory there is a considerable amount of feed-back, and possible non-linear behavior, in that, for example, J depends on χ_c which in turn depends on J through electron-local moment fluctuation coupling.

Transport coefficients depend on the behavior of J at large wavevectors, since scattering phenomena are dominated by the coupling of heavy electrons to large wavevector moment fluctuations. A test of this hypothesis, and of the overall model, is obtained by examining the changes in the resistivity as a function of pressure and magnetic field, in an approach which attributes such changes either to changes in θ_{loc} which proceed à la Kondo, $[\theta^2_{loc}(H) = \theta^2_{loc} + \mu^2_{loc} H^2]$, and/or to changes in J. Batlogg[8] has recently found that such scaling arguments work quite well for the resistivity and magnetization of UBe_{13} in quite large magnetic fields.

Finally, we argued that the mass enhancement of heavy fermions arises from their coupling to local moment fluctuations. As was the case for transport phenomena, the calculated state density will depend primarily on the coupling of the conduction electrons to the large wavevector local moment fluctuations.

Recently Norman[9] has carried out a microscopic model calculation for UPt_3 which serves as a test of our proposed phenonemological approach. He has assumed that quasiparticles on a Fermi surface which is consistent with the de-Haas van Alphen results of Taillefer et al.[10] are coupled to moment fluctuations whose spectra are those measured by Aeppli et al.[7]; he finds both a mass enhancement and a superconducting transition temperature which are in good qualitative agreement with experiment.

Our physical picture and phenomenological description may be rich enough to make possible an understanding of the extraordinary, and diverse, sensitivity of various heavy fermion physical phenomena to pressure and to the presence of impurities. For example, impurities can alter χ by changing either χ_μ and/or J. Either, or both, of these quantities may in turn be quite sensitive to changes in density; moreover, the introduction of impurities can give rise to local changes in density. As a result, a natural explanation may emerge for the fact that in heavy fermion systems the thermal expansion, most often negative, is some four orders of magnitude larger than that of an ordinary metal; the observed values of magnetostriction exceed those of transition metals by two or more orders of magnitude, and, finally, the introduction of impurities can bring about changes in the resistivity which can be two orders of magnitude greater than the value obtained from an estimate based on using for the quasiparticle-impurity scattering amplitude the unitarity limit in a single partial wave.

HEAVY FERMION SUPERCONDUCTIVITY

A fundamental question concerning superconductivity in heavy electron systems is whether it is the heavy electrons that become superconducting. Clear evidence for the pairing of the heavy electrons is provided by measurements which show that the jumps in the specific heat at the transition temperature, T_c, to the superconducting phase, are comparable to the specific heat in the normal phase.

A second fundamental question is whether the superconducting energy gap has nodes on the Fermi surface, and, if so, what their character is. Experimentally, no equilibrium or transport properties in the heavy fermion superconductors exhibit the exponential behavior expected for states with a non-zero energy gap everywhere on the Fermi surface; rather both specific heat and transport measurements display the power-law behavior characteristic of states with gaps which vanish at points or along lines on the Fermi surface. Specific heat measurements at low temperature, which reflect the density of quasiparticle states at energies of order $k_B T$, give direct evidence about the nodes of the gap. At low temperature, the only quasiparticles excited will be those in the vicinity of nodes of the gap. These states possess an energy less than $k_B T$ and lie within an angle $\sim T/\Delta$ of a node, where Δ is the maximum value of the energy gap on the Fermi surface. A simple geometric argument shows that the density of quasiparticles varies as T^2 for nodes at points and as T for nodes on lines, and the corresponding variation of the specific heat is as T^3 and T^2, respectively. In this way the experimental measurement of a T^2-dependence of the specific heat for UPt_3 shows that the energy gap vanishes on a line or lines, while the T^3 dependence found in UBe_{13} is indicative of a gap which vanishes at points. Thus heavy-fermion systems possess at least two superconducting states. Since UBe_{13} possesses cubic symmetry, while UPt_3 is hexagonal, it is possible that crystal structure plays a role in determining the nature of the superconducting state. Evidence that suggests the possible existence of two superconducting states in a single system is provided by specific heat and critical field experiments on $U_{1-x} Th_x Be_{13}$, where x, the concentration of Th impurities, lies between 2 and 4 percent.

A third question of interest is where the nodes lie on the Fermi surface. Information about this is contained in measurements of transport coefficients such as acoustic attenuation. In UPt_3 the attenuation, α, of transverse ultrasound propagating in the basal plane, measured by Shivaram et al.,[11] shows a different temperature dependence according to whether the sound wave is polarized in the basal plane ($\alpha \propto T$) or perpendicular to it ($\alpha \propto T^2$). These results suggest that quasiparticles move more freely in the basal plane than perpendicular to it, which would be consistent with a quasiparticle gap having nodes on lines on the Fermi surface perpendicular to the hexagonal axis. Further evidence for this behavior of the gap is provided by the recent tunneling measurements of Batlogg et al.,[12] which give no evidence for a gap when quasiparticles are injected across crystal faces with normals perpendicular to the hexagonal axis, but show a distinct gap when quasiparticles are injected across faces with moments parallel to the hexagonal axis.

A considerable amount of effort has gone into trying to understand transport in the superconducting states. Under circumstances in which scattering by impurities is the dominant process, as is the case in UPt_3 at temperatures of the order of T_c and lower, the temperature dependence of the transport coefficients seems to disagree with calculations for any anisotropic superfluid state if the scattering is treated in the Born approximation. In this approximation the lowest order s wave scattering by a single impurity is considered; the calculated mean free paths increase with decreasing temperature, and one finds results for the thermal conductivity, κ, and acoustic attenuation, α, which are much larger than those observed experimentally. We have shown[13] that if one takes into account the multiple scattering of quasiparticles by impurities, and if one is near the unitarity limit characterized by a phase shift, $\delta \sim \pi/2$, the mean free path for electron-impurity scattering shows remarkably little dependence on temperature, so that both α and κ/T fall off with decreasing temperature, in agreement with experiment. The transport data for UPt_3, including the anisotropies observed by Shivaram et al.[11] in the attenuation of transverse sound, can be accounted for qualitatively if, as noted above, one has a polar state in which the superconducting gap has nodes on lines on the Fermi surface which are parallel to the c axis of the crystal, and the mean free path is independent of temperature.[14] In our calculations, we did not take pair-breaking into account. Pair-breaking effects are important only at energies close to the gap energy, Δ, and at low energies, $E \sim \hbar/\tau_N$, where τ_N is the lifetime for impurity scattering in the normal state; these have been included in the work of Schmitt-Rink et al.,[14] Hirschfeld et al.,[15] and Scharnberg et al.[16] who find in numerical calculations that with $\hbar/(\tau_N \Delta) \sim 10^{-2}$, pair-breaking effects are important for polar states only at temperatures below $\sim(T_c/10)$, in agreement with the above estimate.

Quite generally features around the nodes are smeared out by impurity scattering. Evidence for this physical affect on the density of states in the superconducting state of UBe_{13} has been found by Ott et al.[17] in experimental measurements of the specific heat at low temperatures $(T \gtrsim 50mK)$; the experimental results are in excellent agreement with theoretical calculations of the state density which assume an axial state, in which the energy gap has point nodes, and electron impurity scattering which is near the unitarity limit.

There can be little doubt that the superconducting states observed in the heavy electron systems are unconventional, when compared to typical metallic superconductors. While there is as yet no theoretical proof or direct experimental demonstration that electron-phonon interactions are essentially irrelevant to heavy fermion superconductivity, in view of the persuasive physical arguments that the physical origin of the large masses is the coupling of conduction electrons to the local moment fluctuations, and that the virtual exchange of such spin fluctuations gives rise to an attractive interaction between heavy electron quasiparticles, and the model calculation of Norman,[9] it would seem overwhelmingly likely that it is the electron-local moment fluctuation coupling which is responsible for heavy electron superconductivity. Whether the resulting pairing state is "p-like" or "d-like" depends on the details of the wavevector dependence of the effective attractive interaction, and present evidence clearly favors the latter possibility.

ARE THE HIGH TEMPERATURE SUPERCONDUCTORS A BRANCH OF THE HEAVY ELECTRON
FAMILY?

Since heavy electron systems provide us with a new mechanism for
superconductivity and new pairing states in metals, it is natural to
inquire whether the physical origin of superconductivity in the ceramic
oxides is similar; thus does superconductivity in these systems arise from
an attractive interaction between electrons or holes induced by their
coupling to spin fluctuation excitations, and are these superconductors
another branch of the heavy electron family? Models in which spin fluctuations induce superconductivity have been proposed by a number of
authors,[18] while the detection of antiferromagnetic ordering in La_2CuO_{4-y}
for non-zero values of y,[19] provides support for this hypothesis. For
$YBa_2Cu_3O_7$, the recent measurements of the charge carrier concentration,[20]
the specific heat jump at T_c,[21] the itinerant carrier magnetic susceptibility,[20,21] the slope of the critical field curve near T_c,[20] together
with the temperature dependence of the London penetration depth,[22] make
possible a self-consistent deduction of the itinerant electron effective
mass (~3.5 me),[4] and demonstrate the existence of substantial exchange
enhancement of the carrier magnetic susceptibility.[4,23] The family
resemblance to heavy electron systems is striking. Quite generally in the
high T_c superconductors, the copper-oxide planes represent a promising
source of spin fluctuations, while if the itinerant carriers belong to a
distinct, but nearby (in energy) band, the basic physics would be remarkably similar, with a spin-fluctuation induced interaction being responsible
for both the heavy carrier mass and superconductivity; it is also possible
that the carriers, which are hole-like for $YBa_2Cu_3O_7$,[20] and spin fluctuations are excitations which belong to the same band.

The reason, then, that one achieves high superconducting temperatures
in the ceramic oxides is that the itinerant carrier density is so small
that the induced spin-spin interaction, J, of Eqtn. (6), plays almost no
role; rather the magnetic behavior is associated with exchange interactions
between spins on nearest neighbor and next nearest neighbor sites, and
their scale is set by the Neel temperature for the copper-oxide layers;
since this temperature can be of the order of room temperature, superconductivity at high temperatures can easily result. Put another way, what
spoils the chances for high T_c in the heavy electron systems is that the
conduction electrons are sufficiently dense to screen the f-electron
moments, so that the scale over which the heavy electron interaction can be
attractive is one or two orders of magnitude smaller than the Curie-Weiss
temperature; that same phenomenon is responsible for the fact that the Neel
temperatures in heavy electron systems are $\lesssim 20K$, rather than being comparable to or greater than room temperature. Since spin-orbit coupling
effects are small in the ceramic oxides, the itinerant carrier long wavelength magnetic susceptibility will be of the Pauli-Landau type, and will
contain no local moment contributions.

It is worth remarking that the measured low itinerant carrier densi-
ties in the ceramic oxides ($n \sim 3 \times 10^{21}$ cm^{-3}) correspond to values of $r_s \sim$
8, where r_s is the interelectron spacing divided by the Bohr radius. As r_s
increases beyond its values at ordinary metallic densities ($r_s \lesssim 3$) the Fermi

liquid parameter F_0^a, associated with direct Coulomb correlations, becomes increasingly negative[24]; thus at $r_s \sim 8$, direct Coulomb correlations favor both the inferred substantial exchange enhancement of the Pauli magnetic susceptibility,[4,23] and p state or d-state superconductivity, depending on the wavevector dependence of that static susceptibility.[2,25] It is possible therefore that direct Coulomb correlations act to enhance the spin-fluctuation-induced interactions associated with the copper-oxide layers, and so further increase the superconducting transition temperatures.

In common with the plasmon-exchange and exciton-exchange mechanisms, the spin-fluctuation mechanism provides a natural explanation for the observed absence of an isotope effect in $YBa_2Cu_3O_7$ and $EuBa_2Cu_3O_7$.[23] Unlike the above mechanisms, it also provides a natural explanation for the extreme sensitivity of the copper-oxide superconductors to substitutions for the copper ions. Such substitutions, it may be argued, can change dramatically the nature of the spin fluctuation excitation in the copper-oxide layers, and easily destroy superconductivity, while it is difficult to see why these substitutions would affect the exchange of virtual plasmons or excitons between carriers, and hence affect any superconductivity arising from that exchange. Spin-fluctuation exchange can give rise, inter alia, to p state pairing with a constant energy gap (as in the Balian-Werthamer phase of superfluid ^3He), and hence yield a temperature dependent London penetration depth consistent with the experimental results of Harshmann et al.[22] Finally, we note that for spin-fluctuation mechanisms, the maximum temperature for a superconducting transition will be of order the Neel temperature, T_N, as has been noted by deGennes;[18] materials with an underlying large Neel temperature, or large energy spin fluctuations, and low carrier concentrations would appear to be promising candidates for superconductors with transition temperatures well in excess of 90K; hence the question-mark in our title.

Much theoretical and experimental work will be required to test the spin fluctuation mechanism for ceramic oxide superconductivity. To cite but two examples, inelastic neutron scattering experiments on single crystals will test whether the superconducting materials possess spin fluctuation excitations of the desired character, while the two-dimensional character of the layers, and anisotropic effects more generally, may well play a special role. It took some three years of intensive experimental and theoretical investigations for the heavy electron community to arrive at a consensus on the physical picture we have set forth in this article; it would not be surprising if a comparable period of time might be required to arrive at a comparable consensus on the new high T_c materials.

INTRINSIC FLUX PINNING IN $YBa_2Cu_3O_7$

The short coherence length, $\xi_o \sim 14A$ which Bedell et al.[4] and Salamon[23] infer from their analysis of experiments on both the normal and superconducting properties of $YBa_2Cu_3O_7$, opens up the interesting possibility of intrinsic flux pinning in this material, i.e. the pinning of magnetic vortices to, for example, Y or Ba atoms in the unit cell, rather than the extrinsic pinning to crystalline imperfections usually found in Type II superconductors. The situation resembles that found in the "other" high temperature superconductors - neutron stars - in which the pinning of vortices in the rotating neutron superfluid to crustal nuclei has been shown to explain glitches and post-glitch behavior in pulsars.[27] Whether flux pinning in the terrestrial high temperature superconductors will correspond to the weak pinning or super-weak pinning situations encountered in neutron stars remains to be determined. It would seem, however, that intrinsic flux pinning both provides a natural explanation for the pinning effects observed by Harshman et al.[22] and makes possible very substantial critical currents in directions parallel to the copper-oxide planes in single crystal defect-free $YBa_2Cu_3O_7$. One would expect that intrinsic pinning phenomena will be highly anisotropic, and there is the further intriguing possibility that although the superconductivity of the 90K superconductors is not affected by the substitution of various rare earth impurities for Y, the resulting pinning phenomena and critical currents might be substantially influenced.

ACKNOWLEDGEMENTS

The present manuscript is adapted, in part, from References 1 and 2. We should like to thank our co-authors on the former manuscript, Zachary Fisk, Daryl Hess, Jim Smith, Joe Thompson, and Jeff Willis, for stimulating conversations on these and related topics, and Kevin Bedell, Stuart Brown, George Gruner, Richard Klemm and Myron Salamon for both stimulating discussions, and for communicating some of their results in advance of their submission for publication. This work was supported in part by a grant from the National Science Foundation, NSF DMR 85-21041; the manuscript was prepared while one of us (DP) enjoyed the hospitality of the Center for Materials Science at Los Alamos National Laboratory, which is supported by the U.S. Department of Energy.

APPENDIX

In this Appendix we give a brief summary of the self-consistent isotropic analysis of the thermodynamic and transport properties of $YBa_2Cu_3O_{7-\delta}$ carried out by Bedell et al.[4] and its possible implications for the pairing mechanism and coupling strengths for the oxide superconductors.[28] Bedell et al.[4] adopt the carrier density of $N_h = 3.5 \times 10^{21}$ cm^{-3} suggested by Hall effect measurements at 100K,[20] and deduce a thermal effective mass of 3.44 m_e from a Fermi liquid analysis of the temperature dependent London penetration depth found by Harshman et al.,[22] an analysis which yields a backflow mass enhancement factor, $(1+F_1^s/3) = 1.46$. Their deduced specific heat is $\gamma = 8.6$mJ/mole YK^2, which, when combined with the experimental results of Inderhees et al.[21] on the specific heat jump at the superconducting transition, yields a coupling constant ratio of $(\Delta C_v/\gamma KT_c) = 4.16$. This is large compared to the BCS week coupling value of 1.43, so that these results suggest the system is a strong coupling superconductor, and, since $(\Delta C_v/\gamma KT_c) > 3.73$ (the maximum value found by Blezius and Carbotte[29] for a phonon mechanism), also argue against a phonon pairing mechanism.

Further insight into the physical origin of the pairing mechanism and strong coupling is obtained from an analysis of the measured magnetic susceptibility. When corrections are made for core and Landau diamagnetism, one finds the electronic (Pauli) contribution to the magnetic susceptibility is $\chi_p = 8.4 \times 10^{-4}$ emu/mole. On combining this value with the Fermi liquid expression for χ_p, and using the density of states obtained above, one obtains an exchange enhancement factor, $(1+F_o^a)^{-1} = 7.35$, corresponding to a Fermi liquid parameter, $F_o^a = -0.864$.

The above results suggest that $YBa_2Cu_3O_{7-\delta}$ may be a member of the heavy electron family, in that spin fluctuations play a significant role and could well be the source of the pairing mechanism; moreover, it belongs to the helium liquid branch, in that the deduced Fermi liquid parameters are remarkably close to those found for ^{3}He. Bedell and Pines are therefore exploring, in work in progress,[28] the possibility that it is a p-state Balian-Werthamer superconductor, with a finite energy gap at the Fermi surface. They adapt the results of strong coupling theory for B-phase (Balian-Werthamer) superfluid ^{3}He, which lead to an energy gap,

$$\Delta_{s.c.} = 1.76 \; KT_c \left[\frac{\Delta C_v}{1.43 \; \gamma \; T_c} \right]^{1/2} = 3.16 \; kT_c \tag{9}$$

and find with this value a coherence length, $\xi_o = 14A$. Assuming that one is in the clean limit (ie $\ell > \xi_o$), they calculate $H_{c1} = 405$ gauss, $H_{c2} = 176$ T, and $-(dH_e/dT)_{T=T_c} = 2.9$ T/K, all of which results are in agreement with experiment. They then use strong coupling theory to calculate the transition temperature,

$$T_c = \alpha(1+F_o^a) \; T_F \; \exp \; 6/1+A_o^a \tag{10}$$

where $\alpha T_F(1+F_o^a)$ is the energy cut off for the spin-fluctuation induced pairing, and $A_o^a \equiv F_o^a/1+F_o^a$; they find $T_c = 126 \; \alpha$ K, so that $\alpha \sim 3/4$, a physically reasonable result.

Given the likely inadequacies of an isotropic model, and the sensitivity of these results to one's choice of the experimental result for n_h (Bardeen et al. take $n_h \cong 9 \times 10^{21}$ cm^{-3} and reach very different conclusions[30]), the above results should be regarded as suggestive, not as definitive; experiments on single crystals and a theory which takes anisotropy into account will evidently be required before one knows whether the coupling is strong or weak, and the nature and physical origin of the pairing mechanism. At present, one can only say that to the extent the number of holes is close to that posited by Bedell et al.,[4] p-wave pairing, strong coupling, and spin-fluctuation exchange represent an attractive way of arriving at a consistent account of experiment.

REFERENCES

1. For a recent review of heavy electron systems see Z. Fisk, D. Hess, C. J. Pethick, D. Pines, J. Smith, J. Thompson, and J. Willis, preprint, June 1987.
2. C. J. Pethick and D. Pines, preprint submitted November 1986 for publication in a Festschrift for V. L. Ginzburg on the occasion of his seventieth birthday (E. Feinberg and L. Keldysh, editors).
3. W. L. McMillan, Phys. Rev. $\underline{167}$, 331 (1968); M. L. Cohen and P. W. Anderson, in Superconductivity in d-and f-band Metals, ed. D. H. Douglas, AIP, New York, 1982; C. M. Varma, S. Schmitt-Rink, and E. Abrahams, Sol. St. Comm. (to be published).
4. K. Bedell, S. Brown, G. Gruner, and D. Pines, in preparation.
5. C. J. Pethick and D. Pines, "Elementary Excitations and Transport in Heavy Fermion Systems; Proceedings of the Fourth International Conference on Recent Progress in Many-Body Theories," ed. P. Siemens and R. A. Smith; Berlin, Springer, 1987 (in press).
6. G. Aeppli, preprint.
7. For UPt_3, see G. Aeppli, A. Goldman, G. Shirane, E. Bucher, and M.-Ch. Lux-Steiner, Phys. Rev. Lett. $\underline{58}$, 808 (1987); for $CeCu_3$, see G. Aeppli, H. Yoshizawa, Y. Endoh, E. Bucher, J. Hufnagl, Y. Onuki, and T. Komatsubara, Phys. Rev. Lett. $\underline{57}$, 122 (1986); for U_2Zn_{17} see C. Broholm, J. Kjems, G. Aeppli, Z. Fisk, J. Smith, S. M. Shapiro, G. Shirane, and H. R. Ott, Phys. Rev. Lett. $\underline{58}$ (in press).
8. B. Batlogg, private communication.
9. M. R. Norman, April 1987 preprint.
10. L. Taillefer, R. Newbury, G. G. Lonzarich, Z. Fisk, and J. L. Smith, J. Magn. Magn. Mat. 63 and 64, 372 (1987).
11. B. S. Shivaram, Y. H. Jeong, T. F. Rosenbaum, and D. J. Hinks, Phys. Rev. Lett. $\underline{56}$, 1078 (1986).
12. B. Batlogg, private communication.
13. C. J. Pethick and D. Pines, Phys. Rev. Lett. $\underline{57}$, 118 (1986).
14. S. Schmitt-Rink, K. Miyake, and C. M. Varma, Phys. Rev. Lett. $\underline{57}$, 2575 (1986).
15. P. Hirschfeld, D. Vollhardt, and P. Wolfle, Solid State Comm. $\underline{59}$, 111 (1986).
16. K. Scharnberg, D. Walker, H. Monien, L. Tewordt, and R. A. Klemm, Solid State Comm. $\underline{60}$, 535 (1986).
17. H. R. Ott, E. Felder, C. Bruder, and T. M. Rice, Europhys. Lett. $\underline{3}$, 1123 (1987).
18. P. W. Anderson, Science $\underline{235}$, 1196 (1987); P.-G. deGennes, preprint; D. Johnston and R. Klemm, private communication, P. A. Lee and N. Read, preprint; Y. Hasegawa and H. Fukuyama, Jap. J. Appl. Phys. 26 L (1987).
19. D. Johnston, D. P. Groshorn, H. Thomann, P. Tindall, and J. P. Stokes, to be published.
20. S.-W. Cheong, S. E. Brown, J. R. Cooper, Z. Fisk, R. S. Kwok, D. E. Peterson, J. D. Thompson, G. L. Wells, E. Zirngiebl, and G. Gruner, preprint, June 1987.
21. S. E. Inderhees, M. B. Salamon, T. A. Friedmann, and D. M. Ginsberg, Phys. Rev. (submitted).
22. D. R. Harshman, G. Aeppli, B. Batlogg, R. J. Cava, E. J. Ansaldo, J. H. Brewer, W. Hardy, S. R. Kreitzman, G. M. Luke, D. R. Noakes, and M. Senba, preprint.
23. M. Salamon, private communication.
24. N. Iwamoto and D. Pines, Phys. Rev. B $\underline{29}$, 3924 (1984).
25. D. Hess, C. J. Pethick, and D. Pines, in preparation.
26. B. Batlogg, R. J. Cava, A. Jayaraman, G. A. Kourouklis, S. Sunshine, D. W. Murphy, L. W. Rupp, H. S. Chen, A. White, A. M. Mujsce, and E. A. Rietman, preprint.

27. For a recent review, see D. Pines and M. A. Alpar, Nature $\underline{316}$, 27 (1985).
28. K. S. Bedell and D. Pines, in preparation.
29. J. Blezius and J. F. Carbotte, preprint.
30. J. Bardeen, D. M. Ginsberg, and M. Salamon, preprint, and these proceedings.

CRITICAL FIELDS OF UBe$_{13}$ FILMS

J. H. Kang, J. Maps, and A. M. Goldman

School of Physics and Astronomy, University of Minnesota
Minneapolis, Minnesota 55455

J. S. Brooks

Department of Physics, Boston University
Boston, Massachusetts

Z. Fisk and J. L. Smith

Los Alamos National Laboratory
Los Alamos, New Mexico

ABSTRACT

The temperature dependences of the parallel and perpendicular
critical magnetic fields of polycrystalline UBe$_{13}$ films have been
measured. The ratio of the critical fields, $H_{c\parallel}/H_{c\perp}$, has been found
to be 1.25 at low temperatures, a value less than would be expected for
s-wave pairing without surface pair-breaking but greater than expected
for any pairing configuration which is a pure angular momentum state
with L $\neq$ 0. Similar efforts directed at the preparation of UPt$_3$ films
were unsuccessful because of the formation of an impurity phase of UPt
which is a ferromagnet.

INTRODUCTION

The nature of the superconducting pairing in heavy-fermion metals[1]
such as UBe$_{13}$ has been the subject of intensive theoretical investigation.[2] The reason for this is that in a number of experimental studies
the results deviate in a qualitative manner from the predictions of the
BCS theory for superconductors with s-wave pairing. Relevant works
include measurements of the temperature dependences of the specific
heat,[3] ultrasonic attenuation,[4] nuclear relaxation rate,[5] critical
magnetic field[6] and superconducting penetration depth.[7] These studies,
together with the observation of peaks in the acoustic attenuation[4]
which have been taken as evidence of collective modes of an anisotropic
order parameter,[8] have led to a picture in which the gap is anisotropic

with points or lines of zeroes on the Fermi surface, but without a
precise identification of the type of pairing. General symmetry
analyses of the order parameter indicate that rather than the use of the
terms singlet and triplet, or s-, p-, and d-wave superconductivity, it
is more appropriate to classify the states according to whether they
have even or odd parity.[9,10] In all odd-parity states, the energy gap
function is either non-zero or vanishes at a set of discrete points on
the Fermi surface. In the even-parity case, lines of zero are possible,
which can never occur in the case of odd-parity.

The classification into odd and even parity states argues against a
single experiment in which a yes or no result determines the symmetry of
the pairing in a heavy fermion material. The discussion of potential
critical experiments has taken place and has been closely connected with
considerations relating to the observability of the ac and dc Josephson
effects between conventional spin-singlet superconductors and
anisotropic superconductors,[11] and the character of the proximity effect
between such systems.[12] Both of these phenomena are sensitive to the
nature of the boundary conditions on the order parameter at the surface
or interface. These boundary conditions in turn are specific to the nature of the pairing. One experiment on the proximity effect has been
interpreted as evidence of triplet pairing in UBe_{13}.[13]

Here we report measurements of both the perpendicular and the
parallel critical fields of the heavy fermion superconductor UBe_{13}. As
was shown some years ago by Saint-James and de Gennes,[14] the nucleation
of the order parameter in a parallel magnetic field usually starts at
the surface in a field higher than that corresponding to the bulk
nucleation field H_{c2}. The latter is the perpendicular critical field
for a film. For an s-wave superconductor the surface critical field,
H_{c3}, which is identified with the parallel critical field, is a factor
1.695 greater than the bulk nucleation field H_{c2} even for a strong-
coupled superconductor.[15] This result follows from a "local" solution
to the Ginzburg-Landau equations and is true as long as the order
parameter has the standard boundary condition. Coating the surface with
a normal metal layer, which acts as a pair-breaker, reduces the ratio to
unity. The observation of a ratio H_{c2}/H_{c3} less than 1.695, at least for
s-wave superconductors, is evidence of pair breaking at the surface.
For pairing in a pure angular momentum state with $L \neq 0$, as we will
argue below, the free surface is a pair-breaker itself. Thus an observation of the ratio of H_{c3}/H_{c2} greater than unity in a heavy fermion
compound would appear to rule out any states not containing some admixture of singlet pairing.

SAMPLE PREPARATION AND CHARACTERIZATION

Thin films of UBe_{13} were fabricated using a static dc sputtering
system which is equipped with ion pumps. Before back-filling with research grade Ar gas, the system was pumped down into the 10^{-9} Torr
range. A sputtering rate of 15-20 Å/minute was used to produce a 3000 Å

thick film on a single-crystal sapphire substrate, which was heated to
800°C during the deposition. The UBe$_{13}$ phase in the films was identified from X-ray diffraction data, which also indicated that the
samples were polycrystalline with randomly oriented crystallites. The
surfaces of the films were highly reflective indicating a certain degree
of smoothness. Investigation of the surface of a film using a scanning
tunneling microscope revealed topological features which varied over
distances of the order of 300-1000 Å.

RESULTS

In Fig. 1 we show the resistive transition of a UBe$_{13}$ film in zero
magnetic field. The transition temperature and transition width are
comparable to those found in bulk material.[1] The sharp rise in the
resistivity observed in bulk material at 2 K above the superconducting
transition was not found in these films.[16] On the other hand, the maximum in the normal state resistivity found near 20 K was observed.

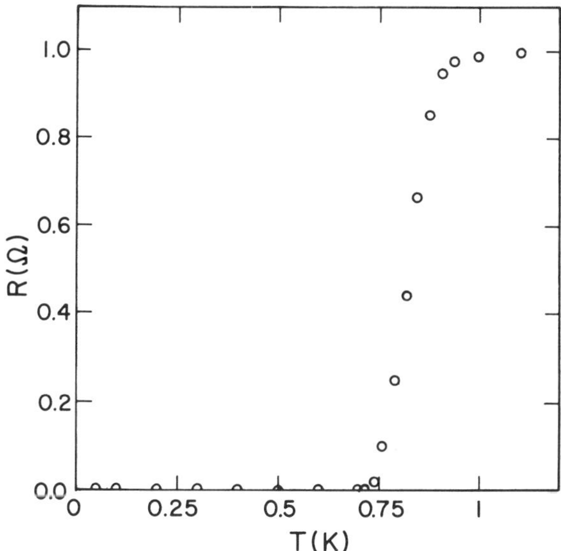

Fig. 1. Resistance vs. temperature of a UBe$_{13}$ film in zero magnetic
         field.

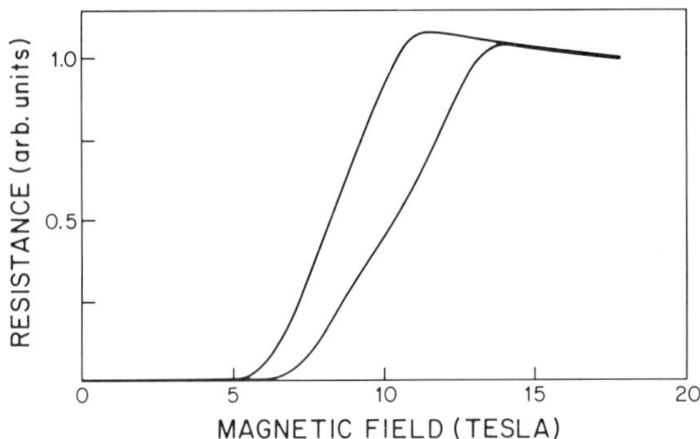

Fig. 2. Resistance vs. magnetic field in Tesla of a UBe_{13} film at 0.05 K.

In Fig. 2 we show the variation of the electrical resistance of a film as a function of applied magnetic field for both the parallel and perpendicular orientations. This data was taken at 0.05 K with the sample cooled in a top-loading ^{3}He-^{4}He dilution refrigerator. The measurements were made at fixed T by sweeping the field which was produced by a Bitter solenoid. The clear shift of the parallel field data relative to the perpendicular field data is evident at all of the temperatures over which measurements were made. In these studies the current densities were 10^{-1} A/cm^2. The normal state film resistance was a linear function of current at the values which were used in the investigations.

Because of the strong negative magnetoresistance in the normal state seen in Fig. 2, there is some ambiguity as to the precise condition which defines H_{c2} or H_{c3}. We have taken the critical fields to be the loci of points in H and T which correspond to the resistance being one half of the peak resistance. The resultant critical field data shown in Fig. 3 are of the same qualitative character as the results of Maple and co-workers for bulk samples of UBe_{13}.[6] The data were not of sufficient detail to compare with specific predictions of any theoretical model of either s-wave or anisotropic superconductivity.

The data taken at the National Magnet Laboratory using a Bitter solenoid were consistent with lower field, higher-temperature data obtained using a small superconducting magnet.

The main feature of Fig. 3 is the fact that the ratio H_{c3}/H_{c2} is about 1.25 in the limit of $T \to 0$. This result appears to be temperature independent below 0.6 K, with the magnitudes of the parallel and perpendicular fields merging above that temperature as T_c is approached from below a dirty-limit s-wave superconductor where the ratio should be 1.695, independent of temperature over the entire range.

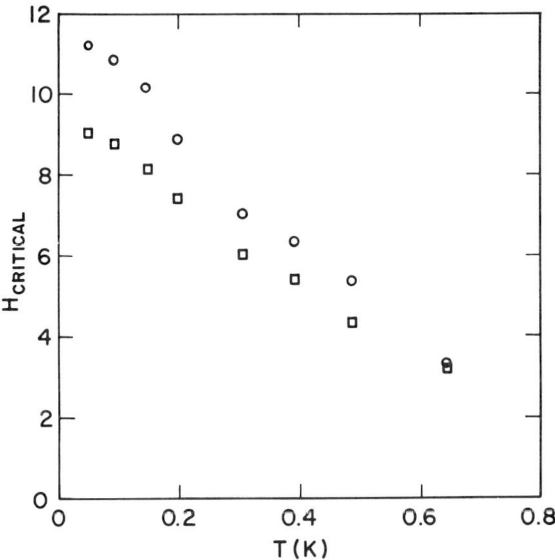

Fig. 3. Critical field in Tesla vs. temperature for a UBe_{13} film. The circles are data taken in a parallel field, the squares in a perpendicular field.

DISCUSSION

The observation of a critical-field ratio 1.695 in UBe_{13} would have categorically ruled out any triplet or anisotropic superconductivity for this material. The fact that the ratio is less than this value does not in of itself preclude pure singlet superconductivity because of the possibility of pair-breaking effects due to some intrinsic or extrinsic character of the surface of an unknown nature which might alter the boundary conditions. Such a pair-breaker could even reduce the critical-field ratio to unity, the value of which would be expected for pairing in any pure angular momentum state with $L \neq 0$. The latter statement follows from the fact that such pair states do not obey Anderson's theorem.[17] Thus nonmagnetic scattering centers and surfaces are both pair-breakers and the critical field ratio might be expected to be unity because of the intrinsic pair-breaking character of the surface for pure states of this type.

Thus the result $H_{c3}/H_{c2} = 1.25$, although ambiguous with regard to ruling out s-wave pairing in UBe_{13}, does preclude the possibility of pure states which are paired with $L \neq 0$, <u>at least at the surface</u> which is the only part of the film actually probed by studies of surface superconductivity. Because surface scattering itself may modify the pairing configuration, it may be hasty to draw conclusions about the interior of the material from the measurement of a surface property such as the surface superconductivity. Such a caveat would also apply to all of the alternative experiments which might reveal the nature of the pairing such as the study of superconducting tunneling and the investigation of the proximity effect.

Although there is no detailed theory for the ratio H_{c3}/H_{c2} for any of the proposed pairing configurations that might describe the superconductivity of UBe_{13}, in principle it is possible to construct a Ginzburg-Landau equation and boundary conditions with the appropriate symmetry and evaluate the ratio for the various cases. Such a theoretical investigation, together with the present experimental results, could greatly illuminate our understanding of pairing in heavy fermion superconductors.

It should be noted that similar preparation techniques were used to fabricate films of UPt_3. Their superconductivity, however, was not impressive, apparently the result of the impurity phase UPt, a material which is a ferromagnet.[18] Surface magneto-optic Kerr effect studies suggest that the UPt is segregated at the surfaces, a fact which if general might explain the absence of a Josephson effect in UPt_3.[19]

In summary, we have fabricated thin films of UBe_{13} and investigated the critical field in both the parallel and perpendicular configuration. The observation of a ratio of 1.25 strongly suggests that the pairing state of UBe_{13} near the surface is not a pure pairing state with $L \neq 0$.

ACKNOWLEDGEMENTS

The authors would like to thank Professor Oriol T. Valls for helpful discussions.

This work was supported by the Air Force Office of Scientific Research under Grant 84-0347. Work at the Francis Bitter National Magnet Laboratory was supported by the National Science Foundation under Grant NSF/DMR-85 11789.

REFERENCES

1. G. R. Stewart, Heavy Fermion Systems, Rev. Mod. Phys. 56:755 (1984).
2. C. M. Varma, Valance Fluctuations, Heavy Fermions and Their Superconductivity, Comments on Solid State Physics XI:221 (1985); P. A. Lee, T. M. Rice, J. W. Serene, L. J. Sham and J. W. Wilkins, Theories of Heavy-Electron Systems, Comments of Solid State Phys. XII:99 (1986).
3. H. R. Ott, H. Rudigier, T. M. Rice, K. Ueda, Z. Fisk and J. L. Smith, p-Wave Suyperconductivity in UBe_{13}, Phys. Rev. Lett. 52:1915 (1984).
4. B. Golding, D. J. Bishop, B. Batlogg, W. H. Haemmerle, Z. Fisk, J. L. Smith and H. R. Ott, Observation of a Collective Mode in Superconducting UBe_{13}, Phys. Rev. Lett. 55:2479 (1985).

5. D. E. MacLaughlin, C. Tien, W. C. Clark, M. D. Lau, Z. Fisk, J. L. Smith and H. R. Ott, Nuclear Magnetic Resonance and Heavy-Fermion Superconductivity in $(U,Th)Be_{13}$, Phys. Rev. Lett. 53:1833 (1984).

6. M. B. Maple, J. W. Chen, S. E. Lambert, Z. Fisk, J. L. Smith, H. R. Ott, J. S. Brooks and M. J. Naughton, Upper Critical Magnetic Field of the Heavy-Fermion Superconductor UBe_{13}, Phys. Rev. Lett. 54:477 (1985).

7. D. Einzel, P. J. Hirschfield, F. Gross, B. S. Chandrasekhar, K. Andres, H. R. Ott, J. Beuers, Z. Fisk and J. L. Smith, Magnetic Field Penetration Depth in the Heavy-Electron Superconductor UBe_{13}, Phys. Rev. Lett. 56:2513 (1986).

8. K. Miyake and C. M. Varma, Landau-Khalatnikov Damping of Ultrasound in Heavy Fermion Superconductors, Phys. Rev. Lett. 57:1627 (1986).

9. G. E. Volovik and L. P.Gor'kov, Superconducting Classes in Heavy Fermion Systems, Sov. Phys. JETP 61:843 (1985).

10. E. Blount, Symmetry Properties of Triplet Superconductors, Phys. Rev. B 32:2935 (1985); K. Ueda and T. M. Rice, p-Wave Superconductivity in cubic metals, Phys. Rev. B 31:7114 (1985).

11. E. W. Fenton, The Josephson Effect in Superconductors with Heavy Fermions, Solid State Commun. 60:347 (1986).

12. A. J. Millis, Proximity Effects Between Singlet and Triplet Superconductors, Physica 135B:69 (1985).

13. Siyuan Han, K. W. Ng, E. L. Wolf, Andrew Millis, J. L. Smith and Zachary Fisk, Observation of Negative s-Wave Proximity Effect in Superconducting UBe_{13}, Phys. Rev. Lett. 57:238: (1986)

14. D. Saint-James and P. G. De Gennes, Onset of Superconductivity in Decreasing Fields, Phys. Lett. 7:306 (1963).

15. Gert Eilenberger and Vinay Ambegaokar, $Bulk(H_{c2})$ and $Surface(H_{c3})$ Nucleation Fields of Strong-Coupling Superconducting Alloys, Phys. Rev. 158:332 (1967).

16. H. R. Ott, H. Rudigier, Z. Fisk and J. L. Smith, UBe_{13}: An Unconventional Actinide Superconductor, Phys. Rev. Lett. 50:1595 (1993).

17. R. Balian and N. R. Werthamer, Superconductivity with Pairs in a Relative p Wave, Phys. Rev. 131:1553 (1963).

18. C. A. Luengo, M. B. Maple and J. G. Huber, Magnon Heat Capacity, Magnon Magnetization and Magnetic Entropy of the Weak Ferromagnet UPt, J. of Magn. and Maget. Mat. 3:305 (1976).

19. U. Poppe, Vacuum Tunneling and Josephson Effect in Heavy Fermion Superconductors, Physica 135B:22 (1985); H. M. Mayer, G. Sparn, N. Grewe, U. Poppe and J. M. Franse, Heavy-Fermions in Kondo Lattice Compounds, J. Appl. Phys. 57:3054 (1985).

KONDO LATTICES : POSSIBLE MECHANISM FOR A NONPHONON SUPERCONDUCTIVITY, MAGNETIC-FIELD-INDUCED SUPERCONDUCTIVITY

Ondrej Hudák

Institute of Experimental Physics
Slovak Academy of Sciences
Solovjevova 47
Košice, CS-04060
Czechoslovakia

INTRODUCTION

In recent years large interest in new mechanisms of super-conductivity appeared due to rising number of new classes of superconducting materials with their properties differing from those of previously known. Let us mention heavy fermion systems and oxide superconductors. In this paper we present results concerning the examination of possible mechanism for a nonphonon superconductivity in Kondo lattices and our suggestion for a new mechanism of magnetic-field-induced superconductive state in heavy fermion systems with appropriate macroscopic symmetry.

POSSIBLE MECHANISM FOR A NONPHONON SUPERCONDUCTIVITY IN KONDO LATTICES

Interpretation of the properties of heavy fermion super-conductors have been proposed in several papers recently, see in /1/. Anderson /1/ and others have called attention to the possibility of a nonphonon mechanism for the effective attraction of electrons in these systems. Our purpose in the present section is to propose mechanism for a nonphonon superconductivity in Kondo lattices. A short summary of our results was published in 1985 ,/2/.

The localization and delokalization of f electrons and the appearance of heavy quasiparticles and a local magnetic moment in heavy fermion systems are conveniently described by the Anderson lattice model. We consider two bands of valence electrons and f electrons, for which we ignore the orbital degeneracy for simplicity

$$H = H_o + H_1 \tag{1}$$

$$H_0 = \sum_{k,j,\sigma} \epsilon_{kj}\, c^{+}_{kj\sigma} c_{kj\sigma} + \sum_{n,\sigma} \epsilon_f\, n_{n,\sigma} + U \sum_{n} n_{n\uparrow} n_{n\downarrow} ,$$

$$H_1 = \sum_{k,j,\sigma,n} \left(V_{knj}\, c^{+}_{kj\sigma} f_{n\sigma} + h.c. \right) , \qquad\qquad n_{n,\sigma} \equiv f^{+}_{n\sigma} f_{n\sigma}$$

where $\epsilon_{kj} \equiv E_{kj} - \mu$ is the energy of the j-th band, j=1,2 , $c^{+}_{kj\sigma}$ and $c_{kj\sigma}$ are the operators that create and annihilate valence electrons, $f^{+}_{n\sigma}$ and $f_{n\sigma}$ are the corresponding operators for f electrons at the n-th f atom, the V_{knj} are mixing parameters, U is the energy of the Coulomb repulsion of the f electrons, and $\epsilon_f \equiv E_f - \mu$ is the energy of their local level. A generalized Schrieffer-Wolff canonical transformation converts this Hamiltonian into the Hamiltonian of a Kondo lattice:

$$H' \equiv exp(S)\, H\, exp(-S) \approx H_0 + \tfrac{1}{2}\,[S, H_1] \qquad\qquad (2)$$

$$S \equiv \sum_{\{\substack{k,j,\\ \sigma,n}\}} \frac{1}{\sqrt{N}}\, V_{kj} \left(\frac{1 - n_{n,-\sigma}}{\epsilon_f - \epsilon_{kj}} + \frac{n_{n,-\sigma}}{\epsilon_f + U - \epsilon_{kj}} \right) \left(f^{+}_{n\sigma} c_{kj\sigma}\cdot exp(i\,k\cdot R_n) - h.c. \right)$$

$$H' = \sum_{k,j,\sigma} \epsilon^{r}_{kj}\, c^{+}_{kj\sigma} c_{kj\sigma} + \tfrac{1}{2} \sum_{\{\substack{j,n,\\ \alpha,\beta}\}} J_j\, \vec{S}_n \cdot \left(c^{+}_{nj\alpha}\, \vec{\sigma}^{\,Pauli}_{\alpha\beta}\, c_{nj\beta} \right)$$

where $J_j \cong 2\,|V_{k_F j}|^2 \cdot [U / |\epsilon_f|(\epsilon_f + U)] > 0$ for momenta at the Fermi surface, and $V_{knj} \equiv V_{kj} \cdot exp(-ikR_n) / \sqrt{N}$. In (2) we ignore the other terms of the new Hamiltonian. Note that we neglected in (2) besides standart terms, /3/, for both sinle bands also interband two fermion terms. Here and below , we assume that the parameter values in (1) are such that localized spins exist at f atoms. The renormalized band energies in (2) are given to second order in mixing parameters, assuming that occupancy of the f level is one electron per f atom and taking into account only diagonal terms in the direct interaction energy between conduction electrons and f electrons, by

$$\epsilon^{r}_{kj} = \epsilon_{kj} + \frac{|V_{kj}|^2}{(\epsilon_{kj} - \epsilon_f)} \cdot \frac{1}{2} - \frac{|V_{kj}|^2}{2U} \; ; \qquad \epsilon^{r}_f(k) = \epsilon_f - \sum_j \left\{ \frac{|V_{kj}|^2}{(\epsilon_{kj} - \epsilon_f)} \left(1 - \frac{N_j}{2}\right) - \frac{|V_{kj}|^2}{(U + \epsilon_f - \epsilon_{kj})} \frac{N_j}{2} \right\} ,$$

where $N_j \equiv \langle c^{+}_{kj\sigma} c_{kj\sigma} \rangle$, $0.5 \lesssim N_j < 1$. Within this approximation we find that for two bands 1 and 2 separated by the energy 2a:

$$\epsilon_{k1} = \frac{k^2}{2m} - a - \mu \qquad\qquad \epsilon_{k2} = \frac{k^2}{2m} + a - \mu$$

the effective masses m_1 and m_2 for the symmetric case $\epsilon_f=-U/2$ and $V_{kj}\equiv V_j$ (independent on k) are given to second order in (μ/U) by

$$\frac{m}{m_1}\approx 1-\frac{J_1}{4U}\;;\qquad \frac{m}{m_2}\approx 1\;;\qquad \frac{m}{m_f}\approx\frac{J_1}{2U}\;.$$

Here we assumed that the mixing energy scales with U in the first band $|V_1|^2=(U\mu/8)$, and in the second band it is a constant independent on U. This assumption we introduce in order to find the exchange energy $J_1=\mu\alpha>0$ a finite constant in the limit of large Coulomb repulsion, $U\to\infty$. The conditions for the Schrieffer-Wollf transformation should be valid and we infere from them that the constant α is small, $\alpha\ll 1$, because $J_1\,N(E_F)\simeq J_1/\mu$. The energy distance between bands 1 and 2 in (2) renormalizes to the value $2a_r = 2a -(J_1/8)+ O(\mu/U)$. The bottom of the first band shifts to higher values by an energy $(J_1/4)$ while that of the second band remains unchanged to $O(\mu/U)$. The chemical potential renormalizes to $\mu_r=\mu -J_1/16$ in the same order. Its value shifts to smaller energies. The f atom energy level also shifts down to $E_{f,r}=E_f -(5J_1/16)$. We see that the halffilled f electron band $(n_f=1)$ contains heavy electrons which are localized here. The first band electrons tend to become heavy for J_1 large enough if U is fixed for a moment. The bottom of the upper (2nd) conduction band remains unchanged , the amount of electrons in this band do not change. The energy gain due to tendency to formate the singlet at the f site is proportional to the exchange energy J_1. The singlets are formed only between f electrons and electrons from the first band.

As it was shown in the ref./4/ and several other papers, in the case of a single band there is a critical exchange energy value $J_c>0$ such that for a system with $J> J_c$ the ground state at low temperatures is a singlet Kondo state. At $0< J< J_c$ this state is antiferromagnetic. We assume here that these results can be extended to the case of the Kondo lattice with two valence electron bands. If the mixing energy for the first band has a magnitude corresponding to $J_1> J_c$, while that for the second band has a magnitude corresponding to $J_2< J_c$, then at low temperatures the effective value of J_1 increases, to $J_1\gg J_c$, while that of J_2 decreases, so that we expect $J_2\ll J_c$. Cosequently , the conduction electrons of the second band are responsible for the magnetic interaction (indirect exchange between localized spins at f atoms) in our model, and the electrons of the first band are scattered by these localized spins. The effective Hamiltonian describing this behaviour of the system is found in the standart way from (2), it is

$$H''=\sum_{k,\sigma}\epsilon^r_{k1} c^+_{k1\sigma} c_{k1\sigma} + (J_1/2)\sum_{n,\alpha,\beta}\vec{S}_n\cdot(c^+_{n1\alpha}\,\vec{\sigma}^{Pauli}_{\alpha\beta}\,c_{n1\beta}) +\qquad (3)$$

$$+ (K/z)\sum_{n,n'}{}'\,\vec{S}_n\cdot\vec{S}_{n'}\quad,$$

where $K \approx \frac{1}{\mu} J_2^2 \cdot f_{RKKY}(2k_F r_{f-f})$. In the last term in (3), we have
considered only the interaction of nearest neighbors (f atoms).
Generalization to magnetic interactions on larger distances is
straightforward. As we have already mentioned, we are assuming
that at low temperatures the first band is in a singlet Kondo
state and that we have $J_1 \gg J_c$. Put in another words, we assume
the itinerant character of first band electrons when they are
screening the localized spins on f atoms. At low temperatures
of all possible states of the system that are described by (3)
only those are effective for which the localized spin $\vec{S}_n$ is

completely screened by the spins of the conduction electrons
of the first band. For these states we can write $\vec{S}_n = (-1/2) \cdot$
$\cdot \sum_{\alpha\beta} c_{n1\alpha}^{+} \vec{\sigma}_{\alpha\beta}^{Pauli} c_{n1\beta}$, and from (3) we find an effective Hamiltonian (here and below, we omit the index 1 specifying the
first band):

$$H''' = \sum_{k,\sigma} \epsilon_k^r c_{k\sigma}^{+} c_{k\sigma} + \frac{1}{2\nu} \sum_{\{k,k',q,\alpha-\delta\}} V^{\alpha\beta\gamma\delta}(q) \, c_{k\alpha}^{+} c_{k+q\beta} c_{k'\gamma}^{+} c_{k'-q\delta} \tag{4}$$

where the interaction energy is

$$V^{\alpha\beta\gamma\delta} \equiv \frac{K \cdot \rho \cdot \gamma(q)}{2} \, \vec{\sigma}_{\alpha\beta}^{Pa.} \cdot \vec{\sigma}_{\gamma\delta}^{Pa.} \quad ; \quad \rho \equiv \nu/N \quad ; \quad \gamma(q) = \frac{1}{k} \sum_{n'}{}' \exp(i\,q \cdot (R_n - R_{n'})) \; .$$

The crystal symmetry of the system is incorporated here in
the particular form of the function $\gamma(q)$. The q-summation
in (4) is restricted to those energies, for which the f-atom
spins are screened by electrons from the first valence band.
The limiting energy for this process is of the order of the
Kondo energy $k_B T_K$.

We have studied the superconductivity in the system described by (4) in the weak-coupling approximation by the Gor'kov
method of Green functions. Let us describe the results of the
isotropic-medium approximation approach firstly. In this
approximation we have $\gamma(q) \rightarrow (1/4\pi) \cdot \int d\Omega_\delta \exp(i\,\delta \cdot q) \equiv \gamma_i(q)$.
We find that a singlet phase occurs for the antiferromagnetic
spin coupling, $K > 0$. If we had $(\gamma_1 \cdot K) < 0$, where γ_1 coefficient
is found from the expansion $\gamma_i(P_1 - P_0) \approx \gamma_0 + \gamma_1 \hat{P}_1 \cdot \hat{P}_0 + \ldots$ on
the Fermi surface, the Anderson-Brinkmann-Morel model phase of
the triplet state would be preferable. The high-temperature
properties of the magnetic susceptibility in heavy fermion superconductors are evidence /1/ for $K > 0$. In the isotropic
approximation, using typical values for the constants of the
materials $CeCu_2Si_2$, UBe_{13}, UPt_3, U_6Fe and U_2PtC_2 from the literature /1/ we find $\gamma_0 > 0$ and $\gamma_1 > 0$. The superconducting gap
for the singlet phase, which according to the isotropic approximation, may realize in mentioned systems is at T=0 given by
$\Delta_{sing.}(0) = 2k_B T_K \cdot \exp(-4/3K\rho \, \gamma_0 \cdot N(E_F))$, where T_K is Kondo
temperature , $\rho = \nu/N$ is the volume per f atom, and $N(E_F)$ is
the state density of electrons of the first band on the Fermi
surface. For nonzero temperatures we find usual BCS-like
expressions for thermodynamic quantities replacing the Debye
energy by the Kondo binding energy and the electron-phonon
interaction constant by the constant $(3K\rho \, \gamma_0/4)$.

The triplet state , which would be realized for $K \cdot \gamma_1 < 0$, is characterized by

$$T_c^t = 1{,}14 \, T_k \exp\left(-12 / |K \gamma_1| \, g \, N(E_F)\right)$$

and other expressions known from the ^{3}He theory of superfluid states.

When the crystal symmetry is taken into account, the solution of the gap equation becomes more complicated. We have used in /5/ the decomposition of the structural factor $\gamma(q)$ into separable forms introduced for the case of the anisotropic electron-phonon interactions by Miyake et al. /6/. Let us consider results for the simple cubic lattice with δ the lattice parameter, and again assume antiferromagnetic interactions between f atoms, $K > 0$. We find that triplet-odd parity pairing is always repulsive. The singlet-even parity pairing is anisotropic. The gap equation has the form

$$\Delta_k = (K/12z) \, \frac{1}{N} \sum_{k'} \left(\phi_k \phi_{k'} + \eta_k \eta_{k'} + \zeta_k \zeta_{k'} \right) \frac{\Delta_{k'}}{2 E_{k'}} \tanh\left(E_{k'} / 2 k_B T\right)$$

where $E_k = (e_k^2 + |\Delta_k|^2)^{1/2}$, c_k being the kinetic energy of the first band electrons, $\phi_k \equiv z \cdot \gamma(k)$, $\eta_k = \sqrt{6} \, (\cos(k_x \delta) - \cos(k_y \delta))$, $\zeta_k = 2(\cos(k_x \delta) + \cos(k_y \delta) - 2 \cdot \cos(k_z \delta))$. The function ϕ has s-like symmetry, η and ζ are $d\gamma$-like. The s-pairing gap is given by the anisotropic function:

$$\Delta_k^s = \Delta_s \cdot \phi_k$$

and its critical temperature T_c^s may be foun from the appropriate equation numericaly. The $d\gamma$-like state gap function is given by

$$\Delta_k^d = \Delta_d \left(\eta_k \pm i \zeta_k \right).$$

The transition temperatures and other properties of the states are analogous to those for the electron-phonon anisotropic interaction superconductivity discussed in /6/, if we put their g constants to $g_2 = g_3 = U_{/6/} = 0$ and $g_s = -g_t = (K/4z)$. Note that effective Hamiltonians similar to our (4) and describing the spin-fluctuation mediated pairing, was recently introduced and discussed in /7/. Within our model and for the antiferomagnetic interaction (K>0) we have found that the triplet p-like superconductivity is not realized at all. In the weak-coupling limit ($k_B T_c^s \cdot N(E_F) \ll 1$) the anisotropic singlet s-like superconductivity may occur only if the inequality $(K/z) \cdot k_B T_K \cdot N^2(E_F) \gtrsim 1.4/z$. The $d\gamma$-like singlet superconductivity has its transition temperature given by

$$T_c^{d\gamma} = 1{,}14 \cdot T_k \exp\left(-33 \, k_B T_k / K\right)$$

Numerical estimations concerning the inequality for singlet s-like superconductivity and for T_c^d are given in the Table one at the end of this paper. Our formula for the critical temperature quite well describes the situation in given systems.

MAGNETIC FIELD INDUCED SUPERCONDUCTIVITY IN THE HEAVY FERMION SYSTEMS

Recently magnetic-field-induced superconductivity in the heavy fermion system $CePb_3$ was observed probably, /8/. The Jaccarino-Peter mechanism[3]/9/ was proposed in /8/ to explain

this phenomenon. We have suggested /10/ recently to employ the
Landau-Ginzburg phenomenological approach developed for systems
with strong spin-orbit coupling. The starting point in our pape
was the free energy density expansion

$$\Delta \tilde{f} = \alpha \, |\vec{\eta}|^2 + \beta_1 \, |\vec{\eta}|^4 + \beta_2 \, |\vec{\eta}\cdot\vec{\eta}|^2 + \beta_3 \, (|\eta_x|^4 + |\eta_y|^4 + |\eta_z|^4) + \tag{5}$$

$$+ \frac{\hbar^2}{2m_1} \, \partial_i^* \eta_j^* \, \partial_i \eta_j + \frac{\hbar^2}{4m_2} \, (\partial_i^* \eta_i^* \, \partial_j \eta_j + \partial_i^* \eta_j^* \, \partial_j \eta_i) +$$

$$+ \frac{\hbar^2}{2m_3} \, (\partial_i^* \eta_i^* \, \partial_i \eta_i) - i\frac{\gamma}{2} \, (\vec{\eta}\times\vec{\eta}^*)\cdot\vec{B} + \frac{1}{8\pi}(\vec{B}-\vec{H})^2 ,$$

where $\partial_j \equiv \dfrac{\partial}{\partial x_j} - \dfrac{2ei}{c} A_j$; $\quad i,j = 1,2,3$; $\quad \vec{\eta} = (\eta_x, \eta_y, \eta_z)$; $\quad \alpha = \alpha_0 \, (T-T_0)>0$

$\gamma>0$, $\vec{B} = \mathrm{rot}\,\vec{A}$.

The three-dimensional order parameter $\vec{\eta}$ is assumed to describe
the ferromagnetic superconducting heavy fermion state with the
magnetization proportional to the vector product $\vec{\eta}\times\vec{\eta}^*$.
Assuming the external field to be oriented along the z-axis,
$\vec{H}$ = (0,0,H), and using the Ansatz $\vec{\eta}$ = $\eta(x)\cdot(1,i,0)$ for modul
tion only in the x-direction we find from (5)

$$\Delta \tilde{f} = \frac{\hbar^2}{4m} \left[\left(\frac{d\eta}{dx}\right)^2 + \frac{4e^2 A^2}{\hbar^2 c^2} \right] + 2\alpha \, \eta^2 + \beta \, \eta^4 - \gamma \eta^2 \left(\frac{dA}{dx}\right) + \tag{6}$$

$$+ \frac{1}{8\pi} \left(\frac{dA}{dx} - H\right)^2 ,$$

where $\dfrac{1}{m} \equiv \dfrac{4}{m_1} + \dfrac{2}{m_2} + \dfrac{2}{m_3}$; $\quad \beta \equiv 4\beta_1 + 2\beta_3 >$; $\eta \equiv \eta^{(x)}$.

Here the vector potential is in the form $\vec{A}$ = (0,A(x),0). In /10
we have solved the Schrödinger-like equation in the linear
approximation, derived from (6) by the same procedure as in the
conventional Ginzburg-Landau case /11,12/, and found descriptio
of the normal-state instability at the critical field. Here we
present our results beyond the linear approximation by using th
variational approach to the equation (6). The total free energy
$\Delta \tilde{F}$ per the sample area S of the crossection perpendicular to
x-direction is found from (6). In the limit where the magnetic
field totaly penetrates the superconducting region (e.i. A(x)=
= H.x):

$$\frac{\Delta \tilde{F}}{S.\sqrt{\pi}} = \frac{\hbar^2}{4m} \left(a^2 /2\xi\right) + \frac{e^2 H^2}{2mc^2} \, a^2 \xi^2 + 2\alpha a^2 \xi + (a^4 \beta \xi/\sqrt{2}) - \gamma H a^2 \xi , \tag{7}$$

assuming the variational form of the order parameter in (6)
is given by $\eta(x)$ = a. exp($-x^2/2.\xi^2$) . Here the amplitude a
and the correlation length ξ are variational parameters. From
equations for the extrema of (7) we have found that there
exists a critical field

$$H_c^* = \left(2.\alpha / (\gamma - (\hbar|e|/2mc))\right)$$

at which a second order transition from the normal state a=0
($H \leq H_c^*$) to the stable inhomogeneous magnetic-field-induced
state with a≠0 (H > H_c^*) occurs. This state is a single vortex
state. Note that the temperature derivative (dH_c^*/dT) is positiv
as it is observed in the experiments. The amplitude a in the
superconducting phase near the phase boundary increases as

$$a^2 \approx \frac{\gamma}{\beta\sqrt{2}}\left(1 - \frac{|e|\cdot\hbar}{2\gamma mc}\right)(H - H_c^*).$$

The free energy (of the single vortex) with respect to the normal state is found

$$\Delta\tilde{F} = -S\cdot\Delta_0\left(\frac{H}{H_c^*} - 1\right)^2, \qquad H \gtrsim H_c^*$$

$$\Delta_0 \equiv \frac{\hbar\cdot c\cdot\gamma\cdot\alpha\cdot\sqrt{2\pi}}{\beta\cdot|e|}\cdot\left(1 - \frac{|e|\cdot\hbar}{2\gamma\cdot mc}\right) > 0.$$

The magnetic moment M_z changes at the transition by amount

$$\Delta M_z = 2\cdot S\cdot\Delta_0\,\frac{H - H_c^*}{H_c^{*2}} \qquad H \gtrsim H_c^*.$$

The susceptibility per volume V changes as

$$\Delta\chi = \frac{2\cdot\Delta_0 S}{V\cdot H_c^*} > 0 \qquad H \gtrsim H_c^*.$$

The change of the heat capacity at constant volume V per unit volume is given as

$$\Delta C_V = \Delta\chi\cdot\left(\frac{dH_c^*}{dT}\right)^2\cdot T\Big|_{H = H_c^*}.$$

We see that the ferromagnetic contribution to the Landau –
–Ginzburg free energy drives the phase transition. In opposite
to the usual (BCS) superconductors the magnetic susceptibility
increases at the transition to the superconducting phase. The
slope and the intercept of the magnetization changes here.
The variational approach, results of which are presented here
for the case of a single one-dimensional vortex , does not
suffer the stability problem which may arise at the approach
using Schrödinger-like equation. Moreover we have found description of the transition for the field values changing in the
neighborhood of the value H_c^*. Note that three-dimensional results for the lattice of vortices are necessary to describe
real situation better. The role of the term linear in the magnetic field in (5) is similar to that in the superfluid
^{3}He case /13/.

The $CePb_3$ material crystallizes in a simple cubic symmetry
class. From the general considerations made by Volovik and
Gorkov /1/ it then follows that a superconducting state existing in a nonzero magnetic field should be identified with one
of the eigth possible states in which the time reversion symmetry is broken and in which net magnetization is nonzero. There
are two three-dimensional irreducible representations of the
group $O{\times}R{\times}U(1)$ which correspond to a ferromagnetic superconductive
state. The crystal symmetry considerations lead to two other
possibilities, either the field-induced magnetization vector is
oriented in the (1,1,1) direction of the crystal (and equivalents)
or in the (1,0,0) direction and its equivalents. The superconductive phase may be of even or odd parity. Degeneracy of the
ground state, the specific heat behaviour at low temperatures
and strong enough fields and the reduction of symmetry should
determine which type of the above-mentioned superconducting
ferromagnetic states occurs in $CePb_3$. The phase transition from
the normal phase (Kondo liquid) to the superconductive phase
occurs in our model through creation of vortex lattice with
single vortices similar to one dimensional described above.
Note that the free energy (5) has the same form for these states.

Table 1

Numerical estimates of $T_c^{d\gamma}$ and $(K/z)kT_K\,N(E_F)^2 \gtrsim 1{,}4/z$.

	T_c^{exp} [K]	$T_c^{d\gamma}$ [K] a.,c.,	$|\Theta|/T_K$ e., exp.;a.,c.,d.	T_m [K] exp.;	$-\Theta$ [K] exp.; b.,
$Ce\,Cu_2\,Si_2$	0,5	0,406	7,5 h.)	20	~150
$U\,Be_{13}$	0,9	0,006 f.)	1,2 ⊖	42	53
$U_2\,Zn_{17}$	0	1,712 g.)	13,9 ⊖	18	250
$U\,Cd_{11}$	0	$1{,}5 \cdot 10^{-3}$	0,23 ⊖	100	23
$Ce\,Cu_6$	0	0,028 ÷ ÷0,0065	3,2 ÷ ÷1,36 ⊖	14 ÷ ÷ 33	45
$Ce\,Al_3$	0	0,0061 ÷ ÷0,0039	1,3 ÷ ÷ 1,11 ⊖	46 ÷ ÷39	35

Remarks:

a., T_K is estimated using Lavagna et al./14/ formula for the temperature of resistivity maximum in Kondo lattices (T_m) and assuming the the 1st band filling n=0,99 , $T_K=10T_m$ T_m is found from /15/

b.,   K is identified from the high-temperature susceptibility temperature Θ given in /15/, $K \cong k_B|\Theta|$

c.,   the number of neighbours is estimated by z≈10, estimated inequality is rewriten into the form $|\Theta|/T_K \gtreqless 70/z$

d.,   the density of states as given by Miyake et. al. /6/ is renormalized to $N(E_F) = N^{s.c.}(E_F) \cdot (t/k_B T_K) = 0{,}14/k_B T_K$, here t is the 1st bandwidth

e.,   ⊖ means that the singlet s-state superconductivity is not allowed , ⊕ allowed

f.,   inappropriate band filling n choosen ?

g.,   the antiferromagnetic phase at 9,7K occurs and superconductivity is not realized

h.,   value on the boundary, for z=6-8 the inequality holds and the singlet s-like superconductivity may also occur

REFERENCES

/1/   P.W.Anderson, Phys.Rev $\underline{B30}$ 1549 (1984)
        G.E.Volovik, L.P.Gor´kov, JETP $\underline{88}$ 1412 (1985)
        P.A.Lee, T.M.Rice, J.W.Serene, L.J.Sham, J.W.Wilkins,
         Comm.Cond.Matt.Phys. $\underline{12}$ 99 (1986)
        P.Fulde,J.Keller, G.Zwicknagl, "Theory of Heavy-Fermion
         Systems" 1987 to appear in "Solid State Physics"

/2/   O.Hudák, JETP Lett $\underline{42}$ 300 (1985)

/3/   J.R.Schrieffer, P.A.Wolff, Phys.Rev. $\underline{19}$ 491 (1966)

/4/   R.Jullien, J.N.Fields, S.Doniach, Phys.Rev $\underline{B16}$ 4889 (1977)

/5/   O.Hudák, 1985, unpublished

/6/   K.Miyake, T.Matsuura, H.Jichu, Y.Nagaoka, Progr. Theor.
         Phys. $\underline{72}$ 1063 (1984)

/7/   K.Miyake, S.Schmitt-Rink, C.M.Varma, Phys.Rev $\underline{B34}$ 6554
         (1986)
        M.Cyrot, Sol.St.Comm. $\underline{60}$ 253 (1986)
        B.Oles, A.M.Oleś, JMMM $\underline{54-57}$ 413 (1986)

/8/   C.L.Lin, J.Teter, J.E.Crow, T.Mihalisin, J.Brooks, A.I.
         Abou-Aly, G.R.Stewart, Phys.Rev.Lett. $\underline{54}$ 2541 (1985)

/9/   V.Jaccarino, M.Peter, Phys.Rev.Lett. $\underline{9}$ 290 (1962)

/10/   O.Hudák, Phys.Lett. $\underline{A119}$ 89 (1986)

/11/   E.M.Lifshitz, L.P.Pitajevskij, "Statistical Physics",
         Part 2, Vol. IX of Theoretical Physics Course,Nauka,
         Moscow 1978 §47

/12/   P.G.de Gennes, "Superconductivity of Metals and Alloys"
         Benjamin, New York, 1966

/13/   V.Ambegaokar, N.D.Mermin, Phys.Rev,Lett. $\underline{30}$ 1981 (1973)

/14/ M.Lavagna,C.Lacroix,M.Cyrot, J.Appl.Phys. $\underline{53}$ 2055 (1982)

/15/ G.R.Stewart, Rev.Mod.Phys. $\underline{56}$ 755 (1984)

HEAVY FERMION PROPERTIES AND THEIR RELATION

TO ELECTRONIC BAND THEORY

Warren E. Pickett

Naval Research Laboratory
Washington, DC 20375-5000

INTRODUCTION

Materials of the "heavy fermion" (HF) class of metals are characterized by extremely large values at low temperature of the linear specific heat coefficient $\gamma(T) \equiv C/T$ and their magnetic susceptibility $\chi(T)$. These highly enhanced thermodynamic properties are interpreted as due to fermionic quasiparticles bearing a very large effective mass m^*. Neither γ nor χ remain constant as the temperature is raised, rather they display unusual temperature dependences before rapidly decreasing to normal values at high temperature. In addition the resistivity, which is large at room temperature, does not begin to decrease monotonically except at low temperatures of the order of 2-20 K. The crossover crudely defines a "coherence temperature" T_0, below which the excitations begin to look like fermionic quasiparticles. For $T \sim T_0$ more complex behavior occurs, indicative of strong many-body interactions which destroy the quasiparticle character of the excitations.

The situation is illustrated in Fig. 1, which shows $\gamma(T)$ for several HF compounds. For several compounds, including UBe_{13}, $CeCu_2Si_2$ and $CeAl_3$, $\gamma(T)$ decreases rapidly (perhaps after an initial rise) on the scale of $T_0 \sim$ 2-5 K. These materials are among the most highly enhanced but do not magnetically order. In UCd_{11} and U_2Z_{17} γ is relatively constant above magnetic transitions at 5 K and 9.7 K respectively. In $NpBe_{13}$ γ increases by an order of magnitude below 12 K before encountering a magnetic transition at 3.4 K, (with similar behavior displayed by UCd_{11}), while in U_2Zn_{17} γ is relatively constant above the magnetic transition and jumps by a factor of two at the transition. Yet another behavior is shown by UPt_3, which is non-magnetic and displays a weak T-dependence which can be fit in the range 0.5-17 K by a $T^2 \log T$ term suggestive of spin fluctuations. These data, which are taken from the review of Stewart[1] and are discussed in more detail there, indicate the wide range of behavior which falls under the "heavy fermion" umbrella.

All known HF compounds contain at least one atom type with a partially filled f shell (e.g. Ce, Yb, U, Nb, Pu) and at least one atom type with no partially filled f shell. Considering the known "narrow band" character of f electrons, this observation suggests that the anomalous behavior arises from the interaction/hybridization of highly correlated f electrons with itinerant conduction electrons. More evidence for this interpretation arises from other data. At high temperature the susceptibilities follow a Curie-Weiss-like behavior with characteristic effective moments (2.5 - 2.7 μ_B for Ce, 2.6-3.5 μ_B for U) before becoming Fermi-liquid-like (constant) at $T \ll T_0$. Resistivities also reflect strong interactions, being very large at room temperature

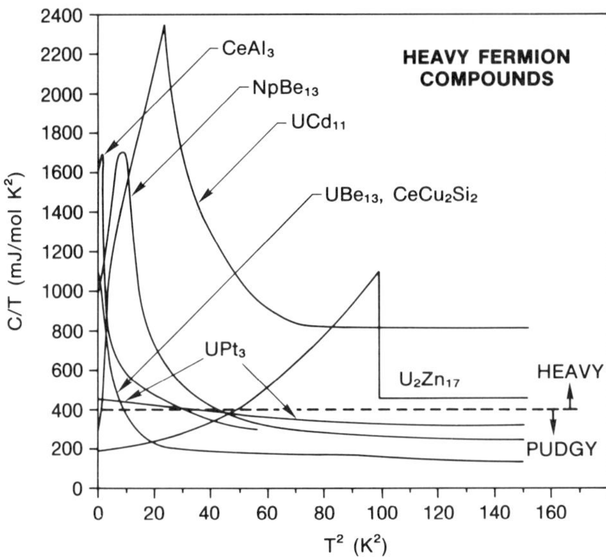

Figure 1 — Behavior of C/T versus T for a number of heavy fermion compounds, illustrating the diversity of behavior which occurs.

and often well below, and commonly (but not universally) showing negative magnetoresistance indicative of spin disorder scattering.

Early reviews of HF properties have been given by Stewart[1] and by Brandt and Moshchalkov,[2] and the work done since these reviews is too extensive and varied to survey here. Since the interest in this volume is centered on superconductivity, and moreover because the discovery of superconductivity was so unexpected and stimulated much interest, we close this Introduction with a synopsis of the history of this area. In 1979 Steglich et $al.$[3] discovered that $CeCu_2Si_2$ became superconducting at $T_c = 0.5$ K, followed by discoveries of superconductivity in UBe_{13} ($T_c = 0.85$ K) and UPt_3 ($T_c = 0.5$ K) by Ott et $al.$[4] and Stewart et $al.$[5], respectively. In metals with such strong many-body interactions the discovery of superconductivity, even at very low temperatures, rapidly led to speculations that a new mechanism was responsible for superconductivity. The discussion of the interactions which may be responsible for the mass enhancement and for the superconductivity remains an active field and has been reviewed by Lee et $al.$[6] and by Fulde et $al.$[7]

The purpose of the present article is to discuss some aspects of modern band theory treatments of HF compounds and identify what they do, and do not, tell us about heavy fermion behavior. To this end it will be useful to compare to and contrast with the other main theoretical approach, that of parameterized model Hamiltonians. These two approaches are essentially distinct and complementary, and one of the recent trends has been to form a combined, or hybrid, attack by incorporating some features of both.

THEORETICAL APPROACHES

Band Theory

One way to view the band theory approach is to write the many-body Hamiltonian for N electrons in the form

$$H = H_0 + \Delta H = \sum_{i=1}^{N} h_i + \Delta H , \tag{1}$$

where H_0 is a sum of one-body operators

$$h_i = -\nabla_i^2 + \sum_l V_{\text{ext}}(r_i - R_l) + V_{\text{hxc}}(r_i) \tag{2}$$

and ΔH contains the many-body Coulomb interactions

$$\Delta H = \frac{1}{2} \sum_{i \neq j}^{N} \frac{e^2}{|r_i - r_j|} - \sum_{i=1}^{N} V_{\text{hxc}}(r_i).$$

Here a mean field V_{hxc} ("Hartree + exchange + correlation") has been added to the "external" potential V_{ext} (due to the nuclei) and subtracted from the many-body term. Formally it is quite arbitrary what one chooses for V_{hxc}, but since an exact solution is not possible and the effects of ΔH are known to be important, it is clearly useful to make an optimal choice of V_{hxc}. This choice can be guided by the desire, say, to make ΔH as small as possible in its effects, or perhaps to make it easy to approximate.

In practice, however, another choice is made for V_{hxc} which is suggested by Density Functional Theory[8] (DFT). This choice, for which V_{hxc} is a functional of the ground state density $n(r)$ alone, leads to a system of non-interacting electrons with the same density as the interacting system. The result is a *parameter-free, self-consistent band theory* for non-interacting electrons in the effective potential $V_{\text{ext}} + V_{\text{hxc}}$, with eigenvalues ϵ_k and wavefunctions ψ_k. This system defines a non-interacting Green's function $G_0(k,\omega) = (\omega - \epsilon_k)^{-1}$, in terms of which the full single-particle Green's function G, which includes the many body effects of ΔH, can be written

$$G^{-1}(k,\omega) = G_0^{-1}(k,\omega) - \Sigma(k,\omega), \tag{4}$$

defining the self-energy Σ. (In these equations ω is understood to include a small imaginary part.) A point of central concern here is the importance and character of Σ when the density functional choice of bands is made. Another important feature of this choice which will not be discussed further is that it can be used to evaluate an *energy functional* which determines the crystal structure, equation of state, phonon frequencies and the like. These properties are characteristic of the electronic *ground state* of HF materials and, with the possible exception of the superconducting state, the ground state properties are not unusual. Rather it is the single particle excitations and their interactions which appear exceptional.

On both theoretical grounds[6,7] and from considerations of transport properties[9] (which are *not* enhanced), it now seems that the energy dependence of Σ may dominate its wavevector dependence in HF metals. When that is the case one may expand the self-energy for low energy to obtain

$$G(k,\omega) = \frac{z_k}{\omega - \tilde{\epsilon}_k} + G_{\text{incoh}} \tag{5}$$

where $z_k = [1 - (d\Sigma/d\omega)_{\omega=0}]^{-1}$ is the wavefunction renormalization factor, and it is also the inverse of the mass enhancement factor in view of the relation

$$\tilde{\epsilon}_k = z_k \left[\epsilon_k + \Sigma(k,0) \right]. \tag{6}$$

From this renormalized spectrum the Fermi surface is determined from the condition $\tilde{\epsilon}_k = \mu$, where μ is the chemical potential. Two important conclusions follow from Eq. (6): first, a purely energy-dependent renormalization does not affect the Fermi surface, and second, it is possible to choose an unenhanced mean field spectrum $\epsilon_k + \Sigma(k,0)$ which describes the Fermi surface of the strongly interacting system. This spectrum can be obtained from an appropriate choice of V_{hxc}, and it now seems clear that the DFT prescription is just the one which gives the Fermi surface (or a very good approximation to it).

It is now well established that the density functional prescription [within the local density approximation (LDA) which is almost universally adopted] gives accurate Fermi surfaces for non-f-band metals. There is now ample evidence that the LDA treatment also gives accurate Fermi surfaces in f electron metals, and even in HF systems. The first strong indication of this came with Koelling and Norman's finding[10,11] that LDA provides an excellent description of the Fermi surface of $CeSn_3$, a mixed valent compound, although earlier there had been successful Fermi surface studies on more nearly normal f electron metals, principally by Koelling and coworkers. A more striking example is that of UPt_3, whose Fermi surface has recently been measured for fields oriented in the basal plane (from $\hat{a}$ to $\hat{b}$) as well as out of the plane ($\hat{b}$ to $\hat{c}$) of the hexagonal lattice, by Taillefer $et\ al.$[12] Wang $et\ al.$[13] have demonstrated that, in spite of the exceedingly complex nature of the bands and Fermi surface, as shown in Figs. 2 and 3, and with as many as eight orbits for a given field direction, two independent LDA calculations provide a very impressive account of the Fermi surface geometry. The implication clearly is that modern LDA band theory provides the spectrum which determines the Fermi surface, that is, $\Sigma(k,0)$ is negligible when the LDA mean field prescription is used.

LDA band theory does not account for the mass enhancement z_k^{-1}. In $CeSn_3$ the measured masses are about five times larger[10,11] than band masses, while in UPt_3 the corresponding mass enhancements are[13] 12-23, with an average[13-16] near 20. Even larger enhancements over the band mass (roughly 90) are predicted[17] for UBe_{13}. These enhancements are described by the self-energy, but below we will discuss some attempts which have been made to describe them by a mean field "renormalized band theory."

Let us enumerate what features of electron-electron interactions band theory does and does not account for. Band theory includes (1) full crystal structure effects, such as anisotropy and appropriate symmetries of band states, (2) orbital and spin degeneracy of the f states, (3) dynamics of the conduction electrons (which are not strongly correlated), (4) hybridization of f states with conduction electrons, (5) spin-orbit coupling, and (6) Luttinger's theorem constraints, which dictate that the Fermi surface contains exactly N states. It is not obvious that any theory which slights any of these aspects can lead to a quantitative description of HF materials. LDA band theory does not describe the dynamics of highly correlated electrons; indeed, it seems clear that such effects may only be described properly by the self-energy. It also does not account for the scattering of conduction electrons at the Fermi surface from a localized spin, (the Kondo effect) or, what is more pertinent here, from a lattice of localized spins.

Model Hamiltonians

The accepted way to try to account for large dynamic many-body effects which dominate the physical properties of mixed valent and HF systems is by studying model Hamiltonians of the Kondo lattice or Anderson lattice type. Since the Anderson lattice has attracted the most attention in the context of HF behavior, we confine these short comments to it. By expanding the electronic states in Bloch functions for the conduction electrons and in localized orbitals for the f electrons one can obtain, by making a series of simplifying approximations, the model Hamiltonian.

$$H^{AL} = \sum_{ks} [\epsilon_k^o c_{ks}^+ c_{ks} + \epsilon_f f_{ks}^+ f_{ks} + (V_k c_{ks}^+ f_{ks} + h.c.)] + U\sum_i f_{i\uparrow}^+ f_{i\uparrow} f_{i\downarrow}^+ f_{i\downarrow}, \qquad (7)$$

where ϵ_k^o is the unhybridized conduction band, ϵ_f is the f orbital energy, V_k is a hybridization matrix element and U is the f state correlation energy which measures the energy cost of occupying a site with two f electrons. Although a full solution to even this limited problem does not appear imminent, several important properties have been inferred as discussed in reviews elsewhere.[6,7] The present discussion will be limited to its relationship to, and implications for, band theory.

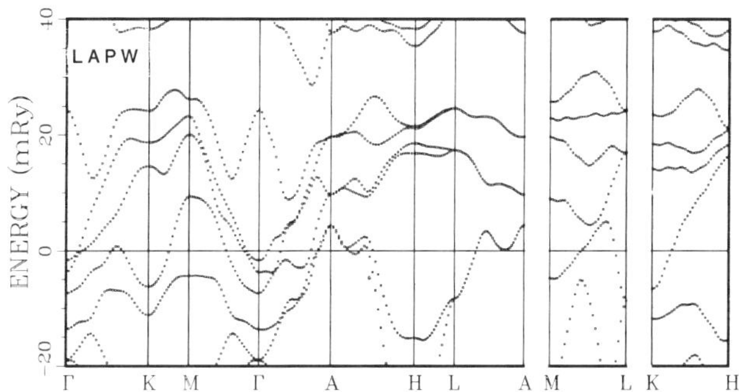

Figure 2 — LAPW band structure of UPt$_3$ near E_F, from ref. 13. The large number of crossings at E_F lead to the complex Fermi surface shown in Figure 3.

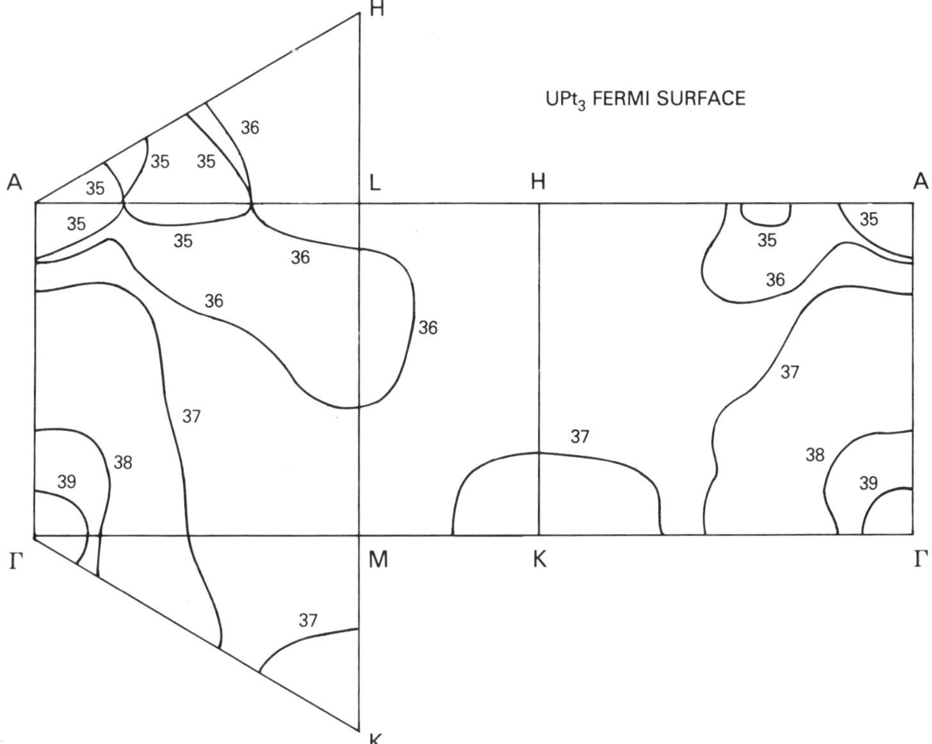

Figure 3 — Fermi surface cross sections of UPt$_3$ in the high symmetry planes. Simple closed surfaces are centered at the Γ, K and A points, with more complex surfaces occurring near the A-H-L plane.

Rather than reconstruct the steps necessary to obtain H^{AL} from H, we present the method proposed by Monnier, Degiorgi and Koelling[18] (MDK) for obtaining the Anderson lattice parameters ϵ_k^0, ϵ_f, V_k and U, which are not directly measurable quantities: (1) calculate the all-electron (conduction +f) LDA band structure, (2) fit this band structure to a parameterized LCAO Hamiltonian, (3) eliminate the f states from the LCAO basis and re-diagonalize, which leads to "conduction states" with spectrum ϵ_k^o which include all effects due to the f electrons (such as the direct Coulomb term) *except for hybridization*, (4) obtain the hybridization terms V_k by taking matrix elements between "conduction" and f states, and (5) apply the

Gunnarsson-Schönhammer[19] (GS) theory (as the best model in existence) to identify the remaining parameters by comparing with experimental spectroscopic data. The GS theory treats a single Anderson impurity to first order in N_f^{-1}, where N_f is the f level degeneracy which is usually extracted from crystal field arguments. From the GS theory ϵ_f and U can be obtained. In addition, U can be calculated within LDA theory by increasing or decreasing the f occupancy on a single site and comparing total energies.[20,21]

The DMK procedure for determining Anderson lattice parameters, or some variation on it, is essential if serious comparison of the Anderson model with experiment is contemplated. DMK carried through the procedure for YbP, with encouraging results. By tying together the band theory and model Hamiltonian approaches, however tenuously, their procedure may serve to illuminate the peculiarities or biases of each approach. The band picture, for example, with its use of Block functions may tend to overemphasize the delocalization of the heavy electron bands, while the Anderson lattice with its strictly localized basis for representing f electrons may tend to underemphasize banding.

Hybrid Approaches

The DMK method of using LDA band theory to identify the Anderson lattice parameters is a novel example of a *hybrid approach* to f electron behavior, which may be defined as any procedure which tries to incorporate the desirable features of the band and model Hamiltonian theories. As mentioned above, LDA theory gives a good description of the Fermi surface. The many-body term in Eq. (7) must not change the Fermi surface, which is to say that the self-energy should be primarily energy (rather than wavevector) dependent. This behavior has been suggested on theoretical grounds by Yoshimori and Kasai[22] and others, and is pictured schematically in Fig. 4.

One approach is to build in, or add, many-body-like effects in a band structure calculation. d'Ambrumenil and Fulde[23] carried out a hybrid calculation by first obtaining a Korringa-Kohn-Rostoker (KKR) band structure of $CeCu_2Si_2$, except that the Ce f electron was treated as a core (localized) state. Then they replaced the KKR LDA phase shifts on the lattice of Ce atoms by a Kondo-like phase shift which varies with energy on the scale of T_0 near the chemical potential μ. This procedure has the effect of concentrating the (hybridized) f bands within T_0 of μ and producing a very large value of the density of states $N(\mu)$ comparable to that seen in specific heat measurements.

A related approach has been tried by Strange and Newns,[24] who suggested shifting the f band position and renormalizing the bandwidth as suggested by Fermi liquid theories. Norman and Koelling[11] find, however, that their application of this approach to $CeSn_3$ degrades the already excellent agreement of the Fermi surface compared to de Haas-van Alphen data. The fact is that it is found now in several f electrons systems that the LDA bands give very good Fermi surfaces, so good that any hybrid method which produces any sizable alteration of the Fermi surface, as the Strange-Newns approach will do in general, will worsen the predicted Fermi surface. Again the point is that what LDA is missing is the *dynamic correlation effects*, which were never expected to be described within LDA.

A somewhat different objective motivated the work of Cooper and coworkers.[25] Similar to the approach of DMK, they begin with a self-consistent LDA calculation. From the resulting bands they obtain parameters which are used in a Schrieffer-Wolf transformation of H^{AL}. They have applied this procedure with considerable success to describe the complex magnetic phases of CeSb and other Ce and Pu pnictides.

Band Theory and the Mechanism for Superconductivity

Much of the interest in HF materials grew out of the discovery of superconductivity in several of the compounds. Since superconductivity involves pairing of electrons at the Fermi surface, and band theory appears to describe both the Fermi surface and the character of the

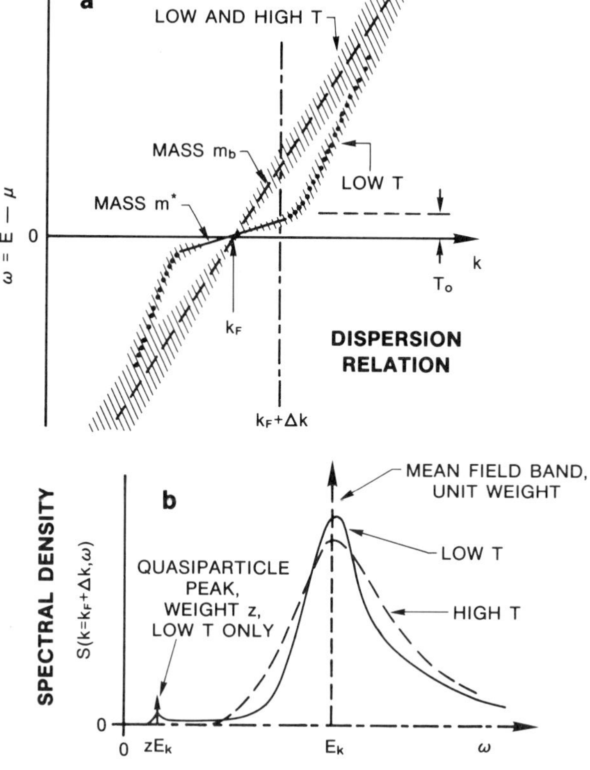

Figure 4 — Schematic representation of quasiparticle features in a heavy fermion material. (a) The dispersion curve ϵ_k of mass m_b from band theory (straight dashed line) and the dispersion relation of mass m* (solid line) which includes dynamic self-energy corrections. The mass enhancement vanishes at high temperature (T) or at high excitation energy $\omega > T_0$. (b) Plot of the associated spectral density near the Fermi surface, along the vertical chain-dashed curve at $k_F + \Delta k$ in (a). The quasiparticle, at energy zE_k relative to the band energy E_k, carries weight $z << 1$, but dominates the thermodynamic behavior below T_0.

quasiparticle wavefunctions, it is instructive to ask what band theory can say *quantitatively* about possible pairing mechanisms.

Perhaps the initial question to ask is whether the *conventional* electron-phonon interaction (EPI) can account for $T_c \sim 0.5 - 1$ K in HF systems. By "conventional" we mean that the screening of the ionic motion is accomplished locally and can be described by standard static dielectric screening theory, in which case it is reasonable to use a rigid-ion type of model[26] to describe the EPI. Jarlborg, Braun and Peter[27] were the first to test this approach, obtaining a coupling strength $\lambda < 0.1$ for $CeCu_2Si_2$ using the Linearized Muffin-Tin Orbital (LMTO) method. Using the McMillan equation[28] with $\mu^* = 0.1$, $\lambda \geq 0.3$ is necessary before a measurable T_c will result. Moreover, the large interactions in HF compounds that are reflected in the large mass enhancements may lead to rather larger values of μ^*, and Jarlborg *et al.* concluded that the EPI was not a viable coupling mechanism.

Razafimandimby, Fulde and Keller (RFK)[29], noting that the $CeCu_2Si_2$ displays Kondo lattice behavior, made an estimate of the strength of λ from the "Kondo volume collapse" mechanism. The possibility of such a mechanism was seen by Allen and Martin,[30] who noted that in a Kondo system the scattering of electrons at E_F is dominated by the Kondo mechanism. Since the Kondo scale T_K has a very strong volume dependence for *homogeneous* volume changes, it was reasoned that the local compression and dilatation due to long wavelength phonons could lead to strong EPI. (By the definition given above this is an *unconventional* EP coupling.) Since no solution of the Kondo lattice problem exists, RFK used a single site approximation for the f phase shift and used experimental data for the Kondo Gruneisen parameter, and found that one could obtain $T_c \sim 1$ K with this procedure.

Pickett, Krakauer and Wang,[17] applying the results of LAPW calculations, evaluated the rigid-ion expression[26] to find $\lambda = 0.035$ in UBe_{13}, much too small to produce $T_c = 0.85$ K even if one assumes the most optimistic limit $\mu^* = 0$. Applying the same approach to UPt_3, Wang, Krakauer and Pickett[15] obtained $\lambda = 0.28$. This value is in the very sensitive range where small changes in the assumed values of μ^* (and λ) can result in calculated T_c's in the range 10^{-5} K to 1 K. They noted, however, that Eliashberg theory describes coupling between phonons and *Coulomb quasiparticles*, so the wavefunction renormalization constant (the reciprocal of the mass enhancement, which is 20 in UPt_3) must be taken into account. When this is done for UPt_3, for which T_0 is the same order of magnitude as the Debye temperature, Θ_D, the effective EP coupling strength is reduced by a factor of roughly $m_{band}/m^* = 1/20$. Similar conclusions have been reached by Fenton[31] in studies of the Eliashberg equations, who concluded that in a HF material with $m^* = 20\, m_{band}$ it would require a value of $\lambda \sim 1.5$ (i.e. similar to that of Nb_3Sn with $T_c \sim 20K$) to produce a $T_c \sim 1$ K.

Oguchi, Freeman and Crabtree[32] however argue that their calculated value of $\lambda = 0.33$ for UPt_3 using the LMTO method (which is consistent with the value of Wang *et al.*[15] given above, considering the different methods used) can be made to provide a consistent picture of $T_c \sim 0.5$ K in this material. Their arguments rely on a highly anisotropic (but s-wave) gap function whose large degree of anisotropy arises from the *small value* of λ. This interpretation must be considered tentative since it uses relations derived in the weak anisotropy limit and neglects the renormalization (discussed above) which affects the solution of the Eliashberg equations.

Although many of these studies have attempted to account for the effects of the HF behavior on the electronic spectrum, none directly addresses the possibly large effects on the coupling *interaction* between quasiparticles. There is now direct evidence from neutron scattering studies of magnetic excitations ("spin fluctuations") in many HF materials[33] including the superconducting members, and these have been investigated theoretically primarily phenomenologically with model Hamiltonian treatments. Very recently Norman[34] has attempted to account for the effects of the observed magnetic behavior on the normal state self-energy, which gives the mass enhancement, as well as on the superconducting state self-energy which leads to the Eliashberg gap equations for T_c. In his approach he combines (1) the Berk-Schrieffer expression for the spin fluctuation coupling to electrons, (2) experimental data for the susceptibility $\chi(q, \omega)$, (thus skirting the most difficult many-body problems), and (3) a two-band model of the Fermi surface which was motivated by LDA calculations. By identifying one parameter from experiment—essentially the Stoner enhancement—he obtains polar ("p-like") solutions to the gap equation and transition temperatures around 0.2-0.4 K, which is the correct order of magnitude. Since he finds the results to be somewhat sensitive to particularities of the two-band model, it is desirable to generalize the study to include the full (very complicated) Fermi surface, but his approach appears to be a fairly realistic procedure for adding self-energy corrections to LDA band structure results and investigating both the normal and superconducting states.

SUMMARY

Modern band theory gives a single particle spectrum and corresponding wavefunctions which are approximations to the quasiparticle excitations. In heavy fermion metals at low temperature it is evident that there are large many-body mass enhancements over and above the band mass which are properly described by the single particle self-energy. Although experimental data on the Fermi surfaces of these systems is still sparse, the implications of studies of UPt_3, on the mixed valent compound $CeSn_3$ and on several other f electron metals indicate that LDA band structures predict Fermi surfaces quite accurately.

There has not been space to review adequately other successes of band theory. For example, Marksteiner *et al*[35] and Wang *et al*[15] have shown that the valence band x-ray photoemission spectra (XPS) of UPt_3 are predicted correctly. Pickett *et al*[17] have found that in UBe_{13} both the XPS and the corresponding unoccupied states measured by inverse photoemission are well described by the LDA bands, although the strong scattering may distribute f-electron spectral density to higher excitation energy than given by the LDA bands. Arko[36] has used angle-resolved photoemission spectroscopy to map a few of the occupied bands within 1.5 eV of ϵ_F, and these appear to be in agreement with the theoretical results as well.

Existing band structure results have very little to say about the unusual T-dependent properties, and are just beginning to be applied in detailed solutions of the Eliashbeng gap equation. Most studies indicate that the *conventional* EPI is much too weak to account for the superconductivity observed in $CeCu_2Si_2$, UBe_{13} and UPt_3.

The encouraging trend which is emerging is an interest in obtaining a more detailed picture of the microscopic processes which lead to, or result from, strong correlations in periodic systems. Band theory provides the foundation upon which more complete pictures can be built and tested. The examples which have been reviewed here represent only the initial stage in a new direction of theoretical study into the origins of heavy fermion behavior.

ACKNOWLEDGMENTS

I have benefitted from discussions with a number of people, especially D. D. Koelling, H. Krakauer, M. R. Norman, and C. S. Wang. This work has been supported partially by the Office of Naval Research Contract No. N00014-84-WR-24055.

REFERENCES

1. G.R. Stewart, Rev. Mod. Phys. **56**, 755 (1984).
2. N.B. Brandt and V.V. Moshchalkov, Adv. Phys. **33**, 373 (1984).
3. F.J. Steglich *et al.*, Phys. Rev. Lett. **43**, 1892 (1979).
4. H.R. Ott, H. Rudigier, Z. Fisk and J.L. Smith, Phys. Rev. Lett. **50**, 1595 (1983).
5. G.R. Stewart, Z. Fisk, J.O. Willis and J.L. Smith, Phys. Rev. Lett. **52**, 679 (1984).
6. P.A. Lee, T.M. Rice, J.W. Serene, L.J. Sham and J.W. Wilkins, Comm. Cond. Matter Phys. **12**, 99 (1986).
7. P. Fulde, J. Keller and G. Zwicknagl, to be published in "Solid State Physics."
8. See, for example, *Density Functional Methods in Physics*, edited by R.M. Dreizler and J. da Providencia (Plenum, New York, 1985).
9. C.M. Varma, Phys. Rev. Lett. **55**, 2723 (1985).
10. D.D. Koelling, Solid State Commun. **43**, 247 (1982).
11. M.R. Norman and Koelling, J. Less-Common Metals **127**, 357 (1986).
12. L. Taillefer, R. Newbury, G.G. Lonzarich, Z. Fisk and J.L. Smith, J. Magn. Magn. Mat. **63&64**, 372 (1987), and unpublished.
13. C.S. Wang *et al.*, Phys. Rev. **B35**, 7260 (1987).
14. T. Oguchi and A.J. Freeman, J. Magn. Magn. Mat. **52**, 176 (1985).
15. C.S. Wang, H. Krakauer and W.E. Pickett, J. Phys. **F16**, L287 (1986).
16. R.C. Albers, A.M. Boring and N.E. Christensen, Phys. Rev. **B33**, 8116 (1986).
17. W.E. Pickett, H. Krakauer and C.S. Wang, Phys. Rev. **B34**, 6546 (1986).
18. R. Monnier, L. Degiorgi and D.D. Koelling, Phys. Rev. Lett. **56**, 2744 (1986).
19. O. Gunnarsson and K. Schönhammer, Phys. Rev. **B28**, 4315 (1983).
20. P.H. Dederichs, S. Blügel, R. Zeller and H. Akai, Phys. Rev. Lett. **53**, 2512 (1984).
21. M.R. Norman *et al.*, Phys. Rev. Lett. **53**, 1673 (1984).
22. A. Yoshimori and H. Kasai, J. Magn. Magn. Mat. **31**, 475 (1983).
23. N. D'Ambrumenil and P. Fulde, J. Magn. Magn. Mat. **47-48**, 1 (1985).
24. P. Strange and D. Newns, J. Phys. **F16**, 335 (1986).
25. J.M. Wills, B.R. Cooper and P. Thayamballi, J. Appl. Phys. **57**, 3185 (1985).
26. G.D. Gaspari and B.L. Gyorffy, Phys. Rev. Lett. **28**, 801 (1972).
27. T. Jarlborg, H.F. Braun and M. Peter, Z. Phys. **B52**, 295 (1983).
28. W.L. McMillan, Phys. Rev. **167**, 331 (1968).

29. H. Razafimandimby, P. Fulde and J. Keller, Z. Phys. **B54**, 111 (1984).
30. J.W. Allen and R.M. Martin, Phys. Rev. Lett. **49**, 1106 (1982).
31. E.W. Fenton, Solid State Commun. **60**, 351 (1986).
32. T. Oguchi, A.J. Freeman and G.W. Crabtree, Phys. Lett. **A117**, 428 (1986).
33. A.I. Goldman, G. Shirane, G. Aeppli, E. Bucher and J. Hufnagl, J. Magn. Magn. Mat. **63 & 64**, 380 (1987).
34. M.R. Norman, private communication.
35. P. Marksteiner, P. Weinberger, R.C. Albers, A.M. Boring and G. Schadler, Phys. Rev. **B34**, 6730 (1986).
36. A.J. Arko, unpublished.

ULTRASONIC INVESTIGATION OF NOVEL SUPERCONDUCTING SYSTEMS[*]

M. Levy, A. Schenstrom, K. J. Sun[†] and B. K. Sarma

Physics Department
University of Wisconsin-Milwaukee
Milwaukee, Wisconsin 53201

ABSTRACT

Ultrasonic attenuation measurements are reported for the superconducting, normal, and mixed states of the heavy Fermion superconductors UPt_3 and URu_2Si_2 and in the alloys of the ternary superconductors $Er_{1-x}Ho_xRh_4B_4$. A peak in attenuation in the mixed state of UPt_3 may be indicative of collective excitations of the flux line lattice. A broad maximum in attenuation is found below the superconducting transition temperature of URu_2Si_2. An increase in attenuation in the superconducting phase of the reentrant ternary superconductors with x = 0.6 and 0.813 is ascribed to increased spin phonon interaction induced by superconducting screening of crystalline electric fields.

I.   INTRODUCTION

Ultrasonic attenuation measurements have been performed in the superconducting, normal and mixed states of the heavy Fermion superconductors UPt_3 and URu_2Si_2 and in the alloys of the ternary superconductors $Er_{1-x}Ho_xRh_4B_4$.

These data will be presented in order to provide some insight into the different interaction mechanisms that can be deduced from ultrasonic attenuation measurements. In none of these systems does the temperature (T) dependence of the attenuation follow the BCS[1] expression $\frac{\alpha_s}{\alpha_n} = \frac{2}{e^{\Delta/kT}+1}$.

Here α_s/α_n is the ratio of the attenuation in the superconducting state to that in the normal state, (Δ is the temperature dependent order parameter and k is the Boltzmann constant.) BCS derived this expression for longitudinal waves in the limit $q\ell > 1$, where q is the sound wave vector and ℓ is the electron mean free path.[1] It has been shown[2] that this expression also holds for longitudinal and transverse waves in the limit $q\ell < 1$. Fig. 1 shows this result for a single crystal sample of V. Figs. 2 and 3 show the temperature dependent data for UPt_3[3] and URu_2Si_2.[4] Fig.

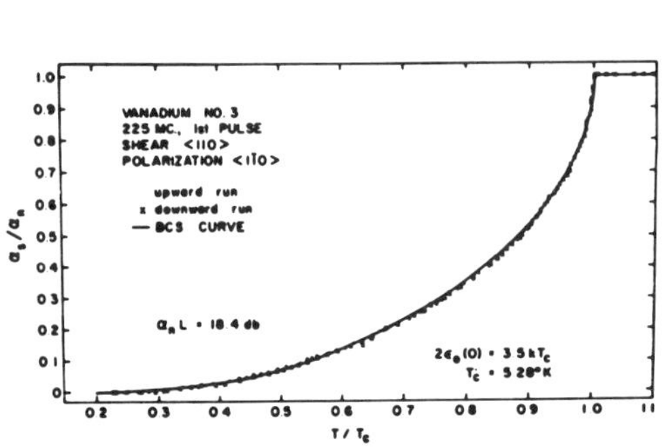

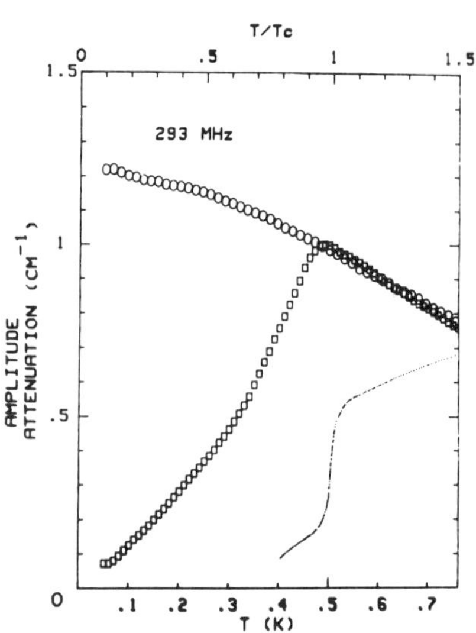

Fig. 1. Ultrasonic attenuation versus temperature in the superconducting state of V. The solid line is plotted according to the BCS expression.

Fig. 2. Temperature dependence of the ultrasonic attenuation in the normal and superconducting states of UPt$_3$.

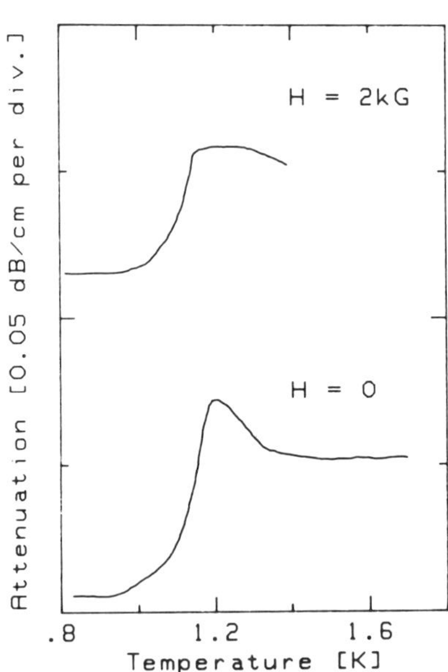

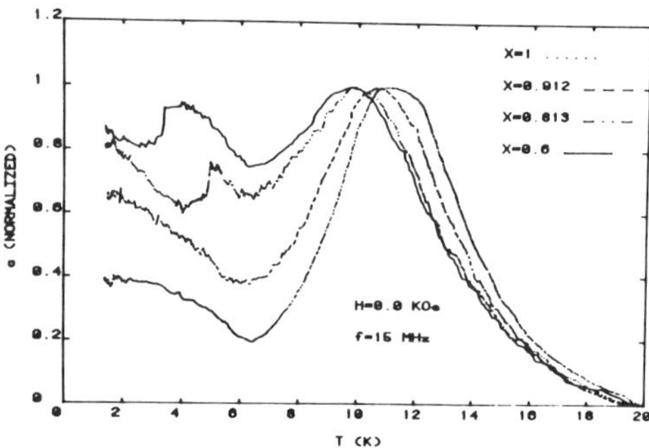

Fig. 3. Ultrasonic attenuation versus temperature for URu$_2$Si$_2$.

Fig. 4. Temperature dependence of the normalized ultrasonic attenuation in the Er$_{1-x}$Ho$_x$Rh$_4$B$_4$ system. Samples with x = 0.6 and 0.813 have superconducting transition temperatures of 6.7 K and 6.0 K respectively; and, ferromagnetic transition temperatures of 3.5 K and 5.0 K respectively. The attenuation increases in the superconducting state. Samples with x = 0.912 and 1 have ferromagnetic transitions at 6.2 K and 6.4 K respectively.

4 shows the normalized data for $Er_{1-x}Ho_xRh_4B_4$[5] with x = 0.6, 0.813, 0.912 and 1; the first two are reentrant superconductors; the last two only exhibit a ferromagnetic transition.

The sharp decrease in attenuation for Figure 1 occurs at the superconducting transition temperature T_c. The solid line is plotted according to the BCS expression with the BCS temperature dependent superconducting energy gap. It is evident that the fit to the data is very good. The ultrasonic data may therefore be inverted in order to obtain the superconducting energy gap, as a function of temperature Fig. 5[6] or it may be plotted as $\ln(\frac{2\alpha_n}{\alpha_s} - 1)$ versus $\frac{1}{T}$ in order to obtain the zero temperature energy gap from the slope of this plot, since the BCS gap is almost temperature independent at low reduced temperatures,[7] Fig. 6.

II. UPt_3

The data displayed in Figure 2 show the attenuation in the normal and superconducting states for UPt_3. The attenuation in the normal state was obtained by measuring it as a function of temperature in a constant magnetic field that was slightly larger than the zero temperature upper critical field $H_{c2}(0)$. The attenuation as a function of field in the normal state was measured at several temperatures below T_c. It was found that α_n was independent of field below about twice $H_{c2}(0)$. The ratio of α_s/α_n is plotted in Figure 7.[8] The attenuation starts to decrease a few millidegrees below T_c as determined from ac susceptibility. Several investigators[9,10,11] have reported that these data may not be fit by the BCS expression, as is plainly evident. However, if the data are plotted vs T^2 or T^3 or $T^2\ln^2 T$ it is possible to obtain regions of the data that satisfy quadratic,[9] cubic[10] or logarithmic[8] temperature dependences. For transverse waves it is possible to find linear[11] temperature dependent regions. At present the most accepted explanation for this deviation from BCS behavior appears to be that the superconducting energy gap of UPt_3 is anisotropic with axial or polar zeros in the energy gap.[9]

The magnetic field dependence of the attenuation in the mixed state of UPt_3 is shown in figure 8. At low fields the attenuation is linearly dependent on the applied field, at high fields it is quadratically dependent and there is a peak in attenuation at the transition between the two behaviors. Before discussing these data further we would like to present the behavior expected for a type II superconductor. In a dirty type II superconductor the electron mean free path ℓ is smaller than the intrinsic coherence length ξ_o and the actual coherence length ξ is given by $\xi = \sqrt{\ell\xi_o}$. Thus ℓ is also shorter than ξ_o. Therefore, electrons in the mixed state of a dirty type II superconductor may experience several collisions inside the normal cores of a flux line and thus produce an

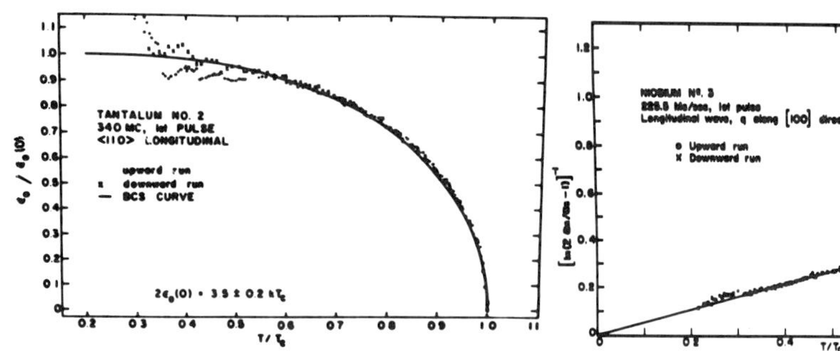

Fig. 5 . Temperature dependence of the superconducting energy gap obtained from α_s/α_n in Ta. The solid line is the BCS gap. The fit is better close to T_c.

Fig. 6. Plot of $\ln(2\alpha_n/\alpha_s - 1)$ versus T for Nb. The value of the zero temperature superconducting energy gap is obtained from the slope of the straight line.

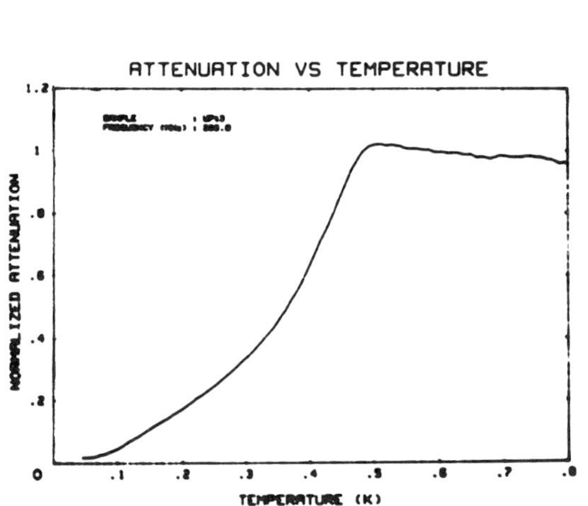

Fig. 7. α_n/α_s versus temperature for UPt$_3$.

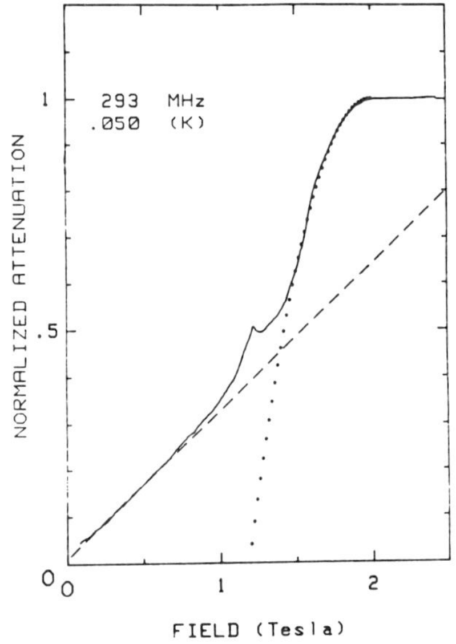

Fig. 8. Magnetic field dependence of the ultrasonic attenuation coefficient in the mixed state of UPt$_3$. The attenuation in the low region is proportional to the field while near H_{c2}, $(\alpha_n - \alpha_s)/\alpha_n \simeq (H_{c2} - H)^2$. The peak between these two regions may be produced by collective excitations of the flux line lattice.

attenuation that is proportional to the density of these normal cores. This is given by $1 - \Delta^2$. However Δ^2 is proportional to the magnetization M which in turn is proportional to $H_{c2}-H$ close to H_{c2}. Therefore $\frac{\alpha_n - \alpha_s}{\alpha_n} \simeq H_{c2}-H$. Figure 9 shows such a linear dependence for Nb_3Sn.[12] These data were obtained with surface acoustic waves.

In the clean limit,[13] the electron mean free path is much larger than the coherence length and an electron traverses through several flux lines before undergoing a collision. Thus the electron takes a spatial average of the order parameter. It is possible to obtain the dependence of the attenuation on H close to H_{c2} by expanding the BCS expression for α_s/α_n for small values of $\langle\Delta\rangle$, where the brackets indicate a spatial average, and since $\Delta^2 \simeq H_{c2} - H$ then $\langle\Delta\rangle \simeq (H_{c2}-H)^{1/2}$. From expansion of the BCS expression we obtain $\frac{\alpha_n}{\alpha_s} = 1 - \frac{\langle\Delta\rangle}{2kT}$, and therefore $\frac{\alpha_n - \alpha_s}{\alpha_n} \simeq \langle\Delta\rangle \simeq (H_{c2} - H)^{1/2}$. Figure 10 shows longitudinal data obtained on a pure sample of Nb which is plotted as a function of $(H_{c2}-H)^{1/2}$ in figure 11a. A similar plot is shown for transverse waves in Fig. 11b. This square root dependence is satisfied up to the transition temperature.[13] The data obtained in the mixed state of UPt_3, Fig. 8, do have a linear dependence on H. But this occurs at low values of H. It may be possible that the same argument still holds in this limit and the change in attenuation will be proportional to the density of the normal cores which is in turn proportional to the applied field. However close to H_{c2} there is a

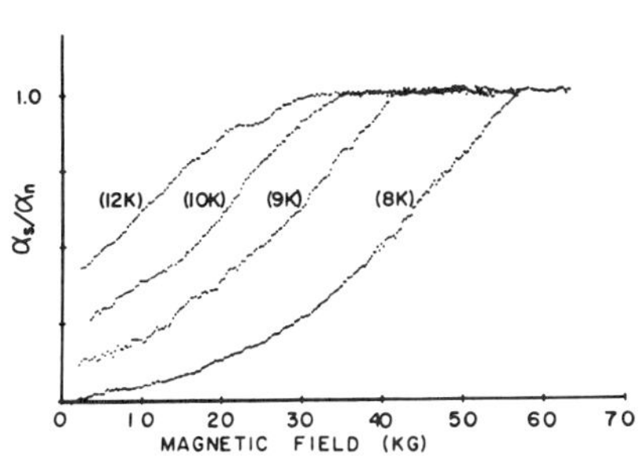

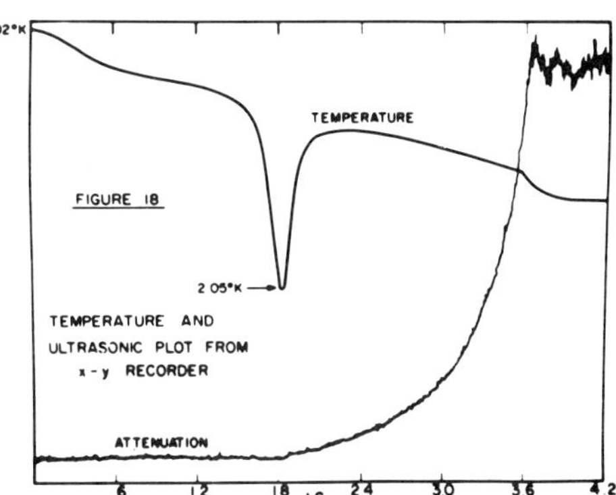

Fig. 9. Magnetic field dependence of the attenuation of surface acoustic waves travelling through a Nb_3Sn film. The attenuation in the mixed state below the upper critical field is linearly dependent on the applied field. This is the expected result for a dirty type II superconductor.

Fig. 10. Magnetic field dependence of the ultrasonic attenuation of longitudinal waves in the mixed state of pure Nb, a clean type II superconductor.

quadratic dependence of the attenuation on the change in field. If we
follow the same reasoning, it would appear that above a certain field,
which is indicated by the peak in attenuation in the mixed state, the flux
lines are more effective in producing attenuation. The extrapolation of
the linear portion of the curve lies below the quadratic portion. It
intersects the normal attenuation curve at a value of H which is higher
than H_{c2}. We would like to propose that in the linear region the flux
lines are independent of each other, but that the screening currents are
not circularly symmetric because their behavior has to reflect the
anisotropy of the superconducting energy gap. Along certain directions
there may be protrusions of the screening currents which extend further
out from the center. In the quadratic region these protrusions have
coalesced and therefore may make the flux lines more effective in
attenuating sound waves. The peak in attenuation occurs at the transition
between the two behaviors which may be indicative of collective
excitations of the flux lines as they start to coalesce.

III. $\underline{URu_2Si_2}$

Preliminary data have been obtained on URu_2Si_2 which are shown in
Figure 3. There is a maximum in the attenuation, more pronounced than the
one evident in UPt_3, Fig. 2, which occurs below the superconducting
transition temperature determined from ac susceptibility. The amplitude

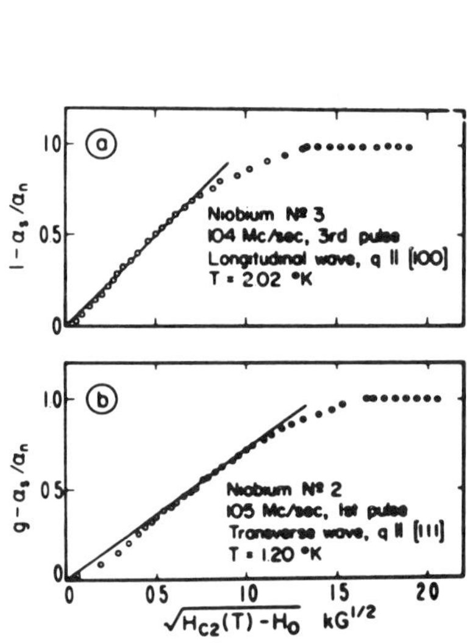

Fig. 11. (a) Data of Fig. 10
plotted as $(\alpha_n - \alpha_s)/\alpha_n$ versus
$(H_{c2} - H)^2$. The straight line
fit near the origin is expected
for a clean type II
superconductor. (b) A similar
plot for transverse waves in a
pure Nb single crystal.

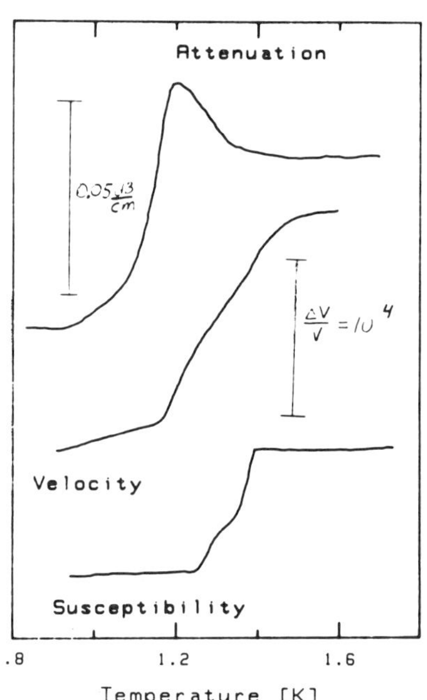

Fig. 12. Sound velocity attenuation and susceptibility versus
temperature for URu_2Si_2. The
susceptibility drops at T_c = 1.4 K.

of this maximum is decreased by a magnetic field. The velocity[4] appears
to decrease in the superconducting state, Fig. 12, by about the same
amount as in a regular BCS superconductor.

IV. $Er_{1-x}Ho_xRh_4B_4$

The temperature dependence of four samples in the $Er_{1-x}Ho_xRh_4B_4$
system is shown in Fig. 4. Those samples with x < 0.89 are reentrant
superconductors. The minima in attenuation at around 6 K correspond to
the superconducting transition temperature as determined by susceptibility
measurements. The sharp drop in attenuation at T = 3.5 K and 5 K
corresponds to the ferromagnetic transition temperature for x = 0.6 and
0.813 respectively. It is evident that the attenuation increases
throughout the superconducting state of these two samples. This increase
in attenuation is quenched by the transition into the ferromagnetic state.
The two samples with x > 0.89 undergo a ferromagnetic transition which
also correspond to the minima in attenuation around 6 K as determined by
susceptibility measurements. Figures 13 and 14 show the magnetic field
dependence of the attenuation in the two reentrant superconductors. The
attenuation is plotted as a function of temperature in various constant

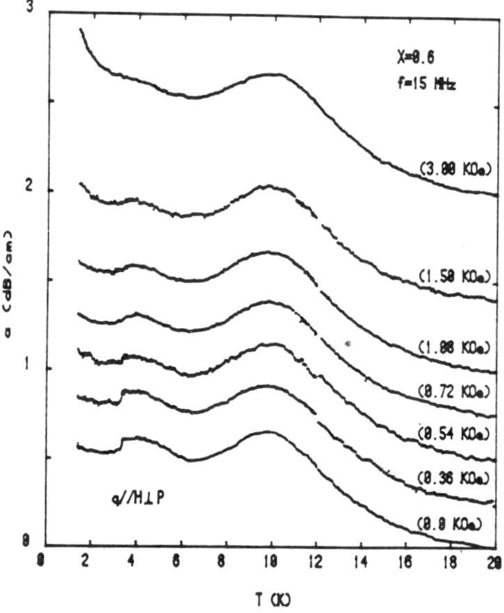

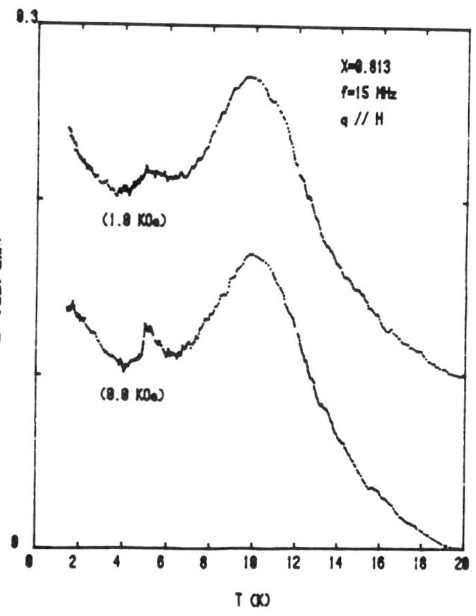

Fig. 13. Attenuation as a
function of temperature for
several constant fields for
$Er_{0.4}Ho_{0.6}Rh_4B_4$. The curves have
been displaced in order of increasing field. In zero field
the superconducting phase lies
between 3.5 K and 6.7 K. The
increase in attenuation present
in the superconducting phase is
decreased by diminishing superconductivity with a magnetic
field.

Fig. 14. Comparison of the
attenuation at zero magnetic
field to that at 1 KOe as a
function of temperature for
$Er_{0.187}Ho_{0.813}Rh_4B_4$. The curves
have been arbitrarily displaced
with respect to each other. In
zero field the superconducting
phase lies between 5K and 6K; the
increase in attenuation present
in the superconducting phase is
decreased by the 1 KOe field,
which is sufficient to destroy
superconductivity over the whole
temperature range.

applied magnetic fields for x = 0.6 in Fig. 13 and x = 0.813 in Fig. 14.
In both figures it is evident that the increase in attenuation associated
with the superconducting state is diminished in a magnetic field. This
decrease is equally evident in Figure 15 where the attenuation is plotted
as a function of magnetic field at various constant temperatures. In the
temperature interval where the sample with x = 0.6 is superconducting, 3.5
K to 6.7 K, there is an initial decrease in attenuation at low fields.
Subsequently, the attenuation increases after the sample enters the
ferromagnetic state.

The data presented in Figs. 12 to 15 demonstrate that there is an
increase in the attenuation in the superconducting state of the reentrant
superconductors which disappears when superconductivity is quenched either
by lowering the temperature or by applying a magnetic field. The samples
with x > 0.89 which only undergo a ferromagnetic transition do not exhibit
these effects. And, in fact, do not show any significant features at the
ferromagnetic transition temperature, such as a peak in attenuation which
would be produced by spin fluctuations associated with a second order
phase transition. The absence of these features is further evidence that
the Ho rich $Er_{1-x}Ho_xRh_4B_4$ samples are mean field ferromagnets. The Ho
spins are aligned along the c axis by the crystalline electric fields
(CEF) and do not undergo spin fluctuations at the ferromagnetic phase
transition. It is suggested[5] that in the superconducting state,
superconducting currents screen the CEF allowing the Ho spins to fluctuate
which then produce significant spin phonon interaction which is reflected
by the increase in ultrasonic attenuation. When superconductivity is
quenched by the transition to the ferromagentic state when the temperature
is reduced, the superconducting screening currents disappear and the Ho
spins are once again aligned along the c-axis and spin phonon interaction
associated with the fluctuations vanishes. A similar result is obtained
when superconductivity is elliminated by a magnetic field.

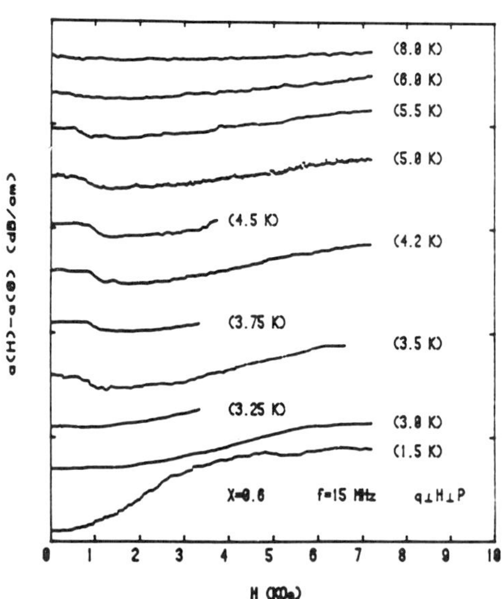

Fig. 15. Attenuation as a function of magnetic field for several constant
temperatures for $Er_{0.4}Ho_{0.6}Rh_4B_4$. The curves have been displaced in order
of increasing temperatures. In zero field the superconducting phase lies
between 3.5 K and 6.7 K. In this temperature region the attenuation
decreases when the field is large enough to destroy superconductivity.

SUMMARY

The attenuation of bulk waves in the heavy Fermion superconductors UPt_3 and URu_2Si_2 and in four alloys of the ternary alloys $Er_{1-x}Ho_xRh_4B_4$ have been reported. None of these samples exhibit BCS attenuation behavior in the superconducting state. The differences seen in the heavy Fermion superconductors may be due to axial or polar zero in the superconducting energy gap, and to collective excitations of the vortex lattice in the mixed state.

The ferromagnetic samples of the ternary system exhibit mean field ultrasonic attenuation behavior. The reentrant superconductors exhibit an increase in attenuation in the superconducting state. This increase is associated with supercurrent screning of the crystalline electric fields. Thus spin fluctuations associated with the HO ferromagnetic transition are enhanced resulting in an incraese in spin phonon interaction.

The insight obtained during these investigations may be applicable to and may help to unraval the ultrasonic data that will be obtained on the new high T_c superconductors.

REFERENCES

*Research supported by the Air Force Office of Scientific Research under grant No. AFOSR 84-0350.

†Present address: NASA-Langley Research Center MS 231, Hampton, Virginia 23665

1. J. Bardeen, L. N. Cooper, and J. R. Schrieffer, Theory of Superconductivity, Phys. Rev. 108:1175 (1957).
2. M. Levy, Ultrasonic Attenuation in Superconductors for $q\ell < 1$, Phys. Rev. 131:1497 (1963), and M. Levy, R. Kagiwada and I. Rudnick, Ultrasonic Attenuation of Transverse Waves in V, Nb and Ta for $q\ell < 1$,, Phys. Rev. 132:2039 (1963).
3. Y. J. Qian, M-F Xu, A. Schenstrom, H.-P. Baum, J. B. Ketterson, D. Hinks, M. Levy and B. K. Sarma, Longitudinal Sound Measurements on UPt_3 in a Magnetic Field, Solid State Communications (to be published).
4. K. J. Sun, A. Schenstrom, B. K. Sarma, M. Levy, J. B. Ketterson, D. Hinks, Longitudinal Sound in URu_2Si_2, Bull. Amer. Phys. Soc. 32:641 (1987).
5. K. J. Sun, M. Levy, M. B. Maple and M. W. Torikachvilli, Induced Spin Phonon Interaction in the Superconducting State of Ho-Rich $Er_{1-x}Ho_xRh_4B_4$ (to be published).
6. M. Levy and I. Rudnick, Ultrasonic Determination in the Superconducting Energy Gap in Tantalum, Phys. Rev. 132:1073 (1963).
7. F. Carsey, R. Kagiwada, M. Levy and K. Maki, Apparent Two Energy Gaps in Pure Niobium, Phys. Rev. B4:854 (1971).
8. M.-F. Xu, Y. J. Qian, A. Schenstrom, H.-P. Baum, J. B. Ketterson, D. Hinks, M. Levy and B. K. Sarma, Ratio of Ultrasonic Attenuation in the Superconducting to the normal state in UPt_3 (to be published).
9. D. J. Bishop, C. M. Varma, B. Batlogg, E. Bucher, Z. Fisk, and J. L. Smith, Ultrasonic Attenuation in UPt_3, Phys. Rev. Lett. 53:1009 (1984).

10. V. Muller, D. Maurer, E. W. Scheidt, Ch. Roth, K. Luders, E. Bucher, and H. E. Bommel, Observation of Lambda-shaped Ultrasonic Attenuation Peak in Superconducting UPt_3, Solid State Comm. 57:319 (1986).

11. B. S. Shivaram, Y. H. Jeong, T. F. Rosenbaum, and D. G. Hinks, Anisotropy of Transverse Sound in the Heavy Fermion Superconductor UPt_3, Phys. Rev. Lett. 56:1078 (1986).

12. H. P. Fredricksen, H. L. Salvo, Jr., M. Levy, R. H. Hammond and T. H. Geballe, Ultrasonic Attenuation of Surface Acoustic Waves in a Superconducting Thin Film of Nb_3Sn in an Applied Magnetic Field, Proceedings of 1979 IEEE Ultrasonics Symposium, 453 (79 CH 1482-9 Eds. J. de Klerk and B. R. McAvoy, IEEE, New York, 1979).

13. R. Kagiwada, M. Levy, I. Rudnick, H. Kagiwada, K. Maki, Ultrasonic Attenuation in a Pure Type II Superconductor Near H_{c2}, Phys. Rev. Lett. 18:74 (1967).

14. R. Kagiwada, Ultrasonic and Thermal Effects in Superconductors, Thesis 1966, UCLA (unpublished).

PHENOMENOLOGY OF SUPERCONDUCTIVITY AND MAGNETIC ORDER

IN HEAVY FERMI LIQUIDS AND NARROW-BAND METALS

L.E. DeLong

Department of Physics and Astronomy
University of Kentucky
Lexington, KY 40506
Materials Science Division
Argonne National Laboratory
Argonne, IL 60439

SUMMARY

Trends for the occurrence of superconductivity and magnetic order in
narrow-band metals are reviewed. Magnetic interactions and strongly nonadiabatic coupling between electrons and the lattice are discussed as important limitations on the superconducting transition temperature T_c. The
quantitative influence of the temperature and magnetic field dependences of
normal state parameters on the upper critical field of heavy fermion and
high-T_c superconductors, including the novel very-high-T_c copper oxides,
are considered. Recent evidence is reviewed for the existence of an unexpectedly low energy scale T_0 that acts as a "hyperstrong" cutoff for the
T_c of heavy fermion, A15 and other narrow-band superconductors.

INTRODUCTION

We have recently presented[1-3] a quantitative correlation between the
observed low-temperature ground state and the values of the electronic heat
capacity C_e and total magnetic susceptibility χ^* of "narrow-band" materials. Our motivation is the desire to understand the instability of a
given material to long-range order such as superconductivity or antiferromagnetism. These instabilities are driven by microscopic interactions that
generally renormalize the magnitudes and temperature dependences of various
physical properties such as the magnetic susceptibility and heat capacity.
We suppose that these effects presage the specific type of ground state
ultimately attained at lower temperatures.

Our previous analysis has focused on two experimental quantities. The
low-temperature heat capacity yields an effective Sommerfeld coefficient
$\gamma^* \equiv C_e T^{-1}$, which is generally temperature-dependent. The uncorrected χ^* and
γ^* can be used to define an effective exchange enhancement ratio R [Ref.4]:

$$R = \lim_{\substack{T \to 0 \\ \text{or } T \to T_c^+}} \frac{1}{3} \left[\frac{\pi k_B}{\mu_B} \right]^2 \frac{\chi^*}{\gamma^*} \tag{1}$$

The limit in Eq. 1 signifies that R is to be evaluated at the lowest possible temperature or just above the onset of long-range order at a critical temperature T_c (magnetic or superconducting).

We have examined magnetic and calorimetric data for a large number of transition metal, actinide and rare earth compounds and elements[4] that can be generally termed "narrow-band" metals. We found that a novel method of plotting R vs γ^* revealed distinct regions of superconducting, magnetically ordered and paramagnetic states[1], as shown in Fig. 1. An examination of Fig. 1 leads to the following observations:

1) Magnetic order is observed only if γ^* exceeds a threshold value $\gamma_M^* \approx 3.5 \times 10^4$ erg/cm^{3}K^2.

2) There are no superconductors observed in the interval $\gamma_M^* \lesssim \gamma^* \lesssim 10^5$ erg/cm^{3}K^2.

3) The "heavy fermion" superconductors (UPt_3, UBe_{13} and $CeCu_2Si_2$) occur for $\gamma^* \gtrsim 10^5$ erg/cm^{3}K^2, and are well isolated from "conventional" superconductors (which we define by $\gamma^* < \gamma_M^*$) by a region of magnetic order.

4) All superconductors have values of $R \leqslant R_c(\gamma^*) \lesssim 4$, a critical value of R above which superconductivity cannot occur.

5) Most materials examined have $R \lesssim 10$, consistent with "almost localized" Fermi liquid models[5].

The existence of trends 1) through 5) is remarkable in view of the extreme simplicity of the definitions of R and γ^*. For example, Eq. 1 ignores the potential influence of spin-orbit coupling, crystalline electric field effects, band orbital magnetism, etc. We have given detailed discussions of these trends and their limitations elsewhere[1-4]. Trends

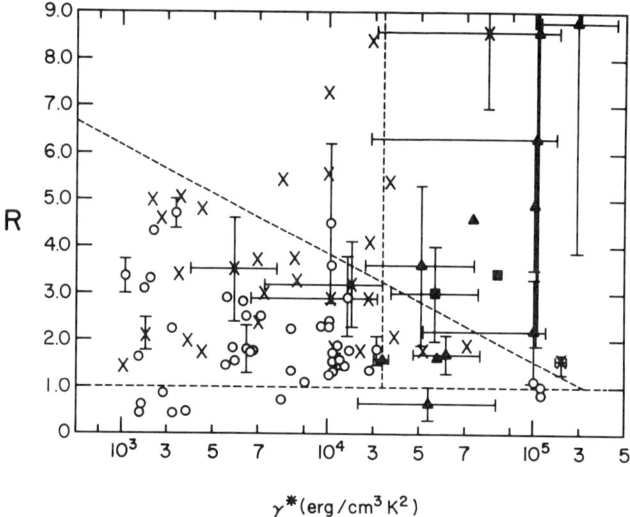

Fig. 1. Exchange enhancement ratio R vs electronic heat capacity coefficient γ^* for a variety of transition metal, actinide and rare earth compounds and elements. Open circles denote superconductors, diamonds antiferromagnets, crosses paramagnets, and squares refer to to materials that order via structural or unknown transformations. The vertical dashed line denotes $\gamma_M^* \approx 3.5 \times 10^4$ergcm^{-3}K^{-2}, the diagonal dashed line $R_c(\gamma^*)$, and the horizontal dashed line $R \equiv 1$ for a free Fermi gas. Note: Cu has $\gamma^* \sim 10^3$erg cm^{-3}K^{-2}, and U_6Fe and UPt_3 have (γ^*,R) values of $(1.8 \times 10^4, 1.5)$ and $(1.1 \times 10^5, 1.1)$, respectively.

1)-3) have been briefly discussed in regard to U compounds by Fisk et al.[6] Below, we will enlarge upon several aspects of these correlations and relate them to magnetoresistance and upper critical field data.

CRITICAL VALUES OF γ^* AND CROSSING ENERGY SCALES

The trends 1) through 3) above can be related to the cross-over of lattice and electronic energy scales[1,3], the Debye temperature θ_D and Fermi temperature T_F^*, respectively. We are forced to adopt an isotropic Fermi liquid model[1,3] in reviewing data for a large group of materials, and we define $T_F^* \equiv \pi k_B^2 n / 2 \gamma^*$, where the fermion density n determines the average Fermi wavevector $k_F = (3\pi^2 n)^{1/3}$. θ_D is obtained from analyses of the lattice heat capacity[2] or from inelastic neutron scattering data, whenever they are available.

We have found that magnetic order occurs for $T_F^*/\theta_D \lesssim 3.5$ within the group of materials surveyed to date[3]. On the other hand, "heavy fermion superconductivity" (HFSC) is thus far observed only for $0.2 \lesssim T_F^*/\theta_D \lesssim 0.85$ (i.e., UPt_3, UBe_{13} and $CeCu_2Si_2$). The disappearance of "conventional superconductivity" (CSC) and onset of magnetic order for $\gamma^* \sim \gamma_M^*$ is evidently related to the breakdown of the Born-Oppenheimer or "rigid lattice" approximation common to much of solid state theory. When $T_F^* \lesssim \theta_D$, we expect the Fermi velocity $v_F^* \sim v_s$, the sound velocity, and the interaction between the heavy quasiparticles and the lattice is highly nonadiabatic. HFSC appears in a regime in which a "fast lattice" must adjust to the slower hopping of heavy quasiparticles, suggesting a polaronic picture[7,8] is a reasonable basis for a description of these materials. Indeed, the correlation of ground-state type with T_F^*/θ_D suggests that lattice excitations play a crucial role in the behavior of heavy fermion materials.

Unfortunately, T_F^* must be determined using auxillary assumptions about the magnitude of n, a parameter not easily extracted from experiment. These difficulties are more fully discussed in Refs. 3 and 9. Our estimates of T_F^* are chosen to generally agree with a heirarchy of energy scales established by neutron scattering, x-ray spectroscopy and deHaas-van Alphen measurements: $k_B T_C \lesssim \Gamma \lesssim k_B T_F^* \lesssim \Delta$, where Γ is the neutron quasielastic line half-width and Δ is the Anderson hybridization parameter extracted from analyses of x-ray data[3]. Note that $\Gamma \sim k_B T_K$, the Kondo or spin fluctuation energy, and for $T \gg$ coherence temperature, is a measure of a <u>single-site</u> magnetic moment lifetime.[3,5] Unfortunately, direct Fermi surface measurements of relevant narrow-band materials are extremely limited so far[10-12]. Nevertheless, low temperature estimates of T_F^* and θ_D extracted from de Haas-van Alphen[11] and neutron scattering data[13] for UPt_3 support the picture presented here[9].

SUPERCONDUCTIVITY AND EXCHANGE ENHANCEMENT

Trends 4) and 5) of the Introduction are relevant to a large literature on the generally antagonistic relationship between superconductivity and magnetism. Jensen and coworkers[14,15] examined trends in the superconductivity of transition metal alloys and compounds and proposed that paramagnons play a decisive role in the behavior of the noble elements and their alloys. T_C was observed to be strongly depressed in noble metal alloys as the "Stoner exchange enhancement" $S \equiv N_\chi/N_\gamma$ increased beyond ~ 1, where N_χ and N_γ are the electronic densities of states determined from susceptibility and heat capacity measurements, respectively.

We have found that these trends can be extended to a much broader data set using our operational definition of "exchange enhancement" given in Eq. 1. It is important to note that R implicitly reflects the generalized susceptibility $\chi(q,\omega)$ at q, $\omega=0$. R is, of course, observed to diverge in <u>ferro</u>magnetic substances, but would not necessarily be large for nearly <u>antiferro</u>magnetic materials, where $\chi(q,\omega)$ is enhanced at finite q. We find, quite generally, that T_c is a monotonically decreasing function of R within a given class (i.e., common structure and similar atomic constituents) of materials. This trend appears to override any other effects in the region $\gamma^* < \gamma_M^*$. Some examples of this trend are given in Fig. 2.

The HFSC, UPt_3, UBe_{13} and $CeCu_2Si_2$, also appear to be constrained by a similar rule, in that $R > 1.0$ for these materials, whereas the nearby compounds $CeCu_6$ ($\gamma^*=2.4 \times 10^5$ erg/cm^3K^2, R~1.6), YbCuAl ($\gamma^* = 8.0 \times 10^4$ erg/cm^3K^2, R=8.6) and $CeA\ell_3$ $\gamma^*=7.1 \times 10^4$ erg/cm^3K^2, R=1.9) are nonsuperconducting. The isolation and confinement of the heavy fermion superconductors to such a small sector in the $R-\gamma^*$ plane suggests that a very delicate balance exists between superconductivity and magnetic order in this region, consistent with recent neutron scattering studies[16] and theoretical work[17].

Another important observation is that a large number of high-T_c A15 superconductors (as well as U_6Fe, URu_2Si_2, U_2PtC_2, $CeRu_2$, etc.) are located in a sector $R<3.0$ and $\gamma^* \lesssim \gamma_M^*$. This behavior implies that magnetic interactions, as opposed to the more accepted constraint of lattice instabilities, are the primary limiting factor for T_c of narrow-band superconductors. However, lattice instabilities may still be

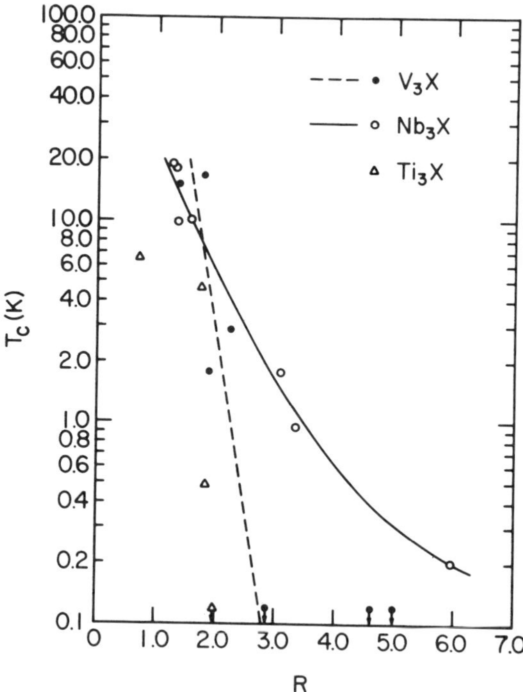

Fig. 2. Superconducting transition temperature T_c <u>vs</u> exchange enhancement ratio R for Nb_3X (open circles), V_3X (solid circles) and Ti_3X (triangles). Lines are guides to the eye and connect materials with similar chemical constituents.

important in high-T_C compounds; Kim[18] has argued that magnetic and lattice instabilities may both compete with superconductivity in high-T_C materials.

MAGNETORESISTANCE AND UPPER CRITICAL FIELD

The large magnitudes and unusual temperature dependences of the upper critical fields H_{c2} of heavy fermion superconductors have been difficult to explain in terms of traditional singlet pairbreaking models such as the WHHM theory[19]. The $H_{c2}(T)$ data for high purity U_6Fe samples exhibit unusual positive curvature[20] that is similar to the anomalous behavior of UPt_3 [Refs. 21,22], as shown in Figs. 3a and 4a.

The WHHM model yields an analytic form for $T_C(H)$ in terms of the orbital pairbreaking parameter $h \equiv (0.281)H/H_{c2}^*(0)$, where, in the dirty limit, the pairbreaking strength is scaled by

$$H_{c2}^*(0) = (2.28 \times 10^{-4}) \frac{T_{CO}}{v_F^* \ell_{tr}} \propto \Delta_0 \gamma^* \rho \quad . \tag{2}$$

ρ is the electrical resistivity, $T_{CO} \equiv T_C(H=0)$, v_F^* is the Fermi velocity, and Δ_0 is the T=0 superconducting gap (in SI units). $H_{c2}^*(T)$ is the maximum possible critical field that is attained in the absence of Pauli limiting. However, the WHHM model, valid for arbitary mean free path ℓ_{tr}, (see Eq. 3.46 of Ref. 23) cannot account for the behavior of U_6Fe and UPt_3 for reasonable choices of T_{CO}, the initial slope $H_{c2}'(T_{CO})$, or arbitrarily large spin-orbit scattering rates $\sim\lambda_{so}$. This is because H_{c2} both exhibits positive curvature at low fields and exceeds the maximum possible field H_{c2}^* at lower T.

These effects were previously observed[23] in the high-field Chevrel phases $Mo_6(Se_{1-x}S_x)_8$, where Decroux and Fischer (DF) have argued that H_{c2}^* must be enhanced by unknown mechanisms. They introduced an empirical function $\beta(T)$ that scaled $H_{c2}^*(T)$ to higher values $H_{c2}^{**}(T) = \beta(T)H_{c2}^*(T)$, and deduced an analytic form $\beta_{DF}(T)$ that brought their data into agreement with the WHHM theory:

$$H_{c2}^{**}(T) \equiv \beta_{DF}(T)H_{c2}^*(T) \equiv \frac{\beta_0 H_{c2}^*(T)}{1+(\beta_0-1)T^2/T_{CO}^2} \quad , \tag{3}$$

where $\beta_0 \geqslant 1$ is a constant.

Recognizing the similarities in $H_{c2}(T)$ data for U_6Fe and Chevrel compounds, DeLong et al.[24-27] employed a similar approach in which the orbital pairbreaking parameter h is rescaled with $H_{c2}^{**}(0) = \beta(T)H_{c2}^*(0)$. This results in excellent agreement between the scaled WHHM model and U_6Fe, U_6Co and UPt_3 data, as shown in Figs. 3 and 4.

DF have considered[23] several possible mechanisms for the enhancement function β_{DF} including strong coupling, "nonlocal" corrections to the gap equation and anisotropy of the pairing interaction. However, DF argued that a "multiple band" model[28,29] (i.e. several bands intersecting the Fermi level with very different v_F^* and scattering times τ_{tr}) provided the best explanation of β and their Chevrel phase data.

On the other hand, DeLong and coworkers[25-27] have observed that the sign of the magnetoresistance $\Delta\rho(T,H) \equiv \rho(T,H) - \rho(T,0)$ correlates with the sign of the curvature of $H_{c2}(T)$ at $T \lesssim T_{co}$ for U_6Fe, U_6Co, UPt_3, UBe_{13} and $CeCu_2Si_2$. $\rho(T)$ [Refs. 30,31], $\Delta\rho$ [Refs. 25-27,31,39] and H_{c2} [Refs. 20-22, 24] for U_6Fe and UPt_3 are remarkably similar, in spite of the approximate five-fold difference in γ^* [Refs. 32,33] for these two materials, and their clear separation in the $R-\gamma^*$ plot of Fig. 1. It is noteworthy that the anomalous behavior of the H_{c2} of UBe_{13} has been interpreted as a manifestation of the field dependences of normal state parameters such as χ^*, γ^* and ρ, or odd-parity superconducting pairing interactions[34-36].

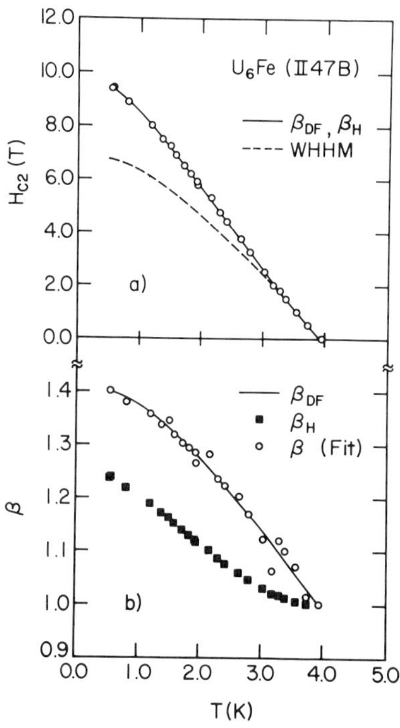

Fig. 3. a) Upper critical field H_{c2} vs temperature T for U_6Fe sample #II47B. The circles are data[20] and the dashed line is the prediction of the standard WHHM model. The solid line is a fit using the scaled WHHM model with $\beta(T) = \beta_{DF}$ of Eq. 3. A fit using $\beta(T) = \beta_H$ of Eq. 5 is nearly indistinguishable from the solid line.
b) Enhancement function $\beta(T)$ vs T for the fits of a), above. The circles are force-fit values of β necessary to exactly fit the measured $H_{c2}(T)$, and the solid line is a fit of these points to the β_{DF} form of Eq. 3. The squares are $\beta_H(T)$ determined from magnetoresistance data and Eq. 5. (After Refs. 25-27)

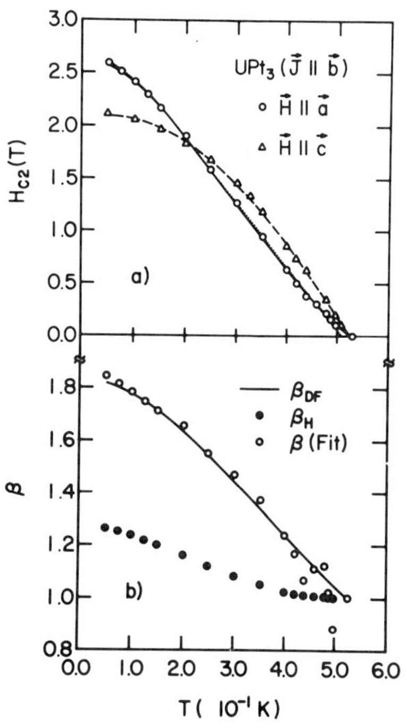

Fig. 4. a) Upper critical field H_{c2} vs temperature T for a single crystal of UPt_3 with current $\vec{J}$ parallel to the hexagonal $\vec{b}$-direction. Open circles and triangles are data from Ref. 22 for the field $\vec{H}$ parallel to the $\vec{a}$-and $\vec{c}$-directions, respectively. The solid line is a fit using the β_{DF} scaling, and the dotted line is for the β_H form. The dashed line represents two indistinguishable fits using either scaling.
b) Enhancement function $\beta(T)$ vs T for selected fits of a), above. The open circles are force-fit values of β necessary to exactly fit the measured $H_{c2}(T)$, and the solid line is a fit of these points to the β_{DF} form of Eq. 3. The solid circles are $\beta_H(T)$ determined from magnetoresistance data. Only data for $\vec{H}||\vec{a}$ are shown for clarity. (After Refs. 25-27)

The above facts and Eq. 2 led DeLong et al.[25-27] to relate $\beta(T)$ to the
field dependences of ρ and γ^*, and they proposed

$$\beta(T) = \left[\eta(T,H) \cdot \frac{\gamma^*(T,H)}{\gamma^*(T,0)} \cdot \frac{\rho(T,H)}{\rho(T,0)} \right]_{H=H_{c2}} \cdot \qquad (4)$$

$\eta(T,H)$ is an initially undetermined factor that reflects potential field and
temperature dependences of the pairing interaction and strong-coupling
effects (note $H_{c2}^* \sim \Delta_0$, by Eq. 2). Most of the enhancement of H_{c2}^*
should be due to $\Delta\rho(T,H)$, since U_6Fe and UPt_3 exhibit only small changes
in γ^* and χ^* for $T \lesssim T_{co}$ and $H \lesssim H_{c2}(0)$ [Refs. 4,20,37,38]. Therefore,
the application of Eq. 4 to these materials is simplified by ignoring
potential variations of $\eta(T,H)$ and defining

$$\beta_H(T) \equiv \left[\frac{\rho(T,H)}{\rho(T,0)} \right]_{H=H_{c2}} \approx \beta(T) \qquad (5)$$

Consequently, the anomalous but similar $H_{c2}(T)$ curves of U_6Fe and UPt_3 are
expected to derive from their similar $\Delta\rho$ behavior.

Indeed, it was found[25-27] that the $\Delta\rho(T,H)$ data for U_6Fe and U_6Co
closely resemble those of UPt_3, in spite of the approximate ten-fold
difference in fermion effective mass m^* between the two types of compound.
At low temperatures, all three compounds exhibit positive $\Delta\rho$ that follows
a novel and perhaps universal form:

$$\Delta\rho(T,H) = \frac{AH^2}{B+H} \qquad (6)$$

A and B are T- and H- independent constants for $T \lesssim T_{co}$ and $H \lesssim H_{c2}(0)$
(see Fig. 5 and Table 1 for details). $\Delta\rho$ for U_6Fe and U_6Co are found to

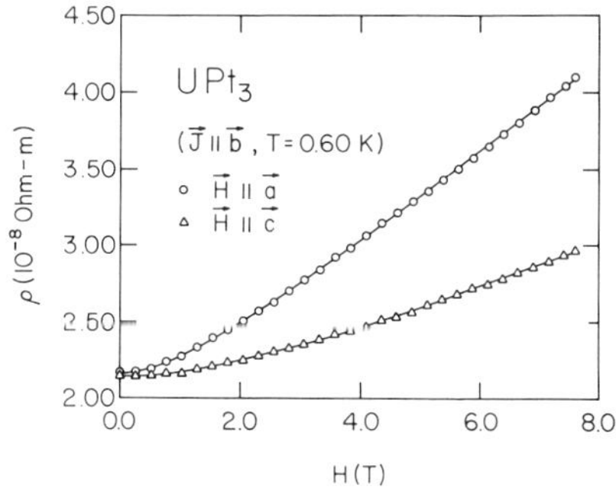

Fig. 5. Resistivity ρ as a function of magnetic field H for a single
crystal sample of UPt_3 at a temperature T=0.60K. Open circles are for
$\vec{H} \parallel \hat{a}$-axis, and closed circles are $\vec{H} \parallel \hat{c}$-axis. The current $\vec{J} \parallel$
$\hat{b}$-axis. The solid lines are fits to $AH^2/B+H$. The $\vec{H} \parallel \hat{a}$ data have been
slightly scaled to correct for geometrical errors in determining ρ. The
excellent quality of the fits of $\Delta\rho$ to Eq. 6 is also attained for U_6Fe
and U_6Co samples. (After Ref. 39)

Table 1: Fit Parameters for $\Delta\rho$ and H_{c2} Data[27,39]

Compound	A ($\mu\Omega cmT^{-1}$)	B (T)	ρ_O ($\mu\Omega cm$)	a ($\mu\Omega cmK^{-2}$)	T_O (K)	T_{CO} (K)	β_O
U_6Fe(II47B)	0.421	4.32	11.4	0.173	5.7	3.94	1.41
U_6Co(II53B)	1.16	3.43	59.9	0.105	5.6	2.53	1.15
$UPt_3(\vec{H}||\vec{a})$	0.311	1.73	1.85	2.36	0.43	0.52	1.84
$UPt_3(\vec{H}||\vec{c})$	0.174	4.68	1.59	1.61	0.66	0.52	1.15

change sign for $T \gtrsim 25K$, similar to the cross-over observed for UPt_3 near 18K [Refs. 31,39]. The T- and H- independence of A and B permit one to extrapolate the values of $\rho(T,0)$ necessary to define β_H in Eq. 5, implying that $\rho(T,0) = \rho_O(0,0) + aT^2$, consistent with measurements of $\rho(T,0)$ for UPt_3 [Refs. 27,31,32], U_6Fe and U_6Co [Ref. 27].

The β_H data for U_6Fe, U_6Co and UPt_3 were found to yield good fits of the $H_{c2}(T)$ data, as illustrated in Figs. 3 and 4. It is therefore easy to conclude that the anomalous H_{c2} behaviors of cleaner samples of U_6Fe, U_6Co and UPt_3 indeed reflect the T- and H- dependencies of ρ that must be included in WHHM model fits. Further, H_{c2} data for very dirty samples (for which $\Delta\rho \approx 0$ and $\beta_H \approx 1$) are increasingly well-fitted by the unscaled WHHM model as ℓ_{tr} decreases, as expected[25-27].

However, DeLong et al.[27] have pointed out that small discrepancies remain between the β_H-scaled WHHM model and H_{c2} data for samples of intermediate purity that exhibit the highest values of $H_{c2}(0)$, yet have very small β_H due to increased $\rho_O(0,0)$, and small $\Delta\rho(T,H)$. They have achieved only approximate fits of these data with the β_H scaling, but still obtain excellent fits using the β_{DF} function[24-27]. The extraordinary positive curvature of H_{c2} $(\vec{H}||\vec{a})$ of clean UPt_3 for $T \lesssim T_{CO}$ is also better described by the β_{DF} scaling. These discrepancies and the obvious differences between β_H and β_{DF} shown in Figs. 3b and 4b make it doubtful that the enhancement of H_{c2}^* is a trivial result of the field dependence of ℓ_{tr}, as suggested by Eqs. 2 and 5.

DeLong et al.[27] have undertaken a detailed analysis of these discrepancies by assuming that the true enhancement function $\beta(T) \approx \beta_{DF} = \eta(T,H) \beta_H$. $\eta(T,H)$ is assumed to parameterize additional factors (other than $\Delta\rho(T,H)$) that influence $H_{c2}(T)$. An analysis of the fitting parameters describing $H_{c2}(T)$ and $\Delta\rho(T,H)$ for U_6Fe, U_6Co and UPt_3 consistently yields $\beta(T,H) \sim \xi^{-1}(T,H) [\xi^{-1}(T,H) + \ell_{tr}^{-1}(T,H)]$ where the inverse Ginzberg-Landau coherence length $\xi^{-1}(T) \sim \Delta(T)$, the order parameter. This intriguing result is unexplained.

The assumption $\beta_{DF} = \eta\beta_H$ leads to a condition on the low-T enhancement factor β_O that is very similar to a strong coupling renormalization[40] of H_{c2}:

$$\beta_O = 1 + \frac{aT_{CO}^2}{\rho_O} \; \frac{AH_{c2}^2(0)[(B+H_{c2}(0))\rho_O]^{-1}}{1+AH_{c2}^2(0)[(B+H_{c2}(0))\rho_O]^{-1}} \tag{7}$$

Note that β_0 diverges as $\rho_0 \to 0$ in the clean limit. Therefore, when ρ_0 decreases to the point at which $\ell_{tr} \gtrsim \xi_0$, the zero-T coherence length, one must set $\rho_0 \to f a T_0^2$ in Eq. 7, where T_0 is a cut-off energy scale analogous to θ_D and f is a numerical factor of order unity[26,27,40].

A crucial point is realized, however, when T_0 is evaluated using experimental data and Eq. 7: $T_{CO} \lesssim T_0 \ll \theta_D$ for U_6Fe, U_6Co and UPt_3! It is extremely unlikely that a physical association can be made between such low values of T_0 and an average phonon frequency, as expected in the case of traditional electron-phonon mechanisms. It is remarkable that $T_0 \lesssim 0.9K$ is the same order as estimates of the "coherence temperature" for UPt_3 (see Table 1).

CONCLUSIONS AND SPECULATIONS

We conclude that the WHHM model must be modified to take into account the T-and H-dependences of normal state parameters. These effects should be particularly important for analyzing H_{c2} data for the very high-T_c copper oxides[41], where the strong linear T-dependence of ρ for $T \lesssim 100K$ should, by itself, result in important reductions of H_{c2}, and reentrant behavior as demonstrated in Fig. 6.

A rather simple extension[25-27] of the WHHM model has yielded the first quantitative fits of $H_{c2}(T)$ for U_6Fe, U_6Co and UPt_3, in spite of the isolation of U_6Fe and U_6Co from UPt_3 by the interval of magnetic order $\gamma_M^* < \gamma^* < 10^5 erg cm^{-3} K^{-2}$ defined in Fig. 1. The novel form of Eq. 6 has been shown to very accurately describe $\Delta\rho$ data for U_6Fe, U_6Co and

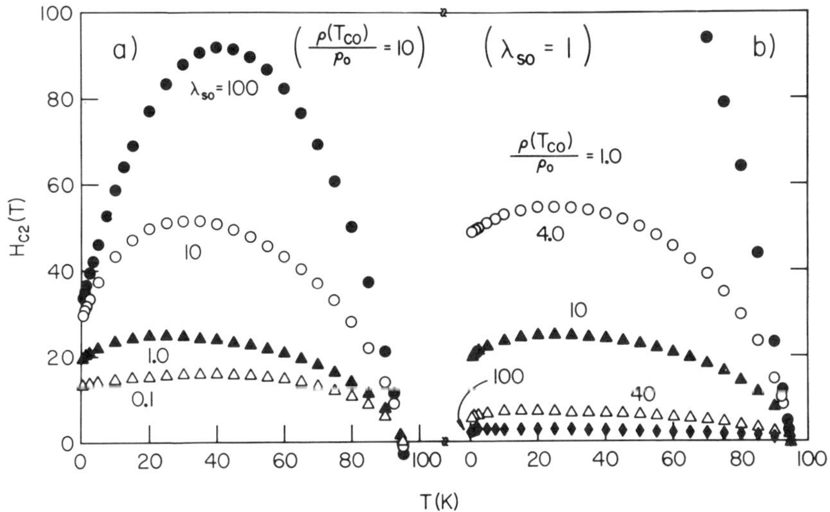

Fig. 6. Upper critical field H_{c2} versus temperature T for a typical ternary copper oxide such as $YBa_2Cu_3O_7$, assuming $\Delta\rho(T,H) \equiv 0$ and $\rho(T) = \rho_0 + bT$, b = constant. Parameters used are $T_{CO} = 95K$, $H'_{c2} = -5T/K$, Maki parameter α = WHHM value[19] = $-0.528 H'_{c2}$. a) $\rho(T_{CO})/\rho_0 = 10$ is fixed and λ_{so} is varied. b) $\lambda_{so} = 1$ is fixed and $\rho(T_{CO})/\rho_0$ is varied. (After Ref. 45).

UPt_3, materials spanning $10 \lesssim m^*/m_e \lesssim 100$. This suggests that Eq. 6 may apply to all heavy fermion materials in the low temperature or "coherent" regime.

The exact nature of the pairing interaction in HFSC is not yet clear. The observed enhancement $\beta(T)$ of H_{c2} may be related to any or all of the mechanisms discussed by DF[23]. More theoretical work and experimentation are needed to assess these possibilities. However, the analyses of DeLong et al.[27] strongly suggest the T^2 dependence of β_{DF} is due to the T^2 coefficient of $\rho(T)$ and/or the T-dependence of ξ. U_6Fe and U_6Co are located in the same region ($R \lesssim 3$, $\gamma^* \lesssim \gamma_M^*$) of the R-γ^* plane as high-T_c A15 compounds, and DeLong[9] has summarized the many other similarities between high-T_c and HFSC. In particular, Webb et al.[42] have suggested the T^2 contribution to ρ observed in high-T_c A15 compounds for $T \lesssim 40K$ is due to a high density of low-lying phonon modes at energies $\lesssim 20$ meV. Positive curvature in $H_{c2}(T)$ of clean A15 samples has also been observed, but dismissed as due to inhomogeneities[43].

These observations suggest[27] that a novel pairing interaction associated with the T^2 term of ρ is present in both HFSC ($\gamma^* \gtrsim 10^5$ erg cm^{-3}K^{-2}) and CSC ($\gamma^* \lesssim \gamma_M$). The similarities in the properties of U_6Fe and UPt_3 suggest that the novel pairing generates an anisotropic component of the order parameter $\Delta(\vec{k},\omega)$. Complete gapping of the Fermi surface in the superconducting state would occur for $\gamma^* \lesssim \gamma_M^*$ due to the coexistence of isotropic and anisotropic pairing; whereas, the isotropic component of the pairing is presumably suppressed by magnetic interactions in HFSC in the regime $\gamma^* \gtrsim 10^5$ ergcm^{-3}K^{-2}.

The cross-over from negative to positive $\Delta\rho$ with decreasing temperature observed for U_6Fe, U_6Co and UPt_3, and the apparent magnetic limitations on T_c (R_c and γ_M^*) derived from Fig. 1 support the existence of an anisotropic pairing cutoff T_O comparable to the coherence temperature. The HFSC may then be regarded as "hyperstrong coupled" materials for which $T_{CO} \sim T_O$. T_{CO} cannot exceed T_O, as the anisotropic pairing interaction would be repulsive at energies above the cutoff T_O; and we know of no HFSC for which T_{CO} greatly exceeds the estimated coherence temperature. We expect that in lower mass materials such as A15 compounds, high values of H_{c2} and T_c may be attributed to phonon softening or "conventional" strong coupling effects, since T_{CO} (and therefore T_O) $\gtrsim 10K$ is one to two orders of magnitude higher than in the case of HFSC. The historical experimental limit $T_c \lesssim 23K$ then implies $T_O \gtrsim 30K$ for narrow-band, high-T_c materials.

The case of UBe_{13} and $CeCu_2Si_2$ can also be analyzed within this basic approach[27]. However, these materials generally exhibit a strong, negative $\Delta\rho$ for $T \lesssim T_{CO}$, and the coefficients governing $\Delta\rho(T,H)$ are themselves T-H- dependent[44] making an accurate analysis of H_{c2} much more difficult. T_O is probably very small and H- dependent in this case, placing these materials outside of a fully coherent state[45]. The apparent differences in the properties of particular materials such as U_6Fe and UBe_{13} then would only be due to differences in energy scales such as the coherence temperature $\sim T_O$, T_K and T_F^*, leading to different magnitudes of "heavy fermion" anomalies in observable properties. The isolating region of magnetic order ($\gamma_M^* \lesssim \gamma^* \lesssim 10^5$ erg/cm^{3}K^2) may only be a result of the crossing of θ_D and T_F^* and a consequent, but temporary, disfavoring of the correlations necessary for superconductivity.

ACKNOWLEDGEMENTS

The author wishes to thank Mr. R.B. Mattingly for his assistance in data
analysis, and several colleagues for their informative discussions and
helpful criticisms: Prof. J. Allen, Dr. K. Bedell, Dr. B. Brandow, Dr. G.
Crabtree, Dr. B. Dunlap, Dr. E. Fenton, Prof. φ. Fischer, Dr. R. Klemm, Dr.
S.H. Liu and Prof. D. Wohlleben. Research at Argonne National Laboratory
was supported by U.S.D.O.E. Contract #W-31-109-ENG-38.

REFERENCES

1. L.E. DeLong, Phys. Rev. B 33, 3556 (1986).
2. L.E. DeLong, J. Less-Common Metals 127, 285 (1987).
3. L.E. DeLong, J. Magn. Magn. Mater. 62, 1 (1986).
4. L.E. DeLong, R.P. Guertin, S. Hasanian and T. Fariss, Phys. Rev. B 31,
     7059 (1985).
5. P.A. Lee, T.M. Rice, J.W. Serene, L.J. Sham and J.W. Wilkins, Comments
     Condensed Matter Phys. 12, 99 (1985).
6. Z. Fisk, J.L. Smith, H.R. Ott and B. Batlogg, J. Magn. Magn. Mater. 52,
     79 (1985).
7. A.S. Alexandrov, J. Ranninger and S. Robaszkiewicz, Phys. Rev. B 33,
     4526 (1986).
8. S.H. Liu, preprint.
9. L.E. DeLong, to appear in the Proceedings of the Fifth Int. Conf. on
     Valence Fluctuations, Bangalore, India, 1987.
10. W.R. Johanson, G.W. Crabtree, A.S. Edelstein and O.D. McMasters, J.
     Magn. Magn. Mater. 31-34, 377 (1983).
11. L. Taillefer, R. Newbury, G.G. Lonzarich, Z. Fisk and J.L. Smith, J.
     Magn. Magn. Mater. 63 & 64, 372 (1987).
12. P.H.P. Reinders, M. Springford, P.T. Coleridge, R. Boulet and D.
     Ravot, Phys. Rev. Lett. 57, 1631 (1986).
13. B. Renker, F. Gompf, J.B. Suck, H. Rietschel and P. Frings, Physica
     136B, 376 (1986).
14. G. Gladstone, M.A. Jensen and J.R. Schrieffer in R.D. Parks, ed.,
     "Superconductivity" (Marcel Dekker, New York, 1969), Vol. 2,
     Chapt. 13.
15. K. Andres and M.A. Jensen, Phys. Rev. 165, 533 (1968); M.A.Jensen and K.
     Andres, ibid., p. 545.
16. G. Aeppli, to appear in the Proceedings of the Fifth Int. Conf. on
     Valence Fluctuations, Bangalore, India, 1987.
17. K. Machida and M. Kato, Phys. Rev. Lett. 58, 1986 (1987).
18. D.J. Kim, Phys. Rev. B 34, (1986).
19. N.R. Werthamer, E. Helfand, and P.C. Hohenburg, Phys. Rev. 147, 295
     (1966); K. Maki, Phys. Rev. 148, 362 (1966).
20. L.E. DeLong, G.W. Crabtree, L.N. Hall, H. Kierstead, H. Aoki, S.K.
     Dhar, K.A. Gschneidner, Jr. and A. Junod, Physica 135B, 81 (1985).
21. J.W. Chen, S.E. Lambert, M.B. Maple, Z. Fisk, J.L. Smith, G.R.
     Stewart and J.O. Willis, Phys. Rev. B 30, 1583 (1984).
22. B.S. Shivaram, T.F. Rosenbaum and D.G. Hinks, Phys. Rev. Lett. 57,
     1259 (1986).
23. M. Decroux and φ. Fischer in "Superconductivity in Ternary Compounds
     II," M.B. Maple and φ. Fischer, eds. (Springer-Verlag, Berlin,
     1982), Chapt. 3.
24. L.E. DeLong, L.N. Hall, S.K. Malik, G.W. Crabtree, W. Kwok and K.A.
     Gschneidner, Jr., J. Magn. Magn. Mater. 63&64, (1987).
25. L.E. DeLong, G.W. Crabtree, L.N. Hall, D.G. Hinks, W.K. Kwok, S.K.
     Malik and K.A. Gschneidner, Jr., Proceedings of the Fifth
     International Conference on Valence Fluctuations, Bangalore, India,
     1987.

26. L.E. DeLong, G.W. Crabtree, L.N. Hall, D.G. Hinks, W.K. Kwok, S.K. Malik and K.A. Gschneidner, Jr., Bull. Am. Phys. Soc. $\underline{32}$, 685 (1987).

27. L.E. DeLong, G.W. Crabtree, L.N. Hall, D.G. Hinks, W.K. Kwok, and S.K. Malik, to be published.

28. P. Entel and M. Peter, J. Low Temp. Phys. $\underline{22}$, 613 (1976).

29. M. Decroux, Ph.D. Thesis, U. Geneva, 1980 (unpublished).

30. Z. Fisk and A.C. Lawson, Solid State Commun. $\underline{13}$, 277 (1973); R.O. Elliott, J.L. Smith, R.S. Finocchiaro and D.A. Koss, Mat. Sci. Eng. $\underline{49}$, 65 (1981).

31. A. de Visser, R. Gersdorf, J.J.M. Franse and A. Menovsky, J. Magn. Magn. Mater. $\underline{54-57}$, 383 (1986); K. Kadowaki, A. Umezawa and S.B. Woods, $\underline{ibid.}$, p. 385.

32. G.R. Stewart, Z. Fisk, J.O. Willis and J.L. Smith, Phys. Rev. Lett. $\underline{52}$, 679 (1984).

33. L.E. DeLong, J.G. Huber, K.N. Yang and M.B. Maple, Phys. Rev. Lett. $\underline{51}$, 312 (1983).

34. U. Rauchschwalbe, U. Ahlheim, F. Steglich, D. Rainer and J.J.M. Franse, Z. Phys. B $\underline{60}$, 379 (1985).

35. M.B. Maple, J.W. Chen, S.E. Lambert, Z. Fisk, J.L. Smith, H.R. Ott, J.S. Brooks and M.J. Naughton, Phys. Rev. Lett. $\underline{54}$, 477 (1985).

36. M. Tachiki, T. Koyama and S. Takahashi, Physica $\underline{135B}$, 57 (1985).

37. H.M. Mayer, U. Rauchschwalbe, F. Steglich, G.R. Stewart and A.L. Giorgi, Z. Phys. B $\underline{64}$, 299 (1986).

38. P.H. Frings, J.J.M. Franse, F.R. deBoer and A. Menovsky, J. Magn. Magn. Mater. $\underline{31-34}$, 240 (1983).

39. L.E. DeLong, G.W. Crabtree, L.N. Hall, D.G. Hinks, W.K. Kwok and R.B. Mattingly, to appear in the Proceedings of the Eighteenth Int. Conf. on Low Temp. Physics, Kyoto, 1987.

40. N.F. Masharov, Sov. Phys.: Solid State Phys. $\underline{16}$, 1524 (1975). See Eq. 27 and note $\ln(1+x) \approx x/1+x$ for $0 < x \ll 1$.

41. M.K. Wu, J.R. Ashburn, C.J. Torng, P.H. Hor, R.L. Meng, L. Gao, Z.J. Huang, Y.Q. Wang and C.W. Chu, Phys. Rev. Lett. $\underline{58}$, 908 (1987).

42. G.W. Webb, Z. Fisk, J.J. Engelhardt and S.D. Bader, Phys. Rev. $\underline{B15}$, 2624 (1977).

43. T.P. Orlando, E.J. McNiff, Jr., S. Foner and M.R. Beasley, Phys. Rev. B $\underline{19}$, 4545 (1979).

44. G. Remenyi, D. Jaccard, J. Flouquet, A. Briggs, Z. Fisk, J.L. Smith and H.R. Ott, J. Physique $\underline{47}$, 367 (1986).

45. B. Batlogg, D.J. Bishop, E. Bucher, B. Golding, Jr., A.P. Ramirez, Z. Fisk, J.L. Smith and H.R. Ott, J. Magn. Magn. Mater. $\underline{63\&64}$, 441 (1987).

46. L.E. DeLong and R.B. Mattingly, unpublished results.

PROXIMITY EFFECTS BETWEEN CONVENTIONAL

AND UNCONVENTIONAL SUPERCONDUCTORS

A. J. Millis

AT&T Bell Laboratories
600 Mountain Avenue
Murray Hill, NJ 07974

D. Rainer

Physikalisches Institut
Universität Bayreuth
Bayreuth D8580
West Germany

J. Sauls

Dept. of Physics
Princeton University
Princeton, NJ 08545

In this article I review recent theoretical work on the possibility of using tunneling and proximity effect experiments to study unconventional superconductors. The basic idea is simple: as shown in Fig. 1 one places a thin layer of some well understood conventional superconductor in good metallic contact with the superconductor one wishes to study, and then measures (e.g. by tunneling at the outer edge the conventional layer) how the superconducting properties of the conventional material are altered by its proximity to the unconventional material. To study this question theoretically one has to solve the gap equation for the inhomogeneous system. Because the symmetry and the physical origin of the pairing interaction may be different for conventional than for unconventional superconductors, it is possible that the proximity effect is different between two conventional superconductors than it is between a conventional and an unconventional superconductor. Our understanding of the nature and observability of these differences is still preliminary. A useful theoretical technique for calculating such proximity effects has only recently been proposed, and only a few calculations in simple model systems have been done. The results from this preliminary work are not encouraging: proximity effects involving unconventional superconductivity seem to differ only in subtle ways from those involving only conventional superconductivity; further, proximity effect experiments do not seem to provide a useful method of distinguishing unconventional singlet (e.g. d-wave) from triplet superconductivity.

The rest of this article is organized as follows: in section II the relevant physics of the proximity effect and of unconventional superconductivity is outlined, and the important problems are posed. In section III some theoretical approaches to the problem are outlined and a few results for an s-wave superconductor in proximity to model d-wave or p-wave superconductors are presented. Josephson coupling between conventional and unconventional superconductors is also briefly discussed. There is a brief conclusion.

This article is a review of the physics of proximity effects involving unconventional superconductors. The emphasis is on the important physics; the reader is directed elsewhere for calculational details.

II

In this section the physics of proximity effects involving anisotropic superconductivity is qualitatively discussed. The mathematical formulation of the problem is given in the next section.

We begin with the proximity effect. Study of the proximity effect concerns itself with the question of how the superconducting properties of one material (e.g. transition temperature, magnitude of gap, penetration depth...) are affected when it is placed in good electrical contact with another superconductor or non-superconducting material as shown e.g. in Fig. 1. Proximity effects involving two conventional s-wave superconductors have been studied for many years. For a review see Refs. 1,2. The basic idea behind the proximity effect is simple: if two metals are in good electrical contact, electrons from one material may pass into the other. Thus Cooper pairs from a superconductor may leak into a normal metal, imparting some superconducting properties to a non-superconductor near an interface. Conversely, normal electrons may leak into a superconductor, weakening the superconductivity near the interface.

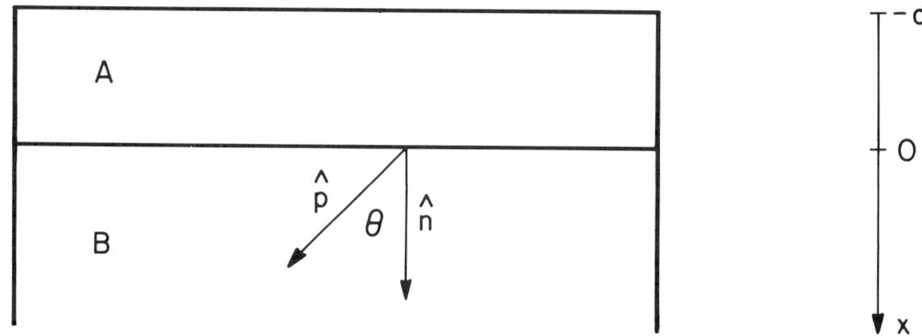

1). Schematic of physical situation considered in this paper. "A" is a thin layer of metal deposited in good electrical contact upon material B, a superconductor to be studied. Coordinates used in section III are also given.

From this physical picture one immediately sees that there are four important parameters governing the proximity effect. These are the bulk superconducting pairing potential and pairbreaking parameters for each of the two materials, the transmissivity of the interface, and a characteristic length ξ_0 for each material. This length, ξ_0, may be defined roughly as follows: the superconducting properties of electrons within a distance ξ_0 of the interface are directly influenced by the presence of the interface and of any material beyond the interface. The superconducting properties of electrons farther than ξ_0 from the interface are, in general, not. We shall make this definition more precise below. The length ξ_0 is usually approximately given by the BCS coherence length: e.g. for a clean material[1] $\xi_0 \sim v_F/\max(T,\Delta)$ (here v_F is the Fermi velocity, T is the temperature and Δ the BCS gap. We use units such that $\hbar = k_B = 1$). Typically $\xi_0 \sim 10^2 - 10^4 \text{Å}$, much larger than a typical atomic dimension; this considerably simplifies the theoretical and experimental problem. Note that it is only if the interface between the two materials is sharp on the scale of ξ_0 that one may parametrize the interface by a transmissivity. Clearly, to maximize the observability of any proximity effect one would like to study films of thickness $d \lesssim \xi_0$, but one would like $d \gg a$, where a is some atomic length (such as the size of a unit cell or a scale determining the sharpness of the interface) so that the intrinsic pairing interaction etc. in the film is the same as in a bulk sample of the same material.

Results that would be obtained from ideal proximity effect experiments involving conventional s-wave superconductivity are sketched in Fig. 2. The system to which these sketches refer is shown in Fig. 1. In Fig. 2a the transition temperature, T_c, of the A-B system is shown as a function of the thickness, d, of material A, for two cases. Curve (1) is for $T_{CB} = 0$, and curve (2) is for $T_{CA} > T_{CB} > 0$. Here $T_{CA,B}$ are the bulk transition temperatures of materials A and B respectively. It is assumed that material B is infinitely thick. In Fig. 2b the superconducting order parameter Δ is plotted as a function of position x (in the direction normal to the A-B interface) for $T_{CB} > T > T_{CA}$. It is assumed that material B is infinitely thick. The length over which Δ varies is set by ξ_0 (for $T > 0$); the discontinuity in Δ is set by the transmissivity of the interface. In Fig. 2 it is assumed that the interface is not pairbreaking (containing, e.g. no magnetic impurities). Then the only role played by the interface is to control the degree to which e.g. the properties of one side influence electrons on the other side near the interface.

The curves in Fig. 2 summarize some of the simplest results of the theory of the proximity effect in conventional superconductivity. From them it is seen that the transmissivity of the interface and the length scale ξ_0 merely fix the relevant length scales and the overall magnitude of the various effects. The important physics is contained in the pairing potentials and pairbreaking parameters of the two materials, which determine the superconducting T_cs. We now consider how these results charge when unconventional superconductivity is involved.

We first define "unconventional superconductivity" and outline some of its properties. Further details and references may be found in[3-4] and in the heavy fermion literature. A homogeneous superconductor is characterized by an order parameter $\Delta_{\alpha\beta}(\vec{k})$ which depends on wave-vector and on two spin variables. The order parameter may be thought of as a bound pair of electrons; $\vec{k}$ is then the Fourier transform of the coordinate describing the relative $(\vec{r}_1 - \vec{r}_2)$ motion, and α

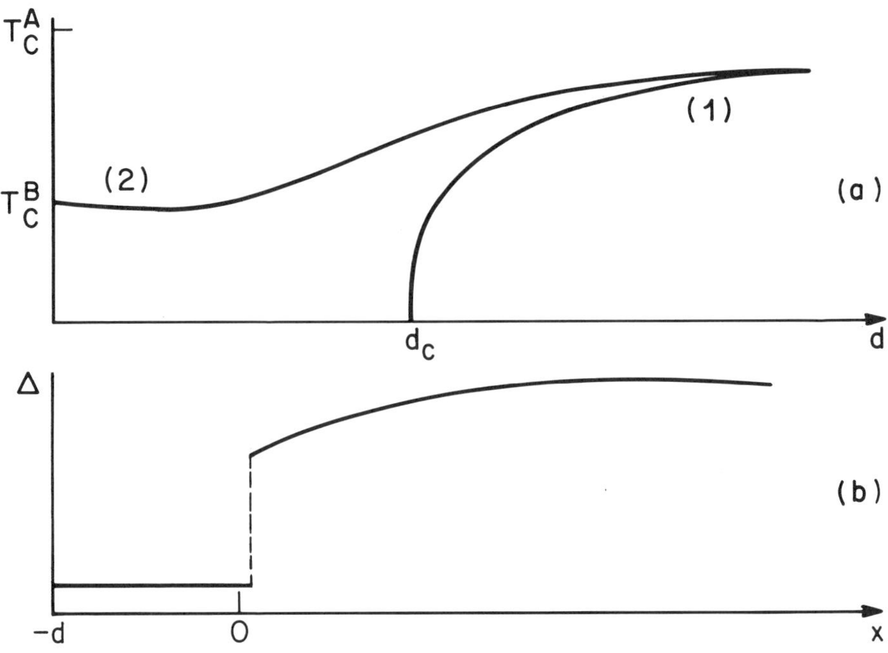

2). Results of proximity effect experiments
a) T_c of A-B system (measured at outer layer of A) for the two cases (1) material B an s-wave superconductor (2) material B a pairbreaker.
b) Gap as function of position for the case $T_{CS} > T > T_{CA}$. It is the thesis of this paper that the curves 1 in 2a and the curve in Fig. 2b apply to the case B=anisotropic ("d-wave" or triplet) superconductor. Only the magnitude of d_c or the discontinuity in Δ distinguish anisotropic from conventional superconductors.

and β are the spins. By the Pauli principle Δ must change sign if we interchange α and β and set $\bar{K} \rightarrow -\bar{K}$. There are then two possibilities: singlet, in which Δ is odd under interchange of α and β and even under $\bar{K} \rightarrow -\bar{K}$ and triplet, in which the reverse is true. In materials with strong spin-orbit coupling (such as the heavy fermion materials), spin is not a good quantum number; however it has been shown by Volovik and Gorkov[5] and by Blount[6] that provided one restricts attention to crystals with inversion symmetry, it is possible to define a "pseudospin" index which is conserved as the electron propagates through the crystal. Pseudospin is related to the conventional spin by a rotation which depends on position on the Fermi surface and thus couples to the magnetic field in a complicated way. But as long as one does not consider magnetic fields, one can, for most purposes, treat pseudospin as ordinary spin.

Thus, a superconducting order parameter may be characterized by its parity under spatial inversion, $\bar{K} \rightarrow -\bar{K}$. Even parity corresponds to spin singlet, odd to spin triplet. One may further characterize a superconducting order parameter by the way it transforms under operations of the crystal symmetry group. If the order parameter transforms into itself under all rotations of the crystal group, one has "s-wave" (and thus, necessarily, singlet) superconductivity, if not, one has "higher partial wave" "anisotropic" or "unconventional" superconductivity. In both s-wave and anisotropic superconductivity one may have an energy gap which vanishes nowhere on the Fermi surface, or on points or (in the case of singlet superconductivity only lines on the Fermi surface.

In an anisotropic superconductor the order parameter must vary in a particular way across the fermi surface. Therefore, any process which does not conserve crystal momentum can be pairbreaking for such a superconductor. For example, the standard techniques for computing the effect of non-magnetic impurities on superconductivity can be used to show that impurity scattering is pairbreaking for higher partial wave superconductors.[8,9] The pairbreaking parameter is of order $(1/\tau T_{co})$, where T_{co} is the transition temperature of a pure crystal and τ the usual impurity scattering time.

Now consider an inhomogeneous situation, e.g. that shown in Fig. 1. The superconducting order parameters will then vary in space, and one must write $\Delta_{\alpha\beta}(\vec{K},\vec{r})$. The coordinate $\vec{r}$ gives the spatial variation of the order parameter and may be thought of as the center of mass of a Cooper pairs. The characteristic scale for variations with $\vec{r}$ is the length ξ_o mentioned above. It is easy to show[9] that up to terms of relative order a/ξ_o, the classification of Δ in terms of symmetry under rotations of spin and $\vec{K}$ applies also to the inhomogeneous case: for this purpose $\vec{r}$ is a dummy variable.

Note that the scattering (reflection from or transmission through) a boundary can change an electron's momentum or spin, and may therefore be pairbreaking for a non s-wave superconductor. Note especially that even an interface with no spin-orbit scatterers or magnetic impurities may have a large amplitude for flipping the "pseudo-spin" of an electron if it connects materials with very different spin-orbit couplings. This pairbreaking effect will be different for different possible anisotropic states and for different orientations of the order parameter in a given state. Thus a boundary will tend to orient and also to change the form of a anisotropic order parameter. This effect has long been known in the context of ^3He[10,11], but the complications induced by a non-spherical fermi surface and by spin-orbit coupling have not so far been considered.

The presence of an interface has another peculiar effect. Essentially because electronic momentum and (pseudo)-spin need not be conserved in interaction with an interface, it is possible in general for an interface to convert e.g. an s-wave spin singlet Cooper pair to a d-wave spin singlet or into a p-wave spin-triplet Cooper pair.[12,14,15]

The result of these considerations is that in general proximity effects involving unconventional superconductivity are more complicated than proximity effects involving only s-wave superconductivity, but there are no qualitative differences, in contrast to a previous claim.[16]

III

In this section we outline various methods of calculating proximity effects, with emphasis on the quasiclassical green function technique, and then give results of a few simple calculations.

The theoretical problem is simple to state: one must solve the BCS-Gorkov (or Eliashberg) equations of superconductivity in the inhomogeneous situation shown in Fig. 1. Even in the familiar s-wave case this is a difficult task because the gap function will vary in space on the scale of the BCS coherence length ξ_o and because one must know something about the normal state electronic wave functions of the system shown in Fig. 1. In practice, two approaches have been taken. One is the Macmillan tunneling model,[17] in which spatial variations of the gap are neglected (except that it may have different values in material A or material B), free electron wave functions are assumed for materials A and B, and phenomenological tunneling matrix element T_{AB} is introduced to provide an amplitude for an electron to go from one material to the other. Because spatial variations of the gap are neglected this method is restricted to situations in which the thicknesses d_A, d_B of the A and B materials satisfy $a < d_{A(B)} < \xi_{A(B)}$. In the MacMillan model, then, the gap in each material has the bulk value appropriate to the material and temperature plus a correction which is of order T_{AB}^2 when T_{AB} is small. The MacMillan model has been applied to proximity effects involving non s-wave superconductors;[18] however this application did not properly include the pairbreaking effect of interfaces on anisotropic superconductivity.[15]

An alternative approach to proximity effects, due to deGennes,[1,2] involves recasting the BCS equations as an integral equation for the gap Δ; one finds for the s-wave case:

$$\Delta_{\alpha\beta}(\hat{p}, x) = \int d^2\hat{p}' \, dx' V(x, \hat{p} \cdot \hat{p}') K_{\alpha\beta;\gamma\delta}(x, x';\hat{p}) \Delta_{\gamma\delta}(\hat{p}', x') \tag{1}$$

Here $V(x)$ is the BCS pairing potential at position x. The p' integral is taken over the Fermi surface. To linear order in the gap function, the kernel K may be expressed in terms of the normal state Green functions of the problem. In the s-wave case (if time reversal invariance applies) deGennes showed it may be expressed in terms of the density-density correlation function of the A-B system. The correct expression for K (including the boundary condition at the A-B interface) may then be deduced from simple physical arguments and related to measurable normal-step quantities. This method may be applied to unconventional superconductivity,[16,19] however, the kernel K turns out to involve higher order correlation functions (current-current, density-current, density-spin current ...) and the correct boundary condition is difficult to deduce, when spin-orbit coupling and a non-spherical Fermi surface are involved. In Refs. 16 and 19 the possibility that passage through even a non magnetic interface could alter the electronic spin as well as momentum was not correctly handled.

A method for computing proximity effects which avoids some of the difficulties of the previously described techniques has recently been devised.[12] It is essentially a way to incorporate interfaces into the technique of quasiclassical Green functions devised by Eilenberger[20] and used extensively to study ^{3}He.[21] In the quasiclassical Green function formalism one eliminates at the outset variations on the atomic length scale a k_F^{-1} and writes equations for quantities that vary on length scales $\sim \xi_0$. These equations contain all of the information needed to compute many physical quantities, including the spatial variation in the gap function, but are substantially less complex than the BCS equations. Because interfaces lead to spatial variations on length scales $\ll \xi_0$, the effect of an interface may be expressed as a boundary condition on the quasiclassical equations.[12,22] The form of this boundary condition has been worked out, in the limit of a weakly transmitting boundary, for an interface which is planar, translationally invariant along the interface and otherwise arbitrary.[12] To compute the boundary condition one must know the S matrix appropriate to the interface. The S matrix gives the amplitude for an incoming particle to be reflected or transmitted with or without its spin being flipped. The S matrix would be difficult to compute, but may be parametrized in a simple way, and the constraints on S imposed by time-reversal, rotation or other symmetries may easily be determined.[12]

Once the S matrix for an interface is given, one may solve the equations for the inhomogeneous system shown in Fig. 1, obtaining the form of the gap function, the superconducting T_c, etc. The simplest case is when the interface is invariant under time reversal and rotations about its normal. One may then parametrize the interface by a transmission matrix of the form[12]

$$S_{\alpha\beta}(\hat{p}) = c(\cos\theta)\delta_{\alpha\beta} + s(\cos\theta)(\hat{n}\times\hat{p})\cdot\vec{\sigma}_{\alpha\beta} \tag{2}$$

Here $\vec{\sigma}$ is the vector of Pauli matrices, $\vec{n}$ is the unit normal to the interface, $\hat{p}$ is a unit vector giving direction on the Fermi surface of material B, $\cos\theta = \hat{n}\cdot\hat{p}$ and c and s are even functions of their argument. In the absence of spin orbit coupling, $s(\theta) = 0$; in heavy fermion materials in which spin-orbit coupling is strong one expects $s \sim c$.

Given an expression for the S-matrix, a general expression for the kernel in the inhomogeneous gap equation (to linear order in the gap functions) may readily be derived.[15] The expression reduces to that given by deGennes' if one restricts to s-wave superconductivity. The full expressions are lengthy; the essential physics is contained in the expression for the case in which x' is in material B and x is in material B:

$$K_{\alpha\beta;\gamma\delta}(x,x') = [|T_{AB}|^2\delta_{\alpha\gamma}\delta_{\beta\delta} + 2\mathrm{Im}(s^{*}c)(\hat{n}\times\hat{p})\cdot\vec{\sigma}_{\gamma\delta}\delta_{\alpha\beta} + ..]e^{x/\xi A}e^{-x'\xi B} \tag{3}$$

Note that $|T_{AB}^2 = |c|^2 + |s|^2$ is related the transmissivity of the interface that would be measured in e.g. a tunneling experiment, while $cs^{*} \leq |T_{AB}|^2$ is difficult to determine by a normal state measurement. If there is spin orbit coupling, $\mathrm{Im}\,s^{*}c \neq 0$. Evidently the kernel contains both terms which do not mix singlet and triplet superconductivity and terms which do. It is therefore possible for e.g. a triplet Cooper pair to cross an interface and convert itself to a singlet pair, or vice versa. The main conclusion of this work follows immediately: in the presence of strong spin-orbit coupling even an interface which is time reversal invariant and rotationally invariant about its normal will couple conventional s-wave superconductivity both to anisotropic singlet (e.g. d-wave) or to triplet superconductivity more or less as strongly as it couples s-wave to s-wave superconductivity.

Very similar arguments have been applied to the study of the Josephson effect between conventional and unconventional superconductors, with similar conclusions.[12,13,14]

In any event, using the kernel given above one may calculate e.g. the form and magnitude of the order parameter induced in the surface layer (material A in Fig. 1) by bulk superconductivity in material B. We assume that material A is then $(d << \xi_A)$ and is an s-wave superconductor with BCS pairing potential λ_A and transition temperature T_{CA} such that $T_{CS} \gtrsim T > T_{CA}$. We linearize in the gap in material B and in the transmissivity of the interface. We consider three cases: (i) B is an s-wave superconductor with gap Δ_B (ii) B is a d-wave superconductor with a gap near the interface of the form $\Delta_{\alpha\beta}(\hat{p}) = (\sigma_2)_{\alpha\beta}[\Delta_{0B}Y_{20}(\hat{p}) + \Delta_{2B}Y_{22}(\hat{p}) + \Delta_{-2B}Y_{2-2}(\hat{p})]$ where σ_2 is a Pauli matrix and Y_{LM} is a spherical harmonic Δ_o and $\Delta_{\pm 2}$ are numbers (iii) B is a p-wave superconductor in an axial[3] state with ℓ vector normal to the interface, so $\Delta_{\alpha\beta}(\hat{p}) = (\hat{p}_x + i\hat{p}_y)((\vec{m}\cdot\vec{\sigma})\sigma_2)_{\alpha\beta}\Delta_B$. The normal to the interface is chosen to be the axis about which spin and orbital angular momentum m is quantized. In (ii) $\vec{m}$ is a vector giving the preferred axis in spin space. The direction $\vec{m}$ is determined by the reflective properties of the interface, and will be discussed in more detail elsewhere.[15] In (ii) and (iii) we have used only those components of the order parameter unaffected by the presence of a perfectly reflecting interface in the xy plane. Other components would (to leading order in T_{AB}^2,) vanish within a distance $\sim\xi_o$ of the interface.

One then finds for $\Delta_A(\bar{p}, x)$, the induced gap in material A, the following expressions:

$$\Delta_A(\hat{p}, x) = \frac{(\lambda_A/\lambda_B)}{(1-\ln T/T_{CA})} C \tag{4}$$

Here C is a constant. For case (i) $C = \int d^2\hat{p}|T_{AB}(\theta)|^2\Delta_B$ for case (ii) $C = \int d^2\hat{p}|T_{AB}(\theta)|^2 Y_{20}(\hat{p}) \Delta_{oB}$; for case (iii) $C = 2\text{Im}(c^*s)f$, where f depends upon the angular average of $\vec{m}$ and $\vec{p}$. Note that the singlet-triplet coupling, in contrast to the singlet-single coupling, involves the relative phase of the spin-independent and spin-dependent amplitudes. One can show that the symmetries of the problem (time reversal, unitarity, rotational invariance) do not require the singlet-triplet term to vanish; however, what determines the phase difference is not currently understood. However, assuming that any term not forbidden by symmetry will occur, one may conclude that except for a numerical factor, the form and magnitude of the induced gap in the layer is independent of the nature of the bulk superconductivity. The constant C is largest in case (i), and may vanish in cases (ii) or (iii) if the bulk order parameter does not have components of the necessary symmetry. C is smaller in case (iii) than in case (ii). Thus, an anomalously weak proximity effect could be a signature of anisotropic superconductivity; however, C is of the same order of magnitude in each of the cases (i)-(iii). Note also that the gap in material A is isotropic in momentum space, provided the pairing interaction in A is isotropic; thus s, p, or d-wave superconductivity in one material may, in the presence of spin-orbit coupling, induce a s-wave gap in another material. Therefore the leakage of Cooper pairs into material A will occur and be measurable regardless of whether A is dirty (and hence pairbreaking for unconventional superconductivity) or not. From this it follows at once that the T_c vs thickness curve (Fig. 2a) for an s-wave superconductor in proximity to an anisotropic superconductor will always be of the form shown in curve (2) in Fig. 2a, independent of the nature of the anisotropic superconductivity, although the numerical values of e.g. d_c will vary. Further results and details of the calculation will be given elsewhere.[15]

In conclusion, proximity effect experiments do not seem to be a useful way of studying unconventional superconductivity, because no qualitative differences from conventional proximity effects are found. However, at present calculations have only been done for simple models; in particular, only specularly reflecting interfaces have been considered. Because diffuse scattering at an interface is to some extent pairbreaking for anisotropic superconductors, one expects in this case the magnitude of e.g. the leakage of Cooper pairs from an unconventional to a conventional superconductor to be reduced relative to that of s-wave to s-wave proximity effects; it is unclear whether this effect is large enough to warrant altering the pessimistic first sentence of the conclusion.

REFERENCES

1) P. G. deGennes, Rev. Mod. Phys. *36*, 225 (1964).
2) G. Deutscher and P. G. deGennes in *Superconductivity* ed. R. D. Parks (Marcel Dekker: New York, 1969).
3) P. W. Anderson and W. F. Brinkman in *The Physics of Solid and Liquid Helium*, ed. K. H. Benneman and J. B. Ketterson (John Wiley and Sons: New York, 1978) p. 177.
4) P. A. Lee, T. M. Rice, J. W. Serene, L. J. Sham and J. W. Wilkins, Comm. Cond. Mat. Phys. *12*, 99, (1986).
5) G. E. Volovik and L. P. Gorkov, Sov. Phys. JETP *61* 843, (1985).
6) E. I. Blount, Phys. Rev. B *32* 2935 (1985).
7) A. A. Abrikosov, L. P. Gorkov and I. E. Dzyaloshinski, *Methods of Quantum Field Theory in Statistical Physics*, trans. R. A. Silverman (Dover: New York, 1963).
8) R. Balian and N. R. Werthamer, Phys. Rev. 131 1553 (1963).
9) A. J. Millis, unpublished.
10) V. Ambegaokar, P. G. deGennes and D. Rainer, Phys. Rev. A 9, 2676 (1974).
11) L. J. Buchholtz and G. Zwicknagl, Phys. Rev. *B23*, 5788 (1981).
12) A. J. Millis, D. Rainer and J. A. Sauls, to be published.
13) J. Sauls, Z. Zou. and P. W. Anderson, unpublished.
14) V. B. Geshkenbein and A. I. Larkin, JETP Lett. 40, p. 395, 1986.
15) A. J. Millis, to be published.
16) A. J. Millis in *Proceedings of the International Conference on Materials and Mechanisms of Superconductivity 1985* eds. K. Gschneider and E. L. Wolf, Physica *135B*, p. 69 (1985).
17) W. L. McMillan, Phys. Rev. *175* 537 (1968).
18) K. Scharnberg, D. Fay, and N. Schopohl, J. de Physique *39*, C6-481 (1978).
19) E. W. Fenton, Sol. State. Comm. *54*, 705 (1985).
20) G. Eilenberger, Z. Phys. *214*, 195. (1968).
21) J. W. Serene and D. Rainer, Phys. Rep. *101*, 221 (1983).
22) A. V. Zaitsev, Zh. Exph. Teor Fiz. *59* 1015 (1984).

CHARGE IMBALANCE RELAXATION AS A PROBE OF

ANISOTROPIC HEAVY FERMION SUPERCONDUCTORS

L.Coffey and T.R.Lemberger

Dept. of Physics
Ohio State University, Columbus Ohio 43210

INTRODUCTION

A quasiparticle charge imbalance [1,2] measures the net charge
associated with the electron and hole-like quasiparticle branches in a
superconductor.A non-zero charge imbalance is created in a
superconductor by injecting a current across the insulating barrier of a
superconductor insulator normal metal (SIN) junction. The charge
imbalance decays by converting into condensate charge with a relaxation
time $\tau_R(T)$ [3,4,5] which is determined by the various pairbreaking
processes in the superconductor and by the tunnelling of excitations
back across the junction into the normal metal [6]. The latter is
described by an inelastic scattering rate $1/\tau_{TUN}$ and results in a rise
in the junction resistance $R(T)$ as the temperature is increased up to
T_C . The excess junction resistance can be studied to yield a variety
of information about the superconductor [7,8,9]

Quasiparticle charge imbalance in the heavy fermion superconductors
UBe_{13} and UPt_3 is the subject of this calculation. $\tau_R(T)$ and $R(T)$ are
calculated using the polar and axial order parameters of superfluid ^3He
as generic models for the order parameters that have been proposed for
the heavy fermion compounds. The primary pairbreaking mechanism
contributing to $\tau_R(T)$, other than $1/\tau_{TUN}$, is quasiparticle scattering
from non-magnetic disorder. Results for $\tau_R(T)$ and $R(T)$ for the polar
case are presented and discussed. A more detailed explanation of the
calculation, along with results for the axial case, will be presented
elsewhere [10].

THEORY AND RESULTS

The total resistance across the SIN junction is given by[8]

$$R(T) = \frac{R_N}{g_{NS}(T) - Q^*/2N_0eV}$$

(1)

where $g_{NS}(T)$ is the usual tunnelling conductance Q^* is the quasiparticle charge imbalance and is defined by [11]

$$Q^* = 2N_0 \int_{-\infty}^{\infty} \int \frac{d\Omega_k}{4\pi} N_1(k,\omega) f^T(k,\omega) \tag{2}$$

where N_1 is the real part of the quasiparticle density of states before averaging over the Fermi surface. The relaxation time $\tau_R(T)$ is defined by

$$\left[\frac{dQ^*}{dt}\right]_i = -\frac{Q^*}{\tau_R(T)} \tag{3}$$

where $[dQ^*/dt]_i$ is the rate at which quasiparticle charge is injected.

The deviation from equilibrium of the quasiparticle distribution function $f^T(\mathbf{k},\omega)$ is calculated in the Keldysh formalism from[11]

$$[\tau_3\omega - \Sigma(k,\omega), g(k,\omega)]_- = 0 \tag{4}$$

where $g(k,\omega)$ is the electronic Greens function matrix[11].

The self energy $\Sigma(k,\omega)$ contains contributions describing the superconducting pairing, the tunnelling of quasiparticles into the superconductor and the proximity induced pairbreaking effect described by $1/\tau_{TUN}$. $\Sigma(k,\omega)$ also contains a contribution describing the pairbreaking due to non-magnetic disorder. The strength of the disorder is characterised by a scattering rate $1/\tau$.

In figures (1) and (2), $R(T)$ and $\tau_R(T)$ are plotted for the polar case respectively for an impurity scattering rate $1/\tau\Delta(0)$ of 0.584 and $1/\tau_{TUN}\Delta(0)$ of 0.0063. As can be seen from figure (1), $R(T)$ increases as the temperature approaches T_C due to the non-equilibrium charge

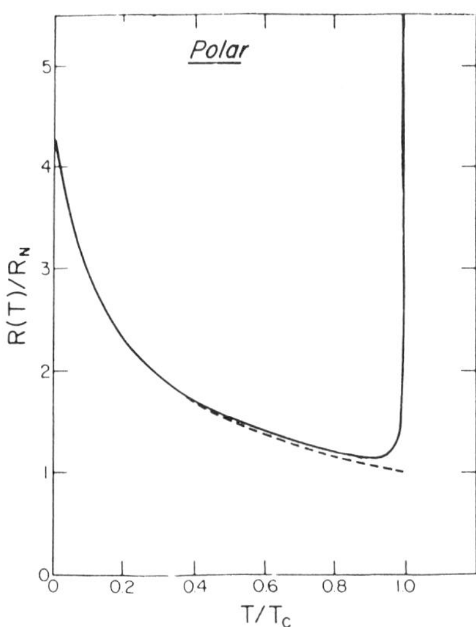

Figure(1): The SIN junction
resistance

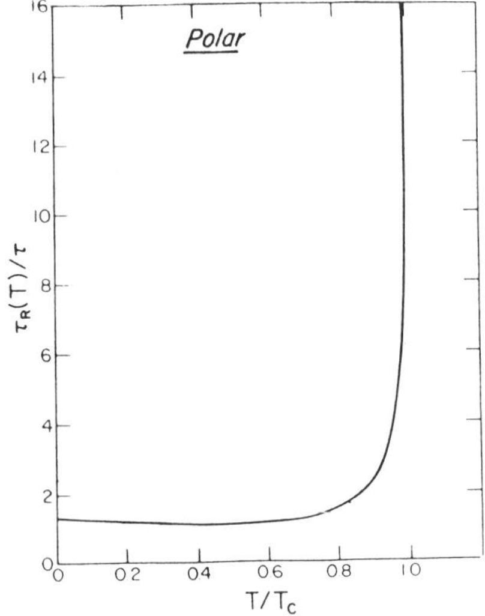

Figure(2):The charge imbalance
relaxation time

imbalance Q^* and it eventually diverges at T_C. For the value of $1/2\tau\Delta(0)$ chosen here $R(T)/R_N \propto (T_C\text{-}T)^{-0.89}$ near T_C. In a conventional BCS superconductor $R(T)/R_N \propto (T_C\text{-}T)^{-0.5}$ near T_C where electron-phonon scattering is the usual pairbreaking mechanism.

At low temperatures $R(T)$ increases in a manner that depends on the low frequency density of states $N_S(\omega)$ and the level of disorder. In the case of a pure polar case , $R(T)$ would diverge as 1/T as the temperature decreases to zero. At low temperatures, $R(T)$ behaves as $R_N(0.2 + 0.8(T/T_c)^{0.74})^{-1}$ for the value of $1/\tau\Delta(0)$ chosen here.

The relaxation time $\tau_R(T)$ diverges at T_c but, unlike $R(T)$, it satur ates to a constant value at zero temperature. This value is comparable to the scattering time τ that characterises the pairbreaking due to the nonmagnetic disorder.

In conclusion, calculating and measuring the temperature and disorder dependences of $\tau_R(T)$ and $R(T)$ provides a novel way to probe the heavy fermion superconducting state. One can also determine the relative importance for $R(T)$ and $\tau_R(T)$ of the various pairbreaking mechanisms such as impurity scattering or inelastic electron electron scattering.

BIBLIOGRAPHY

[1] J.Clarke, Phys. Rev. Lett. 28, 1363 (1972)
[2] J.Clarke and J.L.Paterson, J. Low Temp. Phys. 15, 491 (1974)
[3] M.Tinkham, Phys. Rev. B6, 1747 (1972)
[4] A.Schmid and G.Schon, J. Low Temp. Phys. 20, 207 (1975)
[5] C.J.Pethick and H.Smith, Ann. Phys. (NY) 119, 133 (1979)
[6] T.R.Lemberger, Phys. Rev. B29, 4946 (1984)
[7] T.R.Lemberger, Phys. Rev. Lett. 52, 1029 (1984)
[8] T.R.Lemberger, Y.Yen and S.G.Lee, Phys. Rev. B35, 6670 (1987)
[9] Y.Yen and T.R.Lemberger, in preparation; S.G.Lee and T.R.Lemberger, in preparation.
[10] L.Coffey and T.R.Lemberger, submitted to Phys. Rev. Lett.
[11] J.B.Nielsen, C.J.Pethick and H.Smith, J. Low Temp. Phys. 46, 565 (1982)

EFFECTS OF MASS ENHANCEMENT ON COOPER PAIRING IN HEAVY-FERMION

SUPERCONDUCTORS

E.W. Fenton

Physics Division
National Research Council of Canada
Ottawa, Canada K1A 0R6

INTRODUCTION

With $m^* \approx 200\ m_e$ (including in superconductors $CeCu_2Si_2$, UBe_{13}, and UPt_3) due mainly to many-body enhancement rather than a band structure effect, as confirmed recently by deHaas-van Alphen experiments on $CeCu_6$ and UPt_3[1,2], effects of the mass enhancement taken alone will be shown to be: (a) the coulomb pseudopotential μ^* operating in a frequency range a few times the Debye frequency ω_D is as usual much smaller than μ operating in a much larger range E_F, where E_F is of usual magnitude; (b) "Migdal's theorem" holds for usual electron-phonon interactions and for any interaction operating within a frequency range comparable to ω_D or smaller; and (c) the usual coulomb and phonon-mediated electron-electron interactions cannot be causing the superconductivity. Conditions (a) and (b) occur rather than intuitive results expected if m^* were a band mass because enhancement of the quasiparticle density of states by many-body effects does not mean any enhancement at all of the electron density of states.

With the large mass due mainly to a many-body effect, the quasiparticle trajectory in time and space can be taken as a propagator which has Fourier transform[4,5,6]

$$G(k,\omega) \approx \frac{1}{\omega - \varepsilon_k - \Sigma(\omega)}$$

$$= \frac{1}{Z_{HF}(\omega)(\omega - \varepsilon_k^*)} \tag{1}$$

The frequency renormalization or inverse pole strength factor is $Z_{HF}(\omega) = 1 - \frac{\partial\Sigma(\omega)}{\partial\omega} \gg 1$, at small ω, and $\varepsilon_k^* = \varepsilon_k/Z_{HF}(\omega)$ is the band-structure energy relative to the Fermi level but reduced by many-body effects. The mass m_{band} in ε_k is in effect increased to $m_{band}Z_{HF}(\omega)$ in ε_k^*, and the number of quasiparticle (not electron) states in energy interval $\delta\omega$ at the Fermi level is correspondingly increased.

We turn now to the Eliashberg equations within the framework of the usual theory of superconductivity.[3] The inverse propagator becomes an inverse propagator for a two-component field:

$$\underset{\sim}{G}^{-1}(\underline{k},\omega) = \omega Z(\omega)\underset{\sim}{1} - \varepsilon_{\underline{k}}\tau_3 - \phi(\omega)\underset{\sim}{\tau}_2 \qquad (2)$$

Here $\underset{\sim}{1}$, $\underset{\sim}{\tau}_2$ and $\underset{\sim}{\tau}_3$ are unit and Pauli two-component spinors in the Nambu two-component particle-hole space. $Z(\omega)$ includes $Z_{HF}(\omega)$ and $\delta Z_{phonon}(\omega)$. As usual the contribution to Z from coulomb interactions is frequency-dependent only on the scale of electron volts and can therefore be included as a constant in the band energy $\varepsilon_{\underline{k}}$, the band density of states $N(\varepsilon_{\underline{k}})$, the electron-phonon coupling, and the screened coulomb interaction itself. The Eliashberg equations at zero temperature are[3]:

$$\phi(\omega) = \int_0^\infty \frac{d\omega'}{\pi} \int d\varepsilon_{\underline{k}'} N(\varepsilon_{\underline{k}'}) \mathrm{Im}\left[\frac{\phi(\omega')}{\omega'^2 Z^2(\omega') - \varepsilon_{\underline{k}'}^2 - \phi^2(\omega')}\right] V_\phi(\underline{k},\underline{k}',\omega,\omega') \qquad (3)$$

$$(Z_{HF}(\omega) - Z(\omega))\omega = \int_0^\infty \frac{d\omega'}{\pi} \int d\varepsilon_{\underline{k}'} N(\varepsilon_{\underline{k}'})$$

$$\times \mathrm{Im}\left[\frac{\omega' Z(\omega')}{\omega'^2 Z^2(\omega') - \varepsilon_{\underline{k}'}^2 - \phi^2(\omega')}\right] V_Z(\underline{k},\underline{k}',\omega,\omega') \qquad (4)$$

Here

$$V_\phi = (V_a^{ph}(\underline{k},\underline{k}',\omega+\omega') + V_r^{ph}(\underline{k},\underline{k}',\omega-\omega')) + V_{coul}(\underline{k},\underline{k}')$$

$$V_Z = (V_a^{ph}(\underline{k},\underline{k}',\omega+\omega') - V_r^{ph}(\underline{k},\underline{k}',\omega-\omega'))$$

where a and r refer to advanced and retarded functions for lower and upper halves of the complex frequency plane, and V_{coul} is screened and renormalized by Z_{coul}^{-2} and coulomb three-vertex corrections.[1] The gap function is $\Delta(\omega) = \phi(\omega)/Z(\omega)$.

The full frequency renormalization $Z(\omega)$ is $Z_{HF}(\omega)$ plus a part $\delta Z(\omega)$ due to the electron-phonon interaction. We take[4-6]

$$Z_{HF}(\omega) \gg 1 \quad ; \quad \omega \leqslant E_{HF} \quad ; \quad Z_{HF}\text{real} \qquad (5)$$

$$Z_{HF}(\omega) \approx 1 \quad ; \quad \omega > E_{HF} \qquad (6)$$

For the moment we ignore the imaginary part of Z_{HF}. For the range $\omega \leqslant E_{HF}$, $Z(\omega)$ is dominated by $Z_{HF}(\omega)$ with the condition of inequality 5. With the usual coulomb and phonon-mediated interactions,[1] the integrals in eqs. 3 and 4 are not dominated by the energy range less than E_{HF}, contrary to what would have occurred if m^* were entirely a band mass. In this case we can take advantage as usual of the fact that V^{ph} becomes negligible above a cut-off frequency ω_c which is a few times ω_p. It is easy to show in the usual manner[3] that the upper frequency limit in eq. 3 can be replaced with ω_c if V_{coul} is also replaced by U_{coul} which satisfies the equation:

$$U_{coul}(\underline{k},\underline{k}') = V_{coul}(\underline{k},\underline{k}') + \int\limits_{\omega_c}^{\infty} \frac{d\omega''}{\pi} \int d\varepsilon_{\underline{k}''} N(\varepsilon_{\underline{k}''}) V_{coul}(\underline{k},\underline{k}'')$$

$$\times \text{ Im}\left[\frac{1}{\omega''^2 Z^2(\omega'') - \varepsilon_{\underline{k}''}^2 - \phi^2(\omega'')}\right] U_{coul}(\underline{k}'',\underline{k}') \qquad (7)$$

If $m^* \approx 200\ m_e$ were entirely a band mass, the integral term in eq. 7 with $|\varepsilon_{k''}| \leqslant E_{HF} < \omega_D, \omega_c$ would be essentially zero, and $U_{coul} = V_{coul}$ would occur. However if the large mass is due mainly to a many-body effect, then with $|\varepsilon_{k''}| \leqslant E_F$ where $E_F >> \omega_D, \omega_c$, and if in the wide band the usual approximation is made

$$N(\varepsilon_{\underline{k}}) V_{coul}(\underline{k},\underline{k}') = -\mu \ ; \quad |\varepsilon_{\underline{k}}| \leqslant E_F \ ; \quad E_F >> E_{HF}$$

$$= 0 \ ; \quad |\varepsilon_{\underline{k}}| > E_F$$

then solving eq. 7 leads to the standard result[1]

$$-N(\varepsilon_{\underline{k}}) U_{coul}(k,k') = \mu^* = \frac{\mu}{1 + \mu \ln \dfrac{E_F}{\omega_c}}$$

and $\mu^* << \mu$ as usual. Entirely different results for μ^* are obtained for $m^* \approx 200\ m_e$ due to a many-body effect than if it were a band mass.

MIGDAL THEOREM

The "Migdal theorem" states that in the electron-phonon scattering, corrections of diagrams (b), (c) and so on to diagram (a) in Figure 1 are of order ω_D/E_F and may be neglected.[7,8] For the system with no f electrons and $Z_{ph}(\omega)$ comparable to unity, the three-vertex correction in Fig. 1(b) is

$$\Gamma_o^{(b)} = i \int \frac{d\omega d\underline{k}}{(2\pi)^4} \ |g^2| D(\underline{k}-\underline{k}') G_o(\underline{k}') G_o(\underline{k}'+\underline{q}) \qquad (8)$$

where g and D are the <u>usual</u> bare three-vertex and phonon propagator. For phonons with $qv_F >> \omega$,

$$\Gamma_o^{(b)} \approx \frac{N(0) |g|^2}{\omega_D} \frac{\omega_D}{E_F} << 1$$

When we replace G_o for the system which has no f electrons with G_{HF} for the heavy-Fermion system, then

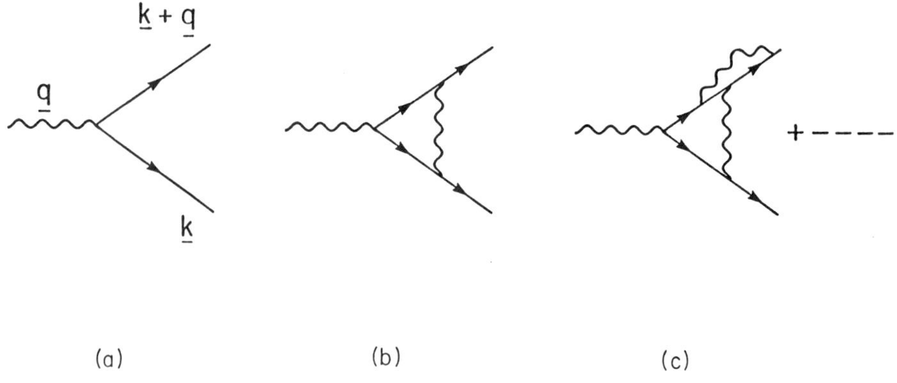

(a)   (b)   (c)

Fig. 1. The bare electron-phonon three-vertex, (a), and corrections (b), (c), ... to the bare vertex which are neglected in the usual theory of superconductivity[6].

$$\Gamma_{HF}^{(b)} \approx \Gamma_o^{(b)} \times \frac{N_{HF}(0)}{N_o(0)} \times \frac{1}{Z_{HF}^2}$$

$$= \frac{\Gamma_o^{(b)}}{Z_{HF}} \ll \Gamma_o^{(b)} \ll 1 \quad ; \quad \omega \lesssim E_{HF} \tag{9}$$

Here $N_{HF}(0)$ is the quasiparticle density of states for heavy-Fermions at the Fermi surface.

The result that $\Gamma_{HF}^{(b)} \propto \frac{\omega_D}{Z_{HF}E_F}$ rather than $\frac{Z_{HF}\omega_D}{E_F}$ suggested by Rice[9] occurs because the <u>electron</u> density of states is unchanged from $N_o(0)$. Only the <u>quasiparticle</u> density of states is enhanced by the factor $Z_{HF} \gg 1$. In the three-vertex diagram of Figure 1(b), the small quasiparticle pole strength $Z_{HF}^{-1} \ll 1$ appears in second order, and the large quasiparticle density of states enhancement $Z_{HF} \gg 1$ appears only in first order.

As discussed by Scalapino[7] and by Engelsberg and Schrieffer[8], Migdal's theorem does not apply to some optical phonons where $\omega \gg qv_F$. The same situation occurs for heavy-Fermion systems. In both cases, for $\omega \gg qv_F$ the electron-phonon three-vertex is given by $\lambda_\omega = 1 - \frac{\delta\Sigma}{\delta w} = Z(\omega)$. In this case, for such phonon states where Migdal's theorem does not apply in usual superconductors, it applies even less-well in heavy-Fermion superconductors. However, for both usual superconductors and heavy-Fermion superconductors, the number of such phonon states constitutes an extremely small part of the phonon spectrum, determined by $\left(\frac{q_{max}}{q_{Debye}}\right)^3 \approx \left(\frac{\omega_D}{E_F}\right)^3$. Absence of Migdal's theorem for such a small number of phonon states corresponds to a correction to the usual superconductivity theory which is even smaller than the neglected $\frac{\omega_D}{E_F}$ three-vertex correction for some average phonon state, even for heavy-Fermion systems.

TRANSITION TEMPERATURE WITH USUAL INTERACTIONS

We take the usual λ_{ph} due to phonons and the usual μ^* due to the coulomb interaction, and investigate the effect of changing $Z(\omega)$ alone to $Z_{HF}(\omega)$ in systems where f electrons are added. We can trivially fold in the frequency range for electron energies (not interaction frequency) from E_F to ω_D as in BCS theory, and in the zero to ω_D range replace $\lambda_{ph}(\omega)$ and μ^* with a simple λ_o defined by T_{co} for the corresponding non-f electron compound,

$$T_{co} = 1.134\, \omega_D e^{-1/\lambda_o} \tag{10}$$

With $Z_{ph} \rightarrow Z_{HF+ph} \approx Z_{HF}$ for $0 < \omega \lesssim E_{HF}$, and $Z = Z_{ph}$ for $E_{HF} < \omega < \omega_D$, we find for these two frequency ranges $\lambda_1 = \lambda_o/A$ (where $A = \frac{Z_{HF}}{Z_{ph}}$) and $\lambda_2 = \lambda_o$.

At $T = 0$ and $\Delta_o \ll E_{HF}$ (and $T_{co} \ll E_{HF}$),

$$\Delta_1 = \frac{\lambda_o \Lambda_1}{A} \ln\left(\frac{2E_{HF}}{\Delta_1}\right) + \frac{\lambda_o \Delta_2}{A} \ln\left(\frac{\omega_D}{E_{HF}}\right) \tag{11}$$

$$\Delta_2 = \lambda_o \Delta_1 \ln\left(\frac{2E_{HF}}{\Delta_1}\right) + \lambda_o \Delta_2 \ln\left(\frac{\omega_D}{E_{HF}}\right) \tag{12}$$

Using $\Delta_2 = A\Delta_1$, we obtain immediately

$$\Delta_{HF} = 2E_{HF} e^{-A/\lambda_o} \left(\frac{\omega_D}{E_{HF}}\right)^A$$

$$= \Delta_o \left(\frac{\Delta_o}{2E_{HF}}\right)^{A-1} \quad ; \quad A = \frac{Z_{HF}}{Z_{ph}} \gg 1 \tag{13}$$

Similar equations for T_c show that $\frac{T_c}{\Delta}$ for the heavy-Fermion system is the same as the ratio $\frac{T_{co}}{\Delta_o}$ for the corresponding compound with no f electrons. T_c for heavy-Fermion systems is exponentially small and essentially zero when T_{co} for the corresponding compound with no f electrons satisfies $T_c \ll E_{HF}$.

Since the phonon-mediated interaction becomes small at low interaction frequencies because the phonon density of states become proportional to ω^2, the electron-electron interaction becomes repulsive at very small ω, as well as when $\omega \gg \omega_D$. This can be represented by taking a repulsive-interaction $-\lambda_1$ for $0 < \omega \leqslant \Omega_1$, and an attractive interaction λ_2 for $\Omega_1 < \omega \leqslant \omega_D$, with $\Omega_1 \ll \omega_D$. With E_{HF} comparable to Ω_1, the possibility exists that when the total $\lambda_o < 0$ and $T_{co} = 0$ for the corresponding compound with no f electrons, then with $Z_{ph} \rightarrow Z_{HF} \gg Z_{ph}$ for $0 < \omega \lesssim E_{HF}$ effectively eliminating the repulsive $-\lambda_1$, then $\lambda_{HF} > 0$ and $T_c \neq 0$ might occur for the heavy Fermion system. However solving gap equations for Δ_1 and Δ_2 in the two frequency ranges shows that <u>both</u> $-\lambda_1$

and λ_2 interactions are reduced by the factor $A = Z_{HF}/Z_{ph}$. In this case, when $\lambda_0 < 0$ so that $T_{co} = 0$, then $\lambda_{HF} < 0$ and $T_c = 0$. When $\lambda_0 > 0$, the situation becomes exactly the same as we have discussed above for the case where the electron-electron interactions contributing to λ_0 are positive in all frequency ranges below ω_D, i.e. T_c for the heavy-Fermion system becomes exponentially small and is essentially zero.

When $\Delta_0 = E_{HF}$, gap equations analogous to eqs. 11 and 12 lead immediately to $\Delta_1 = \frac{\Delta_0}{A}$ for $0 < \omega \leqslant \Delta_0$ (with $\Delta_0 = E_{HF}$), and $\Delta_2 = \Delta_0$ for $\Delta_0 < \omega < \omega_D$. As we have shown earlier,[6]

$$T_c = \frac{T_{co}}{Z_{HF}(0)/Z_{ph}(0)} \quad ; \quad \Delta_0 \approx E_{HF} \tag{14}$$

for this case where the gap function Δ_0 of the corresponding compound with no f electrons is comparable to the energy scale E_{HF} for the heavy-Fermion system. At least before the advent of high-T_c oxide superconductors, eqs. 11-13 meant that with $T_{co} < 24$ K, the maximum transition temperature that could occur for heavy-Fermion superconductivity caused by the usual phonon-mediated interaction would have been roughly 1 K when $Z_{HF} \approx 20$.

Since the compounds with no f electrons which are analogous to $CeCu_2Si_2$, UBe_{13}, and UPt_3 have either unobservably low T_c or $T_c = 0$, eqs. 11-13, mean unambiguously that heavy-Fermion superconductivity must occur because of some special electron-electron interaction or "Kondo boson" in the f electron compounds. This Kondo boson may be, among other possibilities, a phonon-mediated interaction strongly enhanced by a Kondo volume effect, or a magnon-mediated interaction.

We thank H. Rietschel for an informative discussion of the
limitations of the usual superconductivity theory with the usual coulomb
pseudopotential, limitations which occur even for usual
superconductors.[10,11]

REFERENCES

1. P.H.P. Reinders, M. Springford, P.T. Coleridge, R. Boulet and
    D. Ravot, Phys. Rev. Lett. $\underline{57}$, 1631 (1986); and J. Mag. & Mag. Mat.
    $\underline{63\&64}$, 297 (1987).
2. L. Taillefer, R. Newbury, G.G. Lonzarich, Z. Fisk and J.L. Smith, J.
    Mag. & Mag. Mat. $\underline{63\&64}$, 372 (1987).
3. D.J. Scalapino, in "Superconductivity", ed. R.D. Parks (Dekker, New
    York) 1969, p. 449.
4. C.M. Varma, Phys. Rev. Lett. $\underline{55}$, 2723 (1985).
5. H. Fukuyama, Solid State Sciences $\underline{62}$, 209 (1985).
6. E.W. Fenton, Solid State Comm. $\underline{60}$, 351 (1986).
7. Ref. 3, pp. 473-477.
8. S. Engelsberg and J.R. Schrieffer, Phys. Rev. $\underline{131}$, 993 (1963).
9. For $Z_{HF} \rightarrow Z$, T.M. Rice, Solid State Sciences $\underline{52}$, 178 (1984), see
    especially p. 182.
10. H. Rietschel, private communication.
11. H. Rietschel and L.J. Sham, Phys. Rev. B $\underline{28}$, 5100 (1983).

PARAMETERS AND EXOTIC PROPERTIES OF HIGH T_c SUPERCONDCUTORS

V.Z. Kresin

Materials and Chemical Sciences Division
Lawrence Berkeley Laboratory
University of California
Berkeley, CA 94720, U.S.A.

S.A. Wolf

Naval Research Laboratory
Washington, DC 20375

ABSTRACT

A method of determining the values of main parameters such as the
effective mass, the Fermi energy, and the coherence length for the new
high T_c superconductors is developed. The method is based on specific
heat data. The new T_c materials are low dimensional systems and this
feature plays a crucial role in the analysis. Particularly interesting is
the small value of the Fermi energy. The result of the analysis shows
that we are dealing with unusual systems. Energy gap appears to be comparable with E_F, and the coherence length is small. Contrary to the usual
case, the large fraction of carriers is paired.

INTRODUCTION

Since the recent discovery of very high transition temperature superconducting ceramics[1] there has been great activity associated with
characterizing these materials. In order to carry this analysis forward
one must be sure to take into account the dimensionality of the carriers
that are a consequence of the constraints of the structure. In the course
of determining the structure and affects of the structure on superconductivity it has been concluded that these superconducting oxides are lower
dimensional systems. In fact, the superconductors based on the K_2NiF_4
structure such as $La_{2-x}Sr_xCuO_4$ ($x \simeq 2$) are highly two dimensional[2] while
those based on the $Y_1Ba_2Cu_3O_7$ structure contain one dimensional
chains.[3,4] This paper is therefore concerned with the determination of
the materials parameters when the dimensionality is reduced. Expressions
for effective mass, Fermi energy, Fermi velocity and coherence length will
be derived for both one and two dimensional systems.

We will attempt to evaluate many of the materials parameters based on
some recent experimental data.[5-7] Since many of the important parameters can be estimated using only structural information and knowledge of
the Sommerfeld constant γ, the most reliable source for this constant is

electronic heat capacity measurements. For isotropic superconductors, γ can be estimated from the slope of the upper critical field H_{c2} at T_c.[8] However for polycrystalline samples containing anisotropic crystals this is not very reliable. In this paper we focus on the determination of the main parameters from the experimental data on heat capacity. It will be shown that we are dealing with an exotic system with unusual properties.

The paper consists of two parts. In Sec. 2 the expressions allowing us to determine the parameters directly from heat capacity measurements will be derived. We considered both the 2D and 1D cases. It will be shown that the parameters are very sensitive to the dimensionality, and the analysis should be carried out in a consistent way. A set of parameters for the La-Sr-Cu-O and Y-Ba-Cu-O systems is obtained (Sec. 3).

THEORY

2D Case

Consider a model containing 2D sheets. Each of them contains a 2D Fermi gas of carriers. There are a small number of interlayer transitions which are important if we are concerned with 2D fluctuations. But for our present purpose we can consider isolated 2D subsystems. Superconducting pairing occurs between two electrons (or holes) belonging to the same sheet. The energy in the normal state is equal to:

$$E = \sum_i \int \phi_i \, f_i \, d\epsilon_i \, , \tag{1}$$

where $\phi_i = \epsilon_i \, \nu_i^{2D}$, $\nu_i^{2D} = 2(dp_x \, dp_y/d\epsilon_i)S_i \, (2\pi \hbar)^{-2}$ is the 2D density of states (DOS), S_i is the area, $f_i = [\exp(-(\epsilon_i - \epsilon_{Fi})/T + 1)]^{-1}$, ϵ_{Fi} is the Fermi level for the i-th sheet; the summation in (1) is taken over all 2D sheets. The DOS in the 2D case can be written in the form

$$\nu_i^{2D} = m_i \, (2\pi \hbar^2)^{-1} \, , \tag{2}$$

where m_i is the effective mass defined by the relation: $m_i = \int dl v_i^{-1}$; here $v_i = \partial \epsilon_i / \partial p_i$ and the integration is over the Fermi curve. For a simple quadratic dispersion relation $m_i = p_i / v_i$. Calculating the energy with the use of the expression $E = E_o + \sum_i (\partial \phi / \partial \epsilon_i)_F \, (k_B T)^2$, we arrive at the following expression for the specific heat: $C = \gamma T$, where

$$\gamma = (\pi/3 \, \hbar^2) \, k_B^2 \sum_{i=1}^{k} S_i \, m_i \, . \tag{3}$$

Consider the case when all the 2D sheets are equivalent and equidistant. Then we obtain

$$\gamma = (\pi/3 \, \hbar^2) \, m^* \, k_B^2 \, a^{-1} \quad erg/K \, cm^3 \, . \tag{4}$$

Here a is the interlayer distance, $m^* = m_i$, k_B is the Boltzmann's constant. It is essential that in the 2D case γ does not depend on the electron concentration. The situation is entirely different in the 1D (see below) and 3D cases.

The quantity γ can be measured experimentally; then the effective mass can be determined directly from the equation

$$m^* = (3 \hbar^2/\pi) \, k_B^{-2} \, a \, \gamma \ . \tag{5}$$

We will use this expression in order to evaluate m^* for the La-Sr-Cu-O system (see Sec. 3).

We have noted that γ for a 2D system does not depend on the carrier concentration. The latter affects noticeably the Fermi momentum which is equal to (see e.g., ref. [9]):

$$p_F = (2\pi \, N_s)^{1/2} \, \hbar \ , \tag{6}$$

where N_s is the surface concentration. In our case $N_s = na$, where n is the usual volume concentration. The values of the Fermi energy $\epsilon_F = p_F^2/2m^*$ and the Fermi velocity can be evaluated with the use of Eqs. (5), (6). Then the coherence length $\xi_o = (\hbar \, v_F/2\pi kT_c)$ can be also calculated (see below, Sec. 3).

<u>1D Case</u>

Consider the model which is a set of 1D channels (lines). The energy can be described by Eq. (1) with the substitution $\nu^{2D} \to \nu^{1D}$, where ν^{1D} is DOS in the 1D case. As a result we obtain the following expression for γ:

$$\gamma = (\pi/12 \ \hbar) \, k_B^2 \sum_i^k l_i \, v_i^{-1} \ ; \tag{7}$$

where l_i is the length of the i-th line. If all lines are equivalent, we obtain:

$$\gamma = (\pi/12 \ \hbar) \, k_B^2 \, \sigma^{-2} \, v_F^{-1} \ . \tag{8}$$

Here σ^{-2} is the number of lines per unit area. Note that in the 1D case the Fermi momentum is equal to (see e.g. Ref. 9) $p_F = 2\pi \, \hbar \, N_L$, where N_L is the linear carrier concentration.

Consider the special case when the system is a combination of 2D and 1D structures (probably, it is related to $Y_1 \, Ba_2 \, Cu_3 \, O_7$ structures, see Ref. 3,4). Namely, the system is layered, similar to the case studied in p.1, but each layer consist of a set of 1D parallel chains. In this case,

$$p_F = 2\pi \, \hbar nab \tag{9}$$

where b is the distance between neighboring chains (within one layer). The effective mass is described by the expression

$$m^* = 12(\hbar \, k_B^{-1} \, ab)^2 n \, \gamma \ . \tag{10}$$

THE PARAMETERS OF THE La-Sr-Cu-O AND Y-Ba-Cu-O SYSTEM

Eqs. (5) and (6) can be used in order to calculate the main parameters of high T_c 2D superconductors. In this section we apply the method described in Sec. 2 for a calculation of the parameters for $La_{2-x} Sr_x CuO_4$, $x \simeq 0.2$. Our method is based on the heat capacity measurements. These measurements are difficult because of a large lattice contribution. The experiments have been carried out by several groups[5-7] by various methods. Different values of γ have been reported. For example, according to refs. [5,6] $\gamma \simeq 7.5$ mJ/mole K^2. The value $\gamma \simeq 12$ mJ/mole K^2 was obtained in [7]. The value n depends on the quality of the sample, on the concentration x; according to experimental data[10,11] n is within the interval 10^{21} cm^{-3} < n < 5 × 10^{21} cm^{-3}. For the purpose of an estimate, we can put $\gamma \simeq 10$ mJ/mole K^2, n $\simeq 5 \times 10^{21}$ cm^{-3}.

The effective mass m^* can be calculated with the use of Eq. (5); a = 6.6 Å [2]. Then we obtain $m^* \simeq 5.5$ m_e. The value of p_F can be calculated with the use of Eq. (6), and we obtain $p_F = 5 \times 10^{-20}$ g-cm/sec. It is interesting to note that despite the relatively small value of n, the value of p_F is close to that in some metals. This is due to the 2D structure of the material. The Fermi energy is small ~ 0.15 eV, because of the large value of m^*.

Superconductivity in materials with such a small Fermi energy is quite unusual. The fact that E_F and the energy gap Δ are comparible is unprecedented. The small value of E_F might have a strong impact on lattice instability. This problem will be discussed in detail elsewhere.

The Fermi velocity is equal to $v_F \simeq 8 \times 10^6$ cm/sec, and the coherence length $\xi_o = \hbar v_F/2\pi kT_c$ appears to be about 25 Å.

We have estimated the parameters by putting $\gamma \simeq 10$ mJ/mole K^2, n $\simeq 5 \times 10^{21}$ cm^{-3}. It is important to stress that the exact values of the parameters can be obtained from Eq. (5); their accuracy can be improved with more precise measurements of γ and n.

A detailed analysis has been carried out in [5]; the magnetic field dependence of the specific heat was studied. According to [5], the value of γ is within the interval 6.5 mJ/mole K^2 < γ < 8.5 mJ/mole K^2 (the spread is due to the uncertainty in the value of H_{c2} which determines the amount of the normal phase present). A close value of γ was obtained in [6] by a different method. The parameter values obtained from Eq. (5) with $\gamma = 7.5$ mJ/mole K^2 and n = 3 × 10^{21} cm^{-3} are given in the table.

The measurements of heat capacity for $Y_1 Ba_2 Cu_3 O_7$ have been performed [13,14]. Our analysis in this case is based on Eqs. (9), (10). We use the values $\gamma \simeq 20$ mJ/mole K^2 [13], n $\simeq 5 \times 10^{21}$ cm^{-3}. The obtained values of parameters are presented in the table.

CONCLUSION

The values of the parameters depend strongly on the dimensionality of the system. The new high T_c superconductors are low dimensional materials and the evaluation of the parameters should be carried out with considerable care. In many instances, 3D expressions have already been used, leading to incorrect results. In this paper we derived the expressions for such parameters as m^*, p_F, v_F, ϵ_F and ξ_o based on specific heat data. We calculated the values of the parameters. In particular, we would like to stress the small value of the Fermi energy. The low dimensionality plays a crucial role in this analysis.

We would like to stress two important results. First of all, the ratio Δ/T_c is much larger than for conventional superconductors. It means that a large fraction of carriers in the new materials are paired. Note that the expression $\xi_o = \hbar v_F(2\pi T_c)^{-1}$ is valid, if $\Delta \ll \epsilon_F$ (see, e.g. [15]). The smallness of the coherence length also makes the systems unique.

Table 1.

	m^*	p_F	ϵ_F	v_F	ξ_o
$La_{1.8}Sr_{0.2}CuO_4$	$4\ m_e$	3.7×10^{-20} gm × cm/sec	0.12 eV	9.25×10^6 cm/sec	23.5 Å
$Y_1Ba_2Cu_3O_7$	$\sim 10^2\ m_e$	1.5×10^{-19} gm × cm/sec	0.07 eV	1.2×10^6 cm/sec	*

*Expression $\xi_o = \hbar V_F(2\pi kT_c)^{-1}$ is not applicable for this case for such small ϵ_F and such large T_c.

Acknowledgments

The authors are grateful to K. Müller and A. Stacy for valuable and interesting discussions. This work was supported in part by the Office of U.S. Naval Research under Contract No. N00014-86-F0015 and carried out at the Lawrence Berkeley Laboratory under contract No. DE-AC03-76SF00098.

REFERENCES

1. J. Bednorz and K. Müller. Z. Phys. B66, 189 (1986).
2. M. Tahagi et al. Jpn. J. Appl. Phys. 26, L123 (1987).
3. F. Beech et al. (preprint).
4. L. Toth et al. Phys. Rev. Lett. (submitted).
5. N. Phillips et al. (preprint).
6. B. Batlogg et al. Phys. Rev. B35, 5340 (1987).
7. S. Tanaka et al. Proc. of MRS meeting (Anaheim, 1987; in press).
8. T. Orlando et al. Phys. Rev. B35, 5347 (1987).
9. V. Kresin, Phys. Rev. B25, 157 (1982); B34, 7587 (1986).
10. A. Panson et al. Appl. Phys. Lett. 50, 1104 (1987); M. Tonouchi et al. Jpn. J. of Appl. Phys. 26 L519 (1987); N. Ong et al. (preprint).
11. W. Kwok et al. Phys. Rev. B35, 5343 (1987); S. Uchida et al. Jpn. J. of Appl. Phys. 26, L443 (1987).
12. H. Junod et al. (preprint).
13. N. Phillips et al. (preprint).
14. O. Fisher et al. (preprint).
15. P. G. DeGennes, "Superconductivity of Metals and Alloys," W. A. Benjamin, Inc., New York (1966).

THE GINZBURG CRITERION IN HIGH Tc OXIDES

G. Deutscher

Department of Physics and Astronomy
Tel Aviv University
Ramat Aviv, Tel Aviv University, Isreal

Abstract

The application of the Ginzburg criterion yields for the new high Tc
oxides a broad critical region, with a width of order unity. This
width could be reduced in the event of large strong coupling corrections

Most of the recent theoretical interest in the newly discovered
superconducting oxides has been directed towards the various mechanisms
that may be responsible for their high critical temperature. In contrast
we wish here to point out to a basic feature that differentiates these
oxides from ordinary superconductors, irrespective of the mechanism
involved. Based upon available data, we show that the Ginzburg criterion
yields for the critical region, with a width of order unity. This makes
mean field theories inapplicable to the new superconductors. However,
this conclusion may be to some extent altered in the event of a very
strong coupling correction.
The Ginzburg criterion essentially states that the width of the critical
region in a second order phase transition is determined as the range of
temperature where the condensation energy per coherence volume is of
order unity. It is known that this range is extremely small for
superconductors. Expressed in terms of the reduced temperature
$\varepsilon = |T_C-T|/T_C$, it is typically to the order of 10^{-10}. This constitutes
the basis for the validity of the mean field theories of superconductivity
BCS and Landau Ginzburg.
The width of the critical Region can be calculated from experimental data
when cast into the form:

$$[H_c2 \ (\varepsilon)/8\pi] \ \xi^3 \ (\varepsilon) < k_D T_C \tag{1}$$

From an analysis of the available data (jump of the heat capacity at T_C,
critical fields, Hall effect and conductivty), Bardeen concludes that
$H_C(T=0) = 12,000$ Oe and $\xi(T=0) = 12A$. We note that the value of H_C can
be directly obtained from the jump in the heat capacity at T_C, and that
the value quoted is consistent with the field mean approximation. Using
$H_C(\varepsilon) = 2 \ H_C(T=0)\varepsilon$ and the clean limit result $\xi(T)=.74\xi(0)\varepsilon^{-1/2}$, one
obtains the width of the critical region $\varepsilon<0.7$. Thus, the critical
behavior of the new high T_C oxides should be similiar to that of
superfluid He^4, rather than to that of conventional superconductors.
The calculated critical width depends of course on the values used for

H_c and ξ, which might have to be modified if more accurate data becomes available. But we expect non mean field behavior to be observable in any case near T_c.

Finally, we note that the width of the critical region is sensitive to strong coupling corrections. Using $\xi(0) = 0.18 h v_F/k_B T_c$, $H_{c2}(0)/8\pi = 1/2\, N(0)\Delta^2(0)$ and $2\Delta(0) = 3.5\eta k_B T_c$, eq.1 can be rewritten in the clean limit as:

$$\varepsilon < [(13/\eta)(k T_c/E_F)]^4$$

If strong coupling effects turn out to be important, i.e. if the heat capacity at T_c has been actually underestimated by current experiments, the width of the critical region is significantly reduced. Thus, the observation of a broad or narrow critical region might be indicative respectively of weak or strong coupling behavior. As outlined by Bardeen, this has an immediate bearing on the mechanism of the high T_c in the high T_c oxides.

I wish to thank Vladimir Kresin for encouraging me to write this note, and it is a pleasure to acknowledge the hospitality of the Department of Applied Physics at Stanford University during the preparation of this manuscript.

<u>References</u>

1) J. Bardeen, these Proceedings

RVB THEORY OF HIGH T_c SUPERCONDUCTIVITY*

Philip W. Anderson

Department of Physics: Joseph Henry Laboratories
Jadwin Hall, Post Office Box 708
Princeton, New Jersey 08544

The following is a contribution from the entire Princeton group
(listed at the end of the paper) as well as benefiting from a number of
friends and helpers who are working independently. More experimentalists
than I can name have helped us.

I intend here to be neither contentious nor poetic, both of which I
have been accused of being. I want to describe what we know about the RVB
theory. This theory, which is founded in the grubby realities of the
physics of magnetic oxides, contains no wishful thinking as to parameter
values and unusual mechanisms, unlike any other that I know of; in fact,
we are desperately trying to figure out why T_c is so low--if it is. On
the other hand, in the end it runs up against some extraordinarily thorny
theoretical difficulties, but before it does so it explains a remarkable
range of experimental facts, some quite unexpected and strange, and even
better, leaves the door open to explanations for a great number of others.

We start from the premise that the CuO_2 layer lattice (and, if it is
relevant, as I believe it is not, the CuO chain) is appropriately described
by the simplest Hubbard model. Because of the large Jahn-Teller distortion
the 2p - $d_{x^2-y^2}$ admixture stabilizes the square planar structure while
providing a relatively large hopping integral between the $d_{x^2-y^2}$-based
Wannier functions, which are isolated from all other relevant bands. This
is a good description of the band calculations, and when coupled with a
reasonable U-value also explains the fairly large exchange integral
$J = t^2/U$ of order 1 - 2000°K. (I say it is large because it must be much
higher than the La_2CuO_4 antiferromagnetic T_c of 250°K, which results from
higher-order effects such as interplane coupling and anisotropy.)

Thus most of the physics must be a result of the Hubbard model
Hamiltonian

$$-t \sum_{<ij>} c_{i\sigma}^{+} c_{j\sigma} + U \sum_{i} n_{i\uparrow} n_{i\downarrow} \tag{1}$$

____

*Work was supported by NSF Grant No. DMR 8518163.

and its canonically transformed "superexchange" version appropriate for
large U and doping with acceptors:

$$- t \sum (1 - n_{i\,-\sigma}) \; c^+_{i\sigma} \; c_{j\sigma} \; (1 - n_{j-\sigma}$$

$$+ J \sum_{<ij>} (\sigma_i \cdot \sigma_j - 1) \qquad\qquad (2)$$

The parameters t and J are not dynamically affected by spin-Peierls coupling
to breathing modes; unlike my first paper[1] and Kivelson et al.,[2] we believe
we can show this coupling does not change anything. Even if it does, the
parameters here are no more subject to an isotope effect than those of any
other magnetic transition. The zero isotope effect only confirms what the
sound velocity data of Bishop[3] already told us: the basic parameters are
electronic energies. What is often frustrating is that the above is true
even in the unlikely case that some other mechanism intervenes, yet in all
but one or two of the myriads of theories it is totally ignored. In our
opinions, the only other interaction playing an important role is the
longer-range Coulomb effect, which is of course not small in a near-
insulator.

When the Cu lattice is stoichiometrically Cu^{++}, the hopping part of
(2) is zero, and we are left with a Mott insulating crystal of S = 1/2.
14 years ago I observed that these are peculiar beasts, which often do not
exhibit antiferromagnetism even when they should, but may, under some
circumstances, exhibit a new kind of state in which the zero-point fluctuations have destroyed antiferromagnetism and the state is a featureless
singlet spin liquid of singlet pairs. This is the famous "RVB" state. We
believe it exists at least in La_2CuO_4 (with small doping), (1 2 3) $O_{6.5}$,
$NaTiO_2$, and probably $BaBiO_3$. So far we have three theories of this state,
all in agreement if none conclusive, and they all lead to about the picture
I now give you.

The most physical is the idea of a projected BCS pairing:

$$\Psi = P_{(n_i=1)} \Pi_k \; (u_k + v_{k\uparrow} \; c^+_{k\uparrow} \; c^+_{k\downarrow}) \; \Psi_{vac}$$

with $u_k/v_k = \pm 1$ depending on whether

$$E_k = \Delta_0 (\cos k_x \pm \cos k_y) \gtrless 0$$

Δ_0 is a number of order J, and E_k is the quasiparticle excitation energy.
These excitations are fermions with no charge, no true kinetic energy, and
no gap; but the state has an enormous amount of pairing with a typical
binding energy of 1000°K. Unfortunately the material is an insulator.

We have a Fermi liquid argument and a full-fledged bells-and-whistles
gauge field theory argument for this state, not to mention a theorem which
says it is gapless; this is the solid and clean part of our theory. Fortu-
nately at least one measurement of $C = \gamma T$ has been taken in this region.
The gauge field theory argument is very firm: no really genuine broken
symmetry and no macroscopic quantum coherence can occur, because our theory
has a local gauge symmetry.

It is when we dope this state with acceptors and make Cu^{+++} that we run into really hard questions. In fact, the physics is so obvious that without doing any theory we could point out that the preexisting pairing would become coherent at some T_c which is roughly proportional to doping δ. But how?

To see how, we have to examine the rather interesting excitation possibilities of the RVB state. One might think that the effect of doping would be just to make our insulating pseudo Fermi sea into a regular charged one which then went superconducting, and that was the way we originally envisaged it; but actually, the thing it does is very much more interesting than that. In our RVB state we have already seen that the only low-energy excitation is a peculiar object which looks like a fermion but has no charge. This we recognize to be a spin soliton made up by the following process: keeping all the other sites bound up in singlet pairs, take site i _out_ of the pair liquid and put a single, isolated, unbound spin on it. Thus the spinon is a true soliton in that it implies a change in the entire background liquid as well as a local excitation; it cannot be expressed in terms of local operators.

Of course, it is possible to remove an electron directly, leaving its unpaired partner behind; the resulting state is a true charged fermion and can be created by a local operator.

Finally, we can imagine the product of the above two: a _hole soliton_, where we pair up the odd spin with one from a spin soliton, leaving a hole in a paired spinon fluid. This excitation is a boson--the product of two fermions--and was introduced by Kivelson, Rokhsar, and Sethna. It is easy to visualize if we think in terms of a bound pair fluid as simply a bookkeeping mechanism for keeping track of the motion of the underlying pairs: a motion of a hole soliton (perhaps we should call it a kivelson) by two spaces, for instance, can move a nearest-neighbor bond one space. But it is important to recognize that our spin fluid (unlike KRS's) has bonds of all lengths and that is only a rough visualization. Nonetheless, a kivelson by itself cannot tunnel out of a piece of material because that changes the total spin and charge by unphysical values. Whether it can interfere with itself after going round a loop has the jury still out; on this depends the size of the fluxon in a flux array.

Our theory seems to tell us that doping produces kivelsons (hole or holons is easier to say) and not electrons, because it is cheaper to leave all the remaining electrons paired. The first few are bound to defects or potential fluctuations, but when they get free they Bose-condense and that is the superconductive mechanism. Their masses are quite light--should be of the same order as spinons, or $\sim 10\ m_c$ at most, so the Bose condensation temperature $[(h^2/m^*)n_h$ in 2d$]$ should be quite high--this is why we find "low" T_c's so hard to understand.

The mass is light because (as Hsu showed us) as holes hop they do _not_ leave behind a trail of damage as they do in a true antiferromagnet; in fact, true antiferromagnetic fluctuations will make them heavier and may be the reason for low T_c's. Holes move almost as fast as they would in a ferromagnet, so the presence of holes strongly favors RVB, in order to minimize their kinetic energy. The holes need not attract each other at all, and we assume that they actually see mostly the repulsive Coulomb interaction. This immunity to repulsive interactions means that T_c can be almost as high as we like.

All of this can be formalized in a theory due to Zou Zhou, which started as a "slave boson" theory and became a "slave fermion" one. He shows that one can formally separate charge and spin degrees of freedom with a set of two charged bosons and one fermion e_i^*, d_i^*, $s_{i\sigma}^*$ and one constraint $\sum n_e + n_d + n_s = 1$ which makes half of the fermions redundant.

Now we have a superconductor, the question becomes what are its excitations and what is its structure? Our theorems convinced us that it had to have some kind of low-energy excitation, but this <u>cannot</u> be the bosons, because they mix with the plasma oscillations in a way that has been known since 1958 or so and leads to a complete wipe-out of the long-wavelength bosons. We reluctantly came to the conclusion that the gapless spectrum was still the spinons: that the PFS is still there! Conveniently, just about then I talked to Norman Phillips, who told me that $C_{sp} = \gamma T$ was still there too; and Zou's theory also shows that it should still be there.

Thus the basic nature of our superconductivity is a condensation of holes in the presence of a Fermi gas of spinons. Thus the real answer to the question "what is the energy gap?" is "there <u>is</u> no gap," and both specific heat and, to a lesser extent, T_1 measurements confirm this as far as magnetic excitations are concerned.

The next interesting question is how the three kinds of excitations meld and interact? Are there two Fermi surfaces or one--free bosons or not? To these questions our primary answer is "we don't know yet." But I want to make the following radical suggestions.

(1) There may well not be a second Fermi surface: instead, the spinons may acquire charge gradually as the energy increases. At $E = T = 0$, they are purely magnetic; but at high energy, they give us a finite tunneling density of states and electromagnetic absorption in the gap. Once we acquire a hole amplitude, this automatically implies boson pairs $\langle e_k^+ e_{-k}^* \rangle$, which in turn automatically leads, via the hole-hole interaction (whether attractive or repulsive), to an electron pair amplitude and a true "gap function"

$$\Delta^*(k) = \sum_k V(k - k') \langle c_{k'\uparrow} c_{-k'\downarrow}^* \rangle$$

This quantity may or may not be related to singularities in the density of states. What the electron pair amplitude is useful for is tunneling from grain to grain and possibly even from layer to layer, since the single hole amplitude cannot give broken symmetry by Yang's theorem and interparticle tunneling has to occur in pairs.

Let me close with a list of experimental facts that this theory has been able to deal with, or at least seems reasonably compatible with, whereas most of them are quite hard for other theories to deal with. I have not given full discussions of many of these, which will appear elsewhere.

(1)   Insulating property of stoichiometric (nonantiferromagnetic) Cu^{++} layers. (La_2CuO_4, 1 2 3 $O_{6.5}$ ($BaBiO_3$? $NaTiO_2$?) [predicted]

(2) $T_c(\delta)$

(3)   Antiferromagnetism of La_2CuO_4 vs. doping (and $NaTiO_2$?)

(4)   Large linear specific heats in insulators, La - Sr, (1 2 3). [predicted]. Low-energy magnetic states.

298

(5) Anomalous T-dependence of ρ_{Norm} (not complete)

(6) Hall-thermopower: number of carriers (see (1))

(7) Nature of high-T "twitch" in La_2CuO_4-based system (incomplete)

(8) T_1, IR, tunneling--gaps--if any--do not agree (some predicted; some incomplete).

(9) Relative weakness of intergrain coupling.

As you see, there is much to do; the theory is incomplete but in principle we feel it is a fairly well-posed problem; it is the answers that are hard to get.

In closing I want to list the many people to whom I am indebted: Princeton Group: P. W. Anderson, G. Baskaran, Z. Zhou, E. Abrahams, I. Affleck, J. Sauls, Ted Hsu, S.-D. Liang, R. Kan, T. Wen, J. Wheatley, Phuan Ong, S. Coppersmith.

Friends:

    Santa Barbara: S. Kivelson, D. Rokhsar, J. Sethna
    MIT: P. A. Lee, G. Kotliar, N. Read
    La Jolla: A. Ruckenstein
    Elsewhere: T. M. Rice, T. V. Ramakrishnan, B. S. Shastry, P. Fazekas

Experimental acknowledgments: N. E. Phillips, T. Geballe, B. Batlogg, H. R. Ott, G. Aeppli, D. Bishop, G. Thomas, L. Greene, D. McLachlan, C. W. Chu, A. Ramirez, and many preprint senders.

References

1. P. W. Anderson, Science $\underline{235}$, 1196 (1987).
2. S.A. Kivelson, D.S. Rokhsar, and J.M. Sethna, Phys. Rev. $\underline{B35}$, 8865 (1987).
3. D. Bishop et al., preprint.

ELECTRONIC FLUCTUATION AND PAIRING*

N. W. Ashcroft

Laboratory of Atomic and Solid State Physics
Cornell University
Ithaca, NY 14853

<u>Abstract</u>

From one-electron theory it is argued that a starting point for a description of
the electronic structure of the high T_c compounds is a set of almost filled bands
separated by a significant gap from unoccupied bands. But it is also argued that
correlation effects are important, and these are included by separating the electronic
charge into itinerant and quasilocalized, the latter possessing polarization-waves of
exciton character that provide a pairing mechanism for the electrons. It is also
suggested that for low enough densities fluctuation effects in the itinerant electrons
may significantly modify the standard averaged repulsive contribution to the total
pairing interaction.

I. INTRODUCTION

The notion of electron pairing, which is central to the theory of superconductivity, evidently had its origins in the ill-behaved conducting behavior of rapidly
quenched metal-ammonia solutions.[1] Forty years later extraordinary transport behavior is again seen in a remarkable class of chemically complex systems which share
with the metal ammines the property that the carrier densities are relatively low
and also share, in some instances, chemical instability. In the intervening years
the mechanism giving the necessary substance to Ogg's idea of a "di-electron" has
been identified and almost universally accepted as originating from the exchange of
phonons in a superconducting material. This picture begins with Fröhlich[2] and has
culminated in the highly successful BCS theory.

Other mechanisms leading to net attraction between electrons have been proposed over the years, as reviewed by Ginzburg and Kirzhnits.[3] These include the
exchange of spin fluctuations of various kinds, the exchange of charge density waves,
the exchange of excitons, and the exchange of plasmons. With the advent of the
new class of superconducting oxides the energy scales offered by these and other
non-phonon based mechanisms look increasingly attractive. They carry with them a
clear implication that electron correlation effects are important and that approaches
that incorporate them only in an average way—the standard band theory route, for

example—may be discarding essential physics. The systems have interesting magnetic characteristics, and this hints at the possible appearance of spin fluctuation mechanisms. But it is important to note that the systems are often inhomogeneous, and even when not, what is perhaps the key control constituent, namely oxygen, is itself endowed with interesting magnetic properties.[4] It is, moreover, highly mobile, and while the attributes of the companion elements are of interest in their own right, it seems evident that the acute sensitivity of the properties of the compounds to oxygen content is a strong indication of its central importance in the overall electronic structure. To take an example, it may be noted that with <u>conventional</u> valences (Y: 3; Ba: 2; Cu: 1) and with oxygen assigned a valence of 6, the Bloch one-electron view would lead to the possibility, <u>in principle</u>, of exact band filling with precisely 7 oxygens for each unit of $(Y\,Ba_2Cu_3)$. In view of the strength of bonding of the oxides, it is not unreasonable to suppose that in the ideal configuration (O_7) there is a gap to the next set of bands. Optical evidence[5] suggests that this band-gap might be in the electron-volt range. It follows that with oxygen deficiency $(O_{7-\delta})$ the electronic structure is characterized by an uppermost band that is incompletely filled and separated by an energy gap from the next band. This feature is quite uncommon in metals; nevertheless, the photoemission results of Fujimori et al[6] support a picture with a very low density of states at the Fermi level (ϵ_F), with strong structure several volts <u>below</u> ϵ_F, and also with very little agreement with the calculated band structures.[7]

The difficulties of band theory have already been mentioned; a conclusion to draw is that even if the broad aspects of the electron structure described above are correct, it is not necessarily permissible to assume that it can be correctly described in one-electron terms. Magnetic behavior in the normal state seems to support this point of view. In seeking a model to describe the essential effects of electron correlation, there is a further clue. The polarizabilities of the <u>ions</u> of the elements Y, Ba, and Cu in their conventional valence states are all rather large, and in the absence of serious overlap of these ions they would, at the density appropriate to the oxide superconductor, contribute to a significant static background dielectric constant. This is again compatible with the optical data.[5][8] Accordingly, in order to establish the simplest description, and one that for the present does not specifically involve electron spin, the charge will be divided into bound and itinerant. In a one-electron picture the appropriate description of the bound charge would proceed from the tight-binding approach. But it is necessary to go beyond this. To do so, imagine developing the interactions between the bound electronic charge in each cell in terms of multipoles. The essential physics can be illustrated with the lowest surviving (dipole) terms;[9] if the lattice sites are $\{\vec{R}\}$ and the point dipole operator at each site is $\hat{d}_R(\vec{r})$, then electron correlations are manifested in the interaction

$$\sum_{\vec{R},\vec{R}'}' \int d\vec{r} \int d\vec{r}\,'(\hat{d}_{\vec{R}} \cdot \nabla_r)(\hat{d}_{\vec{R}'} \cdot \nabla_{r'})\frac{e^2}{|\vec{r}-\vec{r}'|} \tag{1}$$

Later it will be necessary to incorporate the fact the sites $\vec{R}$ are disturbed by phonons.

Omitting the free charge for a moment, the physics of this problem can be described by a coupled periodic array of fluctuating electronic dipoles; the corresponding Hamiltonian can be straight-forwardly diagonalized.[10] The excitations are collective polarization-waves, and the summation of their energies leads in real space to the familiar Van der Waals attractive interactions between the dipole centers. Because of the lattice symmetry, the waves are periodic in reciprocal space.

They have a dispersion across the zone in multiple branches, and the variation is typically $(4\pi e^2 \rho_a/m)^{\frac{1}{2}}$ where ρ_a is the atomic number density.[11] In the absence of interactions between dipoles there is a characteristic excitation energy defined in the one-particle problem described above by the state at the top of the gap. Inclusion of interactions lowers the excitation energy as expected from the standard theory of excitons. This minimum characteristic energy, occuring at $q = 0$ will be denoted by $\hbar\omega_o$. If it lies in a gap, then the collective polarization-wave excitations will not suffer damping from confluence with the particle-hole spectrum. Damping can certainly occur, however, at higher wave-vectors.

II. ELECTRON-PAIRING

The problem now is to establish the qualitative features of an effective electron-electron interaction appropriate to a model system whose single-particle characteristics consist of a partially filled valence-band separated by a gap from an empty conduction band. The occupied states are of both localized and itinerant types, and correlation effects are to be included for each class. Finally, the ions themselves are dynamic; in a simplified model they will be described by a single ionic plasma frequency ω_i. Given this description, the dielectric constant $\epsilon(q, \omega)$ can be written down immediately: let $\epsilon_e = \epsilon_e(q)$ be the dielectric function for the itinerant carriers. It can always be written in the form $\epsilon_e = 1 + f(q)/q^2$ where in the simplest approximation $f(q) = k_o{}^2$ with $k_o{}^2 = 4\pi e^2 g(\epsilon_F)$, $g(\epsilon_F)$ being the one-electron density of states at the Fermi level. The importance of a proper choice for $f(q)$, especially in the context of dilute carrier systems, is discussed below. Next, let $\omega(q)$ be the dispersion of the polarization-waves. Then if they are not damped[12]

$$\epsilon(q, \omega) = \epsilon_e - \frac{\omega_i{}^2}{\omega^2} + \frac{\Omega^2(q)}{\omega^2(q) - \omega^2}. \tag{2}$$

The Lorentz-oscillator form of the last term in (2) is argued on general physical grounds. Note in particular that at $q \to 0$, $\Omega^2(0) \to \omega_o{}^2(\epsilon_b - 1)$, where ϵ_b is the static background dielectric constant attributable to the bound charge. Otherwise for the present it suffices to know that $\Omega(q)$ and $\omega(q)$ are quite comparable in magnitudes and that both are much larger than ω_i.

The roots of (2) give the dispersion relation for the combined sources of polarization; let $\omega_1(q)$ be the lower (phonon-like) branch and $\omega_2(q)$ the branch from polarization-waves. Then

$$\omega_1{}^2(q) = (\omega_i{}^2/\epsilon_e)/(1 + \Omega^2(q)/\omega_q{}^2\epsilon_e) \tag{4}$$

and

$$\omega_2{}^2(q) = \omega_q{}^2 + \Omega^2(q)/\epsilon_e \tag{5}$$

with the limits $\omega_1 \sim q$ and $\omega_2 \sim \omega_0$ as $q \to 0$. Given these, the inverse of $\epsilon(q, \omega)$ is

$$\frac{1}{\epsilon(q, \omega)} = \frac{1}{\epsilon_e}\left[1 + \frac{1}{\epsilon_e(\omega_2{}^2 - \omega_1{}^2)}\left\{\frac{\omega_2{}^2(\Omega_q{}^2 + \omega_i{}^2) - \omega_i{}^2\omega_q{}^2}{\omega^2 - \omega_2{}^2} + \frac{\omega_i{}^2\omega_q{}^2 - \omega_1{}^2(\Omega_q{}^2 + \omega_i{}^2)}{\omega^2 - \omega_1{}^2}\right\}\right] \tag{6}$$

$$\simeq \frac{1}{\epsilon_e}\left[1 + \frac{\Omega_q{}^2/\epsilon_e}{\omega^2 - \omega_2{}^2(q)} + \frac{\omega_1{}^2(q)}{\omega^2 - \omega_1{}^2(q)}\right] \qquad (7)$$

The third term in (7), from the phonon branch, will not be discussed further, except to note that ω_1 is actually lower than $\omega_i/\epsilon_e^{\frac{1}{2}}$ because of the influence of the second branch. (This should result in an enhancement of the usual electron-phonon-electron coupling). Instead, for wave-vector q and frequency ω the effective interaction arising from the first two terms of (7) is

$$\frac{4\pi e^2}{q^2 \epsilon_e(q)}\left[1 + \frac{q^2 \Omega_q{}^2/(q^2 + f(q))}{\omega^2 - \omega_2{}^2(q)}\right]. \qquad (8)$$

With the repulsive and attractive terms thus identified, the superconducting transition temperature is given by[13]

$$k_B T_c = <\hbar\omega_2> \exp -1/(\lambda^* - \mu^*) \qquad (9)$$

where $\lambda^* = \lambda/(1+\lambda)$ with λ being the standard average over the attractive term but including the selection rule $\vec{q} = \vec{k}' - \vec{k} + \vec{K}$ ($\vec{K}$ a reciprocal lattice vector and $\vec{k}$ and $\vec{k}'$ the wave vectors of the itinerant carriers). In (9) $\mu^* = \mu/(1 + \mu\, ln(\epsilon_F/ <\hbar\omega_2>))$ with μ arising from the average of the direct interaction. This term will be discussed in the next section.

To obtain an estimate for λ, a simple scaling argument can be used by writing the second term in (8) as

$$\left(\frac{\Omega_q}{\omega_i}\right)^2 \left\{ \frac{4\pi e^2}{\left(q^2 + f(q)\right)^2} \frac{\omega_i{}^2 q^2}{\left(\omega^2 - \alpha^2 \omega_{ph}{}^2(q)\right)} \right\} \qquad (10)$$

where α^2 is defined by the scaling assumption $\omega_2(q) = \alpha\omega_{ph}(q)$ with $\omega_{ph}(q)$ a typical phonon dispersion. This is a simple extension of the idea of scaling the Bohm–Staver relation $\left(\omega_{ph}(q \to 0) = v_F q \sqrt{Z^* m/3M}\right)$ in order to obtain a realistic phonon relation.[14] In this case the contribution { } in (10) receives a factor α^{-2} relative to a corresponding phonon calculation in which electron-phonon-electron term is found using a Debye model with a Bohm-Staver sound velocity (as for example, in Ref. 13). Since large q dominate in the calculation of λ, we can obtain α^2 from (5); combining with (10) we find the required average to be

$$\left\langle \frac{\Omega_q{}^2}{\Omega_q{}^2 + \omega_q{}^2} \right\rangle \lambda_{ph}$$

where λ_{ph} is a typical phonon value. As noted, the averages are dominated by high q, so that the scaling factor is not far removed from unity. Values of λ_{ph} are typically calculated to be in the range 0.2–0.4; it follows that if the Coulomb term is not unusually large (see below), then the possibility of high temperature superconductivity is determined by the prefactor $< \hbar\omega_2 >$, which as noted earlier, is in the eV range. It should be emphasized that this conclusion arises from a consideration of a mechanism that is entirely electronic in origin. For this reason the characteristic energy $\hbar\omega_o$ should not be so large that high frequency excitations are unfavorably weighted in the Eliashberg equation nor so low that there is difficulty with the Migdal theorem and separation of time scales. The possibility that the phonon mechanism and the exciton-like mechanism may <u>both</u> act in concert should not be overlooked.[15]

III. THE DIRECT INTERACTION

Rietschel and Sham[16] have noted that because of plasmon exchange, even the direct Coulomb interaction average, μ^*, is not invariably positive. It is worth pointing out that if one goes beyond linear response in calculating the contributions to $f(q)$, then for sufficiently low densities similar behavior can also occur. Again, the origin of these effects may be found in fluctuation terms whose character is similar to those just discussed in the context of the <u>inhomogeneous</u> electron system. To give a specific example, let

$$\Lambda^{(3)} = \Lambda^{(3)}(\vec{q} + \vec{q}', i\omega; \vec{q}', -i\omega; -\vec{q}; 0)$$

be the irreducible 3 point function for the homogeneous electron gas. Next, consider the contribution to the effective interactions between electrons defined by

$$\Delta v_{eff}(q) = v_{sc}{}^2(q,0) \int \frac{d\omega}{(2\pi)^3} \int \frac{d\vec{q}'}{(2\pi)^3} v_{sc}(\vec{q}', i\omega) v_{sc}(q + q', i\omega)(\Lambda^{(3)})^2. \tag{11}$$

where the dynamically screened Coulomb interaction is defined by

$$v_{sc}(q, \omega) = 4\pi e^2 / q^2 \epsilon(q, i\omega)$$

The numerical importance of these fluctuation contributions was first pointed out by Rasolt and Geldart.[17] The essential physics rests with the observation that while the perfect screening sum rule efficiently constrains the <u>non-fluctuating</u> terms, there is no constraint operating in dynamic screening. As a consequence, terms such as (11) can have a numerical significance considerably above what might be expected from their formal order (in the pertubation sense). In a recent paper[18] Maggs and Ashcroft have shown that in real space the interaction resulting from (11) actually has <u>attractive</u> Van der Waals (i.e., power law) behavior at long range. This is confirmed by Vosko and Langreth,[19] who show that the interaction has the form $-a \log br/r^6$ where a and b are both density dependent. The overall density dependence is roughly as $r_s{}^{\frac{9}{2}}$ where r_s is the usual electron spacing parameter. Though some of the expansion methods leading to these results are only rigorously defensible at small r_s, the general success of density functional techniques in their application to systems with highly variable densities suggests that contributions from (11) (and related terms) may be of some considerable importance in low density homogeneous electron systems. The fact that the terms can be attractive points to the possibility of an intrinsic pairing instability;[20] at the very least it suggests that their presence may lead to a reduction in μ. Thus, returning to (9), if $< \hbar\omega_2 >$ is in the eV range, a transition temperature of $O(10^2)^\circ K$ is entirely consistent with the notions of an attractive coupling originating with polarization waves. Gap structure in the one-particle bands seems, however, to be the most favorable situation since this will ensure strong spectral weight in the polarization-wave branch.

IV. COMMENTS

From the structure of the form of the electron phonon coupling, it can be inferred that the electron polarization-wave interaction is

$$|g_{\vec{k},\vec{k}'}|^2 = \frac{4\pi e^2}{(q^2 + f(q))} \cdot \frac{q^2}{(q^2 + f(q))}(\hbar\Omega_q \hbar\omega_2(q)). \tag{12}$$

In the scattering of electrons by phonons, Umklapp processes are very important $(\vec{k}\,' - \vec{k} = \vec{q} + \vec{K})$; typical scattering lengths at high temperatures are $O(10^2)\,\AA$. The scattering of electrons by polarization-waves will be largely elastic; nevertheless there can be significant momentum exchange, especially from the Umklapp processes. Since (12) exceeds an electron-phonon coupling by $(\hbar\omega_2/\hbar\omega_{ph})(\Omega_q/\omega_q)^2 \approx 10^2$, it appears that scattering lengths can be as short as $1-2\,\AA$. But by itself this argument would not lead to any significant temperature dependence. The point, however, is that much of the localized charge adiabatically follows the motion of the ions as they are disturbed by phonons. The fluctuating multipoles are thus not fixed in space but in fact vibrate with mean square displacements ultimately rising as T. From the argument given in the introduction the Fermi surface dimensions will be small so it may not require especially large temperatures in order to achieve large angle scattering even if it were not already provided by the Umklapp processes. A standard Born-approximation argument will result in a linear dependence of resistivity on temperature from a combination of strong coupling to polarization fluctuations which in turn ride on thermally displaced ions.

In this approach to the problem of high temperature superconductivity (which does not exclude the possibility of further spin dependent processes), one can ask what features of the oxide superconductors are especially important with respect to fluctuation effects? To answer this, it is worth noting that the ionization energies of Cu are $7.72, 20.29, 36.83, \ldots \ldots$ eV (to be compared with $6.38, 12.23, 20.5 \ldots$ in Y, $5.21, 10.00, 35.5$ in Ba, and $13.61, 35.11$ in O). Thus, while it is true that Madelung energy gain will partially compensate the energy penalty suffered in increasing the charge state of Cu, it does seem likely that in a <u>metallic</u> state with attendant screening, Cu would tend to favor its normal valence. In these circumstances, the ions Cu^+, Ba^{2+}, Y^{3+} will constitute sources of quasilocalized and quite-polarizable electron charge.[21] It is fluctuations in this system which are thereby considered important, and the point of view is then to include correlation effects via the introduction of polarization-waves.[9] Note that electrons need not be actually transferred as in the model proposed by Varma et al[22] which in other respects bears some similarities to the present picture. But, once again, an overall energy gap will ensure significant weight in the polarization-wave spectrum.

The fact that these oxides possess structural instabilities deserves comment. It suggests that the pair-and-higher-center interactions may be more complex than simple screened ion-ion potentials. As shown by Maggs and Ashcroft,[18] this situation indeed prevails when significant screened Van der Waals interactions are present in a metal. The balance between these two separate contributions to ion-ion interactions depends very much on carrier concentration, which, in turn, is sensitive to oxygen content. For this reason it is entirely conceivable that in sintered samples arrested oxygen diffusion can result in "boundary" phases that border a bulk phase of a different kind. The phases that result are likely to have their own characteristic values of Ω_q, ω_q, etc., and hence have differing superconductive behavior. It should also be noted that in oxygen deficient states the magnetic properties are not at all straightforward to assess. Incomplete occupancy requires the consideration of Van Vleck paramagnetism, among other possibilities. Finally, the mobility of oxygen in those compounds is an interesting phenomenon and not unrelated to what has been discussed so far. Again, the point is that the diffusion of oxygen is governed by the effective pair and multicenter interactions that oxygen possesses in the presence of its neighbors. The fact that ion-ion interactions are both screened and augmented by screened Van der Waals interactions, with rather different length scales,[23] may lead to a net interaction that is rather weak.

REFERENCES

*Work supported by the National Science Foundation under Grant DMR-8415669.

1. R. A. Ogg, Phys. Rev. 69, 243, 544 (1946).

2. H. Fröhlich, Phys. Rev. 79, 845 (1950).

3. V. L. Ginzburg and D. A. Kirzhnits "High Temperature Superconductivity" Consultants Bureau, N.Y. (1982).

4. As an atom, as a molecule, and even as ionized molecules.

5. P. E. Sulewski, T. W. Noh, J. T. McWhirter, A. J. Sievers, S. E. Russek, R. A. Buhrman, C. S. Jee, J. E. Crow, R. E. Salomon, and G. Myer (to be published).

6. A. Fujimori, E. Takayama-Muromachi, and Y. Uchida, Solid State Communications (submitted).

7. L. F. Mattheis and D. R. Hamann, Solid State Communications (to appear).

8. N. W. Ashcroft (unpublished).

9. The higher multipole terms are most conveniently included by using the symmetrized Kubic harmonics.

10. A. Lucas, Physica 35, 353 (1968). See also N. W. Ashcroft in "The Liquid State of Matter: Fluids Simple and Complex" (eds. E. W. Montroll and J. L. Lebowitz) North Holland, 1982 p. 141.

11. If the system is conducting, and Z^* is the mean valence, then this variation can also be written as $Z^{*-\frac{1}{2}}\omega_p$ where ω_p is the free carrier plasma frequency.

12. It is clear from the form of (2) that damping can be included in principle. The effect will be to shift spectral weight from the exciton (or polarization-wave) branch to the single particle modes.

13. See, for example, P. Morel and P. W. Anderson, Phys. Rev. 125, 1263 (1962).

14. N. W. Ashcroft, Phys. Rev. Letts 21, 1748 (1968).

15. In principle the direct and excitonic mechanisms could be attractive, but very weakly so, over an energy scale fixed by $\hbar\omega_2$. The subsequent addition of the phonon mechanism might then have a significant effect.

16. H. Rietschel and L. J. Sham, Phys. Rev. B 28, 5100 (1983); see also M. Grabowski and L. J. Sham, Phys. Rev. B 29, 6132 (1984).

17. M. Rasolt and D. J. W. Geldart, Phys. Rev. Letts 35, 1234 (1975); Phys. Rev. B 13, 1477 (1976).

18. A. C. Maggs and N. W. Ashcroft (submitted for publication).

19. D. C. Langreth and S. H. Vosko (to be published).

20. The depth of the minimum in the $-log\ ar/r^6$ exceeds in magnitude the corresponding Thomas-Fermi screened Coulomb interaction at the same location when r_s is greater than about 5.5. This is the neighborhood of the well known compressibility instability of the electron gas.

21. One notes that Th is even more polarizable than Y!

22. C. M. Varma, S. Schmitt-Rink, and E. Abrahams, Solid State Communications (to appear).

23. See K. K. Mon, N. W. Ashcroft, and G. V. Chester, Phys. Rev. B $\underline{19}$, 5103 (1979).

NON-PHONON MECHANISMS OF SUPERCONDUCTIVITY IN HIGH T_c SUPERCONDUCTING
OXIDES AND OTHER MATERIALS AND THEIR MANIFESTATION

Vladimir Z. Kresin

Materials and Chemical Sciences Division
Lawrence Berkeley Laboratory
University of California, Berkeley
Berkeley, California 94720

ABSTRACT

Low dimensionality and the unusual parameter values in the high
T_c materials lead to a key contribution of the plasmon mechanism of
superconductivity. In addition, these systems provide a unique opportunity to observe a multigap structure. The problem of the lattice
instability is discussed. A manifestation of non-phonon mechanisms
(NPM) in Nb_3Ge and the contribution of the intramolecular vibrations
are analyzed. Proximity systems containing high T_c superconductors
are promising from the point of view of possible applications.

I. INTRODUCTION

Recently discovered high T_c materials[1,2] are characterized by
"exotic" properties. This paper is concerned with the description of
properties and with the analysis of the mechanisms of high T_c superconductivity. In addition we are going to analyze the appearance of
the non-phonon mechanism in some conventional systems. The structure
of the paper is as follows. Section II contains analysis of the
properties of high T_c materials. The problems of the lattice
instability, the appearance of a multigap structure and the influence
of the proximity effect will be discussed. We are going to discuss in
detail the plasmon mechanism and its coexistence with strong electron-
phonon interaction. Section III contains an analysis of some conventional systems.

II. HIGH T_c MATERIALS

A recent exciting development, the discovery of new high T_c
superconducting oxides[1,2] brought up the problem of mechanisms of
superconductivity in these materials. An analysis of their structure
and parameters leads to the conclusion that their superconducting
state is greatly affected by NPM, namely, by exchange of 2D (two-
dimension) plasmons.

1. Low Dimensionality and "Exotic" Properties of High T_c Superconductors

Main parameters. Lattice instability. The new high T_c materials
are low dimensional systems. For example, La_{2-x}, $Sr_x CuO_4$ ($x \simeq 2$) is two
dimensional (the interlayer distance $d \simeq 6.5 \overset{\circ}{A}$), while $Y_1 Ba_2 Cu_3 O_7$ contains
one dimensional chains. S. Wolf and the present author[3] have carried out
our evaluation of the parameters of high T_c materials based on specific
heat data.[4] We think that these data are the most reliable source; they
can be used with high accuracy for analysis even polycrystalline
samples.

Low dimensionality is taken into account in a consistent way and
plays a key role in the analysis.[3] According to,[3] $La_{2-x} Sr_x CuO_4$
is characterized by a large value of the effective mass: $m^* \simeq 5\ m_e$.
A most striking feature is the small value of the Fermi energy: $\varepsilon_F \simeq$
0.12 eV.

The situation with such a small value of ε_F along with large,
comparable value of the energy gap Δ is unique. In connection with it
I would like to stress that the superconducting transition affects the
state of the lattice and this influence is determined by the parameter
$\sim (\Delta/\varepsilon_F)^2$ (see refs. 5 and 6). This parameter is usually small.
However, the situation is entirely different in high T_c superconducting oxides. A large value of (Δ/ε_F) leads to a drastic change
of the phonon spectrum. The following increase in T_c (and Δ) leads
to the lattice instability. We think that the upper limit of T_c is
determined by this factor.

Coherence length. Multigap structure. According to the analysis[3]
the coherence length ε_0 appears to be very small ($\sim 20 \overset{\circ}{A}$). Such small
value of ε_0 leads to the unique opportunity to observe a multigap
stucture.[7] The appearance of a such structure is connected with the
presence of the overlapping energy bands.

The two-gap model has been introduced by Suhl, Matthias and Walker.[8]
Afterwards it has been studied by Geilikman, Zaitsev, and the present
author[6,9] and by the present author in ref. 10. The difficulty of
observing multigap effects, as well as effects caused by gap anisotropy
are due to the Anderson theorem.[11] Namely, the inequality $\ell \ll \xi_0$ (ξ_0
is the coherence length, ℓ is the mean free path) results in the gap
averaging into a single one. Interband transitions are the main
mechanism of this averaging.

The new high T_c materials provide a unique opportunity to observe,
under certain conditions, effects due to the presence of several gaps.
For relatively clean samples, the criterion $\xi_0 \ll \ell$ can be met.

Note that the Anderson criterion allows one to determine whether
one is dealing with a multigap case. Indeed, additional doping of
these materials will result in a decrease of and, subsequently, in
becoming less than ξ_0, when a transition to the one-gap picture will

take place. Such a transition can be observed experimentally, because
the tunneling spectrum and the temperature dependences, e.g., of the
kinetic coefficients are different in the one gap and multigap cases.

High T_c superconductivity and the proximity effect. The
proximity effect allows one to induce the superconducting state in
materials which are not superconductors by themselves. If, for
example, this material is a semiconductor, then, as a result, one can
take advantage of both superconductivity and semiconducting properties.
An important example of such an application of the proximity effect is
the tunneling system Nb-InAs-Nb, studied experimentally.[12] An
externally applied electric field changes noticeably the amplitude of
the flowing Josephson current, which is promising from the point of
view of making a three-terminal device. A theoretical analysis[13a] shows
that the sharpness of the field effect depends strongly on the
temperature and increases with increasing T. That is why the use of
high T_c superconductors, namely, the systems S_h-InAs-S_h, or

S_h-S-InAs-S-S_h, where S_h is a high T_c superconductor, and S is
a conventional material (e.g., Nb or NbN) is promising for the field
effect.

Another interesting proximity system is S_α-S_β consisting of two

superconductors (assume that $T_c{}^\alpha > T_c{}^\beta$). Such as system is charac-

terized by a single T_c with $T_c{}^\alpha > T_c > T_c{}^\beta$ (a general expression has

been obtained by the present author in ref. 13b. As a result of the

proximity effect, one effectively increases the T_c of the S_β

superconductor.

Consider the case with S_β is an A-15 compound with high values of
such critical parameters as the critical current and the critical
field. If S_α is a high T_c superconductor, one can use the proximity
effect in order to increase T_c of the A-15 film and to take advantage
of its high values of the critical parameters.

2. Mechanisms of High T_c Superconductivity. Plasmon Mechanism.

The low dimensionality along with the presence of several overlapping energy bands and a small value of ε_F makes the appearance of
the plasmon mechanism of superconductivity very favorable. This
mechanism has been proposed by the present author (see refs. 14 and
15) and then developed by H. Morawitz and the author.[16] Later the
plasmon mechanism in high T_c oxides has been studied in.[17] The
new materials are characterized by a relatively small carrier
concentration n,[18] and it is important that the intensity of the
electron-plasmon interaction increases with decreasing n.

As is known in the three dimension (3D) case the plasmon branch
has a gap ω_0 at momentum q=o, and the plasma frequency is very high.
In the 2D case the situation is entirely different. Namely, the
plasmon dispersion relation does not contain an energy gap and in the
region of small q has the form (see, e.g., ref. 19) $\omega \sim q^{1/2}$. As a

matter of fact, there are several plasmon branches. The existence
of the $\omega \sim q^{1/2}$ branch is connected with the low dimensionality of
the system and is present even for a single 2D group of carriers. The
presence of several overlapping energy bands lead to the appearance of

an additional acoustic plasmon branch. In the 3D case this acoustic branch was introduced in;[20] its contribution to the superconducting state was studied in.[21-24]

The presence of the plasmon branches results in electron-electron attraction which appears to be large for systems with small carrier concentration.

The effect at the plasmon branch $\omega \sim q^{1/2}$ on the superconducting properties of an inversion layer was studied in.[25] Our approach is based on the method of the thermodynamic Green's function. The plasmon mechanism is affected by a number of factors. The high T_c oxides do not contain just one 2D sheet. They have a layered structure and, strictly speaking, one should take into consideration the interlayer interaction. One can show (see below) that the main contribution comes from the short wavelength region and the interlayer interaction does not play an important role. Moreover, it is necessary to take into account the presence of several energy bands.

The order parameter $\Delta(\omega_n, \vec{\kappa})$ describing the pairing in a 2D layer is described by the following equation, which is a generalized Eliashberg's equation:

$$\Delta(\omega_n, \vec{\kappa}) = \frac{T}{(2\pi)^2 Z} \sum_{n'} \int d\vec{\kappa}\; \Gamma(\omega_n - \omega_{n'}, \vec{\kappa} - \vec{\kappa}')\; F^+(\omega_n, \vec{\kappa}) \tag{1}$$

Here $\omega_n = (2n+1)\pi T$, $\vec{\kappa}$ is the 2D momentum, F^+ is the anomalous Green's function, and Γ is the total vertex. The vertex Γ can be written as a sum of the plasmon and phonon terms: $\Gamma = \Gamma_{p\ell} + \Gamma_{ph}$.

Consider the vertex $\Gamma_{p\ell}(\omega, \vec{\kappa})$. Its poles correspond to collective excitations, i.e., plasmons.

Let us study the properties of a single 2D sheet. Consider the general case of overlapping energy bands. The structure of the vertex in the 3D case was studied by Geilikman[22a] and by Geilikman and the author.[26] The case of 2D bands has been studied by Tavger and author.[27] According to[22a,27], the vertex $\Gamma_{11} \equiv \Gamma_{11;11}$ is equal to:

$$\Gamma_{11} = (V_{11} + \Pi_{22} R)\, S^{-1} \tag{2}$$

where

$$S = 1 + V_{11}\Pi_1 + V_{22}\Pi_2 + \Pi_1 \Pi_2 R \tag{3}$$

$$R = V_{12}^2 - V_{11} V_{22}.$$

Here V_{11}, V_{12}, V_{12} are Coulomb matrix elements and Π_1, Π_2 are polarization operators. The case of a single 2D energy band is described by Eqs. (2) and (3) with $\Pi_{22} = 0$. Equations (2) and (3) are written in the random phase approximation. The quantities Π_1 and Π_2 are given by

$$\Pi_i(\vec{q},\omega) = -\frac{m_i}{2\pi^2} \int_0^{2\pi} d\phi \; \frac{\cos\phi}{\alpha_i - \cos\phi + i\delta\cos\phi} \qquad i = \{1,2\} \qquad (4)$$

Here $\alpha_i = (\omega/V_{Fi}q)$, V_{Fi} is the Fermi velocity for i-th band; the polar axis is chosen along the 2D vector $\vec{q}$. Consider the region in the (ω,q) plane which corresponds to $\alpha_1 \gg 1$, $\alpha_2 \gg 1$. Then $\Pi_i = -(\varepsilon_{iF}/\pi)q^2/\omega^2$. The equation S=o determines the plasmon branch $\omega_{p\ell;b}$ which has the dispersion relation $\omega \sim q^{1/2}$ in the region of small q. Contrary to the usual (see below) acoustic plasmon branch, the branch $\omega_{p\ell;b}$ exists even in the case of a single energy band. The dispersion relation $\omega \sim q^{1/2}$ and the absence of an energy gap at q o is a consequence of the low dimensionality. The analysis of a more general case $\alpha_i > 1$ will be given separately. If $\alpha_1 \ll 1$, $\alpha_2 \ll 1$, we obtain

$$\Pi_1 \simeq m_1/\pi; \; \Pi_2 = -(\varepsilon_{2F}/\pi)\, q^2/\omega^2 \qquad (5)$$

As a result, we obtain the acoustic plasmon branch $\omega_{p\ell,a} \sim q$. For 3D system this branch, which is due to the presence of the overlapping bands was obtained in,[19] see also.[20-24] The vertex $\Gamma(\omega_n,\vec{\kappa})$ obtained by substituting $\omega \to -i\omega_n$, can be written in the form:

$$\Gamma(\omega_n,\vec{\kappa}) = \Gamma_o + D_{eff}(\omega_n,\vec{\kappa}) \qquad (6)$$

where $\Gamma_o = V_1^{src}$ describes the Coulomb repulsion and

$$D_{eff} = \Gamma^o \frac{\omega_{p\ell,a}^2(q)}{\omega_{p\ell,a}^2(q) + \omega_n^2} \qquad (7)$$

has the form of the usual phonon Green's function with the dispersion relation $\omega_{q\ell,a}(q)$ and describes the electron–electron attraction via plasmon exchange.

Consider the effect of the interlayer interaction. One has to introduce quantity $\Gamma_i(\vec{\kappa},z;\omega)$; this function is the Fourie component of the vertex $\Gamma(\vec{r},\omega)$ with respect to $\vec{\rho}$ (the axis z is chosen to be perpendicular to the layer). Let layer "a" be located at z=0 and let us evaluate the quantity $\Gamma_i(\vec{\kappa},0;\omega)$. The equation for this quantity contains the same terms as Eq. (2), but, in addition, we should take into account its Coulomb interaction with other sheets. For example, the presence of sheet b at z=d leads to the appearance of the terms $V_1^{ab}(q_\perp,d)\,\Pi^b(q_\perp;d;\omega)\,\Gamma^{ba}$, etc. These additional terms contain $(V_1^{ab})^2$ in the lowest order (V_1^{ab} describes the Coulomb interaction between carriers, in layers "a" and "b"). It is easy to see that the additional contribution of Γ_1 due to the presence of a layer at z=d is proportional to $e^{-2q_\perp d}$. Hence, in the region of small $q_\perp$ it is necessary to take into account the interlayer interaction and we are dealing with a 3D problem. But this interaction can be neglected in the short wavelength region ($2q_\perp d \gg 1$). For the high temperature oxide $La_{1.8}Sr_{0.2}CuO_4$, $d \simeq 6.5\text{\AA}$.[1] The Fermi momentum $p_F \simeq 3.7 \times 10^{-20}$ gm x cm sec^{-1} (see ref. 3), and therefore, if $q_\perp \gtrsim 0.5\ p_F$, we can consider the 2D sheet only.

The above analysis was carried out in the RPA (see e.g., refs. 28 and 29). One can show[16] that the main conclusions concerning the different plasmon branches the effective electron-electron attraction via plasmon exchange are valid for high T_C systems and play an important role in the understanding of the basic mechanisms of high T_C superconductivity.

<u>Strong electron-phonon coupling</u>. Consider the phonon part of the total vertex Γ (see Eq. (1)). Speaking of the electron-phonon interaction (EPI), one should stress that the BCS theory, based on an analysis of EPI, does not restrict the values of T_C to the low temperature region. The Eliashberg equation (see below, Eq. (8)) is valid if $\widetilde{\Omega} \ll \varepsilon_F$, where Ω is the characteristic phonon frequency ($\widetilde{\Omega} \sim \Omega_D$) and has a solution with a high $T_C \gtrsim \widetilde{\Omega}$ (strong EPI). We are going to describe the mechanism of the appearance of strong electron-phonon coupling λ and the problem of describing a state with arbitrary λ.

The low dimensionality implies the necessity to analyze the properties of a 2D gas of carriers. Such a system is characterized by a Fermi curve $\varepsilon(\vec{k}) = \varepsilon_F$ instead of a Fermi surface ($\vec{k}$ is the two-dimensional momentum). If the 2D system of carriers contains a subgroup with a high DOS near the Fermi level (its presence in the superconducting high T_C oxides is due to the mixed valence state of Cu), then the Fermi curve has sections which are almost linear nesting states. Such a situation has been studied by the author.[30] In[30] the properties of a size-quantizing Bi film were studied. Although Bi films and the layered superconducting oxides are entirely different systems, there is a strong analogy in some aspects of their behavior (anisotropy of the Fermi curve, small carrier concentration, etc.). The method developed[30] can be applied to study the low dimensional superconductors.

One can show by analogy with[30] that the presence of linear sections (nesting state) of the Fermi curve leads to lattice instability. This instability comes from the interaction of phonons with electronic states attached to these linear sections and manifests itself in the appearance of an imaginary pole in the phonon Green's function. The transition (at some $T = T_p$) to the charge density wave state becomes favorable. A decrease is temperature in the region $T > T_p$ is accompanied by a decrease in the phonon frequency (phonon softening). If $T_C > T_p$ (such situation is perfectly realistic for the high T_C materials), then a low phonon mode with finite momentum appears, and the smallness of the phonon frequency makes EPI strong ($\lambda \sim \widetilde{\Omega}^{-2}$).

The T_C for an arbitrary value of the electron-phonon coupling can be determined[31] from the usual Eliashberg equation which can be written in the form (at $T = T_C$):

$$\Delta(n)Z = \sum_{n'} [K_{n-n'} - 2\mu^*]\ \Delta(n')\ \left|2n'+1\right|^{-1}\Big|_{T_C} \tag{8}$$

where

$$K_{n-n'} = 2 \int d\Omega \, g(\Omega) \Omega \left[\Omega^2 + (n-n')^2 (2\pi T)^2 \right]^{-1} \qquad (9)$$

and Z is the renormalization function

$$Z = 1 + (2n+1)^{-1} \sum_{n'} K_{n-n'} (2n'+1) / \left| 2n'+1 \right|_{T_c}^{-1}; \qquad (10)$$

$$g(\Omega) = \alpha^2(\Omega) \, F(\Omega).$$

Equations (8) and (10) can be solved by the matrix method developed by Owen and Scalapino.[32] The solution for any λ is obtained by the author;[31] with high accuracy, it can be written in the form (if $\mu^* = 0$)

$$T_c = 0.25 \, \tilde{\Omega} \, [e^{\frac{2}{\lambda}} - 1]^{-\frac{1}{2}}. \qquad (11)$$

$$\tilde{\Omega} = <\tilde{\Omega}^2>^{1/2}, \quad <\Omega^2> = (2/\lambda) \int d\Omega \, g(\Omega) \Omega$$

If $\lambda \lesssim 1$, we obtain $T_c \, 0.25 \tilde{\Omega} \exp(-1/\lambda)$, $\lambda = \int d\Omega \, g(\Omega)\Omega^{-1}$, see ref. 33; one can show that (see ref. 31) that the expression

$$T_c = 1.14 \, \tilde{\Omega} \exp\{(1+0.5\rho)\rho^{-1}\} = 0.25 \, \tilde{\Omega} \exp(-1/\lambda)$$ does not differ noticeable from the well-known expression obtained in ref. 34; here $\rho = \lambda(1+\lambda)^{-1}$. In the opposite limit we obtain from Eq. (11) $T_c = 0.18\lambda^{1/2} \, \tilde{\Omega}$, in accord with refs. 34-36.

If $\mu^* \neq 0$, we obtain

$$T_c = 0.25 \, \tilde{\Omega} \, [\exp(2/\lambda_{eff}) - 1]^{-\frac{1}{2}}. \qquad (12)$$

where $\lambda_{eff} = (\lambda - \mu^*)(1 + 2\mu^* + \lambda\mu^* t(\lambda))^{-1}$, the function $t(\lambda)$ is defined.[31]

Strong EPI has been found in organic superconductor (see ref. 37). A major manifestation of the strong coupling is the difference $\beta - \beta_{BCS}$, where $\beta = \varepsilon_0/T_c$, and ε_0 is the energy gap at T=0. In the weak coupling approximation, $\beta \equiv \beta_{BCS} = 1.76$. The organic superconductor $\beta - (ET)_2 AuI_2$ is characterized by the value $\beta \simeq 4\beta_{BCS}$, obtained from tunneling spectroscopy.[37] Tunneling data show large value $\beta \gg \beta_{BCS}$ for the new high T_c superconductors.

Recent experimental data (D. Morris and A. Zettl, private communication) show the presence of the isotope effect in La-Sr-Cu-O. This means that the electron-phonon interaction contributes to superconductivity. However, the coupling constant λ_{ph} is not large enough to provide high T_c. Indeed, according to neutron data, $\tilde{\Omega} \simeq$ 120K, and hence in superconducting oxides $\pi T_c \sim \tilde{\Omega}$. In this case, the electron-phonon interaction could provide high T_c if λ_{ph} were large enough (according to Eq. (12) this would require $\lambda_{ph} \simeq 5$).

But if this were so, the ratio $2\epsilon_o/T_c$ would have to be large.[7] At present, a lot of data indicate that $2\epsilon_o/T_c \simeq 5$ which corresponds to intermediate coupling ($\lambda_{ph} \simeq 2$).

Hence, the electron-phonon interaction plays an important role, but in order to provide high T_c, it is necessary to have an additional mechanism. We think that 2D plasmons (this type of excitations exists in the materials of interest) provide this additional attraction.

We came to the conclusion that our concept of a coexistence of phonon and non-phonon mechanisms proposed in[14-15] is receiving experimental support. In the next section we are going to discuss the problem of the coexistence of the phonon and non-phonon mechanisms.

Coexistence of the Plasmon and Electron-Phonon Mechanisms. The Possibility of Experimental Observation of the Phasmon Mechanism. Based on the generalized Eliashberg equation (1) one can evaluate T_c and the order parameter. It is important that the part of $\Gamma_{p\ell}$ which provides the electron-electron attraction via exchange of 2D plasmons can be written in the form of the usual D-function. As a result, Eq. (1) can be written as a usual Eliashberg equation:

$$\Delta(\omega_{n'})\, Z = \pi T \sum_{\omega_{n'}} \int d\Omega \, [g(\Omega) \, D(\omega_n - \omega_{n'};\Omega) - 2\mu^*] \frac{\Delta(\omega_{n'})}{|\omega_{n'}|} \tag{13}$$

Here $D(\omega_n,\Omega) = \Omega^2/\Omega^2 + \omega_n^2$ is a D-function, and $g(\Omega) = g_{ph}(\Omega) + g_{p\ell}(\Omega)$ where $g_i(\Omega) = \alpha_i(\Omega)\, F_i(\Omega)$, $i = \{ph; p\ell\}$. F_i is the phonon (plasmon) density of states, α_i describe the electron-phonon and the electron-plasmon interactions, respectively. In addition, one can introduce the coupling constant $\lambda = 2 \int d\Omega\, g(\Omega)/\Omega$ which can be written as a sum $\lambda = \lambda_{e,ph} + \lambda_{e;p\ell}$, $\lambda_i = 2 \int d\Omega\, g_i(\Omega)/\Omega$.

The critical temperature in the presence of both the electron-phonon and the plasmon mechanisms can be evaluated from Eq. (13) (see ref. 15). We assume weak electron-plasmon coupling (a more general case will be described elsewhere). Then we obtain

$$T_c = T_c^{ph}(\Omega_{p\ell}/T_c^{ph})^h \tag{14}$$

Here T_c^{ph} is the critical temperature in the absence of the plasmon mechanism, $\Omega_{p\ell} \simeq 0.5\,\epsilon_F$, $h = \lambda_{p\ell}(\lambda_{ph} + \lambda_{p\ell})$. Note that the large value of the ratio $\Omega_{p\ell}/T_c^{ph}$ makes the contribution of 2D plasmons crucial even for small $\lambda_{p\ell}$. For example, if $\lambda_{ph} = 1.5$, $\Omega_{p\ell}/T_c^{ph} = 15$, we obtain $T_c \simeq 2T_c^{ph}$.

The very important question arises of how to detect the presence of the non-phonon plasmon mechanism. Such a separation can be carried out experimentally because the plasmons are excitations of carrier system whereas the phonons involve ionic motion. It would be important to carry out a tunneling and neutron scattering experiment. Tunneling spectroscopy based on inversion of the Eliashberg equation will display all modes, including plasmons. As for neutron scattering, it will show

the function $F_{ph}(\Omega)$ only, because neutron scattering is not affected
by the carriers subsystem. Usually $F_{ph}(\Omega)$ and $g_{ph}(\Omega)$ have a similar
structure (position and number of peaks, the value of Ω_{max}). If the
plasmons play an important role (this is the case for the high T_c
superconductors), then comparison of the neutron and tunneling data
would allow one to detect the presence of the additional (plasmon)
mode. It would be particularly important to compare the frequency
ranges.

It is essential to stress two points. First of all, the smallness
of the Fermi energy (e.g., for $La_{1.8}Sr_{0.2}CuO_4$ the value $\varepsilon_F \simeq 0.12$ eV,
see ref. 3),leads to the plasmon edge within the region suitable for
tunneling spectroscopy.

In addition, the electron-plasmon coupling constant $\lambda_{p\ell}$ depends
on the carrier concentration ($\lambda_{p\ell} \sim V_F^{-1} \sim n^{-1/2}$, see, e.g.,
ref. 15 and increases with decreasing n. This is important because
the new high T_c materials are characterized by small values of n.
This fact makes the plasmon contribution crucial for explaining high
T_c in these materials.

III. NON-PHONON MECHANISMS OF SUPERCONDUCTIVITY IN CONVENTIONAL
     SUPERCONDUCTING SYSTEMS

Despite the considerable theoretical progress and support for the
existence of a non-phonon mechanism (NPM) of superconductivity, (see
e.g., refs. 38-40, 21-27) the situation with NPM remains peculiar.
Strictly speaking, it is impossible to point out a single superconductor and state that the superconductivity in this material is
caused by NPM. We do not have any definite experimental evidence of a
non-phonon mechanism. Unusual properties of high T_c superconductors
make the appearance of the non-phonon mechanism very favarable (see
above). NPM does not necessarily lead to high T_c. On the other
hand, it is known that the BCS theory based on an analysis of the
electron-phonon interaction (EPI) is not restricted to small T_c
values. It is difficult to imagine a situation in which EPI would not
play any role. Rather, it is more realistic that the phonon and
non-phonon mechanisms coexist, although their relative contributions
may be different. One can synthesize materials with the desired
structure favorable for appearance of NPM. But there are also many
existing superconductors which might benefit greatly from NPM.
Substances containing non-uniform structures with spatial separation
between different groups of electrons, or those with complex band
structures, can be expected to have a significant contribution from
NPM. We should be able to prove experimentally the presence of a
non-phonon mechanism. In other words, it should be possible to
separate the contributions of NPM and EPI.

It has been noted (see above, Sec. II) that the analysis of the
tunneling and neutron data will allow one to determine the presence of
the plasmon mechanism in the high T_c oxides. In this section we
consider the usual low T_c superconducting system.

<u>Non-phonon contribution from a high frequency peak.</u> Consider the
case when the non-phonon mode is located higher then the tunneling

region. In the paper[41] the present author proposed a method
allowing one to carry out a separation of this mode. The method is
based on tunneling spectroscopy and on measurements of electronic heat
capacity, or on the temperature dependence of the effective mass.

The powerful technique of tunneling spectroscopy allows one to
determine the function $g(\Omega) = \alpha^2(\Omega) F(\Omega)$ ($F(\Omega)$ is the phonon density
of states, $\alpha^2(\Omega)$ describes EPI). This function can be obtained by an
inversion procedure (see, e.g., refs. 42-43) based on the Eliashberg
equation

$$\Delta(\omega) = Z^{-1}(\omega) \int d\omega' \{ \int d\Omega\, g(\Omega)\ [D(\omega'+\omega) + D(\omega'+\omega) -$$

$$\mu^*]\ \mathrm{Re}\{\Delta(\omega')\ [\omega'^2 - \Delta^2(\omega')]^{-1/2}\} \tag{15}$$

where $\Delta(\omega)$ is the order parameter, D is the phonon Green's function,
μ^* is the Coulomb pseudopotential, and Z is the renormalization
function. It is important that the Eliashberg equation (15) is
written under the assumption that the superconducting state is cause
by EPI only; this interaction corresponds to the energy range suitable
for tunneling spectroscopy.

The same function $g(\Omega)$ affects the behavior of the electronic heat
capacity $Ce(T)$. EPI leads to a deviation of $Ce(T)$ from the linear
law. The analysis of the tunneling data and the behavior $Ce(T)$ allows
one to determine the presence of the NPM (see ref. 41).

Recently a detailed analysis of the properties of Nb_3Ge aimed at
the search for NPM has been carried out by Kihlstrom, Hovda, Wolf, and
the present author.[44] The obtained results manifest a major contribution of NPM to the superconducting state Nb_3Ge.

Nb_3Ge has the highest T_c among A-15 compounds ($T_c\ Nb_3Ge = 22.3K$).
Band structure calculations[45] show that Nb_3Ge is an unusual material
among A-15 superconductors, and the usual EPI is not sufficient to
provide its high T_c. The density of states at the Fermi level is relatively small (see e.g., ref. 46). On the other hand, the band structure
of Nb_3Ge is favorable for an NPM.[24,45] The presence of overlapping
bands might result in pairing in one band via virtual transitions to
another band.[22a] In addition, the acoustic plasmon branch (see above)
can also provide electron-electron attraction.

The selection of Nb_3Ge was motivated by these reasons. An analysis
based on the method[41] (see ref. 44) has resulted in a picture entirely
different from those obtained for Pb and V_3Si. According to[44] one
can state that the non-phonon mechanism plays the key role in Nb_3Ge.

<u>Pairing via molecular excitations.</u> In the previous section, we
studied the case when the energy of virtual transitions $\Delta\varepsilon_{virt.}$ exceeds
greatly the tunneling region. Let us discuss now a different case when

$\Delta\varepsilon_{virt.}$ lies within this region. For concreteness consider the system studied by the author.[47] If the superconductor contains complex molecules (e.g., if the molecules are placed on the surface of a thin film), then additional electron-electron attraction arises via vibrational excitation of the molecules. This might result in an increase in T_c (see ref. 47). This change of T_c can be treated on the basis of the interesting theory of local modes developed in.[48] From this point of view, aromatic molecules are best because their vibrational spacing is relatively small ($\sim 10^2 K$) and the contribution to coupling is notable. This mechanism of superconductivity based on intramolecular virtual excitations can be detected by the tunneling technique and will manifest itself as an additional peak. The position of the peak can be obtained from the second derivative of the tunneling characteristic.

It is important that the molecular frequencies are known independently from molecular spectroscopy. If the position of the peak coincides with the molecular frequency, this will manifest a new mechanism of superconductivity, namely the effect of intramolecular degrees of freedom on pairing.

SUMMARY

In this paper we consider several superconducting systems which are greatly affected by non-phonon mechanisms. The main results can be summarized as follows:

1. The low dimensionality and the small value of the carrier concentration in new high T_c oxides lead to unusual values of the main parameters such as ε_F, m^*, ξ_o. As a result, one can observe a multigap structure.

2. The state of the lattice is greatly affected by the superconducting transition.

3. Exchange of 2D plasmons plays a key role in high T_c superconductivity. Its manifestation can be determined experimentally.

    The superconducting state in high T_c materials is due to the coexistence of the phonon and non-phonon mechanisms.

4. Superconducting state Nb_3Ge is due to non-phonon interaction. Intramolecular excitations can provide additional attraction which can be detected by molecular spectroscopy and by tunneling.

5. Effective increase of T_c of A 15 superconductors can be achieved with the use of the proximity effect.

ACKNOWLEDGMENTS

The author is grateful to M. L. Cohen, D. Gubser, K. Kihlstrom, W. A. Little, H. Morawitz, and S. Wolf for many valuable discussions. This work was supported by the U. S. Office of Naval Research under Contract No. N00014-86-F0015 and carried out at the Lawrence Berkeley Laboratory under Contract No. DE-AC03-76SF00098.

REFERENCES

1. J. Bednorz and K. Müller, Z. Phys. B66, 189 (1986).

2. S. Uchida, et al., Jpn. J. Appl. Phys. Lett. 26, L1 (1987); C. Chu, et al., Phys. Rev. Lett. 58, 405 (1987); R. Cava, et al., Phys. Rev. Lett. 58, 408 (1987); M. Wu, et al., Phys. Rev. Lett. 98, 908 (1987).

3. V. Z. Kresin and S. Wolf, Solid State Comm. (in press).

4. N. Phillips, et al., (preprint); R. Battlogg, et al., Phys. Rev. B35, 5340 (1987); S. Tanaka, et al., Proc. of MRS meeting (Anaheim, 1987), in press.

5. J. Bardeen and M. Stephen, Phys. Rev. 136, 1485 (1964).

6. B. Geilikman and V. Z. Kresin, in Kinetic and Non-Stationary Phenomena in Superconductors, Wiley, New York (1974), p. 87.

7. V. Z. Kresin, Solid State Comm. (in press); Proc. of MRS meeting (Anaheim, 1987), in press.

8. H. Suhl, B. Mattias, and L. Walker, Phys. Rev. Lett. 3, 552 (1959).

9. B. Geilikman, R. Zaitsev, and V. Kresin, Proc. of LT-X, vol. IIA, Moscow (1967), p. 173; Sov. Phys. - Solid State 9, 642 (1967).

10. V. Z. Kresin, J. Low Temp. Phys. 11, 519 (1973).

11. P. Anderson, J. Phys. Chem. Sol. 11, 26 (1959).

12. H. Takayanagi and T. Kawakami, Phys. Rev. Lett. 54, 2449 (1985).

13. a) V. Z. Kresin, Phys. Rev. B34, 7587 (1986); b) V. Z. Kresin, Proc. LT-17, ed. by U. Eckern, A. Schmid, W. Weber, and H. Wuhl, North-Holland, Amsterdam (1984), p. 1029.

14. See Physics Today 40, 22 (1987); presented at Special Sessions at APS (March, 1987, New York) and MRS (April 1987, Anaheim).

15. V. Z. Kresin, Phys. Rev. B35, xxx, (1987).

16. V. Z. Kresin and H. Morawitz (preprint).

17. J. Ruvalds, Phys. Rev. B35, xxx, (1987).

18. A. Panson, et al., Appl. Phys. Lett. 50, 1104 (1987).

19. T. Ando, A. Fowler, and F. Stern, Rev. Mod. Phys. 54, 437 (1982).

20. D. Pines, Can. J. Phys. 34, 1379 (1956).

21. H. Fröhlich, J. Phys. C1, 544 (1968).

22. a) B. Geilikman, Sov. Phys.-Usp. 8, 2032 (1966); 16, 17 (1973); b) E. Pashitskii, Sov. Phys.-JEPT 28, 1267 (1969).

23. J. Ihm, M. L. Cohen, and S. Tuan, Phys. Rev. B23, 3258 (1981).

24. J. Ruvalds, Adv. in Phys. $\underline{30}$, 677 (1981).

25. Y. Takada, J. Phys. Soc. Japan $\underline{45}$, 786 (1978); $\underline{49}$, 1713 (1980).

26. B. Geilikman and V. Z. Kresin, Sov. Phys.-Semiconductors $\underline{2}$, 639 (1968).

27. V. Z. Kresin and B. Tavger, Sov. Phys.-JETP $\underline{23}$, 1124 (1966); Phys. Lett. $\underline{20}$, 595 (1966).

28. D. Pines, Elementary Excitations in Solids (Benjamin, 1963).

29. P. Vashishta and K. Singwi, Phys. Rev. $\underline{B6}$, 875 (1972).

30. V. Z. Kresin, J. Low Temp. Phys. $\underline{57}$, 549 (1984).

31. V. Z. Kresin, Phys. Lett. (in press).

32. C. Owen and D. Scalapino, Physica $\underline{55}$, 691 (1971).

33. B. Geilikman, V. Z. Kresin, and N. Masharov, J. Low Temp. Phys. $\underline{18}$, 241 (1975).

34. R. Dynes, Solid State Comm. $\underline{167}$, 331 (1968).

35. P. Allen and R. Dynes, Phys. Rev. $\underline{B12}$, 905 (1975); C. Leavens, Solid State Comm. $\underline{17}$, 1499 (1975); S. Louie and M. L. Cohen, Solid State Comm. $\underline{22}$, 1 (1977).

36. V. Z. Kresin, H. Gutfreund, and W. A. Little, Solid State Comm. $\underline{51}$, 339 (1984).

37. M. Hawley, et al., Phys. Rev. Lett. $\underline{57}$, 629 (1986).

38. W. A. Little, Phys. Rev. $\underline{134A}$, 1416 (1964); H. Gutfreund and W. A. Little, in Highly Conducting One ·Dimensional Systems, ed. by J. Derreese, R. Errard, and V. van Doren (Plenum, New York, 1979), p. 305.

39. High Temperature Superconductivity, ed. by V. Ginzburg and D. Kirzhnits (Plenum, New York, 1982).

40. D. Allander, J. Bray, and J. Bardeen, Phys. Rev. $\underline{B37}$, 1020 (1973).

41. V. Z. Kresin, Phys. Rev. $\underline{B30}$, 450 (1984).

42. W. McMillan and J. Rowell, in Superconductivity, ed. by R. Parks (Dekker, New York, 1969), vol. 1, p. 561.

43. E. Wolf, Principles of Electron Tunneling Spectroscopy, Oxford University Press, New York, 1985.

44. K. Kihlstrom, P. Hovda, V. Z. Kresin, and S. Wolf (preprint).

45. B. Klein, L. Boyer, D. Papaconstantopoulos, and L. Matteis, Phys. Rev. $\underline{B18}$, 641 (1978).

46. G. Stewart, in Superconductivity in d- and f- Band Metals, ed. by W. Buckel and W. Weber, Kernforschungszentrum, Karlsruhe (1982), p. 81.

47. V. Z. Kresin, Phys. Lett. $\underline{49A}$, 117 (1974).

48. H. Schuttler, M. Jarrell, and D. Scalapino, Phys. Rev. $\underline{58}$, 1147 (1987).

INTERACTIONS AT INTERFACES IN LAYERED SYSTEMS

Gerald B. Arnold

Department of Physics
University of Notre Dame
Notre Dame, IN

INTRODUCTION

Artificially layered systems have received a great deal of attention in recent years. One major impetus for the study of such systems was the search for alternative mechanisms of superconductivity. The discovery of superconductivity above nitrogen temperature in a naturally layered system[1] has elevated the awareness of the potentially wide range of properties available in layered systems, and renewed the interest of theorists in alternative mechanisms of superconductivity connected with interactions between layers.

The system which I will discuss in this article consists of two types of layers, weakly-coupled to each other. In one layer, electronic motion parallel to the interface is of the nearly free type in a two dimensional band which is only partially full. In the other layer, at energies near the Fermi Energy, there is an energy gap, so that there are no states which freely propagate parallel to the interface in this layer. The electronic overlap integral between states in these layers, T, is assumed to be small relative to overlap integrals within each layer.

The single particle properties of this system can be obtained easily by using the Bloch-Wannier basis.[2] The interactions between electrons in neighboring layers are also best described in this basis, which is ideally adapted to treating weakly-coupled two dimensional layers. The superconducting mechanism to be discussed involves the exchange coupling of electrons in the conducting layer to <u>singlet</u> particle-hole excitations in the neighboring layer - essentially, the exciton mechanism.[3] Much effort was expended in looking for this mechanism at an interface between an ultra-thin film and a bulk layer.[4] Interestingly, the failure to observe it may have been due in part to a generic instability related to phonons at interfaces, which I will discuss in the first portion of this article. The second portion deals with interactions between two weakly-coupled layers, and the proposed mechanism for superconductivity.

INDEPENDENT PARTICLE GREEN'S FUNCTIONS IN LAYERED SYSTEMS

A functional integral representation of the retarded Green's function for independent non-interacting electrons is (sum over repeated indices and assume that E contains a positive infinitesimal imaginary part)

$$G(ij \mid E) = \langle i \mid (E-H)^{-1} \mid j \rangle = \langle \phi_i \phi_j^* \rangle = \int \frac{D\phi^* D\phi}{N} \exp\left[-\phi_n^*(E-H)_{nm}\phi_m \right] \{\phi_i \phi_j\}$$

The labels i and j refer to individual layers, and H is the one-electron Hamiltonian, which is evaluated in the Bloch-Wannier basis. N is a normalization, equal to the integral in the numerator without the factors in curly backets, and

$$D\phi^* D\phi \equiv \prod_i d(\mathrm{Re}\phi_i)\, d(\mathrm{Im}\phi_i)$$

In the Bloch-Wannier representation as applied to a layered system, the single particle Hamiltonian is diagonalized in directions parallel to the interfaces by Bloch waves, which are characterized by a band index, μ, and a two dimensional wave vector k. This converts the three dimensional problem into a quasi-one dimensional problem.

For lattice vibrations, one can write an analogous expression (ω contains a positive infinitesimal imaginary part)

$$G(ij \mid \omega^2) = \langle i \mid (\omega^2-\Phi)^{-1} \mid j \rangle = \langle u_i u_j \rangle = \int \frac{Du}{N} \exp\left[-\frac{1}{2} u_n(\omega^2-\Phi)_{nm} u_m \right] \{u_i u_j\}$$

The real field u_i represents the deviation of the ith site from its equilibrium position, and Φ is the dynamical matrix.

The conservation of k in this system means that all of the above are evaluated at the same value of k. The local density of states in the nth layer is obtained from summing over all allowed k values:

$$\rho_{nn}(x) = -\frac{D}{\pi} \sum_{\vec{k}} \mathrm{Im}\, G(nn \mid x)$$

where $x = E$, $D = 1$ for electrons and $x = \omega^2$, $D = 2\omega$ for phonons. The k dependence present in both Green's functions is not explicitly indicated, for notational simplicity. The calculations performed in this work will assume that only nearest neighbor planes are coupled. This effectively reduces these independent particle problems to comparatively trivial one-dimensional tight-binding model problems.

INTERFACE PHONON MODES

Consider the interface between two semi-infinite lattices. Let

$$\Phi_{nm} = \delta(n, m)\, \Omega_{L,R} + \delta(n+1, m)\, T_{L,R} + \delta(n-1, m)\, T_{L,R}$$

for the bulk left hand (L) and right hand (R) lattices. Take the interface between planes $n = 0$ and $n = 1$, and let the coupling of these two planes be T ($\Omega_{L,R}$, $T_{L,R}$, and T all depend on k). The method for performing the functional integrals is discussed in the

appendix. Using the method discussed there, integrate over all fields u_n having $n \neq 0$ or 1.
This yields an exponential with argument

$$\frac{1}{2}\left[-T_L e^{i\theta_L} u_0^2 - 2T u_0 u_1 - T_R e^{i\theta_R} u_1^2 \right]$$

where

$$\sin\theta_{L,R} = \begin{cases} \left[1 - \chi_{L,R}^2 \right]^{1/2} & \chi_{L,R} \leq 1 \\[2ex] -i\,\mathrm{sign}\left(\chi_{L,R}\right)\left[\chi_{L,R}^2 - 1\right]^{1/2} & \chi_{L,R} > 1 \end{cases}$$

and

$$\chi_{L,R} \equiv (\omega^2 - \Omega_{L,R})/2T_{L,R}.$$

Integration over u_1 yields an exponent with argument

$$-\frac{1}{2}\left[T_L e^{i\theta_L} - T^2 e^{-i\theta_R}/T_R \right] u_0^2$$

so that a final integration over u_0 produces

$$< u_0 u_0 > = \left[T_L e^{i\theta_L} - T^2 e^{-i\theta_R}/T_R \right]^{-1}$$

According to standard arguments, for fixed k, <u>poles</u> of this function represent modes which
are confined in the direction perpendicular to the interface, but propagate in the plane of the
interface with wave vector k. These are <u>interface</u> <u>modes</u>.

For poles to appear, both exponentials in the denominator must be real, so both χ_L
and χ_R must be greater than unity so that the exponentials are real:

$$\exp(i\theta_{L,R}) = \chi_{L,R} + \mathrm{sgn}(\chi_{L,R})\left[\chi_{L,R}^2 - 1\right]^{1/2}$$

In addition,

$$e^{i(\theta_L + \theta_R)} = T^2/\left(T_L/T_R\right)$$

has solutions only when the right hand side is greater than one, because χ_L and χ_R are
greater than one. When $T^2 = T_L T_R$, I will say that T has its "impedance match" value, so
that interface modes occur only for T <u>greater</u> than its impedance match value.

The <u>stability</u> of such an interface can now be addressed. If there exist interface
modes having zero frequency at <u>finite</u> k, then it costs no energy to produce a lattice
distortion at the interface with wave vector k, so the interface is <u>unstable</u> to such a
distortion. <u>Stability therefore requires that there be no interface modes with zero frequency
and non-zero k.</u>

Set $\omega = 0$, so that $\chi_{L,R} = -\Omega_{L,R}/(2T_{L,R})$. I assume that $|\chi_{L,R}| > 1$, so that
an interface mode is possible. Because $|\chi_{L,R}|$ is an <u>increasing function</u> of k, there will
always exist a solution to

$$\left(\chi_L + \mathrm{sgn}(\chi_L)\left[\chi_L^2 - 1\right]^{1/2}\right)\left(\chi_R + \mathrm{sgn}(\chi_R)\left[\chi_R^2 - 1\right]^{1/2}\right) = T^2/(T_L T_R) > 1$$

at $\omega = 0$ for finite k. In other words, whenever interface modes exist, they imply an unstable interface. This in turn implies that stable interfaces will not tolerate interface couplings which are greater than the impedance matching value.

As an example of such a system, consider cubic lattices, with identical lattice constants, having an interface perpendicular to the (100) direction, and include nearest and next-nearest neighbor couplings. Then

$$\Omega_L = 2t_{1L}\left(3 - \cos(k_y) - \cos(k_z)\right) + 4t_{2L}\left(3 - \cos(k_y)\cos(k_z)\right)$$

$$T_L = t_{1L} + 2t_{2L}\left[\cos(k_y) + \cos(k_z)\right]$$

with analogous expressions for lattice R, and $T = t_1 + 2t_2\{\cos(k_y) + \cos(k_z)\}$.

It appears that, in general, stable interfaces between semi-infinite lattices must have interfacial coupling which is less than or equal to the impedance match value. This implies that interfaces between bulk systems are prone to instabilities, which presumably lead to reconstruction at the interface, with inevitable alteration in the local electronic properties. This fact may explain the failure to observe the exciton mechanism at an interface between bulk materials. The designation "bulk", as used here, really means any system consisting of more than four or five atomic layers of the same material. In the next section, I will show that in the extreme case of a 1x1 superlattice, any amount of interface coupling is tolerated, without the appearance of an instability. Thus the 1x1 superlattice is inherently a much more stable system for realizing interface related phenomena.

PHONON MODES IN SUPERLATTICES

A superlattice is characterized by cells which are repeated ad infinitum. Let each cell consist of $M + 1$ layers, labelled 0 through M. Assume that layer 0 is one material, layers 1 through M are another material. Suppose $<u_{N,0}, u_{N,0}>$ is required, where u_{N0} is the field in the Nth cell, 0th layer in that cell. Integrating over all fields except those in the Nth cell and those in layers neighboring the cell yields an exponent with argument

$$-\frac{1}{2}\left[\beta_{N-1,M}u_{N-1,M}^2 + 2T\, u_{N-1,M}u_{N,0} + A_0 u_{N,0}^2 + 2T\, u_{N,0}u_{N,1} + \alpha_{N,1}u_{N,1}^2\right]$$

where T is the coupling between cells and α and β are obtained by solving recursion relations:

$$\alpha_{N,i} = A_i - T_{i,i+1}^2 / \alpha_{N,i+1} \;;\; \beta_{N,i} = A_i - T_{i,i-1}^2 / \beta_{N,i-}$$

for $0 \leq i \leq M$. By periodicity

$$\alpha_{N,M+1} = \alpha_{N+1,0} \;;\; \beta_{N,-1} = \beta_{n-1,M}$$

A_i is the factor multiplying $u_{N,i}^2$ and $T_{i,i+1}$ is the coupling between the layers i and i+1 $(T_{0,1} = T = T_{M,M+1})$. For an infinite system, α and β are independent of cell label, e.g., $\alpha_{N,1} = \alpha_1$. Integration over the remaining fields gives

$$< u_{N,0}u_{N,0}> = < u_0 u_0> = \left[A_0 - T^2/\beta_M - T^2/\alpha_1\right]^{-1}$$

For M = 1, a "1x1 superlattice"

$$< u_0 u_0 > = \left[(A_0)^2 - 4T^2 A_0 / A_1 \right]^{-1/2}$$

There are no longer any poles, only integrable singularities. This mathematical feature persists for any "1xM" superlattice. However, as a practical matter, for $M \geq 5$ identical layers, there can exist modes which are <u>essentially</u> confined to the zeroth layer in each cell, and instabilities may reassert themselves.

The properties at long wavelength are most immediately affected by interfacial coupling of two dimensional layers. This is reasonable, on physical grounds. On the other hand, at critical points lying between the extremes of the local density of states, a small amount of coupling has <u>very</u> <u>little</u> effect.

ELECTRONIC STATES

If only one band is accounted for in a tight-binding model, then all of the results above for phonons can be converted to results for electrons merely by introducing <u>one</u> extra parameter into the layer-diagonal portion of the Hamiltonian: the difference in energy, V, between the bottoms of the bands in the two materials.[5] Setting $\omega^2 = E$ in the results above then converts the solution for phonons into that for electrons.

The method of calculation is easily adapted to treating multiple electronic bands. For example, neglecting spin, a material with s and p bands is described by a 4x4 Hamiltonian matrix with a 4x4 matrix, $T_{i,i+1}$ coupling layers i and i+1. For a superlattice of M + 1 layers in each cell:

$$\alpha_i = A_i - T_{i,i+1} \, \alpha_{i+1}^{-1} \, T_{i,i+1} \qquad \beta_i = A_i - T_{i,i-1} \, \beta_{i-1}^{-1} \, T_{i,i-1}$$

where all quantities are matrices, and, according to periodicity

$$\alpha_{M+1} = \alpha_0 \qquad\qquad \beta_{-1} = \beta_M$$

The discussion of stability is irrelevant for the electronic case. One can examine the persistence of two dimensional features in the local density of states as a function of interface coupling, however. Just as in the phonon case, the critical points which lie between the extremes in the local density of states are essentially unaltered by a small amount of coupling. This is especially true if one considers a 1x1 superlattice wherein the critical point of interest happens to fall within an energy range where adjoining layers contain no states propagating parallel to the interface, i.e., the two dimensional band gap of the adjoining layers. The appearance of such two dimensional phenomena within a band is important because the Fermi energy may fall at or near the characteristic logarithmic singularity in a two dimensional electron band and lead to an enhancement of the superconducting critical temperature.[6]

INTERACTIONS IN LAYERED SYSTEMS

Using the Bloch-Wannier basis and second quantization, a two particle interaction becomes:

$$\frac{1}{2} \sum_{1,2,3,4} < 12 \mid V \mid 34 > C_1^+ C_2^+ C_4 C_3$$

where the integers J(= 1, 2, 3, 4) represent n_J, k_J, with n_J the layer index and k_J a two dimensional wave vector (band and spin indices have been suppressed, for the moment).

The matrix element is proportional to a Kronecker delta which requires $k_1 + k_2 = k_3 + k_4$. The system of interest consists of two types of layers, arrayed in a 1x1 superlattice.

The matrix element decomposes into pieces involving interactions within a layer and those involving interactions between layers. The assumption of weak electronic coupling between layers allows the neglect of all interlayer matrix elements except those coupling nearest neighbor layers. The coupling between layers is of three types:

1. Coupling between the electron density in layer n and that in layer n + 1:

$$< n, k_1; n+1, k_2 \mid V \mid n, k_3; n+1, k_4 > C^+_{n,\sigma}(k_1) \, C^+_{n+1,\sigma'}(k_2) C_{n+1,\sigma'}(k_4) C_{n,\sigma}(k_3)$$

2. Exchange coupling between electrons in layer n and those in layer n+1:

$$< n, k_1; n+1, k_2 \mid V \mid n+1, k_3; n, k_4 > C^+_{n,\sigma}(k_1) \, C^+_{n+1,\sigma'}(k_2) \, C_{n,\sigma'}(k_4) \, C_{n+1,\sigma}(k_3)$$

3. Coupling between electron pairs in layer n and pairs in layer n+1:

$$< n, k_1; n, k_2 \mid V \mid n+1, k_3; n+1, k_4 > C^+_{n,\sigma}(k_1) C^+_{n,\sigma'}(k_2) \, C_{n+1,\sigma'}(k_4) \, C_{n+1,\sigma}(k_3)$$

All of these matrix elements are proportional to T^2, where T is the (small) one-particle coupling between layers. The pair coupling will not be relevant to the phenomena discussed below, so I will ignore it.

I will discuss the effects of these interactions in the context of the functional integral formalism[7], to be consistent with the rest of this article. One could also employ the operator formalism and Hubbard-Stratonovich procedure, which is equivalent.

The functional integral involves integration over anticommuting Grassmann fields which are functions of imaginary time τ ($0 \le \tau \le \beta$), position within the layer, ρ (a two dimensional vector), spin σ, and layer index. Consider a 1x1 superlattice consisting of two distinct layers: a, wherein the Fermi Energy lies within a 2d band, and b, wherein the Fermi Energy falls within a gap between two bands. I will convert from a description in terms of two dimensional wave vectors to one in terms of two dimensional position vectors describing location within a layer, ρ, because this is more convenient. I will use $a_{\rho\sigma}$ [$\rho \equiv (\rho, \tau)$] and its complex conjugate (c.c.) for fields in layer a, $b_{\rho\sigma}$ and its c. c. for fields in layer b. The functional integral integrates exp(-A) over these fields, where A is the action (sum over all repeated indices):

$$A \equiv a^*_{\rho\sigma} \left(\frac{\partial}{\partial\tau} + H_a \right)_{\rho\rho'} a_{\rho'\sigma} + \frac{1}{2} U^{(a)}_{\rho\rho'} a^*_{\rho\sigma} a^*_{\rho'\sigma'} a_{\rho'\sigma'} a_{\rho\sigma}$$

$$+ a^*_{\rho\sigma} T_{\rho\rho'} b_{\rho'\sigma} + b^*_{\rho\sigma} T_{\rho\rho'} a_{\rho'\sigma} + V_{\rho\rho'} a^*_{\rho\sigma} b^*_{\rho'\sigma'} b_{\rho'\sigma'} a_{\rho\sigma} + J_{\rho\rho'} a^*_{\rho\sigma} b^*_{\rho'\sigma'} a_{\rho'\sigma'} b_{\rho\sigma}$$

$$+ b^*_{\rho\sigma} \left(\frac{\partial}{\partial\tau} + H_b \right)_{\rho\rho'} b_{\rho'\sigma} + \frac{1}{2} U^{(b)}_{\rho\rho'} b^*_{\rho\sigma} b^*_{\rho'\sigma'} b_{\rho'\sigma'} b_{\rho\sigma}$$

The single particle Hamiltonian within the ith layer is designated by H_i. The interactions are all local in τ, e.g., $V_{\rho\rho'} \propto \delta(\tau, \tau')$.

In order to simplify mattters further, assume that all interactions are spatially local within the layers, so that each is proportional to $\delta(\rho, \rho')$. In this approximation the two-particle coupling between planes becomes

$$V a^*_{\rho\sigma} a_{\rho\sigma} b^*_{\rho\sigma'} b_{\rho\sigma'} - J a^*_{\rho\sigma} a_{\rho\sigma'} b^*_{\rho\sigma'} b_{\rho\sigma}$$

$$= (V\!-\!J)\, a^*_{\rho\sigma} a_{\rho\sigma} b^*_{\rho\sigma} b_{\rho\sigma} + V a^*_{\rho\sigma} a_{\rho\sigma} b^*_{\rho-\sigma} b_{\rho-\sigma} - J a^*_{\rho\sigma} a_{\rho-\sigma} b^*_{\rho-\sigma} b_{\rho\sigma}$$

The last term couples electrons in layer a to <u>singlet</u> particle-hole excitations in layer b. Note that if J is positive this coupling is attractive even in lowest order, like the electron-phonon interaction, and is <u>not</u> directly diminished by the repulsive interaction given by V. The other two terms describe coupling to <u>triplet</u> electron-hole excitations in layer b.

The interaction $U^{(b)}$ may now be combined with the term proportional to J to yield:

$$-\frac{1}{2} U^{(b)} M^*_{\rho\sigma} M_{\rho\sigma} + \frac{1}{2}\frac{J^2}{U^{(b)}} a^*_{\rho\sigma} a_{\rho-\sigma} a^*_{\rho-\sigma} a_{\rho\sigma} \qquad M_{\rho\sigma} \equiv b^*_{\rho-\sigma} b_{\rho\sigma} + \frac{J}{U^{(b)}} a^*_{\rho-\sigma} a_{\rho\sigma}$$

The singlet particle hole excitation may be represented by a field $\phi_{\rho\sigma}$ introduced by the functional integral identity

$$\exp\left(\frac{1}{2} U^{(b)} M^*_{\rho\sigma} M_{\rho\sigma}\right) = \int \frac{D\phi^* D\phi}{N} \exp\left\{-\frac{1}{2}\phi^*_{\rho\sigma}\phi_{\rho\sigma}/U^{(b)} + \frac{1}{2}\phi^*_{\rho\sigma} M_{\rho\sigma} + \frac{1}{2}M^*_{\rho\sigma}\phi_{\rho\sigma}\right\}$$

where N is the integral in the numerator with $M_{\rho\sigma} = M_{\rho\sigma}{}^* = 0$. The result of all of these manipulations is to produce an action which is quadratic in the b fields, with linear coupling to the ϕ, ϕ^* fields, and an action for the a fields which includes attractive interactions:

$$-\left[\frac{1}{2}\frac{J}{U^{(b)}} a^*_{\rho-\sigma} a_{\rho\sigma} \phi^*_{\rho\sigma} + \text{h.c.}\right] - \frac{J^2}{2U^{(b)}} a^*_{\rho\sigma} a_{\rho-\sigma} a^*_{\rho-\sigma} a_{\rho\sigma}$$

The first term is similar to that appearing as a result of the electron phonon interaction. If the action for the ϕ fields is approximated by a quadratic form

$$\frac{1}{2}\, \phi^*_{\rho\sigma}(D^{-1})_{\rho\sigma}\phi_{\rho\sigma}$$

then completing the square in the ϕ fields and integrating them after a shift of functional variable from $\phi^*_{\rho\sigma}$ to

$$\left[\phi^*_{\rho\sigma} - \frac{J}{U^{(b)}} a^*_{\rho'\sigma'} a_{\rho'-\sigma'} D_{\rho'\sigma',\rho\sigma}\right]$$

with the corresponding change in the variable $\phi_{\rho\sigma}$, where D is the propagator for the ϕ fields (obeying $D^{-1}D = 1$), we find that the action in layer a acquires the contribution:

$$-\frac{1}{2}\left[\frac{J}{U^{(b)}}\right]^2 a^*_{\rho\sigma} a_{\rho-\sigma} D_{\rho\sigma,\rho'\sigma'} a^*_{\rho'-\sigma'} a_{\rho'\sigma'}$$

This is of the same form as the conventional interaction due to exchange of phonons, with D as the analogue of the phonon propagator. This propagator may be extracted in the usual way from the quadratic portion of the action for the ϕ fields.

Using the two interactions between the electrons in layer a derived above, one may now approximate the complete interaction by a <u>local</u> attractive interaction and implement the conventional BCS analysis to obtain T_c for the electrons in layer a. Because the relevant energy scale for electron-hole pair excitations is of the order of the 2d energy gap in layer b, it is evident that the energy range over which this interaction is attractive may be an order of magnitude greater than that over which the electron phonon mechanism is attractive. Hence, given the same coupling strength, the BCS T_c expression implies an order of magnitude increase in T_c.

The basic idea of the approach outlined here is to use integral identities (such as that invoked above to introduce the ϕ fields) to replace the terms in the action which are of fourth order in the fields, yielding an action which is quadratic in the basic fields, with linear coupling between them and the fields introduced via the integral identities (auxiliary fields, such as ϕ). The basic fields may then be integrated out of the problem, leaving only the action for the auxiliary fields. While the latter action is highly non-linear, in general, often only the portions quadratic in the auxiliary fields need be retained. In the case at hand, one of these auxiliary fields will be the pairing field in layer a. The equation obtained for this field yields the superconducting critical temperature for this layer.

The theory presented here is just an outline of this more detailed theory, which is not completely worked out, as of this writing. Nonetheless, the results obtained here are highly suggestive that two-particle coupling of the exchange type to singlet particle-hole excitations can produce superconductivity at high temperatures.

APPENDIX

Coupled Gaussian integrals of the type

$$< x_o x_o > = \int \frac{Dx}{N} \, \{x_o x_o\} \, \exp\left[-\frac{1}{2} \sum_i (A x_i^2 + 2B x_i x_{i+1}) \right] \quad (Dx \equiv \prod_i dx_i)$$

are required (N is the integral without the product in curly brackets). Each index i labels a plane. Consider a set of 2M+1 planes. Integrate out all fields x_i having $i < 0$, in order to get $< x_0 x_0 >$. Start at $i = -M$. Because only nearest neighbor planes are coupled, each integration may be done by completing the square, viz.:

$$\int dx_{-M} \exp\left[-\frac{1}{2} A x_{-M}^2 - B x_{-M} x_{-M+1} \right] = I_{-M} \exp\left[B^2 x_{-M+1}^2 / 2A \right]$$

The factor I_{-M} also appears in the normalization, so it is irrelevant. The integral over x_{-M+1} is

$$\int dx_{-M+1} \exp\left[-\frac{1}{2} (A - B^2/A) x_{-M+1}^2 - B x_{-M+1} x_{-M+2} \right]$$

Setting $\alpha_{-M+1} = A - B^2/A$ we find that successive values of α_{-M+i}, defined as the coefficients of x_{-M+i}^2, satisfy a recursion equation:

$$\alpha_{-M+i} = A - B^2/\alpha_{-M+i-1}$$

For large M, $\alpha_{-M+i} = \alpha_{-M+i-1} = \alpha$ and

$$\alpha = A/2 \pm [(A/2)^2 - B^2]^{1/2}$$

For $B = 0$, $\alpha = A$ requires the solution

$$\alpha = B \left\{ A/2B \pm \mathrm{sgn}(A)[(A/2B)^2 - 1]^{1/2} \right\} \equiv B \exp(i\theta)$$

which defines θ.

REFERENCES

1. M. K. Wu, J. R. Ashburn, C. J. Torng, P. H. Hor, R. L. Meng, L. Gao, Z. J. Huang, Y. Q. Wang, C. W. Chu, Phys. Rev. Lett. <u>58</u>, 908 (1987).
2. D. Kalkstein and P. Soven, Surf. Sci. <u>26</u>, 85 (1971).
3. W. A. Little. Phys. Rev. <u>134</u>, A1416 (1964); V. L. Ginzburg, Phys. Lett. <u>13</u>, 101 (1964), Sov. Phys.-Uspekhi <u>13</u>, 335 (1970); D. Allender, J. Bray, and J. Bardeen, Phys. Rev. B<u>7</u>, 1020 (1973).
4. D. L. Miller, M. Strongin, O. F. Kammerer, and B. G. Streetman, Phys. Rev. B<u>13</u>, 4834 (1976).
5. Madhu Menon and Gerald B. Arnold, Phys. Rev. B<u>27</u>, 5508 (1983).
6. J. E. Hirsch and D. J. Scalapino, Phys. Rev. Lett. <u>56</u>, 2732 (1986).
7. H. Kleinert, Fortschritte der Physik <u>26</u>, 565 (1978).

EXCITONIC SUPERCONDUCTIVITY IN LAYER STRUCTURES

John Bardeen, D. M. Ginsberg and M. B. Salamon

Department of Physics
University of Illinois at Urbana-Champaign
Urbana, Illinois   61801

INTRODUCTION

Many different theories have been presented at this Conference to try
to account for the high transition temperatures, T_c, of the recently
discovered copper oxide compounds. Present evidence is that the properties
are consistent with a normal BCS superconductor, but with the attractive
interaction that accounts for the pairing mediated by something other than
phonons. Non-phonon mechanisms were suggested not long after the BCS
theory was first presented. Two of the most prominent were those of
Little,[1] who suggested linear metallic chains surrounded by polarizable
ligands, and by Ginzburg,[2] who suggested a sandwich-like layer structure of
metallic layers between which there are polarizable layers.

Allender, Bray and one of the authors[3] gave the basis for a more
complete microscopic theory of layer structures. We used a model in which
the polarizable layer is similar to a narrow gap semiconductor. The
virtual transitions that mediate the attractive interactions are virtual
quasi-particle excitations across the semiconducting gap. These
calculations indicated that high T_c's should be possible with reasonable
values of the parameters involved if the right structures could be found.

In this paper we will discuss thermodynamic and magnetic evidence on
the 90K compounds, based in part on work done at Illinois[4] on measurements
of the specific heat jump at T_c. The thermodynamic data are consistent
with BCS in the weak coupling limit. The superconductors are strongly type
II, with a small Pippard coherence distance, $\xi_o \sim 10^{-7}$ cm and a penetration
depth $\lambda \sim 10^{-5}$ cm. This agreement supports the BCS quasiparticle excitation

spectrum and appears to rule out theories in which pairs are present in the normal state and undergo a Bose transition to a superfluid condensate.

A qualitative discussion will be given of the exciton mechanism and of possible excitonic transitions that might be responsible for the attractive interaction. A discussion is given of experiments that might help distinguish between the various suggestions that have been made. An important question is whether or not collective excitations (plasmons, acoustic plasmons, spin waves) are involved.

THERMAL PROPERTIES

As pointed out, particularly by Batlogg,[5] when T_c's are plotted against the Sommerfeld γ, the oxides appear to form a new class of superconductors with T_c's about three times the upper limit of metals with the same γ's. Cava et al.[6] and others have interpreted magnetic data of $YBa_2Cu_3O_{7-\delta}$ and similar compounds with T_c's in the range of 90K in terms of the Ginzburg-Landau theory. Large values of $\kappa = \lambda_{GL}/\xi_{GL}$, in the range 60 – 100, are required, with $\lambda_{GL} \sim 1500\text{Å}$ and $\xi_{GL} \sim 20\text{Å}$.

Recent measurements at Illinois[4] have shown that there is a specific heat jump at T_c in $YBa_2Cu_3O_{7-\delta}$ that is in good agreement with the BCS prediction of 1.43 γT_c. The value of $\gamma = 3\times10^3$ erg/cm^{3}K^2 is estimated from the magnetic susceptibility. With corrections for diamagnetic contributions the Pauli susceptibility is about $1.37 \times 10^{-9}\gamma$. M. B. Salamon and coworkers[7] have found that when this value of γ is used with the experimental resistivity $\rho = 200\mu\Omega\text{cm}$ in the expressions for the Ginzburg-Landau parameters in the dirty limit, good agreement is found with the magnetic data. They find a value $\xi_{GL} \sim 12\text{Å}$, that is considerably less than the value derived by Cava et al.[6] as well as with other estimates.

They have given a microscopic derivation of the Ginzburg-Landau parameters on the basis of an isotropic free electron model with an effective mass adjusted to give $\gamma = 3\times10^3$ erg/cm^{3}K^2. A similar calculation with slightly different numbers is outlined in Tables I, II and III. The carrier density (holes) of 9×10^{21}/cm^3 is taken from recent unpublished ac Hall data of C. F. Gallo and K. V. Rao.[8] This value corresponds about one hole for every two Cu atoms. The value of k_F is about 6.5×10^7cm^{-1}.

The normal state properties are outlined in Table I. The density of states for one spin at the Fermi surface is proportional to the effective mass ratio m^*/m. The value $m^*/m \simeq 9$ is chosen to give $\gamma = 3\times10^3$. The

Table I. Normal State Properties: $YBa_2Cu_3O_{7-\delta}$

Carrier density: $n = k_F^3/(3\pi^2) = 9\times10^{21}/cm^3$ (ref. 4)
Fermi wave vector: $k_F = 6.5\times10^7 cm^{-1}$
Density of states/spin: $N(0) = m^*k_F/(2\pi^2\hbar^2) = 3\times10^{34}/cm^3 erg$
Sommerfeld const: $\gamma = 2\pi^2 k_B^2 N(0)/3 = 3\times10^3 erg/cm^3 K^2$ (refs. 4, 7)
Effective mass ratio: $m^*/m = 9$
Resistivity (90K): $\rho = 200\mu\Omega cm$ (ref. 6)
Mean free path (90K): $\ell = \hbar k_F/(ne^2\rho) = 17\text{Å}$

Table II. Superconducting Properties: $YBa_2Cu_3O_{7-\delta}$
(Theory, weak coupling, $k_B T_c \ll \hbar\omega_{ex}$)

Transition temperature	$T_c = 90K$
Gap parameter	$\Delta = 1.76 k_B T_c = 2.15\times10^{-14} ergs$
Specific heat jump	$\Delta C_e = 1.43\gamma T_c$
Pippard coherence length	$\xi_o = \hbar v_F/(\pi\Delta) = 12\text{Å}$
London penetration depth	$\lambda_L = (m^*c^2/4\pi ne^2)^{1/2} = 1700\text{Å}$
Critical field	$H_c(0) = [4\pi N(0)]^{1/2}\Delta = 12kOe$

Table III. Ginzburg-Landau Parameters: $YBa_2Cu_3O_{7-\delta}$

	Experiment (ref. 6, 9, 10)	Theory
Penetration depth	$\lambda_{GL} = 1400\text{Å}$	$\lambda_L = 1700\text{Å}$
Coherence distance	$\xi_{GL} = 14\text{Å}$	$\xi_o = 12\text{Å}$
Ratio	$\kappa = 100$	
Critical fields (refs. 9, 10)	$H_c(0) = 12kOe$	$12kOe$
$H_{c1} = H_c(0)\ln\kappa/(\sqrt{2}\kappa)$	$H_{c1} = 400\text{ Oe}$	400 Oe
$dH_{c1}/dT = -(H_{c1}/H_c)(4\pi\Delta C_e/T_c)^{1/2}$	-7 Oe/K	-9 Oe/K
$H_{c2} = \sqrt{2}\kappa H_c$	$80 - 320T$	$170T$

m.f.p. for $\rho = 200\mu\Omega cm$ is about 17Å. The parameters required for the
superconducting properties are outlined in Table II.

The critical field, $H_c(0) = 12kOe$, is in agreement with
experiment.[9, 10] The Pippard coherence distance is about 12Å, less than
the m.f.p., so that one is not in the dirty limit, although with anisotropy
scattering may be sufficient to reduce the penetration depth, λ, below the

London value, 1700Å. Ginzburg–Landau parameters (Table III)
of $\xi_{GL} = \xi_o = 12$Å and $\lambda_{GL} = 1400$Å are therefore reasonable. Because of
the uncertainties involved, we have chosen $\kappa = 100$, with $\xi_{GL} = 14$Å. Values
of H_{c1}, dH_{c1}/dT and H_{c2} are all consistent with experiment. It is in fact
remarkable how well the simple isotropic model fits the data.

The fact that ordinary weak–coupling BCS theory applies shows that the
energies of the excitations $\hbar\omega_{ex}$ that mediate the attractive interaction
between electrons are very large compared with $k_B T_c$. In an analysis of the
Eliashberg equation, Marsiglio, Akis and Carbotte[11] have shown that
for $\Delta C_e/(\gamma T_c)$ to be of the order 1.5 or less, the dominant $\hbar\omega_{ex}$ must be
larger than the order of $20k_B T_c$, or of order ~ 0.2eV. Similar restrictions
apply to the relation $H_c(0) = 2.4\gamma^{1/2}T_c$. This provides severe restrictions
on possible theories.

The product $k_F\xi_o = 2E_F/(\pi\Delta)$ is of order 8. This ratio gives a measure
of the ratio of the pair correlation distance to the interparticle
spacing. A value of $k_F\xi_o \gg 1$ implies that ordinary BCS theory is valid and
that the superconducting state is not a Bose condensation of preformed
pairs.

THE EXCITON MECHANISM

The polarization diagrams[3] that mediate the attractive interaction may
be similar to those used by Louie and students[12] to get improved values for
the bandgaps in semiconductors. The Coulomb interaction between electrons
may be written $4\pi e^2/(q^2\varepsilon(q))$ where $\varepsilon(q)^{-1}_{K,K'}$ must be regarded as a matrix in
reciprocal lattice vectors. What is involved is virtual excitation from an
electron in the valence band of a semiconductor across a narrow gap to the
conduction band. The intermediate state differs by a wave vector, Q , from
the initial, where in general Q lies outside of the first Brillouin zone.
It may be written $Q = q+K$, where q is in the first zone and K is a
reciprocal lattice vector. Scattering between two electrons in states $k\uparrow,-$
$k\downarrow$ to $k'\uparrow,-k'\downarrow$ results when one electron excites a virtual transition with
wave vector Q and another absorbs it. Migdal's theorem that higher order
diagrams are unimportant should apply if the energy of the gap $\hbar\omega_g$ is small
compared to the Fermi energy defined by the average electron density.

Physically, one can think of one electron polarizing the lattice and
another taking advantage of it to give an attractive interaction. To put
it another way, the polarization energy varies as the square of the
charge. If two electrons are in the same vicinity, the energy is

proportional to $-(2e)^2$ as compared with an energy $-2e^2$ of the separate charges.

Band structure calculations[13] on the copper oxide compounds suggest gap transitions of the order of 2eV. This is roughly the Fermi energy of an electron in a gas of density $10^{22}/cm^3$ and effective mass $m*/m = 1$. The total density corresponding to two or three states per Cu atom would be perhaps eight times larger, with a Fermi energy four times larger, $\sim$ 6-8eV, consistent with band calculations. The ratio $\hbar\omega_g/E_F$ thus may be sufficiently small for Midgal's theorem.

Yu et al.[13] have given a band structure calculation of $YBa_2Cu_3O_{7-\delta}$ in which they suggest possible transitions for the excitonic mechanism. They involve charge-transfer excitations in the quasi-1D CuO_3 chains that lie between the Ba ions in the crystal structure. The carriers are in the CuO_2 sheets adjacent to the Y ions. Many other suggestions for excitations that mediate the attractive interaction have also been made, including plasmons and spin fluctuations. It is too early to suggest which may be the correct one.

The band gaps that give the virtual transitions may also give rise to absorption in the normal state. Orenstein et al. [14] have observed anomalies in the reflectivity spectra of both $La_{2-x}Sr_xCuO_4$ and $YBa_2Cu_3O_{7-x}$ that indicate absorption peaks. The latter is in the range 1-2eV, consistent with what is required for the exciton mechanism.

CONCLUSIONS

A complete theory must give both the ground state and the spectrum of elementary excitations and should include effects of anisotropy. The data discussed here involve thermodynamic and magnetic properties. It would be very valuable to have data on the transport properties. These include response to electromagnetic radiation, acoustic attenuation, thermal conductivity and nuclear spin relaxation times. It is hoped that these will be measured when better specimens become available. Such experiments should determine whether or not the usual BCS theory applies or will require revision to account for the high T_c oxides.

REFERENCES

1. W. A. Little, Possibility of synthesizing an organic superconductor, Phys. Rev. 134, A1416–A1424 (1964).

2. V. L. Ginzburg, Concerning surface superconductivity, Zh. Eksp. Teor. Fiz. 47, 2318–2320 (1964) [Sov. Phys.-JETP 20, 1549–1550 (1965)] and The problem of high-temperature superconductivity, Ann. Rev. Mater. Sci. 2, 663–696 (1972).

3. D. Allender, J. Bray, and J. Bardeen, Model for an exciton mechanism of superconductivity, Phys. Rev. B7, 1020–1029 (1973).

4. S. E. Inderhees, M. B. Salamon, T. A. Friedmann, and D. M. Ginsberg, Measurement of the specific heat anomaly at the superconducting transition of $YBa_2Cu_3O_{7-\delta}$, Phys. Rev. (submitted).

5. B. Batlogg, A. P. Ramirez, R. J. Cava, R. B. van Dover, and E. A. Rietmann, Electronic properties of $La_{2-x}Sr_xCuO_4$ high-T_c superconductors, Phys. Rev. B35, 5342 (1987).

6. R. J. Cava, B. Batlogg, R. B. van Dover, D. W. Murphy, S. Sunshine, T. Siegrist, J. P. Remeika, E. A. Rietman, S. Zahurak, and G. P. Espinosa, Bulk superconductivity at 91 K in single-phase oxygen-deficient perovskite $Ba_2YCu_3O_{9-\delta}$, Phys. Rev. Lett. 58, 1676 (1987).

7. M. B. Salamon, J. Bardeen, and D. M. Ginsberg, Comment on "Bulk superconductivity at 91 K in single-phase oxygen-deficient perovskite $YBa_2Cu_3O_{9-\delta}$," Phys. Rev. B (submitted).

8. C. F. Gallo and K. V. Rao, private communication.

9. P. M. Grant, R. B. Beyers, E. M. Engler, G. Lim, S. S. P. Parkin, M. L. Ramirez, V. Y. Lee, A. Nazzal, J. E. Vazquez, and R. J. Savoy, Superconductivity above 90 K in the compound $YBa_2Cu_3O_x$: structural, transport, and magnetic properties, Phys. Rev. B35, 7242 (1987).

10. D. R. Harshman, G. Aeppli, B. Batlogg, R. J. Cava, E. J. Ansaldo, J. H. Brewer, W. Hardy, S. R. Kreitzman, G. M. Luke, D. R. Noakes, and M. Senba, Temperature dependence of the magnetic penetration depth in the high-T_c superconductor $YBa_2Cu_3O_{9-\delta}$: evidence for conventional s-wave pairing, Phys. Rev. Lett. (submitted).

11. F. Marsiglio, R. Akis, and J. P. Carbotte, Thermodynamics in very strong coupling: a possible model for the high T_c oxides, (preprint).

12. M. S. Hybertsen and S. G. Louie, Electron correlation in semiconductors and insulators: band gaps and quasiparticle energies, Phys. Rev. B34, 5390 (1986).

13. J. Yu, S. Massidda, A. J. Freeman, and D. D. Koelling, Bonds, bands, charge transfer excitations and superconductivity of $YBa_2Cu_3O_{7-\delta}$, Phys. Rev. Lett. (submitted).

14. J. Orenstein, G. A. Thomas, D. H. Rapkine, C. G. Bethea, B. F. Levine,
     R. J. Cava, E. A. Rietman, and D. W. Johnson, Jr., A normal-state gap
     transition in Cu-O superconductors, (preprint).

THE EXCITON INTERACTION: ITS POSSIBLE ROLE

IN HIGH TEMPERATURE SUPERCONDUCTIVITY

W.A. Little

Physics Department
Stanford University
Stanford, CA 94305

The recent remarkable developments in superconductivity[1,2]
has forced the group of physicists in the main stream of
superconductivity research to re-examine the possible role of
what has been referred to in the conference as novel
mechanisms of superconductivity. The exciton mechanism is one
such. While the many studies and developments in this subject
are relatively well known to those involved in studies of
organic superconductors and superconductors of reduced
dimension, it appears that it is not well known to that large
body of physicists involved in the more conventional
"mainstream" of superconductivity. In view of this I have been
asked to review the salient features of the mechanism and to
discuss what it can and cannot do. I will base my remarks on
the most recent and most comprehensive review of the subject
published in 1979 by H. Gutfreund and myself [3] plus a few
key papers since that time[4,5]. I would also like to
acknowledge the major role our colleagues in theoretical
physics in the Soviet Union have played in contributing to the
resolution of many of the problems in this field which had
been raised by our colleagues in the West. Works by Ginzburg,
Gorkov, Migdal, Dzyaloshinskii, Kresin, Kirzhnits, Khomski to
name a few, permeate the field and have been crucial in the
development of the subject.

The possibility of obtaining superconductivity at
temperatures approaching ambient by utilizing an electronic
excitation rather than a phonon mechanism was first raised in
1964[6]. While it is now generally accepted that the
replacement of the Debye frequency in the BCS expression for
T_C, $kT_C = 1.14\ \hbar\omega\ \exp(-1/\lambda)$ by an electronic transition of
much higher energy could lead to high T_C this was by no means
accepted in the first several years after its proposal. The
key point which is now agreed upon is that if one ignores the
Coulomb interaction then the coupling constant λ which
appears in this expression for T_C does not contain the mass of
the particles which mediate the electron-electron interaction.
The only factor which does contain this mass m is the the
prefactor $\hbar\omega$, which is proportional to $m^{-1/2}$ and it is this
which gives rise to the Isotope Effect.

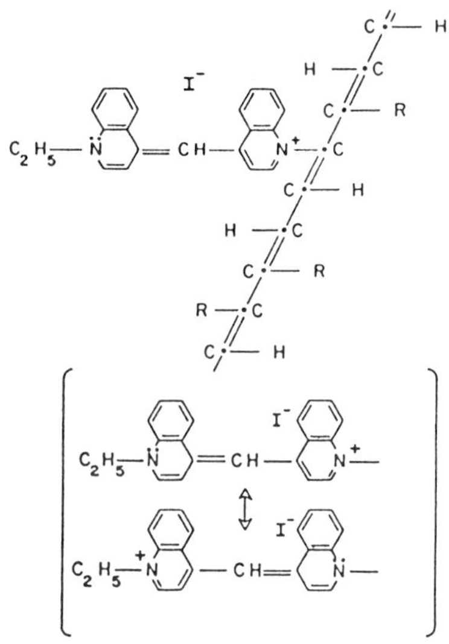

Figure 1. Exciton coupled Organic Superconductor

If a sufficient attraction can be obtained by the virtual
movement of electrons rather than ions then substantially
higher values of T_C should be obtainable. Such an interaction
has been called the "exciton" interaction in reference to the
electronic _excitations_. It does not imply the use of the more
limited concept of excitons described by Mott-Wannier or
Frenkel excitons.

Role of Limited Dimensionality

The original proposal was based on the concept of
deliberately designing a polymeric system with certain
electronically polarizable side-chains to provide the
electronically mediated attraction. See figure 1. This appears
to have been the first time that superconductivity occurring
in a system of limited dimensionality was considered. It was
suggested at that time that the arguments which prohibited
classical phase transitions in one-dimensional (1.D) systems
where condensation occurs in real space, might not apply to a
quantum phase transition in which condensation occurs in
momentum space. This was soon shown not to be the case[7.8].
As a result interest rapidly shifted to the possibility of
realizing the same attractive mechanism in a 2-D system[9].
where fluctuations would tend to play a less crucial role.
Ginzburg[10] suggested a 2-D analogue of the organic
superconductor consisting of a sandwich of an atomically thin
metal layer between layers of dielectric as illustrated in
Figure 2.

This was followed a several years later by the proposal of a
more specific example of this by Allender, Bray and
Bardeen[10] in 1972.

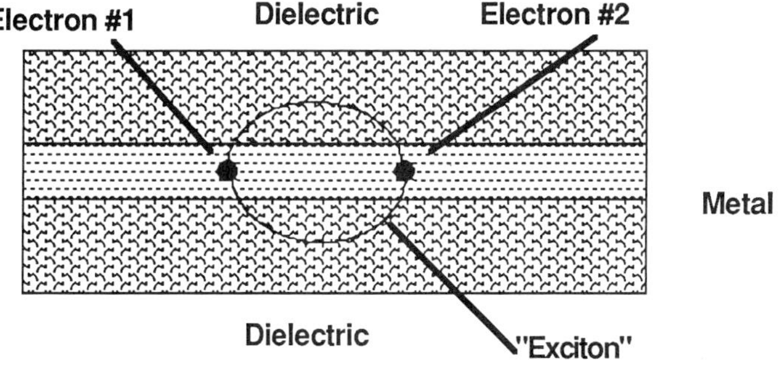

Figure 2. Ginzburg's 2-D Exciton Superconductor.

Fluctuations

While fluctuations in a 2-D system are smaller than those in a
1-D system, nevertheless, they are still sufficient to destroy
the long-range-order. This was shown by Hohenberg[11] in 1967,
who proved that a true pase transition to the superconducting
state was prohibited just as in 1-D. These arguments were
based on the proof that the order parameter is destroyed by
fluctuations in a large sample so that no long range order can
occur. This can be pictured as in Figure 3. If one imagines a
long thin "superconducting" sample then the order parameter is
described by a complex number at each point. Fluctuations in
the local density will be accompanied by fluctuating currents.
Such currents cause, or are represented by, a variation in the
phase of the order parameter, consequently, the relative phase
of the order parameter at two points a distance r apart will
fluctuate. As r gets larger, so will the relative phase
fluctuations. So that the average projections of the order
parameter at one point on that of another a long distance away
approaches zero. This is the condition of the absence of
off-diagonal-long-range and prohibits a true phase transition
in a 1-D system. A similar argument applies in 2-D. However,
we pointed out that such fluctuations do not, of themselves
cause resistance[12]. This can be seen by imagining our 1-D
sample bent in to the form of a ring. Then, if at one time the

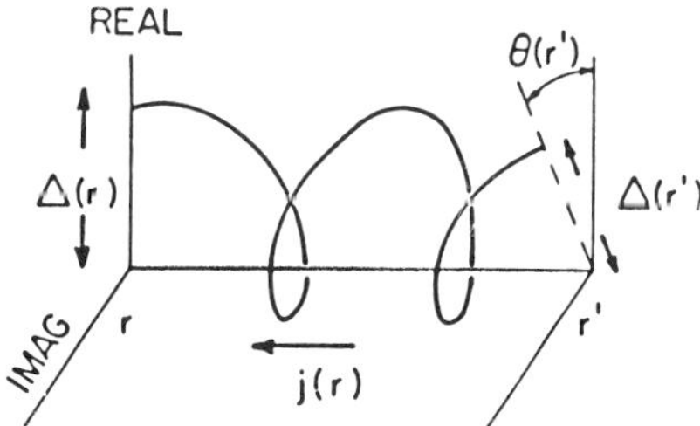

Figure 3. Illustration of Fluctuations of the Order
Parameter which destroy the Off-Diagonal-Long-Range-Order in
1-D and 2-D systems.

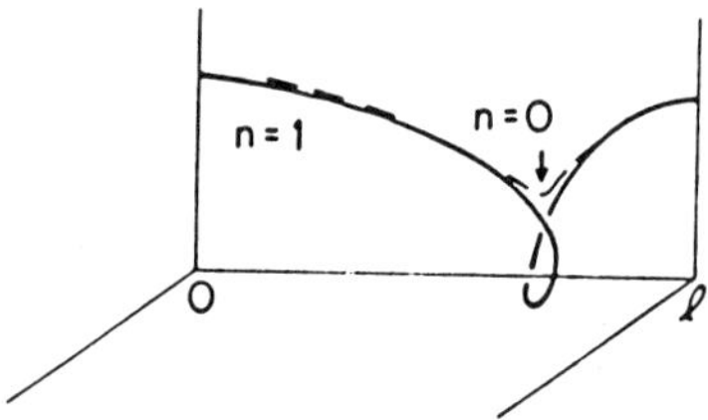

Figure 4. Type of Fluctuation which gives rise to Resistance.

ring has a current circulating round it this will be described
by an order parameter which, as in Figure 3, spirals round the
axis in the complex plane and joins back on itself after going
round the ring. As before fluctuations cause the phase of the
order parameter at two different points on the ring to
fluctuate relative to one another but these do not change the
total winding number of the spiral, which is proportional to
the magnitude of the circulating current. This number is a
topological invariant and hence the current cannot decay by
this means. The only way it can decay is by pulling a loop of
the spiral through the axis as illustrated in Figure 4. This
requires the order parameter to be driven to zero in a small
region of space and to do so requires a finite amount of free
energy. Consequently, the resistance becomes exponentially
small for values of kT less than this. These arguments
provided for the first time the correct explanation of the
decay of persistent currents in small samples[12]. This was
developed more fully by Langer and Ambegaokar[13].

This together with the realization that in the real world
such 1-D or 2-D systems would have weak interactions with
other chains or planes adjacent to them which would partially
suppress these fluctuations, led to the recognition that these
fluctuations were something of a red herring. This was
confirmed by the discovery by Gamble et al.[14] of 2-D
superconductivity in TaS_2 and intercalated derivatives of this
is 1970, and the even more remarkable discovery of
superconductivity in the quasi-1-D polymeric system $(SN)_x$
containing no metal atoms, by R. L. Greene et al.[15] in our
laboratories in 1975.

The exciton interaction raised a number of other problems
in addition to those associated with the role of
dimensionality. These must also be considered.

<u>Vertex Corrections</u>

Of particular importance is the question whether the
interaction with any electronic degrees of freedom can be
described adequately in lowest order of perturbation theory.
This is possible for the phonon interaction because of the,
so called, Migdal approximation[16], which shows that the
vertex corrections such as those of Figure 5 are proportional

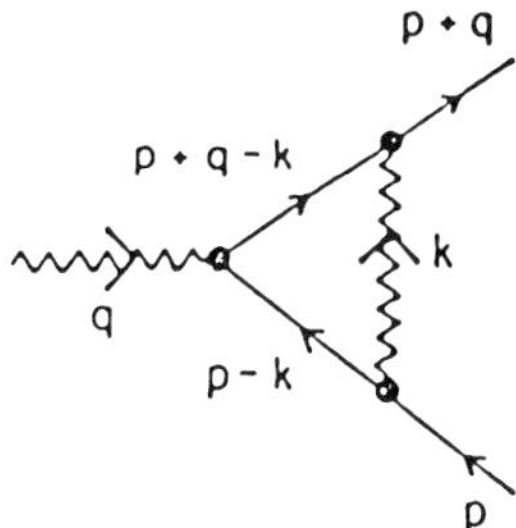

Fig.5. Lowest order correction to the
electron-phonon vertex.

to $(m_e/M)^{1/2}$, where M is the mass of the ions. The question
is, if one replaces the phonons by an electronic excitation,
are these corrections then of order unity i.e. $(m_e/m_c)^{1/2}$? We
have examined this point[3,17] in detail and have shown that
if the exciton system is physically separated from the
conduction electron system by a distance of the order of b,
then the effective coupling constant will be strongly peaked
at q=0 and will drop off for momenta in excess of q=1/b. This
momentum dependence strongly reduces the magnitude of any
vertex corrections. If, on the other hand, the exciton and
conduction systems co-exist in the same region of space then
this is not necessarily so and one should view with caution
results obtained using the simple electron-electron vertex.

<u>Exchange</u>

The exciton interaction differs from the phonon interaction in
another, fundamental way.[18] The electrons and the ions which
participate in the phonon modes are distinguishable. This is
not so for the case of the electron-exciton interaction. The
conduction electrons and the exciton electrons are identical
and consequently in computing their interaction, exchange must
be considered. Such matrix elements are given by an expression
of the form $2J_{ik} - K_{ik}$, where J_{ik} is the Coulomb interaction
between the conduction and exciton electrons and K_{ik} is the
exchange term. In general, if both i and k are drawn from the
same set of orbitals one finds that K_{ik} and J_{ik} are of the
same order of magnitude. So the neglect of K_{ik} would over
estimate the total element by as much as a factor of 2. This
element appears quadratically in the expression for the
effective interaction and consequently a gross error is
committed if exchange is not treated with reasonable care. If,
once more, the conduction and exciton systems are physically
separated in space, as we have considered in several
models[6,17], K_{ik} vanishes and one obtains the simple Coulomb
term. On the other hand in calculations of the
plasmon-mediated interaction[19,20] where the plasmon and the
electrons coexist in the same region of space the exchange
term will be of major importance.

<u>Stability</u>

A number of authors have argued that excitonic superconductivity would be impossible because, for an interaction sufficiently strong to lead to superconductivity, the electron-lattice system would become unstable. This argument was based on the examination of the electron-electron interaction, which can be described in terms of the dielectric function, $\varepsilon(q,\omega)$:

$$V(q,\omega) = \frac{4\pi e^2}{q^2\,\varepsilon(q,\omega)}$$

Here $\varepsilon(q,0)$ is the dielectric constant resulting from the phonon and all electronic excitations. This expression can be decomposed into its phonon and electron components and the interaction averaged over the Fermi surface to give $N(0)V = \mu - \lambda$. It was argued[21,22] that $\varepsilon(q,0)$ would have to be positive in order to prevent an instability in the conduction electron gas and this would require $\mu - \lambda$ to be positive and thus $\lambda < \mu$, prohibiting superconductivity. This conclusion is incorrect. First, the above would apply only if the electron gas was uniformly distributed in the unit cell. The excitonic systems considered[6,17] were specifically designed to have a very limited region within which the electrons and excitons could interact. It is said that this requires the inclusion of Umklapp processes. This is true - but one must not ignore the physics of the real systems!

Second, as we will show shortly, it is not $V(q,\omega)$ which enters the integral equation for the BCS gap function but a better behaved, relative of V, $U(q,\omega)$. The constraints on U are different than those on V.

Third, the energy of the exciton interaction generally will be different from that of the Fermi energy. This allows one to transform away some of the Coulomb term μ to yield a μ^* given by $\mu^* = \mu(1 + \mu\ln(E_f/\omega))^{-1}$, where, as for the phonon case ω represents the excitation energy and E_F the Fermi energy. For the phonon case the logarithmic factor is typically of order 5 but for excitons it can be of order 1 or 2, in which case a significant reduction in μ will still occur, allowing negative values for $\lambda - \mu^*$.

<u>Effects of Phonons</u>

<u>Virtual Phonons</u> In any real physical system the exciton system would have to coexist with the phonon degrees of freedom. It had been argued that the energy region where the excitons would act, the virtual phonons would generate a strongly repulsive term which would nullify any contribution from the excitons. This was based on the use of the Bohm-Pines expression for the electron-electron interaction,

$$V(q,\omega) = \frac{g^2 E(q)}{\omega^2 - E^2(q)}$$

This becomes positive for values of ω larger than the phonon energies $E(q)$. That such repulsion does not occur was shown by Ginzburg[23] and Kirzhnits, Maksimov and Khomskii[24]. They showed that if one transforms the Eliashberg equations to the form of the BCS gap equation, it is not $V(q,\omega)$ which appears in the integral equation but a smooth function $U(\xi,\xi')$ involving the energies ξ and ξ' of the electrons where,

$$U(q,\ \omega) = \frac{-g^2}{E(q) + |\xi| + |\xi'|}$$

This gives an attractive interaction for all excitations, regardless of their energy.

<u>Real Phonons</u> It had also been argued[25] that just as for the Mossbauer effect, the coherent superconducting state would be destroyed by real phonons and this would prevent the occurrence of high T_C superconductivity for temperatures approaching the Debye temperature. Early approximate treatments supported this conjecture[26,27]. However, careful use of the full Eliashberg equations show that the phonon contributions can only help to raise T_C. The underlying formal reason can be seen from the Eliashberg expression for the gap function $\Delta(\omega)$.

$$\Delta(\omega) = \frac{1}{Z(\omega)} \int \frac{d\omega'}{\omega'} \operatorname{Re}[\Delta(\omega')] \int_0^{\omega_0} d\omega_q\, \alpha^2(\omega_q) F(\omega_q)$$

$$\times \{[N(\omega_q) + f(-\omega')][(\omega' + \omega_q + \omega)^{-1} + (\omega' + \omega_q - \omega)^{-1}]$$

$$- [N(\omega_q) + f(\omega')][(-\omega' + \omega_q + \omega)^{-1} + (-\omega' + \omega_q - \omega)^{-1}]\}$$

The major contribution to this comes from those regions of small ω'. If one neglects ω' relative to ω_q in the two last brackets one sees that the phonon occupation numbers cancel. The only contribution from the phonon occupation which remain come from the terms neglected in this approximation and the terms in $Z(\omega)$. These are of the same order of magnitude and are small.

The subject was finally laid to rest by a rigorous proof by Bergmann and Rainer[28] in 1973, that all phonon (or exciton) modes contribute to an increase in T_C regardless of their energy. To this in 1984 we added a minor twist by closing a subtle loophole in the argument[28].

<u>The Strength of the Interaction</u>

The above arguments show that there is no fundamental principle which can prohibit the attainment of high temperature superconductivity using an exciton interaction. The problem is purely one of finding or devising a system with a sufficiently strong interaction. In 1973[29] we derived a set of criteria for obtaining a strong exciton interaction in 2-D systems. These are the following:

(a) The Fermi momentum must be "small" in the sense of (b).
(b) The separation between the electron and exciton system should be as small as possible and in any case should be less than the Fermi wavelength.
(c) The transition density of the exciton system must be large close to the conduction system.
(d) The exciton system must be densely packed. Exciton "impurities" in general, will give only a very small increase in T_C.
(e) The exciton system must have a low energy of excitation to give an adequate coupling term.

Following a suggestion of H. Keller, we[17,3,4] found that a very strong exciton interaction could be obtained by using transition metal complexes in which the exciton and electron systems share the same atomic site. We considered a 1-D system in which a set of linked d_{z^2} orbitals form a 1-D conduction band and the $d_{xz,yz}$ orbitals of the same atom overlap the π-electron orbitals of an organic ligand to form the exciton system. In this case the conduction and exciton electrons come within an Angstrom of one another so their interaction can be large. In this case, however, one must consider carefully the problems of exchange and vertex corrections before drawing any conclusions on the efficacy of the interaction leading to superconductivity. Later [5] we showed that the coupling constant was strongly dependent on the relative energies of the metal and ligand states prior to hybridization. These must be degenerate to give the strongest interaction.

We will show that almost all the above conditions can be satisfied in the recently discovered high transition-
-temperature cuprates. We will argue that the exciton interaction could be responsible for the 90K transition temperature of the $YBa_2Cu_3O_{6.7}$ compound.

EXCITON SUPERCONDUCTIVITY IN THE CUPRATES

The discovery of superconductivity at temperatures above 30 K in the Ba-La-Cu-O system[1] and the subsequent observation of T_C above 90 K in $YBa_2Cu_3O_{6.7}$[2] led us to ask whether the exciton interaction rather than the phonon mechanism might be responsible. The question one might ask then is, which electronic excitations might be involved? From the above arguments I would argue against free-electron-like plasmons being responsible because these are built ,in general, of the same set of orbitals as the conduction electrons and coexist in the same region of space. If this is the case then the arguments about exchange and vertex corrections enumerated above come into play and weaken the interaction. Also, while

2-D plasmons have interesting low energy excitations, there are many other 2-D systems like graphite, TaS_2, etc., which do not exhibit high T_C superconductivity. We believe the inter-band or intra-band excitations are responsible. The states involved in these transitions are characteristic of the particular atoms involved and the precise stoichometry of the compound. This is consistent with the observed sensitivity of T_C to these properties.

$La_{2-x}(Ba, Sr)_xCuO_4$

The superconducting phase in the $La_{2-x}(Ba, Sr)_xCuO_4$ systems has been identified as a tetragonal K_2NiF_4 structure· The copper atoms are surrounded by a distorted octahedron of oxygen atoms with the four in-plane (x,y) CuO bonds shorter than the axial (z) ones. The oxygen octahedra share corners forming planes of CuO_6 separated by (La,Sr,Ba)-O layers. Figure 6A. In our model we assume that the conduction electrons lie in the sheet formed from the overlap of the $d_{x^2-y^2}$ orbitals of the Cu with the p_x and p_y orbitals of the four adjacent oxygen atoms in the x-y plane. Figure 6B. Substitution of small amounts of Sr (or Ba) for La in these systems results in a partially filled, mixed-valence band.

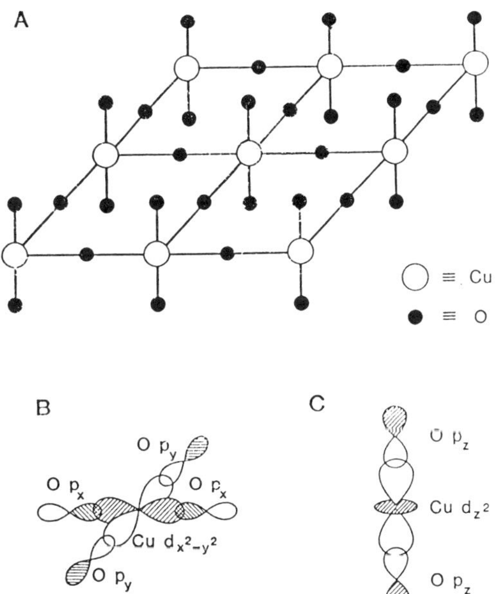

Figure 6. La_2CuO_4 showing square planar array(A) of conduction band orbitals(B) and proposed exciton orbitals(C).

We define a possible exciton system in these compounds as
that formed from the Cu (d_z2) orbitals which overlap with the
two oxygen p_z orbitals of the atoms above and below the x-y
plane. Figure 6C. These form bands of bonding, anti-bonding
and non-bonding orbitals. The bonding and anti-bonding
bands form a two level band of states which constitute the
polarizable exciton system. Interaction with the conduction
electrons will induce virtual excitations to the anti-bonding
level (provided it is empty) causing a breathing-mode type of
oscillation of *electronic charge* from the two oxygen atoms
to the Cu d_z2 orbital and back. This could provide a strong
exciton attraction between electrons in the conduction band.
The interaction is strong because the transition charge moves
from the remote oxygen atoms, where the interaction with
conduction electrons is weak, to the d_z2 orbital where the
interaction with the $d_x2 \, _y2$ electrons is strong, and back to
the remote oxygens again. With reasonable estimates for the
density of states, electron screening and Coulomb integrals,
values of Tc well above 90K would be obtained. However,
Matheiss[30] has recently completed a detailed band
calculation for La_2CuO_4 which shows that both bonding and
antibonding bands are full. If this remains valid for the
$La_{2-x}(Ba, Sr)_xCuO_4$ compounds then these virtual transitions
would be forbidden and no exciton interaction of this form
could contribute. One would conclude that the phonon
interaction must be responsible for the superconductivity as
described by Weber[31]. However, another possibility is the
following. Certain of the conduction band states which are
nearly degenerate with those orbitals responsible for the
triplet of states forming the antibonding band will hybridize
with them. Transitions from these to conduction band states
similarly hybridized, above but near the Fermi surface, would
provide a weaker exciton contribution. This could contribute
something towards the superconductivity of these compounds.

$YBa_2Cu_3O_7$

In the $YBa_2Cu_3O_7$ the situation is different. The structure
as determined by Siegrist et al [32] is shown in figure 7. It
consists of two slightly puckered, square planar Cu - O sheets
similar to the one in La_2CuO_4, but separated by
one-dimensional chains of Cu - O with tightly bonded O's on
each Cu pointing at the Cu atoms in the planes above and below
the chain. We assume as before, that the conduction electrons
lie in the planes. We treat the one dimesional chains as the
exciton system. These chains may distort to spin- or
charge-density states but this does not affect our
conclusions. Matheiss and Hamann[33] have calculated the band
structure for this compound. They find that the Fermi energy

cuts across the bands for the electrons in the planes and for
those in the chains. We focus on the one dimensional band. We
have calculated the hybridization of the states of this band
and find a weak admixture of in-plane Cu d_z2, but a strong
admixture of the oxygen p_z orbitals in these states both above
and below E_F. One can construct exciton states ($\mathbf{q}$) from
products of states below E_F, ($\mathbf{k}$) and above E_F ($\mathbf{k} + \mathbf{q}$). These
describe oscillations of electronic charge on and off the
oxygens adjacent to the planes. Through the screeened Coulomb
interaction this couples to the electrons in the planes to
give an electron-exciton coupling constant. Rough estimates
indicate that the exciton contribution to T_C is significant,
however, our final conclusion awaits a better determination of
the out-of-plane screening and a better estimate of the
hybridization parameters from band calculations. Due to oxygen
deficiencies we expect these chains will have an average
length of only three or four unit cells and this will affect
the hybridization parameters and the energies of the states
involved in these excitations.

Recently[34], superconductivity at 90 K has been observed in
the "336" compound, $La_{3-x}Ba_{3+x}Cu_6O_{14+y}$ in which the
one-dimensional chains of the $YBa_2Cu_3O_7$ compound are missing.
However, in this case there are similar, short one-dimensional
chains of Cu d_z2 and O p_z orbitals perpendicular to the planes
which could play the same role in providing an exciton
interaction.

ACKNOWLEDGEMENTS

Much of the work on the exciton mechanism was done in
collaboration with Professor Hanoch Gutfreund. The most recent

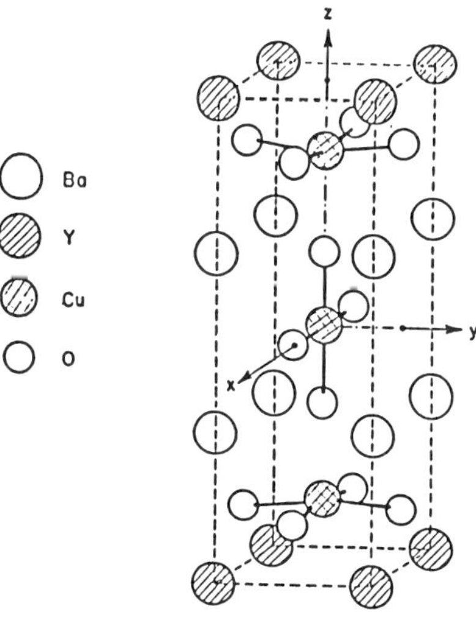

Figure 7. Structure of $YBa_2Cu_3O_7$

work on the cuprates was done in collaboration with Professor J. Collman and John McDevitt of the Chemistry Department, Stanford University. I wish to thank them for their input, suggestions and criticism. I would also like to acknowledge valuable discussions with Vladimir Kresin, Aaron Kapitulnik, Walter Harrison, L. F. Mattheiss, Gordon T. Yee and Matthew Zisk. In addition I would like to thank Len Mattheiss for a preview of his band calculations. This work was supported by the Department of Energy (DEFG03-86-45245).

REFERENCES

1. J. G. Bednorz and K. A. Müller, Z. Phys. **B64,** 189 (1986)

2. C. W. Chu, P. H. Hor, R. L. Meng, L. Gao, Z. J. Huang, and Y. Q. Wang, Phys. Rev. Lett. 58, 405 (1987)

3. H. Gutfreund and W. A. Little, "Prospects of Excitonic Superconductivity" in <u>Highly Conducting One Dimensional Solids,</u> J. T.Devreese, R. P. Evrard, and V. E. Doren, Eds (Plenum , New York 1979) p305.

4. W. A. Little, Int. Jour. Quantum Chem. **15,** 545 (1981)

5. W. A. Little, Jour. de Physique, Colloque C3, **44,** 819 (1983)

6. W. A. Little, Phys. Rev. **134,** A1416 (1964)

8. T. M. Rice, Phys. Rev. **140A,** 1189 (1965)

9 V. L. Ginzburg, Zh. Eksp. Teor. Fiz. **47,** 2318 (1964); [Sov.Phys.-JETP **20,** 1549 (1965)]

10. V. L. Ginzburg, Contemp. Phys. **9,** 355 (1968)

11. P.C. Hohenberg, Phys. Rev.**158,** 383 (1967)

12. W. A. Little, Phys. Rev. **156,** 396 (1967)

13. J. S. Langer and V. Ambegaokar, Phys. Rev. **164,** 498 (1967)

14. F. Gamble, F. J. DiSalvo, R. A. Klemm and T. H. Geballe, Science **168,** 568 (1970)

15. R. L. Greene, G. B. Street and L. J. Suter, Phys. Rev. Lett. **34,** 577 (1975)

16. A. B. Migdal,Sov.Phys.-JETP,**7,**996 (1958).

17. D. Davis, H. Gutfreund, and W.A. Little, Phys. Rev. **13,** 4766 (1976)

18. W. A.Little, J. Polymer Sci. Pt.C**29,**17 (1970).

19. V. Kresin (preprint)

20. J. Ruvalds (preprint)

21. M. L. Cohen and P. W. Anderson, in Superconductivity in d-, and f-Band Metals, D. H. Douglass (Ed.), AIP, New York (1972)

22. J. C. Phillips, Phys. Rev. Lett. **29**, 1551 (1972)

23. V. L. Ginzburg, JETP Lett. **14**, 396 (1971)

24. D. A. Kirzhnits, E. G. Maksimov, and D. I. Khomskii, J. Low Temp. Phys. **10**, 79 (1973)

25. P. W. Anderson, Editorial Comment, Physics **2**, 151 (1966).

26. J. Appel, Phys. Rev. Lett. **21**,1164 (1968)

27. P. B. Allen, Solid State Comm. **12**,379 (1973)

28. G. Bergmann and D. Rainer, Z. Phys. **263**,59 (1973); V. Z. Kresin, H. Gutfreund and W. A. Little, Solid State Comm. **51**, 339 (1984)

29. W. A. Little, J. Low Temp. Phys.**13**, 365 (1973).

30. L. F. Mattheiss, Phys. Rev. Lett. **58**, 1028 (1987)

31. Werner Weber, Phys. Rev. Lett. **58**, 1371 (1987)

32. T. Siegrist,S. Sunshine, D.W. Murphy, R.J. Cava, and S.M. Zahurak,Phys. Rev. Lett.(to be published)

33. L. F. Mathiess and D.R. Hamann, Solid State Commum., in press.

34. D.B.Mitzi, A. F. Marshall, J. Z. Sun, D. J. Webb, M. R. Beasley, T. H. Geballe, and A. Kapitulnik.(preprint)

CHARGE TRANSFER RESONANCES AND SUPERCONDUCTIVE PAIRING

IN THE NEW OXIDE METALS

C. M. Varma and S. Schmitt-Rink

AT&T Bell Laboratories
Murray Hill, New Jersey 07974

Elihu Abrahams

Serin Physics Laboratories
Rutgers University
Piscataway, New Jersey 08855

INTRODUCTION

All theoretical discussions of the superconductivity in the new oxide metals[1-6] have been within the BCS framework of two-particle pairing. The issue in question is what is the pairing mechanism? Do the Cooper pairs have the usual BCS s-wave symmetry? Is the pairing so strong that the Blatt-Schafroth type ideas of Bose condensation of tightly bound pairs are realized? The points of view proposed may be classified as follows:

I. Pairing through electron-phonon interactions

    (i)   conventional[7]

    (ii)   bipolaron formation and Bose condensation[8]

II. Pairing through magnetic correlation of the electrons

    (i)   exchange of antiferromagnetic (AFM) spin fluctuations[9,10]

    (ii)   resonating valence bond (RVB) or spin-bipolaron formation and Bose condensation[11,12]

III. Pairing through exchange of electronic polarization resonances[13]

ELECTRON-PHONON MODELS

Estimates were given earlier of the upper limit on superconducting transition temperatures expected of the new compounds based on their electronic structure.[13] An optimistic upper limit is 30 K. The recent experiments yielding only a very small isotope shift imply that the contribution of electron-phonon interactions to T_c is in fact much smaller.[14,15] Although there are calculations with electron-phonon interactions which give T_c in $La_{2-x}Sr_xCuO_4$ in agreement with experiments, they also predict structural transformations which are not observed.[7] This means there is something crucial missing in the electron-phonon plus band structure models for these materials.

In the bipolaron models, pairs of electrons interact so strongly via their induced lattice deformation that they form real space pairs.[16] This may be modelled by a local effective electron-electron attraction U, $|U| \gg t$, where t is the electron hopping amplitude. The kinetic energy of a pair (which in the strong binding limit is a boson) is $t^2/|U|$ corresponding to an effective mass $|U|/t^2$. Bose condensation occurs at a temperature characteristic of such an effective mass, i.e., $T_c \sim t^2/|U|$. In Fig. 1 we have sketched T_c/t as a function of $t/|U|$.[17] Actually, the renormalization of the hopping amplitude puts an even stronger limit on T_c due to bipolaron condensation. The weak coupling portion of the curve ($t \gg |U|$) corresponds to the BCS result $T_c \sim t \exp(-t/|U|)$. In the bipolaronic region, the Bose line for T_c/t lies below the BCS extrapolation which, as remarked above, has a T_c of at most 30K for optimistic parameters of the materials in question.

MAGNETIC MODELS FOR SUPERCONDUCTIVITY

The starting inspiration of such models is that the stoichiometric compound La_2CuO_4, which in band theory should be metallic with a half-filled band, is in fact an insulator and, below about 220 K, an antiferromagnet.[18,19] It is then quite natural to regard this compound as an antiferromagnetic Mott-Hubbard insulator and seek magnetic mechanisms for the superconductivity away from half-filling.

It is absolutely vital that any model for superconductivity accounts as well for the properties at half-filling. We shall see that this does not mean one is irrevocably lead to the magnetic models of superconductivity. The magnetic models all assume that these materials are describable by the Hubbard model. We shall soon discuss that this is not reasonable.

A theory for superconductivity in heavy fermions through exchange of AFM spin fluctuations is quite successful in explaining their properties.[9,20] It predicts an

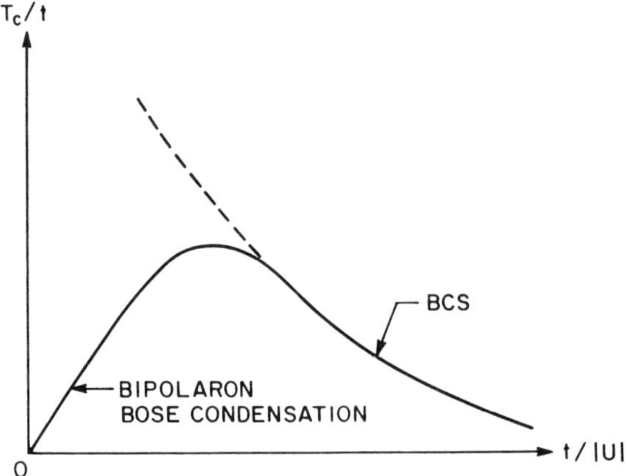

Fig. 1.   Critical temperature T_c as a function of electron hopping amplitude t and on-site attraction U for an attractive Hubbard model.

anisotropic gap and a density of states of quasiparticles which unlike BCS is linear in temperature. The quasiparticle properties and the gap in the new superconductors appear to us to be quite conventional. This would appear to rule out such a magnetic mechanism. Experimentally, the evidence for such a BCS-like gap is emerging. The temperature dependence of the London penetration depth is also consistent with BCS-type pairing.[21] The unusual linear specific heat in the superconducting state as well as the linear resistivity in the normal state can be understood as arising from tunneling states for which there is now independent evidence from very low temperature acoustic measurements.[22,23]

MODEL FOR THE OXIDES

Let us first consider the simple transition metal oxides. The Mott insulating phase, the Hubbard model and the super-exchange process leading to antiferromagnetism were developed to understand their properties. A tight-binding model for these is:

$$H = \sum_i (\epsilon_A n_{Ai} + \epsilon_B n_{Bi} + U_A n_{Ai\uparrow} n_{Ai\downarrow} + U_B n_{Bi\uparrow} n_{Bi\downarrow})$$

$$+ \sum_{<i,j>} [V n_{Ai} n_{Bj} + t \sum_\sigma (C^\dagger_{Ai\sigma} C_{Bj\sigma} + h.c.)] \tag{1}$$

where A stands for the metal ion say and B for the oxygen. V is derived from the Coulomb interaction; it leads to the stability of ionic configurations.

In many cases a truncated version of (1), the Hubbard model, is adequate. The considerations leading to it are as follows: consider the mean field solution of (1) or equivalently some self-consistent one-electron theory, such as Hartree-Fock or density functional theory, etc. This will include the ionic and the covalent interactions among the ions and give rise to a self-consistent band structure which usually has the form of a set of filled bands which are mostly oxygen p with a small amount of covalent d admixture and unfilled bands which are mostly transition metal d with a small amount of oxygen p admixture (see Fig. 2). The formal O^{--} state has of course been stabilized by the electrostatic interaction between the atoms.

When we now begin to consider corrections beyond one-electron (mean field) theory, it is enough to consider an effective Hamiltonian describing the upper band only

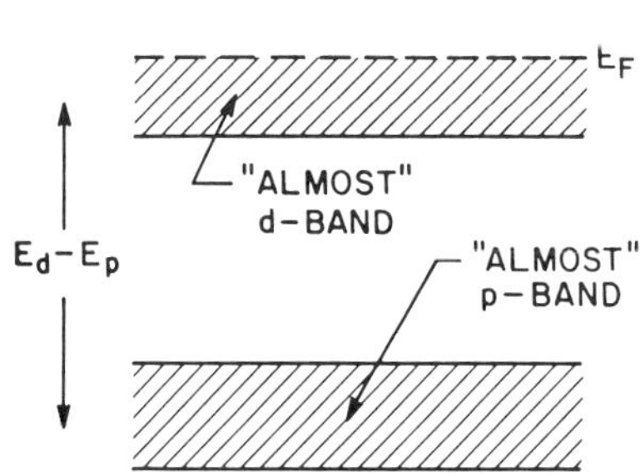

Fig. 2.   Electronic structure of transition metal oxide compounds.

provided no term in the Hamiltonian causes significant resonant transitions between the lower and the upper bands. Thus a kinetic energy term in the space of the "d" orbitals and a local repulsion $\overline{U}$ in this subspace suffices:

$$H = \sum_i \left[\overline{t} \sum_\sigma (C_{i\sigma}^+ C_{i+1\sigma} + h.c.) + \overline{U} n_{i\uparrow} n_{i\downarrow} \right] \tag{2}$$

where $\overline{t} \approx \tilde{t}^2/(\tilde{\epsilon}_A - \tilde{\epsilon}_B)$. If $\tilde{t}/(\tilde{\epsilon}_A - \tilde{\epsilon}_B)$, $U/(\tilde{\epsilon}_A - \tilde{\epsilon}_B)$ and $V/(\tilde{\epsilon}_A - \tilde{\epsilon}_B)$ are much smaller than 1, this picture is adequate. If they are $\gtrsim 1$, it is not. Here, $\tilde{\epsilon}_A$, $\tilde{\epsilon}_B$ and t are the mean field renormalized levels and Cu-O hopping amplitude, respectively.

We shall see that for the oxide superconductors it is not adequate. This can be shown starting either from a band point of view or a localized ionic point of view. Self-consistent band structure calculations have been done for both $La_{2-x}Sr_xCuO_4$ and $YBa_2Cu_3O_7$. A tight-binding fit to the band structure in the plane (as well as along the chains in $YBa_2Cu_3O_7$) in terms of effective atomic levels E_d, E_p and a transfer integral t yields $E_d \approx E_p$![24,25] The band structure has the schematic form shown in Fig 3. The covalency of the d-p orbitals at the self-consistent ionic state is 100%, there is no gap at all and a dispersionless non-bonding $p_x - p_y$ band lies right at the foot of the conduction band. Any fluctuations involving U or the Cu-O Coulomb interaction V dynamically mix the p and d states. Restricting attention to a Cu-like conduction band alone to deduce a Hubbard model is quite useless.

This point of view is reinforced by looking at the problem from a localized point of view following Zaanen, Sawatzky and Allen.[26] Consider the two parameters U_A and E_x, where E_x is the charge transfer excitation energy corresponding to $A^n B^m \rightarrow A^{n+1} B^{m-1}$. The relevant E_x for NiO for example is the energy difference $E(N_i^+ O^-) - E(N_i^{++} O^{--})$. In the band picture, Fig. 3, E_x corresponds approximately to the average energy needed

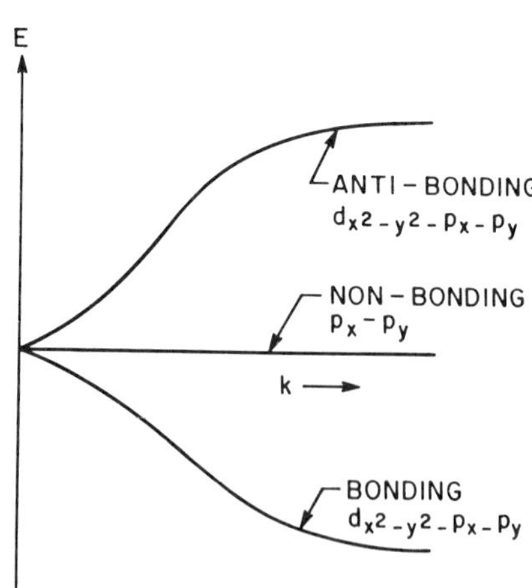

Fig. 3.   Self-consistent one-electron energies for planar CuO_2 networks.

to create a hole in the lower bands and an electron in the conduction band taking into account the effective attraction between the electron and the hole, i.e., the excitonic effect.

For $E_x \gg U_A$, see Fig. 4, the Hubbard model is adequate. Within this sector, for $U_A \gg W$, the Mott-insulator is the ground state (W is the conduction band width); the lowest particle-hole excitation corresponds to $2d^n \rightarrow d^{n+1} + d^{n-1}$. But for $E_x \ll U_A$, completely different behavior occurs: for $E_x \gg W$ in this regime, the material is again an insulator, but with the lowest excitation of the charge transfer variety $M^{++} O^{--} \rightarrow M^+ O^-$. For $E_x \ll W$ the material is a metal even for very large U_A since fluctuations $M^{++}O^{--} \rightarrow M^+O^-$ are mixed quantum-mechanically. For $E_x \gg W$, if M^{++} has local moments, the material will doubtless order magnetically at a temperature of order $(t^2/E_x)^2/U_A$. (In a simple cubic lattice, because of nesting, this will always be the case, no matter how large W.)

As we move to the right of the periodic table, the ionization potential of the transition metals drops making E_x smaller. Even for NiO, the classic Mott-insulator, Sawatzky and Allen have presented strong evidence that a description purely in terms of a Hubbard model is inadequate.[27] The charge transfer excitations are then even more important in CuO. Some experimental evidence for this is available. This means that the conductivity gap is dominated by the process $Cu^{++}O^{--} \rightarrow Cu^+O^-$, whereas in a conventional Mott insulator O^{--} would change only virtually and the conductivity process would be $2Cu^{++} \rightarrow Cu^{+++} + Cu^+$. Imagine now that we have a material in which, as in the new oxide superconductors, some of the Cu is in the Cu^{3+} state. The excitonic energy for the Cu^{3+} ion is considerably lower than for Cu^{2+} since the electron

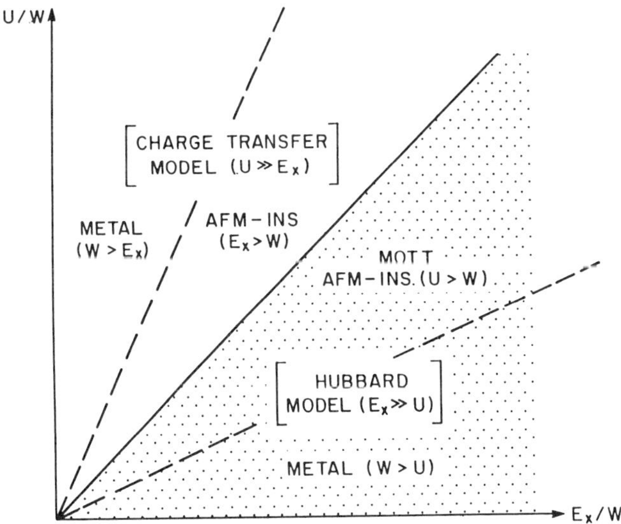

Fig. 4.   Ground state phase diagram of transition metal oxide compounds.

359

affinity energy of Cu^{3+} is much larger than that of Cu^{2+}. There is then the possibility of the material being metallic.

In fact, there is also the real possibility of doping producing O_2^{2-} rather than Cu^{3+}; the peroxide ion is known to occur for *excess* oxygen in La_2NiO_4. But La_2CuO_4 forms with oxygen deficiency — in this situation the formation of the peroxide ion on doping is unlikely. The Pauling-Zachariasen rules, when applied to the measured lattice constants, give strong evidence for Cu^{3+}, as does chemical analysis. The charge transfer energy $Cu^{3+}O^{--} \rightarrow Cu^{2+}O^{-}$ is however bound to be much smaller than U making the Hubbard model quite a wrong starting point for the discussion of these materials. If the energy E_x is very small compared to U, U may be dropped altogether out of the problem (away from half-filling) simultaneously discarding any matrix elements for $2d^n \rightarrow d^{n+1} + d^{n-1}$ in the kinetic energy. Metallic conduction now occurs through the process

$$
\begin{array}{cccc}
& +++ & -- & ++ \\
\rightarrow & ++ & - & ++ \\
\rightarrow & ++ & -- & +++
\end{array}
$$

etc., in which only intermediate states costing energy E_x are used. E_x here is characteristic of $(+++, --) \rightarrow (++, -)$ and not $(++ --) \rightarrow (+ -)$ which would be much higher in energy.

It is worth noting that neither the band description emphasizing strong covalency nor the ionic description are likely to be completely correct. They are useful only as starting points for discussion. In particular, different experiments — EXAFS, photoemission, lattice constants etc. — are likely to give different values for "valence" or the ionic state of Cu and oxygen if interpreted too literally. We talk here of "Cu^{+++}" merely to illustrate a point; our arguments are essentially unchanged if we talk of "O^{-}" instead or a linear combination of various valences.

CHARGE TRANSFER RESONANCES IN THE METALLIC STATE

Given the presence of nominally 3+ Cu ions, not only can metallic behavior occur but also a resonance will appear in the excitation spectrum due to the excitonic transition $Cu^{3+}O^{--} \rightarrow Cu^{2+}O^{-}$ at relatively low energy (several tenths of eV) on the electronic scale but large on the scale of lattice vibrations. Under certain conditions this resonance can be fairly well-defined. Then, since the metallic behavior also

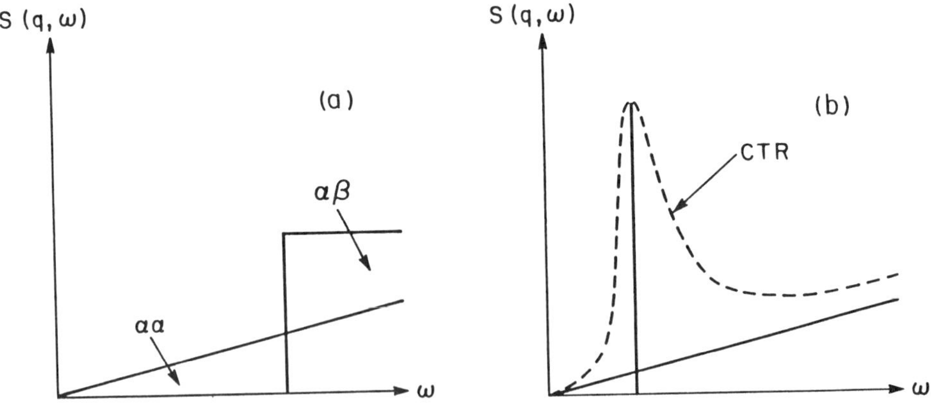

Fig. 5. Particle-hole excitation spectrum for large momentum transfer q as a function of frequency ω (a) without, (b) including Coulomb interactions.

involves the same particle-hole transition, the quasiparticles near the Fermi surface will acquire a large renormalization. We believe this effect is the source of the high-temperature superconductivity. We shall see below that the electronic structure is propitious for a reasonable resonance.

For high-temperature superconductivity not only must this resonance be at a high energy (on the phonon scale), it must also have a large oscillator strength and be fairly localized in real space. The last condition is necessary for the quasiparticle renormalization due to the interaction with the resonance to be large due to local-field effects. Based on the considerations above, we had predicted that a charge transfer excitonic resonance (CTR) at an energy scale of about 1/2 eV must be present in the superconducting materials and its oscillator strength must be related to T_c.[13]

A resonance at about 1/2 eV with a large oscillator strength (about .75 electron per Cu^{3+} ion) has indeed been observed both in $La_{2-x}Sr_xCuO_4$ and in $YBa_2Cu_3O_7$.[28-31] In the insulating La_2CuO_4, the absorption edge is at about 2 eV, corresponding to $Cu^{++}O^{--} \rightarrow Cu^+O^-$ transitions, in accord with our considerations above.

Let us consider the excitation spectrum starting from a situation such as in Fig. 3. Keeping only one non-bonding or bonding band (β) and the anti-bonding (conduction) band (α) and neglecting any interactions, the particle-hole excitation spectrum (at some large momentum transfer) looks then like in Fig. 5a. Let us first ignore the mixing of the $\alpha\alpha$ and the $\alpha\beta$ excitations due to the Coulomb interaction and consider its effect only on the $\alpha\beta$ excitations. For large enough V, a sharp excitonic resonance with nearly zero excitation energy carrying a large fraction of all the oscillator strength of the $\alpha\beta$ transitions appears then as in Toyazawa's calculation in a different context (see Fig. 5b).[32] An important role in such a calculation, if carried out for a realistic model, will doubtless be played by the fact that r_s for the oxide metals is ≈ 3.5 Å, i.e., about twice the Cu-O separation. So V is not screened by metallic processes.

An important (and hard) question is the width of such a resonance due to i) decay into particle-hole transitions in the $\alpha\alpha$ channel and ii) scattering off the latter. Process i) ("Landau damping") is represented diagrammatically in Fig. 6. Note that the vertex includes direct as well as exchange scattering. The other process ii) is due to the inherent three-body nature of the problem and described by such diagrams as shown in Fig. 7. As is well-known from the x-ray problem and discussed elsewhere, it cannot be estimated in conventional perturbation theory, even for finite hole mass.[33]

SUPERCONDUCTIVITY THROUGH CHARGE TRANSFER RESONANCES

Two kinds of charge transfer resonances should be distinguished — longitudinal and transverse. We shall see that it does not appear possible for the former which couple to electrons through their scalar potential to lead to an attractive electron-electron interaction. The latter which couple to electrons through their vector potential in general lead to an attractive electron-electron interaction which is significant only if there are large local-field effects. The transfer resonances must therefore be fairly localized in space (and in energy) and have large oscillator strengths. The electrons must also be of the tight binding variety.

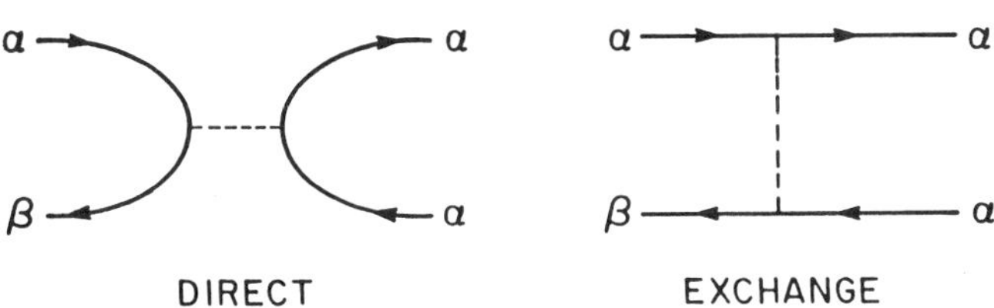

Fig. 6.   Processes leading to Landau damping of the charge transfer resonance.

LONGITUDINAL RESONANCES

In general, the screened Coulomb interaction is given by

$$V_S = V(\epsilon_\infty - V\Pi)^{-1} = V\epsilon_L^{-1} \tag{3}$$

where Π is the irreducible polarizability and ϵ_∞ arises from the neglected (high-frequency) transitions. All objects above have four indices corresponding to the bands of incoming and outgoing particles. For the effective electron-electron interaction, we want V_{S1}, where '1' denotes $\alpha\alpha - \alpha\alpha$ particle-hole pair interactions. Denoting $\alpha\alpha - \alpha\beta$ and $\alpha\beta - \alpha\beta$ particle-hole pair interactions by '2' and '3', respectively, and for simplicity using $V_1 V_3 = V_2^2$, it is easy to show that

$$V_{S1} = \left(\frac{V_1}{\epsilon_1}\right) + \left(\frac{V_2}{\epsilon_1}\right)^2 \frac{(\Pi_{\alpha\beta,\alpha\beta} + \Pi_{\beta\alpha,\beta\alpha})}{1 - \left(\dfrac{V_3}{\epsilon_1}\right)(\Pi_{\alpha\beta,\alpha\beta} + \Pi_{\beta\alpha,\beta\alpha})} \tag{4}$$

where

$$\epsilon_1 = \epsilon_\infty + V_1\Pi_{\alpha\alpha,\alpha\alpha} \tag{5}$$

is the metallic part of the longitudinal dielectric function. The CTR appears as a pole in the second term on the r.h.s. of (4), which may be parametrized as

$$V_{S1} = \left(\frac{V_1}{\epsilon_1}\right)\left(1 + \frac{\tilde{\omega}_o^2}{\omega^2 - \omega_T^2 - \tilde{\omega}_o^2 + i\omega\gamma}\right) \tag{6}$$

Here, $\tilde{\omega}_o^2 = \omega_o^2/\epsilon_1$ gives the screened oscillator strength and $\omega_L^2 = \omega_T^2 + \tilde{\omega}_o^2$ and γ the center frequency and width of the resonance, respectively. ω_T is the frequency of the transverse CTR. (V_1/ϵ_1) may be likened to the usual Coulomb parameter μ, for ω_L, ω_T large enough, there is no point in defining μ^*. An examination of (6) now reveals that it is attractive roughly only between ω_T and ω_L, which is too high a frequency range to be effective for superconductivity. At low frequencies, below ω_T, V_{S1} is always repulsive due to the self-screening of the CTR (see Fig. 8).

TRANSVERSE RESONANCES

Consider now the transverse dielectric function, which including the high-frequency contribution and the effect of the CTR may be written as

$$\epsilon_T = \epsilon_\infty - \frac{\omega_o^2}{\omega^2 - \omega_T^2 + i\omega\gamma} \tag{7}$$

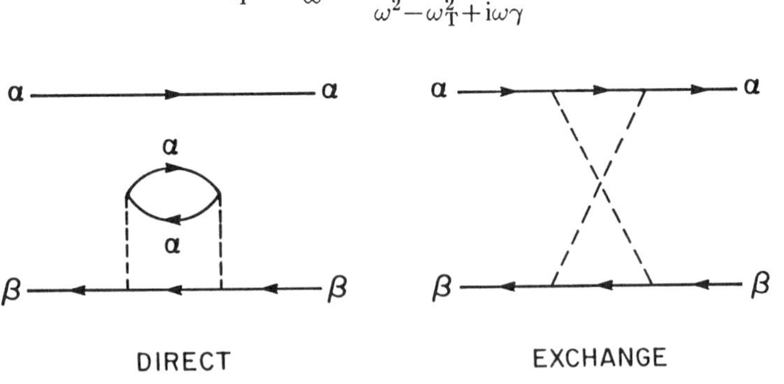

Fig. 7.   Three-body processes leading to broadening of the charge transfer resonance.

Since there is no direct transverse interaction between electrons (except in QED), we need consider only the second term in (7) to obtain the induced interaction. Obviously, it is attractive for all frequencies below ω_T, the frequency of the transverse CTR. In any calculation, both the repulsive longitudinal and this attractive transverse part must be considered.

The conclusion that longitudinal excitonic resonances do not appear to lead to attractive pairing is arrived at by arguments similar to those used by Inkson and Anderson[34] in connection with proposals such as those of Little,[35] Ginzburg[36] and Allender, Bray and Bardeen.[37]

The transverse resonances appear, to our knowledge, not to have been suggested before, as a source of superconductive pairing. They do not have the screening problems which make longitudinal resonances unsuitable. Transverse resonances do not however couple to free electrons. Their coupling to electrons is however as important as that of longitudinal resonances for tightly bound electrons.

The general idea of pairing through electronic resonances has been discussed before.[35-37] Our ideas differ from these in some important respects. Past proposals for excitonic superconductivity have relied on there being two types of electrons, one to give band to band transitions and another to give free electrons. Moreover, only longitudinal processes have been discussed. Using this idea, Little has suggested that in $La_{2-x}Sr_xCuO_4$ superconductivity is not due to electronic resonances and that in $YBa_2Cu_3O_7$ his ideas are realized due to the bonding anti-bonding transition in the Cu-O chains.[38] According to our ideas both in $La_{2-x}Sr_xCuO_4$ and $YBa_2Cu_3O_7$ pairing is caused predominantly by the transverse charge transfer resonances. Chains are not required, nor is the presence of alternate metal and semiconductor layers. We have also presented a model in which the antiferromagnetic insulator ground state of the materials under discussion naturally arises. It is a vital part of this model that slightly away from half-filling charge transfer resonances have large oscillator strengths. This aspect is crucial to the overall understanding of these materials.

RELATIONSHIP TO EXPERIMENTS

The predicted charge transfer resonances have been observed in optical experiments at around 0.5 eV in $La_{2-x}Sr_xCuO_4$ and at around 0.7 eV in $YBa_2Cu_3O_7$.[28-31] They have half-widths of about 0.5 eV as well. They dominate the optical properties — about 2/3 of the spectral weight up to about 2 eV is contained in these resonances. They are in fact responsible for the dull black color of these materials which by the Drude term alone would have been golden.

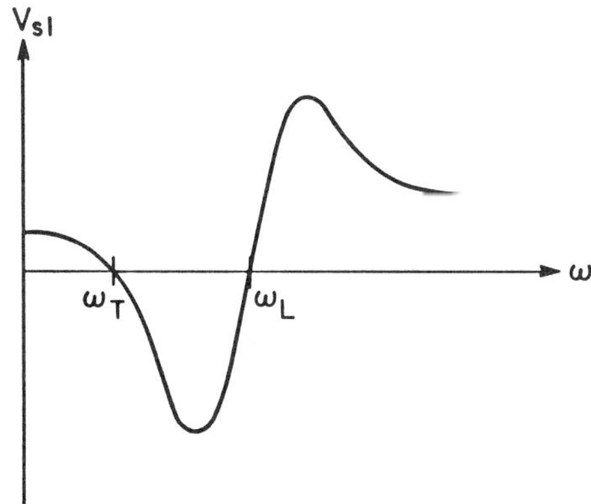

Fig. 8.   Screened Coulomb interaction as a function of frequency ω.

We have also suggested experiments as a function of doping to verify the predicted correlation of the oscillator strength of the resonance with T_c.[13] These have now been done — the results are quite consistent with our ideas. For low ($x < 0.1$) and high ($x > 0.2$) concentration, $La_{2-x}Sr_xCuO_4$ shows little oscillator strength in the CTR. At these concentrations, these particular samples were non-superconducting. At the intermediate concentration $x \approx 0.175$, the highest T_c and the highest oscillator strength in the CTR have been observed.[29] The nonlinear dependence of T_c on this oscillator strength, to which the coupling constant λ is proportional, suggests $\lambda < 1$, i.e., weak coupling. In another set of experiments, a rough measure of the oscillator strength in various samples is found proportional to the fraction of ideal Meissner effect observed in those samples with the onset temperature always near 40K.[30] In $YBa_2Cu_3O_7$, a high $T_c \approx 90K$ is observed with a large oscillator strength for the CTR, while in $YBa_2Cu_3O_{6.2}$, a non-superconducting compound, the CTR is absent.[31] This correlation appears to us to be a strong confirmation of the ideas presented by us.

In $YBa_2Cu_3O_7$ the average valence is 7/3 while in $YBa_2Cu_3O_{6.5}$ the Cu in the planes is in 2+ state (and that along lines is in 1+ state). There should be a general trend of correlation in the oscillator strength of the CTR with the average number of 3+ ions. This is probably loosely followed, but the non-monotonic decrease in T_c with oxygen content in this compound needs further investigation. Similarly, initial addition of Sr in $La_{2-x}Sr_xCuO_4$ must lead to increased average Cu^{+++} configuration. The decrease in the oscillator strength (and T_c) with x beyond 0.2 is doubtless associated with the orthorhombic to tetragonal phase transition. We are led to suspect that at this transition the average number of Cu^{+++} decreases leading to the formation of O^- or more likely the peroxide ion $(O_2)^{--}$.

The quantitative extraction of parameters from the optical data must await experiments with polarized light on single crystals. The highly anisotropic nature of these materials leads to some difficulties in the quantitative analysis of results from experiments in polycrystals.

With such *caveat emptor*, we may proceed to extract λ from the data of Orenstein et al.[28,29] with the additional and rather strong assumption that the resonances observed are almost q-independent. We then find $\lambda \underset{\approx}{<} 1$ for both $La_{1.82}Sr_{0.18}CuO_4$ and $YBa_2Cu_3O_7$ with $\omega_T \approx 0.5$ eV for the former and $\omega_T \approx 0.7$ eV for the latter. This yields a rather large value of T_c compared with experiments. Two things must be borne in mind — such estimates give an upper limit on λ and the usual expression for T_c is not valid for such a large range of the attractive interaction. Migdal's theorem breaks down and the corrections are doubtless in the direction of reducing T_c. This has been discussed by Grabowski and Sham.[39] Another problem well worth looking into is the ratio Δ/T_c for $\omega_T/E_F \approx O(1)$, when λ is in the weak coupling regime. We suspect Δ/T_c will be higher than the BCS value.

With Cu in a mixed valent (2+, 3+) state and a given atom changing from one to the other configuration, it is not possible to sustain magnetic order, if, as we have taken for granted, the 3+ state in these structures is a magnetic singlet. The situation is somewhat similar to the metallic rare-earth mixed valence compounds which also do not magnetically order.

REFERENCES

1. J. G. Bednorz and K. A. Mueller, Z. Physik B *64*, 189 (1986).
2. S. Uchida, H. Tagaki, K. Kitazawa, and S. Tanaka, Jap. J. Appl. Phys. Lett. *26*, L1 (1987).
3. C. W. Chu, P. H. Hor, R. L. Meng, L. Gao, Z. J. Huang, and Y. Q. Wang, Phys. Rev. Lett. *58*, 405 (1987).

4. R. J. Cava, R. B. Van Dover, B. Batlogg, and E. A. Rietmann, Phys. Rev. Lett. *58*, 408 (1987).

5. M. K. Wu, J. R. Ashburn, C. J. Torng, P. H. Hor, R. L. Meng, L. Gao, Z. J. Huang, Y. Q. Wang, and C. W. Chu, Phys. Rev. Lett. *58*, 908 (1987).

6. R. J. Cava, B. Batlogg, R. B. Van Dover, D. W. Murphy, S. Sunshine, T. Siegrist, J. P. Remeika, E. A. Rietmann, S. Zahurak, and G. P. Espinosa, Phys. Rev. *58*, 1676 (1987).

7. W. Weber, Phys. Rev. Lett. *58*, 1371 (1987).

8. P. Prelovsek, T. M. Rice, and F. C. Zhang, J. Phys. C *20*, L229 (1987).

9. K. Miyake, S. Schmitt-Rink, and C. M. Varma, Phys. Rev. B *34*, 6554 (1986).

10. D. J. Scalapino, E. Loh, and J. Hirsch, Phys. Rev. B *34*, 8190 (1986).

11. P. W. Anderson, Science *235*, 1196 (1987); G. Baskaran, Z. Zou, and P. W. Anderson, Solid State Commun., in press.

12. A. E. Ruckenstein, P. J. Hirschfeld, and J. Appel, Phys. Rev. B, in press.

13. C. M. Varma, S. Schmitt-Rink, and E. Abrahams, Solid State Commun. *62*, 681 (1987).

14. B. Batlogg, R. J. Cava, A. Jayaraman, R. B. Van Dover, G. A. Kourouklis, S. Sunshine, D. W. Murphy, L. W. Rupp, H. S. Chen, A. White, A. M. Mujsce, and E. A. Rietman, Phys. Rev. Lett. *58*, 2333 (1987).

15. L. C. Bourne, M. F. Crommie, A. Zettl, H. C. Zur Loye, S. W. Keller, K. L. Leary, A. M. Stacy, K. J. Chang, M. L. Cohen, and D. E. Morris, Phys. Rev. Lett. *58*, 2337 (1987).

16. B. K. Chakraverty and J. Ranninger, Philos. Mag. B *52*, 669 (1985).

17. P. Nozieres and S. Schmitt-Rink, J. Low Temp. Phys. *59*, 195 (1985).

18. P. Ganguly and C. N. R. Rao, J. Solid State Chem. *53*, 193 (1984).

19. D. Vaknin, S. K. Sinha, D. E. Moncton, D. C. Johnston, J. Newsam, C. R. Safinya, and H. E. King, Jr., Phys. Rev. Lett. *58*, 2802 (1987).

20. S. Schmitt-Rink, K. Miyake, and C. M. Varma, Phys. Rev. Lett. *57*, 2575 (1986).

21. D. R. Harshman, G. Aeppli, B. Batlogg, R. J. Cava, E. J. Ansaldo, J. H. Brewer, W. Hardy, S. R. Kreitzman, G. M. Luke, D. R. Noakes, and M. Senba, Phys. Rev. B, in press.

22. B. Golding, N. O. Birge, W. H. Haemmerle, R. J. Cava, and E. Rietmann, preprint.

23. E. Abrahams and C. M. Varma, preprint.

24. L. F. Mattheiss, Phys. Rev. Lett. *58*, 1028 (1987).

25. L. F. Mattheiss and D. R. Hamann, Solid State Commun., in press.

26. J. Zaanen, G. A. Sawatzky, and J. W. Allen, Phys. Rev. Lett. *55*, 418 (1985).

27. G. A. Sawatzky and J. W. Allen, Phys. Rev. Lett. *53*, 2339 (1984).

28. J. Orenstein, G. A. Thomas, D. H. Rapkine, C. G. Bethea, B. F. Levine, R. J. Cava, E. A. Rietmann, and D. W. Johnson, Jr., Phys. Rev. B, in press.

29. J. Orenstein, G. A. Thomas, D. H. Rapkine, C. G. Bethea, B. F. Levine, B. Batlogg, R. J. Cava, D. W. Johnson, Jr., and E. A. Rietmann, preprint.

30. S. Etemad, D. E. Aspnes, M. K. Kelly, R. Thompson, J. M. Tarascon, and G. W. Hull, preprint.

31. K. Kamaras, C. D. Porter, M. G. Doss, S. L. Herr, D. B. Tanner, D. A. Bonn, J. E. Greedan, A. H. O'Reilly, C. V. Stager, and T. Timusk, preprint.

32. Y. Toyozawa, M. Inoue, T. Inui, M. Okazaki, and E. Hanamura, J. Phys. Soc. Japan *21*, 208 (1966); ibid, 209 (1966).

33. A. E. Ruckenstein and S. Schmitt-Rink, Phys. Rev. B *35*, 7551 (1987).

34. J. C. Inkson and P. W. Anderson, Phys. Rev. B *8*, 4429 (1973).

35. W. A. Little, Phys. Rev. *134*, A1416 (1964).

36. V. L. Ginzburg, Usp. Fiz. Nauk. *101*, 185 (1970) [Sov. Phys. - Usp. *13*, 335 (1970)].

37. D. Allender, J. Bray, and J. Bardeen, Phys. Rev. B *7*, 1020 (1973); Phys. Rev. B *8*, 4433 (1973).

38. J. P. Collman, J. T. McDevitt, and W. A. Little, preprint.

39. M. Grabowsky and L. J. Sham, Phys. Rev. B *29*, 6132 (1984).

BONDS, BANDS, CHARGE TRANSFER EXCITATIONS AND SUPERCONDUCTIVITY:

$YBa_2Cu_3O_{7-\delta}$ vs. $YBa_2Cu_3O_6$

Jaejun Yu, A.J. Freeman and S. Massidda

Dept. of Physics and Astronomy and
Materials Research Center
Northwestern University, Evanston, IL 60201

It is generally recognized that highly precise electronic structure calculations provide important information about the properties of the superconducting materials and provide insight into the possible mechanism for their observed high temperature superconductivity. We here report briefly some of our results for the Y-Ba-Cu-O materials[1] - focussing on the role of chains vs planes in terms of state-of-the-art all-electron local density calculations for $YBa_2Cu_3O_{7-\delta}$ (the high T_c superconductor with chains and planes) and for $YBa_2Cu_3O_{6+x}$ (planes but no chains and no superconductivity). The results obtained highlight the role of charge transfer excitations (exciton) which we proposed earlier for the high T_c in $YBa_2Cu_3O_{7-\delta}$.

Important structural features which affect all the properties of the high T_c $YBa_2Cu_3O_{7-\delta}$ compounds[2] arise from the fact that $(2+\delta)$ oxygen atoms are missing from the perfect triple perovskite, $YCuO_3.(BaCuO_3)_2$. These vacancies arise from a total absence of O atoms in the Y-O planes (which seems to separate the Cu-Cu interactions across the Y plane) and an ordered absence of O atoms in the Cu planes between the Ba-O planes (which leads to the formation of linear chains of Cu-O-Cu). As a result, there are two Cu ions (called Cu2) in five-coordinated positions and one Cu ion (called Cu1) in a four coordinated position - as shown in Fig. 1(a). Since the interatomic distance Cu2-O4 (2.303 Å) is much larger than Cu1-O4 (1.850 Å)[3], the Cu1 ions have a rather weak interaction with the Cu2 ions. The Cu2 ions are in a locally very strong tetragonal distortion and this yields a 2D structure for these planes similar to that of $La_{2-x}M_xCuO_4$. The additional distortions of the O2 and O3 ions (the so-called "dimpling")[3] arises from the absence of O ions in the adjacent Y-O plane.

Recently, several neutron experiments[4] showed that the oxygen vacancies, which mainly concentrate on the O1 sites, change the composition and symmetry from orthorhombic (in $YBa_2Cu_3O_7$) to tetragonal (in $YBa_2Cu_3O_6$). Due to the missing oxygens at the O1 site, the 1D chain structure in the $YBa_2Cu_3O_7$ compounds is completely absent from the $YBa_2Cu_3O_6$ compounds as shown in Fig. 1(b). The additional oxygen vacancies, therefore, change the local symmetry as well as the electronic configuration around Cu1 sites.

One notable consequence of the change of structures is that the
Cu1-O4 distance in $YBa_2Cu_3O_6$ is shorter than in $YBa_2Cu_3O_7$.[4] In this
geometry, each Cu1 ion would be completely isolated from the other Cu1
ions in the Cu1 plane (having no oxygens lying between Cu1's) and
remain as Cu^{+1} with completely filled d-shell. Hence, the d-orbital
states of Cu1 are expected to be very localized in the Cu1 plane.

The local density band structure calculations were carried out for
both the Pmmm structure[3] of $YBa_2Cu_3O_7$ and the P4/mmm structure[4] of
$YBa_2Cu_3O_6$, using the highly precise full-potential linearized augmented
plane wave (FLAPW) method[5] with the Hedin-Lundqvist form for the
exchange-correlation potential. The details of the calculations are
presented in a previous paper[6].

The calculated band structures (near E_F) of tetragonal $YBa_2Cu_3O_6$
and orthorhombic $YBa_2Cu_3O_7$ in the basal planes of the tetragonal or
orthorhombic Brillouin zone (BZ) are shown in Fig. 2. The bands at the
top plane of the BZ's have the same features as the basal plane due to
the two dimensional nature of the structure. A remarkable simple band
structure near E_F emerges from the Cud-Op complex of bands. There are
four bands considered to be important for the electronic structure near
E_F of both $YBa_2Cu_3O_7$ and $YBa_2Cu_3O_6$. (Each of them is labelled
according to their symmetries.) Two strongly dispersed bands crossing

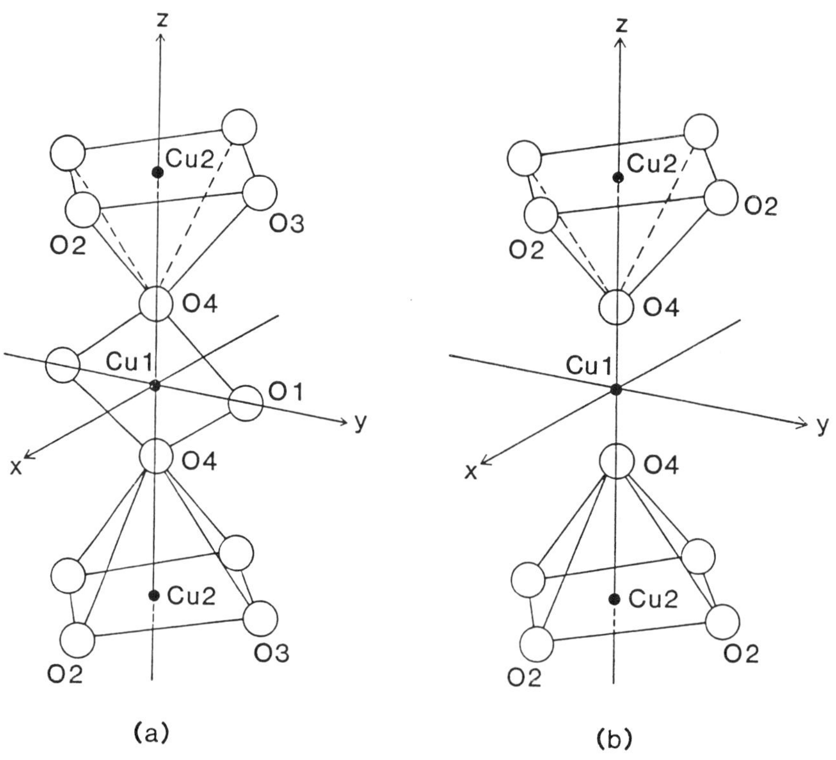

Fig. 1 Local environments for the Cu1 and Cu2 atoms in (a) $YBa_2Cu_3O_7$
and (b) $YBa_2Cu_3O_6$, following the Y-Cu2-Ba-Cu1=Ba-Cu2-Y
ordering along Z.

E_F consist of $Cu2(d_{x2-y2})$ - $O2(p_{x,y})$ or $O3(p_y)$ orbitals in the 2D
Cu2-O planes and show a 2D character which proved to be so important
for the properties of $La_{2-x}M_xCuO_4$.[7] In contrast to the case of
$La_{2-x}M_xCuO_4$, however, the symmetry allowed interactions of the Cu2 dpσ
bands with the Cu1 π-bonding (dpπ) bands result in a complicated
dispersion for the occupied parts of the Cu2 bands along Γ-X (or Γ-Y).
In the case of $YBa_2Cu_3O_6$ and $YBa_2Cu_3O_7$, the dpπ bands originating from
Cu1 lie nearer to E_F than the d_{z2} orbital bands in $La_{2-x}M_xCuO_4$.[7]

In $YBa_2Cu_3O_{7-\delta}$, the $Cu1(d_{z2-y2})$ - $O1(p_y)$ - $O4(P_z)$ anti-bonding
(dpσ) band (labelled as D_1-C_1-Δ_1 in Fig. 2(a)) shows the (large) 1D
dispersion expected from the Cu1-O1-Cu1 linear chains but is almost
unoccupied. This band is in sharp contrast to the π-bonding band
(formed from the $Cu1(d_{zy})$ - $O1(p_z)$ - $O4(p_y)$ orbitals) which is almost
entirely occupied in the stoichiometric (δ = 0) compound and becomes
fully occupied for the superconducting materials ($\delta \geqslant 0.1$). We will
soon see that since for δ = 0 this almost flat π-bonding band (the
state C_2-Δ_4 in Fig. 2(a)) lies just below and crosses E_F along S-Y, it
gives rise to peaks in the DOS near E_F making the DOS at E_F sensitive
to the position of E_F (i.e., to δ).

The dominant 1D electronic structure, coming from the linear chain
Cu1-O1 atoms in $YBa_2Cu_3O_{7-\delta}$ is completely absent from the band
structure of $YBa_2Cu_3O_6$ in Fig. 2(b). Instead of the 1D electronic
structure, there are two degenerate states (d_{yz} and d_{zx}) of Cu1 at Γ
and M. These orbitals form π-bonding (dpπ) bands with O4 atoms.

The prominent change of the 1D electronic structure from $YBa_2Cu_3O_7$
to $YBa_2Cu_3O_6$ is strongly correlated to the lowering or absence of
superconductivity observed in several experiments[4] in the tetragonal
to $YBa_2Cu_3O_6$ is strongly correlated to the lowering or absence of
superconductivity observed in several experiments[4] in the tetragonal
phase of $YBa_2Cu_3O_{7-\delta}$ where δ is usually greater than 0.5. In fact, we
had earlier discussed the importance of the 1D feature in the
electronic structure near E_F,[8] pointing out the possible important role
played by charge transfer excitations ("excitons") of occupied Cu1-O

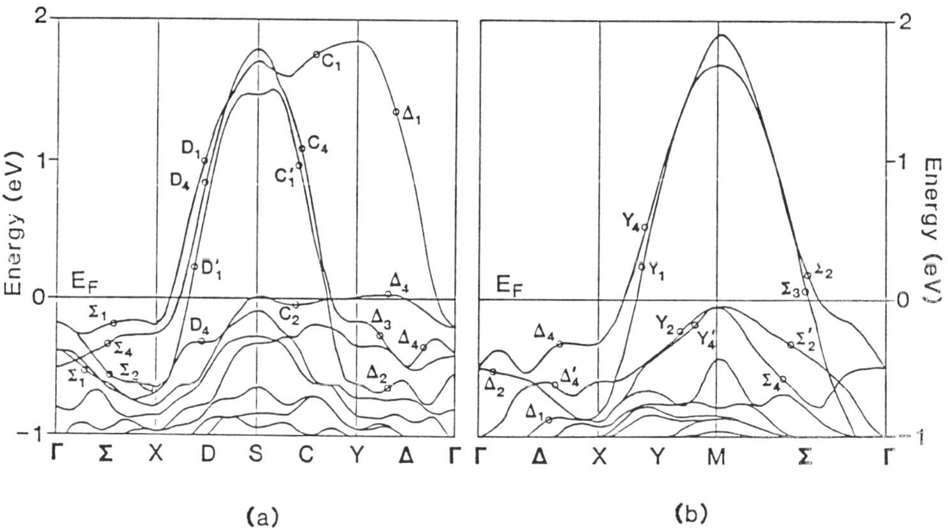

Fig. 2 Energy bands of (a) $YBa_2Cu_3O_7$ and (b) $YBa_2Cu_3O_6$ near E_F.

369

dpπ (localized) orbitals into their empty Cu1-O dpσ anti-bonding
(extended) states. These Cu^{3+}-Cu^{4+}-like charge fluctuations[9] may
induce attractive interactions both in the chains and in the 2D
conduction (Cu2) bands - thereby promoting the high T_c via exchange
of these "excitons".

In the band structure of $YBa_2Cu_3O_6$ in Fig. 2(b), it is also
remarkable that the symmetry allowed interactions of bands play a
dominant role in the resulting band features near E_F. Since the dpπ
bands of Cu1-O are close to E_F, they will inevitably have a crossing or
anti-crossing with the strongly dispersed dpσ bands of Cu2-O. The
two bands Σ_2 and Σ'_2 along ΓM have the same symmetry and anti-cross
each other. This happens because both the odd (under z- reflection)
$Cu2d_{x^2-y^2}$ - $O2(p_{x,y})$ dpσ band and the odd (under x-y reflection)
$Cu1d_{yz}(d_{zx})$ - $O4(p_{y,x})$ belong to the same symmetry representation.
Since the dpπ bands of Cu1-O4 are very localized in the Cu1-plane and
the dpσ bands of Cu2-O2 are well extended states, the strong
hybridization of the two bands near E_F may result in a complicated and
correlated electronic structure near E_F.

In addition to the electronic structures, a number of surprising
features emerged from these calculations, including the low DOS at E_F
and its related properties. In the superconducting $YBa_2Cu_3O_{7-\delta}$
compounds, the DOS at E_F for $\delta > 0.1$ is much lower per Cu atom than in
$La_{2-x}Sr_xCuO_4$, in agreement with recent experiments[2]. This has a number
of important consequences for DOS derived properties and
superconductivity including: reduced screening, an increased role for
the polarization of ionic constituents, lowered conductivity (and
reduced superconducting current carrying capacity), etc.. In
considering magnetic properties we found a surprisingly large Stoner
factor for both $YBa_2Cu_3O_7$ and $YBa_2Cu_3O_6$. In $YBa_2Cu_3O_7$, the largest
contributions to the Stoner factor, $S = N(E_F)I$, come from the O1 and O4
atoms where the localized dpπ states of Cu1 (in the chain) originate.
Whereas $S = 1.12$ for $\delta = 0$ (indicating a possible magnetic
instability), S drops sharply for increased oxygen vacancies to 0.86
for $\delta = 0.1$. This gives an enhancement factor $(1-S)^{-1} = 7$ which is the
same order as Pd metal and may be significant in view of Curie-Weiss
behavior of the magnetic susceptibility for the normal state.[2]

On the contrary, however, the calculated Stoner factor for
$YBa_2Cu_3O_6$ turns out to be greater than one, $S \sim 1.38$, even though the
DOS at E_F is smaller than in $YBa_2Cu_3O_7$. Significantly, the major
contributions to the Stoner factor comes from the Cu2 and O2 in the 2D
conducting planes. This may be a result of an increase in the exchange-
correlation interactions parameter, I, due to greater localization of
the 2D conduction electrons within the planes. In addition to the
highly magnetic instability in the 2D Cu2-O2 planes, the band Σ_3 along
ΓM in Fig. 2(b) suggests possible Fermi surface (FS) nesting along the
(110) direction. From these results of the FS nesting effects and a
large Stoner factor, antiferromagnetic ordering in the conduction plane
of $YBa_2Cu_3O_6$ is expected.

In conclusion, we have dealt with an intriguing aspect of the
origin of the high T_c superconductivity in $YBa_2Cu_3O_{7-\delta}$ namely the role
played by the Cu-O chains vs the Cu-O planes which make up this
structure. The lower T_c ($\approx$ 40K) superconductor, $La_{2-x}M_xCuO_4$, consists
only of Cu-O planes and there is evidence in $YBa_2Cu_3O_{7-\delta}$ that
introducing additional O vacancies into the chains (increasing δ to 0.5
or greater) results in the onset of a tetragonal phase and the lowering
of T_c and its eventual absence[4,10]. We have reported on detailed high
precision local density energy band study of the two structures:

orthorhombic $YBa_2Cu_3O_7$ and tetragonal $YBa_2Cu_3O_6$. The differences found
and discussed indicate that superconductivity is less likely and
magnetism more likely in the O_6 case. From crude rigid ion
calculations[8], the standard electron-phonon mechanism was found to yield
too low a T_c for the $YBa_2Cu_3O_7$ system. Comparisons with results for
the quasi-2D $La_{2-x}M_xCuO_4$ and $YBa_2Cu_3O_6$ indicate that the Cu1-O1 chains
in $YBa_2Cu_3O_{7-\delta}$ may play a key role in excitonic contributions to the
observed high T_c.

ACKNOWLEDGEMENTS

Work supported by the National Science Foundation (DMR grant No. 85-
20280) through the Northwestern University Materials Research Center
and DMR grant No. 85-18607, and a computing grant from its Division
for Advanced Scientific Computing.

REFERENCES

1. Wu, M.K., Ashburn, J.R., Torng, C.J., Hor, P.H., Meng, R.L., Gao, L.,
    Z.J. Huang, Wang, Y.Q., and Chu, C.W., 1987, Superconductivity at
    93K in a New Mixed-Phase Y-Ba-Cu-O Compound System at Ambient
    Pressure, Phys. Rev. Lett., 58:908.
2. Cava, R.J., Batlogg, B., van Dover, R.B., Murphy, D.W., Sunshine,
    S., Siegrist, T., Remeika, J.P., Rietman, E.A., Zahurack, S., and
    Espinosa, G.P., 1987, Bulk Superconductivity at 91K in Single
    Phase Oxygen-Deficient Perovskite $Ba_2YCu_3O_{9-\delta}$, Phys. Rev. Lett.
    58:1676.
    Hinks, D.G., Soderholm, L., Capone II, D.W., Jorgensen, J.D.,
    Schuller, I.K., Segre, C.U., Zhang, K., and Grace, J.D., 1987,
    Phase Diagram and Superconductivity in the Y-Ba-Cu-O System,
    Appl. Phys. Lett. (in press).
3. Beno, M.A., Soderholm, L., Capone II, D.W., Hinks, D.G.,
    Jorgensen, J.D., Schuller, I.K., Segre, C.U., Zhang, K., and
    Grace, J.D., 1987, Structure of the Single Phase High
    Temperature Superconductor $YBa_2Cu_3O_7$, Appl. Phys. Lett. (submitted).
4. K. Ogawa et al. (Natl. Res. Inst. for Metals, Tokyo, Japan).
    (Private Communication).
    Santoro, A., Miraglia, S., Beech, F., Sunshine, S.A. Murphy, D.W.,
    Schneemeyer, L.F., and Waszczak, J.V., 1987, The Structure and
    Properties of $Ba_2YCu_3O_6$, Mats. Res. Bulletin (submitted).
5. Jansen, H.J.F. and Freeman, A.J., 1984, Total-Energy Full-
    Potential Linearized-Augmented-Plane-Wave Method for Bulk
    Solids: Electronic and Structural Properties of Tungsten,
    Phys. Rev. B30:561.
    Wimmer, E. Krakauer, H., Weinert, M., and Freeman, A.J., 1981,
    Full-Potential Self-Consistent Linearized-Augmented-Plane-Wave
    Method for calculating the Electronic Structure of Molecules and
    Surfaces. O_2 Molecule, Phys. Rev. B24:864.
6. Massidda, S., Yu, Jaejun, Freeman, A.J., and Koelling, D.D.,
    1987, Electronic Structure and Properties of $YBa_2Cu_3O_{7-\delta}$ - A
    Low Dimensional, Low Density of States Superconductor, Phys. Lett.
    (to appear).
7. Yu, Jaejun, Freeman, A.J., and Xu. J.-H., 1987, Electronically
    Driven Instabilities and Superconductivity in the Layered
    $La_{2-x}Ba_xCuO_4$ Perovskites, Phys. Rev. Lett. 58:1035.
    Mattheis, L.F., (1987),
    Phys. Rev. Lett. 58:1028.
8. Yu, Jaejun, Massidda, S., Freeman, A.J., and Koelling, D.D.,
    1987, Bonds, Bands, Charge Transfer Excitations and

Superconductivity of $YBa_2Cu_3O_{7-\delta}$, Phys. Lett. (to appear).

9. Fu, C.L. and Freeman, A.J., (submitted Feb. 26, 1987),
    Optic breathing Mode, Resonant Charge Fluctuations and High
    T_c Superconductivity in the Layered Perovskites, Phys. Rev.,
    (to appear).

10. Schuller, I.K. Hinks, Beno, M.A., Capone II, D.W., Soderholm, L.,
    Locquet, J.-P., Bruynseraede, Y., Segre, C.U., and Zhang, K.,
    1987, Structural Phase Transition in $YBa_2Cu_3O_{7-\delta}$: The Role of
    Dimensionality for High Temperature Superconductivity, Solid
    State Commun. (submitted).

POSSIBLE ROLE OF OXYGEN VACANCIES AND EXCITONIC MECHANISM IN HIGH T_C

SUPERCONDUCTING OXIDES

William Y. Hsu and Robert V. Kasowski

Central Research and Development Department*
E. I. du Pont de Nemours and Company
Experimental Station, Wilmington, DE 19898

INTRODUCTION

Superconductivity with very high transition temperatures, T_c, has been observed in pure and doped La_2CuO_{4-y} and cuprate perovskites $ABa_2Cu_3O_{7-y}$ recently.[1] These materials not only have unprecedented T_cs but also other intriquing properties. For example, undoped La_2CuO_{4-y} could be metallic, semimetallic, semiconducting, antiferromagnetic, or super-conducting and likewise $YBa_2Cu_3O_{7-y}$ could be antiferromagnetic semiconducting, metallic with localized moments and superconducting, depending on how the samples are prepared. The key controlling factor appears to be the oxygen content of the materials. In a systematic investigation of the relationship between physical properties and oxygen vacancies in La_2CuO_{4-y}, Johnston *et al* observed the development of antiferromagnetic ordering from a non-magnetic ground state at y = 0 into a state with a Neel temperature of 290 K at y = 0.03.[2] We study here theoretically electronic properties of oxygen vacancies in these cuprates and explore their possible contribution to T_c.

To date, all band structure calculations suggest that La_2CuO_4 and $YBa_2Cu_3O_7$ should have broad bands (about 2 eV wide) at the Fermi energy, E_F. These broad bands would in principle favor Pauli paramagnetism and preclude the existence of localized moment and magnetic ground state since usual models of magnetism, such as the Hubbard and Anderson models, require narrow bands and large correlation energies. We show here from first principles that oxygen vacancies alter the electronic band structure significantly.[3] In particular, the broad conduction bands at E_F for the stoichiometric compounds have been broken up and narrowed down considerably by the vacancies due to the disappearing of some O 2p bands and additional filling of partially occupied Cu 3d states. The new narrow bands are about 0.1 eV wide and could conceivably support localized moments and an antiferromagnetic ground state. These bands are also desirable for the excitonic mechanism[4] and could contribute to the high T_c.

---

* Contribution No. 4454

The electronic structure of O vacancies is computed from first principles by the pseudofunction method[5] using the supercell approach. The supercell approach, which should provide accurate local information of the electronic structure of random vacancies, is adopted for the ease of computation. We focus our attentions on the effects of O vacancies in the linear Cu-O chains of $YBa_2Cu_3O_{7-y}$ and vacancies in planar Cu-O of La_2CuO_{4-y}.

The supercell chosen for $YBa_2Cu_3O_{7-y}$ has two unit cells in the direction of the linear Cu-O chains and every second O atom is missing from the chains. The formal composition corresponding to our supercell with vacancy is $YBa_2Cu_3O_{6.5}$, i.e. y = 0.5. To insure against errors, we have also computed the electronic structure for the stoichiometric y = 0 case with the identical doubled unit cell. The energy bands of $YBa_2Cu_3O_7$ are well known and thus serve as a check on the accuracy of our calculation.

For La_2CuO_{4-y}, we have constructed a supercell with one O vacancy for every 4 formula units, or equivalently, one O vacancy for every 16 oxygen atoms, which formally corresponds to $La_2CuO_{3.75}$, i.e. y = 0.25. The supercell contains 4 tetragonal unit cells, i.e. doubling the cells in the planar directions. An O vacancy is created by removing one of the oxygens in the Cu-O plane. Again to check against possible errors, we have computed the y = 0 case with the same quadrupled unit cell also. Although we are keenly aware of the critical role of the orthorhombic symmetry as we were the first to point out that the orthorhombic instead of the tetragonal phase should be the host of superconductivity in doped La_2CuO_{4-y} and that undoped La_2CuO_{4-y} could superconduct under appropriate conditions[6], the simpler tetragonal symmetry is used here for the ease of computations. In addition, our results for the tetragonal case should hold qualitatively for the orthorhombic case since the crucial conduction bands under consideration are similar and of the same width. Therefore, if spin-polarized band splitting could occur in the tetragonal phase, it should occur more easily in the orthorhombic phase due to the pre-existence of energy gaps in substantial portions of the Brillouin zone.

For the present model calculations we chose not to allow the atoms to relax in the super-cell when an O is removed because we did not want to unnecessarily complicate the essential physics by introducing additional degrees of freedom at this stage.

Each of our pseudofunction basis function is constructed from the radial solution of the spherical part of an appropriate muffin-tin potential by the following recipe. In the atomic core region, the radial solution is orthogonized to the core states; in the interstitial bonding region, the full numerical solution is retained; in the tail region, an appropriate Neuman function is used. But we are careful to keep the basis functions and their slopes continuous everywhere. The most significant feature of our method is that, in the critical bonding region between atoms, the most faithful radial solution to the instantaneous muffin-tin Hamiltonian is used. Since the basis set is so optimally chosen, a small set (< 10 functions per atom typically) is sufficient for accurate total-energy computation. The pseudofunction method is thus tailored to study the band properties of complex structures with large unit cells such as the present layered perovskites. During computation, the basis functions and the total potential are iterated to self-consistency.

For the present computations, our basis sets consist of s functions for Ba and La, s and d functions for Y and Cu, and s and p functions for O. The muffin-tin radii are 1.49 Å for Y, Ba and La, 0.97 Å for Cu and 0.86 Å for O. To compute the nonspherical terms of the Hamiltonian, the basis functions are expanded to 2197 plane waves and the nonspherical part of the potential to 15629 plane waves for La_2CuO_{4-y} and to 3519 and 24245 plane waves respectively for $YBa_2Cu_3O_{7-y}$. Exchange and correlation are treated via the Hedin-Lundquist formalism of the local-density approximation.

RESULTS AND DISCUSSIONS

Figure 1 compares the energy bands for the y = 0 and 0.5 cases of $YBa_2Cu_3O_{7-y}$ and Figure 2 for the y = 0 and 0.25 cases of La_2CuO_{4-y}, respectively. Only the bands near the Fermi energy, E_F, are shown. The electronic bands for the stoichiometric compounds without O vacancies agree very well with published results. In particular, all the additional degeneracies required by the higher symmetry of the no vacancy cases are reproduced in our supercell calculations using the lower symmetry of the vacancy cases.

The introduction of vacancies has a dramatic effect on the electronic structure: new bandgaps were created to break up and narrow significantly the broad and overlapping anti-bonding Cu-O bands near E_F. The dramatic changes arise from the removal of 3 O p-bands and additional filling of partially occupied Cu d-bands, or chemically, from the creation of lower valence Cu states by the O vacancies. Some of the dramatic changes are anticipated. For example, the creation of O vacancies in a linear Cu-O chain would disrupt its connectivity and thus destroy its contribution to the broad metallic band. A Hubbard type anti-ferromagnetic ground state would likely occur in oxygen deficient samples due to the presence of these narrow bands (about 0.1 eV) and a large density of Cu 3d states at E_F. We believe that spin polarized calcualtions, which we plan to do, will yield a magnetic semiconducting state with sufficient amount of O vacancies and a non-magnetic metallic state without any vacancies.

The calculations reported here represent the end members of the interesting compositional range. The experimental structures apparently have fewer vacancies especially in the superconducting region. Then the vacancy-induced narrow band, which should exist locally even for lower vacancy concentrations, could contribute to the high T_c by the excitonic mechanism. However, the excitonic states considered here are not virtual excitations between the Cu-O bonding and antibonding orbitals.[4] Instead they are excitations between the occupied Cu-O antibonding states and the unoccupied antibonding states split off by the O vacancies. Virtual transitions corresponding to charge transfer from the Cu-O bands to the O vacanies would create a breathing-mode type oscillation of the electronic charge. A strong excitonic attraction between conduction electrons could result which would in turn produce the high T_c superconducting state. Computation of the necessary matrix elements to assess this possibility is in progress.

CONCLUSION

The electronic properties of O vacancies in La_2CuO_{4-y} and $YBa_2Cu_3O_{7-y}$ have been investigated by *ab initio* pseudofunction calculations. We find that vacancies alter the electronic structure significantly due to the disappearance of some O 2p bands and additional filling of the partially occupied Cu 3d bands. The net result is the appearance of new bandgaps to break up and considerably narrow the broad anti-bonding Cu-O band near the Fermi energy. The new narrow bands could support local moment or antiferromagnetism and might also contribute to the high T_c by the excitonic mechanism.

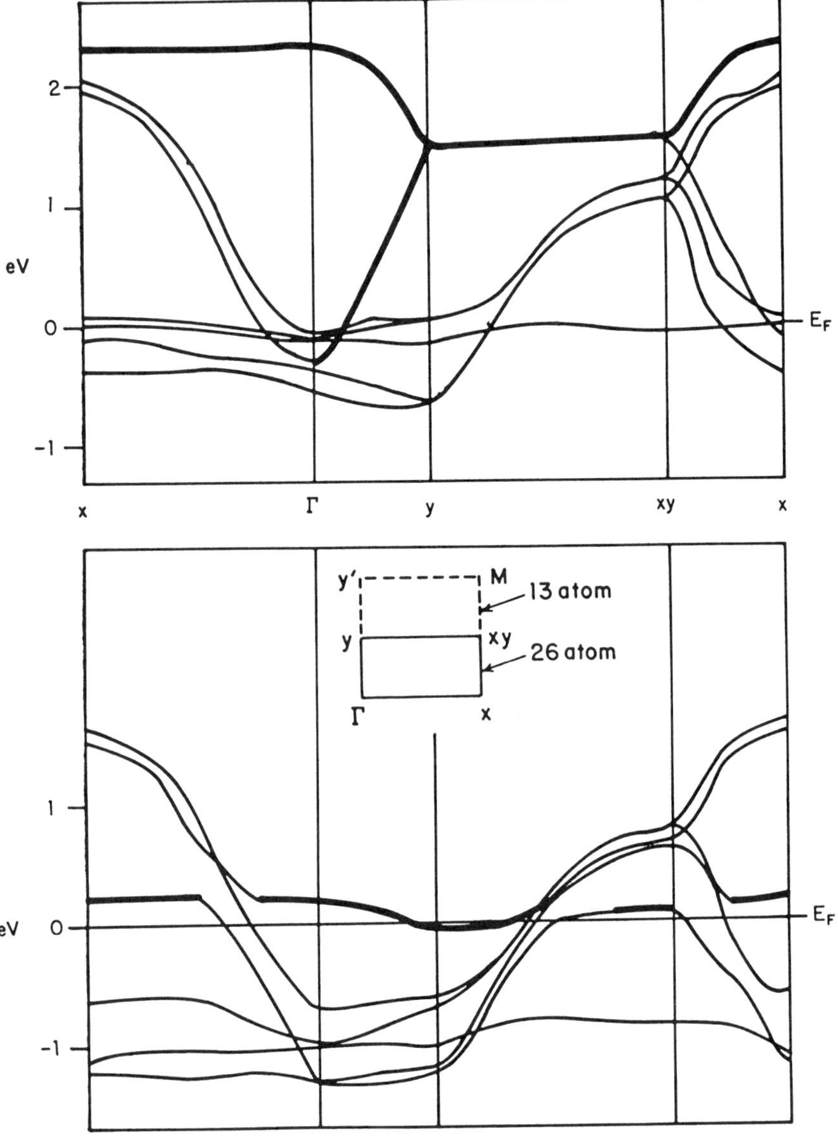

Fig. 1.   Band structures of $YBa_2Cu_3O_{7-y}$: (1) top panel $y = 0$ (no O vacancies) and (2) bottom panel $y = 0.5$ (with O vacancies). The Brillouin zones for the supercell and normal unit cell are shown in the bottom panel. The heavy line indicates how the 1D Cu-O band (top panel) has been broken up and significantly narrowed by O vacancies (bottom panel).

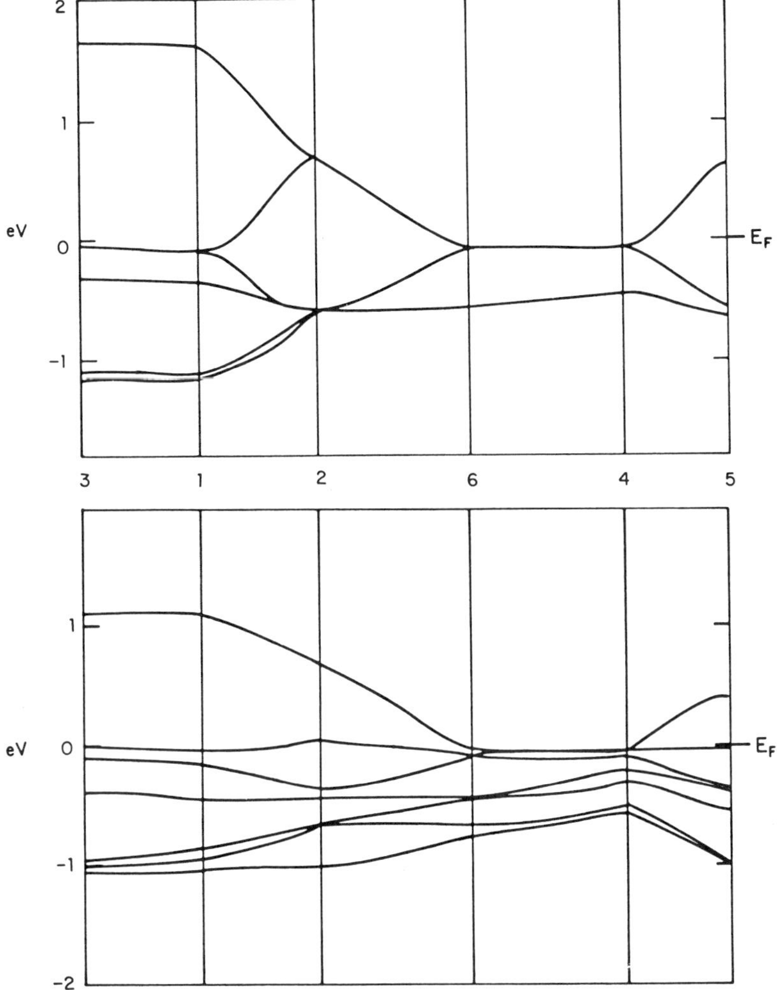

Fig. 2.   Band structures of La_2CuO_{4-y}: (1) top panel y = 0 (no O
vacancies) and (2) bottom panel y = 0.25 (with O vacancies).
The **k** points labelled 1 to 6 along the horizontal axis are:
3 = (.0,.0,.2878), 1 = (.0,.0,.0) = Γ, 2 = (.25,.0,-.0719),
6 = (.25,.25,-.1438), 4 = (.25,.25,.2438) and 5 = (.0,.25,
-.0179). The broad 2 eV band at E_F has been broken up and
significantly narrowed by O vacancies. Note, in particular, the
degeneracy along 2 to 6 direction (top panel) has been lifted and
a bandgap has appeared (bottom panel). Similarly, the
degenerate band just below E_F along 3 to 1 direction has also been
split.

REFERENCES

1. Appropriate references are numerous and well known. See other papers and their citations in this Proceeding.
2. D. J. Johnston, J. P. Stokes, D. P. Goshorn and J. T. Lewandowski, preprint.
3. The $YBa_2Cu_3O_{7-y}$ results have been reported previously: R. V. Kasowski, Superlattices and Microstructures, in press. Details of the La_2CuO_{4-y} calculations will be reported elsewhere: R. V. Kasowski, W. Y. Hsu and F. Herman, to be published.
4. See, for example, J. P. Coleman, J. T. McDevitt and W. A. Little, preprint.
5. R. V. Kasowski, M.-H. Tsai, T. N. Rhodin and D. D. Chambliss, Phys. Rev. B 34, 2656 (1986).
6. R. V. Kasowski, W. Y. Hsu and F. Herman, Solid State Commun., in press.

AN EXCITONIC MODEL FOR THE NEW HIGH TEMPERATURE

SUPERCONDUCTORS

T.C. Collins

Vice Provost for Research, University of Tennessee
Knoxville, TN 37996

A. Barry Kunz

Dept. of Physics, Michigan Technological University
Houghton, MI 49931

J.J. Ladik

Director of the Institute for Theoretical Chemistry
Universitat Erlangen, D-8520 Erlangen - Egerland strasse 3
Erlangen, Federal Republic of Germany

ABSTRACT

An excitonic model is given which can explain the experimental
results of the new high temperature superconductors, $AB_{42}CU_3O_{6+\delta}$ (A =
LA, Nd, Sm, Eu, Gd, Ho, Er, Lu and Y). The excitonic model is an
extension of the one used to describe the temperature-dependent surface
shielding effect in Cu and the anomalously large diamagnetism
measurements found in CuCl and CdS at temperature above that of liquid
nitrogen. Both singlet and triplet paired spins are included in the
description. The parallel paired states contribute a paramagnetic term
to the susceptibility and make possible no change in T_C with magnetic
ion replacements for Y.

I. Introduction

It is very interesting to note that copper associated with oxygen
has shown anomalous effects for at least the last 20 years. Witteborn
and Fairbank[1,2] measured a net force on the electron of zero to within
$\pm$ 6 x 10^{-12} V/m. This value was the result of the combined effects of
gravity on the test electron and an abient electron field of
~ -5 x 10^{-11} v/m. Theoretical predictions and room temperature contact-
potential results indicated that an ambient field of 10^{-6} V/m should be
present. Also the low ambient field was not measured in all
experiments. Measurements made at WPAFB, Ohio found that the tubes
which had the anomalous effects had formed an oxcide layer on the inside
of the tube.

In 1978 and 79 anomalously large diamagnetism and its possible
relation to the Meissner effect in CuCl and CdS at temperatures above
that of liquid nitrogen were reported[3-8]. In CuCl the diamagnetic
susceptibility was reported[3-5] to be anywhere between 7% and 80% of the
ideal Meissner value for superconducting metals. The diamagnetic

anomaly[4] was accompanied by a sharp drop in the resistivity, as well as
by two peaks in a simultaneous differential thermal analysis measurement
at the onset and completion of the anomaly. These effects were observed
over a temperature range of 10 to 20K around a mean temperature of 240K,
and at hydrostatic pressure 5 to 25 kbars. The anomalies were observed
only when the samples were rapidly cooled or warmed.

In pressure-quenched CdS large diamagnetism approaching 100% flux
exclusion was reported at 77K by Rown et. al.[6] The samples were
pressurized to 40 kbars or more and explosively released. At 30 kbars
CdS undergoes a wurtzite-to-NACl phase transition[9], and the pressure-
quenched samples transform from the wurtzite structure to a mixture of
zine-blende and NaCl structures. The magnetic anomaly disappeared after
a day or two of temperature cycling.

While several different groups have repeated these measurements and
observed the above described properties, the lack of consistency of
results from sample to sample plaques their interpretation. This is
likely due to important differences in the sample properties.

In 1986, Bednorz and Muller[10] open the flood gates to
superconductivity above 30K in A-Ba-Cu-O layered compounds with A being
Y, LA, Nd, Sm, Eu, Gd, Ho, Er and Lu.[11] "In particular, the high T_c of
$A-Ba_2Cu_3O_{6+\delta}$ is attributed mainly to the quasi two-dimensional assembly
of the $CuO_2-Ba-CuO_{2+\delta}-Ba-CuO_2$ layers sandwiched between two A layers.
The comparison of the magnetic properties of the superconductive
materials $Ho\ Ba_2Cu_3O_{6+\delta}$ and $Y\ Ba_2Cu_3O_{6+\delta}$ shows experimental evidence
for a nearly complete decoupling of the magnetic effects on the
superconductivity. Also the isotope effect of replacing O_{16} by O_{18} did
not cause a shift in the superconducting transitions temperature
This, of course, tends to rule out that this superconductivity is
related to phonon coupling.

In the next section the excitonic model which explains the results
of CuCl and CdS is presented[15]. This takes in account that impurities
have to be presence. Then in Section III, using the outlined model of
Section II, the comparison of the model with the experimental results of
the $ABa_2Cu_3O_{6+\delta}$ layer compounds is given.

II. Excitonic Model for CuCl and CdS

The proposed model is based on self-consistent field Hartree-Fock
calculations of the band structures of $CuCl^{16}$ and CdS^{17}. In CuCl this
includes the effect of impurities, defects, and pressure.[17] According
to the calculated band structure of CuCl at atmospheric pressure, the
fundamental gap is direct at k=0 and has an energy of 4 eV, in
reasonable agreement with Cardona[18] measurement of 3.4 eV. The valence
band consists mainly of the d bands of Cu and is relatively narrow. The
line of most physical interest is the Γ-X line. The X point in the
lowest conduction band (CB) is about 1 eV higher than the bottom of the
CB at Γ. The direct gap at X is about 7.5 eV, in reasonable agreement
with Goldman's analysis.[19] As pressure is applied the CB at X begins to
move down in energy relative to the CB at Γ. This is calculated by
decreasing the lattice constant by 1% and 5% of its normal value. A
linear interpolation of band energies versus lattice constant indicates
that Γ and X become degenerate at a lattice-constant reduction of 0.2%
which may be consistent with the pressure of 6 kbar that Chu et al.[4]
used. Thus, though an indirect gap is present, its value is not near
0.35 eV as suggested by Abrikosov.[20] As the pressure is increased the
band at X continues to decrease in energy.

The O^{2-} impurity level is calculated to be very shallow.[17] The calculations show that, as a function of hydrostatic pressure with or without uniaxial stress, the O^{2-} impurity level rapidly approaches the conduction band. A 1% reduction in the lattice constant brings the O^{2-} level with an exciton's bringing energy from the conduction band. Electrons could then pressure ionize from O^{2-} and can become superconducting via an excitonic mechanism. Pressure-ionization process is possible for an impurity-defect combination as well. Namely, these carriers can couple to the exciton field which forms an electronic polaron, as formulated by Devtreese et al.[21] The electron is clothed in about one virtual longitudinal exciton. The exciton coupling reduces the average e^- -e^- repulsion in the CB by 0.54 Ry (in the static limit). The critical temperature T_c at which superconductivity might occur is

$$T_c = 1.14 \frac{\hbar w_{ex}}{k_B} \exp\left[-\frac{1}{UD(E_F)} \right],$$

where hw_{ex} is the binding energy of the free exciton, k_B is Boltzmann's constant, U is the coupling between the exciton and the electron, and $D(E_F)$ is the density of states at the Fermi surface. T_c was then calculated for several values of conduction-electron density N(E) assuming (a) that the conduction band at Γ alone contributed to $D(E_F)$ and (b) that both electronic states at Γ and X points contribute to $D(EF)$ [when the Γ and X points are degenerate, this produces the largest $D(E_F)$]. It was found that for carrier densities of 10^{-1} e^-/unit cell, T_c for a nondegenerate CB was 38K while it was 1745 K for the degenerate case. For N(E) = 10^{-2} the corresponding numbers are 10^{-2} and 55 K, respectively. Therefore it is possible that with $10^{-2} e^-$/unit cell superconductivity close to or above liquid-nitrogen temperatures can be achieved.

A crucial point of this model is that there is an optimum range of electron densities for which superconductivity is possible at moderately high temperature. Too small an electron density causes a low density of states at the Fermi surface, and consequently a low T_c. On the other hand, the number of carriers should not be so large that it quenches out the exciton field. In other words, the density of free carriers in the CB should be such that the screening length is greater than the radius of the electronic polaron. The free-carrier density that would quench the exciton field is given by the Thomas-Fermi approximation

$$k_{TF} = \left[\frac{6\pi e N}{E_F} \right]^{1/2},$$

where N is the Free-carrier density and E_F is the Fermi energy. With the use of the band parameters, a cutoff density $N - 10^{-2}$ was obtained.

The model based on the above calculations predicts the following features: Owing to the heavy mass (narrow widths) of the valence band, the exciton cannot follow the motion of the conduction electrons. As long as there are not too many conduction electrons, the repulsion between them will be screened by the virtual excitations of the d-band electrons causing a local polarization field. This polarization field will have a positive local field and a negative part which is less localized (the hole arising from the d band and the electron from the s band).

A pressure-ionized electron from O^{2-} which is now in the conduction band will scatter more strongly on polarized d-band electrons

of like spin. This acts as a hole that will attract a second electron
in the conduction band that will be of the same spin as the first
electron. Thus the polarization field is generated by the same kind of
particles as the scattered conduction electrons. Therefore, the
physical situation is similar to that of ^{3}He, which has the lowest-
energy state by coupling particles of parallel spins.

This model clearly shows that high-temperature superconductivity is
possible through an impurity-based mechanism. It does not arise merely
from the intrinsic band structure of CuCl but by a combination of the
band structure under stress [which shows a degeneracy of the conduction
band at the X and Γ points at moderate pressures (0.2% lattice
contraction)] with an impurity such as O^{2-}, whose levels lie within an
exciton's binding energy of the CB. The impurities donate electrons
both to the Γ and X points in the conduction band and therefore create
a sufficiently large density of states at the Fermi surface in the CB
which is necessary for a superconducting mechanism.

So far there have been no published theories that explain the
diamagnetic anomaly in CdS. Kunz, Weidman, and Collins,[17] however,
propose that an impurity mechanism similar to that in CuCl may be
responsible for effects in CdS. Nam et al[8] have observed the anomaly in
a sample of CdS doped with Li but not in pure CdS. Cote, Capsimalis,
and Homan[22] analyzed samples in which they observed the anomaly and
found that these samples had a metastable rocksalt phase after pressure
quenching, a phase that was pronounced in samples rich in Cl. Kunz et.
al.[17] have calculated that the rocksalt structure has an indirect band
gap at X, which contributes to a high enough density of electronic
states. Therefore, with a large number of impurities, the NaCl phase of
CdS is host to the diamagnetic anomaly in a manner similar to CuCl.

III. Excitonic Model for the A-Ba-Cu-O Layered Compounds

The model of Section II can be applied to the layered compounds by
taking into account the changes of the electronic structure. First
assumption is that the layers can be treated as being almost
independent. Then the metal A layer forms a normal conducting sheet
with a Fermi level near the bottom of the CB of the Cu-O layers. Thus
the interchanging of the tri-valent atoms (LA, Y, ...) has the effect of
changing the Fermi level by a relatively small amount. The copper atoms
in the crystal most closely resembles the $(4s^2, 3d^9)$ configuration. The
d^{10} state is at much higher energy and do not play a role in the
superconducting process. The valence band of the CuO_2 layers are made
up of the $4s^2$ of Cu and the $2P_{x,y}$ states of oxygen along with the 3d
bands of copper.

In the tetragonal structure the oxygen p-band split into the 4-fold
P_{xy} band and the 2-fold P_z band. The $P_z$band forms the conduction band
for the Cu-O layer which in turn remains near the valence band (the
material is black). The A layer Fermi level is very near in energy to
that of the conduction energy levels of CuO_2. This results in a
transfer of a small number of electrons into the copper oxide layer.
These electrons play the same role as the impurity electrons did in the
case of CuCl and CdS.

Another important consequence of this model is that one can expect
to have paired spin 1 states as outlined in section II. Thus the
introduction of magnetic scatters will have a greatly reduced effect
since the two spins are scattered the same. As pointed out
experimentally, the layered A-Ba-Cu-O compounds show little or no
effect due to using Ho for A.[12]

There should also be a difference in making Josephson junctions
between the BCS type superconductors and those of the excitonic model.

In the excitonic model the electors pairs are localized much more than
in BCS pairs. Experimentally[23] a point contact Josephson junction was
made between a bulk Y-Ba-Cu-O compound and a Nd needle but reference
(23) did not measure a Y-Ba-Cu-O — Y-Ba-Cu-O junction.

In order for the excitonic model to have a high critical current
requires that the superconducting electrons in the Cu-O layer polarizes
the electron cloud in the A layer (mirror potential effect). This
process would be analogous to that of giant ferromagnetic or
antiferromagnet spin calculation.

IV. Conclusion

A theoretical model has been proposed for the superconductivity
found in A-Ba-Cu-O layer compounds. The explanation is a simple
extension of the excitonic model applied to the experimental results of
CuCl and CdS. A pressure experiment should change the number of
electrons that are transferred from the A-layer to the Cu-O layer. This
should cause a shift in the critical temperature of the superconductor
which could help in supporting the theory.

References

1)   F.C. Witteborn and W.M. Fairbank, Phys. Rev. Lett. $\underline{19}$, 1049
     (1967).

2)   J.M. Lockhart, F.C. Witteborn and W.M. Fairbank, Phys. Rev.
     Lett. $\underline{38}$ 1220 (1977).

3)   N.B. Brandt, S.V. Kuvshinnikov, A.P. Rusakov, and V.M.
     Smeronov, Zh. Eksp. Teor. Fiz. Pis'ma Red. 27,37 (1978) [JETP
     Lett. 27, 33 (1978)].

4)   C.W. Chu, A.P. Rusakov, S. Huang, S. Early, T.H. Geballe, and
     C.Y. Huang,
     Phys. Rev. B$\underline{18}$, 2116 (1978).

5)   I. Lefkowitz, J.S. Manning, and P.E. Bloomfield, Phys. Rev.
     B$\underline{20}$, 4506 (1979).

6)   E. Brown, C.G. Hornan, and R.K. MacCrone, phys. Rev. Lett. $\underline{45}$,
     478 (1980).

7)   C.G. Hornan, D.P. Kendall, and R.K. MacCrone, Solid State
     Commun. $\underline{32}$, 521 (1979).

8)   S.B. Nam, Y. Chung, and D.C. Reynolds (unpublished).

9)   B.A. Samara and H.G. Drickamer, J. Phys. Chem Solids $\underline{23}$, 457
     (1962).

10) J.G. Bednorz and K.A. Muller, Z. Phys. B$\underline{64}$, 189 (1986).

11) e.g. P.H. Hor, R.L. Meng, Y.Q. Wang, L. Gao, Z.J. Huang, J.
     Bechtold,
     K. Forster, and C.W. Chu, Phys. Rev. Lett. $\underline{58}$ 1891 (1987).

12) J.R. Thompson, D.K. Christen, S.T. Sekula, B.C. Sales, and L.A.
     Boatner Phys. Rev. Lett. (to be published).

13) L.C. Bourne, M.F. Crommie, A. Zettl, Hans-Conrad zur Loye, S.W.
     Keller, K.L. Leary, Angelica M. Stacy, K.J. Chang, Marvin L.
     Cohen, and Donald E. Morris, Phys. Rev. Lett. $\underline{58}$, 2337 (1987).

14) D. Datlogg, R.J. Cava, A. Jayaraman, R.B. van Dover, G.A.
     Kourouklis, S. Sunshine, D.W. Murphy, L.W. Rupp, H.S. Chen, A.
     White, K.T. Short, A.M. Mujsce and E.A. Rietman, Phys. Rev.
     Lett. $\underline{58}$, 2333 (1987).

15) T.C. Collins, M. Seel, J.J. Ladik, M. Chandrasekhar and H.R.
     Chardraschkar, Phys. Rev. B $\underline{27}$ 740 (1983).

16) A.B. Kunz and R.S. Werdman, J. Phys. C. $\underline{12}$, L 371 (1979).

17) A.B. Kunz, R.S. Weidman, and T.C. Collins, J. Phys C $\underline{14}$, L 581
     (1981).

18) M. Cardona, Phys. Rev. $\underline{129}$, 69 (1963).

19) A. Goldman, Phys. Status Solidi B $\underline{81}$, 9 (1977).

20) A.A. Abrikosov, 7h. E ksp. Teor Fiz. Pis'ma Red. $\underline{27}$ 235 (1978). [JETP Lett. $\underline{27}$, 219 (1978)].
21) J.T. Devreese, A.B. Kunz and T.C. Collins, Solid State Commun. $\underline{11}$, 673 (1972).
22) P.J. Cote, G.P. Capsimalis, ad C.G. Homan, Appl. Phys. Rev. Lett. $\underline{38}$, 927 (1981).
23) J.S. Tasi, Y. Kubo, and J. Tabuchi, Phys. Rev. Lett. $\underline{58}$, 1979 (1987).

EXCITONIC THEORY OF HIGH TEMPERATURE OXIDE SUPERCONDUCTORS

AND THE LACK OF AN ISOTOPE SHIFT

C.F. Gallo, L.R. Whitney and P.J. Walsh*

3M Center   St. Paul MN 55144
*Fairleigh-Dickinson Univ. Teaneck, NJ 07666

ABSTRACT

The high temperature superconductivity[1] observed in $Y_1Ba_2Cu_3O_7$[2] is analyzed in terms of the "phonon + exciton" theory of Allender, Bray and Bardeen[3] (ABB) applied to an atomically microscopic interfacial model of this anisotropic layered material. An approximate equation is presented that is conceptually transparent and allows one to visualize the relative interplay between the exciton and phonon contributions. Recent data[4,5] on the lack of an isotope shift on the transition temperature of $Y_1Ba_2Cu_3O_7$ is analyzed. The equation gives conditions for a lack of an isotope shift when the exciton mechanism dominates over the phonon mechanism. From ABB theory, numerical calculations are presented that demonstrate the plausibility of a dominant exciton mechanism in $Y_1Ba_2Cu_3O_7$. It will be very interesting to perform the isotope experiment on other oxide superconductors to further explore the possibility of excitonic superconductivity indicated by our calculations. Superconducting transition temperatures near room temperature appear attainable at reasonable carrier densities and within ABB's excitonic theory and our data extrapolations.

INTRODUCTION

In a recent publication[6], we presented literature data[7] on the superconducting transition temperature (T_c) as a function of free carrier density for known oxide superconductors. To this graph (Figure 1) we added data on the latest high T_c superconductors [doped La_2CuO_4 and $Y_1Ba_2Cu_3O_7$]

where the carrier concentrations were calculated from mixed valence
concepts and confirmed by AC Hall effect measurements[8]. We analyzed the
data in terms of the "phonon + exciton" superconductor theory of Allender,
Bray and Bardeen[3] and concluded that all these oxides could <u>not</u> be treated
as a coherent class with a <u>phonon mechanism alone,</u> and that an additional
mechanism appeared necessary to explain the high T_c oxides. We
quantitatively showed that the excitonic mechanism may supply the binding
for the electron-electron Cooper pairs necessary to achieve high
superconducting transition temperatures.

Since that time, two intriguing publications[4,5] have appeared in which
O^{18} was substituted for O^{16} in $Y_1Ba_2Cu_3O_7$ and <u>no</u> corresponding change in
the transition temperatures was observed although Raman measurements show
the expected phonon frequency shift. The results will be discussed in the
context of ABB's excitonic theory.

EXCITONIC THEORY AND ISOTOPE EQUATIONS

ABB[3] treat superconductivity at a metal-semiconductor interface where
the "electron-electron Cooper pairs" can be bound through a phonon and/or
an exciton. We point out that the latest high T_c oxide superconductors are
sandwich structures[1] with alternating layers of metal-semiconductor-metal-
semiconductor, etc. which we propose can be described by the theory of ABB,
but now on an atomic scale with carriers penetrating the semiconducting
layers from the metallic layers on both sides. From ABB (Eq. 4.10) we have

$$T_c \simeq (0.7)\,\theta_0\,\exp\left(-\,1/g_{eff}\right) \tag{1}$$

where θ_0 is related to the Debye temperature and given by

$$\theta_0 = \omega_{po}/k_B$$

where k_B is the Boltzmann constant, and ω_{po} is the maximum phonon frequency.

Here

$$g_{eff} = \lambda_{ph}^* + \frac{\left(\lambda_{ex}^* - \mu'\right)}{1 - \left(\lambda_{ex}^* - \mu'\right)\ln\left(\omega_g/\omega_{po}\right)} \tag{2}$$

where (3)

$$\lambda_{ph}^{*} = \frac{\lambda_{ph}}{1 + \lambda_{ph}}$$

and λ_{ph} is the electron-phonon coupling constant, and

$$\lambda_{ex}^{*} = \frac{\lambda_{ex}}{1 + \lambda_{ex}} \qquad (4)$$

and λ_{ex} is the electron-exciton coupling constant. In addition

$$\mu' = \frac{\mu}{1 + \mu \ln(\omega_F/\omega_g)} \qquad (5)$$

where μ is the density of states times an average of the screened coulomb interaction, ω_g is the average semiconductor energy gap, and ω_F is the Fermi energy given by

$$\omega_F = \left(\hbar^2/2m^*\right)\left(3\pi^2 N\right)^{2/3} \qquad (6)$$

for a simple parabolic band. Note that N is the carrier density, $\hbar$ is Plancks constant divided by 2π, and m^* is the effective electron mass. We note that Eqs. 3, 4, and 5 represent renormalized interaction parameters that make Eq. 1 reasonably conservative for quantitative estimates.

In order to make these equations more transparent for our subsequent discussion, we introduce the approximation

$$\lambda_{ph}^{*} \ln(\omega_s/\omega_{po}) \ll 1 \qquad (7)$$

into Eqs. 1 and 2 to obtain

$$T_c \simeq (0.7)\,\theta_o \left(\frac{\theta_{ex}}{\theta_o}\right)^{\left(\frac{\lambda_{ex}^{*} - \mu'}{\lambda_{ph}^{*} + \lambda_{ex}^{*} - \mu'}\right)^2} exp\left(-\frac{1}{\lambda_{ph}^{*} + \lambda_{ex}^{*} - \mu'}\right) \qquad (8)$$

We believe this expression will be very useful in understanding super-conductors in which both phonon and exciton mechanisms are active. This expression has the reasonable limits and seems to yield realistic quantitative estimates. Also note that we have introduced

$$\theta_{ex} = \omega_g/k_B$$

as the "excitonic temperature" which is a more accurate description of the exciton formation energy than the bandgap (ω_g).

Equation 8 is obtained from Eq. 1 which was renormalized twice by ABB for maximum accuracy. The Russians[9] have derived a similar expression to

Eq. 8 that also has desirable conceptual transparency, but seems to yield
unrealistically high quantitative estimates. But the Russian expression
has <u>not</u> been renormalized twice, and we think that our Eq. 8 is superior in
that regard.

In order to examine the isotope effect, we use the usual approximate
proportionality

$$\theta_o \propto M^{-1/2} \tag{9}$$

in Eq. 8 and note that all other terms in Eq. 8 are independent of isotopic
mass (M). Thus

$$T_c \propto M^{-\frac{1}{2}\left[1-\left(\frac{\lambda^*_{ex}-\mu'}{\lambda^*_{ph}+\lambda^*_{ex}-\mu'}\right)^2\right]} \tag{10}$$

In the absence of excitonic effects ($\lambda^*_{ex} = 0$), note that Eq. 10 reduces to

$$T_c \propto M^{-\frac{1}{2}\left[1-\left(\frac{\mu'}{\lambda^*_{ph}-\mu'}\right)^2\right]} \tag{11}$$

which is a known relationship that was used in references 4 and 5 to
analyze the data on the isotope effect, but restricted to the case of a
phonon mechanism alone.

Proceeding with the excitonic mechanism, from ABB (eq. 1.5), we take

$$\lambda_{ex} = b\, a\, \mu \left(\omega_p/\omega_g\right)^2 \tag{12}$$

where ω_p is the plasma frequency of the bound valence electrons in the
semiconductor, b is the fraction of time the free carriers spend in the
semiconductor, and a is a reduction factor of order 1/3 - 1/5 which
includes the dielectric screening factor. In favorable cases at
metal-semiconductor interfaces described by ABB, they suggest values

$$\omega_p \simeq 10 \text{ ev}$$
$$\omega_g \simeq 2 \text{ ev}$$
$$q \simeq 1/3 - 1/5$$
$$\mu \simeq 1/3 - 1/2$$
$$b \simeq .2$$

which yields values of

$$\lambda_{ex} \simeq 0.3 - 0.8 \quad . \tag{13}$$

Our estimates for $Y_1Ba_2Cu_3O_7$ indicate that higher values are reasonable.

$$\lambda_{ex} \simeq 1.0 - 1.5 \quad . \tag{14}$$

We propose the dependence of λ_{ex} on free carrier density (N) can be considered more explicitly as it enters through both b and μ as follows

$$b \propto N^{1/3}$$
$$\mu \propto N^{-1/9}$$

which finally yields

$$\lambda_{ex} \equiv (\delta N)^{.22} \quad . \tag{15}$$

ABB considered a single metal-semiconductor interface where free carriers from the metal penetrate into the semiconductor and this affects the factor b. For the <u>layered</u> ceramic oxides with high T_c, note that b is higher than considered by ABB because free carriers tunnel into the semiconductor layer from both adjacent metallic (Cu-O) layers. In fact, Bardeen has recently suggested that all the relevant phenomena (phonon + exciton) occur within the Cu-O layers and we will discuss this in more detail later.

ISOTOPE EXPERIMENTS AND ANALYSIS

In references 4 and 5, the superconducting isotope effect was studied by substituting O^{18} for O^{16} in $Y_1Ba_2Cu_3O_7$ and <u>no</u> corresponding change in the transition temperature was observed although Raman measurements show the expected phonon frequency shift. In these publications, the results were analyzed in terms of Eq. 11

$$T_c \propto M^{-\frac{1}{2}\left[1 - \left(\frac{\mu'}{\lambda^*_{ph} - \mu'}\right)^2\right]} \tag{11}$$

Here the isotope effect is completed reduced under the following three possible sets of conditions:

(1) $\lambda^*_{ph} = 0$. This implies the phonon mechanism is not active and that some other mechanism must be invoked to explain the superconductivity.

(2) $\mu' \gg \lambda_{ph}^*$. As analyzed in references 4 and 5, this situation is untenable within the known framework of phonon mediated superconductivity. In addition, we point out that this situation is inconsistent with the attainment of high temperature superconductivity which will tend to be largest when $(\lambda_{ph}^* - \mu')$ is large.

(3) $\lambda_{ph}^* = 2\mu'$. As analyzed in references 4 and 5, this seems inconsistent with the known properties of $Y_1Ba_2Cu_3O_7$ and is not consistent with high temperature superconductivity.

We are in agreement with the above analyses which leads to two possible generic conclusions.

(1) Either there is some unusual and/or unknown phonon mediated superconducting mechanism involved in $Y_1Ba_2Cu_3O_7$, or

(2) there is some other non-phonon mechanism involved in the high T_c $Y_1Ba_2Cu_3O_7$ superconductor, presumably an electronic mechanism that is not dependent upon isotopic mass. This possibility is mentioned in references 4 and 5 and we shall pursue this viewpoint further as an extension of our previous presentation.

ISOTOPE ANALYSIS WITHIN THE EXCITONIC THEORY

To analyze the isotope experiments within ABB's "phonon + exciton" theory, we rewrite our approximate Eq. 10 here.

$$T_c \propto M^{-\frac{1}{2}\left[1-\left(\frac{\lambda_{ex}^* - \mu'}{\lambda_{ph}^* + \lambda_{ex}^* - \mu'}\right)^2\right]} \tag{10}$$

It is clear that minimal isotopic effect can be obtained when the excitonic mechanism is dominant

$$\tag{16}$$

to yield

$$\lambda_{ex}^* \gg \lambda_{ph}^*$$

$$T_c \propto M^{-\frac{1}{2}\left[1-\left(\frac{\lambda_{ex}^* - \mu'}{\lambda_{ex}^* - \mu'}\right)^2\right]} \propto M^0 \tag{17}$$

In other words, the transition temperature is independent of isotopic mass. This means that an excitonic mediated mechanism may be primarily responsible for the high transition temperature superconductivity in $Y_1Ba_2Cu_3O_7$ rather than a phonon mechanism.

For quantitative analysis, we note that the published experimental results for the exponent in Eq. 10 is $0 \pm .03$. To estimate the maximum possible value of λ_{ph}^{*} consistent with the experiments, we set the exponent in Eq. 10 to the experimental tolerance limit of $-.03$ which yields quite low values of λ_{ph} . Although our Eqs. 8 and 10 are convenient for conceptual clarity, for quantitative analysis we return to ABB's Eq. 1 which is more accurate. The Debye temperature was taken approximately $600^{\circ}K$. Other parameters were estimated from ABB[3]. From our previous efforts, we found that large values of effective carrier mass ($m^{*}=3$) fit the data and was consistent with the polarizable ionic lattice of these materials. Further parameter searches to fit the data in <u>Figure 1</u> yielded the excellent results shown in <u>Figure 2</u> with the following reasonable parameter set for use with ABB Eqs. 1 and 2.

$$\omega_g = 0.16 \text{ ev}$$
$$m^{*} = 3$$
$$\mu = \tfrac{1}{3}$$
$$\lambda_{ex} = \left[(0.85)(10)^{-21} cm^3 \, N \right]^{.22}$$
$$\lambda_{ph} = 0.015$$

For these parameters, we find that λ_{ex} varies up to approximately 1.7 as shown in <u>Figure 3</u>. Values in this range are quite consistent with our calculations from Eq. 12 with our estimate of parameters appropriate for the Y_{123} compound. Realize that this is only a typical parameter set for Y_{123} that also describes the other oxide superconductors as exciton dominant. In reality, we expect that the lower temperature oxides with lower carrier densities will have higher proportionate phonon contributions[6]. Further isotope experiments will help separate the relative contributions of excitons and phonons for the other oxide superconductors.

EXCITONIC TRANSITIONS

The energy gap parameter (ω_g) deserves more discussion. Note that it is really the formation energies of the excitons ($k_B \Theta_{ex}$) which is more relevant than the gap, and there are many possible electronic excitations[10,11]. For example, published energy level calculations[10] in $Y_1Ba_2Cu_3O_7$ show that this high temperature superconductor is metallic along the Cu-O-Cu-O directions while simultaneously possessing highly anisotropic energy gaps that vary from zero to 1.8 ev in other directions. The formation energy $k_B \theta_{ex}$ for virtual electronic excitations (which act to attract electrons to form the superconducting electron pairs) is a complex quantity related to an "average gap energy" taken over the complicated Brillouin zone of these highly anisotropic materials. Note that the "average" will be preferentially weighted to the low energy excitons. Our value of $\omega_g \simeq .2ev$ derived from our curve fitting may act as a guide in identifying the relevant excitons and evaluating energy level calculations.

It appears that a key factor in the high temperature oxide superconductors is the presence of an anisotropic set of energy gaps along with a partially filled conduction band. This situation provides low energy valence excitations by mobile electrons as the excitonic mechanism for high temperature superconductivity.

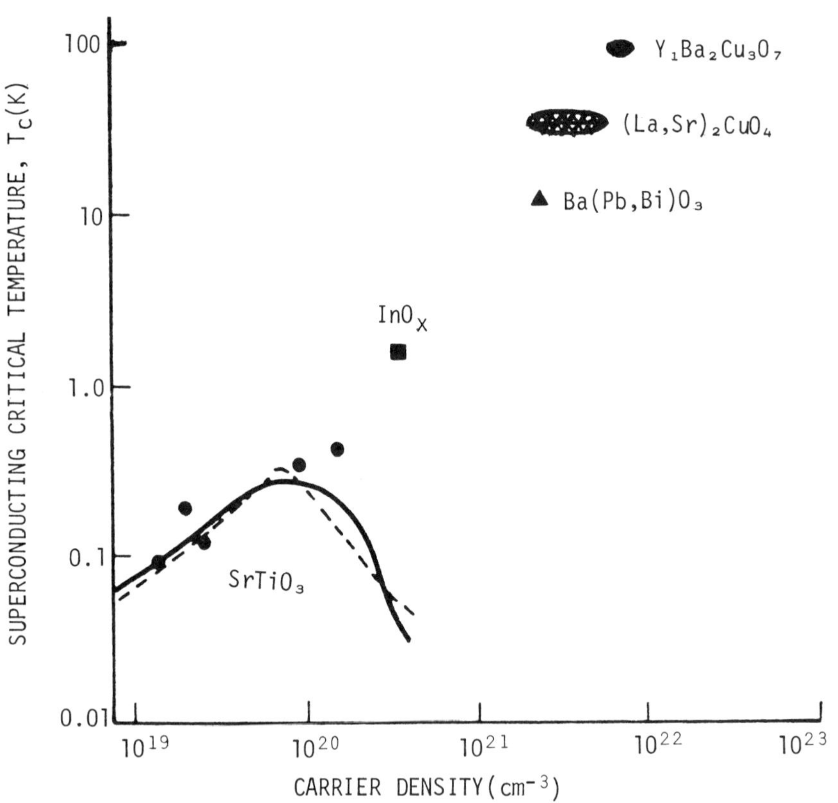

Figure 1: Data on superconducting transition temperature (T_c) vs carrier concentration (N) for the oxide superconductors.

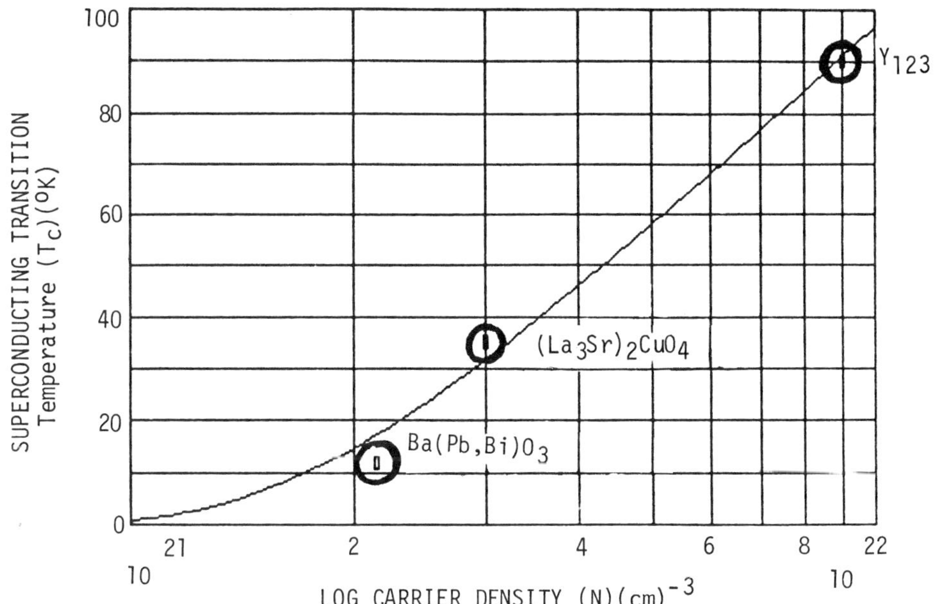

Figure 2: Comparison of excitonic theory and data on transition temperature (T_c) vs carrier concentration (N) for the oxide superconductors.

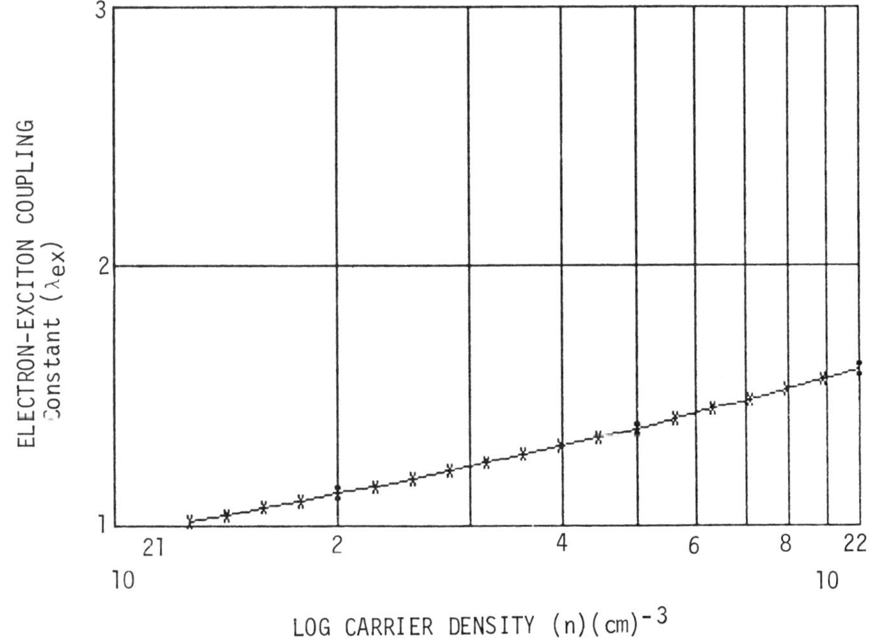

Figure 3: Theoretical values of electron-exciton coupling parameter (λex) vs carrier concentration.

SUMMARY AND CONCLUSIONS

From ABB's[3] "phonon + exciton" theory of superconductivity, new
approximate expressions have been presented which display (1) the relative
interplay of the exciton and phonon mechanisms, and (2) the isotope effect
in terms of "exciton + phonon" mechanisms. Recent data[4,5] on the lack of
an isotope shift in $Y_1Ba_2Cu_3O_7$ is analyzed in terms of an exciton dominant
contribution to the high temperature superconductivity using quantitative
estimates from ABB theory. These high temperature oxide semiconductors are
characterized by: (1) high carrier densities, (2) high carrier effective
masses, (3) anisotropy, and (4) small excitonic excitation energies. It
will be very interesting to perform the isotope experiment on other oxide
superconductors to further explore the possibility of excitonic
superconductivity indicated by our calculations. Superconducting
transition temperatures near room temperature appear attainable at
reasonable carrier densities and within existing excitonic theories and
data extrapolations.

ACKNOWLEDGEMENT

We are delighted to acknowledge many years of dicussions and counsel
from Prof. J. Bardeen (Univ. of IL) on high temperature superconductivity.

REFERENCES

1. For a review of the effort on high temperature superconductors, see A.
    Khurana, Physics Today 40, 17 (1987).
2. M. Wu, et al, Phys. Rev. Lett. 58, 908 (1987).
    S. Hikami, et al, Japan J. Appl. Phys. 26, L314 (1987).
    L. Bourne, et al, Phys. Lett. 120, 494 (1987).
3. D. Allender, J. Bray and J. Bardeen, Phys. Rev. B7, 1020 (1973).
4. L. Bourne, et al, Phys. Rev. Lett. 58, 2337 (1987).
5. B. Batlogg, et al, Phys. Rev. Lett. 58, 2333 (1987).
6. C. Gallo, L. Whitney and P. Walsh, "Excitonic Theory of Mixed Valent
    Doped Superconducting Semiconductors", presented at Material Res.
    Soc. Meeting on April 24-25, 1987 and to be published in
    proceedings.
7. M. Beasley and T. Geballe, Phys. Today 37, 60 (1984).
8. AC Hall Effect Measurements performed in collabortion with K. V. Rao
    of Swedish Royal Institute.
9. V. Ginzburg and D. Kirzhnits, "High Temperature Superconductivity"
    (Consultant Bureau, New York, 1977) (Translated in 1982), Pgs 283-
    289.
10. J. Yu, S. Massidda, A. Freeman and D. Koelling, Phys. Rev. Lett.,
    submitted (1987).
11. W. Little, J. Collman and J. McDevitt, "Role of the Exciton
    Interaction in the New High T_c Superconductors", presented at
    Material Res. Soc. Meeting on April 24-25, 1987 and to be published
    in proceedings.

SUPEREXCHANGE MEDIATED SUPERCONDUCTIVITY IN THE SINGLE BAND
HUBBARD MODEL

S. Doniach[a], P. J. Hirschfeld[a,b], M. Inui[c], and A. E. Ruckenstein[b]

Dept of Applied Physics, Stanford University, Stanford CA 94305[a]
Dept of Physics, University of California at San Diego, La Jolla CA 92093[b]
Dept of Physics, Stanford University, Stanford CA 94305[c]

Abstract We discuss the physics of BCS pairing in the strongly correlated single
band Hubbard model close to half-filling. In contrast to spin-fluctuation mediated pairing of importance in ^3He and heavy Fermion systems, we argue that the superexchange
mediated pairing proposed by Anderson is stabilized by short range antiferromagnetic
spin ordering. A mean field calculation shows that coexistence of antiferromagnetism
and superconductivity is a stable solution for a doped Mott-Hubbard system.

The recent discovery of high T_c superconductivity in the layered copper perovskites
[1] has motivated a number of theoretical ideas attempting to explain the origin of the
pairing interaction in these systems [2]. Although some researchers argue that variants
of the conventional electron-phonon mechanism can account for the high T_c in these
materials, most of the theoretical discussions have appealed to electron-electron correlations, ultimately originating from Coulomb repulsion as the main pairing mechanism.
Our work was motivated in part by Anderson's suggestion [3] that superconductivity
occurs as a result of spin correlations induced by the superexchange interaction between
spins on nearest neighbor Cu sites.

The discovery of antiferromagnetism in nearly stochiometric La_{2-x} Cu O_{4-y} raises
the question of the interplay between superconductivity and magnetism in the high
T_c superconductors [4]. We have focused on the nature of this antiferromagnetic state
using a mean field approach to describe the Neel ground state. This is in contrast
to the work of Anderson and collaborators who argued in terms of a ground state for
stochiometric composition (i.e., for half filling in the language of the Hubbard model)
as a paramagnetic resonant valence bond (RVB) fluid with a Hubbard gap for charge
excitations. Subsequent mean field theories of RVB like pairing [5,6] concentrated on
the possibility of superconductivity without discussing the precise nature of the ground
state at half filling.

Our approach has been formulated [7] in the context of the single band Hubbard
model, defined by the Hamiltonian:

$$H = -t \sum_{<i,j>\sigma} c_{i\sigma}^{\dagger} c_{j\sigma} + U \sum_i n_{i\sigma} n_{i-\sigma} \tag{1}$$

Operators $c_{i\sigma}$ ($c_{i\sigma}^\dagger$) denote hybridized Cu-$3d_{x^2-y^2}$ and O-$2p_{xy}$ "quasi-particle" orbitals [8], localized at the Cu sites. The properties of the model are characterized by just two dimensionless parameters: t/U and the filling of the band, $\delta = 1 - n$, where n is the average numbers of electrons per Cu site. The on site Coulomb repulsion, U, and the hopping amplitude, t, are in principle determined from the knowledge of the "quasi-particle" wave functions.

The Hamiltonian (1) is clearly a highly simplified model for the Cu-oxide superconductors. Its use requires justification in terms of more realistic models of charge transfer and Coulomb energies in the transition metal oxides [9].

Theoretical estimates [6] as well as photoemission [10] and neutron scattering [4] experiments, suggest strong onsite correlations at the Cu sites. We limit ourselves to the strong correlation limit, $t/U \ll 1$, and make use of a canonical transformation [11] to eliminate doubly occupied sites to lowest order in t/U, resulting in the effective Hamiltonian

$$H = H_{single} + H_{superexchange} + H_{pair-hopping}$$

$$H_{single} = -t \sum_{<ij>\sigma} f_{i\sigma}^\dagger f_{j\sigma} b_i b_j^\dagger + \sum_i \lambda_i (b_i^\dagger b_i + \sum_\sigma n_{i\sigma} - 1) + \mu \sum_{i\sigma} n_{i\sigma}$$

$$H_{superexchange} = \frac{t^2}{U} \sum_{<ij>} (\vec{\sigma}_i \cdot \vec{\sigma}_j - \sum_\sigma n_{i\sigma} n_{j\sigma}) \tag{2}$$

$$H_{pair-hopping} = \frac{2t^2}{U} \sum_{<ijl>\sigma} b_i (f_{i\sigma}^\dagger f_{j-\sigma}^\dagger f_{j-\sigma}^\dagger f_{l\sigma} - f_{i\sigma}^\dagger f_{j\sigma} f_{j-\sigma}^\dagger f_{l-\sigma}) b_l^\dagger$$

Here, $f_{i\sigma}^\dagger$ ($f_{i\sigma}$) create (annihilate) chargeless spin-1/2 Fermions which are related to the original electron operators through $c_{i\sigma} = b_i^\dagger f_{i\sigma}$. The "slave Boson" operators, $b_i^\dagger$ (b_i) carry the charge degrees of freedom and satisfy the no double occupancy constraint $b_i^\dagger b_i + \sum_\sigma n_{i\sigma} = 1$, enforced at each site on the lattice by the Lagrange multipliers, λ_i [6]. In (2) $< ij >$ and $< ijl >$ imply, respectively, unrestricted sums over the nearest and next to nearest neighbors (of every site, i); and σ_i are spin operators; $\sigma_i^\alpha = \sum_{\mu\nu} f_{i\mu}^\dagger \sigma_{\mu\nu}^\alpha f_{i\nu}$.

It is easy to see that both $H_{superexchange}$ and $H_{pair-hopping}$ in (2) can be rearranged to appear as an attractive interaction between pairs of electrons, and that this raises the possibility of superconductivity mediated by virtual (high frequency) charge fluctuations, involving empty and doubly occupied nearest neighbor sites.

It seems rather paradoxical that an attractive pairing interaction can be derived from the purely repulsive Hubbard Hamiltonian. In fact, in the low density limit the above argument must break down as can be seen by simply considering the scattering of two electrons in the original Hubbard model. The resulting pseudopotential [12]

$$V_{eff} = \frac{U}{1 + KU} \tag{3}$$

(where K represents a two particle operator) is always repulsive, so that no pairing can occur for sufficiently small electron densities.

396

With increasing band filling, pairing does occur, at least in principle, through the exchange of spin fluctuations. The latter mechanism is, of course, well established as the origin of superfluidity in ^3He [13]. In the case of charged particles the pair breaking effect of the direct Coulomb repulsion must be treated on equal footing with the pairing mechanism, and s-wave superconductivity may occur only at extremely low temperatures [14], if at all. What remains is the possibility of pairing in higher angular momentum states for which the Coulomb repulsion is substantially decreased.

The physics of the large U Hubbard model close to half filling is qualitatively different. The system obtained by doping the $\delta = 0$ antiferromagnetic state may be described in terms of a small concentration, $\delta = 1 - n$, of holes. In analogy with the motion of vacancies in solid ^3He [15], the motion of a hole in an almost half filled Hubbard band leads to spin exchanges which, in turn, have a strong effect on the spin correlations of the system. The interaction between two holes then involves the solution of a many-body problem, in contrast with Eq.(3). An extreme limit of this interaction may be obtained by ignoring all but the Ising part of the interaction terms in (2). In this case, the tunneling of a pair of holes away from one another leads to a chain of reversed spins [16], the energy of which grows linearly with the distance between holes. The resulting pairing mechanism is thus analogous to the confinement of quarks in quantum chromodynamics.

Of course this picture will not be expected to survive the effects of quantum fluctuations (through the transverse part of the superexchange coupling), and in particular single holes can be expected to tunnel independently through the Gutzwiller narrow band mechanism. Thus we do not expect true confinement in the full quantum problem.

In the mean field BCS-like approximation, the hole pairing is then the result of a balance between the single particle kinetic energy which is considerably reduced relative to that in the low density limit, and the superexchange induced pairing. It appears physically plausible that this will be a reasonable approximation for small δ where the holes are still largely confined (i.e., for $t\delta \leq t^2/U$) but will start to break down (and perhaps be replaced by spin-fluctuation induced pairing) as δ becomes $\geq t/U$.

In order to get a more quantitative picture of (i) the possibility of superexchange mediated superconductivity and (ii) the interplay between superconductivity and antiferromagnetism we have studied a generalized Hartree-Fock mean field theory of Hamiltonian (2). In addition, the motion of holes was modeled by the saddle point approximation for the auxiliary boson, b_i, well known in the context of heavy Fermion systems [17]. In this approximation $b_i^\dagger b_i$ was replaced by $< b^2 >= 1- < n >= \delta$ (the hole concentration), so that the single hole tunelling matrix element in H_{single} became $t\delta$. In the paramagnetic state the hole pairing results whenever the superexchange induced pairing interaction is large compared to the renormalized band width, $8t\delta$. The critical temperature is finite at half filling, reflecting the crude treatment of the double occupancy constraint. Once long range antiferromagnetic order is allowed for, we find that the superconducting order parameter exactly vanishes at half filling; while away from $\delta = 0$ superconductivity coexists with antiferromagnetism, with $T_c/T_N \approx \sqrt{\delta}$. The details of these calculations, and further interpretation of the results can be found in Refs.[6,7].

We believe that the observed disappearance of magnetism at 3% doping [4], to be contrasted with the approximately 30% found in our mean field theory, is in part due to spin frustration effects induced by hole motion and accompanying spin exchanges. However, hole pairing should survive even in the absence of AF long range order, provided the spin coherence length is sufficiently large. We note that the physics of the pairing mechanism is substantially different from that in liquid ^3He: in the latter the pairing is due to exchange of spin fluctuations, whereas the pairing of holes in the large U Hubbard model at low doping is due to (localized) virtual charge excitations, stabilized by the strong antiferromagnetic spin correlations of the underlying lattice. Moreover, the latter mechanism implies a rather short superconducting coherence length and it thus involves states far from the Fermi surface.

The picture given above should lead to observable effects. In particular, the coexistence of antiferromagnetism and superconductivity in the mean field calculation is likely to indicate that, even in the absence of long range AF order, short range antiferromagnetic correlations should be present in the superconducting state. Depending on the detailed nature of the excitation spectrum, the magnetic fluctuations should strongly influence spin-lattice relaxation times, and, of course, should be directly observable in inelastic neutron scattering experiments.

Finally, we note that the superexchange-induced pairing mechanism should be very sensitive to the Coulomb repulsion. The nature of retardation effects in a narrow band Hubbard model is not clear at this time. Thus it is not clear whether a finite frequency scale for charge fluctuations (excitons) [18] could come into play to reduce the depairing effects of the direct Coulomb repulsion, or whether a spreading out of the pair wave function would in itself be sufficient to do this.

References

[1] J.G. Bednorz, and K.A. Muller, Z.Phys. **B64**, 189 (1986).

[2] see recent review by T.M. Rice, Preprint

[3] P.W. Anderson, Science **235**, 1196 (1987).

[4] D. Vaknin, S.K. Sinha, D.E. Moncton, D.C. Johnston, J. Newsam, and H. King, preprint; S. Mitsuda, G. Shirani, S.K. Sinha, D.C. Johnston, M.S. Alvarez, D. Vaknin, and D.E. Monkton, preprint; T. Freltoft, J.P. Remeika, D.E. Moncton, A.S. Cooper, J.E. Fischer, D. Harshman, S.K. Sinha, and D. Vaknin, preprint; D.C. Johnston, J.P. Stokes, D.P. Goshorn, and J.T. Lewandowski, preprint.

[5] G. Baskaran, Z. Zou, and P.W. Anderson, to be published in Solid State Comm.

[6] A.E. Ruckenstein, P.J. Hirschfeld, and J.Appel, Phys.Rev.B July 1, 1987.

[7] M. Inui, S. Doniach, P.J. Hirschfeld, and A.E. Ruckenstein, submitted to Phys.Rev. Lett.

[8] P.W. Anderson, Phys.Rev. **115**, 2 (1959).

[9] J. Zaanen, G.A. Sawatzky, and J.W. Allen, Phys.Rev.Lett. **55**, 418 (1985).

[10] J.A. Yarmoff, D.R. Clarke, W. Drube, U.O. Karlsson, A. Taleb- Ibrahimi, and F.J. Himpsl, Phys.Rev.B, Rapid Communications, to be published.

[11] J.E. Hirsch, Phys.Rev.Lett. **54**, 1317 (1985); C. Gros, R. Joynt, and T.M. Rice to be published.

[12] J.Kanamori, J. Theoret. Physics (Kyoto) **30**, 275 (1963).

[13] W.F. Brinkman, J. Serene, and P.W. Anderson, Phys.Rev.A **10**, 2386 (1974).

[14] Grabowski and L.J.Sham, these proceedings.

[15] Y.Nagaoka Phys.Rev. **147**, 392 (1966).

[16] J.E. Hirsch, preprint; B. Shraiman, and E.D. Siggia, private communication; A.E. Ruckenstein, unpublished.

[17] G. Kotliar, and A.E. Ruckenstein, Phys.Rev.Lett. **57**, 1362 (1986).

[18] C.M. Varma, S. Scmitt-Rink, and E. Abrahams, see these proceedings.

ON SPIN-DENSITY WAVE STATE IN $(La_{1-x}M_x)_2CuO_{4-\delta}$

Yasumasa Hasegawa and Hidetoshi Fukuyama

Institute for Solid State Physics
University of Tokyo
7-22-1, Roppongi, Minato-ku, Tokyo 106, Japan

ABSTRACT

The Fermi surface instability in quasi-two-dimensional electrons is
investigated as a possible model of oxides superconductors, $(La_{1-x}M_x)_2CuO_{4-\delta}$. The spin density wave (SDW) transition temperature within the
mean field apploximation is obtained as a function of band filling.   It
is found that SDW is stable only in a small region around the half
filling of the band.

INTRODUCTION

Since the high T_c superconductivity was discovered, [1] the semiconducting state in $(La_{1-x}M_x)_2CuO_{4-\delta}$, (M=Sr, Ba or Ca) for small x has
attracted much interest [2-9]. The characteristics of the semiconducting
state will be closely related to the mechanism of the high T_c superconductivity.   It has been clear by susceptibility measurement [2-6],
neutron diffraction [7,8], and NMR [9] that the ground state for very
small x is magnetically ordered state.   The transition temperature, T_s,
for x=0 is about 230K, but it is sensitive to the oxygen deficiency, δ.
The observed structural transition [10], tetragonal to orthorhombic at
T 530K for x=0, does not affect the electrical property.

The Fermi surface instability against SDW has been pointed out by
the present authors [11] based on the model of two dimensional square
lattice with half-filled band, which results in the perfect nesting of
the Fermi surface.   In the mean time several band structure calculations [12-14] have been reported, which demonstrate the existence of the
almost square-like Fermi surface. The Fermi surface instability has also
been discussed by Jorgensen et al. [15], but they proposed the charge
density wave with the lattice distortion, which has been shown not to
lead to a gap at the Fermi surface [12,14]. We calculate the x dependence of T_s including three dimensionality [16] and the transfer between the
next nearest sites.

MODEL

We investigate the model in the body centered tetragonal lattice with the transfer integrals between the nearest neighbors (t) and next nearest neighbors (t_2) in the basel plane and that between layers (t'). Then the energy band is given by

$$\varepsilon(k) = -2t(\cos k_x a + \cos k_y a) - 4t_2 \cos k_x a \cos k_y a$$

$$- 8t' \cos k_z a/2 \cos k_x a/2 \cos k_y a/2 ,\qquad\qquad (1)$$

where a and c are the sizes of the unit cell in the basal plane, and along the c-axis, respectively. In the rigid band approximation the electron density is given by $n=1-2x+2\delta$. In the following we assume $\delta=0$ and calculate T_S as a function of x.

SDW TRANSITION TEMPERATURE

The SDW transition temperature in the mean field approximations is given by $1=U\chi(Q,T_S)$, where U is the electron-electron repulsive interaction and $\chi(Q,T_S)$ is the correlation function with the wave vector Q.

First, we investigate the case $t_2=t'=0$. The result is shown in Fig.1, where we take $V/4t=0.214$, 0.266, 0.494 so that T_{S0}, which is the maximum of T_S for varying x, is given as $T_{S0}/4t=0.005$, 0.01 and 0.05, respectively. Since 4t is estimated to be 1.6eV according to band structure calculations, $T_{S0}/4t=0.01$ is thought to correspond to the case in $(La_{1-x}M_x)_2CuO_4$. The SDW is seen to be stable only in the region of very small x. We have also seen that this feature is not changed even if Q is varied to yield the maximum $\chi(Q,T)$ for given x (chain and broken lines in Fig.1.) The z component of Q, Q_z, is not determined in the two-dimensional model.

Next, the effect of the interlayer transfer integral is investigated. Then Q_z should be determined to give the maximum $\chi(Q,T)$. We found that $Q=(\pi/a, \pi/a, 0)$ gives the maximum $\chi(Q,T)$ at least for x=0. The x dependences of T_S is calculated for this fixed value of Q as shown in Fig.2, where U is taken to be $U/4t=0.315$. Since variation of Q_z as well as Q_x and Q_y will not affect the phase boundary drastically, we conclude that the SDW state is also confined in the region of small x even in the presence of relatively large three dimensionality.

Lastly the effect of the transfer integrals between next nearest neighbors in the basal plane is investigated. For a choice of $t_2/t'=-0.1$ and t'=0 the x dependence of the density of state is ploted as a broken line in Fig.3. The van Hove singularity is located x=0.041. The x dependence of T_S is also shown as a solid line where Q is fixed to be $Q_x=Q_y=\pi/a$ and $U/4t=0.354$. It is seen that the maximum of T_S and the van Hove singularity are located at different x and no anomaly is present in T_S at x=0.041, where the density of states diverges. Note that for $0.025<x<0.041$ T_S is decreasing as x increases, whereas the density of states is increasing. This feature is due to the fact that T_S is mainly determined by the nesting condition of the Fermi surface but not the density of states only.

DISCUSSIONS

We have shown that SDW state can exist only in the small region of x in the mean field approximation. The similar result has also been discussed by Schulz [17] based on scaling theory. The three dimensinality is shown to have littel effect on the phase diagram. These features seem to be consistent with experiments in $(La_{1-x}M_x)_2CuO_{4-\delta}$. The maximum of T_s and van Hove singularity are shown to be located at different value of x when the transfer between next nearest sites are taken into account. The magnetic susceptibility measurements [4-6] indicate that the densiyt of states at the Fermi surface increases with x up to x~0.1. In our model this feature can be reproduced by taking $t_2<0$. In Fig.3 the density of states and T_s are shown for a choice of $t_2/t=$ -0.1. Although the value of t_2 should be larger to account for the increase of the density of states up to x=0.1, this choice of t_2 is already too large to stabilize SDW at x=0 as seen in the figure. This implies that the x dependences of both T_s and density of states cannot be explained consistently in terms of t_2 only. The observed increase of T_s by the oxygen deficiency [8] also seems to incompatible with the observed magnetic susceptibility in the rigid band approximation where oxygen deficiency corresponds to negative x. This will indicate that oxygen deficiency is not described by the rigid band approximation.

The similarity of the present system to $(TMTSF)_2X$ should be noticed, where the SDW transition temperature is also determined by the nesting condition which is controlled by pressure [18,19]. The superconductivity in the latter case, which is stabilized once the SDW is destroyed, is identified by NMR measurement [20] as anisotropic singlet [21]. There exists a difference, however, between these two superconductors in their phase diagrams plotted as temperature vs. the nesting condition (x or pressure). While superconductivity transition temperature, T_c, is maximum at the boundary with SDW in $(TMTSF)_2X$, T_c appears to be vanishingly low at the boundary waith SDW in the case of $(La_{1-x}M_x)_2CuO_4$. Althouogh the last feature many depend on sample quality such as oxygen deficiency and may not be intrinsic, it is possible that the types of these superconductivities can be different.

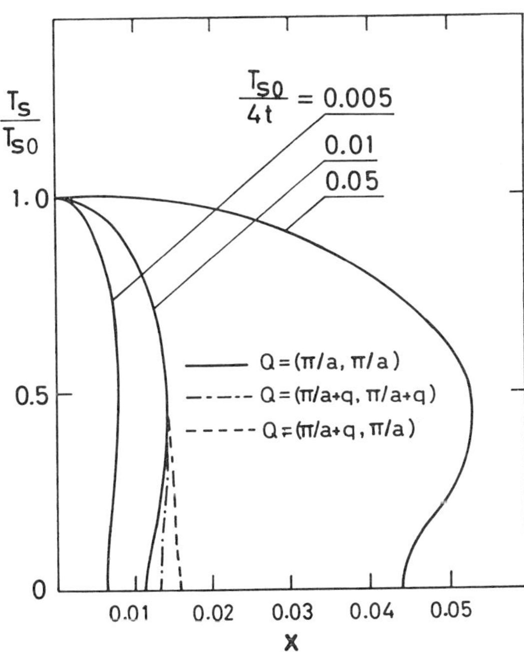

Fig. 1 SDW transition temperature as a function of x. The coupling constant is chosen to give $T_{S0}/4t$=0.005, 0.01 and 0.05 at x=0. The solid lines are the result with $Q=(\pi/a,\pi/a)$. The chain and broken lines are the envnelopes with $Q_0=(\pi/a+q,\pi/a+q)$ and $Q=(\pi/a+q,\pi/a)$ with varying q.

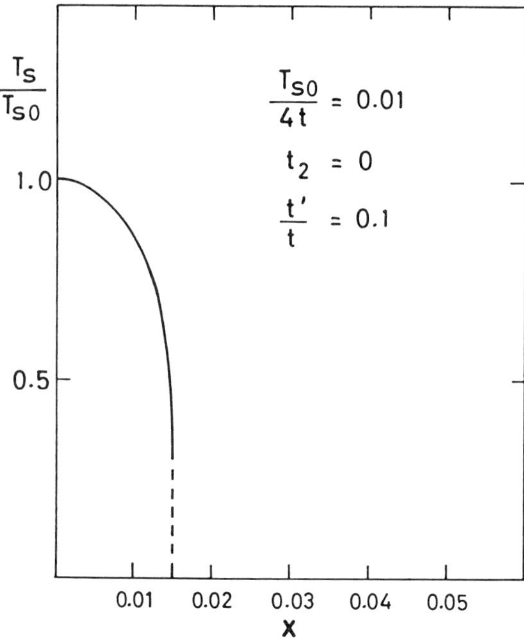

Fig. 2 SDW transition temperature as a function of x in the presence of the interlayer transfer integral, t'.

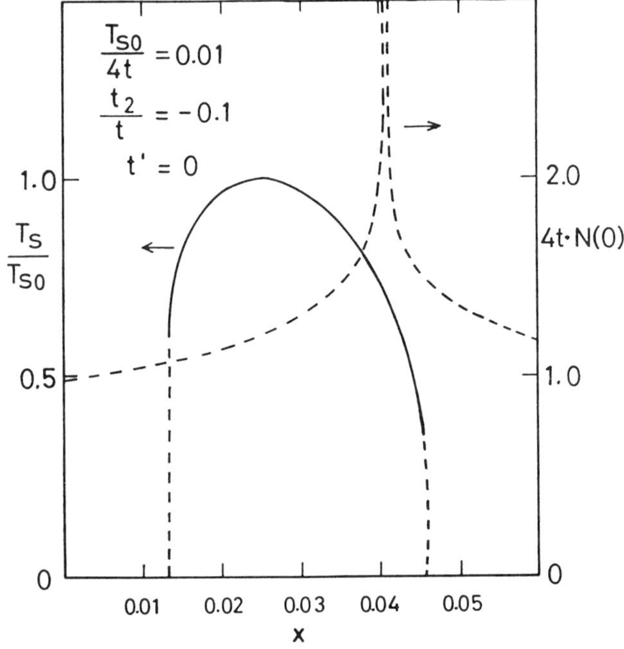

Fig. 3 SDW transition temperature (solid line) and the density of states at the Fermi surface (broken line) as a function of x in the presence of the transfer integral between the next nearest neighbors in the basal plane.

REFERENCES

1. T.G. Bednorz and K.A. Muller, Z. Phys. B64: 189 (1986).
2. T. Fujita, Y. Aoki, Y. Maeno, J. Sakurai, H. Fukuba and H. Fujii, Jpn. J. Appl. Phys. 26: L368 (1987).
3. S. Uchida, H. Takagi, H. Yanagisawa, K. Kishio, K. Kitazawa, K. Fueki and S. Tanaka, Jpn. J. Appl. Phys. 26: L445 (1987).
4. R.L. Greene, H. Maletta, T.S. Plaskett, J.G. Bednorz and K.A. Muller, to be published in Solid State Commun.
5. K. Fukuda, M. Sato, S. Shamoto, M. Onoda and S. Hosoya, preprint.
6. S. Uchida, S. Tajima, H. Takagi, K. Kishio, T. Hasegawa, K. Kitazawa, K. Fueki and S. Tanaka, to be published in Proc. of LT18.
7. D. Vaknin, S.K. Sinha, D.E. Moncton, D.C. Johnston, J. Newsaw, C.R. Safinya and H.E. King, preprint; B.X. Yang, S. Mitsuda, G. Shirame, Y. Yamaguchi, H. Yamauchi and Y. Syono, preprint.
8. T. Freltoft, J.P. Remeika, D.E. Moncton, A.S. Cooper, J.E. Fischer, D. Harshman, G. Shirane, S.K. Shinha and D. Vaknin, preprint.
9. Y.Kitaoka, S. Hiramatsu, T. Kohara, K. Asayama, K. Oh-ishi, M. Kikuchi and N. Kobayashi, Jpn. J. Appl. Phys. 26: L397 (1987) and Y. Kitaoka, private communications; H. Yasuoka, private communications.
10. J.M. Longo and P.M. Raccah, J. Solid State Commun. 6: 526 (1973).
11. H. Fukuyama and Y. Hasegawa, J. Phys. Soc. Jpn. 56: 1312 (1987).
12. L.F. Mattheiss, Phys. Rev. Lett. 58: 1028 (1987).
13. J. Yu, A.J. Freeman and J.H. Xu, Phys. Rev. Lett. 58: 1035 (1987).
14. K. Takagahara, H. Harima, A. Yanase, Jpn. J. Appl. Phys. 26: L352 (1987).
15. J.D. Jorgensen, H.B. Schutler, D.G. Hinks, D.W. Capone, K. Zhang and, M.B. Brodsky, Phys. Rev. Lett 58: 1024 (1987).
16. Y. Hasegawa and H. Fukuyama, Jpn. J. Appl. Phys. 26: L322 (1987).
17. H.J. Schulz, preprint.
18. K. Yamaji, J. Phys. Soc. Jpn. 51: 2787 (1982).
19. Y. Hasegawa and H. Fukuyama, J. Phys. Soc. Jpn. 55: 3978 (1986).
20. M. Takigawa, H. Yasuoka aond G. Saito, J. Phys. Soc. Jpn. 56: 873 (1987).
21. Y. Hasegawa and H. Fukuyama, J. Phys. Soc. Jpn. 56: 877 (1987).

CRITICAL TEMPERATURE OF SUPERCONDUCTIVITY CAUSED BY STRONG CORRELATIONS

Hidetoshi Fukuyama, Yasumasa Hasegawa and Kei Yosida[*]

Institute for Solid State Physics, University of Tokyo
7-22-1 Roppongi, Minato-ku, Tokyo 106, Japan
*Faculty of Science and Technology, Science University of
Tokyo, Noda, Chiba 278

ABSTRACT

Detailed numerical calculations have been performed of the critical
temperature of superconductivity caused by strong correlations based on
the mechanism of Anderson combined with the coherent-potential-approximation. The results confirm our former approximate estimate of the carrier
number dependence reported in Jpn. J. Appl. Phys. __26__ L371 (1987).

Recently the possible high T_C superconductivity in the presence of
strong correlation has attracted much attention.[1] It will conveniently
be described by the Hubbard model with strong Coulomb interactions and
its effective form, H_{eff}, retaining only two sites term is[2]

$$H_{eff} = K + J \, \Sigma' \, (\vec{s}_i \cdot \vec{s}_j - \frac{1}{4} n_i n_j) \quad , \tag{1}$$

where $J=4t^2/U$ is the exchange interaction and K represents the kinetic
energy projected onto the subspace without the double occupancy. By
treating the second term in the mean field approximation and by assuming
that the order parameter is proportional to $\gamma(k) \equiv \cos k_x a + \cos k_y a$ (we
confine ourselves to the case of two-dimensional square lattice) we see
that the critical temperature, T_C, is given by

$$1 = \frac{J}{N^2} \sum_{k,p} \gamma(k) \gamma(p) \pi(k,p;T_C) \quad , \tag{2}$$

where $\pi(k,p;T_C)$ is the particle-particle correlation function to be
evaluated with respect to K in eq.(1). In the treatment by Baskaran
et al.[3] and by Ruckenstein et al.[4] this $\pi(k,p;T_C)$ is evaluated for free
electrons but with reduced band width . On the other hand we employed
the coherent-potential-approximation (CPA) and eq.(2) reads as follows
for each electron density, n,[5]

$$1 = \frac{J}{2\pi t} \, \text{Im} \int_{-\infty}^{\infty} dx \, \tanh \frac{x}{2\tau_C} \, [\{(v-u)(v^2+x^2) - 2vx(u+v) - nx\} \, /$$

$$\{\frac{n}{2}(\frac{1}{u}-\frac{1}{v}) + 2x\} + \frac{n}{8x} (u+v)^2 (u-v) \, / \, (uv-4)] \quad , \tag{3}$$

$$\equiv \frac{J}{2\pi t}(K_1 + K_2) \quad . \tag{4}$$

Here $\tau_c = T_c/w$, $v = \mu/w$, $2w = 8t$ being the band-width, and u and v are defined by

$$u = 2\left[v + x - i\sqrt{1 - \frac{n}{2} - (v+x)^2}\right] \quad , \tag{5.a}$$

$$v = 2\left[v - x + i\sqrt{1 - \frac{n}{2} - (v-x)^2}\right] \quad . \tag{5.b}$$

Equation (5.a) and (5.b) are derived by the following CPA equations[6,7] for the self-energy function $\Sigma(i\varepsilon_n)$ and the Green function $\mathscr{G}(k,i\varepsilon_n)$

$$\Sigma(i\varepsilon_n) = -n/2F(i\varepsilon_n) \quad , \tag{6.a}$$

$$F(i\varepsilon_n) = \frac{1}{N}\sum_k \mathscr{G}(k,i\varepsilon_n) = \frac{2}{w}\left[Z + i\sqrt{1 - Z^2}\right] \quad , \tag{6.b}$$

where $Z = (i\varepsilon_n - \Sigma(i\varepsilon_n) + \mu)/w$, μ being the chemical potential, and the parabolic density of state with the half width, w, is assumed.

The first term on r.h.s. of eq.(3) result from the process shown in Fig.1(a), whereas the second from Fig.1(b), where the broken lines with crosses represent the effect of the excluded volume due to infinite Coulomb interaction[8] treated within the CPA, which is the same as the alloy analogy of Hubbard.[9]

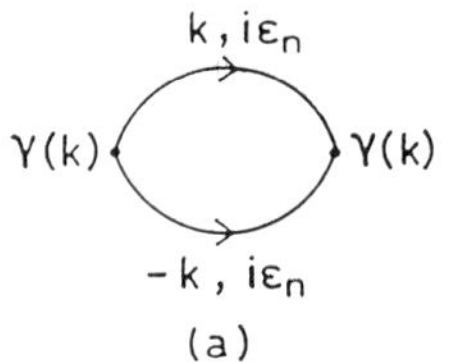

Fig.1(a) Process contributing to K_1.

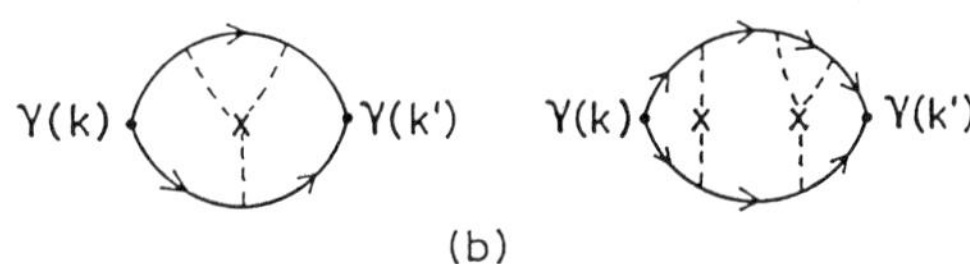

Fig.1(b) Examples of processes contributing to K_2.

The numerical evaluations of eq.(3) together with eqs.(5.a) and (5.b) have been performed. The result for the critical temperature T_c in the case of $J/\pi w = 0.1$ is shown in Fig.2 as a function of the carrier density, n. In Fig.3, K_1 and K_2 defined by eq.(4) are shown as a

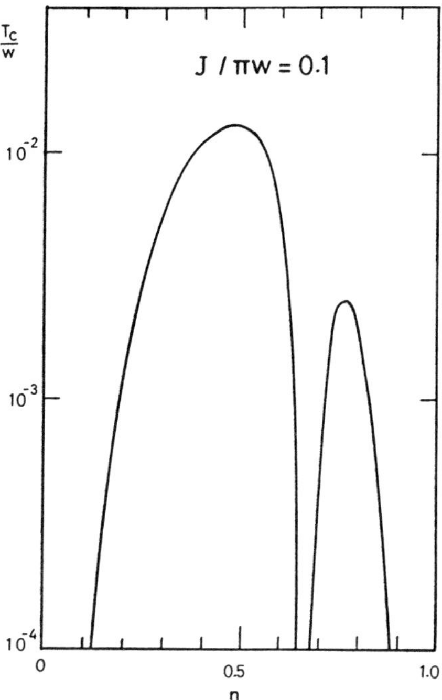

Fig.2 The carrier number, n, dependence
of the critical temperature T_c in unit
of w, 2w being the band width.

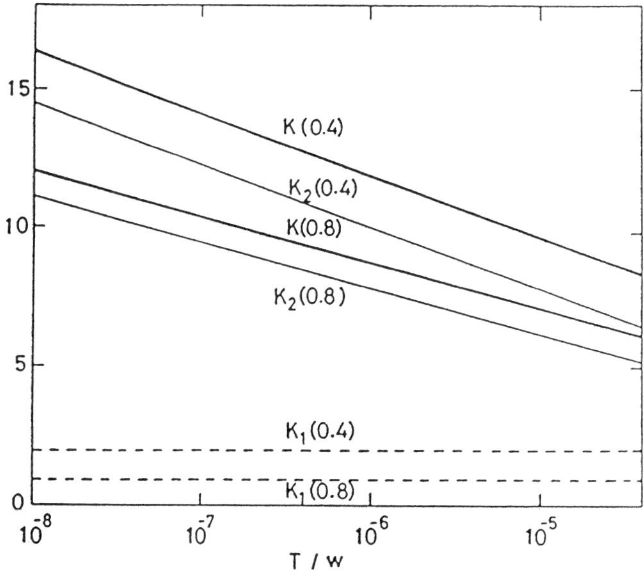

Fig.3. The temperature dependences of K_1, K_2
and $K=K_1+K_2$ for n=0.4 and 0.8.

function of the temperature for two choices of n=0.4 and 0.8. In our former publication only K_2 is retained and evaluated with further approximation for the integrand. The two results, however, yield essentially the same n-dependence of T_c for the present choice of J/t. (For large values of J/t, the present scheme yields finite T_c in the limit of small n, which will be artifact of the approximation.) Although CPA can not be trusted literally, we expect the qualitative aspect of the result obtained in the present scheme will hold.

The very low T_c at around n=2/3 is due to the vanishingly small average value over the Fermi surface of the order parameter characterized by $\gamma(k)$. Hence this structure in T_c-n curve will not be present in the d-wave state where the order parameter is proportional to ($\cos k_x a - \cos k_y a$). Such a d-wave state will actually be more enhanced in the present circumstances of strong correlations[10,11] as in the weakly correlated systems.[12,14] The T_c for this state can also be examined in the present framework but has not be pursued so far because of the complexity in numerical calculations. In the d-wave state the order parameter always has lines of zeros on the Fermi surface by symmetry, whereas in the presnt anisotopic s-wave state it is not always the case: rather it will have finite gaps at values of n, where T_c is high.

From Fig.2 we see that the maximum of T_c will not locate so close to the half-filling, n=1, and in the limit of t/U$\rightarrow$0 T_c is reduced appreciably. Hence we expect that the superconductivity in the Hubbard model will be most enhanced for moderate values of 1-n when J/t$\approx$0(1), i.e. U$\approx$t. The plausibility argument of such intermediate strength of correlation in oxides has been given by Lee and Read.[14]

REFERENCES

1. P.W. Anderson, Science 235: 1196 (1987).
2. J.E. Hirsch, Phys. Rev. Lett. 54: 1317 (1985).
3. G. Baskaran, Z. Zou and P.W. Anderson, Solid State Commun.
4. A. Ruckenstein, P. Hirschfeld and J. Appel, preprint.
5. H. Fukuyama and K. Yosida, Jpn. J. Appl. Phys. 26: L371 (1987).
6. For review, R.J. Elliott, J.A. Krumhansl and P.L. Leath, Rev. Mod. Phys. 46: 465 (1974).
7. H. Fukuyama and H. Ehrenreich, Phys. Rev. B7: 3266 (1973).
8. For detailed derivations, see e.g. D. Yoshioka and H. Fukuyama, J. Phys. Soc. Jpn. 54: 2996 (1985).
9. J. Hubbard, Proc. Roy. Soc. A281: 401 (1964).
10. J.E. Hirsch, preprint.
11. Y. Kuramoto, preprint.
12. D.J. Scalapino, E. Loh and J.E. Hirsch, Phys. Rev. B34: 8190 (1986).
13. J. Miyake, S. Schmitt-Rink and C. Varma, Phys. Rev. B34: 6554 (1986).
14. P.A. Lee and N. Read, preprint.

SUPERCONDUCTING ENERGY GAP AND PAIRING INTERACTION

IN HIGH T_c OXIDES

S. Maekawa, H. Ebisawa,[†] and Y. Isawa[††]

Institute for Materials Research, Tohoku University
Sendai 980, Japan
[†] Department of Engineering Science, Tohoku University
Sendai 980, Japan
[††]Research Institute of Electrical Communication
Tohoku University, Sendai 980, Japan

Abstract
We propose two characteristic features in high T_c oxides, La-Sr-Cu-O and Y-Ba-Cu-O: (i) The apparent discrepancy of the superconducting energy gap between far-infrared and quasi-particle tunneling measurements can be an indication of anisotropic Cooper pairing with spin singlet. (ii) The Coulomb interaction induces spin wave mode with the wave number of the Brillouin zone boundary and the energy close to $2\Delta_0$, Δ_0 being the maximum value of the superconducting energy gap. This mode may be observed by neutron inelastic scattering and tunneling experiments.

1. Introduction
In addition to their high transition temperatures (T_c), various unusual physical properties of the oxide superconductors, La-(Ba,Sr,Ca)-Cu-O and Y-Ba-Cu-O, have raised much theoretical interest. Soon after the discovery of La-Sr-Cu-O systems, the superconducting energy gap was examined by both far-infrared and quasi-particle tunneling experiments. Although both experiments revealed the existence of the gap, it turned out that the observed $2\Delta_0/k_B T_c$ ratio was very different between these measurements: In the tunneling measurements,[1-5] the ratio is $4 \sim 18$. This ratio is larger than the BCS value, 3.5. On the other hand, the ratio was obtained to be $1.3 \sim 2.7$ from the far-infrared measurements.[6-9] The similar discrepancy between far-infrared and tunneling measurements was also obtained in Y-Ba-Cu-O.[10-15]
Here, we would like to propose[16,17] that these observations suggest anisotropic Cooper pairing with spin singlet[18] in the oxide superconductors. In Sec. 2, we show the theoretical results of the far-infrared absorption spectra and the tunneling conductance in poly-crystalline samples with anisotropic Cooper pairing with spin singlet.
Recent observation of the absence of isotope effect in Y-Ba-Cu-O[19,20] suggests that the Coulomb repulsive interac-

tion between electrons is strong and/or non-phonon mechanism is responsible to the superconductivity. It will be conceivable that the Coulomb interaction may act as a pairing mechanism. Anderson[21] has proposed an anisotropic singlet superconducting state in the system with the strong Coulomb repulsive interaction. Monte Carlo simulations by Hirsch[22] suggest the state. The antiferromagnetic spin fluctuation with wave number $\mathbf{Q}$ in the Brillouin zone boundary causes anisotropic Cooper pairing with spin singlet.[23,24] In Sec. 3, we show that the anisotropic Cooper pairing induces a spin wave mode with $\mathbf{Q}$ and energy close to $2\Delta_0$, where Δ_0 is the maximum value of the superconducting energy gap. This is a natural consequence of the coherence factor which appears in the dynamical spin susceptibility with the wave number $\mathbf{Q}$. The direct observation of the mode may be done in the neutron inelastic scattering experiments. Finally, the conclusion and discussion are given in Sec. 4.

2. Far-Infrared Absorption and Tunneling Conductance

The electromagnetic absorption in a superconductor is dependent on the directions of the propagation and electric field relative to the crystal axis. Here, we take a polycrystalline sample so that the angular average of the absorption with respect to the crystal direction is observed. Assuming that the penetration depth of the wave is much smaller than the superconcucting coherence length, the absorption of electromangetic wave with frequency ω is written as[17]

$$\frac{\sigma_{1s}}{\sigma_{1n}} = \int \frac{d\Omega}{4\pi} \left\{ \frac{2}{\hbar\omega} \int_{|\Delta_{\mathbf{p}}|}^{\infty} (f(E) - f(E+\hbar\omega)) \; L(E;\mathbf{p}) \; dE \right.$$

$$\left. - \frac{1}{\hbar\omega} \int_{|\Delta_{\mathbf{p}}|-\hbar\omega}^{-|\Delta_{\mathbf{p}}|} (1-2f(E+\hbar\omega)) \; L(E;\mathbf{p}) \; \Theta(\hbar\omega-2|\Delta_{\mathbf{p}}|) \; dE \right\} , \quad (1)$$

where

$$L(E;\mathbf{p}) = \frac{E^2 + |\Delta_{\mathbf{p}}|^2 + \hbar\omega E}{\varepsilon_1 \varepsilon_2} , \quad (2)$$

$$\varepsilon_1 = \sqrt{E^2 - |\Delta_{\mathbf{p}}|^2} , \qquad \varepsilon_2 = \sqrt{(E+\hbar\omega)^2 - |\Delta_{\mathbf{p}}|^2} , \quad (3)$$

$$\Theta(x) = \left\{ \begin{array}{lll} 1 , & x > 0 \\ 0 , & x < 0 \end{array} \right. \quad (4)$$

$f(x)$ is the Fermi distribution function, and $\sigma_{1s(n)}$ is the real part of the optical conductivity in the superconducting (normal) states. Here, $\Delta_{\mathbf{p}}$ is the order parameter with momentum $\mathbf{p}$.

Let us examine the absorption in the following three types (I, II, and III) of anisotropic superconductors:

$$\Delta_{\mathbf{p}} = \frac{1}{2} \Delta_0 (\hat{p}_x^2 + \hat{p}_y^2 - 2\hat{p}_z^2), \qquad \dots \dots \text{I} , \quad (5)$$

$$\Delta_p = 2 \ \Delta_0 \ \hat{p}_x \hat{p}_y \ , \qquad\qquad\qquad \cdots \cdot \ \text{II} \ , \quad (6)$$

$$\Delta_p = \frac{\Delta_0}{2 + \cos\alpha} \ (\cos\alpha\hat{p}_x + \cos\alpha\hat{p}_y + \cos\alpha\hat{p}_z), \ \cdots \ \text{III} \ , \quad (7)$$

where $\hat{p}_i$ is the i-th component of the unit vector $\hat{p}$ relative to the crystal axis, Δ_0 is the maximum value of the gap, and α is a parameter given by $\alpha = p_F a$, p_F and a being the Fermi wave number and the lattice constant, respectively. When the Fermi surface is sherical, the gap vanishes on the lines with $\hat{p}_z = \pm 1/\sqrt{3}$ for Type I and on the lines with $\hat{p}_x = 0$ and $\hat{p}_y = 0$ for Type II. For Type III,[26] the gap does not vanish for $\alpha < \sqrt{3}\pi/2$, whereas the gap vanishes on lines for $\alpha > \sqrt{3}\pi/2$ when the Fermi surface is spherical. The gap has maxima (and local maxima) on points for Types I, II, and III. The gap has also a local maximum on the line with $\hat{p}_x = 0$ for Type I.

In Fig. 1, the numerical results of the absorption spectra in the anisotropic superconductors are shown.[17]

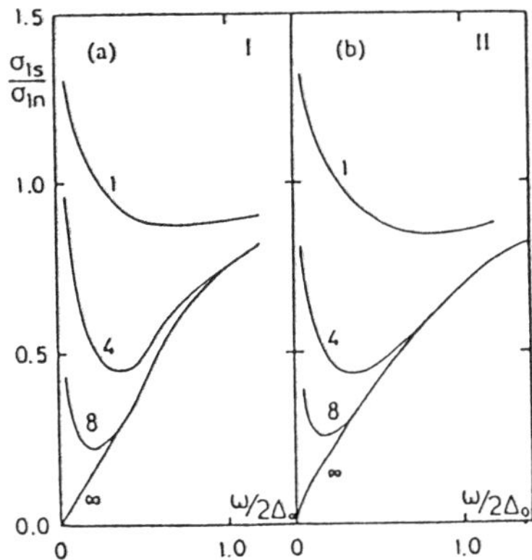

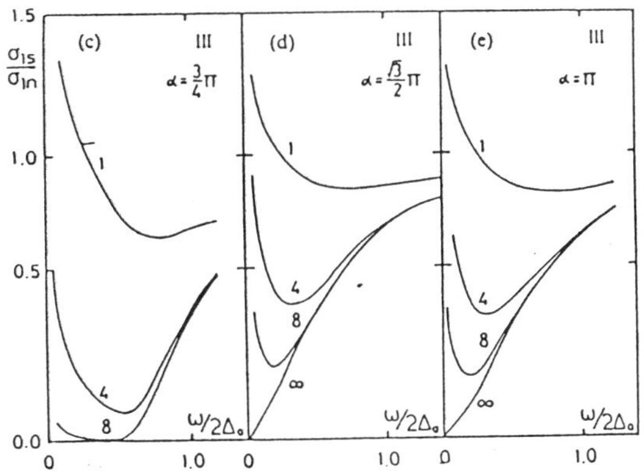

Fig. 1 The optical conductivity at finite temperatures for various types of anisotropy. The parameters on curves are $\Delta(T)/k_B T$. The curves at $T = 0$ are specified by the parameter ∞. The Types of anisotropy are: (a) Type I, (b) Type II, (c) Type III, $\alpha = 3\pi/4$, (d) Type III, $\alpha = \sqrt{3}\pi/2$ and (e) Type III, $\alpha = \pi$. (After ref. 17)

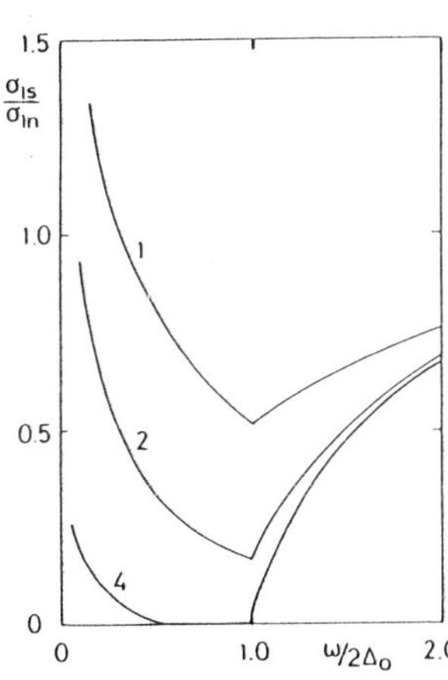

Fig. 2   The optical conductivity by Mattis-Bardeen[27] at finite temperatures in the isotropic BCS superconductors.

The spectra by Mattis–Bardeen[27] in the isotropic BCS superconductors is shown in Fig. 2 for comparison. As seen in
these figures, the absorption edge is strongly reduced in
the superconductors with anisotropic energy gap. It is also
interesting to see that the absorption edge moves toward zero
in Types I and II as temperature decreases, and there is not
any specific structure at $2\Delta_0$.
    The quasi-particle tunneling conductance in a poly-
crystalline sample is given by

$$\frac{\sigma_{Ts}}{\sigma_{Tn}} = \int \frac{d\Omega}{4\pi} \left\{ \int_{-\infty}^{\infty} dE \; \mathrm{Re}\left[\frac{|E|}{\sqrt{E^2 - |\Delta_{\boldsymbol{p}}|^2}} \right] \left[-\frac{\partial f(E+eV)}{\partial eV} \right] \right\} , \qquad (8)$$

where $\sigma_{Ts(n)}$ is the tunneling conductance between superconducting (normal) and normal electrodes, and V is the
voltage between the electrodes. The numerical results of the
tunneling conductance[17] for Types I, II, and III are given

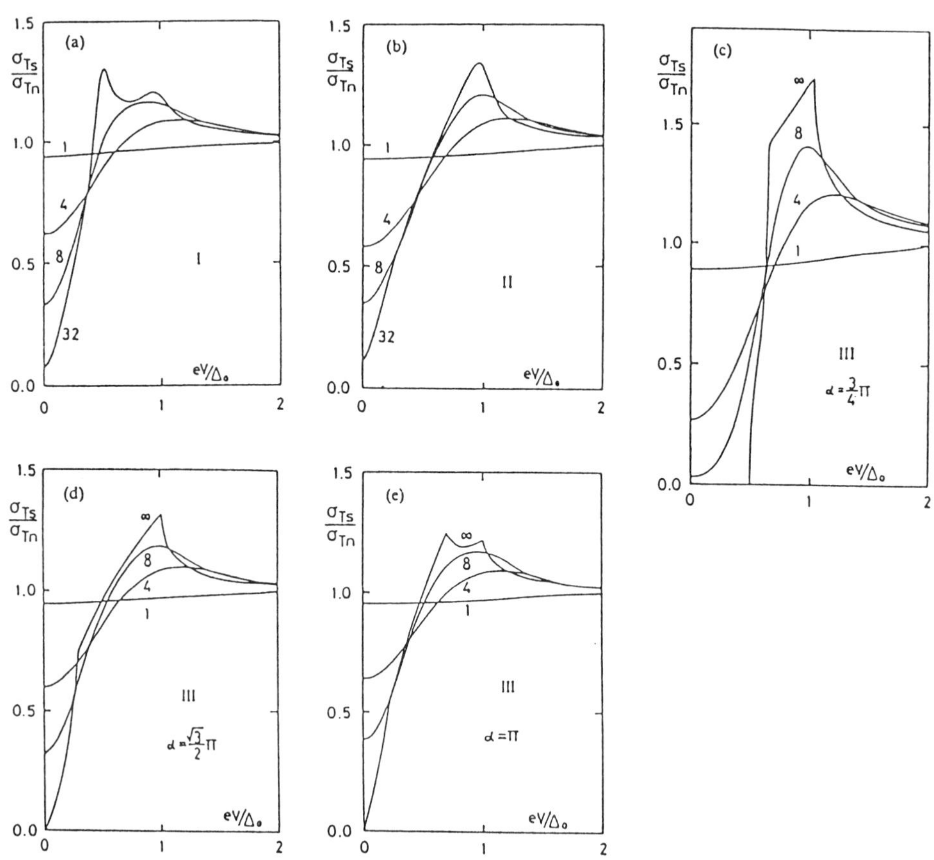

Fig. 3   The tunneling conductance as a function of the voltage
        at finite temperatures.   The parameters on curves are
        $\Delta(T)/k_B T$.   The curves at $T = 0$ are specified by the
        parameter ∞ .   The Types of anisotropy are: (a) Type I,
        (b) Type II, (c) Type III, $\alpha = 3\pi/4$, (d) Type III,
        $\alpha = \sqrt{3}\pi/2$ and (e) Type III, $\alpha = \pi$. (After ref. 17)

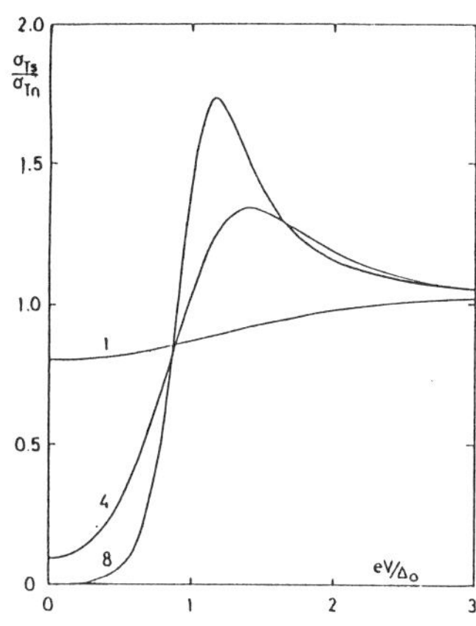

Fig. 4 The tunneling conductance as a function of the voltage at finite temperatures in the isotropic BCS superconductors.

in Fig. 3. The conductance for Types I and II has been discussed by Tachiki et al.[28] at $T = 0$. In Fig. 4, the tunneling conductance in the isotropic BCS superconductors is shown for comparison.[29] Comparing these results, we find that the energy gap defined by the peak-position in the tunneling conductance is much larger than that by the minimum-position in the far-infrared absorption spectra. This fact is in accord with the experimental observations in oxide superconductors, La-Sr-Cu-O and Y-Ba-Cu-O.

In the numerical calculations, we assumed that the order parameter, Δ_p, is real. Therefore, the zeros of the gap were emphasized in the tunneling and far-infrared absorption spectra. However, we note that it is possible for the order parameter to be complex without much change of the discussion in this section. For the complex order parameter, it is possible for the gap to be non-zero at any value of p even for the d-wave pairing. This was also pointed out by Lee and Read.[30] The discussion about this will be given in Sec. 4.

3. Spin Wave Mode in Superconducting States

In this section, we show that the strong Coulomb interaction induces a characteristic spin wave mode with wave number $\mathbf{Q}$ in the Brillouin zone boundary and energy close to $2\Delta_0$. We take, for simplicity, a two dimensional square lattice with the band energy,

$$\xi_{\mathbf{p}} = -2t (\cos p_x a + \cos p_y a) - \mu , \tag{9}$$

μ being a parameter to determine the position of the Fermi energy. The essential features in this section will not depend on the dimensionality.

The dynamical spin susceptibility in the normal state in the random phase approximation is written as

$$\chi (\mathbf{Q}, \omega) = \frac{\chi_0 (\mathbf{Q}, \omega)}{1 - U\chi_0 (\mathbf{Q}, \omega)} , \tag{10}$$

415

where $\chi_0(\mathbf{Q}, \omega)$ is the susceptibility with wave number $\mathbf{Q}$ and frequency ω of the non-interacting electrons and U is the on-site Coulomb interaction ($U > 0$). In the almost half-filled band ($\mu \sim 0$), $\chi_0(\mathbf{Q}, \omega)$ becomes very large near $\mathbf{Q} = (\pi/a)(\pm 1, \pm 1)$ with $\omega = 0$.

In the superconducting states, the susceptibility, $\chi_0(\mathbf{Q}, \omega)$, in eq. (10) is expressed at $T = 0$ as

$$\chi_0(\mathbf{Q}, \omega) = \frac{1}{2} \sum_{\mathbf{p}} \left[\frac{1}{E_{\mathbf{p}} + E_{\mathbf{p}+\mathbf{Q}} - \omega} + \frac{1}{E_{\mathbf{p}} + E_{\mathbf{p}+\mathbf{Q}} + \omega} \right]$$
$$\times \frac{1}{2} \left(1 - \frac{\xi_{\mathbf{p}} \xi_{\mathbf{p}+\mathbf{Q}} + \Delta_{\mathbf{p}} \Delta_{\mathbf{p}+\mathbf{Q}}}{E_{\mathbf{p}} E_{\mathbf{p}+\mathbf{Q}}} \right), \tag{11}$$

where

$$E_{\mathbf{p}} = \sqrt{\xi_{\mathbf{p}}^2 + |\Delta_{\mathbf{p}}|^2} . \tag{12}$$

In eq. (11), the term in the parenthesis is called the coherence factor of the susceptibility.

In the almost half-filled band, the following super-conducting state may be stabilized at T_c,[24,30)]

$$\Delta_{\mathbf{p}} = \frac{1}{2} \Delta_0 (\cos p_x a - \cos p_y a). \tag{13}$$

This is because the order parameter has the relation,

$$\Delta_{\mathbf{p}+\mathbf{Q}} = -\Delta_{\mathbf{p}} , \tag{14}$$

with $\mathbf{Q} = (\pi/a)(\pm 1, \pm 1)$. We note that when the relation (14) is introduced into eq. (11), the coherence factor changes from that in the usual BCS superconductors.

To understand the role of the coherence factor, let us assume, for a moment, that $\Delta_{\mathbf{p}}$ in the quasi-particle energy, $E_{\mathbf{p}}$, is constant, Δ_0. Then, using eq. (14) and the relation $\xi_{\mathbf{p}+\mathbf{Q}} = -\xi_{\mathbf{p}}$, we obtain the imaginary part of $\chi_0(\mathbf{Q}, \omega)$,

$$\chi_0''(\mathbf{Q}, \omega) = \begin{cases} 0, & \text{for } \omega < 2\Delta_0, \\[2ex] \dfrac{1}{4\pi t a^2} \dfrac{\omega}{\sqrt{\omega^2 - 4\Delta_0^2}} \ln\left(\dfrac{8t}{\sqrt{\omega^2 - 4\Delta_0^2}} \right), & \\[2ex] & \text{for } \omega > 2\Delta_0. \end{cases} \tag{15}$$

The singularity of $\chi_0''(\mathbf{Q}, \omega)$ at $\omega = 2\Delta_0$ comes from the density of states of the superconducting quasi-particles (the logarithmic factor is due to the density of states in the normal states). The real part of $\chi_0(\mathbf{Q}, \omega)$, $\chi_0'(\mathbf{Q}, \omega)$, is calculated to be

$$\chi_0'(\mathbf{Q}, \omega) \sim \ln^2 |2\Delta_0 - \omega|. \tag{16}$$

Inserting eqs. (15) and (16) into eq. (11), we obtain that

$\chi(\mathbf{Q}, \omega)$ have a pole at $\omega \lesssim 2\Delta_0$. This pole indicates a spin wave mode induced by the Cooper pairing.

We have numerically calculated $\chi_0(\mathbf{Q}, \omega)$ in the case of eq. (13). We took $\mu = 0$ for simplicity. In Fig. 5(a), the real and imaginary parts of $\chi_0(\mathbf{Q}, \omega)$ are shown. In the vicinity of $2\Delta_0$, the real part shows dispersive behavior, although not shown explicitly. The dynamical susceptibilities, $\chi_0(\mathbf{Q}, \omega)$, in the normal states are shown in Fig. 5(b). From Fig. 5(a), it is easily seen that $\chi(\mathbf{Q}, \omega)$ in eq. (11) gives a spin wave mode at $\omega \lesssim 2\Delta_0$, when the Coulomb interaction is strong. We note here that $\chi_0'(\mathbf{Q}, \omega)$ in Fig. 5(a) results in the life time of the spin wave, since there exist zeros of the energy gap in the case of eq. (13). We expect that this mode may be observed in the neutron inelastic scattering experiments.

The pairing interaction due to the antiferromagnetic spin fluctuation with $\mathbf{Q}$ is calculated to be

$$V(\mathbf{Q}, \omega) \simeq \frac{U^2 \chi_0(\mathbf{Q}, \omega)}{1 - U\chi_0(\mathbf{Q}, \omega)} + \frac{1}{2}\frac{U}{1 - U\chi_0(\mathbf{Q}, \omega)}. \tag{17}$$

When we use the strong coupling formalism by Scalapino et al.[31] and Berk-Schrieffer,[32] we find that the spin wave contributes to the pairing interaction. Although we will leave the detailed calculation in a separate paper, the spin wave mode obtained above may cause peaks of the tunneling conductnance at $\sim 3\Delta_0$ and $\sim 5\Delta_0$ in addition to the peak at Δ_0. We consider that these peaks are the origin of the multiple peak structure observed experimentally in Y-Ba-Cu-O[12-14] and La-Sr-Cu-O.[5]

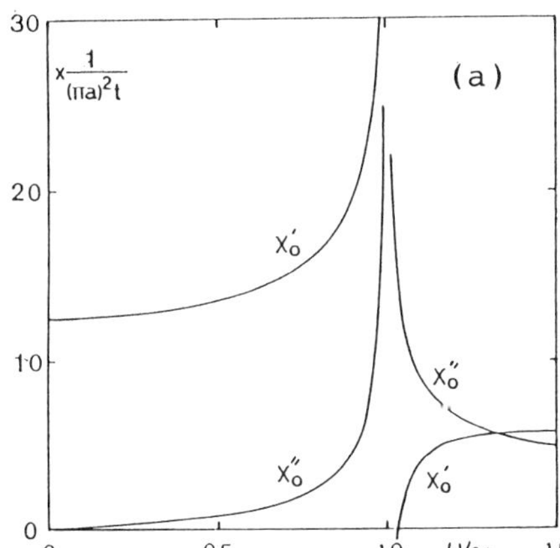
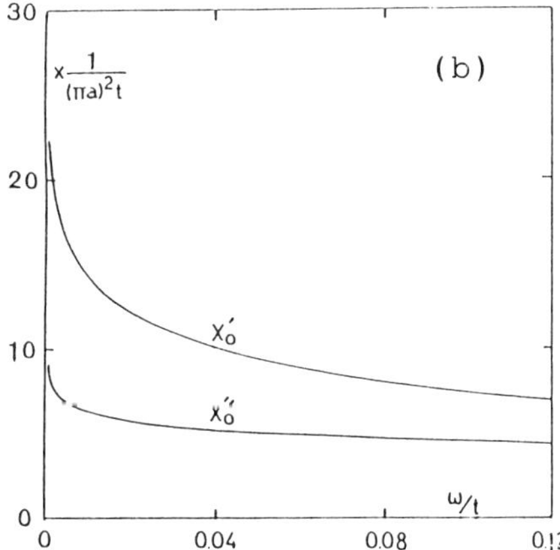

Fig. 5 (a) The real and imaginary parts of the dynamical spin susceptibility, $\chi_0(\mathbf{Q}, \omega)$, in the case of eq. (13) are shown by solid lines. The value of Δ_0/t is chosen as 0.01. (b) The real and imaginary parts of $\chi_0(\mathbf{Q}, \omega)$ in the normal state are shown.

4. Conclusion and Discussion

In the almost half-filled band with the strong Coulomb interaction, the antiferromagnetic spin fluctuation with wave number Q near the Brillouin zone boundary causes the Cooper pairing. In this case, the superconducting states with the relation of the order parameter,

$$\Delta_{p+Q} = -\Delta_p, \tag{14}$$

will be stabilized at T_c. We have shown that this relation of the order parameter changes the coherence factor of the dynamical spin susceptibility in the superconducting states. As a result, a characteristic spin wave mode with Q and energy close to $2\Delta_0$ is induced.

The most probable order parameter at T_c is considered to be that given in eq. (13) in two dimension.[24,30] However, it is not certain what the ground state is. In more general, the order parameter may be of complex form:

$$\Delta_p = \Delta_p^0 \exp(2i\hat{\phi}_p). \tag{18}$$

In this case, the anisotropy of the energy gap can be very different from that of eq. (13). Even so, we believe that the spin wave mode exists, and that the observation of this mode will give us much information of the ground states and the pairing mechanism.

In Sec. 3. we took the on-site Coulomb intaeraction and the anisotropic gap with zeros. Since the type of the coherence factor is of crucial importance for the spin wave mode, It will also be possible to obtain the mode even in the states with isotropic gap if the Coulomb interaction is of finite range. This discussion will be given elsewhere. The effects of impurities[25] on the spin wave mode will be discussed in a separate paper. Finally, the spin wave mode in the superconducting states may also be observed in the heavy fermion systems.

Acknowledgements

The authors wish to express their thanks to Prof. M. Tachiki, Prof. Y. Muto and Prof. I. Iguchi for their valuable discussions.

References

1) T. Ekino, J. Akimitsu, M. Sato and S. Hosoya: to be published in Solid Stat. Commun.
2) J. R. Kirtley, C. C. Tsuei, Sung I. Park, C. C. Chi, J. Rozen and M. W. Shafer: to be published in Phys Rev. B (1987).
3) J. Morland, H. F. Clark and H. C. Ku: preprint.
4) M. E. Hawley, K. E. Gray, D. W. Capone II and D. G. Hinks: to be published in Phys Rev. B (1987).
5) M. Naito, D. P. E. Smith, M. D. Kirk, B. Oh, M. R. Han, K. Char, D. B. Mitzi, J. Z. Sun, D. J. Webb, M. R. Beasley, O. Fisher, T. H. Geballe, R. H. Hammond, A. Kapitulnik and C. F. Quate: preprint.
6) Z. Schlesinger, R. L. Greene, J. G. Bednorz and K. A. Muller: to be sublitted to Phys. Rev. Lett.

7) P. E. Sulewski, A. J. Sievers, S. E. Russek, H. D. Hallen, D. K. Lathrop and R. A. Buhrman: preprint.

8) U. Walter, M. S. Sherwin, A. Stacy, P. L. Richards and A. Zettl: preprint.

9) K. Nagasaka, M. Sato, H. Ihara, M. Tokumitsu, M. Hirabayashi, N. Terada, K. Senzaki and Y. Kimura: Jpn. J. Appl. Phys. **26** (1987) L645; in this work, a larger value, 5, is reported.

10) J. R. Kirtley, W. J. Gallegher, Z. Schlesinger, R. L. Sandstrom, T. R. Dinger and D. A. Chance: preprint.

11) C. C. Tsuei, P. P. Freitas, J. R. Kirtley, S. I. Park, W. J. Gallagher, C. C. Chi, T. S. Plassket, M. W. Shafer, T. R. Dinger, R. L. Sandstrom, D. A. Chance, J. R. Rozen, Z. Schlesinger, R. L. Green and R. T. Collins: to be published.

12) I. Iguchi, H. Watanabe, Y. Kasai, T. Mochiku, A. Sugishita and E. Yamaka: Jpn. J. Appl. Phys. **26** (1987) L615.

13) H. Watanabw, Y. Kasai, T. Mochiku, A Sugishita, I. Iguchi and E. Yamaka: to be publishied to Proc. of LT18.

14) D. P. E. Smith, M. D. Kirk, D. B. Mitzi, J. Z. Sun, D. J. Webb, K. Char, M. R. Hahn, M. Naito, B. Oh, M. R. Beaseley, T. H. Geballe, R. H. Hammond, A. Kapitulnik and C. F. Quate: preprint.

15) G. A. Thomas, H. J. Ng, A. J. Millis, R. N. Bhatt, R. J. Cava, E. A. Rietman, D. W. Jphnson, Jr., G. P. Espinosa, J. M. Vandenberg: preprint.

16) S. Maekawa, H. Ebisawa and Y. Isawa: Jpn. J. Appl. Phys. **26** (1987) L468.

17) H. Ebisawa, Y. Isawa and S. Maekawa: Jpn. J. Appl. Phys. **26** (1987) No. 6.

18) We presume the spin singlet pairing since the Josephson effect between oxide superconductors and Nb is clearly seen.

19) B. Batlogg, R. J. Cava, A. Jayaraman, R. B. van Dover, G. A. Kourouklis, D. W. Murphy, L. W. Rupp, H. S. Chen, A. White, K. T. Short, A. M. Mujsce and E. A. Rietman: Phys. Rev. Lett. **58** (1987) 2333.

20) L. C. Bourne, M. F. Crommie, A. Zettl, H. zur Loye, S. W. Keller, K. L. Leary, A. M. Stacy, K. J. Chang, M. L. Cohen and D. E. Morris: Phys. Rev. Lett. **58** (1987) 2337.

21) P. W. Anderson: Science **235** (1987) 1196.

22) J. E. Hirsch: Phys. Rev. Lett. **54** (1985) 1317.

23) K. Miyake, S. Schmitt-Rink and C. M. Varma: Phys. Rev. **B34** (1984) 6554.

24) D. J. Scalapino, E. Loh Jr. and J. E. Hirsch: Phys. Rev. **B34** (1986) 8190.

25) S. Maekawa, Y. Isawa and H. Ebisawa: Jpn. J. Appl. Phys. **26** (1987) L771.

26) F. J. Ohkawa and H. Fukuyama: J. Phys. Soc. Jpn. **53** (1984) 4344.

27) D. C. Mattis and J. Bardeen: Phys. Rev. **111** (1958) 412.

28) M. Tachiki, M. Nakahara and R. Teshima: J. Mag. Mag. Mat. **52** (1985) 161.

29) I. Giaever, H. R. Hart Jr. and K. Megerle: Phys. Rev. **126** (1962) 941.

30) P. A. Lee and N. Read: preprint.

31) D. J. Scalapino, J. R. Schrieffer and J. W. Wilkins: Phys. Rev. **148** (1966) 263.

32) N. F. Berk and J. R. Schrieffer: Phys. Rev. Lett. **17** (1966) 433.

ARE SPIN DENSITY WAVES INVOLVED IN HIGH-TEMPERATURE SUPERCONDUCTIVITY?

E.W. Fenton

Physics Division
National Research Council of Canada
Ottawa, Canada K1A OR6

INTRODUCTION

In the $La_{2-x}Ba_xCuO_4$, $La_{2-x}Sr_xCuO_4$, and $M_1Ba_2Cu_3O_y$ compounds, where M is a transition metal such as yttrium and y is apparently between 7 and 6, the quasi-two-dimensional metallic planes of repeated CuO_2 have a perfectly-nested square Fermi surface when there are exactly two extra electrons per CuO_2, in a tight-binding band structure. In this case one expects formation of an electron charge density wave (CDW) or spin density wave (SDW). The Fermi surface is significantly different from a simple square in more sophisticated band-structure theory,[1,2] however in this case saddle-points (Van Hove density of states singularities) in E versus $\underline{k}$ may be responsible for a density wave, as suggested by Rice and Scott[3] for the 2H crystal form of several transition metal di-chalcogenides. In this case, when the number of extra electrons per CuO_2 unit is somewhat less than two, e.g. 1.85 for x = 0.15 in the lanthanum compounds, or somewhat more than two for $M_1Ba_2Cu_3O_y$, density wave formation may be controlled by the Van Hove singularities, with no increase or even a decrease in the electrical resistivity of the CDW or SDW state from the normal-metal state[3]: the high-velocity parts of the Fermi surface are not gapped by the density wave.

It thus appears that the question of whether or not an electron density wave occurs, and whether it is a CDW or a SDW, is probably important to understanding just why T_c has increased from less than 24 K to a present maximum of 100 K for $Y_1Ba_2Cu_3O_y$[4] and probably much higher values in the future. At the time of this writing (1 May 1987), there are many rumours that antiferromagnetism has been verified in lanthanum and "one-two-three" compounds by neutron scattering experiments, but to our knowledge no preprints yet exist on this, and moreover it is not known whether the antiferromagnetism if it occurs involves at least partially some SDW formation (characterized by reduced free energy among metallic electrons due to the SDW). As we will discuss later, the Stanford electron tunnelling study[5] of LaSrCuO may show evidence of an incommensurate SDW in the by-now-famous dI/dV versus voltage curve which shows three peaks on the positive voltage side and similar structure on the negative voltage side, instead of the single peak on each side expected for a BCS superconductor. However at the present time it is not known whether a CDW or SDW state occurs or not in the high-T_c superconductors.

From earlier studies of organic metals, it has been found that
formation of a density wave has a variety of effects on the phonon states.
It is well known of course that a Peierls-distortion CDW causes at least
one phonon state to have significantly lowered frequency compared to the
same state for the non-density-wave metal at higher temperatures[6].
However it is not so well known, and was only recently discovered, that
when the condensation energy of the density wave has a net-positive
contribution from the coulomb interaction between metallic electrons, then
a second raised-frequency charge-vibration mode also occurs, where the
raised frequency is still small compared to usual plasma frequencies.[8-11]
(Plasma frequencies are too high to be effective in formation of Cooper
pairs since polarization caused by passage of an electron has already
relaxed when a second electron appears later at the same point. This is
not true at low wave vector for a single metallic plane, but is true for a
repeated and closely-spaced sequence of metallic planes where an
effective-average plasmon frequency is nearly the value for a
three-dimensional metal.) The raised frequency for the SDW with a
second-harmonic CDW may be as much as five or ten times higher than the
corresponding phonon frequency for the non-density-wave metallic state.
This mode must exist for the SDW, where the coulomb exchange energy term
dominates the condensation energy, but may exist for the CDW, where the
Peierls phonon-mediated interaction must be larger than the coulomb effect
in the direct interaction energy (since otherwise the SDW would occur) and
the total contribution to condensation energy from coulomb exchange and
direct interactions may be positive or negative. Moreover, for the CDW
the raised-frequency is raised too much, right to the edge of the CDW gap
for quasiparticle excitations, in which case it has very small spectral
weight and is strongly over-damped. Existence of two modes, rather than
just one softened mode as for the simple Peierls CDW,[6] occurs because the
density wave would still exist and still vibrate in a rigid host lattice.

Coupling of both lowered-frequency and raised-frequency
charge-vibrational modes in a SDW system to electrons is much stronger
than for the corresponding phonon state in the non-density-wave metal. In
this case either mode or both may be involved in high-temperature
superconductivity. At the present time, only the vibrational modes with
wave vector close to the density-wave wave vector Q have been
studied,[6,8-11] and so it is still premature to attempt to estimate T_c for
superconductivity occurring in a SDW system. As for the coexistence of
superconductivity with antiferromagnetism or a SDW state, a subject of
long and spirited controversy, a careful treatment of all
magnetic-Umklapp-vector scattering shows that in the superconductivity gap
equation, the only change from the non-SDW state is to replace the
Bloch-wave density of states of a non-magnetic metal with the actual
density of states for electrons in the presence of the antiferromagnetism
or SDW.[7]

CHARGE VIBRATIONS IN A SDW SYSTEM

For the SDW system with CDW second harmonic, the result of the
theoretical development of refs. 6 and 8-11 is an expression for a
composite propagator at each wave vector describing $n+1$ charge vibrations
resulting from density-wave mixing of n host-lattice $2Q+q$ vibrations
(independent in the normal metal) with each other and with charge
vibration of the density wave itself; e.g.

$$D(2\underline{Q},\omega) = \frac{\lambda(1 + \frac{A^2}{4})^2(\lambda_c\omega_c + \sum_{i=1}^{n} \frac{A^2\lambda_i^{ph}\omega_{i,2\underline{Q}}^{ph}}{(1 + \frac{A^2}{4})^2})^{-1}}{D_o^{-1}(2\underline{Q},\omega) + 1 + \lambda(\frac{\omega^2}{4\Delta^2})f(\frac{\omega}{2\Delta}) - \alpha} \tag{1}$$

where

$$D_o(2\underline{Q},\omega) = \frac{\lambda_c}{\lambda}\frac{\omega_c^2}{\omega^2-\omega_c^2} + \sum_{i=1}^{n}\frac{A^2\lambda_i^{ph}}{(1 + \frac{A^2}{4})^2\lambda}\frac{\omega_{i,2\underline{Q}}^{ph2}}{\omega^2-\omega_{i,2\underline{Q}}^{ph2}+i\omega\gamma} \tag{2}$$

and where

$$\Delta \approx 1.75\ T_{cSDW} \approx E_F e^{-1/\lambda} \tag{3}$$

with $\lambda = \left(\lambda_c + \frac{A^2}{(1 + \frac{A^2}{4})^2\lambda}\sum_{i=1}^{n}\lambda_i^{ph}\right)$

$$A = \langle 2\underline{Q}CDW\rangle/\langle 1\underline{Q}SDW\rangle \quad \text{and} \quad \lambda_i^{ph} = \frac{g_{phi}^2}{\omega_{i,2\underline{Q}}^{ph}}.$$

A is less than one. The function $f(\omega)$ is given by

$$f\left(\frac{\omega}{2\Delta}\right) = \left\{\pi i + \ln[(1-S)/(1+S)]\right\}/2\left(\frac{\omega}{2\Delta}\right)^2 S \quad \text{in which} \tag{4}$$

$$S = \left(1 - \frac{4\Delta^2}{\omega^2}\right)^{\frac{1}{2}}$$

We have taken the sum of exchange and direct coulomb terms with a single frequency ω_c representing some effective average for the plasmon frequency. This procedure does not change the physics in any significant way because ω_c is comparable to E_F in magnitude and is several orders of magnitude larger than 2Δ, T_c, or ω_{Debye} for phonons in the non-density-wave metal above T_{cSDW}. α in eq. 1 represents pinning of the density wave due to high-order commensurability and/or plasma effects of the "incommensurate" density wave.

Coupling of the mixed vibrational modes of the lattice and density wave system to Fermi surface electrons, for the part gapped by the density wave, is given for the n=1 case by

$$g^2 N_{DW}(0) = A^2\lambda_{ph}\omega_{2\underline{Q}}^{ph}(1 + \frac{A^2}{4})^{-2} + \lambda_c\omega_c$$

$$\gg g_{ph}^2 N(0) \tag{5}$$

$N_{DW}(0)$ is the Fermi surface density of states removed by the density-wave energy gap and $N(0)$ is the density of states of the normal metal state for $T > T_{cSDW}$. The inequality occurs because $N_{DW}(0)$ is a significant fraction of $N(0)$ or the SDW would not have occurred, $\omega_c >> \omega_{2Q}^{ph}$, and $\lambda_c \gtrsim A^2 \lambda_{ph}$ since the SDW would not otherwise occur. The second-order (g^2) coupling to the electrons involves the function $f\left(\frac{\omega}{2\Delta}\right)$ as well, squared.

$D(2Q,\omega)$ is actually $D(0,\omega)$ because $2Q$ is a reciprocal lattice vector in the density-wave system. The resonant frequencies for $D(q,\omega)$ change from those of $D(0,\omega)$ by plus $v_F^2 q^2$ times a constant of order unity, for $q << k_F$. In general, the term $\lambda\left(\frac{\omega^2}{4\Delta^2}\right) f\left(\frac{\omega}{2\Delta}\right)$ in the denominator of $D(0,\omega)$ is replaced by a function of ω and q in $D(q,\omega)$. We are at present extending the theory for $D(0,\omega)$ to $D(q,\omega)$ with arbitrary q for a model sufficiently simple to proceed with but at the same time sufficiently representative of the charge-vibrational modes of a 1Q SDW-plus-2Q CDW system in $La_{2-x}Sr_xCuO_4$ or $Y_1Ba_2Cu_3O_y$ crystals. The fraction of phonon states in the normal metal which become the composite vibrational function of eq. 1 with electron to vibrational-mode coupling of eq. 5 is roughly comparable to $N_{DW}(0)/N(0)$.

With electron mean free path less than the size of a Cooper pair, we use the Anderson theorem and pair time-reversed electron eigenstates in the presence of impurities. In this case the enhanced electron-vibrational mode coupling $g^2 N_{DW}(0)$ of eq. 5, to the gapped part of the Fermi surface (and using $f\left(\frac{\omega}{2\Delta}\right)$ twice in a manner analogous to the theory for frequency-dependent conductivity of refs. 8-11) is also coupled to the non-gapped part of the Fermi surface via impurity scattering. Each electron eigenstate of the dirty metal then couples both gapped and non-gapped parts of the Fermi surface to mixed lattice-density wave vibrational modes with the enhanced coupling $g^2 N_{DW}(0)$.

In Figure 1 we show the spectral density for $q = 0$ charge vibration with $\omega << E_F$ in a 1Q SDW-plus-2Q CDW system, for several values of the harmonic amplitude ratio A. Because the maximum value of $\hbar\omega^{ph}$ is about 80 meV for copper-oxygen breathing modes in $La_{2-x}Sr_xCuO_4$, and the width of the approximately half-filled upper Cu_d-O_p electron band is a few eV, with $2\Delta << E_F$ we can expect a raised frequency ω_2 with $\omega_{2Q}^{ph} < \omega_2 < 10\ \omega_{2Q}^{ph}$, or $\hbar\omega_2$ with a maximum value which is several times $1000\ k_B$. In Figure 1 we have taken $\dfrac{\omega_{2Q}^{ph}}{2\Delta} = 0.1$, $\lambda = 0.3$, $\lambda^{ph} = 0.2$, and $\lambda_c = \lambda - \dfrac{A^2 \lambda_{ph}}{1 + \dfrac{A^2}{4}}$. This value of λ means that with $E_F \approx 5$ eV, the density-wave gap in the electron spectrum is $2\Delta \approx 350$ meV ($T_c \approx 1100$ K) and $\omega_{2Q}^{ph} \approx 35$ meV. We have taken the high-order (and therefore small) commensurability-energy parameter α of eq. 1 as 0.005 for an "incommensurate SDW". With $A = 0.4$, the spectral weight of the higher-frequency peak at $\omega_2 \approx 190$ meV is about the same as the spectral weight of the low-frequency peak at $\omega_1 \approx 1.75$ meV, and about 20% of the unity spectral weight of the 2Q phonon in the metal with $T > T_{cSDW}$. (The theoretical frequency ω_1 is of course purely arbitrary since it would be zero except for $\alpha \neq 0$, and α has just been guessed at.)

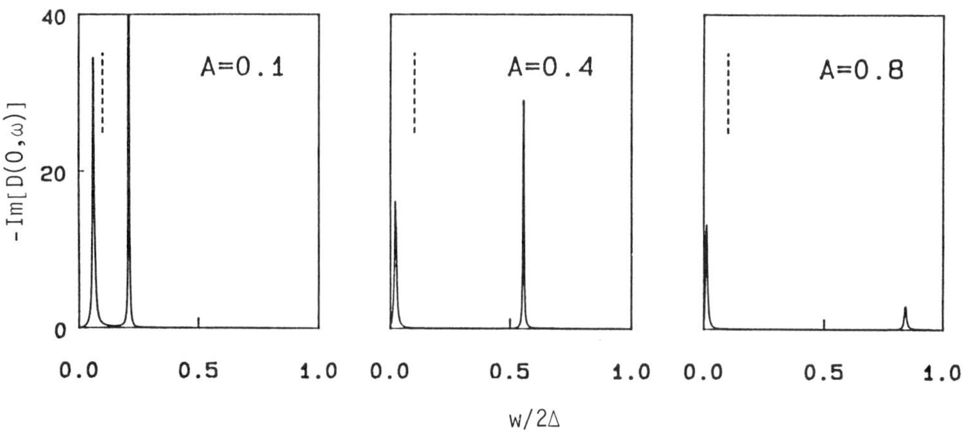

Fig. 1. Spectral weight of the composite charge-vibration mode for phase-mode $2Q$ vibration of a lattice mixing with phase-mode $2Q$ vibration of a $1Q$ SDW-plus-$2Q$ CDW. The dashed line at $\frac{\omega}{2\Delta} = 0.1$ is the position of ω_{2Q}^{phonon} for the $2Q$ phonon state of the normal metal when $T > T_{cSDW}$.

STANFORD ELECTRON TUNNELLING STUDIES

Naito et al[5] at Stanford have recently obtained a highly-unusual dI/dV versus voltage curve in electron tunnelling studies into thin films of high-T_c superconducting LaSrCuO. A tracing of this curve is shown in Fig. 2, for $La_{2-x}Sr_xCuO_4$ with x near 0.15. Instead of the usual BCS-like density of states structure, where a 2Δ gap occurs between two steeply rising and thereafter more-slowly-falling peaks, with the gap centred at E_F, three peaks occur on the positive-V side, and a very similar structure occurs on the negative-V side. The most often suggested explanation is that three peaks on the positive-V side and counterparts on the negative-V side represent 1Δ, 3Δ and 5Δ peaks for a superconductor, with 3Δ and 5Δ peaks caused by some multiple tunnelling transitions. One difficulty with this explanation is that Δ is determined not by the center of the first peak near E_F, but by halfway up the rise on this peak. This means that 3Δ and 5Δ appear to occur well before the second and third peaks, as shown in Figure 2. A second difficulty is that with $T_c \approx 35$ K, the gap $2\Delta \approx 37$ meV means that $\frac{2\Delta}{k_B T_c} \approx 13$, far larger than 3.5 for a weak-coupling

superconductor, or than 4.7 for strong-coupling Pb. Although several electron tunnelling and infrared experiments have attempted to determine $\frac{2\Delta}{k_B T_c}$, a large range of values between about 2 and 6 have been found. (It has been found that with a careful Kramers-Kronig analysis in an infrared measurement, $\frac{2\Delta}{k_B T_c} \approx 3.5$ is obtained.[12]) The value $\frac{2\Delta}{k_B T_c} \approx 13$ in the Stanford studies has been suggested by Naito et al[5] to be due to a very-high-T_c layer near the surface, or due to large anisotropy of the gap.

We suggest that an incommensurate spin density wave (SDW) with T_{cSDW} comparable to 125 K $\left(\frac{2\Delta}{k_B T_{cSDW}} \approx 3.5\right)$ could possibly be causing the multi-peak structure on each voltage side of the $\frac{dI}{dV}$ vs. V curve in Figure 2. Because of the incommensurability of the SDW, the structure

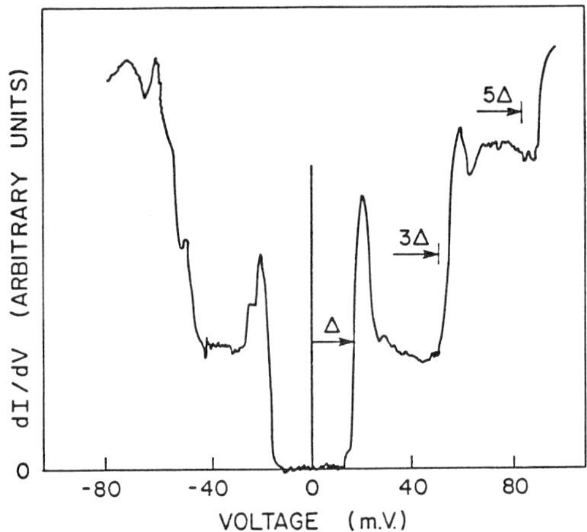

Fig. 2   dI/dV versus V from Figure 3 of Naito et al.[1]

with two peaks at the edge of the $-\Delta$ to Δ gap centred at E_F (and V=0) is replicated again away from E_F, with the centre of the replicated structure at $\pm 2v_F\delta Q$, where $2\delta Q$ is the difference between $2Q$ of the SDW wave vector and a commensurate value which is equal to a reciprocal lattice vector of the host lattice.   The situation is illustrated schematically in the repeated Brillouin zone scheme of Figure 3.   We suggest a SDW rather than a charge density wave (CDW) because the density-of-states peaks would more likely be detected for the SDW because its non-locality length ξ_{SDW} is much larger than ξ_{CDW} when $T_{cSDW} \approx T_{cCDW}$.   This is because

$$\xi_{SDW} \approx \frac{\hbar v_F}{\pi k_B T_{cSDW}}, \quad \text{whereas} \quad \frac{1}{\xi_{CDW}} \approx \left(\frac{\hbar v_F}{\pi k_B T_{cSDW}}\right)^{-1} + \frac{1}{a} \quad \text{and} \quad \xi_{CDW} \approx a, \text{ the lattice}$$

constant.

For an incommensurate density wave state, the crystal unit cell in $\underline{r}$-space becomes the entire volume of the crystal, and every repeated zone obtained by translation from the first zone by some reciprocal lattice vector $\underline{G}_{\ell mn}$ of the host crystal is inequivalent in the repeated zone scheme.   Density-of-states peaks at the edge of the $-\Delta$ to Δ gap centred at E_F are replicated at positions where the two-peak structure is centred at $n2v_F\delta Q$, where n is a positive or negative integer.   The magnitude of the contribution of an Umklapp-replicated structure to the electron density of states is determined by an Umklapp coefficient, i.e. by one of the coefficients $\rho_{\ell mn}$ in the expansion of the electron density of the host crystal, corresponding to the Umklapp vector $-\underline{G}_{\ell mn}$ which translates the repeated zone back to the first Brillouin zone:

$$\rho(\underline{r}) = \sum_{\ell,m,n} e^{i\underline{G}_{\ell mn}\cdot\underline{r}} \rho_{\ell mn} \tag{6}$$

For $La_{2-x}Sr_xCuO_4$, a SDW instability may be occurring with

$$\underline{Q} = \left(\frac{\pi}{a}\,\hat{a} + \frac{\pi}{b}\,b\right) - \delta\underline{Q}(x)$$

$$\equiv \frac{1}{2}\,\underline{G}_{11} - \delta\underline{Q}(x) \tag{7}$$

426

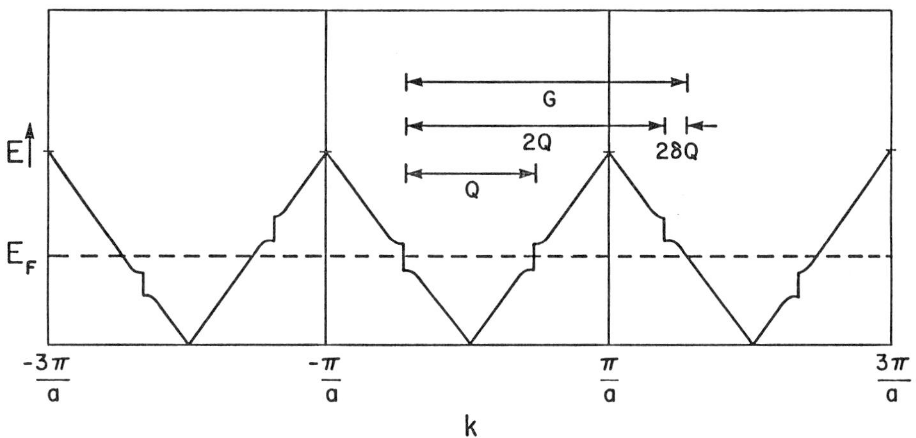

Fig. 3   Schematic diagram of density wave gaps in the repeated zone
scheme, for an incommensurate density wave where $2Q = \frac{2\pi}{a} - 2\delta Q$.
When $\delta Q = 0$, every repeated zone is identical to the central zone.

(The third c-component is probably either zero or $\frac{\pi}{c}$ c.) For $x = 0$,
$\delta Q(0) = 0$ is expected. In the $\delta Q(0) = 0$ case, only two peaks at the edges
of a $-\Delta$ to Δ gap centred at E_F (V=0) would occur. However with e.g.
$x \approx 0.15$, 1.85 extra electrons on each CuO_2 unit should occur in the
metallic planes, rather than 2 extra electrons for La_2CuO_4, and we expect
$\delta Q(0.15) \neq 0$.

In a rigid band model where k_F in the two-dimensional k-space is
proportional to the filling factor in the highest $3d_{x^2-y^2}-2p$ band,
$2\delta Q(x) \approx \frac{x}{2} \underline{G}_{11}$ when $x \ll 1$. Using the band structure calculations of
Mattheiss[1] and Yu et al[2] in $2v_F \delta Q$, the mid-point voltage $\delta E \approx 75$ meV
between second and third positive-V peaks in Figure 2 would in this case
mean that $x \approx 0.04$. This value could in fact be occurring for the one
probe spot among four where Naito et al. observed the unusual dI/dV versus
V curve of Figure 1. However in the band structure calculations,[1,2] a
Van Hove singularity in the electron density of states occurs at
50-100 meV below the $x = 0$ Fermi level for La_2CuO_4. When the filling
factor of the $3d_{x^2-y^2}-2p$ band is reduced with $x \neq 0$, this density of
states singularity probably controls $\underline{Q}$ for sufficiently large x, say
$x \approx 0.05$. When E_F lowers to the level of the Van Hove singularity,
reducing the filling factor through a certain range thereafter does not
lower E_F significantly. In this case, for $x \approx 0.15$ the incommensurability
δQ would result in Umklapp replicas of the usual BCS density of states,
displaced from E_F by roughly ± 100 meV, and consistent with the dI/dV
versus V curve of Figure 2.

The triple-peak structure in the Stanford electron tunnelling studies
of LaSrCuO may possibly be caused by an incommensurate SDW. If this is
so, then experiments extending the voltage range should show the BCS-like
Fermi level density of states structure replicated more than once on each
voltage side, so that more than three peaks in dI/dV occur on each voltage
side.

REFERENCES

1. L.F. Mattheis, Phys. Rev. Lett. $\underline{58}$, 1028 (1987).
2. J. Yu, A.J. Freeman and J.-H. Xu, Phys. Rev. Lett. $\underline{58}$, 1035 (1987); and A.J. Freeman, J. Yu and C.L. Fu, preprint.
3. T.M. Rice and G.K. Scott, Phys. Rev. Lett. $\underline{35}$, 120 (1975).
4. C. Politis (KFK Karlsruhe) and H.L. Luo (San Diego), reported at American Physical Society meeting in New York City, 18 March 1987.
5. M. Naito, D.P.E. Smith, M.D. Kirk, M.R. Hahn, B. Oh, K. Char, D.B. Mitzi, D.J. Webb, J.Z. Sun, M.R. Beasley, O. Fisher, T.H. Geballe, R.H. Hammond, A. Kapitulnik, C.F. Quate, preprint, "Electron Tunnelling Studies into Thin Films of High-T_c Superconducting LaSrCuO".
6. P.A. Lee, T.M. Rice and P.W. Anderson, Solid State Comm. $\underline{14}$, 703 (1974).
7. E.W. Fenton, Solid State Comm. $\underline{54}$, 633 (1985); ibid $\underline{56}$, 1033 (1985); ibid $\underline{57}$, 241 (1986).
8. M.J. Rice, Solid State Comm. $\underline{25}$, 1083 (1978).
9. E.W. Fenton and G.C. Psaltakis, Solid State Comm. $\underline{47}$, 767 (1983).
10. E.W. Fenton and G.C. Aers, Mol. Cryst. Liq. Cryst. $\underline{119}$, 201 (1985).
11. E.W. Fenton and G.C. Aers, Proc. NATO Advanced Study Institute on Low-Dimensional Metals and Superconductors, Mt. Orford, Quebec (Plenum, 1987).
12. T. Timusk, private communication.

THE EFFECT OF STRUCTURAL AND ANTIFERROMAGNETIC

PHASE TRANSITIONS ON SUPERCONDUCTIVITY

A. A. Gorbatsevich, V. Ph. Elesin,
and Yu. V. Kopaev

P. N. Lebedev Physics Institute
Academy of Sciences of the USSR
Moscow, USSR

Band calculations for La_2CuO_4 [1] and $YBa_2Cu_3O_7$ [2, 3] show that there
exists a nesting of Fermi surfaces which seems to be associated with structural (CDW) and antiferromagnetic (SDW) transitions in $La_{2-x}Me_xCuO_4$ [4, 5].
In $La_{2-x}Me_xCuO_4$ the nesting vector corresponds to the dimerization of the
period. The observed CDW period, due to the tilting mode, opens a dielectric
gap only if the degeneracy of the valence orbitals of the Cu-O complex [6]
is taken into account, contrary to the ordinary Peierls situation.

There is no experimental evidence for CDW or SDW phase transitions
in $YBa_2Cu_3O_{7-\delta}$. However, it should be kept in mind that the nesting vector
[2, 3] corresponds to an incommensurate structure generated in the Cu-O
layer plane. Such a structure often tends to form clusters. This prevents
detection of CDW or SDW transitions in X-ray and neutron diffraction experiments. The diffuse scattering technique seems preferable in this context.

It has been shown in [7, 8, 9] that peculiarities in state densities
at CDW transitions may result in considerable enhancement of the superconducting transition temperature. These works report on the spectrum dielectrization due to the CDW effect on superconductivity (studied) within the weak
interaction limit, assuming that the dielectric gap is isotropic and opens
on the whole Fermi surface. In real systems, e.g., in superconducting ceramics, one can neither consider these interactions weak, nor the dielectric
gap isotropic. With increasing doping, the nesting of Fermi surface sections is dominant only in certain directions of k-space. Under this condition the dielectric gap opens only on part of the Fermi surface, meeting
the requirements of "nesting." The remaining part of the Fermi surface acts
as a reservoir for charge carriers. The location of the Fermi level in a
rearranged phase is found from the equation for the number of extra particles
n:

$$\tilde{n} = \frac{n}{2N_D} = \sqrt{\mu^2 - \Delta_D^2} + P(\mu - \mu_0) \tag{1}$$

where Δ_D is the dielectric gap, N_D the density of states on "nesting" sections of the Fermi surface, $\mu_0 = \tilde{n}$ the chemical potential at $\Delta_D = 0$, $P =
N_\rho/N_D$ the power of the reservoir, and N_ρ the state density in the reservoir.
From (1) it follows that when the reservoir power P exceeds the critical
value $P_c = \tilde{n}/(\Delta_D - \tilde{n})$, the chemical potential level is inside the dielectric

gap. This situation was studied in [10]. And the spectrum dielectrization was shown to suppress the superconductivity. Quite a different situation occurs at $P < P_c$, with the chemical potential level located in the
region of singular density of states above the gap ($\mu > \Delta_D$). First we consider the case of CDW. Within the weak interaction limit the equation for
the superconducting transition temperature T_S has the form

$$\frac{1}{\lambda} = \frac{1}{1 + P} \int_0^{\tilde{\omega}} \frac{d\xi}{2E} \left(\tanh \frac{E + \mu}{2T_S} + \tanh \frac{E - \mu}{2T_S}\right) +$$

$$\frac{\mu}{1 + P} \int_0^\infty \frac{d\xi}{2E} \left(\frac{\tanh \dfrac{E - \mu}{2T_S}}{E - \mu} - \frac{\tanh \dfrac{E + \mu}{2T_S}}{E + \mu}\right) + \frac{P}{1 + P} \ln \frac{2\gamma\tilde{\omega}}{\pi T_S} \tag{2}$$

Here $E = \sqrt{\xi^2 + \Delta_D^2}$ and $\tilde{\omega}$ is the cutoff (Debye energy for phonon mechanism
of superconductivity) energy. However, with equal justification, λ could
be attributed to a different mechanisms, to an excitonic mechanism, for
instance. The last term in (2) describes the contribution to the superconducting condensate of reservoir electrons. When $P \leq P_c$ ($\mu \geq \Delta_D$), we
get from (2)

$$T_S = \Delta_D \left(\frac{\lambda}{1 + P}\right)^2 \alpha, \quad \alpha = 2\pi(2\sqrt{2} - 1)^2 \zeta^2 (-\tfrac{1}{2}) \tag{3}$$

This result shows that the isotope effect is strongly suppressed in the
system under consideration because the Debye frequency (cutoff energy) does
not enter into the equation for T_S explicitly. When $P \geq P_c$ ($\mu < \Delta_D$), the
transition temperature jumps down to the value $T_{SO} \ll T_S$, obtained in [10],
which approximately agrees with the BCS theory for reservoir electrons
($T_{SO} \approx (2\gamma/\pi)\tilde{\omega} \exp(-1/\lambda)$). From the selfconsistency equation for the superconducting order parameter $\Delta_S(0)$ at $T = 0$ in model (2), it follows that

$$\frac{2\Delta_S(0)}{T_S} = \frac{K(\pi/4)^2}{2\pi((2\sqrt{2} - 1)\zeta(-1/2))^2} \approx 3.7 \tag{4}$$

where $K(\alpha)$ is the elliptic integral. Equation (4) is not dependent on P
and approximately equals the value of $2\Delta_S(0)/T_S$ in the BCS theory. However,
model (2) allows a wide range of values of this ratio. As a result of (4),
the dielectric order parameter Δ_D was not changed in the interval from
$T = 0$ to $T = T_S$. If we reject this assumption, both the increase of Δ_D
with reduction of T and it decreases below a certain temperature $T' < T$
become possible, depending on the parameter describing the deviation from
perfect nesting [11]. In the former case, the ratio $2\Delta_S(0)/T_S$ is less than
the value given by (4), in the latter, it is greater. Nonmonotonic behavior
of Δ_D accompanied by a temperature change qualitatively agrees with the observed nonmonotonic dependence of the orthorhombic splitting parameter in
$La_{1-x}Ba_xCuO_4$ in a wide range of temperatures. Taking into account the interelectron interaction nonlocalities leads to superconducting gap anisotropy (the value of Δ_S is minimal in the reservoir area) and decreases the
ratio (4). The interpretation of kinetic (NMR, ultrasonic attenuation) and
optical experiments should take into account the complexity of the interference of superconducting and dielectric order parameters which results
in renormalization of the kinetic coherence factors and generates several
thresholds in the infrared spectrum (see references in [9]).

Up to now it was assumed that the system is uniform and its parameters,
including the order parameter, are constants. But it is also known that
the presence of carriers in the region above the dielectric gap initiates
instability with respect to the transition to a nonuniform state [12].

This situation possibly takes place in the Y-ceramics as stated before. The density of states dependence on energy, which in the uniform case contains square-root singularity at the edge of the band, is characterized by the dimensionality of the system in the nonuniform phase. If the initial spectrum is quasi-one-dimensional, there is the square-root singularity at the band edge in the nonuniform case as well. This confirms the exact solution of quasi-one-dimensional models with "nesting." In the two-dimensional case, this singularity is smeared. In the three-dimensional case this smearing is more pronounced. If superconducting and dielectric order parameters co-exist, a system of equations can be obtained and the nonuniform problem can be integrated exactly only for the simplest co-ordinate dependence of the parameters

$$\Delta_D(\vec{r}) = \Delta_D e^{i\vec{q}\cdot\vec{r}}, \quad \Delta_S(\vec{r}) = \Delta_S e^{i\vec{q}\cdot\vec{r}} \tag{5}$$

For a three-dimensional initial spectrum the density of states in a dielectric phase with the order parameter (5) is

$$N(E) = \frac{N_0}{2V_F q} \left(\sqrt{(E + V_F q)^2 - \Delta_D^2} - \sqrt{(E - V_F q)^2 - \Delta_D^2} \right). \tag{6}$$

The direct calculation shows that the superconducting transition temperature decreases quadratically according to the parameter $V_F q / T_S$. If $V_F = 10^7$ cm/s, $T_S = 5\cdot10^{-3}$ eV, the characteristic scale of the nonuniformity at which smearing of the densities of states only slightly changes the temperature of the superconducting transition as compared to the nonuniform case should exceed 10^{-6} cm. In reality there always exist excitations spoiling exact integrability. In this situation the spectrum is no longer final-band. This can be simulated by an order parameter of the standing wave kind:

$$\Delta_D(\vec{r}) = \Delta_D \cos \vec{q}\cdot\vec{r}. \tag{7}$$

Thus the order parameter (5) mixes the state involving $\vec{k}$ only with the state involving $\vec{k} + \vec{q}$, while the order parameter (7) mixes states which differ in vector $\vec{q}$ ($\vec{k} + \vec{q}$, $\vec{k} + 2\vec{q}$, $\vec{k} + 3\vec{q}$, ...). In a most physically real case, when $V_F q / \Delta_D \ll 1$, the calculation of the densities of states gives the following:

$$N(E) = N_0 \eta \begin{cases} \dfrac{E}{\Delta_D} K\left(\dfrac{E}{\Delta_D}\right) & E < \Delta_D \\[2ex] K\left(\dfrac{\Delta_D}{E}\right) & E > \Delta_D \end{cases} \tag{8}$$

here η is a numerical coefficient which is defined by the system dimensions, K the elliptic integral. The nonzero value of $N(E)$ with $E < \Delta_D$ is due to the contribution of the areas with $\Delta_D(r) \to 0$ (7). With $E \to \Delta_D$ the equation logarithmically diverges as compared to (6). However, if (6) approaches the value of the prior example at an energy scale of the order of $V_F q$, (8) reaches the same values at a scale of the order of Δ_D. Thus, the smearing of the density of states in a nonuniform system of a periodic type (7) can considerably suppress the superconducting transition temperature as compared to the uniform system.

There is also one more possibility of realizing a nonuniform structure, i.e., cluster formation. The cause of cluster formation can be associated with incommensurate structure pinning on impurities, and with commensurate potential of the lattice, as well as with attraction of solitons (domain boundaries). In the process of clustering the generally uniform phase areas with a dielectric gap (the short-range order) are separated by metallic areas. If the cluster dimension of the dielectric phase considerably exceeds the radius of the Cooper pair, the dielectrization effect on superconductivity is similar to that of the uniform dielectric phase. When $\xi(T)$ $(T \to T_S)$ approaches the cluster dimensions, the latter act as a periodical structure of the (7) kind with respect to the superconductivity, which suppresses the state density. Because of the increase in superconducting energy, the system tends to change lattice parameters within a narrow region near T_S in just such a way, which increases the degree of dielectrization. Similar behavior has been observed experimentally [13].

The results obtained above could be systematized for strong interactions. When $\Delta_D \gg \max(\mu - \Delta_D, \Delta_S \ln \tilde{\omega}/\Delta_S)$, the fundamental contribution to the right-hand side of (2) is made by the term containing the singular denominator $(E - \mu)$. In this approximation it is possible to restrict our consideration to a single-band model with a singularity in the density of

states of a dimensional type: $N(\varepsilon) = N_0\sqrt{E_0/\varepsilon}$. This approximation also allows us to solve the Eliashberg equation.

In the case of intermediate coupling $(\lambda \sim 1)$, the damping appears to be small. Regarding the smallness of the damping and the decrease of the density of states with energy, the following equation for T_S can be obtained:

$$1 = \frac{\lambda\sqrt{E_0}}{(1 + P)(1 + \tilde{\lambda})} \int_0^\infty \frac{d\varepsilon}{\varepsilon^{3/2}} \tanh \frac{\varepsilon}{2T_S}$$

$$\tilde{\lambda} = \pi\sqrt{E_0} \int_0^\infty \frac{d\Omega\alpha^2(\Omega)F(\Omega)}{\Omega^{3/2}} \tag{9}$$

The solution of Eq. (9) coincides with (3) if $\lambda \to \lambda/(1 + \tilde{\lambda})$ is substituted. It is interesting to compare the value T_S derived from (9) with the McMillan result $T_S^{(M)}$ (supposing $\tilde{\Omega} = E_0$, $\tilde{\lambda} = \lambda$):

$$T_S/T_S^{(M)} = 1.45\alpha z^2 \exp(1/z), \quad z = \lambda/(1 + \tilde{\lambda})$$

The minimal value is obtained when $\lambda = 1$ $(z = 1/2)$ and approximately equals 20. The gap $\Delta_S(0)$ at $T = 0$ can be found in a similar way. The ratio $2\Delta_S(0)/T_S$ coincides with (4).

In the case of very large values $(\lambda \gg 1, T_S > \bar{\Omega})$, when renormalization effects of the spectrum and the electron damping become very significant, it is possible to demonstrate by the scaling approach, following [14], that

$$T_S \simeq (\lambda)^{2/5}(E_0^{1/5} \bar{\Omega}^{4/5}).$$

The problem of the coexistence of SDW and superconductivity has some pecularities. In general, there are two types of superconducting solutions in the presence of a dielectric order parameter: the symmetric solution $\Delta_{s\vec{k}} = \Delta_{s\vec{k} + \vec{Q}}$ (here $\vec{Q}$ = nesting vector), and the antisymmetric one: $\Delta_{s\vec{k}} = -\Delta_{s\vec{k} + \vec{Q}}$. In the case of CDW, the latter was shown to be suppressed by dielectric ordering [8]. The former is enhanced and was considered above. The solution of the self-consistency equations shows that in the presence of SDW the situation is reversed, i.e., the symmetric solution is suppressed and the antisymmetric one is enhanced. In a single-band model the antisymmetric superconducting order parameter must have nodes in $\vec{k}$-space. So in the limit of isotropic s-like interaction, it vanishes. It

can only occur if the interaction potential also has nodes. For singlet
superconducting pairing it must be d-like. All the results obtained above
hold with λ in (2) being replaced by the effective anisotropic constant
and describe the enhancement of anisotropic superconducting pairing by SDW.

The problem of coexistence of three types of ordering, i.e., CDW, SDW,
and superconductivity, can be studied systematically in our model as well.
It can be shown that coexistence is absent in the case of a symmetric superconducting solution, and if the coupling constant for triplet dielectric
pairing (SDW) is smaller than for singlet pairing (CDW), and vice versa
in the case of an antisymmetric superconducting solution.

REFERENCES

1. L. F. Mattheis, Phys. Rev. Lett., $\underline{58}$, 1028 (1987).
2. L. F. Mattheis and D. R. Haman, Preprint.
3. J. Yu, S. Massida, A. J. Freeman, and D. D. Koeling, Preprint.
4. D. McK. Paul, G. Balakrishnan, N. R. Bernhoeft, W. I. F. David, and
    W. T. A. Harrison, Phys. Rev. Lett., $\underline{58}$, 1976 (1987).
5. T. Fujita, Y. Aoki, Y. Maeno, J. Sakurai, H. Fukuba, and H. Fujii,
    Japan J. Appl. Phys., $\underline{26}$, L 368 (1987).
6. Yu. V. Kopaev, N. M. Plakida, and B. A. Volkov, Trieste Conference
    on High-T_S Superconductivity, July (1987).
7. Yu. V. Kopaev, JETP, $\underline{58}$, 1012 (1970).
8. A. F. Rusinov, Do Chan Kat, and Yu. V. Kopaev, JETP, $\underline{65}$, 1984 (1973).
9. Yu. V. Kopaev and A. I. Rusinov, Phys. Lett., $\underline{121}$, 300 (1987).
10. G. Bilbro and W. L. McMillan, Phys. Rev., $\underline{B14}$, 1887 (1976).
11. Yu. V. Kopaev, Trudy FIAN, $\underline{86}$, 3 (1975).
12. T. M. Rice, Phys. Rev., $\underline{B2}$, 3619 (1970).
13. A. I. Golovashkin et al., this volume.
14. J. P. Carbotte, F. Marsiglo, and B. Mitrovic, Phys. Rev., $\underline{B33}$, 6135
    (1986).

EFFECTS OF COULOMB INTERACTION ON SUPERCONDUCTIVITY

Yasutami Takada

Institute for Solid State Physics, University of Tokyo
7-22-1, Roppongi, Minato-ku, Tokyo 106, Japan

I. INTRODUCTION

In recent high-T_c oxide[1] and heavy-fermion[2] superconductors, electron (or hole) correlations are important in understanding their mechanisms
of superconductivity. At present, however, we have only a little knowledge
about the effects of the Coulomb interaction on superconductivity.[3] In
the conventional theory, all the Coulomb effects are pushed into a single
parameter μ^*, called the Coulomb pseudo-potential[4] which is regarded as a
phenomenological parameter.

In this paper, we aim to make a detailed investigation of the Coulomb
effects by considering a possibility of superconductivity in the electron
gas in which only the Coulomb interaction is included. By using the RPA for
the electron-electron interaction, the present author[5] discussed the same
problem in 1978 in the weak-coupling KMK scheme[6]. He obtained the result
that superconductivity appeared when the electronic density parameter r_s is
larger than 6. Later, a similar result was obtained by Rietschel and Sham
[7] even when the strong-coupling Eliashberg equation[8] was solved instead
of the KMK scheme, as long as the RPA was used. It is clear, however, that
the RPA is not a good approximation for $r_s > 1$. Grabowski and Sham[9] tried
to go beyond the RPA, but their calculation was far from a first-principle
one. Probably the ordinary Green's function method which they employed will
not provide reliable results for this problem, because there are no good
expansion parameters in the theory for $r_s > 1$. We will approach the problem
in the method of effective-potential expansion[10] which is known to provide
a very accurate description of the normal state of the electron gas[11].
The results for the correlation energy ε_c in the Green's function Monte
Carlo (GFMC) method[12] are reproduced by the method with relative errors
less than 1% for $1 \leq r_s \leq 10$.

We describe the method briefly in Sec.II and give the results for the
normal-state properties in Sec.III. A gap equation is derived and relation
with the conventional theory of superconductivity is accounted for in Sec.
IV. We show calculated results for T_c in Sec.V and discuss the possibility
to explain high-T_c oxide superconductors in terms of a model of the (anisotropic) electron gas in Sec.VI.

II. CORRELATED BCS STATE

The Hamiltonian for the electron gas in a uniform positive background is written in second quantization as

$$H = \sum_{k\sigma} \varepsilon_k C_{k\sigma}^{\dagger} C_{k\sigma} + \frac{1}{2} \sum_{q\neq 0} \sum_{k\sigma} \sum_{k'\sigma'} V(q)\, C_{k+q\sigma}^{\dagger} C_{k'-q\sigma'}^{\dagger} C_{k'\sigma'} C_{k\sigma} \, , \qquad (1)$$

with $\varepsilon_k = k^2/2m^*$ and $V(q) = 4\pi e^2/\kappa q^2$ where m^* is the band mass and κ is the dielectric constant. For the ground-state wave function, we consider the following trial function[10]:

$$|\Phi\rangle = \sum_{n=0}^{\infty} \frac{1}{n!} \left[\sum_{m=1}^{\infty} U_m \right]^n |0\rangle \, , \qquad (2)$$

where $|0\rangle$ is the BCS state[13], defined by the vacuum state of the Bogoliubov operator[14] $\alpha_{k\sigma}$ which is transformed from $C_{k\sigma}$ as

$$\alpha_{k\uparrow} = u_k C_{k\uparrow} - v_k C_{-k\downarrow}^{\dagger} \qquad \text{and} \qquad \alpha_{-k\downarrow} = u_k C_{-k\downarrow} + v_k C_{k\uparrow}^{\dagger} \, . \qquad (3)$$

The correlation operator U_m is defined with the use of the long- and short-range parts of the effective potential, V_ℓ and V_s, as

$$U_m = (-i)^m \int_{-\infty}^{0} e^{0^+ t_1} dt_1 \cdots \int_{-\infty}^{0} e^{0^+ t_m} dt_m \{ T[e^{iH_0 t_1} V_\ell e^{-iH_0 1} \cdots$$
$$e^{iH_0 t_m} V_\ell e^{-iH_0 t_m}]_L /m!$$
$$+ T[e^{iH_0 t_1} V_s e^{-iH_0 t_1} e^{iH_0 t_2} V_\ell e^{-iH_0 t_2} \cdots e^{iH_0 t_m} V_\ell e^{-iH_0 t_m}]_L /(m-1)! . \qquad (4)$$

where the symbol T and the subscript L represent the T product and the instruction to collect only terms described by linked graphs. We assume the following forms for H_0, V_ℓ, and V_s:

$$H_0 = \sum_{k} e_k \left(\alpha_{k\uparrow}^{\dagger} \alpha_{k\uparrow} + \alpha_{-k\downarrow}^{\dagger} \alpha_{-k\downarrow} \right) , \qquad (5)$$

$$V_\ell = \frac{1}{2} \sum_{q} \sum_{kk'} V_\ell(q)\, \Gamma_{k+q,k} \Gamma_{k'+q,k'} \left(\alpha_{k+q\uparrow}^{\dagger} \alpha_{-k\downarrow}^{\dagger} + \alpha_{-k\downarrow} \alpha_{k+q\uparrow} \right)$$
$$\times \left(\alpha_{-k'-q\uparrow}^{\dagger} \alpha_{k'\downarrow}^{\dagger} + \alpha_{k'\downarrow} \alpha_{-k'-q\uparrow} \right), \qquad (6)$$

and

$$V_s = \frac{1}{2} \sum_{q} \sum_{kk'} V_s(q)\, \Gamma_{k+q,k} \Gamma_{k'+q,k'} (\alpha_{k+q\uparrow}^{\dagger} \alpha_{-k\downarrow}^{\dagger} \alpha_{-k'-q\uparrow}^{\dagger} \alpha_{k'\downarrow}^{\dagger} + c.c), \qquad (7)$$

where

$$V_\ell(q) = V(q)\exp(-q^2/q_c^2), \quad e_k = |k^2 - k_F^2|/2m^*, \quad \text{and} \quad \Gamma_{k+q,k} = u_{k+q} v_k + u_k v_{k+q} .$$

The parameter q_c will be determined variationally in the region $0 \leq q_c \leq 0.1 k_F$ in which k_F is the Fermi momentum.

The expectation value for the operator $H-\mu N$ with respect to $|\Phi\rangle$ can be calculated by using the linked-cluster theorem[10], where μ is the chemical potential and N is the total number operator. We will write $\langle H-\mu N\rangle$ in power series of V_s as

$$\langle H- \mu N \rangle = \sum_{n=0}^{\infty} \Omega^{(n)} \, . \qquad (8)$$

The n-th order term in V_s, $\Omega^{(n)}$, is still expanded in V_ℓ up to infinite order. On the assumption that q_c is small, important contribution will come only from ring terms in the expansion. As a result, we obtain an expression for $\langle H-\mu N\rangle$ which is almost the same as in ref.11. The only difference is that the effective potential $\tilde{V}$ in the paper is replaced by $\bar{V}_S$ which is defined by

$$\bar{V}_S(q) \equiv V_S(q)/[1+V_\ell(q)\ \Pi(q,0)]^2 \ , \tag{9}$$

with the static polarization function in the RPA $\Pi(q,0)$. According to the results in ref.11, $\bar{V}_S(q)$ is small for all q even when r_S is as large as 10. Therefore, we can neglect all terms higher than second order in $\bar{V}_S$. Terms considered in $\Omega^{(0)}$ are represented diagrammatically in Fig.1 in which E_k is defined as

$$E_k = [\varepsilon_k-\mu-\sum_{k'} V(k-k')v_{k'}^2]\ (u_k^2-v_k^2) - 2\sum_q V(q)u_{k+q}v_{k+q}u_k v_k. \tag{10}$$

Similarly, diagrams for $\Omega^{(1)}$ and $\Omega^{(2)}$ are shown in Figs.2 and 3, in which $\bar{V}(q)$ is defined in the same way as in Eq.(9) with the replacement of V_S with V, and the symbol, $\times\varepsilon$, indicates that we should multiply $1+V_\ell\Pi$ in the final expression to avoid double counting of the screening factor.

III. NORMAL STATE PROPERTIES

In the normal limit, i.e., $v_k^2=n_k$ and $u_k^2=1-n_k$ with $n_k=1$ for $k<k_F$ and 0 otherwise, we have determined q_c and $\bar{V}_S(q)$ variationally. Irrespective of

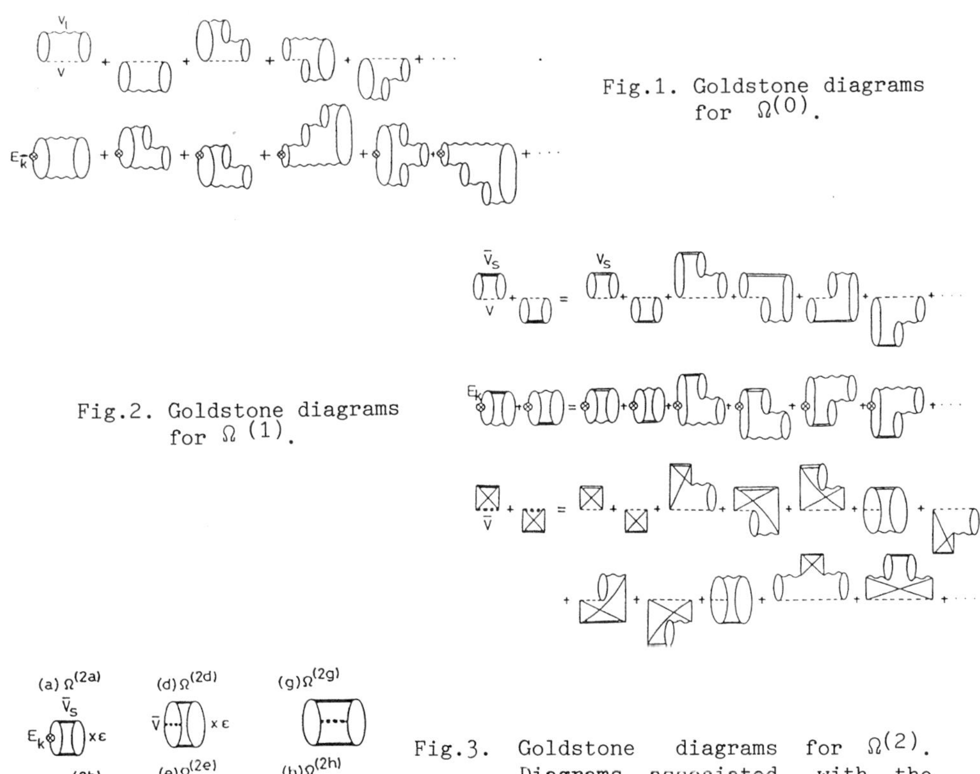

Fig.1. Goldstone diagrams for $\Omega^{(0)}$.

Fig.2. Goldstone diagrams for $\Omega^{(1)}$.

Fig.3. Goldstone diagrams for $\Omega^{(2)}$. Diagrams associated with the symbol, $\times\ \varepsilon$, indicate that we should multiply $1 + V_\ell\Pi$ in the final expression to avoid double counting of the screening factor.

437

r_s, $\langle H - \mu N \rangle$ has been found to have a local minimum at $q_c = 0.06 k_F$. Obtained results for $\bar{V}_s(q)$ depend on r_s, but $\bar{V}_s(q)$ is always much smaller than the bare potential $V(q)$ in the entire region of q as shown in Fig.4. (It must be noted that when $V_s(q)$ is taken to be zero from the outset, $\langle H - \mu N \rangle$ is reduced to the result in the RPA and has a minimum at $q_c = \infty$.) The present results for ε_c agree with those in the GFMC with errors less than 8% for $1 \leq r_s \leq 20$. An example of the calculated results for $N_k = \langle C_{k\sigma}^\dagger C_{k\sigma} \rangle$ is given in Fig.5, in which we have also plotted the renormalization factor z_k^{-1}, defined by

$$z_k^{-1} = (2N_k - 1)(2n_k - 1) \quad \text{for} \quad k \neq k_F \quad . \tag{11}$$

At the Fermi surface, there is a discontinuity in N_k as well as in z_k^{-1}, but the magnitude of the discontinuity in N_k is equal to the average of z_k^{-1} at $k = k_F \pm 0^+$ which defines the renormalization factor at the Fermi surface. In Fig.6, the calculated results for z_k^{-1} at $k = k_F$ are shown as a function of r_s. Results obtained by Rice[15] and Hedin[16] are indicated by the solid circles and crosses, respectively. At $r_s = 14.3$, z_k^{-1} becomes zero at $k = k_F$, which suggests that a metal-insulator transition occurs at this density.

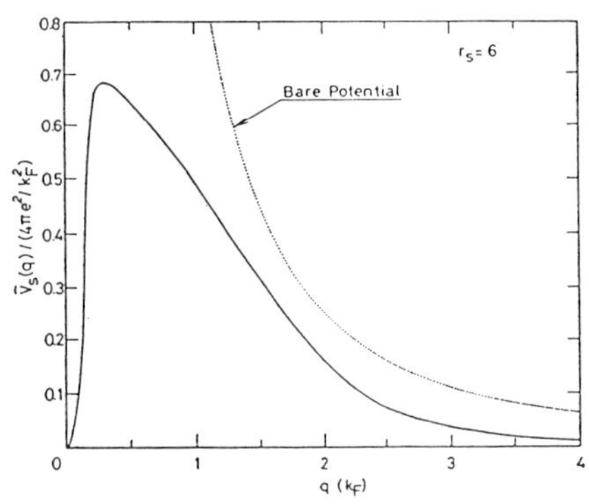

Fig.4. Short-range part of the effective potential $\bar{V}_s(q)$ normalized by $4\pi e^2/\kappa k_F^2$ for $r_s = 6$.

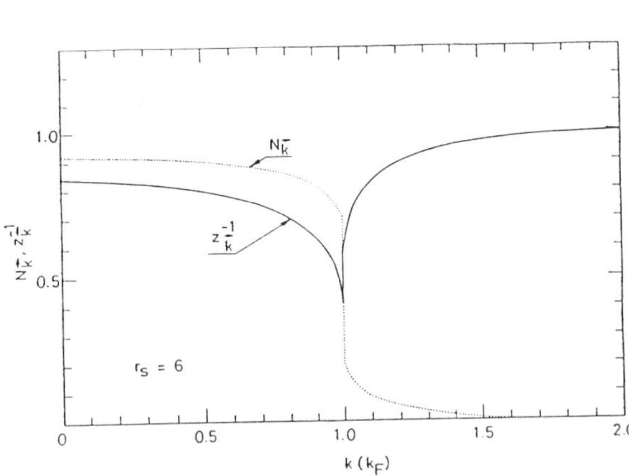

Fig.5 The occupation number N_k (the dotted curve) and the renormalization factor z_k^{-1} (the solid curve) as a function of k for $r_s = 6$.

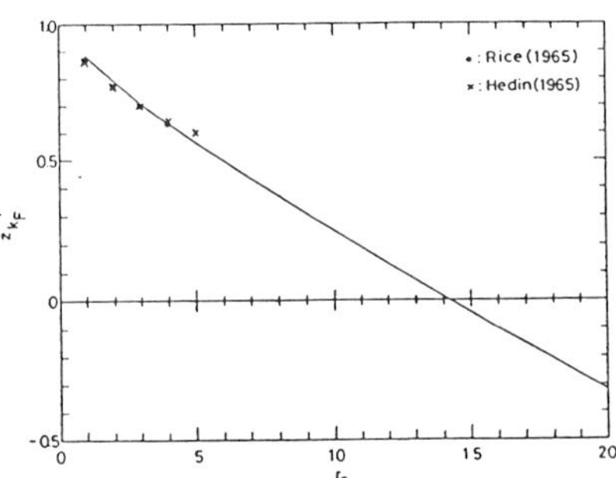

Fig.6 Renormalization factor at the Fermi surface as a function of r_s. The results given by Rice[15] and Hedin[16] are indicated by the solid circles and the crosses, respectively.

IV. GAP EQUATION

For given V_ℓ and V_s, a gap equation can be obtained by the optimization condition for v_k, i.e., $\delta\langle H-\mu N\rangle/\delta v_k=0$. We can express the equation for a gap Δ_k at T=0 as

$$\Delta_k = -\sum_{k'} V_p(k,k')\, \Delta_{k'}/2\sqrt{\tilde\varepsilon_{k'}^2+\Delta_{k'}^2} \quad. \tag{12}$$

Here, both the pairing potential $V_p(k,k')$ and the single-particle energy $\tilde\varepsilon_k$ are expressed in power series of V_ℓ and V_s. They have the following structure:

$$V_p(k,k') = V_p^*(k,k')/z_k z_{k'}, \quad \text{and} \quad \tilde\varepsilon_k = \varepsilon_k^*/z_k, \tag{13}$$

where in the normal limit, ε_k^* and V_p^* are represented diagrammatically in Figs. 7 and 8, respectively. In Fig.8, a term which has repulsive (attractive) contribution to V_p^* is indicated by the sign +(-). In the RPA in which $V_s(q)=0$ and $V_\ell(q)=V(q)$, the present potential V_p^* is reduced to the KMK pairing potential derived in the RPA[5,6]. Thus in view of Eq.(13), our present theory may be regarded as an extension of the KMK theory to the strong-copling region.

Except at $k=k'=k_F$, neither V_p nor V_p^* is a physical electron-electron interaction in this variational theory of superconductivity. The same is true for the single-particle energy. For $k\neq k_F$, both $\tilde\varepsilon_k$ and ε_k^* are different from a physical excitation energy. We can calculate the gap at the Fermi surface more easily by introducing these unphysical quantities than using physical ones. This situation reminds us of the density-functional theory[17] in which we use unphysical Kohn-Sham single-particle energy eigenvalues to obtain the physical charge density easily[18]. Though they do not represent the real energy levels, the Kohn-Sham eigenvalues often give a right structure of the density of states near the Fermi surface qualitatively. Thus by evaluating $(\tilde\varepsilon_k^2+\Delta_k^2)^{1/2}$, we may obtain a qualitative structure of the density of states in the superconducting state.

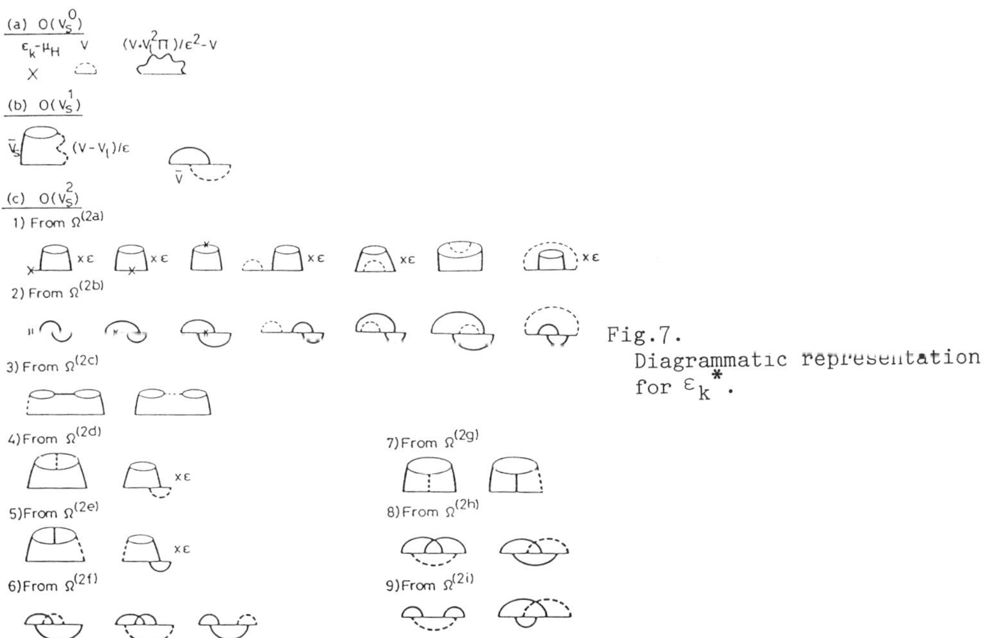

Fig.7.
Diagrammatic representation for ε_k^*.

V. NUMERICAL RESULTS

Numerically, we can perform rigorous integrations for ring terms, but the computation time compels us to make approximate calculations for ring-exchange, ladder, and ladder-exchange terms. For example, the contribution to ε_k^* from the first-order exchange term is given by

$$\varepsilon_k^{(1e)*} = 2 \sum_{qk'} \overline{V}(q) \; [-\overline{V}_s(k-k')n_{k+q} + \overline{V}_s(q+k+k')(1-n_{k+q})]n_{k'}(1-n_{k'+q})$$

$$/ \; (e_k + e_{k'} + e_{k+q} + e_{k'+q}) \; . \tag{14}$$

By replacing $\overline{V}_s(k-k')$ and $\overline{V}_s(q+k+k')$ with $\overline{V}_s(\sqrt{k^2+<k>^2})$ and $\overline{V}_s(\sqrt{(k+q)^2+<k>^2})$, respectively, we can calculate Eq.(14) with the same efforts as for ring terms. Here, $<k>$ is considered to give an average value of $|k'|$ in the k'- integral. Other terms in ε_k^* and V_p^* are also calculated in the same way. Namely we treat the Fermi factors like $n_{k'}(1-n_{k'+q})$ and the energy denominators rigorously, but we use some average values for the interactions like $V_s(k-k')$ in the integrand by introducing $<k>$. For Eq.(14), the best choice for $<k>$ is found to be $0.78k_F$. However for other terms, the optimum value for $<k>$ is probably different. Thus we will calculate ε_k^* and V_p^* for several values of $<k>$ and show its dependence.

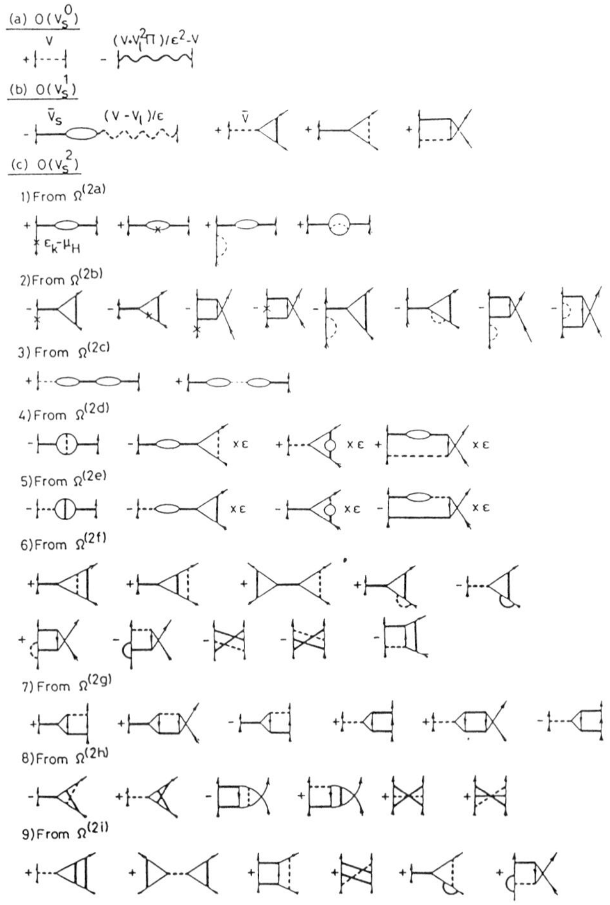

Fig.8.
Diagrammatic representation for $V_p^*(k,k')$. The sign $+$ $(-)$ indicates repulsive (attractive) contribution.

The correlation part of the chemical potential μ_c, obtained from ε_k^* at $k=k_F$, is given in Fig.9 in units of $Ry^*(=m^*e^4/2\kappa^2)$ as a function of r_s for three different values of $\langle k \rangle$. The results for $\langle k \rangle=0.85k_F$ reproduce those in the GFMC with relative errors less than 2.5% for $1\leqq r_s\leqq 10$. Although μ_c depends rather strongly on $\langle k \rangle$, $\tilde{\varepsilon}_k$ itself is virtually independent of it, as can be seen in Fig.10 for $r_s=8$. For $0.965k_F<k<k_F$, $\tilde{\varepsilon}_k$ becomes positive, but such a behavior in $\tilde{\varepsilon}_k$ cannot be seen for $r_s<6$.

We can measure the strength of instability for superconductivity described in Eq.(12) by calculating T_c in the following equation:

$$\Delta_k = - \sum_{k'} V_p(k,k') \frac{\Delta_{k'}}{2\tilde{\varepsilon}_{k'}} \tanh \frac{\tilde{\varepsilon}_{k'}}{2T_c} . \tag{15}$$

This equation can be solved numerically by the division of the integral for $|k'|$ into small intervals, (k_i,k_{i+1}). In Fig.11, an example of the kernel, $K(\bar{k}_i,\bar{k}_j)$, defined by

$$K(\bar{k}_i,\bar{k}_j) \equiv \frac{1}{k_{j+1}-k_j} \int_{k_j}^{k_{j+1}} dk' \int \frac{d\Omega_{k'}}{(2\pi)^3} V_p(k,k') \Big|_{|k|=\bar{k}_i} , \tag{16}$$

is shown at $\bar{k}_i=0.9999998k_F$. In Eq.(16), $d\Omega_{k'}$ is the integral over angular variables of k'. As $\langle k \rangle$ is decreased, the kernel hardly changes its shape but shifts upwards. In particular, for $\langle k \rangle=0.78k_F$ (the solid curve), the calculated kernel at the Fermi surface is about the same as that in the RPA(the dotted curve). As $\bar{k}_j$ goes away from the Fermi surface, the present kernel is much more repulsive than that in the RPA. The difference comes primarily from the repulsive parts in the first-order terms $V_p^{(1+)*}$ in Fig.8. Although superconductivity does not occur in the RPA for $r_s=5$, we

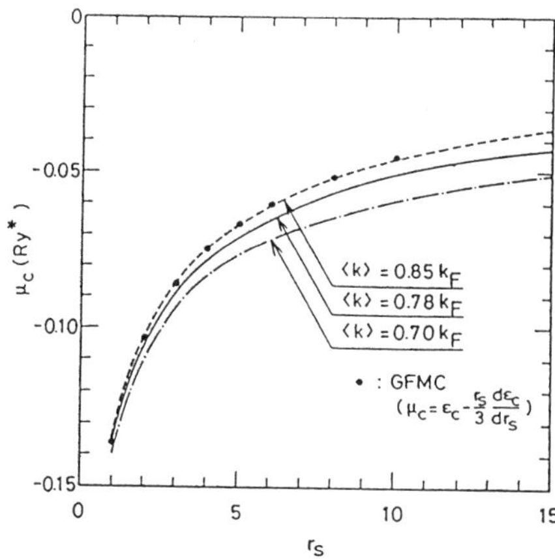

Fig.9. Correlation part of the chemical potential in Ry^*. Solid points provide the results in the GFMC.

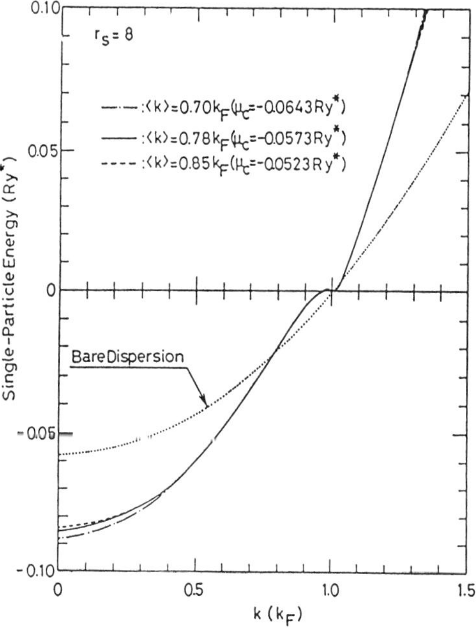

Fig.10 Single-particle energy $\tilde{\varepsilon}_k$ in Ry^* as a function of k for $r_s=8$.

have obtained a finite T_c even for $\langle k \rangle = 0.70 k_F$. (Calculated T_c is given in Fig.10 in units of K^* ($= m^*/m_e \kappa^2$ degrees Kelvin), where m_e is the mass of a free electron.) This example illustrates several important points: (1) Superconductivity occurs even if the kernel is always positive. (2) We cannot discuss the occurrence of superconductivity only by the examination of the pairing potential at the Fermi surface. (3) It is not true that $V_p^{(1+)*}$ is always harmful to superconductivity. Actually, the contribution of $V_p^{(1+)*}$ in the regions $\bar{k}_j < 0.95 k_F$ and $\bar{k}_j > 1.05 k_F$ is favorable for super-conductivity, because Δ_k changes its sign in these regions and $V_p^{(1+)*}\Delta_k$ becomes negative. Thus we can conclude that although the plasmon (the attractive part in the first-order term $V_p^{(1-)*}$) plays a primary role in the occurrence of superconductivity as stated in previous publication[5], both the vertex corrections to the plasmon-mediated interaction and the potential coming from the exchange of spin fluctuations (the part $V_p^{(1+)*}$) are also important, especially for the processes having the momentum transfer of the order of k_F which contribute to the kernel for $\bar{k}_i = k_F$ and $\bar{k}_j < 0.95 k_F$ or $\bar{k}_j > 1.05 k_F$.

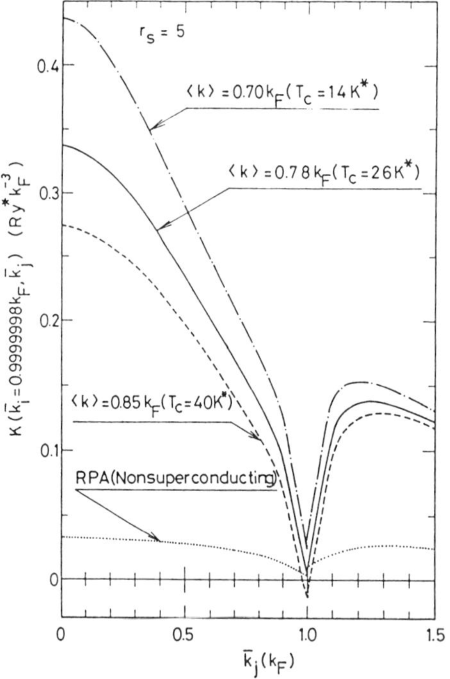

Fig.11. The kernel $K(\bar{k}_i, \bar{k}_j)$ at $\bar{k}_i = 0.9999998 k_F$ in units of $Ry^* k_F^{-3}$ for $r_s = 5$.

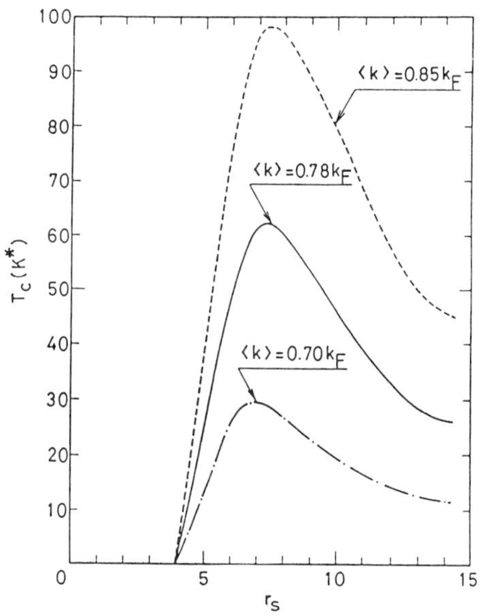

Fig.12. Calculated results for T_c in units of K^* for the electron gas as a function of r_s.

442

We have given the results for T_c as a function of r_s in Fig.12. Super-conductivity appears for $r_s > r_{sc} = 3.9$. As r_s increases, T_c increases first, reaches its maximum, T_{cm}, at $r_s = 7.5$, and then decreases. The value T_{cm} depends on $<k>$ but is in the range 10-100K*.

VI. DISCUSSION

Since relative errors for ε_c and μ_c between the present theory and the GFMC are less than 10% for $1 \leq r_s \leq 8$, the kernel K is also expected to have a relative error less than 10%. When we have changed the kernel by 10% artificially, r_{sc} is found to become as large as 6 and T_{cm} changes as much as a factor of three. We have also checked errors in r_{sc} and T_{cm} by changing the parameters q_c and $<k>$. For example, for $q_c = 0.4 k_F$, r_{sc} is about 6 and T_{cm} is in the range 1-10K*.

There are only a few metals whose r_s values are larger than 3.9. The most familiar examples are the alkali metals, Na, K, Rb, and C_s, but these metals do not show superconductivity experimentally. Discrepancy between experiment and the present theory may be explained in various ways: First, a model of the electron gas without phonons may not be suitable for superconductivity in those metals. When we include phonons with the Coulomb interaction simultaneously, there is a possibility that phonons work destructively. Second, these metals may already be in other ordered states like the CDW one[19]. Thirdly, r_{sc} may be larger than 5.62 which is the r_s value for C_s. Lastly, the usual r_s values for those metals are defined with $m^* = m_e$ and $\kappa = 1$. However, κ may be larger than unity to make r_s smaller than 3.9 even for C_s.

Another system to which we have to pay attention is the recent high-T_c oxide superconductors[1]. Although the r_s values for those materials are not known yet, they are probably larger than r_{sc}. At present, it is a general trend to emphasize one or two dimensionality of the system. However, to have bulk superconductivity without fluctuations, we have to consider the system in three dimensions. Thus there is a standpoint to regard these oxides as an anisotropic three-dimensional electron gas. The band mass m^* , estimated by a geometrical average over three axes will be much larger than m_e. Besides κ will be of the order unity, because any ferroelectric transition has not reported so far in contrast to the case of $SrTiO_3$[20] in which κ becomes as large as 10^4. Thus we expect the factor $m^*/m_e \kappa^2$ to be of the order of unity, which predicts that T_c is in the range 10-100K.

Such a simple model as the electron gas is probably insufficient to explain all the details of experiments. However, we cannot find any apparant inconsistency with experiments. On the contrary, there are several experiments which are consistent with our model: (1) Bell's group[21] has found the absence of an isotope effect, indicating a mechanism other than the usual phonon one. (2) Both Tsukuba's[22] and Stanford's[23] groups have found three gap structures in tunneling spectroscopy. In our theory, the density of states has three peaks at T=0 for $r_s > 6$, reflecting the fact that $\tilde{\varepsilon}_k$ becomes zero at two different k's near the Fermi surface. (3) Ishikawa's group[24] in our institute has found that the high-T_c phases of these oxides show a weak Pauli paramagnetism, which is consistent with our view of the electron gas. (4) Mori's group[25] in our institute has measured the pressure dependence of T_c and found that T_c increases rather sharply with the increase of the pressure for La-Sr-Cu-O but that T_c does not change so much for Y-Ba-Cu-O. In our view, Y-Ba-Cu-O has the r_s value near 7.5 and T_c does not change so much even if r_s is decreased by the application of the pressure. On the other hand, La-Sr-Cu-O may have the r_s value larger than 7.5 (and consequently is closer to the metal-insulator transition point). Thus with the decrease of r_s, T_c is expected to increase.

REFERENCES

1. See, for example, in :"Physics Today", Vol 40, No.4, p.17 (1987).
2. G.R. Stewart, Rev. Mod. Phys. 56 :755 (1984).
3. See, for example, D. Rainer, in :"Progress in Low Temp. Phys." ed. by D.F. Brewer, Vol. 10, p. 371 (1986).
4. P. Morel and P.W. Anderson, Phys. Rev. 125 : 1263 (1962).
5. Y. Takada, J. Phys. Soc. Jpn. 45: 786 (1978).
6. D.A. Kirzhnits, E.G. Maksimov, and D.I. Khomskii, J. Low Temp. Phys. 10: 79 (1973).
7. H. Rietschel and L.J. Sham, Phys. Rev. B28: 5100 (1983).
8. G.H. Eliashberg, Sov. Phys. -JETP 11:696 (1960).
9. M. Grabowski and L.J. Sham, Phys. Rev. B29: 6132 (1984).
10. Y. Takada, Phys. Rev. A28:2417 (1983).
11. Y. Takada, Phys. Rev. B35:6923 (1987).
12. D.M. Ceperley and B.J. Alder, Phys. Rev. Lett. 45: 566 (1980); S.H. Vosko, L. Wilk, and M. Nusair, Can. J. Phys. 58: 1200 (1980).
13. J. Bardeen, L.N. Cooper, and J.R. Schrieffer, Phys. Rev. 108: 1175 (1957).
14. N.N. Bogoliubov, Sov. Phys. -JETP 7: 41 (1958).
15. T.M. Rice, Ann. Phys. (N.Y.) 31: 100 (1965).
16. L. Hedin, Phys. Rev. 139: A796 (1965).
17. P. Hohenberg and W. Kohn, Phys. Rev. 136: B864 (1964).
18. W. Kohn and L.J. Sham, Phys. Rev. 140; A1133 (1965).
19. A.W. Overhauser, Adv. Phys. 27: 343 (1978); T.M. Giebultowicz, A.W. Overhauser, and S.A. Werner, Phys. Rev. Lett. 56: 1485, 2228 (1986).
20. See, for example, Y. Takada, J. Phys. Soc. Jpn. 49: 1267 (1980).
21. B. Batlogg et at., preprint.
22. I. Iguchi et al., to appear in Jpn. J. Appl. Phys.
23. D.P.E. Smith et al., preprint; M. Naito et al., preprint.
24. Y. Nakazawa et al., to appear in Jpn. J. Appl. Phys. Lett.
25. S. Yomo et al., preprint; S. Yomo et al., Jpn. J. Appl. Phys. 26:L 603 (1987).

THE IMPORTANCE OF DIFFERENT TWO-DIMENSIONAL PLASMON BRANCHES FOR HIGH T_c SUPERCONDUCTORS

V. Z. Kresin

Lawrence Berkeley Laboratory
Berkeley, California

H. Morawitz

IBM Almaden Research Center
San Jose, California

ABSTRACT

The plasmon mechanism of superconductivity in the new high T_c superconductors is characterized by the appearance of two physically distinct branches. They arise from the presence of several partially occupied carrier bands and the reduced dimensionality. Dispersion relations of these branches are derived in RPA and the effect of deviation from RPA is discussed. The possible hybridisation with polar phonon modes and the consequences on contributions to superconductivity and optical properties are presented. Additionally the implications for tunneling and neutron scattering experiments are sketched.

INTRODUCTION

This paper is concerned with several interrelated problems involving the role of plasmons in the high T_c materials. We show that within simple effective mass models of several two-dimensional carrier groups, one obtains two types of plasmon branches, both of which are expected to potentially contribute to superconductivity. In this paper we are going to evaluate the dispersion relations for these branches in the random phase approximation (RPA), apply a method for going beyond RPA and include the hybridisation of the plasmons with polar optical modes of the system. Implications of the contributions of these plasmons to relevant experiments such as tunneling and infra-red absorption are also given.

TWO-DIMENSIONAL PLASMON DISPERSION RELATIONS

We consider the general case of Fermi carriers arising from several overlapping bands. This is a plausible assumption for the new superconducting oxides because of their complicated band structure.[1] For simplicity, we restrict ourselves to the case of two bands in the effective mass approximation. Specifically, we assume $m_2 \gg m_1$. The superconducting order parameter Δ is descibed by the generalized Eliashberg equation:

$$\Delta(\vec{\kappa}, \omega_n) = \pi T \sum_{\omega_n'} \int d\vec{\kappa} \Gamma(\vec{\kappa} - \vec{\kappa}', \omega_n - \omega_n') F^+(\vec{\kappa}', \omega_n'). \tag{1}$$

Here κ is a 2D momentum, F^+ is an anomalous Green's function, $\omega_n = (2n + 1)\pi T$ and Γ is the total electron-boson vertex: $\Gamma = \Gamma_{PL} + \Gamma_{PH}$. In this paper we focus on Γ_{PL}.

Within the RPA the vertex Γ_{PL} is given by

$$\Gamma_{11} = V_{11}/\varepsilon \qquad (2)$$

where V_{11} is the bare electron-electron interaction between carriers of type 1 in the two-dimensional system and

$$\varepsilon = 1 + V_{11}\Pi_1 + V_{22}\Pi_2 + R \qquad (3)$$

with $R = (V_{12}^2 - V_{11}V_{22})\Pi_1\Pi_2$ and

$$\Pi_i = -\frac{m_i}{(2\pi)^2} \int_0^{2\pi} d\phi \; \frac{\cos\phi}{\alpha_i - \cos\phi + i\delta\cos\phi} \quad i = 1, 2 . \qquad (4)$$

The indices (1,2) correspond to the different bands, $\alpha_i = \omega(v_{F_i}q)^{-1}$.

Based on eqs (3) and (4) one can obtain several branches. The case of a single two-dimensional group of carriers corresponds to $\Pi_2 = 0$.

Consider first the case $\alpha_1 < < 1$, $\alpha_2 > > 1$. Then we obtain $\Pi_1 = m_1/\pi$ and $\Pi_2 = -(\varepsilon_{F2}/\pi)(q/\omega)^2$. The root of the equation $\varepsilon = 0$ determines the corresponding dispersion relation and we obtain

$$\omega_d(q) = aq \qquad (5)$$

where

$$a = \left(\frac{\varepsilon_{F2}}{\pi} \; \frac{V_{22}}{1 + V_{11}\Pi_{11}} \right) . \qquad (5a)$$

One can see that even for a two-dimensional system in the presence of overlapping bands, the " demon" branch[2] occurs.

In addition to these "demon" states there exists an additional branch even in the absence of overlapping bands. Its appearance is due to to the low dimensionality only. Such modes have been studied extensively for the 2 Dimensional Electron Gas (2DEG) (for an excellent review see reference 3). This corresponds to $\alpha_i > 1$ ($i = 1,2$). We obtain for this case

$$\Pi_i = m_i/\pi\left[\left(1 - \alpha_i^{-2}\right)^{-1/2} - 1\right] . \qquad (6)$$

To simplify the discussion here, we consider the case of a single band (the generalisation to two bands is straightforward). After some algebraic steps , we obtain

$$\omega = v_F q\rho(q)\left[\rho^2(q) - 1\right]^{-1/2} \qquad (7)$$

with

$$\rho(q) = 1 + q/2e^2 m . \qquad (7a)$$

One can see directly from eq (7) that $\omega(q) \sim q^{1/2}$ in the region of small q. Increasing q leads to a cross-over in the functional dependence of $\omega(q)$ to a linear regime $\omega(q) \sim q$. A more detailed analysis will be given elsewhere.[4]

As a result of our analysis we have found two distinct and different branches: one of them arising from the reduced dimensionality and the other is due to the presence of overlapping bands of sufficiently different masses.

PHENOMENOLOGICAL TREATMENT BEYOND THE RPA

Equations 3 to 7, have been obtained in the RPA. A small value of the carrier concentration n in the system may lead to a large value of the electron gas parameter r_s. Even in that case the analysis based on RPA is still qualitatively valid.[5]

A phenomenological method of going beyond the RPA was described in[5] based based on interpolation of the correlation energy between the high and low density limits. The basic idea is in substituting $\Pi_i \rightarrow \Pi_i^*$, where $\Pi_i^* = \Pi_i(1 + f(q)\Pi_i)^{-1}$ and $f(q) \rightarrow 0$ as $q \rightarrow 0$.

One may use , for example the approximation proposed by Hubbard[6] or the approximation proposed by P. Vashishta and K. Singwi.[7]

We would like to stress at this point that going beyond the RPA in the manner discussed above only modifies the dispersion relations in terms of numerical coefficients of the $q^{1/2}$ and q dependent terms. The effect of this more accurate treatment of electron-electron interactions on the superconducting transition arises from differences in the plasmon term $g_{PL}(\omega)$ in the Eliashberg equation. The function $g_{PL}(\omega)$ can be determined by tunneling spectroscopy.

The inclusion of exchange may affect the numerical value of the coupling constant.[8] On the other hand , analysis of a plasmon mechanism as well as the usual phonon mechanism should not be concerned with the problem of wave-function overlap because both subsystems - the superconducting and the mediating pairing modes - are located in the same spatial region.

ACOUSTIC PLASMON-POLAR PHONON MIXING

In addition to the presence of several low-lying electronic excitations of collective nature such as the "demon"-like branches and the lower-dimensional layer plasmons discussed in the last two sections, we want to draw attention to the potential hybridization of electronic charge density oscillations with ionic charge motion such as polar modes.

As a result of this hybridization, the two branches acquire characteristics of each other as in standard polariton theory. The modes do not cross, but are split-off from each other by an amount depending on the strength of the coupling. The lower branch strongly disperses in the intermediate frequency region ($100 \text{ cm}^{-1} < \omega < 1000 \text{ cm}^{-1}$). As a consequence the plasmon density of states behaves similarly to the usual phonon density of states (Fig. 1).

The function $g_{PL}(\omega)$ is therefore characterized by a strong peak in the intermediate frequency range. This peak has a structure similar to the usual phonon peak and can be determined by tunneling spectroscopy.[9] Because of the mixing of electronic and ionic degrees of freedom, it seems plausible that optical measurements in this frequency region will be enhanced over purely ionic oscillator strength. There is recent experimental evidence for very strong absorption in the phonon region in both the La-Ba-Cu-O and Y-Ba-Cu-O systems.[10]

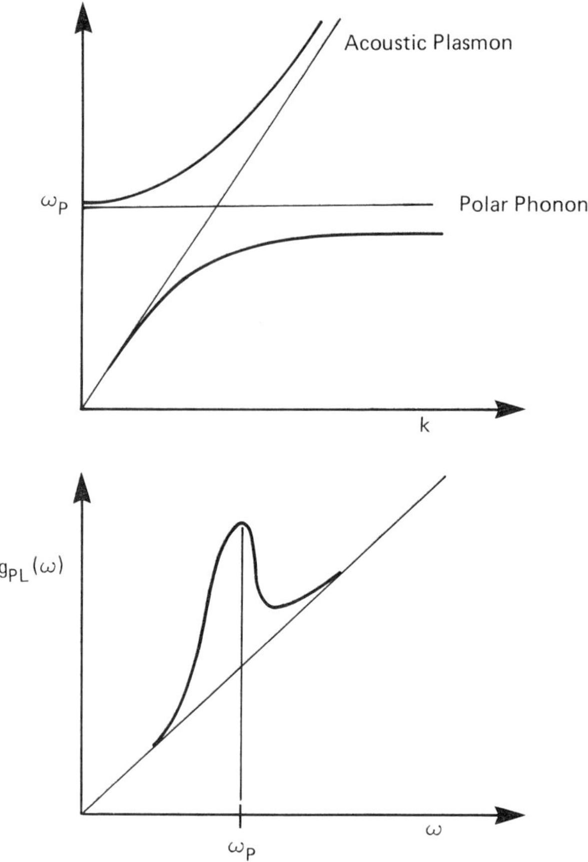

Fig. 1. Schematic plot of the coupled acoustic plasmon-polar phonon dispersion relations (top) and the resulting change in the plasmon density of states $g_{PL}(\omega)$ at the polar phonon frequency ω_p (bottom.

SUMMARY AND CONCLUSION

In the search for collective excitations (bosons) to explain the high T_c in the cuprate oxides, the existence of the two-dimensional plasmons discussed is a certainty. In addition, the coupling strength for plasmons in 2D increases with decreasing carrier concentration n.[11]

We believe that high T_c superconductivity is caused by the coexistence of both the phonon and plasmon mechanism, and that the plasmon contribution is the key difference. The phonon mechanism by itself is only compatible with high T_c for very large values of λ. At the same time, experimental data indicate the intermediate coupling case. The presence of a partial isotope effect in addition to the argument about coupling strength is a strong point for the coexistence of both phonon and plasmon mechanisms. Experiments sensitive to phonon structure only such as neutron scattering should be able to discriminate between plasmon-enhanced phonon structure in the tunneling and infra-red absorption measurements. Such differences would manifest the presence of the plasmon mechanism unambiguously.

REFERENCES

1. L. F. Mattheiss, *Phys. Rev. Lett.* 58:1028 (1987); J. Yu, A. J. Freeman and J. H. Xu, *Phys. Rev. Lett.* 58:1035 (1987); R. V. Kasowski, W. S. Hsu and F. Herman, *Sol. State Comm.* xx:xx (1987).
2. D. Pines, *Can. J. Phys.* 34:1379 (1956).
3. T. Ando, A. B. Fowler and F. Stern, *Rev. of Modern Physics* 54:437 (1982).
4. V. Z. Kresin and H. Morawitz, to be published.
5. D. Pines, 'The Many Body Problem,' W. A. Benjamin, Inc., New York (1961).
6. J. Hubbard, *Proc. Roy. Soc. (London)*, 539:A240 (1957).
7. P. Vashista and K. S. Singwi, *Phys. Rev.* B6:875 (1972).
8. We are grateful to Dr. H. Gutfreund for this remark.
9. E. L. Wolf, "Principles of Electron Tunneling Spectroscopy," Oxford University Press, 1985.
10. D. A. Bonn, J. E. Greedan, C. V. Stager, T. Timusk, M. G. Doss, S. L. Herr, K. Kamaras and D. B. Tanner, *Phys. Rev. Lett.* 58:2249 (1987).
11. V. Z. Kresin (in these proceedings).

BASIS AND CONSEQUENCES OF THE TWO-BAND

MODEL FOR HIGH T_C OXIDE SUPERCONDUCTORS

J. Ihm

Department of Physics, Seoul National University
Seoul, Korea
Bell Communications Research, Red Bank, NJ

D. H. Lee

IBM T.J. Watson Research Center
Yorktown Heights, NY

For high T_C (La, Y, Rare Earth)-(Sr, Ba)-Cu-O system, we
propose that two different bands, one with a light mass and the
other with a heavy mass, straddle the Fermi level E_F. In
$YBa_2Cu_3O_7$, definite evidence exists for this model. In La_{2-x}
Sr_xCuO_4 where band structure calculations give a single d_{x2-y2}
band at E_F, this condition is presumably satisfied by lifting
of the heavy copper d_{z2} band (by electron-electron correlation)
to the level of the light copper d_{x2-y2} band. This unusual
band structure results in not only the interband scattering of
the Cooper pair between the d_{x2-y2} and d_{z2} bands but also the
attractive interaction between the partners of the d_{x2-y2} electron pair via acoustic plasmon and interband electron-hole pair
exchange. Both of these effects are shown to enhance T_C of the
system dramtically.

Since superconductivity around 30K was first reported for the La-Ba-
Cu-O system[1], a great deal of progress has been made on the material
development and the T_C in the 90-100 K range is now routinely achieved in
(Y, Rare Earth)-Ba-Cu-O systems[2]. It seems that the Y-Ba-Cu-O system
abounds in meta-stable structures at higher temperatures, which sustains
hopes for realization of even higher T_C. Furthermore, it is conccivable
that related materials (e.g., O replaced by F or Cl) may also have high T_C
and the search for new oxide superconductors continues all over the world.

On the theoretical side, many interesting ideas have been presented
so far[3-7]. However, it is fair to say that no concensus has been reached
on the major mechanism responsible for the unusually high T_C in this system. We previously proposed a two-band model[8] to explain the high T_C in
$La_{2-x}(Ba, Sr)_xCuO_4$. That paper contained formalism constituting the basis

of the two-band model. In the present short article, we would like to
state the physics of the two-band model in a more intuitive terms, providing motivations and consequences of the model. It is our belief that, even
if details may differ, the basic mechanism for the high T_c in all the (La,
Y, Rare Earth)-(Ba,Sr)-Cu-O systems can be called the "interband interaction" as described below.

In $YBa_2Cu_3O_7$, the wide (light mass) band probably corresponds to the
copper d_{x2-y2}-oxygen 2p sigma band orbitals in copper II plane (there are
two of them, actually) and the narrow band may possibly correspond to pi
band orbitals in copper I chain. In $La_{2-x}Sr_xCuO_4$, on the other hand, we
pay attention to two bands; band 1 consisting predominantly of the copper
d_{x2-y2} and oxygen 2p orbitals, which is usually assumed to be the only band
that includes the Fermi level, and band 2 consisting predominantly of the
copper d_{3z2-r2} and oxygen 2p orbitals, which is usually assumed to be completely below the Fermi level due to the Jahn-Teller splitting. For simplicity in presentation, we will focus on $La_{2-x}Sr_xCuO_4$ in the discussion
below, and band 1 and 2 are called the copper d_{x2-y2} and d_{z2} band, respectively. The same physics holds, however, for $YBa_2Cu_3O_7$ as well. Our
fundamental assumption is that band 2 is lifted in energy by the strong
electron-electron correlation in that narrow band so that both band 1 and
2 straddle the Fermi level. Earlier band structure calculations for the
tetragonal phase La_2CuO_4 show that the Fermi level crosses the wide d_{x2-y2}
band only and the flat d_{z2} band lies completely below the Fermi level (Of
course, the d_{x2-y2} band splits along with the orthorhombic distortion in
real La_2CuO_4). However, the fact that the d_{z2} band is very flat means d_{z2}
electrons are highly localized, indicating inapplicability of the band
picture for these elctrons. The elctron-electron correlation (the Hubbard
U) becomes important in this system, and this is the motivation for assuming the total Hamiltonian as

$$H = H_0 + H_1 + H_2 \, , \qquad (1)$$

where H_0 contains the on-site energy and the hopping terms to neighbors (
which are typical for band electrons), H_1 corresponds to the positive correlation energy at each site, and H_2 consists of the electron-phonon coupling (which was missing by mistake in reference 8) and the phonon
Hamiltonian.

After a series of approximations mentioned elsewhere[8], we arrive at
the following effective pairing Hamiltonian

$$H_p = \sum_{k\sigma} \tilde{\varepsilon}_{1k} c_{k\sigma}^+ c_{k\sigma} + \sum_{k\sigma} \tilde{\varepsilon}_{2k} f_{k\sigma}^+ f_{k\sigma} - V_{11} \sum_{kk'} c_{k\uparrow}^+ c_{-k\downarrow}^+ c_{-k'\downarrow} c_{k'\uparrow}$$
$$- V_{12} \sum_{kk'} (c_{k\uparrow}^+ c_{-k\downarrow}^+ f_{-k'\downarrow} f_{k'\uparrow} + f_{k\uparrow}^+ f_{-k\downarrow}^+ c_{-k'\downarrow} c_{k'\uparrow}) \qquad (2)$$

where $\tilde{\varepsilon}_1$ and $\tilde{\varepsilon}_2$ are the renormalized energies of band 1 and 2, c^+ and c^+
are electron creation and annihilation operators, and f^+ and f are hole
creation and annihilation operators, respectively. The description of the
band 2 in terms of the hole is based on the conjecture of the more than
half-filling of the band: it is straight forward to formulate the Hamiltonian in terms of holes alone or electrons alone as desired. $V_{11}(\;0)$
and V_{12} are the averaged intraband and interband interactions of the Cooper
pair. The BCS theory for two-band models has been worked out by Suhl et.
al[9]. The result is a dramatic enhancement of the T_c due to the interband
interaction V_{12}, compared to the case of the single band with the same V_{11},
irrespective of the sign of V_{12}. The intraband interaction in band 2, V_{22},
is assumed to be zero, indicating that the Cooper pair of pure d_{z2} character
could be absent or negligibly weak (But mixing to the d_{x2-y2} pair via V_{12}
makes it significantly stronger). A more accurate result can be obtained

if one solves the Eliashberg equation[10] for the two-band model. Such calculations are in progress.

In addition to introducing the Cooper pair scattering term V_{12}, the two-band mechanism gives rise to another important process which enhances the T_c. The collective motion of band 2 carriers(plasmons) is renormalized by the presence of band 1 to become acoustic, and it overscreens the electron-elctron repulsion in band 1. Furthermore, the polarization due to interband transition creating a hole in one band and an electron in another gives rise to attractive (negative) contributions to the dielectric screening function. We have previously[8] identified this as the "excitonic mechanism" in a broad sense. It is noted that the excitonic mechanism is realized in the two-band model in the absence of insulating or semiconducting region. Both acoustic plasmons and interband excitons contribute to the Cooper pair binding V_{11}. All these effects are automatically included in actual calculation of the total dielectric function, irrespective of how people call them. A general formalism for this calculation has been worked out previously[11]. It is not clear whether the traditional phonon-mediated attraction gives a negligible contribution to V_{11}, but the calculation of the dielectric function including above effects[8],[11] shows that the new contributions are essential to get high T_c.

Kresin[7] also mentioned the improtance of the plasmons in enhancing the T_c and emphasized the 2-dimensional plasmons. Our calculations show that existence of two bands is more crucial than the 2-dimensional nature of the band in plasmon-aided enhancement of T_c for the following reasons. Firstly, one can always get acoustic (very low frequency) plasmons, which are essential in enhancing T_c, in two-band (one light band and one heavy band) systems in both 2 and 3 dimensions. Secondly, in order for the plasmons to contribute to superconductivity, they must scatter electron pairs. The frequency and the wavevector of the plasmons occur at the right place to scatter elctron pairs very effectively only if there are two or more bands around the Fermi level. In this respect of frequent scattering event, it is unlikely that one can observe well-defined (long-lived) acoustic plasmons independently.

Our model has already explained number of experimental data[8], including the orthorhombic to tetragonal structural transition for $La_{2-x}Sr_xCuO_4$, normal state resistivity and susceptibility in terms of coexistence of light and heavy mass bands. It also predicts some important properties which can be tested experimentally. First of all, existence of two kinds of carriers (among them, it was concluded previously[8] that the heavy carriers are always holes) is a direct consequence of the model. Rather intriguing Hall effects seem to indicate that the relative lineup of two bands may change as the temperature varies or there is a compensation effect between holes and electrons. A careful study of photoemission (hopefully angle-resolved) and Auger spectroscopy for a fresh sample without change in oxygen content should be able to tell us whether a more localized and a less localized bands coexist at the Fermi level as claimed in our model. Another important prediction of the model is the existence of two gaps and a significant gap anisotropy. This effect can be somewhat washed out by scattering by defects and impurities abundant in the system. But if the two-band mechanism is truly responsible for high T_c, two gaps should manifest themselves in some way. The larger gap should correspond to states with a smaller mass (confined predominantly in the xy plane) whereas the smaller gap should correspond to states with a larger mass. A detailed comparison with experiment will be published elsewhere.

Finally, we note that the apparently important role of oxygen content in the system is a puzzling problem. Whether oxygen plays a more active role than to stabilize the crystal structure and determine the amount of charge carriers is still an open question.

REFERENCES

1. J.G. Bednorz and K.A. Muller, Z. Phys. B64, 189 (1986)
2. M.K. Wu, J.R. Ashburn, C.J. Torng, P.H. Hor, R.L. Meng, L. Gao, Z.J. Huang, Y.Q. Wang, and C.W. Chu, Phys. Rev. Lett. 58, 908(1987)
3. J.D. Jorgensen, H.B. Schuttler, D.G. Hinks, D.W. Capone, K. Zhang, M.B. Brodsky, and D.J. Scalapino, Phys. Rev. Lett. 58, 1024(1987)
4. L.F. Mattheiss, Phys. Rev. Lett. 58, 1028(1987)
5. J. Yu, A.J. Freeman, and J.H. Xu, Phys, Rev. Lett. 58, 1035(1987)
6. P.W. Anderson, Science, 235, 1196(1987)
7. V.Z. Kresin, Phys. Rev. B(to be published)
8. D.H. Lee and J. Ihm, Solid State Commun.(to be published)
9. H. Suhl, B.T. Matthias, and L.R. Walker, Phys. Rev. Lett. 12, 552 (1959)
10. G.M. Eliashberg, Soviet Physics JETP 11, 696 (1960)
11. J. Ihm, M.L. Cohen, and S.F. Tuan, Phys. Rev. 23, 3258(1981)

SPECTRA OF PLASMONS IN SUPERCONDUCTING CUPRATES

John Ruvalds[*]

Lyman Laboratory of Physics
Harvard University
Cambridge, MA 02138

INTRODUCTION

Plasmons may provide a promising mechanism for superconductivity providing that certain restrictive conditions on the electronic structure are satisfied. However, these favorable situations were presumed to be quite rare until recently, when the exciting discoveries of high temperature superconducting oxides rejuvenated interest in novel mechanisms of superconductivity.

In the search for theoretical clues to alternate processes which may strongly bind electrons and thereby allow their pairs to condense into a superconducting state, it is constructive to recall the original motivation of the pioneers who discovered the copper oxide superconductors.

Bednorz and Müller[1] were intrigued by studies of other oxides whose transition temperatures were not exciting, but were remarkably high in cases where the density of electron states was rather low. Thus their idea was to find new oxides whose electron concentration may be improved with the hope of maintaining the extraordinary attractive coupling strength. Their original find of T_C near 35K in La-Ba-Cu-O alloys set the stage for a series of discoveries of new materials in this class. An important consideration which emerged from these alloys is the unusual extreme sensitivity of the transition temperature T_C to the Ba concentration; for example, substitution of $Ba_{.05}$ for the equivalent number of La atoms produces no benefit whereas a replacement of $Ba_{.15}$ triggers a superconducting state up to $T_c \simeq 40\,K$.

A second key development was achieved by C.W. Chu and his colleagues by applying pressure to the La-Ba-Cu-O alloy and thereby elevating T_C. Furthermore, by remarkable intuition, Chu attempted to imitate the influence of external pressure by substitution of yttrium, and thus he discovered the Y-Ba-Cu-O series of 90 K superconductors which surpassed the technological goal of liquid nitrogen refrigeration and decisively stimulated theorists to look for new explanations.

---

[*] On leave from the University of Virginia, Charlottesville, VA 22901.

The conventional BCS [3] approach, which relies on phonons to generate
an electron pairing interaction, are difficult to reconcile with the
absence of an isotope effect in the Y-Ba-Cu-O superconductor, which was
established by B. Batlogg *et al.* [4] and independently by L.C. Bourne. [5] The
Bell Labs group [4] determined the large Raman scattering shifts associated
with the O^{18} isotope substituted in place of O^{16}, which would indicate an
expected shift of T_C by 4 K in the BCS theory. By contrast, their experimental data on the resistivity and magnetic susceptibility rule out changes
in T_C in excess of 0.2 K. Hence their measurements suggest minimal contribution by the oxygen-related phonons to the superconductivity of the
yttrium based oxides.

In view of the above developments theoretical options have been significantly narrowed. Furthermore, a wealth of data is becoming available
which provides further clues, as well as constraints, regarding the viability of mechanisms that have been proposed for the superconducting oxides.

A key feature of the present work is an attempt to focus on other
physical measurements, in addition to the determination of T_C, which may
have a bearing on the theoretical proposals. Thus we emphasize the structure which may be observable in the infrared spectroscopy, electron
tunneling and neutron scattering whose origin may be traced to an electronic
mechanism such as the presence of acoustic plasmons.

Although we cannot do justice in this space to the large number of
theories which have recently arisen, it should be noted that the highly
original ideas developed by P.W. Anderson, [6] on the basis of resonating
valence bonds, are particularly intriguing in that they explain several
anomalous features such as the specific heat and susceptibility measurements in addition to the very high transition temperatures.

Our analysis follows the acoustic plasmon mechanism that we envisioned
for the copper oxides quite recently. [7] The basic requirement for the
superconductivity in this approach is the existence of two plasmon branches,
which are well separated in energy. Thus the higher acoustic branch extending to conventional energies of order $\Omega^\ell \sim 1$ eV represents the oscillations of the majority hole carriers which should occur in both superconducting and normal metal varieties of the copper oxide alloys. However,
the distinguishing feature of those alloys which become superconducting
at high temperatures, is that the Fermi energy is postulated to intersect
a second minority hole band whose oscillations yield a lower plasmon energy
$\Omega^h \lesssim 0.1$ eV; The "heavy mass" hole carriers from a band with energy
$e_h \cong k^2/2m_h$, thus provide a Boson mode which may provide the attractive
interaction between the majority hole carriers having energy $E_\ell \simeq k^2/2m_\ell$.
This type of coincidence requires a modification of the electronic structure calculated for La_2CuO_4 by Mattheiss [8] and independently by the Freeman
group, [9] and our conjecture was that alloying with Ba, Sn or Y may shift
the Fermi energy (or the E_h band) to achieve the desirable intersection
shown in Figure 1.

In the case of Y-Ba-Cu-O superconductors, the more recent band structure calculations [10,11] in fact exhibit the two-band spectrum which is
qualitatively similar to the desired superconducting alloy situation shown
in Figure 1. Also, resonant photoemission experiments [12] on Y-Ba-Cu-O
alloys provide evidence for the two-band structure near the Fermi energy.

Given the favorable electronic situations described above, it is now
of interest to determine if the plasmons exist as well-defined excitations
and which of their characteristics, if any, correlate with high transition
temperatures.

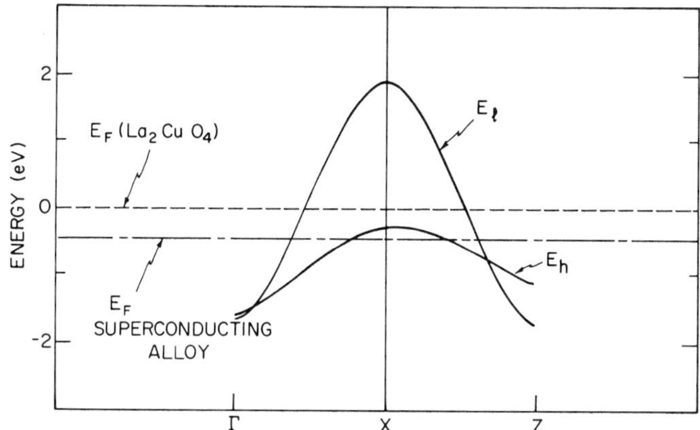

Fig. 1. Representation of the electron energy as a function of momentum
for the copper oxides. The shapes of E_ℓ and E_h are
patterned after the band structure calculations of Refs. 8 and 9,
with the Fermi energy E_F intersecting only the majority hole
band. The desired situation for the plasmon-induced superconductivity is the partial occupation of the E_h band as well,
which may be achieved by shifting the Fermi energy (or E_h) in
the alloys.

The concept of acoustic plasmons in metals has been discussed for
some thirty years, and yet experimental evidence for their existence was
lacking until a few months ago. However, analogous cases of two component
plasmas are known to occur in astrophysics and in semiconductors. Also,
the acoustic plasmon modes generated by the modified Coulomb potential in
a two-dimensional electron gas on a helium surface have been verified.

Pines originally suggested that metals containing two groups of
electrons, with the heavy mass m_h carriers being strongly screened by
the lighter mass m_ℓ electrons, may exhibit an acoustic plasmon branch[13]
whereas the majority light electrons would give an ordinary "optical"
plasmon. His analogy to the screening of ions to yield sound waves
indicates the feasibility of acoustic plasmons providing that their lifetime is not too limited by Landau decay into the light electron-hole
continuum. The desirability of acoustic plasmons as a replacement for
phonons as a mediating source of electron pairing in the BCS theory was
realized by Radhakrishnan[14] and by Fröhlich,[15] and investigated by many
theorists. Yet, there has been a reluctance to assign the plasmon
mechanism to a particular substance, and in this connection, it is worth-
while to mention the review[16] of earlier work.

The allure of the plasmon mechanism for superconductivity is evident
from the strong-coupling version of the BCS equation, which yields the
transition temperature

$$T_c \simeq 0.7\theta \exp\left[-\frac{1+\lambda}{\lambda-\mu^*}\right] , \qquad (1)$$

where λ is the electron pairing interaction, μ^* is the modified
Coulomb repulsion and θ is a characteristic Boson temperature. In the
phonon induced pairing, $\mu^* \simeq 0.1$, $\lambda \sim 1$, and $\theta \cong 300\,K$ gives $T_c \leqslant 25\,K$.
However, if a plasma energy of order $\theta_{p\ell} \simeq 100,000\,K$ is substituted in
Eq. (1), the evident enhancement should make room temperature superconductivity readily available *if* the coupling parameters λ and μ^* remain
in a reasonable range. Unfortunately, the latter constraints are difficult

to achieve, as evidenced by the abundance of ordinary nonsuperconducting
metals which have high plasma frequencies. The problems with a very high
energy boson mode show up in the electron pairing interaction

$$\lambda = 2 \int \frac{\alpha^2 F(\omega)}{\omega} \, d\omega \quad , \tag{2}$$

which decreases as the peak in the density of states $F(\omega)$ moves to
higher energies. Also, the Coulomb repulsion between electrons, which is
reduced by screening and retardation effects, is of the form

$$\mu^* = \frac{\mu}{1 + \mu \, \ell n \, (E_F/\theta)} \quad , \tag{3}$$

where μ is the screened Coulomb coupling. Hence if the boson energy θ
becomes comparable to E_F, the Coulomb repulsion can overcome the attractive pairing interaction, i.e., $\lambda \leqslant \mu^*$, resulting in a normal metal.

Another extreme of very low plasmon energies can be achieved in
doped semiconductors, and acoustic plasmon modes may be realized in thin
films for semiconductor devices. However, in these situations the number
of electron (or hole) carriers is low and the corresponding transition
temperatures are expected to be very low, $T_c \leqslant 1 \, K$.[18]

In the case of superconducting oxides, the plasmon structure should
resemble the two-dimensional electron dynamics in the Cu-O planes. There
are two different plasmon models, in addition to our two-band model, which
have been proposed to explain the high temperature superconductivity.
Kresin has invoked a single component two-dimensional plasmon which acts
in conjunction with a soft phonon mode,[19] and Ashkenazi *et al.*[20] have
suggested that an interlayer plasma oscillation may provide the pairing of
electrons in the Cu-O planes.

It should be possible to distinguish the viability of these independent theories by infrared absorption and electron tunneling measurements
which we discuss in the following order. In the next section, the alloy
concentration is used as a guide to correlate the Fermi energy, plasmon
structure and T_c. The plasmon damping is also analyzed, and the prospects
for infrared spectroscopy measurements are discussed next.

Some comments on electron and neutron scattering spectra are also
presented. Finally, conclusions with regard to the search for new
materials are considered.

PLASMON STRUCTURE

For the two-dimensional electronic structure idealized in Figure 1,
it is reasonable to use an effective mass approximation of the light
majority electron energy $E\ell$ and the minority band with the "heavy" mass
label E_h. The relevant scaling of the momenta is to measure wave vectors
in units of k_F^ℓ and k_F^h respectively, and to define energies in units of
the Fermi energy E_F^ℓ measured from the top of the majority hole band
shown in Fig. 1. A crucial element in the physical basis of the plasmon
mechanism, is the large separation in energy between the two bands at the
symmetry point X shown in Fig. 1: This allows the lower plasmon branch
of the h-hole oscillations to appear in the low energy region $\omega << E_F^\ell$,
where Landau damping by the ℓ-electron-hole continuum is very much reduced.
In fact, this energy separation is more vital to the existence of two
well-defined plasmon branches than changes in the effective masses m_ℓ

and m_h. The smaller splitting of two degenerate bands in the $Rb_x WO_3$ may account for the smaller T_c values in that group of oxides.[17]

For convenience, the analysis of the plasmon structure can be represented in the dielectric formulation using the two-dimensional functions[7]

$$\varepsilon_1 \equiv \text{Re } \varepsilon(q,\omega) = 1 + \frac{2\pi\alpha}{k^2}\left[k + \text{sgn}(\nu_+)H(\nu_+^2 - 1)(\nu_+^2 - 1)^{1/2} \right.$$
$$\left. + \text{sgn}(\nu_-)H(\nu_-^2 - 1)(\nu_-^2 - 1)^{1/2} \right] , \tag{4}$$

and

$$\varepsilon_2 \equiv \text{Im } \varepsilon(q,\omega) = \frac{2\pi\alpha}{k^2}\left[H(1 - \nu_+^2)(1 - \nu_+^2)^{1/2} - H(1 - \nu_-^2)(1 - \nu_-^2)^{1/2} \right], \tag{5}$$

where $\alpha = 0.22\, r_s$, $\nu_+ = \pm \omega/k - k/2$, and $H(x)$ is the Heaviside step function. The total dielectric function is

$$\varepsilon_{TOT} = \varepsilon_\ell + \varepsilon_h , \tag{6}$$

where the individual contributions of the ℓ and h bands naturally includes appropriate r_s^ℓ and r_s^h screening parameters.

In the strictly two-dimensional case, even a single band will yield an "acoustic" plasmon with dispersion

$$\Omega^\ell \cong (aq + b^2 q^2)^{1/2} , \tag{7}$$

where the coefficients a and b are determined by the screening r_s^ℓ, and the mode is well defined since $\varepsilon_2 = 0$ in the region of the zeroes for $\varepsilon_1(q,\omega)$ which yield $\Omega^\ell(q)$. Interactions between the layers will of course yield an anisotropic plasmon with a nonvanishing $\Omega^\ell(q=0)$ as in graphite. However, it is interesting to note that the first evidence for a two-dimensional plasmon in a metal has just been reported for single crystals of La_2NiO_4:[21] These infrared measurements show a dramatic change in the reflectivity which indicates strong plasmon features only when the electric field of the light is polarized parallel to the plane of the Ni-O atoms; this data confirms an anisotropy in excess of 250 for the plasmon response.

Analogous experiments on the superconductor oxides using polarized light would be of great interest. Preliminary infrared data on polycrystalline samples of the Ba-La-Cu-O alloys yields the majority plasmon response near $\Omega^\ell \sim 1\,\text{eV}$ in La_2CuO_4 as well as in various Ba (and Sr) concentration alloys, and it is intriguing that this plasmon remains at roughly the same energy in the superconducting cases as well as in the normal metal alloys.[22]

Our calculations of T_c using the one-component plasma with $\Omega^\ell \sim 1\,\text{eV}$ yield $T_c = 0$ as expected because the Coulomb repulsion μ^* overwhelms the attraction λ associated with exchange of a plasmon of relatively high energy.

On the other hand, if the Fermi energy can be moved by alloying to intersect the minority hole band, as in Fig. 1, then a lower energy plasmon with $\Omega^h \sim 0.1$ eV, can provide the electron pairing mechanism. By explicit calculation of the two-component plasmons using Eqs. (5) and (6), and parameters appropriate to the band structures, i.e., $r_s^\ell = 0.5$,

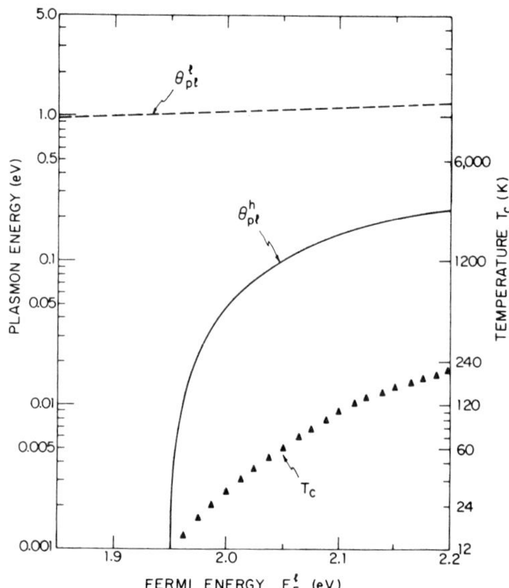

Fig. 2.   Plasmon energies as a function of the Fermi energy measured from the top of the E^{ℓ} hole band. Note that the calculated T_c values scale with lower plasmon mode $\theta_{p\ell}^{h}$, which is the maximal value of $\Omega^{h}(q)$.

$r_s^h = 5.0$ and $\Delta E_F^h = 0.1$ eV measured from the top of the E^h band, we find $\Omega^h \simeq 0.15$ eV and $T_c \simeq 200$ K using the screened static Coulomb potential for α^2 in Eqs. (1)-(3). As the Fermi energy energy moves slightly, dramatic changes in the calculated plasma energy and T_c are obtained as seen in Figure 2.

Experimental evidence for low frequency plasmon structure in the super-conducting alloys has just been announced by several groups, but the poly-crystalline nature of their samples rules out a specific determination of the plasmon parameters at the moment.[24,25]

An ideal probe of the lower plasmon branch would be the infrared spectroscopy with polarized light on single crystal samples. Hence we discuss the type of structure that may be observable by these techniques.

INFRARED RESPONSE

A complicating feature of the infrared spectrum is the presence of high frequency phonons, principally originating from the Cu-O vibrations, and the hybridization of longitudinal optical phonons with the acoustic plasmons. For simplicity we demonstrate this effect for one phonon having a constant energy ω_{LO}, coupled only to the lower energy h plasmon mode. If we introduce a hybridization strength parameter g, we obtain the mixed mode eigenvalues

$$\omega_{\pm} = \frac{1}{2}\left\{\omega_{LO} + \Omega^h(q) \pm \left[(\omega_{LO} - \Omega^h(q))^2 + 4g^2\right]^{1/2}\right\}. \qquad (8)$$

The resulting dispersion is shown in Figure 3, along with the majority ℓ-hole plasmon with a higher sound velocity (its coupling to the phonon is neglected solely for aesthetic purposes).

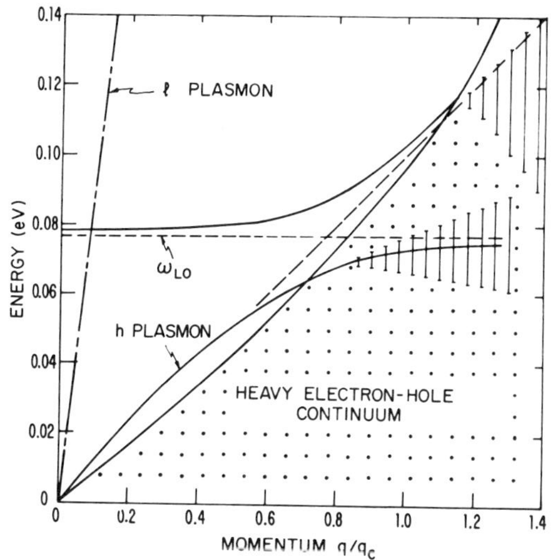

Fig. 3. Excitation spectrum proposed for the superconducting copper
oxides. The high frequency ℓ plasmon continues to $\Omega^\ell \sim 1$ eV,
whereas the minority band h-plasmon is an order of magnitude
lower in energy. The hybridization with an optical phonon near
ω_{LO} is shown for g = 0.01 eV. Damping of these branches is
severe within the h-electron-hole continuum, as shown by shaded
region. However, outside the h-electron hole-continuum we find
$\Omega^h \tau^h \simeq 30$, for $r_s^\ell = 0.5$ and $r_s^h = 2.0$.

Landau damping of the low energy plasmon by the majority electron
hole continuum is very sensitive to the electronic structure. A major
benefit of the superconducting oxides is the substantial separation of
the energy bands, i.e., $E^\ell(o) - E^h(o) \simeq 2$ eV, which allows the lower
branch Ω^h to fall in a region of the ℓ electron-hole continuum where
the density of allowable decay states is small. Nevertheless, the damping
is quite sensitive to the effective masses and particularly to the placement
of the Fermi energy. For example, with $r_s^\ell = 0.5$, we find that
$M_s^h = 2.0$ gives $\Omega^h \tau^h \sim 30$, whereas $r_s^h = 5.0$ yields $\Omega^h \tau^h \sim 10$.
These r_s values yield the plasmon spectrum with $\Omega^\ell \sim 1$ eV using realistic energy bands. In addition the finite electron mean free path, caused
by impurity and electron scattering, will also limit the viability of the
low energy plasmon as a well-defined excitation. Thus, this electronic
mechanism of superconductivity manifests an extreme sensitivity to alloy
content, sample quality and, in some instances, to external pressure.

Infrared absorption by these acoustic modes is not described by the
conventional Drude model which works well for optical plasmons, Rather,
the required momentum conservation for the light can be satisfied by analogy to the Holstein process which involves the simultaneous creation of
an electron-hole pair as well as an acoustic mode. Such a process
describes very well the acoustic phonon spectrum of lead, and provides a
direct and accurate measure of phonons in that superconductor.[26-28]

By analogy to the phonon analysis,[26,27] we expect the infrared
absorption to reveal the acoustic plasmon density of states $F(\omega)$. For a
two-dimensional electron gas it is reasonable to expect $\Omega^h \simeq b\, q$, and
then the simplest approximation gives

$$F(\omega) \;\widetilde{=}\; c\omega, \qquad\qquad \text{for} \quad \omega \leqslant \theta^h \;. \qquad\qquad (9)$$

This wedge shape will be modified by a finite plasmon width Γ and coupling to the phonons as in the ω_{LO} case shown in Fig. 3. With these modifications, the density of plasmon states becomes:

$$F(\omega) \;=\; \frac{\pi\Gamma}{b^2}\,\ell n\left|\frac{b^2 Q^2 - 2bQ\,[\omega - \sigma(\omega)] + [\omega - \sigma(\omega)]^2 + \Gamma^2}{[\omega - \sigma(\omega)]^2 + \Gamma^2}\right|$$

$$+ \;\frac{\pi}{b^2}\,[\omega - \sigma(\omega)]\left[\tan^{-1}\left(\frac{bQ - \sigma(\omega)}{\Gamma}\right) - \tan^{-1}\left(\frac{-\sigma(\omega)}{\Gamma}\right)\right], \qquad (10)$$

where

$$\sigma(\omega) \;=\; \frac{g^2}{\omega - \omega_{LO}} \;, \qquad\qquad\qquad (11)$$

represents the phonon-induced structure, and Q is a cut-off moment which indicates the limiting momentum range for well-defined acoustic plasmons.

The corresponding plasmon density of states is shown in Fig. 4, using typical values for the plasmon and phonon energies and a damping width $\Gamma \simeq 0.05\,\theta^h$.

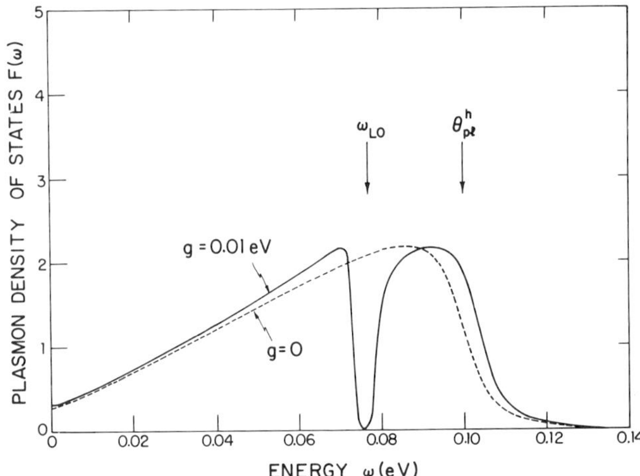

Fig. 4. Plasmon density of states $F(\omega)$ as a function of energy for the lower energy Ω^h plasmon. The dotted curve shows the rounding of the wedge shape by a finite damping width $\Gamma = 0.005$ eV. Hybridization with an LO phonon yields an antiresonance near ω_{LO} as shown by the solid curve.

Another likely absorption process involves the creation of two acoustic plasmons with zero total momentum. In this case the spectrum is similar in form to the F function, but the density of states refers to a pair of plasmons and extends to $\omega \leqslant 2\theta^h$.

Single crystal samples should exhibit the two-dimensional nature of the plasmons as well as the frequency variation shown in Fig. 4. Hence the use of polarized light, as in the experiments[21] on La_2NiO_4 may reveal

further clues to the plasmon structure and its relation to superconductivity.

ELECTRON TUNNELING AND NEUTRON SCATTERING

Now that considerable progress has been achieved in making thin
film single crystals, the tunneling spectroscopy which has pinpointed
the phonon spectrum and the electron-phonon coupling in ordinary superconductors, should provide an ideal method to investigate plasmons as
well. Following the derivation of McMillan and Rowell, it is reasonable
to expect structure proportional to the plasmon density of states $F(\omega)$
in Eq. (10), providing, of course, that the coupling strength is not
strongly energy dependent.

In the superconducting oxides, our calculations predict a smooth
linear energy dependence of the tunneling current arising from the
majority of ℓ - hole plasmon which extends to $\Omega^{\ell} \sim 1$ eV, and an
additional structure from the minority h - hole plasmon extending
roughly to 100 meV.

Preliminary data[29] on Y-Ba-Cu-O samples in fact shows considerable
structure near 100 meV, and the plasmon structure may provide an alternative explanation for the observed data. Excitation of a pair of plasmons by an electron may also be observable. Observation[30] of the
Josephson effect and the Shapiro step structure in Y-Ba-Cu-O provides
evidence for singlet electron pairing.

Neutron scattering may reveal plasmon contributions to the phonon
structure factor arising from the hybridization process shown in Fig. 3.
However, these experiments face difficulties in the high energy region
$\omega_{LO} \simeq 77$ meV (for Y-Ba-Cu-O) where the plasmon contributions are
expected to be most prominent. Nevertheless, as the sample preparation
techniques improve to a point where large single phase targets, or perhaps even single crystals, become available, the neutron spectrum may
exhibit the pronounced features noted above, which occur at energies
away from the peaks found in Raman scattering experiments.

CONCLUSIONS

The plasmon spectra proposed by three groups should be distinguishable by infrared absorption and electron tunneling measurements. Our
proposal of a two-band[7] structure which yields a low energy plasmon only
in the superconducting alloys, can be probed by varying the alloy content.
The interlayer plasmon contribution,[20] also at low energies, should have
a distinctive response to polarized light. The combined influence of a
soft phonon and a single component plasmon proposed by Kresin[19] should be
examined by neutron scattering in addition to the other probes.

Another electronic mechanism for superconductivity based on the
exchange of excitons has also been proposed. Originally the idea was
proposed for organics by W.A. Little and for metal-semiconductor interfaces by J. Bardeen *et al*. The exciton mechanism has been invoked
recently for the copper oxides[31,32] and is another possible explanation
which is compatible with the absence of an isotope effect on T_c.

Recently there have been several articles on spin fluctuations and
electron pairing induced by a variety of spin interactions. Many of
these are discussed elsewhere in the present book.

Our goal has been to emphasize those spectral features of the plasmon mechanism which may be observable and hopefully may provide a guide to the desirable properties for elevating T_c. In principle, room temperature superconductors are within the realm of the plasmon mechanism. Thus the prospect of predicting new materials with desirable electronic structure presents a crucial challenge.

It is a pleasure to thank Q.G. Sheng for technical assistance. We have enjoyed stimulating discussions with H. Ehrenreich, B. Halperin, P. Martin, M. Stephen, Z. Tesanovic and many other colleagues at Harvard. We are indebted to P. Sulewski, S. Perkowitz and S. Wolf for communicating data prior to publication. Our research is supported by the U.S. Department of Energy Grant No. DEF 605-84-ER45113.

REFERENCES

1. T.G. Bednorz and K.A. Müller, Z. Phys. B64:189 (1986).
2. M.K. Wu, J.R. Ashburn, C.J. Torng, P.H. Hou, R.L. Meng, L. Gao, Z.J. Huang, Y.Q. Wang and C.W. Chu, Phys. Rev. Lett. 58:908 (1987).
3. J. Bardeen, L.N. Cooper and J.R. Schrieffer, Phys. Rev. 106:162 (1957).
4. B. Batlogg, R.J. Cava, A. Jayaraman, R.B. van Dover, G.A. Kourouklis, S. Sunshine, D.W. Murphy, L.W. Rupp, H.S. Chen, A. White, K.T. Short, A.M. Mujsce and E.A. Rietman, Phys. Rev. Lett. 58:2333 (1987).
5. L.C. Bourne, M.F. Crommie, A. Zettl, H.C. zur Loye, S.W. Keller, K.L. Leary, A.M. Stacy, K.J. Chang, M.L. Cohen and D.E. Morris, Phys. Rev. Lett. 58:2337 (1987).
6. P.W. Anderson, Science 235:1196 (1987). An interesting overview of recent theoretical developments may be found in P.W. Anderson and E. Abrahams, Nature 327:363 (1987).
7. J. Ruvalds, Phys. Rev. B35, No. 16 (1987).
8. L.F. Mattheiss, Phys. Rev. Lett. 58:1028 (1987).
9. Jaeyun Yu, A.J. Freeman and J.H. Xu, Phys. Rev. Lett. 58: 1035 (1987).
10. L.F. Mattheiss and D.R. Hamann, Sol. St. Comm. (1987).
11. Jaeyun Yu, S. Massida, A.J. Freeman and D.D. Koelling, preprint.
12. R.L. Kurtz, R.L. Stockbauer, D. Mueller, A. Shih, L.E. Toth, M. Osofsky and S.A. Wolf, Phys. Rev. B35, No. 16 (1987).
13. D. Pines, Can. J. Phys. 34:1379 (1956).
14. V. Radhakrishnan, Phys. Lett. 16:247 (1965).
15. H. Fröhlich, J. Phys. C1:544 (1968).
16. J. Ruvalds, Adv. Phys. 30:677 (1981).
17. L.M. Kahn and J. Ruvalds, Phys. Rev. B19:5641 (1979).
18. Y. Takada, J. Phys. Soc. Japan 45:786 (1978); ibid 49:1713 (1980).
19. V. Kresin, Phys. Rev. B35, No. 16 (1987).
20. J. Ashkenazi, C.G. Kuper and R. Tyk, preprint.
21. J.M. Bassat, P. Odier and F. Gervais, Phys. Rev. B35:7126 (1987).
22. S. Tajima, S. Uchida, S. Tanaka, S. Kanbe, K. Kitazawa and F. Fueki, Jpn. J. Appl. Phys. 26:4 L432 (1987).
23. K. Ohbayashi, N. Ogita, M. Udagawa, Y. Aoki, Y. Maeno and T. Fujita, Jpn. J. Appl. Phys. 26:4 L420 (1987).
24. S. Perkowitz, G.L. Carr, B. Lou, S.S. Yom, R. Sudharsanan and G.S. Ginley, preprint.
25. P. Sulewski and A.J. Sievers, private communication.
26. R.R. Joyce and P.L. Richards, Phys. Rev. Lett. 24:1007 (1970).
27. P.B. Allen, Phys. Rev. B3:305 (1971).
28. B. Farnworth and T. Timusk, Phys. Rev. B14:5119 (1976).
29. M.D. Kirk, D.P.E. Smith, D.B. Mitzi, J.Z. Sun, D.J. Webb, K. Char, M.R. Hahn, M. Naito, B. Oh, M.R. Beasley, T.H. Geballe, R.H. Hammond, A. Kapitulnik and C.F. Quate, preprint.
30. J.S. Tsai, Y. Kubo and J. Tabuchi, Phys. Rev. Lett. 58:1979 (1987).
31. C.M. Varma, S. Schmitt-Rink and E. Abrahams, Sol. St. Comm. (1987).
32. J.P. Collman, J.T. McDevitt and W.A. Little, preprint.

THE ROLE OF SPATIAL SEPARATION IN

PAIRING INDUCED BY ELECTRONIC MODES

H. Gutfreund

The Racah Institute of Physics
The Hebrew University
Jerusalem, Israel

INTRODUCTION

Various non-phonon mechanisms of superconductivity have been
suggested since the pioneering work by Little on the excitonic mechanism.[1]
Such mechanisms are now attracting considerable attention in view of the
discovery of high T_c superconductors. The naive logic which has sometimes
been applied in the discussion of these mechanisms proceeds as follows:
a) identify a boson-like electronic excitation with a typical energy E_{ex} ;
b) the scattering amplitude for exchanging such an excitation between two
electrons with momenta $\underline{k}$, $-\underline{k}$ is negative for $\omega < E_{ex}$; c) assume that the
coupling constant λ exceeds the parameter μ^* which measures the repulsive
Coulomb interaction; d) apply directly the BCS theory and conclude the
existence of superconductivity at the temperature

$$T_c = E_{ex} \ \exp \left[\frac{1}{\lambda - \mu^*} \right] \tag{1}$$

This reasoning is oversimplified and frequently misleading because the
parameters appearing in this equation depend on each other in a
complicated way. Moreover, additional problems of conceptual nature arise
in the context of non-phonons mechanisms. These problems were discussed
in detail in a review article by Gutfreund and Little[2] and briefly in
Little's contribution at this conference.

In the present paper I shall elaborate on three points which
distinguish between the coupling of an electron to an electronic excitation
mode and the electron-phonon coupling: the question of Migdal's theorem

and vertex corrections, the importance of local fields, and the role of the exchange interaction. The latter will be demonstrated for the case of pairing via acoustic plasmons. The discussion of these points leads to an emphasis on the confinement of the electrons of the excitonic medium and those which are expected to form Cooper pairs to non-overlapping orbitals. Such a spatial separation between the two types of electrons is a characteristic feature of the exciton mechanism proposed by Little[1] and Ginzburg,[3] and of the specific models considered by Allender et al.[4] and by Davis et al.[5]

PROBLEMS OF PAIRING BY ELECTRONIC MODES

a) Vertex Corrections

The formulation of the theory of superconductivity both in the weak and strong coupling regimes depends on the validity of Migdal's theorem, which asserts that vertex corrections to the electron-phonon interaction are small. For a phonon of phase velocity ω/q much smaller than the Fermi velocity, the lowest correction is of the order of $\omega_D/E_F \approx 10^{-2}$. Most of the phonons involved in conventional superconductivity have momentum $q \approx p_F$ and hence a small phase velocity. If this argument is carried over to any of the non-phonon mechanisms, one has to replace ω_D by an electronic frequency which results in large vertex corrections.

Indeed, it was demonstrated by Rietschel and Sham[6] and by Grabovski and Sham[7] (see also Sham's contribution at this conference) that if one neglects vertex corrections then the electron-plasmon interaction alone leads to superconductivity in the alkali metals. A strong effect of vertex corrections on T_c was found by these authors even when $\omega_D/E_F \approx .1$.

On the other hand, it was shown in ref. 2 that if the electron-exciton interaction is restricted to low momentum transfers then the vertex corrections are small. Such momentum dependence of the interaction is due to the separation in space between the excitonic medium and the conduction electrons.

b) Limitation on Values of $\lambda - \mu^*$

The total electron-electron interaction can be represented in the form

$$V(q,\omega) = \frac{4\pi e^2}{q^2 \epsilon(q,\omega)}, \tag{2}$$

where the dielectric function contains the direct Coulomb repulsion and the effect of all the excitations in the system. This interaction may be separated into the Coulomb part and into a part due to the exchange of a virtual excitation. In particular, at $\omega = 0$, $N(0)V(q,0) = \mu - \lambda$. ($N(0)$ is the density of states at the Fermi surface). In a simple model of an electron liquid immersed in a uniform positive background, causality and stability[8] arguments imply that $\epsilon(q,0) \geq 0$, namely $\lambda \leq \mu$. This does not exclude superconductivity in such a model because the relevant parameter is μ^* and not μ, but it restricts severely the value of $\lambda - \mu^*$.

However, in ordinary materials λ is frequently significantly larger than μ. This fact is commonly assigned to local field corrections or, synonymously, to umklapp processes. These corrections are due to the fact that the local fields on the localized ions are different from the average field sensed by the extended electrons. It has been argued,[9,10] that such corrections are not effective for non-phonon mechanisms, because the electrons which now replace the ions occupy a large part of the unit cell and there is hardly any difference between the local field seen by these electrons and by those participating in the paired condensate. This argument is correct and should be taken seriously. However, in the models proposed originally for the realization of the exciton mechanism, the electrons of the excitonic mode and their interaction with the conduction electrons is restricted to a small and localized part of the unit cell. When this is not the case the coupling strength is drastically reduced.

c) The Exchange Interaction

The most obvious difference between an electron-phonon and an electron-exciton or electron-plasmon interaction, is that in the latter case the interaction vertex involves two identical particles. This results in exchange terms in the interaction which have an opposite sign to the direct interaction integrals and may reduce substantially the interaction strength. In view of the delicate balance between the Coulomb repulsion and the pairing attraction, this reduction may have a disastrous effect. It was found[2] that there is an optimal separation, of the order of a chemical bond length, between the excitonic and conducting electrons, at which the short-range exchange interaction is strongly suppressed and the long-range direct interaction is less affected.

This argument cannot be applied to justify the neglect of exchange interactions in pairing by plasmon modes. In the following paragraph we shall derive the effect of these interactions in the case of pairing by acoustic plasmons.

ACOUSTIC PLASMONS

The collective motion of a system of two types of electrons of widely different effective masses in a solid may, under suitable conditions, result in a well defined low-frequency mode referred to as an acoustic plasmon. Let us label the "light" electrons by the index "s", and the heavy electrons by the index "d", then the frequency of the acoustic plasmons is given by

$$\omega(q) = \frac{1}{\sqrt{3}} \frac{\omega_d v_s}{\omega_s} q \ , \tag{3}$$

and their width

$$\gamma(q) = \frac{\pi \ \omega_d^2}{4 \ k_s^2 \ v_s} q, \tag{4}$$

where ω_s, ω_d are the plasmon frequencies of the s- and d-electrons, respectively, and v_s, k_s are the Fermi velocity and inverse screening length of the s-electrons.

The existence of these modes was first suggested by Pines[11] who called them "demons". It was later proposed[12,13] that they could replace the phonons and provide a new mechanism of superconductivity. This idea was pursued in detail by Ruvalds and his co-workers, who discussed the possibility of acoustic plasmons playing an important role in the superconductivity of the A15 compounds. This work is summarized in a review article by Ruvalds.[14] The coupling of electrons to acoustic plasmons was studied by Ihm et al.[15] Acoustic plasmons were also proposed as a possible explanation of the mechanism of superconductivity in the new high T_c materials.[16,17,18]

In all these studies the exchange interaction was not included. The effect of the latter on the electron-acoustic plasmon coupling was studied by Entin-Wohlman and Gutfreund.[19] Let me summarize here the main results of this study.

The starting point is a two band Hamiltonian in which only two-particle scattering processes which conserve the two band indices are allowed. These are the two intraband terms V_s (**q**), V_d (**q**) and the interband term V_{sd} (**q**). It can be shown that the other terms are small.[15,20] To each of these, there corresponds an exchange term V_i (**p** + **q** − **p'**) (i = s, d, sd), where **p**, **p'** are the momenta of the incoming electrons. We adopt the exchange approximation suggested by Schrieffer,[21] and replace V_i (**p** + **q** − **p'**) by a short-range, momentum independent potential V_i. It is then possible to apply the generalized random phase approximation introduced by Pines and Nozieres[8] to derive the effective interaction between two s-electrons, which like in eq. 2 can be represented by a dielectric function of the medium. The inverse dielectric function may be conveniently written as a sum of four terms

$$\frac{1}{\epsilon} = \frac{1}{\epsilon_c} + \frac{1}{\epsilon_{sd}} + \frac{1}{\epsilon_s^p} + \frac{1}{\epsilon_{sd}^p} \tag{5}$$

The first term represents the Coulomb interaction within the s-band and leads to the parameter μ. The second term gives the contribution of the acoustic plasmon mode, which is now modified by the exchange interaction. In the long wavelength limit this term is

$$\frac{1}{\epsilon_{sd}} = \frac{1}{1 + 2I_{sd} - I_s} \cdot \frac{\omega^2(q)}{\omega^2 - \omega^2(q) + i\gamma(q,\omega)} \tag{6}$$

where $I_s = N(0)V_s$, $I_{sd} = N(0)V_{sd}$, and the acoustic plasmon frequency is given by eq. 3 multiplied by $(1 + 2I_{sd} - I_s)^{1/2}$. The width γ is not affected by the exchange interaction.

The third term in eq. 5 represents the contribution of spin-density fluctuations in the s-band, and in principle it exists also in the electron-phonon system. The last term is the additional contribution of interband spin-density fluctuations. The latter is proportional to U_{sd}^2, and therefore is generally expected to be small. It should be pointed out that when all the exchange and spin-density terms are omitted then eq. 5 reduces to the dielectric function derived in ref. 15.

One can now proceed to calculate the electron-acoustic-plasmon coupling constant λ,

$$\lambda = -2 < V_c \ (\mathbf{p},\mathbf{p'}) \int \frac{p(\mathbf{p},\mathbf{p'},\omega)}{\omega} \ d\omega > . \tag{7}$$

where the spectral density p is given by

$$p = - \frac{1}{\pi} \ \mathrm{Im}(\frac{1}{\varepsilon_{sd}} + \frac{1}{\varepsilon_s^p}) \ , \tag{8}$$

and the average over the Fermi surface is essentially an integration over the angle between $\mathbf{p}$ and $\mathbf{p'}$. When the exchange interaction parameters I_{sd} and I_s are small, and when ε_{sd} and ε_s are taken in the long wavelength limit, one gets

$$\lambda \approx 1/2 - I_{sd} - I_s \tag{9}$$

These are leading order estimates which only emphasize that exchange interactions should be taken into account, and demonstrate how they appear in the theory. A quantitative estimate of their effect requires a more realistic calculation of the effective interaction and an evaluation of the parameters I_s and I_{sd} based on the actual band structure.

CONCLUSIONS

The main purpose of this paper is to emphasize the role of the spatial separation into non-overlapping orbitals of two types of electrons participating in an exciton mechanism of superconductivity. When this is the case, one can apply qualitative arguments, which are supported by actual calculations, to show that the problems raised in the preceeding discussion do not limit drastically the electron-exciton coupling. Such qualitative arguments cannot be applied directly to the plasmon mechanisms. This does not exclude the possibility of superconducting pairing induced by the exchange of an acoustic plasmon or a two-dimensional plasmon in a layered structure. All it means is that one still needs a reliable estimate of the coupling strength of electrons to these plasmons, to show if reasonable coupling constants can be obtained in spite of the apparent limitations due to the vertex corrections and the exchange interactions.

REFERENCES

1.   W.A. Little, Phys. Rev. 134, A1416 (1964).
2.   H. Gutfreund and W.A. Little, "Prospects of Excitonic Superconductivity" in Highly Conducting One-Dimensional Solids, J.T.

Devreese, R.P. Evrard and V.E. van Doren, Eds. (Plenum, New York 1979), p. 305-372.

3.   V.L. Ginzburg, Zh. Eksp. Teor. Fiz $\underline{47}$, 2318 (1964) [Sov. Phys. JETP $\underline{20}$,.

4.   D. Allender, J. Bray and J. Bardeen, Phys. Rev. $\underline{B7}$, 1020 (1973).

5.   D. Davis, H. Gutfreund and W.A. Little, Phys. Rev. $\underline{B13}$, 4766 (1976).

6.   H. Rietschel and L.J. Sham, Phys. Rev. $\underline{B28}$, 5100 (1983).

7.   M. Grabowski and L.J. Sham, Phys. Rev. $\underline{B29}$, 6132 (1984).

8.   D. Pines and P. Nozieres, The Theory of Quantum Liquids, W.A. Benjamin, New York, 1966.

9.   M.L. Cohen and P.W. Anderson, in Superconductivity in d- and f-Band Metals, D.H. Douglass ed., AIP, New York, 1972.

10.   J.C. Phillips, Phys. Rev. Lett. $\underline{29}$, 1551 (1972).

11.   D. Pines, Can. J. Phys. $\underline{34}$, 1379 (1956).

12.   B.T. Geilikman, Soviet Phys. JETP $\underline{21}$, 796 (1965).

13.   H. Fröhlich, J. Phys. $\underline{C1}$, 544 (1968).

14.   J. Ruvalds, Advances in Physics $\underline{30}$, 677 (1981).

15.   J. Ihm, M.L. Cohen and S.F. Tuan, Phys. Rev. $\underline{B23}$, 3258 (1981).

16.   V. Kresin, preprint and present proceedings.

17.   J. Ruvalds, preprint and present proceedings.

18.   Y. Ashkenazi and C. Kuper, preprint.

19.   O. Entin Wohlman and H. Gutfreund, J. Phys. $\underline{C17}$, 1071 (1984).

20.   H. Gutfreund and I. Unna, J. Phys. Chem. Solids $\underline{34}$, 1523.

21.   J.R. Schrieffer, J. Appl. Phys. $\underline{39}$, 642 (1968).

ELEMENTARY REMARKS ON HIGH TEMPERATURE SUPERCONDUCTORS

H. Fröhlich

Department of Physics, Oliver Lodge Laboratory,
The University of Liverpool, Oxford Street, P.O. Box 147
Liverpool, L69 3BX England

These materials are based on metal-oxides, including perovskites
and certain atoms with incomplete inner shells, e.g., rare earths. It
should be noted that more than 90% of all ferroelectrics are oxides,
based on 0^{--}. In these cases the 0^{--} electrons are displaced arising
from mode softening. In our case the mode is not softened, but a polar
oscillation ω of the 0^{--} electrons should exist. Its frequency would be
expected to be higher than that of an ionic vibration.

The incomplete inner shells will give rise to a narrow electron
band, with high effective mass m^* relatively small number density n.

The Fermi energy

$$\zeta = \hbar^2 \ (3\pi^2 n)^{2/3} \ /m^*$$

will then be relatively small.

Applying BCS theory with ω replacing the ion frequency, yields the
transition temperature

$$K \ T_c = \hbar W \ \exp(-^1/F) \ .$$

Note that the interaction parameter F which is proportional to

$$^1/\zeta \propto m^*/ \ n^{2/3}$$

will be high. Thus both ω and $\exp(-1/F)$ will be
high.

Diluting the number of incomplete shell atoms, i.e., reducing n
leads to an increase in T_c. At the same time, however, reducing n
reduces the current that can be carried.

POLARON EFFECTS IN HIGH–T_c OXIDE SUPERCONDUCTORS

D.J. Scalapino and R.T. Scalettar

Department of Physics
University of California
Santa Barbara, CA 93106

N.E. Bickers

Institute for Theoretical Physics
University of California
Santa Barbara, CA 99106

ABSTRACT

We argue that measurements of the specific heat, static susceptibility, dc and ac conductivity, and thermopower are consistent with a polaron model for normal–state conduction in the high–T_c superconductors $La_{2-x}Sr_xCuO_{4-y}$ and $YBa_2Cu_3O_{7-y}$. Strong electron–phonon coupling is also suggested by measurements of elastic constants and thermal Debye–Waller factors. We investigate a model for polaron formation in these systems and a possible mechanism for polaron superconductivity.

The discovery of high–temperature superconductivity[1,2] in the layered perovskite systems $La_{2-x}Sr_xCuO_{4-y}$ and $YBa_2Cu_3O_{7-y}$ has stimulated new interest in unconventional pairing mechanisms. In this paper, we argue that a number of normal–state experiments are consistent with conduction by small polarons. We investigate a polaron pairing mechanism[3] which could lead to superconductivity and argue that the resulting transition temperature may be reasonably high.

Measurements of elastic constants[4,5] and oxygen Debye–Waller factors[6] suggest that the electron–lattice interaction plays an important role in the new systems. In particular, measurements of the sound velocity,[4] ultrasonic attenuation,[4] and Young's modulus[5] indicate that the lattice in $La_{2-x}Sr_xCuO_{4-y}$ is anomalously soft in the temperature range from 200 K down to the superconducting transition. Furthermore, a strong electron–lattice interaction would provide a natural link between the new layered perovskites and the 13 K perovskite superconductor[7] $BaPb_{1-x}Bi_xO_3$: magnetic interactions probably cannot provide such a link, since $BaPb_{1-x}Bi_xO_3$ contains no transition elements.

A number of experiments seem consistent with conduction by small polarons, rather than by electrons weakly coupled to phonons. First of all, measurements of the normal–state specific heat coefficient and magnetic susceptibility[8–11] show that the high–T_c superconductors are strongly–coupled systems. While the Sommerfeld χ/γ ratio appears to be of order unity,[9] the measured thermodynamic density of states is enhanced by an order of magnitude over values from band structure.[12,13] So large a discrepancy probably cannot be explained in terms of a conventional $(1 + \lambda)$–enhancement from weak–coupling phonons. (On the other hand, the approximate equality of χ and γ argues against some magnetic mechanisms, which predict a Stoner–like enhancement of the susceptibility.) One explanation for the uniform enhancement of χ and γ is band narrowing by small–polaron formation.

A second argument against the conventional electron–phonon mechanism is the absence of an oxygen isotope effect[14,15] in $YBa_2Cu_3O_{7-y}$. While it is possible to envisage phonon modes which lead to superconductivity yet do not involve large–scale oxygen motion, the coupling to such modes is believed to be weak.[16] The polaron–pairing mechanism investigated below exhibits an anomalous isotope effect, whose details remain to be investigated.

Additional evidence for polaron formation in the normal state is provided by recent measurements of the frequency–dependent electrical conductivity in single crystals of $La_{2-x}Sr_xCuO_{4-y}$.[17] The conductivity $Re\,\sigma(\omega)$ within the CuO_2 planes exhibits the following qualitative features: (a) a prominent peak near 2 eV, (b) a peak at approximately 0.5 eV, which grows in intensity with increased doping, and (c) a narrow Drude–like feature at zero energy. The integrated weight of the Drude peak is estimated at only 5% of the total oscillator strength. Experiments on polycrystalline samples of $YBa_2Cu_3O_{7-y}$ also suggest the presence of a high–energy peak in $Re\,\sigma(\omega)$ and drastically reduced weight in the Drude peak. These measurements are consistent with conduction by small polarons[18]: when a polaron hops from site i to $i+1$ (see Figure 1), it alters the potential energy surface for the local ionic coordinates. A ground–state to ground–state transition is exponentially suppressed by the Franck–Condon factor $\exp(-E_P/\omega_0)$, where E_P is the polaron binding energy and ω_0 is the characteristic phonon frequency. These "zero–phonon" hopping processes lead to formation of a polaron band, whose width is narrowed from the original electronic bandwidth by the factor $\exp(-E_P/\omega_0)$. Scattering processes within this narrow band can account for the Drude peak in the optical conductivity: since the probability for hopping without phonon emission is small, the weight in this feature is expected to be only a small fraction of the total oscillator strength.

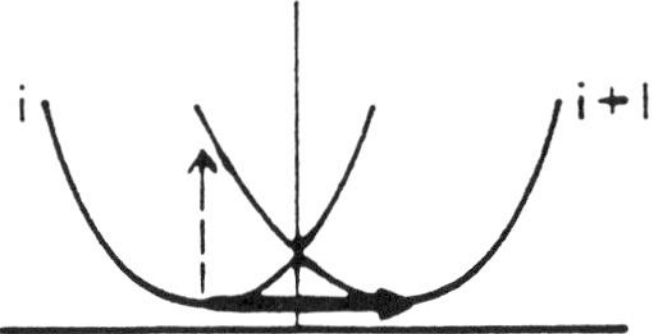

Figure 1. Schematic local ion potential for a polaron at sites i and $i+1$. When the polaron hops, the background ions may undergo a "zero–phonon" transition (solid arrow) or, more likely, be left in an excited state (dashed arrow). Only the first process contributes to the formation of a coherent polaron band.

On the other hand, the most favorable hopping process is one in which the local ionic coordinates do not rearrange (see Figure 1). In the simplest approximation, such a process requires an excitation energy of order $2E_P$.[18] A peak is expected in the optical conductivity at this energy. Since the hopping process requires that an unoccupied site be adjacent to an occupied site, the peak should scale with $x(1-x)$, where x is the polaron concentration. In the scenario discussed below, x is just the divalent dopant concentration in $La_{2-x}M_xCuO_{4-y}$. Experimental results are consistent with this dependence in the low doping limit.[17] The value of the dc conductivity is limited by impurity scattering and need not correlate with the height of the peak at energy $2E_P$.[19]

Note that the polaron bandwidth D_P is drastically reduced from the rigid–lattice value. For a two–dimensional tight–binding model with hopping matrix element $\tilde{t}$, the rigid–lattice bandwidth is just $8\tilde{t}$. Assuming $E_P = 0.25$ eV (the rough value indicated by optical measurements in $La_{2-x}Sr_xCuO_{4-y}$), $\omega_0 = 0.1$ eV, and $\tilde{t} = 0.1$ eV gives

$$m^*/m = e^{E_P/\omega_0} = 12$$
$$D_P = 8\tilde{t}(m/m^*) = 800 \text{ K}.$$

(1a)

For low dopant concentration x, the polaron Fermi energy is roughly

$$\epsilon_F = \tfrac{1}{2}xD_P = x(400 \text{ K}).$$

(1b)

For such small band energies, unusual thermal effects are possible. In particular, at temperatures of order the bandwidth, the electrical resistivity due to impurity scattering becomes linear in T, rather than constant. Approximate linearity may persist down to temperatures of order one–tenth the bandwidth.[20] A nearly linear resistivity above T_c has been observed in both $La_{2-x}Sr_xCuO_{4-y}$ and $YBa_2Cu_3O_{7-y}$.[21]

The measured thermopower $S(T)$ of $La_{2-x}Sr_xCuO_{4-y}$ also seems inconsistent with conduction by nearly free carriers, but consistent with polaron conduction.[22] As shown by Cooper et al.,[22] the thermopower in $La_{2-x}Sr_xCuO_{4-y}$ and $La_{2-x}Ba_xCuO_{4-y}$ is large (of order 100–300 $\mu V/K$ for low doping levels) and nearly constant with temperature in the range 100–300 K. Further, the high–temperature thermopower is well represented by the formula[18,23]

$$S = -\frac{k_B}{|e|}\ln 2 - \frac{k_B}{|e|}\ln\frac{x}{1-x} \qquad (2)$$

for x between 0 and 0.15. This is the appropriate expression for conduction by spin-$\frac{1}{2}$ polarons with no double site occupancy (the $\ln 2$ term represents the transport of spin entropy).

Transport and thermal properties may be modeled qualitatively by assuming that conduction occurs in an *uncorrelated* narrow band. The presence of correlations to inhibit double site occupancy is quantitatively important: for example, in the absence of correlations, the high–temperature thermopower becomes

$$S = -\frac{k_B}{|e|}\ln\frac{x}{2-x} \ . \qquad (3)$$

Nevertheless, the present calculation for an uncorrelated band illustrates the general temperature and dopant dependences of ρ and S. As shown in Figure 2, the high–temperature slope of the resistivity and the saturation limit of the thermopower decrease monotonically with increasing doping. Such a dependence has been observed in $La_{2-x}M_xCuO_{4-y}$.[22,24] Alternate explanations for the linear resistivity are based on scattering by tunneling centers,[25] electron–electron scattering near a van Hove singularity,[26] and impurity scattering of Bose charge carriers.[27]

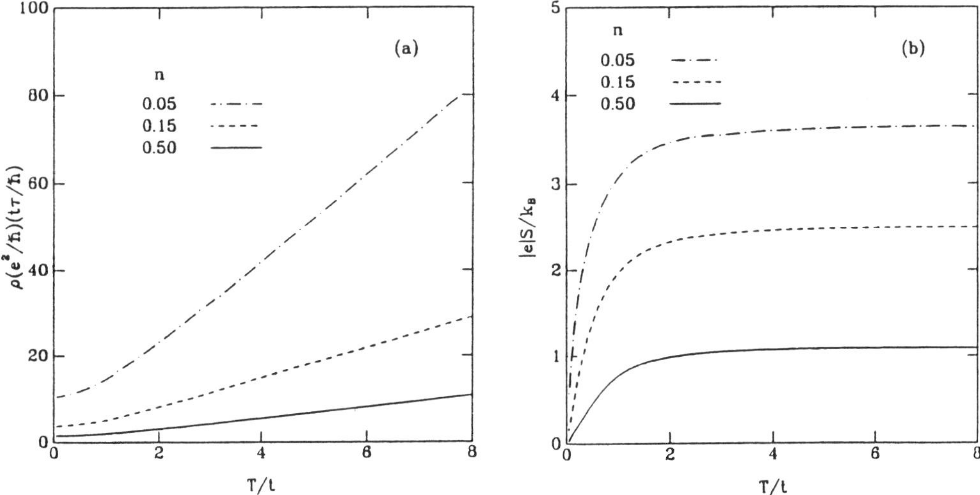

Figure 2. Resistivity (a) and thermopower (b) for a two–dimensional hole band with $\epsilon_k = -2t(\cos k_x + \cos k_y)$. A constant hole scattering rate τ^{-1} is assumed, and the chemical potential $\mu(T)$ is adjusted to keep n, the number of holes per site, fixed. Hole–hole correlations are ignored.

A strongly temperature–dependent specific heat coefficient $\gamma(T) = C(T)/T$ and a high–temperature Curie susceptibility also follow from the narrow band model.[20] Note that the zero–temperature mass enhancement m^*/m may be significantly larger than the value deduced from $\chi(T)$ or $\gamma(T)$ at high temperature.

Motivated by the experimental results cited above, we suggest a simple scenario for polaron formation. We assume that, at least in $La_{2-x}M_xCuO_{4-y}$, attention may be restricted to the two–dimensional CuO_2 layers. Mounting experimental evidence suggests that doping with divalent ions introduces holes in the oxygen 2p orbitals. A large intra–atomic Coulomb repulsion U_{dd} prevents double hole occupancy of the copper 3d orbitals. Hopping of 2p holes may proceed through direct overlap of near–neighbor oxygen orbitals, or by an indirect transfer through the copper sites. In either case, an O–O tranfer matrix element $\tilde{t}$ of order 0.1 eV seems reasonable. To discuss the coupling between 2p holes and lattice deformations in a simple approximation, we assume the heavy copper ions form a rigid lattice and that only the oxygen ions are mobile. If only a near–neighbor Cu–O spring constant is retained, there is, by symmetry, no tendency for bond relaxation when a 2p hole is introduced on an oxygen ion. For this reason, the near–neighbor O–O bond provides the dominant coupling to the lattice. The hole–phonon interaction takes the form

$$H_{\mathrm{h-ph}} = \tilde{V} \sum_{\langle ij\rangle} (n_i + n_j) x_{ij} = V \sum_{\langle ij\rangle} (n_i + n_j)(a_{ij} + a_{ij}^+), \tag{4}$$

where n_i counts the number of 2p holes on site i; x_{ij} is the O–O bond displacement; and a_{ij}^+ creates a bond phonon. Here, the coupling constant

$$V = \tilde{V} \sqrt{\frac{\hbar}{2M\omega_0}}, \tag{5}$$

with ω_0 and M the phonon frequency and reduced mass. The complete Hamiltonian takes the form

$$\begin{aligned}
H &= H_{\mathrm{ph}} + H_{\mathrm{h-ph}} + H_{\mathrm{h}} \\
H_{\mathrm{ph}} &= \omega_0 \sum_{\langle ij\rangle} a_{ij}^+ a_{ij} \\
H_{\mathrm{h}} &= -\tilde{t} \sum_{\langle ij\rangle} (c_i^+ c_j + c_j^+ c_i) \,.
\end{aligned} \tag{6}$$

Solving $H_{\mathrm{ph}} + H_{\mathrm{h-ph}}$ for a given configuration of holes generates a pattern of static displacements in the oxygen lattice. The resulting static contribution to the polaron Hamiltonian is

$$\begin{aligned}
H_{\mathrm{stat}} &= -\Delta \sum_{\langle ij\rangle} (n_i + n_j)^2 \\
&= -z\Delta \sum_i n_i - 2\Delta \sum_{\langle ij\rangle} n_i n_j,
\end{aligned} \tag{7a}$$

where

$$\Delta = \frac{V^2}{\omega_0} = \tilde{V}^2 \cdot \frac{\hbar}{2M\omega_0^2} \tag{7b}$$

and $z = 4$ is the oxygen–lattice coordination number. Since $(M\omega_0^2)^{1/2}$ is the static O–O spring constant, Δ is (at least nominally) independent of the oxygen ion mass. The second term in (7a) is a near–neighbor polaron attraction of the type first considered by Chakraverty.[3] The matrix element for polaron hopping between two lattice ground states is just

$$t = \tilde{t} e^{-(z-1)\Delta/\omega_0} \,. \tag{8}$$

Taking into account the Coulomb repulsion between holes, the full polaron Hamiltonian becomes

$$\begin{aligned}
H_P &= -t \sum_{\langle ij\rangle} (c_i^+ c_j + c_j^+ c_i) + U_c \sum_i n_{i\uparrow} n_{i\downarrow} - V_{\mathrm{eff}} \sum_{\langle ij\rangle} n_i n_j \\
V_{\mathrm{eff}} &= 2\Delta - (V_c + \delta V),
\end{aligned} \tag{9}$$

where U_c and V_c are on–site and near–neighbor Coulomb repulsions; and δV ($\propto t^2/\omega_0$) is an additional repulsion which emerges in second–order perturbation theory.[28] For V_{eff} positive and larger than the polaron bandwidth, this Hamiltonian is expected to lead to bipolaron formation at high temperature and possible Bose condensation at low temperature.[29] For smaller V_{eff}, the model should instead lead to BCS–like superconductivity. The parameter regime for the high–T_c oxides may lie between these two extremes. Note that V_{eff} is an instantaneous coupling which incorporates Coulomb repulsion. As usual, the effect of U_c on superconductivity may be minimized by constructing a pseudopotential for the pair wave function.[30]

Oxygen isotope dependence enters through the hopping matrix element t (and through δV). Since the polaron bandwidth serves as the BCS energy cutoff, a detailed calculation is required to establish the isotope dependence of T_c for arbitrary band filling. (The conventional ladder–graph summation to find T_c is likely to be invalid, since no equivalent of Migdal's Theorem is available. We are currently carrying out Monte Carlo simulations for the model in equation (6).) The isotope dependence of the polaron bandwidth leads to interesting predictions for normal–state properties as well: decreasing the phonon frequency should increase the effective mass m^* and the high–temperature slope of the resistivity $d\rho/dT$.

Throughout this discussion, we have not referred explicitly to the role of the copper lattice. A more complete treatment, taking into account the hybridization of $Cu^{2+}O^{2-}$ and Cu^+O^- states in the absence of doping,[31] seems essential for an explanation of magnetic properties. Our intent is simply to point out that polarons may play a role in determining the properties of the high–T_c oxides.

ACKNOWLEDGMENTS

We would like to thank J.E. Hirsch for correcting some of our earlier ideas regarding the Holstein model and for discussing with us the results of reference 28. We would also like to acknowledge useful discussions with S. Sachdev and J.W. Wilkins. This work has been supported in part by the National Science Foundation under Grants DMR86–15454 and PHY82–17853 and by funds from NASA.

REFERENCES

1. J.G. Bednorz and K.A. Müller, Z. Phys. B64, 18 (1986).

2. M.K. Wu, J.R. Ashburn, C.J. Torng, P.H. Hor, R.L. Meng, L. Gao Z.J. Huang, Y.Q. Wang and C.W. Chu, Phys. Rev. Lett. 58, 908 (1987).

3. B.K. Chakraverty and C. Schlenker, J. Phys. (Paris) Colloq. 37, C4–353 (1976); B.K. Chakraverty, J. Phys. (Paris) Lett. 40, L–9 (1979).

4. K. Fossheim, T. Laegreid, E. Sandvold, F. Vassenden, K.A. Müller and J.G. Bednorz, Solid State Commun., in press (1987).

5. L.C. Bourne, A. Zettl, K.J. Chang, M.L. Cohen, A.M. Stacy and W.K. Ham, Phys. Rev. B35, in press (1987).

6. D. McK. Paul, G. Balakrishnan, N.R. Bernhoeft, W.I.F. David and W.T.A. Harrison, Phys. Rev. Lett. 58, 1976 (1987).

7. A.W. Sleight, J.L. Gillson and P.E. Bierstedt, Solid State Commun. 17, 27 (1975).

8. E. Zirngiebl, J.O. Willis, J.D. Thompson, C.Y. Huang, J.L. Smith, P.H. Hor, R.L. Meng, C.W. Chu and M.K. Wu, preprint (1987).

9. A. Junod, A. Bezinge, T. Graf, J.–L. Jorda, J. Muller, L. Antognazza, D. Cattani, J. Cors, M. Decroux, O. Fischer, M. Banovski, P. Genoud, L. Hoffmann, A.A. Manuel, M. Peter, E. Walker, M. Francois, K. Yvon, Europhysics Lett., in press (1987).

10. A. Junod, A. Bezinge, D. Cattani, J. Cors, M. Decroux, O. Fischer, P. Genoud, L. Hoffmann, J.–L. Jorda, J. Muller and E. Walker, Proc. of LT–18, in press (1987).

11. H. Maletta, R.L. Greene, T.S. Plaskett, J.G. Bednorz and K.A. Müller, Proc. of LT–18, in press (1987).

12. L.F. Mattheiss, Phys. Rev. Lett. 58, 1028 (1987).

13. J. Yu, A.J. Freeman and J.-H. Xu, Phys. Rev. Lett. **58**, 1035 (1987); S. Massidda, J. Yu, A.J. Freeman and D. Koelling, preprint (1987).

14. B. Batlogg, R.J. Cava, A. Jayaraman, R.B. van Dover, G.A. Kourouklis, S. Sunshine, D.W. Murphy, L.W. Rupp, H.S. Chen, A. White, K.T. Short, A.M. Mujsce and E.A. Rietman, Phys. Rev. Lett. **58**, 2333 (1987).

15. L.C. Bourne, M.F. Crommie, A. Zettl, H.-C. zur Loye, S.W. Keller, K.L. Leary, A.M. Stacy, K.J. Chang, M.L. Cohen and D.E. Morris, Phys. Rev. Lett. **58**, 2337 (1987).

16. W. Weber, Phys. Rev. Lett. **58**, 1371 (1987).

17. J. Orenstein, private communication.

18. See, e.g., I.G. Austin and N.F. Mott, Adv. in Physics **18**, 41 (1969); or G.D. Mahan, *Many-Particle Physics* (Plenum, New York, 1981).

19. An alternate interpretation of the optical conductivity has been provided by C.M. Varma, S. Schmitt-Rink and E. Abrahams, Solid State Commun., in press (1987).

20. The magnetic susceptibility should exhibit Curie–law behavior $(\chi^{-1} \propto T)$ in the same temperature range. A Curie law has been observed in some $YBa_2Cu_3O_{7-y}$ samples (see, e.g., Refs. 9 and 21), but may be due to contamination with $BaCuO_2$.

21. R.J. Cava, B. Batlogg, R.B. van Dover, D.W. Murphy, S. Sunshine, T. Siegrist, J.P. Remeika, E.A. Rietman, S. Zahurak and G.P. Espinosa, Phys. Rev. Lett. **58**, 1676 (1987).

22. J.R. Cooper, B. Alavi, L.-W. Zhou, W.P. Beyermann and G. Grüner, Phys. Rev. **B35**, in press (1987).

23. P.M. Chaikin and G. Beni, Phys. Rev. **B13**, 647 (1976).

24. J.M. Tarascon, L.H. Greene, B.G. Bagley, W.R. McKinnon, P. Barboux and G.W. Hull, Proc. of the International Workshop on Novel Mechanisms of Superconductivity (this conference) (1987).

25. E. Abrahams and C.M. Varma, preprint (1987).

26. P.A. Lee and N. Read, Phys. Rev. Lett. **58**, 2691 (1987).

27. P.W. Anderson, G. Baskaran, Z. Zou and T. Hsu, Phys. Rev. Lett. **58**, 2790 (1987).

28. J.E. Hirsch and E. Fradkin, Phys. Rev. **B27**, 4302 (1983).

29. A. Alexandrov and J. Ranninger, Phys. Rev. **B23**, 1796 (1981).

30. N.F. Berk and J.R. Schrieffer, Phys. Rev. Lett. **17**, 433 (1966).

31. See, e.g., reference 19; and C.M. Varma, S. Schmitt-Rink and E. Abrahams, Proc. of the International Workshop on Novel Mechanisms of Superconductivity (this conference) (1987).

SUPERCONDUCTIVITY DUE TO NEGATIVE-U IMPURITIES

H.-B. Schüttler

Materials Science Division
Argonne National Laboratory
Argonne, IL 60439

M. Jarrell and D. J. Scalapino

Department of Physics
University of California
Santa Barbara, CA 93106

ABSTRACT

We review our recent work on induced superconductivity due to negative-U impurities. Results have been obtained by a new method that combines Monte Carlo simulation with diagrammatic techniques. We discuss
their possible relevance for the copper oxide high-T_c materials.

It was suggested some time ago[1,2] that negative-U centers[3] could
enhance the superconducting transition temperature T_c of a metallic host
lattice. We have recently carried out an extensive study[4-6] of this
effect, using a novel approach which combines quantum Monte Carlo methods
with diagrammatic techniques. We have shown that, with relatively small
impurity concentrations c, substantial T_c-enhancements $(dT_c/dc|_{c=0})$ may be
achieved and that these enhancements occur at coupling strengths where the
ordinary weak coupling perturbation techniques[2] break down. We have also
studied the possibility of induced superconductivity, sustained solely by
the presence of a small concentration of these impurity centers in an
intrinsically non-superconducting host. It was subsequently proposed[7,8]
that such negative-U impurities might be responsible for the large T_c-
values observed in the recently discovered copper oxide superconductors.
Here, we will review our results for induced superconductivity due to
negative-U impurities and discuss their possible relevance[7] for the high-
T_c materials.

The negative-U center consists of a localized orbital which is
hybridized with the host lattice conduction band and coupled to a local,
bosonic degree of freedom. Its Hamiltonian has the form

$$H = \sum_{ps} \varepsilon_p c^\dagger_{ps} c_{ps}$$

$$+ \sum_{ps} N^{-1/2} V_p \left(c^\dagger_{ps} d_s + d^\dagger_s c_{ps} \right) + H_i \tag{1}$$

where, $d^\dagger_s$ and $c^\dagger_{ps}$ create electrons (with spin $s = \uparrow,\downarrow$) in the impurity orbital and in a conduction band state p with energy ε_p, respectively. V_p denotes the hybridization matrix element and N is the number of host lattice sites. The Fermi level is set to $\mu = 0$ in the following. H_i describes the internal structure of the impurity. We will consider here the case of a local exciton model with[4,9]

$$H_i = \lambda \sigma_x \left(n_{d\uparrow} + n_{d\downarrow} \right) + \frac{\Omega}{2} \sigma_z + \varepsilon_d \left(n_{d\uparrow} + n_{d\downarrow} \right) \tag{2}$$

Here $n_{ds} = d^\dagger_s d_s$ denotes the occupation number of the impurity d_s-orbital with a bare on-site energy ε_d. Via the first term in (2), it is coupled to a two-level excitonic system, described by Pauli matrices σ_x and σ_z. The two-level system has an excitation energy Ω set by the second term in (2). By polarizing the two-level system, d-electrons can lower their effective on-site energy. In this way, an attractive interaction between the d-electrons is mediated: In the absence of hybridization ($V_p \equiv 0$), for example, the ground state energy E_0 of the single impurity (H_i) depends on the occupation $n_d = n_{d\uparrow} + n_{d\downarrow}$ of the impurity orbital,

$$E_0(n_d) = \varepsilon_d n_d - \left[(\Omega/2)^2 + (n_d - 1)^2 \lambda^2 \right]^{1/2} , \tag{3}$$

from which we can extract an effective on-site interaction[4,9]

$$-U \equiv E_0(0) + E_0(2) - 2E_0(1)$$

$$= -2\left[(\Omega/2)^2 + \lambda^2 \right]^{1/2} + \Omega < 0 \tag{4}$$

that is obviously attractive. If we choose the on-site energy (measured from the Fermi level $\mu \equiv 0$) $\varepsilon_d = 0$, the unoccupied and the doubly occupied impurity orbital configurations are degenerate while the singly occupied orbital has $E_0(1) = E_0(0) + U > E_0(0)$. In this case, the hybridization will give rise to strong electron-pair fluctuations between the conduction band and the impurity which allows the conduction electrons to participate in the attractive on-site interaction and may hence induce superconductivity.

Formally, at finite impurity concentration c, these pair fluctuations give rise to a selfenergy Σ_c and an effective interaction V_c of the conduction electrons. Our basic approach in calculating the induced T_c is now to take disorder averages and expand Σ_c and V_c in powers of the impurity concentration.[10-12] As shown in Fig. 1, Σ_c and V_c are, to leading order in c, given by

$$\Sigma_c(p,i\nu) = c|V_p|^2 G_d(i\nu) + O(c^2) \tag{5}$$

$$V_c(p',i\nu'|p,i\nu) = -cT|V_p|^2|V_{p'}|^2 \Gamma_d(i\nu'|i\nu) + O(c^2) \tag{6}$$

where ν,ν' denote odd Matsubara frequencies and p,p' are conduction electron momenta. $G_d(i\nu)$ and $\Gamma_d(i\nu'|i\nu)$ are, respectively, the single and two-particle d-electron Green's functions of a <u>single</u> impurity, as defined in Ref. 4, Eqs. (7) and (8). They are obtained from the Hamiltonian (1) by the Monte Carlo simulation techniques,[13] described previously.[4-6] With Σ_c and V_c, we carry out the usual Eliashberg-type ladder summation to obtain the conduction electron particle-particle t-matrix t_c. We then search for the superconducting instability at which t_c diverges.[10-12] Since the time consuming single impurity simulations require the temperature T (but not the concentration c) as an input, it is numerically convenient to specify a certain transition temperature $T = T_c$ and then vary c such that the instability occurs at that T. We thus calculate the concentration c needed to achieve a given T_c, rather than T_c as a function of c.

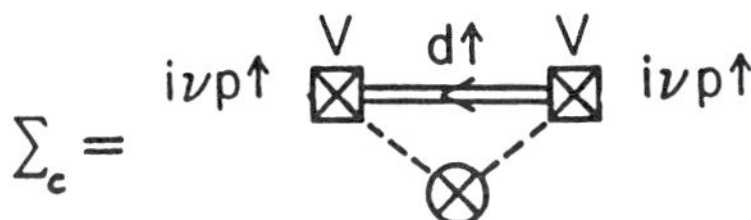

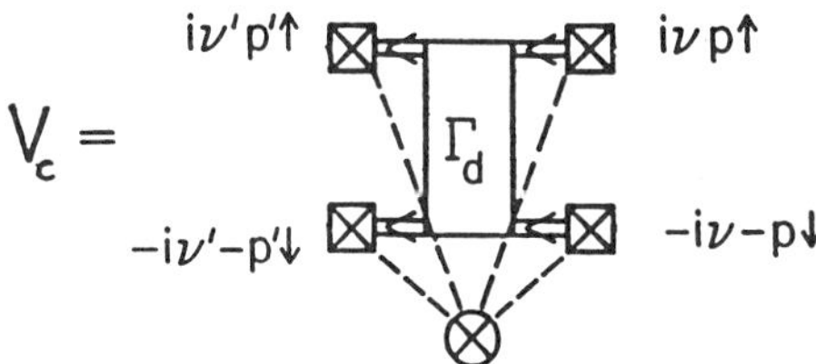

Fig. 1. Disorder averaged conduction electron selfenergy Σ_c and effective interaction V_c expressed in terms of the single impurity d-electron Green's functions G_d and F_d to lowest order in the impurity concentration c.

We confine ourselves here to the simplest case of a momentum independent hybridization $V_p = V$ = const which we measure in terms of the reso-
nance width $\Delta \equiv \pi\rho(0)|V|^2$ (where $\rho(0)$ is the conduction electron density
of states at the Fermi level per site and spin).[14] Assuming furthermore
that Δ is small compared to the bandwidth $D \sim 1/\rho(0)$, we find that, for a
given Δ, the impurity concentration and the conduction electron bandstruc-
ture enter the Eliashberg equation only via the ratio $\tilde{c} = c/(\pi\rho(0))$ which
is thus a universal function of T_c, Δ, ε_d, Ω, and U, independent of band-
structure details or $\rho(0)$. The inverse of this quantity is displayed as
a function of T_c, U, Δ, and ε_d in Figs. 2-4, respectively. The unit of
energy has been chosen such that Ω = 1. Notice that $1/\tilde{c}$ is a convenient
measure for the strength of the induced superconductivity. Its dependence
on the model parameters is very similar to that of the linear enhancement
coefficient $(dT_c/dc)_{c=0}$ studied in Refs. 4,5:

$1/\tilde{c}$ increases with decreasing T_c and increasing U (Fig. 2). As a
function of Δ (Fig. 3), it exhibits a maximum when $\Delta \sim \Omega$, i.e., when the
average time spent by an electron on the impurity orbital, $\tau_\Delta \sim \Delta^{-1}$,
becomes comparable to the characteristic time scale $\tau_\Omega \sim \Omega^{-1}$, on which the
excitonic polarization can respond to the presence of a d-electron. As a
function of the on-site energy ε_d (Fig. 4), $1/\tilde{c}$ shows a sharp maximum at
$\varepsilon_d = 0$ where the unoccupied and the doubly occupied impurity configurations are "in resonance," i.e., $E_0(0) = E_0(2)$. If $\varepsilon_d > 0$ or $\varepsilon_d < 0$, it

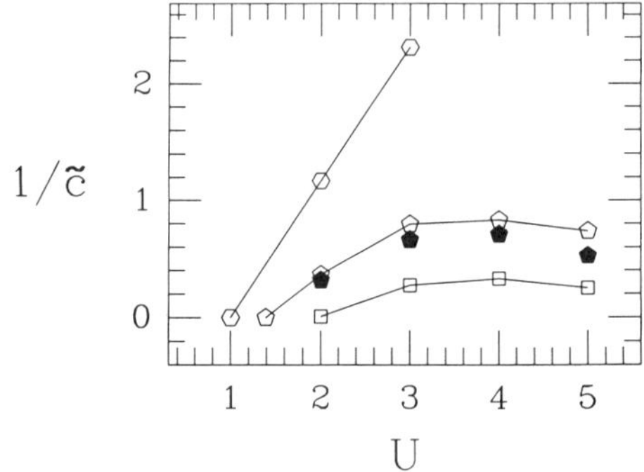

Fig. 2. $1/\tilde{c}$ vs. U for $T_c = 0.025$ (⬡), $T_c = 0.05$ (⬡),
$T_c = 0.1$ (⬜), $\Delta = \Omega = 1$, and $\varepsilon_d = 0$. The open
and full symbols indicate data obtained from
Monte Carlo runs with imaginary time slices
$\Delta\tau = 0.25$ and $\Delta\tau = 0.125$, respectively
(c.f. Ref. 4 for details)

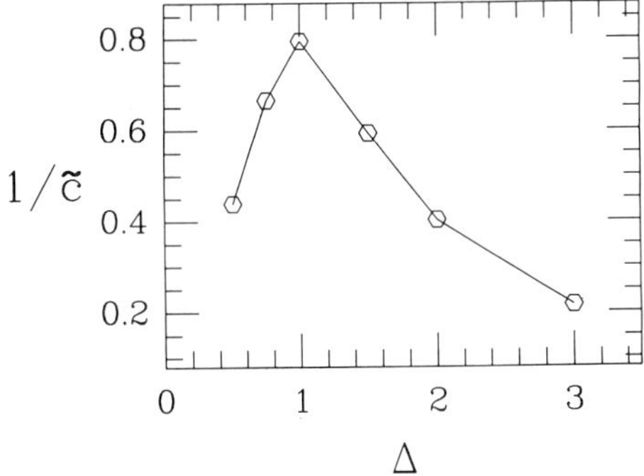

Fig. 3. $1/\tilde{c}$ vs. Δ for $T_c = 0.05$, $U = 3$, $\varepsilon_d = 0$, $\Omega = 1$

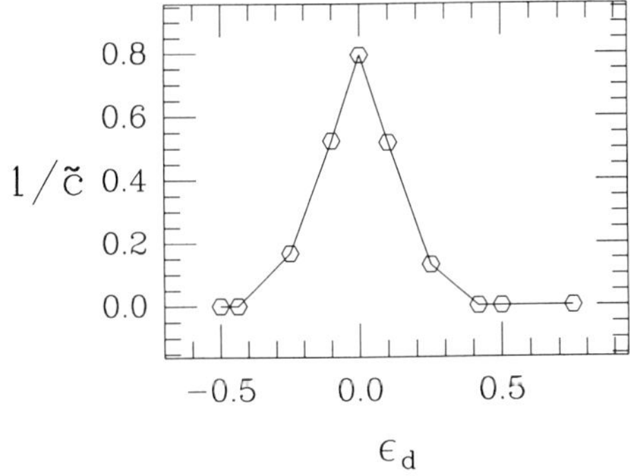

Fig. 4. $1/\tilde{c}$ vs. ε_d for $T_c = 0.05$, $U = 3$, $\Delta = \Omega = 1$

becomes energetically unfavorable for an electron pair to tunnel into or out of the d-orbital, respectively. Pair fluctuations are thus strongly suppressed if $|\varepsilon_d| \gtrsim \Gamma/2$.

To estimate typical orders of magnitude for the T_c's that can be obtained with this mechanism, let us assume a host lattice with a bandwidth $\rho(0)^{-1} = 3$ eV, a typical value for the high-T_c oxides.[15-20] We assume further a negative-U impurity with optimal on-site energy $\varepsilon_d = 0$, an exciton energy $\Omega = 0.1$ eV, a resonance width $\Delta = 0.1$ eV $= \Omega$ (corresponding to a hybridization strength $|V| \sim 0.3$ eV) and an attractive $U = 0.3$ eV $= 3\Omega$. Then according to Fig. 1, with a $\tilde{c} \cong (2.3)^{-1} \Omega \cong 43$ meV, corresponding to an impurity concentration $c = \pi\rho(0) \tilde{c} \cong 4.5\%$, we would get $T_c = 2.5$ meV $= 27$ K. Similarly, with $c \cong 15\%$, a $T_c = 54$ K could be obtained and, extrapolating to larger concentrations, with $c = 40\%$ we would reach $T_c \cong 100$ K.

It is interesting to compare these parameters with those of a conventional BCS superconductor such as aluminum, say. Al has[21] a $T_c = 1.2$ K, a Debye energy $\Omega = 0.034$ eV, an inverse density of states $\rho(0)^{-1} \cong 2.6$ eV and thus, according to the BCS formula, $T_c = 1.13 \, \Omega \, \exp[-1/U\rho(0)]$, an attractive $U = 0.46$ eV, about 50% larger than the 0.3 eV assumed above. If we rescale these values for $\rho(0)$, Ω and U to those assumed in our hypothetical "negative-U material," $\rho(0)^{-1} = 3$ eV, $\Omega = 0.1$ eV, and $U = 0.3$ eV, the BCS formula gives an even smaller transition temperature, $T_c = 0.05$ K. Thus, in the BCS superconductor, T_c is over 2 orders of magnitude smaller than in the 4.5% doped negative-U material with identical Boson frequency Ω, density of states $\rho(0)$, and attractive strength U. The results for the negative-U material are therefore quite surprising, particularly in view of the fact that, in the BCS superconductor, <u>all</u> lattice sites contribute to the attractive interaction whereas, in the negative-U system, only a small fraction (c) of the sites contributes.

One might suspect that these results are a consequence of the particular excitonic mechanism, Eq. (2), that mediates the attraction. However, we have also studied a model H_i where the d-electron is coupled to a local phonon mode.[5,6] We find that, for comparable values of the Boson frequency Ω and the effective on-site attraction U, and the resonance width Δ, the results in both models are quite similar. Apparently, the crucial difference between the BCS superconductor and the negative-U system is that in the former, the virtual Boson is exchanged directly between the conduction electrons, whereas in the latter, the exchange can take place only after the electrons have tunneled into the impurity orbital. If one estimates perturbatively to lowest order in the coupling the effective attraction [Eq. (6)], mediated between conduction electrons by a local phonon impurity,[5,6] one finds that indeed this attraction is enhanced relative to the on-site U by a factor $(\pi\rho(0)\Delta)^{-2} \gg 1$.[6] This is roughly the square of the occupancy time ratio τ_Δ/τ_ρ where $\tau_\Delta \sim \Delta^{-1}$ is the average time span for which an electron remains at an impurity site after tunneling into the d-orbital and $\tau_\rho \sim \rho(0)$ is the (much shorter) time it spends at any lattice site while propagating in the conduction band.

The foregoing results show that the negative-U center mechanism may well be capable of producing quite large superconducting T_c's even with a quite small on-site coupling strength. We should caution, however, that this is achieved only if the model parameters are carefully tuned such that $\Delta \sim \Omega$ and, most importantly, $|\varepsilon_d| \sim 0$ ($|\varepsilon_d| < \Delta$). For example, if we shift ε_d off resonance by only a small fraction of Δ, $\varepsilon_d = \pm 0.015$ eV, we would have to increase the impurity concentration from c = 15% to over 30% (c.f. Fig. 4) in order to maintain a T_c of 54 K (assuming again $\Delta = \Omega = 0.1$ eV, U = 0.3 eV, $\rho(0)^{-1} = 3$ eV). If we shift ε_d by more than ± 0.04 eV, we get, according to Fig. 4, $1/\tilde{c} = 0$, indicating that the specified T_c = 54 K cannot be reached even at the highest doping levels. In the high-T_c materials, considerable variations in their stoichiometry and a concomitant shift of their Fermi level μ are possible. Assuming that their large T_c is indeed due to the presence of negative-U centers, small variations in their stoichiometry should thus give rise to dramatic changes in T_c, since ε_d (which is the on-site energy measured <u>relative</u> to μ) would be shifted. Although the dependence of T_c on sample stoichiometry is well documented in those materials,[22] the variations do not appear to be as drastic as the results in Fig. 4 would lead us to expect. On the other hand, one might argue that such variations can easily be masked by sample inhomogeneities. They could also be suppressed by any mechanism that would "pin" μ and/or ε_d to a constant value. Nevertheless, we feel that the strong variation of T_c with Δ and ε_d predicted by the negative-U center model may be difficult to reconcile with the experimental observation that high transition temperatures in the oxide superconductors survive considerable changes in stoichiometry and chemical composition. Thus, further experimental and theoretical work is needed to investigate how strongly crucial system parameters, such as the position of the Fermi level, electronic hybridization strengths and on-site energies, are affected by the compositional changes.

We wish to acknowledge helpful discussions with D. G. Hinks, J. D. Jorgensen, E. Loh, I. K. Schuller, R. L. Sugar, and D. Toussaint. This work was supported by the National Science Foundation under Grant No. DMR-83-20481 and by the U.S. Department of Energy, Basic Energy Science-Materials Sciences, under Contract No. W-31-109-ENG-38. Support from E. I. du Pont de Nemours and Company and the Xerox Corporation is gratefully acknowledged.

REFERENCES

1. E. Simanek, Solid State Commun. <u>32</u>, 731 (1979).
2. C. S. Ting, D. N. Talwar, and K. L. Ngai, Phys. Rev. Lett. <u>45</u>, 1213 (1980).
3. P. W. Anderson, Phys. Rev. Lett. <u>34</u>, 953 (1975).
4. H.-B. Schüttler, M. Jarrell, and D. J. Scalapino, Phys. Rev. Lett. <u>58</u>, 1147 (1987); and to be published.
5. H.-B. Schüttler, M. Jarrell, and D. J. Scalapino, J. Low. Temp. Phys. (in press).
6. M. Jarrell, H.-B. Schüttler, and D. J. Scalapino, to be published.
7. W. Y. Lai, C. S. Ting, and D. Y. Xing, preprint.
8. J. Yu, S. Massida, A. J. Freeman, and D. D. Koelling, Phys. Lett., submitted.
9. J. E. Hirsch and D. J. Scalapino, Phys. Rev. B <u>32</u>, 5639 (1985).
10. A. A. Abrikosov and L. P. Gofkov, JETP <u>12</u>, 1243 (1961).
11. T. Matsuura, S. Ichinose, Y. Nagaoka, Prog. Theor. Phys. <u>57</u>, 713 (1977).

12. A. Sakurai, Phys. Rev. B **17**, 1195 (1978).

13. J. E. Hirsch and R. M. Fye, Phys. Rev. Lett. **56**, 2521 (1986).

14. In view of the discrepancy between values for the superconducting gap that has been observed by different experimental methods in the high T_c oxides, it is worth mentioning that an anisotropic p-dependence of the hybridization matrix element p, in this model, would lead to an anisotropic superconducting gap function, varying like $|V_p|^2$ as a function of p. (For a review of the experiments, see, for example, the contribution by K. E. Gray et al., in these proceedings, and D. A. Bonn, J. E. Greedan, C. V. Stager, T. Timusk, M. Doss, S. Herr, K. Kamaras, G. Porter, D. B. Tanner, J. M. Tarascon, W. R. McKinnon, L. H. Greene, preprint.)

15. L. F. Mattheiss, Phys. Rev. Lett. **58**, 1028 (1987); L. F. Mattheiss and D. R. Hamann, Solid State Commun., submitted.

16. J. Yu, A. J. Freeman, and J.-H. Xu, Phys. Rev. Lett. **58**, 1035 (1987); J. Yu, S. Massida, A. J. Freeman, and D. D. Koelling, Phys. Lett., submitted.

17. M.-H. Whangbo, M. Evain, M. A. Beno, and J. M. Williams, Inorg. Chem. (Comm.), submitted.

18. W. Weber, Phys. Rev. Lett. **58**, 1371 (1987).

19. W. E. Pickett, H. Krakauer, D. A. Papaconstantopoulos, L. L. Boyer, preprint.

20. R. V. Kasowski, W. S. Hsu, F. Herman, preprint.

21. R. Meservey and B. B. Schwartz, in _Superconductivity_, Vol. I, R. D. Parks, ed., Marcel Dekker, Inc., New York (1969).

22. For a review, see the contribution by I. K. Schuller, D. G. Hinks, J. D. Jorgensen, L. Soderholm, M. Beno, K. Zhang, C. U. Segre, and J.-P. Locquet, in these proceedings.

PREDICTION OF ANISOTROPIC THERMOPOWER OF $La_{2-x}M_xCuO_4$

Philip B. Allen,* Warren E. Pickett, and Henry Krakauer**

Condensed Matter Physics Branch, Naval Research Laboratory
Washington, DC 20375-5000, *Department of Physics, State
University of New York, Stony Brook, NY 11794-3800
**Department of Physics, College of William and Mary
Williamsburg, VA 23185

INTRODUCTION

The thermopower tensor $S_{\alpha\beta}(T)$ of a tetragonal or orthorhombic crystal
is diagonal in the crystallographic axes and equals, according to
Boltzmann theory[1]

$$S_{\alpha\alpha}(T) = -\frac{k_B}{e} \int d\epsilon \left(\frac{\epsilon-\mu}{k_BT}\right) \sigma_{\alpha\alpha}(\epsilon) \left(-\frac{\partial f}{\partial \epsilon}\right) / \int d\epsilon \sigma_{\alpha\alpha}(\epsilon) \left(-\frac{\partial f}{\partial \epsilon}\right) \tag{1}$$

$$\sigma_{\alpha\alpha}(\epsilon) = e^2 N(\epsilon) v_\alpha^2(\epsilon) \tau(\epsilon) \tag{2}$$

where $N(\epsilon)$, $v_\alpha^2(\epsilon)$, and $\tau(\epsilon)$ are the density of states, mean square
velocity (α-component) and scattering lifetime, averaged for states with
energy ϵ. If a Sommerfeld expansion of $\sigma_{\alpha\alpha}(\epsilon)$ is valid, the integrals in
Eq. (1) yield a result linear in T and proportional to $\sigma'_{\alpha\alpha}(\mu)/\sigma_{\alpha\alpha}(\mu)$.
However, the integrands in Eq. (1) do not become vanishingly small until
$\pm 10k_BT \sim .02Ry$ (at 300K) above and below ϵ_F, so the Sommerfeld expansion
fails and S is not linear.

We have used the band structure of Pickett et $al.$[2] and the rigid band
approximation to calculate $S_{xx}(T)$ and $S_{zz}(T)$ for $La_{2-x}M_xCuO_4$ where M is
Ca, Sr, or Ba. The values of $N(\epsilon)$ and $v_\alpha^2(\epsilon)$ come directly from the
calculated eigenvalue spectrum, whereas $\tau(\epsilon)$ requires an additional
calculation. If electron-phonon scattering dominates (which provides a
reasonable and conservative explanation[3] for the measured $\rho_{\alpha\alpha}(T)$ in
samples with linear T-dependence and rrr≥4) then $\tau(\epsilon)$ is given by

$$1/\tau(\epsilon) = 2\pi\lambda(\epsilon)k_BT/\hbar \tag{3}$$

and $\lambda(\epsilon)$ can be approximated by $\sum_a \eta_a(\epsilon)/M_a <\omega^2>$ where $\eta_a(\epsilon)$ has already
been calculated[2] in the rigid muffin tin model. Alternately, if impurity
scattering dominates, one might assume $1/\tau(\epsilon) \propto N(\epsilon)$, or possibly $1/\tau(\epsilon) =$
const. The results for $\sigma_{\alpha\alpha}(\epsilon)$ and $S_{\alpha\alpha}(T)$ for all three models are shown
in Figs. 1 and 2. All three models predict moderately large positive
("hole-like") behavior of S_{zz} and smaller, negative ("electron-like")
behavior of S_{xx}. In polycrystalline samples, Maeno et $al.$[4] have reported

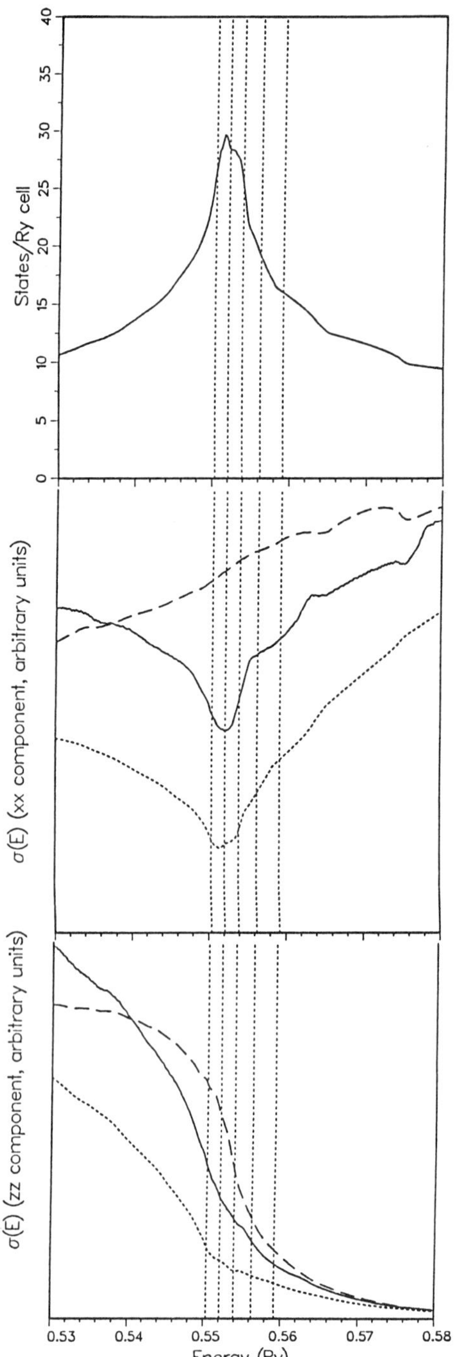

Figure 2. $S_{zz}(T)$ (positive values) and $S_{xx}(T)$ (negative values) versus T for concentration x=0.15. The solid, dotted, and dashed lines follow the code of Fig. 1.

Figure 1. Upper panel: $N(\epsilon)$ vs ϵ; middle and lower panels: $\sigma_{xx}(\epsilon)$ and $\sigma_{zz}(\epsilon)$ assuming: (solid line: $1/\tau \propto \sum_a \eta_a/M_a$); (dotted line: $1/\tau \propto N(\epsilon)$); (dashed line: $1/\tau = $ const.) The vertical dotted lines denote $\epsilon_F(x)$ for concentrations x of divalent atoms equal to 0.2, 0.15, 0.1, 0.05, and 0 in order of increasing ϵ_F.

a positive and nearly T-independent $S(T) \approx 25\mu V/K$ at x=0.15 and T=300 K. We find a fairly weak x-dependence of S, as shown in Table I. It is amusing that the "sign of the carriers" predicted in our thermopower calculation switches depending on crystallographic orientation, just as our prediction of the Hall tensor,[3] but in an opposite way: the Hall coefficient is "hole-like" if the orbits are in the xy plane, whereas $S_{xx} = S_{yy}$ is "electron-like".

ACKNOWLEDGMENTS

P. B. Allen acknowledges a sabbatical stay at the Naval Research Laboratory, with support under the Intergovernmental Personnel Act. This work was supported by the Office of Naval Research Contract No. N00014-84-WR-24055 and the National Science Foundation (NSF) Grant No. DMR-84-16046. We also acknowledge SDIO/IST and DARPA for their partial support of our research. Computations were done with NSF support on CRAY X-MP's at the University of Pittsburgh and the University of Illinois.

Table I. Thermopower (in $\mu V/K$) at 300 and 600 K for various concentrations x (the crystal structure is assumed tetragonal K_2NiF_4-type; the scattering rate $1/\tau$ is taken proportional to $\sum_a \eta_a/M_a$).

x	$S_{xx}(300K)$	$S_{xx}(600K)$	$S_{zz}(300K)$	$S_{zz}(600K)$
0.00	-15.	-20.	59.	111.
0.05	-17.	-19.	61.	105.
0.10	-15.	-14.	60.	97.
0.15	- 6.	- 7.	57.	87.
0.20	6.	1.	50.	76.

REFERENCES

1. N.W. Ashcroft and N.D. Mermin, Solid State Physics, (Holt, Rinehart, and Winston, New York, 1976, pp 250-8.
2. W.E. Pickett, H. Krakauer, D.A. Papaconstantopoulos, and L.L. Boyer, Phys. Rev. B35, 7252 (1987).
3. P.B. Allen, W.E. Pickett, and H. Krakauer, to be published.
4. Y. Maeno, Y. Aoki, H. Kamimura, J. Sakurai, and T. Fujita, Japan. J. Appl. Phys. 26, L402 (1987).

BAND STRUCTURE AND ELECTRON-PHONON INTERACTION CALCULATIONS FOR PROPOSED
HIGH-T_C SUPERCONDUCTING OXIDES: $MCUO_3$ (M = LA, BA, CS, Y) IN THE
PEROVSKITE STRUCTURE

D. A. Papaconstantopoulos and L. L. Boyer*

Metal Physics Branch and *Condensed Matter Physics Branch
Naval Research Laboratory
Washington, DC 20375-5000

The crystal structures reported for the recently discovered high-T_c
superconductors containing copper and oxygen have one thing in common;
structural similarity to ideal perovskite. The systems which superconduct
with $T_c \sim$ 35 K, e.g., $La_{2-x}Ba_xCuO_4$ with x $\sim$ 0.1 - 0.2, either have the
tetragonal K_2NiF_4 structure or small orthorhombic distortions there-
from.[1-4] The K_2NiF_4 structure can be described qualitatively as alternating layers of perovskite ($KNiF_3$) and rock salt (KF) structures. The
structures reported for the Y-Ba-Cu-O systems, which have $T_c \sim$ 90 K, are
described approximately as an ordered superlattice of oxygen deficient
perovskite cells.[5-7] The common elements in all the superconductors
discovered so far with $T_c >$ 30 K are copper and oxygen. These results
suggest that the local environment of the copper and oxygens in the
perovskite structure is somehow conducive to high temperature superconductivity.

In an effort to explore these ideas theoretically we have carried out
electronic band structure calculations for $MCuO_3$ (M = Cs, Ba, La and Y) in
the ideal perovskite structure. Band structure quantities were calculated
using the self-consistent APW method in the muffin-tin approximation with
a local density approximation[8] for exchange and correlation. From these
quantities we determine values for the McMillan-Hopfield parameter η
(total and site decomposed), using the rigid muffin-tin approximation,[9] to
provide some indication of the strength of the electron-phonon interaction
in these idealized systems.

The calculations for M = La, Ba, and Cs were carried out for a
lattice constant a = 7.8234 a.u., which is two times the average copper-
oxygen separation in the La_2CuO_4 compound. A smaller value, a = 7.2943
a.u., was selected for the M = Y compound to correspond approximately to
the difference between the ionic radii of Y and La. The muffin-tin radii
used in the calculation were chosen to minimize the interstitial volume:
R_M = 3.5761 a.u. and R_{Cu} = R_O = 1.9558 a.u. for M = La, Ba, and Cs; R_Y =
3.3343 a.u. and R_{Cu} = R_O = 1.8236 a.u. for the yttrium compound. States
falling within $\sim$ 1.5 Ry of the Fermi energy (E_F) were treated as bands in
a semirelativistic approximation[10]. These include, of course, the valence
bands, which are predominantly Cu-3d and O-2p, and O-2s, La-5p, Ba-5p, Cs-
5p and Y-4p states as well. Deeper core levels were allowed to relax self
consistently using a fully relativistic atomic code.

The band structures and total densities of states N(E) are shown in Figs. 1-4 along with selected contributions to N(E), $N(Cu-t_g)$, $N(Cu-e_g)$ and N(O-p), which reveal the predominant character of the valence bands. The La-f decomposed density of states (DOS) is also shown since this is a prominent feature in the conduction bands for the La compound. For the La, Ba and Y compounds, $N(E_F)$ is composed mainly of Cu-3d states with e_g symmetry and O-2p states, while for $CsCuO_3$, there is a large contribution from the t_g symmetry copper states as well. This can be understood as a nearly rigid band shift of the Fermi level into the $Cu-t_g$ bands, which are fully occupied for the La, Ba and Y compounds. The additional contribution from the $Cu-t_g$ bands approximately triples the total DOS at E_F for $CsCuO_3$. The band structure of $CsCuO_3$ differs from the La and Ba compounds in another respect: the M-5p states (not shown for M = La and Ba) shift from ~-0.5 for La to ~-0.35 for Ba to ~-0.1 for Cs, where for Cs they hybridize with the valence bands.

The electron-phonon interaction parameter λ, which governs T_c in the conventional theory of superconductivity, is given by $\lambda = N(E_F)<I^2>/F$, where F is a force constant which gives an average stiffness of the crystal against phonon modes, and $<I^2>$ is an average electron-phonon matrix element. We have calculated values for $\eta_s = N(E_F)<I^2>_s$ using the rigid muffin-tin approximation to evaluate the average electron-phonon matrix elements $<I^2>_s$ for each site s. These results are listed in Table 1 along with the total and decomposed DOS at E_F, which enter the calculation of η_s.

The calculated values of η_s for the Y compound are qualitatively similar to those for La and Ba compounds, the latter of which were reported and discussed in a previous paper[11]. In Ref. 11 it was argued that T_c could be calculated as a function of a single parameter representing an average Cu-O force constant. The argument was based on the result that η_M was nearly zero for these materials. Calculations of T_c as a function of this parameter showed that T_c's of 30 to 40 K could be achieved with a modest amount of phonon softening. The same conclusion is probably true for the Cs compound, although a similar quantitative analysis would not be valid owing to the substantial contribution of η_{Cs}. In spite of the large increase in $N(E_F)$ for $CsCuO_3$, η is about the same as for the Y, La, and Ba compounds. Unfortunately, the large increase in $N(E_F)$ is approximately cancelled by a corresponding decrease in the size of the matrix elements for the Cu and O sites.

Our calculations suggest that the ideal perovskite analogs of the high-T_c oxide superconductors would themselves be high-T_c materials if they were to form in the ideal (no defects) perovskite structure. It seems unlikely that this conclusion would change if the calculations were carried out using more reliable estimates of the lattice constants. We plan to carry out total energy calculations on these idealized materials in the future to answer conclusively this question and to investigate the stability of the perovskite structure.

We acknowledge ONR, SDIO/IST and DARPA for their partial support of this research.

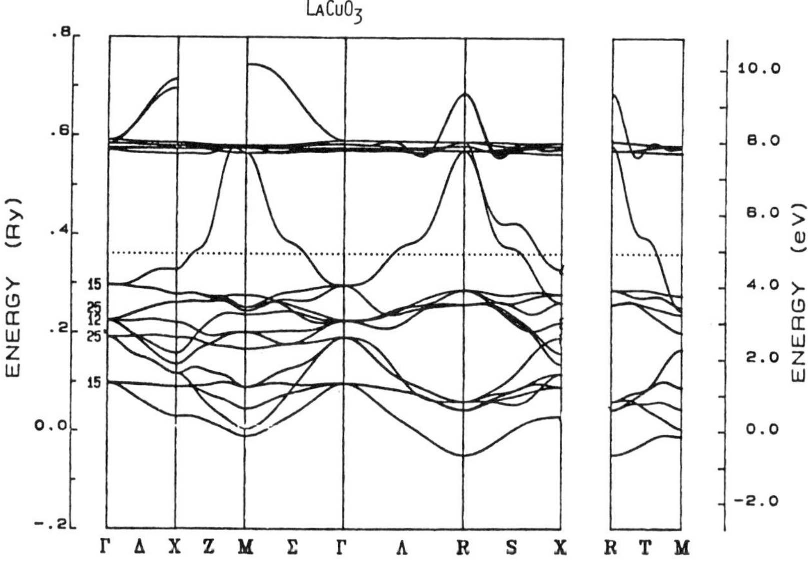
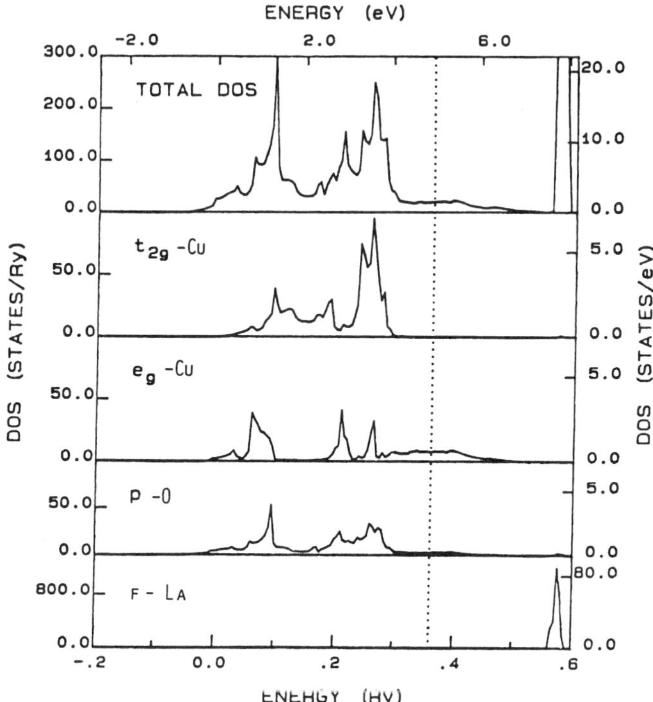

Figure 1. Band structure, density of states (DOS) and selected decomposed DOS for LaCuO$_3$.

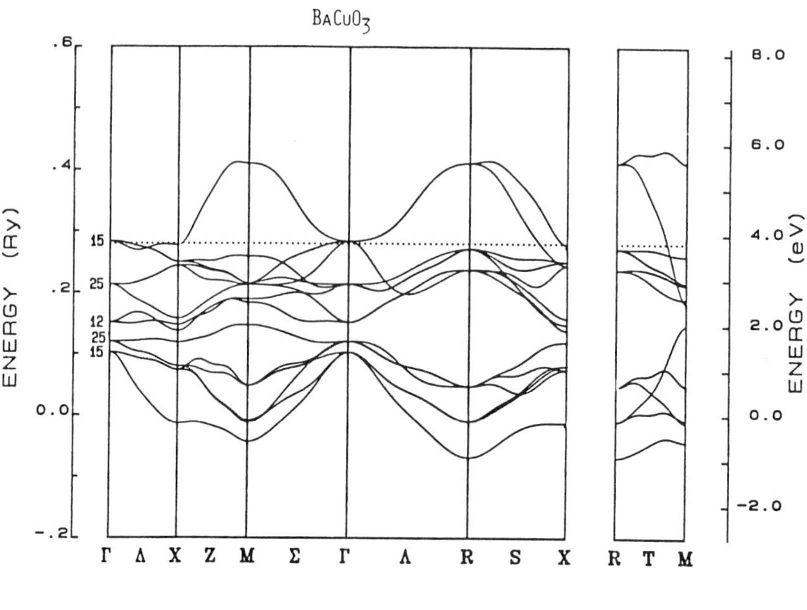

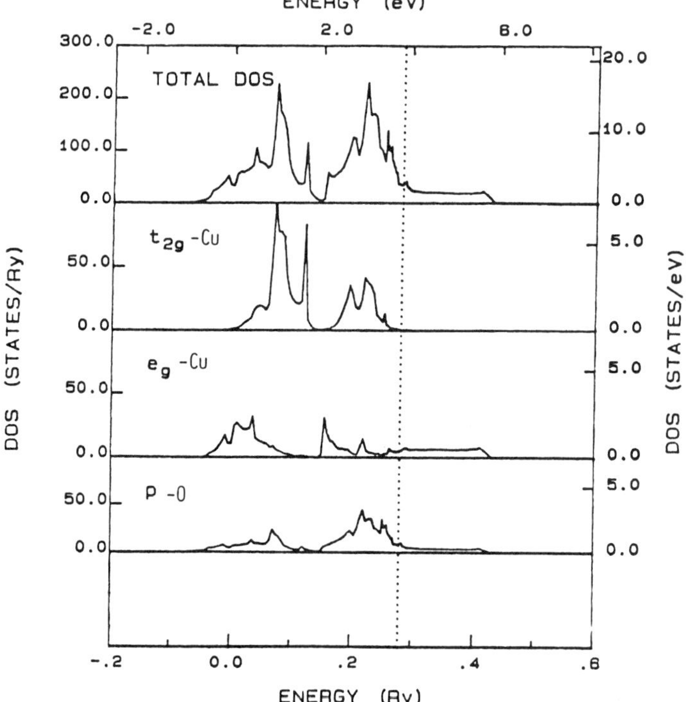

Figure 2. Band structure, density of states (DOS) and selected
decomposed DOS for $BaCuO_3$.

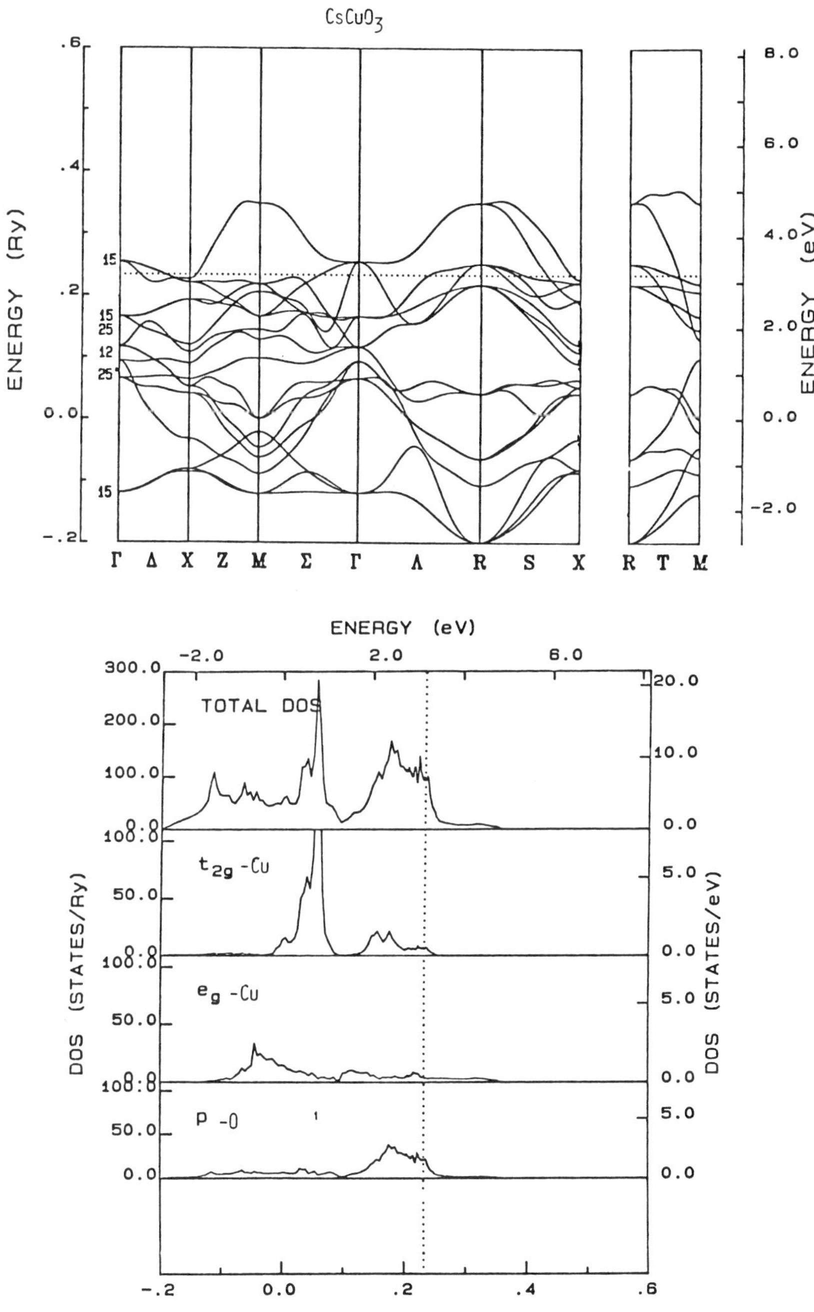

Figure 3. Band structure, density of states (DOS) and selected decomposed DOS for CsCuO3.

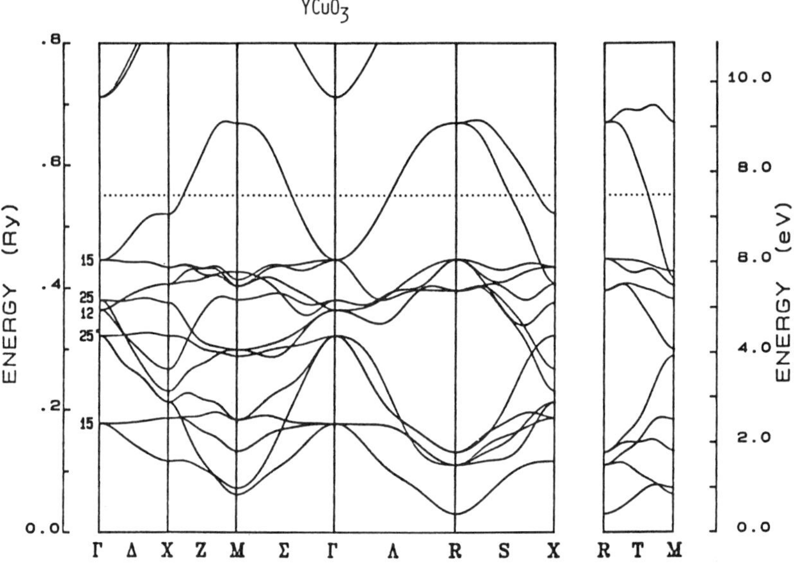

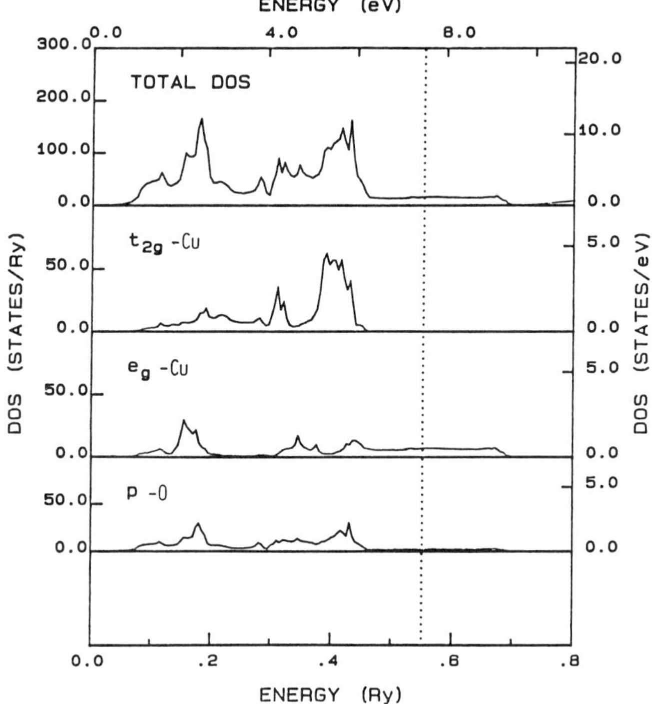

Figure 4. Band structure, density of states (DOS) and selected decomposed DOS for YCuO$_3$.

Table 1. Fermi energy E_F (Ry), decomposed and total densities of states (states/Ry) at E_F and contributions to η (eV/A^2) from the M, Cu, and O sites in MCuO$_3$.

	M=La	M=Ba	M=Cs	M=Y
E_F	0.382	0.280	0.233	0.551
N(M-s)	0.0042	0.0071	0.0645	0.0050
N(M-p)	0.0189	0.4968	15.7506	0.0095
N(M-t_{2g})	0.0442	0.1601	0.3147	0.0307
N(M-e_g)	0.0034	0.0013	0.1767	0.0060
N(M-f)	0.0956	0.0899	0.4303	0.0244
N(Cu-s)	0.1219	0.1205	0.0743	0.1205
N(Cu-p)	0.2607	0.4087	0.2578	0.1923
N(Cu-t_{2g})	0.0754	0.7294	6.6017	0.0156
N(Cu-e_g)	7.3835	4.9116	3.6857	7.0666
N(Cu-f)	0.0181	0.0890	0.0628	0.0119
N(O-s)	0.2175	0.1614	0.1058	0.2733
N(O-p)	9.6388	26.4336	61.3162	6.8178
N(O-t_{2g})	0.0144	0.0135	0.0238	0.0136
N(O-e_g)	0.0868	0.0765	0.0581	0.0982
N(O-f)	0.0229	0.0099	0.0190	0.0277
N(E_F)	19.3697	36.7274	94.9097	15.8271
ηM	0.000	0.012	0.485	0.000
ηCu	0.894	1.650	0.815	1.127
ηO	1.433	1.810	1.528	2.139

REFERENCES

1. E. F. Skelton, W. T. Elam, D. U. Gubser, S. H. Lawrence, M. S. Osofsky, L. E. Toth, and S. A. Wolf, Phys Rev. B 35:7140 (1987).
2. J. D. Jorgensen, H.-B. Schuttler, D. G. Hinks, D. W. Capone II, K. Zhang, M. B. Brodsky, and D. J. Scalapino, Phys. Rev. Lett. 58:1024 (1987).
3. R. M. Fleming, B. Batlogg, R. J. Cava. and E. A. Rietman, Phys. Rev. B 35:7191 (1987).
4. D. McK. Paul, G. Balakrishnan, N. R. Bernhoeft, W. I. F. David, and W. T. A. Harrison, Phys. Rev. Lett. 58:1976 (1987).
5. R. M. Hazen, L. W. Finger, R. J. Angel, C. T. Prewitt, N. L. Ross, H. K. Mao, C. G. Hadidiacos, P. H. Hor, R. L. Meng, and C. W. Chu, Phys. Rev. B 35:7238 (1987).
6. Y. LePage, W. R. McKinnon, J. M. Tarascon, L. H. Green, G. W. Hull, and D. M. Hwang, Phys. Rev. B 35:7245 (1987).
7. T. Siegrist, S. Sunshine, D. W. Murphy, R. J. Cava, and S. M. Zahurak, Phys. Rev. B 35:7137 (1987).
8. L. Hedin and B. I. Lundqvist, J. Phys. C 4:2064 (1971).
9. G. D. Gaspari and B. L. Gyorffy, Phys. Rev. Lett. 28:801 (1972).
10. D. D. Koelling and B. N. Harmon, J. Phys. C 10:3107 (1977).
11. W. E. Pickett, H. Krakauer, D. A. Papaconstantopoulos, and L. L. Boyer, Phys. Rev. B 35:7252 (1987).

CHARACTER OF STATES NEAR THE FERMI LEVEL IN $YBa_2Cu_3O_7$

H. Krakauer* and W. E. Pickett**

*College of William and Mary, Williamsburg, VA 23185
**Condensed Matter Physics Branch, Naval Research
Laboratory, Washington, DC 20375 5000

INTRODUCTION

The discovery of very high temperature superconductivity in copper oxide materials[1,2] raises a number of fundamental questions. Many of these are related to the fact that these oxides are qualitatively different than nearly all other high temperature superconductors (they are similar in some ways to the Ba-(Pb-Bi)-O system). Preparation requires techniques which are used in the preparation of ceramics, and their properties are similar to many ceramics - brittle rather than malleable, strongly tending to crack rather than bend. Yet, at high temperature they show conductivity characteristic of low density of states metals, and become superconducting in the range T_c = 30-100 K.

Band structure calculations[3-5] indicate these materials are metals with bands crossing the Fermi level which have Cu-O d-p character. They are relatively broad bands, nearly 10 eV wide, which arise from the strong Cu-O overlap due to their small separation. In the La-Ba-Cu-O system the bands at E_F are $dp\sigma$ antibonding bands, and several studies[3,5,6] have shown that these states are strongly modulated by the relative motion of the atoms, that is, there is a strong electron-phonon interaction (EPI). These studies and others[7] indicate that the EPI is strong enough to account for the observed T_c = 30-45 K in this system.

In the $YBa_2Cu_3O_7$ system (including materials obtained by the replacement of Y by a number of rare earth elements), there are several indications that the EPI may not be responsible for the superconductivity. The high values of T_c, in the 90-100 K range[2], alone are enough to suggest that a novel mechanism is responsible, and recently two laboratories[8,9] have found no detectable dependence of T_c on the mass of oxygen. The replacement of Y by a number of rare earth elements which have masses up to twice as large and T_c over 90 K also suggests that superconductivity is independent of the mass of the trivalent atom in this structure. Preliminary studies[10] of the calculated EPI in this material (using the rigid atom approximation) also suggests it is no larger than that in the La-(Ba,Sr)-Cu-O system.

Whatever the mechanism giving rise to superconductivity in $YBa_2Cu_3O_7$, there seems to be little doubt that it is due to pairing of electron states at the Fermi level E_F. Therefore it is of central interest to investigate the character of the states at and near E_F and their energy- and real-space distribution. In this paper we present results of detailed Linearized Augmented Plane Wave calculations of these properties.

CALCULATIONAL RESULTS

The atomic geometry of $YBa_2Cu_3O_7$ has been determined by several groups.[11] The unit cell is orthorhombic with space group Pmmm, containing one formula unit per unit cell. The structure can be pictured as a defect perovskite structure, with a superlattice-like ordering of the Y and two Ba atoms along the c direction serving to define the unit cell. In addition, the oxygen atom is missing from the Y layer, and in the Cu-O layer sandwiched between the Ba planes one of the two oxygen atoms is missing. The structure then consists of the Y and two symmetry-related Ba atoms, Cu(1) and O(1) atoms in the Cu-O chain with O(4) oxygen atoms above and below the Cu(1) atom, and two Cu-O corrugated planes with Cu(2), O(2) and O(3) atoms in each. We find that, although the interactions between the two planes and between the planes and the chains are small, they are larger than the interactions between the Cu-O planes in the La_2CuO_4 system.

The details of the calculations will be given elsewhere. We note here only that although this material has a very large unit cell with low symmetry, we have carried out very well converged calculations. The charge density and potential were expanded in the interstitial region in 14600 planes waves and inside the inscribed spheres around each atom in lattice harmonics through l=12. The secular equation dimension was about 850. The densities of states and band structure plots given below were obtained from disciplined Fourier series representations based on 165 first-principles points in the irreducible zone. The present calculations are much more highly converged than those of Mattheiss and Hamann[12] (who used a geometry which has since been shown to be incorrect) and are also more extensive than those of Massidda et al.[13]

The density of states (DOS) of $YBa_2Cu_3O_7$ is shown in Fig. 1. The same DOS is also shown on an expanded scale in Fig. 2 for energies within 0.4 eV of the Fermi energy. The values of the projected DOS at E_F are summarized in Table I and compared to those of $La_{2-x}(Ba,Sr)_xCuO_4$. Among the notable features seen in Figs. 1 and 2 is the nearly identical character of the O(2) and O(3) DOS, reflecting the nearly identical chemical environment of these two atoms due to the small size of the orthorhombic distortion. By contrast, the spectral weight of the O(1) and O(4) atoms is shifted to higher energies, and at E_F, the DOS of the O(1) and O(4) atoms is more than twice as large as that of the O(2) and O(3) atoms and comparable to that of the Cu atoms. The linear chain Cu(1) atom is four-fold planar coordinated with the O(1) (along the b axis) and O(4) (along the c-axis) atoms, and this accounts for the overall similarity of the spectral densities of the O(1) and O(4) atoms. There is a sharp peak 0.1 eV below E_F on the O(1) atom and, slightly reduced, also on the O(4) atom. E_F itself falls on a rather smooth region of spectral density.

The DOS at the Fermi energy $N(E_F)$ of the O(1) and O(4) atoms is much larger than that of the O atoms in the La-Ba-Cu-O compounds. For the ideal compounds ($\delta=0$ and x=0 in Table I), the DOS of the O(1) and O(4) atoms is more than four times larger than on the O(x,y) atom in La_2CuO_4. Highest T_c's have been achieved in Y-Ba-Cu-O compounds with oxygen vacancies as low as $\delta=0.03$,[14] while x=0.15 seems to yield the highest T_c

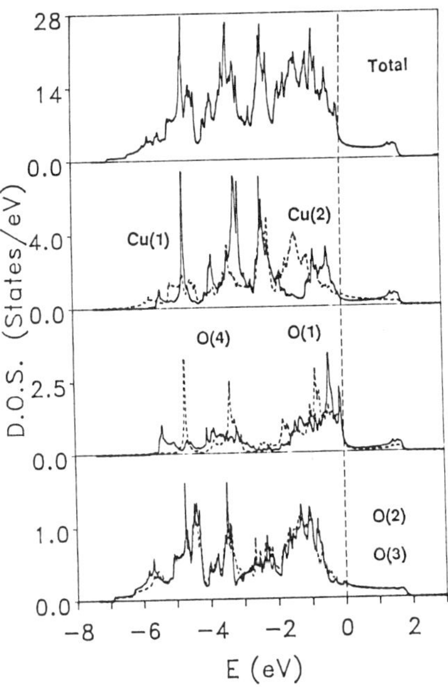

Figure 1. Density of states for YBa$_2$Cu$_3$O$_7$. The total DOS is shown in the top panel and muffin-tin projected DOS per atom are shown in the other panels (see text for atom identification). Note the change of scale from panel to panel.

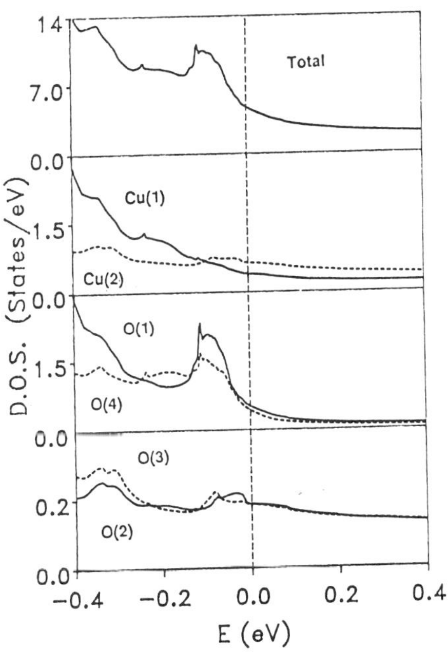

Figure 2. Same as Fig. 1, but for an expanded energy scale near E$_F$. Note the change of scale from panel to panel.

in the La-Sr based systems. For these compounds, the rigid band values in
Table I still give O(1) and O(4) N(E_F)'s which are more than twice as
large as that on that on the La-(Ba,Sr)-Cu-O compound. Furthermore, N(E_F)
on the Cu, O(1) and O(4) atoms are more nearly comparable to each other
than are those in the La-based compounds.

The "rigid band" values presented in Table I have been determined
using the standard procedure whereby the oxygen is assigned the nominal
O^{2-} character, and the creation of an oxygen vacancy implies the loss of 6
O p states below E_F while the removal of the neutral O atom subtracts only
4 p electrons. Thus the creation of δ O vacancies provides 2δ extra
electrons for the band states above E_F. This picture is certainly not
strictly correct when O-related bands cross E_F, both because the O^{2-}
designation is not correct and because the O p states which are removed in
forming the vacancy do not lie entirely below E_F. Other problems with the
rigid band model for oxygen vacancies are discussed in the next section.
On the other hand, the rigid band model should be more appropriate for
substitutional fluorine in place of oxygen.[15] For the T_c=150 K compound
in Ref. 15, $YBa_2Cu_3F_2O_y$ (assuming the same structure and y=5) this would
increase E_F by about 1 eV into a region of <u>reduced</u> N(E_F).

The band structures near E_F are shown in Figs. 3 and 4. Since it is
essential to identify the contributions of the chains and the planes
separately, the bands which are strongly associated with the chains, that
is, the Cu(1), O(1) and O(4) atoms, are emphasized in Fig. 3. In Fig. 4
the bands associated with the layers, comprised of the Cu(2), O(2) and
O(3) atoms, are emphasized. There are four bands which cross E_F (numbers
33 to 36, to accommodate the 68 electrons per unit cell). The layers give
rise to two of the bands crossing E_F, corresponding to one band for each
layer. These bands are similar to those crossing E_F in La_2CuO_4, but
differ here in that they are not precisely half filled. The Cu-O chain
gives rise to one steep band crossing E_F, which is strongly dispersing
only in the chain direction, that is, along X-S and Γ-Y. In addition
there is a <u>flat chain-related band just at</u> E_F <u>along the</u> Y-S <u>direction</u>.
Here we show the bands only in the k_z=0 plane, but there is some small but
important dispersion along the c direction which will be discussed
elsewhere.

From Fig. 2 the peak in N(E_F) 0.1 eV below E_F is seen to be associated solely with the chain-related O(1) and O(4) atoms, and does not
show up at all on the Cu(1) chain atom. Therefore this peak arises from
states which lie along the chain but <u>avoid</u> the Cu(1) atom, and apparently
relating primarily to O(1)-O(4) ppσ interactions. Figures 3 and 4 show
there are mostly chain derived states, rather than layer-derived, within
0.5 eV below E_F, and the peak is associated with the flat portion of the
uppermost filled band midway between Γ and S.

DISCUSSION

The presence of a more complex unit cell here than in the La_2CuO_4
compound leads to bands which are not half filled and have more complex
Fermi surfaces. The chain-related bands crossing E_F may lead to qualitatively new processes related to its strong one-dimensional character, and
due to the two dimensional planes and one dimensional chains the electronic properties have strongly anisotropic character in all three
directions.

In the La-(Ba,Sr)-Cu-O system we have shown that a rigid band model
should be realistic for moderate (15-20%) substitutions of La. In the
present system one would expect because of the similarity of the materials
that a rigid band model should also work for substitution of Y on Ba
sites, and vice versa. The results given in Table I should be applicable

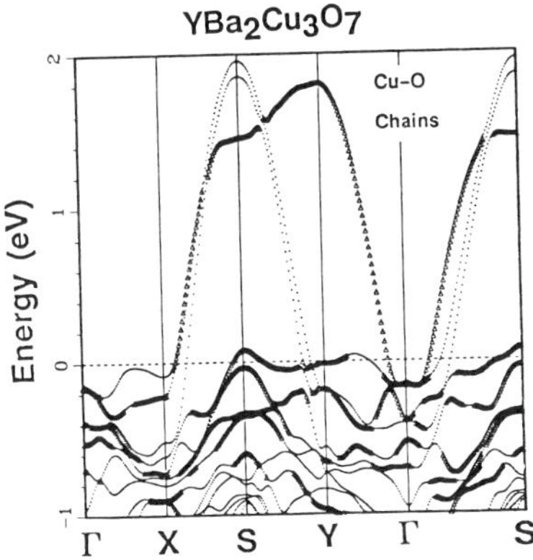

Figure 3. Band structure of YBa$_2$Cu$_3$O$_7$ in the k_z=0 plane of the orthorhombic zone, near the Fermi level E_F=0. States with more than 60% of their charge on the Cu(1)-O(1)-O(4) chains are emphasized with the large symbols.

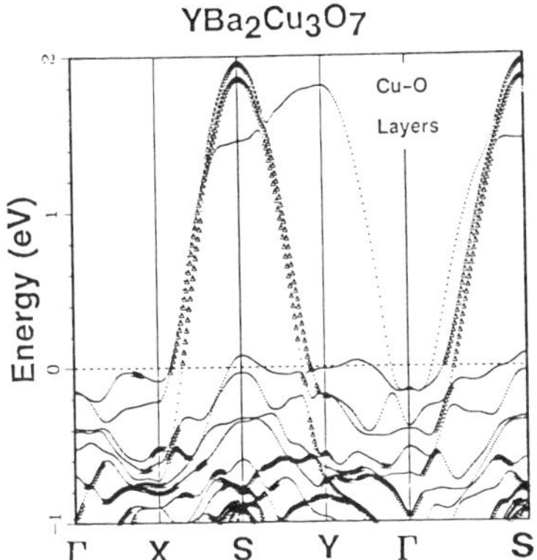

Figure 4. As in Fig. 4, except the emphasized states are those with more than 80% charge on the two Cu(2)-O(2)-O(3) layers.

for this case, although substantial variation away from the Y:Ba ratio of 1:2 leads to the formation of other phases.[2] One is tempted to apply a rigid band model to the case of oxygen vacancies (which occur on the O(1) sites along the Cu-O chains under non-optimum conditions), but caution is necessary in this situation. The change in the on-site potential due to the removal of an oxygen atom is drastic, and several possibilities arise. The extra electrons originally on the O ion when the neutral O atom is removed may go equally into the bands crossing E_F, but they may also perhaps remain in part on the empty site, stabilized by the Madelung potential as in the F-center in ionic oxides. Not only is the on-site change in potential large, which will alter the band structure, but the coordination of the neighboring copper atoms decreases from 4 to 3. As a result, oxygen vacancies will result in both a change in the Cu(1) states as well as a disruption of the one-dimensional chain structure.

It is evident that the disruption of the chain, such as by the removal or displacement of an O(1) atom, can have serious consequences for transport properties if the one dimensionality is strong, because an electron cannot avoid a defect in one dimension. If, as seems to be the case in La$_2$CuO$_4$, that the Cu has a tendency to become magnetic, the lowering of its coordination will suppress the banding of the d-states and enhance magnetic tendencies. [Vacancies are a necessary precondition of the antiferromagnetic state in La$_2$CuO$_4$.] Although there have been studies in $YBa_2Cu_3O_{7-\delta}$ of the effects of oxygen vacancies, from near $\delta=o$ to $\delta=1$ per formula unit, no magnetic transitions have been reported. It has been found, however, that superconductivity is strongly depressed as δ is increased, and a metal-to-insulator transition occurs at $\delta=0.5$. Such a transition is not predicted by the rigid band model, again suggesting it is not applicable in describing oxygen defect compounds.

This work has been supported by the National Science Foundation (NSF) Grant No. DMR-84-16046 and the Office of Naval Research Contract No. N00014-84-WR-24055. We also acknowledge SDIO/IST and DARPA for their partial support of this research. Computing was done under the auspices of the NSF at the supercomputing centers at Cornell University and at the University of Pittsburgh, as well as the Naval Research Laboratory.

TABLE I. Comparison of Fermi level densities of states $N(E_F)$ (in states /eV-atom). Also shown are rigid band values of $N(E_F)$ corresponding to oxygen vacancies in Y-Ba-Cu-O and alloying with Sr or Ba in La-Cu-O.

$YBa_2Cu_3O_{7-\delta}$

$E_F(Ry)$	$E_F(eV)$	δ	holes	N(E)	Cu(1)	Cu(2)	Cu-avg	O(1)	O(4)	O(2)	O(3)
.4412	.000	.00	.00	4.78	.352	.602	.519	.518	.452	.175	.175
.4446	.047	.10	-.20	3.77	.316	.571	.486	.319	.246	.171	.168
.4490	.107	.20	-.40	3.02	.257	.501	.420	.195	.155	.155	.152
.4542	.177	.30	-.60	2.62	.229	.451	.377	.143	.117	.143	.141

$La_{2-x}Sr_xCuO_4$

$E_F(Ry)$	$E_F(eV)$	x	elect	N(E)		Cu	O(z)	O(x,y)
.5590	.000	.00	.00	1.24		.540	.039	.113
.5530	-.082	.14	-.14	2.03		1.045	.108	.181

REFERENCES

1. J.G. Bednorz and K.A. Muller, Z. Pys. B64, 189 (1986).
2. M.K. Wu *et al.*, Phys. Rev. Lett. 58, 908 (1987).
3. L.F. Mattheiss, Phys. Rev. Lett. 58, 1028 (1987).
4. J. Yu and A.J. Freeman, Phys. Rev. Lett. 58, 1035 (1987).
5. W.E. Pickett, H. Krakauer, D.A. Papaconstantopoulos, and L.L. Boyer, Phys. Rev. B35, 7252 (1987).
6. W.Weber, Phys. Rev. Lett. 58, 1371 (1987).
7. P.B. Allen, W.E. Pickett, and H. Krakauer, preprint.
8. B. Batlogg *et al*. Phys. Rev. Lett. 58, 2333 (1987).
9. L.C. Bourne *et al.*, Phys. Rev. Lett. 58, 2337 (1987).
10. H. Krakauer and W.E. Pickett, unpublished.
11. S.B. Qadri, L.E. Toth, M. Osofsky, S. Lawrence, D.U. Gubser, and S.A. Wolf, Phys. Rev. B35, 7235 (1987); T. Siegrist, S. Sunshine, D.W. Murphy, R.J. Cava, and S.M. Zahurak, Phys. Rev. B35, 7137 (1987); P.M. Grant *et al.*, Phys Rev. B35, 7242 (1987); R.M. Hazen *et al.*, Phys. Rev. B35, 7238 (1987); Y. LePage, W.R. McKinnon, J.M. Tarascon, L.H. Greene, G.W. Hull, and D.M. Hwang, Phys. Rev. B35, 7245 (1987).
12. L.F. Mattheiss and D.R. Hamann, preprint.
13. S. Massidda, J. Yu, A.J. Freeman, and D.D. Koelling, preprint.
14. J.J. Rhyne, D.A. Neuman, J.A. Gotaas, F. Beech, L.E. Toth, S. Lawrence, S. Wolf, M. Osofsky, and D.U. Gubser, preprint; L.E. Toth, private communication.
15. S.R. Ovshinsky, R.T. Young, D.D. Allred, G. DeMaggio, and G.A. Van der Leeder, Phys. Rev. Lett. 58, 2579 (1987).

ELEMENTARY THEORY OF THE PROPERTIES OF THE CUPRATES

Walter A. Harrison

Department of Applied Physics
Stanford University
Stanford, CA 94305-4090

INTRODUCTION

Cuprates have proven to be superconducting at high temperatures.[1,2]
Mattheiss[3] and Yu, Freeman, and Xu[4] have already provided self-consistent
LAPW calculations of the bands of La_2CuO_4, and Mattheiss has interpreted
them in terms of a tight-binding description. Here we provide a preliminary account of an independent tight-binding analysis which provides a
self-contained semiquantitative description of the bands, roughly consistent with those given earlier. It also allows estimates of the other
properties of this system. In particular we analyze the vibrational and
antiferromagnetic properties, finding in the nonmagnetic metallic state an
electron-paramagnon coupling constant which becomes arbitrarily large as
one approaches the antiferromagnetic state. It may thus account for the
high-temperature superconductivity.

PARAMETERS

The first step in a tight-binding analysis is to obtain the free-atom
term values for the elements La, Cu, and O. Hartree-Fock values[5,6] for
La are ε_s = 4.63 eV and ε_d = -7.31 eV. For Cu they are ε_s = -6.49 eV and
ε_d = -16.38 eV (Ref. 5 gives a value -13.36 eV for a configuration $d^{10}s^1$
but this was lowered by $\frac{1}{2}$ U for the appropriate $d^{9.5}$ configuration using
a value U = 6.04 eV), and for O they are ε_p = -16.72 eV and ε_s = -34.02 eV.
The three electrons from each of the La atoms and the one from the copper
s-state fill all but one of the oxygen p-states if we leave the d-states
full. Those empty metallic levels are inessential to the band structure
as in the perovskites (Ref. 6, p. 438ff), and the essential bands arise
from the copper d-states and oxygen p-states. We discard all other basis
states. In fact, since the energies of the copper d-states and oxygen
p-states we have given are almost identical, we neglect the difference,
taking ε_p = ε_d.

The crystal structure[8] of a plane containing the copper atoms in the
orthorhombic structure (D_{4h} symmetry) of La_2CuO_4 is shown in Fig. 1. The
predominant coupling will be the nearest-neighbor coupling between Cu d-
states and O p-states. For these we take[9]

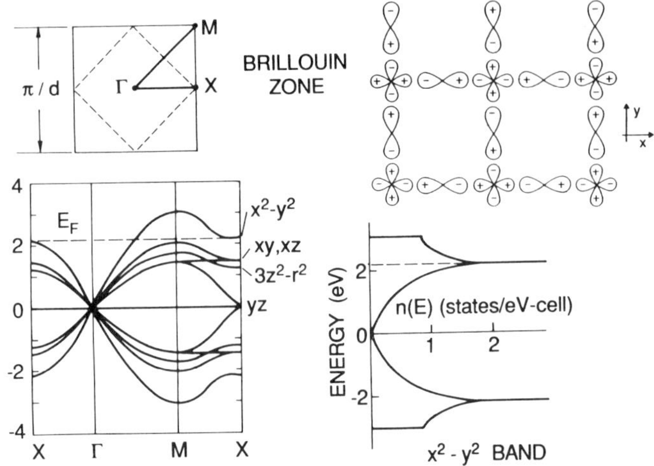

Fig. 1.   The square Brillouin zone for the two-dimensional La_2CuO_4 structure shown to the right; x^2-y^2 orbitals are on Cu and x- and y-
orbitals are on oxygen. Below to the left are the tight-binding
bands. The Fermi energy is shown as E_F and the corresponding
Fermi surface is the dashed line in the Brillouin zone above.
To the right of the bands is shown the density of states for
the x^2-y^2 band.

$$V_{pd\sigma} = -\frac{3\sqrt{15}}{2\pi}\frac{h^2}{m}\frac{\sqrt{r_p r_d^3}}{d^4} \tag{1}$$

and $V_{pp\pi}$ is positive and smaller by a factor of $1/\sqrt{3}$. The parameter r_p =
4.41Å for oxygen[9] and r_d = 0.67 Å for copper.[6] At the equilibrium spacing
of 190 Å this gives $V_{pd\sigma}$ = -1.24 eV. There are oxygen atoms above and
below the plane of the figure by 2.40 Å at each copper site. These are
more weakly coupled than those shown in Fig. 1 because of the larger spacing and in fact have no coupling to the particular d-state that we find to
be important (it has two units of angular momentum around the internuclear
axis), so we neglect it completely.

THE BANDS

     The remaining six oxygen p-levels and five copper d-levels per cell
broaden out into eleven bands in the square Brillouin zone. They may be
directly calculated using the parameters given. At the point M in the
Brillouin zone, shown in Fig. 1, only a single d-state from each atom
enters each band, the state x^2-y^2 for the highest band, and the calculation is immediate. If we use only the single d-state for each band, and
the coupled p-states, the calculation for the entire band is elementary,
giving

$$\varepsilon_k = \varepsilon_d + V_2 \sqrt{\frac{\sin^2 k_x d + \sin^2 k_y d}{2}} \tag{2}$$

for the highest band [a similar form given by Jorgensen et al.[11] appears
to be based upon wavenumbers measured from M rather than Γ], where the
parameter

$$V_2 = -\sqrt{6}V_{pd\sigma} = 3.04 \text{ eV} \tag{3}$$

is called the <u>covalent energy</u>. This, along with the Coulomb repulsion U,
is the principal parameter of the theory. The corresponding approximation
for all the bands is shown in Fig. 1.

There are enough electrons to fill all but one-half of a band. We
see from Fig. 1 that the corresponding Fermi surface lies entirely in the
top band given by Eq. (2), so that band will suffice for our analysis. As
noted by Mattheiss,[3] the constant-energy surface for a half-filled band,
appropriate to La_2CuO_4, forms a square Fermi surface for this system, as
shown in Fig. 1. The energy is given by $\varepsilon_F = \varepsilon_d + \frac{1}{2}\sqrt{3}V_2$. This comes
at a saddle point in the bands at X, which is responsible for a divergent
density of states at the Fermi energy (in this undoped compound with one
hole per cell), as seen in the density of states for the highest band
plotted in Fig. 1.

THE SPECIAL POINTS METHOD FOR BAND INTEGRATIONS

In calculating properties it will be necessary to average band energies over the occupied states. In fact the average over each complete
bonding band and its corresponding antibonding band is just the d-state
energy (equal to the p-state energy), ε_d. Thus it will be sufficient just
to obtain the average over the empty portion of the highest band. That
can be done with the special-points, or Baldereschi-point, method.[12] In
this method a wavenumber $\mathbf{k}^*$ is selected as representative, such that the
leading Fourier components of the bands cancel out and only the average is
left. The special point for a square lattice of edge a is[12] $\mathbf{k}^* =
[1,1]\pi/2a$, or $[1,1]\pi/4d$ for the cuprate lattice, and the average over the
entire band could be approximated as $\varepsilon(\mathbf{k}^*)$.

Here we wish an average only over half of the band, the region outside
the Fermi surface in Fig. 1. We may do that by distinguishing alternate
copper atoms, as we shall do for the antiferromagnetic state. Then the
primitive cell doubles and the new Brillouin zone becomes congruent with
the Fermi surface shown in Fig. 1. The new special point, for the unfilled
portion of the band becomes $\mathbf{k}^* = [\frac{1}{2},1]\pi/2d$. The corresponding average
for the band in Eq. (2) is $\varepsilon_d + \frac{3}{4}V_2$.

THE NATURE OF THE ELECTRONIC STRUCTURE

We are of course principally interested in the properties of the doped
cuprates, such as $Ba_xLa_{2-x}CuO_4$, because of their superconductivity. However, the properties of the pure compound give clues as to the important
properties. The bands we have given would indicate metallic behavior, but
the pure compound is insulating.

This could be explained by a Peierls distortion, as suggested by Mattheiss,[3] which would suggest[13] that soft phonons are responsible for the superconductivity. Indeed distortions of the structure are observed[14] but they appear not to be of the type which would cause a Peierls distortion. (A glide-plane symmetry can again reduce the gap to zero.) Further, the superconductor does not show the isotope effect[15,16] that would be expected if the phonons were responsible for the superconductivity. We shall nevertheless consider the vibrational properties.

The pure compound is found to be antiferromagnetic,[17] but the antiferromagnetism disappears with doping. It has been widely suggested that the proximity of the antiferromagnetic instability is important to the superconductivity, and we consider also the antiferromagnetism.

PHONON PROPERTIES

The bonding properties can be estimated[6,9] in terms of the electronic structure using the tight-binding parameters we have given. In particular, we may directly estimate the vibrational frequencies. Of particular interest are the optical modes in which the oxygen atoms vibrate against the heavier copper lattice. It is reasonable to treat these modes in an Einstein approximation.

The calculation of the bonding properties requires consideration of the coupling between the oxygen p-states and the copper s-states,[18] as well as the pd-coupling we have been discussing. Considering both has led us to (we plan to publish details separately) an optical mode frequency ω_0 given by

$$M_0 d^2 \omega_0{}^2 = 8[8V_{sp\sigma}{}^2/(\varepsilon_s - \varepsilon_p) + V_2] = 74.44 \text{ eV} \tag{4}$$

with M_0 the mass of oxygen. This corresponds to $\omega_0/2\pi c = 589$ cm^{-1}, in fair accord with an observed[19] vibrational mode at 677 cm^{-1}. More convenient parameters are $\hbar\omega_0 = 74$ meV and $T_0 = \hbar\omega_0/k_B = 855°$K.

We are treating the optical mode frequencies in the Einstein approximation, and in the same spirit we may look at the coupling between electronic states arising when a single oxygen is moved along the oxygen-copper axis by a distance u. This modifies the coupling with the two neighboring d-states and produces a matrix element between Bloch states of the band. If we average the square of the coupling matrix element over the bands, the square root of the average is found to be $-2\sqrt{2}V_{pd\sigma} u/Nd$, the average coupling between two Bloch states due to the displacement of one oxygen by u along a copper-oxygen axis.

For the purposes of calculating properties, we should rewrite[20] u in terms of annihilation and creation operators (a_j and $a_j{}^+$ for displacement of the jth oxygen) for the corresponding Einstein mode of frequency ω_0. Then the electron-phonon interaction becomes $H_{e\phi} = (1/N) \Sigma_j V_{e\phi}(a_j + a^+{}_j)$ with

$$V_{e\phi} = (2\hbar/3M_0\omega_0 d^2)^{1/2} V_2 \tag{5}$$

The most useful parameter representing the electron-phonon interaction
is the dimensionless electron-phonon coupling constant given for a metal
by[21]

$$\lambda_\phi = \frac{2\Omega}{(2\pi)^3} \int \frac{V_q^* V_q}{\hbar\omega_q \, \partial\varepsilon_k/\partial k} \, dA_{FS} \tag{6}$$

where the wavenumber q is from one point on the Fermi surface to some other
point, and in the integral the other point varies over the Fermi surface.
Here $V_q^* V_q$ is taken constant at $V_{e\phi}^2/N$ and ω_q is taken constant at ω_0.
The remaining factors and integral, if divided by N, are simply the density
of states per copper per unit energy. We have

$$\lambda_\phi = \frac{n(E_F) V_{e\phi}^2}{\hbar\omega_0} = \frac{2V_2^2 n(E_F)}{3M_0\omega_0^2 d^2} \tag{7}$$

The subscript ϕ indicates the contribution from the optical phonons. There
is an additional contribution from acoustical phonons and any other relevant excitations.

If we take the density of states averaged over the band, $n(E_F) = 2/V_2$,
and the other parameters we have given, we obtain $\lambda_\phi = 0.050$. This is quite
weak coupling, though we note that it varies inversely as the square of
the frequency. If there _were_ strong mode softening, as suggested by
Weber,[13] λ_ϕ could become as large as one wishes. At present there appears
not to be any.

This parameter, including all contributions to it, enters three
interesting properties. First is the superconducting energy gap parameter
Δ at $T = 0$. For weak coupling, appropriate for λ well below one, Δ and λ
are related by the BCS theory, which gives[18,19]

$$\Delta = \hbar\omega_0/\sinh(1/\lambda) \approx 2\hbar\omega_0 e^{-1/\lambda} \tag{8}$$

for small λ. If λ becomes large because of mode softening, the relation
becomes $\Delta \approx \lambda\hbar\omega_0 = 4\hbar V_2/3M_0\omega_0 d^2$ and could become large. In that case the
BCS theory becomes inaccurate and one should really use the Eliashberg
theory.[22]

A second interesting property is enhancement of the density of states
(for the nonsuperconducting metal) at the Fermi energy, relevant to the
specific heat at low temperatures; $n(E_F)_{enhanced} = (1+\lambda)n(E_F)$. A third
property is the scattering rate for electrons by phonons at high tempera-
tures, given by $1/\tau = \lambda k_B T/\hbar$, leading to $\rho = m/Ne^2\tau = m\lambda k_B T/Ne^2\hbar$. This
applies only at temperatures above the Debye temperature, in this case
$T_0 = \hbar\omega_0/k_B$, which we have indicated is 855°.

ANTIFERROMAGNETISM AND UNRESTRICTED HARTREE-FOCK

Our Hamiltonian is based upon two p-states and one d-state per cell, all with the same energy ε_d. We may now explicitly add the intraatomic electron-electron repulsion on the copper, $\sum_j U c^+_{j+} c_{j+} c^+_{j-} c_{j-}$, and seek an approximate solution as an antiferromagnetic state. We do this in the unrestricted Hartree-Fock approximation, in which we allow the occupation of the spin-up and spin-down electrons to differ on each copper in an antiferromagnetic pattern, giving a net spin per atom of $\mu = \langle c^+_{1+} c_{1+} \rangle - \langle c^+_{1-} c_{1-} \rangle$. This is a variational calculation, just as the band calculation we described, but now with a more general state. We may in fact require that the variational state have a moment μ and then plot the resulting total energy as a function of this μ. We see that if U is small, the resulting energy as a function of μ is minimum at $\mu = 0$, the nomagnetic state. If U is sufficiently large, the energy at $\mu = 0$ becomes a local maximum and two minima occur at finite μ. Then the antiferromagnetic state is predicted and the value of μ at the minimum is the predicted moment. Using the special-point method, and adding back Coulomb energies counted twice, we obtain

$$E(\mu) = -\sqrt{\tfrac{3}{4}V_2^2 + \tfrac{1}{4}\mu^2 U^2} - (1 - \mu^2)U/4 \qquad (9)$$

Minimizing this with respect to μ we obtain $\mu = (1 - 3V_2^2/U^2)^{1/2}$. For our parameters antiferromagnetism is predicted with a moment near $\mu = 0.5$. This value of μ also leads to a gap of $\tfrac{1}{2}\mu U = 1.48$ eV over the entire Fermi surface.

We also may at this point note that if we change the occupation of the band by doping, for example with barium in $Ba_x La_{2-x} CuO_4$, we may retain the antiferromagnetic bands but now remove x electrons per copper which had contributed to the potential $\tfrac{1}{2}\mu U$. This has the effect of replacing U in Eq. (9) by $U^* = U(1 - x)$, and with our parameters the criterion for antiferromagnetism is no longer satisfied if x exceeds 0.128. We turn next to antiferromagnetic fluctuations in the doped, metallic, nonmagnetic state.

PARAMAGNONS AND ELECTRON-PARAMAGNON COUPLING

The great formal similarity of the antiferromagnetic transition to the Peierls transition discussed earlier in this paper would suggest a model for the excitations similar to the soft-Einstein vibrational-mode picture discussed there. We have already given in Eq. (9) the energy as a function of μ, which we may expand for small μ as $E(0) + (U^*/4)(1 - U^*/\sqrt{3}V_2)\mu^2$. Here we have assumed sufficient doping, $U^* < \sqrt{3}V_2$, and the ground state is nonmagnetic. We may expect that if a small moment μ is induced in some way, it will oscillate with a magnetic frequency ω_m. If we know that frequency we may treat the system as a set of such paramagnetic oscillators.

We may in fact obtain this frequency as for a two-level system using the special-point method. We write the spin-up state as u_1 times a Bloch sum on half the atoms and u_2 times a Bloch sum on the other half. The Hamiltonian in the time-dependent Schroedinger equation for that state will contain a term $\pm V_3 \cos\omega_m t$, with V_3 equal to the amplitude of $\tfrac{1}{2}U^*\mu(t)$, due to the oscillating spin-down electrons, and it will contain a coupling term $\tfrac{1}{2}\sqrt{3}V_2$ as seen from Eq. (9). We may then substitute the form $u_1 = 1/\sqrt{2} + \alpha\cos\omega_m t + i\beta\sin\omega_m t$, and for u_2 the same form with α and

β of opposite sign; both are of order V_3. Substituting these in the time-dependent Schroedinger equation, we find that it is satisfied to order V_3 with real α and β only if

$$(\hbar\omega_m)^2 = 3V_2^2(1 - U^*/\sqrt{3}V_2) \tag{10}$$

This is the required result.

We may now complete the Hamiltonian for these modes by adding a "kinetic" energy proportional to $(d\mu/dt)^2 = -\mu^2\omega^2$, which must average to the potential energy, fixing the coefficient. It is then straightforward to treat these as Einstein modes, as we did the optical phonons. The electron-magnon interaction becomes[20] $H_{em} = (1/N) \Sigma_j V_{em}(m_j + m^+_j)$, where the m_j and $m_j{}^+$ are magnon annihilation and creation operators, and $V_{em} = (3V_2^2/U^*\hbar\omega_m)^{1/2}U^*/2$. Then just as in Eq. (7) for phonons we obtain

$$\lambda_m = \frac{3V_2^2 U^* n(E_F)}{4(\hbar\omega_m)^2} \tag{11}$$

The paramagnon frequency, given in Eq. (10), goes to zero as we approach the antiferromagnetic transition (as x approaches 0.128 with our parameters), and the electron-paramagnon coupling constant can become arbitrarily large. In terms of the weak-coupling theory of superconductivity Eq. (11) leads, as after Eq. (8), to an energy-gap parameter of $\Delta = 3V_2U^*/2\hbar\omega_m$. This leads also to an enhanced density of states of approximately $3U^*/\hbar\omega_m)^2$ per copper atom and a high-temperature resistivity of $\rho = 3mV_2U^*k_BT/2Ne^2\hbar(\hbar\omega_m)^2$. This should apply at temperatures above $T = \hbar\omega_m/k_B$, which is below the superconducting critical temperature by a factor of order $1/\lambda$, and may therefore explain the large observed linear resistivity in the normal state of the cuprates.[23]

ACKNOWLEDGMENT

The author is indebted to W. C. Herring, J. S. Langer, P. Lee, and W. Weber for fruitful discussions. This work was supported by the National Science Foundation under Grant No. DMR-84 14126.

REFERENCES

1. J. G. Bednorz and K. A. Müller, Z. Phys. B64, 189 (1986).
2. M. K. Wu, J. R. Ashburn, C. J. Torng, P. H. Hor, R. L. Meng, L. Gao, Z. J. Huang, Y. Q. Wang, and C. W. Chu, Phys. Rev. Letters 58, 908 (1987). Other observations have been made by a number of other workers in Phys. Rev. Letters this year.
3. L. F. Mattheiss, Phys. Rev. Letters 58, 1028 (1987).
4. Jaejun Yu, A. J. Freeman, and J.-H. Xu, Phys. Rev. Letters 58, 1035 (1987).
5. Values schrom J. B. Mann, "Atomic Structure Calculations, 1. Hartree-Fock Energy Results for Elements Hydrogen to Lawrencium." Distributed by Clearinghouse for Technical Information, Springfield, VA 22151, are listed in Ref. 6 on page 534. Values listed for the noble-metal s-states are in error.
6. W. A. Harrison, "Electronic Structure and the Properties of Solids," W. H. Freeman (New York, 1980).

7.   S. Froyen, Phys. Rev. $B22$, 3119 (1980).

8.   J. M. Longo and P. M. Raccah, J. Solid State Chemistry $\underline{6}$, 526 (1973).

9.   W. A. Harrison and G. K. Straub, submitted to Phys. Rev. $\underline{B}$.

10. J. C. Slater and G. F. Koster, Phys. Rev. $\underline{94}$, 1498 (1954); listed also in Ref. 6, p. 481.

11. J. D. Jorgensen, H. B. Schüttler, D. G. Hinks, D. W. Capone, II, K. Zhang, M. B. Brodsky, and D. J. Scalapino, Phys. Rev. Letters $\underline{58}$, 1024 (1987).

12. A. Balderschi, Phys. Rev. $\underline{B7}$, 5212 (1973), described also in Ref. 6.

13. Werner Weber, Phys. Rev. Letters $\underline{58}$, 1371 (1987).

14. V. B. Grande, Hk. Muller-Buschbaum, and M. Schweizer, Z. Anorg. Allg. Chem. $\underline{428}$, 120 (1977).

15. B. Batlogg, R. J. Cava, A. Jayaraman, R. B. van Dover, G. A. Kourouklis, S. Sunshine, D. W. Murphy, L. W. Rupp, H. S. Chen, A. White, K. T. Short, A. M. Mujsce, and E. A. Rietman, Phys. Rev. Letters $\underline{58}$, 2333 (1987).

16. L. C. Bourne, M. F. Crommie, A. Zettl, Hans-Conrad zur Loye, S. W. Keller, K. L. Leary, Angelica M. Stacy, K. J. Chang, Marvin L. Cohen, and Donald E. Morris, Phys. Rev. Letters $\underline{58}$, 2337 (1987).

17. Yamaguchi quoted from a conference in Nature, $\underline{326}$, 638 (1982).

18. W. A. Harrison, Phys. Rev. $\underline{B24}$, 5835 (1981).

19. M. Stavola, R. J. Cava, and E. A. Rietman, Phys. Rev. Letters $\underline{58}$, 1571 (1987).

20. See, for example, W. A. Harrison, "Solid State Theory," McGraw-Hill (New York, 1970); reprinted by Dover (New York, 1979), p. 410.

21. P. B. Allen, in "Dynamical Properties of Solids," edited by G. K. Horton and A. A. Maradudin, North Holland (Amsterdam, 1980), Vol. 3, p. 95.

22. A simple analysis leading to this result is given by J. Callaway, "Quantum Theory of the Solid State," Academic Press (New York, 1976).

23. See, for example, J. Z. Sun, D. J. Webb, M. Naito, K. Char, M. R. Hahn, J. W. P. Hsu, A. D. Kent, D. B. Mitzi, B. Oh, M. R. Beasley, T. H. Geballe, R. H. Hammond, and A. Kapitulnik, Phys. Rev. Letters $\underline{58}$, 1574 (1987).

MEAN FIELD APPROACH TO DOUBLE-OCCUPANCY INDUCED PAIRING IN OXIDE SUPERCONDUCTORS

D. M. Newns

IBM Thomas J. Watson Research Center
P. O. Box 218
Yorktown Heights, NY 10598

The double-occupancy induced pairing mechanism (DOM for short) of superconductivity[1-7] applies to systems with strongly correlated narrow energy bands in which an on-site repulsive electron-electron interaction U plays an important role. Imagine first a band with only spin degeneracy and an electron occupation, written as $(1 - x_h)$, assumed less than unity per atom. x_h is then the number of holes per site. Assume $U = \infty$, so double occupancy is projected out. If the system is not magnetic, i.e. a Fermi liquid, it's energy scale, i.e. the band width for quasiparticles, is proportional to x_h times the $U = 0$ band width[8-11] . This is because electrons can only hop into empty sites, whose concentration x_h then enters the intersite hopping rate and thus band width. The reduction factor can be as large as .01 in heavy fermion materials.

If now U is reduced to a large but finite value, thus allowing some double occupation, the possibility that two quasiparticles may scatter into a site together increases their phase space, i.e. lowers their energy, and thus acts as an attractive interaction. This DOM can lead to superconductivity, as has been proposed by the author[2] for heavy fermion systems.

A very similar mechanism has been proposed by Anderson and coworkers[3,5] and taken up by others[4,6] for oxide superconductors. The Anderson theory is formulated for a Hubbard model, in contrast to Ref.2 which is based on the Anderson lattice. We find that, at least within our approach, the Hubbard model behaves rather singularly near half-filling, leading to a very small critical temperature there. This may be seen to be unphysical behaviour, and be circumvented, by taking into account the important role played by the oxygen 2p -band, and working with an Anderson lattice model, as in Ref.2.

Experimentally, we mention two key features. Firstly, the hole-like nature of conduction in the nearly half-filled $Cu\,d_{x^2-y^2}/O_{2p}$ band of $La_{2-x}Sr_xO_4$ systems is shown by Hall effect[12] and thermopower[13] measurements. For a nearly half-filled band, this is hard to understand without assuming a very large electron-electron interaction U. Secondly, T_c is found to be a nearly linear function of x_h, when this is directly measured[12], i.e. it scales with renormalized band width. This is closely similar to the scaling of T_c with $T_K^{3/4}$ found in the application of the DOM to heavy fermion materials[2], motivating extension of the DOM concept to oxide superconductors.

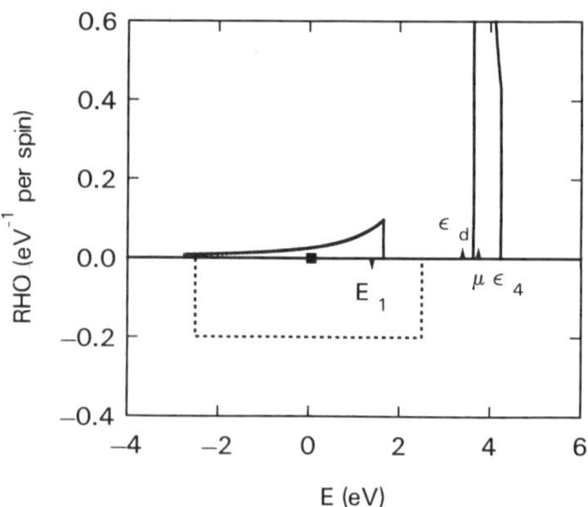

Fig. 1.   DOS generated by mean field Hamiltonian (4); $x_h = 0.4$, other parameters as text. Lower side of abscissa illustrates unhybridized O-band and original d-level E_1. Above abscissa are shown the renormalized d- and O-band densities of states and effective d-level ϵ_d .Fermi level μ lies in d-band. ϵ_4 is uppermost band edge.

The Hamiltonian which we use to describe in particular $La_{2-x}Sr_xCuO_4$ involves an O-band, whose width 2D is controlled by O-La interactions and is not renormalized by the Copper-U, interacting with Cu $d_{x^2-y^2}$ orbitals of energy E_1 via matrix elements V. A sketch of the energy levels is found in Fig. 1. The Hamiltonian is written in terms of slave bosons, which are convenient for describing strongly correlated electron systems[9-11]:

$$H = E_0\sum_i b_i^+ b_i + E_1\sum_{i\sigma} d_{i\sigma}^+ d_{i\sigma} + E_2\sum_i a_i^+ a_i$$

$$+ \sum_{k\sigma} \varepsilon_k c_{k\sigma}^+ c_{k\sigma} + \sum_{i\sigma k} V_k\left[b_i d_{i\sigma}^+ c_{k\sigma} e^{ik\cdot R_i} + hc\right] \tag{1}$$

$$+ \sum_{i\sigma k} V_k\left[a_i d_{i\sigma}^+ c_{k-\sigma}^+ e^{ik\cdot R_i} sgn\sigma + hc\right] - \mu\hat{N} .$$

In (1), the first three terms define explicitly the energies E_0, E_1, E_2 of the d^8, d^9 and d^{10} states of a Cu ion at site i. The two singlet states d^8 and d^{10} are represented by scalar bosons b_i and a_i respectively. The spin-1/2 state d^9 is represented by a fermion $d_{i\sigma}$, where σ is spin. The $c_{k\sigma}$ are conventional Fermi operators for the oxygen band states k of energy ε_k, spin σ, with which the d orbitals hybridize. V_k is the matrix element $<0 \mid V \mid k>$ for hopping into site 0 in the lattice from band state k. When the hopping is into a d^8 site, a b-boson is annihilated along with a k-fermion state of spin σ, and a d-fermion of spin σ is created (fifth term in (1)). When hopping is into a d^9 site, a spin σ f-fermion and a k-fermion of spin σ are annihilated and an a-boson created, (sixth term in (1)). Finally, a chemical potential term is added involving a number operator $\hat{N}$, which commutes with H:

$$\hat{N} = \sum_{i\sigma} d^+_{i\sigma} d_{i\sigma} + 2\sum_i a^+_i a_i + \sum_{k\sigma} n_{k\sigma}. \tag{2}$$

Expression (1) is only equivalent to the electron problem in question if a constraint applied to every lattice site i ('local constraint') is satisfied:

$$Q_i = b^+_i b_i + \sum_\sigma d^+_{i\sigma} d_{i\sigma} + a^+_i a_i = 1 \quad . \tag{3}$$

The operator Q_i on *any* site i commutes with H, in principle enabling (3) to be exactly satisfied for all i. Eq. (3) ensures that one, and only one, of the states d^8, d^9 and d^{10}, can be occupied in any configuration.

Applying the mean field theory developed in heavy fermion systems[2, 9–11] to (1), we take expectation values $< a^+_i > \; = \; < a_i > \; = \; a, < b^+_i > \; = \; < b_i > \; = \; b$, and include the constraint (3) in an essentially mean-field like way by adding a Lagrange multiplyer term $\lambda\sum_i (Q_i - 1)$. Taking E_0 as energy zero we obtain:

$$H_{mf} = \sum_{k\sigma}(\varepsilon_k - \mu)c^+_{k\sigma}c_{k\sigma} + (\varepsilon_d - \mu)\sum_{k\sigma}d^+_{k\sigma}d_{k\sigma}$$

$$+ b\sum_{k\sigma}V_k(d^+_{k\sigma}c_{k\sigma} + hc) + a\sum_{k\sigma}V_k(d^+_{k\sigma}c^+_{-k-\sigma}\mathrm{sgn}\sigma + hc) \tag{4}$$

$$+ N_s\Big[(\varepsilon_d - E_i)b^2 + (E_2 - E_1 + \varepsilon_d - 2\mu)a^2 + E_1 - \varepsilon_d\Big].$$

Here λ is incorporated into a new effective d-level $\varepsilon_d = E_1 + \lambda$, and the parameters a, b and ε_d are to be determined by making the expectation value of (4) stationary with respect to them.

In this paper, we restrict ourselves to a weak-coupling approximation in which a<<b, i.e. $T_c<<$ 'renormalized bandwidth' is assumed. Also for simplicity the k-dependence of V_k is neglected, and an 'ionic' limit of the theory in which the upper d-like band (see Fig 1) is assumed narrow with respect to its distance from the lower, O-like, band is taken. ε_d is then related to E_1 by $\varepsilon_d \approx E_1 + V^2 D^{-1} \ln | \varepsilon_d/(\varepsilon_d - D) |$, the renormalized Fermi level DOS is $\rho_d(\mu) = \varepsilon_d^2/2DV^2 b^2$, and the renormalized d-band width is $\sim b^2 V^2 D/(\varepsilon_d^2 - D^2)$ (ε_d measured from O-band center). The renormalized DOS is shown in Fig. 1. The proportionality between d-band width and x_h has its origin in site blocking.

The gap equation is

$$U_{eff} = \frac{1}{2N_s} \sum_k \frac{v^2_k \tanh[\frac{\beta}{2}(E^2_k + \Delta^2_k)^{1/2}]}{(E^2_k + \Delta^2_k)^{1/2}}, \tag{5}$$

where $U_{eff} = (E_2 - 2E_1) + (E_1 + \varepsilon_d - 2\mu)$, the gap $\Delta_k = 2abV^2/[(\varepsilon_d - \varepsilon_k)^2 + 4b^2V^2]^{1/2}$, and $v_k = \Delta_k/a$. $\Delta = \Delta_{k_F}$ is the energy gap for excitations.

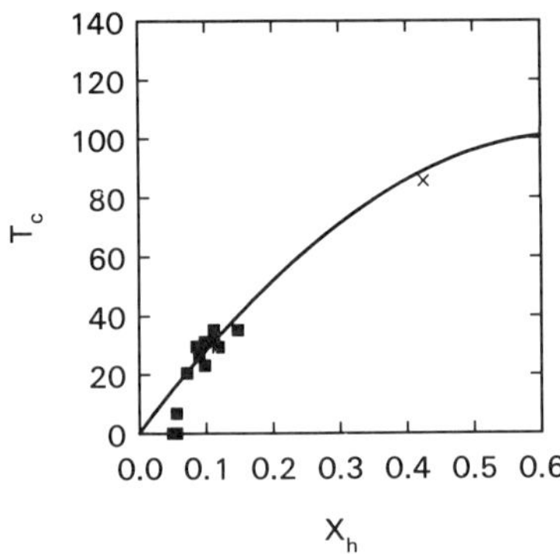

Fig. 2. Plot of T_c vs. hole concentration x_h. Squares are experimental data for $La_{2-x}CuO_4$ samples, after Ref. 12, Cross is data[12] for $YBa_2Cu_3O_{6.5+x/2}$ (one sample), interpreted assuming holes go only into the 1/3 of Cu's which are in square planar configuration. Curve is Eq. (12) for U = 12.1 eV, other parameters as text.

T_c is given by

$$T_c = 1.14 \frac{x_h\sqrt{1 - x_h^2}}{\varepsilon_d + x_hD} \frac{V^2D}{\sqrt{\varepsilon_d^2 - D^2}} \exp[-U_{eff}/4\rho_0V^2], \qquad (6)$$

where $\rho_0 = (2D)^{-1}$.

Parameters chosen to optimize the fit to the Matheiss band structure calculation[14] on La_2CuO_4 are V = 1.93 eV, D = 2.5 eV, ε_d = 3.4 eV, yielding the DOS in Fig. 1 for $x_h = 0.4$. A one-parameter fit to the data of Ref. 12 is now possible, and is shown in Fig. 2, with U_{eff} = 12.1eV . This probably too large value would be reduced on taking into account repulsive interparticle interactions. $YBa_2Cu_3O_{6.5+x/2}$ lies on the curve if we assume that the holes are only on the square planar-coordinated Cu ions which should have the larger Jahn-Teller splitting of the e_g orbitals.

Calculation of γ with our parameters yields γ = 24 mJ mole^{-1}K^{-2} at $x_h = 0.12$, about twice the value from a weighted average of experimental observations[16].

The zero-temperature gap Δ and critical field $H_c(0)$ are predicted to be BCS-like $\Delta = 1.75T_c$, $H_c(0) = [4\pi\rho_d(\mu)v^{-1}]^{1/2}\Delta$, where v is volume per Cu atom. Some measurements yield nearly BCS-like values for the gap[16,17].

In conclusion, we have adapted our theory of heavy fermion superconductivity to oxide superconductors, arriving at a theory more general than the Anderson-Hirsch approach[1,3-7] which fits much of the available data to within a factor of about two.

518

References

1. J. E. Hirsch, Phys. Rev. Lett. **54**, 1317 (1985).

2. D. M. Newns, Phys. Rev., in press.

3. P. W. Anderson, Science **235**, 1196 (1987).

4. H. Fukuyama and K. Yosida, Jap. J. Appl. Phys., **26**, 160 (1987).

5. G. Baskaran, Z. Zou and P. W. Anderson, preprint.

6. A. E. Ruckenstein, preprint.

7. G. Kotliar, preprint.

8. T. M. Rice and K. Ueda, Phys. Rev. Lett. **55**, 995; **55**, 2093 (1985).

9. N. Read, Ph.D. thesis, University of London, (1985); N. REad and D. M. Newns, J. Phys. C. **16**, 3273, L1055 (1983); N. Read, J. Phys. C 2651 (1985); D. M. Newns, N. REad and A. C. Hewson in *Moment Formation in Solids* Ed. W. L. Buyers, Plenum 1984; N. Read and D. M. Newns Solid State Commun, **52**,, 993 (1984).

10. P. Coleman, Phys. Rev. B28, 5255 (1983); Phys. Rev. B29, 3035 (1984); Phys. Rev. B35, 5072 (1987).

11. A. Auerbach and K. Levin, Phys. Rev. Lett. **57**, 877 (1986).

12. M. W. Shafer, B. L. Olson and T. Penney, reprint.

13. J. R. Cooper, B. Alavi, L. W. Zho, W. P. Beyermann and G. Gruner, preprint.

14. L. F. Mattheiss, Phys. Rev. Lett. **58**, 1028 (1987); L. F. Mattheiss and D. R. Hammann, preprint.

15. R. L. Greene, private commun.

16. J. R. Kirtley, private commun.

17. Z. Schlesinger, private commun.

THEORETICAL MODEL FOR OXYGEN VACANCY DEPENDENCE
OF T_c IN THE $YBa_2Cu_3O_x$ SYSTEM ($6 \leq x \leq 8$)

Frank Herman[‡]

IBM Almaden Research Center, San Jose CA 95120-6099
Max-Planck Institute for Solid State Research
D-7000 Stuttgart 80, Federal Republic of Germany

Robert V. Kasowski and William Y. Hsu

Central Research and Development Department[§]
E.I. du Pont de Nemours and Company
Experimental Station, Wilmington, DE 19898

INTRODUCTION

With a view to understanding the relationship between oxygen content, oxygen vacancy ordering, and T_c in the $YBa_2Cu_3O_x$ system, we have calculated the electronic structure of these materials for $x = 6$, 7, and 8 using the first-principles pseudofunction method.[1] Employing a simple interpolation scheme, we have estimated the electronic density of states at the Fermi level, $N(E_F)$, for materials containing varying numbers of oxygen vacancies and varying degrees of oxygen vacancy disorder. Adopting a simplified BCS model according to which T_c depends only on $N(E_F)$, we find that T_c increases monotonically as the oxygen content, x, increases from 6 to 8. Moreover, for a fixed x, the more highly ordered the oxygen vacancies, the higher is T_c. In this paper we will summarize the essential features of our theoretical model and use it to interpret recent experimental results reported at this conference by Batlogg.[2,3] A more detailed account of the present work will be published elsewhere.[4]

CRYSTAL STRUCTURE

The crystal structure of $YBa_2Cu_3O_7$ with ordered vacancies is shown in Fig. 1. The atomic positions were obtained from neutron scattering experiments.[5] This triply-layered oxygen-deficient perovskite is composed of 2-D Cu O networks involving the Cu2, O2, and O3 sites, and 1-D Cu-O ribbons involving the Cu1, O1, and O4 sites. There are two O vacancies in the unit cell, one at Oz (in the yttrium plane), and another at Oy (in the basal plane). In the ideal structure, the O1 and Oy sites are full and empty, respectively. By virtue of the asymmetrical filling of these two sites, the crystal is orthorhombic. It is possible to disorder, partially or fully, the O atoms in the basal plane by suitable heat treatment. The orthorhombic distortion decreases from about 2 percent to zero as the vacancy arrangement changes from fully ordered to fully disordered. In the limiting case of total disorder, the crystal is tetragonal, with the probabilities of occupancy of the O1 and Oy sites both equal to 0.5.

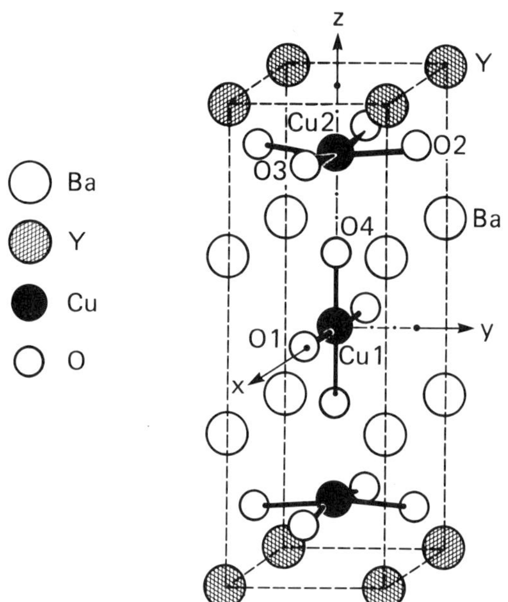

Fig. 1. Crystal structure of orthorhombic $YBa_2Cu_3O_7$ (ordered vacancy model).[5] The vacant sites in the Cu1 and Y planes are denoted by Oy and Oz, respectively. In $YBa_2Cu_3O_6$, the O1 site is also vacant. In $YBa_2Cu_3O_8$, the only remaining vacancy is at Oz. See also Table I.

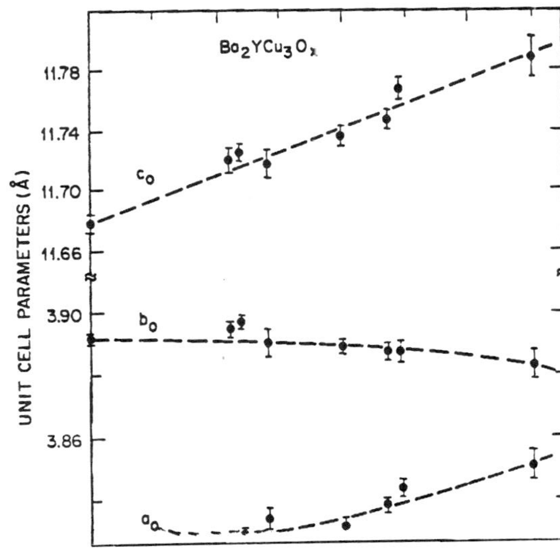

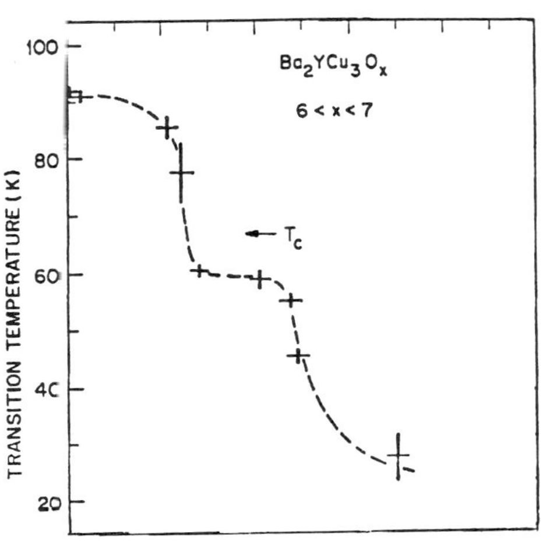

Fig. 2. Dependence of resistive superconducting transition temperature (left panel) and crystallographic cell parameters (right panel) on oxygen content in $YBa_2Cu_3O_x$ Based on Refs. 2 and 3. We are grateful to Dr. B. Batlogg for permission to reproduce these drawings.

If oxygen is removed from $YBa_2Cu_3O_7$, O atoms are removed from the O1 (chain-link) position, leading eventually to $YBa_2Cu_3O_6$, which is tetragonal, containing no O atoms in the basal plane. In $YBa_2Cu_3O_6$, the 1-D Cu-O ribbons have been reduced to linear arrays of isolated O4-Cu1-O4 molecules. This crystal model of $YBa_2Cu_3O_6$ is consistent with neutron scattering experiments, and with extensive experimental evidence indicating that O atoms are removed from the O1 rather than the O4 positions (or from the O2 or O3 positions for that matter), as one goes from $YBa_2Cu_3O_7$ to $YBa_2Cu_3O_6$. Experimental references for these and subsequent remarks will be found in Ref. 4.

Although nearly all experimental work on the $YBa_2Cu_3O_x$ system has focused on the range of x between 6 and 7, it is instructive to consider the extended range between 6 and 8. Bearing in mind that there is no evidence for O atoms occupying site Oz, and that O atoms are known to populate the O1 and Oy positions in partially or fully disordered $YBa_2Cu_3O_x$ (with x between 6.5 and 7), we will assume that in the hypothetical crystal $YBa_2Cu_3O_8$ the O1 and Oy sites are both fully occupied, and that site Oz remains empty. According to this model, $YBa_2Cu_3O_8$ is tetragonal.

BAND STRUCTURE CALCULATIONS

Using the first-principles pseudofunction method,[1] we calculated the band structures of orthorhombic and tetragonal $YBa_2Cu_3O_6$, $YBa_2Cu_3O_7$, and $YBa_2Cu_3O_8$. In the case of orthorhombic $YBa_2Cu_3O_7$, we used the atomic coordinates for the ordered vacancy model obtained by neutron scattering experiments.[5] For simplicity, we assumed that the atomic coordinates are not changed (on the average) by increasing or decreasing the O content. Thus, we used the same atomic coordinates for orthorhombic $YBa_2Cu_3O_6$ and $YBa_2Cu_3O_8$ as for $YBa_2Cu_3O_7$, for which we had experimental information. We then studied the tetragonal counterparts of these three crystals, using the same atomic coordinates as before, except that the lattice constants in the basal planes were averaged.

We then calculated the electronic density of states at the Fermi level, $N(E_F)$ for each of these structures. In the case of $YBa_2Cu_3O_6$, $N(E_F)$ was virtually identical for the tetragonal and orthorhombic structures, and similarly for $YBa_2Cu_3O_8$, indicating that local strains are not likely to affect $N(E_F)$ significantly. The orthorhombic distortion itself was also found to have a considerably weaker influence on the band structure of $YBa_2Cu_3O_7$ than the asymmetrical occupancy of O1 and Oy. Having thus summarized the structural background, we will concern ourselves beyond this point only with tetragonal $YBa_2Cu_3O_6$, $YBa_2Cu_3O_7$, and $YBa_2Cu_3O_8$, and orthorhombic $YBa_2Cu_3O_7$.

DEALING WITH INTERMEDIATE STRUCTURES

As reported in Ref. 4, the band structure changes dramatically as we increase the O content by unit amounts. These band structure changes are very complex indeed, involving not only the shifting and distortion of existing bands, but also the creation and interposition of new bands arising from the newly introduced O atoms. The simplest way to deal with an O content intermediate between 6 and 8 is to consider the electronic density of states, $N(E)$, which we can interpolate, as opposed to energy bands, which we cannot because of their changing number.

As shown in Table I, $N(E_F)$ becomes progressively larger as we proceed from $YBa_2Cu_3O_6$ to $YBa_2Cu_3O_7$ to $YBa_2Cu_3O_8$. To deal with intermediate cases, let us assume that the O1 and Oy sites are progressively filled, on an equal basis, as we move from $YBa_2Cu_3O_6$ to $YBa_2Cu_3O_8$. Thus, the occupancy of each of these sites is (x-6)/2, as x ranges from 6 to 8. With these simple considerations in mind, we can estimate $N(E)$ and $N(E_F)$ for intermediate values of x, and for varying degrees of O vacancy ordering, by forming suitably weighted averages of $N(E)$ for the three crystals that we have studied.

The results of such interpolations are also presented in Table I. It is clear that $N(E_F)$ increases with increasing oxygen content, x, and that for a fixed value of x, $N(E_F)$ is larger the more ordered the vacancies are. Assuming that T_c depends primarily on $N(E_F)$, these calculated results parallel recent experimental observations[6-8] that T_c increases with increasing O content as x increases from 6.5 to 6.9, and that T_c at x = 6.9 is higher the greater the orthorhombic distortion, and, by implication, the more ordered the vacancies.

TABLE I. Electronic density of states at the Fermi level, (E_F), for $YBa_2Cu_3O_x$ as a function of oxygen content x and occupancy of sites O1, Oy, and Oz. Values for $N(E_F)$ were obtained from crystal calculations and from interpolations between the cases indicated. The units are states/eV-cell.

Case	$N(E_F)$	x	O1	Oy	Oz	Crystal	Interpolation
(a)	5.04	6.0	0	0	0	$YBa_2Cu_3O_6$	
	5.54	6.5	1/4	1/4	0		(a) and (f)
(b)	5.81	6.5	1/2	0	0		(a) and (d)
(c)	6.03	7.0	1/2	1/2	0		(a) and (f)
(d)	6.61	7.0	1	0	0	$YBa_2Cu_3O_7$	
	6.53	7.5	3/4	3/4	0		(a) and (f)
(e)	6.81	7.5	1	1/2	0		(d) and (f)
(f)	7.02	8.0	1	1	0	$YBa_2Cu_3O_8$	

THEORETICAL MODEL FOR T_c

In view of this correlation between theory and experiment, we are encouraged to go a step further and attempt to make the correlation more quantitative. Accordingly, let us assume that changes in the O content and vacancy ordering affect the lattice vibrational spectrum and electron-lattice interaction much less strongly than they do changes in $N(E_F)$, so that we can simplify the problem by assuming that T_c depends only on $N(E_F)$. In this spirit, let us adopt the simplified BCS expression

$$T_c = 1.14 <\omega> \exp[-1/N(E_F)V],$$

where $<\omega>$ is an average excitation energy (expressed in degrees K) and V is an interaction parameter. We will not attempt to draw distinctions between weak and strong coupling. We merely note that $<\omega>$ could turn out to be an excitation energy involving exciton and plasmon as well as phonon effects. It will be sufficient for our purposes to evaluate $<\omega>$ and the coupling coefficient $N(E_F)V$ for various choices of T_c for $YBa_2Cu_3O_6$ and $YBa_2Cu_3O_7$.

In order to establish the scale, let us set T_c equal to 100 K for the ordered vacancy model for $YBa_2Cu_3O_7$. Since the 1-D ribbons are totally absent in $YBa_2Cu_3O_6$, we are left only with 2-D Cu-O networks, not unlike the situation in La_2CuO_4-type superconductors. Accordingly, let us set T_c equal to 40 K for $YBa_2Cu_3O_6$, this being the upper range of measured transition temperatures in La_2CuO_4-type superconductors. With these two adjustments, we can obtain values for $<\omega>$ and $N(E_F)V$, and thereby estimate T_c for the remaining cases, as indicated in Table II.

In this way we find that as x is raised from 6 to 7, T_c increases from 40 to 100 K (by design), and that at x = 6.5, T_c is 57 K if the vacancies are disordered, and 67 K if they are partially ordered. These estimates bear a resemblance (most probably fortuitous) to the experimental results of Schuller[7] and Tarascon,[8] both of whom reported values of T_c of about 55 K at roughly x = 6.5. Referring to Table II, we see that at x = 6.5, 7.0, and 7.5, the more ordered the vacancies, the higher the values of T_c. This theoretical conclusion is consistent with the experimental results of Beyers,[6] who found that for values of x just below 7, the greater the orthorhombic distortion, and by implication the more ordered the vacancies, the higher the value of T_c . The interested reader is referred to Ref. 4 for a discussion of the other fits shown in Table II and their theoretical implications.

Taking this model at face value, we would predict that T_c can be increased by about 19 percent (relative to ordered $YBa_2Cu_3O_7$) if we could synthesize $YBa_2Cu_3O_8$. Since this may not be feasible, we could at least try to increase the O content as much as possible, keeping the O vacancies as nearly ordered as possible, since both of these strategies would tend to enhance T_c.

TABLE II. Estimates for the superconducting transition temperature, T_c, based on the BCS expression $T_c = 1.14 <\omega> \exp[-1/N(E_F)V]$. Adjusted values of T_c are indicated by asterisks. Values of NV (at 100 K) and $<\omega>$ in various units are given at bottom.

Case		T_c (K)				x	Vacancies
(a)	30*	40*	50*	60*	70*	6.0	ordered
	47	57	65	73	80	6.5	disordered
(b)	59	67	74	80	85	6.5	partially ordered
(c)	69	75	81	86	90	7.0	disordered
(d)	100*	100*	100*	100*	100*	7.0	ordered
	95	96	97	98	99	7.5	disordered
(e)	112	109	107	105	103	7.5	partially ordered
(f)	125	119	114	110	107	8.0	ordered
$N(E_F)V$	0.26	0.34	0.45	0.61	0.87	(at 100 K)	
$<\omega>$	4184	1661	820	515	314	(K)	
$<\omega>$	2908	1154	570	358	218	(cm^{-1})	
$<\omega>$	0.36	0.14	0.07	0.04	0.03	(eV)	

RANGE OF VALIDITY OF MODEL

Our (spin-unpolarized) electronic structure calculations clearly show that $YBa_2Cu_3O_x$ (x = 6, 7, 8) are all metals, whether the structure is tetragonal or orthorhombic. Although we have not dealt directly with intermediate compositions, we would expect $YBa_2Cu_3O_x$ to be metallic over the entire range of x between 6 and 8 within the framework of spin-unpolarized one-electron band theory. It has been reported by various investigators[9-11] that $YBa_2Cu_3O_x$ is a semiconductor and not superconducting in the range of x between 6 and 6.5. Assuming this to be true, one would have to explore the possibility that the band gap arises from an antiferromagnetic spin arrangement, or from electron correlation effects. In either case, semiconducting behavior below x = 6.5 would limit the applicability of our theory to values of x larger than 6.5.

Assuming that the semiconducting band gap has an antiferromagnetic origin, it would be interesting to find ways of suppressing the antiferromagnetism experimentally. One would then be able to see whether T_c still falls to zero below $x = 6.5$, or whether T_c approaches values appropriate to the 2-D Cu-O networks in the La_2CuO_4-type superconductors, an idea we used to fit T_c to 40 K for $YBa_2Cu_3O_6$ earlier.

INTERPRETATION OF RECENT AT&T BELL RESULTS

At the present conference, Batlogg[2,3] reported new results for the oxygen-vacancy dependence of T_c in the $YBa_2Cu_3O_x$ system based on studies of materials prepared by a gettered annealing technique which results in samples of uniform oxygen deficiency. These materials were found to be single-phase orthorhombic compounds over the entire range of x studied. It was found that T_c is approximately constant near 90 K for $6.8 \leq x \leq 7.0$, then drops sharply to about 60 K between $x = 6.8$ and 6.7, and then remains constant for $6.6 \leq x \leq 6.7$. Finally, T_c drops again, but less sharply, to about 30 K at roughly $x = 6.3$. The AT&T group suggest that a short or long range ordered oxygen vacancy phase with T_c of about 60 K exists near the composition $x = 6\ 2/3$. That is to say, when the relevant O sites are about 1/3 empty, T_c drops by about 1/3, from 90 K to 60 K. These results are significantly different from earlier ones because the present materials are prepared differently.

Let us now examine these striking new results, which are shown in Fig. 2, in the light of the present model. For this purpose let us return to Table II, and consider the two fits for $T_c = 30$ and 40 K for $YBa_2Cu_3O_6$. (Both of these are also fitted to $T_c = 100$ K for ordered $YBa_2Cu_3O_7$.) These two fits come closest to describing the AT&T results, on the average, if we multiply all our T_c values by 0.9. This amounts to reinterpreting the T_c entries in Table II as percentages of the experimental value of T_c for $YBa_2Cu_3O_7$ with ordered vacancies.

The slow decrease in T_c just below $x = 7$ can be construed as being due principally to the decrease in x, since this would have a less drastic effect on T_c than would disordering of the O vacancies in the basal plane. This is consistent with the right side of Fig. 2, where the basal lattice constants, a_o and b_o, are seen to change only slightly just below $x = 7$.

Concerning the rapid fall of T_c in a narrow range of x near $6\ 2/3$ in the AT&T experiments, let us first explore the possibility that the O atoms in the basal plane rearrange, so that the O1 and Oy sites become roughly equally occupied. Although this could lead to a reduction in T_c of about 20 to 30 K, as suggested by the numerical results in Table II, we must reject this mechanism because it would lead to a tetragonal structure, contrary to experiment. In fact, the experimental orthorhombic distortion (cf. Fig. 2) falls only gradually with decreasing oxygen content, there being no sharp discontinuity suggestive of an order-disorder transition. In the light of these remarks, we must look for an alternate explanation for the rapid fall in T_c at about $x = 6\ 2/3$.

Since the AT&T materials were prepared in a fashion quite different from that used in earlier studies,[9-11] some of the structural assumptions on which our model are based must be re-examined in the light of these new findings. It is conceivable, for example, that due to the special method of fabrication, O atoms can be removed not only from the O1 (chain-link) sites, but also from the O4 (ribbon-edge) sites, or even from the 2D-Cu-O networks (O2 and O3 sites).

Although we have not yet carried out explicit calculations involving the removal of O atoms from the O2, O3, or O4 sites, we can see from our partial DOS that the O4 sites contribute more strongly to $N(E_F)$ than do the O2 and O3 sites. Accordingly, we suggest that the rapid fall in T_c at about $x = 6\,2/3$ may be due to the preferential removal of atoms from the O4 (ribbon-edge) sites. This would lead to a sufficiently large reduction in $N(E_F)$ to account for the large reduction in T_c. Moreover, depleting O4 would not change the orthorhombic distortion drastically, as demanded by experiment.

We can now go on and interpret the experimental plateau in T_c as a redistribution of O atoms between the O4 sites on the one hand, and the O1 and Oy sites on the other. That is to say, the fall in T_c due to the reduction in oxygen content is offset by the rise in T_c due to an increase in $N(E_F)$, which is due in turn to atoms favoring the O4 over the O1 and Oy sites. The further fall in T_c below $x = 6.5$ can again be accounted for by a preferential removal of atoms from the O4 sites, followed possibly by a gradual redistribution between the O4 and O1,Oy sites.

CONCLUDING REMARKS

We are able to account for many features of the oxygen content and oxygen-vacancy ordering dependence of T_c in the $YBa_2Cu_3O_x$ system by adopting a simple BCS model and assuming that T_c depends only on $N(E_F)$. In this way we can demonstrate that T_c increases with increasing oxygen content, and that for a fixed value of x, the higher the oxygen-vacancy ordering (in the basal plane), the higher the value of T_c. These ideas apply only to the metallic phase, and would not carry over to a non-superconducting semiconductor phase.

In this paper we have summarized the essential features of our model. For a more detailed account of the present work, see Ref. 4. In addition, we have offered an interpretation of recent experimental results by the AT&T group,[2,3] which indicate that samples prepared in a novel manner exhibit a non-monotonic decrease of T_c with decreasing oxygen content. It is noteworthy that samples prepared in this manner remain superconducting over a wider range of x than samples prepared by more conventional techniques. According to our interpretation, the overall trend in the AT&T results for T_c vs. x can be explained by the reduction in T_c due to the reduction in $N(E_F)$ associated with the reduction in oxygen content and the degree of oxygen vacancy ordering. Moreover, we suggest that the non-monotonic variation in T_c with x is due to staged redistributions of oxygen vacancies between the O1 (chain-link) and O4 (ribbon-edge) sites.

ACKNOWLEDGEMENTS

The authors are grateful to many authors for sending us preprints in advance of publication. We are particularly grateful to Drs. B. Batlogg, A. J. Freeman, D. E. Cox, J. B. Jorgensen, W. A. Little, L. F. Mattheiss, M. W. Shafer, and J. M. Tarascon for preprints and stimulating discussions. We also thank Dr. S. S. P. Parkin for a critical review of the manuscript. One of us (F. H.) wishes to express his appreciation to O. K. Andersen and his group at the Max-Planck-Institute Stuttgart for their hospitality and scientific stimulation during his stay as an Alexander von Humboldt Senior U. S. Fellow. The work at IBM was supported in part by the Office of Naval Research.

REFERENCES

‡   Permanent address: IBM Almaden Research Center, 650 Harry Road, San Jose, CA 95120-6099.

§   Contribution No. 4488.

1.   R. V. Kasowski, W. Y. Hsu, T. N. Rhodin, and D. D. Chambliss, *Phys. Rev. B* **34**, 2656 (1986).

2.   B. Batlogg, in *Novel Mechanisms of Superconductivity,*, S. A. Wolf and V. Z. Kresin, eds., (Plenum Press, New York, 1987), this volume.

3.   R. J. Cava, B. Batlogg, C. H. Chen, E. A. Rietman, S. M. Zahurak, and D. Werder, submitted to *Phys. Rev. B1.*

4.   F. Herman, R. V. Kasowski, and W. Y. Hsu, submitted to *Phys. Rev. B1.*

5.   D. E. Cox, A. R. Moodenbaugh, J. J. Hurst, and R. H. Jones, *J. Phys. Chem. Solids*, in press.

6.   R. Beyers, G. Lim, E. M. Engler, V. Y. Lee, M. L. Ramirez, R. J. Savoy, R. D. Jacowitz, T. M. Shaw, S. La Placa, R. Boehme, C. C. Tsuei, S. I. Park, M. W. Shafer, W. J. Gallagher, and G. V. Chandrasekhar, *Appl. Phys. Lett.*, in press.

7.   I.K. Schuller, D. G. Hinks, M. A. Beno, D. W. Capone II, L. Soderholm, J.-P. Locquet, Y. Bruynseraede, C. U. Segre, and K. Zhang, *Solid State Commun.*, in press.

8.   J. M. Tarascon, L. H. Greene, W. R. McKinnon, G. H. Hull, and T. H. Geballe, paper at MRS Meeting, Anaheim, California, April 30, 1987; R. J. Cava, ibid; J. M. Tarascon, W. R. McKinnon, L. H. Greene, G. W. Hull, B. G. Bagley, E. M. Vogel, and Y. LePage, preprint.

9.   P. K. Gallagher, H. M. O'Bryan, S. A. Sunshine, and D. W. Murphy, preprint.

10.   J. van den Berg, C. J. van der Beek, P. H. Kes, G. J. Nieuwenhuys, J. A. Mydosh, H.W. Zandbergen, F. P. F. van Berkel, R. Steens, and D. J. W. Ijdo, *Europhys. Lett.*, in press.

11.   J. D. Jorgensen, M. A. Beno, D. G. Hinks, L. Soderholm, K. J. Volin, R. L. Hitterman, J. D. Grace, I. K. Schuller, C. U. Segre, K. Zhang, M. S. Kleefisch, submitted to *Phys. Rev. B*; also J. D. Jorsensen, paper at European MRS Meeting, Strasbourg, France, June 3, 1987;

COMPLEX HAMILTONIANS: COMMON FEATURES OF

MECHANISMS FOR HIGH-T_c AND SLOW RELAXATION

K.L. Ngai, R.W. Rendell and A.K. Rajagopal

Naval Research Laboratory
Washington, D.C. 20375-5000

INTRODUCTION

The discoveries of the new ceramic superconductors $La_{2-x}A_xCuO_{4-z}$[1] and $YBa_2Cu_3O_{7-x}$[2] with T_c of about 40K and 90K respectively have given new impetus to the discussions of high-T_c mechanisms. The specific mechanism by which electrons lower their energy by pairing in these superconducting ceramics is at issue. Conventional lattice vibration mechanisms may be inappropriate in view of experimental evidence of the absence of an isotope effect[3], and inadequate on theoretical grounds for the high transition temperature. For these reasons, alternative models have been seriously considered. By now, a number of mechanisms for high-T_c superconductivity have either been proposed specifically for these new materials or suggest themselves to be possible candidates. These include the resonating valence bond (RVB)[4], the negative-U models[5-7], the Bose condensation of bipolarons models[8,9], the d-wave pairing near a spin-density-wave instability of a positive-U Hubbard model[10], the surface or two-dimensional plasmon models[11,12], acoustic plasmon models[13], the excitonic models[14,15], the dynamic instability model[16], and many others that possibly we are not aware of. The answer to which one of these many theoretical ideas is ultimately responsible for the superconductivity in these ceramics will have to wait for further experimental investigations of both the normal and superconducting states. Arguments have been given that existing experimental data including the linear specific heats at low temperatures seems to eliminate even some unconventional versions of BCS theory of coupling between electron pairs and phonons, and indicates some models are favored.[17] The possibility should also be considered that the linear specific heat observed[36,37] at low temperatures in these new high-T_c materials may contain contributions from excess modes such as two-level systems.

It has been recognized that in these high-T_c oxides, defects are present. Some of the defects are definitely related to vacancies caused by deficiency in oxygen. Studies of $La_{2-y}Sr_yCuO_{4-x}$ perovskite have shown that T_c is dramatically affected by subtle changes in oxygen content. Similarly, changes in oxygen content in the 90K series $REBa_2Cu_3O_{7-x}$ also drastically affect their physical properties and T_c[18]. Even in undoped La_2CuO_{4-y}, the occurrence of trace and filamentary superconductivity is extremely process dependent[19]. It can be reversibly removed or restored by heating the material in argon or oxygen. Moreover, the antiferromagnetism found[20] in La_2CuO_{4-y} is quite sensitive to the oxygen concentration parameter y. The

antiferromagnetic ordering temperature determined from neutron scattering
rose from 50K to 150K with annealing in a vacuum to remove oxygen. The
existence of a small oxygen vacancy concentration ($y>0$) appears to be crucial for promoting antiferromagnetism. Changes in the oxygen content may
affect the average copper valence via corresponding changes in the oxidation states of copper atoms. The average valence of copper increases with
the oxygen content. For example[18], it was found in $YBa_2Cu_3O_7$ that the oxygen content can change from 6.7 to 6.2 resulting in a change in the average
valence of copper from 2.2 to 1.8. The valence implies the formation of
Cu^{+1}. This change of the oxidation state has a number of implications
including the possible charge fluctuations between $Cu^{++}O^{--}$ and $Cu^{+}O^{-}$ in the
CuO_{1+x} planes[21], the spawning of negative-U centers[5], and the possible
modification of the suggested Cu^{3+} - Cu^{4+} - like charge fluctuations which
induce attractive interactions (-U centers) both in the chains and to the
2-dimensional (Cu2) bands.[7] Oxygen vacancies may also modify the RVB or
spin-liquid model in the present form. The occurrence of "unsynchronized"
resonance in the RVB model due to oxygen vacancies has also been pointed
out[22].

The discussions in the previous paragraph serve the purpose of emphasizing that the real high-T_c materials are more complicated than the representations given by the model Hamiltonians of the proposed mechanisms.
Some interactions are neglected and provisos are missing in the description
of the realistic superconducting ceramics in the context of any one of
these proposed models. In writing down the Hamiltonian for any of these
models with a specific mechanism, a simplified form of it is usually
adopted. The actual Hamiltonian may be more complex. For example, the
negative-U Hamiltonian of a double-valence fluctuating molecule hybridized
with the conduction-electron band is often drastically simplified to have
the manageable form as given in References 5 & 6 for calculations of
enhancement of T_c by use of perturbative techniques and Monte Carlo simulation techniques. A negative-U term $Un_{j\uparrow}n_{j\downarrow}$ is used to describe the
attractive interaction when conduction electrons are transferred by hybridization onto the negative-U centers. The actual picture may be more
truthfully described by orbitals, potentials, lattice distortions and
kinetic degrees of freedom of both the conduction electron and the impurity
electron. For the purpose of addressing the problem of enhancement of
superconductivity, the simplification is not a problem and the approach
adequate. However, a full representation of all the details and complicated aspects of the physics corresponding to any of these mechanisms would
require a Hamiltonian more complex than is given. A large class of complex
Hamiltonians possesses the generic feature of low-lying near degenerate
states. The latter are particularly prominent if electron-electron correlations are important which appears from experimental inference[20] to be the
case in La_2CuO_{4-y}. Low-lying near degenerate states are characteristics of
metastability and the glassy state. However the existence of these near
degenerate states does not necessarily imply metastability. In either
case, a subset of these near degenerate states are responsible for the
slowing down of long time relaxation processes resulting in slow nonexponential relaxations of the type commonly observed in equilibrium viscous
liquids as well as in nonequilibrium glasses[23]. The existence of these
near-degenerate states for quantum Hamiltonians in which there are strong
correlations between different degrees of freedom has been documented to be
a generic property. We will refer to systems with this property as complex
Hamiltonians. In the corresponding classical limit, Hamiltonians are
generally nonintegrable and possess a characteristic structure of regular
and chaotic orbits in phase space. In either the quantum or the classical
context, we have provided a framework to show how the respective generic
property manifests itself in slowing down of the primary relaxation rate W_o
to have the self-similar time-dependent form of $W(t) = W_o (\omega_c t)^{-n}$, $0<n<1$,
for $\omega_c t>1$. From this, the fractional exponential relaxation (Kohlrausch)

function $\phi(t) = \exp[-(t/\tau^*)^{1-n}]$ along with additional predictions follow.[24] Another approach based on classical mechanics with time-dependent Dirac constraints and statistical mechanics of irreversible processes has also led to the same conclusions.[25]

From the previous discussions, it is clear that slow relaxation may naturally arise from the complex and non-integrable (in the classical mechanics sense) nature of the total unabridged Hamiltonian for many of the proposed high-T_c mechanisms if realistic interactions, conditions such as the existence of clusters and their different phase coherences, and imperfections in material are taken into account. If any of these proposed mechanisms turn out to be correct, then both high-T_c superconductivity and slow relaxation are two accompanying consequences of the same complex Hamiltonian. We are led to expect that in some of the high-T_c superconducting ceramics we may be able to observe experimentally a slow relaxation phenomenon. This turns out to be indeed the case and will be discussed in the next section. By contrast, the conventional BCS phonon mechanism for superconductivity in simple metals does not involve a complex Hamiltonian and slow relaxation is not expected there. Slow relaxation will provide, in addition to superconductive and magnetic properties, another handle on the physics of the high-T_c materials. A parallel study of superconductivity and slow relaxations and a comparison of their variations with conceivable modification of sample structure and defect concentrations by various means including heat treatment and sample preparation method, etc. may reveal correlations of their variations. The result may be able to provide some guidance in the search for the origin of high-T_c in these ceramic materials.

INFERENCES FROM EXPERIMENTS

In the previous section, we discussed that any alternative models to the BCS theory of phonon mediated superconductivity to account for the high-T_c superconductivity in the ceramics are likely to involve a complex Hamiltonian. Whichever it and the mechanism it represents may be, slow relaxation will be an accompanying feature as expected from our general considerations of relaxations in complex systems. In fact, Müller, Takashige and Bednorz[26] have found signatures for both a glass state and slow nonexponential decays from susceptibility and magnetic moment measurements in powder samples of La_2CuO_{4-y}:Ba. The experimental data and the sample properties suggest the picture of coupled superconducting clusters or grains each with size small compared with the London penetration depth and are weakly coupled into closed loops. The picture[27] is similar to a spin glass. This state in La_2CuO_{4-y} has been called the superconductive-glass state[26]. An estimate of the single phase area and the grain size has led to the conclusion that the superconductive-glass state is present in the La_2CuO_{4-y}:Ba grains. The Hamiltonian describing the entire array of weakly coupled grains[27] involves the sum over all distinct pairs of grains of terms $J_{ij}\cos(\phi_i-\phi_j-A_{ij})$ where ϕ_i and ϕ_j are phases of the complex energy gaps and A_{ij} is a magnetic field phase factor. Slow relaxation is attributed to correlation, cooperativity and coupling between grains. It is argued that "the phase coherence for a certain cluster has to become close enough to neighboring ones that it can relax. This becomes less probable when the system evolves in time, and large volumes with different phase coherences are present". This qualitative argument for slow relaxation is consistent with our view of the slowing down of the relaxation rate by the coupling between clusters with Hamiltonian $\sum_{ij}J_{ij}\cos(\phi_i-\phi_j-A_{ij})$.

Measurements of the real and imaginary parts of the ac susceptibility of the Y-Ba-Cu-O system have been made over a frequency range of a few decades[28]. The susceptibility exhibits frequency dependence at temperatures below 70K. The complicated temperature and frequency dependences of the susceptibility are attributed to inter-grain boundary effects. In the future, we can expect more investigations of relaxation properties in the high-T_c ceramics to be reported. The results will be useful for correlation with the superconductivity properties and their relation to the microstructure and chemical constituents of these materials.

COMPLEX HAMILTONIANS

In addition to the complex Hamiltonian discussed in the previous section that arises from coupled superconducting clusters that are present in some though perhaps not all of the high-T_c oxides, another source of complex Hamiltonian comes from the electron correlation effects that are important in any of the mechanisms for high-T_c superconductivity. Again let us consider the negative-U models or bipolaron models. The interaction for zero and double occupancies at the negative-U center usually involves significant lattice distortions and hence nonlinear potentials in the dynamical trapping of the pair of electrons. There are examples of such nonlinear Hamiltonians which in the classical context gives rise to chaos in phase space, and in the quantum context to near degenerate levels with a distribution conforming to so-called Gaussian Orthogonal Ensemble (GOE) of random matrix theory. These include several Hamiltonians with two degrees of freedom coupled by nonlinear potentials in the coordinates.

$$H = \sum_i (p_i^2/2m_i) + \alpha V(q_1, q_2) \tag{1}$$

For a large class of potentials V, the quantum energy levels are found to exhibit level repulsion and their fluctuation properties conform to well-defined statistical distributions as α increases and the correlations between degrees of freedom become important. An example is the Morse potential, $V = [1-\exp[-a(q_2 - q_1)]]^2$, which is often used to represent molecular bonds. This is also equivalent via canonical transformation to a Hamiltonian with coupling in the kinetic energy. The quantized energy spectra of the Morse potential system has been studied numerically.[29] In particular, the nearest-neighbor level-spacing distributions were constructed for different values of the Hamiltonian parameters. In the limit of strong coupling, where the Hamiltonian is expected to be complex, the calculated distributions are closely described by $P(S/\bar{S}) = (\pi S/2\bar{S}^2) \exp(-\pi S^2/4\bar{S}^2)$, where S is the nearest-neighbor level spacing and $\bar{S}$ is the mean spacing. This is the Wigner distribution which also closely describes the GOE of random matrix theory. The same limit of strong coupling in this system corresponds to classical motion which is nonintegrable and predominantly chaotic as exhibited by trajectories in Poincare sections of the Hamiltonian. As α decreases and the interparticle correlations become less important, $P(S/\bar{S})$ tends instead to a simple Poisson distribution and the corresponding classical motion becomes dominated by regular trajectories. These quantum and classical results are borne out by explicit solutions of many other nonlinear potentials.

In addition to these explicit numerical solutions, arguments have been proposed by Pechukas, and improved upon by Berry and Yukawa, to understand the spectral properties of Hamiltonians in a more general context as α varies in strength.[30] These are based on the observation that for complex Hamiltonians, there are no strong selection rules and the off-diagonal matrix elements of V between nearby states are large and of the same average strength. These matrix elements also fluctuate rapidly with α. However, for simple Hamiltonians, with strong selection rules, off-diagonal matrix elements are small compared to the mean level spacing. Within this context, an explicit analytical construction of the level spacing distribution functions were developed by supplementing the Hamiltonian $H = H_o + \alpha V$ with statistical mechanical arguments. For Hamiltonians representing quantum systems with time-reversal symmetry, the results (in arbitrary number of dimensions) are again in accord with GOE when the Hamiltonian is complex and α is large and makes a transition to the Poisson distribution as α decreases. More recently, Nakamura and Lakshmanan[30] have proposed a rigorous basis for these developments based on Lax forms and their solutions. Berry[31] has also discussed the generic occurrence of level repulsion in connection with the Wigner-von Neumann theorem for energy level degeneracies. Thus there is increasing understanding of how specific features of complex Hamiltonians influence the generic features of level spectra. Others have suggested the use of Lanczos-Haydock-Heine tridiagonalization methods to study the spectra of electronic bands in imperfect solids. A simple example of this was actually studied previously by Berry[31] for motion in a 2-d lattice of hard scatterers (i.e. quantum billiard). Solution of the KKR determinant of this system revealed a GOE spectrum. The corresponding classical behavior is strongly chaotic.

Consider next the RVB model with inclusion of possible oxygen vacancies and frustrations. The RVB or spin singlet liquid state has been suggested as being the appropriate ground state for $S = \frac{1}{2}$ Heisenberg antiferromagnets. The characterization of quantum and classical generic features in lattice spin systems has recently been addressed by Muller.[32] Spin systems containing N spins can be studied in the limit of N finite, $S \rightarrow \infty$ (classical limit) as well as in the limit of S finite, $N \rightarrow \infty$ (thermodynamic limit). The classical limit was examined by Nakamura and Bishop[33] by explicit calculations on a three-spin triangular lattice with antiferromagnetic coupling:

$$H = J\sum(\vec{S}_i \cdot \vec{S}_{i+1} + \sigma S^z_i S^z_{i+1}) \tag{2}$$

where $J>0, -1 \leq \sigma < 0$ and $i = 1,2,3$. Boundary conditions are periodic. The triangular lattice with antiferromagnetic coupling is frustrated in the sense that there are competing interactions, especially for small values of S. For $S = \frac{1}{2}$, the presence of frustration will enhance the formation of the RVB. For $S > \frac{1}{2}$, the classical dynamics of this system appear to be non-integrable (excluding the isotropic limit $\sigma = 0$) and the semiclassical quantum spectra exhibits level repulsion. A sufficient number of levels to explore the statistical properties of the spectrum were available even for the three-spin cluster by using spin values as large as $S = 36\frac{1}{2}$. From these results, the authors expect studies with even larger S will reveal a Poisson distribution for $\sigma = 0$ and a Wigner distribution for $\sigma < 0$. Muller expects this situation to be generic for the classical limit of spin systems. He also discusses the work of Frahm and Mikeska on a driven one-spin system who found both a Wigner distribution and nonintegrable classical dynamics in the $S \gg 1$ regime. Thus the behavior of spin systems in the classical and semiclassical regimes is accessible by study of small spin clusters of large but finite S. Here the generic level spacing structure is a precursor of classical dynamical chaos. For the situation representative of the high-T_c ceramics, the Hamiltonian is far more complicated than the simple Heisenberg system in Eq. (2). Of more immediate interest is the study of the appropriate Hamiltonian in the complementary limit of small S (e.g. $S = \frac{1}{2}$) and a large spin cluster, i.e. the thermodynamic limit. This is more difficult to analyze theoretically but Muller[32] has suggested that the spectrum of quantum spin systems is generically irregular in this limit with strong level repulsion due to the lack of a sufficient number of conservation laws (i.e. selection rules). He discusses and presents evidence for the emergence in this limit of an irregular spectrum which is not constrained by a piece-wise smooth energy density with isolated van Hove singularities. Muller presents the $N \rightarrow \infty$ limit of quantum spin systems as a regime of quantum chaos for which the long time dynamical behavior will be modified. Although the corresponding spectrum has not yet been completely studied, it can be expected that large numbers of quantum spins with complicated and frustrated interactions will introduce some type of generic complexity into the Hamiltonian.

Let us finally return to the other class of complex Hamiltonians describing the coupling between superconducting clusters in some samples of the high-T_c ceramics. The diamagnetic response of weakly linked superconducting clusters was modeled by Ebner and Stroud[27] in terms of coupling between the host material and the grains according to the Hamiltonian

$$H = \sum J_{ij} \cos(\phi_i - \phi_i - A_{ij}) \tag{3}$$

where J_{ij} is the coupling energy between grains i and j, and the phase factors A_{ij} are due to the presence of a static external magnetic field. This "Josephson tunneling" model is characterized by frustration in that any cluster of grains has numerous competing ground states with nearly equal energy. Such frustration is familiar from previously studied systems such as spin glass models. Randomness can enter through the Josephson coupling energy J_{ij} and also through the grain sizes. The existence of many local energy minima in classical XY systems and planar spin models with random J_{ij} similar to Eq. (3) are known from numerical studies. For the purpose of relaxation, this situation leads to a picture in which some clusters must wait to relax until the phase coherence becomes close enough to neighboring ones. Such cooperativity due to correlations between degrees of freedom has many of the ingredients present in our discussion of relaxation of complex Hamiltonians. It is possible that a system described by a Hamiltonian such as Eq. (3), supplemented by other additional interactions and randomness which will inevitably be present in a realistic representation of the ceramics, may constitute a complex Hamiltonian in this sense. Efetov[34] has given an example of an electron in which the presence of a random distribution of impurities induces a complex level spectra described by the GOE distribution. Simple driven spin systems in the regime in which the driving force acts as a random interaction also exhibit the generic GOE level spectra. The more complicated granular superconductor system must then be seriously considered as a possible complex Hamiltonian.

SLOWING DOWN OF RELAXATION RATE

We have seen that it is likely that the complex Hamiltonians proposed for high-T_c will give rise to low-lying degenerate levels. A possible source of the latter is the strong electron correlation effects common to a number of these complex Hamiltonians. If we pick out a dynamical variable (say the spin) and consider its relaxation, then the electron correlation effects present would make the relaxation process difficult to describe. Basically, we have to deal with a many-body problem in an irreversible process for which there is no standard method of solution. We have recognized the possibility that the dynamical variable is being coupled to others and its relaxation rate is slowing down with time by the constraints imposed by the others. The idea that the relaxation rate that enters into a master equation can be a time-dependent function is a central idea of our models and this is physically reasonable for relaxation in complex systems with correlations. Rates for kinetic equations are known to be generally time-dependent and constant relaxation rates have been derived only for simple integrable Hamiltonians under some simplifying assumptions of the heat bath such as the weak-coupling limit. Intuitively, a time-dependent relaxation rate $W(t)$ can be motivated by cooperativity which entails dynamics of other variables sequential to the initial relaxation of a primary variable. It is also consistent with general properties of relaxation functions. Remarkably, the only time-dependence of W(t) which guarantees that a change of temperature will change only the timescale but not the functional form of the relaxation function of a relaxation process (i.e. thermorheological simplicity)[25] has the monomial form of

$$W(t) = W_o(\omega_c t)^{-n} \tag{4}$$

where W_o is the primary relaxation rate, ω_c is a characteristic frequency and n is a fraction of unity. We can arrive at the same result by another requirement for the stability of the relaxation spectrum in forming the macroscopic relaxation spectrum by superposition of microscopic relaxation spectra. In addition to these general considerations, W(t) can be derived on the basis of Hamiltonian dynamics where the generic properties of complex Hamiltonians are taken into account. We have modeled the sequential dynamics of the other variables by the transitions among the low-lying near degenerate levels of the complex Hamiltonian. If the initial (at short times) relaxation rate of the primary variable is W_o, then the accompanying transitions modify it to have the time-dependent form of Eq. (4). This can also be derived[35] dircotly in terms of the corresponding classical generic property, classical chaos. The current understanding of the relation between kinetic equations for the approach to thermodynamic equilibrium and microscopic equations of motion are based on the methods developed by van Hove, Prigogine, and others. A Hamiltonian H_S interacting with a heat bath will systematically produce irreversible kinetic equations with constant relaxation rates W_o. Previous workers have only derived kinetic equations corresponding to integrable Hamiltonians H_S, whereas it has become known that Hamiltonians are generically nonintegrable. We have recently derived[35] kinetic equations directly from the equations of motion for classical nonintegrable H_S's interacting with a heat bath. In the weak-coupling limit, we find that the effect of nonintegrability for a large class of systems is generally to modify the constant relaxation rates by a multiplicative time-dependent factor: $W_o f(t)$. The form of f(t) has been investigated both analytically and numerically and its origins can be traced to the mechanism of classical chaos. The results are again consistent with: $f(t) = 1$, $\omega_c t<1$ and $f(t) = (\omega_c t)^{-n}$, $\omega_c t>1$ where $0<n<1$. This leads directly to nonexponential relaxation $\exp[-(t/\tau^*)^{1-n}]$ along with the modified relaxation time $\tau^* = [(1-n)\omega_c^n/W_o]^{1/(1-n)}$ as has been experimentally documented for glasses and polymers. These are examples of our continuing effort to improve our understanding of relaxations in complex systems which may include some of the high-T_c materials.

This work is supported in part (K. L. Ngai and R. W. Rendell) by ONR Contract No. N0001487WX24039.

REFERENCES

1. J. G. Bednorz and K. A. Müller, Z. Phys. B64:189 (1986).
2. M. K. Wu, J. R. Ashburn, C. T. Torng, P. H. Hor, R. L. Meng, L. Gao, Z. J. Huang, Y. Q. Wang, and C. W. Chiu, Phys. Rev. Lett. 58:908 (1987).
3. B. Batlogg, R. J. Cava, A. Jayaraman, R. B. van Dover, G. A. Kourouklis, S. Sunshine, D. W. Murphy, L. W. Rupp, H. S. Chen, A. White, K. T. Short, A. M. Mujsie, and E. A. Rietman, to be published (1987).
4. P. W. Anderson, Science 235:1196 (1987).
5. C. S. Ting, D. N. Talwar and K. L. Ngai, Phys. Rev. Lett. 45:1213 (1980); C. S. Ting, K. L. Ngai, and C. T. White, Phys. Rev. B22:2318 (1980).
6. H. B. Schuttler, M. Jarrell, and D. J. Scalapino, Phys. Rev. Lett. 58:1147 (1987).

7. J. Tu, S. Massidda, A. J. Freeman, and D. D. Koelling, to be published.

8. A. S. Alexandrov, J. Ranninger, and S. Robaszkiewicz, Phys. Rev. B33:4526 (1986); L-Y. Zhang, Solid State Comm. 62:491 (1987).

9. N. Mott, Nature 327:185 (1987).

10. D. J. Scalapino, E. Loh, and J. E. Hirsch, Phys. Rev. B34:8190 (1986).

11. E. N. Economou and K. L. Ngai, Solid State Comm. 17:1155 (1975).

12. V. Kresin, to be published.

13. J. Ruvalds, to be published.

14. J. Bardeen, in: Superconductivity in d- and f-Band Metals, D. Douglas, ed., Plenum, NY (1976).

15. C. M. Varma, S. Schmitt-Rink, and E. Abrahams, to be published.

16. K. L. Ngai and T. L. Reinecke, Phys. Rev. B16:1077 (1977).

17. P. W. Anderson and E. Abrahams, Nature 327:363 (1987).

18. J. M. Tarascon, W. R. McKinnon, L. H. Greene, G. W. Hull and E. M. Vogel, to be published.

19. P. M. Grant, S. S. P. Parkin, V. Y. Lee, E. M. Engler, M. L. Ramirez, J. E. Vazquez, G. Lim, R. D. Jacowitz, and R. L. Greene, Phys. Rev. Lett. 58:2482 (1987).

20. D. Vaknin, S. K. Sinha, D. E. Moncton, D. C. Johnston, J. Newsam, C. R. Safinya, and H. E. King, Jr., to be published, Science 236:780 (1987).

21. P. H. Hor, R. L. Meng, Y. Q. Wang, L. Gao, Z. J. Huang, J. Bechtold, K. Forster, and C. W. Chu, Phys. Rev. Lett. 58:1891 (1987).

22. L. Pauling (preprint).

23. Dynamic Aspects of Structural Change in Liquids and Glasses, C. A. Angell and M. Goldstein, eds., Annals, New York Acad. Sci. 484 (1986).

24. For a review, see K. L. Ngai, R. W. Rendell, A. K. Rajagopal, and S. Teitler, in Ref. 23, pp. 150-184; pp. 321-323.

25. K. L. Ngai, A. K. Rajagopal, and S. Teitler, to be published.

26. K. A. Müller, M. Takashige and J. G. Bednorz, Phys. Rev. Lett. 58:1143 (1987).

27. C. Ebner and D. Stroud, Phys. Rev. B31:165 (1985).

28. K. V. Rao, D. X. Chen, J. Nogues, C. Politis, C. Gallo, and J. A. Gerber, to be published.

29. T. Terasaka and T. Matsushita, Phys. Rev. A32:538 (1985).

30. K. Nakamura and M. Lakshmanan, Phys. Rev. Lett. 57:1661 (1986) and references therein.

31. M. V. Berry, Ann. Phys. 131:163 (1981).

32. G. Muller, Phys. Rev. A34:3345 (1986).

33. K. Nakamura and A. R. Bishop, Phys. Rev. B33:1963 (1986).

34. K. B. Efetov, Adv. Phys. 32:53 (1983).

35. R. W. Rendell and K. L. Ngai, to be published.

36. M. Decroux, A. Junod, A. Bezinge, D. Cattani, J. Cors, J. L. Jorda, A. Stettler, M. Francois, K. Yvon, O. Fischer and J. Muller, Europhys. Lett. 3:1035 (1987).

37. M. E. Reeves, T. A. Friedmann and D. M. Ginsberg, Phys. Rev. B35:7207 (1987).

SUPERCONDUCTIVITY DUE TO LOCALIZED BIPOLARONS IN METAL-OXIDE COMPOUNDS

C. S. Ting and D. Y. Xing

Department of Physics
University of Houston
Houston, Texas 77004

INTRODUCTION

In this article, the effect on superconductivity of a wide band of
electrons due to localized bipolarons is studied by Nambu's formalism. The
bipolarons are from electrons of a narrow d band which interact strongly
with local phonons and become localized polarons. In the presence of a
small mixing term between the wide band electrons and the localized d
electrons, we show that the superconducting transition temperature T_C of
the system can be drastically enhanced. The order parameters, energy gap
and specific heat jump at $T=T_C$ have also been investigated.

The idea of bipolaron formation in a crystal with large local lattice
deformation which could lead to an attractive on-site interaction between
two electrons with opposite spins was first proposed by Anderson[1] to
explain the pinning of Fermi level in amorphous semiconductors.
Chakraverty[2] generalized this idea to the pairing of electrons between two
nearest neighboring sites and discussed the possible electronic phases of
an oxide compound as the electron-phonon interaction λ varies.[2] At low
temperature and for λ less than a critical value λ_1, the electron-phonon
interaction is weak and the system is a metal. For moderate λ such that
$\lambda_2 > \lambda > \lambda_1$, the phonon mediated attractive electron-electron interaction can
overcome the repulsive interaction between electrons. The system is a BCS
superconductor.[3] However, for very strong electron-phonon interaction, the
oxide compound becomes a bipolaronic insulator in which the bipolarons are
locally trapped and can be regarded as nearest neighboring electron pairs
with opposite spins coupled with lattice deformations. These bipolarons
have been considered[2] as localized Cooper pairs in the limit of large λ.
The formation of localized bipolarons is favored if the electrons come from
a narrow d band of a metal ion. These pre-existing localized Cooper pairs
do not contribute to superconductivity. In the crystal there may be conduction electrons with a wide band in presence. If a hybridization term
exists between the wide band and the localized electrons, the localized
Cooper pairs can be transferred to and become delocalized in the conduction
band via the proximity effect.[4] With proper concentration of bipolarons

and other parameters, the superconductivity transition temperature T_c of the conduction electrons can be dramatically enhanced.

When the on-site repulsive Coulomb interaction u between two electrons with opposite spins is weak, the formation of an on-site bipolaron is possible. Each of the bipolaron sites can be regarded as a negative U-center, and its influence on the superconducting transition temperature of a wide band electron has been studied previously.[5,6] In this paper we shall assume u to be large; the bipolaron can only be formed on the neighboring metal ion sites.[2] In Section II, the effective Hamiltonian of the system will be given. In Section III the Nambu's formulation of superconductivity will be employed to study the transition temperature, order parameters and energy gap of the system. In Section IV, thermodynamic property of the superconductor, including the specific heat jump at T_c will be studied. In the final section we discuss the obtained results and their possible connection with the recently discovered high T_c oxide compounds, particularly $La_{2-x}(Ba,Sr)_xCuO_4$-type compounds.[7-10]

EFFECTIVE HAMILTONIAN

A. Normal State

The system under consideration is an oxide compound and consists of the following compositions: the localized bipolarons are made of electrons from a narrow d-band, and a wide band of "s" electrons which may in fact come from s, p or other d-orbitals. The "s" electrons of the wide band have a superconducting critical temperature T_{co} and are described by the following BCS Hamiltonian[3]:

$$H_s = \sum_{k,\sigma} \varepsilon_k \, a_{k\sigma}^+ a_{k\sigma} - g\sum_{k,k'} a_{k\uparrow}^+ a_{-k\downarrow}^+ a_{k'\downarrow} a_{-k'\uparrow} \cdot \tag{1}$$

ε_k is the kinetic energy of the "s" electron and $-g$ is the phonon mediated electron-electron interaction. $a_{k\sigma}^+$ and $a_{k\sigma}$ are the conduction electron operators. The narrow d band electrons are strongly interacting with local phonons. By using Holstein's transformation[11], the Hamiltonian for d electrons can be written as

$$H_d = \sum_{i,\sigma} \varepsilon_d (n_{ia,\sigma} + n_{ib,-\sigma}) + U \sum_{i,\sigma} n_{ia,\sigma} n_{ib,-\sigma} \quad , \tag{2}$$

$$\varepsilon_d = \varepsilon_0 - \lambda^2 \sum_q \frac{1}{\omega_q} \quad ,$$

$$U = U_{ab} - J - 2\lambda^2 \sum_q \frac{1}{\omega_q} e^{i\vec{q}\cdot(\vec{a}-\vec{b})} \quad ,$$

where $n_{ia,\sigma} = d_{ia\sigma}^+ d_{ia\sigma}$ is the occupation number of the d or localized electron at the a-site of the i-th unit cell with spin σ. $\vec{a}$ and $\vec{b}$ represent the two nearest neighboring positions of metal-ions in i-th bipolaron site (or unit cell). ε_0 is the unperturbed single electron energy of the d electron, and ε_d is regarded as the energy of a locally trapped polaron. ω_q is the local phonon frequency. U_{ab} is the repulsive Coulomb interaction between electrons on the nearest neighbors, and its value is reduced by the screening due to the wide band electrons. J is the superexchange or antiferromagnetic coupling[2,12] between electrons on sites a and b via the p-orbital of an in-between oxygen atom. Here a large on-site Coulomb repulsive interaction u is implicitly assumed. For large electron-phonon interaction λ, the renormalized hopping term T_{ij} for a narrow d-band polaron[11] can be extremely small and is thus neglected in Eq. (2). When U<0, Eq. (2) can be used to describe the formation of a localized bipolaron or Cooper pair. The Hamiltonian which describes the hybridization of the s and d electrons can be written as

$$H_{sd} = \sum_j \sum_{k,\sigma} \left[V_{kd} a_{k\sigma}^+ (d_{ja\sigma} + d_{jb\sigma}) e^{i\vec{k}\cdot\vec{R}_j} + c.c. \right] , \tag{3}$$

V_{kd} is the mixing term between the localized and extended electrons and its value is also much reduced by λ. For the sake of simplicity we have taken $V_{kda} = V_{kdb} = V_{kd}$, here V_{kda} and V_{kdb} are respectively the mixing terms between the extended electrons with localized electrons at a-site and b-site in the j-th unit cell. With this approximation, the mathematical structure of the present Hamiltonian

$$H = H_s + H_d + H_{sd} , \tag{4}$$

becomes similar to that of negative U centers in Ref. 6. The summation over j in Eqs. (2) and (3) is extended over all the localized bipolaron sites, which are randomly distributed and most likely are pinned near the (Ba,Sr) sites in the case of $La_{2-x}(Ba,Sr)_x CuO_4$ compound.

B. Superconducting State

Let us introduce the superconducting order parameter Δ for the s band electrons and local bipolaron order parameter Δ_d:

$$\Delta = g \sum_k \langle a_{k\uparrow} a_{-k\downarrow} \rangle , \tag{5}$$

$$\Delta_d = -U \langle d_{ja\uparrow} d_{jb\downarrow} \rangle , \tag{6}$$

with $\langle d_{ja\downarrow} d_{jb\uparrow} \rangle = \langle d_{ja\uparrow} d_{jb\downarrow} \rangle$. In the self-consistent mean field approximation, the effective Hamiltonians for H_s and H_d of Eq. (4) in superconducting state reduce to

$$H_s = \sum_{k,\sigma} \varepsilon_k a_{k\sigma}^+ a_{k\sigma} - \sum_k (\Delta a_{k\uparrow}^+ a_{-k\downarrow}^+ + c.c.) + \frac{\Delta^2}{g} , \tag{7}$$

$$H_d = \sum_{j,\sigma} E_{d\sigma} (d_{ja\sigma}^+ d_{ja\sigma} + d_{jb\sigma}^+ d_{jb\sigma}) - \sum_{j,\sigma} (\Delta_d d_{ja\sigma}^+ d_{jb-\sigma}^+ + c.c.) , \tag{8}$$

where $E_{d\sigma} = \varepsilon_d + U \langle n_{ja-\sigma} \rangle$ and $\langle n_{jb-\sigma} \rangle = \langle n_{jb\sigma} \rangle = \langle n_{ja\sigma} \rangle = \langle n_{ja-\sigma} \rangle$. When a-site and b-site in the j-th unit cell reduce to a single site, it corresponds to

the negative U-center discussed by Anderson.[1] The negative U-centers not
only can be formed from the local lattice deformations,[1,5] but also may be
originated from local exciton centers.[13,14] The enhancement of T_c due to
these negative U centers has been studied by using the various mean field
theory[5,6] and the Monte Carlo simulation.[14]

NAMBU'S FORMALISM

The Green's function for the localized electrons in a bipolaron site
can be represented by a matrix

$$\hat{G}_d(a,b,t) = -i \begin{pmatrix} \langle T(d_{ja\uparrow}(t)d^+_{ja\uparrow}(0))\rangle \, , & \langle T(d_{ja\uparrow}(t)d_{jb\downarrow}(0))\rangle \\[12pt] \langle T(d^+_{jb\downarrow}(t)d^+_{ja\uparrow}(0))\rangle \, , & \langle T(d^+_{jb\downarrow}(t)d_{jb\downarrow}(0))\rangle \end{pmatrix} , \quad (9)$$

where $\langle\ldots\ldots\rangle$ is the statistical average over the grand canonical
ensemble defined by the effective Hamiltonian in superconducting state.
Similarly, one can define the s electron Green's function matrix $\hat{G}_s(k,t)$.
Its explicit expression in Nambu's formulation of superconductivity[15] is
well known and needs not be repeated here. The Fourier transforms of
$\hat{G}_d(a,b,t)$ and $\hat{G}_s(k,t)$ in ω space can be represented as
$\hat{G}_d(\omega) = \hat{G}_d(a,b,\omega) = \hat{G}_d(b,a,\omega)$ and $\hat{G}_s(k,\omega)$

$$\hat{G}_s^{-1}(k,\omega) = \omega Z_s(\omega)\,\hat{\tau}_o - \varepsilon_k\hat{\tau}_3 - \Phi_s(\omega)\hat{\tau}_1 \, , \quad (10)$$

$$\hat{G}_d^{-1}(\omega) = \omega Z_d(\omega)\hat{\tau}_o - E_d\hat{\tau}_3 - \Phi_d(\omega)\hat{\tau}_1 \, , \quad (11)$$

with $\hat{\tau}_1=\hat{\sigma}_x$, $\hat{\tau}_3=\hat{\sigma}_z$ as the Pauli matrices and $\hat{\tau}_o$ as a 2 x 2 unit matrix.
$Z_s(\omega)$, $Z_d(\omega)$, $\Phi_s(\omega)$ and $\Phi_d(\omega)$ in Eqs. (10) and (11) need to be obtained
from the self-energy diagrams. When the hybridization H_{sd} is not con-
sidered, $Z_s(\omega) = Z_d(\omega) = 1$, $\Phi_s(\omega) = \Delta$ and $\Phi_d(\omega) = \Delta_d$. It is interesting to
note that under this condition ($H_{sd}=0$) and from Eq. (6), the order para-
meter Δ_d of the localized Cooper pair or bipolaron satisfies the following
equation

$$\Delta_d = -UT \sum_n \frac{\Delta_d}{\omega_n^2 + E_d^2 + \Delta_d^2} \, , \quad (12)$$

with $\omega_n = (2n+1)\pi T$. The summation over n can be easily carried out, Eq.
(12) becomes

$$1 = -\frac{U}{2\sqrt{E_d^2 + \Delta_d^2}} \tanh\left(\frac{\sqrt{E_d^2 + \Delta_d^2}}{2T}\right) . \quad (13)$$

The transition temperature T_b for which the localized bipolaron begins to
form can be obtained by putting $\Delta_d=0$. For $E_d \ll T_b$, $T_b \simeq -U/4$. The value of
T_b may reach 500K if $-U \approx 0.2$ eV. On the other hand, the superconducting

transition temperature T_{co} of the s electron is much smaller than T_b. However, when the hybridization term H_{sd} of Eq. (3) is turned on, we shall show that the onset temperatures for the bipolaron formation and the superconducting state of the s-electrons begin at the same temperature T_c and the value of T_c lies between T_{co} and T_b.

In the presence of H_{sd}, the self energies $\Sigma_s(\omega)$ and $\Sigma_d(\omega)$ for the s electrons and the localized d electrons due to H_{ds} need to be considered. They are derived as

$$\Sigma_s(\omega) = 2N_I \left| V_{kd} \right|^2 \hat{\tau}_3 \hat{G}_d(\omega) \hat{\tau}_3 \ , \tag{14}$$

$$\Sigma_d(\omega) = \sum_k \left| V_{kd} \right|^2 \hat{\tau}_3 \hat{G}_s(k,\omega) \hat{\tau}_3 \ . \tag{15}$$

N_I is the number of localized bipolarons per unit volume. Using the above equations, $Z(\omega)$ and $\Phi(\omega)$ in Eqs. (10) and (11) can be shown to satisfy the following self-consistent equations:

$$Z_s(\omega) = 1 + \left(\frac{2n_I\Gamma}{\pi N(0)}\right) \frac{Z_d(\omega)}{B(\omega)} \quad , \quad \Phi_s(\omega) = \Delta + \left(\frac{2n_I\Gamma}{\pi N(0)}\right) \frac{\Phi_d(\omega)}{B(\omega)} \quad , \tag{16}$$

$$Z_d(\omega) = 1 + \frac{\Gamma}{\pi} \int d\xi \frac{Z_s(\omega)}{A(\xi,\omega)} \quad , \quad \Phi_d(\omega) = \Delta_d + \frac{\Gamma}{\pi} \int d\xi \frac{\Phi_s(w)}{A(\xi,\omega)} \quad , \tag{17}$$

where $B(\omega) = -\omega^2 Z_d^2(\omega) + \Phi_d^2(\omega) + E_d^2$, $A(\xi,\omega) = -\omega^2 Z_s^2(\omega) + \Phi_s^2(\omega) + \xi^2$, and $\Gamma = \pi N(0) \langle \left| V_{kd} \right|^2 \rangle$ is the lifetime broadening of the d electrons, $n_I = N_I/N_s$ with N_s as the number of s electrons per unit volume, $N(0)$ is the density of states for s electron at the Fermi level. The above equations are similar to those obtained by Kiwi and Zuckermann[16] for magnetic impurities with positive U. The gap functions for s and localized d electrons are defined as

$$\Delta(\omega) = \frac{\Phi_s(\omega)}{Z_s(\omega)} = \frac{\Delta + n_I \alpha(\omega)\Delta_d}{1 + n_I \alpha(\omega)} \ , \tag{18}$$

$$\Delta_d(\omega) = \frac{\Phi_d(\omega)}{Z_d(\omega)} = \frac{\Delta_d \sqrt{\Delta^2(\omega) - \omega^2} + \Gamma\Delta(\omega)}{\sqrt{\Delta^2(\omega) - \omega^2} + \Gamma} \ , \tag{19}$$

with $\alpha(\omega) = 2\Gamma/\pi N(0) B(\omega)$. The gap function $\Delta(\omega)\left(\Delta_d(\omega)\right)$ does not equal to the order parameter $\Delta(\Delta_d)$ except for $\Gamma=0$. They, in general, are complex quantities and can be self-consistently solved from Eqs. (18) and (19). The order parameters Δ and Δ_d on the other hand should be obtained from Eqs. (5) and (6).

A. Superconducting transition temperature T_c

At finite temperature T, the frequency ω which appeared in previous equations needs to be replaced by $\omega = i\omega_n$ with $\omega_n = (2n+1)\pi T$ and n is an integer. For example, $\Delta(\omega = i\omega_n) = \Delta(\omega_n)$. The order parameters Δ and

Δ_d according to Eqs. (5) and (6) can be shown to have the following forms:

$$\Delta = 2gN(0)T\sum_n \frac{\Delta(\omega_n)}{\sqrt{\omega_n^2+\Delta^2(\omega_n)}} \tan^{-1} \frac{\omega_D}{\sqrt{\omega_n^2+\Delta^2(\omega_n)}} \quad , \tag{20}$$

$$\Delta_d = -UT \sum_n \frac{\Phi_d(\omega_n)}{B(\omega_n)} = -U_{eff}\, \Gamma T \sum_n \frac{\Delta(\omega_n)}{B(\omega_n)\sqrt{\omega_n^2+\Delta^2(\omega_n)}} \quad , \tag{21}$$

and

$$U_{eff} = U/\left(1+UT \sum_n 1/B(\omega_n)\right) \quad . \tag{22}$$

ω_D here is the cut-off energy or Debye frequency in BCS theory. In deriving Eq. (21) we have assumed that the band width of the s electrons is much larger than T_c and $\Delta(\omega_n)$. Near the transition temperature T_c, both Δ and Δ_d are small. Eqs. (20) and (21) are expanded to the linear order in Δ and Δ_d, and we obtain

$$(1-a)\Delta - b\Delta_d = 0 \quad \text{and} \quad c\Delta - (1-d)\Delta_d = 0 \quad , \tag{23}$$

where

$$a = 4gN(0)\, T_c \sum_{n>0} \frac{1}{\omega_n\left(1+n_I\alpha(w_n)\right)} \tan^{-1}(\omega_D/\omega_n) \quad , \tag{24}$$

$$b = 4gN(0)T_c \sum_{n>0} \frac{n_I\alpha(\omega_n)}{\omega_n\left(1+n\,\alpha(\omega_n)\right)} \tan^{-1}(\omega_D/\omega_n) \quad , \tag{25}$$

$$c = -2U_{eff}\,\Gamma T_c \sum_{n>0} \frac{1}{\omega_n B(\omega_n)\left(1+n_I\alpha(\omega_n)\right)} \quad , \tag{26}$$

$$d = -2U_{eff}\,\Gamma T_c \sum_{n>0} \frac{n_I\alpha(\omega_n)}{\omega_n B(\omega_n)\left(1+n_I\alpha(\omega_n)\right)} \quad . \tag{27}$$

It is understood that, in the coefficients a, b, c and d, the order parameters Δ and Δ_d are put to zero. T_c can easily be obtained from Eq. (23) or

$$(1-a)(1-d) - bc = 0 \quad . \tag{28}$$

When $n_I=0$, $b=d=0$. Eq. (28) yields the result from BCS theory. For $n_I\neq0$, the superconducting transition temperature T_c has already been discussed[6] in various limits and can be determined numerically from the above equation. In Fig. 1 we display the variation of T_c as a function of E_d/Γ (solid curves) for several different values of U. The pure phonon mediated BCS transition temperature is chosen to be $T_{co}\approx10K$ (the dashed curve). For larger values of $-U$, T_c is enhanced drastically as compared with T_{co}. For small values of $-U$, T_c is depressed. The randomly distributed bipolaron centers have two effects.[6] One corresponds to the attractive interaction between conduction electrons mediated by the localized centers which

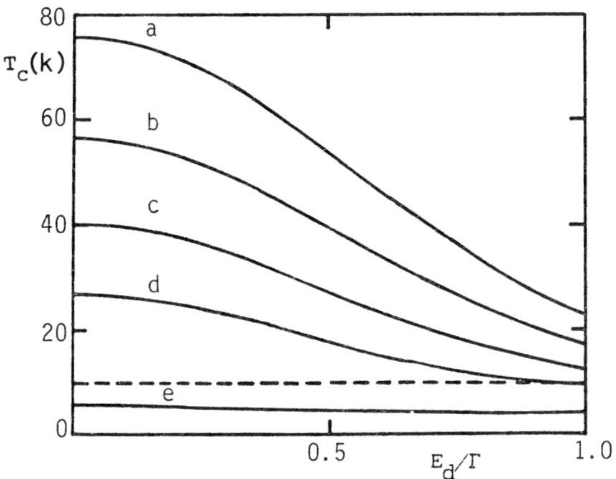

Fig. 1. Superconducting transition temperature T_c as a function
of E_d/Γ for several values of U, with $\Gamma=0.1$ eV,
$N(0)=0.15$ eV^{-1} and $n_I=0.025$. The values of U for curves
a, b, c, d and e respectively are -0.16, -0.14, -0.12,
$-.10$ and -0.05 eV. The dashed line corresponds to the
BCS transition temperature without U-center ($n_I=0$).

enhances T_c. The second effect corresponds to the resonance scattering of
conduction electrons from these centers, which decreases T_c. Therefore,
for large $-U$ the first effect is dominant, and for small $-U$ the second
effect may become important.

B. <u>The energy gap Δg and density of states of the extended and the
localized electrons.</u>

From Eqs.(18), (20) and (21), the order parameters Δ and Δ_d have only
real values and can be self consistently solved with the following
expression for $B(\omega_n)$

$$B(\omega_n) = \omega_n^2 + \Gamma^2 + E_d^2 + \Delta_d^2 + \frac{2\Gamma[\omega_n^2 + \Delta_d\Delta(\omega_n)]}{\sqrt{\Delta^2(\omega_n)+\omega_n^2}} . \tag{29}$$

When Δ and Δ_d are known, the complex gap functions $\Delta(\omega)$ and $\Delta_d(\omega)$ can be
obtained from Eqs. (18) and (19). It is easy to see that as $\omega < \Delta(\omega)$, $\Delta(\omega)$
and $\Delta_d(\omega)$ have only real roots. Otherwise, complex roots appear. The
energy gap Δ_g is defined[17] as the value of $\omega=\Delta_g$ such that complex values
begin to appear in $\Delta(\omega)$ and $\Delta_d(\omega)$. In Fig. 2, we show the real and imagi-
nary parts of $\Delta(\omega)$ and $\Delta_d(\omega)$ at T=0K and $E_d=0$ as functions of ω. For $\omega \geqslant \Delta_g$,
the imaginary parts of $\Delta(\omega)$ and $\Delta_d(\omega)$ become nonzero, reflecting the decay
processes due to the hybridization term. $\Delta(\omega)$, to a good approximation,
is real and constant except for ω near Δ_g. The absolute value of $Im\Delta_d(\omega)$
becomes large as ω increases, which indicates that the localized bipolaron
is unstable in the large ω limit.

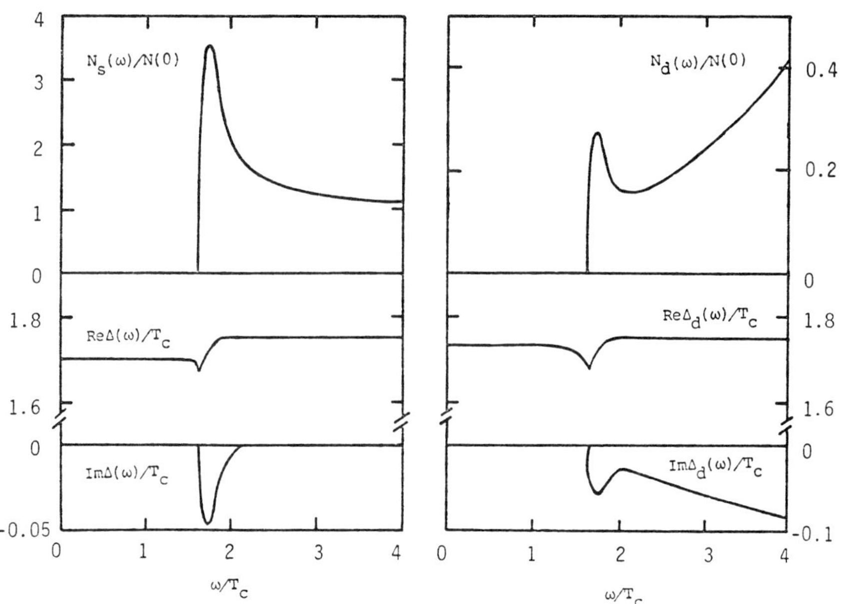

Fig. 2. Real and imaginary parts of $\Delta(\omega)$ and $\Delta_d(\omega)$, and electronic densities of state for s and d electrons with $E_d=0$, $U=-0.12$ and $T_c=40$ K.

According to the Green's function defined in Eqs. (10 and (11), the densities of state spectra for the wide band s and localized d electrons at T=0K are given by

$$N_s(\omega) = N(0) \ \mathrm{Re} \left[\frac{\omega}{\sqrt{\omega^2 - \Delta^2(\omega)}} \right] \ , \tag{30}$$

$$N_d(\omega) = \frac{1}{\pi} \ \mathrm{Im} \left[\frac{\omega Z_d(\omega) + E_d}{B(\omega)} \right] \ . \tag{31}$$

Fig. 2 gives the typical result of $N_s(\omega)$ and $N_d(\omega)$ as functions of ω with $E_d=0$. It is interesting to note that the energy gap Δ_g in the excitation spectrum is the same for both s and d electrons regardless of the value for E_d. $N_s(\omega)$ has a large peak at ω near Δg, while $N_d(\omega)$ at $\omega \simeq \Delta_g$ has a small peak, and its main peak is near $\omega \simeq (E_d^2 + \Gamma^2 + \Delta_d^2)^{1/2}$. In Fig. 3A, we show the value of Δ_g/T_c at T=0K and $E_d=0$ as a function of T_c. There the bipolaron concentration $n_I=0.025$ is fixed. T_c increases as U increases. In general the value of Δ_g/T_c is smaller than the value of 1.76 derived from the BCS theory.[3] This result seems to be consistent with the gap of $La_{2-x}Sr_xCuO_4$ measured by far-infared absorption.[18] In Fig. 3B the energy gap Δ_g has been plotted as a function of T with $E_d=0$ for one set of parameters. The general behavior $\Delta_g(T)$ is similar to that BCS result, but with $\Delta_g(0)$ being considerably smaller than that of BCS theory. In the same figure Δ_d and Δ are also plotted as functions of T. It is easy to see that $\Delta_d > \Delta_g > \Delta$.

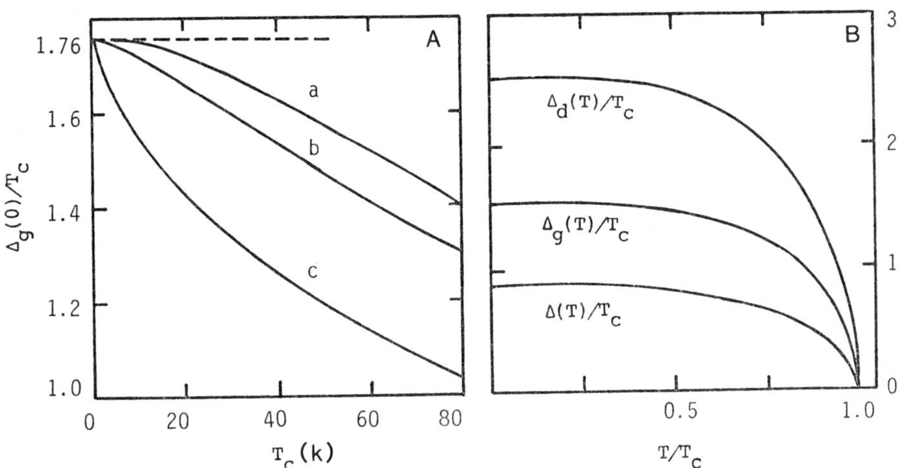

Fig. 3.   (A) $\Delta_g(0)/T_c$ vs. T_c for several values of BCS transition temperature T_{co}: 10 K (a), 2.5 K (b) and $T_{co}=0$ (c). The dashed line corresponds to $n_I=0$. (B) Curves showing Δ_d/T_c, Δ_g/T_c and Δ/T_c vs. T/T_c with $U=-0.14$ eV, $E_d=0$, $T_c=56.4$ K, and other parameters being the same as those in Fig. 1.

THERMODYNAMICS NEAR T_c

To find the difference in thermodynamic potential between the superconducting phase and the normal state, the following relation is employed

$$\Omega_s - \Omega_n = \int_0^\Delta \frac{\partial g^{-1}}{\partial \Delta} \Delta^2 \, d\Delta + 2n_I \int_0^{\Delta_d} \frac{\partial |U|^{-1}}{\partial \Delta_d} \Delta_d^2 d\Delta_d \ . \tag{32}$$

The first term on the right-hand side of the above equation is well known in BCS theory. The second term which comes from the localized bipolaron order parameter can be derived by a similar procedure. When $(T_c-T)/T_c\ll1$ and the order parameters Δ and Δ_d are small, we can expand the right hand side of Eqs. (20) and (21) to the third order in Δ and Δ_d respectively.

$$\Delta = g\,[S_1(T)\Delta + S_2(T)\Delta^3], \tag{33}$$
$$\Delta_d = -U[D_1(T)\Delta_d + D_2(T)\Delta_d^3] \ , \tag{34}$$

where the coefficients $S_1(T)$, $S_2(T)$, $D_1(T)$ and $D_2(T)$ are listed in the Appendix. Using Eq. (28) it is straightforward to show that

$$1 - gS_1(T_c) = 1 - a - \frac{bc}{1-d} = 0 \ , \tag{35}$$

$$1 - |U|\,D_1(T_c) = \left(1 - d - \frac{bc}{1-a}\right)\left(1 + UT_c \sum_n 1/B(\omega_n)\right) = 0 \ . \tag{36}$$

$B(\omega_n)$ here is given by Eq. (29) with $\Delta=\Delta_d=0$ and $T=T_c$, a, b, c and d are defined in Eqs. (24) to (27). From the above relations and Eqs. (33) and (34), Δ and Δ_d near $T<T_c$ can be written as

$$\Delta = \left[\frac{1}{S_2(T_c)} \frac{dS_1(T_c)}{dT_c}\right]^{1/2} (T_c-T)^{1/2} \ , \tag{37}$$

$$\Delta_d = \left[\frac{1}{D_2(T_c)} \frac{dD_1(T_c)}{dT_c}\right]^{1/2} (T_c-T)^{1/2} . \tag{38}$$

The behavior of Δ and $\Delta_d \propto (T_c-T)^{1/2}$ as $T \rightarrow T_c$ is identical to that of BCS theory. If we define

$$-K = \frac{1}{S_2(T_c)} \left[\frac{dS_1(T_c)}{dT_c}\right]^2 + 2n_I \frac{1}{D_2(T_c)} \left[\frac{dD_1(T_c)}{dT_c}\right]^2 \ , \tag{39}$$

the thermodynamic potential difference of Eq. (32) near $T \simeq T_c$ can be derived as $\Omega_s - \Omega_n = -K(T_c-T)^2/2$. From which the specific heat jump at T_c is obtained as

$$\Delta C = \left[-T \frac{d^2}{dT^2} (\Omega_s - \Omega_n)\right]_{T=T_c} = KT_c \ . \tag{40}$$

In Fig. 4, the quantity $\Delta C/\alpha T_c$ is plotted as a function of n_I for several different values of U. $\alpha = 2\pi^2 N(0)/3$ is the coefficient of the linear T specific heat in the normal state. It is interesting to note that the calculated $\Delta C/\alpha T_c$ is in general higher than the BCS value 1.43. For $n_I \ll 1$, the contribution to $\Delta C/\alpha T_c$ comes mainly from the wide band electrons, and the value of $\Delta C/\alpha T_c$ is closer to that of BCS theory. When n_I increases, the contribution due to the localized bipolarons becomes important so that $\Delta C/\alpha T_c$ can be larger than the BCS value. This result seems to be consistent with the experimental measurements of $La_{2-x}Sr_xCuO_4$ with x=0.15.[19]

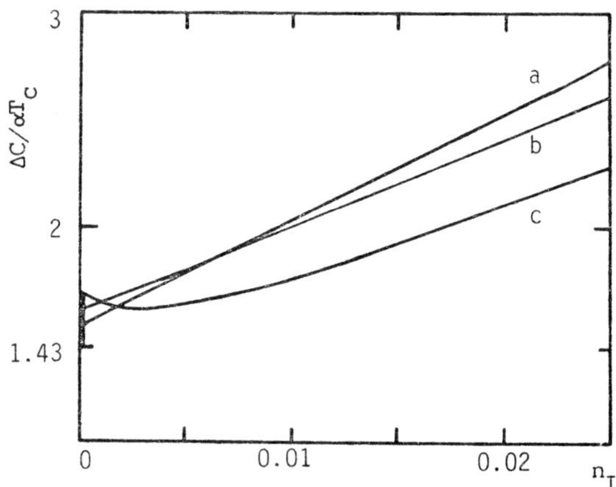

Fig. 4. The ratio of the electronic specific heat jump ΔC to αT_c as a function of n_I for several values of U: -0.06 (a), -0.1 (b) and -0.14 eV (c).

DISCUSSION

In this section we wish to make a possible connection of the present model to $La_{2-x}(Ba,Sr)_xCuO_4$ type compounds. In these systems, the electron phonon has been shown to be rather strong.[20] The current carriers come from the extra holes in the Cu^{+3} ions as compared with the Cu^{+2} background ions. The wide band carriers come from $3d_{x^2-y^2}$ orbitals, and the narrow band carriers are from $3d_{z^2}$ orbitals.[21] With small n_I, the localized bipolarons are expected to be distributed sparsely and randomly among pairs of neighboring Cu ions. Comparing with the host La ions, the dopant Ba (Sr) can be regarded as an impurity carrying a negative charge. It is likely that the bipolaron which is made of a pair of holes with opposite spin may be pinned near one of the negatively charged Ba (Sr) sites. Our numerical value for Δ_g/T_c is, in general, less than that of BCS value 1.76. This result seems to agree with the infared measurements.[18] However, it does not agree with the result from the tunneling experiments.[22] The former[18] measures the averaged bulk property of the sample, while the result of the latter experiment[22] may critically depend on the sample's surface condition. However, for large n_I ($n_I \lesssim 0.2$), the value of Δ_g/T_c can become slightly larger than that of BCS theory. The specific heat jump at T_c has also been calculated. The obtained value for $\Delta c/\alpha T_c$ is usually larger than that of BCS value 1.43. This result is consistent with the specific heat measurements[19] of $La_{2-x}Sr_xCuO_4$ with x=0.15. As was mentioned in Section II, the present model is almost identical to that of the negative U centers studied in Ref. 6. The contribution to the strength of $-U$ not only can come from the local lattice deformations, but may also be from the local exciton centers[13] which are results due to impurities[14] and charge fluctuations[23,24] between the same and different host ions. The results of the present paper are based upon a mean field theory in Nambu's formulation of superconductivity. The effect due to the localized bipolarons is treated on the basis of self-consistent Born approximation, which is valid only for $n_I \ll 1$. For larger values of n_I, the standard coherent potential approximation should be applied to study the problem. Among various mechanisms[25] proposed for the high T_c oxide compounds, the effect due to oxygen defects (vacancies) on the Cu basal plane has not been carefully investigated. Whether these oxygen vacancies could act as a source of large local deformations or exciton centers needs to be answered. It is apparent that further theoretical work in this area is necessary in order to explain the various properties associated with these oxide superconductors, particularly $YBa_2Cu_3O_7$-type compounds[26] with T_c above 90K.

ACKNOWLEDGEMENT

We would like to thank C. W. Chu, P. H. Hor, C. Y. Huang, W. Y. Lai, K. L. Ngai, W. P. Su and M. K. Wu for helpful discussions and useful comments. This work is partly supported by a grant from the Texas Advanced Technological Research Program and supplemented by the Space Vacuum Epitaxy Center at the University of Houston via a NASA grant.

APPENDIX: Expressions for $D_1(T)$, $D_2(T)$, $S_1(T)$ and $S_2(T)$

We expand the right-hand side of Eqs. (20) and (21) to the third order in Δ and Δ_d, respectively. The procedure is simple but tedious. The coefficients in Eqs. (32) and (33) are given below.

$$D_1(T) = 2T \sum_{n \geqslant 0} \left[1 + \Gamma E(\omega_n)/\omega_n\right]/B_0(\omega_n) \ , \tag{A1}$$

$$D_2(T) = 2T \sum_{n \geqslant 0} \left[(1 + \Gamma E(\omega_n)/\omega_n)B_1(\omega_n)/B_0(\omega_n)^2 \right.$$
$$\left. - \Gamma\left(F(\omega_n)-E(\omega_n)^3/2\right)/\left(B_0(\omega_n)\,\omega_n^3\right)\right] \ , \tag{A2}$$

$$S_1(T) = 4N(0)T \sum_{n \geqslant 0} \left[\left(\xi(T)E(\omega_n)/\omega_n\right)\tan^{-1}\left(\omega_D/\omega_n\right) \ , \tag{A3}$$

$$S_2(T) = 4N(0)T \sum_{n \geqslant 0} \left[\frac{H(\omega_n) - \xi(T)^3 E(\omega_n)^3/2}{\omega_n^3}\right]\tan^{-1}(\omega_D/\omega_n)$$
$$- \frac{\omega_D \xi(T)^3 E(\omega_n)^3}{2\omega_n^2(\omega_n^2+\omega_D^2)}\right] \ , \tag{A4}$$

where

$$E(\omega_n) = 1 + \gamma(\omega_n)\left(1-\xi(T)\right)/\xi(T) \ , \tag{A5}$$

$$F(\omega_n) = \omega_n^2\left[\left(1-\gamma(\omega_n)\right)\left(E(\omega_n)-1\right)B_1(\omega_n)/B_0(\omega_n) + \gamma(\omega_n)\eta(T)\right] \ , \tag{A6}$$

$$H(\omega_n) = F(\omega_n)\xi(T)^3 - \omega_n^2 E(\omega_n)\xi(T)^4 \eta(T) \ , \tag{A7}$$

with

$$B_0(\omega_n) = \left(\omega_n + \Gamma\right)^2 + E_d^2 \ , \tag{A8}$$

$$B_1(\omega_n) = 1 + \left(\Gamma/\omega_n\right)E(\omega_n)\left(2 - E(\omega_n)\right) \ , \tag{A9}$$

$$\gamma(\omega_n) = 1/\left[1+n_I\alpha_0(\omega_n)\right] \ , \tag{A10}$$

$$\alpha_0(\omega_n) = 2\Gamma/\pi N(0)B_0(\omega_n) \ , \tag{A11}$$

$$\xi(T) = c/(1-d) \ , \tag{A12}$$

$$\eta(T) = \frac{4gN(o)T}{1-a} \sum_{n > 0} \left\{\left[\frac{\left(1-\gamma(\omega_n)\right)B_1(\omega_n)}{\omega_n B_0(\omega_n)}\left(E(\omega_n)-1\right) - \frac{E(\omega_n)^3}{2\omega_n^3}\right]\tan^{-1}\left(\frac{\omega_D}{\omega_n}\right)\right.$$
$$\left. - \frac{E(\omega_n)^3\omega_D}{2\omega_n^2(\omega_n^2+\omega_D^2)}\right\} \ . \tag{A13}$$

Here c, d and a in Eqs. (A12) and (A13) have been given by Eqs. (24) to (27).

REFERENCES

1. P. W. Anderson, _Phys. Rev. Lett._ 34:953 (1975).
2. B. K. Chakraverty, _J. Physique Lett._ 40:L-99 (1979) and _J. Physique_ 42:1351 (1981).
3. J. Bardeen, L. N. Cooper and J. R. Schrieffer, _Phys. Rev._ 108:1175 (1957).
4. W. L. McMillian, _Phys. Rev._ 175:537 (1968).
5. E. Simanek, _Solid State Comm._ 32:731 (1979).
6. C. S. Ting, D. N. Talwar and K. L. Ngai, _Phys. Rev. Lett._ 45:1213 (1980).
7. J. G. Bednorz and K. A. Muller, _Z. Phys._ B64:189 (1986).
8. C. W. Chu, P. H. Hor, R. L. Meng, L. Gao, Z. L. Huang and Y. Q. Wang, _Phys. Rev, Lett._ 58:405 (1987).
9. R. J. Cava, R. S. Van Dover, B. Batlogg and E. A. Reitman, _Phys. Rev. Lett._ 58:408 (1987).
10. Z. Zhao, L. Chen, C. Cui, Y. Huang, J. Liu, G. Chen, S. Lu, S. Gao and Y. He, to appear in _Science Report_, China (1987).
11. T. Holstein, _Ann. Phys._ 8:325, 343 (1959).
12. P. W. Anderson, _Science_ 235:1196 (1987).
13. B. D. Nguyen and E. Simanek, _Solid State Comm._ 42:671 (1982).
14. H. B. Schüttler, M. Jarrell and D. J. Scalapino, _Phys. Rev. Lett._ 58:1147 (1987).
15. Y. Nambu, _Phys. Rev._ 117:648 (1960).
16. M. Kiwi and M. J. Zuckermann, _Phys. Rev._ 164:548 (1967).
17. A. A. Abrikosov, L. P. Gorkov, _Zh. Eksperim. i Teor. Fiz._ 39:1781 (1961) and _Soviet Phys.-JETP_ 12:1243 (1961).
18. U. Walter, M. S. Sherwin, A. Stacy, P. L. Richards and A. Zettl, _Phys. Rev._ B35:5327 (1987); P. E. Sulewski, A. J. Sievers, S. E. Russek, H. D. Hallen, D. K. Lathrop and R. A. Buhrman, _Phys. Rev._ B 35:5330 (1987); Z. Schlesinger, R. L. Greene, J. G. Bednorz and K. A. Müller, _Phys. Rev._ B 35:5334 (1987).
19. B. D. Dunlap, M. V. Nevitt, M. Slaski, T. E. Klippert, Z. Sungaila, A. G. McKale, D. W. Capone, R. B. Poeppel and B. K. Flandermeyer, _Phys. Rev._ B35:7210 (1987).
20. W. Weber, _Phys. Rev. Lett._ 58:1371 (1987).
21. L. F. Mattheiss, _Phys. Rev. Lett._ 58:1028 (1987).
22. J. R. Kirtley, C. C. Tsuei, Sung I. Park, C. C. Chi, J. Rozen and M. W. Shafer, _Phys. Rev._ B35:7216 (1987).
23. C. M. Varma, S. Schmitt-Rink, and E. Abraham, preprint.
24. C. L. Fu and A. J. Freeman, preprint.
25. V. Z. Kresin, (preprint), and Refs. 5, 12, 13, 20, 23 and 24.
26. M. K. Wu, J. R. Ashburn, C. J. Torng, P. H. Hor, R. L. Meng, L. Gao, Z. J. Huang, Y. Q. Wang and C. W. Chu, _Phys. Rev. Lett._ 58:908 (1987).

DILUTE FERMI LIQUID OF HEAVY POLARONS IN COPPER OXIDE SUPERCONDUCTORS

R.B. Laughlin

Department of Physics
Stanford University
Stanford, California 94305
Lawrence Livermore National Laboratory
University of California
P.O. Box 808, Livermore, California 94550

C.B. Hanna

Department of Physics
Stanford University
Stanford, California 94305

In this paper we point out that a large body of data on the newly discovered copper oxide superconductors points to the existence in these materials of a new type of dilute fermi liquid composed of heavy polarons, which we believe acquire their mass magnetically. Each polaron is a hole doped into the host material, a Mott insulator, self-trapped in a well, much the way a hole added to an alkali halide is trapped. The polaron is mobile, has a mass m^* of approximately 6 electron masses, and has an ionization energy, an upper limit on its binding energy, of approximately 0.15 eV. The fermi energy of the polaron liquid in $La_{2-x}Sr_xCuO_4$ with $x = 0.15$ is roughly 80 meV. The smallness of this energy is unprecedented in the phenomenology of superconductivity and is probably the key to understanding these materials.

Stoichoimetric La_2CuO_4 is a magnetic insulator[1], despite the fact that local density calculations[2] find it to be a metal with a large fermi velocity. This behavior is seen in a class of materials[3] exemplified by NiO and is our practical definition of a Mott insulator. Like NiO, La_2CuO_4 is antiferromagnetically ordered[4] at low temperatures. While the precise definition of a Mott insulator in solid state physics is somewhat controversial, the basic concept is not. One usually imagines hydrogen atoms on a cubic lattice. If the near-neighbor distance in this lattice is comparable to the bohr radius, it acts like an alkali metal. If this distance is one meter, on the other hand, it acts like isolated hydrogen atoms, and in particular does not conduct electricity. The difference between these two extremes is the relative size of T, the Hamiltonian matrix element for an electron to tunnel between adjacent hydrogen $1s$ orbitals, compared with U, the energy cost to put two electrons on the same atom. Unless the atoms are sufficiently close together to make T greater than U, the system is an insulator with an energy gap of approximately U. The energy gap of La_2CuO_4 is not presently known. Estimates range from 0.5 eV to 2.0 eV.

The most compelling evidence for the existence of heavy polarons is the presence of optical impurity bands of the type seen in organic metals. Fig. 2 shows the optical reflectivity of doped and undoped $La_{2-x}Sr_xCuO_4$ as reported by Orenstein et al[5]. We show this data primarily because it proves that the oscillator strength below 1 eV **grows with doping**. It also shows, consistent with other researchers' results[6], that the oscillator strength is correctly accounted for by one carrier per dopant, just as it would if one were doping silicon. This is reflected in the plasma edge, marked ω_{ps} in the figure. One carrier per dopant gives a carrier density of

$$n = \frac{x}{\Omega} = 1.84 \times 10^{-3} \text{ Å}^{-3} \quad ,$$

where $x = 0.175$ is the doping fraction and $\Omega = 95$ Å^3 is the formula unit volume, and thus a bare plasma frequency of

$$\hbar\omega_p = \hbar \sqrt{\frac{4\pi n e^2}{m}} = 1.59 \text{ eV}.$$

The observed plasma frequency is then

$$\hbar\omega_{ps} = \frac{\hbar\omega}{\sqrt{\epsilon_\infty}} = 0.83 \text{ eV} \quad ,$$

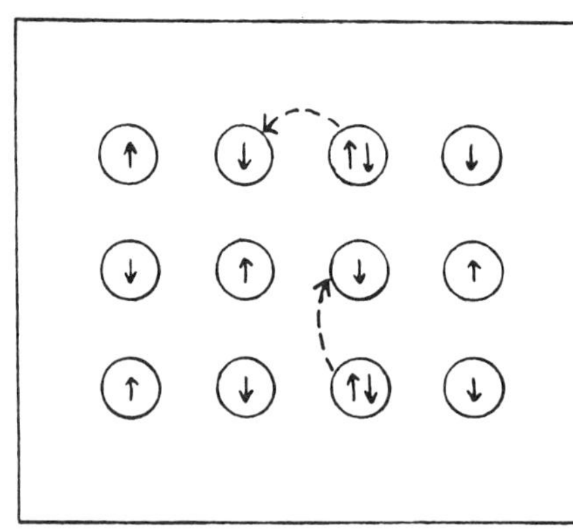

Fig. 1. Illustration of Mott insulator. Gap formation is adequately understood, but transport of donated carriers is not.

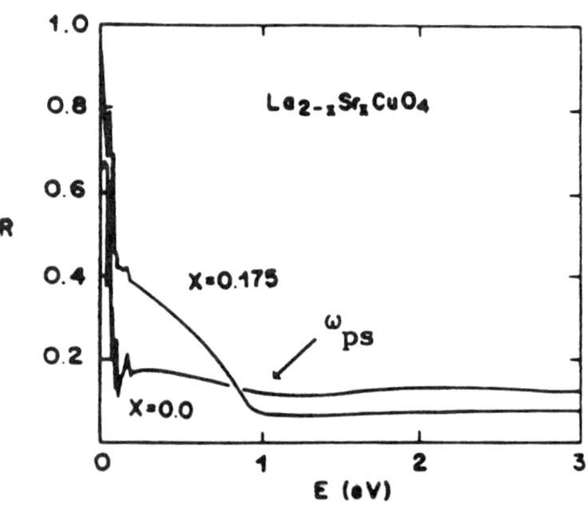

Fig. 2. Optical reflectivity data of Orenstein, $et.$ $al.$ showing doping dependence and "observed" plasma frequency ω_{ps} in $La_{2-x}Sr_xCuO_4$.

554

where $\epsilon_\infty \approx 3.7$ is deduced from the observed reflectivity of 0.1 at 3 eV through the relation

$$\epsilon = \left[\frac{1 + \sqrt{R}}{1 - \sqrt{R}} \right]^2 \ .$$

In our opinion this agreement is not coincidental. It indicates that each dopant contributes one carrier to the lower Hubbard band. We note particularly that the oscillator strength does not **slide** down from high frequency, but rather **materializes** at low frequency as the dopant is added.

Orenstein *et. al*[5]. also point out that **95%** of the oscillator strength added by the dopant appears as a broad absorption band centered near 0.5 eV. This feature can be seen more clearly in the data of Herr *et. al.*[7], which is reproduced in Figs. 3 and 4. Fig. 3 is the reflectivity of $La_{2-x}Sr_xCuO_4$ with x = 0.15 measured over a range sufficient to see both the broadband absorption and phonons. Fig. 4 is the optical conductivity as deduced from this reflectivity via a Kramers–Kronig analysis. The broadband absorption is the feature in the conductivity that peaks at 3500 cm^{-1}. Note that raising the sample temperature to $300^\circ K$ reduces the oscillator strength in this band by approximately 30%. This has the following simple interpretation. The band appears to have a threshold near 1200 cm^{-1} which is reasonably interpreted as the polaron ionization threshold. A good guess[8] for the polaron binding energy E is half this value, or 74 meV. If the lower Hubbard band contained one state per formula unit, a crude estimate of the fraction of the polarons un–ionized at $300^\circ K$ would be

$$f = \cfrac{1}{1 + \cfrac{1}{x} \exp(- \cfrac{E}{k_B T})} = 0.72 \ ,$$

which is the observed value. Elevating the temperature also increases the d.c. conductivity, as would be expected if carriers were liberated from

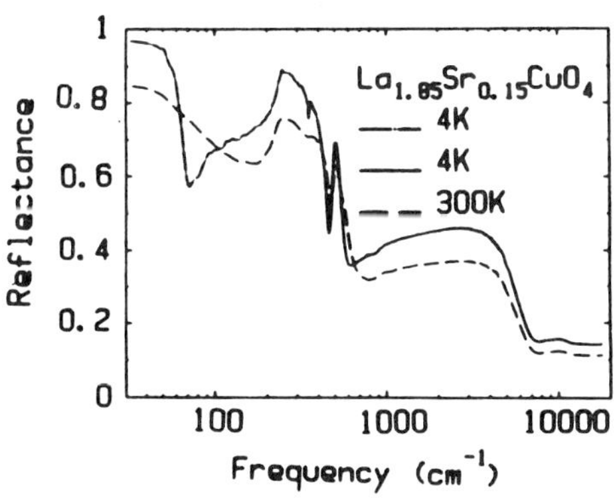

Fig. 3 Optical reflectivity data of Herr *et al.* analogous to Fig. 2.

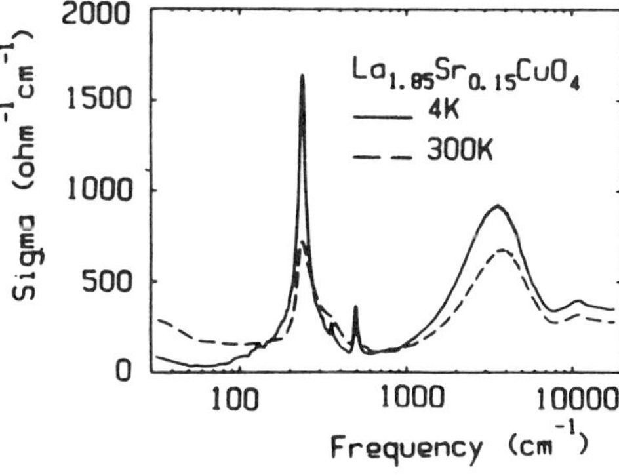

Fig. 4 Optical conductivity of $La_{2-x}Sr_xCuO_4$ obtained by Kramers–Kronig analysis of data in Fig. 3.

their polaronic wells. We note that the phonon-like feature at 230 cm^{-1} also loses oscillator strength as the temperature is increased, as if it were also caused by the polaron. This identification is supported by the observations of Herr *et. al.*[7] and of Bonn *et. al.*[9] that the 230 cm^{-1} feature contains 7 times the oscillator strength expected of a phonon. Taking the height σ of the feature to be 1600 Ω^{-1}cm^{-1} and its width $\Delta\omega$ to be 50 cm^{-1}, one obtains

$$\hbar^2 \, \sigma\Delta\omega = 5700 \text{ meV}^2 \quad .$$

The value expected for a phonon is

$$\hbar^2 \, \frac{\pi}{2} \, \frac{4\pi e^2}{\Omega m} = 790 \text{ meV}^2 \quad ,$$

where M is the oxygen mass. We have word-of-mouth confirmation[10] that the size of this feature increases with doping, but in any event we feel sufficiently confident to predict it. The identification of this feature as **electronic** in origin is supported by the absence of anomalous behavior in the raman data of Sugai *et. al.*[11] and the lack of doping-induced changes in the phonon density of states as measured by inelastic neutron scattering by Renker *et. al.*[12] We note that Bonn *et. al.*[9] report that the strength in this feature does **not** change when the temperature is raised only to 77°K, a result consistent with our assumed binding energy.

It was pointed out to us by S. Kivelson that these three features in the optical spectrum, namely

1. A broad absorption band containing most of the available oscillator strength.
2. A "phonon" with an anomalously large oscillator strength.
3. Drude conductivity. (The doped material conducts electricity.)

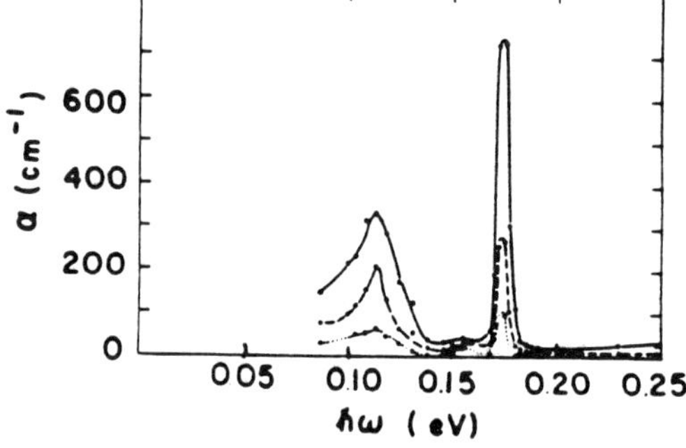

Fig. 5 Change to optical absorption coefficient of polyacetylene resulting from increased doping with AsF$_5$. Taken from Fincher *et. al.*

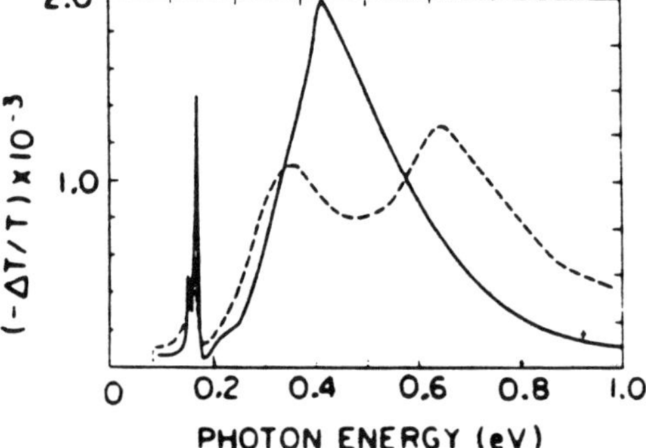

Fig. 6 Change to optical absorption coefficient of polyacetylene resulting from illuminating with 2.4 eV photons. Taken from Vardeny *et. al.*

have a striking physical precedent in polyacetylene. Polyacetylene is an organic semiconductor which becomes a metal when doped. Doping also induces two "phonons" with anomalously large oscillator strengths. Fig. 5 shows the change to the optical absorption spectrum of cis-polyacetylene resulting from doping, as reported by Fincher $et.\ al.$[13] These features grow with doping, have a position unrelated to the dopant, and are unusually intense. They are thought to be associated with solitons, the particular type of polaron occurring in polyacetylene. Also associated with the presence of solitons is a broadband absorption centered near 0.5 eV. Fig. 6 shows the optical absorption spectra of cis- and $trans$-polyacetylene induced by irradiating with 2.4 eV light, as reported by Vardeny $et.\ al.$[14] One can see clearly that the "phonons" induced by doping are also induced by photoexcitation and that a broadband absorption is also present. The ability to photoexcite these features supports the evidence from Fig. 5 that they are associated with a polaron and not with a particular dopant. The theory of solitons has been worked out by a number of authors and is, in our opinion, on solid ground. In particular, Mele and Rice[15] have shown that the expected fraction of the carrier's oscillator strength in the broadband absorption is 65%, that the remainder is shared between the "phonon" and drude conductivity, and that the polaron mass m^* is of the order of 6 electron masses.

The doping-induced features in $La_{2-x}Sr_xCuO_4$ must be associated with **mobile** polarons, as opposed to trapped ones, because the carrier density measured in the hall effect is equal to the doping density. Assuming one hole per dopant, the hall coefficient is expected to be

$$R_H = \frac{1}{ne} = \frac{5.9 \times 10^{-10}}{x} \text{ m}^3/\text{coulomb} \quad .$$

The hall coefficient of $La_{2-x}Sr_xCuO_4$ as measured by Ong $et.\ al.$[16] is reproduced in Fig. 7. The above equation is the solid line in the figure. One can see that it agrees with experiment perfectly for $x \leq 0.15$, the doping level at which the superconductivity begins to degrade.

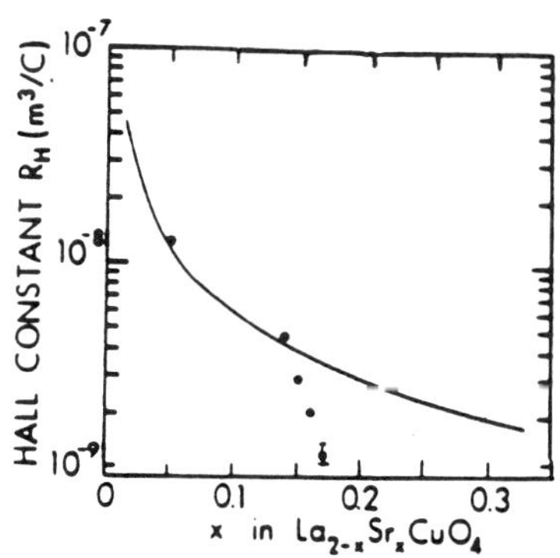

Fig. 7 Hall coefficient of $La_{2-x}Sr_xCuO_4$ reported by Ong $et.\ al.$ The solid line is the theoretical curve discussed in the text.

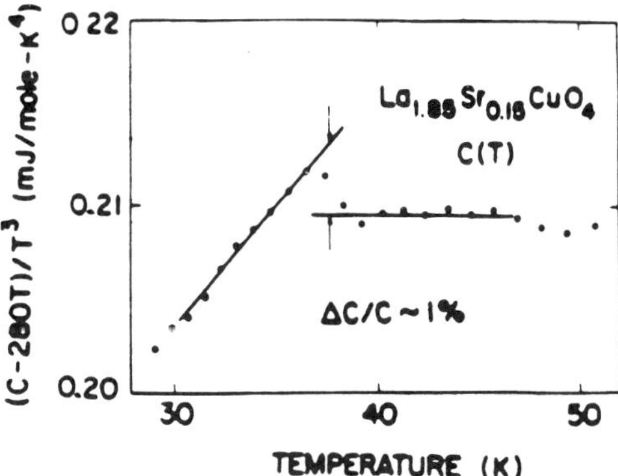

Fig. 8 Heat capacity of $La_{2-x}Sr_xCuO_4$ reported by Batlogg $et.\ al.$

Given that the polarons are mobile, they must be **massive**, since most of their oscillator strength is exhausted in the broadband absorption. The drude, "phonon", and broadband oscillator strengths must add up to the total, in the manner

$$[f_D + f_P + f_B] \frac{\pi}{2} \frac{4\pi n e^2}{m} = 1 \quad .$$

The drude contribution, however, is related to the polaron effective mass m^* by

$$f_D = \frac{m}{m^*} \quad .$$

Thus, if f_B is 95% of the total, the polaron mass can be no smaller than 20 electron masses. This is not precisely correct because the f-sum rule confined to energies less than 3 eV should reflect the "band" effective mass, which is not known. A reasonable guess would be 0.5 electron masses, considering the large fermi velocity predicted by the local density calculations[2], but this is very uncertain. This estimate would lead to an m^* of roughly 10 electron masses.

Let us now examine the known properties of $La_{2-x}Sr_xCuO_4$ to see if they are compatible with a dilute fermi liquid of polarons. In making this comparison, we shall use an m^* of 6 electron masses, as this is the value appearing in many experimental papers.

The fermi energy of $La_{2-x}Sr_xCuO_4$ is predicted to be tiny. Even if the fermi surface is spherical, which it probably is not, the fermi energy at x = 0.15 is of the order

$$\epsilon_f = \frac{\hbar^2}{2m^*} (3\pi^2 n)^{2/3} = 83 \text{ meV} \quad ,$$

a value comparable to the **debye frequency**. If the polarons do not disperse along the c-axis, one has, with d the interlayer spacing, the even smaller value of

$$\epsilon_F = \frac{\hbar^2}{2m^*} (2\pi n d) = 41 \text{ meV} \quad .$$

One immediate consequence of a small fermi energy is that the weak-coupling gap formula

$$\Delta = \hbar\omega_D \, e^{-1/\lambda}$$

is totally inappropriate even for truly weak-coupled superconductors because the physically important cutoff is the fermi energy rather than the debye frequency. Provided that the polaronic mass is **not** due to phonons, the isotope shift in this limit will be smaller than that of an ordinary superconductor[17]. This is true even if pairing is mediated by phonons. It is our understanding that a substantial isotope shift has now been observed in $La_{2-x}Sr_xCuO_4$, in contrast to to YBaCuO, which exhibits less than 5% of the shift expected in weak coupling.

The linear specific heat coefficient of the dilute polaron fermi liquid is expected to be comparable to that of a normal metal because the large mass counteracts the reduced carrier density. For the case of three-dimensional dispersion we have, with N_A denoting Avogadro's number,

$$\gamma = \frac{k_B^2 \, m^*}{3\hbar^2} \, \Omega \, (3\pi^2 n)^{1/3} \, N_A = 6.1 \text{ mJ/mol/}^\circ\text{K}^2 \quad .$$

This compares well with the value obtained experimentally by a number of authors. We reproduce in Fig. 8 the heat capacity data of Batlogg $et.$ $al.$[18], from which these authors infer a value of 5.3 mJ/mol/$^\circ$K^2 using the weak-coupling heat capacity anomaly formula

$$\frac{\Delta C}{T_c} = 1.43 \, \gamma \quad .$$

They also estimate γ from the temperature derivitives of H_{c1} and H_{c2}, and report a compromise value of 6 mJ/mol/$^\circ$K^2.

The theoretical London penetration depth, given by

$$\lambda_L = \sqrt{\frac{m^* \, c^2}{4\pi n e^2}} = 3200 \text{ Å}$$

for x = 0.15 agrees less well with experiment. Kossler $et.$ $al.$[19] and Aeppli et $al.$[20] report values of 2000 Å and 2500 Å determined from muon spin relaxation measurements. It should be remarked that the uncertainty in these numbers is large. Both of these groups needed to assume the vortex structure of a traditional type-II superconductor in order to extract λ_L from the data.

Another experiment supporting the existence of a small fermi energy is the thermopower measurement of Hundley $et.$ $al.$[21], reproduced in Fig. 9. Rather than finding the linear temperature dependence typical of a metal they find the thermopower to be roughly constant. This is exactly the behavior one would expect of an electron gas heated past its degeneracy temperature. If the elastic collision rate is insensitive to carrier energy, the thermopower is related to the heat capacity in the manner

$$Q = \frac{C}{3ne} \quad .$$

When the temperature of an electron gas exceeds its fermi temperature the heat capacity is that of an ideal gas, and one has

$$Q = \frac{k_B}{2e} = 43 \text{ } \mu\text{V/}^\circ\text{K} \quad .$$

In the opposite limit, the carrier heat capacity freezes out, and one has

$$Q = \frac{\pi^2}{3} \frac{k_B T}{e} \frac{m^*}{\hbar^2 (3\pi^2 n)^{1/3}} = 0.17 \ \mu V/^\circ K^2 \ T \ .$$

The solid line in Fig 9. the complete noninteracting electron gas thermopower appropriate for $x = 0.15$ and an m^* of 6 electron masses. It correctly predicts the magnitude of the data and the roll-off from linear behavior near $200^\circ K$. Note that it contains no adjustable parameters.

A large m^* is also consistent with the apparent discrepancy between the optical conductivity of $La_{2-x}Sr_xCuO_4$ and its measured static conductivity. The data of Bonn $et.\ al.$[9], which are similar to those of Herr $et.\ al.$[7] in Fig. 4 except that they are measured at 77 $^\circ K$, extrapolate to a zero-frequency conductivity of approximately 100 $\Omega^{-1}cm^{-1}$, a number considerably lower than the measured conductivity of 2500 $\Omega^{-1}cm^{-1}$. This latter value is the correct one. Values of order 1000 $\Omega^{-1}cm^{-1}$ are commonly reported for optimally doped $La_{2-x}Sr_xCuO_4$. Since the materials are grainy, it is not unreasonable that the true conductivity might be as high as $10^4 \ \Omega^{-1}cm^{-1}$, the saturation resistivity of an ordinary metal. If one assumes such a value for σ, the elastic scattering rate becomes

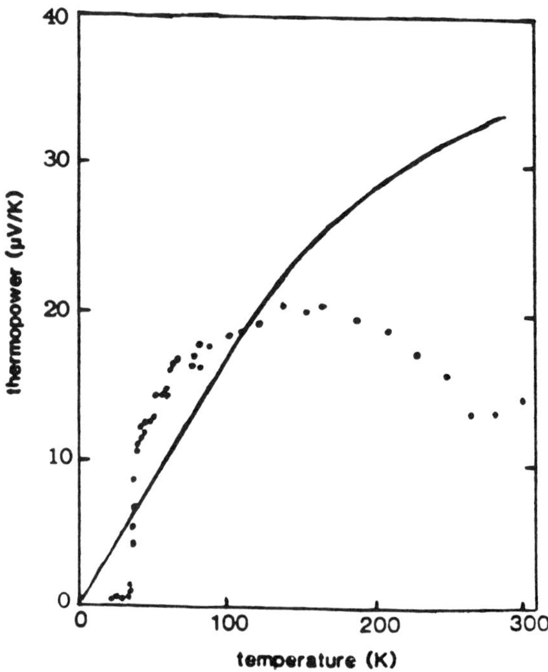

Fig. 9 Thermopower of $La_{2-x}Sr_xCuO_4$ reported by Hundley $et.\ al.$ The solid line is the theory discussed in the text.

$$\frac{\hbar}{\tau} = \frac{\hbar n e^2}{m^* \sigma} = 4.8 \text{ meV} \quad ,$$

or 39 cm^{-1}, a value below the resolution of the experiment. Thus even though the conductivity is low by normal metal standards, the collision time is extremely long. This occurs because the fermi velocity is low. For three-dimensional dispersion we obtain

$$v_F = \frac{\hbar \, (3\pi^2 n)^{1/3}}{m^*} = 6.9 \times 10^6 \text{ cm/sec} \quad ,$$

and thus an elastic mean free path of

$$l = v_F \tau \approx 130 \text{ Å} \quad ,$$

a resonable value for this material.

Another consequence of the small fermi energy is that the tunneling current should be asymmetric, with a larger tunnel conductance when the sample is biased **negative**. This is because the fermi sea, which is made of holes, is small. There are many states above the fermi surface into which holes can be injected, but only a small number below the fermi surface which can receive electrons. Asymmetric tunneling of the proper polarity is apparently commonly seen[22].

We do not believe that the data enable us as yet to distinguish between an ordinary polaron, one caused by phonons, and a magnetic polaron. However, we lean toward the notion of a magnetic polaron because, as Anderson[23] has repeatedly stressed, the energy gap in this material is inherently magnetic. The physics of carrier transport in Mott insulators is not understood. To appreciate the problem, imagine, as shown in Fig. 1, that spins of the Mott insulator are ordered antiferromagnetically and that an extra electron is added. The extra electron has no preferred atom on which to sit and can tunnel with matrix element T from one atom to the next. It is therefore expected to be mobile. However, it cannot tunnel to a neighboring atom unless the spin **already there** has the correct orientation. This means that quantum transport in a Mott insulator cannot be understood without first solving a spin problem. This fundamental difficulty is the impetus behind Anderson's invocation of the "resonating valence bond" state to understand these materials, and it is of concern to us as well. It is likely, at any rate, that the mechanism by which the carriers acquire mass is related to the mechanism by which they pair.

ACKNOWLEDGEMENTS

Because the field of high-T$_c$ superconductivity is moving so quickly, we have found it necessary to discuss data and the ideas, of other researchers before they are published through formal channels. We wish to express special thanks to J. Orenstein, D. Tanner, and N. Ong for allowing us to reproduce their data, and to S. Kivelson and P. Littlewood for numerous useful discussions. We also wish to acknowledge the help of T. Geballe, M. Beasley, J.R. Schrieffer, and E.J. Mele. This work was supported by the National Science Foundation under Grant No. DMR-85-10062 and by the NSF-MRL Program through the Center for Materials Research at Stanford University.

REFERENCES

1. The most recent reference on this subject is P.M. Grant, S.S.P. Parkin, V.Y.Lee, E.M. Engler, M.L. Ramirez, J.E.Vazquez, G. Lim, R.D. Jacowitz, and R.L. Greene, Phys. Rev. Lett. 58, 2482 (1987). The superconductivity discussed in this paper is due to deviations from perfect stoichiometry.
2. L.F. Mattheiss, Phys. Rev. Lett. 58, 1028 (1987); J. Yu, A.J. Freeman, and J.-H. Xu, Phys. Rev. Lett. 58, 1035 (1987).
3. B.H. Brandow, Adv. Phys. 26, 651 (1977); N.F. Mott, Proc. Roy. Soc., London, A62, 416 (1949).
4. D. Vaknin, S.K. Sinha, D.E. Moncton, D.C. Johnston, J. Newsam, C.R. Safinya, and H.E. King, Jr., Phys. Rev. Lett. 58, 2802 (1987).
5. J. Orenstein, G.A. Thomas, D.H. Rapkine, C.G. Bethea, B.F. Levine, R.J. Cava, E.A. Rietman, and D.W. Johnson, Jr., (preprint).
6. S. Tajima, S. Uchida, S. Tanaka, S. Kanbe, K. Kitizawa, and K. Kueku, Jpn. J. Appl. Phys. 26, (1987).
7. S.L. Herr, K. Kamaras, C.D. Porter, M.G.Doss, D.B. Turner, D.A. Bonn, J.E.Greedon, C.V. Stager, and T. Timusk, (preprint).
8. J. Appel, Solid State Phys. 21, 193 (1966).
9. D.A.Bonn, J.E. Greedon, C.V. Stager, T. Timusk, M.G. Doss, S.L. Herr, K. Kamaras, C.D. Porter, D.B. Tanner, J.M. Tarascon, W.R. McKinnon, and L.H. Greene, Phys. Rev. B35, 8843 (1987).
10. G.A. Thomas, private communication.
11. S. Sugai, M. Sato, and S. Hasoya, Jpn. J. App. Phys. 26, L495 (1987).
12. B. Renker, F. Gompf, E. Gering, N. Nucker, D. Ewert, W. Reichardt, and H. Rietschel, (preprint).
13. C.R. Fincher, Jr., M. Ozaki, A.J. Heeger, and A.C. MacDiarmid, Phys. Rev. B19, 4140 (1979).
14. Z. Vardeney, J. Orenstein, and G.L. Baker, Phys. Rev. Lett. 50, 2032 (1983).
15. E.J. Mele and M.J. Rice, Phys. Rev. Lett. 45, 926 (1980).
16. N.P. Ong, Z.Z. Wang, J. Clayhold, J.M. Tarascon, L.H. Greene, and W.R. McKinnon, Phys. Rev. B35, 8807 (1987).
17. C.S. Koonce and M.L. Cohen, Phys. Rev. 177, 707 (1969).
18. B. Batlogg, A.P. Ramirez, R.J. Cava, R.B. van Dover, and E.A. Rietman, Phys. Rev. B35, 5340 (1987).
19. W.J. Kossler, J.R. Kemton, X.H. Yu, H.E. Shone, Y.J. Uemura, A.R. Moodenbaugh, M. Suenaga, and C.E. Stronach, Phys. Rev. B35, 7133 (1987).
20. G. Aeppli, R.J. Cava, E.J. Ansaldo, J.H. Brewer, S.R. Kreitzman, G.M.Luke, D.R. Noaker, and R.F. Kiefl, Phys. Rev. B35, 7129 (1987).
21. M.F. Hundley, A. Zettl, A. Stacy, and M.L. Cohen, Phys. Rev. B35, 8800 (1987).
22. R.C. Dynes and W.J. Gallagher, private communication.
23. P.W. Anderson, Science 235, 1196 (1987); P.W. Anderson, G. Baskaran, Z. Zou, and T. Hsu, Phys. Rev. Lett. 58, 2790 (1987).

MOLECULAR ORBITAL BASIS FOR SUPERCONDUCTIVITY

WITH APPLICATIONS TO HIGH-T_c MATERIALS

K. H. Johnson, M. E. McHenry, C. Counterman, A. Collins,
M. M. Donovan, R. C. O'Handley, and G. Kalonji

Department of Materials Science and Engineering
Massachusetts Institute of Technology
Cambridge, MA 02139

INTRODUCTION

In a paper published in 1983,[1] a "real-space" molecular-orbital basis
for superconductivity was proposed. There it was shown that the occurrence
of superconductivity in a broad range of materials is correlated with the
nature of the molecular-orbital topology at the Fermi energy. This topology
leads to a formula for the superconducting transition temperature (T_c) which
(1) depends on two chemical bond parameters at the Fermi energy, (2) reduces
to the well known BCS[2] formula for T_c at certain limits of these parameters,
and (3) produces values of T_c in remarkable agreement with experiment. This
molecular-orbital theory furthermore explains in simple chemical-bonding
language why many materials are superconducting, why superconductivity and
magnetism are generally complementary phenomena, and why some materials are
neither superconducting nor magnetic. We will first review key features of
this molecular-orbital model of superconductivity. Then it will be
demonstrated that the theory accurately describes the recently discovered
high-T_c superconductivity in La-Sr-Cu-O and Y-Ba-Cu-O oxides.

THE MOLECULAR-ORBITAL THEORY OF SUPERCONDUCTIVITY

The basis for a molecular-orbital theory of superconductivity lies in
the observation of a persistent correlation between the occurrence of
superconductivity and the existence of molecular orbitals at the Fermi
energy which are **bonding** within a "network" of spatially extended **"layers"**
or **"tubes"** of overlapping atomic orbitals. Prototypes of these molecular
orbitals for several well known types of superconductors are shown in Figs.
1 and 2. In Fig. 1 we display a computer-generated three-dimensional
contour map of one of the degenerate "layered" $p\pi$-bonding orbitals
characteristic of the real-space electronic structure of the superconductor,
FCC aluminum, at the Fermi energy (E_f). Fig. 2 illustrates schematically
the "tubular" $d\delta$-bonding orbitals at E_f for superconducting transition
metals (e.g. niobium) and their compounds (e.g. Nb_3Sn).

From these types of molecular orbitals one can construct an electron-
pair ("Cooper-pair") wavefunction by forming a **"valence-bond-like"** electron
occupation of the spatially extended layered or tubular orbitals, with the

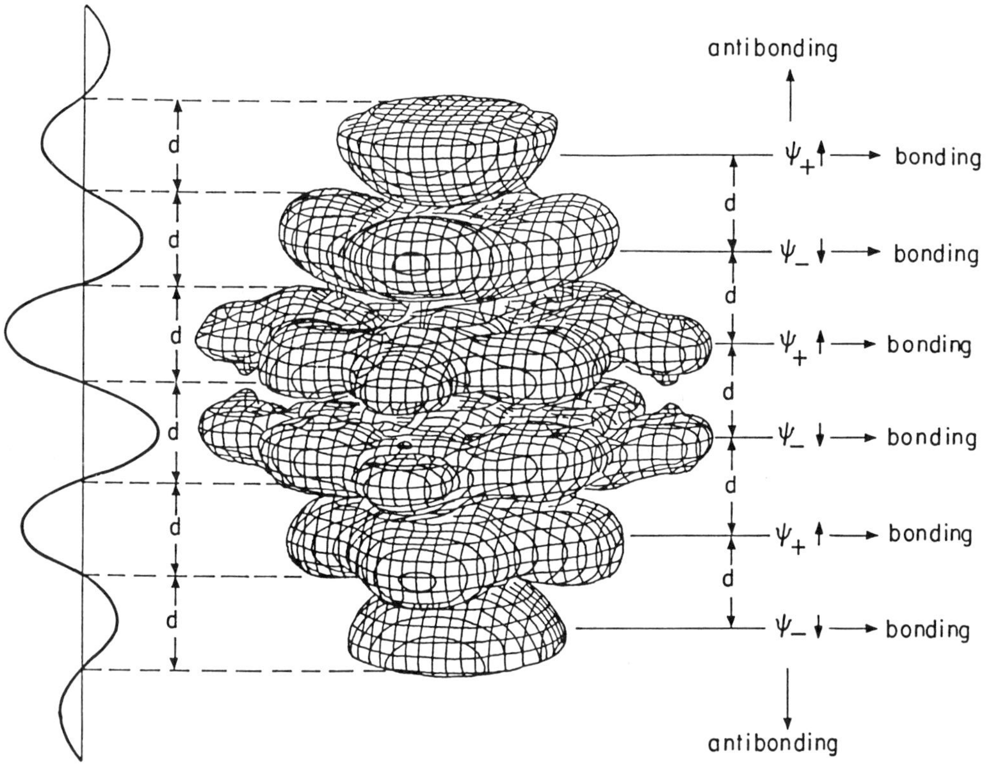

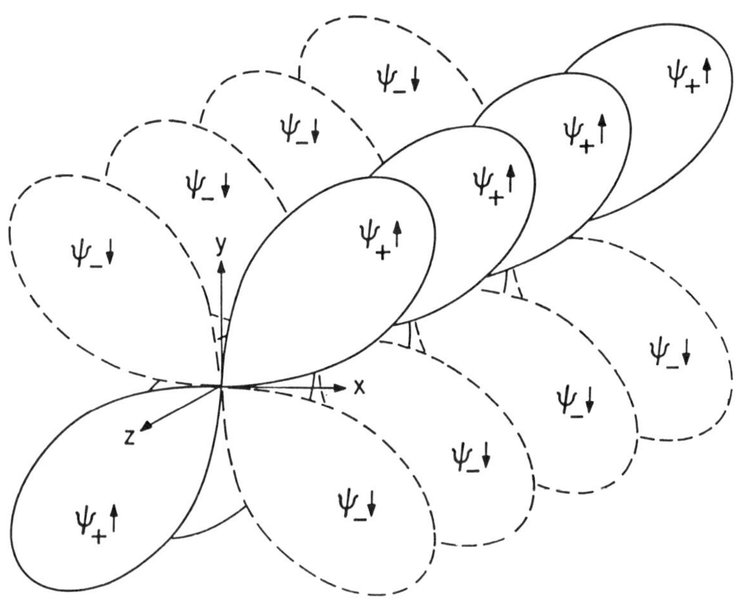

Fig.1

$$d\delta\ (d_{xy})$$

Fig. 2

correlation of "spin-up" and "spin-down" electrons in the positive- and negative-phase regions of the wavefunction. This correlation results in a commensurate antiferromagnetic conduction electron spin density wave. The superconducting electron-pair wavefunction can then be described by the **singlet** state

$$\Psi(1,2) = [\psi_+(1)\psi_-(2) + \psi_-(1)\psi_+(2)][\alpha(1)\beta(2) - \beta(1)\alpha(2)], \qquad (1)$$

where ψ_+ and ψ_- represent the positive- and negative-phase regions of the spatially extended "layered" and "tubular" molecular orbitals (see Figs. 1 and 2), and α and β denote "spin-up" and "spin-down" eigenfunctions, respectively.

The binding energy or "energy gap" of the this pair function with respect to the normal state is defined as

$$\Delta = 2E_f - E(d), \qquad (2)$$

where $E(d)$ is the pair-function energy, which depends on the distance d between neighboring ψ_+ and ψ_- "layers" or "tubes" (scc Fig. 1). The significance of the ψ_+-ψ_- distance d is that it represents the distance beyond which the Coulomb repulsion of an electron pair is screened out by the chemical bonding at E_f. $E(d)$ is less than $2E_f$ because of the correlation of "spin-up" and "spin-down" electrons in the ψ_+ and ψ_- layers or tubes. **This correlation is driven by dynamic Jahn-Teller-induced relative displacements of the lattice ions and electrons in response to the degeneracy or near-degeneracy of the molecular orbitals at the Fermi energy.** The Jahn-Teller coupling can be quantified through the bond overlap integral $S(E_f)$ at the Fermi level and the normalized amplitude of the Jahn-Teller displacements, δ, by the following approximate expression[1]

$$\delta/d \sim S(E_f) \sim (m/M)^\beta \qquad (3)$$

where m and M are the electron and lattice ion masses, respectively. The quantity β is the Jahn-Teller coupling parameter which, through the above expression, can be described as a measure of the strength of the delocalized bonding at E_f. The values of β are constrained by two well-defined limits imposed by the harmonic oscillator problem. The "harmonic limit" $\delta = (m/M)^{1/2}d$ represents the small oscillations characteristic of a classical oscillator, whereas the limit $\delta = (m/M)^{1/4}d$ is the "zero-point" vibrational amplitude of a quantum oscillator, beyond which the system is likely to be unstable with respect to a static Jahn-Teller distortion. These β-values imply two domains of Jahn-Teller-induced effects:

$$1/4 \lesssim \beta \lesssim 1/2 \qquad \text{the dynamic Jahn-Teller effect}$$

$$0 \lesssim \beta \lesssim 1/4 \qquad \text{the pseudo Jahn-Teller effect}$$

Within the domain of the pseudo Jahn-Teller effect the electron pair wavefunction (expression 1) may still be bound, resulting in an energy gap Δ, through a correlation driven by a static strain and an electronic ("excitonic") "resonance" between two nearly degenerate molecular orbitals around E_f of different symmetries and spatial characters.

On the basis of the superconducting state pair wavefuncion in Eq. 1 and the above dynamic Jahn–Teller-driven electron–lattice displacements, it is possible to write down an effective Schrödinger wave equation for such a pair and to solve this equation for E(d) in Eq. 2, thereby obtaining Δ. This has been carried out in detail in Ref.1 so we repeat only the results:

$$\Delta \sim [h^2 (m/M)^\beta /2\pi md^2]\exp[-h^2/\{2me^2 d[1-(m/M)^\beta]\}] \tag{4}$$

where e is the electron charge and h is Planck's constant. If we assume here the validity of the traditional relationship[2]

$$\Delta \sim 2k_B T_c \quad (\text{or, more precisely, } k_B T_c = 0.57\Delta) \tag{5}$$

between the energy gap Δ, the Boltzmann constant k_B, and the superconducting transition temperature T_c, we can write down the following approximate formula for T_c:

$$k_B T_c \sim [h^2 (m/M)^\beta /4\pi md^2]\exp[-h^2/\{2me^2 d[1-(m/M)^\beta]\}] . \tag{6}$$

In the limit of very small bond overlap at E_f where, according to expression 3, $\beta = 1/2$, formula 6 reduces to

$$\lim_{\beta \to 1/2} k_B T_c \sim [h^2/4\pi (mM)^{1/2}d^2]\exp(-h^2/2me^2 d) . \tag{7}$$

It can be shown[1] that this expression is equivalent to the well known BCS formula[2]:

$$k_B T_c \sim \overline{h}\omega_D \exp[-1/N(E_f)V_{ep}], \tag{8}$$

where ω_D is the Debye frequency, $N(E_f)$ is the electronic density of states at the Fermi energy, and V_{ep} is the electron–phonon interaction potential. Formulae 7 and 8 agree if

$$\omega_D \sim h/2(mM)^{1/2}d^2 , \tag{9}$$

an approximate relation originally proposed by Weisskopf[3], and if

$$N(E_f)V_{ep} \sim 2me^2 d/h^2 . \tag{10}$$

While the molecular–orbital model for superconductivity reduces to the conventional BCS theory in the limit of the dynamic Jahn–Teller coupling parameter $\beta = 1/2$, it complements and indeed goes beyond BCS theory in a number of important respects:

(1) It allows one to predict which materials are likely to be superconductors and which are not entirely on the basis of the real–space molecular–orbital topology at the Fermi energy, which must typically be of the spatially delocalized pπ- or dδ-bonding characters shown in Figs. 1 and 2 (f–electron superconductors are also within the scope of the theory). In contrast, localized antibonding (e.g. dσ^*, dπ^*, or dδ^*) character at E_f is a precursor to ferromagnetism in transition metals (see Section 6 of Ref. 1).

(2) It permits one to estimate T_c via formula 6 from two simple parameters derived from the real-space molecular-orbital topology at the Fermi energy, namely the distance d between neighboring ψ_+ and ψ_- "layers" or "tubes" (the distance beyond which the Coulomb repulsion is screened out by the chemical bonding at E_f) and a Jahn-Teller coupling parameter β. For example, from the aluminum "layered" $p\pi$-bonding molecular orbitals shown in Fig. 1, it follows that $d \sim 1.8$Å and $\beta \sim 1/2$, the latter value being deduced from the overlap criterion, Eq. 3. Substitution of these two values into formula 6 yields $T_c \sim 1.2°$K, which is in excellent agreement with the experimental value $(T_c)_{exp} = 1.18°$K. As a bonus the Debye temperature, $\theta_D \sim 375°$K, calculated from formula 9 and the definition $k_B \theta_D = \hbar\omega_D$, is also in excellent agreement with the value $(\theta_D)_{exp} \sim 375°$K measured for aluminum.

(3) It is also readily applicable to more strongly Jahn-Teller-coupled transition-metal superconductors (including transition-metal compounds and alloys), where the molecular-orbital topology at E_f is generally of the "tubular" $d\delta$ topology shown in Fig. 2. The usually larger bond overlap of such $d\delta$ orbitals typically results in values of $\beta < 1/2$. For example, values of $\beta \sim 0.35$ and 0.25 are appropriate for BCC crystalline niobium and vanadium, respectively. The "intertubular" $(\psi_+ - \psi_-)$ distance parameter d for the "tubular" $d\delta$-bonding orbitals (Fig. 2) at E_f for these metals has the corresponding values of $d \sim 2.3$Å and 1.5Å for Nb and V, respectively. Substituting these quantities in formula 6, we obtain values of $T_c \sim 8.6°$K and $5.2°$K for Nb and V, respectively, in good agreement with the corresponding measured values of $(T_c)_{exp} = 9.2°$K and $5.4°$K.

(4) For even greater amounts of bond overlap at E_f such that the Jahn-Teller coupling parameter β falls into the range $0 \lesssim \beta \lesssim 1/4$, a **static** Jahn-Teller (or Peierls) distortion of the lattice can remove the molecular-orbital degeneracy at E_f, inducing a lattice instability and structural transformation (e.g. from cubic to tetragonal symmetry). An intrinsically low crystal symmetry or amorphous structure can also remove the orbital degeneracy at E_f. If the energy separations between the resulting nondegenerate orbitals are small, then electron pairing and the superconducting state can still occur through dynamic **"pseudo-Jahn-Teller" coupling** of the electrons and the lattice ions, induced by the "mixing" of the nearly degenerate but spatially different molecular orbitals at E_f or through "excitonic" fluctuations in orbital occupancy. The "competition" between the static and dynamic Jahn-Teller effects for larger bond overlaps at E_f $(\beta \lesssim 1/4)$ supports the conjecture that the interactions responsible for lattice instabilities and structural phase transformations are similar to those responsible for superconductivity in the same materials (e.g. A15 compounds). For example, in A15 compounds such as Nb_3Sn, the Fermi energy coincides with **$d\delta$-bonding** molecular orbitals, resulting in **"tubular"** $d\delta$ orbitals extending along the Nb "chains", as illustrated in Fig. 2. The $d\delta$-bond overlap at E_f in Nb_3Sn is substantial, implying a relatively small value of $\beta \sim 0.15$. For the derived interorbital distance $d \sim 1.9$Å (the distance between neighboring tubular orbital components ψ_+ and ψ_- in Fig. 2), formula 6 yields a value of $T_c \sim 18.4°$K, in excellent agreement with the measured value $(T_c)_{exp} = 18.0°$K for Nb_3Sn. This value of β is also consistent with the nearly **vanishing isotope effect** in Nb_3Sn (see discussion below) and with the tendency of this material to undergo a lattice transformation from the cubic to tetragonal phase.

M = 16
Tc
200
100
0
d = 7
d = 6
d = 5
d = 4
d = 3
d = 2
0.0   0.1   0.2   0.3   0.4   0.5
β

Fig. 3

M = 100
Tc
200
100
0
d = 7
d = 6
d = 5
d = 4
d = 3
d = 2
0.0   0.1   0.2   0.3   0.4   0.5
β

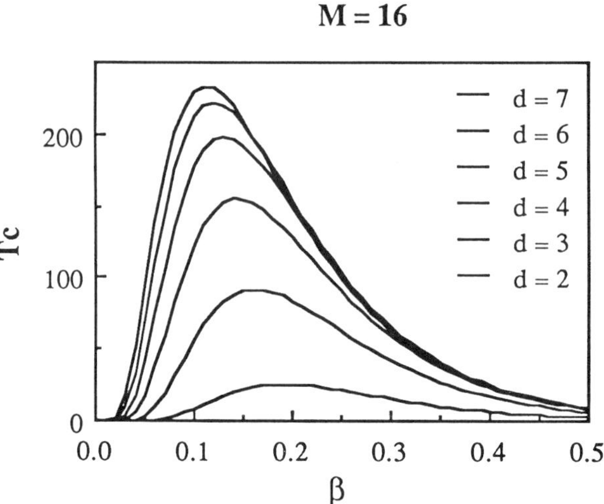

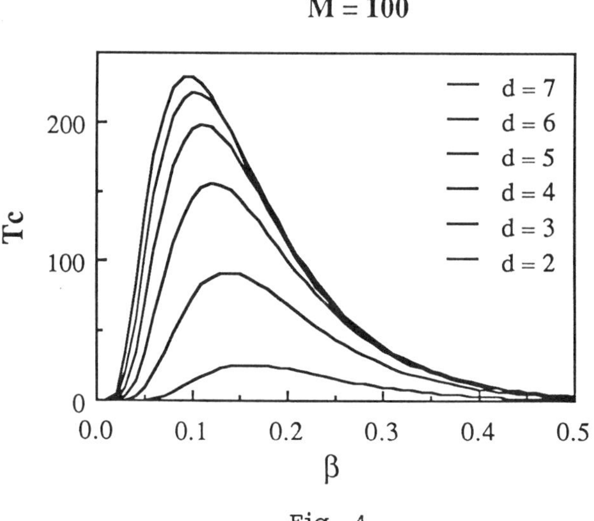

Fig. 4

(5) Figs. 3 and 4 illustrate the functional dependence of T_c on the Jahn-Teller coupling parameter β for various values of the $\psi_+ - \psi_-$ distance parameter d. Several key features are worthy of discussion. First, the increase in d up to ~ 8Å is seen to have dramatic effects on the T_c vs β mappings, indicating that the highest T_c's are attainable only with relatively large d-values. As will be shown in the discussion of our recent molecular-orbital studies of **high-T_c oxide superconductors** (see following section), the primary distinction between these new oxide superconductors and their metallic predecessors is their increased $\psi_+ - \psi_-$ distances d and Jahn-Teller coupling constants in the "pseudo-Jahn-Teller" range $\beta \lesssim 1/4$. Also evident from Fig. 5 is the fact that the T_c vs β curves are double valued functions with a single maximum at the value $\beta = \beta_{crit}$. The approximate relationship for the exponent of the isotope effect is

$$z \cong - \frac{\partial \ln T_c}{\partial \ln M} = - \frac{M}{T_c}\left(\frac{\partial T_c}{\partial \beta}\right)\left(\frac{\partial \beta}{\partial \gamma}\right)\left(\frac{\partial \gamma}{\partial M}\right) \tag{11}$$

where $\gamma \equiv (m/M)^\beta$ and $(\partial \beta / \partial \gamma)(\partial \gamma / \partial M)$ is always positive. Therefore, we find that **a vanishing isotope effect can be observed in approaching the maximum T_c at β_{crit} for any given d-value.** This is consistent with the observation of vanishing isotope effects in Nb_3Sn and in more recently discovered high-T_c oxide superconductors (see following section), in which the β-values approach β_{crit}. It is also interesting to note from Figs. 3 and 4 and from Eq. 11 that, although seldom attained, **$\beta < \beta_{crit}$ leads to a negative isotope effect.** This may useful in explaining the negative isotope effects measured in materials such as uranium and palladium hydride. Because the molecular-orbital theory of superconductivity can straightforwardly explain these anomalous or vanishing isotope effects for $\beta < 1/4$, yet reduces to the BCS theory in the $\beta = 1/2$ limit, there is no need for esoteric non-phonon mechanisms beyond the Jahn-Teller effects elucidated in this paper and in Ref. 1.

(6) The mutual exclusivity of superconductivity and magnetism throughout the Periodic Table and in most alloys and compounds, can be explained simply in terms of the corresponding real-space molecular-orbital topologies at E_f. **While spatially delocalized bonding (e.g. $p\pi$ or $d\delta$) along "layers" or "tubes" at E_f is a precursor to superconductivity, localized antibonding (e.g. $d\sigma^*$ and $d\pi^*$ in transition metals) is a precursor to ferromagnetism**[1]. In those rare cases where superconductivity and magnetism coexist in the same material, the antibonding regions around the magnetic ions are spatially removed from the extended tubes or layers at E_f associated with the superconductivity.

(7) In the molecular-orbital model the superconducting "coherence length" can be estimated from the approximate formula[1]

$$1 \sim (M/m)^\beta d. \tag{12}$$

In aluminum, for example, this yields a value of 400Å, whereas in the recently diiscovered high-T_c oxide superconductors the calculated coherence lengths range from 18Å to 59Å (see following section).

(8) The molecular-orbital model permits a simple description, analogous to that originally suggested by Schafroth[4], of Cooper pairs as "quasimolecules" undergoing Bose-Einstein-like condensation to the many-

Vanishing isotope effect
Tc max
- isotope effect
+ isotope effect
Tc
βcrit
β

Fig. 5

Neutral Atom Electronegativities
-2
-4
-6
-8
-10
-12
Mulliken Electronegativity
χ (eV)
s-orbital
d-orbital
Oxygen 2p
Sc Ti V Cr Mn Fe Co Ni Cu
Transition Metal Element

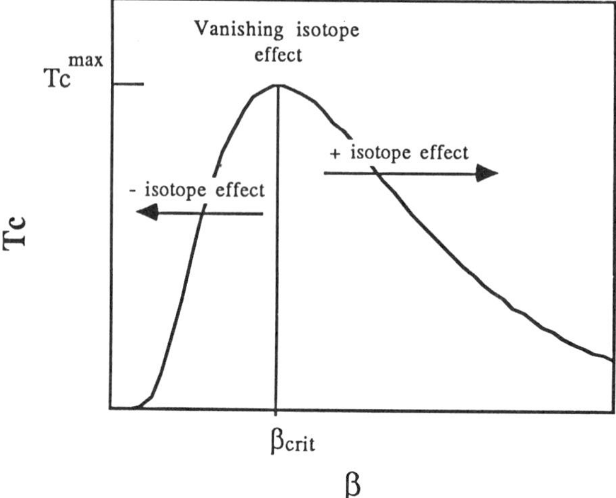

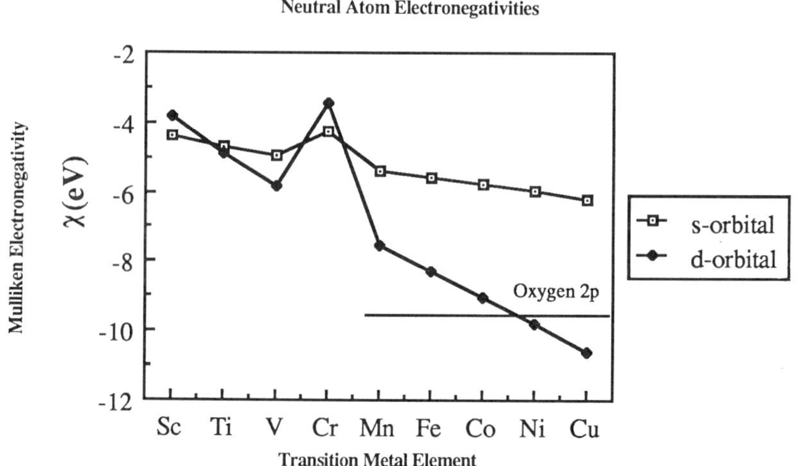

Fig. 6

electron superconducting state. The spatially delocalized "valence-bond-like" correlation (Eq. 1) of an electron pair at E_f over many "layered" or "tubular" molecular-orbital components ψ_+ and ψ_- (Figs. 1 and 2) uses up only a very small fraction of the available phase space defined by the layered or tubular "cells". Thus there is ample phase space left to put many more electron pairs into the same valence-bond-like pair function (Eq. 1).

APPLICATIONS OF THE MOLECULAR-ORBITAL THEORY TO HIGH-T_c SUPERCONDUCTORS

The recent discovery[5,6] of high-T_c superconductivity in metal oxides such as doped La_2CuO_4 and $YBa_2Cu_3O_7$ presents a challenge to the above-described molecular-orbital theory. The **self-consistent-field xalpha scattered-wave (SCF-Xα-SW) method**[7] has been employed to calculate the real-space molecular-orbital energy levels and wavefunction topologies for clusters containing up to thirty-eight atoms representing the local chemical environments in these materials. The results indicate that for both the La- and Y-based materials, the molecular-orbital topologies at E_f are dominated by the presence of "tubular" pπ bonds running parallel to the copper oxygen planes. Important differences between the "intertubular" ψ_+-ψ_- distances d and Jahn-Teller coupling parameters β, related to the relative amounts pπ-bond overlap at E_f and unique crystal structures of the La- and Y-based materials, are believed to be responsible for the wide range of measured T_c-values.

The first indication of the possibility of oxygen-oxygen pπ-bonding at E_f comes from an examination of the **"orbital electronegativities"** of the transition-metal atomic d-states as compared with the oxygen 2p states, derived from local-density (Xα) eigenvalues.[8] In Fig. 6 we show the s- and d-orbital electronegativities for 3d transition metals as compared with that of the oxygen 2p states. It will be observed in this figure that moving to the "right" in the 3d transition series decreases the d-orbital electronegativity significantly, while the gap between the s- and d-orbitals increases. The large electronegativity difference between transition-metal s-states and the oxygen p-states results in a strongly ionic interaction between the two, leading to deep lying TM-s,O-p bonding states in solids along with high lying unoccupied antibonding states. In contrast, the TM-O d-p electronegativity difference is decreasing in traversing the 3d series to the right, implying a change from ionic to increasingly covalent character of the p-d interactions. In Fig. 6 it will also be noticed that copper-oxygen d-p interactions are unique in that they imply moderate covalency while having oxygen p-character significantly above the d-character for the first time in the transition series. The covalent Cu-O d-p interaction, coupled with the filling of the Cu d-orbtals, implies localized Cu-O d-p antibonding character around E_f. In solids having such localized nearest-neighbor antibonding character at E_f, it is natural to find more delocalized second-nearest-neighbor bonding character at E_f. Since the oxygen atoms are second-nearest neighbors in the copper-oxide-based superconductors, it follows that delocalized oxygen-oxygen pπ-bonding is likely at the Fermi energies.

The **local** molecular-orbital wavefunction topology at E_f for Sr-doped La_2CuO_4, resulting from the SCF-Xα-SW cluster calculations, is shown

O(p)—O(p)
"Pi" bonding
"tubes"
d
(2.7A)
La(d)—O(p)
"delta"
bonding
Sr(d)—O(p)
"delta"
bonding
Cu(d)—O(p)
antibonding
Direction of "tubes"
<110>

Fig. 7.

O(p)—O(p)
"Pi" bonding
"tubes"
(3.88A)
Ba(d)—O(p)
"delta"
bonding
Cu(d)—O(p)
antibonding
Ba(d)—O(p)
"delta"
bonding
Direction of "tubes"

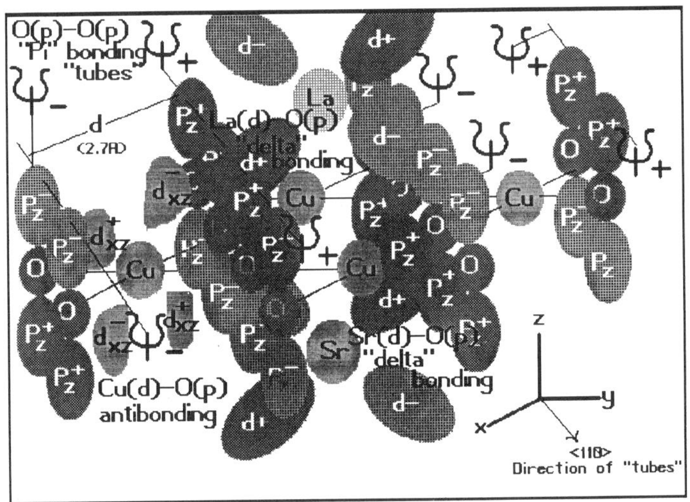

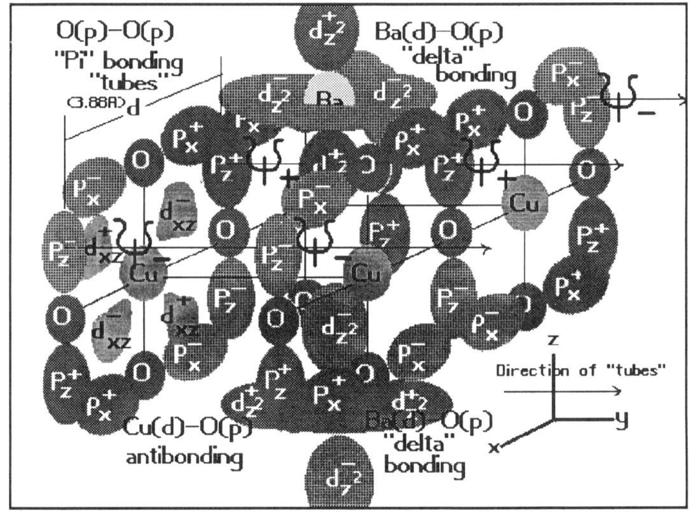

Fig. 8.

schematically in Fig. 7. The most striking feature of this diagram is the oxygen p_z orbitals which couple in **oxygen-oxygen π-bonding** ψ_+ and ψ_- **"tubes"** parallel to the Cu-O (xy) plane along the <110> direction of the crystal lattice. Significant **copper-oxygen d-p antibonding character** at E_f "compresses" the oxygen p_z orbitals so as to promote π-bond overlap. Thus the above conjectures about the Cu-O/O-O composite chemical antibonding/bonding characters at the Fermi energy are confirmed by the detailed molecular-orbital calculations. Furthermore, quantitatively small but qualitatively important **delocalized La and Sr dδ-bonding** character couples in phase with the oxygen $p\pi$-bonding "tubes" so as to further enhance their bond overlap and to "lock" in their phase with that of the tubular $p\pi$-bonds lying within the "upper" and "lower" oxygen planes (not shown in the diagram). This molecular orbital corresponds to one column of a degenerate e_g representation (of the D_{4h} point group), so it does not break symmetry in the tetragonal phase of Sr-doped La_2CuO_4. However, the intrinsic anisotropy in the oxygen-oxygen $p\pi$-bonding arising from the partial electron occupancy of this molecular orbital is the driving mechanism for a Jahn-Teller instability. The $p\pi$-bond overlap is consistent with a Jahn-Teller coupling parameter $0.25 \lesssim \beta \lesssim 0.30$ (see formula 3). The $\psi_+-\psi_-$ $p\pi$ "intertubular" distance $d - 2.7$Å, which is simply the square root of two over two times the lattice constant. Substitution of these parameters along with the oxygen ion mass into formula 6 yields $34°K \lesssim T_c \lesssim 49°K$, which is in good agreement with the range of superconducting transition temperatures measured for optimally doped La_2CuO_4 superconductors. It will be noted here that the parameter β lies within the range associated with the **dynamic** Jahn-Teller coupling mechanism of superconductivity described above (analogous to the BCS electron-phonon mechanism). However, the β-values are perilously close to the $\beta = 0.25$ threshold for a static Jahn-Teller lattice instability and, in conjunction with expression 11, predict only a very small if not negligible isotope effect. These oxide superconductors are indeed subject to such lattice instabilities. The Debye temperature and coherence length of La_2CuO_4-based superconductors predicted from formulae 9 and 12, respectively, are $\theta_D \sim 221°K$ and 35Å $\lesssim 1 \lesssim 59$Å for $0.25 \lesssim \beta \lesssim 0.30$. These values are in reasonable agreement with available experimental data.

The above molecular-orbital analysis of La_2CuO_4-based superconductors is qualitatively consistent with the results of band-structure calculations that suggest the presence of significant oxygen p character (along with antibonding Cu d character) at the Fermi energy and the "two-dimensional" nature of the electronic states involved in superconductivity. Where the present model differs markedly from the band-structure conclusions is the observation that the electronic-structure features of importance to the superconductivity lie **between** the Cu-O and La(Sr) planes and **not** within the Cu-O planes. Moreover, significant delocalized $p\pi$-bonding between oxygen atoms that are second-nearest neighbors in the doped La_2CuO_4 structure is present in the electronic states at E_f. Such an interaction is missed in the decomposition of reciprocal-space band structures using a typical "tight-binding" analysis where only nearest-neighbor interactions are considered.

We consider next the molecular-orbital analysis of the high-T_c superconductor, $YBa_2Cu_3O_7$. SCF-Xα-SW cluster calculations again indicate the presence of oxygen-oxygen $p\pi$-bonding orbitals at E_f, enhanced by localized Cu(d)-O(p) antibonding and by small amounts of highly delocalized

Ba(d)–O(p) bonding character. However, in contrast to the La_2CuO_4 system, the ψ_+ and ψ_- "tubes" at E_f, resulting from the oxygen–oxygen $p\pi$–bond overlap along the "one–dimensional" Cu–O chains (parallel to the x-axis), now are oriented parallel to the y-axis of the Cu–O basal plane, as illustrated schematically in Fig. 8. Note that in this figure the x-, y- and z-axes correspond the a(=3.884Å), b(=3.822Å), and c(=11.675Å) crystallographic axes, respectively. The larger one–dimensional O–O $p\pi$–bond overlap together with the lower symmetry (as compared to La_2CuO_4) implies from Eq. 3 a Jahn–Teller coupling parameter $\beta \lesssim 0.25$. This is the "pseudo–Jahn–Teller" region described above, in which electron pairing is associated with "excitonic" fluctuations in the electron occupancy of nearly–degenerate but spatially different molecular orbitals around E_f, rather than dynamic–Jahn–Teller–induced oxygen ion displacements. Very close in energy to the molecular orbital displayed in Fig. 8 is one in which the barium–oxygen interaction is $d_{x^2-y^2}$–p_z π–bonding rather than d_{z^2}–p_z δ–bonding. The near degeneracy of these states is not surprising from the point of view of crystal-field theory, since the d_{z^2} and $d_{x^2-y^2}$ orbitals of transition–metal ions are usually degenerate in oxides where the oxygen coordination number is greater than or equal to six. Because the $Ba(d_{z^2})$–$O(p_z)$ $d\delta$–bonding states shown in Fig. 8 are localized half way between parallel oxygen chains, while the $Ba(d_{x^2-y^2})$–$O(p_z)$ $d\pi$–bonding states (not shown) are localized closer to the oxygen chains, pseudo–Jahn–Teller–induced "resonance" between these two nearly–degenerate states at E_f implies a dynamic fluctuation δ (parallel to the y-axis) in the average spatial location of an electron relative to the oxygen ion positions (viewed here as fixed). The upper limit to the value of δ deduced from Fig. 8 is approximately equal to b/4, suggesting a maximum $\delta \sim 0.95$Å. The distance between the $p\pi$ "tubular" orbitals ψ_+ and ψ_- shown in Fig. 8 is d = a = 3.88Å. Substituting these values of δ and d into expression 3 (assuming again M=16 for the oxygen mass) gives a value of $\beta \sim 0.14$ for the pseudo–Jahn–Teller coupling parameter. This may be considered to be the lower limit of this parameter in the range $0.14 \lesssim \beta \lesssim 0.25$. Substitution of this range of β-values along with d = a = 3.88Å into formula 6 yields the following range of superconductivity transition temperatures: $87^\circ K \lesssim T_c \lesssim 145^\circ K$. These values are in reasonable agreement with the range of T_c's reported for Y–Ba–Cu–O–based superconductors. It should be noted that the above range of β-values together with the pseudo–Jahn–Teller electron "resonance" ("excitonic") mechanism for electron pairing provide an explanation of the vanishing isotope effect in this superconductor. Moreover, the $\beta = 0.14$ value deduced above corresponds almost exactly to β_{crit} for d = 3.88Å $\sim$ 4Å in Fig. 5 (see also Fig. 3). The Debye temperature calculated for this material from Eq. 9 is $\theta_D \sim 101^\circ K$. The "coherence length" estimated from formula 12 falls in the range $16Å \lesssim l \lesssim 51Å$ for $0.14 \lesssim \beta \lesssim 0.25$.

It will be noticed that no mention of the yttrium coordination region of the $YBa_2Cu_3O_7$ crystal has been made in explaining the origin of high–T_c superconductivity in this material. In other words, the superconducting current is viewed as running along the oxygen–oxygen $p\pi$–bonding "tubes" (parallel to the y-axis) between the basal Cu–O and Ba planes (see Fig. 8). Thus the high–T_c superconductivity is "one–dimensional" in $YBa_2Cu_3O_7$, being spatially confined to the regions between the one–dimensional Cu–O chains and Ba sites. This may be contrasted to our findings above for La_2CuO_4–based superconductors, namely that the supercurrents are effectively "two–

dimensional", flowing along the oxygen-oxygen pπ-bonding tubes that run along eqivalent <110> directions between the square-planar Cu-O planes and La (or Sr,Ba) planes (see Fig. 7). The yttrium regions of $YBa_2Cu_3O_7$, which are somewhat similar locally to the La_2CuO_4 structure (except for the "puckering" of the associated Cu-O planes), could also possibly provide superconductivity along corresponding $O(p)-O(p)$ <110> π-bonds, if the one-dimensional pπ-bonds at E_f shown in Fig. 8 were "broken", e.g. through the formation of oxygen vacancies along the one-dimensional Cu-O chains. The remaining superconductivity associated with the yttrium regions should then occur at the lower transition temperatures $34°K \lesssim T_c \lesssim 49°K$ calculated above for La_2CuO_4-based superconductors. Recent experimental findings for oxygen-deficient $YBa_2Cu_3O_{7-x}$ superconductors may be explained by these conjectures.

PREDICTING NEW SUPERCONDUCTORS

In view of the success of the above molecular-orbital theory in explaining and correlating the observed properties of high-T_c oxide superconductors, along with those of many other types of superconductors[1], one might expect the theory to have truly **predictive** value. Because of the close dependence of this theory on the nature of real-space chemical bonding, it can be applied in conjunction with established principles of coordination chemistry (such as the orbital-electronegativity concept discussed above) to look for novel superconducting materials. For example, how high a transition temperature can one ultimately expect to observe with the right choice of crystal chemistry parameters? The present molecular-orbital theory is quite specific in this regard. If we accept the conventional fixed relationship between the superconductor energy gap Δ and T_c (as given above in Eq. 5), then by optimizing the parameters d and β in formula 6 it can be proven that the theoretical maximum attainable transition temperature is $(T_c)_{max} \sim 240°K$. The conditions for this limit are that $[(m/M)^\beta]_{opt} = 1/3$ and $d_{opt} = 3D/4$, where $D = 10.4Å$. For the oxygen mass $M = 16$ the parameters giving $(T_c)_{max}$ are $\beta = 0.10$ and $d = 7.8Å$, respectively. It is interesting to note that $d = 7.8Å$ is almost exactly twice the lattice constant $a = 3.88Å$ of $YBa_2Cu_3O_7$ and that T_c-values in the vicinity of $240°$ have been reported in a few laboratories. The bad news is that $\beta = 0.10$ is well below the threshold $\beta = 0.25$ discussed above for the static Jahn-Teller effect, which can produce lattice instabilities. Indeed, the oxides which have been reported to show signs of superconductivity around $240°K$ also seem to be structurally unstable, with the result that such high-T_c is short lived. Comparing Figs. 3 and 4, one sees that materials with atoms of relatively low mass at interatomic distances $3Å \lesssim d \lesssim 7Å$, producing spatially delocalized second-nearest-neighbor pπ-bonding at E_f, are the most likely candidates for producing more stable superconductors ($0.20 \lesssim \beta \lesssim .25$) with transition temperatures in the range $100°K \lesssim T_c \lesssim 150°K$. The substitution of the neighboring elements of oxygen (e.g. fluorine and nitrogen) in this class of materials could possibly result in more optimal pπ-bond overlap at E_f and therefore to higher T_c's. According to arguments presented above, optimal second-nearest-neighbor pπ-bond overlap at E_f (which determines β through Eq. 3) will most likely be obtained if the nearest-neighbor interaction (typically with a noble or transition metal) at E_f is antibonding. The orbital electronegativities, such as those shown in Fig. 6, can be used as a guide in choosing the atomic composition that ensures these interactions at E_f. Finally, if the standard BCS relation 5 between Δ and T_c is not correct as Δ increases, then it is possible for the present theory to admit transition temperatures above 240°K. Some experiments have indeed suggested that the BCS proportionality factor between Δ and T_c should be realistically changed by as much as 1.5. This would shift the maximum transition temperature predicted by the molecular-orbital theory to well above room temperature.

ACKNOWLEDGEMENTS

We are grateful to the following agencies for sponsoring this research: the Office of Naval Research (KHJ); the National Science Foundation, Grant No. DMR 8318829 (RCO); the Department of Energy, Grant No. DE-FG-02-ER 45174 (GK).

REFERENCES

1. K. H. Johnson and R. P. Messmer, _Synth. Metals_ 5:151 (1983).
2. J. Bardeen, L. N. Cooper, and J. R. Schrieffer, _Phys. Rev._ 108: 1175 (1957).
3. V. F. Weisskopf, _Contemp. Phys._ 22:375 (1981).
4. M. R. Schafroth, _Phys. Rev._ 130:1244 (1958).
5. J. G. Bednorz and K. A. Müller, _Z. Phys._ B7:189 (1986).
6. M. K. Wu, J. R. Ashburn, C. J. Torng, P. H. Hor, R. L. Meng, L. Gao, Z. J. Huang, Y. Q. Wang, and C. W. Chu, _Phys. Rev. Lett._ 58:908 (1987).
7. J. C. Slater and K. H. Johnson, _Phys. Rev._ B5:844 (1972).
8. M. E. McHenry, R. C. O'Handley, and K. H. Johnson, _Phys. Rev._ B35:3555 (1987).

SUPPRESSION OF HIGH TEMPERATURE SUPERCONDUCTIVITY BY DISORDER

IN THE RESONANT VALENCE BOND MODEL

L. Coffey and D.L. Cox

Department of Physics
The Ohio State University
Columbus, OH

INTRODUCTION

The recently proposed resonant valence bond (RVB) model for high
temperature superconductivity generates a pairing mechanism via the
super-exchange interaction between strongly correlated electrons on
different sites[1,2]. The need to consider such an unconventional,
electronic pairing mechanism is supported by the lack of a measurable oxygen
isotope effect in the 1-2-3 (T_c=90K) materials[3] and apparent absence of
any mode softening as a function of temperature or doping in the
$(La_{1-x},Sr_x)_2CuO_{4-y}$ alloys[4]. These data cast doubt on any explanation of
high transition temperatures relying upon exchange of high frequency phonons
associated with the light oxygen atoms.

The mean field treatment of superconductivity within the RVB model
yields even-parity singlet states (both s-wave and d-wave). The extended
singlet s-wave state considered previously[1,2] is suppressed by non-
magnetic disorder unlike conventional BCS superconductivity which has a
uniform gap function in momentum space.

In this article, we summarize preliminary considerations on the
suppression of superconductivity by non-magnetic disorder in the RVB model.
A brief discussion is presented of the relevance of these results to other
proposed theoretical models for high T_c superconductivity and to some of the
recent experimental data.

RESULTS OF PAIRBREAKING CALCULATIONS

In the absence of disorder, the following results have been obtained in
the RVB model[1,2]: i) At half-filling (where $\delta \to 0$, δ the mean number of
unoccupied sites) for the extended singlet state T_c tends to $J/4$, J being
the nearest neighbor super-exchange interaction. As δ increases, T_c
decreases rapidly. (Recently, it has been pointed out that the d-wave order
parameter is in fact more stable away from half-filling[5].) ii) The
superconducting state quasiparticle energies E_k are given by

$$(1) \qquad E_{\vec{k}} = \sqrt{\xi_{\vec{k}}^2 + (\Delta + \Delta'\xi_{\vec{k}})^2}$$

where Δ is the value of the extended singlet gap function at the fermi
level and Δ' is the slope of the gap function with respect to $\xi_{\vec{k}}$, the normal

state quasiparticle energy. The latter quantity is given by
$-2t\delta(\cos k_x a + \cos k_y a) - \tilde{\mu}$, where a is the lattice constant and $\tilde{\mu}$ the effective
chemical potential of the quasiparticles.

Disorder is included by inserting a one particle level broadening
$\Gamma = 1/2\tau$ in the quasiparticle spectra. The broadening is assumed to be due
to potential scattering, and is proportional to the concentration of
scattering defects. Away from half-filling (and the logarithmic singularity
in the density of states), standard analysis[6] yields a Maki pairbreaking
parameter ζ given by

$$(2) \qquad \zeta = \frac{1}{2\tau\Delta}\frac{\Delta'^2}{1 + \Delta'^2}.$$

Note that ζ is explicitly proportional to the derivative of the gap function
with respect to $\xi_{\vec{k}}$, so that Anderson's theorem holds in the absence of
momentum dependence. The excitation gap $\tilde{\Delta}_{ex}$ is reduced from its bare value
$\Delta_{ex}=\Delta/\sqrt{1 + \Delta'^2}$ according to

$$(3) \qquad \tilde{\Delta}_{ex} = \Delta_{ex}(1 - \zeta^{2/3})^{3/2}.$$

Clearly gapless superconductivity is possible for ζ exceeding the critical
value of unity.

Because ζ vanishes at T_c, there are no pair-breaking effects in the BCS
logarithmic (in T_c) term appearing in the transition temperature equation.
However, pairbreaking effects do appear in the additional terms associated
with the momentum dependence of the gap function. These terms dominate the
T_c equation near half filling. Expanding systematically in powers of δ near
half-filling, the reduction of T_c is given to the leading two orders by

$$(4) \qquad \frac{\delta T_c}{T_c} \simeq -\frac{1}{J\tau}[1.08 + 8.85(\frac{t\delta}{J})^2].$$

A similar analysis holds for the d-wave gap for which the coefficient of the
δ^2 term is reduced to 0.98. (Corrections of order $(J/t)^2$ with t the hopping
integral have been omitted from the above result.)

It should be noted that as $\delta \rightarrow 0$, the Born approximation for the
scattering rate diverges as $1/\delta$. Thus, the full superconducting state
t-matrix must be used in the equations for the renormalized frequency and
gap functions. This is the subject of future studies. The full normal
state t- matrix yields a scattering rate which <u>vanishes</u> as δ, so that T_c
suppression should set in only as one moves away from half-filling.

RELATION TO OTHER THEORY AND EXPERIMENT

The above results clearly apply to other models of superconductivity
which involve extended singlet order parameters, including those mediated by
antiferromagnetic spin fluctuations[7] and localized phonon modes[8]. It is
well known that for order parameters involving higher angular momentum
states, sensitivity to disorder will result. Interestingly, within the RVB
model the d-wave order parameter is less sensitive to disorder than the
s-wave order parameter.

The application of the RVB model to the Sr doped La_2CuO_{4-y} system is
plausible, particularly in view of the straightforward interpretation of the
Sr atoms as electron acceptors[9]. There is no dopant atom playing a
similar role in the 1-2-3 90K superconductors, and the applicability of the
RVB model is unclear. The only tuning of the occupancy is controlled by the

oxygen defect content away from ideal stoichiometry (O_7), which should serve
to _increase_ the filling. Nevertheless, rapid quenching of the high
temperature tetragonal phase shows significant suppression of the transition
temperature to as low as 20-30K[10]. This tetragonal structure apparently
has disordered oxygen vacancies in the plane between the Ba atoms. Thus, it
appears that the orthorhombic order is necessary to restore the full
transition temperature of 90K, a result consistent with the present
calculations.

ACKNOWLEDGEMENTS

We thank C. Jayaprakash and D. Stroud for useful discussions, and we
thank J.W. Wilkins in particular for a conversation on the small δ limit.
We thank the several authors for sharing preprints prior to publication.
D.L.C. acknowledges research support from an Ohio State University Seed
Grant, and L.C. acknowledges support from General Motors.

References

[1] P.W. Anderson, Science _235_, 1196 (1987); G. Baskaran, Z. Zou and P.W.
    Anderson, preprint (submitted to Sol. St. Comm.).

[2] A. Ruckenstein, P.J. Hirschfeld, and J. Appel, preprint.

[3] B. Batlogg et. al., Phys. Rev. Lett. _58_, 2333 (1987), and L.C. Bourne
    et. al., Phys. Rev. Lett. _58_, 2337 (1987).

[4] B. Renker et. al., preprint.

[5] G. Kotliar, preprint.

[6] K. Maki, in _Superconductivity_, ed. R.D. Parks (Marcel-Dekker, New York,
    1969) p. 1036.

[7] J. Hirsch, Phys. Rev. Lett. _54_, 1317 (1985).

[8] J. Hirsch, preprint.

[9] N.P. Ong et. al., preprint.

[10] R. Beyers, Special Proceedings of the Materials Research Society
     Meeting, Anaheim, April 1987 (to be published).

DISCOVERY AND PHYSICS OF SUPERCONDUCTIVITY ABOVE 90K

C. W. Chu, P. H. Hor, R. L. Meng, L. Gao, Z. J. Huang,
Y. Q. Wang and J. Bechtold

Department of Physics and Center for Space Vacuum Epitaxy
University of Houston

I.   THE DISCOVERY OF 90K-SUPERCONDUCTIVITY

The search for superconductivity with a practically high transition T_c,
i.e. above 77K-the liquid nitrogen temperature, has been one of the most
challenging tasks of physicists and material scientists ever since the
discovery of superconductivity in 1911. In spite of the great efforts over
the last three quarters of a century, until 1986 T_c had been raised by only
20K, to 23.2K first observed[1] in 1973. In view of the slow progress, many
non-conventional approaches have been proposed and tested. They fall into
the general categories of synthesizing and investigating novel materials
under normal and extreme conditions, and the search for novel
superconducting mechanisms. There exist, to date, no definitive guidelines
to predict material systems that will exhibit high T_c and/or novel
superconducting mechanisms. In this respect, an empirical search for new
high-T_c materials remains the most effective approach. It has been
suggested[2] that it may be fruitful to investigate superconductivity under
various conditons in compounds generally not favorable for conventional
superconductivity. The superconducting oxides are examples of such systems,
since the great majority of the known superconductors are inter-metallic
compounds.

Oxidation is known to degrade, in general, the metallic and, in
particular, the superconducting characteristics of matter. However,
superconductivity has been observed in some highly oxidized perovskite and
related compounds[3] such as $SrTiO_{3-x}$, Na_xWO_3, $Li_{1+x}Ti_{2-x}O_4$ and $BaPb_{1-x}Bi_xO_3$,
with a T_c exceeding 13K for the last two oxides. Such a high T_c is
especially unusual for $BaPb_{1-x}Bi_xO_3$, which consists of no transition metal
elements and displays no large electron density of states[4], both being
essential for conventional high T_c intermetallic compounds. Therefore,
$BaPb_{1-x}Bi_xO_3$ the predecessor of the current high-T_c oxides, has been
one[2,5] of the focal points in the search for high-T_c materials at Houston
immediately after its discovery in 1975. Other groups carried out extensive
studies in this unusual $BaPb_{1-x}Bi_xO_3$ compound system including AT&T Bell

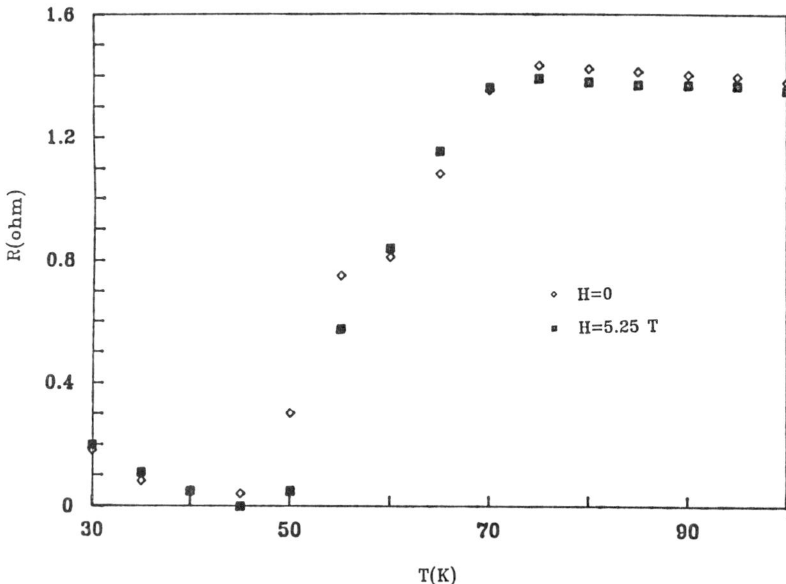

Figure 1. R vs. T for a LaBCO sample.

Labs and the University of Tokyo. Non-conventional superconducting mechanisms have been suggested[2] to account for the observations in $BaPb_{1-x}Bi_xO_3$.

To raise the T_c of the oxides, the Zürich group carried out a series of investigations in an attempt to enhance the electron-phonon interaction through polaron formation in a Jahn-Teller system. In 1986, they achieved[6] superconductivity in the 30K range in the La-Ba-Cu-O (LaBCO) system. Independently, the Tokyo group[7] and Houston group[8] reproduced the observation in late 1986. Furthermore, the Tokyo group[9] unambiguously attributed the superconductivity observed in the 30K-range to be the single layer K_2NiF_4 phase of LaBCO, i.e., $(La_{0.85}Ba_{0.15})_2CuO_{4-\delta}$.

To explore the nature of this unusally high temperature superconductivity, we subjected the LaBCO compounds to high pressure. The T_c was found to be enhanced at an unprecedentedly high rate of 1K/kbar, about 100 times that for superconductors previously known. An onset temperature T_{c0} up to 57K was achieved[10] in LaBCO, with the K_2NiF_4 structure under 12 kb. This suggests that a non-conventional superconducting mechanism may be in operation in these oxides. To simulate the pressure effect, we replaced Ba with the smaller Sr and found $(La_{1-x}Sr_x)_2CuO_{4-\delta}$ to superconduct with a T_{c0} ~42.5K at ambient pressure. Similar results were

observed by other groups including AT&T Bell Labs[11], Beijing[12], Belcore[13],
Argonne[14] and Naval Research Lab.[15] Unfortunately, we observed that
$(La_{1-x}Ca_x)_2CuO_4$ only has a T_c ~20K, although Ca has a smaller radius than
Sr. This immediately showed that the mere reduction in inter-atomic
distance cannot continue to raise T_c to above 60K in oxides with the
K_2NiF_4 structure.

By examining the large volume of data accumulated by the end of 1986,
we found that 1) the T_{c0} is always reduced when the LaBCO sample is made a
pure single-layer K_2NiF_4 structural phase which displays a sharp transition
at ~35K, and 2) a sharp resistance drop, as shown in Fig. 1, indicative of a
superconducting transition above 70K, occurs randomly but only in mixed
phase LaBCO samples consisting of structures in addition to K_2NiF_4 (a
similar observation by the Beijing group was also reported on January 16,
1987). Therefore, we concluded that superconductivity above 70K can be
found only in compounds with structure different from K_2NiF_4. Many mixed
phase samples were subsequently made under different conditions and tested
for superconductivity. Some of these were purposely synthesized in an
oxygen-deficient environment in the hope that a wide range of phases could
be formed in one sample. One of these samples was a Ba-rich LaBCO coated
with a dull-pink insulating layer. Without perturbing the sample, we
decided to measure its magnetic properties. On January 12, 1987, we
observed a large diamagnetic shift in this sample, starting at ~100K and

Ba-La-Cu-O J-31

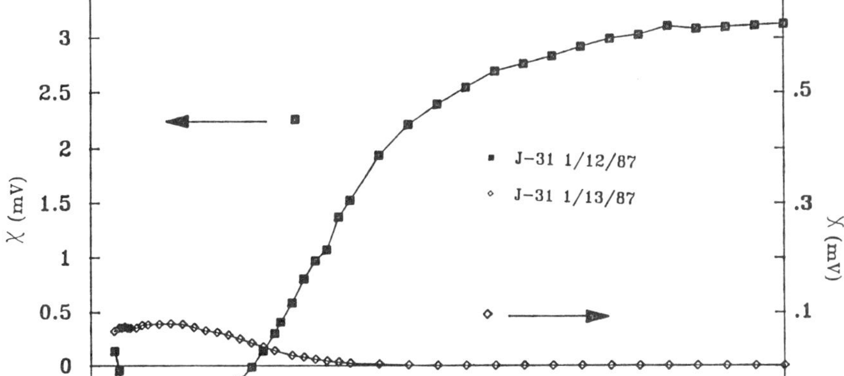

Figure 2. The diamagnetic shift of a mixed phase LaBCO
sample below 100K, and its disappearance a day later.

reaching at ~4K a value of ~40% of that for a bulk superconductor, as shown
in Fig. 2. Unfortunately, the diamagnetic signal disappeared the next day.
These observations strongly convinced us that superconductivity above 77K
must exist and the only question left was how to stabilize the high
T_c phase.

From our high pressure results, we found that smaller atoms tend to
form a high temperature superconducting phase more easily. La (in Ba-rich
LaBCO) is the largest among all rare-earth elements. It was, therefore,
very natural for us at Houston and Alabama to try to synthesize mixed phase
A-Ba-Cu-O compounds with A=Y, Yb and Lu which all occupy the same column in
the periodic table and are not magnetic (only later did we learn[16] that
magnetism in A does not affect the superconductivity). On January 29, 1987,
we detected in the mixed Y-Ba-Cu-O (YBCO) compounds a resistance (R) drop,
starting at 93K and completing at 80K. Meissner effect was clearly evident
in < 24% of the samples below ~90K in these compounds. Consequently,
superconductivity above 77K was finally unambiguously and reproducibly
achieved.[17] Subsequent magnetic field effect measurements up to 7T, made by
us in Houston, indicated a record-high upper critical field H_{c2} ~180T at 0°K
for these compounds. Within a few days, the superconducting transition was
sharpened to between 93K and 90K, and later enhanced[18] to between 98K and
94K. Preliminary pressure and x-ray studies showed[19] that the newly
discovered compound was different from the K_2NiF_4 structural compound giving
rise to 30K-superconductivity. Subsequent dc-magnetization measurements
caried out in collaboration with the Lockheed and Los Alamos group
showed[20] a large Meissner effect and unusual glassy behavior in our samples.
Part of the results were received on February 6, 1987 by Physical Review
Letters (PRL) for publication. A press release was made on February 16,
1987 following the acceptance of our papers by PRL on February 11, 1987. It
is interesting to point out that the decisive Y-element used by us to
stabilize the first 90K-superconductor YBCO appeared in the local newspaper,
the Houston Chronicle, on both February 16 and 17. The next observation of
superconductivity above 70K in YBCO came on February 23, from a Japanese
group[21] at Tokyo. The People's Daily reported the similar observation on
February 25, 1987 by the Physics Institute group[22] at Beijing and on
February 28 by the Berkeley group.[23] Also in February, superconductivity
starting at 90K was detected[24] in the Lu-Ba-Cu-O compound by the Brookhaven
and Ames group. More than 30 copies of three of our preprints on YBCO were
sent to various laboratories by Federal Express on February 25, 1987. An
avalanche of reports reproducing the results followed the next week.

After superconductivity above 90K was firmly established by us in the
mixed phase YBCO, we decided to identify the phase responsible for the
observation in collaboration with the group at Geophysical Laboratory at
Washington, D.C. By February 27, both the chemical formulas of
$YBa_2Cu_3O_{6+\delta}$[25] and the $Y-CuO_2-BaO-CuO_{2-\delta}-BaO-CuO_2-Y$ layer structure were
determined.[25] Similar structural results were also obtained the following
week by the AT&T[26] and IBM groups.[27] However, the exact location of the
oxygen vacancies were not determined until later by the Argonne group using
neutron-scattering.[28] Linear Cu-O chains along the b-axis were found to
exist due to the complete absence of oxygen along the a-axis between the
BaO-layers.

II. THE NEW CLASS OF 90K-SUPERCONDUCTORS: $ABa_2Cu_3O_{6+\delta}$

With the exact stoichiometry and the general structure of the
superconducting phase determined, we immediately replaced Y by the rare-
earth elements to examine the role of Y in high-temperature
superconductivity. To our great surprise, the complete replacement of Y
even with the most magnetic Gd does not affect the T_c of these compounds. A
new class of superconductors, $ABa_2Cu_3O_{6+\delta}$ with A = Y, La, Nd, Sm, Eu, Gd,
Ho, Er, and Lu, with T_c above 90K was therefore discovered.[16]
Independently, the groups at Los Alamos,[29] Tokyo,[30] and AT&T[31] had also made
the similar observation prior to the Special High T_c Session of the March
American Physical Society Meeting on March 18, 1987. The results
show[32] that the "A" elements are used only to stablilize the so-called
three-layer structure to sustain the 90K superconductivity between the Y-
layers and that superconductivity must be confined to the three Cu-O layer
assembly sandwiched between the Y-layers. By comparing the single-layer
30K-superconductor $(La_{1-x}Ba_x)_2CuO_{4-\delta}$ and the three-layer 90K-superconductor
$LaBa_2Cu_3O_{6+\delta}$, one may not be surprised to observe high T_c in compounds with
more Cu-O layers in the future. It should be noted that due to the large
size of La, the three-layer structure is marginally stable and the
transition is broad from 95K to 75K. However, this also provides a unique
opportunity for us to explore many possibilities of enhancing the T_c and
improving the superconducting properties of these and related compounds. To
bring La back to the mainstream of 90K-superconductivity is important[32] for
large scale application because it is inexpensive and plentiful.

However, it should be pointed out that, as of yet, no effort of ours
has succeeded in synthesizing superconducting $ABa_2Cu_3O_{6+\delta}$ with A=Ce or Pr,
or $ASr_2Cu_3O_{6+\delta}$, with A = Y, La, and Eu. Whether this is due to our failure
in achieving the optimal synthesis conditions or the mismatch of the atomic
sizes of the elements invovled remains to be determined. We have observed
the greatest stability of the compounds when A lies in the middle of the
rare-earth series. Careful examinations of the above will undoubtedly lead
to a better understanding in high temperature superconductivity.

III. THE PHYSICS OF SUPERCONDUCTIVITY ABOVE 90K

As mentioned earlier,[10] oxidation is known to degrade the
superconducting properties of matter, and yet superconducting has been
observed in a few highly oxidized perovskite related compounds, in addition
to the newly discovered high T_c oxides. In spite of the great variation of
their T_c's, these compounds share some characteristics in common. For
instance, $BaPb_{1-x}Bi_xO_3$, $(La_{1-x}Ba_x)_2CuO_{4-\delta}$ and $YBa_2Cu_3O_{6+\delta}$ all have very low
electron density of states at the Fermi surface $N(0)$. Superconductivity
occurs over a narrow x (or δ)-range, with the highest T_c near the metal-
insulator phase boundary. All of them possess oxygen-metal bonds of
particular configuration and mixed valence sites. Due to the small $N(0)$,
many mechanisms have been proposed to enhance the superconducting pairing
interaction. They include soft modes,[33] interfaces,[34] negative
U-centers,[35] bipolarons,[36] plasmons,[37] spin fluctuations[38] and charge
transfer fluctuations.[39] Because of the close interplay between the metal-
insulator transition and superconductivity, a reversed approach from a

highly correlated or even localized state to an itinerant superconducting
state has been suggested. The resonating valence bond model[40] falls into
this category. Many of these models depend critically on the
dimensionality, valence state of the Cu-ions, and the scaling energy of the
newly discovered high-temperature superconductors which are not yet
completely known. Very often, the question of if these superconductors
belong to the so-called BCS superconductors is also raised. In the absence
of a complete picture of this unusually high temperature superconductivity
in oxides, we would like to discuss the several points mentioned above:

A. BCS or Non-BCS Superconductors

There are three ingredients in the BCS theory: 1) superconducting
electron pair formation, 2) mathematical framework developed to describe the
pair-formation and superconducting properties and 3) the interaction
responsible for the electron pairings. They will be addressed separately.

First, electron pair formation in the superconducting state of
these high T_c oxides has been firmly established by flux quantization[41] and
micro-wave experiments on these materials.[42] The temperature dependence of
the penetration depth obtained from a muon-experiment[43] is in agreement with
BCS prediction.

Most of the models proposed to date still fall into the framework
of BCS. The BCS universal gap equation predicts a $2\Delta/kT_c$ at 0°K of 3.5, in
contrast to the experimental values[44] ranging from 2.2 to 16. The short
coherence length and the poor surface quality of these high T_c oxides do
pose great challenges to experimentalists to determine the gap. The small
and poorly-shaped specific heat anomaly[45] at T_c and the linear term[46] of the
specific heat below T_c observed recently lead to some of the
suggestions[47] of a gapless superconducting state in these oxides. This
would be consistent with predictions of the bipolaron model, according to
which superconductivity results from a Bose-Einstein condensation of the
bipolarons.[48] However, it is not clear if the above observations on
specific heat are intrinsic properties of the superconducting phase. It
appears to us that a glassy superconducting state[49] due to the porous nature
of the sample may give rise to a cusp-anomaly at T_c and linear term below
T_c in its specific heat, similar to spin-glasses (although cusp-anomaly has
been observed in the magnetic susceptibility but not in the specific heat at
the spin-glass transition temperature). On the other hand, the glassy
superconducting behavior can also arise from the unusal intrinsic structure
of the materials.

Finally, we consider the superconducting pairing interaction.
Although any attractive interaction between electrons is sufficient for
electron-pair formation by BCS theory, traditionally, such an attraction has
been attributed to an electron-phonon interaction. One may argue
qualitatively that a system with a strong electron-phonon interaction is
intrinsically unstable. As a result, such a system will undergo a
structural transformation upon cooling to a structure no longer favorable to
strong electron-phonon interaction before high temperature superconductivity
can be realized. Several estimations were made[39] based on such an
interaction and could not give a T_c above ~40K. However, it was also

proposed that higher T_c might be possible in the strong-coupling limit.[50]
For these oxides, it has been suggested that the strong electron-phonon
interaction is associated with the soft phonon modes of the oxygen. This
would predict an isotope effect of $T_c \sim M^{-1/2}$, where M is the isotopic mass
of the oxygen. However, no isotope effect was detected.[51] Therefore, it
seems that phonons cannot play the same dominant role here in the high
T_c oxide as that in the conventional low T_c intermetallic superconductors.
We believe that the combined effects of phonons and other excitations of
electronic nature mentioned earlier will be important.

<u>B. Dimensionality</u>

Almost all models[35-40] proposed and developed after the obervation
of superconductivity above 30K are based on the low dimensional (1 or 2)
character of the oxides. Structural studies[25-28] show Cu-O layers and even
Cu-O chains in these compounds. Magnetic field measurements on epitaxial
films[52] and single crystals[53] did demonstrate that these compounds are
highly anisotropic. Band calculations[54] indicated that these layers and
chains are only weakly coupling to each other. Low dimensional systems are
the theorists' dream to obtain singularities and instabilities which will
provide the much needed large density of states and soft modes for high T_c.
They also form the proper environment for two branches of plasmons.[37,55]
However, experimentally, the $N(0)$ determined from low temperature specific
heat studies[45] is low, and little difference in the phonons spectra[56] exists
between the superconducting and non-superconducting $(La_{1-x}S_x)_2CuO_{4-\delta}$. On
the other hand, if one considers these oxides as composite systems
consisting of conducting Cu-O layers separated by insulating layers of Ba-O
or Y, the $N(0)$ so determined can be underestimated.

Now let us consider the dimensionality problem more specifically
from an experimentalist viewpoint:

<u>1. 1D linear Cu-O chains</u>

From neutron scattering,[28] Cu-O forms linear chains along the
b-axis between the two Ba-O layers. The high-temperature superconductivity
observed has been attributed to these chains. In order to maintain the
integrity of these claims, δ has to be 1 in $ABa_2Cu_3O_{6+\delta}$. We have measured
both the magnetic and electrical properties of Eu-Ba-Cu-O, Y-Ba-Cu-O and
La-Ba-Cu-O three-layer compounds, heat treated under vacuum condition, or
180 atmospheres of oxygen. A T_c-maximum was found to exist in the vacuum
annealed samples.[57] More experiments are needed to determine if δ is 1 when
T_c is maximum. However, the samples prepared in <1 atomosphere of oyxgen
are known[28] to have $\delta \sim 0.8$, with a $T_c \sim 90K$. One may argue that since the
coherence length is so short ($\sim 15A$) for these oxides, chains of comparable
length might be sufficient to achieve their full superconductivity.

On the other hand, there exist many organic linear chain
superconductors[58] in nature. To obtain a complete superconducting
transition one invariably needs a single crystal. Polycrystalline samples
behave like semiconductors. However, polycrystalline superconducting oxides
always display a complete transition when properly treated.

Presently, no a-b anisotropy has been detected in either
epitaxially grown films[52] or single crystalline samples[53] in contrast to a

linear chain conductor. However, the twin phase boundary due to the tetragonal-orthohombic transition[58] at ~750°C may mask the anisotropy. A definitive answer cannot be made without results on a single crystalline sample with no twinning.

2. 2D Cu-O layers

The fact that the T_c of $ABa_2Cu_3O_{6+\delta}$ is almost independent of "A" suggests[16,32] that superconductivity in this class of compounds must be confined to the CuO_2-BaO-$CuO_{1+\delta}$-BaO-CuO_2 layer assemblies sandwiched between the A-layers, and Josephson-type coupling must exist between these layer-assemblies. This is consistent with the broadening of the superconducting transition observed in polycrystalline samples in magnetic fields. We have carried out several experiments to determine the exact location of superconductivity in these compounds. Careful magnetization measurements[59] were made on $ABa_2Cu_3O_{6+\delta}$ with A=Y, Gd, Sm , and Eu as a function of magnetic field (H) up to 0.1T from T_c down to 4K. For A = non-magnetic Y, the diamagnetic moments (-M) vary linearly with H when H is smaller than the lower critical field H_{c1}. However, for A = the magnetic Gd, Sm and Eu, there exists an anomally at ~0.01T as shown in Fig. 3. The slope below ~0.01T is greater than above, suggesting that there exists two superconducting components, one with H_{c1} smaller than ~0.01T and the other greater. ESR measurements[60] on the Gd-resonance line in a Gd-doped $YBa_2Cu_3O_{6+\delta}$ show a g-shift of 2.3, suggesting a rather strong coupling between the Gd-atoms and their immediate vicinity, i.e. CuO_2-layers adjacent to the Y-layers. Specific heat measurements[61] on $GdBa_2Cu_3O_{6+\delta}$ display a large lambda anomaly at 2.2K resulting from an antiferromagnetic ordering, consistent with its Curie-Weiss magnetic behavior above T_c with a negative Curie-Weiss temperature of ~4K. The magnetic studies[62] on a non-superconducting $GdBa_2Cu_3O_{6+\delta}$ after vacuum annealing show that the magnetic coupling is strongly enhanced, suggesting the excess oxygen in the center $CuO_{2+\delta}$ layers between the BaO-layers affects the magnetism of the Gd-layers. From the above experiments, there apparently exists some coupling between the Cu-O layer assemblies and the A-layers. It is possible[32] that the outer two CuO_2-layers act (with lower H_{c1}) like a shield to protect the center $CuO_{2+\delta}$ layers (with greater H_{c1}). However, for superconductivity to occur, three Cu-O layers between the A-layers are required simultaneously, i.e. the outer two CuO_2-layers by themselves cannot support superconductivity, as evident from the absence of superconductivity when oxygen is removed from the center layers by vacuum heat-treatment.

The deterioration of the 90K-transition in $LaBa_2Cu_3O_{6+\delta}$ by oxygenation at high pressure[63] is puzzling. Whether this is caused by the occupation of the vacant sites in the La-layers by oxygen-atoms, thus reducing the 2D character of the compounds is yet to be determined. This seems plausible in view of the large atomic size of La.

3. Valence state (states) of Cu ions

As pointed out earlier, all high T_c oxides appear to possess mixed valence sites by charge balance. Many of the models proposed depend critically on the mixed valence state of the Cu ions in $ABa_2Cu_3O_{6+\delta}$. However, the experimental situation is far from being clear at the present time. Some conflicting reports are cited below.

The X-ray near edge absorption experiments showed +2 and +3-states in $(La_{1-x}Sr_x)_2CuO_{4-\delta}$ by the Argonne group. However, only +2-states were reported by the Brookhaven group.[64] Apparently, the absorption peak at ~20eV originally identified as the +3-state remains even for the non-superconducting La_2CuO_4-sample where only the +2-state can exist.

The XPS experiments[65] also show different results for Cu-ions: +1, +2 and +3 in one case, and +2 and +3 in another.

The ESR results on Cu-sites are not definitive either. It is known that only +2-Cu (not +1 or +3-Cu) can give rise to an ESR line. The Illinois group found no Cu-resonance line present in their pure $(La_{1-x}Sr_x)CuO_4$ samples. We observed[60] an ESR line in $YBa_2Cu_3O_{6+\delta}$ but not in $EuBa_2Cu_3O_{6+\delta}$ up to 300K, while both are superconducting above 90K. It is not yet clear if the Cu-resonance line in $YBa_2Cu_3O_{6+\delta}$ is caused by the presence of minute Y_2BaCuO_5 where Cu is +2-valent, or the absence of the

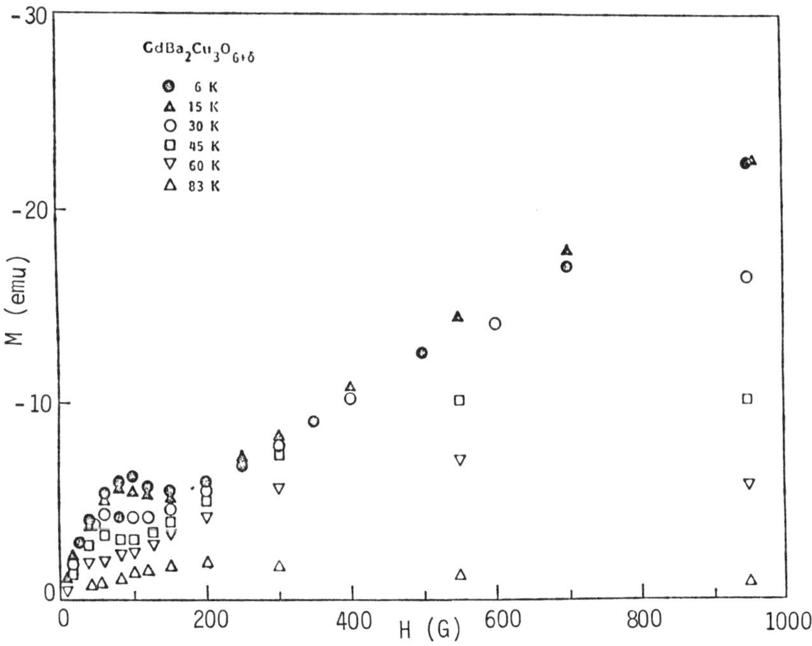

Figure 3. M vs. H for $GdBa_2Cu_3O_{6+x}$.

Cu-line in $EuBa_2Cu_3O_{6+\delta}$ is due to possible magnetic broadening by the Eu atoms. However we do know that $EuBa_2Cu_3O_{6+\delta}$ appears to be a more stable compound than $YBa_2Cu_3O_{6+\delta}$, and that Eu is generally not expected to become magnetic until cooled to low temperature.

Since +2-Cu ions have spin = 1/2, they exhibit magnetic moments. The existence of these magnetic moments should result in a Curie or Curie-Weiss behavior. However, such a behavior has been observed in some but not all 90K-superconducting compounds. Much attention has been given to the antiferromagnetic ordering[66] in La_2CuO_4, associated with the +2-Cu ions in the compound with the exact stoichiometry given by the above formula. However, it has been shown recently that the antiferromagnetic ordering occurs only in oxygen-deficient samples.[67] A fully oxidized compound becomes superconducting[63,68] below 40K. Information about the exact amount of oxygen will be extremely important in clarifying the role of the valence state of Cu in these compounds.

The above experimental results, even though they are incomplete, indicate that questions concerning the role of +2-Cu ions in high temperature superconducting oxides, and the interplay between antiferromagnetic and superconducting interaction deserve re-examination.

4. Scaling Energy

Recently, a sharp drop in the e^+-lifetime was detected[69] at T_c in the superconducting oxides but not in the non-superconducting ones as shown in Fig. 4. Since the lifetime of e^+'s depends on the density of all electrons not just those near the Fermi surface, the observation implies that a substantial fraction of electrons must be involved in the superconducting process. This is in agreement with more recent results[43] that the carrier density in the superconducting state obtained from muon experiments is similar to that in the normal state determined from optical studies. The fraction of electrons (or holes in the present case) participating in the electron-pairing depends on Δ/E_F, where Δ is the energy gap and E_F the Fermi energy. The results imply a small E_F for the high T_c oxides. This is consistent with the small coherence length observed and the prediction[70] based on low dimensionality of the compounds. The large fraction of electrons participating in the superconducting state may also be related to a large cutoff energy in the energy spectrum in contrast to the Debye energy for conventional superconductors. A cutoff energy larger than the Debye energy may imply excitations of electronic nature responsible for the superconductivity observed in oxides.

The total electron density of the compound is not expected to change at T_c without undergoing a structural transformation as in the high T_c oxides. The drastic drop in the e^+-lifetime may be caused by the variation in the annihilation rates or the electrical current paths above and below T_c. It is therefore interesting to examine the possible difference between the boson (Cooper pair)-fermion (e^+) interaction below T_c and the fermion (e^-)-fermion (e^+) interaction above T_c; and to investigate the possible change in anisotropy above and below T_c.

There have been a few newspaper reports of detecting superconductivity
above room temperature, 300K. It is, therefore, helpful to summarize the
general situation of superconductivity above 100K. We all know that the two
necessary and sufficient criteria for the establishment of superconductivity
are: 1) zero resistivity and 2) Meissner effect in a large fraction of the
samples. For the convenience of discussion and for practicality, we would
like to add two more constraints: 3) stability and 4) reproducibility. A
superconductor is considered to be stable only if it can survive the thermal
cycling and the associated disturbance for thorough testings. It is
reproducible only if one can synthesize the samples with proper
superconducting properties with a large yield probability. For all four
criteria to be satisfied, the highest T_c remains in the 90–100K range. If
one wants to relax criterion 2) to Meissner effect for a small fraction of
the samples and abandon 3) and 4), T_c at 155K[71] and 225k[72] have been
reported, although the 225K sample survived a week for resistance
measurement and three weeks for magnetic studies. In case one is satisfied
with only criterion 1), T_c at 285K[73] was detected. Indications of
superconductivity at 240K and higher have been detected[23,74] as sharp drops
in resistance without truly achieving the zero–resistance state. Such
resistance drops were too unstable and too difficult to reproduce in a
systematic fashion for diagnosis. However, strong evidence of
superconductivity, the appearance of the reverse Josephson effect, has been
reported by the Wayne State group[75] below 240K and by an Indian group at
302K.

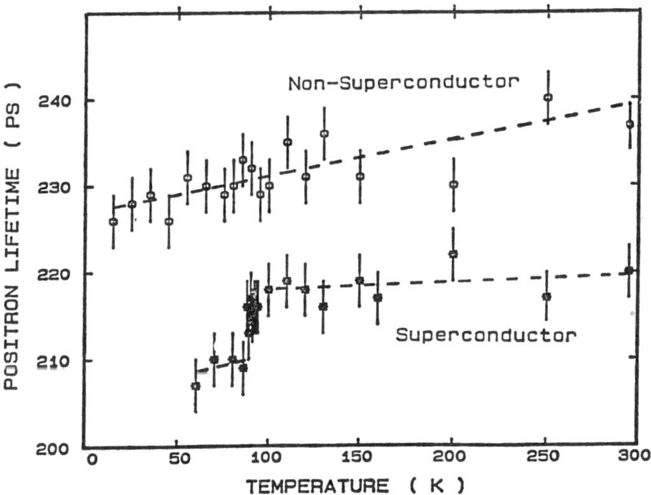

Fig. 4. Temperature dependence of the e^{+}-lifetime in
superconducting and non-superconducting
$YBa_2Cu_3O_{6+x}$.

If one accepts all reports of superconductivity substantially
higher than 100K, the immediate question is: what is the location of
superconductivity in the samples? Based on our studies, we conclude that it
cannot be associated with a random distribution of superconducting
dispersions because a zero resistance path would have required a minimum of
16% of the volume fraction to be superconducting which is substantially
larger than the magnetic anomaly observed. It must be filamentary and
located in the grain boundaries or in the twin phase boundaries due to the
tetragonal-to-orthohombic transition at ~750°C. The superconducting
filaments may then be extremely fragile physically due to difference in

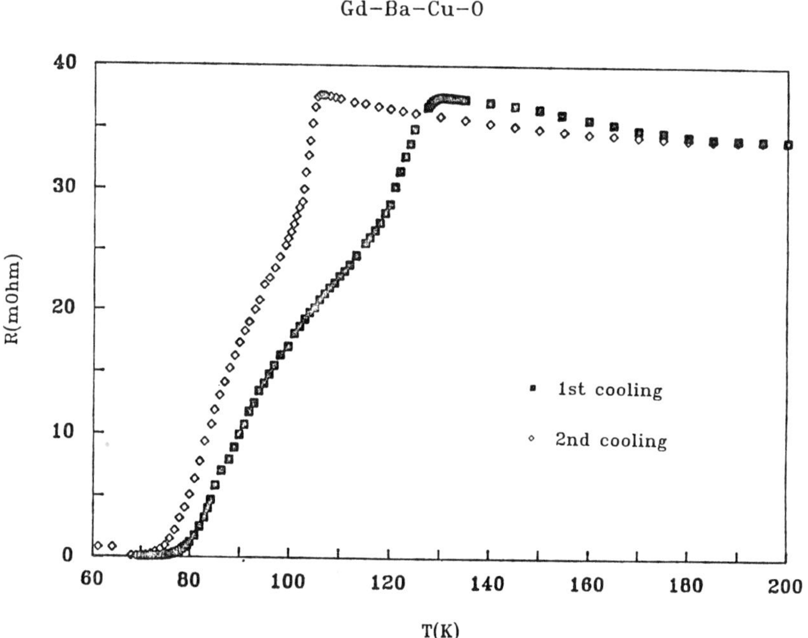

Figure 5. R vs. T for a specially treated Gd-Ba-Cu-O
sample, displaying the effect on T_C due to thermal
cycling.

thermal expansions between them and the surroundings, or unstable chemically
due to their strong interaction with the surrounding. Thermal cycling can
also perturb the oxygen content at the grain boundaries and in turn cause a
change in chemistry. It would not be surprising to conjecture that there
might exist a competition between the 90K-superconducting phase and the
higher T_c filamentary phase. It also appears to us that $ABa_2Cu_3O_{6+\delta}$ may
just be loosely-jointed "compounds" of $AO_{1.5}$, BaO and CuO_2 with some O
"glue". We found that the joining force varies with the atomic size of "A".
To exploit the relative stabilities of the 90K-phase and the filamentary
components may therefore prove fruitful.

In the face of the above stability and reproducibility problems, we at
Houston are trying to control the uncontrollable. We found that by the
method of evacuation, application of electrical field and variation of
synthesis conditions, the reproducibility of achieving an onset
superconductivity between 110 and 130K has tremendously improved. As shown
in Fig. 5, the sharp superconducting transition starts at 130K on the first
cooling but is lowered to 110K on subsequent warming and finally stays at
110K. An unusually large temperature-dependent resistance has been observed
by us in a La-Ba-Cu-O compound starting at a temperature above 360K, in
strong constrast to that for ordinary 90K-superconductors above T_c.

V.   CONCLUSIONS

Great progress has been made in the understanding of the high
T_c compounds. However, many basic problems remain, such as the valence
state of Cu ions, the role of +2-Cu ions, interplay between magnetic and
superconducting interactions, the exact role of dimensionality, and the
mechanisms responsible for the unusally high temperature superconductivity
in oxides.

Stable and reproducible superconductivity has been unambigously
establshed only up to the 90-100K range in the class of compounds
$ABa_2Cu_3O_{6+\delta}$ with A= Y, La. Evidence of superconductivity, although strong
but plagued by poor stability and poor reproducibility, has been reported at
temperatures above 300K. However vague, it is interesting to point out that
similar problems existed in the 30K and then 90K-superconductivity prior to
the identification of the correct phases. Whether history will repeat is
yet to be seen.

VI.   ACKNOWLEDGEMENTS

Discussion with C.Y. Huang, M.K. Wu, H.K Mao and C.Y. Jean are greatly
appreciated. The work is partially supported by NSF grant DMR 8616537 and
NASA Grants NAGW-977 and NAG8-51. It should be noted that references cited
here are far from being complete. They only represent a very small fraction
of the work done to date in this exciting field.

REFERENCES

1. J. R. Gavaler, Appl. Phys. Lett. $\underline{23}$, 480 (1973); L. R. Testardi, J. H. Wernick and W. A. Roy, S. S. Comm. $\underline{15}$, 1 (1974).

2. See for example, C. W. Chu and M. K. Wu, in High Pressure Science and Technology, ed. by C. Homan, R. K. MacCrone and E. Whalley (North-Holland, New York, 1983) Vol. 1, P. 3; C. W. Chu, T. H. Lin, M. K. Wu, P. H. Hor and X. C. Jin, in Solid State Physics under Pressure, ed. by S. Minimura (Terra Scientific, Tokyo, 1985) P. 223.

3. $SrTiO_{3-x}$: J. F. Schooley, W. R. Hosler, M. L. Cohen, Phys. Rev. Lett. $\underline{12}$, 474 (1964); Na_xWO_3: A. R. Sweedler, C. J. Raub and B. T. Matthias, Phys. Lett. $\underline{15}$, 108 (1965); $Li_{1+x}Ti_{2-x}O_4$: D. C. Johnston, H. Prakash, W. H. Zachariasen and R. Viswanathan, Matr. Res. Bull. $\underline{8}$, 777 (1973); $BaPb_{1-x}Bi_xO_3$: A. W. Sleight, J. L. Gibson and F. E. Bielstedt, S. S. Comm. $\underline{17}$, 27 (1975).

4. C. E. Methfessel, G. R. Stewart, B. T. Matthias and C. K. Patel, Proc. Natl. Acad. Sci. U.S.A., $\underline{77}$, 6307 (1980).

5. C. W. Chu, S. Huang, A. W. Sleight, S. S. Comm. $\underline{18}$, 977 (1976).

6. J. G. Bednorz and K. A. Müller, Z. Phys. B $\underline{64}$, 189 (1986).

7. S. Uchida, H. Takagi, K. Kitazawa and S. Tanaka, Jpn. J. Appl. Phys. $\underline{26}$, L1 (1987).

8. C. W. Chu, P. H. Hor, R. L. Meng, L. Gao, Z. J. Huang and Y. Q. Wang, Phys. Rev. Lett. $\underline{58}$, 405 (1987).

9. H. Takagi, S. Uchida, K. Kitazawa and S. Tanaka, Jpn. J. Appl. Phys. $\underline{26}$, L123 (1987).

10. C. W. Chu, P. H. Hor, R. L. Meng, L. Gao and Z. J. Huang, Science $\underline{235}$, 567 (1987); and unpublished.

11. R. J. Cava, R. B. Van Dover, B. Batlogg and E. A. Rietman, Phys. Rev. Lett. $\underline{58}$, 408, (1987).

12. Z. X. Zhao, L. Q. Chen, C. G. Cui, Y. Z. Huang, J. X. Liu, G. H. Chen, S. L. Li, S. Q. Guo and Y. Y. He, Kexue Tongbao $\underline{32}$, 522 (1987)

13. J. M. Tarascon, L. H. Greene, W. R. McKinnon, G. W. Hull and T. H. Geballe, Science $\underline{235}$, 1373 (1987).

14. D. W. Capone, D. G. Hinks, J. D. Jorgensen, Appl. Phys. Lett. $\underline{50}$, 543 (1987).

15. D. Gubser, R. A. Hein, S. H. Lawrence, M. S. Osofsky, D. J. Schrodt, L. E. Toth and S. A. Wolf, Phys. Rev. B $\underline{35}$, 5350 (1987).

16. P. H. Hor, R. L. Meng, Y. Q. Wang, L. Gao, Z. J. Huang, J. Bechtold, K. Forster and C. W. Chu, Phys. Rev. Lett. $\underline{58}$, 1891 (1987).

17. M. K. Wu, J. R. Ashburn, C. J. Torng, P. H. Hor, R. L. Meng, L. Gao, Z. J. Huang, Y. Q. Wang and C. W. Chu, Phys. Rev. Lett. $\underline{58}$, 908 (1987).

18. C.W. Chu, P.H. Hor, R.L. Meng, L. Gao, Z.J. Huang, Y.Q. Wang, J. Bechtold, D. Campbell, M.K. Wu, J. Ashburn and C.Y. Huang, to appear in Phys. Rev. B (1987).

19. P. H. Hor, L. Gao, R. L. Meng, Z. J. Huang, Y. Q. Wang, K. Forster, J. Vassiliou, C. W. Chu, M. K. Wu, J. R. Ashburn and C. J. Torng, Phys. Rev. Lett. $\underline{58}$, 911 (1987).

20. E. Zirngiebl, J. D. Thompson, C. Y. Huang, P. H. Hor and C. W. Chu, preprint.

21. S. Hikami, T. Hirai and S. Kagoshima, Jpn. J. Appl. Phys. $\underline{26}$, L314 (1987).

22. Z. X. Zhao, L. Q. Chen, Q. S. Yang, Y. Z. Huang, G. H. Chem, R. M. Tang, G. R. Liu, G. C. Cui, L. Chen, L. Z. Wang, S. Q. Guo, S. L. Li and J. Z. Bi, Kexue Tongbao $\underline{33}$, 661 (1987); Peoples' Daily, February 25, 1987.

23. L. C. Bourne, M. L. Cohen, W. N. Creager, M. F. Crommie, A. M. Stacy and A. Zettl, Phys. Lett. A $\underline{120}$, 494 (1987)

24. A. Moodenbaugh, M. Suenaga, T. Asano, R. N. Shelton, H. C. Ku, R. W. McCallum and P. Klavins, Phys. Rev. Lett. $\underline{58}$, 1885 (1987).

25. R. M. Hazen, L. W. Finger, R. J. Angel, C. T. Prewitt, N. L. Ross, H. K. Mao, C. G. Hadichiacos, P. H. Hor, R. L. Meng and C. W. Chu, Phys. Rev. B $\underline{35}$, 7238 (1987).

26. T. Siegrist, S. Sunshine, D. W. Murphy, R. J. Cava and S. M. Zahurak, Phys. Rev. B $\underline{35}$, 7137 (1987).

27. P. M. Grant, R. Beyers, E. M. Engler, G. Lim. S. S. P. Parkin, M. L. Ramirez, V. Y. Lee, H. Nazzal, J. E. Vasquez and R. J. Savoy, Phys. Rev. B $\underline{35}$, 7242 (1987).

28. M. A. Beno, L. Soderholm, D. W. Capone II, D. G. Hinks, J. D. Jorgensen, I. K. Schuller, C. U. Segre, K. Zhang and J. D. Grace, to appear in Appl. Phys. Lett.

29. Z. Fisk, J. D. Thompson, E. Zirngiebl, J. L. Smith and S. W. Cheong to appear in S. S. Comm.

30. S. Hasoya, S. I. Shamoto, M. Onoda and M. Sato, Jpn. J. Appl. Phys. $\underline{26}$, L325 (1987).

31. D. W. Murphy, S. Sunshine, R. B. van Dover, R. J. Cava, B. Batlogg, S. M. Zahurak and L. F. Schneemeyer, Phys. Rev. Lett. $\underline{58}$, 1888 (1987).

32. C. W. Chu, to appear in Proc. Natl. Acad. Sci. (USA) (1987).

33. L. F. Mattheiss, Phys. Rev. Lett. $\underline{58}$, 1028 (1987).; J. J. Yu, A. J. Freeman and J. H. Xu, Phys. Rev. Lett. $\underline{58}$, 1035 (1987).

34. See for example, J. Bardeen, in <u>Superconductivity in d- and f-Band Metals</u>, ed. by D. Douglas (Plenum, New York 1978) p. 1.

35. C. S. Ting, D. N. Talwar and K. L. Ngai, Phys. Rev. Lett. $\underline{45}$, 1213 (1980).

36. A. S. Alexandrou, J. Ranninger and S. Robaszkiewiez, Phys. Rev. B. $\underline{33}$, 4526 (1986).

37. J. Ruwald, Phys. Rev. B. $\underline{35}$, 8869 (1987); V. Kresin, Phys. Rev. B. $\underline{35}$, 8716 (1987).

38. P. A. Lee, N. Reed, Phys. Rev. Lett. $\underline{58}$, 2691 (1987); V. J. Emery, Phys. Rev. Lett. $\underline{58}$, 2794 (1987).

39. C. M. Varma, S. Schmitt-Rink and Elihu Abrahams, Phys. Rev. Lett. $\underline{58}$, 2691 (1987)

40. P. W. Anderson, Science $\underline{235}$, 1196 (1987).

41. C. E. Gough, M. S. Colclough, E. M. Gorgan, R. G. Jordan, M. Keene, C. M. Muirhead, A. I. M. Rae, N. Thomas, J. S. Abell and S. Sutton, Nature $\underline{326}$, 855 (1987).

42. W. R. McGrath, H. K. Olsson, T. Claeson, S. Eriksson and L. G. Johansson, submitted to Europhysics Lett.

43. W. J. Kossler, J. R. Kempton, X. H. Yu, H. E. Schonne, Y. J. Uemura, A. R. Moodenbaugh, M. Suenage and C. E. Stronach, preprint.

44. See for example, U. Walter, M. S. Sherwin, A. Stacy, P. L. Richards and A. Zettl, preprint; J. Moreland, A. F. Clark, L. F. Goodrich, H. C. Ku and R. N. Shelton, submitted to Phys. Rev. B. Rapid. Comm.

45. B. D. Dunlap. M. V. Nevitt, M. Slaski, T. E. Klippert, Z. Sungaika, A. G. McKale, D. W. Capone, R. B. Poeppel and B. K. Flandermeyer, preprint.

46. Neuman Phillips et al. Proc. International Workshop on Novel Mechanisms of Superconductivity, June 22-26, Berkeley, CA (U.S.A) (1987).

47. P. W. Anderson, private communication.

48. C. S. Ting and D. Y. Xing, Proc. International Workshop on Novel Mechanisms of Superconductivity, June 22-26, Berkeley, CA (U.S.A.) (1987).

49. K. A. Müller, M. Takashige, and J. G. Bednorz, Phys. Rev. Lett. $\underline{58}$, 1143 (1987); P. H. Hor, R. L. Meng, C. W. Chu, M. K. Wu, E. Zirngiebl, J. D. Thompson and C. Y. Huang, Nature $\underline{326}$, 669 (1987).

50. P. B. Allen and R. C. Dynes, Phys. Rev. B $\underline{12}$, 905 (1975).

51. B. Batlogg, R. J. Cava, A. Jayaraman, R. B. Van Dover, G. A. Kourouklis, S. Sunshine, D. W. Murphy, L. W. Rupp, H. S. Chen, A. White, K. T. Short, A. M. Mujsce and E. A. Rietman, Phys. Rev. Lett. $\underline{58}$, 2333 (1987); L. C. Bourne, M. F. Crommie, A. Zettl, H.-C. Zurloye, S. W. Keller, K. L. Leary, A. M. Stacy, K. J. Chang, M. L. Cohen and D. E. Morris, Phys. Rev. Lett. $\underline{58}$, 2337 (1987).

52. P. Chandhari, R. H. Kock, R. B. Laibowitz, T. R. McGuire and R. J. Gambino, preprint.

53. Y. Iye, T. Tamegai, H. Takeya and H. Takei, preprint.

54. J. Yu, S. Massidda, A. J. Freeman and V. Koelling, to appear in Phys. Lett.; L. F. Mattheiss and D. R. Hamann, to appear in S. S. Comm.

55. J. Ashkenzai, C. G. Kuper and R. Tyk, preprint.

56. A. Masaki, H. Sato, S.-I. Uchida, K. Kitazawa, S. Tankaa and K. Inone, Jpn. J. Appl. Phys. $\underline{26}$, L405 (1987).

57. R. L. Meng, P. H. Hor, Z. J. Huang, J. Bechtold, C. W. Chu and D. Mao, to be published.

58. D. Jerome, A. Mazand, M. Ribault and K. Bechgaard, J. Phys. (Paris) Lett. $\underline{41}$, L95 (1980).

59. P. H. Hor, R. L. Meng, C. W. Chu, C. Y. Huang and R. B. Frankel, preprint.

60. C. Y. Huang, Z. J. Huang, P. H. Hor, R. L. Meng, C. W. Chu, L. Kevan, preprint.

61. J. C. Ho, P. H. Hor, R. L. Meng, C. W. Chu and C. Y. Huang, to appear in Solid State Comm.

62. L. Gao, P. H. Hor, R. L. Meng and C. W. Chu, to be published.

63. R. L. Meng, P. H. Hor, Z. J. Huang, J. Bechtold, C. W. Chu and H. K. Mao, to be published.

64. J. M. Tranguada, S. M. Heald, A. R. Moodenbaugh and M. Suenaga, preprint.

65. A. Fujimari, E. Takayama-Muromachi, Y. Uchida and B. Ukai, to appear in Phys. Rev. B. Rapid Comm.

66. D. C. Johnston, D. P. Goshorn, H. Thamann, P. Tindall and J. P. Stokes, preprint.

67. T. Freltoft, J. P. Reimerka, D. E. Moncton, A. S. Cooper, J. E Fischer, D. Harshwan, G. Shirane, S. K. Sinha and D. Vaknin, preprint.

68. P. M. Grant, S. S. P. Parkin, V. Y. Lee, E. M. Engler, M. L. Ramirez, J. E. Vazquez, G. Lim, R D. Jacowitz and R. L. Greene, preprint.

69. Y. C. Jean, S. J. Wang, H. Nakanishi, W. N. hardy, M. E. Hayden, R. F. Kiefl, R. L. Meng, P. H. Hor, J. Z. Huang and C. W. Chu, preprint.

70. V. Kresin and S. Wolf, preprint.

71. S. R. Ovshinsky, R. T. Young, D. D. Allred, G. DeMaggio, and G. A. Van der Leeden , Phys. Rev. B. $\underline{58}$,2579 (1987).

72. C. Y. Huang, L. Dries, R. L. Meng, P. H. Hor, C. W. Chu and R. Frankel, preprint

73. A. Zettl and M. Cohen, private communication.

74. C. W. Chu, et al., NSF News Release, February 16, 1987

75. J. T. Chen, L. E. Wenger, C. J. McEwan and E. M. Logothetis, Phys. Rev. Lett. $\underline{58}$, 1972 (1987).

MIXED VALENCE COPPER OXIDES, HIGH Tc SUPERCONDUCTORS :

STRUCTURAL STUDY AND ELECTRON TRANSPORT PROPERTIES

Bernard Raveau and Claude Michel

Laboratoire de Cristallographie et Sciences des
Matériaux, ISMRa, Campus 2, Bd du Maréchal Juin
Université de Caen, 14032 Caen Cedex

INTRODUCTION

Perhaps more than hundred papers have been published about the subject of mixed valence copper oxides related to the pervoskite since the
first signs of superconductivity shown by Müller and Bednorz in the system Ba-La-Cu-O[1,2]. The rush in this area of research results from the
fact that most of the physicists were convinced that the critical
temperature of 23.3K observed for Nb_3Ge was a limit difficult to overtop
Infact a great mumber of those oxides were synthesized and studied for
the first time in Caen from 1980 to 1986. In the first part of the present paper synthesis, crystal chemistry and electron transport properties
of the oxides A-La-Cu-O (A = Ba, Sr, Ca) carried out in Caen are
reported. The second part of this study deals with our recent studies of
the second generation of high Tc superconductors, $YBa_2Cu_3O_7 \pm \epsilon$.

GENERATION OF METALLIC OR SEMI-METALLIC ANISOTROPIC CONDUCTORS : THE OXY-
GEN DEFICIENT MIXED VALENCE COPPER PEROVSKITES

It is well known, since the discovery of the oxygen tungsten bronzes
by Magneli[3], that the synthesis of oxides with a metallic or
semi-metallic conductivity requires a possibility of superexchange due to
the overlapping from the d orbitals of the transition metal M with the p
orbitals of the oxygen atoms forming the MO_n framework. Thus a first
condition appears for conductivity element M. In order to induce an anisotropy of the conductivity, we have to generate a mixed framework MO_n
built up at least from two different polyhedra. Consequently it appears
that the transition metal M should be able to take several oxidation
states and several coordinations simultaneously. In this respect, the
elements belonging to the first transition series, Fe, Mn, Ni, Cu are
potential candidates. Among them, copper was chosen because of its greater number of coordinations, its particular Jahn Teller effect, and its
ability to present weaker magnetic interactions. However, the unusual
Cu(III) valence is generally difficult to stabilize. It must indeed be
pointed out, that at the beginning of this study, the only tervalent
copper oxide which was known was the perovskite[4] $LaCu^{III}O_3$ which could
only be synthesized under high pressure (65 kbars).In order to favour the
partial oxidation of divalent copper into tervalent copper, it is possible to insert into the MO_n framework a basic A ion such as an alkaline
earth cation (Ba, Sr, Ca) ; exotic upper oxidation states like Bi(V) or
Pb(IV) are indeed stabilized under normal oxygen pressure by such cations

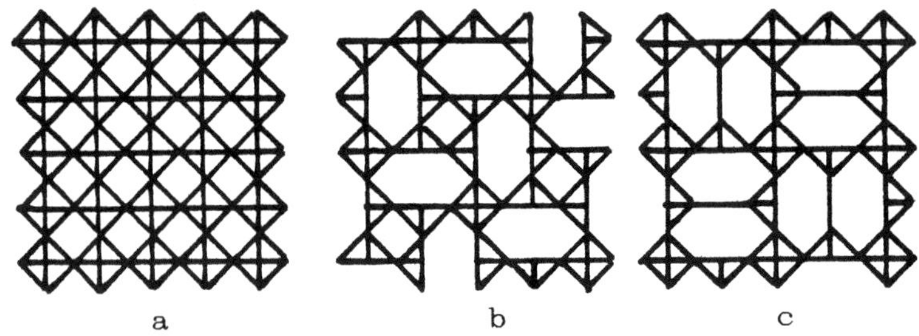

FIG. 1. Projection on to the (001) plane of the stoichiometric perovskite host lattice (a) and of the oxygen deficient perovskites $BaLa_4Cu_5O_{13}$ (b) and $La_{8-x}Sr_xCu_8O_{20}$ (c)

as shown for instance for the superconducting perovskite[5] $Ba(Pb_{1-x}Bi_x)O_3$.
The great flexibility of the ReO_3-type framework[6] suggests the possibility to create anisotropic frameworks by an ordered elimination of oxygen
atoms or of rows of oxygen atoms parallel to a particular direction.

The elimination of rows oxygen atoms parallel to $\langle 100 \rangle$ in the basal
plane of the octahedra of the perovskite host lattice MO_3 (Fig. 1a) leads
to the formation of hexagonal tunnels running along $[100]_P$. This is illustrated by the structure of the oxide $La_4BaCu_5O_{13+\delta}$ (Fig. 1b) which can
be described as formed of corner-sharing CuO_5 pyramids and CuO_6
octahedra[7]. A very closely related structure has been recently observed[8]
for the oxide $La_{8-x}Sr_xCu_8O_{20-\epsilon}$ (Fig. 1c) ; in this latter case one can
also observe infinite hexagonal tunnels running along $[100]_P$, but the
greater number of anionic vacancies in that oxide, leads to the formation
of CuO_4 square planar groups sharing their corners with the CuO_5 and CuO_6
polyhedra. It results that the tunnels form pairs, sharing their edges.
Both oxides exhibit metallic properties[9,10] (Fig. 2) with a p-type conduction in agreement with the mixed valences Cu(III)-Cu(II), which made
one of them - the barium oxide - attractive for the investigation of
superconductivity[1]. No sign of superconductivity was found in fact in
those oxides, likely owing to their three-dimensional character.

Row of oxygen atoms parallel to the $\langle 110 \rangle$ direction of the perovskite
can be eliminated in an ordered way. This is the case of the oxygen deficient perovskite $La_3Ba_3Cu_6O_{14+\delta}$[11] whose framework is built up from

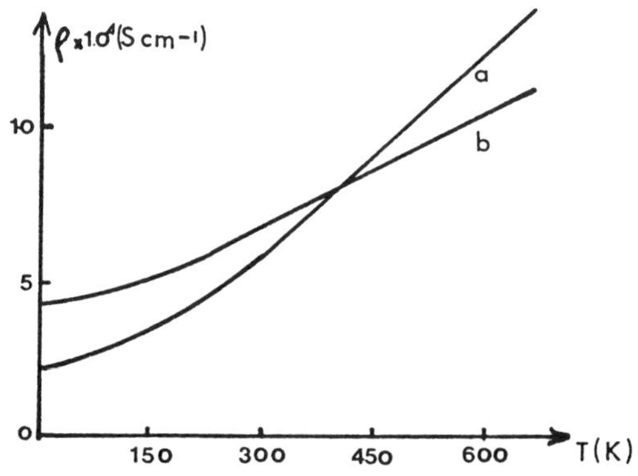

Fig. 2. ρ = f(T) for the oxide $BaLa_4Cu_5O_{13.1}$
(a) and $La_{6.4}Sr_{1.6}Cu_8O_{19.8}$ (b)

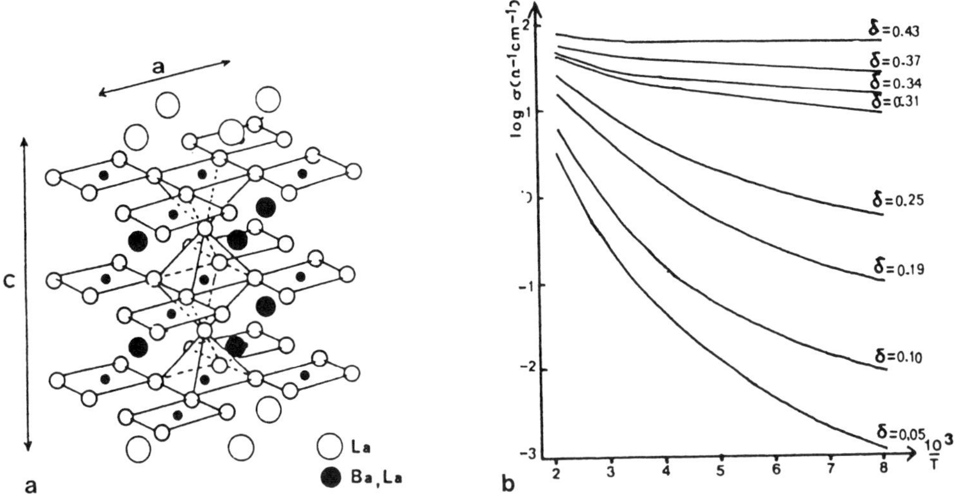

Fig. 3. a) Structure of the oxygen deficient perovskite $Ba_3La_3Cu_6O_{14}$ (δ = 0) ; b) Variation of the electrical conductivity versus temperature for some δ values.

corner-sharing CuO_6 octahedra, CuO_5 pyramids and CuO_4 square planar groups (Fig. 3a). It is worth pointing out that the limit $La_3Ba_3Cu_6O_{14}$ (δ = 0) corresponds to a layer structure. Curiously this latter composition (in fact δ = 0.05 for the oxide prepared at 1050°C in air and quenched at room temperature) is only semiconducting (Fig. 3b) in spite of the exis- tence of the mixed valence of copper ; one can also notice that the con- ductivity increases as the intercalated oxygen amount increases, tending towards a semi-metallic behaviour for δ = 0.43 which corresponds to the oxide annealed under one oxygen atmosphere at 450°C. It must also be noticed that no sign of superconductivity was observed at low temperature for $La_3Ba_3Cu_6O_{14}$, in spite of the bidimensional character of this oxide, perhaps because of the presence of residual oxygen between the layers, leading to a tridimensional framework.

The orthorhombic oxide $YBa_2Cu_3O_{8-x}$ which was recently isolated by two groups almost simultaneously[12,13] has its structure[14,15] closely related to the one of $La_3Ba_3Cu_6O_{14}$. It can be described (Fig. 4a) as formed of triple layers $[Cu_3O_7]_\infty$ of corner-sharing CuO_5 pyramids and CuO_4 square

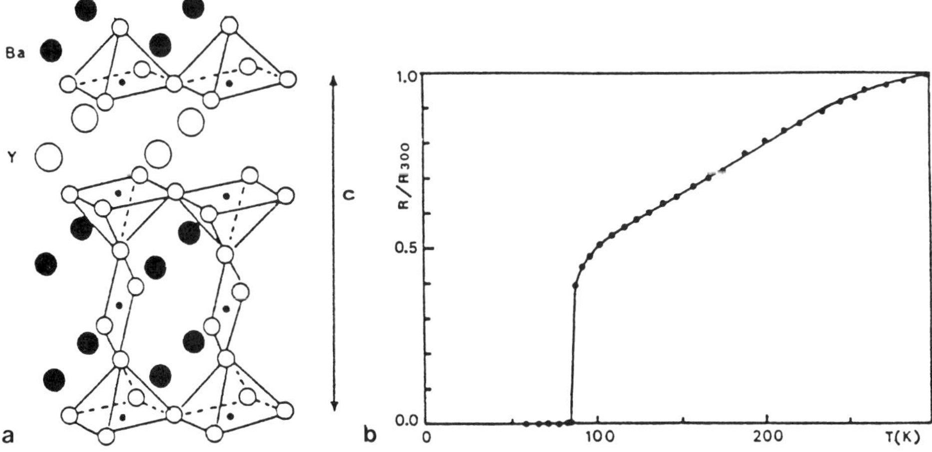

Fig. 4. a) Structure of the superconductor $Ba_2YCu_3O_7$; b) Resistivity ratio versus temperature for the oxyde $Ba_2YCu_3O_7$.

601

planar groups. Each triple layer is built up from two $[CuO_{2.5}]_\infty$ pyramidal
layers connected through one $[CuO_2]_\infty$ layer composed of CuO_4 square planar
groups. In this oxide, barium and yttrium ions are ordered in such away
that two barium planes alternate with one yttrium plane along $\vec{c}$: the
barium ions are located in the triple layers $[Cu_3O_7]_\infty$, whereas the yttrium ions ensure the cohesion between those layers. This structure can
be described from that of the perovskite by an ordered elimination of
rows of oxygens parallel to the $[010]_P$ direction i.e. at the same level
along $\vec{c}$ as the yttrium ions, and as the copper ions of the $[CuO_2]_\infty$
layers. This oxygen deficient ordered perovskite exhibits a metallic
conductivity and a zero resistance below 91K (Fig. 4b) in agreement with
the signs of superconductivity observed for the first time in a mixture
of oxides with composition $Y_{1.2}Ba_{0.8}CuO_{4-y}$ by Chu et al.[16]. Recently we
have also isolated a similar superconductiviting oxide in the case of
lanthanum $LaBa_2Cu_3O_{8-x}$[17], characterized by a lower critical temperature
$T_c = 75K$. Nevertheless it must be pointed out that for this latter oxide,
the examination of the X-ray powder diffraction pattern suggests a tetragonal symmetry. A careful study by electron microscopy and will be necessary to understand this difference with $YBa_2Cu_3O_{8-x}$ and to correlate it
with superconductivity.

Another way to introduce anisotropy in the electron transport properties of copper oxides consists in juxtaposition of insulating layers and
conducting layers. In this respect, the intergrowths of the perovskite
structure with the sodium chloride type structure, observed several years
ago[18] in the strontium titanates $SrO(SrTiO_3)_n$ are interesting frameworks.
The consideration of the K_2NiF_4-type structure La_2CuO_4, first member of
the series, found as a semi-conductor by many physicists, led us to study
the possible substitution of an alkaline-earth ion (Ca, Sr, Ba) for
lanthanum ; this substitution induces in a first step anionic vacancies

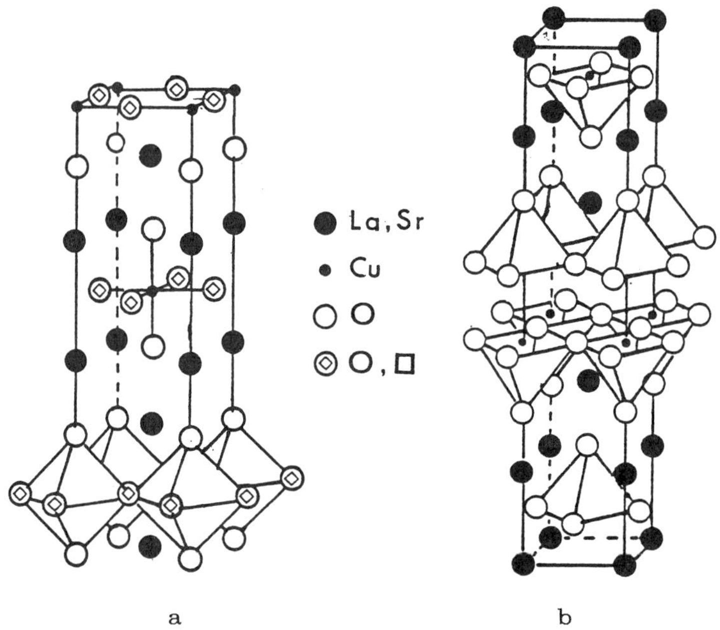

Fig. 5. Structure of the mixed valence copper oxides $La_{2-x}Sr_xCuO_{4-x/2+\delta}$
(a) and $La_{2-x}Sr_{1+x}Cu_2O_{6-x/2+\delta}$ (b)

in the structure according to the formulation $La_{2-x}A_xCu^{II}O_{4-x/2}$; but those oxygen vacancies are partly filled in oxidizing atmosphere, according to the formulation $La_{2-x}A_xCu^{II,III}O_{4-x/2+\delta}$, due to the presence of alkaline earth ions which favours the partial oxidation of Cu(II) into Cu(III). Thus a first series of oxides was isolated with $0 \leq x \leq 0.20$ for barium and calcium[19] and $0 \leq x \leq 1.33$ for strontium[20]. The superconducting properties of those oxides with critical temperatures ranging from 30K for barium[1,2] to 37K for strontium[21] take their origin in the oxygen deficient single perovskites slabs (Fig. 5a) separated by insulating stoichiometry SrO-type layers in agreement with the model developped by Labbé and Bok[22,23]. The second member of the series $La_{2-x}Sr_{1+x}Cu_2O_{6-x/2+\delta}$ could also be isolated[24]. Its structure (Fig. 5b) consists of double perovskite layers intercaled with SrO-type layers. The double perovskite layers are oxygen deficient and may tend to form single layers of corner-sharing pyramids as shown for instance for the limit oxide $La_2SrCu_2^{II}O_6$ which is characterized by the only presence of Cu(II). It is also worth pointing out that the second series of oxides, have not found to be superconducting, in spite of the mixed valence of copper. Further exploration of those compounds will be necessary to rule out the possibility of superconductivity.

THE OXYGEN DEFICIENT ORDERED PEROVSKITE, $YBa_2Cu_3O_{7\pm\epsilon}$, A COMPLEX SOLID STATE CHEMISTRY

Everybody is now able to synthesize the oxide $YBa_2Cu_3O_{7\pm\epsilon}$. However the oxygen stoichiometry of this oxide is variable and depends upon the thermal treatment used during synthesis. Moreover it is now well established that two forms are available : the orthohombic form which is super-conducting and the tetragonal form which does not exhibit superconductivity any more. DTA and thermogravimetric measurements performed in air up to 1100°C[25] show that the transition orthorhombic to tetragonal appears at 750°C, whereas a pronounced weight loss due to oxygen gas being evolved, is observed above 930°C. Consequently a tetragonal oxide $YBa_2Cu_3O_6$ can be prepared by heating $YBa_2Cu_3O_7$ under vacuum or argon[26,27]. The structure of this latter compound (Fig. 6) is deduced from the one of $YBa_2Cu_3O_7$ by a simple ordered elimination of the rows of oxygen atoms parallel to a in the $[CuO_2]_\infty$ layers (Fig. 4a). It results that this structure consists of $[CuO_{2.5}]_\infty$ layers built up of corner sharing CuO_5 pyramids involving Cu(II), connected through $[CuO]_\infty$ layers containing Cu(I) in twoofold coordination.

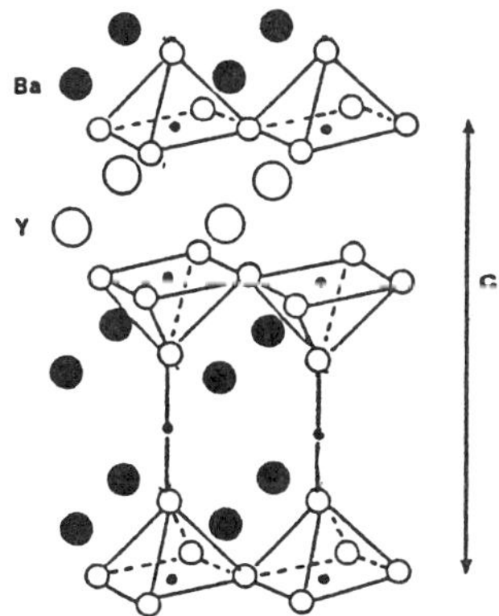

Fig. 6. Structure of the tetragonal oxide $Y_2BaCu_3O_6$.

The orthorhombic superconducting oxide $YBa_2Cu_3O_7$, is in fact complex from the structural point of view. Systematic investigations of the crystals by X-ray diffraction[14,15,28] as well as by electron microscopy[25,29,30] showed that not a single one turned out to be untwinned. It is now sure that the twinning is associated with the structural transformation from the tetragonal high temperature to the orthorhombic low temperature phase[25]. Those crystals exhibit an electron diffraction pattern (Fig. 7) in which one can observe a splitting of reflection hh0 (and $\bar{h}h0$) into two peaks, a small splitting of reflections h00 and 0k0, and no splitting of the reflections $\bar{h}h0$ (and $h\bar{h}0$). The dark field image (Fig. 8), obtained by selecting reflection (220) shows that the size of the twin domains is variable ranging from 100 Å to 1000Å. Thus it appears that the orthorhombic phase obeys a twin law where the tetragonal (110)-mirror plane is the twin element. Several structural models based on the consideration of the change of orientation of the ions of oxygen vacancies along the [010] direction can explain the formation of such domains. One of them (Fig. 9a) involves the presence of additionnal CuO_6 octahedra and CuO_5 pyramids at the domain boundary, whereas a second model (Fig. 9b) would lead to the formation of CuO_4 distorted thetrahedra. The first one appears as more probable since it leads to a much a smaller distortion of the polyhedra and avoids a displacement of the metallic ions contrary to the second model.

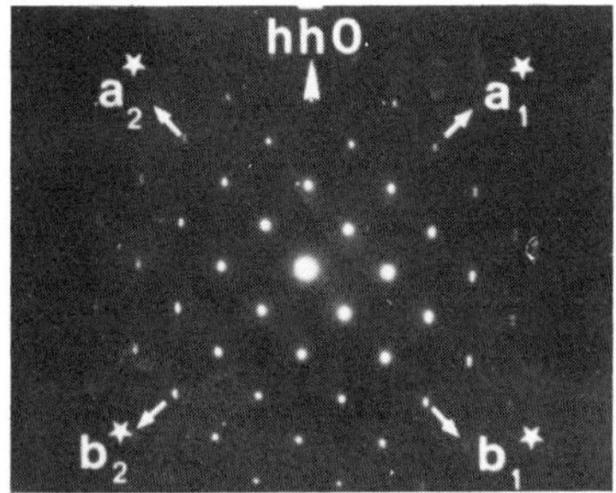

Fig. 7. Typical [001] electron diffraction pattern of microcristals of the oxide $YBa_2Cu_3O_7 \pm \epsilon$.

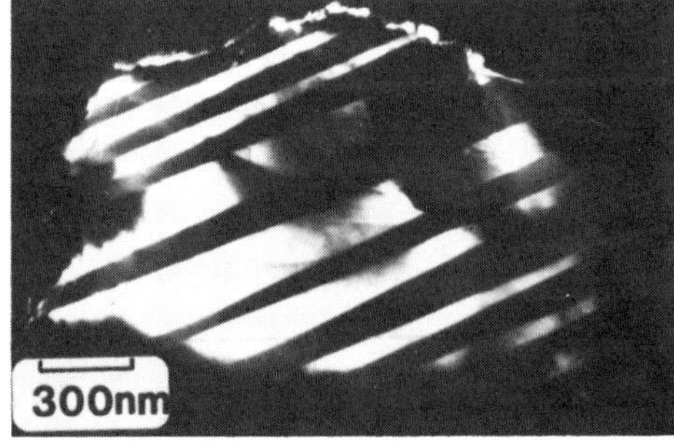

Fig. 8. Dark field image [001] (selected reflection (220) of a microcrystal of $YBa_2Cu_3O_7 \pm \epsilon$.

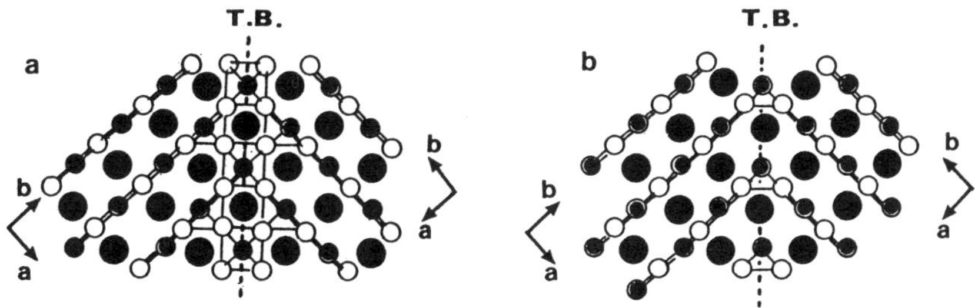

Fig. 9. Orientation of the CuO₄ groups through the quasi perpendicu-
lair twin boundaries involving the formation of CuO₅ pyramids and CuO₆
octahedra (a) or CuO₄ tetrahedra (b).

Besides the twin domains a second family of domains is often observed
which corresponds to the formation of orientated slices perpendicular to
the c axis (Fig. 10a). Such domains are shown on the HREM image (Fig.
10b) which corresponds to an orientation of the c axis parallel to the
edge of the thin crystal. Area labelled 1 exhibits a contrast characteristic of a [100] image whereas the area labelled 2 is characteristic of
a [010] image. These two different views of the structure in the same
crystal could be determined only by a careful simulation of the image
for all the defocus series[31] using Skarnulis programm. Thus an idealized
model of the orientated domains can be proposed (Fig. 10c) which shows
that the CuO₄ groups belonging to the [CuO₂]∞ layers are turned out of
90° form one domain to the other. It must be pointed out that those domains are characterized by a jonction involving a juxtaposition of two
different parameters "a" and "b" respectively at the domain wall. It results that the domain interfaces are particularly disturbed.

Frequently, one observes a misorientation of the crystals as shown
from thin E.D. patterns (Fig. 11a) where we can see the contribution of
each component. In such crystals, the c axis of the domains do not coincide any more, making a variable angle. This is due to the fact that the a
and b parameters are different, and take a different orientation in order
to adjust one to each other. The low resolution image of those crystals
(Fig. 11) shows that both types of domains are observed perpendicularly
to c̄ : twinned domains (area labelled 1) and orientated domains (area
labelled 2). But besides those domains, spectacular strains and bending
of the crystals are clearly visible. Such a feature is likely connected
to the formation of microfractures which has just started on the edges of
the crystal but is not yet achieved. The existence of "unachieved
microfractures" could explain that the angle between the domains varies
from one crystal to the other.

The investigation of the defects by high resolution electron microscopy of the orthorhombic phase has just started. However it must be
pointed out, that a great number of extended defects are observed in all
the crystals. Two examples will only be presented here, which deal with
the existence of oxygen overstoichiometry[32] and oxygen substoichiometry[33]

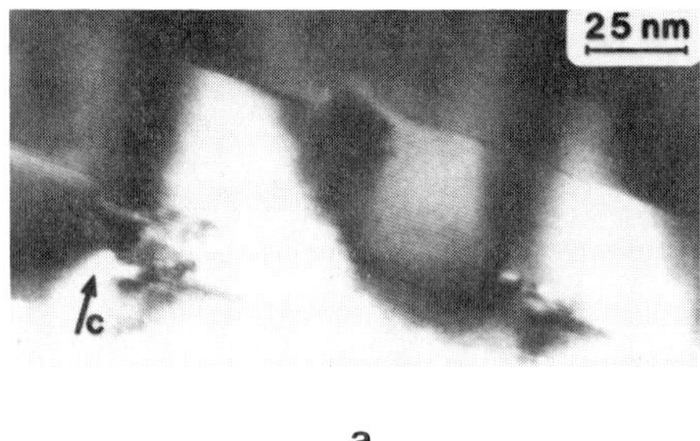

a

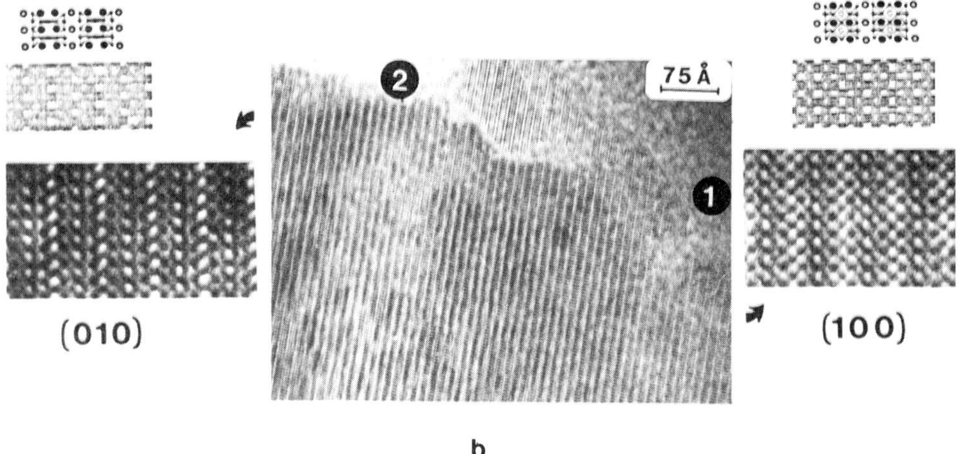

b

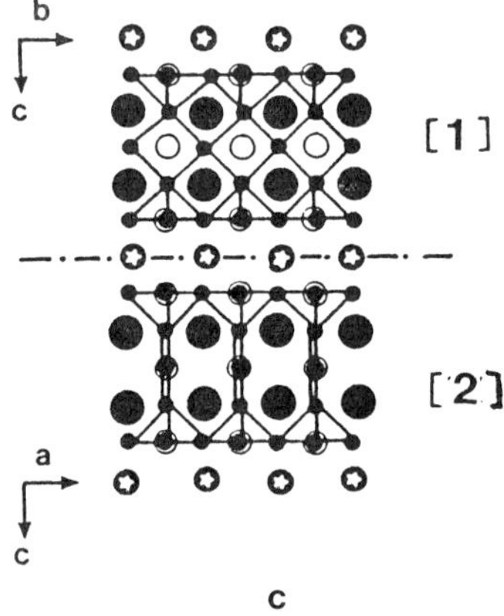

Fig. 10. a) Orientated slices perpendicular to the c axis. b) High resolution image of the boundary : enlargement of regular zone on each side
of the boundary area and corresponding calculated images. c) Idealized
drawing of two orientated domains (Labelled 1 and 2).

606

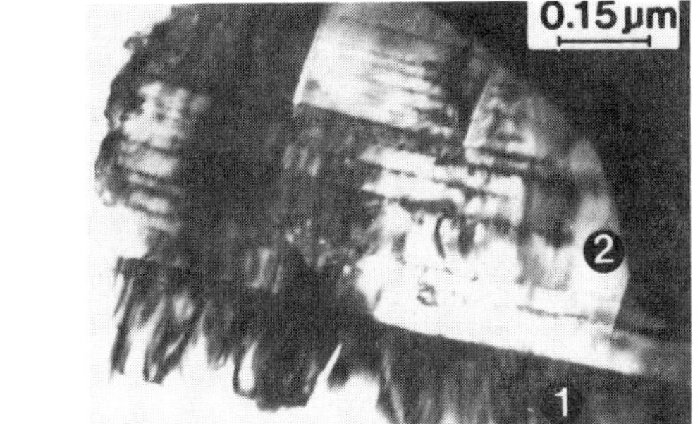

Fig. 11. a) Electron diffraction pattern of a crystal exhibiting misorientated domain. b) low resolution image of such domains.

a

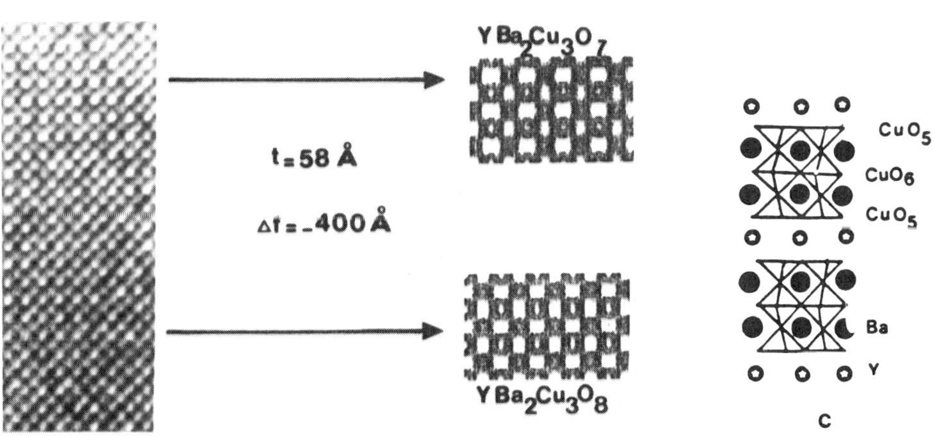

Fig.12. a) [001] image of the oxide $YBa_2Cu_3O_{7\pm\epsilon}$ showing the contrast variation in more or less large domains. b) High magnification. c) Idealized drawing [100] of $YBa_2Cu_3O_8$.

respectively. The [001] high resolution image shows the presence of domains of oxygen overstoichiometry whose size varies from about 30 Å² to some hundred of Å², and characterized by a variation of contrast (Fig. 12a). This variation can better be seen on an enlargement of the HREM image (Fig. 12b). From a careful simulation of the image[31] it could be proved that the upper region corresponds to the $YBa_2Cu_3O_7$ structure, whereas the lower part corresponds to the insertion in this latter structure of additionnal oxygen atoms in the $[CuO_2]_\infty$ layers, leading to the formulation $YBa_2Cu_3O_8$.

The structure of this latter region can then be described as a succession of triple layers built up from two $[CuO_{2.5}]_\infty$ layers of corner sharing octahedra (Fig. 12c) as previously described for the hypothetical limit $YBa_2Cu_3O_8$[12].

On the opposite oxygen substoichiometry can be observed in different regions of the crystals[33], as shown for instance on the [100] HREM image (Fig. 13a). From this micrograph it can be seen that the rows of white spots move apart, implying a bending of the adjacent double-row and that an extra row of small spots appears in the middle. The correlation between the image and the projected potential allowed a structural model of this defect to be established. Such a feature results from the formation of a double row of edge-sharing CuO_4 groups (Fig. 13b) similar those observed in the $SrCuO_2$-type structure[34].

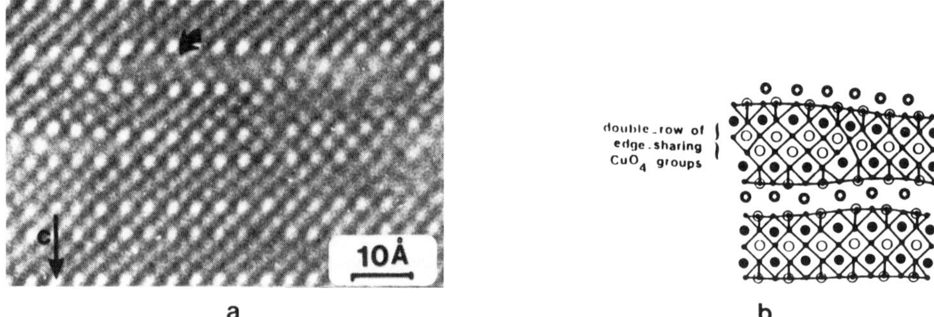

Fig. 13. a) High resolution [100] image of a typical defect.
b) Model of the defect.

As a conclusion, a great number of physical measurements have been performed on those high T_C superconductors which show their interest without any ambiguity. However a careful study will now be absolutely necessary in order to determine what is the role played by the oxygen stoichiometry the twin and orientated domains, and by the different extended defects in the superconducting properties of these compounds.

REFERENCES

1. J.G. Bednorz and K.A. Muller, Possible high T_C superconductivity in the Ba-La-Cu-O system, Z. Phys. B., 64:189 (1986).
2. J.G. Bednorz, M. Takashige and K.A. Muller, Susceptibility measurements support high T_C superconductivity in the Ba-La-Cu-O system, Europhysics Lett., 3:379 (1987).

3. A. Magneli and B. Blomberg, Contribution to the knowledge of the alkali tungsten bronzes, <u>Acta Chem. Scand.</u>, 5:372 (1951).

4. G. Demazeau, C. Parent, M. Pouchard and P. Hagenmuller, Two new oxygenated phases of copper (III). $LaCuO_3$ and $La_2Li_{0.5}Cu_{0.5}O_4$, <u>Mat. Res Bull.</u>, 7:913 (1972).

5. A.W. Sleight, J.L. Gilson and P.E. Bierstedt, High temperature superconductivity in the barium plumbate bismuthate $(BaPb_{1-x}Bi_xO_3)$ systems, <u>Solid State Comm.</u>, 17:27 (1975).

6. B. Raveau, Numerous structures of oxides can be built up from the ReO_3-type framework, <u>Proc. Indian natn. Sci. Acad.</u>, 52:67 (1986).

7. C. Michel, L. Er-rakho, M. Hervieu, J. Pannetier and B. Raveau, $BaLa_4Cu_5O_{13+\delta}$, an oxygen-deficient perovskite built up from corner sharing CuO_6 octahedra and CuO_5 pyramids, <u>J. Solid State Chem.</u>, to be published.

8. L. Er-rakho, C. Michel and B. Raveau, $La_{8-x}Sr_xCu_8O_{20}$, an oxygen deficient perovskite built up from CuO_6, CuO_5 and CuO_4 polyhedra, <u>J. Solid State Chem.</u>, submitted.

9. C. Michel, L. Er-rakho and B. Raveau, The oxygen defect perovskite $BaLa_4Cu_5O_{13.4}$, a metallic conductor, <u>Mat. Res. Bull.</u>, 20:667 (1985)

10. C. Michel, L. Er-rakho and B. Raveau, $La_{8-x}Sr_xCu_8O_{20-\epsilon}$: a metallic conductor belonging to the family of the oxygen deficient perovskites, <u>J. Phys. Chem. Sol.</u>, submitted.

11. L. Er-rakho, C. Michel, J. Provost and B. Raveau, A series of oxygen defect perovskites containing Cu^{II} and Cu^{III} ; the oxides $La_{3-x}Ln_xBa_3[Cu^{II}_{5-2y}Cu^{III}_{1+2y}]O_{14+y}$, <u>J. Solid State Chem.</u>, 37:151 (1981).

12. C. Michel, F. Deslandes, J. Provost, P. Lejay, R. Tournier, M. Hervieu and B. Raveau, The oxide $YBa_2Cu_3O_{8-y}$: a novel mixed valence copper perovskite, with oxygen defect, superconductive below 91K, <u>C.R. Acad. Sci.</u>, 304 II:1059 (1987).

13. R.J. Cava, B. Battlog, R.B. Vandover, D.W. Murphy, S. Sunshine, T. Siegrist, J.P. Remeika, E.A. Reitman, S. Zahurak and G.P. Espinosa, Bulk conductivity at 91K in single phase oxygen deficient perovskite $Ba_2YCu_3O_{9-\delta}$, <u>Phys. Rev. Lett.</u>, 58:1676 (1987).

14. Y. Lepage, W.R. McKinnon, J.M. Tarascon, L.H. Greene, G.W. Hull and D.M. Hwang, Room temperature structure of the 90K bulk superconductor $YBa_2Cu_3O_{8-x}$, <u>Phys. Rev. B</u>, 35:7245 (1987).

15. J.J. Capponi, C. Chaillout, A.W. Hewat, P. Lejay, M. Marezio, N. Nguyen, B. Raveau, J.L. Soubeyroux, J.L. Tholence and R. Tournier, Structure of the 100K superconductor $Ba_2YCu_3O_7$ between 5 and 300K by neutron powder diffraction, <u>Europhysics Lett.</u>, to be published.

16. M.K. Wu, J.R. Ashburn, C.J. Torng, C.J. Hor, R.L. Meng, L. Gao, Z.J. Huang, Y.Q. Wang and C.W. Chu, Superconductivity at 93K in a new mixed-phase Y-Ba-Cu-O compound system at ambient pressure, <u>Phys. Rev. Lett.</u>, 58:908 (1987).

17. C. Michel, F. Deslandes, J. Provost, P. Lejay, R. Tournier, M. Hervieu and B. Raveau, $LaBa_2Cu_3O_{8-y}$, a mixed valence copper oxygen deficient perovskite superconductive below 75K, <u>C.R. Acad. Sci.</u>, 304 II:1169 (1987).

18. S.N. Ruddlesden and P. Popper, The compound $Sr_3Ti_2O_7$ and its structure, <u>Acta Crystallogr.</u>, 11:54 (1958).

19. C. Michel and B. Raveau, Oxygen intercalation in mixed valence copper oxides related to the perovskites, <u>Rev. Chim. Miner.</u>, 21:407 (1984).

20. N. Nguyen, J. Choisnet, M. Hervieu and B. Raveau, Oxygen defect K_2NiF_4 type oxides : the compounds $La_{2-x}Sr_xCuO_{4-x/2+\delta}$, <u>J. Solid State Chem.</u>, 39:120 (1981).

21. R.J. Cava, R.B. van Dover, B. Battlogg and E.A. Rietman, Bulk super-
    conductivity at 36K in $La_{1.8}Sr_{0.2}CuO_4$, Phys. Rev. Lett., 58:408
    (1987).
22. J. Labbé and J. Bok, Europhysics Lett., to be published.
23. J. Bok and J. Labbé, Specific heat and Pauli susceptibility of
    $La(Ba, Sr)CuO_4$ compounds, Europhysics Lett., to be published.
24. N. Nguyen, L. Er-rakho, C. Michel, J. Choisnet and B. Raveau, Inter-
    croissance de feuillets perovskites lacunaires et de feuillets type
    chlorure de sodium : les oxydes $La_{2-x}A_{1+x}Cu_2O_{4-x/2}$ (A = Ca, Sr),
    Mat. Res. Bull., 15:891 (1980).
25. G. Roth, D. Ewert, G. Heger, M. Hervieu, C. Michel, B. Raveau, F.
    d'Yvoire and A. Revcolevski, Phase transformation and microtwinning
    in crystals of the high T_C superconductor $YBa_2Cu_3O_{8-x}$, x = 1.0, Z.
    Phys., submitted.
26. G. Roth, D. Renker, G. Heger, M. Hervieu, B. Domenges and B. Raveau,
    On the structure of the non superconducting $YB_2Cu_3O_{6+\epsilon}$, Z. Phys.,
    submitted.
27. P. Bordet, C. Chaillout, J.J. Capponi, J. Chevanas, M. Marezio, The
    crystal structure of $Ba_{2.1}Y_{0.9}Cu_3O_6$ a compound related to the high
    T_C superconductor $Ba_2YCu_3O_7$, Nature, submitted.
28. T. Siegrist, S. Sunshine, D.W. Murphy, R.J. Cava and S.M. Zahurak,
    Appl. Phys. Soc., March Meeting in N.Y.
29. G. van Tendeloo, H.W. Zandberger and S. Amelinckx, Electron diffrac-
    tion and electron microscopy study of Ba-Y-Cu-O superconducting ma-
    terials, Solid State Com., submitted.
30. M. Hervieu, B. Domenges, C. Michel, G. Heger, J. Provost and B.
    Raveau, Twins and orientated domains in the orthorhombic supercon-
    ductor $YBa_2Cu_3O_7 \pm \epsilon$, Phys. Rev. Lett., submitted.
31. M. Hervieu, B. Domenges, C. Michel, J. Provost and B. Raveau, Ortho-
    rhombic superconductivity $YBa_2Cu_3O_7 \pm \epsilon$, HREM study. I. Structure of
    the ordered oxygen deficient perovskite and the problems of twins
    and domains, J. Solid State Chem., submitted.
32. M. Hervieu, B. Domenges, C. Michel and B. Raveau, HREM study of the
    defects in the orthorhombic superconductor $YBa_2Cu_3O_7 \pm \epsilon$. I. Oxygen
    overstoichiometry, Europhysics Lett., submitted.
33. B. Domenges, M. Hervieu, C. Michel and B. Raveau, HREM study of the
    defects in the orthohombic superconductor $YBa_2Cu_3O \pm \epsilon$. II. Oxygen
    substoichiometry and cationic disorder, Europhysics Lett., submitted
34. L. Teske and H.K. Muller Buschbaum, Zur Kenntnis von Sr_2CuO_3, Z.
    anorg. allg. Kem., 371:325 (1969).

TUNNELING SPECTROSCOPY OF NOVEL SUPERCONDUCTORS

K. E. Gray, M. E. Hawley, and E. R. Moog

Materials Science Division

Argonne National Laboratory, Argonne, Illinois   60439

INTRODUCTION

Recent discoveries of exciting new superconductors have led to further
exciting speculations about novel mechanisms and/or pairing. Tunneling
spectroscopy can again play an important role in establishing the applicability of these ideas to specific superconductors. In addition to the
traditional role of verifying in detail the electron-phonon coupling
through $\alpha^2 F$, in many cases the magnitude of the gap compared to the BCS
prediction or the crystalline gap anisotropy can reveal direct information
about novel mechanisms and/or pairing.

Since many of these new materials have only been available as bulk
samples, or bulk single-crystal studies are desired, the technique of
vacuum tunneling spectroscopy, pioneered by Poppe (1981) for superconductors, is most appropriate. However, thick, nonconducting surface layers
are often found which prevent true vacuum tunneling. For these samples,
mechanical contact of the tunneling tip is required to break through the
surface layer to the superconductor below. The resulting point-contact
tunneling (Blonder and Tinkham, 1983 and Hawley, et al., 1986) can, however, emulate many of the results of true tunneling through a vacuum or
insulator.

In this paper, we shall briefly review relevant tunneling techniques
and some recent experiments on magnetic, organic, heavy fermion and high-T_c
oxide superconductors. Connections are made to theoretical ideas,
especially regarding novel mechanisms and/or pairing.

TUNNELING

Giaever (1960) first pointed out that for tunnel junctions the energy
gap, Δ, can be measured with a voltmeter. Actually Δ is determined from
the current-voltage characteristic I(V) of a tunnel junction by various
practical methods which depend on whether the counterelectrode is nonsuperconducting (NS) or superconducting (SS). For NS junctions, one can fit the
I(V), for $V > \Delta$, by the standard (T = 0) BCS expression $RI(V)=(V^2 - \Delta^2)^{1/2}$,
where R is the high-voltage ($V \gg \Delta$) junction resistance. Corrections for

finite temperature must be made for small values of $\Delta/k_B T$. Note that this procedure works even for strong coupling Pb which has $\pm 5\%$ deviations from the BCS density of states. This procedure avoids the weakness of assigning Δ to the peak in dI/dV or the intersection of dI/dV with the normal state dI/dV, each of which could be affected more by gap smearing and possible leakage current, and even in ideal cases require 5%-10% corrections to Δ due to finite temperature.

An SS junction has a jump in current at $V = 2\Delta/e$ which results in a peak in dI/dV with less serious thermal smearing problems, but intrinsic gap smearing due to inhomogeneities will broaden this peak. For an $S_1 S_2$ junction, I(V) has a sharp jump at $V = (\Delta_1 + \Delta_2)/e$ and a cusp, whose magnitude is temperature dependent, at $V = |\Delta_1 - \Delta_2|/e$. From these points, which correspond to a peak and a zero respectively in dI/dV, both Δ_1 and Δ_2 can be determined individually.

The reduced density of electron states, $\rho(E)$, gives information related to the coupling mechanism responsible for the superconducting condensation. For traditional, BCS superconductors one obtains $\alpha^2 F(\omega)$ where α^2 represents the electron-phonon coupling strength and $F(\omega)$ is the density of phonon states. Direct confirmation of phonon coupling comes from a comparison of $\alpha^2 F(\omega)$ from tunneling with $F(\omega)$ from neutron scattering. For NS tunneling, $\rho(E)$ comes directly from the difference between dI/dV, measured at sufficiently low temperature to reduce thermal smearing, and dI/dV taken in the normal state of the superconductor under study. Thermal smearing at high temperature is reduced in SS tunneling, but a knowledge of $\rho(E)$ in the counterelectrode is necessary. In either case the inversion of McMillan and Rowell (1969) is used to obtain $\alpha^2 F(\omega)$ from $\rho(E)$.

In addition to the above, many other effects, unwanted for present purposes, can be found in tunnel junctions, mostly related to imperfect barriers. For example, a low barrier height leads to positive curvature of I(V) at high voltages, and states in the barrier can open new tunneling channels at voltages corresponding to their excitation energy.

For greater detail on all the above points see Wolf (1985).

TUNNELING TECHNIQUES

A brief discussion of the three usual counterelectrodes is followed by comments on their advantages and disadvantages in particular situations.

<u>Thin film</u> Thin film counter electrodes are usually used with thin film samples and can be either N or S. The barriers include native or plasma-grown oxides or artificial barriers like Mg or amorphous Si which are later oxidized. Use of thin-film counterelectrodes with bulk samples is difficult due to the possibility of electrical shorts resulting from the uneven surface of the bulk (however, some examples are given below).

<u>Sharp tip</u> Sharp tip counterelectrodes are associated with either point-contact or vacuum tunneling and are usually used with bulk samples. Sharpened tips of either N or S approach the surface of the superconductor using a sophisticated arrangement of piezioelectric and/or mechanical devices. Considerable efforts to reduce vibration are required. In most systems, feedback of the tunnel current to the piezioelectric transducers is required to maintain a constant junction resistance. Because the tunnel junction area is small, true vacuum tunneling is only possible using uncontaminated samples and tips. Even normal oxide thicknesses of 20-40 Å may reduce the current below the detectable limit. Samples with a nonconducting contamination layer require mechanical touching to break through the

layer in a process we call point-contact tunneling. Such point-contact
tunneling was analyzed experimentally and theoretically by Blonder and
Tinkham (1983) and by Blonder, et al. (1982).

A very powerful potential of vacuum tunneling is to study the gap
anisotropy in single-crystal samples since tunneling occurs preferentially
along the direction perpendicular to the sample surface. However, for
point-contact tunneling, and probably the fracture surfaces of break-
junctions (see below), it is not clear which direction is being probed.

<u>Superconductor under study</u>   The counterelectrode can also be the
actual superconductor being studied. For samples other than thin films,
Poppe and Schröder (1984) first did this in vacuum tunneling. Recently
Moreland and Ekin (1985) used the idea to form a break-junction. A superconducting sample (e.g., a Nb_3Sn multifilamentary wire) was stretched past
the yield point in a controlled manner resulting in a fracture in liquid
He. Upon slowly relaxing the stretch, tunneling could be observed across
the fracture.

<u>Analysis of Technique</u>   Two features of point-contact, break-junction
or vacuum tunneling reduce their usefulness for obtaining $\rho(E)$, and hence
$\alpha^2(F(\omega)$ or the corresponding function for non-phononic mechanisms. Both
the inherent instability caused by vibration and the low current due to the
small tunneling area lead to significantly greater noise than for thin film
sandwiches. Secondly, the necessary determination of the normal state
dI/dV must be done without disturbing the junction. However, thermal
expansion and residual magnetic interactions preclude changing to higher
temperatures or fields to quench the superconductivity. This latter point
also makes it difficult to study the temperature or field dependence of Δ
which are of considerable interest. It is also often more difficult than
with thin films to measure Δ with these techniques. The ultra-small
tunneling area leads to a necessarily narrow barrier which results in
greater curvature of the normal state $I(V)$ at high voltages.

However, for studying bulk samples, especially single crystals, these
techniques are distinctly more straightforward than using thin film
counterelectrodes (or single crystal films). Note that of these techniques, only true vacuum tunneling can be used reliably to determine gap
anisotropy.

MAGNETIC SUPERCONDUCTORS

Historically, the first vacuum tunneling into a superconductor was
reported by Poppe (1981) in the magnetic superconductor $ErRh_4B_4$ which is
superconducting below 8.5 K, and re-enters into a normal, ferromagnetic
state below $T_M \sim 1$ K. The dI/dV curves showed a well-defined gap structure
with $\Delta \approx 1.4$ meV at 1.55 K and $\eta = 2\Delta(0)/k_B T_c \sim 3.8$. Unfortunately
temperatures below T_M were unattainable. Earlier studies by Rowell, et al.
(1980) on a thin film of $ErRh_4B_4$ with an aluminum oxide artificial barrier
and a Pb counterelectrode showed that Δ dropped to zero as T decreased from
1.05 K to $\sim$0.95 K. However η for these films was only about one-half the
BCS value of 3.5. Umbach, et al. (1981) reported $\eta \sim 4.2$ in thin-film
junctions using an erbium oxide artificial barrier and a normal metal (Mg)
as counterelectrode. They likewise did not go below T_M. Josephson
tunneling critical current between thin films of $ErRh_4B_4$ and In was shown
by Umbach et al. (1982) to drop to zero below T_M accompanied by a splitting
of the Fraunhofer pattern.

Tunneling studies of coexistent antiferromagnetic superconductors were reported by Poppe and Schroeder (1984). They used the vacuum tunneling technique on crystals of $TbMo_6S_8$ and $GdMo_6S_8$ to show that Δ was preserved to a large extent below the magnetic ordering temperature T_N. Josephson tunneling by Vaglio et al. (1984) showed that the superconducting order parameter was larger below T_N than the value extrapolated from above T_N, which was interpreted as a decrease in magnetic pair-breaking in the ordered antiferromagnetic state.

HEAVY FERMION SUPERCONDUCTORS

The first single-particle tunneling measurements on a heavy fermion superconductor were reported recently by Iguchi, et al. (1987a). A polycrystalline ingot of $CeCu_2Si_2$ was polished, plasma oxidized and a Pb counter electrode evaporated to complete the junction. They found an assymetric dV/dI in a magnetic field indicative of a valence-fluctuation compound. Below the superconducting T_c, a series of structures were seen near the Pb gap, i.e., at the minimum of dV/dI. Preliminary interpretation indicates a singlet d-wave superconductor with $\eta \approx 4.6$.

Several point-contact Josephson tunneling experiments have been performed to test whether the pairing is singlet or triplet. Poppe (1985) observed a supercurrent between $CeCu_2Si_2$ and Al and concluded that this heavy fermion did not exhibit triplet pairing. However, Han, et al. (1987) observations based on a proximity effect model led them to conclude that UBe_{13} does exhibit triplet pairing, although this conclusion has been disputed (Kadin and Goldman, 1987).

ORGANIC SUPERCONDUCTORS

Many superconducting organics contain no metallic elements and exhibit lower dimensionality leading to anisotropic band structure. These properties alone have led to a great deal of interest and the natural desire to obtain detailed tunneling information on, e.g., gap anisotropy or the mechanisms of superconductivity. On the basis of Schottky barrier tunneling into bulk $(TMTSF)_2PF_6$ under pressure, More, et al. (1981) inferred a "pseudo-gap" which was 12 times that predicted from weak-coupling BCS theory and concluded the existence of one-dimensional fluctuations was necessary to explain this. Additional tunneling studies of the ambient pressure superconductor $(TMTSF)_2ClO_4$ have not confirmed this observation; however, only one study (Bando, 1984) included temperatures well below T_c, and in that case the large gap of the superconductor Pb counterelectrode could have obscured the determination of the gap in the organic metal. For example, no variation of gap with temperature nor magnetic field was reported.

More recently point-contact tunneling has been reported by Hawley, et al. (1986) on bulk single crystals of $\beta\text{-}(BEDT\text{-}TTF)_2AuI_2$, or $\beta\text{-}(ET)_2AuI_2$, which has the highest T_c ($\sim$5 K) for an ambient pressure organic superconductor. A counterelectrode of Au was used and this guaranteed that any measured gaps are entirely due to the organic superconductor. In order to avoid distortion of the very non-linear $I(V)$, no feedback to the piezoelectric transducers was used. Considerable efforts to eliminate vibration resulted in very stable $I(V)$ without feedback.

614

Figure 1 shows a collection of I(V) curves for a single sample under various experimental conditions. It should be noted that while such curves showing distinct gap structure were reproduced many times on different low-temperature runs, they could not be obtained routinely, nor maintained indefinitely. Thus a complete set of curves versus temperature was impossible with the present apparatus, since changing the bath temperature, and hence pressure, led to irreconcilable movements of the tip. It is also important to realize that several qualitatively different I(V) curves could be obtained at different tip "positions" (caused by changes of the piezo-electric voltage and hence the force on the point contact). A variety of point-contact I(V) curves, complied by Blonder and Tinkham (1983) have been observed, as well as a higher-resistance region for low voltages, but without the sharp rise at $V = \Delta$ and bending back at higher voltages that one expects for tunneling. We have no explanation for this, but it may be related to the nonsuperconducting behavior of $\beta-(ET)_2AuI_2$, since it was also seen above T_c.

We have analyzed only those data which are symmetric and show bending back of I(V) at high voltage. It is most convenient to make plots of I^2 against V^2, as shown in Fig. 2. In that case the slope is R^{-2} and the intercept is Δ^2. Deviations are to be expected for small currents because of finite temperature and possible leakage current. The effect of finite-temperature smearing leads to an over-estimate of the actual Δ, but by no more than 7% in the worst case.

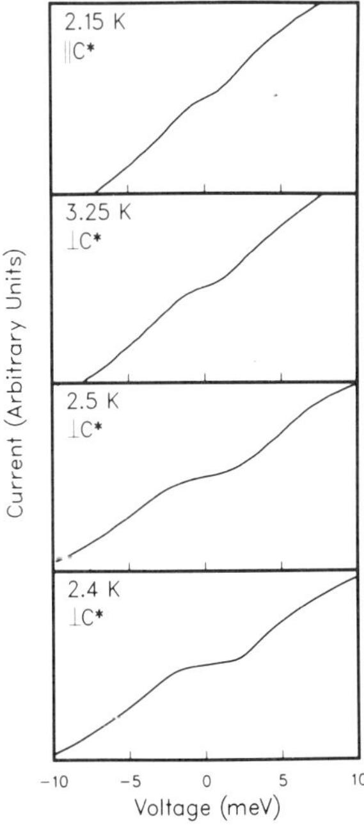

Fig. 1. Various I(V) for $\beta-(ET)_2AuI_2$ point-contact tunneling with Au tip from Hawley, et al. (1986).

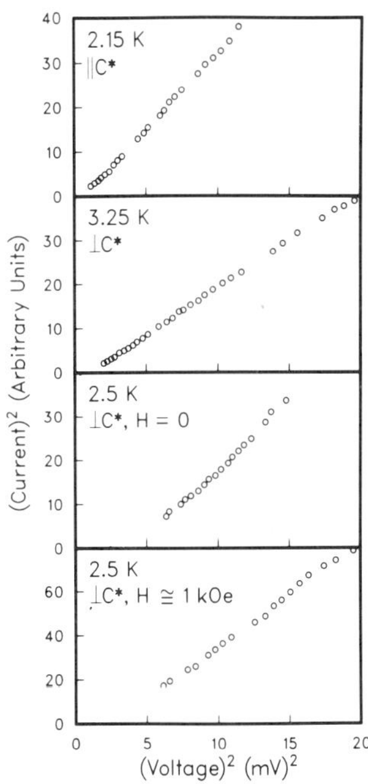

Fig. 2. Plots of I^2 vs. V^2 from Hawley, et al. (1986). Intercepts at $I^2 = 0$ give $(\Delta/e)^2$.

A summary of the Δ values, with error bars reflecting the latitude of fitting the data of Fig. 2, are shown in Fig. 3. The data for zero field, with the tip movement perpendicular to the c^* axis, but at an unknown azimuthal angle, are consistent with a BCS temperature dependence, but with a magnitude well in excess of the weak-coupling value at $T = 0$ of $1.76\,k_B T_c \simeq 0.6$ meV. An additional datum point shows that application of a small magnetic field (~1 kOe) reduced the gap by about 10-20%, which is reasonably consistent with expectations based on inductive measurements on this material. The above measurements give a strong indication that we are measuring a BCS superconducting gap, in spite of its large value ($n \sim 14$). We see a considerably smaller Δ (about 30% as big) for the tip movement parallel to the c^* axis, which is suggestive of large gap anisotropy, however, for the present case of point-contact tunneling, one cannot be sure of the tunneling direction.

The large gap value deserves further discussion, since experimental artifacts usually result in too small a gap value. We can, however, consider whether tunneling may be occurring between two grains of β-(ET)$_2$AuI$_2$, which would yield structure in I(V) at 2Δ. Although a possibility, it seems unlikely for several reasons. First, the I(V) for SS tunneling shows a sharp jump at $V = 2\Delta$ rather than the smooth increase calculated for NS tunneling, with which our data agree quite well. Also, the similar values of Δ measured at ≈ 2.4 K were taken on three separate low-temperature runs for which the precise locations of the tunneling areas, on both the tip and sample, must have been significantly different. It is unlikely that they all resulted in the unusual SS tunneling. After completion of the tunneling measurements, the tip was observed under an SEM and revealed no evidence for chips of β-(ET)$_2$AuI$_2$ nor damage. One can also consider whether the large Δ results from an enhanced T_c due to the pressure exerted by the point-contact tip. However, SEM studies of the sample reveal no evidence of damage by the tip for the surface on which the large gap was measured, and pressure studies (Shirber, et al. 1986) of β-(ET)$_2$AuI$_2$ show only decreases in T_c.

In Fig. 4 we show conductance curves for our data, and for comparison the BCS dependence (Bermon, 1964) for the same gap value and reduced temperature. The much sharper features for tunneling into the a-b plane

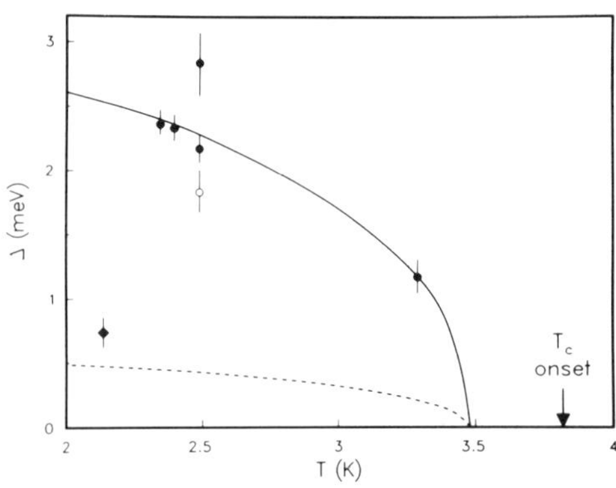

Fig. 3. Various Δ determined from plots like Fig. 2, from Hawley, et al. (1986). Open circle: field of ~1 kOe. Diamond: tip movement parallel to c^* axis.

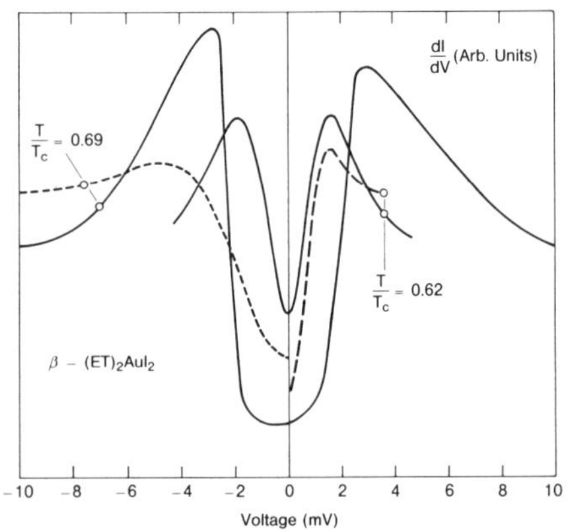

Fig. 4. dI/dV for top and bottom curves of Fig. 1 (T/T$_c$ = 0.62 and 0.64 respectively). Dashed curves are BCS from Bermon (1964).

are only consistent with a gap significantly greater than BCS, whereas for tunneling perpendicular to the c^*-axis, the data are reasonably close to the BCS prediction.

We offered a speculation on the origin of the large Δ. Strong-coupling superconductors are usually associated with very low-frequency phonons (e.g., Pb and Hg). Analysis of the Eliashberg equations by Bergmann and Rainer (1973) and Mitrovic, Leavens, and Carbotte (1980) indicate that high phonon frequencies ($\omega \sim 9\ k_B T_c$) affect T_c most, while $\Delta(0)$ and η are affected most by lower phonon frequencies ($\sim 4\ k_B T_c$ and $\sim 1.3\ k_B T_c$, respectively). Low-frequency libration and/or intramolecular (especially torsional and bending) modes may couple strongly to the conduction electrons in these organic superconductors, since the electron transmission probability from molecule to molecule may be strongly affected by such modes. Subsequently, several entirely different theoretical ideas have been proposed which may explain the large energy gaps found in β-(ET)$_2$AuI$_2$. However, since similarly large gaps have been found in the high-T_c oxide superconductors, discussion will be deferred.

Recently, evidence for very low frequency modes have resulted from ir absorption (Hamberg, et al., 1987) in β-(ET)$_2$AuI$_2$ (21 cm^{-1} or 2.6 meV) and in β-(ET)$_2$I$_3$ (32 cm^{-1} or 4 meV). The latter mode at 4 meV is confirmed in measurements of $\alpha^2 F$ in β-(ET)$_2$I$_3$ by point-contact tunneling (Nowack, et al., 1987) which also show a mode at about 1.3 meV which is very strongly coupled to the electrons.

Nowack, et al., (1987) also present tunneling measurements of the energy gap in β-(ET)$_2$AuI$_2$ using both (a) Au and (b) another crystal of β-(ET)$_2$AuI$_2$ as counterelectrodes. By fitting the conductance peak they find values for η of 4-5 (8-10) if SS (NS) tunneling is assumed for both cases. In our experience using Au counterelectrodes, less than 10% of the curves showed gap structure, indicating that nonsuperconducting regions are more usual. Hence it may be more realistic to assume NS tunneling in case (b) above as well as case (a).

In conclusion, the best results have come from point-contact tunneling, but difficulties of junction stability have prevented full temperature dependence, and anisotropy measurements must await better surfaces so true vacuum tunneling is possible. It seems reasonable to conclude that the energy gap in β-(ET)$_2$AuI$_2$ is significantly greater than the BCS prediction.

HIGH-T$_c$ OXIDE SUPERCONDUCTORS

It is impossible to completely and fairly report all the work worldwide in this fast moving field. We can only summarize the research of which we are aware. Tunneling studies have used, almost exclusively, two superconductors, La$_{1.85}$Sr$_{0.15}$CuO$_4$, with T_c up to ~ 40 K and denoted in the following by LSCO, and YBa$_2$Cu$_3$O$_{6.85}$, with T_c up to ~ 100 K and denoted by YBCO. Although these materials may share many of the unusual properties responsible for high T_c, the tunneling studies to date are different. Most researchers report relatively usual I(V) curves with a gap for LSCO, while more varied and difficult-to-interpret results are found for YBCO.

(La,Sr)$_2$CuO$_4$ The results from the various laboratories for single-particle tunneling into LSCO are summarized in Table I. In most cases the results are for point-contact tunneling into bulk, sintered pellets (Moreland, et al. (1987a) use the break-junction technique and Naito, et al. (1987) measured thin-film samples both with point-contact and with a

Table I. Compendium of tunneling results on $(La,Sr)_2CuO_4$

References all (1987)	Tip	Δ (meV)	$\dfrac{2\Delta(0)}{k_B T_c}$	Assymetry	R* (MΩ)
Kirtley, et al.(a)	PtIr	4.4–7	4.5	weak	~1
Naito, et al.	W	10–30	8–18	strong	~100
Ekino, et al. (a)	Aℓ	7	4.8	weak	?
Moreland, et al. (a)	LSCO	7–9	4.5–6	weak	~1
Hawley, et al.	Au	7–14	5–9	weak	~1
Pan, et al.	PtRh	9–15	8.7	yes	~1

*Approximate normal state junction resistance.

Pb-film counterèlectrode). With the exception of Naito, et al. (1987) the results are reasonably similar. The I(V) for junction resistances of ~1 MΩ show slight asymmetries, and gap values range from 7 to 15 meV, resulting in η values from 4.5 to 9. Some of these are illustrated in Figs. 5–9. The methods of determining Δ from I(V) or dI/dV vary, but different methods would have little effect on the general trends.

For very high resistance junctions, both Kirtley, et al. (1987a) and Moreland, et al. (1987a) found similarities with the ZG model, first proposed by Zeller and Giaever (1969), which consists of a <u>distribution</u> of small superconducting particles embedded in the tunneling barrier. The results shown in Fig. 5 agree qualitatively with this model which predicts a quadratic I(V) and hence a linear dI/dV. For quantitative agreement, one must assume the spatial inhomogeneities to be on a size scale much smaller than the grain sizes ($\gtrsim$ 1 μm) seen in TEM analysis.

The I(V) of Naito, et al. (1987) shown in Fig. 8 is strikingly different from others. Its derivative, shown in Fig. 9 as "spot 3", is very similar to the data of Barner and Ruggiero (1987), shown in Fig. 10, who sandwiched Cu particles, with a relatively narrow distribution of sizes, between artificial barriers separating Pb and Cu electrodes. The

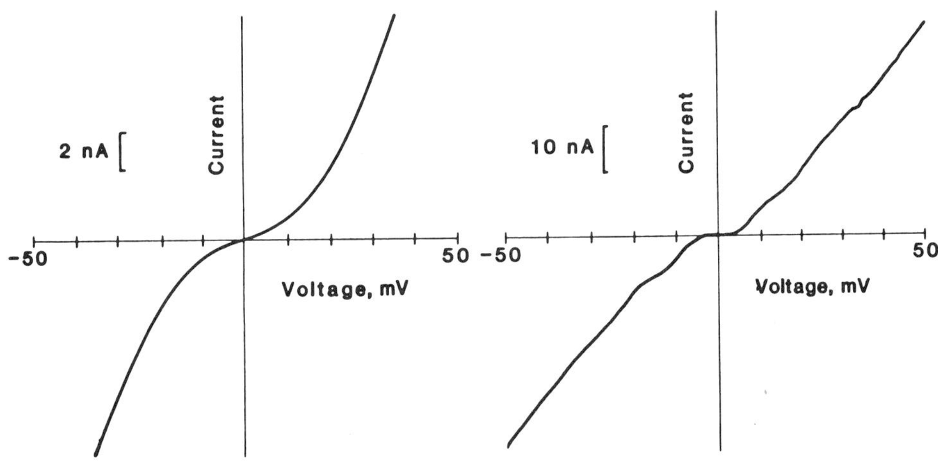

Fig. 5. I(V) of break junction of LSCO showing ZG effect, from Moreland, et al. (1987a).

Fig. 6. I(V) of break junction of LSCO showing gap structure, from Moreland, et al. (1987a).

explanation of the peaks in dI/dV is that a new tunneling channel is available at voltages corresponding to the charging energy of the Cu particles, and at multiples 1, 3, 5, etc. Barner and Ruggiero (1987) found this behavior for junction resistances of ~100 MΩ (just like Naito, et al. (1987), suggesting that the process is strong enough to explain their data as well) and not for much lower resistances. Further work is needed to clarify this point, because such structure in I(V) may have profound implications on the mechanism of high-T_c superconductivity.

$YBa_2Cu_3O_{6.85}$ The results from various laboratories are summarized in Table II. Not only are the variations of Δ greater than for LSCO, but so are the assymetries and deviations from standard I(V). Two preliminary studies of single crystals (Kirtley, et al., 1987c and Moog, et al., 1987b) indicate results similar to polycrystalline samples, and in the former work, little or no anisotropy. Iguchi, et al. (1987b) obtained a very

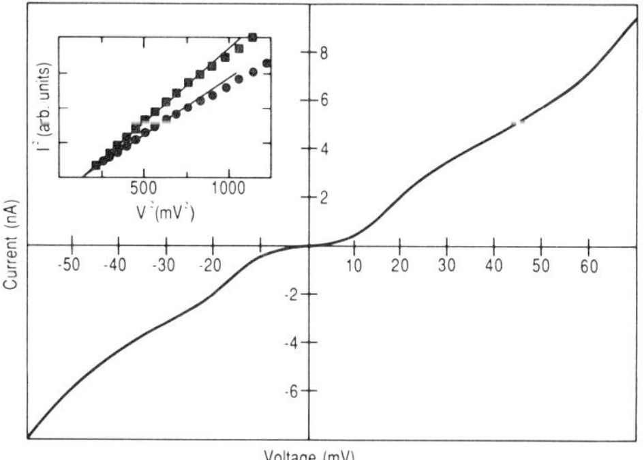

Fig. 7. I(V) for $La_{1.85}Sr_{0.15}CuO_4$ point-contact tunneling with Au tip from Hawley, et al. (1987). Inset: I^2 vs V^2, intercepts give $\Delta \sim 11$ meV.

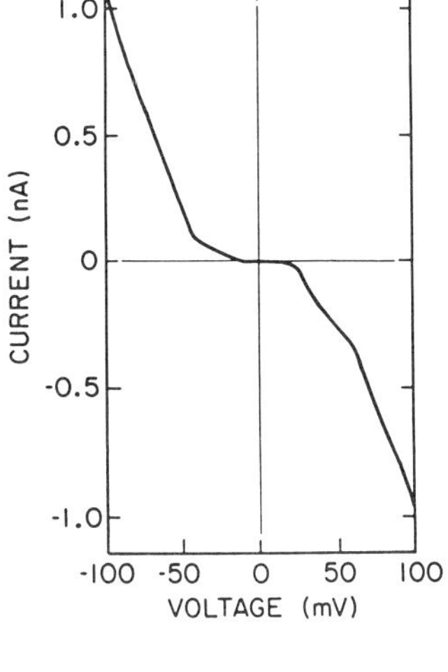

Fig. 8. I(V) for point-contact tunneling with W tip onto film of LSCO from Naito, et al. (1987).

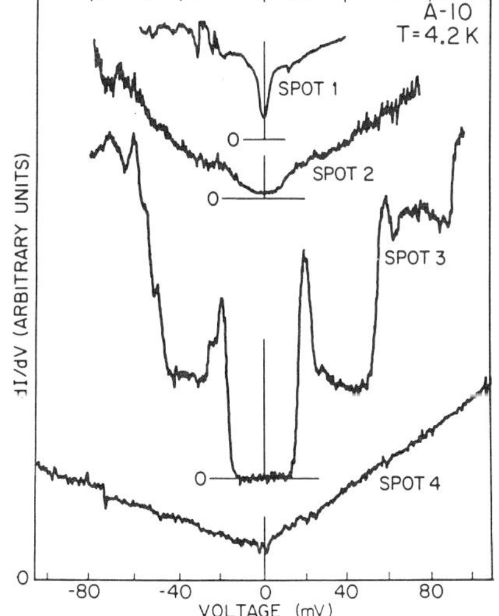

Fig. 9. dI/dV for point-contact tunneling for four spots on LSCO film of Naito, et al. (1987). Spot 3 corresponding to I(V) of Fig. 7.

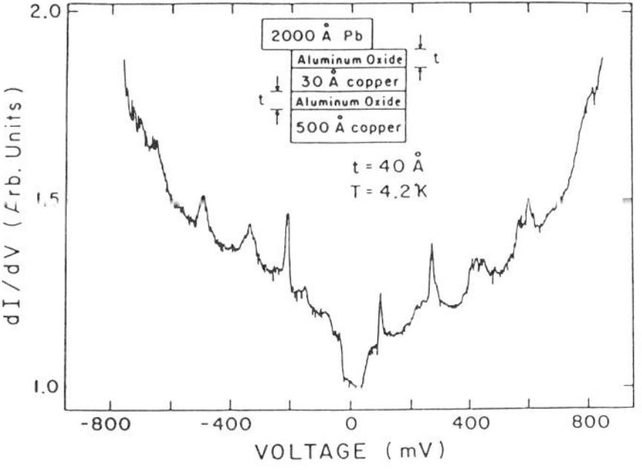

Fig. 10. dI/dV for 30 Å Cu particles imbedded in tunnel barrier, from Barner and Ruggiero (1987).

Table II. Compendium of tunneling results on $YBa_2Cu_3O_{6.9}$

References all (1987)	Tip	Δ (meV)	$\dfrac{2\Delta(0)}{k_BT_c}$	Assymetry	R* (MΩ)
Kirtley, et al. (b)	PtIr	13–20	3.4–5	strong	~10
Smith, et al.	W	50–170	12–44	strong	~100
Ekino, et al. (b)	Al	13–16	2.8–5.5	yes	~10^{-5}
Moreland, et al. (b)	YBCO	20–40[a]	5–10[a]	weak	~1
Moog, et al. (a)	Au	25–35[b]	6–9[b]	weak	~0.01
Ng, et al.	Al	45–85	11–21	weak	~5
Iguchi, et al. (b)	Pb	20–25[d]	10–12[d]	none	?
Kirtley, et al. (c)	PtIr	15–23[e]	4–6[e]	yes	~100
Moog, et al. (b)	Au, W	18–20[e]	~5[e]	weak	10–50

*Approximate normal state junction resistance.

a)Assumes NS tunneling. b) Pure metallic bridge. c) Thin-film counterelectrode. d) Note T_c ~ 48 K. e) Single crystal.

important result: the temperature dependence of Δ. This was possible because they evaporated a Pb-film counterelectrode onto polished, sintered samples, thereby avoiding the thermal expansion problem inherent in point-contact measurements. Although T_c was reduced to ~48 K by the polishing, Δ was found to decrease towards zero as $T \rightarrow T_c$ in a BCS-like fashion (see Fig. 11). This clearly confirms that the I(V) structure is related to superconductivity and gives strong evidence that it is also related to Δ. However, they also see similar structures at voltages of 2Δ/e and 3Δ/e, and these are not easily explained. Further work is needed to clarify the multiple structure.

In our studies of polycrystal YBCO (Moog, et al., 1987a), the most common dI/dV generally display a variety of quite asymmetrical structures which are difficult to interpret in terms of Δ. Smith, et al. (1987) report very strong peaks in dI/dV in the polarity with the sample negative with respect to the tip. We find similar shaped dI/dV, but for different tip positions they can occur for either polarity. Their explanation based on Schottky barriers would otherwise seem reasonable. By increasing the force on the Au tip, lower junction resistances with negligible assymetry were obtained. For a 10 kΩ junction, peaks in dI/dV occurred at ±25 mV and for a 3 kΩ junction we found the I(V) shown in Fig. 12 of a pure metallic bridge (Blonder and Tinkham, 1983) resulting in Δ ~ 30–35 meV.

Interestingly, Ng, et al. (1987) using a soft Al tip found symmetrical peaks at 85 mV and less symmetrical peaks at 45 meV which they analyzed in terms of their depairing model. The results of Kirtley, et al. (1987b) show no peak in dI/dV, seem to be dominated by Zeller and Giaever (1969) tunneling like their LSCO results, and are fit with a much smaller gap than others find. A possible explanation is that the tunneling is into very small particles, necessary for the ZG model, which have reduced T_c and Δ. The I(V) shown in Fig. 13 (Moreland, et al., 1987b) shows structure at about 39 meV, but there is uncertainty in the interpretation between NS and SS tunneling.

For studies of a single crystal of YBCO (Moog, et al., 1987b), both Au and W tips were used. For the W tip, junction resistances were in excess

of 10 MΩ, presumably because the oxide coating and because W does not
deform like Au to increase the junction area. For one of the lowest resistance junctions, a rather poor I(V) could be interpreted to show
Δ~15-20 meV. For very high resistances (~100 MΩ), some data are almost
identical to the I(V) found by Naito, et al. (1987) on LSCO (see Fig. 7),
and reinforces the contention that it is related to very high resistances.

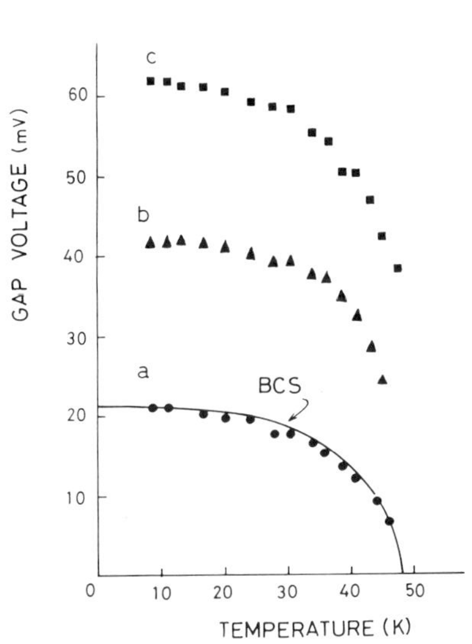

Fig. 11. Temperature dependence of
the resistance minima in tunneling
from YBCO to Pb film, from Iguchi,
et al. (1987b).

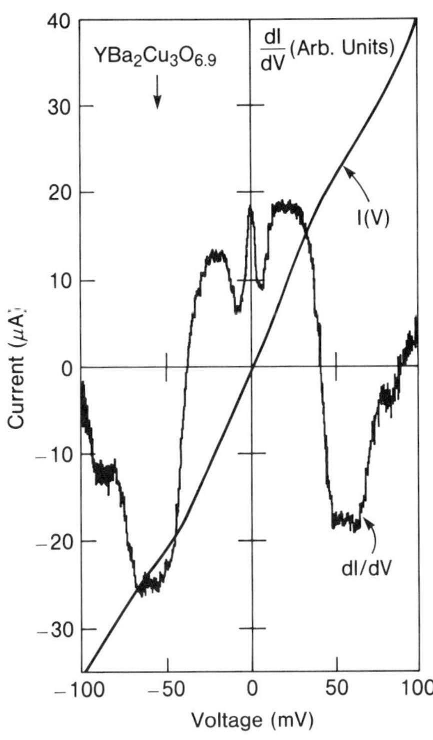

Fig. 12. I(V) and dI/dV for pure
metallic bridge between YBCO and
Au tip, from Moog, et al. (1987).
Inferred gap is 30-35 meV.

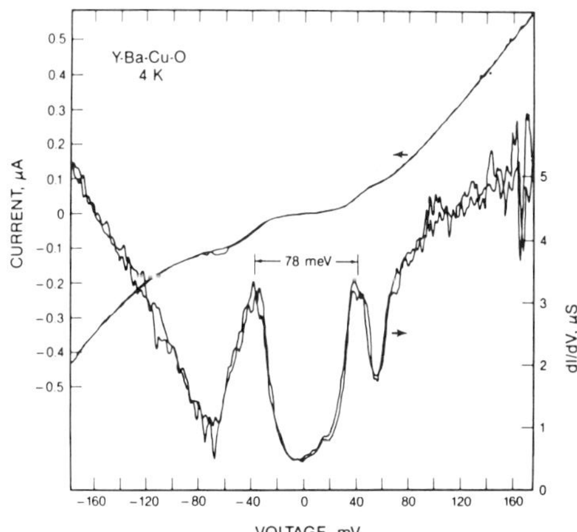

Fig. 13. I(V) and dI/dV for break
junction, from Moreland, et al.
(1987b).

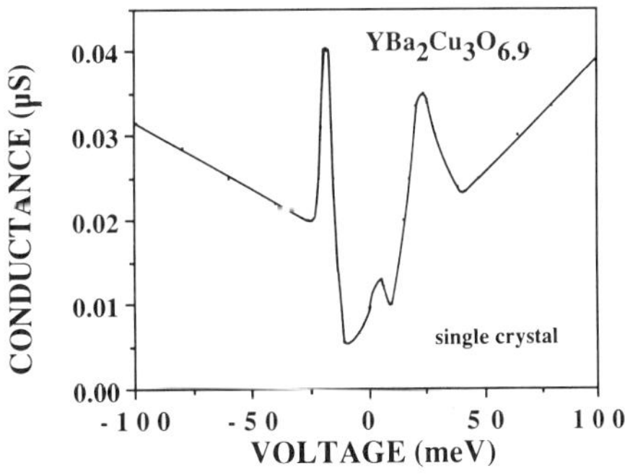

Fig. 14. dI/dV for point-contact
tunneling to a single crystal,
with Au tip movement parallel to
c-axis, from Moog, et al. (1987b).

The Au tip yielded a wider range of results. The polarity-dependent peak in dI/dV occurs for either polarity, as in the polycrystal case above. However, the peak voltage was observed to shift as a function of time (~25% in 20 min.) without readjusting the force on the tip (see also the variation in peak position in Smith, et al., 1987). Symmetrical I(V) are found occasionally with a range of "gaps". One of these is similar to that of Moreland, et al. (1987b) shown in Fig. 13, but gives $\Delta \sim 45$ meV. Other curves have poorer definition of the gap structure and yield smaller $\Delta \sim 20$-25 meV, while occasional peaks in dI/dV at 80-100 mV are found. For a somewhat high resistance junction, the dI/dV shown in Fig. 14 gives clear indications of single-particle tunneling gap at about 20 meV with a background conductance closely following the ZG model. It is similar to the single-crystal data of Kirtley, et al. (1987c) shown in Figs. 15 and 16, which also give values of $\Delta \sim 18$-20 meV.

    <u>Conclusions</u>   Although our understanding of tunneling in the high-T_c oxide superconductors is far from complete, we can made some tentative observations. Excluding results where the Zeller and Giaever (1969) or Barner and Ruggiero (1987) tunneling dominates, the gap values fall in a range of 7-15 meV for LSCO and 20-45 meV for YBCO. The large values (5-12) of $\eta \equiv 2\Delta(0)/k_B T_c$ will be addressed in the Discussion section below. The reasons for the variations of Δ cannot be ascertained at this time, but intrinsic anisotropy and sample inhomogeneities seem to be leading candidates. For YBCO the results are more ambiguous. In cases where background from the ZG model are clearly seen, the gaps are ~ 20 meV; whether Δ is depressed in the small particles necessary for the ZG model, due to variations off stoichiometry, is unclear, but the smallest reported values of Δ in LSCO are found in this case (Kirtley, et al., 1987a). Iguchi's impressive temperature dependence for $T_c \sim 48$ K, would imply $\Delta \sim 40$ meV for $T_c \sim 95$ K, and the results of Moreland, et al. (1987b) could be justifiably interpreted as NS tunneling, like their LSCO results, leading to $\Delta \sim 40$ meV also. The results of Moog, et al. (1987b) and Ng, et al. (1987), showing symmetrical peaks in dI/dV at 45 meV are clouded by the fact that much higher symmetrical peaks (80-100 meV) are sometimes found. Further experimental progress is called for.

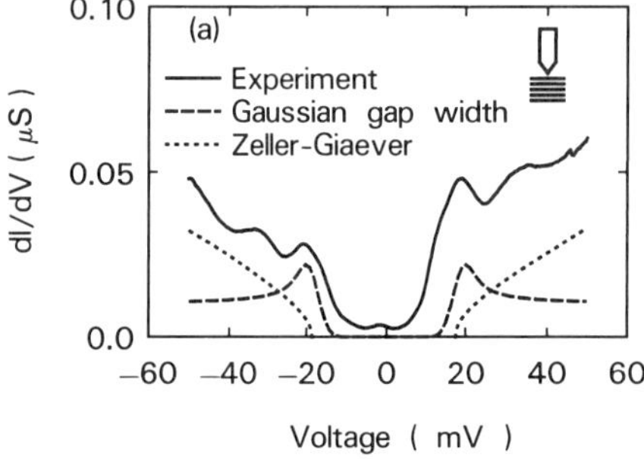

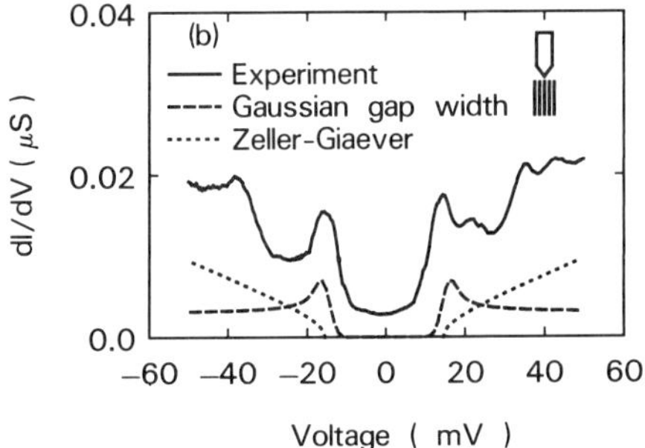

Fig. 15. dI/dV for point-contact tunneling to single crystal of YBCO with tip movement parallel to c-axis, from Kirtley, et al. (1987c).

Fig. 16. dI/dV for point-contact tunneling to single crystal of YBCO with tip movement perpendicular to c-axis, from Kirtley, et al. (1987c).

DISCUSSION

It is likely that the high-T_c of oxide superconductors cannot be entirely accounted for by electron-phonon coupling. Tunneling provides an important experimental tool to help answer this question through measurements of Δ, anisotropy of Δ and the density of states, $\rho(E)$, through which the coupling mechanism can possibly be extracted. At present, $\rho(E)$ is unattainable, evidence for anisotropy is weak, at best, but the large value of $\eta \equiv 2\Delta(0)/k_B T_c$ seems well substantiated at least in LSCO. This property is shared with the organic superconductor, β-$(ET)_2 AuI_2$, but so is the two dimensional conductivity and the relatively low carrier density. Both exhibit a hopping-type conductivity which is strongly modulated, and therefore coupled, to specific lattice vibrations which should result in a particularly strong electron-phonon coupling constant λ. This is especially true for β-$(ET)_2 AuI_2$ in which very low frequencies are expected for these modes. However, the high-T_c of the oxide superconductors has prompted suggestions of new non-phononic mechanisms such as two-dimensional plasmons (Kresin, 1987a), charge density waves (Fenton, 1987), resonant valence bond (Anderson, 1987), excitons, etc.

Lacking definitive tunneling evidence for such mechanisms, we shall confine our final remarks to the large values of η found in both β-$(ET)_2 AuI_2$ and high-T_c oxide superconductors. Three entirely different explanations have been proposed: extremely strong-coupled superconductivity; extremely high inelastic scattering rate; and anisotropy.

<u>Extremely strong coupling</u>   This idea, first proposed by Hawley, et al. (1986), was placed on firm theoretical grounds by the work of Kresin (1987b). He showed that in the limit of extremely strong coupling $(\lambda \rightarrow \infty)$ the Elaishberg equations can be solved for the gap and give $\Delta(0) \approx 1.2 \, \lambda^{1/2} \, \tilde{\Omega}$, where $\tilde{\Omega}$ is a characteristic phonon energy. When this is coupled to the known result that $k_B T_c \cong 0.18 \, \lambda^{1/2} \, \tilde{\Omega}$ in the same limit (e.g., Allen and Dynes, 1975), one obtains the limiting value of $\eta \equiv 2\Delta(o)/k_B T_c = 13.4$ for $\lambda \rightarrow \infty$. This is very close to the value of ≈ 14 reported by Hawley, et al. (1986) for the organic superconductor, β-$(ET)_2 AuI_2$ and is at the upper limit of values reported in Tables I and II for the high-T_c oxide superconductors. Therefore, large η itself cannot exclude the phonon mechanism.

<u>Extremely high inelastic scattering rate, τ_{in}^{-1}</u>   When $\hbar\tau_{in}^{-1} > k_B T$, then Lee and Read (1987) show that the relevant energy scale is $\hbar\tau_{in}^{-1}$, not $k_B T$ and that $\Delta \approx \hbar\tau_{in}^{-1}(T_c)$ instead of the usual BCS value. The fact that superconductivity can survive such strong inelastic scattering (pair-breaking) indicates that the pairing energy is very large. The normal state resistivity of the high-T_c (organic) superconductors is proportional to T(to T^2) from T_c right up to room temperature (Cava, et al., 1987 and Tanner, et al., 1987). Lee and Read (1987) estimate $\hbar\tau_{in}^{-1} \approx 3 \, k_B T > k_B T$ for LSCO and the measurements of Tanner, et al., (1987) for β-$(ET)_2 AuI_2$ result in $\hbar\tau_{in}^{-1} \approx 6 \, k_B T^2 \gg k_B T$, for $T \sim T_c \sim 5$ K. Therefore, in both cases the energy gap should be greater than BCS.

<u>Anisotropy</u>   In an effort to simultaneously explain the small gaps found in ir measurements and the large gaps found in tunneling, Ebisawa, et al. (1987) propose that the high-T_c superconductors are anisotropic, singlet superconductors. Because of their two-dimensional (2D) nature this is not unreasonable. Here, true vacuum tunneling measurements on single crystals could clearly resolve this question. The organic superconductor, β-$(ET)_2 AuI_2$, is also 2D with a large gap from tunneling and none found in

ir measurements (Hamberg, et al., 1987), and thus could also use anisotropy
as an explanation of the large Δ. Unfortunately, energy gap studies
(Hawley, et al., 1986) on this organic could not unambiguously resolve the
anisotropy because they were the point-contact type.

    <u>Summary</u>   It is clear that there is still alot to be learned about
mechanisms of superconductivity in the exotic magnetic, heavy fermion,
organic and high-T_c oxide superconductors. Tunneling can potentially play
a very important role. Also, one is struck by the many similarities of the
organic superconductor β-(ET)$_2$AuI$_2$ and the high-T_c oxide superconductors,
although the mechanisms of superconductivity may be quite different.

REFERENCES

Allen, P. B. and Dynes, R. C., 1975, <u>Phys. Rev.</u>, B12: 905.

Anderson, P. W., 1987, <u>Science</u>, 235: 1196.

Bando, H., Kajimura, K., Anzai, H., Ishiguro, T. and Saito, G., 1984, <u>in</u>:
    "LT-17", U. Eckern, A. Schmid, W. Weber and H. Wühl, eds.,
    Elsevier Science, Amsterdam.

Barner, J. B. and Ruggiero, S. T., 1987, to appear in: "Proc. TMS-AIME:
    Metallic Multilayers and Epitaxy", M. Hong and S. A. Wolf, eds.

Bergmann, G. and Rainer, D., 1973, <u>Z. Phys.</u>, 263: 59.

Bermon, S., 1964, "The BCS Differential Conductance for a Metal-Insulator-
    Superconductor Tunnel Junction", Technical Report No. 1,
    Department of Physics, University of Illinois, Urbana, IL.

Blonder, G. E., Tinkham, M. and Klapwijk, T. M., 1982, <u>Phys. Rev.</u>, B25:
    4515.

Blonder, G. E. and Tinkham, M., 1983, <u>Phys. Rev.</u>, B27: 112.

Cava, R. J., Batlogg, B., van Dover, R. B., Murphy, D. W., Sushine,
    S., Siegrist, T., Remeika, J. P., Rietman, E. A., Zahwrak, S. and
    Espinosa, G. P. , 1987, <u>Phys. Rev. Lett.</u>, 58: 1676.

Ebisawa, H., Isawa, Y. and Maekawa, S., 1987, preprint.

Ekino, T., Akimitsu, J., Sato, M. and Hosoya, S., 1987a, to appear in <u>Solid</u>
    <u>State Commun.</u>

Ekino, T. and Akimitsu, J., 1987b, <u>Japan J. Appl. Phys.</u>, 26: L452.

Fenton, E. W., 1987, preprint.

Giaever, I., 1960, <u>Phys. Rev. Lett.</u>, 5: 147.

Hamberg, I., Tanner, D., Jacobson, C., Williams, J. M. and Wang, H. H.,
    1987, <u>Bull. Am. Phys. Soc.</u>, 32: 805.

Han, S., Ng, K. W., Wolf, E. L., Millis, A., Smith, J. L. and Fisk, Z.,
    1987, <u>Phys. Rev. Lett.</u>, 57: 238.

Hawley, M. E., Gray, K. E., Terris, B. D., Wang, H. H., Carlson, K. D. and
    Williams, J. M., 1986, <u>Phys. Rev. Lett.</u>, 57: 629.

Hawley, M. E., Gray, K. E., Capone, D. W. and Hinks, D. G., 1987, <u>Phys.</u>
    <u>Rev.</u>, B35: 7224.

Iguchi, I., Yasuda, T., Onuki, Y. and Komatsubara, T., 1987a, <u>Phys. Rev.</u>,
    B35.

Iguchi, I., Watanabe, H., Kasai, Y., Mochiku, T., Sugishita, A. and
    Yamaka, E., 1987b, <u>Japan J. Appl. Phys.</u>, 26: L645.

Kadin, A. M. and Goldman, A. M., 1987 <u>Phys. Rev. Lett.</u>, 58: 2275.

Kirtley, J. R., Tsuei, C. C., Park, S. I., Chi, C. C., Rozen, J. and
    Shafer, M. W., 1987a, <u>Phys. Rev.</u>, B35: 7216.

Kirtley, J. R., Gallagher, W. J., Schlesinger, Z., Sandstrom, R. L.,
    Dinger, T. R. and Chance, D. A., 1987b, preprint.

Kirtley, J. R., Tsuei, C. C., Park, S. I. Chi, C. C., Rozen, J.,
        Shafer, M. W., Gallagher, W. J., Sandstrom, R. L., Dinger, T. R.
        and Chance, D. A., 1987c, to appear in: "LT-18".
Kresin, V. Z., 1987a, preprint.
Kresin, V. Z., 1987b, preprint.
Lee, P. A. and Read, N., 1987, preprint.
McMillan, W. L. and Rowell, J. M., 1969, in: "Superconductivity",
        R. D. Parks, ed., Decker, New York.
Mitrovic, B., Leavens, C. B. and Carbotte, J. P., 1980, Phys. Rev., B21:
        5048.
Moog, E. R., Hawley, M. E., Gray, K. E., Hinks, D. G. and Capone,
        D. W., 1987a, private communication.
Moog, E. R., Gray, Hawley, M. E., K. E., Liu, J. Z. and Downey, J., 1987b,
        private communication.
More, C., Roger, G., Sorbier, J. P., Jerome, D., Ribault, M. and Bechgaard,
        K., 1981, J. Physique Lett., 42: L313.
Moreland, J. and Ekin, J. W., 1985, J. Appl. Phys., 58: 3888.
Moreland, J., Clark, A. F., Goodrich, L. F., Ku, H. C. and Shelton,
        R. N., 1987a, Phys. Rev., B35:
Moreland, J., Ekin, J. W., Goodrich, L. F., Capobianco, T. E.,
        Clark, A. F., Kwo, J., Hong, M. and Liou, S. H., 1987b, preprint.
Naito, M., Smith, D. P. E., Kirk, M. D., Oh, B., Hahn, M. R., Char, K.,
        Mitzi, D. B., Sun, J. Z., Webb, D. J., Beasley, M. R., Fischer,
        O., Geballe, T. H., Hammond, R. H., Kapitulnik, A. and Quate,
        C. F., 1987, Phys. Rev., B35: 7228.
Nowack, A., Poppe, U., Weger, M., Schweitzer, D. and Schwenk, H., 1987,
        preprint.
Ng, K. W., Pan, S., de Lozanne, A. L., Panson, A. J. and Talvacchio, J.,
        1987, to appear in : "LT-18".
Pan, S., Ng, K. W., de Lozanne, A. L., Tarascon, J. M. and Greene, L. H.,
        1987, Phys. Rev., B35: 7220.
Poppe, U., 1981, Physica, 108B: 805.
Poppe, U. and Schroeder, H., 1984, in: "LT-17", U. Eckern, A. Schmid,
        W. Weber and H. Wühl, eds., Elsevier Science, Amsterdam.
Poppe, U., 1985, Physica, 136B: 22.
Rowell, J. M., Dynes, R. C. and Schmidt, P. W., 1980, in: "d and f-Band
        Metals", H. Suhl and M. B. Maple, eds., Academic Press, New York.
Shirber, J. E., Azevedo, L. J., Kwak, J. F., Venturini, E. L.,
        Leung, P. C. W., Beno, M. A., Wang, H. H. and Williams, J. M.,
        1986, Phys. Rev., B33: 1987.
Smith, D. P. E., Kirk, M. D., Mitzi, D. B., Sun, J. Z., Webb, D. J.,
        Char, K., Hahn, M. R., Naito, M. Oh, B., Beasley, M. R., Geballe,
        T. H., Hammond, R. H., Kapitulnik, A. and Quate, C. F., 1987,
        preprint.
Tanner, D. B., Jacobsen, C. S., Williams, J. M. and Wang, H. H., 1987,
        preprint.
Umbach, C. P., Toth, L. E., Dahlberg, E. D. and Goldman, A. M., 1981,
        Physica, 108B: 803
Umbach, C. P. and Goldman, A. M., 1982, Phys. Rev. Lett., 48: 1433.
Vaglio, R., Terris, B. D., Zasadzinski, J. F., and Gray, K. E., 1984, Phys.
        Rev. Lett. 53: 1489.
Wolf, E. L., 1985, "Principles of Electron Tunneling Spectroscopy", Oxford,
        New York.
Zeller, H. R. and Giaever, I., 1969,. Phys. Rev., 181: 789.

THERMODYNAMIC CRITICAL FIELD OF $Y_1Ba_2Cu_3O_7$

D. K. Finnemore, M. M. Fang, J. R. Clem, R. W. McCallum,
J. E. Ostenson, Li Ji, and P. Klavins

Ames Laboratory and Department of Physics
Iowa State University, Ames, Iowa 50011
PAC74.30 Ci; 74.30 Ek

ABSTRACT

The thermodynamic critical field curve has been measured for
$Y_1Ba_2Cu_3O_7$ close to the superconducting transition temperature in order
to determine the magnitude of the specific heat jump at the transition.
The slope of the critical field curve is found to be 165 Oe/K, which
corresponds to a specific heat jump of 680 mJ/moleCuK. If the BCS
theory applies, this would also imply a density of electronic states of
3.39 states per electron volt-unit cell and the Sommerfeld term,
$\gamma = 5.3$ mJ/moleCuK^2.

INTRODUCTION

Magnetization curves for a superconductor can give important
information about the electronic properties of the material.[1] If the
magnetization curves are reversible, the free energy difference between
the superconducting and normal state is given by the area under the
magnetization (M) vs magnetic field (H) curve,

$$\frac{H_c^2}{8\pi} = G_n - G_s = \int_0^{H_{c2}} MdH$$

and the thermodynamic critical field, H_c, is given by $H_c^2/8\pi = G_n - G_s$.
The slope of the critical field is then related to the jump in specific
heat by the thermodynamic relation,

$$\Delta C = \frac{T_c}{4\pi} \left(\frac{dH_c}{dT}\right)^2$$

The goal of this work is to directly measure the thermodynamic critical
field close to the transition temperature, T_c, and to derive various
electronic properties from it.

For the case of the $(La_{1.85}Sr_{0.15})CuO_2$ superconductor the
magnetization was found to be reversible over a wide temperature range.[1]
For the $Y_1Ba_2Cu_3O_7$ samples reported here, however, there is only a small
window in temperature close to the transition temperature, T_c, of about

4K width where the magnetization curves are reversible but this is quite adequate to determine the slope $(dH_c/dT)_{T=T_c}$. At 86K the uncertainty in the area of the magnetization curve becomes greater than 3% of the area due to hysteresis. In this work, a study of the magnetization curves is presented and the data are analyzed for the jump in specific heat at T_c. Direct measurement of ΔC is difficult because the phonon contribution is so large.

EXPERIMENTAL

Samples were prepared in the form of a sintered pellet by procedures similar to those outlined by other workers.[2-6] Proper amounts of Y_2O_3, $BaCO_3$ and CuO were ground in a mortar and pestle, put in a Pt lined aluminum boat and heated in air for 16 hr at 900°C. This powder was then ground again and fired again in flowing O_2 for 16 hr at 950°C. Different samples of this powder were then pressed into pellets and fired in flowing O_2 at temperatures ranging from 950°C to 1020°C for 16 hr. The furnace was turned off in steps with one hour holds at 100°C intervals, with stops at 900, 800, 700, 600, and 500°C and then furnace cooled to 200°C, a process which took about 6 hours. Several different reaction temperatures were used to determine where the sample began to decompose and deteriorate.

Magnetization measurements were made in a commercial apparatus in which the sample is slowly pulled through a gradiometer connected to a superconducting quantum interference device (SQUID).

RESULTS AND DISCUSSION

Electrical resistivity for these samples are typical to those reported by other groups.[2-6] Between room temperature and T_c, the curve is rather linear and drops from 1200 $\mu\Omega$cm at 300K to 700 $\mu\Omega$cm at 95K for

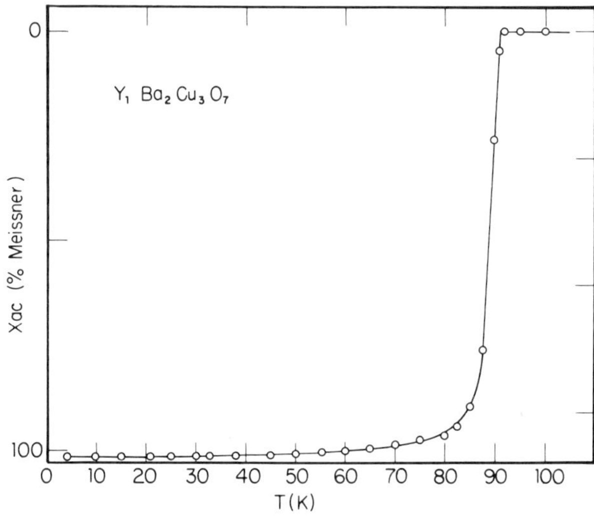

Fig. 1. The ac susceptibility of $Y_1Ba_2Cu_3O_7$ taken at 100 Hz and 1 Oe measuring field.

a typical sample. The temperature interval for the resistive drop at T_c between 90% and 10% of R(95K) was 1.2K. The ac susceptibility at 100 Hz and 1.0 Oe measuring field is shown in Fig. 1. There is a sharp drop at T_c indicating a good quality sample.

Magnetization curves in the normal state shown by the solid line of Fig. 2 are linear in field and give a magnetization of 0.046 emu/cm^3 at 1 Tesla. The curves measured at 95 and 100K are identical to an accuracy of 1%. In the superconducting state, the magnetization rises sharply at the full flux exclusion slope at low temperature. Very close to T_c, the superconducting penetration depth is large so the flux expulsion is not complete.[1]

The magnetization curve for 88K is shown by the solid curve of Fig. 3. The area under the magnetization curve for the magnetic field increasing is the same as that for the magnetic field decreasing to an accuracy of 2%, which is about the accuracy of the magnetization measurements. Using these areas in conjunction with Eq. 1, the critical field is found to be the values shown in Fig. 4. The data give a T_c of 91K, which is the temperature at which the electrical resistivity has dropped below the normal state resistance at 95K by a factor of more than 10,000.

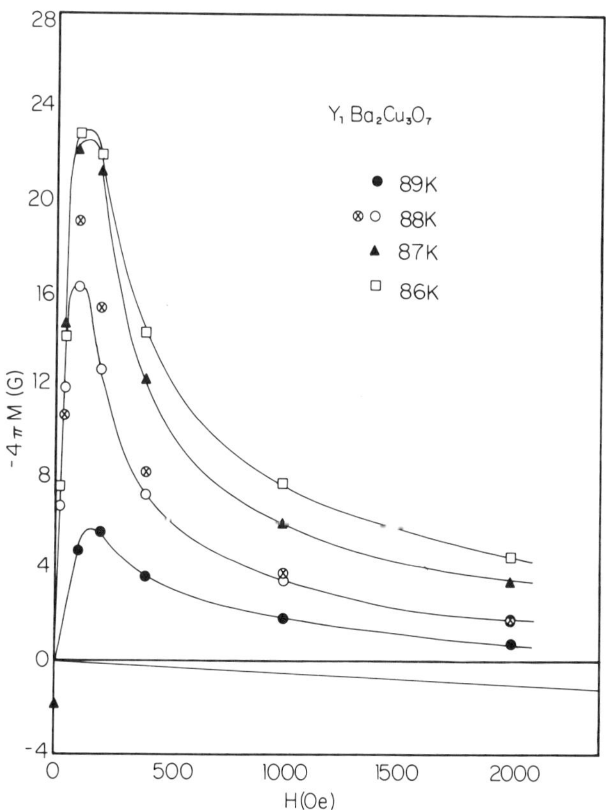

Fig. 2. Low field magnetization data.

The slope dH_c/dT is found to 165 Oe/K, a value rather close to elemental superconductors such as Sn and Pb. Inserting this value into Eq. 2 gives a jump in specific heat at T_c of 19.7 mJ/cm^{3}K or 689 mJ/moleCuK. These values depend only on reversibility of the M vs H curves and the applicability of thermodynamics. There is considerable variation in dH_c/dT for samples with different heat treatment procedures and values range between 102 Oe/K for a sample heat treated at 1020°C and 165 Oe/K for one heat treated 950°C. The higher heat treatment temperatures give much better critical currents, but they have considerable second phase and are not indicative of bulk 123 phase. The sample heat treated at 950°C is indicative of the most pure bulk 123 phase.

An important feature of superconductivity is that there is a great deal of similarity in the critical field curves and there is a law of corresponding states for various materials. A wide variety of elements and compounds show a critical field slope of approximately 170 Oe/K. These measurements show that $Y_1Ba_2Cu_3O_2$ fits well into this pattern. If one assumes that BCS describes the free energy of this material, the density of states can be calculated from the relation:

$$\gamma = 0.0558 \ (dH_c/dT)^2$$

and a value of 0.152 mJ/cm^{3}K^2 or 5.3 mJ/moleCuK^2. Using $N(0) = 3\ /2\pi^2 k^2\gamma$, the density of state is found to be 3.4 states/eV unit cell.

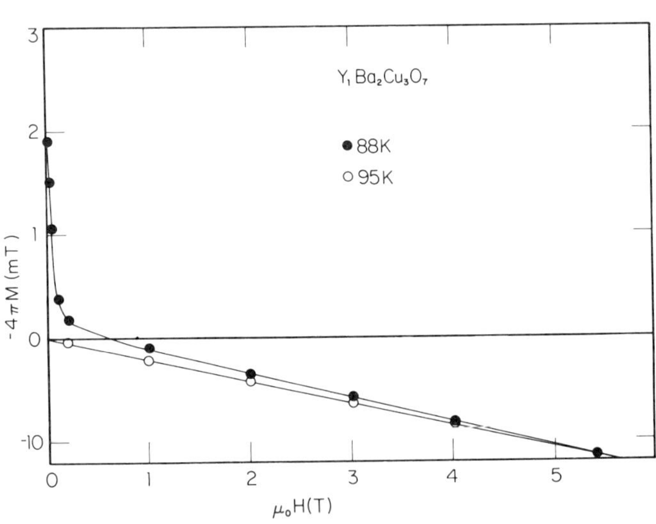

Fig. 3. Reversible magnetization curve shown intersecting the normal state magnetization.

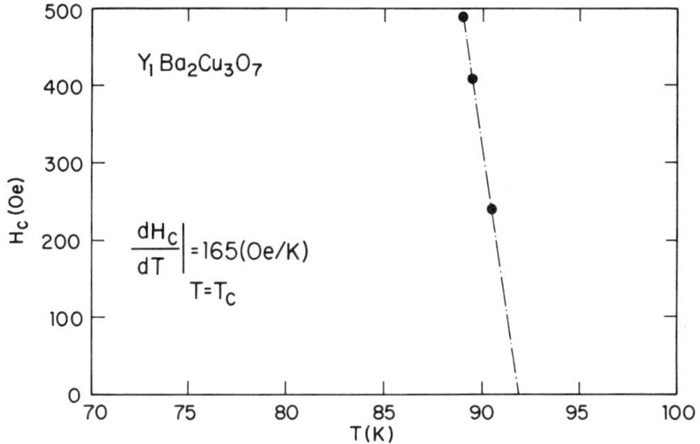

Fig. 4. Thermodynamic critical field.

CONCLUSION

There is a limited range of temperature close to T_c where the
magnetization curves are reversible and one can obtain the free energy
difference between the superconducting and normal states and thus, the
thermodynamic critical field curve. Thermodynamics then give a jump in
specific heat of 1.03 J/moleCuK or $(C_n-C_s)/T_c$=11.4 mJ/moleCuK^2. The
slope of the thermodynamic critical field curve is found to be very
similar to elemental superconductors. The density of states is found to
be N(0)=3.4 states/ev unit cell.

ACKNOWLEDGEMENTS

We would like to thank many authors for providing preprints. Ames
Laboratory is operated for the U.S. Department of Energy by Iowa State
University under Contract No. W-7405-ENG-82. This work was supported by
the Director for Energy Research, Office of Basic Energy Sciences.

REFERENCES

1. D. K. Finnemore, R. N. Shelton, J. R. Clem, R. W. McCallum, H. C.
    Ku, R. E. McCarley, S. C. Chen, P. Klavins, and V. G. Kogan, Phys.
    Rev. B 35 5319 (1987).
2. A. J. Panson, A. I. Braginski, J. R. Gavaler, J. K. Hulm, M. A.
    Janocko, H. C. Phol, A. M. Stewart, J. Talvacchio and G. R. Wagner,
    Phys. Rev. (submitted).
3. A. J. Panson, G. R. Wagner, A. I. Braginski, J. R. Gavaler, M. A.
    Janacko, H. C. Pohl and J. Talvacchio, Appl. Phys. Lett. 50, 1104
    (1987).

4. R. J. Cava, B. Batlogg, R. B. van Dover, D. W. Murphy, S. Sunshine, T. Siegrist, J. P. Remeika, E. A. Reitman, S. Zahwrak and G. P. Espinasa, Phys. Rev. Lett. <u>58</u>, 1676 (1987).

5. D. C. Larbalestier, M. Daeumling, X. Cai, S. Seuntjens, J. McKinnel, D. Hampshire, P. Lee, C. Meingast, T. Willis, H. Muller, R. D. Ray, R. G. Dillenberg, E. E. Hellstrom and R. Joynt, J. Appl. Phy. (accepted).

6. D. Welch, M. Suenaga and H. Saito, Phys. Rev. Lett. (submitted).

7. R. J. Cava, B. Batlogg, R. B. van Dover, D. W. Murphy, S. Sunshine, T. Siegrist, J. P. Remeika, E. A. Rietman, S. Zahurak and G. P. Espinoza <u>58</u> 1676 (1987).

MICROSTRUCTURE AND SUPERCONDUCTIVITY IN HIGH T_c MATERIALS

R. W. McCallum[+], R. N. Shelton[+*], M. A. Noack[+],
J. D. Verhoeven[+†], C. A. Swenson[+*], M. A. Damento+,
K. A. Gschneidner+†, Jr., E. D. Gibson+ and
A. R. Moodenbaugh**

+ Ames Laboratory, USDOE
* Department of Physics
† Department of Materials Science and Engineering
   Iowa State University
   Ames, IA 50011

** Brookhaven National Laboratory, USDOE
   Department of Applied Science
   Upton, NY 11973

ABSTRACT

This paper considers a number of aspects relevant to the preparation and superconducting properties of high purity $YBa_2Cu_3O_7$. Evidence for the incongruent melting of the material and its implications for producing large grained samples are discussed. Thermal expansion data for a sintered sample is presented and preliminary measurements on single crystalites are given.

INTRODUCTION

Since the discovery of high temperature superconductivity in the Y-Ba-Cu-O system[1] literally hundreds of groups have produced samples of the material where the details of the sample preparation have varied over a wide range. In this paper we will attempt to show the limits of the temperature range in which high quality samples may be produced and show that it is necessary to tightly control the temperature at which the material is produced. Following the discussion of the phase diagram, the results of magnetization measurements on a small single crystal will be discussed. Finally preliminary measurements of the thermal expansion of a sintered pellet exhibit behavior not normally associated with superconductivity.

EXPERIMENTAL TECHNIQUES

Two batches of $YBa_2Cu_3O_x$ were made by completely independent processing. Both materials were made by mixing Y_2O_3, CuO and $BaCO_3$ powders. In batch 1, 100 gm lots of the three powders were mixed and then ground in a motorized agate mortar and pestle. Fifty gram lots were given three 24 hours reactions in air at 900°C each followed by regrinding, one 24 hour reaction in air at 920°C followed by regrinding

and a final air reaction for 24 hours at 940°C. The x-ray diffractometer trace on this powder indicated single phase $YBa_2Cu_3O_x$. Pellets
pressed from this powder were fired at 900°C for 16 hours in air and
furnace cooled. Optical and SEM metallographic examination indicated
the presence of three impurity phases Y_2BaCuO_5, $BaCuO_2$ and CuO, all at
levels on the order of 2-5%, and a porosity of roughly 30%. Phase
identification was verified using electron probe microanalysis (EPMA) in
an SEM. Direct current magnetization measurements [8] in a field of 100
Oe indicated a T_c onset of between 90 and 91 K. At 5 K the sample
exhibited 70% screening and 40% Meissner flux expulsion on cooling in
the applied field of 100 Oe.

In batch 2, 10-15 gram lots were ground in a mortar and pestle for
1 hour; lightly pelletized and then reacted in air at 890°C for 36
hours. After this same process was repeated, the sample was again
reground, pelletized and fired at 950 to 960°C for 2.5 days in pure O_2,
followed by furnace cooling to 700°C, holding for 4 hours, and furnace
cooling to 200°C. X-ray diffraction studies of this material revealed
only the $YBa_2Cu_3O_x$ phase. Optical metallographic studies revealed a
dramatic improvement over batch 1 material. The porosity was reduced
to around 15% as measured by automatic image analysis, the Y_2BaCuO_5
phase was difficult to detect anywhere, and the amount of CuO and $BaCuO_2$
both appeared at less than roughly 1%. Careful SEM work, however, using
backscattered electron imaging revealed all three impurity phases more
abundant than indicated by optical microscopy. The DC magnetization
measurements showed a superconducting transition temperature of 91K.
Measurements on a sample cooled in a field of 100 Oe showed a 60% flux
exclusion, while zero field cooling measurements showed 170% screening.
No correction to these numbers has been made for demagnetization factors. The high value of the screening is indicative of a large degree
of connectivity in the sample. Single crystals of $YBa_2Cu_3O_7$ were
produced in a manor described elsewhere.

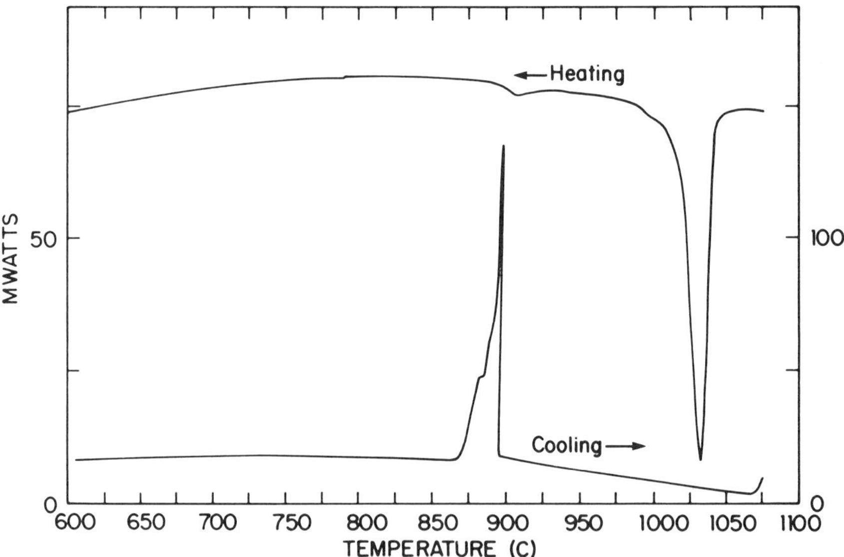

Fig. 1 Differential Thermal Analysis of $YBa_2Cu_3O_7$. Sample weight =
58.9 mg, scan rate 5.0 deg/min. Atmosphere 20% oxygen in Argon
at 30 cc/min.

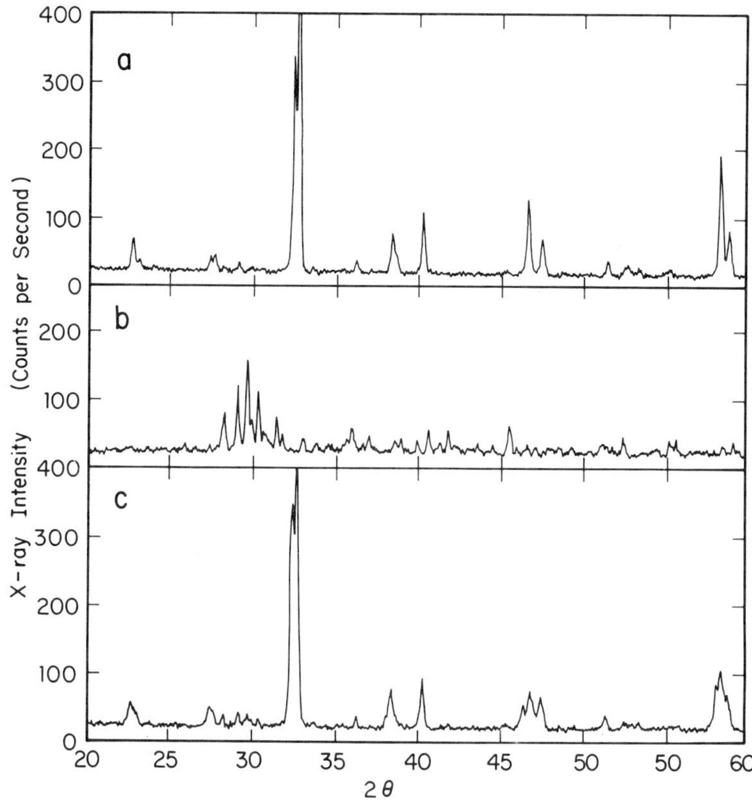

Fig. 2 X-ray diffraction patterns for $YBa_2Cu_3O_7$ using Cu Kα radiation
a) as prepared b) as quenched follow 20 minutes at 1020°C in
air c) material in (b) after 16 hours and 940°C in air.

INCONGRUENT MELTING

In order to determine the upper limit at which the $YBa_2Cu_3O_7$
material may be prepared, a sample of batch 2 material was examined
using differential thermal analysis (DTA). An atmosphere of flowing
Argon-20% oxygen was used. In Fig. 1 the heating and cooling curves
are shown. Upon heating the material exhibits a large melting endo-
therm at 1010°C the sharpness of the event is characteristic of a
single-phase material. Aside from a small < 2-3% impurity melting peak
there are no other features in the curve up to 1075 C. Upon cooling the
solidification is depressed to the $BaCuO_2$-CuO eutectic temperature. In
order to study the exact nature of the melting a large sample of batch 1
material was heated in air to just above the melting temperature and
quenched. In Fig. 2a the x-ray diffraction pattern of the material
prior to melting is shown. The material is predominately the 1-2-3
compounds with possible 2nd phase impurities at the threshold of
detection for x-ray diffraction. After melting (Fig. 2b) the 1-2-3

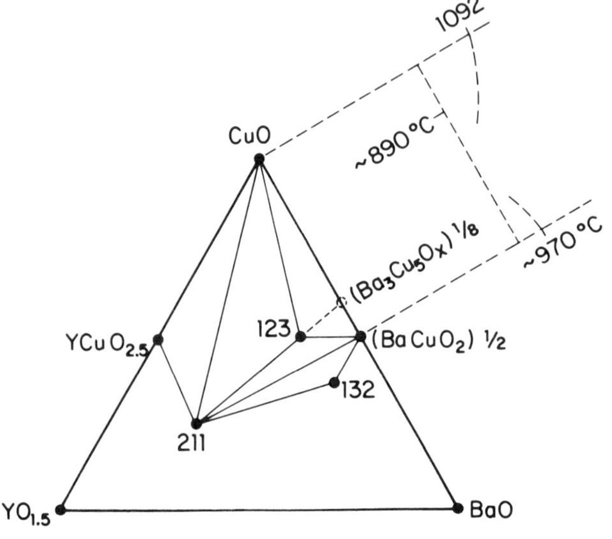

Fig. 3 The Cu rich part of the CuO-BaO-YO$_{1.5}$ 950°C isotherm
(Ref. 2, 3, 4).

phase has totally disappeared. The x-ray pattern may be analyzed in
terms of Y_2BaCuO_5, $BaCuO_2$, CuO and Cu_2O. This analysis was confirmed by
optical and SEM studies of the sample. From the shape of the melted
sample it was clear that the entire sample had not melted. This is
consistent with the peritectic decomposition of the sample into Y_2BaCuO_5
which melts above 1150°C and a liquid of $BaCuO_2$ and CuO. If this is
indeed the case the 1-2-3 compound should be recoverable by annealling
below the melting temperature. After a 16 hr. anneal at 940°C in air
the x-ray diffraction pattern shown in Fig. 2c was obtained. This
clearly shows the 1-2-3 structure. The detectable impurity level at
this point is ascribed to the loss of some $BaCuO_2$-CuO eutectic which wet
the platinum crucible during melting of the sample.

Based on these measurements we propose that the pseudo binary
phase diagram along the Y_2BaCuO_5 - "$Ba_3Cu_5O_8$" line of the pseudo ternary
phase diagram of Fig. 3 is. as shown in Fig. 4. In this phase diagram
$YBa_2Cu_3O_7$ is formed at 1010°C from the reaction of Y_2BaCuO_5 and a melt
of composition $Ba_3Cu_5O_y$. If care is taken to retain the liquid phase,
very large grains of $YBa_2Cu_3O_7$ may be grown. A second consequence of
this phase diagram is that a deviation from stoichiometry in the Y poor
direction results in a liquid phase which coats the particle boundaries
and solidified at 890°C. The presence of this phase is detrimental to
the conducting properties of the sample.

SINGLE CRYSTAL MAGNETIZATION MEASUREMENTS

Single crystals of $YBa_2Cu_3O_7$ were grown by a technique described in
detail in a previous paper[5]. The magnetization as a function of temperature for a collection of these crystals was measured to determine the

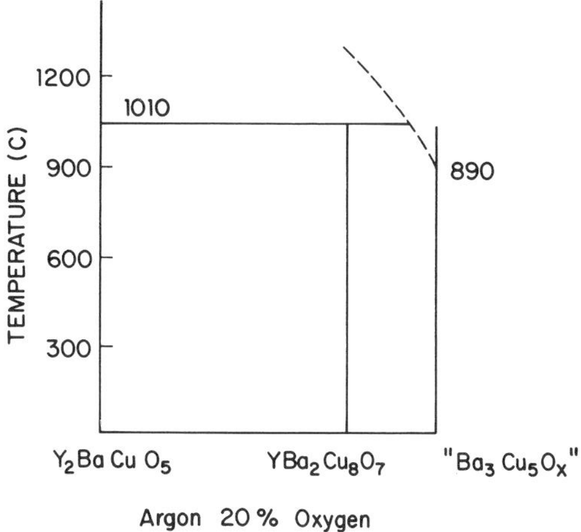

Fig. 4   Pseudo binary phase diagram along the Y_2BaCuO_5 - "$Ba_3Cu_5O_x$" cut of Figure 3.

onset of superconductivity and the magnitude of the magnetic flux expulsion. For the detailed investigation of the effect of the anisotropy, we selected a crystal of rectangular cross-section with approximate dimensions 0.50 by 0.24 by 0.08 mm^3. From single crystal x-ray diffraction data, we determined that the c-axis of the orthorhombic unit cell was perpendicular to the face with the largest area. Metallographic studies using both optical microscopy and SEM revealed that the crystals contained both voids and inclusions of a second phase. The magnetic properties of the crystals were measured in a Quantum Design SQUID magnetometer equipped with a 5.5 T superconducting solenoid and automatic temperature control.

In Fig. 5 we present the low field magnetization of a collection ($\approx$24 mg) of single crystals taken from the same preparation which shows a superconducting onset between 88 and 89 K. The orientation of the crystals is random with respect to the applied field in this experiment. When the sample is cooled in zero applied field, then measured with increasing temperature, the diamagnetic signal is essentially 100% (Fig. 5, open circles). The Meissner flus expulsion expressed as a percentage of this diamagnetic shielding is smaller, but still significant at about 15% (Fig. 5, crosses). This reduced Meissner effect has been reported previously for single crystals of other high T_c superconducting oxides[6,7]. It is most likely associated with flux trapping within the sample. Such trapping may result from defects within the sample such as inclusions, voids or slight variation in oxygen concentration.

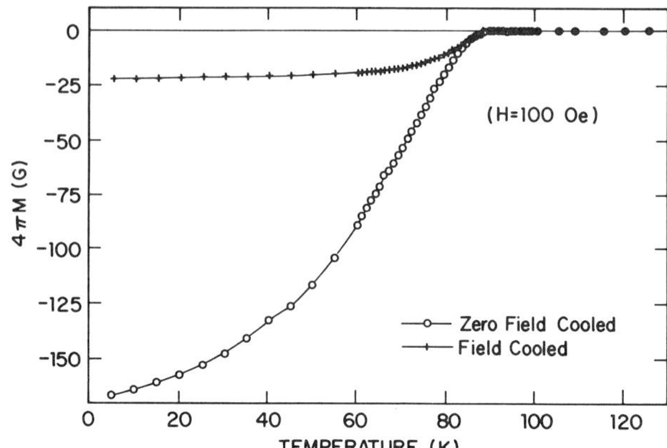

Fig. 5 Magnetization versus temperature for a collection of single
crystals of $YBa_2Cu_3O_7$. Both dc screening (open circles) and
Meissner effect (crosses) curves as shown.

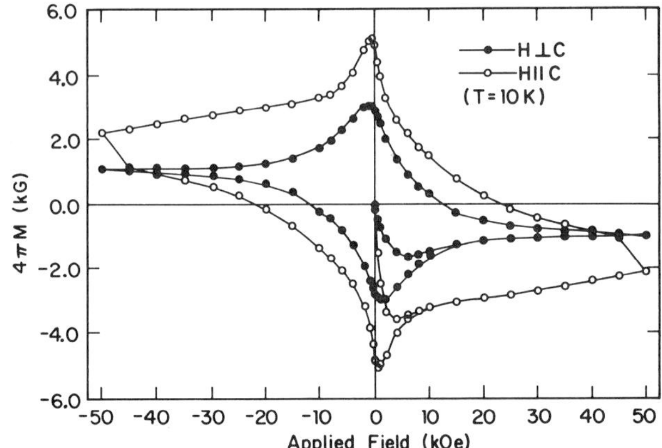

Fig. 6 Magnetization hysteresis loops at 10 K for a single crystal of
$YBa_2Cu_3O_7$. The orientation of the applied field (H) with
respect to the crystallographic c-axis is indicated ⌐ ⌐
curve.

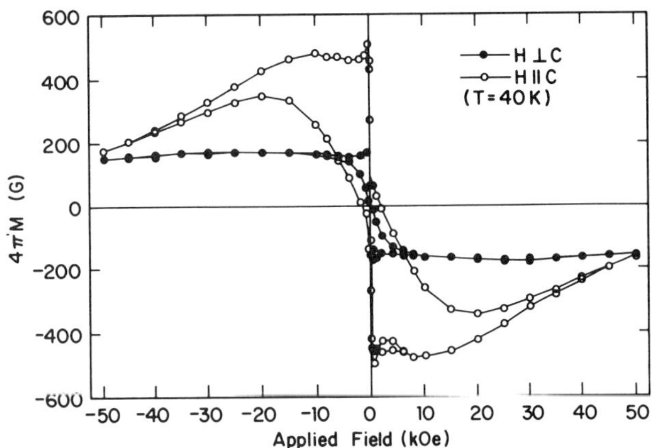

Fig. 7 Magnetization hysteresis loops at 40 K for a single crystal of
$YBa_2Cu_3O_7$. The orientation of the applied field (H) with
respect to the crystallographic c-axis is indicated for each
curve.

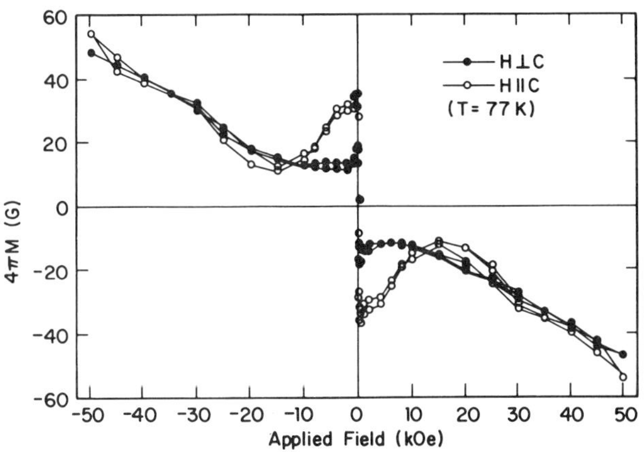

Fig. 8 Magnetization hysteresis loops at 77 K for a single crystal of
$YBa_2Cu_3O_7$. The orientation of the applied field (H) with
respect to the crystallographic c-axis is indicated for each
curve.

The magnetization loops for the rectangular single crystal chosen from the collection are shown for three distinct temperatures in Figs. 6-8. At each temperature, we present data for two orientations of the crystal; namely, with the applied field parallel to the c-axis and perpendicular to the c-axis. There is a high degree of symmetry for each of the curves. In Fig. 8, the diamagnetic background of the sample support is evident as a negative slope in the data at higher fields. This contribution does not affect the analysis at 10 or 40 K presented below. In each orientation, the lower critical fields, $H_{c1}^{\parallel}$ and $H_{c1}^{\perp}$, can be identified as the point at which the virgin magnetization curve deviates from a linear dependence on the applied field, provided corrections for the demagnetization factors are applied. In our data collection above H_{c1}, we noted a time-dependent change in the magnetization, similar to that described by other authors[6,8]. To minimize the effect on our measurements, after establishing a new field, we performed a series of ten measurements over a period of 20 to 30 minutes. During this series, the magnetization stabilized to within 1/2%; therefore, these final values are plotted in each figure. We note that during the first few minutes following a field change the change in the sample magnetization which is attributed to flux creep was sufficiently large to interfere with the computer analysis of the squid data. It is clear that the method of measurement of hysteresis loops will significantly effect the results obtained and that very slow field sweeps must be used to obtain equilibrium data.

Focusing on Fig. 6 which shows the magnetization data taken at a temperature of 10 K, we note that marked difference in the shape and size of the two hysteresis loops. As a measure of the anisotropy of the crystal, the lower critical fields obtained directly from the graph are $H_{c1}^{\parallel} = 3$ kϕe and $H_{c1}^{\perp} = 0.3$ kϕe. After correcting for the demagnetization factors, these become $H_{c1}^{\perp} = 6$ kϕe and $H_{c1}^{\perp} = 0.4$ kϕe, yielding an anisotropy ratio of 15 which is similar to that reported by Dinger and coworkers[6]. The difference in the magnitude of the width of the two loops is most pronounced at the higher applied fields. Applying a critical state model to a cylindrically shaped sample of radius R, Bean[9] related the magnitude of the critical current density to the magnetization by the formula: $J_c = 30M/R$, where J_c is the critical current density in A/cm^2, R is in centimeters, and M is the magnetization in emu/cm^3. Following Dinger and coworkers[6], we approximate R as the geometric mean of one-half of the length of each side of the rectangular face perpendicular to the applied field. Using the difference in the magnetization at an applied field of 40 kϕe, we estimate the critical current density for each of the two crystal orientations. For H parallel to the c-axis so that the currents are induced in the ab planes, we calculate $J_c^{\parallel} = 2.1 \times 10^5$ A/cm^2. In contrast, for H perpendicular to the c-axis so that the currents are induced perpendicular to the ab planes, we calculate $J_c^{\perp} = 4.6 \times 10^4$ A/cm^2.

The ratio $J_c^{\parallel}/J_c^{\perp} = 4.6$ is a measure of the anisotropy of the superconducting properties. This critical current anisotropy is greater at higher temperatures and at increased fields. For example, the current densities obtained at 10 K and 20 kϕe are $J_c^{\parallel} = 3.8 \times 10^5$ and $J_c^{\perp} = 2.1$ z 10^5 A/cm^2 for an anisotropy of 1.8, while the values at 40 K (Fig. 4) and 20 kϕe are $J^{\parallel} = 1.1 \times 10^4$ A/cm^2 and $J_c^{\perp} = 1.8 \times 10^3$ A/cm^2 which yield an anisotropy of 6.1. The magnitudes of our critical current densities are somewhat lower than those reported in an earlier study on single crystals with volumes comparable to that of our specimen[6]. Our lower values could be due to a difference in oxygen concentration or the overall ordering of oxygen vacancies in the crystals. Both of these parameters could have an effect on flux pinning and the critical current density.

II. THERMAL EXPANSIVITY EXPERIMENTS

The thermal expansivities were measured with respect to copper using a differential parallel plate capacitance dilatometer of the type developed by White[10]. This "White-type" capacitance dilatometer has been modified so that an irregularly-shaped sample of arbitrary length pushes against a flat copper plate which is kept parallel to the capacitor plate with three wires. The alignment of this plate can be adjusted to give a gap which varies only by a few percent across the width of the capacitor plate. The set-up was calibrated against a copper standard.

The Grueneisen model[11] for a single-component isotropic system gives the following relationship between the volume thermal expansivity, β, the heat capacity per unit volume, C_V/V or C_p/V, and the bulk moduli, B_T or B_S,

$$\beta = \gamma C_V/B_T V = \gamma C_p/B_S V \qquad (1)$$

The differences between C_p and C_V, or between B_S and B_T are not significant for our experiments. In Eq. (1),

$$\gamma = -d\ln\Phi/d\ln V, \qquad (2)$$

is the logarithmic volume-dependence of the characteristic energy, Φ, for the system. Φ, for example, is Θ for the Debye model, is the ordering temperature for a magnetic material, and is the Fermi temperature for a free electron gas. Since the bulk modulus, the volume and γ in practice vary slowly with temperature, the temperature-dependence of β can be expected to reflect closely the temperature-dependence of C_V.

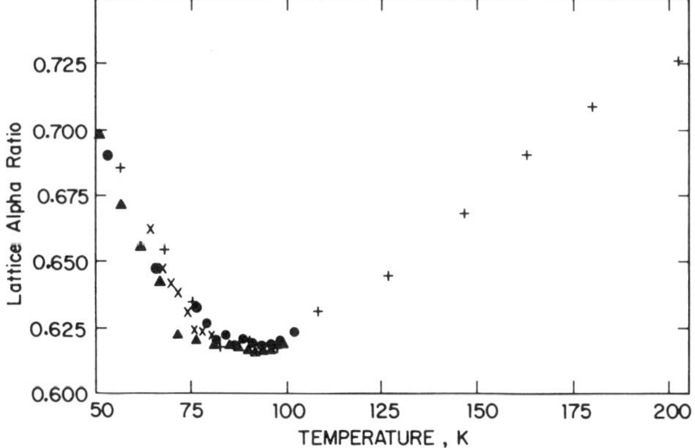

Fig. 9 Ratio of coefficients of linear expansion for $YBa_2Cu_3O_7$ and Cu $\alpha(T)/\alpha_{Cu}(T)$

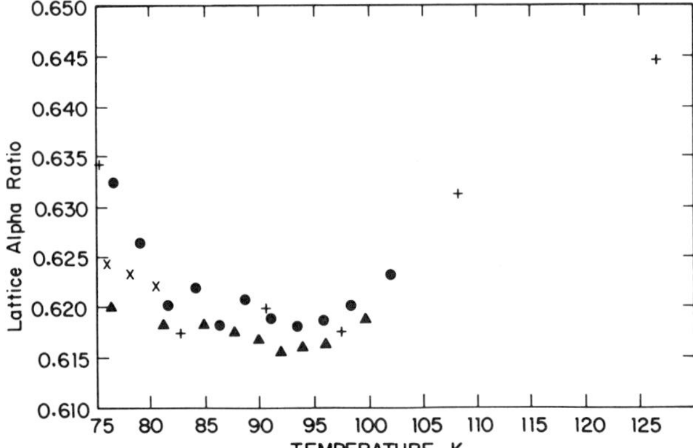

Fig. 10 Ratio of $\alpha(T)/\alpha_{Cu}(T)$ near the superconducting transition.
+ cooling   x, ▲ warming   ● warming from 4 K

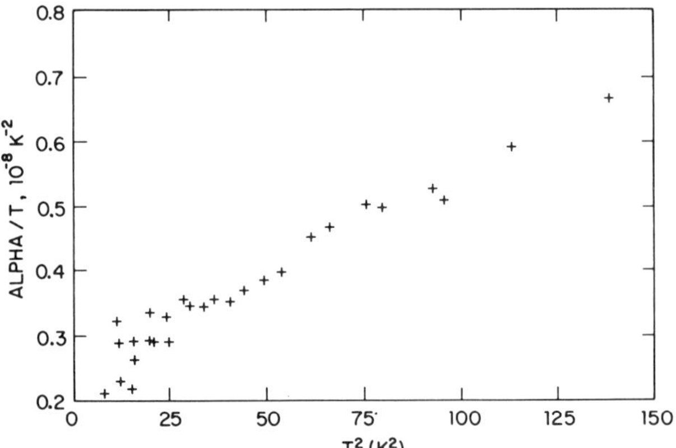

Fig. 11 Coefficient of linear expansion over temperature vs square of
temperature.

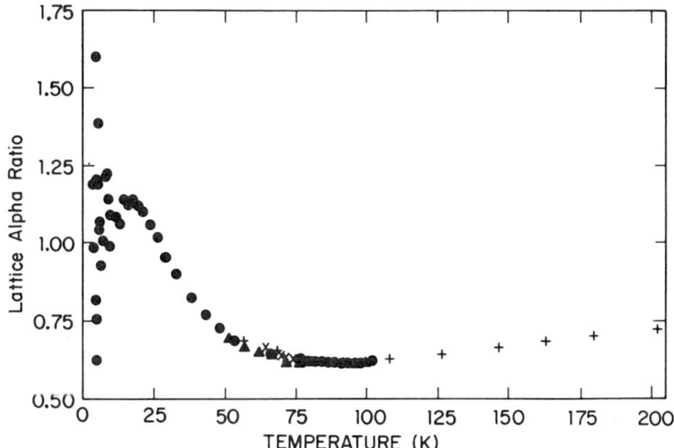

Fig. 12 Ratio of $\alpha(T)/\alpha_{Cu}(T)$ after removal of linear term.

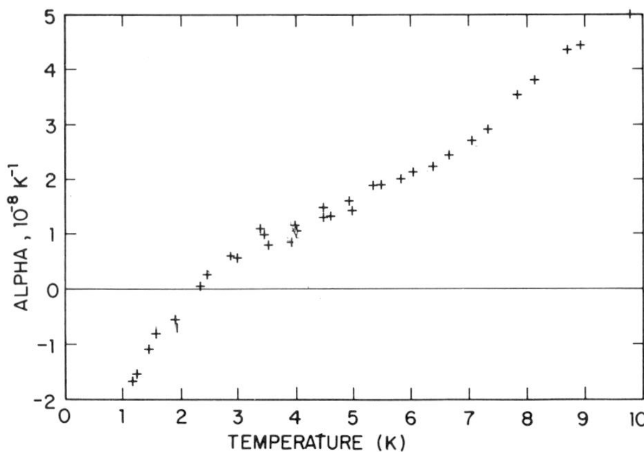

Fig. 13 Low temperature behavior of $\alpha(T)$ showing the change of sign.

The linear expansivity data at high temperature (above 15 K or so) are presented in terms of their ration to that of copper at the same temperature in order to remove the rather strong temperature dependence and to detect small effects. The data from 15 to 240 K are given in Fig. 9, where a monotonic decrease in the ratio is observed down to roughly 95 K. This linear dependence is indicative of a greater temperature dependence of the expansivity that is found for copper, and, therefore, of a greater Debye temperature for the compound than for copper. The data in the region of the superconducting transition are given in Fig. 10, where the results of several coolings and warmings are shown, with appreciable (0.5%) scatter in the data. There is no evidence in the data taken for a monotonic warming for a discontinuity at the superconducting transition. The deviations from the linear relationship appear to begin at about 98 K, well above the 92 K (or so) superconducting transition. The total effect at 92 K is only one percent, however, so this may not be significant when compared with less precise Meissner effect results.

Fig. 11 gives an α/T vs T^2 plot for the low temperature region. Here, the data above 4.5 K to (not shown) roughly 16 K can be represented by

$$10^9 \alpha = 2.5T + 0.00302T^3 \qquad (3)$$

The linear term is approximately ten times that for copper, while the cubic (lattice) term is comparable with that for copper. Figure 12 shows the data below 5 K on an α vs T plot, to emphasize that α becomes negative below 2.5 K, reminiscent of the high temperature side of an magnetic transition. The estimated uncertainty for these results is shown on this plot.

The relatively large value for the linear contribution to the expansivity suggests that Fig. 9 be redone with only the lattice contribution included -- that is, as $(\alpha - A_1 T)/(\alpha_{Cu} - A_{1Cu} T)$. This is done in Fig. 13, where the violent low-temperature temperature-dependence of α is modified to a great extent. The change in the character of α at T_c is quite pronounced, with the Debye temperature appearing to become smaller (that is, the lattice becoming softer) than that for copper. Any such conclusion must be tempered by aneed to determine the temperature dependence of the Grueneisen parameter for the sample.

CONCLUSIONS

In this paper we have shown evidence for the incongruent melting of the high temperature superconductor $YBa_2Cu_3O_7$ and indicated that the nature of the melting has observable consequences on the electrical behavior of the sample. We have demonstrated the existence of magnetic anisotrophy in single crystals of $YBa_2Cu_3O_7$. Finally we have shown that in a sintered pellet of $YBa_2Cu_3O_7$ the thermal expansion indicates that there may be significant lattice softening occurring below 100°F.

ACKNOWLEDGEMENTS

Ames Laboratory is operated for the U.S. Department of Energy by Iowa State University under contract No. W-7405-ENG-82. Brookhaven National Laboratory is supported by the U.S. Department of Energy under contract No. DE-AC02-76CH00016. This work was supported by the Office of Basic Energy Sciences.

REFERENCES

1. M. K. Wu, J. R. Ashburn, C. J. Torng, P. H. Hor, R. L. Meng, L. Gao, Z. J. Huang, Y. Q. Wang, and C. W. Chu, Phys. Rev. Lett. **58**, 908 (1987).
2. K. G. Frase, E. G. Kiniger and D. R. Clarke (submitted Communication of Amer. Cer. Soc., April 1987).
3. G. Wang, S.-J. Hwu, S. N. Song, J. B. Ketterson, L. D. Marks, K. R. Poeppelmeir and T. O. Mason (submitted J. Solid State Chem. Lett., 1987).
4. R. W. McCallum, J. D. Verhoeven, M. A. Noack, E. D. Gibson, F. C. Laabs and D. K. Finnemore (to be published).
5. M. A. Damento, K. A. Gschneidner, Jr., and R. W. McCallum, Appl. Phys. Lett. (submitted).
6. T. R. Dinger, T. K. Worthington, W. J. Gallagher, and R. L. Sandstrom, Phys. Rev. Lett. **58**, 2687 (1987).
7. R. B. Beyers, G. Lim, E. M. Engler, R. J. Savoy, T. M. Shaw, T. R. Dinger, W. J. Gallagher, and R. L. Sandstrom, Appl. Phys. Lett. (to be published).
8. K. A. Mueller, M. A. Takachige, and J. G. Bednorz, Phys. Rev. Lett. **58**, 1143 (1987).
9. C. P. Bean, Phys. Rev. Lett. **8**, 250 (1962).
10. G. K. White, Cryogenics **1**, 151 (1961).
11. E. Grueneisen Handbuch der Physik (Julius Springer, Berlin, 1926), Vol. X, pp. 1-52.

COPPER OXIDATION STATES, VACANCY ORDERING AND THEIR EFFECT ON HIGH

TEMPERATURE SUPERCONDUCTIVITY

Ivan K. Schuller, D.G. Hinks, J.D. Jorgensen, L. Soderholm,
M. Beno, K. Zhang,* C.U. Segre,* Y. Bruynseraede,** and
J.-P. Locquet**

Materials Science and Chemistry Divisions, Argonne National
Laboratory, Argonne, IL 60439

We briefly review our comprehensive work on high temperature oxides
which are isostructural with $YBa_2Cu_3O_{7-\delta}$. X-ray and neutron diffraction,
together with superconductive and normal state measurements show that the
Cu-O chains present in the structure play an important role in the
superconductivity. Whether the oxygen ordering or the presence of copper
in a 3^+ oxidation state is responsible for the high temperature
superconductivity cannot be ascertained at the present time.

In recent months many experiments have been performed on high temperature oxide superconductors,[1-2] however the superconducting mechanism still
remains elusive.[3] In a systematic series of experiments we have determined
the thermodynamic phase diagram,[4] the structure,[5] and the effect of oxygen
stoichiometry[6,7] and rare-earth substitutions[8] on the superconducting
transition temperature of the high temperature superconductor in the
Y-Ba-Cu-O system. Our results show that a single stoichiometric compound
$YBa_2Cu_3O_{7-\delta}$ is responsible for the superconductivity and that its structure
consists of dimpled Cu-O planes weakly linked by "fence" like Cu-O chains.

By quenching processes it is possible to prepare samples with varying
oxygen stoichiometries and varying degrees of ordering in the Cu-O chains.
As the oxygen concentration is reduced and the Cu-O chains disappear the
superconducting transition temperature is depressed. At the present time
it is not possible to ascertain uniquely whether the decrease in oxygen
concentration (possibly responsible for a decrease in the Cu^{3+} concentration) or the destruction of the Cu-O chains is the principal cause
for the destruction of superconductivity.

High temperature superconducting oxide samples were prepared using a
variety of synthetic methods with conventional powder metallurgy. X-ray
and neutron diffraction, resistivity, critical temperatures and currents
were used to characterize the samples in a conventional way.

The thermodynamic phase diagram in the ternary (Y-Ba-Cu) phase space
projected onto the zero oxygen plane, was determined from a large number of
samples. Figure 1 shows the phase diagram determined in this fashion. The
superconducting compound is labeled $YBa_2Cu_3O_{6.5}$, with the oxygen
stoichiometry hypothesized solely from charge balance considerations,

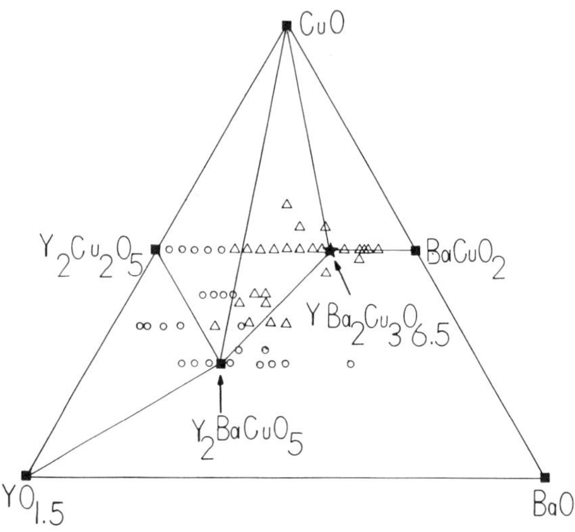

Fig. 1. Phase diagram of the Y–Ba–Cu system projected onto a constant
"preparation" plane. Note that no tie-line exists between Y_2BaCuO_5 and
$BaCuO_2$ because we have evidence for an additional unidentified compound in
this region.

assuming only Cu^{2+} to be present. Note that no tie-line is drawn between
Y_2BaCuO_5 and $BaCuO_2$ because we have evidence for an additional unidentified
compound in this region.

A detailed neutron diffraction study, in the 10–1200 K temperature
range, using Rietveld refinement techniques shows that below an oxygen
stoichiometry dependent temperature ($T_s \sim 950$ K) the structure is
orthorhombic with Pmmm symmetry (Fig. 2a) whereas above this temperature it
is tetragonal with P4/mmm symmetry (Fig. 2b). The structure consists of
ordered perovskite cubes with Cu on the cube corners, oxygens on the sides
of the cubes and Y or Ba in the center of the cube. The Y plane is totally
devoid of oxygen and the sublattice below the Ba planes has an ~50% oxygen
occupancy. In the orthorhombic phase the oxygens are ordered along the <u>b</u>
axis so as to form a one dimensional "chain" whereas in the tetragonal
phase the oxygens are disordered in this sublattice. Besides the slight
change in lattice spacing, the main changes that occur in going from the
orthorhombic to the tetragonal phase are: a) a disordering of the oxygen
vacancies (as shown in Fig. 2) and b) a decrease in total oxygen concentration, as shown in Fig. 3, for experiments conducted under various oxygen
atmospheres. The change in lattice parameters as well as the change in
oxygen occupancy of the various sites show that the orthorhombic to

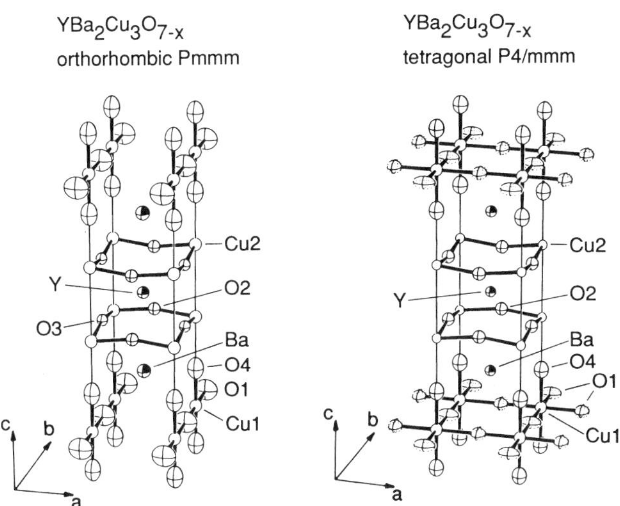

Fig. 2. Orthorhombic and tetragonal phases of YBa$_2$Cu$_3$O$_{7-\delta}$. Note the disordering of the chains and that the oxygen are drawn with dotted lines indicating partial (<50%) occupancy in the tetragonal phase.

tetragonal phase transition occurs for an oxygen stoichiometry of 6.5 (dashed lines in Fig. 3). Note that this is precisely the oxygen stoichiometry for which the presence of Cu^{3+} is expected to vanish from charge balance considerations.

If the high temperature phase is quenched-in to low temperature the superconducting properties are drastically affected. Figure 4 shows two samples; a slowly cooled YBa$_2$Cu$_3$O$_{7-\delta}$ sample (A) and the same sample quenched from high temperature (B). Note that the quenched sample has a much higher resistivity and that the T_c is considerably depressed. A systematic study[9-10] in a large number of samples has shown that T_c decreases monotonically with increasing δ and that it vanishes if the quenched samples remain in the tetragonal phase. Moreover, our neutron diffraction studies[7] show that up to the structural phase transition or down to an oxygen stoichiometry of 6.5 most of the vacancies are formed in the sublattice containing the Cu-O chains with very little change on the oxygen stoichiometry of the dimpled planes around the Y ion, where presumably the conduction occurs. Based on an analysis of the Cu-O distances it was claimed that most of the Cu^{3+} atoms are located on the Cu-O chains.[11] All these measurements indicate that the sublattice containing the Cu-O chains play an important role in the high temperature superconductivity.

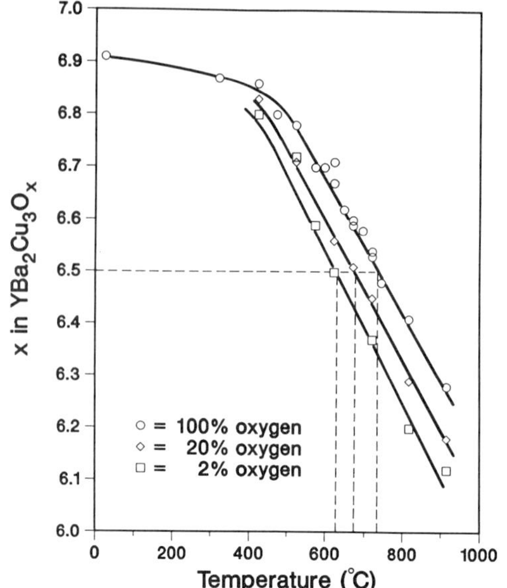

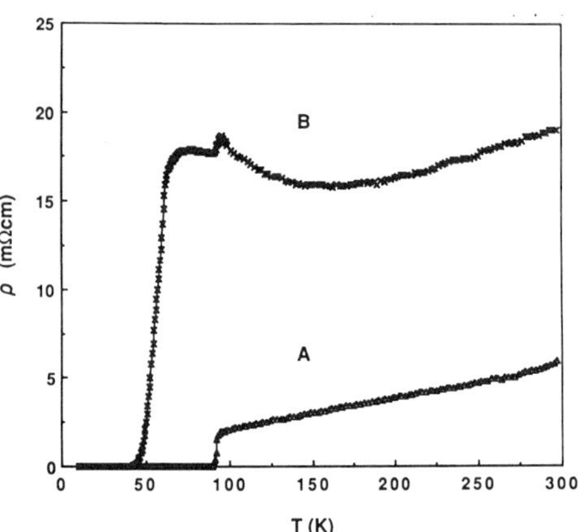

Fig. 3. Change in total oxygen stoichiometry as a function of temperature for three different oxygen atmospheres. The orthorhombic to tetragonal phase transition occurs close to a stoichiometry of 6.5 for oxygen.

Fig. 4. Resistivity versus temperature for A) a slowly cooled sample and B) the same sample quenched from high temperature.

A similar behavior occurs if Y is substituted with Pr, a rare-earth ion which can be found occasionally in a tetravalent state.[8] Figure 5 shows the transition temperature and room temperature resistivity (inset) as a function of x in $Y_{1-x}Pr_xBa_2Cu_3O_{7-\delta}$. Presumably, introducing the tetravalent ion Pr in the structure changes the Cu oxidation state and decreases the concentration of Cu^{3+} in a monotonic fashion with x. Substitutions with other, <u>trivalent</u> rare-earths (Nd, Eu, Gd, Tb, Dy, Ho, Yb, Lu^{12-14}) has very little effect on the transition temperature. The common feature of both experiments is that as the oxygen stoichiometry decreases (i.e., increasing δ) or as the presumably tetravalent Pr ion concentration increases (i.e., increasing x) the transition temperature decreases and the normal state resistivity increases in a monotonic fashion. We should point out the absence of symmetry around the oxygen concentration of 6.5 or Pr concentration x $\cong$ 0.5; i.e., the superconductivity is not recovered for an oxygen stoichiometry below 6.5 or x $\gtrsim$ 0.5.

At this stage, we should stress that it seems that independent of the cause, changes in the Cu^{3+} density is accompanied by a destruction of the Cu–O chains and therefore we cannot ascertain uniquely which effect is responsible for the high temperature superconductivity. However, since up to the structural phase transition no major changes occur in the vacancy density in the dimpled Cu–O planes and all the vacancies occur in the Cu–O chains we can uniquely ascertain that the Cu–O chains play a very important role in the high temperature superconductivity of metal oxides which are isostructural with $YBa_2Cu_3O_{7-\delta}$.

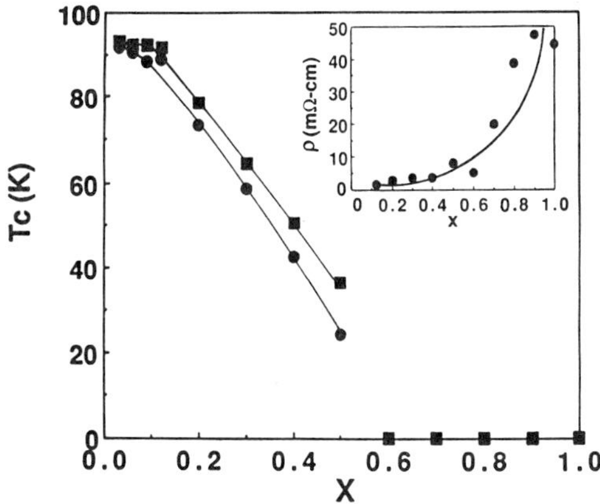

Fig. 5. Transition temperature versus x for $Y_{1-x}Pr_xBa_2Cu_3O_{7-\delta}$, (■) shows the onset of the transition and (●) shows the zero resistance point. The inset shows the resistivity versus x.

In summary, we have performed a detailed study of the structure and superconductivity of high temperature oxides which are isostructural with the $YBa_2Cu_3O_{7-\delta}$ compound. Our results indicate that the Cu-O chains present in these materials play a crucial role in the superconductivity.

This work was supported by the U.S. Department of Energy, BES-Materials Sciences and Chemical Sciences, under Contract #W-31-109-ENG-38. International travel was provided by NATO grant #RG85-0695.

REFERENCES

  * Also at the Physics Department, Illinois Institute of Technology, Chicago, Illinois 60618.
 ** Instituut voor Natuurkunde, Katholieke Univesiteit, Celestijnenlaan 200D, Leuven B-3030, Belgium.
1. J. G. Bednorz and K. A. Müller, Z. Phys. B64, 189 (1986).
2. M. K. Wu, I. R. Ashburn, C. J. Torng, D. H. Hor, R. L. Meng, L. Gao, Z. J. Huang, Q. Wang and C. W. Chu, Phys. Rev. Lett. 58, 908 (1987).

3. For a review see, T. M. Rice, to be published.

4. D. G. Hinks, L. Soderholm, D. W. Capone II, J. D. Jorgensen,
    I. K. Schuller, C. U. Segre, K. Zhang and J. D. Grace, Appl. Phys.
    Lett. 50, 1688 (1987).

5. M. A. Beno, L. Soderholm, D. W. Capone II, D. G. Hinks,
    J. D. Jorgensen, I. K. Schuller, C. U. Segre, K. Zhang and J. D.
    Grace, Appl. Phys. Lett. (in press).

6. I. K. Schuller, D. G. Hinks, M. A. Beno, D. W. Capone II,
    L. Soderholm, J.-P. Locquet, Y. Bruynseraede, C. U. Segre and
    K. Zhang, Sol. St. Comm. (in press).

7. J. Jorgensen, M. A. Beno, D. G. Hinks, L. Soderholm, K. J. Volin,
    R. L. Hitterman, J. D. Grace, I. K. Schuller, C. U. Segre,
    K. Zhang and M. S. Kleefish, Phys. Rev. B (submitted).

8. L. Soderholm, D. G. Hinks, M. A. Beno, J. D. Jorgensen, C. U. Segre,
    K. Zhang and I. K. Schuller, (to be published).

9. J. Jorgensen, B. Veal, G. Crabtree, et al. (to be published).

10. J. van den Berg, C. J. van der Beek, P. H. Kes, G. J. Nieuwenhuys,
    J. A. Mydosh, H. W. Zandbergen, F. P. F. van Berkel, R. Steens and
    D. J. W. Ijdo, (to be published).

11. W. I. F. David, et al., Nature 327, 310 (1987).

12. S. Kagoshima, K.-I. Koga, H. Yasuoka, Y. Nogami, K. Kubo and
    S. Hikami, Jap. Jour. Appl. Phys. 26, 1355 (1987).

13. E. M. Engler, V. Y. Lee, A. I. Nazzal, S. S. P. Parkin, M. L. Ramirez,
    J. E. Vasquez and R. J. Savoy, J. Amer. Chem. Soc. (in press).

14. Z. Fisk, J. D. Thompson, E. Zirngiebl, J. L. Smith and S. Cheong, (to
    be published).

BULK SUPERCONDUCTIVITY AT 60 K IN OXYGEN-DEFICIENT $Ba_2YCu_3O_{7-\delta}$

AND OXYGEN ISOTOPE EFFECT IN $La_{1.85}Sr_{0.15}CuO_4$

B. Batlogg, R. J. Cava, C. H. Chen, G. Kourouklis*, W. Weber**,
A. Jayaraman, A. E. White, K. T. Short, E. A. Rietman,
L. W. Rupp, D. Werder, and S. M. Zahurak

AT&T Bell Laboratories
Murray Hill, NJ 07974

ABSTRACT

Oxygen deficient polycrystalline samples of $Ba_2YCu_3O_{7-\delta}$ have been prepared by a Zr-gettered annealing technique. A distinct bulk superconducting phase with a T_c of 60 K is found for $0.3<\delta<0.4$, which has partial ordering of the O vacancies in the Cu-O chains. This ordering is seen in electron diffraction studies and also in both the value and the temperature dependence of the electrical resistivity. Isotope effect studies on $La_{1.85}Sr_{0.15}CuO_4$ have been performed by partially substituting ^{18}O for ^{16}O. Raman spectroscopy confirms the expected frequency shift of the 428 cm^{-1} phonon line and T_c is found to decrease. The isotope effect exponent α is 0.16 ± 0.02, in contrast to $Ba_2YCu_3O_7$ where α is 0 ± 0.02.

ORDERED OXYGEN VACANCIES IN $Ba_2YCu_3O_{7-\delta}$:
A BULK SUPERCONDUCTOR WITH $T_c \sim 60$ K

A natural way to change the average valence state of Cu in $Ba_2YCu_3O_7$ is given by removal of oxygen, which can be achieved easily by heating the samples to elevated temperatures.[1-9] In this study we chose a somewhat different approach that allows extraction of oxygen at lower temperatures[10], and thus reduces possible complications due to temperature induced disorder. Indeed, this technique leads to distinctly different results: (1) a new 60K T_c bulk superconductor is found with metallic R (T) characteristic and (2) the orthorhombic symmetry is preserved for all oxygen contents ($\delta < 0.7$).

A master batch of $Ba_2YCu_3O_7$ was prepared[11] and the pellets were finally annealed at 500° C in flowing O_2. Part of each pellet was then sealed in an evacuated quartz tube (15 mm diam., 150 mm length) together with a piece of degreased Zr foil (~ 2.5 x 2.5 cm^2). These tubes were then placed for 48 hours into furnaces between 360° C and 520° C, at 20-30° C

---

*   Permanent address: National Technical University, Athens, Greece
** Permanent address: Kernforschungszentrum Karlsruhe, INFP, D-7500 Karlsruhe, FRG

intervals. More oxygen was removed at higher temperature. The tubes were air quenched to room temperature, and resistivity and magnetic measurements were done on identical or neighboring sections of the samples. The lattice parameters were determined from powder X-ray diffraction, and the oxygen content was analyzed thermogravimertically.

The superconducting transition temperatures are shown in Fig. 1. as function of oxygen content. The most prominent observation is a step-like decrease of T_c with a plateau at $\sim$ 60 K at x = 6.6-6.7. The transitions are narrow as indicated by the symbols, except where T_c varies rapidly with x. Additional evidence for the non-monotonic variation of the physical properties as function of x comes from resistivity measurements. In Fig. 1. the room temperature values of ρ are given and the behavior is characterized by the rapid initial increase when oxygen is removed from the samples. Further increase of the vacancy concentration, however, leads to a reduction of ρ in the range of x = 6.6-6.7, and finally an increase of ρ towards x = 6. Even more pronounced is the distinct character of the x = 6.6-6.7 range when the temperature dependence of the resistivity is studied. (See Fig. 2) The $\rho(T)$ characteristic is metallic only for x close to 7.0 and for the samples with a sharp transition at $\sim$ 60 K. All other samples have a negative $d\rho/dT$ over at least part of the temperature range. Thus, both the relatively low resistivity and the metallic $\rho(T)$ indicate a higher degree of order in the 60 K T_c in samples. The diamagnetic susceptibility and Meissner effect data also support this conclusion. The 60 K T_c samples have a transition as sharp as the 90 K samples, whereas significant broadening is found for x = 6.76. The Meissner effect is the same within experimental uncertainly and amounts to $\sim$ 90% of the estimated full value.

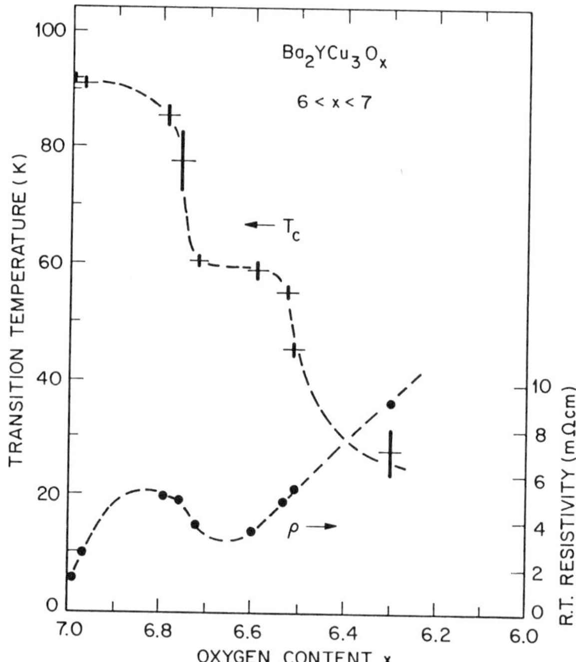

Fig. 1   Non-monotonic dependence of physical properties on the oxygen concentration in Ba$_2$YCu$_3$O$_x$ prepared by a gettered annealing technique. The width of the superconducting transition is given by the length of the symbols. The room temperature resistivity is reduced in the range of the 60 K superconductor.

Powder X-ray diffraction revealed orthorhombic symmetry for the whole composition range studied. Both the a_o and c_o cell parameters increase with decreasing x, whereas b_o slightly decreases. These measurements did not reveal any additional scattering intensity which could be associated with ordering in the 60 K T_c materials. Electron diffraction studies where thus performed in search of vacancy ordering.

In samples with a sharp 60 K transition, additional diffraction intensity was found. When the beam is focused on a single domain, diffuse scattering is seen in the a^*b^* plane in the form of narrow steaks located along b^* halfway between Bragg reflections and elongated in the b^* direction. This pattern is similar to one reported in Ref. 12 for oxygen deficient $Ba_2YCu_3O_7$, and suggests vacancy ordering. The full 3-dimensional arrangement of the vacancies will have to be identified in the ongoing investigations. We note here that approximately 1/3 of the oxygen atoms in the Cu-O chains along b are missing in the 60 K superconductors, but the diffuse scattering indicates a (short range) doubling of the b-periodicity. It should be added also that the 60 K phase is observed in a range of x where the numbers of occupied and empty O sites in the chains includes simple ratios such as 5/8, 2/3 and 3/4.

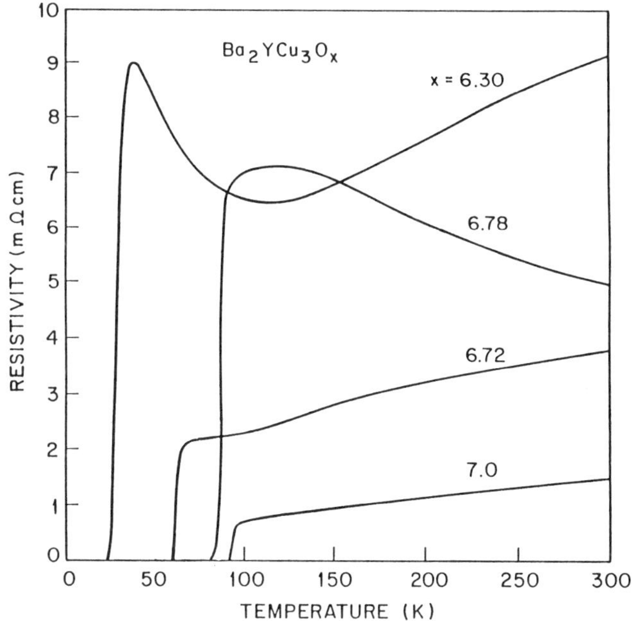

Fig. 2 Temperature dependence of the electrical resistivity for various $Ba_2YCu_3O_x$ samples. The metallic R (T) characteristic for the 60 K T_c superconductors is noteworthy.

The gettered annealing technique employed here produces oxygen deficient $Ba_2YCu_3O_7$ with physical properties that are different from the ones obtained by high-temperature treatment. The absence of the orthorhombic-tetagonal transformation is simply explained by the temperatures used to extract oxygen, which are too low to promote occupation of the ideally vacant O-sites in the chains along the a direction. The existence of a 60 K T_c phase with vacancy ordering clearly points to the particular importance of the Cu-O chains. It is not only the average concentration, but also the spatial arrangement of the vacancies that determines the superconducting transition temperature.

OXYGEN ISOTOPE EFFECT IN $La_{1.85}Sr_{0.15}CuO_4$

One of the outstanding problems in the field of high T_c superconductivity is the identification of the pairing mechanism. With bulk transition temperatures reaching 100 K,[13,14] a number of models have been proposed as alternatives or extensions to the conventional phonon - mediated pairing. Thus the study of the isotope effect appeared most appropriate. When we found no shift[15] of T_c upon substitution of ^{16}O by ^{18}O in $Ba_2EuCu_3O_7$ and $Ba_2YCu_3O_7$ (for the latter compound see also Ref. 16), the result was interpreted[17] as strongly favoring non-phonon mechanisms. Here we report on a similar isotope effect study in $La_{1.85}Sr_{0.15}CuO_4$, where we find a dependence of T_c upon the oxygen mass, with the corresponding exponent α of 0.16 ± 0.02. (A full account is given in Ref.18)

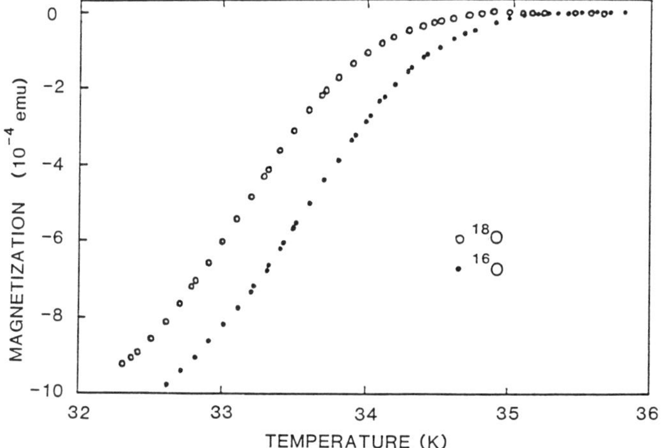

Fig. 3   Shift of the superconducting transition temperature due to partial (70-75%) substitution of ^{18}O for ^{16}O in $La_{1.85}Sr_{0.15}CuO_4$. (dc magnetization in a field of 6.5 Oe).

The same concept of gas phase exchange of oxygen[15] was also applied in this experiment, and the principle of identical thermal treatment of sample and reference was also followed. Slower exchange rates required heating to higher temperatures (750° C) for longer time periods (up to 3 weeks) than in the case of the 90 K T_c materials. The $^{16}O{:}^{18}O$ ratio was measured by Rutherford-Backscattering-Spectroscopy. After one and two cycles, the ^{18}O concentration is 52 ± 5 and 70-75%, respectively. The shift of phonon frequencies involving only the motion of oxygen atoms can be calculated from the average mass and compared to Raman spectroscopy results. In the high frequency range only one mode could be measured (428 cm^{-1}) and it is found to shift to lower frequencies by the expected amount. (See Table) This is an important result since this mode involves only one set of the two inequivalent O sites in this structure, and thus one concludes that O-exchange is not site selective. The superconductive transitions were measured magnetically in a SQUID magnetometer, and representative data are shown in Fig. 3 for a sample with ~ 70-75% ^{18}O. The T_c's for the ^{18}O containing samples are consistently lower than for the ^{16}O ones. The results of the various measurements are compiled in Table I and are summarized in the isotope exponent $\alpha = 0.16 \pm 0.02$.

TABLE I

^{18}O	$\left[\dfrac{\Delta m}{m}\right]^{1/2}$	$\Delta\omega$	$\dfrac{\Delta\omega}{\omega}$	δT_c	α
(%)	(%)	(cm^{-1})	(%)	(K)	
52 ± 5	3.1	14.5 ± 1	3.4	0.39 ± 0.03	0.16 ± 0.02
73 ± 5	4.3	18 ± 1	4.2	0.46 ± 0.04	0.16 ± 0.02

Summary of the experimental results for two sets of samples with different ^{18}O concentration. Compared are the measured frequency shift $\Delta\omega$ of the 428 cm^{-1} Raman line with the average change in Oxygen mass Δm. The isotope effect exponent α is defined as $T_c \sim M^{-\alpha}$.

Because not only oxygen vibrations will contribute to electron phonon coupling, a value less than the classical 1/2 is to be expected for α. This point can be made more quantitative by a calculation of the full phonon spectra and their dependence on the mass of the individual constituents. In extension of the work in Ref. 19, the partial isotope exponents were calculated and for a choice of a particular Cu-O spring constant to reproduce the experimental T_c, the exponent α for oxygen was found to be ~ 0.3. The difference between the calculated and measured α can have various reasons, some of which are discussed in Ref. 18.

The presence of an isotope effect in $La_{1.85}Sr_{0.15}CuO_4$ thus indicates that superconducting pairing in the "40 K" superconductor is at least partly caused by electron-phonon interaction.

REFERENCES

1. The number of papers dealing with oxygen-deficient $Ba_2YCu_3O_{7-\delta}$ grows rapidly and the following list is therefore bound to be imcomplete.

2. P. Strobel, J. J. Capponi, C. Chaillout, M. Marezio and J. L. Tholence, Nature *327*, 306 (1987).

3. J. M. Tarascon, W. R. McKinnon, L. H. Greene, G. W. Hull, B. G. Bagely, E. M. Vogel and Y. LePage, Proc. of the Materials Research Society 1987 Spring Meeting, Symposium S. (D. V. Gubser and M. Sclüter, eds.) p. 65; and also these Proceedings.

4. P. K. Gallagher, H. M. O'Bryan, S. A. Sunshine and D. W. Murphy, Mat. Res. Bull., (submitted).

5. A. Santoro, S. Miraglia, F. Beech, S. A. Sunshine, D. W. Murphy, L. F. Schneemeyer and J. V. Waszczak, Mat. Res. Bull. (submitted).

6. I. K. Schuller, D. G. Hinks, M. A. Beno, D. W. Capone II, L. Souderholm, J. P. Locquet, Y. Bruynseraede, C. U. Segre and K. Zhang, Solid State Comm., 1987.

7. R. Beyers, G. Lim, E. M. Engler, R. J. Savoy, T. M. Shaw, T. R. Dinger, W. T. Gallagher and R. L. Sandstrom, Appl. Phys. Lett. 1987 (submitted).

8. A. W. Hewat, J. J. Capponi, C. Chaillout, M. Marezio and E. A. Hewat, Nature (submitted).

BULK MODULUS ANOMALIES AT THE SUPERCONDUCTING TRANSITIONS OF SINGLE-PHASE
$YBa_2Cu_3O_7$ AND $La_{1.85}Sr_{.15}CuO_4$

D. J. Bishop, P. L. Gammel, A. P. Ramirez, B. Batlogg,
R. J. Cava, and A. J. Millis

AT&T Bell Laboratories, Murray Hill, N.J. 07974

We have measured the changes in sound velocity in the two high-T_c
superconductors $(La,Sr)_2CuO_4$ and $YBa_2Cu_3O_7$ from room temperature through
their superconducting transitions. Both systems exhibit anomalous behavior
at T_c. In a conventional superconductor at T_c there is a small drop in
the sound velocity typically of the order of a few ppm and the sound velocity at $T = 0$ stays below its value at T_c. In $La_{1.85}Sr_{.15}CuO_4$ there is a
large drop ($\sim$150 ppm) at T_c and then a large hardening below T_c of
$\sim$ 1000 ppm.[1] In $YBa_2Cu_3O_7$ there is no drop at T_c[2] but there is a large
hardening below T_c of $\sim$1000 ppm. In both cases the hardening seems to
scale with the superconducting gap squared. This behavior is shown in
Fig. 1 for $La_{1.85}Sr_{.15}CuO_4$ and Fig. 2 for $YBa_2Cu_3O_7$.

The standard thermodynamic analysis of the changes in sound velocity
at the superconducting transition suggests that there are two contributions. The first is a drop which occurs at T_c and is related[3] to the
specific heat jump ΔC and bulk modulus B by

$$\frac{v_s^2 - v_n^2}{v_n^2} = -\frac{\Delta C}{T_c} B \left(\frac{\partial T_c}{\partial P}\right)^2 \tag{1}$$

This may be rewritten as

$$\frac{\Delta B}{B} = \frac{-B}{4\pi}\left[\left(\frac{\partial H_c}{\partial P}\right)^2 + H_c\frac{\partial^2 H_c}{\partial P^2}\right]$$

In $La_{1.85}Sr_{.15}CuO_4$ this drop is seen at T_c and its magnitude ($\sim$150 ppm)
agrees well with Eq. (1) using the independently determined values of
$\partial T_c/\partial P$.[4] For $YBa_2Cu_3O_7$, as shown in Fig. 2 there is no measurable drop in
the sound velocity at T_c. This is not surprising since the measured
value of $\partial T_c/\partial P = 0.07$ K/kbar[5] and Eq. (1) would predict a change of
$\sim$1 ppm, which is outside our resolution. Equation (1) therefore seems to
be valid for both $La_{1.85}Sr_{.15}CuO_4$ and $YBa_2Cu_3O_7$.

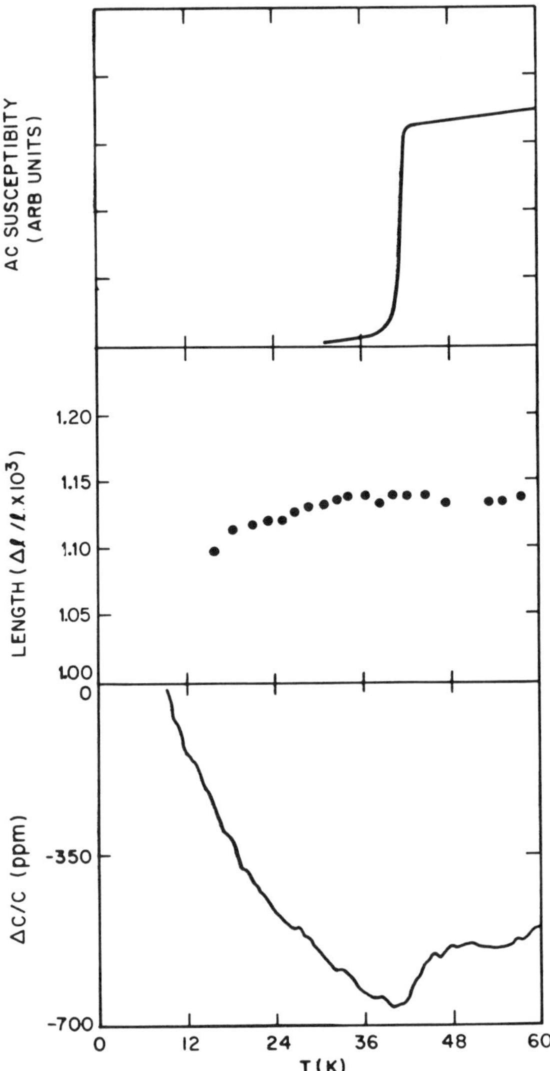

Fig. 1.   The AC susceptibility, thermal expansion, and sound velocity are
shown near T_c for $(La,Sr)_2CuO_4$.

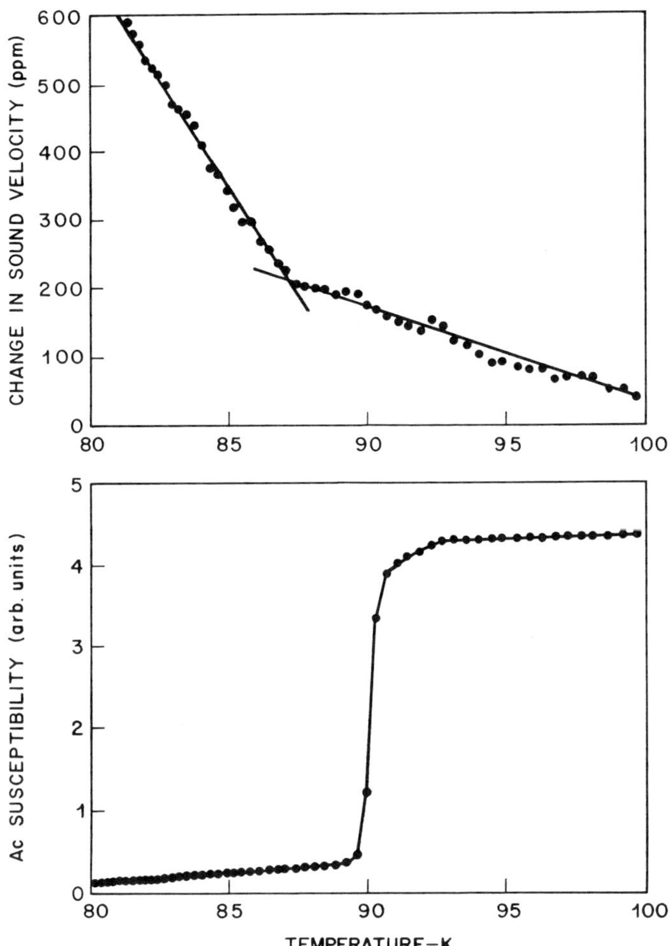

Fig. 2. The change in sound velocity (top) and Ac susceptibility (bottom) are shown for $YBa_2Cu_3O_7$.

One may extend the calculation in Eq. (1) away from T_c by using a BCS model to compute the changes in the properties of the conduction electrons due to the superconductivity in combination with the Bohm-Staver expression for the conduction electron contribution to the sound velocity. Assuming that $\partial\Delta/\partial P = 1.76\ \partial T_c/\partial P$ and also that the other contributions to the lattice stiffness do not change with temperature, one finds for the change in sound velocity at $T = 0$,

$$\frac{V_s^2 - V_n^2}{V_n^2} = \frac{(1.76)^2\ \gamma}{B}\left[\frac{1}{9}\ T_c^2 + \frac{4}{3}\ T_c B\ \frac{\partial T_c}{\partial P} - \left(B\ \frac{\partial T_c}{\partial P}\right)^2\right] \qquad (2)$$

Substituting the known values for T_c, dT_c/dP, and γ into Eq. (2) and using a reasonable estimate for $B = \rho_m V^2$ (where ρ_m is the mass density and $V \sim 5 \times 10^5$ cm/sec) yields a velocity change of at most 10 ppm, which is a factor of ~100 smaller than we see. We conclude that this significant

increase of the sound velocity below T_c which varies like the gap squared
and is seen in both $(La,Sr)_2CuO_4$ and $YBa_2Cu_3O_7$ is a generic feature of
these materials and, unlike the jump at T_c, cannot be explained within a
simple BCS model. The only assumption which goes into deriving Eq. (2) is
that the background lattice does not change its behavior near T_c. Thus,
the present observations suggest the presence of an electronically driven
structural anomaly in the vicinity of T_c, indicating the need for detailed,
temperature-dependent crystallographic data. Another possibility is that
$(\Delta/E_f)^2$ is a much larger number than the standard expressions would suggest.

In conclusion, we have measured the sound velocity of the high-T_c
superconductors $YBa_2Cu_3O_7$ and $(La,Sr)_2CuO_4$ through their superconducting
transitions. Both show an anomalously strong hardening of the bulk
modulus below T_c. This behavior cannot be explained by the standard
thermodynamics of the BCS model which describe conventional superconductors.

REFERENCES

1. D. J. Bishop, P. L. Gammel, A. P. Ramirez, R. J. Cava, B. Batlogg,
    and E. A. Rietman, Phys. Rev. <u>B35</u> (1987) in press.
2. D. J. Bishop, A. P. Ramirez, P. L. Gammel, B. Batlogg, E. A. Reitman,
    R. J. Cava, and A. J. Millis, Phys. Rev. <u>B35</u> (1987), in press.
3. G. A. Alers, "Physical Acoustics," edited by W. P. Mason, Vol. IVA,
    p. 277 (Academic Press, 1966).
4. J. E. Schirber, E. L. Venturini, J. F. Kwak, D. S. Ginley, and B.
    Morosin, Mat. Res. Bull., Rapid Comm., to be published.
5. J. E. Schirber, D. S. Ginley, E. L. Venturini, and G. Morosin, Phys.
    Rev., submitted.

ABSENCE OF RESISTIVITY SATURATION AND ITS IMPLICATIONS

FOR THE HIGH T_c SUPERCONDUCTORS

M. Gurvitch and A. T. Fiory

A T & T Bell Laboratories
Murray Hil, New Jersey NJ 07974

ABSTRACT

Temperature-dependent resistivity $\rho(T)$ was measured from T_c to 1100K in ceramic samples of $La_{1.825}Sr_{0.175}CuO_4$ (LSCO) and $YBa_2Cu_3O_7$ (YBCO). Measurements above 300K were performed in a flow of O_2. In LSCO $\rho(T)$ is linear to 1100K; in YBCO $\rho(T)$ is linear to 600K, increases faster than linear at $T>600K$, owing to the loss of oxygen, and shows a break in slope at the structural phase transition near 950K. It is argued that the observed absence of saturation in the linear parts of the data implies weak electron-phonon coupling in both superconductors, and, coupled with absolute values of the resistivity, places constraints on band structure parameters. Our results take anisotropy into account explicitly and are combined with recent μSR measurements of the penetration depth to obtain experimental estimates of the plasma energy.

1. INTRODUCTION

In this paper we present first measurements of the normal-state resistivity $\rho(T)$ of the high-T_c oxide superconductors [1], from the T_c to 1100K. In the past, studies of $\rho(T)$ in transition metals and A-15 compounds helped to elucidate the nature and strength of coupling in those materials [2,3]. The peculiar nearly linear temperature dependence of the resistivity below room temperature reported for $La_{1.825}Sr_{0.175}CuO_4$ (LSCO) [4] and $YBa_2Cu_3O_7$ (YBCO) [5] motivated us to measure $\rho(T)$ to the highest possible temperatures. Measurements were performed in O_2 atmosphere to minimize the chemical effects of oxygen desorption.

We find resistivities which are essentially linear to 1100K in LSCO and to 600K in YBCO, as shown in Fig. 1. The sharp upturn of $\rho(T)$ in YBCO above 600K is undoubtedly related to oxygen desorption [6], while LSCO is known to be stable in O_2 to the highest temperature used in this experiment [7]. Furthermore, in YBCO we observe an abrupt change in slope at the orthorhombic-tetragonal phase transition. This behavior of YBCO for $T > 600K$ is of interest in itself, and will be briefly discussed below.

The main emphasis of this work is placed on the implications which the observed linear parts of $\rho(T)$ have for the electron-phonon coupling in the two compounds. In essence, we argue that when resistivity is linear to high temperature, as in Fig. 1, it can be understood only in terms of weakly-coupled electrons and phonons (see Sec. 7). Since in the electron-phonon theory of superconductivity the same phonons are responsible for the T_c [8], resistivity tells us that high T_c must be non-phonon in nature. This assumes that $\rho(T)$ results from scattering on phonons. Clearly, if $\rho(T)$ were to have a non-phonon origin (paramagnons, excitons, etc.) then the phonon coupling would be even weaker yet. From our analysis we derive an *upper-bound* estimate for the strength of the electron-phonon coupling λ [9]. Furthermore, making use of the measured absolute resistivities at high temperatures, we argue that recent calculations for the plasma energy $\hbar\omega_p$ are incompatible with these data, for they would imply a violation of the Mott-Ioffe-Regel (MIR) rule [10] that the minimum mean-free-path is an interatomic spacing. We are thus compelled to expect that $\hbar\omega_p$ is considerably smaller than in band structure calculations. This is corroborated by numbers we deduce independently from recent experimental determinations of the London penetration depth $\lambda_L(0)$ [11,12].

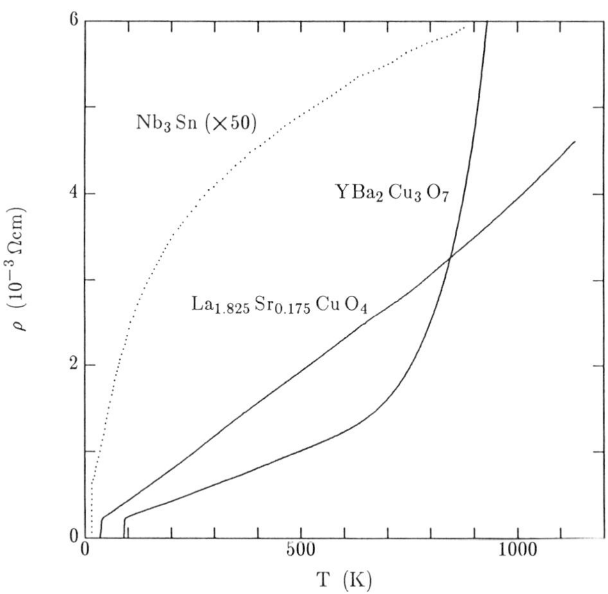

Fig. 1. Temperature dependence of the resistivity $\rho(T)$ for LSCO and YBCO compounds measured in this work. Data for Nb$_3$Sn, scaled up by a factor of 50, taken from Ref. 14, are shown to exhibit classical resistivity saturation behavior.

The paper is organized as follows: we next describe the experiment and discuss the "chemical" behavior of YBCO at T > 600 K, then make our points concerning the linear part of $\rho(T)$, and finally discuss possible problems and some relevant issues such as the absence of the isotope effect in YBCO.

2. EXPERIMENT

Bar-shaped samples, approximately $1\,\mathrm{mm}^2$ cross-section were cut from fully-fired ceramic pellets and reannealed in oxygen on a piece of zirconia, at 975°C for LSCO and at 950°C for YBCO. Four contact wires of either gold or silver were attached with silver cement, a method which limits the maximum temperature to about 900°C. The resistivity data presented in Fig. 1 are a composite of separate runs taken with an O_2 flowing furnace and a liquid-helium cryostat. The T_c values are 38 K and 92 K for LSCO and YBCO, respectively. Temperatures were swept sufficiently slowly to minimize hysteresis associated with oxygen reactions at elevated temperature. Durations of cryostat runs were about 2 hrs. and furnace runs about 4 hrs.

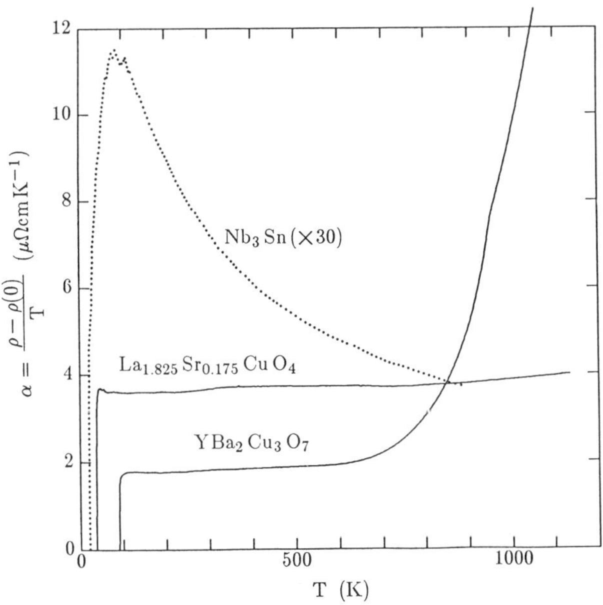

Fig. 2. Resistivity slopes $\alpha = [\rho(T) - \rho(0)]\,/\,T$, where $\rho(0) = 72\,\mu\Omega\,\mathrm{cm}$ for YBCO and LSCO, and $10\,\mu\Omega\,\mathrm{cm}$ for Nb_3Sn. The Nb_3Sn data has been scaled up by a factor of 30.

3. YBCO ABOVE 600 K

Figure 2 gives the resistivity slope, defined as $\alpha = [\rho - \rho(0)] / T$, where $\rho(0)$ is the extrapolated residual resistivity ($72\,\mu\Omega\,\mathrm{cm}$ for both LSCO and YBCO) indicating clearly the regions of linearity and that of oxygen desorption in YBCO. Figure 3 shows the furnace-data portion of the YBCO resistivity for three concentrations of O_2 (pure and diluted with Ar, as indicated), where the derivative curve in the inset emphasizes the sharp feature at the orthorhombic-tetragonal phase transition. The transition temperature has been previously shown to depend on the O_2 partial pressure [6]. We find that these transitions are about 2 K wide, non-hysteretic and continuous. The resistivity is linear up to a temperature T_2, above which the deviation from linearity is caused by the decrease in the oxygen content in the $YBa_2Cu_3O_{7-\delta}$ compound, where $\delta = \delta(T)$ is an increasing function of T [6]. Loss of oxygen produces two effects: it depletes the hole-carrier density, which can be thought of as roughly proportional to $(1-\delta)$, as metallic resistivity disappears at

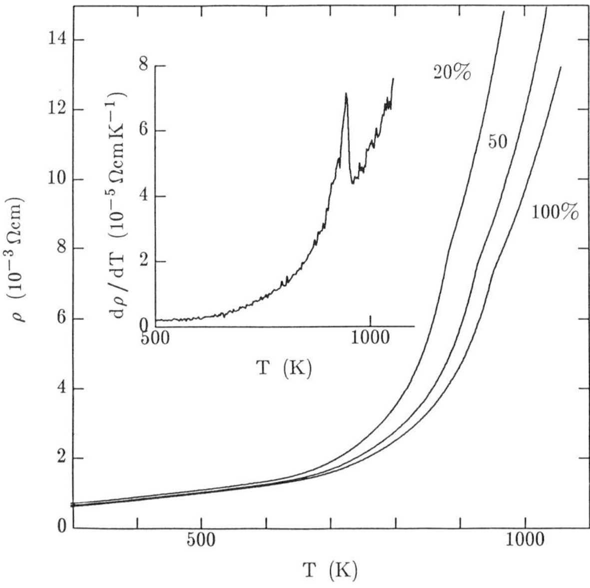

Fig. 3. Resistivity data for the YBCO compound above room temperature at three concentrations of O_2 in the furnace. The break in slope occurs at the temperature of the orthorhombic-tetragonal structural transition, which depends upon O_2. Inset: numerical derivative curve for data in pure O_2.

$\delta \approx 1$, and, at the same time, it creates δ oxygen-vacancy scattering centers per unit cell. These two effects produce static-disorder resistivity which depends on temperature through δ, as $\delta(T)[1 - \delta(T)]^{-1}$, a function in qualitative agreement with the faster-than-linear experimental $\rho(T)$ dependence shown in Fig. 3. It appears that phonon scattering plays a relatively small role in this regime. Indeed, it has been shown [6] that at the structural transition, δ is essentially a constant independent of the partial oxygen pressure, $\delta \approx 0.4$. We see from Fig. 3 that resistivity at this phase transition, occuring at $\delta \approx 0.4$, is almost the same for the three curves, although the transition temperatures themselves are different. Hence, resistivity must be determined mainly by δ, i.e., the number of carriers and disorder, and not by scattering on phonons. Details of this analysis for YBCO at $T > T_2$ will be published elsewhere.

4. LINEAR RESISTIVITIES: ABSENCE OF SATURATION

Naive Estimates of Mean Free Paths

We now turn to the linear part of the $\rho(T)$ data. As can be seen in Fig. 2, the slope $\alpha = [\rho(T) - \rho(0)] / T$ shows no indication of decreasing at high temperature. (The direct slope α agrees to better than 10% with the derivative, $d\rho / dT$, except for the region near T_c and above 600 K in YBCO.) Qualitatively, this reveals at once that mean-free paths ℓ are much longer than interatomic spacings a at temperatures up to 600 K for YBCO and to 1100 K for LSCO. In contrast, metals with $\ell \sim a$ always display strong deviation from linearity at high temperature, the phenomenon known as resistivity saturation [13], with α falling at high temperature. We have included $\rho(T)$ for Nb_3Sn [14] in Fig. 1 and the slope α in Fig. 2 for comparison, as this is a classic example of this saturation behavior. In the parallel-resistor model of resistivity saturation [15], the observed slope $\alpha = [\rho(T) - \rho(0)] / T$ can be expressed in terms of an ideal slope $\alpha_{id} = [\rho_{id} - \rho_{id}(0)] / T$ through [3]

$$\alpha = \alpha_{id} \left(1 - \frac{\rho(0)}{\rho_s}\right)\left(1 - \frac{\rho(T)}{\rho_s}\right) , \tag{1}$$

where ρ_s is the saturation resistivity and $\rho(0)$ is the residual resistivity at $T = 0$. In the following we will compare slopes at two temperatures, at $T_1 = 300$ K and at the highest temperature at which $\rho(T)$ is approximately linear, T_2. Noting that $\alpha_{id} = $ const., and defining the ratio of slopes $\epsilon_S = \alpha(T_2) / \alpha(T_1)$, we find from Eq. (1)

$$\frac{\rho_s}{\rho(T_2)} = \frac{1 - \epsilon_S \dfrac{\rho(T_1)}{\rho(T_2)}}{1 - \epsilon_S} . \tag{2}$$

For LSCO, $T_2 = 1100$ K and for YBCO, $T_2 = 600$ K. From Table I, $\rho(T_1) / \rho(T_2)$ is 0.27 for LSCO and 0.5 for YBCO. According to Fig. 2, between T_1 and T_2 α is constant to approximately 10%; hence, we take $\epsilon_S \geq 0.9$, and find from Eq. (2) $\rho_s / \rho(T_2) \geq 7.6$ for LSCO and $\rho_s / \rho(T_2) \geq 5.5$ for YBCO. In terms of mean free paths, which we write as $\ell = ca$, and where a may be taken (conservatively) as the nearest-neighbor Cu-O distance, $a \approx 2\text{Å}$, the condition on ρ_s means $c \geq 7.6$, and

Table I. Resistivities $\rho(T)$ $[\mu\Omega\,cm]$ and slopes $\alpha = [\rho(T_2)-\rho(0)]/T_2$ $[\mu\Omega\,cm\,K^{-1}]$, measured and derived intrinsic values, for the $La_{1.825}Sr_{0.175}Cu_3O_4$ (LSCO) and $YBa_2Cu_3O_7$ (YBCO) superconductors. T_2 [K] refers to the highest temperature at which $\rho(T)$ is linear in T

	measured values				in-plane intrinsic values			
	$\rho(0)$	$\rho(300\,K)$	T_2	$\rho(T_2)$	$\rho_{\parallel}(0)$	$\rho_{\parallel}(300\,K)$	$\rho_{\parallel}(T_2)$	$\alpha_{\parallel}$
LSCO	72	1185	1100	4400	36	593	2200	1.97
YBCO	72	622	600	1234	36	311	617	0.97

Table II. Band-structure parameters: plasma energy $(\hbar\omega_p)_{\parallel}$ [eV] and Fermi velocity $(v_F)_{\parallel}$ $[10^7\,cm\,s^{-1}]$. Band structure estimates of electron phonon coupling λ and mean-free-path $\ell(T_2)$ [Å]

	band structure			
	parameters		estimates	
	$(\hbar\omega_p)_{\parallel}$	$(v_F)_{\parallel}$	λ	$\ell(T_2)$
LSCO	2.9	2.2	4.1	0.6
YBCO	3.7	2	3.3	1.2

Table III. Experimental estimates for the parameter $c(T_2)$ in the relation $\ell = ca$ and λ, as given by "naive" and "conservative" methodologies discussed in the text, and $(\hbar\omega_p)_{\parallel}$ as determined by resistivity analysis alone and in combination with μSR data

	experimental estimates					
	$c(T_2)$ naive	$c(T_2)$ consv.	λ naive	λ consv.	$(\hbar\omega_p)_{\parallel}$ resis.	$(\hbar\omega_p)_{\parallel}$ μSR
LSCO	≥ 7.6	2.7	≤ 0.16	0.45	≤ 0.96	1.0
YBCO	≥ 5.5	1.7	≤ 0.37	1.2	≤ 2.24	2.2

$c \geq 5.5$ for the two respective compounds at their respective T_2's. This first "naive" estimate may be modified to some extent by other effects such as phonon anharmonicity (thermal expansion) and lifetime broadening of the band structure, which create a counterposing effect of increasing the observed α. We will now consider these effects.

Phonon Anharmonicity

To calculate the effect of phonon anharmonicity on resistivity we note that $\rho \propto T/M\theta_D^2$, where M is an ionic mass and θ_D is the Debye temperature. From this we obtain a correction for the slopes [3]

$$c_{anh} = \left(\frac{\theta_D(T_1)}{\theta_D(T_2)}\right)^2 .$$

Thermal expansion leads to phonon softening, making $\theta_D(T)$ a decreasing function of T; this effect $increases$ the slope at higher T, and hence can counterpose the saturation. With the use of Grüneisen constant $\gamma = -d(\ln\theta_D)/d(\ln V)$ we obtain $\epsilon_{anh} = \exp[2\alpha_v\gamma(T_2-T_1)]$, where α_v is the volume coefficient of thermal expansion, and $\gamma = \alpha_v / (X_o c_v d_o)$, where X_o is the compressibility, c_v the specific heat, and d_o the density. Using the measured parameters for the two compounds we find [16] $\alpha_v = 3.3 \times 10^{-5} \mathrm{K}^{-1}$, $\gamma = 1.88$ in LSCO and $\alpha_v = 3.9 \times 10^{-5} \mathrm{K}^{-1}$, $\gamma = 1.67$ in YBCO, so that anharmonic effects give us $\epsilon_{anh} = 1.1$ and $\epsilon_{anh} = 1.04$ in LSCO and YBCO, respectively.

Lifetime Broadening of the Band Structure

Band-structure broadening due to finite lifetime of electronic states [17] can also produce a trend which counterposes saturation. This effect on the slope α can be estimated by writing [18]

$$\rho(T) - \rho(0) = \frac{4\pi}{\omega_p^2 \tau} \quad , \tag{3}$$

where ω_p is the plasma frequency and τ^{-1} is the scattering rate, which is approximately proportional to the density of states $N_F(0)$ and T [19]. So we obtain for the slope $\alpha = [\rho - \rho(0)] / T$ that $\alpha \propto N_F(0)\omega_p^{-?}$, or, noting that $\omega_p^2 = (8\pi/3)e^2 N_F(0) <v_F^2(0)>$ [20], $\alpha \propto <v_F^2(0)>^{-1}$. The broadening energy $\hbar/\tau$ can be obtained from Eq. (3)

$$\hbar/\tau = 1.33 \times 10^{-4} \rho <(\hbar\omega_p)^2>_\rho \quad , \tag{4}$$

where energies are in eV and resistivity in $\mu\Omega\,\mathrm{cm}$. The band structure (BS) broadening effect on the slopes at T_1 and T_2 can be expressed as

$$\epsilon_{BS} = \frac{<v_F^2(0)>_{\rho(T_1)}}{<v_F^2(0)>_{\rho(T_2)}} \quad ,$$

where the values of $<v_F^2(0)>$ are broadened by $\hbar/\tau$ given in Eq. (4), with $\rho(T_1)$ and $\rho(T_2)$, respectively.

We performed preliminary estimates of this effect for LSCO and YBCO, using band structure calculations reported by Mattheiss [21] and Mattheiss and Hamann [22]. Averaging was performed using Fermi-Dirac distribution functions. It should be emphasized that the procedure we used is not entirely self-consistent. In a self-consistent procedure one would use in Eq. (4) the band structure values of $\hbar\omega_p$. However, as we show below, our experimental estimates of $\hbar\omega_p$ are considerably smaller than band-structure values. We therefore adopted a compromise in which the realistic *experimental* values of $\hbar\omega_p$ are substituted into Eq. (4) to find the broadening effects on the *theoretical* band structure.

We found that broadening will increase the slope α with increasing temperatures (increasing $\rho(T)$), and hence it will, along with anharmonic effects, oppose the saturation. This effect can be rather significant. The approximate factors are $\epsilon_{BS} \cong 1.3$ in LSCO and $\epsilon_{BS} \cong 1.6$ in YBCO. Note that the broadening effect is stronger in YBCO, mainly due to higher $\hbar\omega_p$ which implies higher broadening energy in Eq. (4).

Conservative Estimates of the Mean Free Path

The combined effect of anharmonicity and energy broadening can produce an increase in the slope α given by $\epsilon_{LSCO} = \epsilon_{anh} \cdot \epsilon_{BS} = 1.1 \times 1.3 = 1.43$ and $\epsilon_{YBCO} = \epsilon_{anh} \cdot \epsilon_{BS} = 1.04 \times 1.6 = 1.66$. Saturation acts to lower the slope; experimentally, we observed essentially linear curves. To be on the conservative side in our estimates of mean free path, we assume that these trends are essentially balanced [23]. In this manner we arrive at our estimate for the possible decrease in slope due to saturation: $\epsilon_S = (\epsilon_{anh} \cdot \epsilon_{BS})^{-1}$. With the numbers given above we find $\epsilon_S = 0.7$ in LSCO and $\epsilon_S = 0.6$ in YBCO. Substituting these values into Eq. (2) we obtain $\rho_s / \rho(1100K) = 2.7$ for LSCO and $\rho_s / \rho(600K) = 1.7$ for YBCO, or, in terms of mean free path $\ell = ca$, $c = 2.7$ and 1.7 for the two compounds at their respective maximum temperatures of linearity.

The conclusions of this section are summarized in Table III under the headings "naive" and "conservative" experimental estimates for the parameter $c(T_2)$. These values will be used in our estimates of the strength of the electron-phonon coupling.

5. MEASURED AND INTRINSIC RESISTIVITIES

<u>Absolute Values of ρ and α in Ceramic Samples</u>

The extrapolated residual resistivities $\rho(0)$ of the two compounds are relatively small compared to the $\rho(T)$ at high temperatures ($\rho(0) = 72\mu\Omega$cm in both compounds). Hence, we need not be concerned with the effects of large static disorder in these samples. Further, good ceramic samples (i.e., samples which are within $\sim 15\%$ of the theoretical density, have highest T_c's and sharp transitions) all have similar values of $\rho(300K)$ and α [24], and as discussed below, are within the expected factor of 2 from the resistivity of the best oriented epitaxial films. We list the values we measured in this work in Table I; they are similar to the ones published at 300 K by Cava et al. [4,5].

<u>Correction for Anisotropy</u>

Recent work on epitaxial films of YBCO shows that the resistivity anisotropy $\rho_\perp / \rho_\parallel$ is at least an order of magnitude [25], owing to preferential conduction within the Cu-O planes, confirming anisotropic band structure [22]. Resistivity of a polycrystalline ceramic with randomly-oriented isotropically-shaped grains can be modeled by effective medium theory as a composite of 2/3 by volume conducting grains and 1/3 insulating grains ($\rho_\perp \gg \rho_\parallel$). Neglecting small percolation corrections, this leads to the resulting correction factor $\beta = \rho / \rho_\parallel = 2$ [26]. However, if grains are not well-connected, i.e., if there are dead-end paths for current transport, then it is possible to have $\beta > 2$. Comparing our data with various results for films grown epitaxially on $SrTiO_3$ show [27,28] that $\beta = 2$ is indeed an experimental upper bound. Suzuki and Murakami [27] find for $La_{1.94}Sr_{0.06}CuO_4$ grown with the c-axis perpendicular to the plane of the film, $\rho_\parallel = 700\,\mu\Omega$ cm and $\alpha_\parallel = 1.9\,\mu\Omega$ cm K^{-1} at 300 K. (We note however that this film had a low Sr content and $T_c \approx 18K$). In YBCO, to the best of our knowledge, the lowest $\rho_\parallel$ and $\alpha_\parallel$ values at 300 K were reported by the Stanford group, $\sim 300\,\mu\Omega$ cm and $\sim 1\,\mu\Omega$ cm K^{-1}, respectively [28]. Somewhat higher values are reported on epitaxial films in Ref. 25. Since these values are almost precisely a factor of 2 lower than measured values of polycrystalline samples, we feel that taking $\beta = 2$ is justified.

We cannot rule out, however, that this apparent factor of 2 difference between ceramic samples and oriented films is merely a coincidence, and that the true intrinsic resistivities are indeed smaller yet. The crux of our argument about the strength of the electron-phonon coupling does not really depend on the intrinsic value of resistivity, but rather is based on the linearity of $\rho(T)$ alone, as explained below. However, estimates of $\hbar\omega_p$ depend on the absolute value of the resistivity $\rho_{\parallel}$ or slope $\alpha_{\parallel}$. We return to this issue later.

6. BAND STRUCTURE ESTIMATES: VIOLATION OF MOTT-IOFFE-REGEL RULE

We are now prepared to combine resistivity results with the parameters of the band theory in order to calculate the mean free path and electron-phonon coupling. The standard relations between these quantities come from Eq. (3) where, at $T \geq \theta_D$, $1/\tau$ can be expanded in terms of the electron-phonon coupling parameter λ_{tr} [2,18], which is closely related to the McMillan coupling constant λ appearing in the theory of superconductivity. These relations, in convenient units [3] are

$$\rho\ell = 4.95 \times 10^{-4} \frac{v_F}{(\hbar\omega_p)^2} \ , \tag{5a}$$

$$\lambda = 0.246(\hbar\omega_p)^2 \alpha \ , \tag{5b}$$

where units are $[\alpha] = \mu\Omega\,cm\,K^{-1}$, $[\ell] = \overset{\circ}{A}$, $[v_F] = cm\,s^{-1}$, and $[\hbar\omega_p] = eV$. Previous considerable experience in using these relations to estimate coupling in simple and transition metals [2] and, with additional corrections, A-15's [3] had led to the conclusion that band structure calculations correctly predict the strength of electron-phonon coupling in those materials [2,3]. Before we attempt to couple Eqs. (5) with experimentally-derived quantities, we point out that for the LSCO compound Allen et al. [29] recently presented their calculated basal-plane values, $v_{F\parallel} = 2.2 \times 10^7$ cm s^{-1} and $\hbar\omega_{p\parallel} = 2.9$ eV (for Sr concentration x=0.05), and used them in conjunction with the resistivity data of Suzuki and Murakami [27] to calculate $\lambda = 3.2$. We note that this value is too high for the modest $T_c = 18$ K of Ref. 27. As explained further below, high estimates of λ go hand in hand with small estimates for ℓ; puzzlement over the apparent absence of saturation at room temperature has already been expressed in this regard in Ref. 9 and Ref. 29.

For YBCO, Mattheiss calculated v_{Fx} and v_{Fy} components [21] which give us the average in-plane value of $(v_F)_{\parallel}) = [(v_{Fx}^2 + v_{Fy}^2) / 2]^{1/2} = 2 \times 10^7$ cm s^{-1}. Combining this with $N_F(0) = 7.7$ states/eV-cell [22] and using the expression for the average in-plane component $(\hbar\omega_p)_{\parallel} = [4\pi e^2 \hbar^2 N_F(0) v_{F\parallel}^2 / \Omega_{cell}]^{1/2}$ of Allen et al. [29], where $\Omega_{cell} = 1.74 \times 10^{-22}$ cm^3 is the volume of the unit cell in YBCO, we find $(\hbar\omega_p)_{\parallel} = 3.7$ eV. (The average 3D value is $<\hbar\omega_p> = 3.1$ eV [21].) Using these in-plane band structure values together with in-plane resistivities and slopes from Table I, we calculate ℓ and λ from Eq. (5). The results are shown in Table II under the heading "band structure estimates". We find large values of λ at the expense of completely unrealistic values of $\ell \sim 1\overset{\circ}{A}$. These are not only in contradiction with our conservative lower-bound estimates but they are clearly in violation of the MIR rule, i.e., the uncertainty relation. It is possible that this situation could, at least in part, arise from intrinsic resistivities $\rho_{\parallel}$ smaller than we think, as was mentioned earlier. However, we show in Sec. 8 independent experimental evidence that $(\hbar\omega_p)_{\parallel}$ is indeed smaller than the band structure predictions and find no other evidence at present that intrinsic resistivities are as small as required to justify the band structure values of $\hbar\omega_p$.

7. GENERAL ARGUMENT: ESTIMATES OF λ AND $\hbar\omega_p$

It is not difficult to see that strong coupling and long mean free paths at high temperatures are in general mutually exclusive. If we substitute $\ell = ca \ (c > 1)$

into Eq. (5b) and substitute $\hbar\omega_p$ from Eq. (5a) into Eq. (5b), we find $\lambda = 1.22 \times 10^{-4} v_F / caT$. If c gives the lower bound on ℓ based on the absence of saturation at T_2, the highest temperature of linearity, then we obtain an upper-bound estimate for λ:

$$\lambda_{max} = 1.22 \times 10^{-5} \frac{v_F}{caT_2} \ . \tag{6}$$

Note that here resistivity itself drops out (except for the indirect presence through the value of c) telling us essentially that linear resistivity at high T_2 by itself imposes restrictions on λ. With our conservative estimates of $c(1100K) = 2.7$ in LSCO and $c(600K) = 1.7$ in YBCO and taking $a = 2\text{Å}$, T_2 from Table I, and $(v_F)_{\parallel}$ as given above by the band structure, we find from Eq. (6):

$$\lambda_{max} = 0.45, \quad \text{for LSCO},$$

$$\lambda_{max} = 1.2, \quad \text{for YBCO}.$$

If we used our "naive" values of $c(T_2)$, these estimates would become proportionally smaller: $\lambda_{max}^{naive} = 0.16$ for LSCO and $\lambda_{max}^{naive} = 0.37$ for YBCO. Needless to say, given the knowledge of $\theta_D \lesssim 400\,\text{K}$ for the two compounds[30], these values of λ are too small to account for high T_c's. These estimates further imply that if the values of $\alpha_{\parallel}$ in Table I are indeed intrinsic, then $\hbar\omega_p$ must be lower than the band structure predictions: indeed, in order to satisfy Eq. (5b) with these low values of λ we need to have upper limits of $(\hbar\omega_p)_{\parallel} = 0.96\,\text{eV}$ for LSCO and $2.24\,\text{eV}$ for YBCO. We appeal to other experiments in the next section to corroborate this important result.

8. OTHER EXPERIMENTAL DETERMINATIONS OF $\hbar\omega_p$

Drude plasma frequency values derived from infrared reflectivity measurements were recently reported by Orenstein et al.[31], who obtain values $\hbar\omega_p = 2.0\,\text{eV}$ for ceramic samples of the LSCO compound and correspondingly $3\,\text{eV}$ for the YBCO compound. These quantities were derived from the integral of the optical conductivity. The infrared response exhibits strong peaks, at $0.5\,\text{eV}$ and $0.75\,\text{eV}$, respectively, for the two compounds, which are attributed to gap transitions. The plasma resonance itself appears to be heavily damped as no zero crossing is observed in the real part of the optical dielectric response. These complications makes straightforward interpretation of the infrared data quite difficult.

We suggest that the plasma frequency $(\omega_p^2 = 4\pi e^2 n / m^*)$ can also be computed from measurements of the zero-temperature London penetration depth,

$$\lambda_L(0) = \left[\frac{\tilde{m}^* c^2}{4\pi n e^2} \right]^{1/2} , \tag{7}$$

where $\tilde{m}^* = (1+\lambda) m^*$ is the band-structure effective mass m^* dressed by phonon interactions. We have to concern ourselves with this renormalization because, unlike the Drude frequency, penetration depth is a low-energy, low-temperature quantity. Hence we may write

$$\hbar\omega_p = 1.8 \times 10^3 (1+\lambda)^{1/2} \lambda_L(0) , \tag{8}$$

where $[\hbar\omega_p] = eV$ and $[\lambda_L(0)] = \text{Å}$. This scheme works well for niobium, for example, where $\lambda = 1.0$ [32] and $\lambda_L(0) = 315\text{Å}$ [33], so that Eq. (8) predicts $\hbar\omega_p = 8.9\,eV$, in excellent agreement with band structure $(8.8\,eV)$ [18]. Ideally, one could also combine measurements of $\lambda_L(0)$ and of the optical $\hbar\omega_p$ to find λ. For the cuprate superconductors, probably the most reliable measurements of $\lambda_L(0)$ have been obtained from muon-spin-rotation spectroscopy (μSR), which is a bulk measurement that is inherently insensitive to surface preparation or minor damage to the surfaces of grains in ceramics. Several groups have recently reported values of the μSR linewidth in the low-temperature mixed state of ceramic samples [11,12]: we use the values $\Lambda_{\mu SR} = 0.9\,\mu s^{-1}$ for the LSCO compound and $\Lambda_{\mu SR} = 3\,\mu s^{-1}$ for the YBCO compound. Moreover, it was discovered that the local magnetization is very small compared to that expected for an isotropic 3-dimensional superconductor [12], which we take as independent confirmation that anisotropy dominates the local field distribution. We wish to account for this anisotropy in our interpretation of these raw data, recognizing that the vortex lattice modulates mainly magnetic field components along local c-axes of the grains, since the penetration depth parallel to the c-axes is very large [c.f. large m^* in Eq. (7)], i.e., shielding currents flow mainly in the basal plane. This corrects $\Lambda_{\mu SR}$ by a multiplicative $\sqrt{3}$ factor. We compute the following result for the penetration depth:

$$\lambda_L(0)^2 = 0.0245\frac{\phi_0\gamma_\mu}{\Lambda_{\mu SR}} \quad,$$

where the numerical prefactor was calculated for a triangular vortex lattice at a field appropriate to the experiments — $\lambda_L(0)$ exceeding the vortex-lattice spacing; 2D corrections being unimportant — and γ_μ is the muon gyromagnetic ratio. This gives values of $\lambda_L(0) = 2000\text{Å}$ for LSCO and 1200Å for YBCO, values that differ slightly from those reported [11,12]. Using the "conservative" electron-phonon coupling parameters given in Sec. 7 for the $(1+\lambda)^{1/2}$ correction factor, we obtain plasma energies of $1.0\,eV$ for LSCO and $2.2\,eV$ for YBCO, which for consistency should be interpreted as in-plane experimental values for $(\hbar\omega_p)_\parallel$. These results are presented in Table II and substantiate the conclusions reached in Sec. 7 on the basis of resistivity.

9. DISCUSSION AND CONCLUSIONS

Potential Inconsistencies

Note that there is one inconsistency in the estimates presented in Sec. 7, namely, we arrived at the conclusion that $\hbar\omega_p$ must be smaller than the band-structure values, while at the same time we used band-structure values of v_F in Eq. (3). Generally, if $(\hbar\omega_p)_\parallel = 4\pi\hbar^2 e^2(n/m^*)$ is lower than we thought, then $(v_F)_\parallel = \hbar(2\pi nd)^{1/2}/m^*$ (written here in 2D, with d being the distance between planes) is going to be lower, also. This in turn will have the effect of reducing our estimate of λ through Eq. (6), and lower $\hbar\omega_p$ even further. We do not believe this to be a dramatic effect, since the μSR-derived values seem to bear out our conclusions. In addition, we followed Kresin and Wolf [34] in obtaining 2D estimates of m^* from the measured Sommerfeld constant γ for the two compounds. (The number of carriers n does not enter into the expression for γ in 2D [34].) These m^* values can be then coupled with μSR estimates of $\hbar\omega_p$ to find n, and eventually $v_F = \hbar(2\pi nd)^{1/2}/m^*$. We convinced ourselves that these 2D values of v_F are in reasonable agreement with band structure v_F, which supports our procedure in Sec. 7.

2D Nature and Possible Problems

In what we have done so far, we strived to map the problem into 2D, where it presumably belongs. This has been done with resistivities ($\rho_\parallel$) as well as with band structure quantities. The importance of the 2D analysis for the layered high-T_c perovskites was recently stressed by Kresin [34]. Allen, Pickett and Krakauer [29], in their work on LSCO explicitly exposed the anisotropies in the band-structure quantities, including $\hbar\omega_p$, and accordingly based their estimates of λ on the in-plane values. We have proceeded likewise in this work, although our general conclusions are just the opposite. The question of the applicability of the usual formalism of the Boch-Boltzmann theory of resistivity to an essentially 2D material remains to be addressed, to the best of our knowledge. In particular, it is not obvious to us that the expansion of τ^{-1} in terms of the coupling parameter λ_{tr} will be the same as in 3D; nor is it clear that the difference between λ_{tr} and λ in the theory of superconductivity [2] will not be more serious in 2D. It is somewhat reassuring, however, that in a comparison between 3D and 2D free-electron pictures, i.e., using $v_F^{2D} = \hbar(3\pi^2 n)^{1/3}/m^*$ and $v_F^{3D} = \hbar(2\pi n d)^{1/2}/m^*$ for the in-plane Fermi velocity, with d being the distance between the layers, the numerical results for $\rho\ell$ turn out to be rather similar. From this observation, at least, we should not be too far off in our estimates of ρ_s and mean-free-paths.

Mott-Ioffe-Regel Rule and Absence of Saturation

The MIR rule can be tied to the uncertainty relation — the well-known original argument given by Ioffe [10] — and to the requirement that the uncertainty in the wavevector Δk_F be smaller than k_F itself, leading to the the rule $k_F\ell \gtrsim \pi$. For the oxide superconductors, k_F is smaller than for typical metals, so that the minimum value of ℓ allowed by the MIR rule can be larger than an interatomic spacing. A less restrictive rule follows from the argument of Mott [10], that the minimum distance at over which phase memory of the wavefunction may be lost due to scattering is the distance between atoms and thus the interatomic spacing a is the minimum ℓ. In the context of the present work, this latter view appears to be the correct one.

On the Absence of Sharp Plasma Resonance in the IR Data

According to Eq. (4), using $\hbar\omega_p$ of $1\,\text{eV}$ and $2.2\,\text{eV}$ for the two compounds and $\rho_\parallel(300\text{K})$ from Table I, the broadening energy $\hbar\tau^{-1}$ at room temperature is $\sim 0.08\,\text{eV}$ and $0.2\,\text{eV}$ for LSCO and YBCO, respectively. Hence, overdamping does not seem to be the reason for the absence of a sharp plasma resonance in the infrared reflectivity of Ref. 31.

On the Absence of Isotope Effect

Recently Cohen [35] performed analysis of the Eliashberg equation in terms of the possible absence of the isotope effect and found that the effect can be absent for a combination $\lambda \approx 7$ and $\mu^* = 0.3$. Our analysis excludes large λ on the basis of linear resistivity, and hence excludes the possibility that zero isotope effect [36] in the YBCO may be understood in this context. The recently reported non-zero isotope effect for LSCO [37] contradicts our conclusion of a non-phonon mechanism, since we found $\lambda^{LSCO} \lesssim 0.45$.

Linearity of $\rho(T)$ at Low Temperature

In 3D linearity of $\rho(T)$ at low enough temperatures (above $40\,\text{K}$ in LSCO, above $100\,\text{K}$ in YBCO) would imply low Debye temperatures, since Bloch-

Grüneisen resistivity should have noticeable curvature below $\sim\frac{1}{2}\theta_D$. However, according to the recent work of Micnas et. al [38] in 2D this may not be the case, and so there may be no contradiction between $\theta_D \lesssim 400\,\mathrm{K}$ [30] and the observed linear resistivity.

<u>Conclusions</u>

1. Linear resistivities at high temperatures are inconsistent with strong electron-phonon interaction, irrespective of the absolute values of resistivity, as argued in Sec. 7. This places limits to the possible strength of electron-phonon coupling in oxide superconductors: $\lambda_{\mathrm{LCSO}} \lesssim 0.45$ and $\lambda_{\mathrm{YBCO}} \lesssim 1.2$.

2. These values are inadequate to account for high T_c's; hence, the mechanism of superconductivity must be largely non-phonon in origin.

3. We believe that intrinsic values of resistivities are known from oriented-film work and are in agreement with ceramic values scaled by the factor $\beta \sim 2$ which is consistent with simple effective medium theory argument. These intrinsic resistivities are inconsistent with band structure values of $\hbar\omega_p$ which would predict mean free paths $\sim 1\,\mathring{\mathrm{A}}$ at the highest temperatures of linearity.

4. Our estimates of $\hbar\omega_p$ are in agreement with values we deduced from the recent penetration depth measurements: $(\hbar\omega_p)_\parallel = 1.0\,\mathrm{eV}$ for LSCO and $(\hbar\omega_p)_\parallel = 2.2\,\mathrm{eV}$ for YBCO.

ACKNOWLEDGEMENTS

It is a pleasure to acknowledge stimulating conversations with P. B. Allen, B. Batlogg. R. N. Bhatt, R. B. Cava, R. B. van Dover, G. P. Espinosa, P. K. Gallagher, J. Halbritter, D. R. Harshman, W. J. Kossler, L. F Mattheiss, J. Orenstein, H. Ott, A. Ramirez, B. Raveau, M. Strongin, C. M. Varma, and W. Weber. R. B. Cava, G. P. Espinoza, R. B. van Dover, and A. Ramirez kindly provided ceramic material for our samples.

REFERENCES

1. J. G. Bednorz and K. A. Müller, Z. Phys. b **64**, 189 (1986); M. K. Wu, J. R. Ashbun, C. J. Tong, P. H. Hor, R. L. Meng, L. Gao, Z. J. Huang, Y. O. Wang, and C. W. Chu, Phys. Rev. Lett. **58**, 908 (1987).

2. P. B. Allen, T. B. Beaulac, F. S. Khan, W. H. Butler, F. J. Pinski, and J. C. Swihart, Phys. Rev. B**34**, 4331 (1986); P. B. Allen, "Empirical Electron-Phonon λ Values", preprint.

3. M. Gurvitch, Physica **135B**, 276 (1985).

4. R. J. Cava, B. Batlogg, R. B. vanDover, and E. A. Rietman, Phys. Rev. Lett. **58**, 408 (1987).

5. R. J. Cava, B. Batlogg, R. B. van Dover, D. W. Murphy, S. Sunshine, T. Siegrist, J. P. Remeika, E. A. Rietman, S. Zahurak, and G. P. Espinosa, Phys. Rev. Lett. **58**, 1676 (1987).

6. P. K. Gallagher, to be published in Adv. Ceram. Mater.; K. Kishio, J. Shimoyama, T. Hasegawa, K. Kitazawa, and K. Fueki, preprint.

7. C. Michel and B. Raveau, Rev. Chim. Min. **21**, 407 (1984).

8. This was first realized a long time ago by J. Bardeen, Phys. Rev. **80**, 567 (1950); H. Frölich Phys. Rev. **79**, 845 (1950).

9. This line of thought was not obvious to us until we performed the measurements at high temperatures and observed the absence of saturation. At the APS Meeting in March one of us argued that room-temperature $\rho(T)$ coupled with band structure $\hbar\omega_p$ provides sufficiently large values of λ, at least in LSCO: M. Gurvitch, Bull. Am. Phys. Soc. **32**, 905 (1987).

10. N. F. Mott and E. A. Davis, "Electronic Processes in Non-Crystalline Materials", Clarendon Press, Oxford (1979); A. F. Ioffe and A. R. Regel, Prog. Semicond. **4**, 237 (1960).

11. G. Aeppli, R. J. Cava, E. J. Ansaldo, J. H. Brewer, S. R. Kreitzman, G. M. Luke, D. R. Noakes, and R. F. Kiefl, Phys. Rev. B **35**, 7129 (1987); W. J. Kossler, J. R. Kempton, X. H. Yu, H. E. Schone, Y. J. Uemura, A. R. Moodenbaugh, M. Suenaga, and C. E. Stronach, *ibid.*, p.7133;

12. D. R. Harshman, G. Aeppli, E. J. Ansaldo, B. Batlogg, J. H. Brewer, J. F. Carolan, R. J. Cava, M. Celio, A. C. D. Chaklader, W. N. Hardy, S. R. Kreitzman, G. M. Luke, D. R. Noakes, and M. Senba, Phys. Rev., to be published; W. J. Kossler, private communication; D. R. Harshman, private communication.

13. Z. Fisk and G. W. Webb, Phys. Rev. Lett. **36**, 1084 (1976).

14. D. W. Woodard and G. D. Cody, Phys. Rev. **136**, A166 (1964).

15. H. Wiesmann, M. Gurvitch, H. Lutz, A. K. Ghosh, B. Schwarz, M. Strongin, P. B. Allen, and J. W. Halley, Phys. Rev. **3**, 782 (1977).

16. The detailed account of all the parameters which enter into this calculation will have to be given elsewhere.

17. J. E. Crow, M. Strongin, R. S. Thompson, and O. F. Kammerer, Phys. Lett. **30**, 161 (1969); L.F. Mattheiss and L. R. Testardi, Phys. Rev. B **20**, 2196 (1979).

18. B. Chakraborty, W. E. Pickett and P. B. Allen, Phys. Rev. B**14**, 3227 (1976).

19. P.B. Allen, Phys. Rev. Lett. **37**, 1638 (1976).

20. M. H. Cohen, Phil. Mag. **3**, 762 (1958).

21. L.F. Mattheiss, Phys. Rev. Lett. **58**, 1028 (1987); also private communication.

22. L.F. Mattheiss and D. R. Hamann, Solid State Commun., to be published.

23. The observed small increase in α of LSCO at the highest temperature (Fig. 2) can in fact be attributed to marginal winning of the described mechanisms over saturation. Alternatively, it may be due to incipient oxygen desorption, as in YBCO.

24. R. B. van Dover, private communication.

25. Y. Enomoto, T. Murakami, M. Suzuki, and K. Moriwaki, preprint.

26. R. N. Bhatt, private communication.

27. M. Suzuki and T. Murakami, Jpn. J. Appl. Phys. **26**, L524 (1987).

28. A. Kapitulnik and T. H. Geballe, private communication.

29. P. B. Allen, W. E. Pickett and H. Krakauer, preprint. It is interesting to compare the average parameters derived from the anisotropic values given by Allen et al. in this reference for LSCO with those calculated by Mattheiss, Ref. 21. We find fine agreement for the plasma energy $<\hbar\omega_p> = 2.45\,\mathrm{eV}$ in both cases (at zero Sr concentration). For $<v_F>$ there appears to be a discrepancy: while from this reference we obtain $[(v_x^2+v_y^2+v_z^2)/3]^{1/2} = 2.45\times10^7\,\mathrm{cm\,s^{-1}}$, Mattheiss finds $<v_F> = 4.5\times10^7\,\mathrm{cm\,s^{-1}}$. The discrepancy may be simply in the $\sqrt{3}$.

30. A. Junod, A. Bezinge, and D. Cattani, J. Cors, M. Decroux, O. Fischer, P. Genoud, L. Hoffmann, J.-L. Jorda, J. Muller, and E. Walker, preprint.

31. J. Orenstein, G. A. Thomas, D. H. Rapkine, C. G. Bethea, B. F. Levine, R. J. Cava, E. A. Reitman, D. W. Johnson, Jr., Phys. Rev. B, in press.

32. J. Geerk, M. Gurvitch, D. B. McWhan, and J. M. Rowell, Physica **109-110B**, 1775 (1982).

33. C. Varmazis and M. Strongin, Phys. Rev. B**10**, 1885 (1974).

34. V. Z. Kresin, Proceedings of the Materials Research Society Spring 1987 Meeting; V. Z. Kresin and S. A. Wolf, preprint.

35. M. L. Cohen, this volume.

36. B. Batlogg, R. J. Cava, A. Jayaraman, R. B. van Dover, G. A. Kourouklis, S. Sunshine, D. W. Murphy, L. W. Rupp, H. S. Chen, A. White, K. T. Short, A. M. Mujsce, and E. A. Rietman, Phys. Rev. Lett. **58**, 2333 (1987); L. C. Bourne, M. F. Crommie, A. Zettl, H.-C. zur Loye, S. W. Keller, K. L. Leary, A. M. Stacy, K. J. Chang, M. L. Cohen, and D. E. Morris, *ibid.*, p. 2337.

37. B. Batlogg et al., submitted to Phys. Rev. Lett.

38. R. Micnas, J. Ranninger, and S. Robaszkiewicz, preprint.

EPR, MAGNETIZATION, AND RESISTIVITY STUDIES IN DOPED (4f, 3d IONS)

AND UNDOPED R (OR Y) $Ba_2Cu_3O_{9-x}$ HIGH T_c SUPERCONDUCTORS*

S.B. Oseroff[1], D.C. Vier[2], J.F. Smyth[2], C.T. Salling[2],
S. Schultz[2], Y. Dalichaouch[2], B.W. Lee[2], M.B. Maple[2]
Z. Fisk[3], J.D. Thompson[3], J.L. Smith[3], E. Zirngiebl[3],

[1]San Diego State University, San Diego 92182
[2]University of California, San Diego, La Jolla 92093
[3]Los Alamos National Laboratory, Los Alamos 97545

INTRODUCTION

After years of searching for superconductors with higher transition
temperatures, an incredible success has been achieved in recent months, as
a consequence of the discovery of Bednorz and Müller of a 30 K transition
in $La_{2-y}Ba_yCuO_{4-x}$,[1] followed by the subsequent findings that $YBa_2Cu_3O_{9-x}$
has a T_c well above liquid nitrogen temperatures.[2,3] These discoveries have
resulted in an explosive growth of interest in the field, with the present
challenge being to explain the pertinent superconducting mechanisms.

Superconductivity above 90 K has subsequently been found in the series
of oxide superconductors, $RBa_2Cu_3O_{9-x}$, where R = Y,[2,3] or rare-earths,[4-6]
including those possessing a local moment. Thus, in contrast to most superconductors, where the presence of magnetic moments breaks the Cooper-pairs
thereby depressing T_c, the rare earth ions in these oxides appear to have a
negligible interaction with the superconducting electrons. In addition, the
substitution of Ba by Sr and/or Ca has also been shown to have little effect
on T_c.[7,8]

The high temperature superconducting phase has been identified as
$RBa_2Cu_3O_{9-x}$ with an orthorhombically distorted, oxygen deficient, perovskite-
like structure.[9] It has been suggested that the high temperature superconductivity is associated with a two-dimensional structure in which two CuO_2
layers sandwich one CuO_3 chain, with two Ba^{2+} cations per unit cell.[10] Because the Cu ions seem to be the essential component of the conduction path
in these oxides, we sought to study the effects of doping with several percent of 3d ions, particularly those which also form local moments. We have
measured the suppression of T_c, the changes in magnetic susceptibility, and
the nature of the Electron Paramagnetic Resonance (EPR) signals in compounds
of nominal composition $EuBa_2(Cu_{1-y}M_y)_3O_{9-x}$ where M = Cr, Mn, Fe, Co, Ni, or
Zn and y $\leq$ 0.15. We also have examined the EPR signals in the pure oxide
compounds $RBa_2Cu_3O_{9-x}$, with R = Y, Pr, Nd, Eu, Gd, Ho, Er, or Yb; and in the

compounds $(R_{1-z}R'_z)Ba_2Cu_3O_{9-x}$, with R = Eu or Y, R' = Gd or Er, and z < 0.05. We believe EPR will be a fruitful technique to apply to the superconducting oxides since EPR enables one to study the static and dynamic interactions between added local moments and (hopefully) the conduction electron system, as has been shown for more well known metal systems[11,12] and Type II superconductors.[13-15]

SAMPLE PREPARATION AND EXPERIMENTAL METHODS

Two somewhat different procedures (A or B) were followed in sample preparation. Procedure A consisted of grinding together rare earth oxide, 3d metal oxide, and $BaCO_3$ powders, undried, in appropriate amounts to give the desired (R, R'):Ba:(Cu, M) = 1:2:3 stoichiometric phase. This was followed by sintering in air at 950°C for > 6 hours. The grinding and sintering process was repeated once more. After grinding a third time, samples were pressed into pellets and annealed at 925 - 950°C in flowing oxygen at 1 atm for 12 - 24 hours. Samples were then slow cooled (6 - 12 hours) under oxygen.[5]

Procedure B differed from A in two respects.[4] In procedure B the initial powders were carefully dried (by heating) before weighing. Since the powders are all hygroscopic to some extent, the stoichiometry could differ from that for procedure A by possibly a few percent. Also, in procedure B, oxygen was used instead of air for the initial sinterings. As will be discussed below, these differences apparently had significant effects on some of the observed EPR spectra, even though there appeared to be no difference in sample quality as determined by X-ray diffraction and the resistive transition at T_c. Most samples were characterized by microprobe analysis and X-ray diffraction to determine their overall quality. X-ray powder diffraction patterns were obtained for all of the samples studied. All patterns corresponded to the perovskite-like structure associated with these high T_c oxide superconductors. In most samples no other phases were detected from the X-ray diffraction patterns, except for small amounts ($\leq$ 3%) of the green phase compound R_2BaCuO_5. All samples, with the exception of those containing 3d impurities, were confirmed by X-ray diffraction as having the desired orthorhombically distorted structure. The addition of 3d impurities typically resulted in a less orthorhombic structure as the impurity concentration was increased.

Although the X-ray diffraction, resistivity, and susceptibility results indicated most samples to be of good quality, the microprobe results were less encouraging. Variations in the R:Ba:Cu ratio of ~ 10% were typical. For a few samples, regions of pure Cu or pure Ba were detected. However, the granularity of these superconductors made microprobe analysis difficult. The reliability of the results are suspect as surface topology likely played a strong role. Further investigations are in progress, and will be reported elsewhere.

Dc magnetic susceptibility and ac electrical resistivity measurements were made on samples pressed into pellets. EPR measurements were made on powdered material (grain size ~ 10 μm) filed or ground from the originally prepared pellets and subsequently sealed in Parafilm. The powdering was utilized to allow the microwaves to penetrate a much greater percentage of the sample volume. The EPR spectra were taken from 1.6 K - 300 K with an X-band superheterodyne spectrometer operating at 9.2 GHz. The derivative EPR spectra which are presented were obtained by field modultion and subsequent lock-in detection. Typically the applied modulation field was 20 G and the incident microwave power to the cavity about 0.1 mW. The dc magnetic susceptibility, χ, was measured with a SQUID magnetometer[16] from 6 K - 300 K in magnetic fields of 50 G and 10 kG. Electrical resistivity measurements were performed from 2 K - 300 K at ~ 16 Hz using a 4 probe configuration in which Pt leads were attached to gold-plated spots on the sample using silver epoxy.[5]

RESULTS

A. Resistivity

For all samples, except those containing 3d impurities, the resistance showed a metallic character for $T > T_c$, and exhibited a sharp transition into the superconducting state, with the transition midpoint occurring at 93 ± 2 K. The width of the transitions was ~ 2 K.

Fig.(1) shows the effects on T_c due to substitution of the Cu ions with 3d impurity ions in $EuBa_2Cu_3O_{9-x}$. In Fig.(1a) we present ρ vs T data for

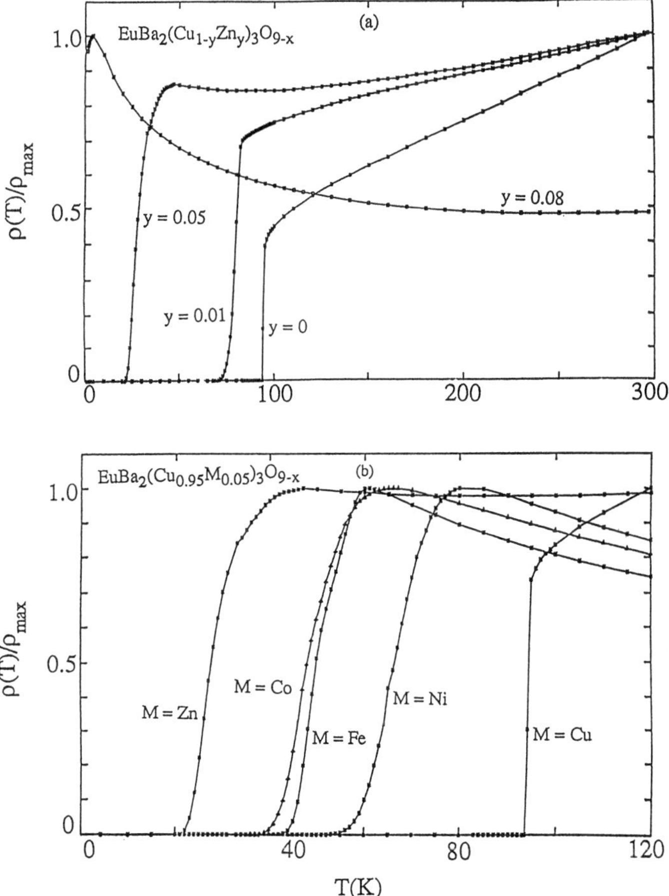

Fig.1: Normalized electrical resistivity as a function of temperature for: (a) $EuBa_2(Cu_{1-y}Zn_y)_3O_{9-x}$ with y = 0, 0.01, 0.05, and 0.08, (b) $EuBa_2Cu_3O_{9-x}$ and $EuBa_2(Cu_{0.95}M_{0.05})_3O_{9-x}$ with M = Fe, Co, Ni, or Zn.

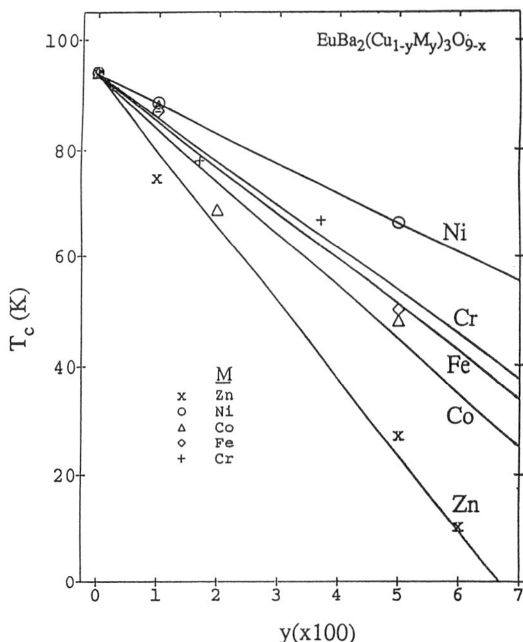

Fig.2:   Critical temperature, T_c, as a function of dopant
          concentration, y, for EuBa$_2$(Cu$_{1-y}$M$_y$)$_3$O$_{9-x}$ with
          M = Cr, Fe, Co, Ni, or Zn. The solid lines are
          drawn as a linear representation of the data
          points.

EuBa$_2$(Cu$_{1-y}$Zn$_y$)$_3$O$_{9-x}$ with $0 \leq y \leq 0.08$. As y increases, the behavior of
the resistivity becomes more semiconductor-like, and T_c shifts to lower

temperatures. X-ray powder diffraction shows that as y increases, the
perovskite-like structure also becomes less orthorhombic as indicated by
a decrease in the splitting of the appropriate peaks. The above behaviors
are typical of all 3d impurities substituted, although each 3d impurity
has a different effectiveness in reducing T_c [as shown in Fig.(1b)] and the
degree of orthorhombicity.

     In Fig.(2) we summarize the effects of the 3d ion substitutions on T_c
for EuBa$_2$(Cu$_{1-y}$M$_y$)$_3$O$_{9-x}$ with M = Cr, Fe, Co, Ni, and Zn. We have also sub-

stituted Mn ions for the Cu and qualitatively find that Mn is less effective
in suppressing T_c than the other 3d ions studied. However, we have not in-

cluded the Mn data in Fig(2) because the change of T_c appears to depend on the

method of sample preparation. Further investigations are underway in an
attempt to clarify the situation. The data of Fig.(2) indicate that there is
no correlation between the free ion magnetic moment of a 3d substitutional
impurity and its effectiveness in reducing T_c. This is in contrast to

standard BCS superconductors, in which magnetic impurities are particularly
effective in suppressing T_c. It is most striking that non-magnetic Zn, which

one would expect to be Zn^{2+}, is more effective in suppressing T_c than the
magnetic 3d ions.

B. Magnetic Susceptibility

     At low 3d ion concentrations (y ~ 0.01), the flux exclusion (Meissner
effect) was similar to that observed for the undoped and 4f doped samples.
For all undoped and 4f doped samples the flux exclusion (Meissner effect)

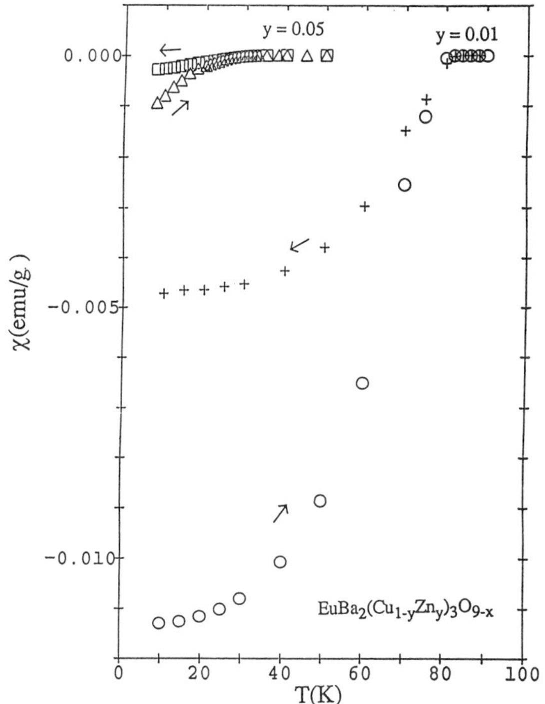

Fig.3: The magnetic susceptibility per gram as a function
of temperature, taken in a dc field of ~ 50 G, for
$EuBa_2(Cu_{0.99}Zn_{0.01})_3O_{9-x}$ and $EuBa_2(Cu_{0.95}Zn_{0.05})_3O_{9-x}$.
The arrows indicate that the data were taken by either
field cooling ($\leftarrow$) or zero field cooling ($\rightarrow$).

was found to be ~ 10 - 40% of the perfect diamagnetism value $-\frac{1}{4}\pi$, as
obtained assuming a macroscopic density of 6 g/cm^3. If the 3d dopant
concentration is increased, a large suppression in the diamagnetism
is observed. As shown in Fig.(3), the flux exclusion is reduced by
more than one order of magnitude when the concentration of Zn is increased from y = 0.01 to y = 0.05.

We have also measured the magnetization above T_c to estimate the
number of Bohr magnetons (μ_B) associated with the 3d elements which
replace the Cu ions. The values determined from our data are approximately 0, 1, 4, 5, and $5\frac{1}{2}$ μ_B for Zn, Ni, Mn, Fe, and Co, respectively.

C. Electron Paramagnetic Resonance

In taking the EPR spectra we typically swept the applied dc magnetic field, H, from ~ 50 G to 6.5 kG and back to ~ 50 G. Below T_c
the microwave impedance of most samples was strongly dependent on both
H and T, giving rise to a "baseline" signal that exhibited significant
variations and hysteresis as a function of dc field magnitude and
direction of sweep. This baseline signal is a consequence of the diamagnetic and hysteretic behavior of the magnetic susceptibility below T_c,
which is typical of Type II superconductors. We note that the field dependence of the sample impedance is such that for H in the vicinity of H_{c1}
(the lower critical field), the baseline can often have sharp changes which

mimic a Lorentzian type of signal. It is important to perform various tests,
such as sweeping H through the signal from both directions, to exclude these
other spurious signals.

In spite of the baseline signal, which is often relatively intense, for
most samples we were able to discern an anti-symmetric (A/B $\cong$ 1.0) Lorentzian
EPR signal below $\sim$ 40 K. We will refer to this EPR signal as the low temperature (LT) line. We have made an extensive study of the LT line, for a large
variety of samples, but we caution that we have not yet made an identification
as to the origins of this signal. In general, since large EPR signals can
arise from phases which might be present in amounts too small to be readily
identified by standard means (e.g., X-ray powder diffraction), it is
necessary to conduct additional experiments and tests to clarify the matter.
Several suitable tests are in progress, but we feel that it is appropriate
to describe the LT signals at this stage because if the LT line does originate from a foreign phase, a detailed knowledge of its properties will be
valuable in future sample preparations designed to ensure its absence.

The peak-to-peak linewidth, ΔH_{pp}, and field for resonance, H_R, of the
LT line are shown as a function of temperature in Fig.(4) for various
$RBa_2Cu_3O_{9-x}$ compounds. The LT line was observed to exhibit similar behavior

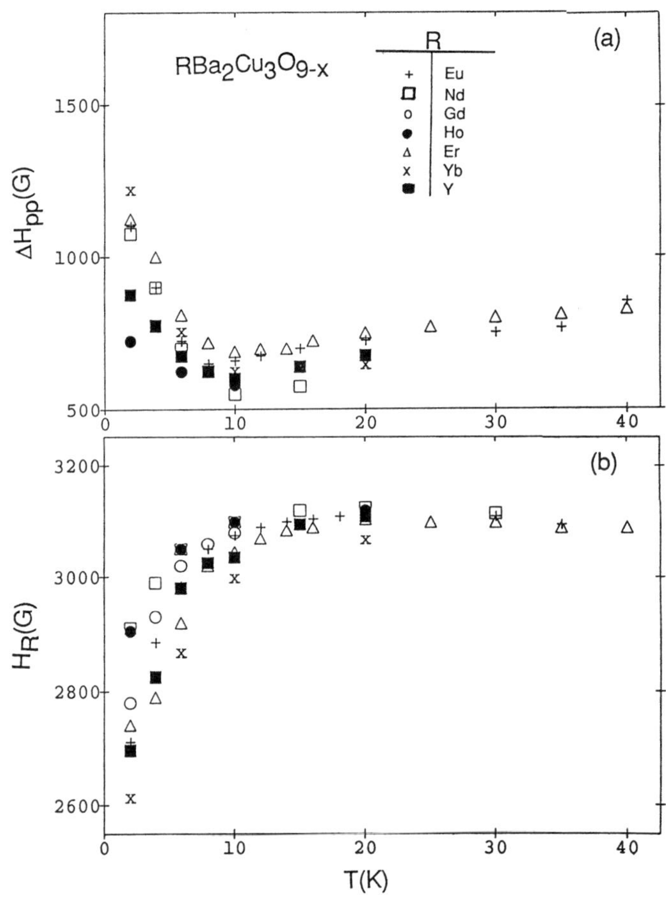

Fig.4: Properties of the low temperature (LT) EPR line as a
function of temperature for $RBa_2Cu_3O_{9-x}$ with R as in-
cated in the figure. (a) The peak-to-peak linewidth
ΔH_{pp}, and (b) the field for resonance, H_R, at a frequency of 9.2 GHz.

in $YBa_2Cu_3O_{9-x}$ and $EuBa_2Cu_3O_{9-x}$ samples doped with Er, and in most, but not all, of the $EuBa_2Cu_3O_{9-x}$ samples doped with Cr, Mn, Fe, Co, Ni, and Zn.[17] As the temperature is decreased from ~ 40 K, the LT line exhibits the following general propoerties: (a) the intensity (defined as the peak-to-peak signal amplitude, A, multiplied by $(\Delta H_{pp})^2$ continually increases down to the lowest temperatures studied (~ 1.6 K); (b) the EPR linewidth, ΔH_{pp}, decreases until ~ 15 K and then rapidly increases below ~ 10 K; (c) the field for resonance, H_R, at 9.2 GHz remains constant at ~ 3100 G down to ~ 15 K and then begins to decrease, with the decrease becoming more rapid at lower temperatures. It is likely that the large shifts in H_R below

~ 10 K are due to the internal rearrangement of an underlying spin system, and that the changes in linewidth are a reflection of the large shifts in H_R, which may be inhomogeneously felt throughout the sample. Further experiments to substantiate this interpretation are in progress and will be reported separately.

To determine if the LT line was only present in the superconducting state, we prepared samples which were non-superconducting by various means: (a) we prepared $PrBa_2Cu_3O_{9-x}$, which has the same crystal structure as the superconducting $RBa_2Cu_3O_{9-x}$ compounds, but is tetragonal instead of ortho-hombically distorted; (b) we substituted 8% of the Cu in $EuBa_2Cu_3O_{9-x}$ with Zn [the $EuBa_2(Cu_{0.92}Zn_{0.08})_3O_{9-x}$ sample of Fig.(1a)]; and (c) a few $RBa_2Cu_3O_{9-x}$ samples with R = Eu and Yb were annealed for several hours at 700°C in 1 atmosphere of Ar to partially deplete the oxygen. For each of the various non-superconducting samples prepared, the LT line was observed, but without the hysteretic baseline typically observed for the supercon-ducting samples. At 2 K the intensity of the LT line, $A(\Delta H_{pp})^2$, for the

non-superconducting samples was about an order of magnitude greater than for the superconducting samples. For the Ar annealed samples, the shifts in H_R and ΔH_{pp} below 10 K were found to be significantly reduced after the

anneal.

In addition to the LT line just described, we observed in many of the doped and undoped samples a partially resolved spectrum similar to that found for Cu^{2+} ions in an orthorhombic host, with g_1 ~ 2.27, g_2 ~ 2.12,

and g_3 ~ 2.05. When present, this spectrum was observable over the full temperature range (2 K - 300 K). We will refer to this as the high tem-perature (HT) spectrum. In most Eu-based samples the intensity of the HT spectrum was either relatively weak or not observable. In Y-based samples, however, the HT spectrum intensity was typically much greater. The inten-sity of the HT spectrum varied greatly from sample to sample, even though most samples were characterized by X-ray diffraction as having only the oxygen deficient, perovskite-like structure present. In some samples the HT spectrum intensity was observed to increase with time. Finally, samples made by procedure B rarely gave an observable HT spectrum, whereas samples made by procedure A frequently exhibited an observable HT spectrum. These observations indicate that the HT spectrum very likely arises from a spurious phase in amounts small enough ($\leq$ 2%) to escape detection by X-ray diffraction. We have attempted to identify this spurious phase without success.[17]

For the undoped $RBa_2Cu_3O_{9-x}$ samples with R = Nd, Pr, Ho, Y, Eu, Er, we did not observe any EPR signals other than those already discussed. However, for R = Yb we did observe an additional, weak EPR line at 77 K with ΔH_{pp} = ~ 1000 G and a g-value of about 3.3 which may correspond to one of the doublet states for Yb^{3+}. We have also observed an additional EPR signal for R = Gd which we attribute to the Gd^{3+} ion. The Gd^{3+} signal is observable from 300 K down to ~ 20 K where it begins to be obscured by the LT line. The linewidth, amplitude, and field for resonance are roughly temperature independent above T_c. At T_c a sharp anomaly occurs in the linewidth and amplitude. The amplitude decreases by ~ 50% and ΔH_{pp} decreases by ~ 500 G upon entering the superconducting state. These decreases take place over a temperature interval of a few degrees K.

We have also initiated investigations in $Y_{1-z}R'_zBa_2Cu_3O_{9-x}$ and $Eu_{1-z}R'_zBa_2Cu_3O_{9-x}$ with R' = Gd or Er, and $z \leq 0.05$. In brief, for R' = Er we observe, in addition to the LT line, an EPR signal with a g-value of ~ 7.6 which is possibly related to one of the eight Er^{3+} doublets. The line was observed for $T \leq 60$ K so we have not been able to determine the effect of the superconducting transition on its properties. For R' = Gd, a complicated EPR spectra has been observed at temperatures both above and below T_c. The intensity of the spectra is stronger at lower temperatures. The spectra consists of several peaks which are presumably due to crystal field splitting of the ground state of Gd. Further investigations of these interesting spectra are in progress and the results will be presented elsewhere.

We did not observe any additional EPR signals in the $EuBa_2(Cu_{1-y}M_y)_3O_{9-x}$ compounds with M = Cr, Fe, Co, Ni, or Zn. For M = Mn, we have observed an additional EPR line in some samples, but not in others. It remains to be determined whether this sample-dependent signal derives from a spurious phase, or is characteristic of Mn ions substituting at the Cu sites.

SUMMARY

In summary, we have studied a wide variety of high T_c oxide superconductors, either undoped, or doped with various 3d or 4f impurities. The 4f impurities have no significant effect on T_c or the diamagnetism below T_c for these oxide superconductors. However, T_c and the diamagnetism are both reduced with increases in 3d impurity concentration. Non-magnetic Zn is found to be the most effective of the 3d impurities studied in this respect. A low temperature (LT) EPR signal was observed in the superconducting state below ~ 40 K for most of the oxide superconductors. We repeat our reservations stated earlier, that we have not yet identified the observed LT signal with the $RBa_2Cu_3O_{9-x}$ phase. As the temperature was lowered the LT signal intensity monotonically increased. The field for resonance was constant down to ~ 15 K then decreased at lower temperatures. The peak-to-peak linewidth, ΔH_{pp}, exhibited a minimum at ~ 10 - 15 K, and rapidly increased below 10 K. Similar LT line behavior was observed in 3 types of non-superconducting samples, but the LT line intensity was about an order of magnitude greater at 2 K than for superconducting samples. A partially resolved spectrum (HT) similar to that found for Cu^{2+} ions in an orthorhombic host was observed for many oxide superconductor samples. It likely arises from the presence of small amounts of another phase and is dependent on the method of sample preparation. We have observed additional EPR signals attributed to

Gd, Er, Yb, and Mn moments in various samples in which these moments are
present. Further investigations are in progress to determine the origins
of the LT and HT EPR signals, and to further characterize the signals observed in samples containing Gd, Er, Yb, and Mn moments.

The authors wish to acknowledge, with thanks, the technical assistance
provided by Mr. Roger Isaacson, University of California, San Diego, with
the EPR spectrometer.

REFERENCES

*   Research supported by NSF-DMR-86-13858, US DOE DE-FG03-86ER45230, and
     US DOE, Los Alamos National Laboratories.
1.   J. G. Bednorz and K. A. Müller, Z. Phys.,B64, 189 (1986).
2.   M. K. Wu, J. R. Ashburn, C. J. Torng, P. H. Hor, R. L. Meng, L. Gao,
     Z. J. Huang, Y. Q. Wang, and C. W. Chu, Phys. Rev. Lett., 58, 908 (1987).
3.   Z. Zhongxian, C. Liquan, Y. Qiansheng, H. Yuzhen, L. Jinxiang,
     C. Genghua, T. Ruming, L. Guirong, C. Changgeng, C. Lie, W. Lianzhong,
     G. Shuquan, L. Shanlin, and B. Jianqing, to be published in Kexue
     Tongbao, No. 6 (1987).
4.   Z. Fisk, J. D. Thompson, E. Zirngiebl, J. L. Smith, and S. N. Cheong,
     to be published in Sol. St. Commun.
5.   K. N. Yang, Y. Dalichaouch, J. M. Ferreira, B. W. Lee, J. J. Neumeier,
     M. S. Torikachvili, H. Zhou, M. B. Maple, and R. R. Hake, to be
     published in Sol. St. Commun.
6.   W. R. McKinnon, J. M. Tarascon, L. H. Greene, and G. W. Hull, to be
     submitted.
7.   E. M. Engler, V. Y. Lee, A. I. Nazzal, R. B. Beyers, G. Lim, P. M.
     Grant, S. S. P. Parkin, M. L. Ramirez , J. E. Vazquez, and R. J. Savoy,
     to be published in J. Am. Chem. Soc.
8.   B. W. Veal, W. K. Kwok, A. Umezawa, G. W. Crabtree, J. D. Jorgensen,
     J. W. Downey, L. J. Nowicki, A. W. Mitchell, A. P. Paulikas, and
     C. H. Sowers, submitted to Appl. Phys. Letters.
9.   M. A. Beno, L. Soderholm, D. W. capine, II, D. G. Hinks, J. D. Jorgensen,
     I. K. Schuller, C. U. Segre, K. Zhang, J. D. Grace, submitted to
     Appl Phys. Letters.
10. Myung-Hwan Whangbo, Michel Evain, Mark A. Beno, and Jack M. Williams,
     submitted to Inorganic Chem.
11. S. B. Oseroff, B. Gehman, S. Schultz, C. Rettori, Phys. Rev. Lett.,
     35, 679 (1975).
12. S. B. Oseroff, M. Passeggi, D. Wohlleben, S. Schultz, Phys. Rev. B,
     15, 1283 (1977).
13. D. Davidov, C. Rettori, and H. M. Kim, Phys. Rev.B, 9, 147 (1974).
14. K. Baberschke, Z. Phys. B, 24, 53 (1976).
15. C. Rettori, D. Davidov, P. Chaikin, R. Orbach, Phys. Rev. Lett.,
     30, 437 (1973).
16. BTI Corporation, 4174 Sorrento Valley Blvd., San Diego CA 92121.
17. S. B. Oseroff, D. C. Vier, J. F. Smyth, C. T. Salling, S. Schultz,
     Y. Dalichaouch, B. W. Lee, M. B. Maple, Z. Fisk, J. D. Thompson,
     J. L. Smith, and E. Zirngiebl, accepted for publication in Sol. St.
     Commun..

UPPER CRITICAL FIELD MEASUREMENTS FOR

$RBa_2Cu_3O_7$ (R = Y,Eu$_{.9}$Y$_{.1}$,Sm)

A. P. Ramirez*, B. Batlogg*, R. J. Cava, L. Schneemeyer,
R. B. van Dover, E. A. Rietman and J. V. Waszczak

AT&T Bell Laboratories, Murray Hill, New Jersey 07974

ABSTRACT

We have measured the upper critical fields for superconducting $RBa_2Cu_3O_7$ (R=Y,Eu$_{.9}$Y$_{.1}$,Sm) up to 27 Tesla, using the midpoint, T_m, of the resistance transition. At the highest fields we find critical field slopes of $|dH_{c2}/dT_m|$ = 2.5, 2.1, and 3.3 Tesla/K for R = Y, Eu$_{.9}$Y$_{.1}$ and Sm, respectively, indicating insensitivity of pairing to local moment magnetism in these compounds. The normal state magnetoresistance is extremely small ($<10^{-2}$).

EXPERIMENTAL

The material used for this study was prepared by ceramics procedures, as described in detail in Ref. 1. Thin bars were shaped and Pt or Au electrical leads attached either with indium or silver epoxy. The resistance was measured by either ac or dc techniques with typical currents of ~1 mA, perpendicular to the applied field. Measurements below 12 Tesla were performed using a superconducting solenoid. The measurements up to 27 Tesla were obtained using the superconducting-Bitter hybrid magnet at Francis Bitter National Magnet Laboratory. For these measurements around 90K, a Pt thermometer was used, the magnetoresistance of which was corrected for following Sample, Brandt, and Rubin (SBR)[2], and using our magnetoresistance at 77K. The corrections involved amount to no more than 1.5% of the temperature in the region of interest with an estimated uncertainty of correction of 10-20% and are considered small for the present requirements.

DISCUSSION

In Fig. 1 we show resistance data for $YBa_2Cu_3O_7$ in fields up to 19T. We note the absence of any noticeable magnetoresistance for the material in the normal state. For instance, for $SmBa_2Cu_3O_7$ we can put an upper limit of ~1% on the magnetoresistance at B = 24 Tesla and T = 100K. The fact that no magnetoresistance greater than 1% is observed up to 24 Tesla raises questions about mechanisms of superconductivity [3] in which pairing is due to magnetic interactions among the conduction electrons.

In Fig. 2 is shown the behavior of the transition temperatures, T_m, here are defined graphically in Fig. 1 to be the intersection of a given R vs. T curve with the line halfway between the extrapolated normal state resistance and zero resistance. The value $|dH_{c2}/dT_m|$ = 2.5 T/K for $YBa_2Cu_3O_7$ is significantly higher than that of 1.6 T/K reported for a similar compound $Y_{1.2}Ba_{1.8}Cu_3O_{6.6}$ in the same field region [4]. We also note the appearance of low field curvature in H_{c2} versus T_m, a feature which might also be associated with the low-dimensional nature of these materials [5]. The present values for the critical fields lead to revised [1] estimates for the Ginzberg-Landau coherence length $\xi(0) \sim 16\text{\AA}$ and specific heat coefficient $\gamma = 8 \pm 2$ mJ/mole(Cu)$-K^2$. The latter is in good agreement with recent direct measurements of the specific heat jump at T_c [6], and adds additional evidence for the empirically-based conjecture [7] that the new oxide compounds form a new class of superconductors with perhaps a novel microscopic pairing mechanism.

Although the zero field T_c of the Sm compound is not significantly different from the Y or $Eu_{.9}Y_{.1}$, the slope dH_{c2}/dT_m is steeper, approaching the value -3.3 Tesla/K in the region 15-27 Tesla. Extrapolating this linearly leads to a value of 47 Tesla at 77K. Sm^{3+} is a Kramers ion with ground state $J = 5/2$, and $g = 2/7$ and antiferromagnetic ordering is known to take place in the compound at 0.6 K [8]. One can speculate that the high critical field slopes might in fact be due to a reduction of internal field for the conduction electrons due to antiferromagnetic correlation is among the local moments.

We would like to thank C. M. Varma for many helpful discussions, and B. Brandt and L. Rubin for much assistance in the measurements carried out at the MIT Francis Bitter National Magnet Laboratory.

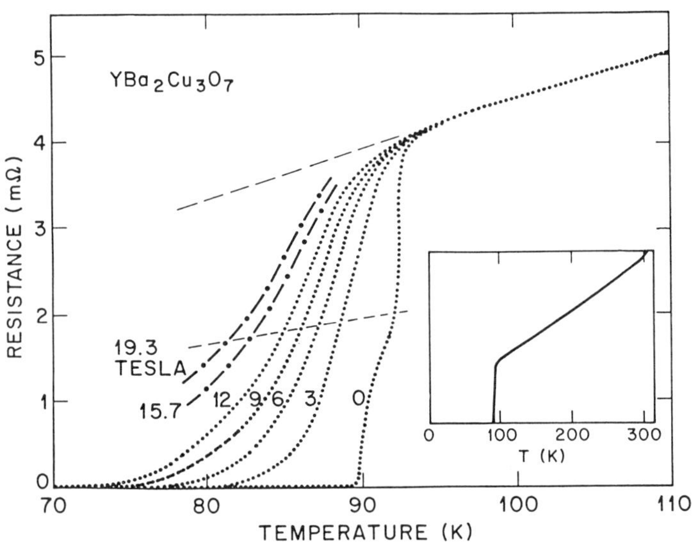

Fig. 1.   Resistance versus temperature of $YBa_2Cu_3O_7$ in the region of the superconducting transition temperature, at several values of applied magnetic field. Note the absence of observable magnetoresistance (<1%) in the normal state. The midpoints of the transition are shown as the intersection of the lower dashed line and an individual resistance curve. The inset shows the zero field resistance over a broader temperature range.

690

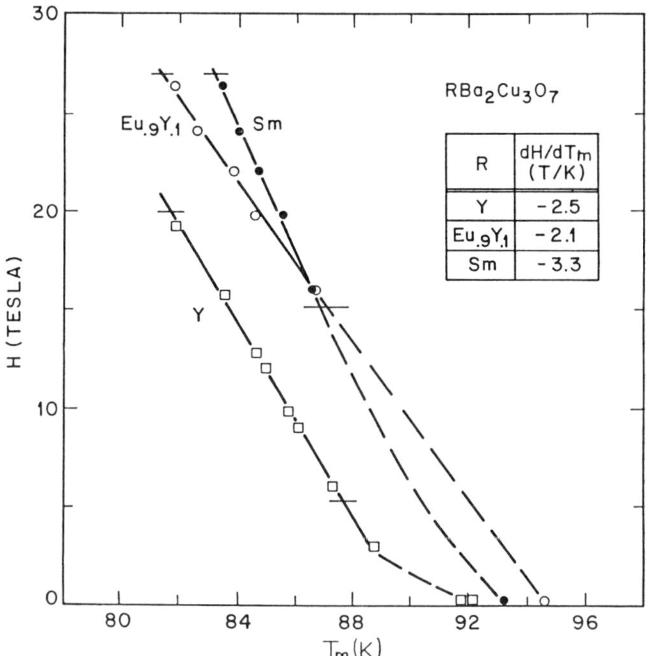

Fig. 2. The magnetic field dependence of the midpoints, T_m, as described in Fig. 1, for three different members of the $RBa_2Cu_3O_7$ family. The table inset gives the limiting dH/dT_m in the high field region, as denoted by the short horizontal bars on the lines drawn through the data.

REFERENCES

*visiting scientists at the MIT Francis Bitter National Magnet Laboratory

1. R. J. Cava, B. Batlogg, R. B. van Dover, D. W. Murphy, S. Sunshine, T. Siegrist, J. P. Remeika, E. A. Rietman, S. Zahurak, and G. P. Espinosa, Phys. Rev. Lett. 58, 1676 (1987).

2. H. H. Sample, B. L. Brandt, and L. G. Rubin, Rev. Sci. Instr. 53, 1129 (1982).

3. P. W. Anderson, Science 235, 1196 (1987).

4. N. Kobayashi, T. Sasaoka, K. Oh-ishi, M. Kikuchi, M. Furuyama, T. Sasaki, K. Noto, Y. Syono, and Y. Muto, unpublished.

5. B. J. Dalrymple and D. E. Prober, J. Low Temp. Phys. 56, 545 (1984).

6. S. Tanaka et al., Mat. Res. Soc. April 1987 Conf., Anaheim.

7. B. Batlogg, A. P. Ramirez, R. J. Cava, R. B. van Dover, and E. A. Rietman, Phys. Rev. B 35, 5340 (1987).

8. D. J. Bishop, A. P. Ramirez, P. L. Gammel, B. Batlogg, L. F. Schneemeyer, and J. V. Waszczak, to be published.

Cu-0 SUPERCONDUCTORS: THROUGH A LENS, BUT DARKLY

Joseph Orenstein, G. A. Thomas, D. H. Rapkine,
C. G. Bethea, B. F. Levine, R. J. Cava, A. S. Cooper,
D. W. Johnson, Jr., J. P. Remeika, and E. A. Rietman

AT&T Bell Laboratories
Murray Hill, NJ 07974

The shock of superconducting transition temperatures (T_c's) greater than 25 K in Cu-O based compounds[1] has loosed a tidal wave of research. We have involved ourselves in a study of the optical properties of these materials, with particular emphasis on $La_{2-x}Sr_xCuO_4$ compounds. Recently we reported [2] reflectivity, R, spectra of $La_{2-x}Sr_xCuO_4$ for x = 0 and 0.175, and of $Ba_2YCu_3O_{9-\delta}$ ($\delta \sim 2.1$) over a broad range of photon energy. The most striking observation was the profound deviation from the expected Drude-like R in the metallic compounds. Fitting R required that $\epsilon(\omega)$ be dominated by a transition at energy 0.45 eV in $La_{1.825}Sr_{.175}CuO_4$ and 0.65 eV in $Ba_2YCu_3O_{9-\delta}$. From our fit we determined that the integrated oscillator strength in the 0.45 eV feature in $La_{1.825}Sr_{0.175}CuO_4$ corresponded to a value of N_{eff} m/m^* of $2.5\times10^{21}cm^{-3}$. This density agrees well with the density of positive carriers (assuming one hole per Sr atom) of $1.9\times 10^{21}cm^{-3}$. Similar conclusions for $La_{1.825}Sr_{0.175}CuO_4$ were reached by Herr et al.[3] based upon Kramers-Kronig (KK) analysis of their R data.

In this work we report measurements of R for single crystal samples of La_2CuO_4 (c-La) and pressed pellets of ceramic $La_{2-x}Sr_xCuO_4$ (LaSr) for x = 0, 0.04, 0.075, 0.175, 0.225. A plot of R vs log ν for c-La is shown in Fig. 1. The samples were cube-like objects approximately 2 mm on a side, which were polished as previously described.[4] Cutting and polishing revealed small holes due to flux inclusions which covered approximately 10% of the reflecting surface. The spectrum labelled $\perp\vec{c}$ was recorded with unpolarized light incident nearly normal to the ab plane of the crystal. The "$||c$" spectrum was taken with light polarized so that the electric field ($\vec{E}$) was parallel to $\hat{c}$ (perpendicular to the ab plane). R is quite different for the two directions, as expected from the crystal structure.[5] R for $\vec{E}$ $||$ $\hat{c}$ is almost constant for $\nu > 600cm^{-1}$, indicating that ϵ_2 (or σ) is near zero in this frequency range.

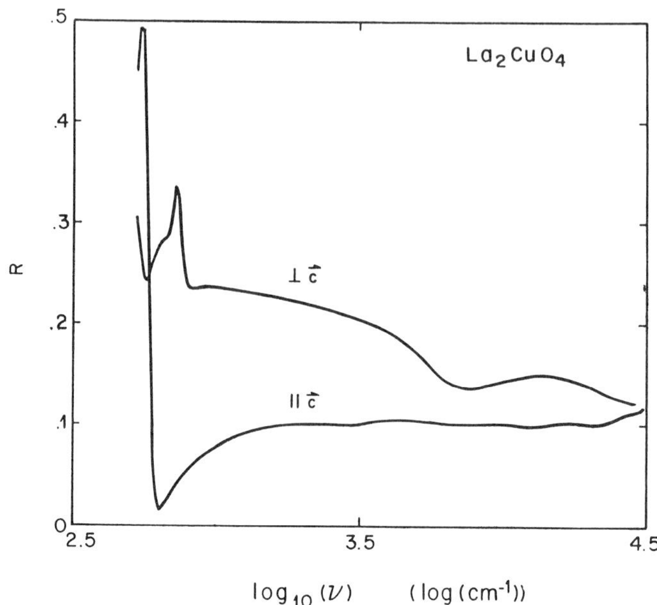

Fig. 1   Reflectivity as a function of log wave number in cm^{-1} for a single crystal of La$_2$CuO$_4$ with the optical electric field aligned along the c-axis and perpendicular to it.

Below 600cm^{-1} there is a sharp dip and rise characteristic of a phonon in an insulating ionic crystal. The spectrum closely resembles data recently reported[6] for La$_2$NiO$_4$. It is clear that R of ceramic pellets for $\nu < 600$cm^{-1} will be strongly influenced by grains aligned so that $\vec{E}$ is nearly parallel to $\hat{c}$. KK analysis in this frequency range will be misleading. Above 600 cm^{-1} where R for $\vec{E} \parallel \hat{c}$ is small and featureless, KK analysis of pressed pellets should be qualitatively correct with somewhat reduced values of $\epsilon(\omega)$ compared with data on single crystals.

The optical conductivity, σ, obtained by KK analysis of the $\vec{E}\perp\hat{c}$ spectrum is shown on a linear scale of energy E in Fig. 2. The derived spectrum of σ vs E was found to be insensitive to the extrapolation to $\nu = 0$, but sensitive to the extrapolation to $\nu = \infty$. In Fig. 2, R was extrapolated at high frequencies according to the relation R $= R_0 + (R_u - R_0)\omega_u^4/\omega^4$, where $R_0 = 0.1$ and ω_u is the upper limit of our experimental range (3.7eV) and R_u is the corresponding value of R. Substituting $R_0 = 0$ yielded a higher σ for energies above about 2 eV by about a factor of two but had little effect at lower energy.

Two transitions are readily observable in this spectrum with peaks at 0.60 eV and at E $>$ 3 eV. To evaluate the oscillator strengths of these features we follow Herr et al. [3] and calculate the quantity f where

$$f(E) \equiv N_{eff}m/N_{Cu}m^* = 2m/\pi e^2 N_{Cu} \int_0^E \sigma d\omega.$$

The significance of f is that (if m $=$ m*) it equals the fraction of electrons which contribute to σ below energy E, normalized to one electron per Cu atom ($N_{Cu} = 1.1 \times 10^{22}$cm^{-3}). From numerical integration of $\sigma(\omega)$ we find f(1 eV) $=$

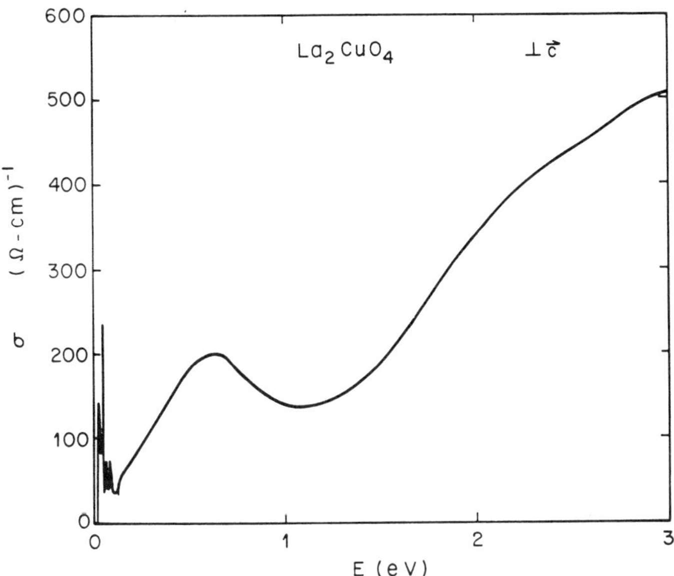

Fig. 2   Conductivity as a function of photon energy in eV for a single crystal of La_2CuO_4 with the optical electric field aligned in the relatively highly conducting a-b crystallographic plane.

0.025 and f (3 eV) = 0.13. We conclude on the basis of these data that c-La is a semiconductor with a large bandgap $E_{gap} \sim$ 2-3 eV. This is incompatible with weak coupling theories of SDW or CDW formation. Rather, it seems that the scale of E_{gap} is set by the short range Coulomb interaction.

The peak we previously reported in LaSr at 0.45 eV is present, but smaller in the nominally undoped c-La (Figs. 1 and 2). In order to study the evolution of this peak with Sr concentration we measured R for ceramic samples with x = 0, 0.04, 0.075, 0.175, 0.225. The results are plotted vs log ν in Fig. 3. The variation of R with x is striking: for $\lesssim$ 1 eV, R increases with Sr concentration until x = 0.175 and then decreases in the sample with x = 0.225. At higher energies, the trend in R is reversed indicating that oscillator strength is transferred from the transition at 2 eV to the 0.45 eV transition with increasing x. This trend is illustrated more clearly in Fig. 4 which shows σ vs E obtained from KK analysis as before. The peak in σ at $\sim$ 0.5 eV increases with x until 0.175 and then decreases. Fig. 5 shows the f(E) vs E obtained by numerical integration of σ. We define the integrated oscillator strength in the 0.45 eV peak by f(1 eV) and plot this quantity vs. x in Fig. 6. The oscillator strength peaks at x = 0.175, the Sr concentration previously identified[4] as possessing the largest superconducting volume fraction.

The transition which peaks at 0.45 eV in LaSr (and at $\sim$ 0.65 eV in $Ba_2YCu_3O_{9-\delta}$) accounts for essentially all of the oscillator strength of carriers in these compounds. This oscillator strength would appear in the Drude mode in a conventional metal. In addition, there appears to be a direct relation between the strength of this transition and the superconductivity. It is important, therefore, to identify the nature of this transition. In an earlier paper we identified a dipole-active transition which would appear out of a resonating valence bond $(RVB)^-$ ground state of the Hubbard model. In the presence of unoccupied sites (Cu^{+++}) a transition is possible which breaks a bond and

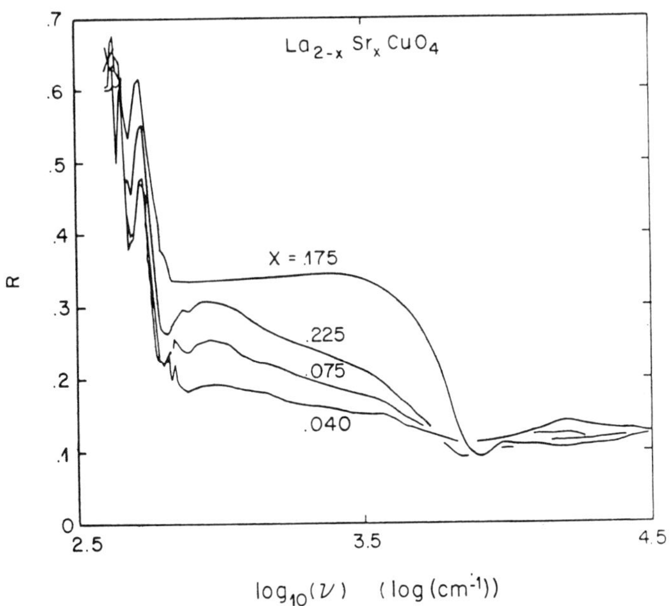

Fig. 3   Reflectivity as a function of log wave number in cm^{-1} for a series of polycrystalline pressed pellet samples of La$_{2-x}$Sr$_x$CuO$_4$, with the values of x as labelled.

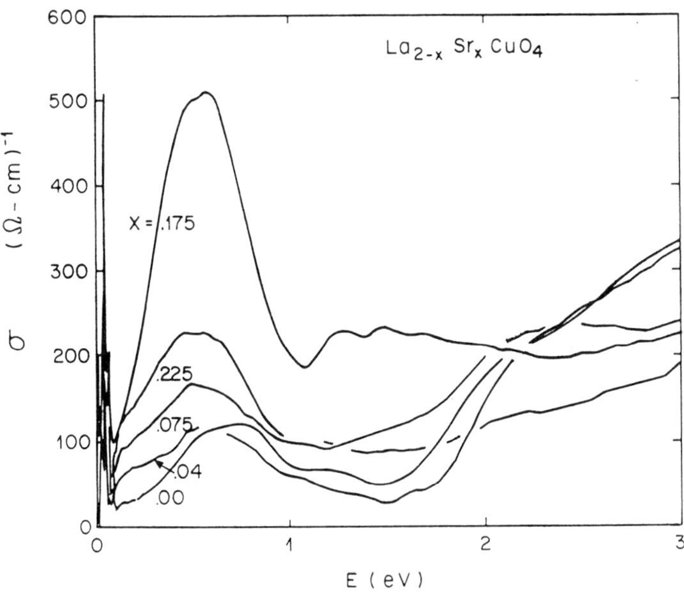

Fig. 4   Optical conductivity as a function of photon energy in eV for a series of polycrystalline pressed pellet samples of La$_{2-x}$Sr$_x$CuO$_4$ with different values of x as labelled. These data are obtained from reflectivity measurements, including those shown in Figures 1 and 3, using a Kramers-Kronig transformation.

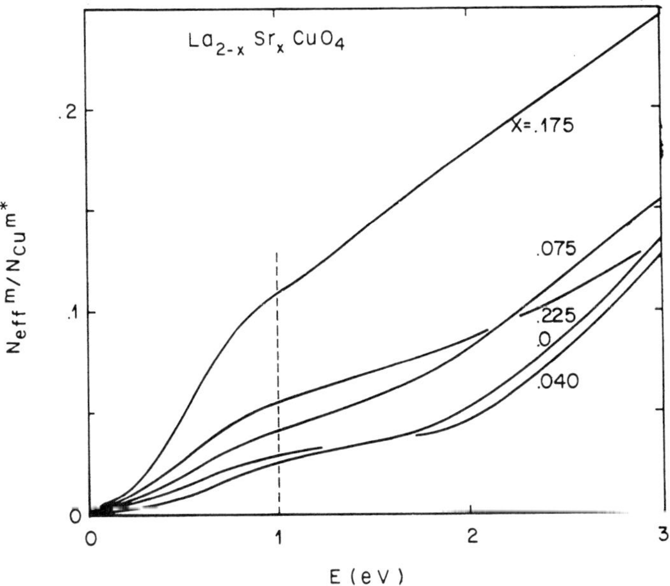

Fig. 5   Normalized effective number of carriers contributing to the optical absorption as a function of photon energy E for a series of polycrystalline pressed pellet samples of La$_{2-x}$Sr$_x$CuO$_4$ for different values of x as labelled. These curves are obtained from an integral of the conductivity as shown in Figure 4.

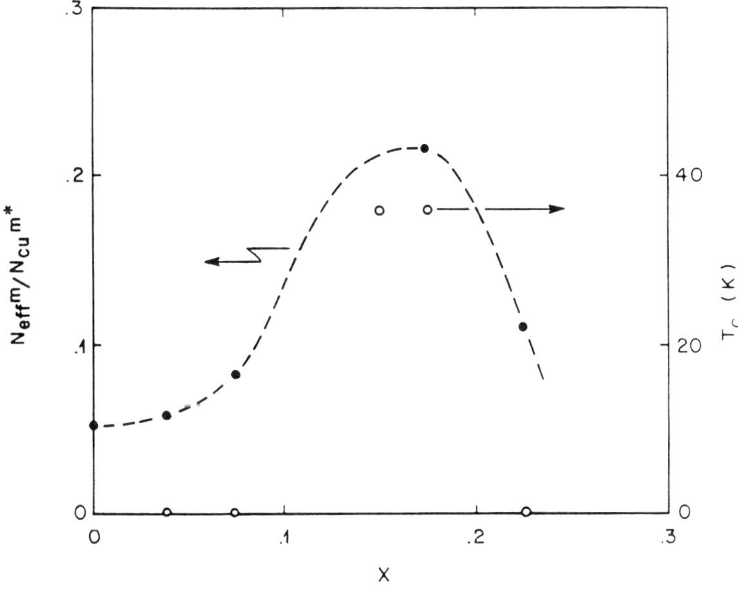

Fig. 6   Normalized effective number of carriers up to 1 eV as a function of Sr concentration x for a series of La$_{2-x}$Sr$_x$CuO$_4$ samples. This number of carriers primarily represents the strength of the conductivity feature near .5 eV.

transfers a formerly bonded electron to a Cu^{+++}. This would have energy $\sim t^2/U$ which is ~ 0.1 eV if $t\sim 0.5$ eV and $U\sim 2$-3 eV. This is somewhat smaller than the energy we observe. Varma et al.[8] have suggested that the superexchange formalization leading to a Hubbard model is inadequate for Cu-O superconductors because the charge transfer exciton energy, Δ, (required for $Cu^{++}O^{--} \rightarrow Cu^{+}O^{-}$) is smaller than the Hubbard U. They suggest that this transition rather than $Cu^{++}Cu^{++} \rightarrow Cu^{+}Cu^{+++}$, is the band gap transition seen at $E \sim 2$ eV. Recently Varma (private communication) has suggested that the 1/2 eV peak results from the transition $Cu^{+++}O^{--} \rightarrow Cu^{++}O^{-}$, and further that the polarization resulting from virtual excitations of this form provide an effective pairing mechanism.

In summary, we have measured $\sigma(\omega)$ in c-La and a series of ceramic samples of LaSr for several values of x. We find that c-La is an anisotropic (quasi-2 D) semiconductor with a gap of ~ 2 eV. A peak in $\sigma(\omega)$ is observed in the gap at 0.6 eV. Measurements of R in ceramic samples show that this peak initially grows with x until the optimum value for superconductivity is reached near $x = 0.175$ and then decreases. The strong correlation of the oscillator strength with T_c may provide a clue to the pairing mechanism in these compounds.

REFERENCES

1. J. G. Bednorz and K. A. Muller, Z. Phys. B *64*, 189 (1986).

2. J. Orenstein, G. A. Thomas, D. H. Rapkine, C. B. Bethea, B. F. Levine, R. J. Cava, E. A. Rietman, and D. W. Johnson, Phys. Rev. B, in press.

3. S. L. Herr, K. Kamaras, C. D. Porter, M. G. Doss, D. B. Tanner, D. A. Bonn, J. E. Greedan, C. V. Stager and T. Timusk, Phys. Rev. B, in press; D. A. Bonn, J. E. Greedan, C. V. Stager, T. Timusk, M. Doss, S. Herr, K. Kamaras, C. Porter, D. B. Tanner, J. M. Tarascon, W. R. McKinnon and L. H. Greene, Phys. Rev. B, in press.

4. R. J. Cava, R. B. van Dover, B. Batlogg and E. A. Rietman, Phys. Rev. Lett. *58*, 408 (1987).

5. R. J. Cava, R. B. Batlogg, R. B. van Dover, D. W. Murphy, S. Sunshine, T. Siegrist, J. P. Remeika, E. A. Rietman, S. Zahurak, and G. P. Espinosa, Phys. Rev. Lett. *58*, 1676 (1987).

6. J.-M. Bassat, P. Odier, and F. Gervais, Phys. Rev. B *35*, 7126 (1987). in press.

7. P. W. Anderson, Science *235*, 1196 (1987). S. A. Kivelson, D. S. Rokhsar, and J. P. Sethna, to be submitted.

8. C. M. Varma, S. Schmitt-Rink, and E. Abrahams, Solid State Comm. *62*, 681 (1987).

SINGLE CRYSTAL SUPERCONDUCTING $Y_1Ba_2Cu_3O_{7-x}$ OXIDE FILMS BY MOLECULAR BEAM EPITAXY

J. Kwo, M. Hong, R. M. Fleming, T. C. Hsieh, S. H. Liou, and B. A. Davidson

AT&T Bell Laboratories
Murray Hill, N.J. 07974

Introduction

Recent breakthrough of superconductivity over 90 K in the Cu oxide based perovskite-type materials is of immense importance in both science and technology.[1,2] The superconducting oxide crystals prepared in thin film form are vital for fundamental physical studies as well as for device applications.[3,4] We report here the structural and superconducting properties of single crystal $Y_1Ba_2Cu_3O_{7-x}$ films[5] prepared by molecular beam epitaxy (MBE) developed for thin film semiconductor technology. The films are epitaxial crystals grown on $SrTiO_3$ (100) face with an orientation of a axis perpendicular to the film plane. Sharp resistive superconducting transitions with T_c (R=0) above 77 K have been reproducibly achieved. The critical current densities are 10^3 A/cm^2 at 77 K, 5×10^4 A/cm^2 at 70 K, and $\sim 10^6$ A/cm^2 at 4.2 K.

MBE Growth

The superconducting films were prepared by the metal molecular beam epitaxy technique in a versatile ultrahigh vacuum deposition system previously described.[6] The films reported here were thermally coevaporated from three separate sources simultaneously; Y from an electron beam heated evaporator, and Ba and Cu from the effusion cells. The oxygen incorporation in the film was made by introducing a molecular oxygen overpressure about 1×10^{-5} torr near the substrate. This pressure is about one order of magnitude higher than the required oxygen flux at the current growth rate of 1.0 Å/sec. The background pressure is less than 2×10^{-10} torr prior to oxygen doping, and is maintained in the low 10^{-7} torr range during growth. For most depositions the growth temperature is kept at 400 $\pm$ 50 C. The total film thickness is about 9000 Å.

The substrates used in this study are primarily of single crystal $SrTiO_3$ (100) face.[7-9] The in-plane lattice constant of this cubic structure is 3.9051 Å, and is within 0.4% match with the lattice spacing of b axis or 1/3 of c axis of the orthorhombic $Y_1Ba_2Cu_3O_{7-x}$ phase. Judging from in-situ high energy electron diffraction (RHEED) pattern, the as-grown film over thousands of angstrom thick remains to be amorphous or microcrystalline. The superconducting $Y_1Ba_2Cu_3O_{7-x}$ phase is formed by further heat treatment in an oxygen flow at 870-900 C for 1 hour, then slowly furnace cooled.

Structural Properties

The chemical composition was determined by the Rutherford backscattering (RBS) technique. Our findings are that the film with the composition closest to stoichiometry has the highest superconducting transition, T_c (R=0). RBS studies also reveal some interdiffusion occuring near the interface between the substrate and the deposits due to high temperature heat treatment.

The crystal structure was analyzed by X-ray diffraction. The sample is of predominantly ($\sim 99\%$) the superconducting $Y_1Ba_2Cu_3O_{7-x}$ phase, with a very small amount of $BaCuO_2$ phase. An X-ray $\theta-2\theta$ scan is shown in Figure 1. The oxygen deficient perovskite $Y_1Ba_2Cu_3O_{7-x}$ phase is grown in the orientation of a axis perpendicular to the film plane, as opposed to the c axis orientation obtained on similar substrates reported recently.[9] The full width at half maximum (FWHM) of the rocking

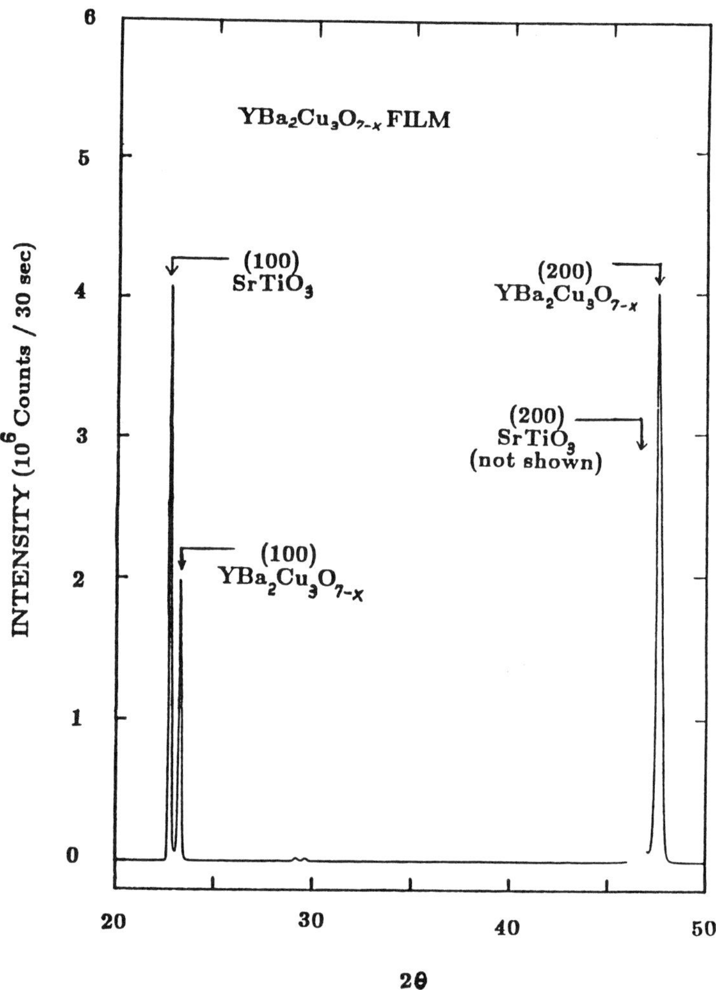

Figure 1. X-ray $\theta-2\theta$ scan for a MBE grown oriented film with a composition of $Y_{0.91}Ba_{2.0}Cu_{3.0}O_{7-x}$ deposited on $SrTiO_3$ (100).

curve of a axis is 0.82°, compared to 0.13° of the $SrTiO_3$ substrate. In the film plane, b and c axes are oriented with a rocking curve of similar magnitude $\sim$0.82°. The correlation length is about 200Å. The lattice spacings were determined from a least square fit among 16 reflections collected from radial scans through (h00), (20ℓ), and (21ℓ). They are a = 3.821(6), b = 3.900(6), and c = 11.68(2)Å, in agreement with the values determined for a stoichiometric, orthorhombic $Y_1Ba_2Cu_3O_{7-x}$ ceramic sample by neutron diffraction.[10]

Superconducting Properties

The superconducting transitions were measured resistively by the standard 4-point measurement using the ac method. Films grown on the $SrTiO_3$ substrates show reproducibly T_c (R=0) above 77 K, and the transition widths are much sharper than those deposited on other types of substrates like Al_2O_3 or MgO. Figure 2 is resistivity vs temperature of one sample with nearly perfect stoichiometry of $Y_{0.91}Ba_{2.0}Cu_{3.0}O_{7-x}$. The room temperature resistivity is $\sim$ 750 $\mu\Omega$–cm, and the resistivity ratio R(300K)/R(90K) is 2.5. The resistive T_c onset is 90 K, and the 10% and 90% transitions are 84-83 K, respectively. The fully zero resistance state is reached at 82 K. No resistance tail was seen.

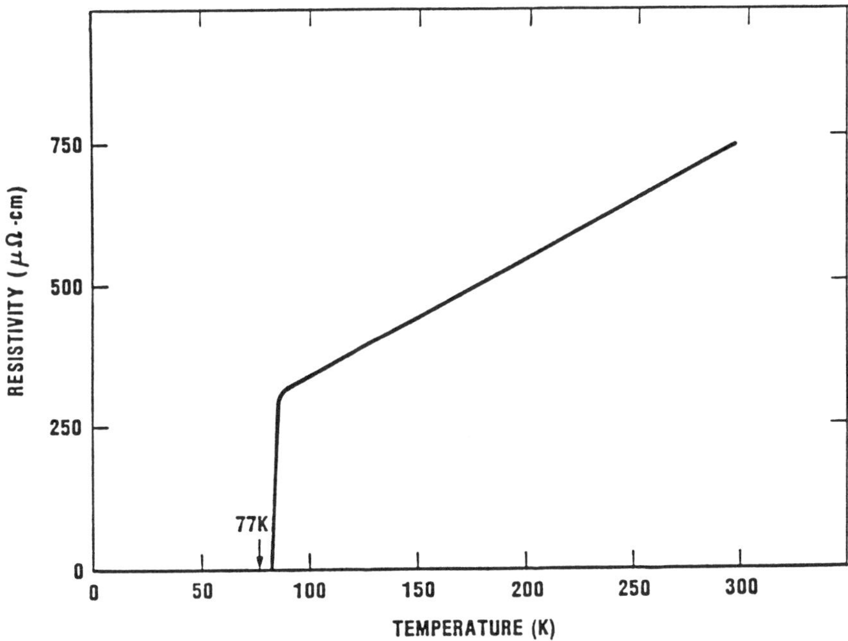

Figure 2.   Resistivity vs temperature for the same film as in Figure 1.

Critical current was measured by the transport method in the van der Pauw configuration without lithographic patterning. Hence the values reported here are likely lower limits of the actual critical current density J_c. At T = 77 K, J_c is $\sim$$10^3$ A/cm^2, and at T = 70 K, J_c rapidly increases to $\sim$5.0 $\times$ 10^4 A/cm^2.

Preliminary magnetization hysteresis measurement showed that J_c at 4.2 K is about 2 orders of magnitude higher than that at 70 K. The results suggest that the present critical current at 77 K is limited by the fact that this temperature is fairly close to the R=0 transition temperature. It is expected that J_c at 77 K can be drastically improved if only a sight increase of T_c (R=0) by ~ 5 K is made.

In a comparative study, we have also prepared the Y-Ba-Cu-Ox films on $SrTiO_3$ (100) by magnetron sputtering from a composite sintered target of $Y_1Ba_2Cu_3O_{7-x}$. The deposition temperature is kept at ambient room temperature. Note that the compositions of the sputtered films usually deviate substantially form that of the target. Following similar heat treatment procedure, the sample also superconducts with the resistive transition shown in Figure 3. The T_c onset is 88 K, and T_c (R=0) is at 70 K. However, X-ray diffraction analysis indicated that the structure is polycrystalline with no string preferential texture. Multiple phases other than $Y_1Ba_2Cu_3O_{7-x}$ were also found. The magnitude of the critical current density is comparable to those obtained on substrates of Al_2O_3, and MgO; namely, J_c at 4.2 K is in the range of $3 \times 10^3 - 1.0 \times 10^4$ A/cm^2.

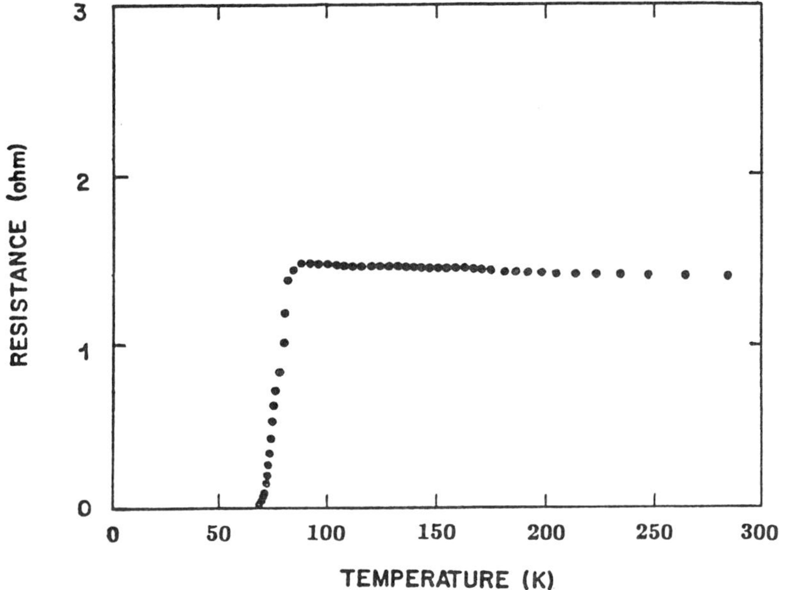

Figure 3.   Resistance vs temperature for a magnetron sputtered Y-Ba-Cu-Ox polycrystalline film on $SrTiO_3$.

Present studies strongly suggest that the attainment of a highly orientation-ordered superconducting layered perovskite with minimal intergrain phases or weak links is crucial for achieving high critical current in this class of superconductors. Hence the MBE produced, epitaxial single crystal superconducting films offer new opportunities in future device applications.

It is a pleasure to acknowledge helpful discussions with R. J. Cava, R. C. Dynes, M. Gurvitch, A. F. Hebard, D. W. Murphy, J. C. Phillips, C. E. Rice, and R. B. VanDover.

References

1. J. G. Bednorz and K. A. Müller, Z. Phys. B64, 189 (1986).

2. M. K. Wu, J. R. Ashburn, C. J. Tong, P. H. Hor, R. L. Wong, L. Gao, Z. J. Huang, Y. Q. Wang, and C. W. Chu, Phys. Rev. Lett. 58, 908 (1987).

3. R. B. Laibowitz, R. H. Koch, P. Chaudhari, and R. J. Gambino, Phys. Rev. B, June (1987).

4. R. H. Hammond, M. Naito, B. Oh, M. Hahn, P. Rosenthal, A. Marshall, N. Missert, M. R. Beasley, A. Kapitulnik, and T. H. Geballe, Extended Abstracts for MRS Symposium on High Temperature Superconductors, Anaheim, CA, April 23-24, 1987.

5. A full account of this work is given in J. Kwo, T. C. Hsieh, R. M. Fleming, M. Hong, S. H. Liou, B. A. Davidson, and L. C. Feldman, to be published.

6. J. Kwo, E. M. Gyorgy, D. B. McWhan, M. Hong, F. J. DiSalvo, and J. E. Bower, Phys. Rev. Lett. 55, 1402 (1985).

7. M. Suzuki and T. Murakami, J. Appl. Phys. 56, 2330 (1984).

8. M. Suzuki, K. Moriwaki, Y. Ecomoto, M. Oda, Y. Hidaka, and T. Murakami, submitted to Jpn. J. Appl. Phys.

9. P. Chaudhari, R. H. Koch, R. B. Laibowitz, T. R. McGuire, and R. J. Gambino, Preprint.

10. J. J. Capponi, C. Chaillout, A. W. Hewat, P. Lejay, M. Marezio, N. Nguyen, B. Raveau, J. L. Soubeyroux, J. L. Tholence, and R. Tournier, Preprint.

11. M. Hong, S. H. Liou, J. Kwo, and B. A. Davidson, Submitted to Appl. Phys. Lett.

CHEMICAL DOPING AND PHYSICAL PROPERTIES OF THE NEW

HIGH TEMPERATURE SUPERCONDUCTING PEROVSKITES

J. M. Tarascon, L. H. Greene, B. G. Bagley, W. R. McKinnon,
P. Barboux and G. W. Hull

Bell Communications Research
331 Newman Springs Road
Red Bank, N.J.   07701-7020

ABSTRACT

The effects of chemical doping on the structural, magnetic, transport and superconducting properties of the new high T_c superconducting oxides are reviewed. The substitution of Sr or rare earths for La in La_2CuO_{4-y} and the replacement for Y by rare earths and for Ba by Sr in the $YBa_2Cu_3O_{7-y}$ system have been investigated, as well as the substitution of Ni for Cu in both systems. In contrast to the rare earth substitutions, a Ni substitution dramatically affects T_c. The importance of the oxygen content (y) in these oxides, which strongly depend on their processing conditions (heat treating time, temperature and ambient), is emphasized and a detailed study of changes in the physical properties as a function of y is reported. The superconducting properties of these perovskites can be destroyed by decreasing their oxygen content. However, full superconductivity can be recovered by reannealing the sample under oxygen or by a low temperature (80°C) plasma oxidation. This later method may be very useful for the technological application of these materials.

1. INTRODUCTION

Until recently, superconductivity has been limited to temperatures below 23K. The announcement in late 1986 by Bednorz and Muller[1] of possible high T_c (30K) superconductivity in the La-Ba-Cu-O system generated much technological interest in potential applications and sparked an intense study of this new family of metallic cuprate oxides by several groups, including our own. Bulk superconductivity was established at 40K by several groups[2-4] in the $La_{2-x}Sr_xCuO_4$ series (at x = 0.15) with the K_2NiF_4 structure and subsequently T_c was increased to temperatures as high as 98K in a multiphase Y-Ba-Cu-O system.[5,6] Superconductivity above liquid nitrogen temperature was independently confirmed by others,[7-8] and the high T_c phase in this system was rapidly isolated $(YBa_2Cu_3O_{7-y})$[9] and its structure established.[10] The structure can be simply viewed as a stacking of Y, CuO_2, BaO, CuO, BaO, CuO_2 and Y planes perpendicular to the c-axis with the main feature being the presence of Cu-O chains parallel to the b-axis in the central copper-oxygen plane (Fig. 1).

The absence of an isotope effect[11] in these materials suggests that the pairing mechanism is not a standard electron-phonon interaction.[12] New theoretical models have been proposed and, interestingly, all of them stress the importance of the oxidation of the copper. For instance, in Anderson's model[13] (RVB) a sufficient concentration of Cu^{+3} is required to destroy the insulating state which results from singlet pairs on adjacent Cu^{+2} ions. Varma et. al.[14] propose that the charge transfer exciton $Cu^{3+}O^{2-} \rightarrow Cu^{2+}O^-$ is the source of the effective electron-electron attraction. Finally, band structure calculations by Mattheis et. al.[15] led to the conclusion that the superconducting properties of these oxides are mainly governed by the Cu-3d and O-2p electrons.

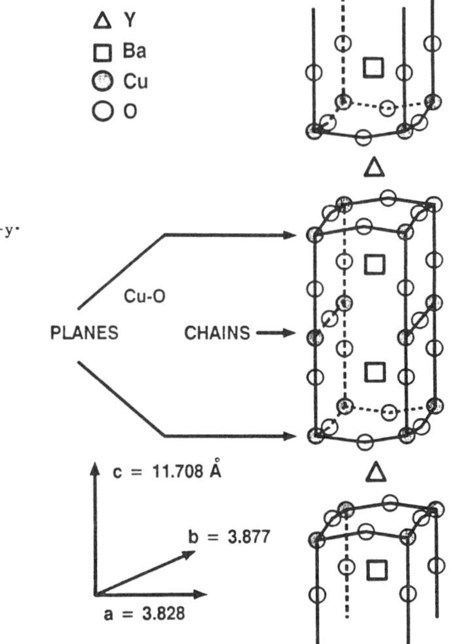

Figure 1. Structure of $YBa_2Cu_3O_{7-y}$.

The physical properties of these new superconducting materials are generally sensitive to deviations from stoichiometry, which in turn are related to their method of preparation. Oxygen non-stoichiometry in the perovskites has been extensively studied through the years because of the interesting ability of these materials to reversibly intercalate the oxygen atoms.[16] During such a process the valence of the transition metal element (e.g., copper) changes in order to balance the electronic charge. We have systematically studied the effect of oxygen stoichiometry on T_c and find that changes in oxygen content produce dramatic changes in the magnetic, electronic transport properties and T_c, and that these changes are completely reversible. We believe from these studies, that the valence of copper is one of the crucial parameters for the occurrence of superconductivity in these materials. In the La-Cu-O system the valence of copper can also be changed through the substitution of divalent ions (Sr) for trivalent ions (La), with a resulting marked effect on the transport properties.

Historically, considerable experimental and theoretical work has been addressed towards the possible coexistence of magnetism and superconductivity. Experimentally, such studies have been complicated by the lack of high T_c materials; thus, when magnetic doping was performed, superconductivity was destroyed before enough of the magnetic component was introduced for magnetic ordering to occur. The new high T_c oxides offer an opportunity for further study on this problem. In most superconductors the interaction between the magnetic moment ion and the superconducting electrons breaks the Cooper pairs responsible for superconductivity thereby suppressing or destroying T_c. In the new perovskite based superconductors the magnetic ions can be placed either on rare earth sites (thus separated from the coppers and oxygens at which the superconducting electrons are believed to be located) or directly on copper sites. We investigated both cases; the effect of having magnetic rare earth ions on lanthanum sites, and of 3d transition metals on copper sites, for both the La-Sr-Cu-O and Y-Ba-Cu-O systems. We find that rare earth ions affect superconductivity only weakly whereas 3d ions destroy it.

In the following, for simplicity, the La-Sr-Cu-O and Y-Ba-Cu-O systems will be denoted as 40K and 90K materials, respectively. We first present the details of our sample synthesis and experimental procedure. Then we present our data (x-ray powder diffraction, magnetic susceptibility, thermogravimetric analysis and resistivity measurements as a function of the temperature and composition), for the effects of strontium, rare earth and nickel substitutions and oxygen doping. In each case the 40K and 90K systems are treated separately. Finally we discuss our results in relation to the large amount of work which already exists for these new high T_c materials.

2. EXPERIMENTAL

The three 40K series $La_{2-x}Sr_xCuO_{4-y}$, $La_{1.6}RE_{0.2}Sr_{0.2}Cu_{4-y}$ and $La_{1.85}Sr_{0.15}Cu_{1-x}Ni_xO_4$ were studied (RE refers to rare earth plus Yttrium). The materials were prepared from stoichiometric amounts of $SrCO_3$, La_2O_3, CuO and the corresponding nickel or rare earth oxide powders, each 99.999% pure. The powders were thoroughly mixed and pressed at 7kbar into a pellet. The pellets were reacted in a platinum boat in a tubular furnace under flowing oxygen. The temperature of the sample was increased to 1120°C, maintained at this temperature for 36 hours, and then slowly cooled (5 hours) to 500°C.

A 90K series of compounds $REBa_2Cu_3O_{7-y}$, $YBa_2Cu_{3-x}Ni_xO_{7-y}$ and $YBa_{2-x}Sr_xCu_3O_{7-y}$ were made from $BaCO_3$, $SrCO_3$ and the corresponding oxides. The mixed powders were placed in an alumina crucible in a furnace. The sample was reacted at 950°C for 48 hours, cooled to room temperature, ground and pressed into a pellet, reheated at a similar temperature for another 48 hours in oxygen, and finally cooled slowly to room temperature taking 6 hours. The resulting pellets were cut into rectangular bars ($.5x3x8mm^3$) for resistivity and critical field measurements and the remainder of the pellet was used for powder diffraction, thermogravimetic analysis (TGA) and magnetic measurements. We observed that depending on the ratio of the starting materials, the annealing temperature, and cooling rate, single crystals (1mm size) with superconducting properties similar to those of the bulk ($T_c=92K$) could be prepared.

The oxygen stoichiometry (7-y) of the 90K materials was obtained by TGA in which the weight loss of a sample heated at 20°C/min in a reducing atmosphere (4 percent H_2 in Ar) was monitored. Although the value of y in the starting material is only known to ±0.1, changes in oxygen content relative to the as-grown material can be determined by TGA with precision of ±0.01. Oxygen removal was also accomplished by annealing the sample under vacuum at 420°C for several hours. Re-intercalation of the oxygen was done either by reannealing the sample at 720°C for 1 hour in oxygen or by a low temperature (80°C) plasma oxidation for a longer period of time.

The samples were characterized by powder x-ray diffraction in a Bragg-Brentano geometry using $CuK\alpha$ radiation (in this paper the designation single phase material will refer to the x-ray powder diffraction characterization). Resistivity measurements were made at constant current by a standard four probe method described elsewhere.[4] Leads were attached to the samples using silver paint and a silicon diode thermometer with $\pm0.2K$ accuracy was used to monitor the temperature. The superconducting transition temperatures were also determined by ac susceptibility measurements. The shielding currents, Meissner expelled flux, and magnetic susceptibility (from above T_c to room temperature) were measured with a SQUID magnotemeter on powder samples.

3. RESULTS

A. <u>Strontium Substitution</u>

Structural and resistivity parameters for the $La_{2-x}Sr_xCuO_{4-y}$ series are summarized in Table I. Note that $x=1$ is the upper limit of the strontium solubility. Beyond this limit the resulting materials are multiphase, as determined by x-ray analysis. Figure 2 shows the variation of the resistivity as a function of temperature for several members of the $La_{2-x}Sr_xCuO_{4-y}$ series ($0<x<3$ in steps of 0.025). The maximum superconducting temperature (midpoint $T_c = 40$ K) is reached at $x=0.15$. For neighboring compositions the superconducting transitions are lower in temperature and broader. AC susceptibility measurements indicated no superconductivity down to 4.2K for a strontium content greater than 0.3. Above T_c the resistivity varies linearly with temperature with a room temperature value, $\rho(300K)$, and a slope $d\rho/dT$ decreasing with increasing x (Table I). Hall measurements show, at least for $x<0.15$, that each Sr dopant atom donates one itinerant hole to the electrical transport in these p-type materials.[17] Bulk superconductivity at 40K for the strontium system was confirmed by a 55% Meissner effect.

La_2CuO_4 exhibits a large paramagnetic signal and evidence for antiferromagnetic ordering at 240K.[18] This magnetic ordering, recently confirmed by powder neutron diffraction,[19] could be affected by a Sr substitution. Figure 3 shows the magnetic data for several members of the 40K series. No evidence of magnetic ordering is observed for the strontium doped samples. In addition, a significant feature is that the susceptibility per gram, χ_g, is lower for

Table 1. Resistivity, transition temperatures, and lattice parameters for materials studied. The onset temperature T_{co} is taken where the resistance noticeably deviates from the extrapolation from high temperature; the midpoint temperature T_{cm} is the temperature where the resistance has fallen to half its extrapolated value. (After Ref. 4).

Sample	x	$\rho(300K)$ ($\mu\Omega$-cm)	$\dfrac{\rho(300K)}{\rho(50K)}$	T_c^{onset} (K)	$T_c^{midpoint}$ (K)	a (Å)	c (Å)	Volume (Å³)
3S15	.1	4400	2.4	32.5	26.0	3.7839(8)	13.211(4)	189.16(12)
1S14	.125	5100	3.4	40.4	36.1	3.7784(8)	13.216(4)	188.68(10)
2S14	.15	2300	4.5	41.2	39.3	3.7771(10)	13.226(3)	188.60(10)
3S14	.175	2200	4.2	40.3	38.2	3.7739(4)	13.232(2)	188.45(6)
4S14	.2	1800	4.3	40.9	35.9	3.7739(8)	13.230(3)	188.54(10)
5S14	.225	1500	4.8	36.4	29.1	3.7708(8)	13.242(4)	188.30(10)
6S14	.25	1300	5.2	41.7	31.2	3.7685(8)	13.247(4)	188.12(9)
7S14	.275	990	5.4	31.7	21.8	3.7666(8)	13.255(3)	188.06(7)
3S8	.3	620	4.9	11.5	5.7	3.7657(6)	13.259(4)	188.02(9)
1S11	.4	4000	0.68	-	-	3.7614(6)	13.228(4)	187.15(8)
	.5					3.7684(4)	13.220(3)	187.75(7)
	.6					3.7644(8)	13.185(5)	186.85(10)
	.7					3.7625(6)	13.156(7)	186.24(10)
	.8					3.7651(10)	13.129(4)	186.12(10)
	.9					3.7616(8)	13.071(4)	184.96(8)
	1.0					3.7626(10)	13.064(8)	184.95(13)

the highest T_c superconducting compound (x=0.15; 0.125) and decreases with decreasing temperature, while for higher strontium content χ_g increases with decreasing temperature.

The ability to prepare the 90K materials with barium but not strontium has been a puzzle. To better understand this chemical problem we studied the $YBa_{2-x}Sr_xCu_3O_{7-y}$ series. We observe that the upper limit of strontium solubility is at $x \simeq 1.4$, beyond which the materials are multiphase. The variation of lattice parameters as a function of x is shown in Figure 4. Note that a, b and c decrease monotonically with increasing strontium concentration. Strontium is smaller than barium and it is possible that the repulsive interactions between the Cu-O layers are not as well screened as they are with barium so that the layers separate and the 123 phase becomes unlikely to form. The effect of the strontium substitution on T_c is shown in Figure 5. Note that T_c decreases monotically from 93K (x=0) to 80K (x$\simeq$1.4). It appears that the decrease in T_c is correlated with the unit cell volume, also shown in Figure 5.

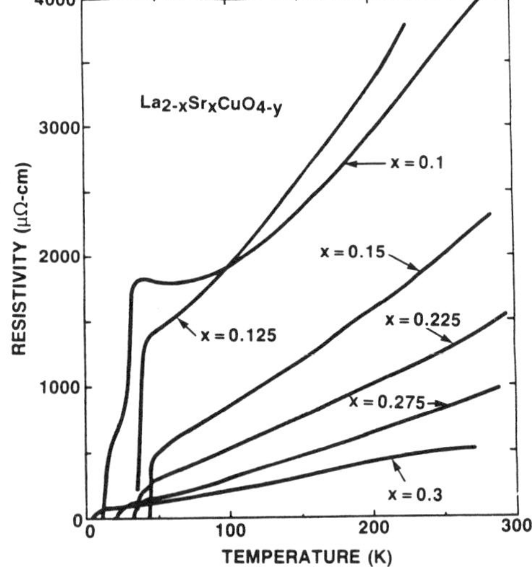

Figure 2. Resistivity as a function of the temperature from below T_c to room temperature from several members of the $La_{2-x}Sr_xCuO_{4-y}$ series (after ref. 4).

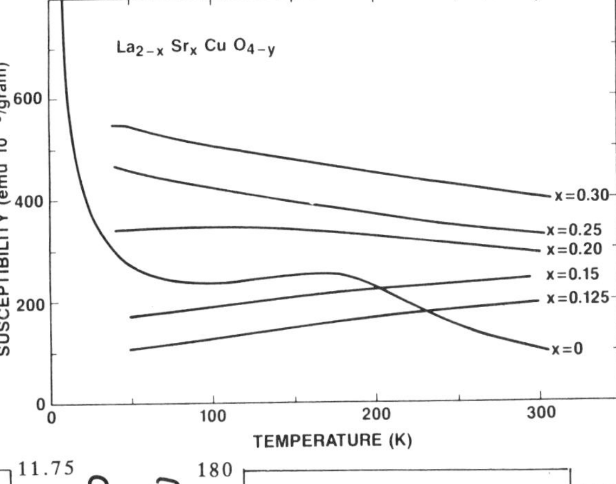

Figure 3. Observed magnetic susceptibility vs. temperature for several members of the $La_{2-x}Sr_xCuO_{4-y}$ series.

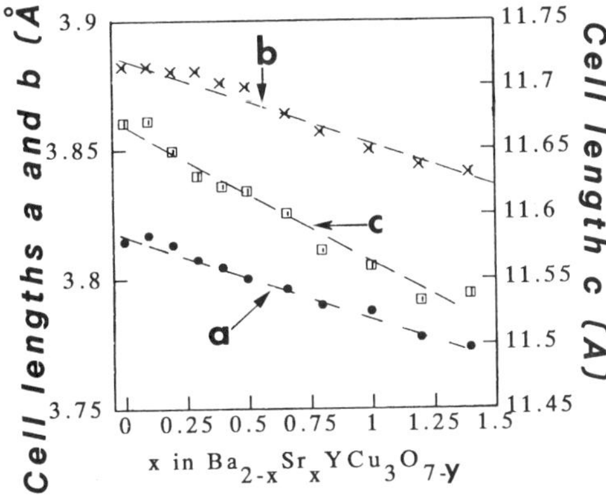

Figure 4. The orthorhombic lattice parameters a,b,c are reported for the $YBa_{2-x}Sr_xCu_3O_{7-y}$. The dashed lines are a guide to the eyes.

Figure 5. Variation of the superconducting critical temperature (determined by ac susceptibility) and of the orthorhombic unit cell volume as a function of x for the $YBa_{2-x}Sr_xCu_3O_{7-y}$ series.

B. Rare Earth Substitution

The substitution of a RE magnetic ion for La or Y, in contrast to that of strontium, does not change the oxidation state of the copper. However, the effect of the RE ions on the superconducting properties of the 40K or 90K phases may help our understanding of the interplay between magnetism and superconductivity and provide insight as to the origin of the superconductivity in these high T_c perovskites.

With the exception of La_2CuO_4, all the rare earth oxides (RE_2CuO_4) show a structure slightly different than the K_2NiF_4 structure[20] formed by the La-Sr-Cu-O system, so that the substitution of RE ions for La is limited to low RE concentrations. We focus on the $La_{1.6}Sr_{.2}RE_{.2}CuO_{4-y}$ series[21] which we observe are single phase for all the rare earths with exception of the heavy ones (Er, Dy, etc.). The x-ray powder diffraction patterns are completely indexed on a basis of a tetragonal unit cell. The cell lengths are summarized in Table 2. Note that a and c decrease continuously with increasing Z of the lanthanide

Table 2. Lattice parameters and superconducting transition temperatures (determined by DC resistance measurements) for the $La_{1.6}Sr_{0.2}Ln_{0.2}CuO_4$ series. (After Ref. 20).

$La_{1.6}Sr_{0.2}Ln_{0.2}CuO_4$ Compounds (Ln =)	ρ 300 k ($\mu\Omega - cm$)	$\frac{\rho 300k}{\rho 50k}$	T_c onset (k)	T_c midpoint (k)	a (Å)	c (Å)	Volume (Å³)
La	1600	4.6	40	39	3.7739(4)	13.238(1)	188.54(4)
Pr	2010	4.0	37	33	3.7678(4)	13.213(2)	187.66(5)
Nd	1840	3.6	33	29	3.7684(4)	13.189(2)	187.15(5)
Sm	2040	3.1	27	23	3.7668(4)	13.189(2)	187.15(5)
Eu	2540	3.0	25	21	3.7669(4)	13.181(1)	187.03(4)
Gd	2340	2.9	23	18	3.7666(3)	13.178(1)	186.94(4)

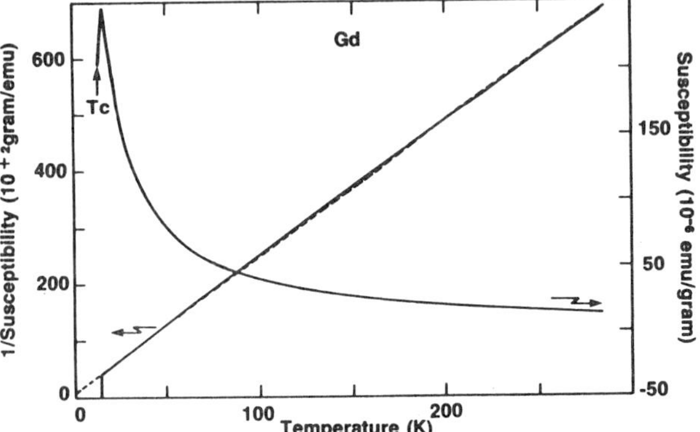

Figure 6. The susceptibility and reciprocal susceptibility are plotted as a function of the temperature for the $La_{1.6}Sr_{0.2}Gd_{0.2}CuO_{4-y}$.

substitution. The smooth and gradual decrease of the unit cell volume V in going from La to Gd suggests that all the rare earths replace La as a trivalent ion.

Magnetic measurements from above T_c to room temperature have been performed on all the members of the series to demonstrate this lack of a magnetic effect more conclusively. As an example, the susceptibility (χ) and reciprocal susceptibility (χ^{-1}) as a function of temperature are shown in Figure 6 for the Gd-doped sample. The reciprocal susceptibility may be fit to a straight line indicating that the data obey the Curie-Weiss law $X = C/(T + \theta_p)$ very well. An effective moment of 8 μ_B per Gd atom is deduced which is in good agreement with the value expected for a free Gd^{+3} (7.94 μ_B). From these experiments the magnetic moment of the rare earth ion can be extracted and is consistent with a +3 valence for all the substituted rare earths.

Figure 7 shows the temperature dependent resistivities for the rare earth doped La-Sr-RE-Cu-O phases. The superconducting transition temperature T_c decreases with the substitution of rare earths of increasing atomic number such that T_c drops from 39K for RE=La to 18K for RE=Gd. For the first members of the series the resistivity above T_c varies linearly as a function of temperature. For the Eu and Gd phases the resistivity goes through a broad minimum above T_c. In the presence of magnetic ions the upturn of the resistivity may be ascribed to Kondo scattering.[22] However, a similar effect observed for the non-magnetic phases $La_{2-x}Sr_xCuO_{4-y}$ (x<0.15) was ascribed earlier to incipient semiconducting behavior.[4]

The depression of T_c upon the introduction of magnetic impurities is generally described by the Arbrikosov and Gor'Kov formula[23] $T_c = n(N(E_f))I_2(g-1)^2J(J+1)$, where n is the concentration of the rare earth ions, $N(E_f)$ is the density of states at the fermi level, I is the exchange integral, and $(g-1)^2J(J+1)$ is the de Gennes factor. In Figure 8 we show that the

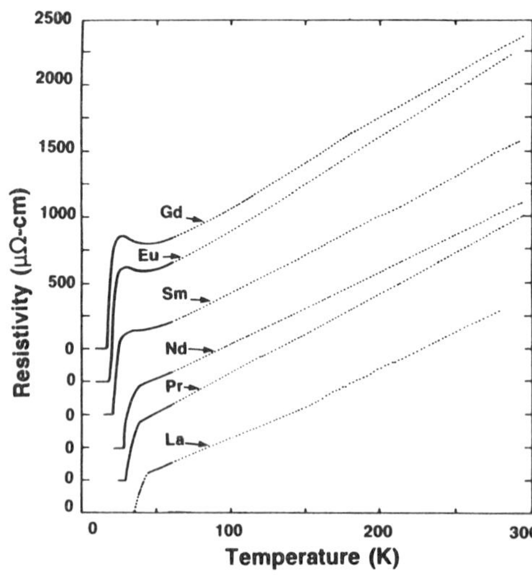

Figure 7. Resistivity as a function of the temperature from below T_c to 300K for the $La_{1.6}Sr_{0.2}RE_{0.2}CuO_{4-y}$ series. (after ref. 20).

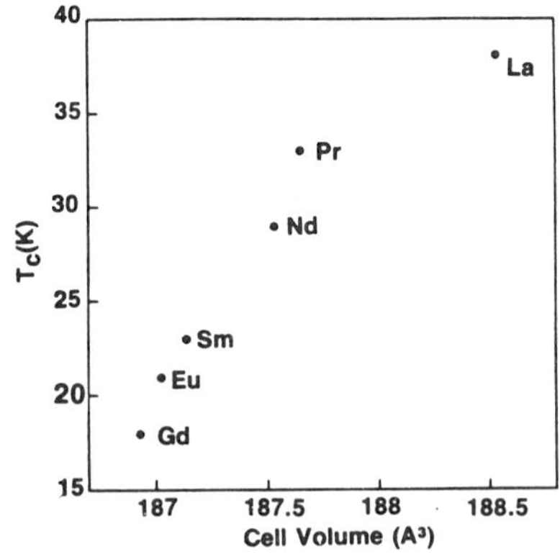

Figure 8. Superconducting critical temperatures and tetragonal unit cell volume are shown as a function of the rare earth for the $La_{1.6}Sr_{0.2}RE_{0.2}CuO_{4-y}$ series. (after ref. 20).

values of T_c, measured at the midpoint of the resistive transition, decreases smoothly through the rare earth series. The de Gennes factor, however, is not monotonic across the series and, in particular, is a minimum and a maximum for the neighboring elements Eu^{+3} ($j=0$) and Gd^{+3} ($j=7/2$), respectively. Thus, we conclude that the decrease in T_c cannot be a magnetic effect. The decrease appears to be correlated with the decrease of the tetragonal unit cell volume, which is also shown in Figure 8. We conclude, therefore, that the observed depression in T_c is essentially a volume effect.

In the 90K material, $YBa_2Cu_3O_{7-y}$, the substitution of all the rare earths for Y has been investigated.[24-25] With the exception of RE=Tb we could prepare compounds for all the rare earths. For RE = La, Pr, Nd, Yb the materials contained a small amount (<5%) of a second phase, but for RE=Lu the second phase was at least 50% of the sample. Crystal data for the rare earth doped 90K materials are reported in Table 3 and shown in Figure 9, together with the data for vacuum annealed samples. With the exception of RE=Pr, the as-grown materials are all orthorhombic. Annealing these samples under a vacuum of 10^{-3} torr at 450°C for 24 hours produces a tetragonal unit cell such that, within our experimental resolution, a becomes equal to b· and c remains constant. In the as-grown material the oxygen atoms in the middle

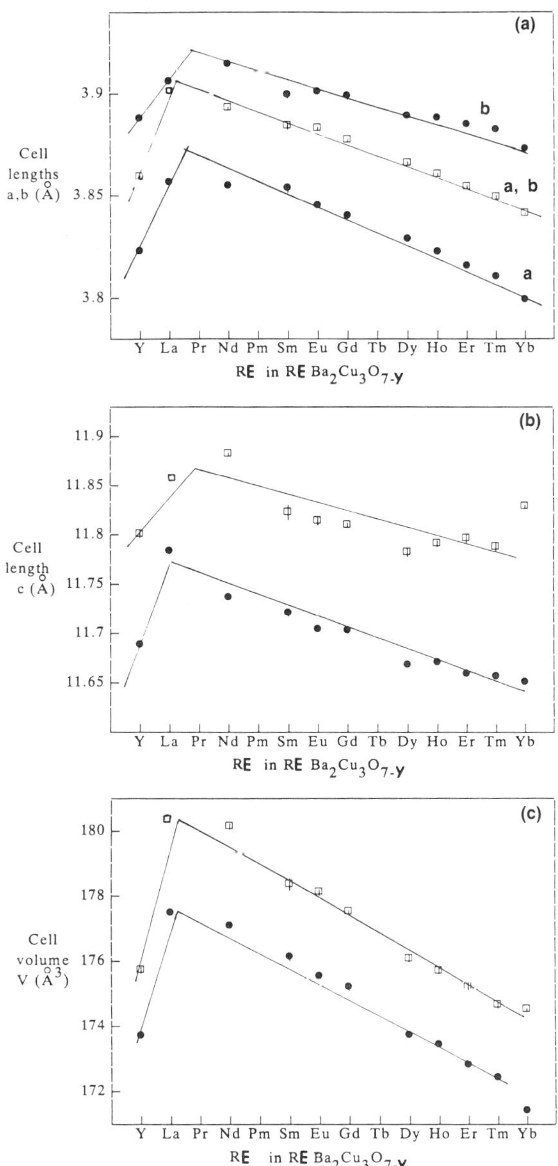

Figure 9. The orthorhombic lattice parameters a, b, c and V are reported as a function of the rare earth for the as-grown $REBa_2Cu_3O_{7-y}$ phases (full circles). The tetragonal unit cell lengths for the vacuum annealed samples are also reported (squares). The dashed lines are a guide to the eye .

Table 3. Crystal data for $REBa_2Cu_3O_{7-y}$, RE = Y or a rare-earth element, for compounds as grown under 1 atm O_2 (orthorhombic) or annealed under vacuum (tetragonal). The oxygen content in the samples as-prepared and the change in oxygen y after annealing in vacuum are determined by TGA as described in the text. a, b, c are the lengths of the unit cell, and V is the volume of the unit cell. (After Ref. 24).

		As prepared (under O_2)					Annealed under vacuum		
RE	y	a (Å)	b (Å)	c (Å)	V (Å³)	Δy	a,b(Å)	c (Å)	V (Å³)
Y	6.72	3.8237(8)	3.8874(8)	11.657(2)	173.28(6)	0.53	3.8589(8)	11.800(4)	175.72(10)
La		3.8562(8)	3.9057(10)	11.783(3)	177.47(7)		3.9007(8)	11.858(3)	180.42(9)
Pr		3.922(2)			180.87(11)/3				
Nd	7.16	3.8546(8)	3.9142(12)	11.736(2)	177.07(7)	0.69	3.8930(6)	11.882(2)	180.13(9)
Sm	7.11	3.855(2)	3.899(2)	11.721(4)	176.11(13)	0.47	3.884(2)	11.822(8)	178.4(2)
Eu	7.10	3.8448(8)	3.9007(10)	11.704(3)	175.53(7)	0.51	3.8829(8)	11.814(4)	178.11(10)
Gd	7.08	3.8397(12)	3.8987(18)	11.703(3)	175.19(11)	0.56	3.8770(6)	11.810(2)	177.51(6)
Dy	6.82	3.8284(8)	3.8888(8)	11.668(2)	173.71(5)	0.54	3.8656(10)	11.783(5)	176.07(11)
Ho	6.71	3.8221(8)	3.8879(8)	11.670(2)	173.42(6)	0.58	3.8601(8)	11.791(3)	175.69(9)
Er	6.82	3.8153(8)	3.8847(12)	11.659(2)	172.80(7)	0.66	3.8540(8)	11.796(4)	175.22(11)
Tm	6.65	3.8101(8)	3.8821(12)	11.656(2)	172.41(7)	0.58	3.8491(10)	11.788(4)	174.65(10)
Yb	6.71	3.7989(8)	3.8727(10)	11.650(2)	171.40(6)	0.70	3.8411(4)	11.828(2)	174.51(5)

Cu-O planes are ordered along the b direction and absent in the a direction hence leaving Cu-O chains along b. For the structure to become tetragonal the oxygen must be disordered. Both the orthorhombic lattice parameters (a, b, c, V) and the tetragonal unit cell (a, c, V) decrease smoothly through the rare earth series without any anomaly, suggesting that the substituted rare earth ions are all trivalent as observed for the 40K system.

The resistivity of the rare earth substituted samples, from room temperature to 4.2K, is shown in Figure 10. The temperature of the resistive midpoint T_c and zero resistance T_{c0} are reported in Table 4 for all the members of the series. With the exception of RE=La, Pr which do not superconduct, the substitution of a rare earth for Yttrium has only a minor effect on T_c.

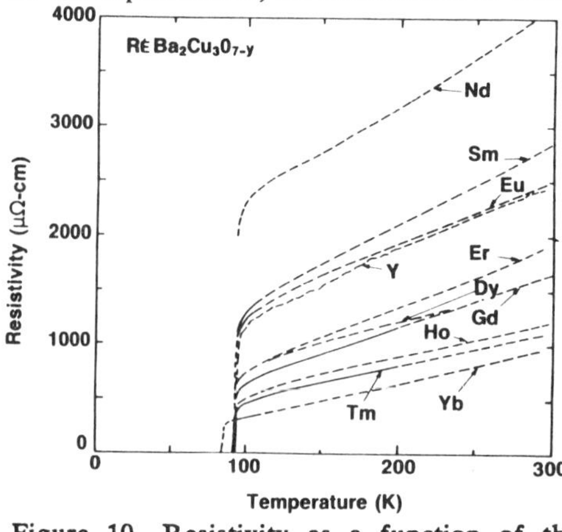

Figure 10. Resistivity as a function of the temperature for the rare earth series $REBa_2Cu_3O_{7-y}$.

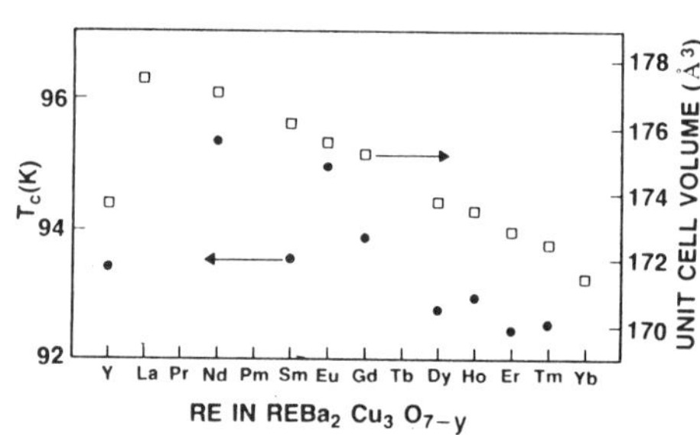

Figure 11. Orthorhombic unit cell volume V and superconducting transition temperatures T_c are plotted as a function of the rare earth for the $REBa_2Cu_3O_{7-y}$ series.

Table 4. Meissner, magnetic effective moments, and paramagnetic Curie temperatures are reported for the $REBa_2Cu_3O_{7-y}$ series, along with their electrical properties (resistivity and transition temperatures). The midpoint (T_{cm}) of the superconducting transition are reported as well. (After Ref. 24).

Compounds	Meissner Effect %	$\mu_{eff\,Th}$	$\mu_{eff\,exp}$	θ_p	T_c midpoint	T_{cO}	$\rho(300)$ $(\mu\Omega-cm)$
$NdBa_2\,Cu_3\,O_{7-y}$	48	3.62	-	-	95.3	92.0	4200
$SmBa_2\,Cu_3\,O_{7-y}$	32	0.85	-	-	93.5	88.3	2800
$EuBa_2\,Cu_3\,O_{7-y}$	35	0	-	-	94.9	93.7	2500
$GdBa_2\,Cu_3\,O_{7-y}$	36	7.94	7.75, 7.85*	-2, -3*	93.8	92.2	1650
$DyBa_2\,Cu_3\,O_{7-y}$	31	10.65	10.45, 10.69*	-7, -8*	92.7	91.2	1550
$HoBa_2\,Cu_3\,O_{7-y}$	33	10.61	10.48, 10.60*	-16, -19*	92.9	92.2	1200
$ErBa_2\,Cu_3\,O_{7-y}$	34	9.58	9.93, 9.63*	-15, -16*	92.4	91.5	1880
$TmBa_2\,Cu_3\,O_{7-y}$	33	7.56	7.53, 7.69*	-35, -38*	92.5	91.2	1110
$YbBa_2\,Cu_3\,O_{7-y}$	49	4.54	-	-	87.0	85.6	1000
$LuBa_2\,Cu_1\,O_{7-y}$	10	0	-	-	89.5	88.2	17,000
$YBa_2\,Cu_3\,O_{7-y}$	35	-	-	-	93.4	91.5	2,500

* Annealed under vacuum

Most of the rare earth compounds have superconducting transition temperatures between 92 and 95K; those of Yb and Lu, however, are at 87 and 89K respectively. As was observed for the 40K materials there is no correlation between T_c and the magnetism of the RE ions. A correlation with the unit cell volume may exist (Figure 11), but it is less obvious than in the 40K system. Since the T_c's and $B_{c2}(T)$'s (upper critical fields) of these materials are all nearly identical,[26] we believe that the most likely explanation for the differences in room-temperature resistivity are due to differences in sample preparation and not intrinsic. Here again, from above T_c to room temperature, the resistivity varies linearly with temperature with no resistance minimum. Resistivity and room temperature Seebeck measurements have shown that the tetragonal vacuum annealed samples do not superconduct and are n-type insulating semiconductors. We note that the orthorhombic phase with both a p-type metallic-like conductivity and superconductivity can be simply restored by reannealing the material under oxygen at 720°C.

Magnetic measurements were performed to characterize the oxidation state of the rare earth ions in these compounds. The susceptibility was measured as a function of temperature for several members of the rare earth series. From the data, using a Curie-Weiss fit as described previously for the 40K material, the magnetic moment can be extracted and compared to the theoretical values expected for a free trivalent rare earth ion. The results of this fit are reported in Table 4. Note that the experimental values agree well with the calculated ones for trivalent ions, thus suggesting a $+3$ valence for all the rare earths. Upon removal of oxygen the materials do not superconduct, so that the magnetic measurements can be extended to 4.2K. For the vacuum annealed samples, the reciprocal temperature dependent susceptibility fits to a Curie-Weiss law demonstrates magnetic moments per rare earth atom similar to those obtained from the as-grown samples. A remarkable result is that the paramagnetic curie temperature θ_p is negative and ranges from -35K for thulium to -2K for gadolinium. These negative values of θ_p reflect antiferromagnetic interactions and suggest the presence of antiferromagnetic ordering in these materials at very low temperatures. To examine this possibility, magnetic measurements were extended down to 1.6K and the results are shown in Figure 12. For most

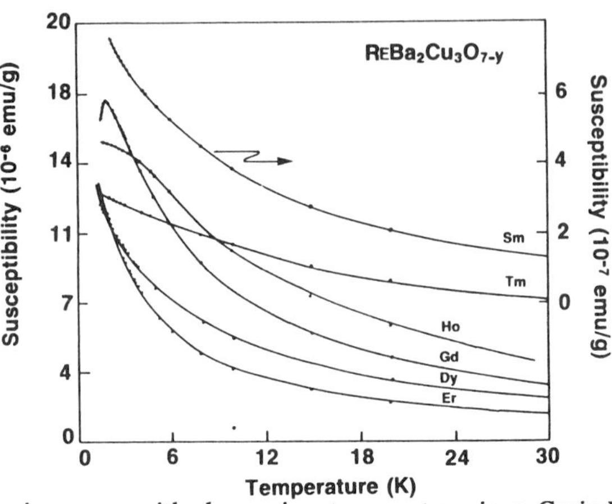

Figure 12. The temperature dependent susceptibility plotted over the range of temperature 1.6, 30K for several members of the REBa$_2$Cu$_3$O$_{7-y}$ series is shown.

of the compounds, the susceptibility increases with decreasing temperature in a Curie-Weiss type manner, the exception is Ho for which a leveling off of χ is observed. A striking feature of this figure is the well pronounced maximum in χ at 2.3K for the gadolinium compound, indicating an antiferomagnetic ordering. The peak at 2.3K in χ is in good agreement with the temperature (2.24K) of the maximum in the specific heat observed by Willis et. al.[27] Magnetic ordering at temperatures of 0.6K have recently been reported for the RE = Dy and Er[28] compounds. We have proposed that the dipole-dipole interactions are responsible for the magnetic ordering in these materials.[24]

As with the 40K materials, we again demonstrate that for the 90K materials there is no correlation between T_c and the magnetism of the rare earth since T_c (and $B_{c2}(T)$) remains nearly constant, independent of the RE, indicating little interaction between the RE and the superconducting electrons. Since the Cu-3d and O-2p electrons are responsible for superconductivity in these materials, a simple way to enhance the interactions of magnetic ions with the superconducting electrons is to introduce 3d magnetic ions (Ni, Co,...) on the copper sites (as we present next).

C. Nickel Substitution

A series of compounds La$_{1.85}$Sr$_{0.15}$Cu$_{1-x}$Ni$_x$O$_4$ with $0<x<0.3$ in steps of 0.025 were investigated.[29] The materials were single phase and were completely indexed based on a tetragonal unit cell. The lattice parameters are reported in Table 5. The lattice parameters "a" increases and "c" decrease with increasing x. The opposite trend was observed (discussed in Section A) through a strontium substitution.

Table 5. Resistivity, transition temperatures, and lattice parameters for both the La$_{0.85}$Sr$_{0.15}$Cu$_{1-x}$Ni$_x$O$_{4-y}$ and YBa$_2$Cu$_{3-x}$Ni$_x$O$_{7-y}$ series. Shown also are the Curie-Weiss constant θ_ρ, the temperature independent susceptibility χ_0, the effective magnetic moment per Ni atom derived from the Curie constant C, the temperature interval over which the fit was determined, and the rms deviation of the data from the Curie-Weiss law (in percent).

Compounds La$_{1.85}$Sr$_{0.15}$Cu$_{1-x}$Ni$_x$O$_4$	a (Å)	b (Å)	c (Å)	V (Å)^3	T$_{cm}$ (K)	P$_{300}$ μ ohm /cm	μ_{eff} per mole of Ni	θ (K)	10^{-4} χ_0 emu/mole Ni	σ %	T fit (K)
x = 0	3.777(3)		13.226(1)	188.60(1)	39.3	2300	-	-	-	-	-
x = 0.025	3.7749(2)		13.222(1)	188.42(2)	22.6	2380	-	-	-	-	-
x = 0.05	3.7761(2)		13.2078(9)	188.33(2)	< 4.2	3500	-	-	-	-	-
x = 0.075	2.7789(4)		13.187(1)	188.31(4)	no	4000	0.74	1	16.1	0.93	15-300
x = 0.1	3.7915(3)		13.1671(3)	188.29(6)	no	4200	0.83	2	16	0.27	20-300
x = 0.125	2.7833(3)		13.1582(3)	188.36(3)	no	9300	1.14	9	16	0.9	20-300
x = 0.2	3.7873(4)		13.1138(2)	188.09(6)	no	-	-	-	-	-	-
x = 0.3	3.7971(3)		13.0460(1)	188.10(3)	no	-	1.72	14	8.47	0.76	10-300
							1.80	20	7.58	0.05	150-300
YBa$_2$Cu$_{3-x}$Ni$_x$O$_7$							Per mole of 3d metal		emu/mole of 3d metal		
x = 0	3.8237(8)	3.8874(3)	11.657(3)	173.28(6)	91.6	1800	0.29	-21	0.88	0.5	100-300
x = 0.25	3.8191(8)	3.8857(8)	11.6571(3)	172.98(7)	63.9	1360	0.62	14	1.17	0.13	80-300
x = 0.5	3.8197(7)	3.8832(4)	11.6470(5)	172.75(8)	52.3	2500	0.87	-4	1.45	0.23	60-300

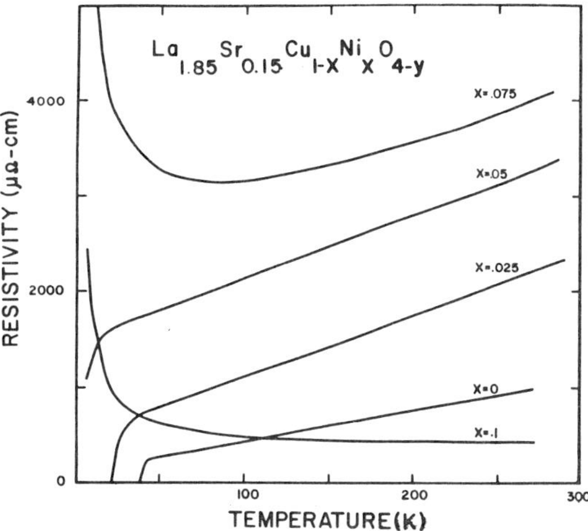

Figure 13. The temperature dependence of the resistivity for several members of the La$_{1.85}$Sr$_{0.15}$Cu$_{1-x}$Ni$_x$O$_{4-y}$ series are shown.

The temperature dependence of the resistivity for the nickel doped 40K materials is shown in Figure 13. The nickel substitution affects T_c dramatically. The superconducting transition temperature decreases from 39K to 21K by just substituting 0.025 of the Cu by Nickel. Increasing the substitution to x=0.05 lowers T_c to 4.2K and for a higher nickel concentration (0.075) the resistivity exhibits a pronounced semiconducting behavior. The substitution for Cu by Ni induces a metal-insulator transition which occurs between $0.05 < x < 0.075$.

Magnetic measurements were performed on the same set of samples from above T_c (or 4.2K) to room temperature (Figure 14). The susceptibility is nearly temperature independent for the compositions x=0; 0.025 and 0.05. At higher compositions the susceptibility increases with decreasing temperature in a Curie-Weiss type manner. The data were fit to a Curie-Weiss law of the form $\chi_g = C_g/(T + \theta) + \chi_0$ having introduced the additional χ_0 term. The results of this fit are summarized in Table 5. Note that for x=0.3 the data yields a magnetic moment of 1.80 μ_B per nickel atom. A similar value per Ni ion was obtained by Singh et. al.[18] for the La$_2$Cu$_{0.75}$Ni$_{0.25}$O$_4$ phase over the same range of temperature. However, at high temperatures their data fit a different Curie-Weiss leading to an effective magnetic moment of 3.0 μ_B as expected for Ni^{+2}, an S=1 system. Thus, we believe that nickel exists as a divalent ion in our materials.

It is important to stress that the dramatic changes in transport properties (metal-insulator transition) and magnetic properties (Pauli paramagnetism to Curie-like) occurs over the same range of composition, $0.05 < x < 0.075$. As nickel replaces copper the carrier density decreases so that the electrons become localized on the magnetic ions and then contribute to the susceptibility.

In contrast to the 40K system, the range of solubility of nickel in the 90K material is quite limited as compounds of nominal composition YBa$_2$Cu$_{3-x}$Ni$_x$O$_{7-y}$ with x greater than 0.5 were multiphase. For x=0.5 and 0.25 the materials were single-phase and their orthorhombic lattice parameters are given in Table 5. The cell lengths a, b, c as well as the unit cell volume slightly

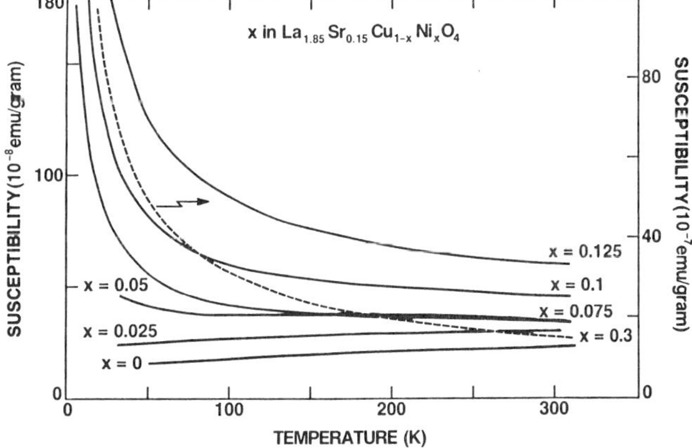

Figure 14. The observed magnetic susceptibility between 4.2K and room temperature is given for the La$_{1.85}$Sr$_{0.15}$Cu$_{1-x}$Ni$_x$O$_{4-y}$ series.

decreases upon nickel substitution. Resistivity measurements on the 90K nickel doped samples have shown[29] that the superconducting transition temperature decreases from 93K at $x=0$ to 50K at $x=0.5$. The metal-insulator transition, observed in the 40K system, cannot be observed here because of the inability to prepare material with a nickel content greater than 0.5. Contrary to the behavior of the 40K system, the susceptibility of the undoped nickel compound is largely paramagnetic which suggests a contribution due to the copper ions. As nickel is introduced the susceptibility is enhanced, and the values of the effective magnetic moment extracted from the Curie-Weiss fit are summarized in Table 5.

We also performed the substitution of Co for Cu in the 40K materials and the results were similar to those for nickel. For instance, the replacement of 0.05 Cu by 0.05 Co destroys T_c and promotes a semiconducting behavior. This is further evidence that the superconducting properties of the new high T_c oxides are extremely sensitive to the presence of 3-d transition metals on the copper sites.

D. Oxygen Doping

It has been shown by Nguyen et. al.[30] that the oxygen content y in the $La_{2-x}Sr_xCuO_{4-y}$ phases can be increased or decreased by annealing the sample at 500°C in oxygen or a vacuum respectively. We determined the effect of oxygen content on the superconducting properties of the 40K materials and found that T_c is enhanced by 1K when the sample is heated at 500°C in oxygen, whereas superconductivity (determined by ac susceptibility) is progressively destroyed by subsequent annealing under vacuum at 500°C.[4] An important observation was that superconductivity can be restored by reheating the sample at 500°C under oxygen. This clearly illustrates that changes in oxygen content and T_c in these 40K materials are completely reversible.

The sensitivity of the resistivity to oxygen content is illustrated in Figure 15. The as-grown material (curve a) has a metallic-like conductivity with a T_c of 38K. After subsequent heating under vacuum (curve b) we observe incipient semiconducting-like behavior with T_c reduced to 29.7K. More interesting here is that the initial T_c and the metallic-like behavior (curve c) is completely recovered by a low temperature plasma oxidation (80°C) confirming that the changes in T_c and oxygen content are reversible.

The plasma oxidation is as efficient as a heat treatment under oxygen to reinsert oxygen in the 40K materials with the advantage that the process occurs near room temperature. This may help in producing phases at room temperature with higher oxygen contents than those formed at elevated temperatures. Recent reports[31-32] have shown that undoped La_2CuO_{4-y} made at temperature of 950°C exhibit signs of superconductivity at 40K. They attributed this effect to parts of the sample being rich in oxygen. None of our samples made at temperatures of 1120°C have shown this effect, likely due to a deficiency in oxygen. Thus, we exposed our as-grown high temperature La_2CuO_{4-y} oxide to an oxygen plasma for periods of 18 hours, 3 days and 6 days. The changes in transport properties are illustrated in Figure 16. The as-grown material has semiconductor-like behavior while the plasma oxidized sample has metallic-like conductivity and superconducts at 40K. The part of the sample superconducting was estimated at 18% by Meissner measurements.

The range of oxygen non-stoichiometry (determined by TGA) for the 40K materials is quite small (0.04), thus preventing a systematic compositional dependence study of its effect on physical properties. In contrast, such a study is possible with the 90K materials which have a large range of non-stoichiometry in oxygen.

<table>
<tr><td>

Figure 15. The effect of oxygen content on the superconducting properties of $La_{1.86}Sr_{0.14}CuO_{4-y}$ is shown. a, b, c refers to the material as-grown, annealed under 10^{-8} torr for 9 days at 550°C and plasma oxidized for 48 hours, respectively.

</td><td>

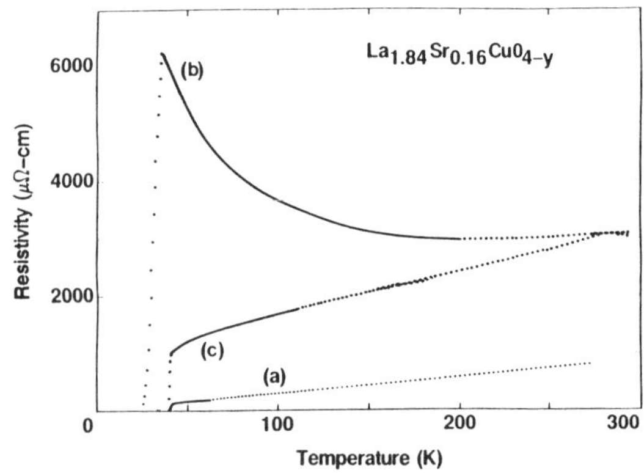

</td></tr>
</table>

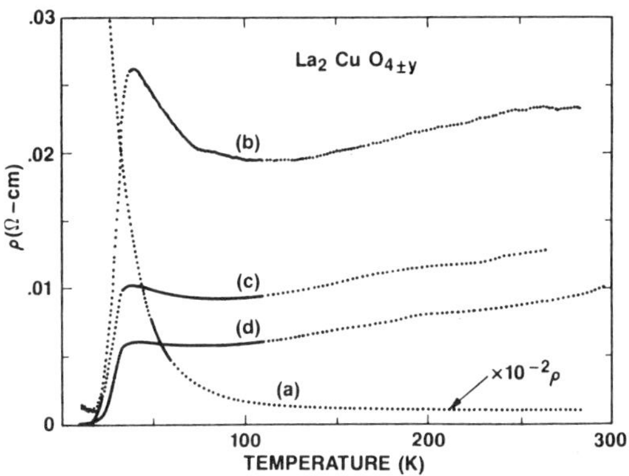

Figure 16. Effect of the plasma oxidation on the electrical properties of pure La$_2$CuO$_4$. a, b, c and d refers to the materials as-grown (1120°C) and after subsequent plasma oxidations.

The superconducting YBa$_2$Cu$_3$O$_{7-y}$ materials are deficient in oxygen. Single crystal and powder neutron diffraction studies have shown that oxygen vacancies are present and ordered in the central Cu-O planes. The oxygen content in our as-grown materials was evaluated by means of thermogravimetric analysis to be about 6.96 and will be taken as 7.0 ± 0.1 in the following. (This is a more accurate value than our previous result of 6.8 ± 0.1) [24] The difficulty in determining this number is due to uncertainties in the final product and, in addition, it depends strongly upon the processing of the sample, namely cooling rate and ambient gas.

The changes in oxygen content as a function of thermal treatment and ambient is clearly illustrated by the thermogravimetric data shown in Figure 17. A sample vacuum annealed at 420°C, i.e., a partial removal of oxygen, is heated and cooled in an oxygen ambient at a rate of 10°C/min (a). The maximum incorporation of oxygen occurs around 500°C and the sample then loses oxygen on further heating. On cooling, the curve starts to follow the heating one but deviates from it below 500°C, at which point the sample continues gaining weight upon cooling leading to a final change in oxygen content per unit formula (y) of 0.53. The removal of oxygen from the 90K material can also be produced by heating it in an inert gas (i.e., under zero partial pressure of oxygen) as shown in Figure 17b where the as-grown material is heated and cooled in argon and then reheated in oxygen. The important point here is the complete reversibility of the oxygen absorption, as the weight loss which occurs on heating in argon is equal to the weight gain on reheating in oxygen. This change corresponds to a y of 0.81 per unit formula. Finally, Figure 17c illustrates the importance of the cooling rate during the processing of these materials. Note a change in oxygen content y as large as 0.33 between the slowly cooled and quenched samples.

Figure 17. Effect of processing (temperature and ambient) on the oxygen content for the YBa$_2$Cu$_3$O$_{7-y}$ compound. Shown are the TGA traces for a) sample annealed under vacuum heated and then cooled under oxygen at a rate of 10°C/min, b) an as-grown sample heated and cooled in argon and then reheated in oxygen and c) for a vacuum annealed sample heated in air, quenched in air, reheated in air and slowly cooled in air.

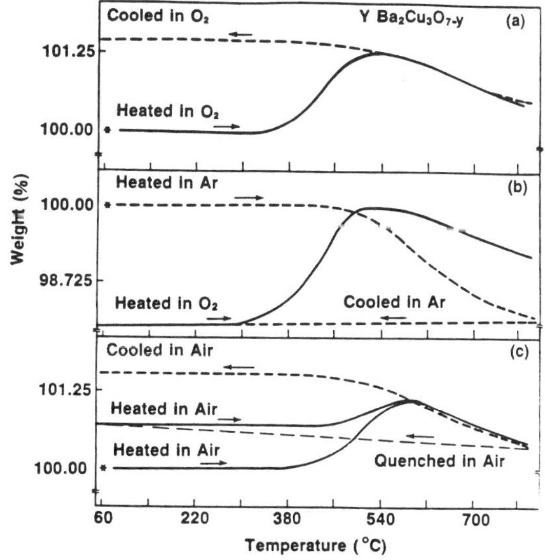

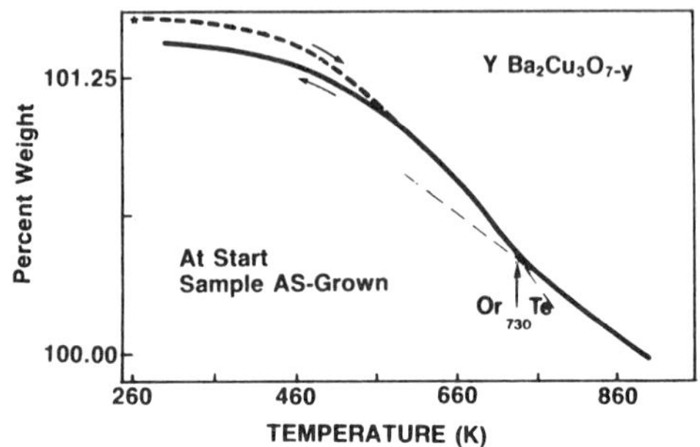

Figure 18. The TGA trace for an as-grown YBa$_2$Cu$_3$O$_{7-y}$ sample first heated, and then cooled in oxygen at a rate of 4°C/min is shown. The arrows distinguish between warming or cooling the sample.

Room temperature x-ray studies[25,33] have shown that the structure evolves from orthorhombic for the as-grown material (Y Ba$_2$Cu$_3$O$_7$) to tetragonal for the vacuum or argon annealed sample (YBa$_2$Cu$_3$O$_6$). One would expect this oxygen induced orthorhombic-tetragonal transition to also occur as a function of temperature, since we had shown that the material was losing oxygen above 500°C. Figure 18 shows the TGA traces for the as-grown material heated and cooled in oxygen. We ascribe the change in slope occurring around 730°C to the orthorhombic-tetragonal phase transition. This was unambiguously confirmed by Gallagher et. al.[34] using high temperature x-ray diffraction. A detailed study on the dependence of this structural transition with the partial pressure of oxygen is described elsewhere.[35]

The effect of oxygen content on the superconducting properties of these materials is shown on Figure 19 for Er Ba$_2$Cu$_3$O$_{7-y}$. The as-grown material has a metallic-like conductivity and superconducts at 93K. After annealing for 15 hours at 450°C in a vacuum of 10^{-3} torr the sample exhibits semiconducting-like behavior. An interesting result is that superconductivity and metallic-like behavior are fully restored by reannealing the sample at 720°C for one hour in oxygen. The loss or absorption of oxygen in these compounds, which induces a superconducting-semiconducting transition (and vice versa), can lead to new devices in which the physical properties will be changed chemically. Therefore the heat treatments mentioned above may limit these applications. To overcome this problem we developed a low temperature process, namely an oxygen plasma, to reintercalate oxygen in an oxygen deficient sample thereby producing marked changes in resistivity and T$_c$.[36]

The efficacy of this technique is illustrated in Figure 20. The as-grown material was annealed under a vacuum of 10^{-3} torr at 420°C and (as noted before) the resistivity evolves from a metallic-like behavior with a T$_c$ (curve a) to a semiconducting-like behavior with no T$_c$ (curve b). The sample is then given 210 and 285 hours in an oxygen plasma and the resistivities measured again (curves c and d, respectively). Note that after such a treatment, both the metallic conductivity and superconductivity at 93K are recovered. To ensure that this is a bulk effect, the Meissner effect and shielding effect were measured for the sample as-grown, after vacuum annealing and after plasma oxidation (Figure 21). Note that after such a process the resulting material shows a 24% Meissner effect suggesting that we are affecting more than the surface of the sample. The exposure time reported here is for a rectangular bar sample (.5x5x12mm). This time is expected to be much shorter to effect similar changes in thin film materials.

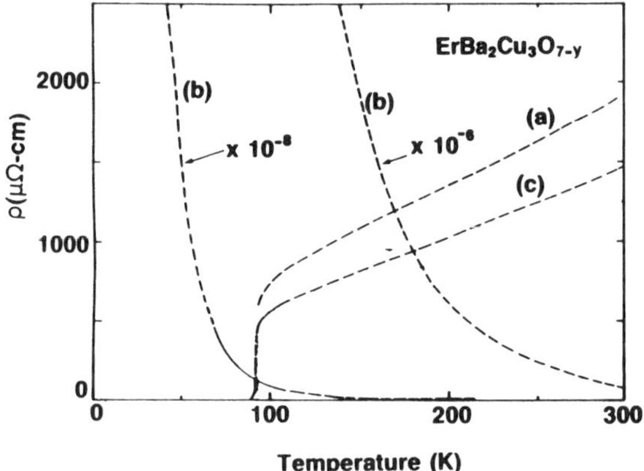

Figure 19. The effect of oxygen content on the temperature-dependent resistivity of ErBa$_2$Cu$_3$O$_{7-y}$, as grown (a), after vacuum anneal (b), and oxygen anneal (c), as described in the text. Note the dramatic increase in resistivity for (b), plotted with two contracted scales of 10^{-6} and 10^{-8} xρ.

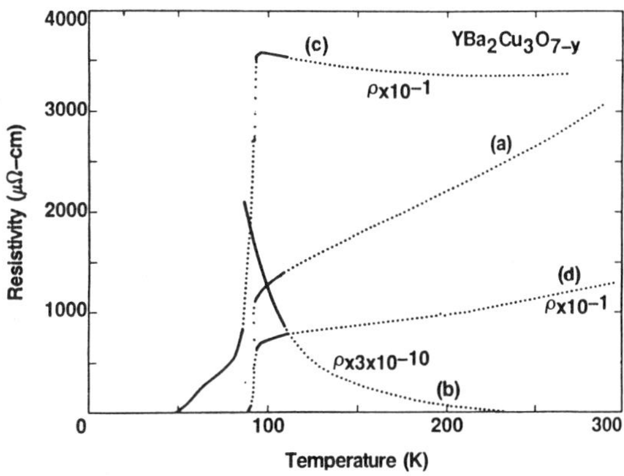

Figure 20. The efficiency of the plasma oxidation process to promote the semiconducting to superconducting transition in YBa$_2$Cu$_3$O$_{7-y}$. The curves a, b, c and d refer to the sample as-grown, vacuum annealed (10^{-3} torr at 450°C) for 15 hours, and after plasma oxidation for 210 and 285 hours, respectively.

Thus far we have dealt mostly with the two extreme cases; the as-grown, orthorhombic, high T$_c$ superconducting material of composition YBa$_2$Cu$_3$O$_7$ and the vacuum (or argon) annealed, tetragonal, non-superconducting composition YBa$_2$Cu$_3$O$_6$. There are, of course, many intermediate states between these two extremes, and we now present data correlating, in more detail, the superconducting, magnetic, and structural behavior of the YBa$_2$Cu$_3$O$_{7-y}$ phase with its oxygen content.

Samples with different oxygen contents were prepared as follows: First a pellet of the as-grown material was sliced into bars of a size 0.25x4x8mm and average weight from 20 to 25mg. These bars, one at a time, were heated up to 500-600°C at a rate of 4°C/min in an argon atmosphere. This step was done in the TGA apparatus so that we were able to directly control the weight change of the sample. When the desired value of y was reached, the sample was cooled to room temperature. In this way changes in oxygen stoichiometry y (with respect to the starting material) can be determined quite accurately (± 0.01).

One major concern was the homogeneity of the resulting material with respect to oxygen. Sample inhomogeneity will produce broadening of the Bragg diffraction peaks. No detectable peak broadening at the intermediate y is observed. Thus, we conclude that, to the accuracy of this technique, our samples are homogeneous. The materials are single phase with sharp diffraction peaks independent of y. The measured lattice parameters are summarized in Table 6 and their variation as function of y is shown in Figure 22. Note that the orthorhombic-tetragonal transition occurs at y = 0.60.

Since resistivity measurements may be misleading, the superconducting properties of these materials, as a function of y, were obtained by measuring the screening and exclusion of magnetic flux. A Meissner effect ranging from 20 to 24% was observed for the superconducting samples. The shielding traces are shown in Figure 23. Note that, upon removal of oxygen, T$_c$ shifts to low temperatures; but more interesting is the width of these superconducting transitions. As y is increased, the transition broadens, but becomes sharp with a T$_c$ of 55K for y = 0.43, and then, becomes broader again at higher y. These results support the existence of a distinct bulk superconducting transition at 55K in the Y-Ba-Cu-O system.

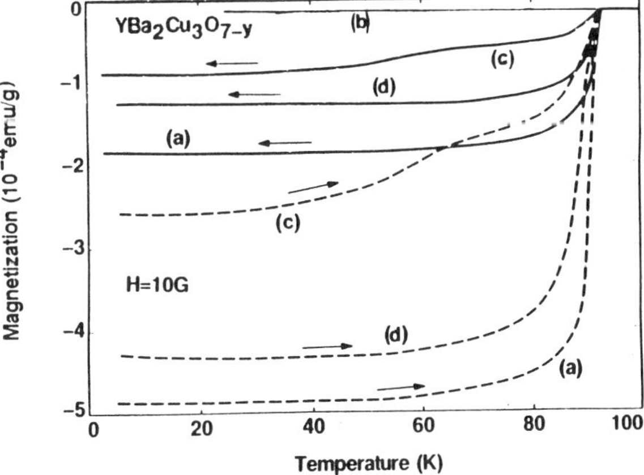

Figure 21. Magnetizaton as a function of temperature on the same sample in the same stages of oxidation, as described in Figure 20. The upper curves are for the sample cooling in a field of 10G (Meissner); the lower curves are for warming in 10G after cooling in zero field (shielding).

Table 6. Crystal data, Meissner, magnetic effective moments μ_{eff}, and paramagnetic Curie temperature θ_ρ as well as the rms deviation of the data from the Curie-Weiss law (in percent) for $YBa_2Cu_3O_{7-y}$. (After Ref. 35)

$YBa_2Cu_3O_{7-y}$	$a(\text{Å})$	$b(\text{Å})$	$C(\text{Å})$	$V(\text{Å})^3$	% Meissner	C μ_d per mole of Cu	$\theta p\,(K)$	rms %	T_c (K)
$y = 0$	3.8146(3)	3.8819(6)	11.670(1)	172.80(3)	35	0.29	-21	0.15	92
$y = 0.11$	3.8221(3)	3.8851(2)	11.701(1)	173.75(3)	28	-	-	-	64
$y = 0.18$	3.8220(3)	3.8872(3)	11.700(1)	173.83(3)	23	-	-	-	-
$y = 0.25$	3.8217(2)	3.8848(3)	11.697(1)	173.66(3)	30	0.30	-30	0.57	57
$y = 0.33$	3.8286(4)	3.8820(3)	11.715(1)	174.12(3)	27	-	-	-	52
$y = 0.435$	3.8347(3)	3.8765(4)	11.720(1)	174.14(3)	27	0.30*	-25	0.50	50
$y = 0.51$	3.8433(5)	2.8573(5)	11.765(1)	174.87(3)	21	0.30	-24	0.47	37
$y = 0.59$	3.8471(5)	3.8654(3)	11.763(1)	174.93(3)	II	0.31	-20	0.19	22
$y = 0.66$	3.8600(5)	-	11.786(3)	175.58(6)	0	0.37	-20	0.8	0
$y = 0.9$	3.8559(3)	-	11.819(1)	175.73(4)	0	0.62	-11	0.39	0

Cava et. al.[9] have shown that the magnetic susceptibility temperature dependence of the as-grown material follows a Curie-Weiss type behavior above the superconducting critical temperature. It was of interest to us to determine how the susceptibility changes with y. Magnetic data were collected, for several members of the series, from above T_c (or 4.2K) to room temperature. The results are shown in Figure 24. The susceptibility varies, independent of y, in a Curie-Weiss type manner. The data were least-square fit to the Curie-Weiss form $\chi_g = C_g/(T + \theta) + \chi_0$ and the results are given in Table 6. Note that the effective moment per mole of Cu ions increases slightly from 0.29 (y = 0) to 0.37 (y = 0.66) and then increases rapidly to 0.62 at (y = 0.9).

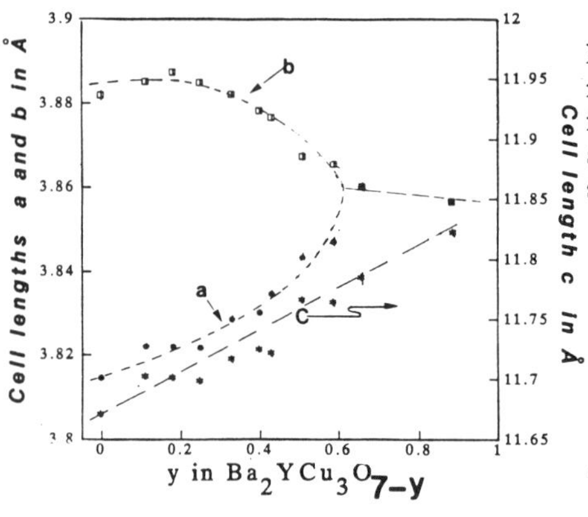

Figure 22. The lattice parameters a, b, c are reported as a function of the oxygen content for the $YBa_2Cu_3O_{7-y}$ series. The dashed lines are a guide to the eye. Note that the structural transition occurs near x = 0.6.

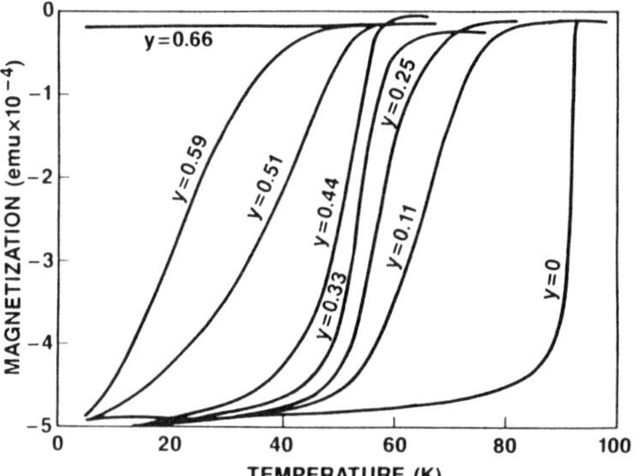

Figure 23. Temperature dependent magnetization for the $YBa_2Cu_3O_{7-y}$ series. The curves shown are for warming the samples in a 10G field after they have been cooled in zero field (shielding effect). T_c is defined at the midpoint (50 percent of the total diamagnetic shift).

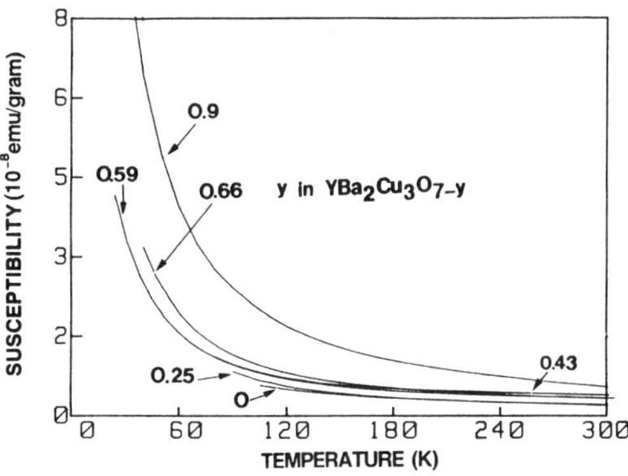

Figure 24. The temperature dependent susceptibility is reported for several members of the $YBa_2Cu_3O_{7-y}$ series from above T_c to room temperature. Note the dramatic change between $x=0$ and $x=0.9$.

4. DISCUSSION

We have shown that a strontium substitution enhances T_c for the 40K materials but decreases it for the 90K materials. Furthermore, for both systems we found that substituting magnetic rare earth ions has little effect on T_c whereas 3d transition metal substitutions destroy it. In all cases the superconducting properties of these new oxide compounds have been found to be extremely sensitive to oxygen content; which is determined by the processing conditions, namely heating time, temperature and ambient. Although the structure of these materials is now known and there has been much theoretical work, the mechanism for superconductivity in these new oxides has yet to be determined. However, the experimental results strongly suggest that the unexpected, high oxidation state of copper in these new materials is associated with their superconductivity. We now discuss our data with respect to the structure of these materials.

Changes in the valence of copper can be achieved either through a strontium substitution or by oxygen doping. We assume all variable valence is assigned to the copper. For the Sr doped 40K system the maximum T_c occurs at $x=0.15$, which corresponds to an average valence of copper (AVC) of 2.15. For the 90K system, T_c is maximum for an oxygen content of 6.9-7 equivalent to an AVC of 2.26-2.33. Thus it appeared that a valence of copper greater than 2 was required for superconductivity in these oxides. This empirical observation was satisfactory until the recent data on La_2CuO_4.[27-28] Indeed, the oxygen deficient La_2CuO_4 (AVC lower than 2) is semiconducting whereas the (ostensibly) stoichiometric phase (AVC=2) superconducts. Although we cannot rule out the possibility of a deficiency in either La or Cu in the La_2CuO_4 compound which would then yield an AVC above 2; a more likely explanation, in view of the plasma oxidation studies, is that the superconducting La_2CuO_4 contains excess oxygen such that the oxygen content per unit formula will be $4+y$ instead of 4, again yielding an AVC greater than 2. Work remains to be done to test this important supposition.

Superconductivity is generally quite affected by the presence of magnetic impurities which interact directly with the superconducting electrons and break the Cooper pairs, thereby suppressing or destroying T_c. The absence of a correlation between the superconducting properties of the rare earth doped compounds and the magnetism of the rare earth, for both the 40K and 90K systems, suggests that the superconducting electrons in these materials do not interact with the magnetic ions. In the structure, the magnetic ions may be situated away from the superconducting electrons (Cu3d-O2p electrons) so that the interactions are weak. In the 90K materials, the rare earth ions are adjacent to the CuO_2 planes ($\sim1.5\text{Å}$), but are about 6Å from the Cu-O chains situated in the middle plane. Thus, the indifference of T_c to the rare earth substitution could be explained by the chains being responsible for superconductivity in the 90K systems. For the 40K materials, the rare earth ions are adjacent to the Cu-O planes responsible for superconductivity, but are situated at a distance of 4.6Å so that the interactions are weak. The presence of 3d magnetic impurities on copper sites rapidly destroys T_c in the 40K system, as expected from the Arbrikovsov and Gor'Kov theory,[23] because of the direct interaction between Ni magnetic ions and the superconducting electrons. In contrast, for the 90K system T_c is only lowered to 50K by a nickel substitution. There are two type of copper sites in the structure and it may be that nickel substitutes for the coppers which do not contribute to the superconductivity, thus T_c is less affected.

We demonstrated that the superconducting properties of these new oxides are extremely sensitive to oxygen content. For instance, upon removal of oxygen the orthorhombic p-type superconducting phase $YBa_2Cu_3O_7$ transforms to the tetragonal n-type semiconducting phase

YBa$_2$Cu$_3$O$_6$. It appears that this structural transition, which occurs near y=0.6, is correlated with the onset of the superconducting-semiconducting transition. Although the value of y is somewhat inaccurate, it is tempting to speculate that at y=0.5 the conduction band is half full thus producing the observed semiconducting behavior. On the other hand, the band is half filled for stoichiometric "La$_2$CuO$_4$," and we have shown bulk superconductivity in this compound.

From a simple valence calculation, a value greater than 2 (Cu^{+3}, Cu^{+2}) and smaller than 2 (Cu^{+1}, Cu^{+2}) has been ascribed to the valence of copper in the YBa$_2$Cu$_3$O$_7$ and YBa$_2$Cu$_3$O$_6$ phases, respectively. This is confirmed by structural studies which have shown that,[37-38] during reduction, the oxygens are removed from the Cu-O chains leaving behind twofold coordinated copper atoms like those in Cu$_2$O, where copper is monovalent. The removal of oxygen from the chains destroys T$_c$ leading to the same conclusion as that established by the rare earth substitution, namely that the Cu-O chains are responsible for superconductivity in these materials. The removal of oxygen from the chains generates some Cu^{+1}, (as observed by photoemission[39]) decreases the number of carriers, and interrupts the path for the superconducting carriers, all of which promote a semiconducting-like behavior. Hall measurements show that this transition occurs rather abruptly with respect to composition.[40] Magnetic measurements show a large increase in the susceptibility, most likely due to electron localization, which is not correlated with this transition. An unexpected result is that the removal of oxygen appears not to change T$_c$ continuously, but instead produces a material with a T$_c$ of 55K $\pm$ 5K. Distinct T$_c$'s of 55K and 90K have been observed in controlled processed materials. We suggested three possible mechanisms to explain the two T$_c$s in these materials; 1) the occurrence of two distinct phases which would correspond to a different ordering of the oxygens within the Cu-O chains (such an ordering if it exists should be detectable by neutron diffraction), 2) different superconductivity mechanisms (superconductivity in the Cu-O chains and in the Cu-O plane) and 3) electronic coupling through the Cu-O plane when required by the presence of oxygen vacancies to produce the 55K material, whereas this coupling is not required for the stoichiometric material. Our experimental data favors superconductivity in the Cu-O chains. On the other hand, recent work[41] has shown superconductivity at 90K in the La$_3$Ba$_3$Cu$_6$O$_{14+y}$ phase, and there are no Cu-O chains in this structure.

We point out that the replacement of Cu by nickel and the loss of oxygen have a similar effect on the superconducting properties of these materials. They each promote a superconducting-semiconducting transition and a considerable increase in the magnetic susceptibility. One possibility is that the decrease in T$_c$ induced by the removal of oxygen may be of a magnetic nature. The origin of such a magnetic effect could result from an electron localization which makes some copper magnetically active, or simply from an oxygen vacancy trapping a single electron (S=1/2) and interacting, like nickel, with the Cu3d-O2p superconducting electrons. Another possibility is that the nickel substitution introduces defects which produce a disordering of the Cu-O network. We found that the removal of oxygen introduces disordering in the middle plane (orthorhombic-tetragonal transition). Thus the suppression of T$_c$ with both oxygen and nickel may be related to an order-disorder effect. Distinguishing between these two possibilities may be resolved by introducing a diamagnetic divalent ion (e.g., Zn) on the copper sites, but again we risk simultaneously introducing oxygen vacancies.

ACKNOWLEDGMENTS

We wish to thank J. M. Rowell, N. G. Stoffel, E. M. Vogel, J. H. Wernick and E. Yablonovitch for valuable discussions, and W. L. Feldmann and W. E. Quinn for technical assistance.

REFERENCES

1. J. G. Bednorz and K. A. Muller, Z. Phys. B 64, 189 (1986).

2. K. Kishio, K. Kitazawa, S. Kanbe, I. Yasuda, N. Sugii, H. Takagi, S. Uchida, K. Fueki and S. Tanaka, Chem. Lett., (Japan) 429 (1987).

3. R. J. Cava, R. B. van Dover, B. Batlogg and E. A. Rietman, Phys. Rev. Lett. 58, 408 (1987).

4. J. M. Tarascon, L. H. Greene, W. R. McKinnon, G. W. Hull and T. H. Geballe, Science 235, 1373 (1987).

5. M. K. Wu, J. R. Ashburn, C. J. Torng, P. H. Hor, R. L. Meng, L. Gao, Z. J. Huang, Y. Q. Wang and C. W. Chu, Phys. Rev. Lett. 58, 908 (1987).

6. P. H. Hor, L. Gao, R. L. Meng, Z. J. Huang, Y. Q. Wang, K. Forster, J. Vassiliou, C. W. Chu, M. K. Wu, J. R. Ashburn and C. J. Torng, Phys. Rev. Lett. 58, 911 (1987).

7. People's Daily, China, 25 Feb 1987.

8. J. M. Tarascon, L. H. Greene, W. R. McKinnon and G. W. Hull, Phys. Rev. B35, 7115 (1987).

9. R. J. Cava, B. Batlogg, R. B. van Dover, D. W. Murphy, S. Sunshine, T. Siegrist, J. P. Remeika, E. A. Rietman, S. Zahurak and G. Espinosa, Phys. Rev. Lett. 58, 1676 (1987).

10. Y. Le Page, W. R. McKinnon, J. M. Tarascon, L. H. Greene, G. W. Hull and D. M. Hwang, Phys. Rev. B35, 7245 (1987).

11. B. Batlogg, R. J. Cava, A. Jayaraman, R. B. van Dover, G. A. Kourouklis, S. Sunshine, D. W. Murphy, L. W. Rupp, H. S. Chen, A. White, K. T. Short, A. M. Mujsce and E. A. Rietman Phys. Rev. Lett. 58, 2333 (1987).

12. J. Bardeen, L. N. Cooper, J. R. Schrieffer, Phys. Rev. 108, 1175 (1957).

13. P. W. Anderson, Science 235, 1196 (1987).

14. C. M. Varma, S. Schmitt-Rink, and E. Abrahams, Solid State Commun. 62 681 (1987).

15. L. F. Mattheis and D. R. Hamann, Solid State Commun. (in press).

16. C. Michel and B. Raveau Revue de Chimie Minerale 21, 407 (1984).

17. N. P. Ong, Z. Z. Wang, J. Clayhold, J. M. Tarascon, L. H. Greene and W. R. McKinnon, Phys. Rev. B36 July 1, 1987 (in press).

18. K. K. Singh, P. Ganguly and J. B. Goodenough, J. Solid State Chem. 52, 254 (1984).

19. Y. Yamaguchi, H. Yamauchi, M. Ohashi, H. Yamamoto, N. Shimoda, M. Kikuchi and Y. Syono, Jpn. J. Appl. Phys. 26, L447, (1987).

20. P. Ganguly and C. N. R. Rao, J. Solid State Chem., 53, 193 (1984).

21. J. M. Tarascon, L. H. Greene, W. R. McKinnon and G. W. Hull, Solid State Commun. 63, 499 (1987).

22. J. Kondo, Prog. Theoret. Phys. 32, 37, (1964); Solid State Physics Vol. 23 (F. Seitz and D. Turnbull eds.), Academic Press, NY, 1969. p. 183.

23. A. A. Abrikosov and L. P. Gor'kov, Zh. Eksp. Teor. Fiz., 39, 1781 (1960); Sov. Phys. -JETP (Engl. Transl), 12, 1243 (1961).

24. J. M. Tarascon, W. R. McKinnon, L. H. Greene, G. W. Hull and E. M. Vogel, Phys. Rev. B36, July 1, 1987, (in press).

25. J. M Tarascon, W. R. McKinnon, L. H. Greene, G. W. Hull, B. G. Bagely E. M. Vogel and Y. LePage, Proceedings of the Materials Research Society 1987 Spring Meeting, Symposium S: High Temperature Superconductors with T_c over 30K. (D. V. Gubser and M. Schluter, eds.) in press; and W. R. McKinnon, J. M. Tarascon, L. H. Greene and G. W. Hull, ibid.

26. T. P Orlando, K. A. Delin, S. Foner, E. J. McNiff, Jr., J. M. Tarascon, L. H. Greene, W. R. McKinnon and G. W. Hull, Phys Rev. Lett., (submitted).

27. J. O. Willis, Z. Fisk, J. D. Thompson, S.-W. Cheong, R. M. Aikin, J. L. Smith and E. Zirngiebl, J. Magn. Mag. Mat. 67, L139 (1987).

28. S. E. Brown, J. D. Thompson, J. O. Willis, R. M. Aikin, E. Zirngiebl, J. L. Smith, Z. Fisk and R. B. Schwarz, Phys. Rev. B. (submitted).

29. J. M. Tarascon, L. H. Greene, W. R. McKinnon, G. W. Hull and P. Barboux, Solid State Commun. (to be submitted).

30. N. Nguyen, F. Studer and B. Raveau, J. Phys. Chem. Solids 44, 6389 (1983).

31. J. Beille, R. Cubanel, C. Chaillout, B. Chevalier, G. Demazeau, F. Deslandes, J. Elourneau, P. Lejay, C. Michel, J. Provost, B. Raveau, A. Sulpice, J. Tholence and R. Tournier, C. R. Acad. Sc. Paris 18, 304 (1987).

32. P. M. Grant, S. S. P. Parkin, V. Y. Lee, E. M. Engler, M. L. Ramirez, J. E. Vazquez, G. Lim, R. D. Jacowitz and R. L. Greene, Phys. Rev. Lett. *58*, 2482 (1987).

33. J. M. Tarascon, W. R. McKinnon, L. H. Greene, G. W. Hull, B. G. Bagley and E. M. Vogel. Proceedings of the American Ceramic Society 1987 Spring Meeting, Special Session on High T_c Superconductor Oxides (D. R. Clarke, ed.) in press.

34. P. K. Gallagher, H. M. O'Bryan, S. A. Sunshine and D. W. Murphy, Mat. Res. Bull., in press.

35. W. R. McKinnon, J. M. Tarascon, L. H. Greene and G. W. Hull Phys. Rev. B (to be submitted).

36. B. Bagley, L. H. Greene, J. M. Tarascon and G. W. Hull, Appl. Phys. Lett. (submitted).

37. J. E. Greedan, A. O'Reilly and C. V. Stager, Phys. Rev. B. *36*, June 1, 1987 (in press).

38. A. S. Santoro, S. Miraglia, F. Beech, S. A. Sunshine, D. W. Murphy, L. F. Schneemeyer and J. W. Waszczak, (preprint).

39. N. Stoffel, J. M. Tarascon, Y. Chang, M. Onellion, D. W. Niles and G. Margarito, Phys. Rev. B (submitted).

40. N. P. Ong, Z. Z. Wang, J. Clayhold, J. M. Tarascon, L. H. Greene and W. R. McKinnon (this conference).

41. D. B. Mitzi, A. F. Marshall, J. Z. Sun, D. J. Webb, M. R. Beasley, T. H. Geballe and A. Kapitulnik, Phys. Rev. B (submitted).

PROPERTIES OF ORIENTED OXIDE SUPERCONDUCTOR THIN FILMS PREPARED BY

PULSED LASER EVAPORATION FROM HIGH T_c BULK MATERIAL

D. Dijkkamp, X. D. Wu[*], S. B. Ogale[+], A. Inam[*], E. W. Chase,
P. Miceli, J. M. Tarascon and T. Venkatesan

Bell Communications Research, Red Bank, NJ 07701-7020

ABSTRACT

We report on the preparation of thin films of Y-Ba-Cu-O superconductors
using pulsed excimer laser evaporation of bulk material. Rutherford backscattering spectrometry showed the composition of these films to be close to
that of the bulk material. Growth rates were typically 1 nm per laser shot.
This new deposition method is relatively simple, very versatile, and does
not require the use of ultrahigh vacuum techniques. After an annealing
treatment in oxygen the films exhibited superconductivity with zero resistance in the range 55 K - 85 K, depending on the annealing conditions and
the type of substrate. In particular, films prepared on $SrTiO_3$<100> showed
transition widths as narrow as 2 K, and a considerable amount of orientation, as observed by X-ray diffraction and ion channeling techniques. There
is evidence for strong interface reaction with in-diffusion of Ba and Cu and
out-diffusion of Ti and Sr, resulting in a thickness dependence of the
superconducting properties of the films.

INTRODUCTION

Since the discovery of superconductivity above 30 K in the La-Ba-Cu-O
system[1], and the subsequent discovery of superconductivity in Y-Ba-Cu
oxides[2], the preparation of bulk superconducting oxide samples, using conventional ceramic techniques has been reproduced by many groups. On the
other hand, preparing thin films of these materials has turned out to be a
more difficult task. Successful deposition using sputtering from bulk materials has been reported for the 40 K superconductors[3,4], including the
preparation of single crystal thin films[5]. Co-evaporation of the metallic
constituents has been used to prepare 90 K films[6,7]. Critical current densities as high as 3×10^6 A/cm^2 at 4 K were shown in single crystal films of
Y-Ba-Cu-oxide[8]. However, for both methods control of the deposition conditions as well as the post-annealing conditions was critical.

We have recently reported the successful deposition of Y-Ba-Cu-O thin
films on sapphire substrates using pulsed excimer laser evaporation from a
bulk target[9]. After an annealing treatment in oxygen these films showed
superconductivity with onsets around 95 K and zero resistivity at 75 K.
In this paper we will briefly describe our preparation method, followed by
recent results of electrical and structural measurements on thin films
deposited on $SrTiO_3$ single crystal substrates.

DEPOSITION METHOD

Pellets with nominal composition $YBa_2Cu_3O_{9-x}$ were prepared in the usual way[10]. Electrical measurements using a four point probe with indium soldered contacts showed onset temperatures of 95 K and transition widths of less than 1 K. A pellet was mounted in a small vacuum system with a base pressure of 5×10^{-7} Torr, and irradiated through a quartz window with a KrF excimer laser (Lambda Physik EMG200E, 30 ns FWHM, 1 J/shot) at 45° angle of incidence. A quartz lens was used to obtain an energy density of approximately 2 J/cm^2 on the target. To avoid texturing of its surface, the pellet was slowly rotated during irradiation. The substrates were mounted at a distance of typically 3 cm from the pellet surface, and were heated to about 450 C during deposition. The laser was fired at a repetition rate of 3-6 Hz, for a total of a few thousand shots. With each shot, a plume of intense white light emission could be observed normal to the sample surface. During deposition, the pressure in the system rose to about $1-2 \times 10^{-6}$ Torr. Using a quadrupole mass spectrometer (Inficon, model Quadrex 200) the following fragments could be detected: O, O_2, O_3, Cu and CuO.

The as deposited films were shiny, dark brown but electrically insulating. Annealing of these films in an oxygen atmosphere for typically one hour at 850 - 900 C, followed by slow cooling to room temperature yielded the desired low resistivity (in the 10^{-3} Ohm cm range). We have tried to optimize the annealing conditions for films on sapphire substrates, by varying the temperatures, times, heating and cooling rates and correlating these with the quality of the resulting superconducting transitions. However, there seems to be some non-reproducibility in the annealing result, though it was clear that a minimum temperature of 850 C for about 30 minutes and slow ([100 C per hour) cooling are essential to obtain superconducting properties. Thus far, the results on $SrTiO_3$ are relatively more reproducible.

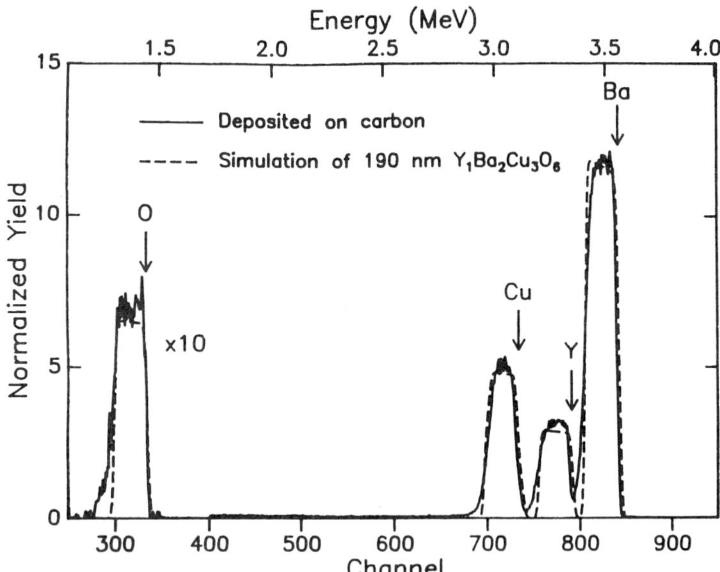

Fig. 1 RBS-spectrum (4 MeV He^{2+}, 175° backscattering angle) of an as-deposited laser evaporated film on a carbon substrate (solid line), overlayed with a simulation of 190 nm $Y_1Ba_2Cu_3O_6$ (dashed line).

The composition of the as-deposited and annealed films was determined
using Rutherford backscattering spectrometry (RBS). Figure 1 shows the RBS
spectrum of a film deposited at 450 C on vitreous carbon, overlayed with a
simulation of a 190 nm thick film with composition $Y_1Ba_2Cu_3O_6$. The simulation was made using the RUMP program[11] and assumes a density of 6.68 g/cm^3.

Obviously, the as-deposited film has almost the same composition as the
starting material, including even the oxygen. Pulsed laser heating is a
highly non-equilibrium process, in which the surface layer is strongly
overheated. This leads to massive expulsion of material, on the order of
100 nm per shot, as determined from mechanical measurements. This material
is most likely ejected in the form of fairly large molecular clusters, which
may explain the preservation of the stoichiometry in the deposited film, in
particular the high oxygen content. Moreover, the rates of heating and
cooling are so fast that phase separation cannot occur in the bulk material.
This was confirmed by RBS spectra of the irradiated area, which were identical to that of the pristine material.

ELECTRICAL MEASUREMENTS

Resistivity vs. temperature measurements were made using the conventional four probe technique. Contacts were made either by indium soldering or
silver ink (Engelhard, Flexible Silver #16). Currents of 1 to 10 micro-amps
were used. The samples were cooled using a closed helium refrigerator
system, which could reach a minimum temperature of about 10 K, as measured
with a calibrated silicon diode (Lake Shore Cryotronics, model DR-500CU).
Although temperature gradients are hard to avoid completely in such a system, we believe that the absolute temperature was determined to better than
3 K, during slow warming up of the system.

We have deposited thick (350 nm) and thin (100 nm) films on
<100> $SrTiO_3$ which is a lattice matched substrate. For the thick film, part
of the substrate was masked to produce a conventional I-V pattern, with
0.2 mm linewidth and 2 mm distance between the voltage leads. After
annealing at 900 C for one hour, followed by slow cooling in an oxygen
ambient, the thick film exhibited a very sharp superconducting transition,

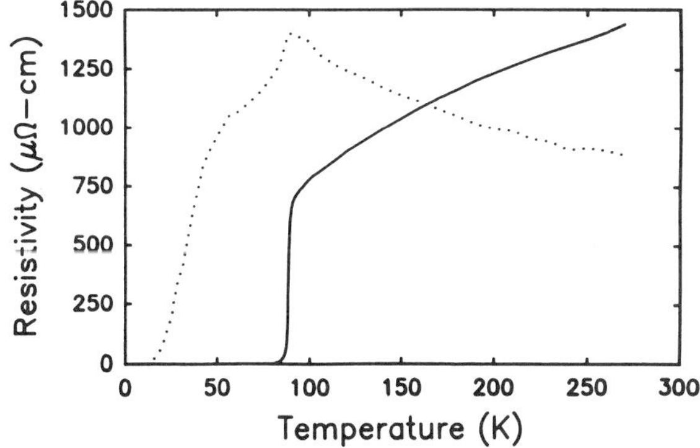

Fig. 2 Resistivity vs. temperature for two films
       on <100> $SrTiO_3$ and annealed for one hour
       at 900 C. Solid line: 350 nm thick film
       with a transition width (90 % - 10 %) of
       about 2 degrees K. Dotted line: 100 nm thick
       film with resistivity in arbitrary units.

as shown in figure 2 (measured on the I-V pattern). The width of the
transition (90 % - 10 %) is about 2 degrees Kelvin. Note that the resistivity above T_c is decreasing towards lower temperature, as is characteristic
for good bulk material. The result on the thin film (also shown in
figure 2, but in arbitrary units of resistivity) is not as good, indicating
contamination effects from the substrate, as was confirmed by the RBS-
spectra discussed later on. We have used ion channeling and X-ray spectrometry to study the films deposited on $SrTiO_3$. The results will be discussed
in the remainder of this paper.

STRUCTURAL MEASUREMENTS

X-ray diffraction performed on films deposited on <100> $SrTiO_3$ substrates show preferential orientation, where the c-axis of the 3-fold
stacked perovskite, $YBa_2Cu_3O_{7-x}$, points normal to the plane of the film. As
shown in figure 3, there are strong (001) reflections corresponding to
l=1,2,5,7, while the l=3 and l=6 are obscured by the (001) and (002) $SrTiO_3$
reflections. A transverse scan (theta scan) across a (001) reflection revealed that the orientation of the c-axis is better than 1^0. We have not
done measurements with a component of the scattering vector in the plane of
the film; thus, at present, we have not determined if there is an azimuthal
orientation of the films with the substrate.

In addition to a large portion of the film which is preferentially
oriented, there is a smaller portion which is powder-like. A transverse scan
on the reflection labeled "P" in figure 1, corresponding to the (110), (103)
and (013) reflections in $YBa_2Cu_3O_{7-x}$, showed no intensity variation and is
characteristic of a powder. The group of peaks near $2\theta = 30^0$ is believed to
be due to a slight presence of $BaCuO_2$.

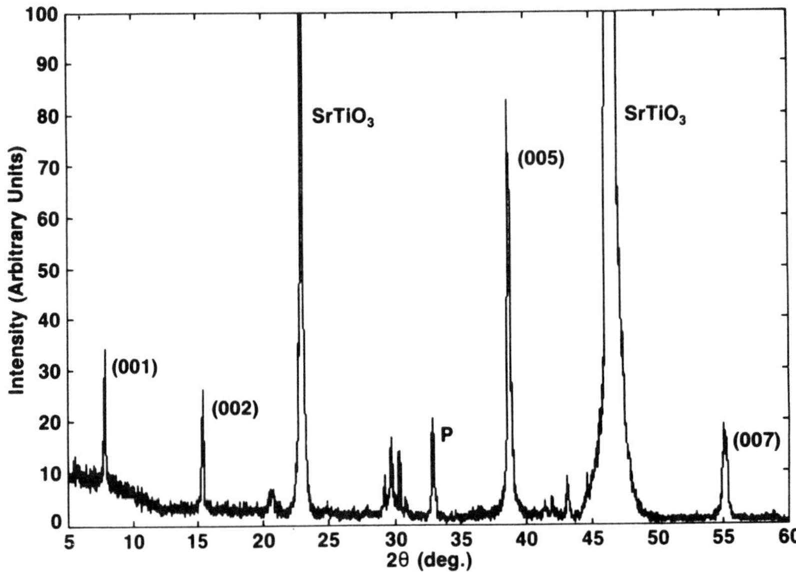

Fig. 3 X-ray diffraction spectrum of an annealed 350 nm thick film on
<100> $SrTiO_3$. The strong presence of (001) $YBa_2Cu_3O_{7-x}$ reflections
indicates preferential orientation along the c-axis.

These experiments were performed on a Scintag Pad V diffraction system using Cu radiation, where a solid state detector is employed to discriminate the K component. The instrumental resolution places a lower limit on the domain size of 75 nm, although we believe that it may be larger. Further studies are being pursued to determine the upper limit on the crystallite size. There is considerable peak broadening at high scattering angles, which may be attributed to a residual strain of about 5 %.

In contrast to the films on $SrTiO_3$, the films deposited on Al_2O_3 substrates showed only a small degree of orientation, with the c-axis parallel to the surface.

Random and channeling spectra were taken on the thin samples. Figure 4 shows the results for the as-deposited film. The random spectrum is well fitted by a simulation of 100 nm $Y_1Ba_2Cu_3O_7/SrTiO_3$, as shown. Channeling is observed in the substrate but not in the film, implying that the as-deposited film is not oriented with respect to the substrate.

The situation differs for the annealed film, as shown in figure 5. In this case, the intensities of all three metal peaks in the channeled spectrum are about 70 % of the random values, which clearly demonstrates the orientation of the film with respect to the substrate. To our knowledge, these are the first channeling measurements on this material.

The random spectrum is best fitted by a simulation of a 100 nm thick film with composition $YBa_{1.5}Cu_2O_7$ on $SrTiO_3$. The apparent enrichment of yttrium may either be due to loss of barium and copper from the film or significant out-diffusion of strontium, which would enhance the yttrium signal, since the two masses are nearly the same. However, in the case of these films on sapphire a clear yttrium enrichment was also observed for similar anneals, which can only be due to loss of barium (and to a lesser extent copper) by evaporation.

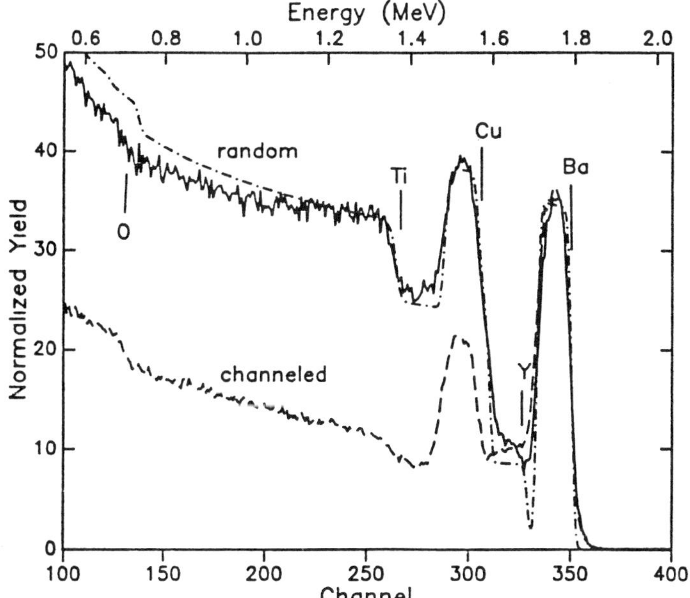

Fig. 4 Random and channeled RBS spectra of an
       as-deposited film on $SrTiO_3$. The dot-
       dashed line is a simulation of 100 nm
       $YBa_2O_3O_7/SrTiO_3$.

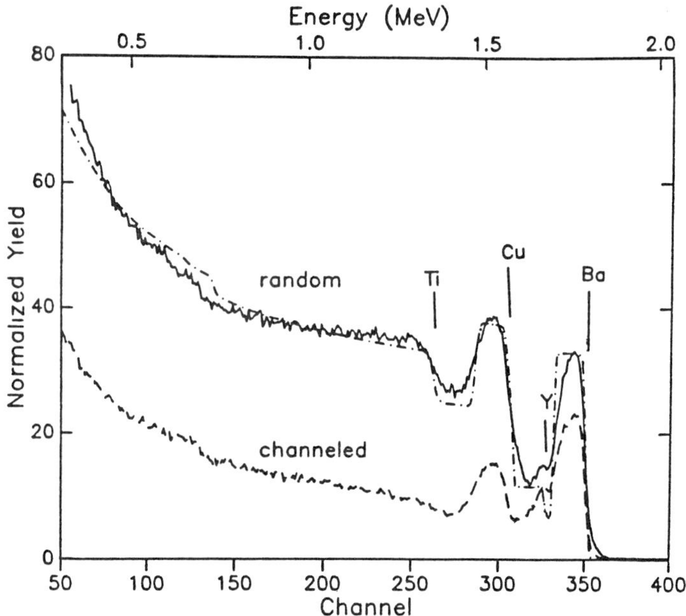

Fig. 5 Random and channeled RBS spectra of annealed
film on $SrTiO_3$, showing a 70 % minimum yield
for the Y, Ba and Cu signals in the channeled
spectra. The dot-dashed line is a simulation
of 100 nm $YBa_{1.5}Cu_{2.3}O_7/SrTiO_3$.

It can be seen that both barium and copper are depleted at the interface, indicating in-diffusion. In addition, outdiffusion of titanium (and
presumably strontium) can be observed. The channeled spectrum clearly shows
pile-up of yttrium at the surface, most likely due to the formation of
yttrium oxide, which was also observed for annealed films on sapphire.

The poor electrical properties of this film must be attributed to
interface mixing and loss of the correct stoichiometry during the annealing
process over a thickness of at least 100 nm. At present, these effects pose
serious limitations for achieving high quality superconducting films with
thickness much less than 300 nm.

CONCLUSION

We have shown that laser deposition is a powerful, versatile and
extremely convenient technique to prepare thin films of superconducting
oxides, from a given bulk material. To obtain the desirable electrical
properties, these films need to undergo a high temperature anneal. We have
shown that films deposited on <100> $SrTiO_3$ are highly oriented after
annealing, and exhibit narrow superconducting transitions. However, a
considerable amount of interface reaction and surface evaporation takes
place during annealing, which puts a lower limit on the thickness of the
films. Presently, studies are being carried out to characterize the superconducting films in more detail.

The authors wish to thank J. M. Rowell and L. H. Greene for many
helpful suggestions in the course of the experiments.

REFERENCES

* Physics Department, Rutgers University, NJ 08855-0849
+ Permanent address: Physics Department, Poona University, India

1. J. G. Bednorz and K. A. Muller, Z. Phys. B **64**, 189 (1986).
2. M. K. Wu, J. R. Ashburn, C. T. Torng, P. H. Hor, R. L. Meng, L. Gao,
    Z. J. Huang, Y. Q. Wang and C. W. Chu, Phys. Rev. Lett. **58**, 908 (1987).
3. M. Kawasaki, M. Funabashi, S. Nagata, K. Fueki and H. Koinuma,
    Jpn. J. of Appl. Phys. **26**, L388 (1987).
4. N. Terada, H. Ihara, M. Hirabayashi, K. Senzaki, Y. Kimura, K. Murata
    and M. Tokumoto, Jpn. J. of Appl. Phys. **26**, L508 (1987).
5. M. Suzuki and T. Murakami, Jpn. J. of Appl. Phys. **26**, L524 (1987).
6. R. Hammond, MRS Spring Meeting, April 1987, Anaheim, CA, to be published.
7. R. B. Laibowitz, R. H. Koch, P. Chaudari and R. J. Gambino,
    MRS Spring Meeting, April 1987, Anaheim, CA, to be published.
8. P. Chaudhari, R. H. Koch, R. B. Laibowitz, T. R. McGuire and
    R. J. Gambino, to be published.
9. D. Dijkkamp, T. Venkatesan, X. D. Wu, S. A. Shaheen, N. Jisrawi,
    Y. H. Min-Lee, W. L. McLean and M. Croft, submitted to Appl. Phys. Lett.
10. J. M. Tarascon, W. R. McKinnon, L. H. Greene, G. W. Hull, B. G. Bagley
    and E. M. Vogel, MRS Spring Meeting, April 1987, Anaheim, CA, to be
    published.
11. L. R. Doolittle, Nucl. Instr. and Meth. in Phys. Res. **B9**, 344 (1985).

OXYGEN ISOTOPE EFFECT IN THE HIGH TEMPERATURE SUPERCONDUCTORS
$YBa_2Cu_3O_{7-\delta}$ and $La_{1.85}Sr_{0.15}CuO_4$, WITH ^{18}O SUBSTITUTED BY DIFFUSION

Marvin L. Cohen[a,b], D. E. Morris[a,c], A. Stacy[b,d], and A. Zettl[a,b]

[a] Department of Physics, University of California, Berkeley, CA 94720
[b] Materials and Chemical Sciences Division, Lawrence Berkeley Laboratory
   Berkeley, CA 94720
[c] Physics Division, Lawrence Berkeley Laboratory, Berkeley, CA 94720
[d] Department of Chemistry, University of California, Berkeley, CA 94720

One of the most suggestive tests for electron-phonon coupling in conventional superconductors is the isotope effect, where T_c is sensitive to ionic mass: $T_c \propto M^{-\alpha}$. The presence of an isotope shift would demonstrate an important role played by the phonons in the superconductivity mechanism. The high-T_c superconducting oxide $La_{1.85}Sr_{0.15}CuO_4$ and $YBa_2Cu_3O_{7-\delta}$ have many structural[1] and electronic[2] features in common. An important feature common to both the layered perovskite K_2NiF_4 structure of $La_{2-x}M_xCuO_4$ (M = Ba, Sr, or Ca) and the distorted oxygen-defect pervoskite structure of $RBa_2Cu_3O_{7-\delta}$ (R = Y or some lanthanide rare-earth elements) is the existence of two-dimensional copper-oxygen planes which are most responsible for the metallic conduction.

Despite the similarities, however, there are important differences between $YBa_2Cu_3O_{7-\delta}$ and $La_{1.85}Sr_{0.15}CuO_4$, including a significantly lower T_c, ($\approx$ 40K $vs.$ 90K), and a lack of one dimensional Cu-O "chains" in the La material. Also, $La_{1.85}Sr_{0.15}CuO_4$ has dramatically different elastic properties[3]. Band-structure calculations[2] on $La_{1.85}Sr_{0.15}CuO_4$ demonstrate that the density of states at the Fermi level is due largely to hybridized Cu $3d$ and O $2p$ states. Calculations of the electron-phonon interactions in $La_{2-x}Sr_xCuO_4$ by Weber[4] suggest that strong coupling of specific oxygen phonons to the conduction electrons can lead, within conventional (phonon mediated) BCS theory, to T_c values of 30-40K. It is therefore expected that vibration of the O ions would be involved if electron-phonon coupling is responsible for the superconductivity in $La_{1.85}Sr_{0.15}CuO_4$.

A non-zero oxygen isotope shift has been observed in the superconducting oxide $La_{1.85}Sr_{0.15}CuO_4$, in sharp contrast to our observation[5,6] of no oxygen isotope shift in $YBa_2Cu_3O_{7-\delta}$. Substitution of ^{18}O for ^{16}O in $La_{1.85}Sr_{0.15}CuO_4$ results in a depression of the superconducting transition temperature T_c. The substitution was carried out in a similar manners as in $YBa_2Cu_3O_{7-\delta}$ (see below), but was more difficult. Magnetic (Meissner effect) and resistive measurements indicated that T_c is depressed by 0.3 - 1.0K in various samples when the isotope ^{18}O is substituted for 65% to 78% of the ^{16}O in the sample. The observed shifts in T_c were extrapolated to 100% ^{18}O. With $T_c \approx M^{-\alpha}$, where M is the oxygen mass, we find $0.10 < \alpha < 0.35$. This non-zero isotope shift indicates that in $La_{1.85}Sr_{0.15}CuO_4$, phonons may play an important role in the superconductivity mechanism. We examine the consequences of this result for phonon mediated electron pairing, within the framework of BCS theory.

We previously searched for an isotope effect in $YBa_2Cu_3O_{7-\delta}$, by substituting the isotope ^{18}O for ^{16}O. The oxygen isotope exchange was carried out by placing a fine grained porous sample in a fused quartz tube connected to an ^{18}O reservoir and heating at 950°C for 12hr. We were able to substitute an estimated 90% of the ^{16}O in the sample by ^{18}O (see below). No shift in T_c was observed in $YBa_2Cu_3O_{7-\delta}$ within 0.3 K, the experimental uncertainty. The expected isotope shift in basic BCS theory ($T_c \propto M^{-1/2}$) would correspond to a reduction in T_c of ~4.6 K on substitution of 90% of the ^{16}O by ^{18}O, if T_c is determined by motion of the O ions. We found[5,6] that the oxygen isotope effect is absent in this material, $\alpha = 0 \pm 0.027$.

In order to determine the appropriate conditions for oxygen isotope substitution, a series of thermogravimetric analysis (TGA) experiments were carried out on $YBa_2Cu_3O_{7-\delta}$ (Ref 6) and on $La_{1.85}Sr_{0.15}CuO_4$. Samples were heated in oxygen or argon and then cooled at various rates. All experiments were carried out at ambient pressure. The change in oxygen content of the sample was measured by the change in sample weight following a change in temperature or of the partial pressure of oxygen in the surrounding space. It was found that the equilibrium oxygen content varies with temperature, and the partial pressure of oxygen in the surrounding space. At temperatures above about 450°C in $YBa_2Cu_3O_{7-\delta}$, or above 700°C in $La_{1.85}Sr_{0.15}CuO_4$, oxygen is lost, but the process is reversible up to at least 1000°C, where the weight loss in O_2 at one atmosphere pressure reaches about 1.3% in $YBa_2Cu_3O_{7-\delta}$. This corresponds to about 9% of the total oxygen content or a change of ~0.5 oxygen per formula unit. When $YBa_2Cu_3O_{7-\delta}$ samples were cooled slowly in O_2, the oxygen content increased until a maximum was reached at 400-450°C. The oxygen loss from $La_{1.85}Sr_{0.15}O_4$ was much smaller.

Heating $YBa_2Cu_3O_{7-\delta}$ in argon nearly doubled the weight loss at each temperature; the loss was only reversible by annealed and cooling in O_2. The effects of heating and cooling in argon and in oxygen at various rates on the superconducting properties were studied.

The time constants for oxygen diffusion into or out of $YBa_2Cu_3O_{7-\delta}$ were found by observing the change in sample weight following a rapid temperature step. The sample weight was found to approach the new equilibrium value by an exponential decay, rather than following a $(time)^{-1/2}$ dependence as would be expected if the diffusion was slow and the interior of the grains did not reach equilibrium. The time constants for oxygen release and uptake were found to be quite short, 10^1 - 10^3 sec., depending on temperature and sample grain size. We concluded that oxygen in this material is mobile at temperatures above 400-500°C, with very high diffusion rates. The time constants for oxygen diffusion into or out of $La_{1.85}Sr_{0.15}CuO_4$ at elevated temperatures were also found to be short, though the changes in oxygen content with temperature in an O_2 atmosphere were much smaller than in $YBa_2Cu_3O_{7-\delta}$.

The very rapid oxygen diffusion in $YBa_2Cu_3O_{7-\delta}$ and in $La_{1.85}Sr_{0.15}CuO_4$ suggested that simple diffusion at elevated temperature at constant O_2 pressure would be an effective method to replace the ^{16}O in a sintered sample by ^{18}O, from $^{18}O_2$ gas placed in the surrounding space. Fine grained porous pellets were used in these experiments. The untreated pellets had sharp resistive transitions to the superconducting state. Several pieces of the $YBa_2Cu_3O_{7-\delta}$ pellets were heated to 950°C in an atmosphere of $^{18}O_2$, held at that temperature for 10 hours and cooled in 6 hours. This temperature and reaction time were chosen to insure maximal ^{18}O exchange with ^{16}O at both weakly bound and tightly bound sites.

In the case of $La_{1.85}Sr_{0.15}CuO_4$ the porous pellets were heated to 940°C-1000°C in an $^{18}O_2$ atmosphere, held at that temperature for 20 hours or more, and then heated to 1050°C-1100°C to complete the sintering. All samples were slow cooled. The extent of isotope substitution was verified by temperature programmed reduction measurements using a quadrupole mass spectrometer. The superconducting resistive transitions remained sharp after the isotope substitution. Pieces broken from the same pellets were given almost identical diffusion treatment in ^{16}O as a control. The T_c of the ^{18}O substituted materials were compared to the controls, and to fragments of the untreated pellets.

Figure 1 shows χ *vs.* temperature for an ^{18}O substituted $La_{1.85}Sr_{0.15}CuO_4$ sample and for the corresponding ^{16}O sample. For the ^{16}O samples the diamagnetic onset temperature is 35 K, while for the ^{18}O substituted sample (65% ^{18}O, 35% ^{16}O) the onset temperature is 34.4 K. The shift in diamagnetic onset temperature is 0.6 K. At lower temperatures the offset between the two curves increases for reasons not completely understood.

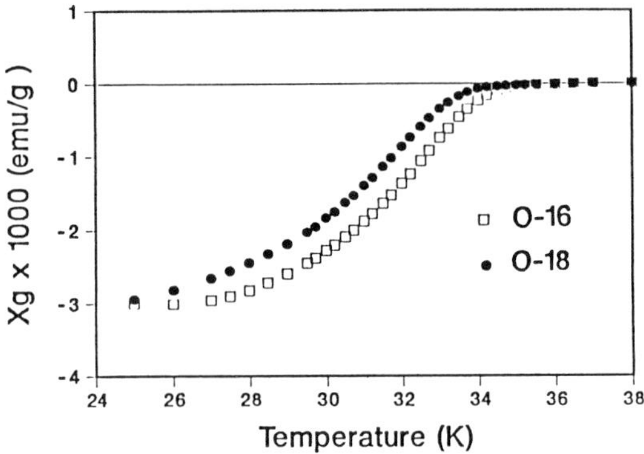

Figure 1: Magnetic susceptibility per gram $La_{1.85}Sr_{0.15}CuO_4$ versus temperature for an ^{18}O isotope substituted portion (65% ^{18}O, 35% ^{16}O) and the ^{16}O portion of the same pellet.

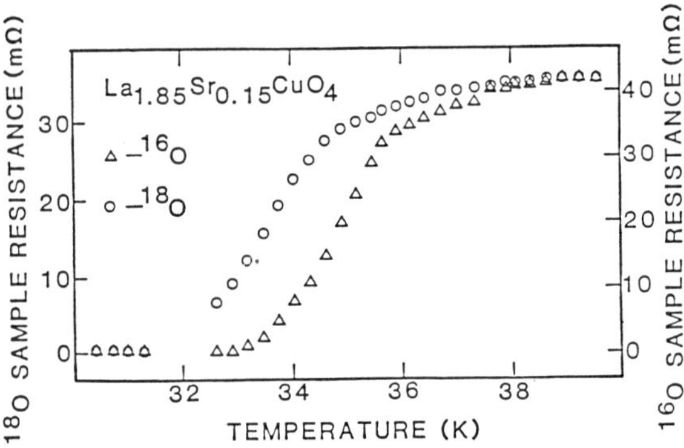

Figure 2: Resistance of $La_{1.85}Sr_{0.15}CuO_4$ *vs.* temperature for the ^{18}O substituted and ^{16}O portions of the same pellet as Figure 1. The sample dimensions and lead spacing vary, so the normal state resistances differ.

Figure 2 shows resistance data near T_c for the same specimen used for the susceptibility measurements of Fig. 1. A well defined break in the R *vs.* temperature curve occurs at $T_c = 35.5$K for the ^{16}O sample and at $T_c = 34.5$K for the ^{18}O sample.

Both our magnetic and resistance experiments demonstrate a consistent and finite isotope effect in $La_{1.85}Sr_{0.15}CuO_4$. With $\alpha = 0.5$ as predicted from phonon BCS theory, one would expect an ^{18}O - ^{16}O shift of T_c of about 2.1 K, much larger than we observed. We find $0.1 < \alpha < 0.3$ for this material, significantly less than the BCS prediction $\alpha = 0.5$, but in sharp contrast to the null isotope effect observed for $YBa_2Cu_3O_{7-\delta}$, where $\alpha = 0.0 \pm 0.027$ (Refs 5 and 6).

The implications of the observed oxygen isotope shift in $La_{1.85}Sr_{0.15}CuO_4$ may be examined within the context of conventional phonon-mediated electron pairing within the BCS theory as we have done previously for $YBa_2Cu_3O_7$ (Ref 5). We have considered the possibility of obtaining an isotope shift in the observed range both within the two square-well model and from numerical solutions of the Eliashberg equations using a model phonon spectrum. Within these models, we have considered the limits on the parameters of the electron-electron interaction that arise from the constraints of the observed T_c and give isotope shifts in the range $0.10 < \alpha < 0.35$. We use a model phonon spectrum derived from the inelastic neutron-scattering data of Renker *et al.*[7] assuming a constant electron-phonon interaction. Within this model, we find solutions with $T_c = 37$K for all $\lambda > 1.1$. The separate conditions $\alpha = 0.1, 0.15, 0.20, 0.25, 0.30,$ and 0.35 can also be satisfied for $\lambda > 3$. An isotope effect of $\alpha = 0.15$ with $T_c = 37$K is found for $\lambda \approx 5.25$ and $\mu^* \approx 0.43$.

Experimentally, values of $\alpha > 0.3$ are possible and further studies are needed to obtain a firm upper limit. If we take an upper limit of $\alpha = 0.35$, then the resulting $\lambda \approx 2.9$ and $\mu^* \approx 0.28$ are consistent with values obtained by Weber[4], when account is taken of the different phonon spectra used. Phillips[8] has predicted an isotope shift $La_{1.85}Sr_{0.15}CuO_4$ of $\alpha \approx 0.2$, which agrees well with the results presented here. Phillips' analysis is based on the role of oxygen vacancies in determining lattice stability and a phonon-induced electron-electron pairing.

We conclude therefore that if we do not consider material properties as Phillips has done but used conventional Eliashberg theory, then $\alpha \approx 0.15$ gives very large values for λ and μ^*. However, $\alpha \approx 0.35$ leads to μ's near the range observed previously for other superconductors.[9] We also note that if parts of the phonon spectrum are shifted independently, then the $\alpha = 0.15$ value can be obtained for smaller λ and μ^*.

In conclusion, a non-zero oxygen isotope shift has been found in the superconducting oxide $La_{1.85}Sr_{0.15}CuO_4$, in sharp contrast to $YBa_2Cu_3O_{7-\delta}$ in which we found that the oxygen isotope effect is absent.

Acknowledgements   The authors thank T. W. Barbee III, L. C. Bourne, K. J. Chang, M. F. Crommie, T. A. Faltens, W. K. Ham, C. S. Hoen, S. W. Keller, K. J. Leary, A. G. Markelz, J. N. Michaels, U. M. Scheven, and H.-C. zur Loye. AZ acknowledges support from the Alfred P. Sloan Foundation. DEM thanks Prof. D. W. Fuerstenau for providing the TGA apparatus, and Profs. R. A. Muller and L. W. Alvarez for encouragement and support during the course of this work. This research was supported by National Foundation Grants DMR-83-19024 (MLC) and DMR-84-00041 (AZ), and by the Director, Office of Energy Research, Office of Basic Energy Sciences, Materials Sciences Division of the U. S. Department of Energy under Contract No. DE-AC03-76SF00098.

References

[1]Cava, R. J. *et al.*, *Phys. Rev. Lett.* **58**, 1676-1679 (1987);   H. Takagi, S. Uchida, K. Kitazawa, and S. Tanaka, *Japan J. Appl. Phys. Lett.* **26**, L1 (1987).

[2]L. F. Mattheiss, *Phys. Rev. Lett.* **58**, 1371 (1987).

[3]L. C. Bourne, A. Zettl, K. J. Chang, Marvin L. Cohen, Angelica M. Stacy, and W. K. Ham (to be published); L. C. Bourne, Marvin L. Cohen, and A. Zettl (to be published).

[4]W. Weber, *Phys. Rev. Lett.* **58**, 1371 (1987).

[5]L. C. Bourne *et al.*, *Phys. Rev. Lett.* **58**, 2337 (1987).

[6]D. E. Morris *et al.*, in *Proceeding of Symposium S of the 1987 Spring Meeting of the Materials Research Society, 21-25 April 1987*, Vol. EA-11.

[7]B. Renker, F. Gompf, E. Gering, N. Nucker, D. Ewert, W. Reichardt, and H. Rietschel, to be published.

[8]J. C. Phillips, *Phys. Rev. B,* in press.

[9]P. B. Allen and R. C. Dynes, *Phys. Rev. B* **12**, 905 (1975).

MAGNETIC FIELD DEPENDENCE OF THE SPECIFIC HEAT

OF SOME HIGH-T_c SUPERCONDUCTORS*

N. E. Phillips, R. A. Fisher, S. E. Lacy, C. Marcenat, J. A. Olsen, W. K. Ham and A. M. Stacy

Materials and Chemical Sciences Division, Lawrence Berkeley Laboratory, University of California, Berkeley, CA 94720

Measurements of specific heat, C, were made on samples of La_2CuO_{4-y} (La1), $La_{1.85}M_{0.15}CuO_{4-y}$ (Ca1, Sr1, Sr2, Ba1 and Ba2) and $YBa_2Cu_3O_{9-y}$ (Y1). Meissner effect measurements were made in 12.5G. The high-temperature onset of a change in magnetic susceptibility, χ, was taken as T_c, and the transition width, ΔT_c, as the temperature interval of the 10-90% change in χ. Calculation of the fractional Meissner effect, $-4\pi\chi_V$, was based on the total sample volume, with χ_V corrected for demagnetizing effects. Figure 1 is a plot of $-4\pi\chi_V$ vs T. Two germanium thermometers were used for the specific heat

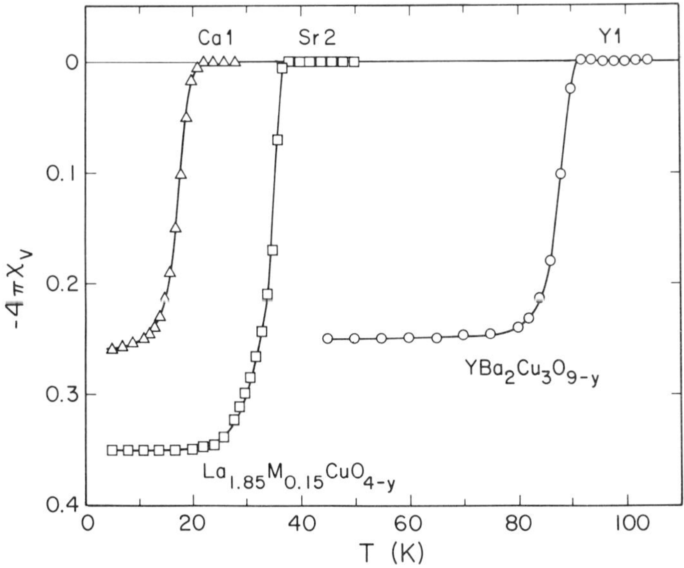

Fig. 1. Fractional Meissner effect, $-4\pi\chi_V$ vs T.

Table 1. Parameters characterizing the high-T_c superconductors $La_{1.85}M_{0.15}CuO_{4-y}$ and $YBa_2Cu_3O_{9-y}$. (All units are in mJ, mole, K and T. ND = not determined.)

	$-4\pi\chi_V$	T_c	ΔT_c	$A(0)$	$\gamma(0)$	$\partial\gamma(H)/\partial H$	$B_3(0)$	$\partial B_3'(H)/\partial H$	B_5	Θ_D
La1	–	–	–	0.16	1.10	−0.096	0.137	0	0.0019	460
Ca1	0.26	22	6	<0.02	3.05	0.035	0.145	0.0019	0.0013	450
Sr1	0.20	37	18	~0	3.9	~0	0.15	~0	ND	450
Sr2	0.35	37	8	0.16	1.54	0.109	0.168	0.0011	0.00085	430
Ba1	0.24	33	11	0.34	3.6	~0	0.16	~0	ND	440
Ba2	~0.02	34	>30	0.36	3.6	~0	0.16	~0	0.0013	440
Y1	0.25	91	9	~4400	20	0.6	0.47	ND	0.0006	380

measurements, a standard $0.3 \leqslant T \leqslant 40K$ thermometer and a high-resolution one for $2.5 \leqslant T \leqslant 40K$ which was used only for Ca1 and Sr2. The precision of the total measured C was 0.1% or 0.01% depending on the thermometer. Samples La1, Ca1, Sr2, Ba2 and Y1 were about 25g, and Ba1 and Sr1 only 5g. For the former samples, C ranged from 30 to 60% of the total measured. Parameters characterizing the samples are given in Table 1.

In the following discussion and analysis of the data, subscripts are used to distinguish the various components of C: e for electronic, l for lattice, h for hyperfine, and i for impurity; additional subscripts n, s and m are used to distinguish the normal, superconducting and mixed states; and a quantity in parentheses following the symbol for a component of C or for the coefficient of one of its terms specifies the value of H. Even for H=0, all samples show a linear term in C, $\gamma(0)T$, which is taken to be C_e for a fraction of the sample, $1-f_s$, that is not superconducting. One expects C_l to be independent of H, and $C_l=B_3T^3+B_5T^5$ in the low-temperature limit. (Θ_D was derived from B_3 for one formula weight.) In the mixed state there are field-dependent terms in T and T^3 for C_e: $C_{em}=\gamma(H)T+B_3'(H)T^3$ [1]. For H≠0 a hyperfine specific heat, $C_h=A(H)/T^2$ is expected. However, for H=0 most samples show deviations from the expected low-temperature limiting be-

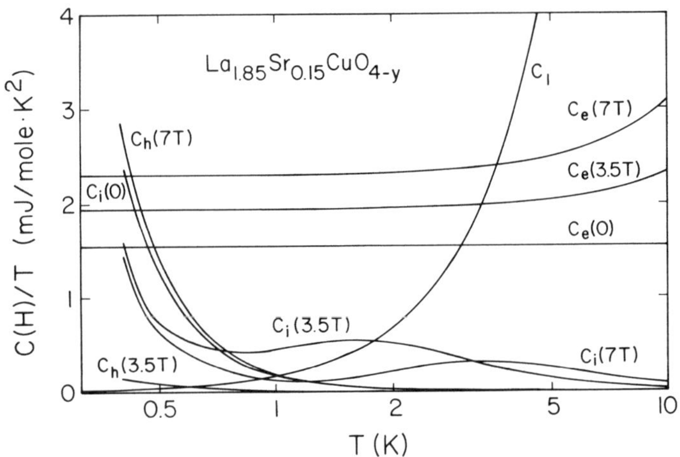

Fig. 2. Components of C for $La_{1.85}Sr_{0.15}CuO_{4-y}$.

havior, $C = C_e + C_1$, of the form $A(0)/T^2$. These deviations are apparently
associated with a magnetic "impurity" that becomes partially ordered for
$0.4<T<1K$. The evidence for this is clearest for the sample Sr2 for which
Schottky anomalies with characteristic temperatures proportional to H are apparent for H=7T and particularly for H=3.5T. The results for all other samples of the La-based compounds are consistent with amounts of the same
impurity that vary from sample to sample in proportion to the observed value
of $A(0)$. The experimental values of $A(H)$ are then accounted for by the sum
of the impurity contributions, $A(0)/T^2$, for that fraction of the sample not
penetrated by flux and the hyperfine contribution, $A(H)/T^2$, calculated for
the interaction of H with the nuclear moments for that part of the sample
penetrated by flux, the penetration being measured by $\gamma(H)$.

For sample Sr2, the analysis of C into its components for T<10K is
represented in Fig. 2. For Ca2, Sr2 and La1, the only samples that show
a measurable dependence of C_e on H, the field dependence is illustrated in
Fig. 3. Within the precision of the data C_1 is nearly the same for all the
La-based compounds. It is shown as C_1/T^3 for sample Ca1 in Fig. 4.
Parameters derived from these analyses are listed in Table 1 for all the
La-based samples. Fig. 5 shows $[C_e(0)-C_e(7T)]/T$ for Ca1 and Sr2. The

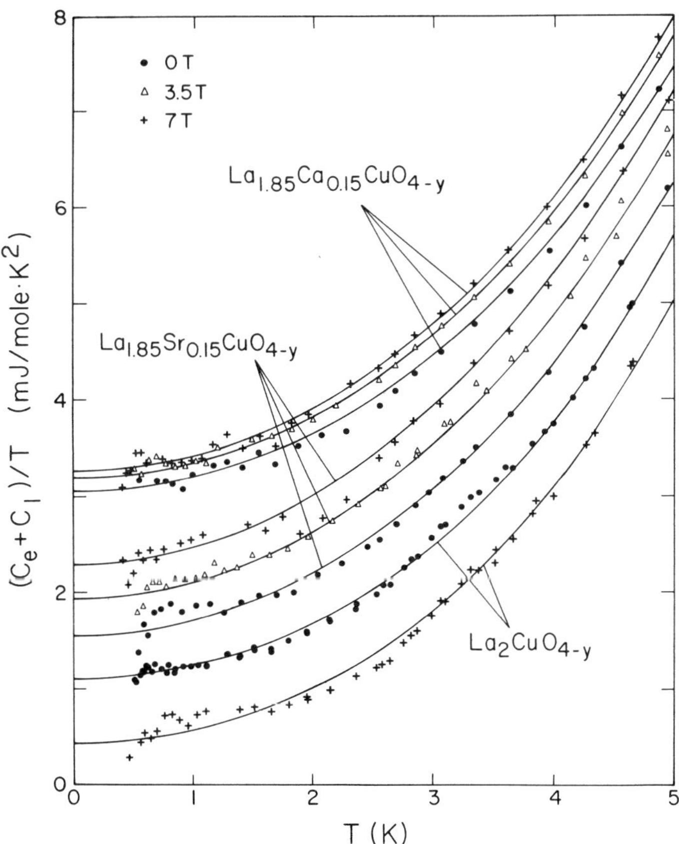

Fig. 3. $(C_e + C_1)/T$ vs T for La$_{1.85}$M$_{0.15}$CuO$_{4-y}$.

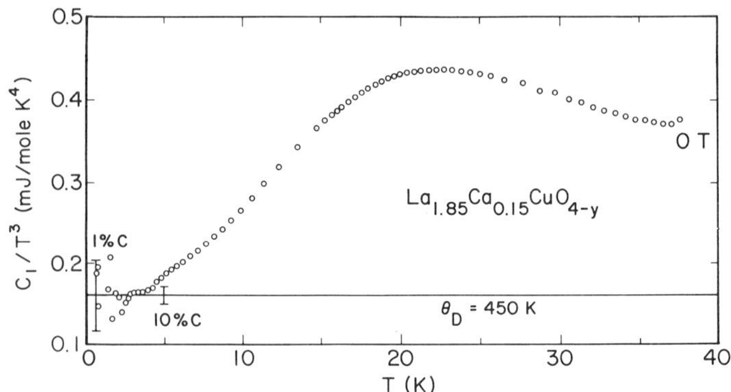

Fig. 4. C_1 for $La_{1.85}Ca_{0.15}CuO_{4-y}$ vs T.

low-temperature behavior is in qualitative agreement with expectation
for the mixed state in H=7T; the dashed lines represent entropy conserving
constructions used to estimate ΔC at T_c; the horizontal bars represent ΔT_c.

    If a fraction f_s of the sample is superconducting, $\Delta C=\beta f_s\gamma T_c$ and
$\gamma(0)=(1-f_s)\gamma$, where β is a numerical coefficient equal to 1.43 in the weak
coupling limit, γ is the coefficient of C_e for the whole sample and $\gamma(0)$ is
the value measured in zero field. If $\beta=1.43$ is assumed, these two relations
can be solved to obtain: for Cal, $f_s=0.31$ and $\gamma=4.4$ mJ/mole$\cdot$K^2; for Sr2,
$f_s=0.82$ and $\gamma=8.6$ mJ/mole$\cdot$K^2. Empirically, $\gamma(H)$ is approximately linear in H
for $H_{c1}\leqslant H\leqslant H_{c2}$ [2]. With this approximation H_{c2} at 0K, extrapolated from
$\gamma(7T)$ is 39T for Cal and 65T for Sr2. These values are in quite reasonable

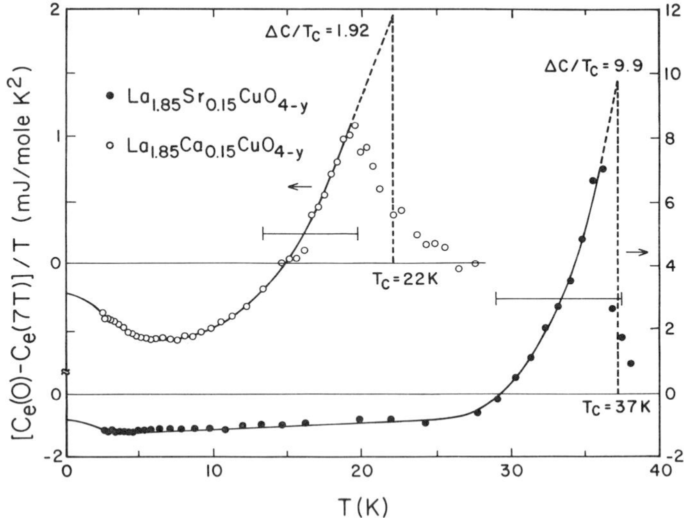

Fig. 5. $[C_e(0)-C_e(7)]/T$ vs T for $La_{1.85}M_{0.15}CuO_{4-y}$.

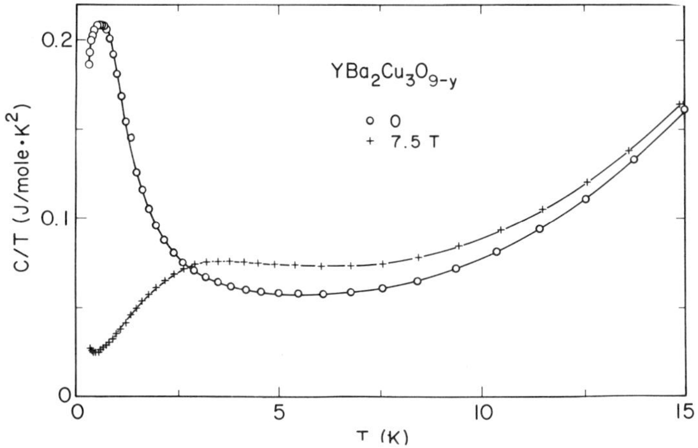

Fig. 6. C/T vs T for YBa$_2$Cu$_3$O$_{9-y}$ at 0 and 7.5T.

agreement with reported values [3] but there is enough latitude to allow $\beta \approx 2$. Other interesting results include the poor correlation of f_s obtained in this way for Cal and Sr2 with the Meissner effect, the obvious discrepancy between the field independent γ values and Meissner effect data for other samples, the nonzero value of γ for La$_2$CuO$_{4-y}$, which is predicted theoretically [4], and its strong field dependence. For La$_{2-x}$Sr$_x$CuO$_{4-y}$ other measurements of C(0) and $\Delta C/T_c$ have been reported [5].

For the Y-based compound, as shown in Fig. 6, analysis of the data is complicated by a large impurity effect. However, rough estimates of γ(0), $\partial \gamma$(H)/∂H and Θ_D are included in Table 1.

REFERENCES

*Work supported by the Director, Office of Energy Research, Office of Basic Energy Sciences, Materials Sciences Division of the U.S. Department of Energy under Contract DE-AC03-76SF00098.
1) K. Maki, Phys. Rev. A3, 702 (1965).
2) cf.: G.R. Stewart and B.L. Brandt, Phys. Rev. B29, 3908 (1984); J.F. DaSilva, N.W.J. Van Duykern and Z. Dokoupil, Physica 32, 1253 (1966); R. Radebaugh and P.H. Keesom, Phys. Rev. 149, 217 (1966); F.J. Morin, J.P. Maita, H.J. Williams, R.C. Sherwood, J.H. Wernick and J.E. Kunzler, Phys. Rev. Lett. 8, 275 (1962).
3) cf.: T.P. Orlando, K.A. Delin, S. Foner, E.J. McNiff, Jr., J.M. Tarascon, L.H. Greene, W.R. McKinnon and G.W. Hull, Phys. Rev. B35, 5347 (1987); D.K. Finnemore, R.N. Shelton, J.R. Clem, R.W. McCallum, H.C. Ku, R.E. McCarley, S.C. Chen, P. Klavins and V. Kagan, Phys. Rev. B35, 5319 (1987).
4) P.W. Anderson, preprint.
5) B. Batlogg, A.P. Ramirez, R.J. Cava, R.B. van Dover and E.A. Rietman, Phys. Rev. B35, 5340 (1987); B.D. Dunlap, M.V. Nevitt, M. Sloski, T.E. Klippert, Z. Sungoila, A.G. McKale, D.W. Capone, R.B. Poeppel and B.K. Flandermeyer, preprint; M. Decroux, A. Junod, A. Bezinge, D. Cattani, J. Cors, J.L. Jorda, A. Stettler, M. Francois, K. Yvon, O. Fischer and J. Muller, preprint.

A SUMMARY OF SOME WORK ON HIGH TEMPERATURE SUPERCONDUCTORS

AT BROOKHAVEN NATIONAL LABORATORY[*]

Myron Strongin

Physics Department
Brookhaven National Laboratory
Upton, New York 11973

The following presentation gives a summary of recent high T_c superconductivity work by the Brookhaven staff. Significant work done by outside users on the major facilities is not discussed here. Theoretical work on mechanisms of superconductivity, will be discussed separately by V. Emery elsewhere in this proceedings. The experimental work can be divided into two major categories; scattering and spectroscopy experiments at the Brookhaven facilities, and measurements of the macroscopic electrical and magnetic properties of the superconductors, as well as an additional contribution on sample preparation. An example of the complimentary nature of these techniques is shown for the case of penetration depth measurements, where the value of λ obtained using mu mesons as a microscopic probe of the internal fields is different than that obtained inductively. This difference leads to some insight into the inter-granular coupling.

Theoretical and experimental work described here shows how the effect of grains, and their anisotropy affects transport properties and critical field measurements. A similar analysis is clearly needed for the penetration depth results. Work at the facilities described here includes neutron and X-ray scattering experiments, and near edge and photoemission measurements. The neutron scattering measurements give insight into the nature of pure La_2CuO_4 and the dependence of the observed anti-ferromagnetism on O content. Near edge measurements show the nature of the Cu valence state in La_2CuO_4, and X-ray scattering results show evidence of structural effects with temperature in this compound. Photoemission measurements in the 1-2-3 compound of Y-Ba-Cu oxide indicate a significant disagreement with band theoretical calculations. All work is ongoing at Brookhaven and although the great value of microscopic probes is already evident, it is expected that they will grow in importance as the quality of the materials increases.

The format of this collection is a two or three page contribution from the authors, including many outside collaborators, involved in each research project.

---

[*]The submitted manuscripts have been authored under contract DE-AC02-7600016 with the Division of Materials Sciences, U.S. Department of Energy.

A NEUTRON POWDER DIFFRACTION STUDY OF $Ba_2YC_3O_7$

D. E. Cox, A. R. Moodenbaugh, J. J. Hurst, and
R. H. Jones

Brookhaven National Laboratory
Upton, New York 11973

The structure of an orthorhombic sample of $Ba_2YCu_3O_7$ with a
superconducting transition at 89K has been determined by Rietveld profile
analysis of neutron powder diffraction data. The space group chosen was
Pmmm (a=3.817Å, b=3.882Å, c=11.671Å). The structure may be viewed as a
heavily-distorted oxygen-defective ordered derivative of a perovskite-
type structure. Two-thirds of the Cu atoms have four near-neighbor
oxygens and one more distant neighbor in a tetragonal pyramidal
arrangement, and form a two-dimensional network in the ab planes linked
through corner shared oxygens about 0.3Å out of the planes. The
remaining one-third of the Cu atoms have fourfold planar coordination in
the bc planes and form chains linked through corner-shared oxygens along
the b axis.

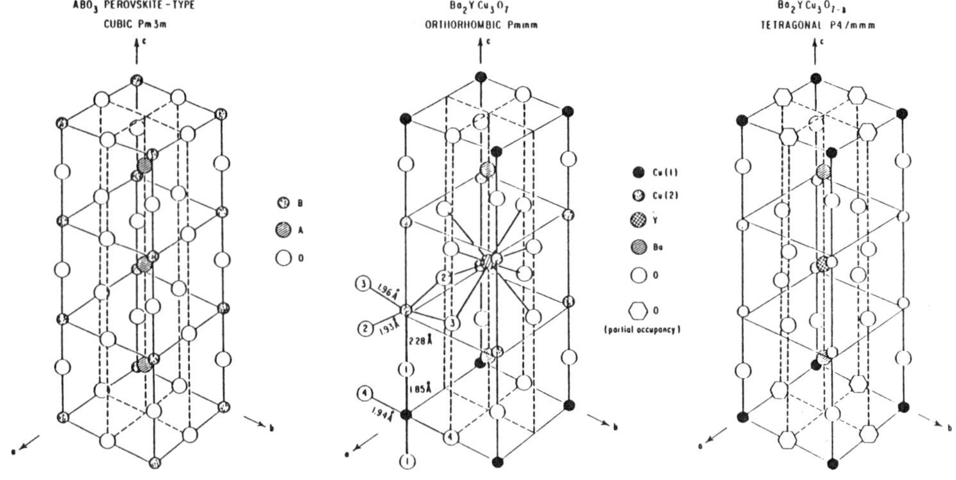

Fig. 1 (a) Three unit cells of the prototype cubic ABO_3 perovskite-
             type structure.
       (b) Orthorhombic structure of $Ba_2YCu_3O_7$ with Pmmm symmetry.
             Heavy lines show the fourfold arrangement of oxygens around
             Cu(1) (small closed circles), the tetragonal pyramidal ar-
             rangement around (Cu(2) (Small stippled circles) and the near-
             cubic coordination of Y (cross-hatched circles).
       (c) Hypothetical tetragonal structure of $Ba_2YCu_3O_7$ with
             oxygen disorder in the Cu(1) basal planes, depicted by open
             hexagons.

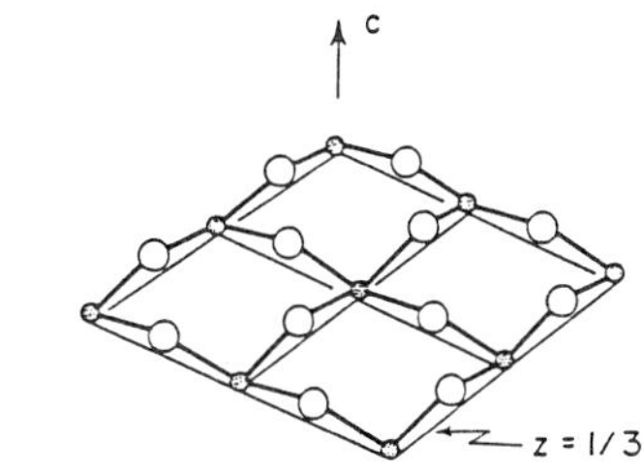

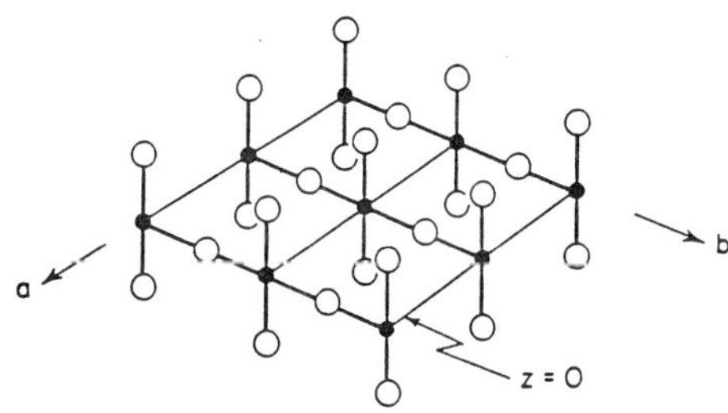

Fig. 2   One-dimensional CuO_4 chains in the
bc planes of $Ba_2YCu_3O_7$ (bottom)
and two-dimensional dimpled CuO_4
sheets in the ab planes (top).

The relationship between the ideal cubic perovskite structure, the
orthorhombic $Ba_2YCu_3O_7$ structure and a disordered form of the latter with tetragonal symmetry; is shown in Fig. 1.   The stereochemistry of
the Cu-O networks is shown in detail in Fig. 2.

ACKNOWLEDGEMENT

This manuscript has been authored under contract DE-AC02-7600016 with the
Division of Materials Sciences, U.S. Department of Energy.

STRUCTURAL PHASE TRANSFORMATIONS AND HIGH-T_c SUPERCONDUCTIVITY

J. D. Axe, H. You, D. Hohlwein, and D. E. Cox
Department of Physics
*National Synchrotron Light Source
Brookhaven National Laboratory
Upton, New York 11973

S. C. Moss and K. Forster
Department of Physics
University of Houston
Houston, TX 77004

P. Hor, R. L. Meng and C. W. Chu
Department of Physics
Space Vacuum Epitaxy Center
University of Houston
Houston, TX 77004

There is an intimate if imprecisely understood connection between
high T_c superconductivity and crystal structure instability. Most of
the highest T_c superconductors ($T_c \approx 20K$) discovered in the past 20
years have the so-called A15-type structure, which is marginally stable
in the sense that many of the A15 compounds spontaneously distort at low
temperatures into a lower symmetry structure. These distortions are
driven by unstable low frequency phonon modes. It is believed that these
"soft" phonon modes enhance the attractive interaction between electrons
and thus tend to increase T_c. On the other hand (the argument goes)
phase transformation which lower the symmetry introduce structural gaps
at the Fermi surface of metals which compete with superconductivity and
tend to decrease T_c.

The recent discoveries of $T_c \approx 40K$ (much higher T_c's are rumored)
in the La_2CuO_4-type systems raise questions concerning the importance
of lattice instabilities to superconductivity in these systems. We have
performed high resolution synchrotron x-ray powder diffraction measurements on the superconductor $La_{1.8}Ba_{0.2}CuO_4$ which show unusual
temperature dependent broadening of certain Bragg peaks, as shown in
Fig. 1. This pattern of peak broadening can be consistently explained as
an unresolved splitting due to a spontaneous reduction of the tetragonal
symmetry of the high temperature structure to a lower monoclinic (or
possibly triclinic) form.

In order to confirm this interpretation of the powder diffraction
data, we were also able to carefully reduce the size of the highly
collimated synchrotron x-ray beam and to isolate and study individual ≈ 70
µm single crystals within the powder aggregate. Fig. 2a shows more
clearly than the powder data the splitting of the reflections upon
cooling. Fig. 2b shows a greatly enlarged image of the diffracted beam
from the same microcrystalline reflection shown in Fig. 2a, using a novel
high resolution Weissenberg camera that we have developed. Again the
splitting is clearly evident.

The relationship of the structural transformation we have discovered
to the superconductivity which occurs at lower temperature ($T_c \approx 36K$) is
unclear. But in view of the interplay between superconductivity and the
structural transformations noted above, further structural studies with
larger single crystals should prove rewarding. It must be pointed out
that the degree of monoclinic distortion that we have found is very
small, and that the brightness of a synchrotron x-ray surce was essential
for high resolution powder and single crystal diffraction experiments
presented here.

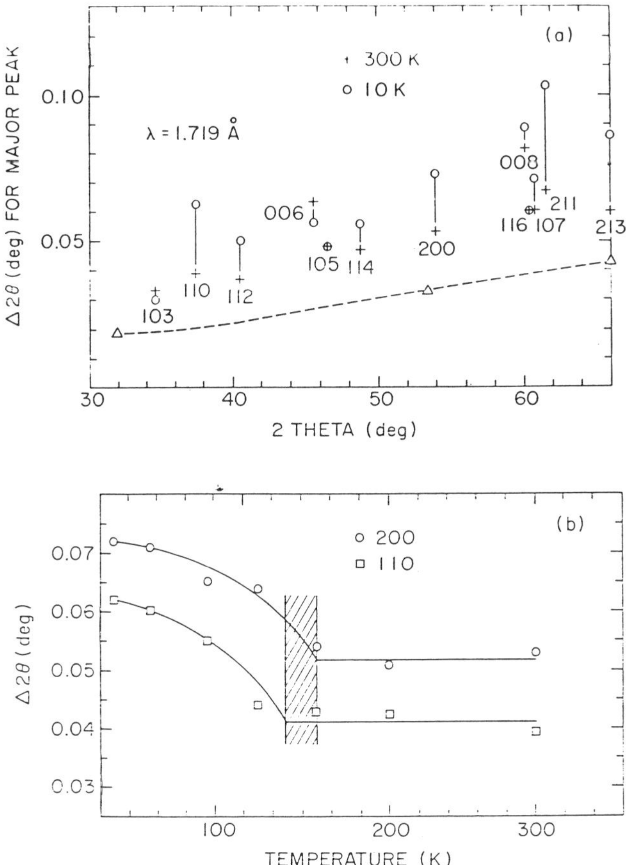

Fig. 1a   Comparison of width of powder diffraction peaks in the high-
T_c superconductor $La_{1.8}Ba_{0.2}CuO_4$ at 300K and 10K. Many
peaks show an unusual broadening as the temperature is lowered.
The triangles represent instrumental resolution.

Fig. 1b   The temperature dependence of the width of two peaks shown
above. The onset of the broadening indicates a phase transformation at $T \approx 130$–150K.

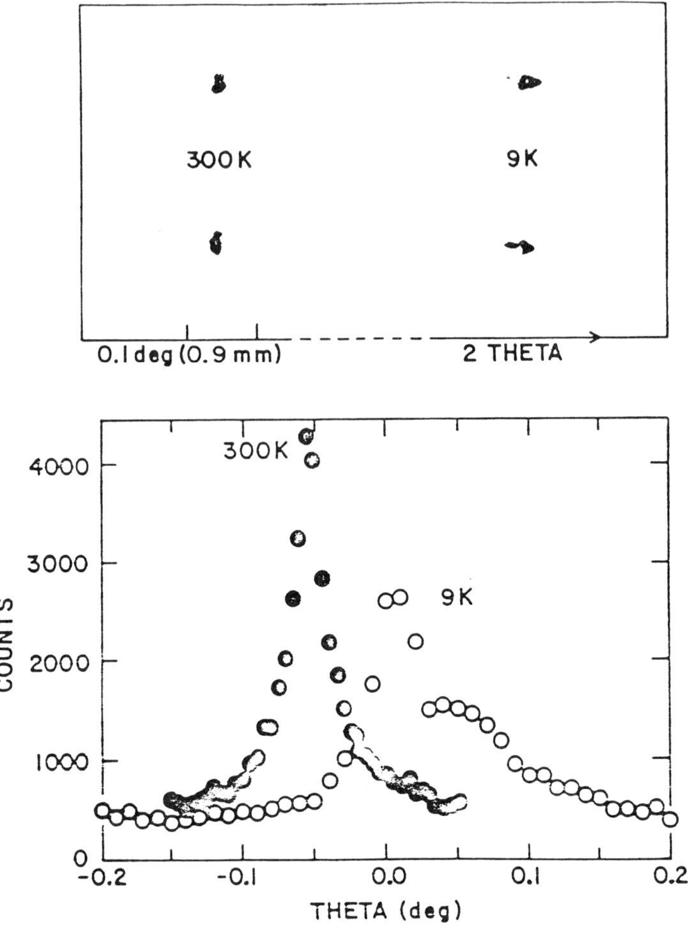

Fig. 2a Splitting of the (110) rocking curve of a small (70 μm) single crystal of $La_{1.8}Ba_{0.2}CuO_4$ at low temperature.

Fig. 2b Magnification of direct image of diffracted beam from crystal studied in Fig. 2a. The horizontal splitting is direct evidence of two slightly different (110) lattic plane spacings at low temperature.

ACKNOWLEDGEMENT

This manuscript has been authored under contract DE-AC02-7600016 with the Division of Materials Sciences, U.S. Department of Energy.

NEUTRON SCATTERING STUDIES OF $La_2CuO_{4-\delta}$

T. Freltoft,[a] J. P. Remeika,[b] D. E. Moncton,[c]
A. S. Cooper,[b] J. E. Fischer,[a,d] D. Harshman,[b]
S. Mitsuda,[a] G. Shirane,[a] S. K. Sinha,[c]
D. Vaknin,[c,d] and B. X. Yang[a]

[a] Brookhaven National, Upton, New York 11973
[b] AT&T Bell Laboratories, Murray Hill, New Jersey 07974
[c] Exxon Research and Engineering Company, Annandale
    New Jersey 08801
[d] University of Pennsylvania, Philadelphia
    Pennsylvania 19104

La_2CuO_4 is the parent compound for some of the newly discovered
high-Tc superconductors, $La_{2-x}B_xCuO_4$, where B is Ba or Sr. In
order to approach an understanding of the superconductivity, structural
and magnetic studies have been conducted on powders and single crystals
of $La_2CuO_{4-\delta}$, where δ is the oxygen deficiency. δ is suggested to
play an important role for the magnetic and superconducting properties of
these materials. La_2CuO_4 has tetragonal structure at high temperatures, but undergoes an orthorhombic distortion at lower temperatures.
The results obtained thus far reveal an antiferromagnetic phase transition in the orthorhombic regime and the magnetic structure in the ordered
phase has been determined.[1] These results which are measured on a
strongly heat-treated powder sample have been reexamined and confirmed
with both high resolution[2] and polarized beam[3] measurements also
conducted on heat-treated powder samples ($\delta \approx 0.03$).

In order to investigate in more detail how the magnetic and structural properties of $La_2CuO_{4-\delta}$ are related to the oxygen deficiency δ,
a series of single crystals was grown and heat-treated in vacuum at
different temperatures.[4] In this way crystals of varying δ are produced. One crystal was heat-treated in oxygen atmosphere to create an
excess oxygen concentration, and also an untreated crystal was investigated.

It is found that the magnetic structure is independent of oxygen
deficiency δ, while the transition temperature T_N and the magnetic
moment μ per Cu atom strongly depend on δ. For the oxygen-treated
crystal no magnetic ordering is found down to T=5K while all the other
samples order antiferromagnetically at higher temperatures. It is also
found that the orthorhombic distortion, measured as the relative
difference of the lattice constants a and c, c/a - 1 is not very
sensitive to the oxygen concentration at low temperature. (a = c in the
high temperature tetragonal phase.)

In the table the results for all samples are summarized. It may be
seen that the magnetic properties T_N and μ are highly dependent on
heat-treatment. A general trend seems to be that a high transition
temperature provides a higher moment and also that heat-treatment at
higher temperature generally produces samples with high μ and T_N.

TABLE

Structural and Magnetic Properties from Different La$_2$CuO$_4$ Samples

Sample	Heat treatment Temperature °C	Lattice const. a (Å) T = 10K	Distortion c/a-1 T = 10K	T**(K) N	Moment* (μ_B)	Ref.
Sgl. Cryst.	Oxygen 800°C	5.334	1.34%	–	–	5
Sgl. Cryst.	None	5.362	1.41%	70	0.17 ± 0.02	4
Sgl. Cryst.	vac. 200°C	5.334	1.44%	85	0.26 ± 0.05	5
Sgl. Cryst.	vac. 250°C	5.334	1.40%	200	0.38 ± 0.02	5
Sgl. Cryst.	vac. 400°C	5.324	1.45%	205	0.23 ± 0.02	4
Powder	950°C	5.333	1.43%	220	0.40	1
Powder	950°C	5.377	1.51%	250	0.40 ± 0.05	2
Powder	950°C			290	0.43 ± 0.13	3

*All moments are calculated assuming the magnetic form factor f(100) = 0.835
T_N is the **onset temperature of magnetic ordering.

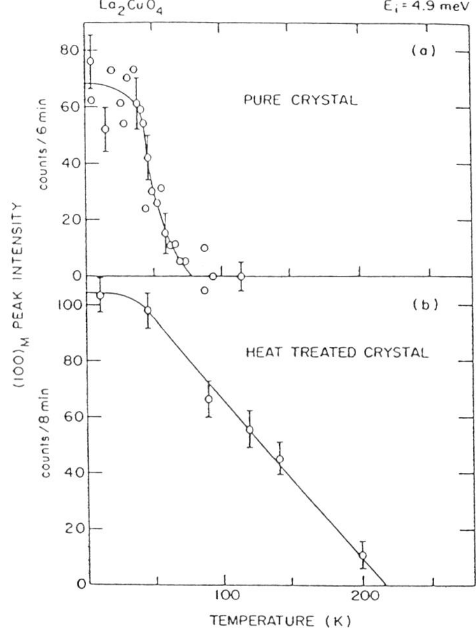

The figure shows the peak intensities of the strongest magnetic reflection (100) in the orthorhombic notation for (a) a untreated single crystal of La$_2$CuO$_{4-\delta}$, and (b) a 400°C heat-treated single crystal with $\delta > 0$. It may be seen that the phase transition in (b) is spread out over the temperature range between 100K and 200K. We suggest that this is caused by inhomogenities of the oxygen vacancies in the crystal.

References

1. D. Vaknin, S. K. Sinha, D. E. Moncton, D. C. Johnston, J. Newsam, and H. King to be published.
2. B. X. Yang, S. Mitsuda, G. Shirane, Y. Yamaguchi, H. Yamauchi, and Y. Syono, J. Phys. Soc., Japan, 56, July 1, 1987.
3. S. Mitsuda, G. Shirane, S. K. Sinha, D. C. Johnston, M. S. Alvarez, D. Vaknin, and D.E. Moncton, Phys. Rev. B, July 1, 1987.
4. T. Freltoft, J. P. Remeika, D. E. Moncton, A. S. Cooper, J. E. Fischer, D. Harshman, G. Shirane, and D. Vaknin, Phys. Rev. B, July 1, 1987.

ACKNOWLEDGEMENT   This manuscript has been authored under contract DE-AC02-7600016 with the Divison of Materials Sciences, U.S. Department of Energy.

752

X-RAY ABSORPTION STUDIES OF $La_{2-x}(Ba,Sr)_xCuO_4$

SUPERCONDUCTORS

J. M. Tranquada, S. M. Heald, A. R. Moodenbaugh,
M. Suenaga, R. F. Garrett, E. D. Johnson, E. Kneedler,
and G. P. Williams

Brookhaven National Laboratory
Upton, NY

Oxides with the general formula $La_{2-x}(Ba,Sr)_xCuO_4$ have been
found to have T_c's of up to 40 K. An important question concerns the
nature of the electronic structure in these systems. The crystal
structure is highly anisotropic, containing planes of copper atoms in a
square lattic with an oxygen atom between each pair of coppers. In a
purely ionic model for x=0, the Cu atoms would have a 2+ valence with a
$3d^9$ configuration, while the oxygen is 2- with filled 2p levels. Band
structure calculations indicate that the Cu 3d and O 2p states hybridize
to form broad bands, the topmost one being half-filled. As La^{3+} ions
are replaced by Sr^{2+} or Ba^{2+}, the number of valence electrons is
reduced and the Fermi energy is lowered through the band. In such a
picture, the superconductivity is due to strong electron-phonon
interactions which are induced by the shape of the Fermi surface. At the
other extreme, many workers have suggested that the reduction in valence
electrons due to doping causes some Cu^{2+} to become Cu^{3+}. The higher
valence would cause the oxygen neighbors to be displaced in order to
screen the charge, resulting in polarons. Superconductivity would then
result from a polaronic or excitonic mechanism.

We have studied[1,2] the unoccupied densities of states in La_{2-x}
$(Ba,Sr)_xCuO_{4-y}$ for x = 0.0-0.3 by measuring x-ray absorption near
edge structure (XANES) at the Cu K, La L, and O K edges. From a
comparison with several reference materials and taking into account
recent cluster calculations on copper complexes, we find that the copper
ions have an integral $3d^9$ configuration (2+ valence) independent of
dopant concentration. Variations in "white line" intensities observed at
La L_2 and L_3 edges indicate that the density of empty states having
d-symmetry with respect to the La site increases with doping. The extra
empty states are assumed to be holes in the O 2p band. Changes near the
quadrupole feature at the Cu K-edge are consistent with linear combinations of O 2p holes having d-symmetry with respect to the Cu site.
Measurements at the O K edge indicate that the edge shifts to higher
energy with increasing x as would be expected for a reduction in 2p
charge.

The XANES results are inconsistent with both the itinerant and mixed
valence pictures. We find instead that the Cu 3d electrons are
localized, with one 3d hole per Cu atom. As the amount of valence charge
is reduced by doping, it is energetically more favorable to remove an
electron from the O 2p band than to put a second 3d hole on a Cu atom.
These results are evidence against the electron-phonon mechanism based on

the band structure calculations and also bipolaronic or excitonic coupling. They are, however, consistent with a model due to Emery[3] in which O 2p holes are paired through exchange interactions via the copper site.

Further evidence against strong electron-phonon interactions has been obtained from temperature-dependent extended x-ray absorption fine structure (EXAFS) measurements at the Cu K edge.[1] The only phonons expected to couple significantly to the valence electrons are breathing modes of the O atoms with respect to the Cu sites in the CuO_2 planes. From the EXAFS measurements, we have determined the rms relative motion between a Cu atom and its O neighbors, a quantity which provides a measure of the interatomic force constant. We find that the Cu-O in-plane bond is quite strong and shows little variation as a function of doping. Model calculations indicate that the EXAFS should be sensitive to the large breathing-mode softening required to explain the high T_c values, and such softening is not observed.

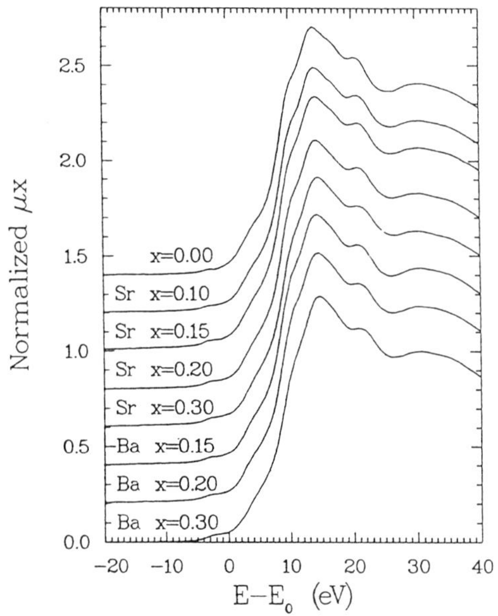

Fig. 1   Cu K edge in $La_{2-x}(Ba,Sr)_xCuO_4$, illustrating the point that there is no shift in the edge position nor significant variation in peak intensities with doping. Copper has a 2+ valence independent of x.

References

1.   J. M. Tranquada, S. M. Heald, A. R. Moodenbaugh, and M. Suenaga, Phys. Rev. B 35, 7187 (1987).
2.   J. M. Tranquada, S. M. Heald, A. R. Moodenbaugh, submitted to Phys. Rev. B.
3.   V. J. Emery, Phys. Rev. Lett., (in press).

ACKNOWLEDGEMENT   This manuscript has been authored under contract DE-ACO2-7600016 with the Division of Materials Sciences, U.S. Department of Energy.

PHOTOEMISSION STUDIES OF THE HIGH T_c SUPERCONDUCTOR $YBa_2Cu_3O_{9-\delta}$

P. D. Johnson[1], S. L. Qiu[1], L. Jiang[1],
M. W. Ruckman[1], M. Strongin[1], S. L. Hulbert[2],
F. R. Garrett[2], B. Sinkovic[3], N. V. Smith[3],
R. J. Cava[3], C. S. Jee[4], D. Nichols[4]
E. Kaczanowicz[4], R. E. Salomon[4], and J. E. Crow[4]

1)   Physics Department, Brookhaven National Laboratory
2)   National Synchrotron Light Source, Brookhaven National
     Laboratory
3)   AT&T Bell Laboratories
4)   Temple University

VUV photoemission spectra (Fig. 1) of the valence bands of the
high-T_c superconductor[1] $YBa_2Cu_3O_{9-\delta}$ $\delta=2.1$, have been measured
using the UV undulator (beam line U5U). Photon energies h ν = 45.0 eV
and 52.0 eV lie within the first harmonic flux output from the undulator,
while h ν = 105.0 lies within the second harmonic. The spectra at h ν =
21.2 eV and 40.8 were taken using a He resonance lamp.

In Fig. 2, the h ν = 21.2 eV spectrum is compared with the total
muffin-tin projected density of states for this material calculated in
the local density approximation (LDA) by Mattheiss and Hamann[2] (to be
published). The bands at binding energies of 2.3 eV and 4.3 eV are
mainly oxygen 2p- and copper 3d- derived, respectively, as confirmed by
the decreasing cross-section of the 2.3 eV peak relative to the 4.3 eV
peak as a function of increasing photon energy. It is to be noted that
these spectra are as-prepared (by scraping) and that the intensity of the
oxygen-derived peak decreases as a function of time in ultra-high
vacuum. Also note that a definite Fermi level (see dashed line in left
figure inset, magnified from the 21.2 eV spectrum) is observed in these
materials.

The question of the surface sensitivity of photoemission
measurements must be emphasized and the particular scraping method that
we have used appears to give somewhat different results than for samples
which are cleaved in vacuum. At the present time these issues are still
unsettled and will undoubtedly be settled as the quality and density of
the samples increase. We emphasize that the measurements reported here
were done on samples prepared in two different laboratories and similar
results were obtained. Scraping was also done below Tc and similar
results were also obtained in the valence band region shown here.
Samples cooled to low temperatures are very sensitive to adsorbed impurities, and a cleaning technique like scraping or cleaving is crucial.

We mention here that the shift of the Cu and O valence band features
to higher binding energy probably implies that the LDA used in the band
structure calculations is not accurate in this case. In fact it is known
that in the case of Cu, LDA calculations[3] give bands that are too low

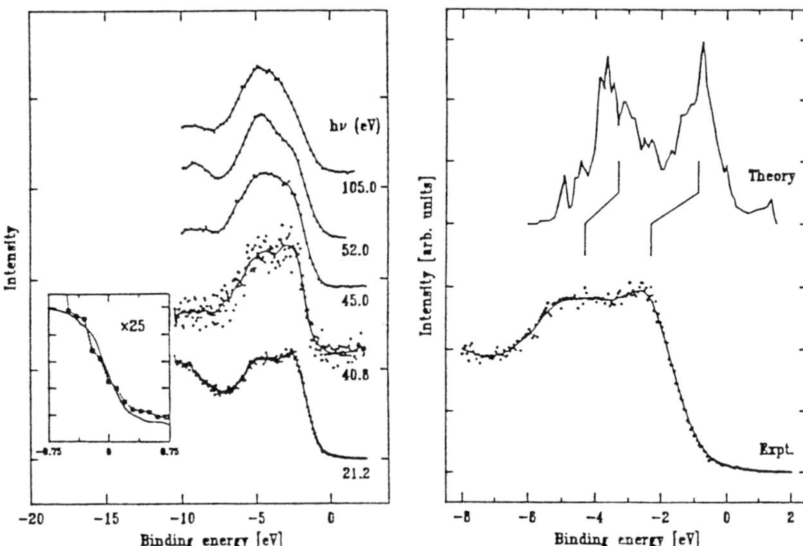

Fig. 1   Photoemission spectra at 45.0, 52.0 and 105.0 eV from
sample 2. Photoemission spectra at 21.2 and 40.8 eV
from sample 1. The inset shows the Fermi level step
scaled by a factor of 25 from sample 1 at 21.2 eV
(dashed line) and the Fermi level from a copper film
(solid line) scaled to the same height.

Fig. 2   Comparison of the 21.2 eV spectrum from sample 1 with
the calculated density-of-states from Mattheiss and
Hamann.

in energy by about 1 eV. It is also well to mention that the photoemission results themselves can be shifted due to the presence of final-
state relaxation or excited state effects. However in a metallic system
these effects are expected to be small.

Finally we mention that we have also studied the 1-2-3 compound with
Gd substituted for Y and the valence band features are similar to those
shown here, indicating that neither Gd or Y are involved in the state
near E_f, and probably do not contribute to the superconductivity except
for determining the structure. Hence we argue that these results provide
evidence that Tc calculations based in the electron phonon mechanism and
LDA band structure calculations can be significant errors.

References

1.   I. Wu, J. R. Ashburn, C. J. Torng, P. H. Hoz, R. L. Ming, L. Gab,
     A. J. Huang, Y. Q. Wang and C. W. Chu, Phys. Rev. Lett., <u>58</u> 908
     (1987).
2.   L. F. Mattheiss and D. R. Hamann (unpublished).
3.   See "The Theory of the Inhomogeneous Electron Gas ", edited by
     S. Lundquist and N. H. March, Plenum, New York, 1983, p. 291.

ACKNOWLEDGEMENT - This manuscript has been authored under contract DE-ACO2-
7600016 with the Division of Materials Sciences, U.S. Department of Energy.

MUON SPIN RELAXATION STUDIES ON HIGH-Tc SUPERCONDUCTORS

W. J. Kossler 1), J. R. Kempton 1), A. Moodenbaugh 2),
D. Opie 1), H. Schone 1), C. Stronach 3), M. Suenaga 2),
Y. J. Uemura 2), and X. H. Yu 1)

1) College of William & Mary, Virginia
2) Brookhaven National Laboratory, New York
3) Virginia State University, Virginia

The AGS proton synchrotron at Brookhaven National Laboratory
facilitates a medium-intensity polarized muon channel. Muon spin
relaxation (μSR) experiments on single-phase specimens of
$(La_{1.85}Sr_{0.15})CuO_4$ and $YBa_2Cu_3O_7$ have been carried out at
this facility. In transverse-field μSR measurements with external
magnetic field (500 - 1000 G) applied perpendicular to the initial muon
spin direction, the magnetic-field penetration depth λ in the superconducting state has been determined.

In the type II superconducting state, the external field penetrates
as flux vortices. Then the magnetic field within the specimen has the
inhomogeneous width $\langle \Delta H^2 \rangle$ which depends on the penetration depth λ.
In transverse-field μ+SR, one observes the spin precession pattern of
positive muons stopped in the specimen by measuring the time spectrum of
muon decay positrons emitted preferentially towards the muon spin
direction. The width of the magnetic field can be determined from the
relaxation rate of muon spins, i.e., the damping of the precession
pattern.

The observed relaxation rate was very small above Tc, while it
increased rapidly with decreasing temperature below Tc in the both
systems. Comparing the observed values of $\langle \Delta H^2 \rangle$ with a model
simulation for the field distribution, the penetration depth λ was
determined as shown in Fig. 1. The temperature dependence of λ for
$YBa_2Cu_3O_7$ was consistent with the prediction for a BCS superconductor which has singlet spin pairing of the superconducting carriers.

The penetration depth at low temperatures is about 2000 A for
$(La_{1.85}Sr_{0.15})CuO_4$ and 1300 A (this number is preliminary and may
be subject to some minor correction) for $YBa_2Cu_3O_7$. Combining
these results on λ with the property proportional to the density of state
(e.g., liner term of the specific heat or Pauli susceptibility above Tc),
one can derive the concentration n of superconducting carriers. The
carrier density thus estimated from the observed penetration depth was n
$= 3 \times 10^{21}/$ cm^3 (n $\sim$ 0.3 / formula unit) for
$(La_{1.85}Sr_{0.15})CuO_4$, and n $\sim 8 \times 10^{21}$ / cm^3 (n $\sim$ 1 / formula
unit).

The results on $(La_{1.85}Sr_{0.15})CuO_4$ have been published in W. J. Kossler, et al., Phys. Rev. B35, 7133 (1987); those on $YBa_2Cu_3O_7$ will be published in forthcoming papers.

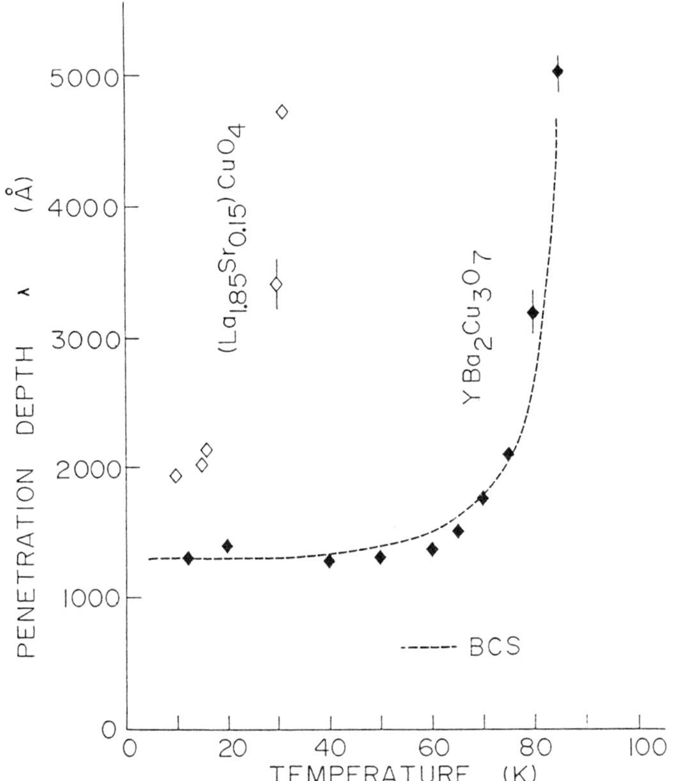

Fig. 1 Magnetic-field penetration depth in the high-Tc superconductors as determined from the muon spin relaxation measurement performed at AGS/BNL. The values for $YBa_2Cu_3O_7$ are still somewhat preliminary and may be subject to minor correction.

ACKNOWLEDGEMENT   This manuscript has been authored under contract DE-AC02-7600016 with the Division of Materials Sciences, U.S. Department of Energy.

INDUCTIVE TRANSITION OF $YBa_2Cu_3O_{7-\delta}$ IN THE 4 MHz FREQUENCY RANGE

R. R. Corderman, H. Wiesmann, M. W. Ruckman and
M. Strongin

Brookhaven National Laboratory
Upton, New York 11973

The inductive transition of superconducting $YBa_2Cu_3O_{7-\delta}$ was
investigated at 4 Mhz. These measurements provide a value of the London
penetration depth, $\lambda_L(0)$, in $YBa_2Cu_3O_{7-\delta}$ which differs by a
factor of $\approx$ 10 from the value obtained from muon spin relaxation
measurements. This discrepancy may be related to the granular nature of
this superconductor.

The magnitude and variation of the penetration depth as a function
of temperature can be predicted from the Bardeen-Cooper-Schrieffer theory
of superconductivity[1], or alternatively, from the phenomenological
Gorter-Casimir two-fluid theory[2]. In the present experiment, a
superconducting sample is placed in the core of a solenoid. Below the
superconducting transition temperature, T_c, the inductance, L, of a
solenoid, and thus the oscillation frequency of the LC section of a
tunnel-diode oscillator[3,4], changes with the penetration depth. By
measuring the frequency change a a function of sample temperature below
T_c, $\lambda_L(0)$ can be determined.

The method of calibrating the apparatus to obtain λ values from the
frequency changes with temperature has been previously discussed by
Schawlow and Devlin[5] and by Varmazis and Strongin[3] and is only
outlined here. The basis of the analysis is an expression, which gives

$$\delta\lambda = -(\delta f/f)\ G$$

where f is the frequency and G is a geometrical factor.

The apparatus can be calibrated either by using specimens of
different diameter or by measuring the normal state skin depth. Assuming
that the penetration depth is a function of the form $\lambda(T) = \lambda(0)Z(T)$,
then $\delta\lambda/\delta Z = \lambda(0)$. $Z(T)$ is a theoretical temperature dependence given
either by the BCS theory or from the Gorter-Casimir phenomenological
theory[2]. The actual measurements give the variation of f vs T and, a
value for $\lambda(0)$ can be obtained from a knowledge of the sample radius R,
the oscillator frequency f, and the geometrical factors.

Figure 1 shows the inductive transiton near 4 MHz for a small
$YBa_2Cu_3O_{7-\delta}$ sample. Following the analysis described above,
$\lambda_L(0)$ is obtained from the slight decrease in oscillation frequency
below T_c, and the superconducting transition temperature ($T_c \approx$ 89.7
°K) is determined from the midpoint of the frequency transition.

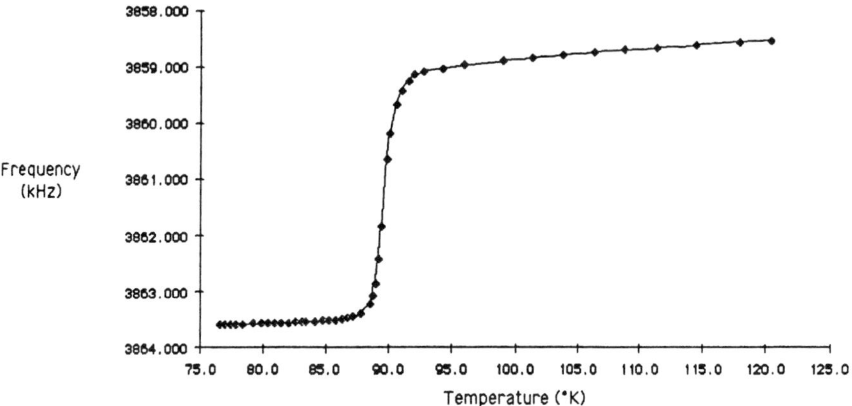

Fig. 1.

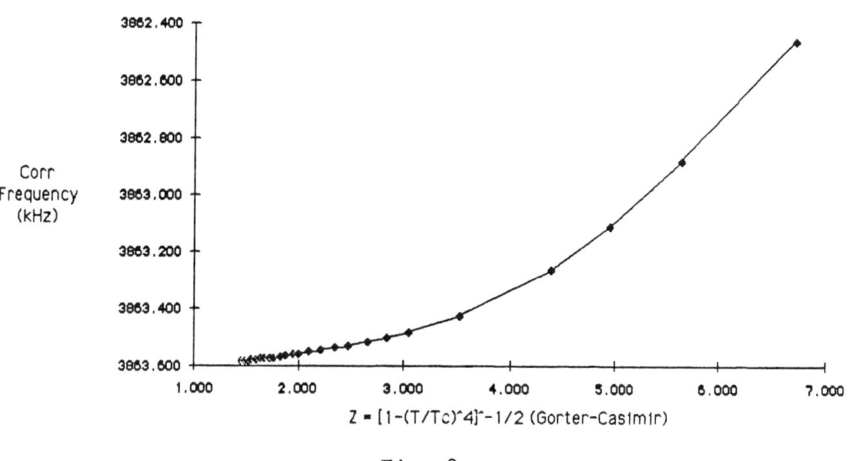

Fig. 2.

Fig. 1. $YBa_2Cu_3O_{7-\delta}$ inductive transition near 4 MHz. $\lambda_L(0)$ is obtained from the slight decrease in oscillation frequency below T_c.

Fig. 2. Corrected frequency vs. Z_{GC} for the $YBa_2Cu_3O_{7-\delta}$ sample. Using the frequency vs. temperature data of Figure 1, $Z_{GC}(T)$ (eq. 3) from the Gorter-Casimir two-fluid phenomenoligical theory was determined and the frequency measurements were corrected by subtracting the frequency measured for the probe with no sample present (see text).

ACKNOWLEDGEMENT   This manuscript has been authored under contract DE-ACO2-7600016 with the Division of Materials Sciences, U.S. Department of Energy.

Using the frequency vs. temperature data of Figure 1, $Z_{GC}(T)$
was determined, and a plot of $Z_{GC}(T)$ vs. frequency is shown in Figure
2. The frequency measurements were corrected by subtracting the
frequency measured for the probe with no sample present. This correction
amounts to $\approx$ 0 - 20 Hz from T $\approx$ 77 - 90 °K and is most likely due to the
presence of the platinum resistance thermometer and leads in the solenoid
coil. While the plot of frequency vs. Z(T) should nominally be a
straight line, the present $YBa_2Cu_3O_{7-\delta}$ sample shows a nonlinear
decrease in frequency with increasing Z. Calculating the slope from the
linear region between $Z\approx1.5$ and $Z\approx2.0$, an apparent penetration depth of
2.4 μm is obtained. This value can be compared to a $\lambda_L(0)$ value of $\approx$
1500 Å obtained for $YBa_2Cu_3O_{7-\delta}$ using muon spin relaxation
techniques and discussed elsewhere in this paper.

The value of $\lambda_L(0)$ obtained from inductive transition measurements
differs by a factor of $\approx$ 10 from the value obtained from muon spin
relaxation experiments. Previous work on roughened superconducting
surfaces and granular superconductors suggests that: 1) surface
roughness, 2) weak-link grain boundary Josephson junctions, and 3) a
distribution of grain T_c's could account for an order of magnitude
difference in the London penetration depth. The additional deviation
when $Z_{GC}>2$ is thought to be due to sample inhomogeneity, however the
large deviation from the intrinsic penetration depth is thought to be due
to the weak coupling between the grains of superconductor. The effect of
anisotropic conduction in these superconductors has not been considered
yet, but it would appear that both methods give some average λ.

Acknowledgments

The authors thank R. Jones and A. Moodenbaugh for preparing the
$YBa_2Cu_3O_{7-\delta}$ samples used in these experiments. This work was
supported by the U. S. Department of Energy, Division of Material
Science, Office of Basic Energy Sciences under Contract No. DE-AC02-
76CH00016.

References

1. J. Bardeen, L. N. Cooper, and J. R. Schrieffer, Phys. Rev. 108,
    1175 (1957).
2. D. Shoenberg, Superconductivity (Cambridge University Press,
    Cambridge, 1952), Chaps. 5 and 6.
3. C. Varmazis and M. Strongin, Phys. Rev. B 10, 1885 (1974).
4. C. Boghosian, H. Meyer, and J. E. Rives, Phys. Rev. 146, 110 (1966).
5. A. L. Schawlow and G. E. Devlin, Phys. Rev. 113, 120 (1959).

INTRINSIC CRITICAL CURRENT AND MAGNETIZATION OF $YBa_2Cu_3O_7$

A. Ghosh, M. Suenaga, and A. Moodenbaugh

Brookhaven National Laboratory
Upton, New York 11973

Dramatic improvements in critical temperature T_c of super-conductors were recently made starting with the discovery of a La-Ba-Cu oxide by Bednorz and Müller [1-3]. Although a large number of reports have been published on both the $La(Ba/Sr)CuO_4$ and $YBa_2Cu_3O_x$ systems, one concern regarding possible uses of these new superconductors in high magnetic fields is the fact that the reported values of critical current density J_c are very low, e.g., the best value of J_c for $YBa_2Cu_3O_x$ is ~ 11 A/mm^2 [4] at 77 K. This value is at least 2 to 3 orders of magnitude lower than what is required for possible high field applications.

In order to investigate the factors which influence the values of J_c in these oxides, we have measured the magnetization for a number of well-characterized specimens of $YBa_2Cu_3O_7$. All of the specimens were prepared by grinding appropriate amounts of the oxides and heat treating them in air or an oxygen atmosphere. T_c was measured by both inductive and resistive techniques. Powder x-ray diffraction techniques and optical microscopy were used for phase identification and measure-ments of grain size. Magnetization at 4 K was measured for bars and disks of the sintered materials as well as for ground powders.

The results of magnetization of the sintered and powdered $YBa_2Cu_3O_7$ indicate that these specimens behave as granular superconductors for H greater than a few tens of millitesla, i.e., each superconducting region is surrounded by a weak superconducting layer. This conclusion is drawn by noting that the reduction in the width of the magnetic hysteresis ΔM for the powders in comparison to that for the bars was substantially less than the value which is expected from a "bulk superconductor." Since $\Delta M \alpha dJ_c$, ΔM should be reduced as the size of the specimen, d. (See Fig. 1.) Samples with varying degrees of granularity were observed. However it is not clear that grain boundaries are the sole source for breaking up the superconducting region.

It is also interesting to note tht it is possible to reduce the "intrinsic" critical current density of $YBa_2Cu_3O_7$ from the magnetization width ΔM of the powdered specimen since the powders can be reduced to the size where each particle is essentially single grains. We found that the magnetization induced $J_c > 10^3$ A/mm^2 at 4.2 K and a few tesla. This result was later confirmed by the IBM report on the single crystal film, $J_c > 10^4$ A/mm^2 at 4.2 K and zero field.

<u>Acknowledgments</u>

The authors appreciate R. H. Jones and J. J. Hurst for preparation
of the specimens and D. O. Welch for stimulating discussions. This work
was performed under the auspices of the U.S. Department of Energy,
Division of Materials Sciences, Office of Basic Energy Sciences under
Contract No. DE-AC02-76CH00016.

<u>References</u>

1.   J. G. Bednorz and K. A. Müller, Z. Phys. B. 64, 189 (1986).
2.   H. Takagi, S. Uchida, K. Kitazawa, and S. Tanaka, Jpn. J. Appl.
      Phys. Lett. (submitted).
3.   M. K. Wu, J. R. Ashburn, C. J. Torng, P. H. Hor, R. L. Meng, L. Gao,
      Z. J. Huang, Y. O. Wang, and C. W. Chu, Phys. Rev. Lett. 58, 908
      (1987).
4.   R. J. Cava, B. Batlogg, R. B. vanDover, D. W. Murphy, S. Sunshine,
      T. Siergrist, J. P. Rameika, E. A. Reitman, S. Zahurak, and G. R.
      Espinoza, Phys. Rev. Lett. 58, 1676 (1987).
5.   D. K. Finnemore, R. N. Shelton, J. R. Clem, R. W. McCallum, H. C.
      Ku, R. e. McCarley, S. C. Chen, P. Klavins, and V. Kogan (to be
      published in Phys. Rev., 1987).

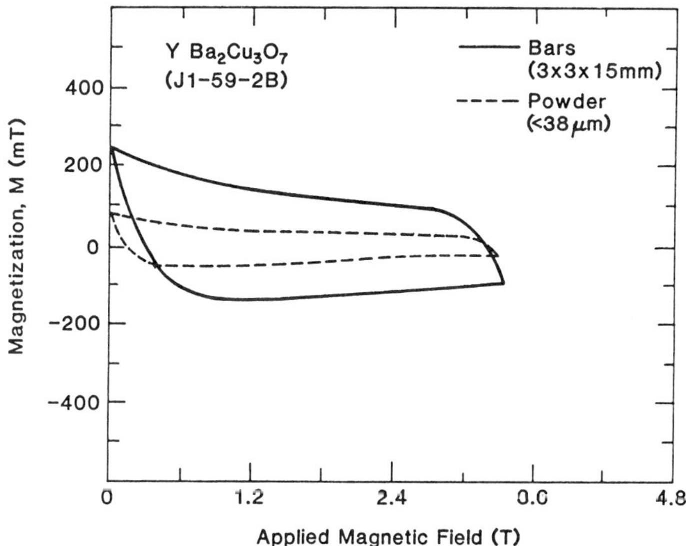

Fig. 1.   Magnetization curves for bulk (2x3x15 mm^3) and poweder (<38
          μm) YBa$_2$Cu$_3$O$_7$ at 4 K.

ACKNOWLEDGEMENT   This manuscript has been authored under contract DE-AC02-
7600016 with the Division of Materials Sciences, U.S. Department of Energy.

THE EFFECT OF ANISOTROPY IN H_{c2} ON THE BREADTH
OF THE RESISTIVE TRANSITION OF POLYCRYSTALLINE
$YBa_2Cu_3O_{7-x}$ IN A MAGNETIC FIELD

D. O. Welch, M. Suenaga, and T. Asano[*]

Materials Science Division
Brookhaven National Laboratory
Upton, NY

[*]on leave from National Research Institute
for Metals, Tsukuba, Ibaraki, Japan

Previous studies have shown that an applied magnetic field severely
broadens the transition to the superconducting state in polycrystalline
$YBa_2Cu_3O_{7-x}$ and La_2CuO_4-based superconductors. Two possible
sources of this broadening are the poor superconducting properties of
the intergranular material in these granular superconductors and the
anisotropy in the upper critical field H_{c2} arising from the layered
nature of their crystal structures. Both factors can cause low values of
the critical current density in magnetic fields for polycrystals, but
whereas the effect of granularity can in principle be removed by proper
processing, the effect of anisotropy is intrinsic and can severely limit
the usefulness of randomly oriented polycrystals if the anisotropy is
large enough.

To investigate the anisotropy we have recently measured the
resistive transition of polycrystalline $YBa_2Cu_3O_{7-x}$ in the
temperature range 83–93K in magnetic fields up to 25 T and analyzed the
results using the Ginzburg–Landau theory for the H_{c2} of axially
symmetric superconductors, together with a percolation theory of the
resistivity, to obtain a measure of the anisotropy ratio
$R \equiv H_{c2\parallel}/H_{c2}$, where $\parallel$ and denote, respectively applied fields
parallel and perpendicular to the copper-oxygen layers.[1]

The assumptions of the analysis are that even though the crystal
structure of $YBa_2Cu_3O_{7-x}$ has orthorhombic symmetry, it is
approximately axially symmetric with the symmetry axis normal to the
planes of the ... $CuO-BaO-CuO_2-Y-CuO_2-BaO$ ... layers comprising the
structure. In this case the orientation dependence of H_{c2} for each
crystallite can be described approximately by the axial Ginzburg–Landau
theory[2], and with this it is readily shown that for a random
polycrystal the average upper critical field is:

$$\langle H_{c2} \rangle = H_{c2}^{max} \frac{\ln[(R^2 - 1)^{1/2} + R]}{[R^2 - 1]^{1/2}} \tag{1}$$

where H_{c2}^{max} is presumably $H_{c2\parallel}$. However, for any field H greater than the minimum value of the orientation-dependent H_{c2}, presumably $H_{c2\perp}$, some fraction of crystallites in the polycrystal will have become normally conducting, and it is readily shown that the superconducting fraction is:

$$f_s(H) = [((H_{c2}^{max}/H)^2 - 1)/(R^2 - 1)]^{1/2} \tag{2}$$

when $(H_{c2}^{max}/R) \leq H \leq H_{c2}^{max}$. (If $H < H_{c2}^{min} = H_{c2}^{max}/R$, f_s is unity; if $H > H_{c2}^{max}$, f_s is zero. Nevertheless, as long as $f_s(H)$ is greater than the percolation threshold $\simeq 1/6$, the polycrystal will still have zero resistance. The onset of resistivity occurs for a field H_{onset} such that f_s falls below the percolation threshold:

$$H_{onset} = H_{c2}^{max} \left(\frac{36}{R^2 + 35}\right)^{1/2} \tag{3}$$

We have described the resistivity of the polycrystal, when the onset field is exceeded, by the effective-medium-percolation-theory-based phenomenological equation of Davidson and Tinkham[3], which for approximately spherical crystallites becomes

$$\frac{\rho(H)}{\rho_N} = \frac{1 - 6 f_s(H)}{1 + 3 f_s(H)} \tag{4}$$

where $\rho(H)$ is the resistivity of the aggregate in a field H and ρ_N is the normal-state resistivity. From the expression, it is easily deduced that the field at the midpoint of the transition, $H_{1/2}$, is given by

$$H_{1/2} = \frac{H_{c2}^{max}}{[(R^2 + 224)/225]^{1/2}} \tag{5}$$

It may be seen in Figure 1 that eqs. 2, 4, and 5 provide a good fit to the experimental data for temperatures in the range 83–93K for a value of R in the range 25–50, and there is an indication that R increases with increasing T, even through Ginzburg–Landau theory does not predict such a variation. Furthermore, we find that the upper critical field slope $(dH_{c2}/dT)_{T_c}$ is about 5T/K for H_{c2}^{max}, 2.6 T/K for $H_{1/2}$, 1.2 T/K for H_{onset}, and 0.2 T/K for H_{c2}^{min}. Note that if the large value of R obtained here is true, then the large breadth of the resistive transition in magnetic fields is an intrinsic property of randomly oriented polycrystals of $YBa_2Cu_3O_{7-x}$, and this, together with limitations on the critical current density caused by the anisotropy, may cause one to place limits on the applicability of this material in high-field magnets and related applications. However, on the beneficial side, a strong anisotropy can enhance critical current densites by providing flux pinning at interfaces between differently oriented grains.

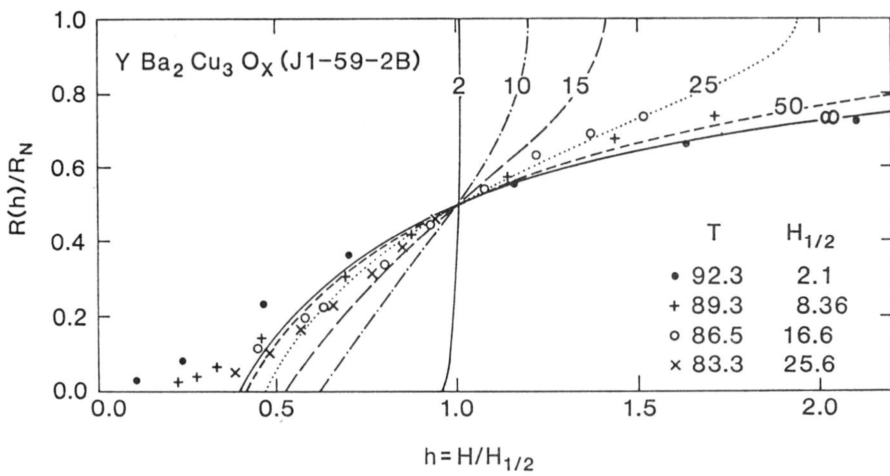

Fig. 1. The magnetic field dependence of the electrical resistance of polycrystalline $YBa_2Cu_3O_{7-x}$ for four different temperatures. The resistance data scaled by the normal state resistance at the appropriate temperature and the applied magnetic field is scaled by the value at which the resistance reaches half the normal state value ($H_{1/2}$). The theoretical curves were calculated for several H_{c2} anisotropy ratios, as noted, using equations 2, 4, and 5.

References

1. D. O. Welch, M. Suenaga, and T. Asano, Phys. Rev. B, to be published in the August 1, 1987 issue.
2. H. Teichler, in Anisotropy Effects in Superconductors, H. W. Weber, Ed., p. 7, Plenum Press, New York, 1977.
3. A. Davidson and M. Tinkham, Phys. Rev. B 13, 3261 (1976).

ACKNOWLEDGEMENT   This manuscript has been authored under contract DE-AC02-7600016 with the Division of Materials Sciences, U.S. Department of Energy.

SUPERCONDUCTING PROPERTIES AND STRUCTURAL CHARACTERIZATION OF High Tc OXIDES

A. R. Moodenbaugh, J. J. Hurst, T. Asano
R. L. Sabatini and M. Suenaga

Brookhaven National Laboratory
Upton, NY

A desire to provide good quality oxide superconducting specimens for
research in our materials laboratory and for collaborators at the
Synchrotron Light Source, High Flux Beam Reactor, and Alternating
Gradient Synchrotron required us to carefully characterize our samples.
Earlier, straightforward methods for preparation $YBa_2Cu_3O_7$[1]
superconductors were described. However, these methods are not
completely defined, and samples prepared nominally according to recipe
may vary in superconducting properties and with respect to the presence
of non superconducting minority phases (primarily observed in x-ray
diffraction work). Here we describe methods of preparation for these
oxides which we utilized, and of characterization to evaluate these
materials.

$YBa_2Cu_3O_7$ was prepared from Y_2O_3, fully oxidized CuO, and
$BaCO_3$ powder. Thorough grinding in an agate mortar was followed by
pelletization before the first firing in Pt (pellets reduce the reaction
with Pt). Samples were put into a cool furnace set to 900 C for 30 hr.
After this firing the sample was promptly reground, pelletized, and
refired under identical conditions. After again grinding and
pelletizing, the sample was heated to approximately 950 C in oxygen flow
for about 60 hr, followed by furnace cooling over several hours to below
200 C. Controlling the furnace at 700 C for several hours as a last
additional step during this furnace cooling appears to produce samples
with exceptionallly narrow diffraction peaks. It has been our experience
that samples closest to being single phase have been heated to a
temperature rear 950 C at which a very small amount of liquid (probably
$BaCuO_2$) runs out of the sample. Longer sintering times at slightly
lower temperatures might produce better specimens.

The temperature of final anneal in oxygen is crucial. A lower
temperature results in a sample with a lower transition temperature, and
often containing a significant amount of minority phases. Higher
temperatues result in a significant partial sample decomposition; the
$BaCuO_2$ runs out of the sample and minority phases again apper in
diffraction diagrams.

Superconducting transitions were measured by mutual inductance
methods. The tendency of these samples to be granular[2] makes the
amplitude of signal a reliable indication of quantity of superconductor.
A better correspondence between superconducting volume and size of signal
occurs when the sample is ground to a powder before measurement. Note
that the volume of superconductor is not a direct function of the

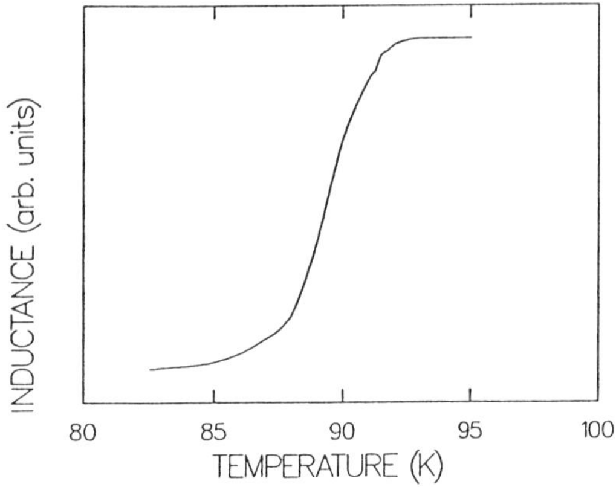

Fig. 1 Superconducting transition of
YBa$_2$CuO$_7$ as measured by mutual
inductance.

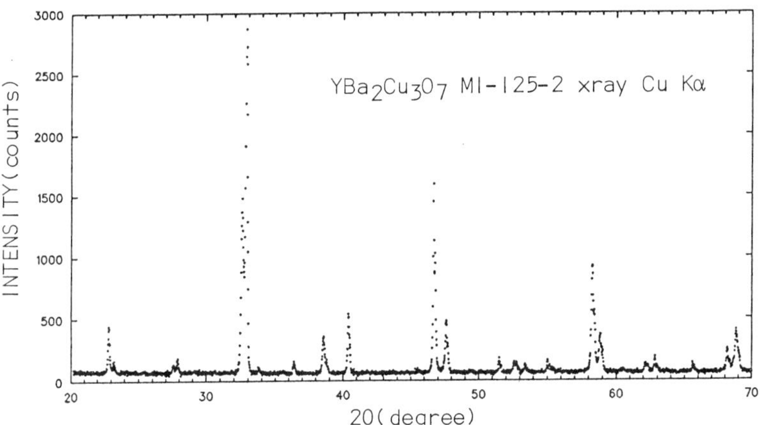

Fig. 2 Portion of X-ray diffraction diagram of
YBa$_2$Cu$_3$O$_7$.

quantity of the "superconducting phase" appearing in x-ray diffraction
diagrams. Hence the size of the transition is an important diagnostic.

Samples of $YBa_2Cu_3O_7$ having mutual inductance superconducting
transition midpoints near 90 K were prepared (see Fig. 1). These samples
are single phase by standard x-ray diffraction (see Fig. 2). Neutron
diffraction measurements of this sample showed an impurity peak at the
0.5% level (d=1.41) that we attribute to BaY_2CuO_5. Energy disperive
spectroscopy[3] performed on a similar sample of $YBa_2Cu_3O_7$
reveals detectable levels of Y_2BaCuO_5, $BaCuO_2$, and CuO. Minority
phases are often present in sufficient quantity to be seen in diffraction. For example, a sample that had undergone a final heat treatment of
900 C in oxygen (rather than the more typical 950 C) had a transition
temperature of 80 K, and contained minority phases BaY_2CuO_5
(Carnegie), $BaCuO_2$, and perhaps $Ba_2Y_2O_5$.

<u>Reference</u>

1.   See, for example R. J. Cava, B. Batlogg, R. B. van Dover, D. W.
     Murphy, S. Sunshine, T. Siegrist, J. P. Remeika, E. A. Reitman,
     S. Zahrate, and G. P. Spinoza, Phys. Rev. Lett., <u>58</u>, 1676 (1987).
2.   M. Suenaga, A. Ghosh, T. Asano, R. L. Sabtini, and A. R.
     Moodenbaugh, Materials Research Society Spring Meeting, 1987,
3.   Araheim, R. W. McCallum, J. D. Verhaven, M. A. Noack, E. D.
     Gibson, F. C. Laabe, D. K. Finnemore, and A. R. Moodenbaugh
     (in preparation).

ACKNOWLEDGEMENT   This manuscript has been authored under contract DE-ACO2-
7600016 with the Division of Materials Sciences, U.S. Department of Energy.

THE VARIATION OF T_c WITH HOLE CONCENTRATION IN $La_{2-x}Sr_xCuO_{4-\delta}$ SUPERCONDUCTORS AND COMPARISON WITH $YBa_2Cu_3O_{7-\delta}$

M. W. Shafer, T. Penney and B. L. Olson

IBM T. J. Watson Research Center
Yorktown Heights, NY 10598

Since the discovery of superconductivity in the La-Ba-Cu oxides by Bednorz and Müller[1] a number of similar copper oxide type materials have been shown to be high temperature superconductors. Presently there are two copper oxide systems which are being thoroughly studied, the $La_{2-x}(Ba,Sr)_xCuO_{4-\delta}$ with a $T_c \simeq 35 - 40K$ and $YBa_2Cu_3O_{9-\delta}$ with a transition temperature $T_c \simeq 90K$.[2]. From a purely scientific point of view, however, the $La_{2-x}(Ba,Sr)_xCuO_{4-\delta}$ series has a major advantage over the 90K system because of the large solubility of divalent ions such as Ba and Sr, whose concentrations can be systematically varied to control the state of the copper - oxygen interactions, and hence the physical properties.

The series $La_{2-x}Sr_xCuO_{4-\delta}$ with the K_2NiF_4 type structure, was studied in considerable detail even before they were known to be superconductors.[3,4,5] Pure La_2CuO_4 is orthorhombic but the substitution of Sr for La stabilizes a tetragonal phase and forces an equivalent number of Cu^{2+} to be converted to Cu^{3+}. It is generally accepted that the transport properties are strongly a function of the Cu^{3+}/Cu^{2+} ratio. The Sr doped samples for $x > 0.06$ were shown to be superconductors with a maximum transition temperature at about 36 - 38K when $x = \sim 0.15$.[6,7,8,9]

This report is an extension of these earlier studies and our previous work.[10,11] We show that the superconductivity is related to the electron deficiency (hole concentration) in the copper–oxygen layer. Our resistivity and Hall data for the La-Sr-Cu-oxide system is then compared with that of $YBa_2Cu_3O_7$. These results are consistent with recent non-phonon theories of oxide superconductivity based on strongly correlated electrons or dynamic peroxide formation.[12,13]

Both the $La_{2-x}Sr_xCuO_{4-\delta}$ and the $YBa_2Cu_3O_7$ materials were prepared from the nitrates. Copper metal (6N), and assayed quantities of the appropriate metal oxides were each dissolved in nitric acid and combined in the desired proportions by standard volumetric techniques. The solutions were evaporated and converted to oxides by heating in oxygen. Although the reaction to the K_2NiF_4 -like phase was essentially complete at $850°C$, they were again reground and pressed into pellets and sintered at $1140\text{-}1180°C$ in oxygen, then annealed at $550\text{-}600°C$. The $YBa_2Cu_3O_7$ material was reacted in oxygen at $920°C$, then pressed into pellets and sintered at $950°C$. They were cooled slowly to $450°C$ and annealed there in oxygen for 16 hrs. In order to avoid crucible contamination during the high temperature sintering, the pellets were supported on powders of identical composition in high-density alumina boats. The resulting materials were characterized as follows: by x-ray diffraction, microprobe analysis and by phase contrast microscopy. The total copper was determined by inductively coupled plasma atomic emission spectroscopy. The electron deficiency, which we call the hole concentration and refer to as a $[Cu-O]^+$ complex rather than Cu^{3+}, was determined by wet chemistry according to the reaction $[Cu-O]^+ + Fe^{2+} = Cu^{2+} + Fe^{3+} + O^{2-}$. For the La-Sr-Cu oxides, the powdered materials were dissolved in $6N\ H_2SO_4$ containing standardized $0.04N\ Fe^{++}$. After the addition of H_3PO_4 the remaining Fe^{+2} was determined by titrating with standard $0.04N\ KMnO_4$.[14]

The fact that these materials are metals, even in the non-superconducting state, implies non-ionic type bonding without defined Cu^{3+} species on any given lattice site. Therefore, we assume the electron deficiency to be distributed between the copper and oxygen ions to form a complex of the type $[Cu^{3+}-O^2-]^+ \leftrightarrow [Cu^{2+}-O^{1-}]^+$. The latter can be thought of as a peroxide complex.[13] This electron deficient complex, which we write as $[Cu-O]^+$, like a Cu^{3+}ion, is a good oxident and easily oxidizes ferrous iron to ferric. On the other hand, the Cu^{+2} ion will not do so. Attempts to determine the accuracy of this analytical procedure were hampered because the lack of a good Cu^{3+} standard. However, based on favorable redox potentials and using stable volumetric solutions, we believe that the actual hole concentration is being measured. Supporting evidence for the reliability of our analysis in determining the concentration of the $[Cu-O]^+$complex is based on a total oxygen determination as done by Nguyen.[5] Here, dry hydrogen is used to reduce the copper in the La-Sr-Cu oxides to the metal. The La and Sr are assumed to remain as their stoichiometric oxides i.e. La_2O_3 and SrO, so the weight loss is due to the oxygen associated with the copper. When we compare these results for total oxygen with our chemical method, there is good agreement. For example, sample No. 23 (x=0.06) had an oxygen content of $3.99\pm.02$ and $3.98\pm.03$ for the solution and H_2 reduction methods respectively.

At room temperature the orthorhombic La_2CuO_4 transforms to tetragonal at $x = \sim .06$, in general agreement with previous work.[5,9] Further, for the tetragonal

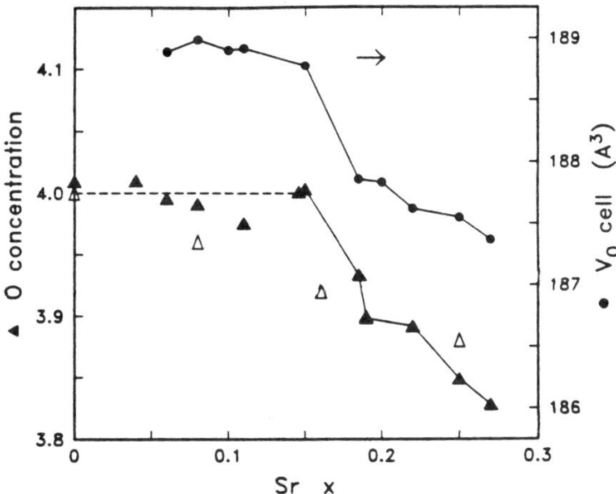

Fig. 1. Cell volume (closed circles) and oxygen content determined from our analysis of $[Cu - O]^+$ (closed triangles) vs.
strontium concentration (x). The open triangles are data from
Nguyen et al[5] as determined by hydrogen reduction.

hase, a_o decreases with x while c_o increases. The small difference between the val-
ies we obtained for the lattice constants and those reported earlier[5,6] are likely due
.o a different preparation procedure and subsequent thermal history. It is clear from
Fig. 1, that the cell volume,V_o, does not change linearly with Sr doping. There is a
sharp decrease around the Sr = 0.15 composition. In Fig. 1,(closed triangles), we
show the variations in oxygen stoichiometry obtained from our compositional analysis. It is seen that an oxygen stoichiometry of O_4 , is essentially maintained, meaning there are few, if any, oxygen vacancies, for compositions from x=0 to about x
= 0.15. For compositions with x > 0.15, the measured decrease in the $[Cu - O]^+$
concentration indicates a decrease in the oxygen content- and therefore an increase
in the number of oxygen vacancies. Also given in Fig. 1 are the data of
Nguyen et al[6] (open triangles). The rather sharp decrease in the cell volume at
x=0.15 - 0.20 (Fig. 1) can be correlated with the oxygen vacancy content, determined from the $[Cu - O]^+$ concentration, which increases sharply above x = 0.15.
This decrease is due to oxygen vacancies and the presence of the $[Cu - O]^+$ complex, whose size is undoubtedly smaller than the Cu^{2+} ion. It appears that when
about 15% of the "octahedral" sites are occupied by $[Cu - O]^+$, in order to maintain the K_2NiF_4 structure, oxygen vacancies are formed, rather than additional
$[Cu - O]^+$ complexes.

The dependence of T_c on Sr concentration x for the $La_{2-x}Sr_xCuO_{4-\delta}$ system
is shown in Fig. 2. Also shown is the quantity $[Cu - O]^+$ which is a measure of the
electron deficiency, that is the hole concentration. In two other papers it is shown
that there is a direct correlation between T_c and electron deficiency. [10,11] The decrease in T_c for x > 0.15 is related to electron deficiency not x.

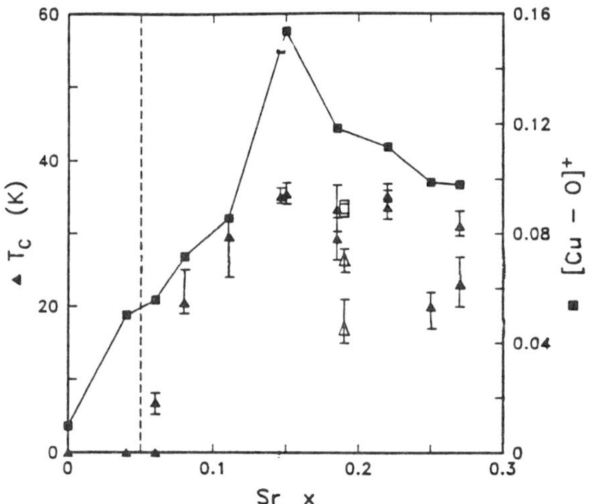

Fig. 2. T_c (closed triangles) and $[Cu - O]^+$ concentration (closed squares) vs. strontium concentration (x). Error bars indicate transition widths. Points without error bar did not show a transition down to 5K.

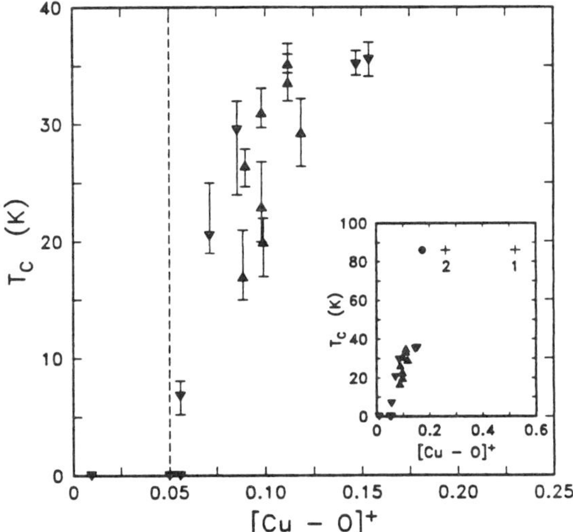

Fig. 3. T_c vs. the hole concentration, $[Cu - O]^+$, as a fraction of total copper. Down triangles are for compositions with x < 0.15, up triangles are for x > 0.15. Inset shows same data plus points for a single $YBa_2CuO_{6.67}$ sample with three normalizations. See text.

In Fig 3, we plot T_c vs. $[Cu - O]^+$ and see a strong dependence of T_c on the $[Cu - O]^+$ concentration with a maximum T_c of 35 degrees for $[Cu - O]^+$ concentration of 15% of the total copper. It is clear that the $[Cu - O]^+$ concentration and not the Sr doping, determines the T_c. T_c increases rapidly above a $[Cu - O]^+$ threshold, i.e. about 5%, and reaches a maximum at roughly 15% . It appears that 15 -16% $[Cu - O]^+$ is the maximum that this structure will accept.

In order to see   if the relationship between T_c and $[Cu - O]^+$ in the $La_{2-x}Sr_xCuO_4$ system can be extended to the $YBa_2Cu_3O_{9-\delta}$ system, we have re-plotted the $La_{2-x}Sr_xCuO_{4-\delta}$ system data together with the results for one $YBa_2Cu_3O_{6.6}$ sample (dot in Fig. 3 inset). In the $YBa_2Cu_3O_{6.6}$ structure there are two different copper layers, one between the barium planes and two between yttrium and barium planes. If only one or two layers are active then the ratio of $[Cu - O]^+$ to active Cu is given by the crosses labeled 1 and 2 respectively in the inset. An extrapolation of the $La_{2-x}Sr_xCuO_{4-\delta}$ data does pass within the range of values for $YBa_2Cu_3O_{6.6}$ indicating that the same relation between T_c and hole concentration might apply in this system too.

For Sr concentration $x<0.15$ the $[Cu - O]^+$ complex concentration follows x, the dashed line in Fig. 4. Each Sr substituted for La creates a hole. The hole concentration, per formula unit, V_O/R_He, determined from the Hall effect also follows the Sr concentration for $x < 0.15$. This agreement shows that the conductivity is dominated by one band, for only in this case does the Hall effect give the true carrier concentration.

It is evident from Fig. 4 that there is a critical Sr concentration x_c of about 0.15 beyond which the system changes drastically. Beyond the critical Sr concentration, the $[Cu - O]^+$ concentration decreases as x increases. On the other hand the Hall number increases far above x, the dashed line, as previously observed by Ong et. al.[15] The well known formulas for the Hall effect for two types of carriers are given below. It is convenient to normalize the hole and electron densities p and n by the

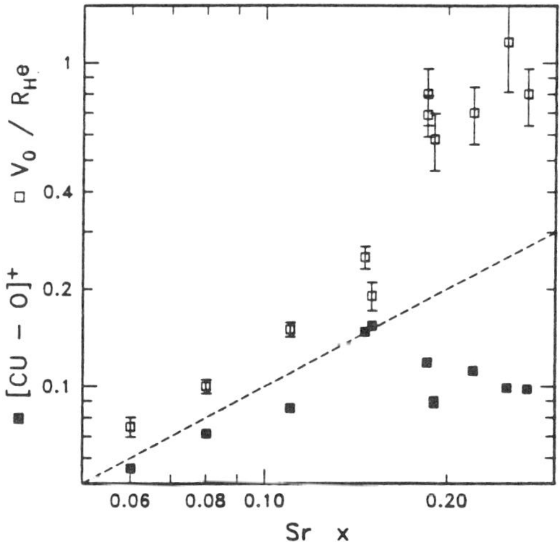

Fig. 4. $[Cu - O]^+$ and Hall number V_0/R_He at 50K vs. strontium concentration (x). The Hall number is the number of holes per formula unit if only one band contributes.

volume per formula unit, V_O (94 $Å^3$ and 174 $Å^3$ for the LaSr and YBa systems respectively) so that $P = pV_O$ and $N = nV_O$. Then the normalized conductivity is

$$\sigma V_0/e \ = \ (P + NQ)\mu_P \tag{1}$$

the Hall number is

$$V_0/R_H e \ = \ (P + NQ)^2/(P - NQ^2) \tag{2}$$

and the Hall mobility is

$$\mu_H \ = \ R_H/\rho \ = \ \mu_p(P - NQ^2)/(P + NQ) \tag{3}$$

where the mobility ratio is

$$Q \ = \ \mu_N/\mu_P \tag{3}$$

From eq. 2 there are two simple explanations for the increase in the Hall number for x above x_c. Either there are still only holes and the hole concentration actually increased or there is a compensation of hole and electron contributions. It is not possible from measurements of Hall constant and resistivity alone to distinguish the two cases unambiguously, nor, in the case of two species of carriers, to determine concentrations and mobilities of both holes and electrons. Since in a simple metal one has a temperature independent Hall number, if one finds a strongly temperature dependent Hall number it is reasonable to ascribe it to changes in the compensation of holes and electrons due to differences in the temperature dependences of the mobilities. In our case, the Hall numbers are only weakly temperature dependent for all of our samples reported in Fig. 4. We find that samples with x > 0.15, the critical concentration, x_c , have Hall numbers about three times larger than those with x = 0.15. It is not possible to support a model of hole and electron compensation on the basis of temperature dependence.

The ambiguity is resolved in our case by the knowledge of the hole concentration, $[Cu - O]^+$, from the wet chemical analysis. Since the holes decrease for $x > x_c$ the large Hall number must come from compensation by electrons. It is still not possible to uniquely determine the four conductivity parameters from three numbers. It is possible to say that the holes still dominate the conductivity. If one arbitrarily sets the mobility ratio Q=1 then the electron number is about 0.7 per formula unit, slightly smaller than the hole number given by $[Cu - O]^+$. The Hall mobility is about 4 cm^2/V sec for the x = 0.15 sample and is the actual hole mobility since only one carrier type is active. The smaller Hall mobility found for the x > 0.15 samples is reduced by the cancellation of electrons and holes and is not the true mobility of either. In addition to the previous study[15] of the Hall effect in this system already noted, several groups have reported Hall measurements on single compositions.[16-19]

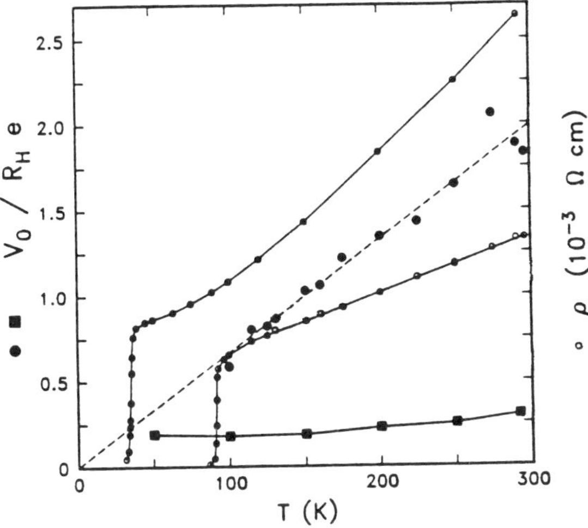

Fig. 5. Resistivity data (open circles) and Hall number (closed squares and circles) vs. temperature for $La_{1.85}Sr_{.15}CuO_4$ and $YBa_2Cu_3O_7$. The closed squares are the Hall data for $La_{1.85}Sr_{.15}CuO_4$

The Hall and resistivity results for $La_{1.85}Sr_{.15}CuO_4$ are compared with those of $YBa_2Cu_3O_7$ in Fig. 5. In both cases linear resistivities (open circles) and positive Hall numbers are seen. The major difference is in the temperature dependence of the Hall number; a strong dependency is seen for $YBa_2Cu_3O_7$(closed circles), while for $La_{1.85}Sr_{.15}CuO_4$ it is essentially independent of temperature (closed squares). In fact, within the error of our measurements, the Hall number is linear in T with little or no intercept. Since one expects a temperature independent carrier concentration in a metal, a likely explanation of this effect is the temperature dependent compensation of hole and electron contributions. This is a valid model, since the four conductivity parameters are undetermined by the resistivity and Hall constant. There is however no simple choice of parameters. In the case of $La_{2-x}Sr_xCuO_{4-\delta}$ there was only weak temperature dependence and the x dependence was satisfied in a simple way. Here the nonlinear eq. 2 must produce a linear result. This unsatisfactory situation may be evidence for some exotic transport mechanism, rather than the usual two band model.

There are a number of all electronic models for high temperature oxide superconductivity. Among them there is one[20] which is based on the coexistence of two d bands and which explains the occurrence of two types of carriers for large Sr concentration and only one type for small concentration, consistent with our observation. Another[12] is based on strong electron electron correlations and is a generalization of a theory of heavy fermion superconductivity. This theory predicts a monotonic increase of T_c with hole concentration, consistent with our results. Still another[13] is based on the occurrence of dynamic peroxide formation which would be consistent with what we write as $[Cu - O]^+$.

In summary, we have prepared a series of samples in the $La_{2-x}Sr_xCuO_{4-\delta}$ system and studied the superconducting and transport properties as a function of composition. A unique wet chemical analytical technique is used to determine the electron deficiency (hole concentration) in the copper-oxygen layer, which we define as a $[Cu - O]^+$ complex. From this analysis we obtain a value for the oxygen stoichiometry and contrary to previously published results, we find few if any oxygen vacancies for $0 \leq x \leq 0.15$. The $[Cu - O]^+$ concentration is approximately equal to the Sr concentration and the Hall number to about $x = 0.15$. Conduction is therefore due to holes in a single band. For $x > 0.15$ the $[Cu - O]^+$ decreases and the number of oxygen vacancies increases. The Hall effect is qualitatively different in this range. It does not measure the number of carriers but rather indicates hole and electron conduction or other exotic behavior. A strong correlation between the $[Cu - O]^+$ concentration and T_c is shown which supports an all-electronic model with strong electron-electron correlations.[12] For $YBa_2Cu_3O_7$, an unusual T dependence is found for the Hall number. The origin can be two carrier conduction with unusual temperature dependences of the conductivity parameters or some novel conduction mechanism related to the novel pairing in these high T_c materials.

We thank R. A. Figat, C. F. Guerci and J. M. Rigotty for technical assistance and A. Malozemoff, R. A. de Groot and D. M Newns for valuable discussions.

References

1. J. G. Bednorz and K. A. Müller, Z. Phys. **B64,** 189 (1986); J. G. Bednorz, M. Takashige and K. A. Müller, Europhys. Letter **3,** 379 (1987).

2. M. K. Wu, J. R. Ashburn, C. J. Torng, D. H. Hor., R. L. Meng, L. Gao, Z. J. Huang, Q. Wang and C. W. Chu, Phys. Rev. Let.**58,** 908 (1987).

3. J. M. Longo and P. M. Raccah, J. Sol. Stat. Chem. **6,** 526 (1973).

4. N. Nguyen, J. Choisnet, M. Hervieu and B. Raveau, J. Sol. Stat. Chem. **39,** 120 (1981).

5. N. Nguyen, F. Studer and B. Raveau, J. Phys. Chem. Solids **49** 6389 (1983).

6. J. M. Tarascon, L. H. Green, W. R. McKinnon, G. W. Hull and T. H. Geballe, Science **235** 1373 (1987).

7. R. J. Cava, R. B. Van Dover, B. Batlogg and E.A. Reitman, Phys. Rev. Lett.**58** 408 (1987).

8. S. Uchida, H. Takagi, K. Kitzawa and S. Tanaka, Jap. J. Appl. Phys. Lett. **26** L1-L2-(1987).

9. R. M. Fleming, B. Batlogg, R. J. Cava and E. A. Reitman, Phys. Rev. B **35** 7191 (1987)

10. M. W. Shafer, T. Penney and B. Olson, to be published

11. T. Penney, M. W. Shafer, B. L. Olson and T. S. Plaskett, Advanced Ceramic Materials, July 1987, Special issue on ceramic superconductors

12. D. M. Newns, to be published.

13. R. A. de Groot, H. Gutfreund and M. Weger, to be published.

14. A slightly different technique was used for the Y-Ba-Cu oxide since they could not be dissolved in H_2SO_4.

15. N. P. Ong, Z. Z. Wang, J. Clayhold, J. M. Tarascon, L. H. Greene and W. R. McKinnon, to be published.

16. M. F. Hundley and A. Zettl, to be published.

17. A. J. Panson, G. R. Wagner, A. I. Braginski, J. R. Gavaler, M. A. Janocko, H. C. Pohl and J. Talvacchio, to be published.

18. M. Suzuki and T. Murakami, to be published.

19. S. Uchida, H. Takagi, H. Yanagisawa, K. Kishio, K. Kitazawa, K. Fueki and S. Tanaka, to be published.

20. D. H. Lee, private communication and D. H. Lee and J. Ihm, Sol. State Comm.,to be published

ANISOTROPY IN SINGLE-CRYSTAL $Y_1Ba_2Cu_3O_{7-x}$

T.K. Worthington, W.J. Gallagher, T.R. Dinger, and R.L. Sandstrom

IBM T.J. Watson Research Center
P.O. Box 218, Yorktown Heights, NY 10598

Considerable activity on high-T_c superconductivity in Cu-O based perovskites has followed the breakthrough discovery of superconductivity in this class of materials by Bednorz and Müller[1]. Nevertheless, the macroscopic nature of the superconductivity in these new materials and the mechanism responsible for their superconductivity are still unclear. The recent availability of single crystals of the new Cu-O based superconductors has led to the observation of considerable anisotropy in resistivity[2], critical current density[3], and upper[2,4-7] and lower[3] critical fields. In this paper we summarize our recent measurements of H_{C2}, H_{C1}, and critical current density as a function of orientation for single crystal $Y_1Ba_2Cu_3O_{7-x}$. We use this data to estimate the anisotropic Ginzburg-Landau parameters. We find that the anisotropic Ginzburg-Landau theory describes our data well near T_C and that, although the superconductivity in $Y_1Ba_2Cu_3O_{7-x}$ is strongly anisotropic, it is three dimensional in nature. Elaborations of available theories are needed to explain lower temperature features of our data.

The $Y_1Ba_2Cu_3O_{7-x}$ single crystals used in this study were obtained from a partially melted pellet made with an off-stoichiometric composition according to a procedure described by Dinger et al.[3] This procedure routinely resulted in highly faceted crystals with dimensions of approximately 200 μm, with occasional crystals approaching 0.5 mm in size. The two crystals used in this study had dimensions of 400 μm by 370 μm by 120 μm and 340 μm by 280 μm by 160 μm. Measurements of the latter in a scanning X-ray diffractometer showed this crystal to be of high quality with unit cell dimensions of a = 3.83 Å, b = 3.89 Å and c = 11.71 Å. The degree of orthorhombicity was b/a = 1.016 . The crystal was macroscopically twinned in the a-b plane. The diffraction peak widths were limited by instrumental resolution indicating domains in excess of 500 Å in size and little compositional variation.

Magnetization hysteresis curves taken as an applied field is swept allow the determination of the lower critical field and provide a noncontact way of determining critical current density from the magnetic moment resulting from induced screening currents[8]. Figure 1 shows magnetic hysteresis loops at 4.5 K for one of the crystals. The crystal was mounted with the CuO planes perpendicular to the field lines for Fig. 1a and with the planes parallel to the field lines for Fig. 1b. The differences in the scale and in the shape of the magnetization for the two orientations is striking. The lower critical fields, $H_{c1}^{\perp}$ and $H_{c1}^{\parallel}$, for crystals oriented such that the applied field is perpendicular and parallel to the Cu-O planes, respectively, can be determined from the point in the initial part of each loop at which the departure from linearity begins. The anisotropy for both crystals was in excess of 10:1. The values for one crystal, corrected for the demagnetizing factors, are listed in Table 1.

The difference in the magnitude of the hysteresis in the two loops in Fig. 1 is also large. In the perpendicular orientation the gradual departure of the magnetization curve from perfect diamagnetism indicates there is strong pinning. In contrast, the sharp break at $H_{c1}^{\parallel}$ in the parallel orientation is indicative of weak pinning. From the magnitude of the magnetization at different fields, the critical current density can be determined[2]. In Fig. 2 we show the field and temperature dependence of the critical current density as determined from the hysteresis curves in Fig. 1 and similar curves taken at higher temperatures for this crystal. (Note that the curves in Fig. 2 are labeled according to the direction of the applied field. Thus, for example, the curves labeled with $H_\perp$ are for fields applied perpendicular to the Cu-O planes such that the induced currents flow along the planes. The critical currents in the table are also listed according to the direction of applied field.) According to Fig. 2, the critical current anisotropy is ~7:1 at 4.5 K and zero field, increasing to ~20:1 at 4 Tesla. The anisotropy increases considerably at higher temperatures and fields. It is significant to note that the critical current density along the Cu-O planes remains substantially in excess of 10^6 A/cm^2 out to 4 T with no sign of a rapid fall off. Magnetization measurements taken on the second crystal confirmed the same general behavior, and actually indicated a stronger anistropy in the critical current density at low temperature and fields (20:1, as given in Table 1).

Above H_{c1}, we observed a slow time dependant change in the magnetization similar to that observed in $La_{2-x}Ba_xCuO_4$ by Müller et al.[9]. The magnetization was seen to change by as much as 20 percent and the time dependence was consistent with logarithmic behavior for times longer than 1000 minutes. (For practical reasons, the magnetization points reported in Fig. 1 were taken a few minutes after establishing the field.) Figure 3 shows a plot of the magnetization versus time for a

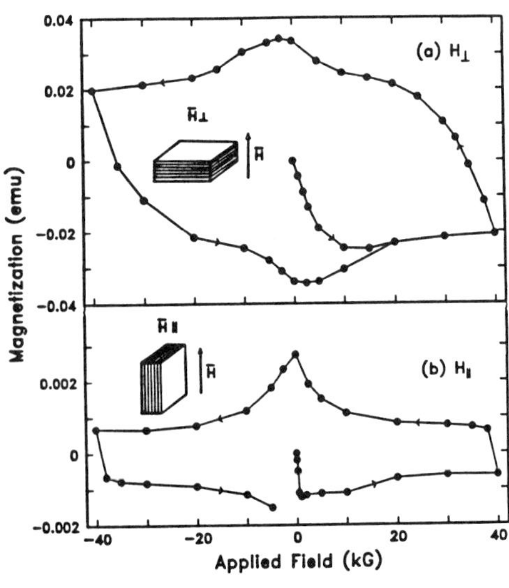

Fig. 1: Magnetization hysteresis loops at 4.5 K for a single crystal with the Cu-O planes oriented (a) perpendicular and (b) parallel to the applied field.

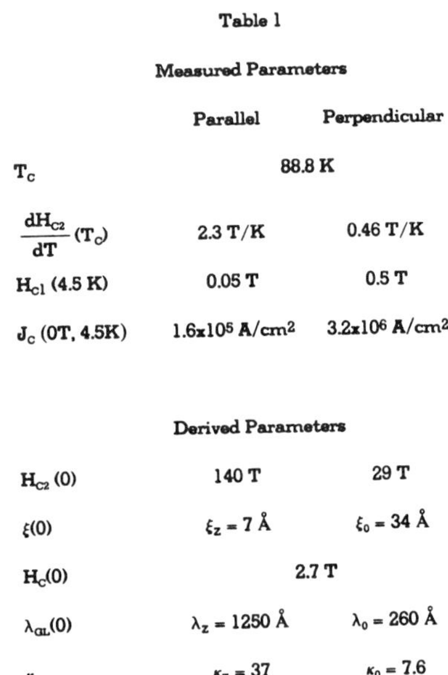

Table 1

Measured Parameters

	Parallel	Perpendicular
T_c		88.8 K
$\dfrac{dH_{c2}}{dT}\,(T_c)$	2.3 T/K	0.46 T/K
H_{c1} (4.5 K)	0.05 T	0.5 T
J_c (0T, 4.5K)	1.6×10^5 A/cm^2	3.2×10^6 A/cm^2

Derived Parameters

	Parallel	Perpendicular
H_{c2} (0)	140 T	29 T
$\xi(0)$	$\xi_z = 7$ Å	$\xi_0 = 34$ Å
$H_c(0)$		2.7 T
$\lambda_{GL}(0)$	$\lambda_z = 1250$ Å	$\lambda_0 = 260$ Å
κ	$\kappa_z = 37$	$\kappa_0 = 7.6$

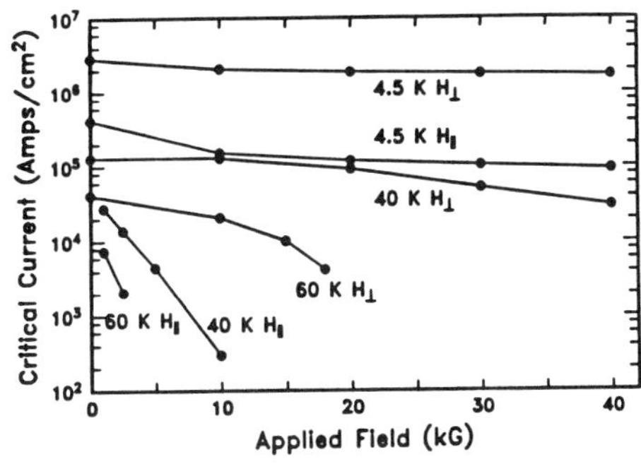

Fig. 2. Induced critical current densities at various temperatures as a function of magnetic field applied parallel and perpendicular to the Cu-O planes.

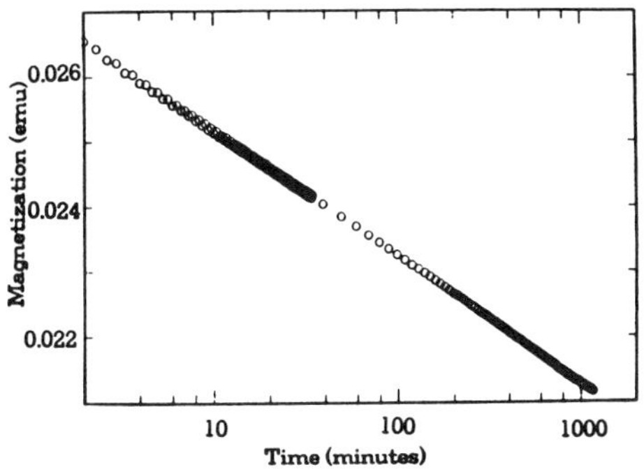

Fig. 3. Magnetization as a function of time at 4.5 K for a 10 kG field applied perpendicular to the Cu-O planes at time t = 0.

field of 1 Tesla applied perpendicular to the Cu-O planes. A similar rate of decrease (~9 %/decade) was also seen in another crystal for identical initial conditions. Logarithmic time dependence has long been associated with flux creep in type II superconductors, as originally observed by Kim et al.[10] It is significant to note when considering the superconducting glass state described by Müller et al., that we have established that the logarithmic behavior exists within a single crystal above H_{C1}.

Our results[7] for the upper critical field, H_{C2}, measured by an inductive technique[11] are shown in Figure 4 for the second crystal mounted with the Cu-O planes oriented parallel and perpendicular to the applied field. The zero field transition temperature extrapolated from both the parallel and perpendicular data is 88.8 K. The temperature dependence of $H_{C2}^{\parallel}$ is consistent with a straight line with a slope of -2.3 T/K. Extrapolation of the $H_{C2}^{\parallel}$ curve back to zero field according to the dirty limit isotropic formula (with no Pauli paramagnetism limiting effects)[12]. gives $H_{C2}^{\parallel}(0) = 140$ T. In contrast, the temperature dependence of the perpendicular field $H_{C2}^{\perp}$ shows a pronounced upward deviation from a linear dependence. Near T_C, $H_{C2}^{\perp}$ has a slope of -0.46 T/K, but below 78 K the data is consistent with a slope of -0.71 T/K. Extrapolating these two curves back to zero temperature using the isotropic dirty limit formula we get estimates for $H_{C2}^{\perp}(0)$ of 29 and 42 T.

We looked for, but did not observe any change in the transition temperature for rotations of 0, 45 and 90 degrees in the a-b plane in a field of 10 Tesla. Even though the crystal is twinned, we would expect to see a variation if there were a strong uniaxial anisotropy associated with the Cu-O chains that exist within the Cu-O planes. We believe that we would have seen an anisotropy in our crystal if it were as large as the ~ 1 K change reported by Hidaka et al.[5] for their resistive measurements of (twinned) single crystal $Y_1Ba_2Cu_3O_{7-x}$. Our observation of no anisotropy within the CuO plane is consistent with our observation of large supercurrents within

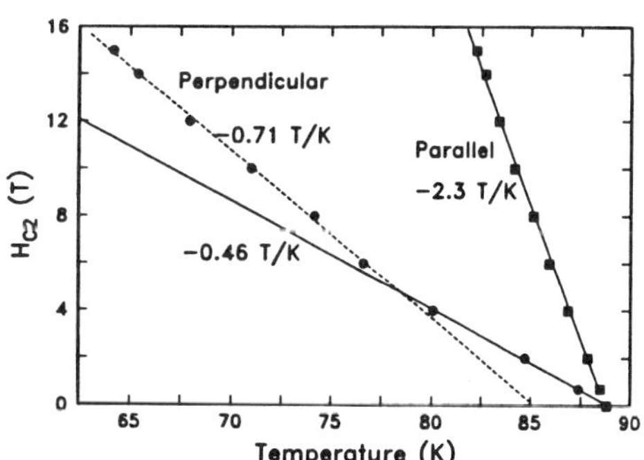

Fig. 4. Temperature dependence of the H_{C2} for crystal oriented with the Cu-O planes parallel and perpendicular to the applied field.

the planes. If only the 1-dimensional chains carried current, it would be difficult to account for such large values of supercurrent.

Our measured values for some of the anisotropic parameters of $Y_1Ba_2Cu_3O_{7-x}$ are given in Table 1 along with various quantities derived from anisotropic Ginzburg-Landau theory. The coherence length in the Cu-O plane, ξ_0, is calculated for the estimate of $H_{C2}^{\perp}(0) = 29$ T and the relation[13] $H_{C2}^{\perp}(0) = \Phi_0/2\pi\xi_0^2$. The ratio of $H_{C2}^{\parallel}/H_{C2}^{\perp} = \xi_z/\xi_0$ results in $\xi_z = 7$ Å. The coherence length perpendicular to the Cu-O layers, ξ_z , is significantly greater than the spacing between the Cu-O layers, 3.9Å, indicating that although the superconductivity is strongly anisotropic, it is still three dimensional in nature. According to the Josephson-coupled layer model[14], the cross-over to two dimensional behavior is expected when $\xi_z = s/\sqrt{2} = 2.8$ Å, where s is the Cu-O interplanar spacing. Our inference that the superconductivity of $Y_1Ba_2Cu_3O_{7-x}$ is three dimensional is consistent with the measurements of Freitas et al.[15] which showed that the fluctuations near T_C were three dimensional in nature. For these calculations we used the anisotropy ratio given by the temperature dependance of H_{C2} near T_C. Obviously, use of the larger anisotropy observed for H_{C1} at lower temperatures would alter the numerical results.

Above 78 K, the linearity of our data indicates agreement with anisotropic Ginzburg-Landau theory. However, some of the lower temperature features of our data do not agree with the most straight-forward expectations from theory. Our experimental results indicate H_{C2} anisotropies of 5:1 near T_C and H_{C1} anisotropies of at least 1:10 at low temperatures. Existing anisotropic theories[13,14,16-18] predict anisotropies for H_{C1} that are equal or less than that for H_{C2}. The pronounced upward bend in $H_{C2}^{\perp}$ below 78 K is another feature of our data that is not expected from simple theories of anisotropic superconductors, although complicated Fermi surfaces[11] could give rise to such a behaviour.

In conclusion, we have established from measurements on single crystals, that the high temperature superconductivity in $Y_1Ba_2Cu_3O_{7-x}$ is strongly anisotropic, though three dimensional in nature. We have established that the anisotropy in the measured properties is associated with the Cu-O planes and that these planes can support large supercurrent densities out to high fields at low temperatures. The larger anisotropy measured for the lower critical field than for the upper critical field, and the temperature dependence of the perpendicular upper critical field indicate that, at least, extensions of the simplest theoretical models for anisotropic superconductors are needed to account for these aspects of the observed properties of $Y_1Ba_2Cu_3O_{7-x}$.

ACKNOWLEDGEMENT

This work was supported in part by the Office of Naval Research Contract No. N00014-85-C-0361. We thank S.J. LaPlaca, P.M. Horn, and D. Keane for single crystal x-ray results, K.F. Etzold, H.R. Lilienthal, T.R. McGuire, J.J. Nocera, S.S.P. Parkin, K.P. Roche, and T.P. Smith for experimental assistance in the magnetic measurements, and S. Foner for advise on high field measurents. We acknowledge informative discussions with R. Schwall, A. Davidson, R.L. Greene, F. Hotzberg, T.N. Jackson, A.W. Kleinsasser, A.P. Malozemoff, M.W. Shafer, T.M. Shaw, and A.R. Williams.

REFERENCES

1. J.G. Bednorz and K.A. Müller, Z. Phys. B 64, 189 (1986).
2. Y. Hidaka, Y. Enomoto, M. Suzuki, M. Oda, and T. Murakami, Jpn. J. Appl. Phys. 26, L377 (1987).
3. T.R. Dinger, T.K. Worthington, W.J. Gallagher, and R.L. Sandstrom, Phys. Rev. Lett. 58, 2687 (1987).
4. S. Shamoto, M. Onoda, and M. Sato, Solid State Commun. (submitted).
5. Y. Hidaka, Y. Enomoto, M. Suzuki, M. Oda, A. Katsui, and T. Murakami, Jpn. J. Appl. Phys. 26, L726 (1987).
6. Y. Iye, T. Tamegai, H Takeya, and H. Takei, Jpn. J. Appl. Phys. Lett. (submitted).
7. T.K. Worthington, W.J. Gallagher, T.R. Dinger, Phys. Rev. Lett. (submitted).
8. C.P. Bean, Phys. Rev. Lett. 8, 250 (1962).
9. K.A. Müller, M Takashige, and J.G. Bednorz, Phys. Rev. Lett. 58, 1143 (1987).
10. Y.B. Kim, C.F. Hempstead, and A.R. Strnad, Phys. Rev. Lett. 9, 306 (1962).
11. B.J. Dalrymple and D. E. Prober, J. of Low Temp. Phys., 56, 545 (1984).
12. N.R. Werthamer, E. Helfand, and P.C. Hohenberg, Phys. Rev. 147, 295 (1966).
13. W.E. Lawrence and S. Doniach, in Proceedings of the Twelfth International Conference on Low Temperature Physics, edited by E. Kanda (Academic Press of Japan, Kyoto, 1971), p. 361.
14. R.A. Klemm, A. Luther, and M.R. Beasley, Phys. Rev. B12, 877 (1975).
15. P.P. Freitas, C.C. Tsuei and T.S. Plaskett, Phys. Rev. B Rapid Communications (to be published July 1, 1987).
16. P. deTrey, S. Gygax, and J.-P. Jan, J. of Low Temp. Phys. 11, 421 (1973).
17. A.V. Balatskii, L.I. Burlachkov and L.P. Gor'kov, Sov. Phys. JETP, 63, 866 (1986).
18. V.G. Kogan and J.R. Clem Phys. Rev. B24, 2497 (1981).

PRESENT STATUS OF HIGH T_c OXIDE SUPERCONDUCTIVITY STUDIES

AT TOHOKU UNIVERSITY

Yoshio Muto, Norio Kobayashi,
and Yasuhiko Syono

Institute for Materials Research*
Tohoku University, Sendai 980, Japan

ABSTRACT

The present status of studies of new high T_c oxide
superconductors held at Institute for Materials Research,
Tohoku University, is described. On single phase
orthorhombic $YBa_2Cu_3O_{6.74}$ and tetragonal $YBa_2Cu_3O_{6.05}$,
results of sample preparation, thermal and X ray analyses,
high resolution electron microscopy study, and neutron
diffraction study are reviewed. The electrical resistivity
measurements are done on $YBa_2Cu_3O_{6.74}$ and $ErBa_2Cu_3O_z$ under
magnetic fields up to 21.6 Tesla. $H_{c2}(0)$ of $YBa_2Cu_3O_{6.74}$
and $ErBa_2Cu_3O_z$ are 150 and 200 Tesla, respectively. In
addition, the resistance of $YBa_2Cu_3O_{6.74}$ begins to appear at
2 Tesla but that of $ErBa_2Cu_3O_z$ does not appear till 5
Tesla.

INTRODUCTION

The discovery of superconductivity in La-Ba-Cu oxides
system with transition temperatures, T_c, more than 30 K by
Bednorz and Müller[1] was a great breakthrough in the
research field for high T_c superconductors. Since we
believed that the high T_c superconductor had to also
exhibit high upper critical fields, H_{c2}, we measured the
electrical resistance of $La_{1.78}Ba_{0.22}Cu_{1.25}O_{4-\delta}$ under
magnetic fields up to 21.1 Tesla between 4.2 and 50 K in
last December and found that the H_{c2} at 0 K, $H_{c2}(0)$,
defined at a mid-point of the resistive transition was
estimated to be 56 Tesla[2,3] according to the dirty limit
theory of Type II superconductivity, which was comparable
to that of the Chevrel phase compound $PbMo_6S_8$, well-known
as the superconductor with the highest H_{c2} till recently.
Since then, we were studying on both the La-Ba-Cu[4] and La-
Sr-Cu oxide[5] systems. The maximum $H_{c2}(0)$ obtained by us
attained to 71 Tesla for a La-Sr-Cu oxide.

---

* Formerly Research Institute for Iron, Steel and Other
  Metals.

Last February, Wu et al[6] in USA and Hikami et al[7] in Japan succeeded independently to get a new system, Y–Ba–Cu oxide with T_c higher than the boiling temperature of liquid nitrogen, 77 K. Immediately after their findings, we could also synthesize a perovskite type oxide $Y_{1.2}Ba_{1.8}Cu_3O_{6.66}$ whose resistivity was zero up to 88 K. We pointed out that this compound had the orthorhombic crystal structure[8], where the unit cell of the cubic perovskite structure was tripled along the c axis, probably due to the ordering of Y^{3+} and Ba^{2+} ions and/or of oxygen vacancy. Furthermore, we found that the $H_{c2}(0)$ of this oxide was estimated to be 96 Tesla[9] by the electrical resistivity measurements in the magnetic fields up to 23.4 Tesla with the slope of $H_{c2}(T)$ at T_c of − 1.6 Tesla/K.

There are many different characteristics which distinguish these systems, although they are both based on the perovskite related structures. One of the most remarkable features lies in the fact that Y–Ba–Cu oxide system essentially comprises a large number of oxygen vacancy which stabilizes the peculiar Ba–Y ordered perovskite-derivative structure, while the La related systems with K_2NiF_4 structure has a rather small amount of oxygen vacancy. Therefore, to control the amount of oxygen vacancy as well as its ordering is the crucial point for the understanding of superconducting characteristics of Y–Ba–Cu oxide compound. Recently we have succeeded to obtain a single phase $YBa_2Cu_3O_{6.74}$. Main part of the present paper is devoted to describe the sample preparation, thermal and X ray analyses[10], high resolution electron microscopy study[11], neutron diffraction study[12] and electrical resistivity measurements under magnetic fields between zero and 21.6 Tesla. In this paper, studies of the sample with tetragonal phase[9] are also given. Furthermore, the H_{c2} data for $ErBa_2Cu_3O_z$ are included. This paper reviews important results of high T_c oxide studies done at Tohoku University.

SYNTHESIS AND CHARACTERIZATION

The starting materials were prepared by mixing, pressing into pellet and firing the mixture of 1:2:3 ratio of 4N $1/2Y_2O_3$, GR grade $BaCO_3$ and 3N CuO at 930°C for more than 20 hrs in air. The sintered material was cooled in air to room temperature in the furnace. The product was confirmed to be solely of an orthorhombic phase by powder X-ray diffraction analysis, as shown in Fig. 1(a). Orthorhombic unit cell dimensions were determined to be a = 3.8180(6)Å, b = 3.8888(7)Å, and c = 11.668(2)Å, where the cubic perovskite unit was tripled along the c axis. These values are practically identical to those of our previously reported specimen whose overall composition is $Y_{1.2}Ba_{1.8}Cu_3O_{6.66}$[8] and also of the orthorhombic phases prepared by other investigators.[13-16]

A tetragonal phase (Fig. 1(b)) was obtained by quenching the sintered specimen from 930°C to room temperature by rapid withdrawal from the furnace. The tetragonal unit cell dimensions were measured to be a = 3.8588(4)Å and c = 11.771(3)Å, which were found to be in agreement with previously reported values of polycrystalline tetragonal phase.[15,16] The oxygen

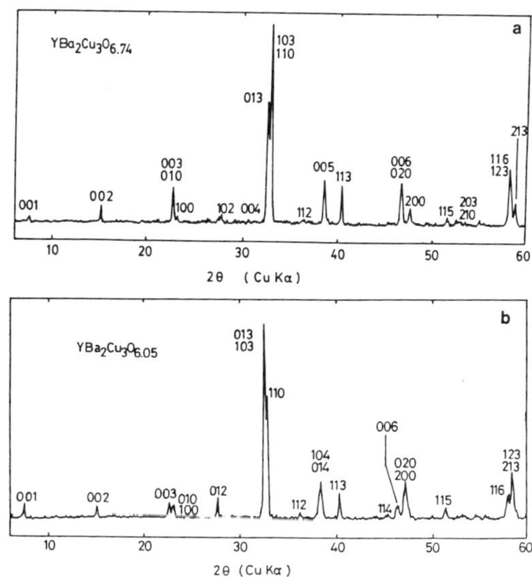

Fig. 1   X-ray powder diffraction patterns of the
orthorhombic phase $YBa_2Cu_3O_{6.74}$ (a) and the
tetragonal phase $YBa_2Cu_3O_{6.05}$ (b).

compositions of the orthorhombic and tetragonal phases were
determinded   to   be $YBa_2Cu_3O_{6.74}$ and $YBa_2Cu_3O_{6.05}$,
respectively, by measuring the weight loss during heat
treatment at 780°C in hydrogen atmosphere, which reduced
only copper ions to copper metal.   The average valence of
copper ions in the orthorhombic and tetragonal phases is
estimated to be 2.16 and 1.70 respectively.
     Resistance measurements for the sintered specimen were
carried out by using a conventional four probe method with
a measuring current of 1 mA.   Temperature was measured by
previously calibrated platinum resistance thermometer.   The
experimental accuracy in temperature measurements was
estimated to be 0.5 %.   The orthorhombic phase showed a
metallic behavior above the onset of superconducting
temperature of 93.1 K, where the resistivity was found to
be less than 1 $m\Omega\cdot cm$, as shown in Fig.2, curve a.   Zero
resistance was realized at 91.3 K, indicating a sharp
superconducting transition.   On the other hand, the
tetragonal phase showed a rather broad transition with the
onset, middle and end points of 79.3, 61.8 and 50.5 K,
respectively (Fig. 2, curve b).   An anomaly at 91.9 K
coincides with the onset temperature of the orthorhombic
phase, indicating the presence of a trace amount of that
phase.

THERMAL ANALYSIS

     Thermogravimetry (TG) and differential thermal analysis
(DTA) were carried out with equal heating and cooling rates
of 5°C/mim in a flow of air.   Figure 3 shows the results of
TG and DTA on the orthorhombic and tetragonal phases.
Significant oxygen desorption in the orthorhombic phase
starts around 400°C and continues with increasing
temperature (Fig. 3 curve a).   The weight loss by heating

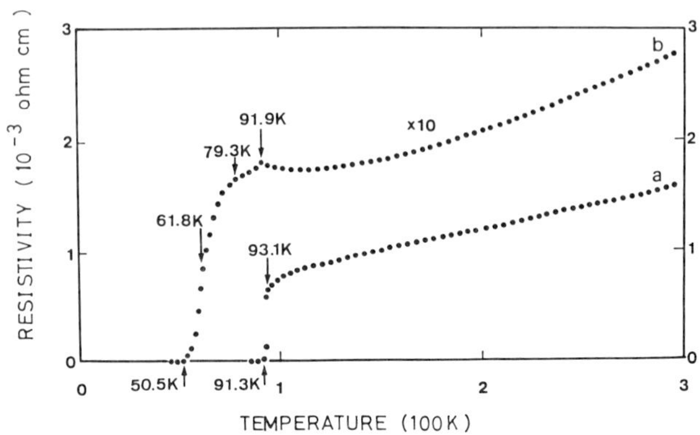

Fig. 2 Temperature variation of resistivity of the orthorhombic phase $YBa_2Cu_3O_{6.74}$ (a) and the tetragonal phase $YBa_2Cu_3O_{6.05}$ (b).

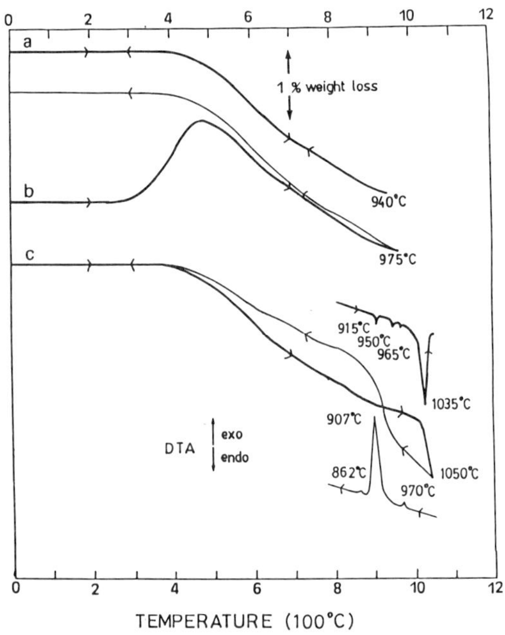

Fig. 3 Thermogravimetry(TG) of heating (thick line) and cooling (thin line) processes. Inserts are recorded DTA signals. Starting materials are the orthorhombic phase synthesized at 930°C and cooled in air to room temperature in the furnace (a) and (c), and the tetragonal phase quenched to room temperature from 930°C (b).

790

the specimen to 930°C was measured to be 2.1 %, which corresponded to 0.9 oxygen atom in the orthorhombic unit cell, giving the composition of $YBa_2Cu_3O_{5.8}$. The average valence of the Cu ions is estimated to be about 1.5, indicating coexistence of Cu^+ and Cu^{2+} ions. The tetragonal phase of $YBa_2Cu_3O_{6.05}$ quenched from 930°C had greater oxygen content than this value, indicating that oxygen is cosiderably accomodated during the quenching process.

A rather broad endothermic reaction took place at around 980°C and was apparently completed around 1050°C. The further weight loss during this reaction was about 0.8 %, which corresponded to 0.3 oxygen atoms in the unit cell, yielding the average composition and valence at 1050°C to be $YBa_2Cu_3O_{5.4}$ and 1.3, respectively. This anomaly might be ascribed to partial melting of the material accompanied by the oxygen loss in the liquid phase. Evidence for such partial melting has been already noticed in sintering the specimen at 950-1000°C.

When the turning point was chosen to be below 950°C before the endothermic anomaly started, the cooling trend was found to be completely <u>reversible</u> (Fig. 3, curve a), indicating that the desorbed oxygen during heating was absorbed again into the specimen during cooling. This implies that no drastic change had occurred by the heat treatment up to 950°C, although a very weak endothermic anomaly was noticed at about 915°C. However, when the specimen was heated up to 1050°C, beyond the temperature range of the remakable endothermic anomaly, reversibility was no longer guaranteed in the cooling cycle (Fig. 3 curve c). The cooling trend revealed a large hystereis: Rapid weight increase due to oxygen absorption was accompanied by a remarkable exothermic signal at a temperature of 130-140°C below the endothermic point in the heating process.

The results of TG and DTA starting from the tetragonal phase are also shown in Fig. 3 curve b. The weight gain due to oxygen absorption commenced around 300°C and merged into the decreasing trend of the orthorhombic phase at around 500°C, showing a broad maximum. The weight difference of 1.55 % between the tetragonal and orthorhombic phase obtained by TG was found to correspond to the oxygen number of 0.64 which was in good agreement with the results of hydrogen reduction experiments.

It is also noticeable that a small, but definite, inflecion in both the absorption and desorption rates was observed at 630°C. This is probably due to the tetragonal-orthorhombic transition as inferred by previous reports.[17,18] Actually the specimen quenched from above 700°C was found to be mostly of tetragonal single phase, while almost single phase material with the orthorhombic symmetry was obtained by annealing at or below 600°C.

ELECTRON MICROSCOPY OF THE ORTHORHOMBIC PHASE

Samples for transmission electron microscopy were prepared by crushing the sintered product[10] and dispersing with ethanol on a micro-grid. High resolution observations have been made with a 400 kV electron microscope (JEM-400EX) having a thereotical resolution of 0.17 nm. The $YBa_2Cu_3O_{6.74}$ compound prepared at 930°C shows high-quality crystals with few defects[11] The image contrast

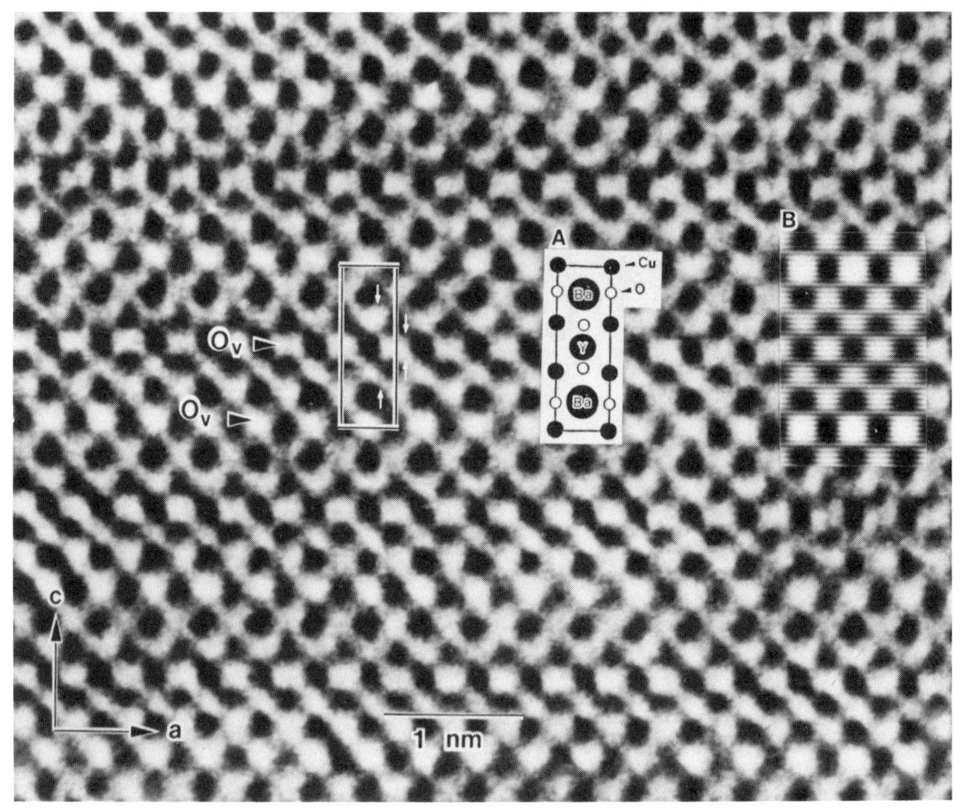

Fig. 4 Enlarged micrograph with a projected structure
model (A) and a calculated image of the structure
model (B).

drastically changed with crystal thickness. The image was
taken with the incident beam exactly parallel to the b axis
of the orthorhombic structure and is shown in Fig. 4,
together with a projection of the structure model proposed
by neutron and X-ray diffractions and an image calculated
from the structure model. The observed image shows a good
correspondence to the calculated one of the structure
model.

In the image, Y and Ba atoms are represented as dark
spots and Cu atoms as weak dark spots and positions of
these dark spots are systematically displaced as indicated
with small arrows. These displacements are consistent with
those of the structure model[12-15], although it is
impossible to make definite estimations of atomic
displacements from the observed image owing to the limited
resolution of the electron microscope used for the
observation. Oxygen atom positions located between metal
atoms correspond to bright regions and deficient oxygen
positions located Y atom planes and sandwiched by Ba atoms,
which are indicated with arrows (O_v), are distinguished as
brighter regions from the other bright regions of oxygen
atom positions. This is clearly seen when the image is
viewed at glancing incidence parallel to the b axis. The
result confirms directly the structure model formed with
perovskite unit cells with ordered arrangement of Y and Ba,

and deficient oxygen atoms, which was proposed by neutron and X-ray diffractions.

NEUTRON DIFFRACTION STUDIES OF BOTH PHASES

Neutron diffraction studies[12] aimed at the oxygen site-occupancy problem in the oxides. Conventional neutron diffractometry was used to obtain reasonable diffraction intensities at low index Bragg reflections in the range $d^{-1} < 0.5$ A^{-1}. Low index Bragg reflections are very important to measure the site-occupancies at each atom site, though they are also necessary to look for magnetic superlattice peaks in the crystals.

Figs. 5(a) and 5(b) show neutron diffraction patterns of the orthorhombic and tetragonal phases at room temperature. Structural parameters were obtained by the Rietveld profile refinement method, using the program RIETAN made by Izumi[19]. Experimental data (dots) and calculated curves (solid lines) fitted reasonably well. Refined structural parameters and residuals, R_{wp}, R_p, R_B and R_F are listed in Tables I and II. R_{wp} and R_p are the residuals for the total diffraction profiles. R_B and R_F correspond to the residuals for the structure factor F. In the present calculation, the temperature factors, B(A^2) are set at values from the TOF data reported by Izumi et al[20] who observed many high index reflections, being crucial to refine the temperature factor. Site-occupancies and

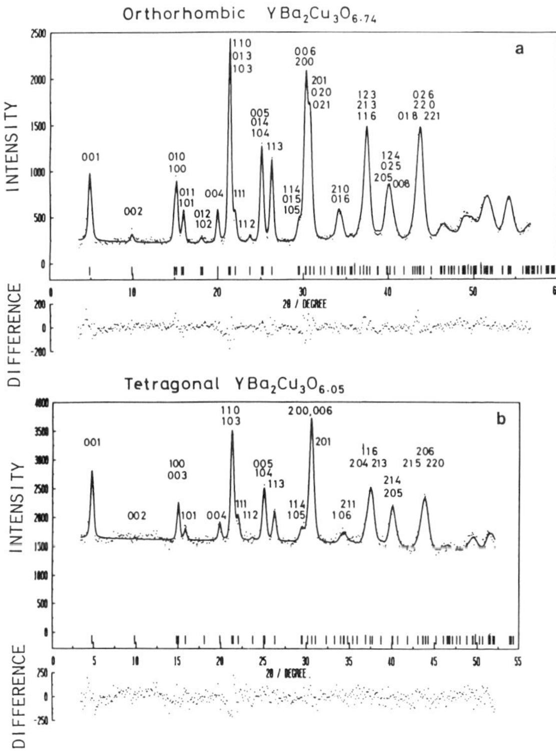

Fig. 5 Observed and calculated neutron diffraction profiles of orthorhombic YBa$_2$Cu$_3$O$_{6.74}$ (a) and tetragonal YBa$_2$Cu$_3$O$_{6.05}$ (b).

Table I   Structural parameters for orthorhombic $YBa_2Cu_3O_{6.74}$. Space group Pmmm. $a=3.8180(6)Å$, $b=3.8888(7)Å$ and $c=11.668(2)Å$. $R_{wp}=0.075$, $R_P=0.056$, $R_B=0.023$ and $R_F=0.013$.

Atom	site	x	y	z	$B(A^2)$	Occupancy
Ba	2t	1/2	1/2	0.186(1)	0.51	1.0
Y	1h	1/2	1/2	1/2	0.80	1.0
Cu_1	1a	0	0	0	0.42	1.0
Cu_2	2g	0	0	0.355(1)	0.42	1.0
O_1	1e	0	1/2	0	0.48	0.91(3)
O_2	2q	0	0	0.156(1)	0.48	1.0
O_3	2r	0	1/2	0.376(2)	0.48	1.0
O_4	2s	1/2	0	0.380(2)	0.48	0.92(2)

Table II   Structural parameters for tetragonal $YBa_2Cu_3O_{6.05}$ at room temperature. Space group P4/mmm. $a = 3.8588(4)Å$ and $c = 11.771(3)Å$. $R_{wp} = 0.036$, $R_P = 0.028$, $R_B = 0.046$ and $R_F = 0.031$.

Atom	site	x	y	z	$B(A^2)$	Occupancy
Ba_1	2h	1/2	1/2	0.192(3)	0.51	0.85(10)
Ba_2	1d	1/2	1/2	1/2	0.51	0.30(5)
Y_1	2h	1/2	1/2	0.192(3)	0.80	0.15(5)
Y_2	1d	1/2	1/2	1/2	0.80	0.70(5)
Cu_1	1a	0	0	0	0.42	1.0
Cu_2	2g	0	0	0.361(2)	0.42	1.0
O_1	2f	0	1/2	0	0.42	0.20(4)
O_2	2g	0	0	0.161(4)	0.42	1.0
O_3	41	0	1/2	0.383(2)	0.42	0.91(2)

location parameters are mainly refined. The chemical compositions of the sample were determined by Kikuchi et al[10]. The sysmmetries Pmmm and P4/mmm are tentatively selected for the refinement.

In the orthorhombic phase, it was found that the oxygen site-occupancies at $(0\ 1/2\ 0)$ and $(1/2\ 0\ z)$, where $z = 0.380(2)$, are, respectively, $0.91(3)$ and $0.92(2)$. These values show better agreement with the reported ones, $0.92(3)$ and $0.95(1)$, by Beno[14] than the values, $0.63(2)$ and 1.0, reported by Izumi et al[20]. The residuals, R_B and R_F are reduced from 4.4 and 2.2 % to 2.2 and 1.3 %, respectively, under the assumption that $(1/2\ 0\ z)$ site, where $z = 0.380(2)$, is fully occupied. The $(1/2\ 0\ 0)$ and $(0\ 0\ 1/2)$ sites are totally deficient. Atom location parameters are similar to the reported values[14,20].

The oxygen site-occupancies in the tetragonal phase showed similar tendencies. The oxygen site-occupancies at $(0\ 1/2\ 0)$ and $(0\ 1/2\ z)$, where $z = 0.383(2)$, are $0.20(4)$ and $0.91(2)$, respectively. In the $(1/2\ 1/2\ z)$ site, where $z = 0.193(2)$, 15 % of Ba atoms were substituted by Y atoms.

In this situation, the residuals, R_B and R_F, are reduced from the values 7.3 and 4.1 % which are obtained, assuming ordering of heavy cations and no oxygen difficiency at the $(0\ 1/2\ z)$ site, where $z = 0.383(2)$, to 4.6 and 3.1 %, respectively. This is due to a distinguishable improvement in the calculated intensities of important reflections such as 110, 111 and 104. No extra reflections corresponding to a magnetic superlattice are observed at room temperature as seen in Fig. 5.

UPPER CRITICAL FIELDS

Figs. 6(a) and (b) are the electrical resistance as a function of temperature under magnetic fields up to 21.6 Tesla, where magnetic fields more than 13 Tesla were produced by the Tohoku hybrid magnet, HM-2[21]. Fig. 6(a) is data for above mentioned $YBa_2Cu_3O_{6.74}$ with the orthogonal phase and Fig. 6(b) data for $ErBa_2Cu_3O_z$, where Y was replaced by Er, while amount of oxygen was not determined. The preparation method was similar to $YBa_2Cu_3O_{6.74}$ in the orthorhombic phase.

While the transition width ΔT_c as defined by 90 and 10 % of the normal state resistance is very sharp at zero field, only about 1 K, ΔT_c under magnetic fields becomes large, exhibiting gradual tail in lower temperature side. The tail of $YBa_2Cu_3O_{6.74}$ attains in lower temperature side when it is compared with that of $ErBa_2Cu_3O_z$. At 77 K, the resistance of $YBa_2Cu_3O_{6.74}$ begins to appear at 2 Tesla, but that of $ErBa_2Cu_3O_z$ does not appear till 5 Tesla. At 21.6 Tesla, the resistance appears for $YBa_2Cu_3O_{6.74}$ at 58 K and for $ErBa_2Cu_3O_z$ at 67 K .

There may be two possible sources for the broading of such a transition in a magnetic field. One is based on the intergranular materials in these ceramic superconductors and the other on the intrinsic property of superconducting material. From the viewpoint of the latter, we can conclude that the $ErBa_2Cu_3$ oxide system is promising for use under the field of several Tesla at liquid nitrogen temperature.

In Fig. 7 the upper critical fields defined both at the midpoint and the end point (zero reistance) of transition curves are shown as a fuction of temperature.

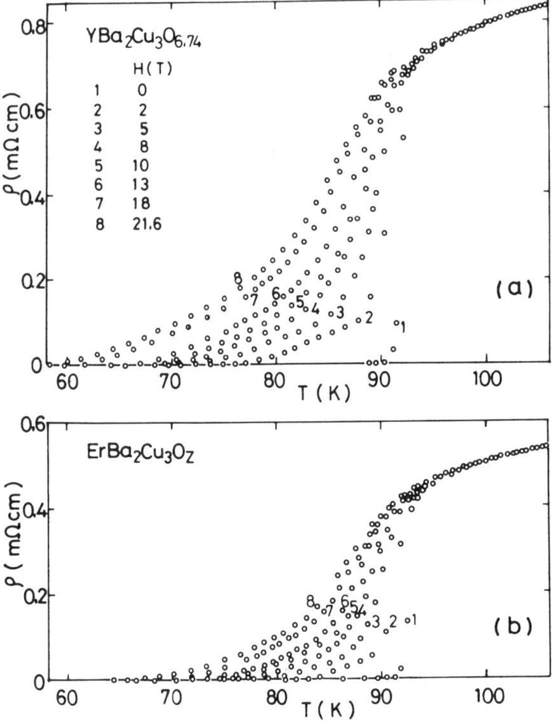

Fig. 6 Electrical resistivity as a fuction of
temperatue under several magnetic fields.
$YBa_2Cu_3O_{6.74}$ (a) and $ErBa_2Cu_3O_z$ (b).

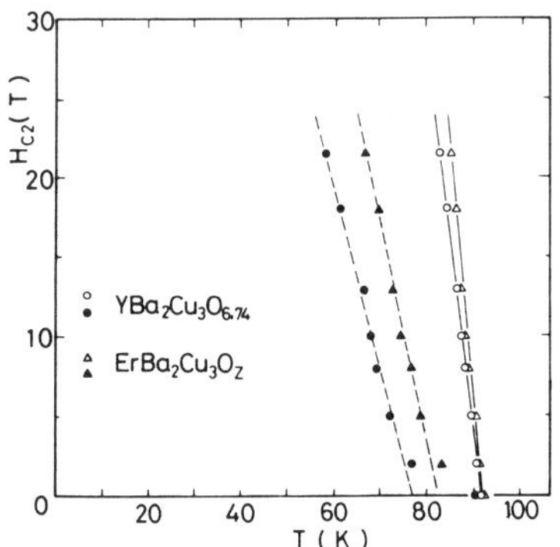

Fig. 7 $H_{c2}(T)$ as a function of temperature. Open and
solid symbols represent H_{c2} defined at mid- and
end-points of the resistive transition.

sample	$T_{c\,end}$ (K)	$T_{c\,mid}$ (K)	$T_{c\,on}$ (K)	$-dH_{c2\,end}/dT$ (T/K)	$-dH_{c2\,mid}/dT$ (T/K)
$YBa_2Cu_3O_{6.74}$	91.3	92	93.1	1.2	2.4
$ErBa_2Cu_3O_z$	91.5	92.6	92.9	1.4	3.2

sample	$H_{c2\,end}(0)$ (T)	$H_{c2\,mid}(0)$ (T)	ξ_{GL} (nm)	ρ_{100} (mΩcm)	γ (mJ/moleK^2)
$YBa_2Cu_3O_{6.74}$	60	150	1.2	0.79	7
$ErBa_2Cu_3O_z$	80	200	1.0	0.51	15

H_{c2} behaves linearly to the temperature except for the low field data below 2 Tesla. The slope of H_{c2} at T_c is shown in Table III. According to WHH theory $H_{c2}(0)$ values are obtained from the slope of H_{c2} and also shown in Table III. $H_{c2}(0)$ value defined at midpoint attains 150 and 200 Tesla for $YBa_2Cu_3O_{6.74}$ and $ErBa_2Cu_3O_z$, respectively. Similar results are recently also obtained by Orlando et al in the MIT group[22]. These values are comparable with the Pauli paramagnetic limiting field, 160 Tesla. The coherence length ξ_{GL} shown in Table I is obtained from the $H_{c2}(0)$ value, which is comparable with the unit cell dimension along c axis.

The electronic specific heat coefficient can be also evaluated from the slope of $H_{c2}(T)$,

$$dH_{c2}/dT \ (T/K) = 4.48 \ \rho_{100}(\Omega cm)\gamma(erg/cm^3K^2)$$

which is also shown in Table III, where γ is the density of states at the Fermi level. Therefore, the γ value is presumed to be about 10 mJ/mole·K^2 for these oxide systems. This value is comparable order of magnitude reported by Kitazawa et al[23].

SUMMARY

In-situ observation of oxygen absorption and desorption during heating and cooling cycle with TG and DTA was done to determine the optimum preparation conditions of high T_c Y-Ba-Cu oxide compounds. The best recipe is to fire the raw mixture of 1/2 Y_2O_3, $2BaCO_3$ and $3CuO$ at 930°C for more than 10 hours, followed by slow cooling in air in the furnace to take as much as oxygen as possible within the specimen. With this method, single phase material of the orthorhombic superconducting oxide $YBa_2Cu_3O_{6.74}$ with T_c = 91.3 K was synthesized.

The crystal structure of $YBa_2Cu_3O_{6.74}$ was studied by high resolution electron microscopy. An observed high-resolution micrograph is consistent with the oxygen deficient perovskite structure model, in agreement with neutron and X-ray diffraction. In the image, Y, Ba and Cu atoms were represented as dark spots with different contrast, depending on their atomic numbers. Unoccpied oxygen atom positions are distinguished as brighter regions from occcupied oxygen atom positions.

Neutron powder diffraction studies together with X-ray and electron microscopy studies give following results. In the orthorhombic phase ($YBa_2Cu_3O_{6.74}$), the occupancies of the $O_1(0\ 1/2\ 0)$ and O_4 sites ($1/2\ 0\ z$, where $z = 0.380(2)$, were found to be 0.91(3) and 0.92(2), respectively. In the tetragonal phase, the occupancies of the $O_1(0\ 1/2\ 0)$ and the $O_3(0\ 1/2\ z)$ sites were 0.20(4) and 0.91(2), respectively. Some disordering of Ba and Y atoms was suggested in the tetragonal phase.

The electrical resistance was measured on $YBa_2Cu_3O_{6.74}$ and $ErBa_2Cu_3O_z$ under magnetic fields up to 21.6 Tesla. According to the WHH theory, $H_{c2}(0)$ values were estimated to be 150 and 200 Tesla for $YBa_2Cu_3O_{6.74}$ and $ErBa_2Cu_3Cu_z$, respectively. In addition, the resistance of $YBa_2Cu_3O_{6.74}$ and $ErBa_2Cu_3O_z$ begins to appear at 2 and 5 Tesla, at 77 K, respectively. It is interesting that the coherence length of the oxide system ξ_{GL} is estimated to be comparable with the unit cell dimension along c axis.

ACHNOWLEDGMENT

The authors are grateful to Profs. M. Hirabayashi, K. Hiraga, M. Kikuchi, M. Tachiki, S. Maekawa, T. Masumoto, K. Fujimori, K. Noto, and A. Inoue, Dr. T. Kajitani , Mrs. H. Arai, Miss A. Tokiwa and Messrs. T. Sasaoka, K. Oh-ishi, O. Nakatsu and S. Terada for their cooperation on these studies. A part of the present work was supported by the Grant-in-Aid for the Special Project Research on New Superconducting Materials from Ministry of Education, Science and Culture of Japan. The experiments for H_{c2} measurements were done at High Field Laboratory of Superconducting Materials, Tohoku University.

REFERENCES

1. J. G. Bednorz and K. A. Müller, Z. Phys. B64 199 (1986).
2. K. Noto, K. Watanabe and Y. Muto, Sci. Rep. RITU A33 393 (1986) .
3. N. Kobayashi, K. Oh-ishi, T. Sasaoka, M. Kikuchi, T. Sasaki, S. Murase, K. Noto, Y. Syono and Y. Muto, J. Phys. Soc. Japan. 56 1309 (1987).
4. K. Oh-ishi, M. Kikuchi, Y. Syono, K. Hiraga and Y. Morioka, Jap. J. Appl. Phys. 26 L484 (1987).
5. N. Kobayashi, T. Sasaoka, K. Oh-ishi, T. Sasaki, M. Kikuchi, A. Endo, K. Matsuzaki, A. Inoue, K. Noto, Y. Syono, Y. Saito, T. Masumoto and Y. Muto, Jpn. J. Appl. Phys. 26 L358 (1987).
6. M. K. Wu, J. R. Ashburn, C. J. Torng, P. H. Hor, R. L. Meng, L. Gao, Z. J. Huang, Y. Q Wang and C. W. Chu, Phys. Rev. Lett. 58 908 (1987).

7. S. Hikami, T. Hirai and S. Kagoshima, Jpn. J. Appl. Phys. $\underline{26}$ L314 (1987).
8. Y. Syono, M. Kikuchi, K. Oh-ishi, K. Hiraga, H. Arai, Y. Matsui, N. Kobayashi, T. Sasaoka and Y. Muto, Jpn. J. Appl. Phys. $\underline{26}$ L498 (1987).
9. N. Kobayashi, T. Sasaoka, K. Oh-ishi, M. Kikuchi, M. Furuyama, T. Sasaki, K. Noto, Y. Syono and Y. Muto, Jpn. J. Appl. Phys. $\underline{26}$ L757 (1987).
10. M. Kikuchi, Y. Syono, A. Tokiwa, K. Oh-ishi, H. Arai, K. Hiraga, N. Kobayashi, T. Sasaoka and Y. Muto, Jpn. J. Appl. Phys. $\underline{26}$ (1987) No. 6, in press.
11. K. Hiraga, D. Shindo, M. Hirabayashi, M. Kikuchi, K. Oh-ishi and Y. Syono, Jpn. J. Appl. Phys. $\underline{26}$ (1987) No. 7, in press.
12. T. Kajitani, K. Oh-ishi, M. Kikuchi, Y. Syono and M. Hirabayashi, Jpn. J. Appl. Phys. $\underline{26}$ (1987) No. 7 in press.
13. R. J. Cava, B. Batlogg, R. B. Dover, D. W. Murphy, S. Sunshine, T. Siegrist, J. P. Remeika, E. A. Rietman, S. Zahurak and G. P. Espinosa, Phys. Rev. Lett. $\underline{58}$ 1676 (1987).14.
14. M. A. Beno, L. Soderholm, D. W. Capone II, D. G. Hinks, J. D. Jorgensen, I. K. Schuller, C. U. Segre, K. Zhang, and J. D. Grace, Appl. Phys. Lett. in press.
15. F. Izumi, H. Asano, T. Ishigaki, E. Takayama-Muromachi, Y. Uchida, N. Watanabe and T. Nishikawa, preprint
16. T. Hatano, A. Matsushita, K. Nakamura, Y. Sakka, T. Matsumoto and K. Ogawa, Jpn. J. Appl. Phys. $\underline{26}$ L721 (1987).
17. K. Yukino, T. Sato, S. Ooba, M. Ohta, F. Okamura and A. Ono, Jpn. J. Appl Phys. $\underline{26}$ L869 (1987).
18. I. K. Schuller, D. G. Hinks, M. A. Beno, D. W. Capone II, L. Soderholm, J-P. Locquet, Y. Bruynseraede, C. U. Segre, and K. Zhang, Solid State Commun. in press.
19. F. Izumi, J. Cryst. Soc. Japan $\underline{27}$ 23 (1985) (in Jananese).
20. F. Izumi, H. Asano, T. Ishigaki, A. Ono and F. P. Okamura, Jpn. J. Appl. Phys. $\underline{26}$ L611 (1987).
21. On Tohoku hybrid magnets, see Sci Rep. RITU A$\underline{33}$ No. 2 (1986).
22. T. P. Orlando, K. A. Delin, S. Foner, E. J. McNiff Jr., J. M. Terascon, L. H. Greene, W. R. MuKinnon and G. W. Hull, submitted to Phys. Rev. Lett.
23. K. Kitazawa, T. Atake, H. Ishii, H. Sato, H. Takagi, S. Uchida, Y. Saito, K. Fueki and S. Tanaka, Jpn. J. Appl. Phys. $\underline{26}$ L748 (1987).

CURRENT CARRYING PROPERTIES IN HIGH T_c Y-Ba-Cu-O SYSTEM

K. Noto, K. Watanabe, H. Morita, Y. Murakami,
I. Yoshii, I. Sato, H. Sugawara,
N. Kobayashi, H. Fujimori and Y. Muto

Institute for Materials Research,
Tohoku University, Sendai 980, Japan

ABSTRACT

The transition temperature, T_c, and the critical current density, $J_c(B)$, at 4.2 K have been studied in high T_c Y-Ba-Cu oxide system. A very large anisotropy, very fast degradations at low fields, and a large hysteresis in $J_c(B)$ have been observed in this system. The cause of the large anisotropy might be the strong c-axis orientation parallel to the direction of the pressing and the low dimensionality of the crystal structure of this material. Although J_c value is very small, the decrease in J_c with increasing magnetic field from 2 to 27 T is estimated to amount to only about 20 %, which shows a high potential for practical use at high fields when J_c is raised by the introduction of strong pinning centers in the Y-Ba-Cu oxide system.

INTROCUDTION

The recent discovery of superconductivity above liquid nitrogen temperature in Y-Ba-Cu oxides[1,2] has generated a large amount of research on these oxides system during very short time. We also have studied the Y-Ba-Cu-O system from the viewpoint of practical applications. In this paper, results of the basic research for practical use, such as an anisotropy and a hysteresis of J_c, are presented.

SAMPLE PREPARATION AND T_c

The sample preparation[3] was performed by powder process of mixture of Y_2O_3, $BaCO_3$ or BaO and CuO in the desired concentration. Some samples were examined by X-ray diffraction analysis with monochromated Cu K_α radiation. Figure 1-a and 1-b show the typical results on $Y_1Ba_2Cu_3O_y$. As seen in Fig. 1-a, the diffraction pattern for a well treated sample indicates that the sample is fully orthorhombic, while one can notice that Fig. 1-b for a

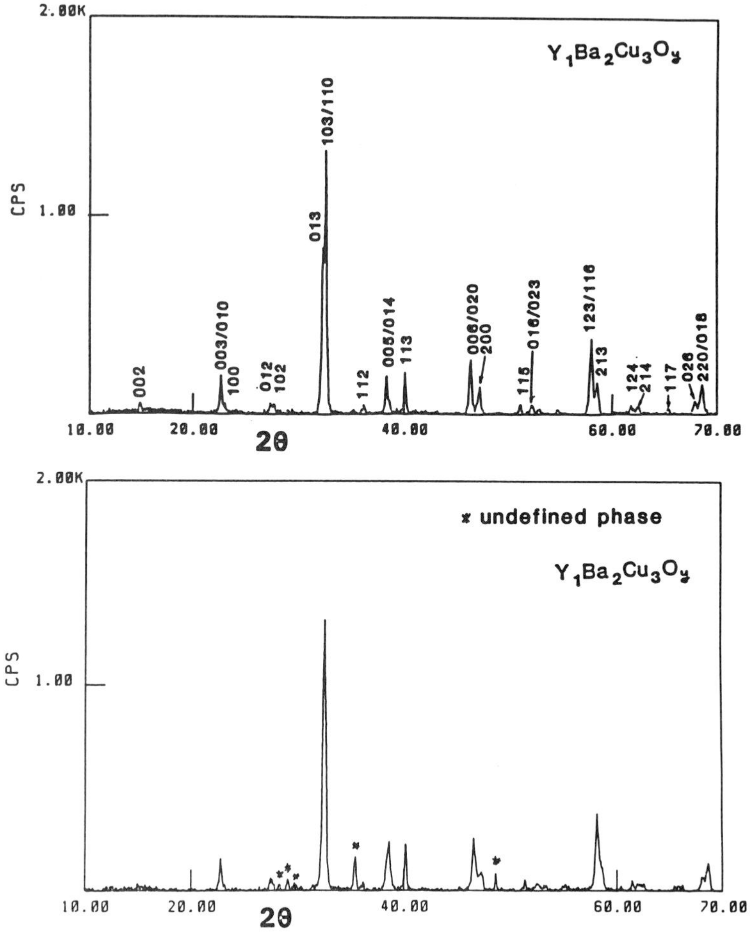

Fig. 1-a. X-ray diffraction patterns for the well treated Y$_1$Ba$_2$Cu$_3$O$_y$ sample with T_c onset of 94 K and T_c offset of 89 K.

Fig. 1-b. The Y$_1$Ba$_2$Cu$_3$O$_y$ sample calicined only once has T_c onset of 92.5 K and T_c offset of 85 K. The undefined phase and the unclear orthorhombic splitting were observed.

badly treated sample shows the appearance of an undefined phase. This Y$_1$Ba$_2$Cu$_3$O$_y$ sample including different phases was prepared only by one calcining without remixed and refired procedure, though it had a superconductive onset of 92.5 K and reached zero resistance at 85 K. In the previous paper[3,4], we observed the two maxima of T_c around x=0.33 and 0.4 in Y$_x$Ba$_{1-x}$CuO$_y$ system. This might be related to the inhomogenous tetragonal phase or undefined another phases as shown in Fig. 1-b, because even in the stoichiometric composition Y$_1$Ba$_2$Cu$_3$O$_y$, X-ray diffraction pattern exhibits sample-to-sample variations. Moreover, if only the stoichiometric $Y_1Ba_2Cu_3O_y$ phase is responsible for the superconductivity, the concentration dependence of T_c in the $Y_xBa_{1-x}CuO_y$ system may become complex due to any extra non-superconducting phases.

CRITICAL CURRENT DENSITY J_c

We measured the magnetic field depencence of J_c by the resistive method for a sample with the current carrying cross section of about 3×3 mm^2. Electrical contacts were made by ultrasonic soldering and the distance between voltage probes was about 5 mm. Magnetic fields were produced by a water cooled magnet up to 17 T and were applied to the sample in the direction perpendicular to the current. Figure 2 shows the magnetic field dependence of J_c at 4.2 K in the $Y_1Ba_2Cu_3O_y$ samples. We also measured the field dependence of J_c in the different magnetic field direction to the pellet surface. As can be seen in this figure, a large anisotropy in the critical current density was observed. This fact seems to imply that the sintered pellet prefers the c-axis orientation parallel to the pressing direction[5]. The orientation structure like texture may introduce strong pinning centers because of its anisotropy of J_c. It looks like the in-situ V_3Ga tape which is well known to have the anisotropic surface pinning coming from columnar structure[7]. It was also found that very fast degradations at low fields and a strange large hysteresis in J_c (B) existed. The very small slope of J_c as a function of the magnetic field in the high field region was confirmed by the previous experiment[3] that the decrease in J_c by the increase in fields from 2 to 27 T was

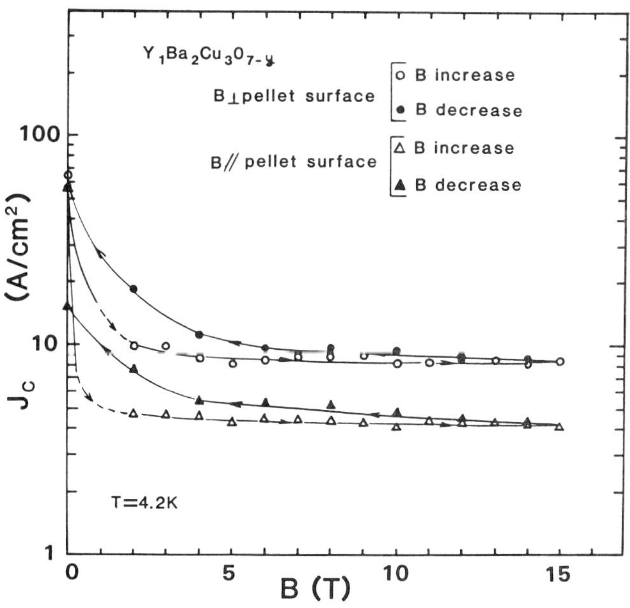

Fig. 2. Magnetic field dependence of J_c at 4.2 K.

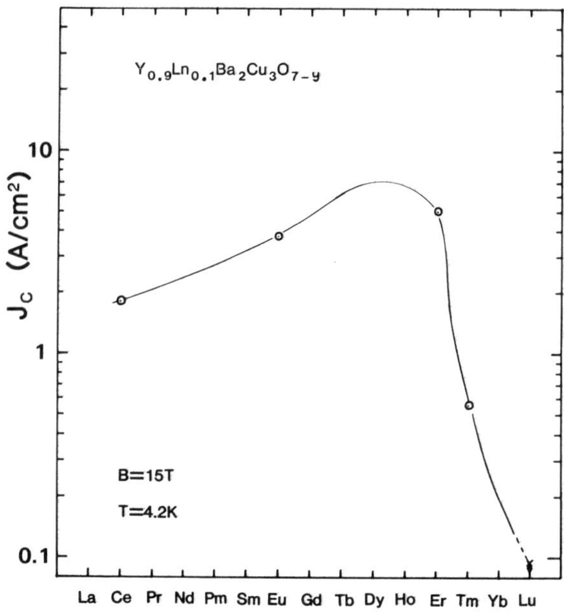

Fig. 3. Lanthanoid dependence of J_c at 4.2 K in the $Y_{0.9}Ln_{0.1}Ba_2Cu_3O_{7-y}$ system. J_c value of the Lu sample was less than 0.1 A/cm^2 at 4.2 K and 15 T.

estimated to be only 20 % by use of the hybrid magnet system[6]. We examined the effect of substitution of Y atom by 10 at. % lanthanoid for the improvement of J_c. The results are indicated in Fig. 3. Although J_c value at 4.2 K is very small due to the low density and the insufficient

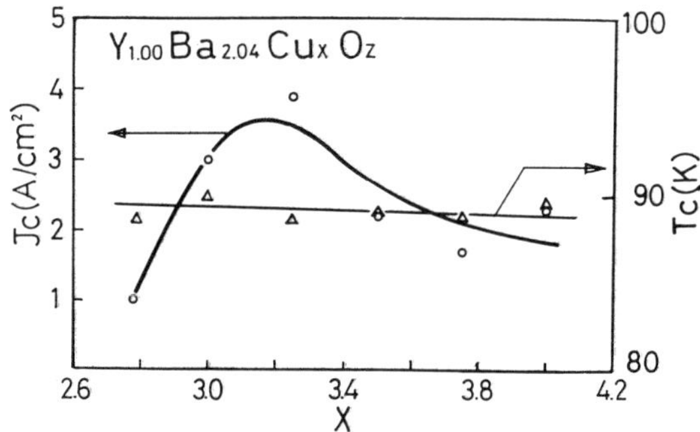

Fig. 4. Copper concentration dependence of T_c and J_c in the $Y_{1.0}Ba_{2.04}Cu_xO_z$ system.

surface contact among the grains, the tendency of the
effect of the substitution is clearly exhibited in the
present study. Figure 3 predicts the most effective
improvement of J_c by the addition of Dy, Ho or Er.
Especially, the substitution of Y by Dy was reported to
show high $H_{c2}(0)$ of more than 340 T for 5.5 T/K using
midpoint[8]. The present data of the lanthanoid dependence
of Jc are in good agreement with that on H_{c2}. Figure 4
shows the Cu concentration dependence of T_c and J_c for the
$Y_{1.0}Ba_{2.04}Cu_xO_z$ system. T_c reduces only a little with Cu
concentration increasing from x=3.0, while the small excess
concentration of Cu rather than stoichiometric $Y_1Ba_2Cu_3O_z$
seems to improve J_c at 4.2K.

ACKNOWLEDGEMENTS

We express our thanks to Drs. S. Miura, Y. Watanabe
and T. Ichikawa for the X-ray diffraction measurements.
This work is supported by Grant-in-Aid for Fusion Research
from the Ministry of Education, Science and Culture, Japan.

REFERENCES

1.   M. K. Wu et al, <u>Phys. Rev. Lett.</u> 58: 908 (1987).
2.   S. Hikami et al, <u>Jpn. J. Appl. Phys.</u> 26: L314 (1987).
3.   K. Noto et al, in: "Proc. LT-18," (1987, Kyoto) to be
     published.
4.   K. Noto et. al, <u>Jpn. J. Appl. Phys.</u> 26: L802 (1987).
5.   H. Watanabe et al, <u>Jpn. J. Appl. Phys.</u> 26: L657
     (1987).
6.   Y. Muto et al, <u>Sci. Rep. RITU</u> A33: 221 (1986).
7.   R. G. Sharma et al, <u>Cryogenics</u> 25: 381 (1985).
8.   K. Takita et al, submitted to <u>Jpn. J. Appl. Phys.</u>
     (1987).

PREPARATION, STRUCTURE, AND MAGNETIC FIELD STUDIES OF

HIGH T_c SUPERCONDUCTORS

M.S. Osofsky[*], W.W. Fuller, L.E. Toth[**], S.B. Qadri[+],
S.H. Lawrence[++], R.A. Hein[+++], D.U. Gubser, S.A. Wolf,
C.S. Pande, A.K. Singh[++], E.F. Skelton and B.A. Bender

Naval Research Laboratory, Washington, DC 20375-5000

We have prepared and characterized high T_c single phase samples of $La_{2-x}M_xCuO_4$ (where M=Sr, Ba, and Ca and x=0.1, 0.2 or 0.3) and $Ba_2YCu_3O_7$. Transmission electron microscopy of the $Ba_2YCu_3O_7$ samples reveals twinning defects in the [110] direction which disappear when the sample is heated above $400°C$. We report on the dc susceptibility, magnetization, and magnetoresistance. We have extracted many of the critical superconducting and normal state parameters from the data.

INTRODUCTION

Superconductivity in La-Ba-Cu-O at $30°K$ was reported by Bednorz and Muller [1] Since then other compounds with a transition temperature, T_c, near $40°K$ have been reported [2], [3]. In the Y-Ba-Cu-O compound a T_c above $90°K$ was reported by Wu et al. [4]. We have prepared $La_{2-x}M_xCuO_4$ (M = Sr, Ba, and Ca) and $Ba_2Y_1Cu_3O_7$, for magnetic field studies of their superconducting transitions.

SAMPLE PREPARATION AND CHARACTERIZATION

The $La_{1.8}Sr_{0.2}CuO_4$, $La_{1.9}Ba_{0.1}CuO_4$, and $La_{1.7}Ca_{0.3}CuO_4$ samples were prepared as described in an earlier paper [5]. This procedure produced nearly single phase samples with the K_2NiF_4 structure as determined by x-ray diffraction.

The $Ba_2Y_1Cu_3O_7$ samples were prepared under processing conditions which produced an orthorhombic structure (as determined by neutron and x-ray diffraction) with two one-dimensional Cu-O-Cu chains on the basal plane of the unit cell [6]. The procedures were: (a) Premix powders of $BaCO_3$, CuO

and Y_2O_3, (b) immerse mixture in alchohol and pulverize loose
agglomerates in an ultrasonic mixer, (c) calcine for 6 hours at 925-950°C
with intermediary grindings, (d) grind and cold press into pellets and
sinter in air at 925-950°C for 12 hours, and (e) move into an oxygen
furnace , heat to 950C for several hours and slow cool (1°/min).
Alternatively, step (e) could be equilibrate in oxygen, reduce the
temperature to 500-700°C, hold for several hours and cool at a moderately
slow rate.

Samples made with these procedures consist of 10-20 micron single
phase grains surrounded by trace amounts of amorphous Y-Ba-Cu-O compounds.
Extensive transmission electron microscopy of the individual grains reveal
parallel 'domains' with an average width of 2000Å (Figure 1). Electron
diffraction shows that the boundaries between these regions are in the
[110] direction. The electron diffraction pattern (Figure 2) shows the
splitting of spots along only one diagonal with increasing separation in
higher order reflections. This can be explained by twinning on the [110]
plane in the orthorhombic crystals confirming earlier work [7]. In situ
heating and cooling experiments showed no changes in the diffraction
pattern down to liquid nitrogen temperatures. The domains disappeared
when the specimen was heated above 400°C. Many of the domains reappeared
when the sample was re-cooled. This behavior is consistent with the many
reports of a tetragonal to orthorhombic transformation above 400°C in
these materials.

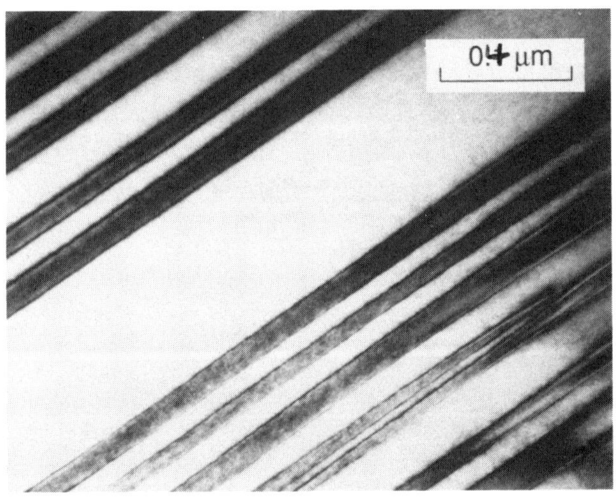

Fig. 1. Transmission Electron Micrograph of a region of
a grain of $Ba_2YCu_3O_7$. The parallel domains average
2000Å in width.

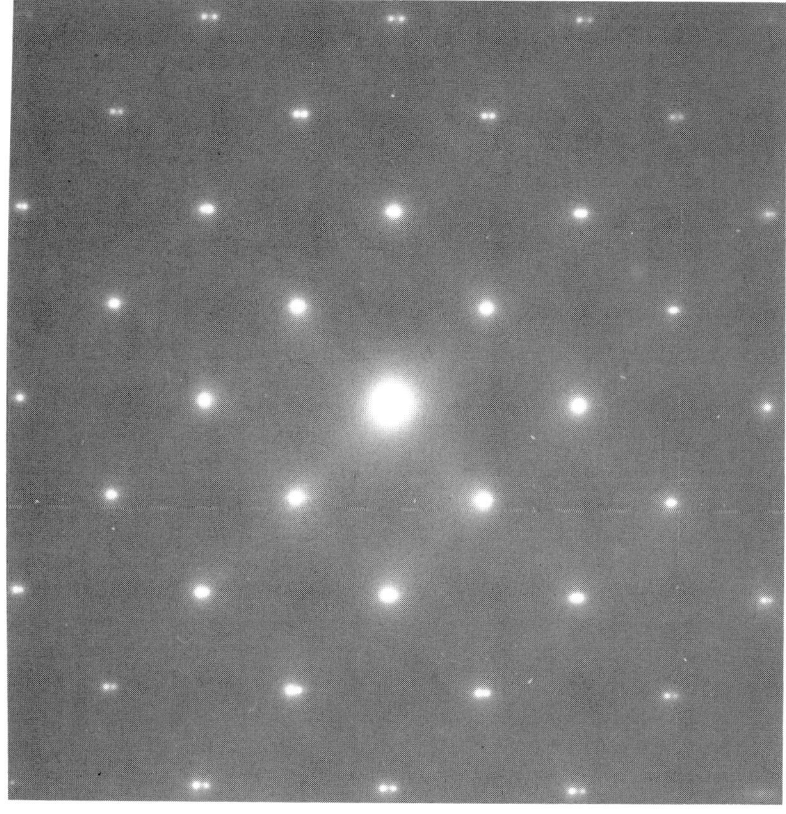

Fig. 2. Electron diffraction pattern of $Ba_2YCu_3O_7$. Spots split
in one diagonal direction with increasing separation in
higher order reflections indicating twinning in the [110]
direction.

EXPERIMENTAL TECHNIQUES AND RESULTS

The samples were first cooled in a SQUID susceptometer in a small
(typically 100 G) magnetic field. The magnetization data (see Table I)
for the $La_{2-x}M_xCuO_4$ (M = Ba and Ca) showed diamagnetic onsets at $20^{\circ}K$ and
$31^{\circ}K$ with 10% and 20% flux expulsion at $10^{\circ}K$ (the lowest temperature
measured) respectively. The curves were not flat at $10^{\circ}K$ indicating that
the samples were still expelling flux. The presence of these broad
transitions with small flux expulsion along with the x-ray data indicating
nearly single phase materials is a manifestation of the oxygen's
importance in the superconductivity (x-rays do not couple strongly to
oxygen). The M = Sr sample had an onset at $36^{\circ}K$ with 40% flux expulsion
at $10^{\circ}K$ and the $Ba_2Y_1Cu_3O_7$ sample showed the onset at $93^{\circ}K$ with 90 - 100%
flux expulsion by $40^{\circ}K$ (Figure 3).

The samples were next warmed above their transition temperatures, cooled in zero field and magnetization, M, versus field, H, measurements were made with fields up to 80 kG. All of the samples have hysteretic M vs. H curves characteristic of type II superconductors with flux pinning centers. The lower critical fields, H_{c1}, for the $La_{2-x}M_xCuO_7$ materials

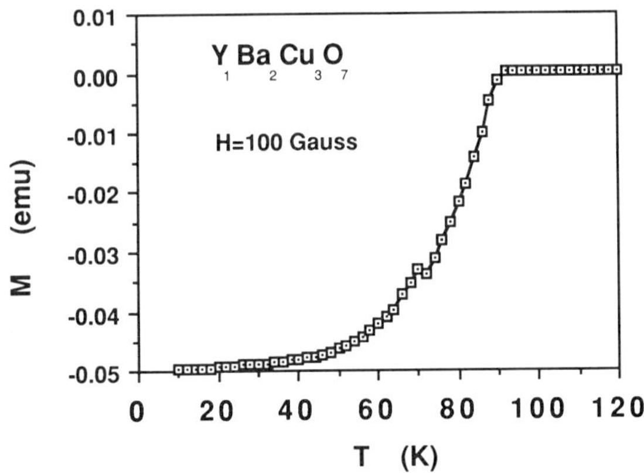

Fig. 3. Magnetization versus temperature for $Ba_2YCu_3O_7$. The low temperature diamagnetic signal shows nearly complete flux expulsion.

are 100 – 300 G at $10^{\circ}K$ (Table I). The critical currents, estimated from the hysteresis in the M vs. H curve [8], are rather low at $10^{\circ}K$. A series of measurements on $La_{1.8}Sr_{0.2}CuO_4$ at several temperatures shows a critical current of 12000 A/cm^2 at $4.2^{\circ}K$ which degrades dramatically at higher temperatures. Figure 4 shows J_c versus H for this sample. This curve is typical of a superconductor with strong pinning centers (insert).

The $4.2^{\circ}K$ magnetization data for the BaYCuO sample was highly hysteretic (Figure 5). The lower critical field was ~0.6 kG (Figure 6). Figure 7 shows critical current versus field for this sample. At H_{c1} the critical current is 2×10^5 A/cm^2 consistent with the results reported by IBM [9]. The shape of the low field region (insert) is typical of a weak pinning superconductor,

The upper critical fields for these materials were measured by monitoring the samples' 4-probe resistance as the temperature was swept in a constant magnetic field. The magnetoresistance of the carbon glass thermometer was taken into account in determining the temperature. The critical temperature in a given field was taken as that temperature where the sample resistance was 50% of the normal value.

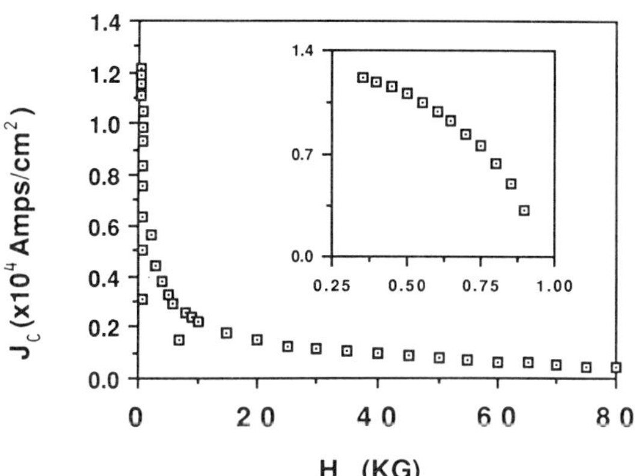

Fig. 4.   Critical current versus magnetic field for $La_{1.8}Sr_{0.2}CuO_4$ (estimated from the hysteresis in the M vs. H curve).   The low field data is typical of a superconductor with strong pinning centers (insert).

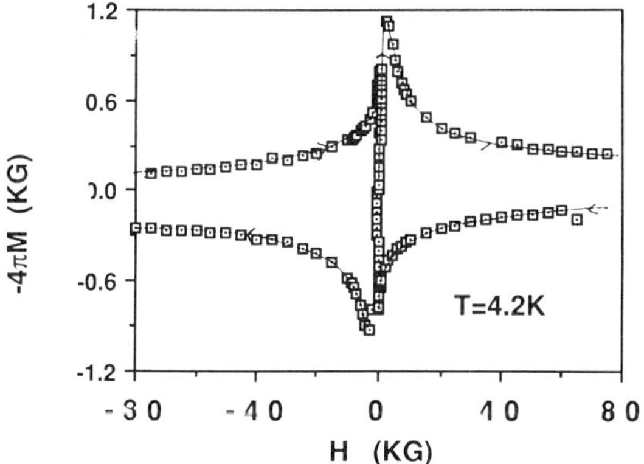

Fig. 5.   Magnetization versus magnetic field hysteresis loop of $Ba_2Y_1Cu_3O_7$ at 4.2°K.   The sample was cooled in zero field.

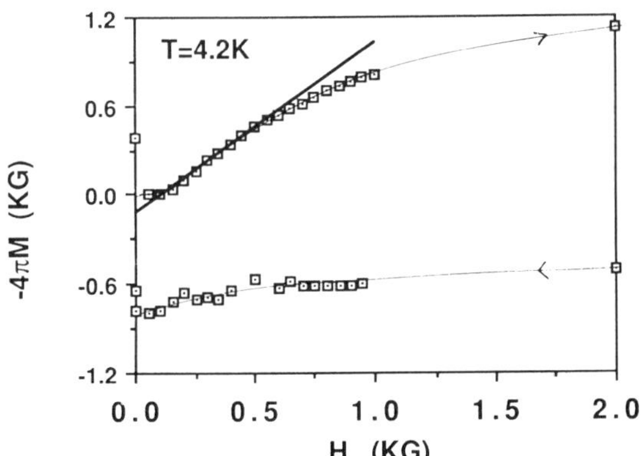

Fig. 6.   Low field data from Figure 5.   The lower critical field, H_{c1}, is 0.6 kG.

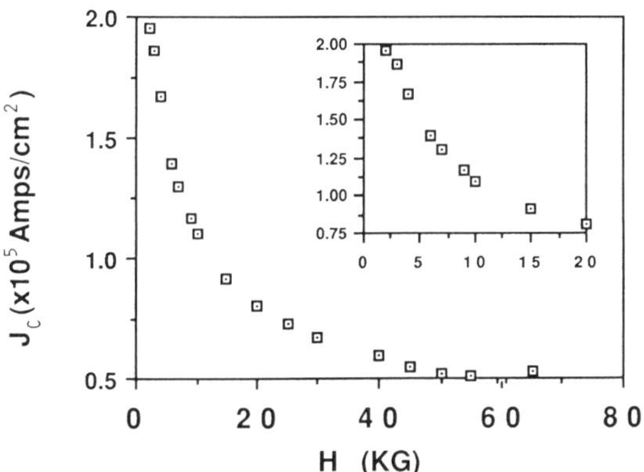

Fig. 7.   Critical current versus magnetic field for $Ba_2Y_1Cu_3O_7$ (estimated from the hysteresis in the M vs. H curve).   The low field data is typical of a superconductor with weak pinning centers (insert).

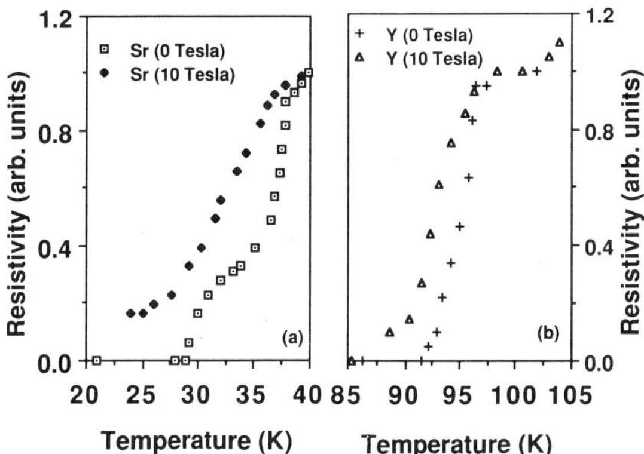

Fig. 8.   Resistance versus temperature of (a)La$_{1.8}$Sr$_{0.2}$CuO$_4$ and (b)Ba$_2$Y$_1$Cu$_3$O$_7$ in 0 and 10 Tesla magnetic fields.

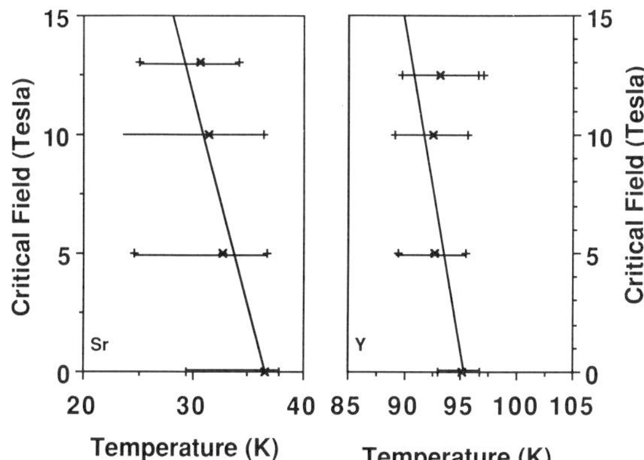

Fig. 9.   The upper critical field as a function of temperature for La$_{1.8}$Sr$_{0.2}$CuO$_4$ (left) and Ba$_2$Y$_1$Cu$_3$O$_7$ (right). The extent of the 10% – 90% points of the transitions are shown. The lines through the data points are guides to the eye.

The magnetic field significantly broadened the transition for the lower T$_c$ materials (Figure 8). This is perhaps due to inhomogeneities in these samples. A field of 100 kG has a negligable effect on the onset temperature. The magnetic field only slightly broadened the transition of

Table I. Parameters calculated from the dc magnetization and critical
field data.

Material	$T_c(50\%)$ (K)	$dH_c/dT\vert_{T_c}$ (kG/K)	$H_{c2}(0)$ (kG)	ξ_{GL} (Å)	γ (ergs/cc – K^2)	$H_c(0)$ (kG)	κ_{GL}	$H_{c1}(0)$ kG calculated	H_{c1}:T kG:K measured	$J_c(4K)$ A/cm^2	T_c onset (K) magnetic	% flux expelled at 10K
$Sr_{0.2}$	34	–13 – –40	330 – 930	19 – 31	840 – 2500	4.5 – 7	76 – 132	0.180 – 0.190	0.35:4.2	1.2×10^4 800^* $1.1\times10^8]$ $1.7\times10^8]$	36	40
$Ba_{0.1}$	27	–9.6 – –15.6	190 – 280	34 – 42	600 – 1000	3.0 – 3.6	65 – 83	0.134 – 0.136	0.2:10	– 57^* $6.4\times10^7]$ $7.6\times10^7]$	31	20
$Ca_{0.3}$	15	–22 – –26	230 – 270	35 – 38	1370 – 1630	2.4 – 2.6	97 – 106	0.079 – 0.081	0.1:10	– 12.8^* $3.8\times10^7]$ $4.2\times10^7]$	20	16
Y	96	–22 – –36	1470 – 2370	12 – 15	2500 – 4030	20. – 26.	75 – 95	0.820 – 0.870	~0.6:4.2	2×10^5 $1.1\times10^9]$ $1.4\times10^9]$	93	100

Note: For the resistivity we used $\rho=350$ $\mu\Omega$–cm for the Sr, Ba, and Ca
compounds and 200 $\mu\Omega$–cm for the Y. This was measured for the Sr and
compounds and assumed for the Ba and Ca.

bracketed values are theoretical values for J_c at 0°K.

$^*J_c$ measured at 10°K.

the higher T_c material. In Fig. 9 the critical field is plotted as a
function of temperature for the $La_{1.8}Sr_{0.2}CuO_4$ and $Ba_2Y_1Cu_3O_7$ samples.

From the slope, $dH_{c2}/dT\vert_{T_c}$, one is able to calculate a variety of the
material parameters [10]. Table I presents the results of these
calculations. The ranges given represent the uncertainty in determining
$dH_{c2}/dT\vert_{T_c}$. For the $La_{2-x}M_xCuO_4$ samples the H_{c1} obtained from $dH_{c2}/dT\vert_{T_c}$
and the resistivity, ρ, agree within a factor of two with the value
obtained from the magnetization data. In $Ba_2Y_1Cu_3O_7$ the agreement is
quite good which is surprising since the calculations assume an isotropic,
3D Fermi surface and these materials are known to be anisotropic, with the
superconductivity believed to be quasi-2D for $La_{2-x}M_xCuO_4$ and possibly
quasi-1D for $Ba_2Y_1Cu_3O_7$ [6].

Our value for γ, the density of states at the Fermi level, is in the
range 2500 – 4030 (ergs/cc –K^2) for $Ba_2Y_1Cu_3O_7$. This is to be compared

with 2150 (ergs/cc $- K^2$) as obtained from the jump in the specific heat
[11]. The agreement is quite good considering the imprecision in
determining $dH_{c2}/dT|_{T_c}$ and ρ for this material.

CONCLUSION

We have prepared, characterized and performed a variety of magnetic
field measurements on the new superconductors. The $La_{2-x}M_xCuO_4$ (M = Ba,
Sr, Ca) materials show incomplete flux expulsion and very broad
transitions, both magnetic and resistive, in magnetic fields. The samples
of $Ba_2Y_1Cu_3O_7$, which have extensive twinning domains due to the crystal's
orthorhombic structure, show negligible broadening of the resistive
transition at 130 kG. This result, along with processing [6] and neutron
studies, shows that good superconducting behavior strongly depends upon
the Cu-O-Cu chains in the basal plane of the structure.
This work at the Naval Research Laboratory could not have been done
without the support of our management and sponsors. We specifically wish
to acknowledge the sponsorship of the Office of Naval Research (ONR), The
Strategic Defence Initiative Organization, Office of Innovative Science
and Technology (SDIO/IST), and the Defence Advanced Research Project
Agency (DARPA).

REFERENCES

* Office of Naval Technology Postdoctoral Fellow
** On sabbatical leave from the National Science Foundation, Washington,
   DC, USA
+ Sachs-Freeman Associates, Landover, MD.
++ Crystal Growth and Material Testing Associates, Lanham, MD.
+++ Catholic University of America, Washington, DC
1) J.G. Bednorz and K.A. Muller: Z. Phys. B64(1986) 189.
2) R.J. Cava, R.B. van Dover, B. Battlogg, and E.A. Reitman: Phys. Rev.
Lett. 58(1987) 408.
3) K. Kishio, K. Kitazawa, S.   Kanbe, I. Yamada, N. Sugii: Jpn. J. Appl.
Phys. (to be published).
4) M.K. Wu, J.R. Ashburn, C.J.Torng, P.H. Hor, R.L. Meng, L. Gao, Z.J.
Huang, Y.O. Wang, and C.W. Chu: Phys. Rev. Lett. 58(1987) 908.
5) D.U. Gubser, R.A. Hein, S.H. Lawrence, M.S. Osofsky, D.J. Schrodt, L.E.
Toth, and S.A. Wolf: Phys. Rev. B35(1987) 5350.
6) L.E.Toth, E.F. Skelton, S.A. Wolf, S.B. Qadri, M.S. Osofsky, B.A.
Bender, S.H. Lawrence, and D.U. Gubser: (submitted to Phys. Rev B)
7) r.Beyers, G.Lim, E.M.Engler, and R.J.Savoy:(to be published in Appl.
Phys. Lett., 29 June 1987)
8) W.A. Fietz and W.W. Webb: Phys. Rev. 178(1969) 657.
9) private communication
10) T.P. Orlando, E.J. McNiff, Jr., S. Foner, and M.R. Beasley: Phys. Rev
19(1979) 4545.
11) A. Junod, A. Bezinge, T. Graf, J.L. Jorda, J. Muller, L. Antognazza,
D. Cattani, J. Cors, M. Decroux, O. Fischer, M. Banovski, P. Genoud, L.
Hoffmann, A.A. Manuel, M. Peter, E. Walker, M. Francois, and K. Yvon:

MAGNETIC BEHAVIOR OF YBaCuO

T. Datta and C. Almasan

Physics and Astronomy Department
University of South Carolina
Columbia, SC 29208

D.U. Gubser, S.A. Wolf, M. Osofsky and L.E. Toth

Naval Research Laboratory
Washington, D.C. 20375

INTRODUCTION

Some of the most intriguing behavior of a superconducting medium
pertain to its magnetic properties. An ideal plasma is just as highly
conducting as a superconductor, but magnetically these are totally different systems. Indeed the statement, that it may be more accurate to
regard a superconductor as a perfect diamagnet than a perfect conductor,
has been attributed to John Bardeen. This pivotal role of magnetism
makes it essential to understand the magnetic properties of superconductors.

EXPERIMENT

We have studied the d.c. magnetization $M(T)$, and susceptibility
$\chi(T)$, of YBaCu-oxide superconductors, as a function of external d.c.
field, temperature and experimental protocols. Specimens of simply connected and multiply connected topology such as a macroscopic ring, were
examined. The samples were sintered in the form of thin disks at several
laboratories, but here attention will be focused on the materials fabricated at NRL. The details of the material preparation have been reported
elsewhere.[1] Particular emphasis was given to the "single phase, 1-2-3"
composition ($Y_1Ba_2Cu_3O_7$) and the mixed phase or the non-stoichiometric
material of the type $Y_{1.4}Ba_{1.6}Cu_2O_{9-y}$. Individual specimens were cut into
irregular-rectangular prisms with approximate dimension $\sim$ 2mmx3mmx6mm.
The ring was of $\sim$ 7mm${}_2$diameter and 3mm high with $\sim$4mm hole. To minimize
demagnetizing effects[2] during the exact diamagnetic susceptibility determination, a thin and long perfect rectangle of $\sim$1mmx1mmx6mm was chosen.
Typical specimen mass was $\sim 10{-}10^2$mg.

The magnetic measurements were performed with a computer-controlled-
superconducting quantum interference-device (SQUID) variable temperature
susceptometer (VTS-SHE/BTI). It is well-known[3] that unless special care

is taken, sizable remnant fields can build up in the superconducting magnet field coils. This trapped field may result in considerable uncertainty in $\chi(T)$. To eliminate such spurious field buildup, the VTS was routinely warmed up to the room temperature. Also, between such procedures, the experiments were queued in increasing order of the magnetic field B. The measurements were made over a field range of $1.0 \leq B \lesssim \cdot 10^3$ Gauss and the range of temperatures was $5K \lesssim T \lesssim 400K$.

Starting with the specimen at the laboratory ambient temperature two distinct experimental protocols were observed[4]: (1) zero-field-cooled-warm up (ZFCW); in this case the specimen was introduced into the VTS cryostat in nominal zero field condition. The sample was allowed to equilibrate for a fixed waiting time ($\lesssim 10^3 s$) at T, the cryostat temperature. T, was chosen to satisfy $T/T_c \ll 1$, were T_c is the targeted phase transition temperature. For example when investigating the purported 240K superconducting transition the ZFCW protocol was initiated with $T \sim 100K$. At the end of the waiting time the desired measuring field was turned on. Since the measurements cannot be done without the field these ZFCW data could only be obtained during sample warm up from $T < T_c$. In the field cooled (FC) and field warm up (FCW) processes, the specimen entered the VTS with B already on. Warm up or cool down data could be collected in this process both above and below room temperature. In both the processes the data were obtained by holding B constant and varying the temperature in a programmed sequence. For $T < T_c$, the ZFCW data represents the diamagnetic shielding response and FC data represents flux expulsion response.

The actual measured quantity is the flux difference between the "top" and "bottom" coil responses of the oscillating sample (vertical speed $\lesssim$ 3mm/s). Prior to the run the sample was carefully centered at the exact middle of these two coils. This is a rather sensitive and critical adjustment. It is of paramount importance in determining the sign of the magnetic response. This flux difference was electronically converted to the appropriate pre-calibrated sample response. The longitudinal dc mass magnetization was calculated from the known specimen mass and the longitudinal mass susceptibility $\chi(T)$ was defined to be

$$\chi(T) = \frac{M(T)}{B} \tag{1}$$

RESULTS

For these specimens qualitatively distinct $\chi(T)$ behaviors were observed, in three temperature ranges relative to the transition temperature, $T_c \sim 90$ K, and are as follows:

i)   In the $T \ll T_c$ domain $X(T) \sim 10^{-3}$ emu/g and was observed not to be a strong function of temperature. At the lowest temperature (for the "best" specimen) we determine that, $\chi = -1/4\pi$ emu/cm^3.

Because the flux is excluded only from the superconducting phase, the field cooled data represents the diamagnetism of the fraction of the sample volume which is superconducting. Thus the ratio of the FC and ZFCW data at any temperature $T(<T_c)$ is less than one i.e., $M(T)_{FC}/M(T)_{ZFCW} < 1$ and is a measure of the fraction of volume of the sample which is superconducting. For the "best" sample the total superconducting volume was $\sim 1/3$. This behavior is exhibited in fig. 1. This means a long range bulk superconducting order, as opposed to a glassy phase was achieved.

An enhanced Meissner effect was observed particularly in the "1-2-3" specimens. This behavior was evidence by a irreversibility between FC and FCW data and often appeared as a negative peak in M(T) or χ(T) for $T \lesssim T_c$. Data showing this effect are displayed in fig. 2. In several runs an enhancement of as much as $\sim$ 40% in the Meissner effect was noticed. Two possible mechanisms are believed to be responsible for this phenomenon, flux detrapping and reentrant short range order in the superconducting phase coherence between the finite clusters.

ii) In the high temperature or $T > T_c$, region $\chi(T) \sim 10^{-7}$ emu/g, temperature dependent and paramagnetic. For the mixed phase specimen χ increased by a factor $\sim$ 2 as T decreased from 400K to 100 K. The sign of this paramagnetic behavior was confirmed by concurrently testing the specimen with a known paramagnet.

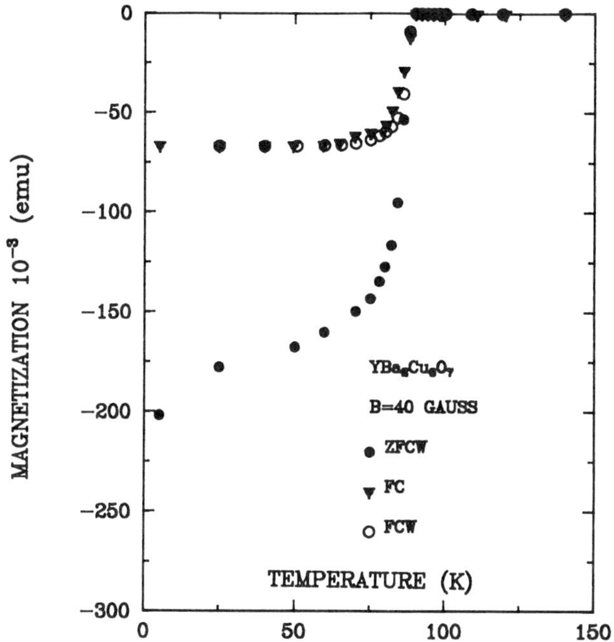

Fig. 1   Diamagnetic shielding and Meissner effect in a "1-2-3" specimen.

iii) The transition region or $T \sim T_c$. In this third domain, χ was adversely affected by the applied field as in spin glass materials[3,4]. Also, χ(T) was observed to be most sensitive to temperature changes. As T_c was approached from above first χ increased to $\sim 4 \times 10^{-7}$ emu/g then χ decreased down to $\sim - 1 \times 10^{-6}$ emu/g just below T_c, as it was driven by a paramagnetic spin-glass to a superconducting transition. This behavior is depicted in fig. 3. The data shown are from a non-stoichiometric specimen, but the "1-2-3" specimen exhibited similar behavior. On further cooling, $|\chi|$ rapidly increased several orders of magnitude to $\sim 10^{-3}$ emu/g in the "1-2-3" specimens and to $\sim 10^{-4}$ emu/g in the non-stoichiometric samples.

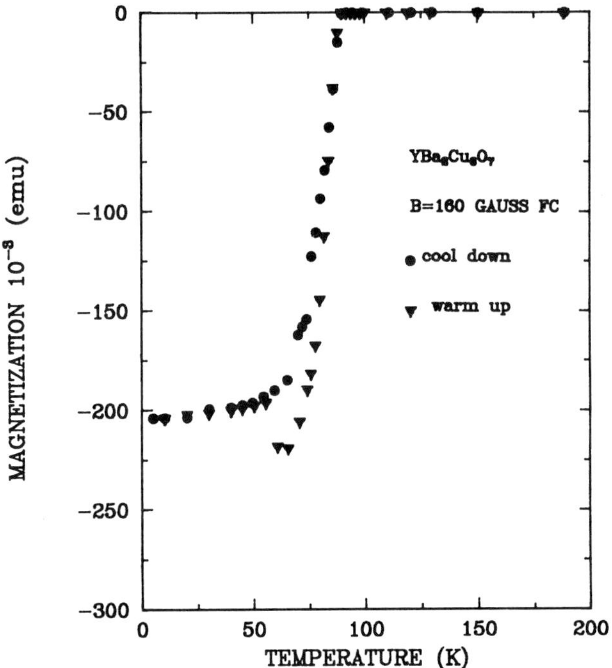

Fig. 2.   Enhanced Meissner effect and FC FCW irreversibility data for a "1-2-3" sample.

DISCUSSION

The magnetic response in the lowest temperature region $T<T_c$ is straight forward.   For the YBa$_2$Cu$_3$O$_7$, the observed value of the low temperature volume susceptibility is $-1/4\pi$, indicating that the material is in a perfect superconducting state.   The "saturation" in χ for low temperatures implies a total bulk superconductivity over the entire sample.   In contrast no such saturation was observed in the nonstoichiometric samples.   For these the $|\chi(ZFC)|$ continues to be increase with decreasing T.   This may be because as the temperature is lowered more and more finite clusters join the percolative supercluster due to increased wave function-phase coherence.

The observation of the Meissner effect confirms gauge-symmetry breaking in the superconducting ground state of these "high-T$_c$" materials.   The ratio $\chi(FC)/\chi(ZFCW)<1$, indicates that microscopically the bulk specimen is an aggrate of non-simply connected regions of normal and superconducting media.

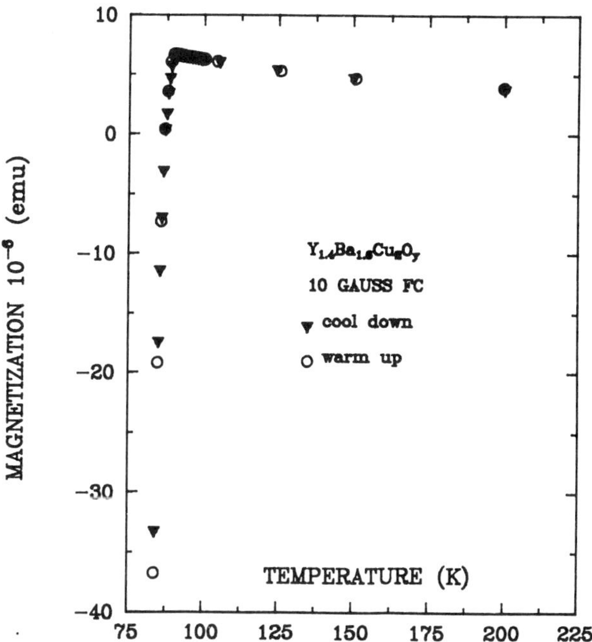

Fig. 3   High temperature paramagnetism and the onset of superconductivity
in the non-stoichiometric material.

The behavior in the intermediate temperature region can only be
clear after of the high temperature phenomenon is understood.

The most prominent aspect of $\chi(T)$ for $T \gg T_c$ is that it is paramagnetic spin-glass. One possible model that is consistent with the measured data is as follows: the behavior is due to two magnetic contributions, such as,

$$\chi(T) = \chi_0 + \chi'(T) \tag{2}$$

where, χ_0 is the temperature independent diamagnetic background[5-8] due
to the core, conduction electrons and the lattice. The second contribution χ' is paramagnetic spin-glass like, with a Curie-Weiss type
temperature dependence that is,

$$\chi'(T) = \frac{C}{T-\theta} \tag{3}$$

The temperature dependence of χ', represents a collection of interacting localized moments. The $\chi(T)$ data fits for eqs. (2) and (3) are
shown in fig. 4. The results of the statistical analysis (SAS) of the
data in the paramagnetic region are (i) For $YBa_2Cu_3O_7$; $\chi_0 = 5.6 \times 10^{-7}$
emu/g, $C = 3.66 \times 10^{-5}$ emu·K/g and $\theta = -38.7K$. (ii) For $Y_{1.4}Ba_{1.6}Cu_2O_y$;
$\chi_0 = 1.0 \times 10^{-7}$ emu/g, $C - 6.42 \times 10^{-5}$ emuK/g and $\theta = -14.17K$. The negative
sign of the Curi-Weiss parameter θ indicates a net antiferromagnetic

interaction between the localized moments. We estimate this localized magnetic moment μ, per formula unit to be $5.68\mu_B$ for the "1-2-3" specimens. Assuming four atoms of oxygen (i.e., y=4) we obtain $\mu = 4.67\mu_B$ per formula unit of the mixed phase samples. Notice, that for the 1-2-3 material χ_0 is larger and C is smaller than in the mixed phase. These indicate a higher density of state $D(E_F)$ and greater de-localization of the states at the Fermi energy and may account for the sharper transition in the 1-2-3 specimens compared with the mixed phase materials.

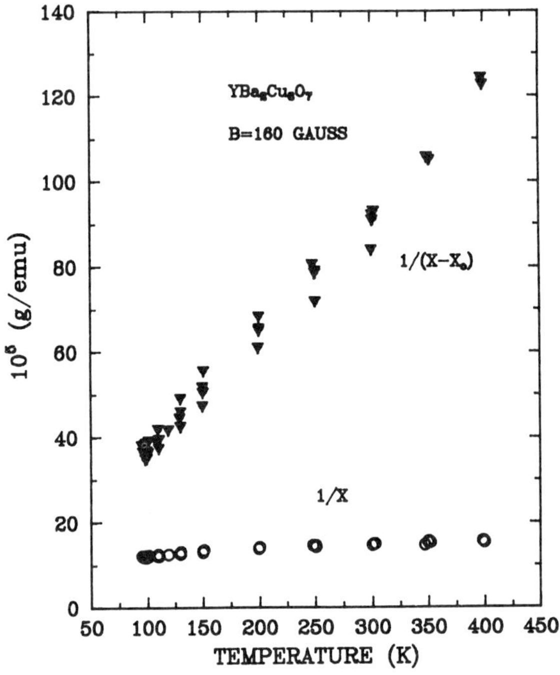

Fig. 4   Comparison of Curie-Weiss ($1/\chi$ vs T) and modified Curie-Weiss ($1/(\chi-\chi_0)$ vs T) behaviors for the 1-2-3 material.

Recently, several[9,10] electronic and magnetic interactions have been discussed as the non-phonon mechanisms for superconductivity in these materials. If indeed the superconductivity is related to the antiferromagnetic behavior then $|\theta|T_c < 1$ indicates a type -1 antiferromagnetic ground state[7] with effective first and second nearest neighbor interactions. Further studies such as neutron measurements are needed to test this assumption.

ACKNOWLEDGEMENTS

This work was partially supported at USC by a grant #1070K101. C. Almasan acknowledges a Zonta Amelia Earhart (graduate) Fellowship. M. Osofsky, acknowledges a ONT-Postdoctoral Fellowship and L.E. Toth

is on a sabbatical leave from the NSF. Timir Datta acknowledges useful
discussion with Professors C.P. Poole, E.R. Jones, Jr., R. Creswick
and Dr. M. Nevitt. Dr. A. Barrientos and J. Estrada assisted with some
of the measurements.

REFERENCES

1.   S.B. Qadri, L.E. Toth, M.S. Osofsky, S. Lawrence, D.U. Gubser, and
     S.A. Wolf, Phys. Rev. $\underline{B35}$, 7235 (1987); E.F. Skelton, W.T. Elam,
     D.U. Gubser, S.H. Lawrence, M.S. Osofsky, L.E. Toth and S.A. Wolf,
     ibid. 7140 (1987).

2.   Soshin Chikazumi, <u>Physics of Magnetism</u>, John Wiley & Sons, NY
     (1964).

3.   T. Datta, D. Thornberry, C. Almasan and E.R. Jones, Sol. St. Comm.
     $\underline{56}$, 523 (1985).

4.   T. Datta, A. Barrientos, J. Amirzadeh and E.R. Jones, Sol. St. Comm.
     $\underline{62}$, 571 (1987).

5.   T. Datta, A. Barrientos. J. Amirzadeh and E.R. Jones, J. Appl. Phys.
     $\underline{61}$, 3555 (1987).

6.   E.R. Jones, T. Datta, C. Almasan, D. Edwards and H.M. Ledbetter,
     Mat. Sc. and Engr., In press.

7.   J.S. Smart. <u>Effective Field Theories of Magnetism</u>, W.B. Saunders
     Philadelphia (1966).

8.   C. Almasan, T. Datta, R.D. Edge, E.R. Jones, J.W. Cable and H.M.
     Ledbetter, preprint.

9.   D. Pines in this conf. proceedings.

10. P.W. Anderson in this conf. proceedings.

A COUPLED STRUCTURAL AND ELECTRICAL TRANSITION IN La_2CuO_4 NEAR 30 K

E. F. Skelton, W. T. Elam, D. U. Gubser, R. A. Hein, V. Letourneau, M. S. Osofsky[†], S. B. Qadri, L. E. Toth[††], and S. A. Wolf

Condensed Matter and Radiation Sciences Division
Naval Research Laboratory, Washington, DC 20375-5000

[†] Supported by the Office of Naval Technology
[††] Permanent address: National Science Foundation
Washington, DC 20550

INTRODUCTION

Recently there has been an explosion of activity focused on the new class of oxygen defect, perovskite related, copper oxide superconductors. All of this was initiated by the discovery of superconductivity at temperatures in excess of 30 K in La-Ba-Cu-oxides[1] and the subsequent discovery of superconductivity in excess of 90 K in Y-Ba-Cu-oxides.[2] The prototypical compound for the lower transition temperature materials is La_2CuO_4 which, at room temperature, crystallizes in an orthorhombic distortion of the K_2NiF_4-structure.[3,4] It has recently been shown that the superconducting compounds $La_{2-x}M_xCuO_4$, where M = Ba, Ca, or Sr, are also orthorhombic above T_c.[5,6]

Based on first principles calculations, Pickett et al. predicted that tetragonal La_2CuO_4 should be superconducting.[7] However, resistance measurements of Jorgensen et al. on La_2CuO_4 indicated that the material is metallic down to about 100 K, below which it behaves like a doped semiconductor.[4] Most recently Grant et al. reported evidence of superconductivity below 40 K of 1 part in 6000 of La_2CuO_4.[8] In consideration of the presence of both metal-like and semiconductor-like behavior in this material, Kasowski et al. performed band structure calculations and group theoretical analyses.[9] From this, they suggested that La_2CuO_4 may undergo a structural phase transition at lower temperatures from the orthorhombic phase to a lower symmetry phase, possibly monoclinic. We have found evidence of such a transition and find that it is correlated with a change in the resistance near 30 K.

EXPERIMENTAL PROCEDURES

Test samples were prepared by mixing appropriate proportions of reagent grade powders of La_2O_3 and CuO to yield a product of La_2CuO_4.

The La$_2$O$_3$ was 99.99% pure and contained 1 ppm of Ca, with no Sr or Ba detected; the CuO contained 1 ppm Ba and less than 1 ppm of Sr. The powders were reacted in air at 950° C for 6 hr, with one grinding after 3 hr. The powders were again ground and pressed at 20 kpsi into pellets, sintered at 950° C in air for 15 hr, and then slowly cooled (2° C/min).

Simultaneous resistance and x-ray diffraction measurements were performed in a temperature controlled cryostat. The resistance measurements were made using standard ac, four probe methods with 100 microamps of current at a frequency of 31 hz. The x-ray data were collected on a computer controlled diffractometer using Cu K-alpha radiation. Additional details are given on our earlier report on the superconductor La$_{1.9}$Ba$_{0.1}$CuO$_4$.[10]

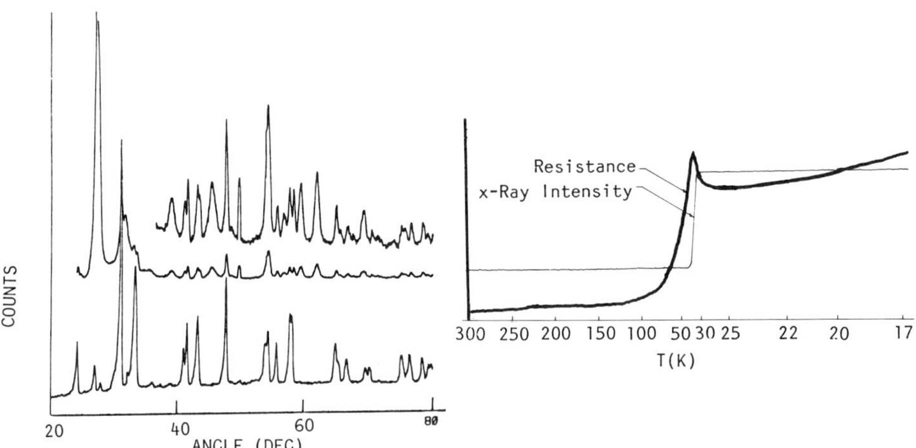

Fig. 1(left): X-ray diffraction spectra recorded at 295 K (lowest curve) and 10-12 K (upper curves). Fig. 2(right): Temperature dependence of the electrical resistance and x-ray intensity at 27.3° 2θ.

RESULTS AND DISCUSSION

A diffraction spectrum taken at 295 K is shown in Fig. 1 (lowest curve). The positions of all the major peaks are in agreement with the orthorhombic lattice reported by Grande et al.[3] and our measured intensities agree with those calculated from their structure. Expressed in the standard setting, the space group is Cmca; a least-squares fit to 37 diffraction peaks between 60° and 120° yields the following unit cell parameters: a = 5.361+0.001; b = 13.151+0.003; c = 5.408+0.001 A.

The temperature dependence of the resistance is plotted in Fig. 2. On cooling from room temperature, there is initially a gradual increase in the resistance, near 100 K a significant increase begins which peaks to a local maximum near 36 K, the resistance then drops until about 25 K below which it monotonically rises. The resistance curves are reversible and seen in every sample. A crystallographic phase transformation accompanies this resistance anomaly. The upper diffraction spectra shown in Fig. 1 are of the same material, but recorded at 10-12 K. There are 17 additional diffraction peaks between 20° and 80° 2θ. We have tentatively indexed these peaks to a monoclinic lattice; attempts to fit higher symmetry lattices were unsuccessful. We are in the process of determining the structure of this new phase.

Magnetization measurements made on a squid susceptometer show a diamagnetic onset at about 35 K. The magnitude of the signal indicates that one part in 5000 of the sample expells flux, if the diamagnetism is due to superconductivity.

To determine the temperature at which the structural transition
occurs, the x-ray detector was centered on the prominent diffraction
peak in the low temperature phase at 27.3° 2θ. The other curve shown
in Fig. 2 represents the temperature dependence of this peak and hence
the relative amount of the low temperature phase. Repetitive cycling
through the transition clearly establishes a connection between the two
phenomena. However, the magnitude of the resistive anomaly near 30 K
shows no measurable relationship to the magnitude of the x-ray peak at
27.3°. But, the magnitude of the low temperature resistance does
increase with increasing magnitude of the x-ray peak. This implies that
this new phase is responsible for the semiconductor-like behavior at low
temperatures. The kinetics of the structural transition are very
sluggish on cooling. Moreover, in addition to the new diffraction peaks
in the specta at low temperatures, remanents of the orthorhombic structure could always be seen. Comparisons of peak intensities in the
vicinity of the transition reveals that the low temperature phase grows
at the expense of the orthorhombic phase.

It is our belief that these observations confirm the predictions of
Kasowski et al.[9], viz. that it is this new phase which is responsible
for the low temperature semiconductive behavior of La_2CuO_4. We further
speculate that this phase transition inhibits the superconducting
transition in most of the material; as noted above, superconductivity
has been reported only in small fractions of these materials.[8] We, as
well as others, have searched without success for structural transitions
in the superconductor $La_{1.9}Ba_{0.1}CuO_4$.[4,10] We believe that one
effect of the introduction of the metal inpurity (Ba, Ca, or Sr) is to
suppress this phase transition and thereby allow a substantial portion
of the orthorhom bic phase to become superconducting. A series of low
concentration alloys is under preparation to test this hypothesis.

REFERENCES

1. J. G. Bednorz and K. A. Müller, Z. Phys. **B64**, 189 (1986)

2. M. K. Wu, J. R. Ashburn, C. J. Torng, P. H. Hor, R. L. Meng, L.
Gao, Z. J. Huang, Y. Q. Wang, and C. W. Chu, Phys. Rev. Lett. **58**, 908
(1987)

3. V. B. Grande, H. Müller-Buschbaum, and M. Schweizer, Z. Anorg.
Allg. Chem. **428**, 120 (1977)

4. J. D. Jorgensen, H. -B. Schuttler, D. G. Hinks, D. W. Capone, II,
K. Zhang, M. B. Brodsky, and D. J. Scalapino, Phys. Rev. Lett. **58**, 1024
(1987)

5. U. Geiser, M. A. Beno, A. J. Schultz, H. H. Wang, T. J. Allen, M.
R. Monaghan, and J M. Williams, Phys. Rev.-B **35**, 6721 (1987)

6. M. Onoda, S. Shamoto, M. Sato, and S. Hosoya, Jpn. J. Appl. Phys.
26, 363 (1987)

7. W. E. Pickett, H. Krakauer, D. A. Papaconstantopoulos, and L. L.
Boyer, Phys. Rev.-B **35**, 7252 (1987)

8. P. M. Grant, S. S. P. Parkin, V. Y. Lee, E. M. Engler, M. L.
Ramirez, J. E. Vazquez, G. Lim, R. D. Jacowitz, and R. L. Green, Phys.
Rev. Lett. **58**, 2482 (1987)

9. R. V. Kasowski, W. Y. Hsu, and F. Herman, Solid State Commun.
-in press

10. E. F. Skelton, W. T. Elam, D. U. Gubser, S. H. Lawrence, M. S.
Osofsky, L E. Toth, and S. A. Wolf, Phys. Rev.-B **35**, 7140 (1987)

A PHOTOEMISSION STUDY OF HIGH T_C OXIDES

Donald Mueller* and Arnold Shih,
Louis E. Toth**, Michael Osofsky*** and
Stuart A. Wolf

Naval Research Laboratory
Washington, D. C. 20375

Richard L. Kurtz and Roger L. Stockbauer

National Bureau of Standards
Gaithersburg, Maryland 20899

INTRODUCTION

A number of ceramic compounds including $YBa_2Cu_3O_7$ have been
reported to superconduct at temperatures in excess of 35K [1-9].
A critical feature in the processing of these compounds that is
required to obtain good superconductive properties appears to
be a final annealing stage in an oxygen containing atmosphere
to increase the oxygen content of the materials [10]. Without
the oxygen anneal step, the superconducting transition
temperature drops and the drop in the temperature versus
resistance curve becomes less defined.

To aid in the development of a better understanding of the
electronic structure in this class of materials, we have
examined changes in the electronic density of states for
$YBa_2Cu_3O_x$ with oxygen content via ultraviolet photoelectron
spectroscopy. Samples with x = 6.95, 6.5, and 6.05 were
examined.

Photoelectron energy distribution curves provide a
semiquantitative measure of the density of occupied electronic
states in a material. They differ from the actual density of
states in two respects. First, the photoionization
cross-sections for different electronic states can be quite
different, and their dependence on photon energy can be
different as well. For the copper 3d and oxygen 2p levels the
photoionization cross sections are approximately equal at a
photon energy of 40eV [11]. Secondly, the position of the peaks
in a photoelectron energy distribution curve may be shifted
slightly with respect to the actual peaks in the density of
states due to interaction of the photoelectron with the hole it
leaves behind.

Our experimental determination of the electronic structure in
$YBa_2Cu_3O_{6.95}$ [12] is in good agreement with similar
determinations made by other research groups [12-14]. The

experimentally determined electronic density of states, however, differs from the calculated valence band density of states [15-17], both in terms of the valence band width and in terms of the peak positions relative to the Fermi level.

Although photoemission is a surface sensitive technique, it has been used successfully to examine the bulk electronic structures of a large variety of both metallic and semiconducting compounds [18].

EXPERIMENTAL

In the experiment, electrons photoemitted from the sample, were energy analyzed by a double pass cylindrical mirror electron energy analyzer. The source of the photons was the SURF-II storage ring. The energy of the incident light was selected by a toroidal grating monochromator whose energy band width at 60eV photon energy was calculated to be 350meV. At lower energies the resolution was better, at higher energies, worse. Spectra presented below have been normalized to the incident light flux as determined by a calibrated in line tungsten mesh photodiode. The resolution of the electron energy analyzer was fixed independent of the photon energy at 240meV.

The $YBa_2Cu_3O_{6.95}$ sample was prepared as described in reference 10. Other samples prepared in the same batch were characterized by X-ray diffraction, neutron diffraction, resistance versus temperature measurement and Meissner effect [19]. T_c was determined to be 93K with a 2K width. The sample was found to have an orthorhombic crystal structure. The $YBa_2Cu_3O_{6.5}$ sample was prepared in the same way as the 93K sample except that the final oxygen anneal step was omitted [10, 20]. Identical samples characterized by resistance measurements and x-ray diffraction were found to have transition temperatures of about 60K with a tetragonal crystal structure. The $YBa_2Cu_3O_{6.05}$ sample was prepared by annealing $YBa_2Cu_3O_{6.5}$ in flowing argon at $750^{\circ}C$ for 15 hours. Samples prepared by an identical procedure were found by neutron diffraction to have a tetragonal crystal structure and 6.05 oxygen atoms per unit cell [19]. Auger analysis verified that the last two samples had an oxygen content ratio of approximately 6.5/6.05.

Photoemission measurements were conducted in UHV at pressures from $1x10^{-10}$ to $3x10^{-10}$ Torr. Rectangular samples were cut in air from pellets sintered to approximately 80% density, and mounted rigidly in indium encased in an aluminium holder. Immediately before photoemission measurements were conducted, the samples were fractured in UHV with a stainless steel blade. The samples were positioned so that the photoemitting face was at an angle of 45° with respect to both the photon beam and the analyzer axis. Fermi level determinations were made on the basis of the Fermi edge observed in photoemission spectra from a sputtered gold foil that was in electrical contact with the samples. The sample holder was mounted on a liquid nitrogen cooled manipulator. Temperatures were determined using a W-5% Re W-26% Re thermocouple imbedded in the sample holder. The measurements were angle integrated throughout the Brillouin zone due to both

the large acceptance cone of the CMA and the polycrystalline
nature of the sample. No evidence for radiation induced damage
was observed in our experiments for any of these samples.
Spectra taken immediately after fracturing the sample to expose
fresh surface appeared identical to spectra obtained after
periods as long as 3.5 hours. This corresponds to a radiation
dose of about 10^4 photons/surface atom at photon energies from
30 to 110eV.

RESULTS AND DISCUSSION

Figure 1 shows a series of photoelectron energy
distribution curves collected from the $YBa_2Cu_3O_{6.95}$ sample at
photon energies from 50 to 108eV. Different features are
enhanced at different photon energies. The features at about 14
and 29eV below the Fermi level (E_F) are due to emission from
the barium 5p and 5s levels, and are enhanced above the Ba 4d
photoionization threshold. The peak at 12.4eV binding energy
prevalent in the 76eV photon energy spectrum is assigned to a
valence band satellite feature that resonanted at the copper 3p
threshold. Resonance satellites have also been observed for
CuO and Cu_2O [22]. The binding energy of the satellite feature
and the photon energy for which it is most intense depend upon
the charge state of copper. The binding energy and photon
energy dependence of the satellite observed here are most
nearly matched by CuO, although oxidation states other than
Cu^{++} cannot definitely be ruled out on the basis of our data.
The features that appear at 28 and 31eV in the 60eV photon
energy spectrum and at 24 and 27eV in the 64eV photon energy
spectrum are due to the barium 4d core levels excited by second
order light. Above 70eV photon energy the intensity of second
order light transmitted by the monochromator is essentially
zero. The features at 20 and 24eV binding energy visible
particularly in the low energy spectra arise from the oxygen 2s
and yttrium 4p levels. The main valence band photoemission
feature occurs with a binding energy 4.5eV below the Fermi
level. Two smaller peaks with binding energies of 9.4 and 2.3eV
also occur. Both these peaks drop more rapidly in intensity
than the 4.5eV feature with increasing photon energy. At photon
energies of approximately 40eV, the photoionization
cross-sections of oxygen and copper are approximately equal. At
photon energies below 40eV, emission from predominantly oxygen
derived levels is more intense than from levels of
predominantly copper character [11]. At photon energies above
40eV, emission from predominantly copper derived levels is more
intense than emission from oxygen derived levels. This
observation together with the behavior of the spectra in figure
1 suggests a greater oxygen contribution for the levels at 2.3
and 9.4eV binding energy than to the main peak. On cooling the
93K superconductor to 88K in UHV, no changes in the valence
band photoemission spectra were noted. The 2.3eV peak was
particularly sensitive to water adsorption. After exposure of
the sample to 1 L water vapor this peak was significantly
reduced in intensity.

Figure 2 shows photoemission spectra of the valence band
region for the three $YBa_2Cu_3O_x$ samples together with a spectrum
from a non-superconducting La_2CuO_4 sample. All four spectra

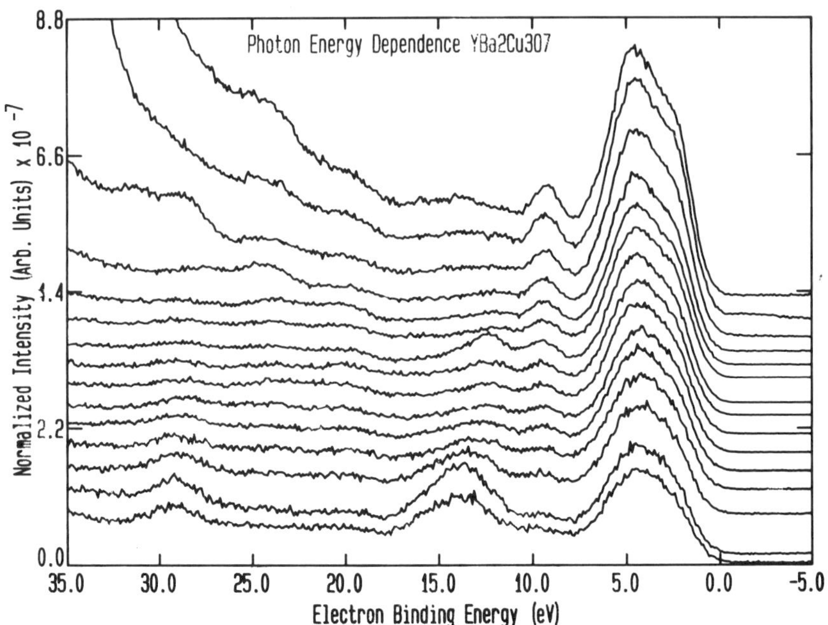

Figure 1. Photoelectron energy distribution curves from $YBa_2Cu_3O_{6.95}$ at photon energies from 50 to 108eV. The electron binding energies are measured with respect to the Fermi energy (0.0eV). The curves have been offset vertically for clarity. Photon energies were 50eV (top curve), 55, 60, 64, 68, 72, 76, 80, 84, 88, 92, 96, 100, 104, and 108eV (bottom curve).

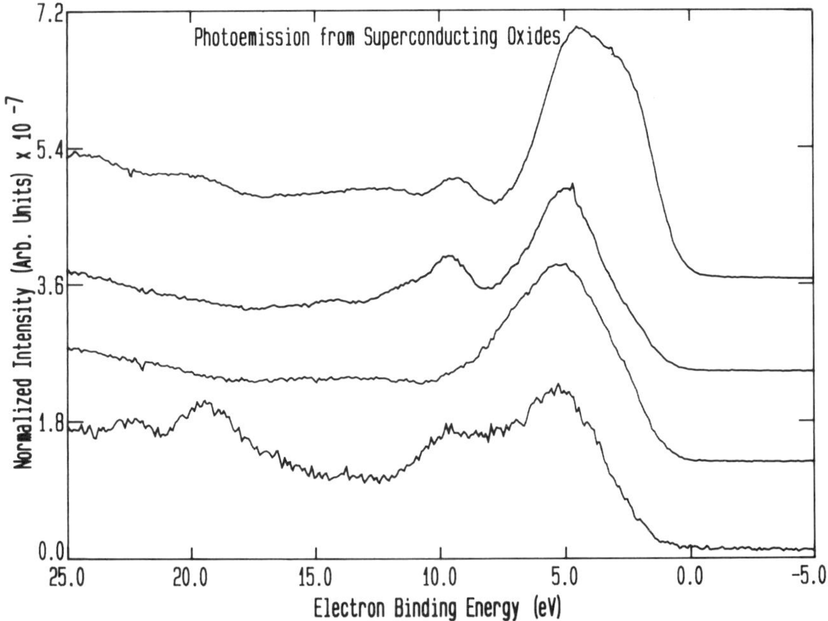

Figure 2. Photoemission spectra (photon energy=60eV) are compared for the compounds $YBa_2Cu_3O_{6.95}$ (top curve), $YBa_2Cu_3O_{6.5}$, $YBa_2Cu_3O_{6.05}$ and La_2CuO_4 (bottom curve).

show the main peak in the electron energy distribution curves
4.5 to 5eV below the Fermi level. This peak is shifted slightly
toward higher binding energies in the oxygen deficient
compounds and in La_2CuO_4. The feature at 9.4eV binding energy
is also present in all the spectra except that of $YBa_2Cu_3O_{6.05}$,
where it appears to be shifted to 8.8eV and reduced in
intensity. The main difference between the 93K superconductor
and other compounds is the presence of the feature at 2.3eV
binding energy. It is much less intense in the $YBa_2Cu_3O_{6.5}$
spectrum, where it appears to be shifted slightly toward higher
binding energy (2.6eV), and has all but vanished in the
$YBa_2Cu_3O_{6.05}$ spectrum (See figure 3). Minor shifts in the O 2s,
Y 4p and Ba 5s and 5p levels also take place with changes
in the oxygen concentrations. These shifts may be due to a
combination of changes in the atomic charge states and changes
in the hole screening. The levels of binding energies are
summarized in table I.

The measurements presented here provide two constraints on
theoretical treatments aimed at understanding the electronic
structure of $YBa_2Cu_3O_x$. (1) The presence of the 2.3eV feature
in spectra from the 93K compound implies a higher charge carrier
concentration at the Fermi level in this compound than in the
related lower T_c and non-superconductive compounds. (2) The
oxygen-like cross sectional dependence of the 2.3eV states
implies a large contribution from oxygen to the density of
states near the Fermi level in the 93K compound. In agreement
with the measurements, calculations by Massidda et. al. [16] and
Krakauer et. al. [17] show that the density of states near the
Fermi level are mainly derived from the oxygen in the Cu-O
chain. More significantly, the observed 2.3eV feature is
intense in the orthorhombic sample, but weak in the tetragonal
samples. The X-ray and neutron diffraction data [10,19] as well
as Cu-O vibrational spectra [22] indicated that the linear Cu-O
chains are removed by the orthorhombic to tetragonal phase
transition. The measurements however disagree with the
calculations in terms of the locations of peaks in the
electronic density of states with respect to the Fermi level.
The measured levels lie roughly 2eV further below E_F than
the calculated peaks. Part of this shift may be due to
interaction between the photoemitted electron and the hole it
leaves behind in the final state. Further investigations are
planned to determine the source (or sources) of this shift.

ACKNOWLEDGMENT

RLK and RLS wish to acknowledge support from the United
States Office of Naval Research.

---

*   National Research Council Postdoctoral Fellow.
**   On sabbatical leave from the National Science
    Foundation.
*** Office of Naval Technology Postdoctoral Fellow.

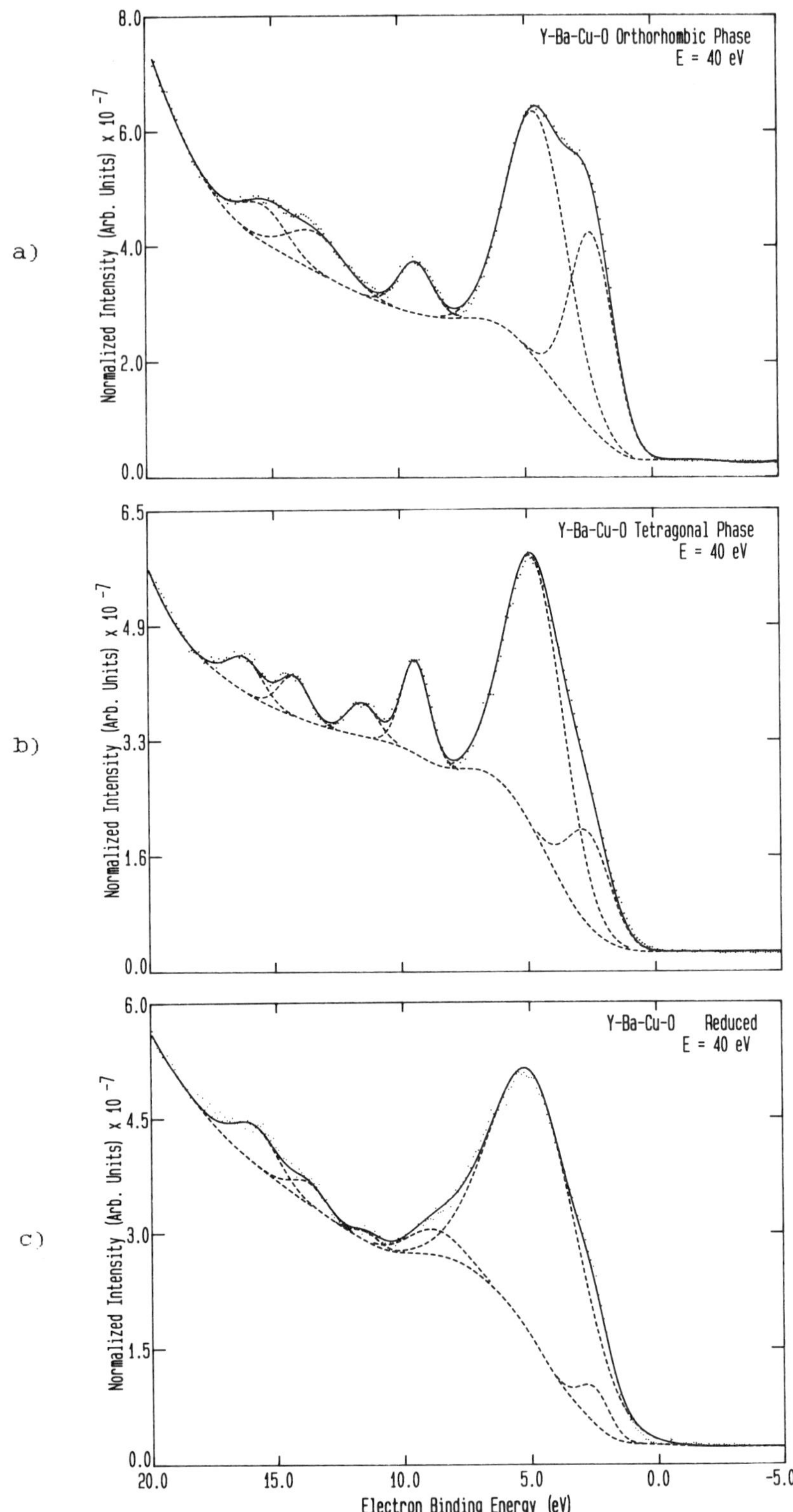

Figure 3. Deconvolution of the 40eV spectra into a background and Gaussian peaks, a)$YBa_2Cu_3O_{6.95}$, b)$YBa_2Cu_3O_{6.5}$, c)$YBa_2Cu_3O_{6.05}$.

Table 1.   Comparison of electronic binding energies

for $YBa_2Cu_3O_x$ Compounds (in units of eV)

Levels	Orthorhombic $x = 6.95$	Tetragonal $x = 6.5$	Tetragonal $x = 6.05$
Valence	2.3	2.6	——
Valence	4.5	4.7	4.8
Valence+ Satellite	9.4	9.4	8.8 (?)
Ba $5p_{3/2}$	13.3	14.2	13.7
Ba $5p_{1/2}$	15.3	16.2	15.8
O 2s	20.1	20.5	21.0
Y 4p	24.0	24.0	24.5
Ba 5s	28.8	28.8	29.8

REFERENCES

1. M. K. Wu, J. R. Ashburn, C. T. Torng, P. H. Hor, R. L. Meng, L. Gao, Z. Huang, Y. Q. Wang and C. W. Chu, Phys. Rev. Lett. 58:908 (1987).
2. P. M. Grant, S. S. P. Parkin, V. Y. Lee, E. M. Engler, M. L. Ramirez, J. E. Vazquez, G. Lim, R. D. Jacowitz and R. L. Greene, to be published Phys. Rev. Lett. (1987)
3. S. Uchida, H. Takagi, K. Kitazawa and S. Tanaka, Jpn. J. Appl. Phys. Lett. 26:L1 (1987).
4. C. W. Chu, P. H. Hor, R. L. Meng, L. Gao, Z. J. Huang, and Y. Q. Wang, Phys. Rev. Lett. 58:405 (1987).
5. D. U. Gubser, R. A. Hein, S. H. Lawrence, M. S. Osofsky, D. J. Schrodt, L. E. Toth and S. A. Wolf, Phys. Rev. B35: 5350 (1987).
6. P. H. Hor, R. L. Meng, Y. Q. Wang, L. Gao, Z. J. Huang, J. Bechtold, K. Forster and C. W. Chu, Phys. Rev. Lett. 58: 1891 (1987).
7. D. W. Murphy, S. Sunshine, R. B. VanDover, R. J. Cava, B. Batlogg, S. M. Zhurak and L. F. Schneemeyer, Phys. Rev. Lett. 58:1888, (1987).
8. D. B. Mitzi, A. F. Marshall, J. Z. Sun, D. J. Webb, M. R. Beasley, T. H. Geballe and A. Kapitulnik (submitted to Phys. Rev.)
9. Shiou-Jyh Hwu, S. M. Song, J. Thiel, K. R. Poeppelmeier, J. B. Ketterson and A. J. Freeman, Phys. Rev. B35:7119 (1987).
10. L. E. Toth, M. Osofsky, S. A. Wolf, E. F. Skelton, S. B. Qadri, W. W. Fuller, D. U. Gubser, J. Wallace, C. S. Pande, A. K. Singh, S. Lawrence, W. T. Elam, B. Bender, and J. R. Spann, submitted to Proceeding of Journal of Chemistry, the American Chemical Society meeting, to be held in New Orlean, Aug. 20, 1987.
11. J. A. Yarmoff, D. R. Clarke, W. Drube, U. O. Karlesson, A. Taleb-Ibrahimi, and F. J. Himpsel, private communication.
12. R. L. Kurtz, R. L. Stockbauer, D. Mueller, A. Shih, L. E. Toth, M. Osofsky and S. A. Wolf, to appear in Phys. Rev. B., Rapid Communication, June 1, 1987.
13. P. D. Johnson et. al., to appear in Phys. Rev. B, Rapid communication, June 1, 1987.
14. F. C. Brown, T. C. Chiang, T. A. Friedmann, D. A. Ginsberg, G. N. Kwawer, T. Miller, M. G. Mason, submitted for publication.
15. L. F. Mattheiss and D. R. Hamann, submitted for publication.
16. S. Massidda, Jaejun Yu, A. J. Freeman, D. D. Koelling, Submitted for publication.
17. H. Krakauer and W. E. Pickett, Noval Mechanisms of Superconductivity, ed. by S. A. Wolf and V. Z. Kresin, this issue.
18. E. W. Plummer and W. Eberhardt in "Angle-Resolved Photoemission as A Tool For Study of Surfaces", Advances in Chemical Physics XLIX:533.
19. J. J. Rhyne, J. Gotaas, F. Beech, L. Toth, S. Lawrence, M. Osofsky, and S. A. Wolf, submitted for publication in Phys. Rev. B, Rapid communication.
20. A. M. Kini, U. Geiser, H-C. I. Kao, K. D. Carlson, H. H. Wang, M. R. Monaghan, and J. M. Williams, submitted for publication in Inorg. Chem., communication, (1987).

21. M. R. Thuler, R. L. Benbow and Z. Hurych, Phys. Rev. B26:
    669 (1982)
22. M. Stavola, D. M. Krol, W. Weber, S. A. Sunshine, A.
    Jayaraman, G. A. Kourouklis, R. J. Cava and E. A.
    Rietman, submitted for publication.

EXPERIMENTS ON HEAVY ELECTRON AND HIGH T_c OXIDE
SUPERCONDUCTORS

M. B. Maple, Y. Dalichaouch, J. M. Ferreira,[*] R. R. Hake,[†]
S. E. Lambert,[‡] B. W. Lee, J. J. Neumeier,
M. S. Torikachvili, K. N. Yang, and H. Zhou
Department of Physics and Institute for Pure and
Applied Physical Sciences,
University of California, San Diego, La Jolla, CA 92093

Z. Fisk, M. W. McElfresh, and J. L. Smith
Los Alamos National Laboratory, Los Alamos, NM 87545

INTRODUCTION

Two classes of superconducting materials, heavy electron superconductors (frequently referred to as "heavy fermion superconductors")[1,2] and superconducting oxides,[3] have attracted a considerable amount of attention in recent years because of their extraordinary superconducting properties which may be associated with an unconventional type of superconductivity. During the past year, two types of oxides containing a rare earth element (or yttrium), an alkaline earth element, and copper, have been found to exhibit superconductivity at high temperatures, one type at temperatures as high as ~100 K! These spectacular developments have brought an unprecedented level of excitement and intensity to research on these new oxides and related materials.

The heavy electron materials, compounds of a rare earth element (usually Ce) or an actinide element (usually U), appear to have an enormous density of states $N(E_F)$ at the Fermi level E_F, as inferred from γ, the coefficient of the electronic specific heat $C_e = \gamma(T)T$, which attains values as large as several J/mole-K^2 for some materials at temperatures $T < 1$ K. Alternatively, the electrons can be viewed as having an immense effective mass as large as several hundred times the mass of the free electron. Since the specific heat jump ΔC at the superconducting transition temperature T_c is of the order of $\gamma(T=T_c)T_c$, the same heavy electrons that are responsible for the large γ in the normal state are also involved in the superconductivity. The heavy electron superconductors have low T_c's ($\leqslant 1$ K) and, for such low T_c's, extremely large upper critical magnetic fields H_{c2}

[*]Departamento de Física, Universidade Federal de Pernambuco 50000, Recife, Brasil.
[†]Department of Physics, Indiana University, Bloomington, IN 47405.
[‡]IBM Almaden Research Center, 650 Harry Rd., San Jose, CA 95120-6099.

with anomalous temperature dependences. Superconducting properties such as the specific heat, thermal conductivity, magnetic field penetration depth, ultrasonic attenuation rate, and nuclear spin lattice relaxation rate have power law temperature dependences $\sim T^n$, where n is an integer, suggesting that these materials may exhibit anisotropic superconductivity in which the superconducting energy gap vanishes at points or lines on the Fermi surface.[1] An especially intriguing possibility is that the anisotropic superconductivity involves triplet-spin pairing of electrons,[2] mediated by paramagnon exchange, in analogy with superfluid ^{3}He.

Compared to the heavy electron superconductors, the oxide superconductors have very low values of $N(E_F)$, as inferred from low values of the electronic specific heat coefficient γ. As a result, the specific heat jump ΔC at T_c is very small and barely discernable in some of these materials [e.g., the compound[3] $Ba(Pb_{1-x}Bi_x)O_3$]. In the '70's, superconductivity at moderately high temperatures was reported in two oxide systems, $Li_{1-x}Ti_{2+x}O_4$ with $T_c \sim 14$ K [Ref. 4] and $Ba(Pb_{1-x}Bi_x)O_3$ with $T_c \sim 13$ K [Ref. 5]. Recently, superconductivity at high temperatures has been observed in two copper oxide compounds, $La_{2-x}M_xCuO_{4-\delta}$ (M = Ca, Sr, Ba) which has a maximum $T_c \approx 40$ K for M = Sr and $x \approx 0.15$ [Refs. 6-9], and $RBa_2Cu_3O_{7-\delta}$ which has a $T_c \gtrsim 90$ K for R = Y or a rare earth element (except for Ce, Pr or Tb) [Refs. 10-14]. The unexpectedly large values of T_c displayed by these materials, especially in view of such low values of $N(E_F)$, has led to the speculation that these materials may display an unconventional type of superconductivity, perhaps similar in nature to that of the heavy electron compounds.

The feature that is common to the heavy electron superconductors and the high T_c superconducting oxides may be a magnetic mechanism that is responsible for the formation of the superconducting electron pairs (Cooper pairs). Such a magnetic pairing mechanism is, for example, incorporated in the "resonating valence bond" (RVB) model that was recently advanced by Anderson[15] to account for the high T_c superconductivity of the new copper oxide materials. In the RVB model, the superconducting electron pairs are identified with nearest neighbor pairs of Cu^{2+} S = 1/2 electrons in spin-singlet states, as a result of the superexchange interaction, which become mobile when a sufficient number of Cu^{3+} ions are present. Lines of zeroes of the superconducting energy gap on the Fermi surface are also suggested which, along with the antiferromagnetic correlations involving the nearest-neighbor spin-singlets, may indicate a relationship between the RVB and heavy electron superconducting states.

In the first part of this paper, we briefly describe some of our efforts to determine the type of pairing and pairing mechanisms that are associated with the superconductivity of heavy electron materials by means of $H_{c2}(T)$ measurements in Gd-doped $(U_{1-x}Th_x)Be_{13}$ alloys. An account of some of our recent work on the new high T_c copper oxide superconductors is given in the second part of the paper.

UPPER CRITICAL MAGNETIC FIELD MEASUREMENTS ON Gd-DOPED $(U_{1-x}Th_x)Be_{13}$ ALLOYS

A striking T_c vs x phase diagram has been reported[1,16] for the heavy electron system $(U_{1-x}Th_x)Be_{13}$; T_c exhibits a nearly linear decrease with x, a sharp minimum at $x_{min} \approx 0.017$, a broad maximum at $x_{max} \approx 0.025$, and a subsequent decrease with x. For compositions x between ~ 0.017 and ~ 0.04, two features have been observed in the specific heat, the upper one associated with the development of the superconducting state and the lower one corresponding to another phase transition that occurs without destroying the superconductivity.[17] Two possible explanations for the lower peak are that (1) it is associated with a second superconducting phase with a different order parameter (in analogy with superfluid ^{3}He), and (2) that it is due to the formation of an itinerant electron antiferromagnetic state that coexists with superconductivity.

We have made measurements of T_c as a function of pressure P on the $(U_{1-x}Th_x)Be_{13}$ system for $0 \leq P \leq 12$ kbar.[18] The results indicate that there are two distinct superconducting phases in the $(U_{1-x}Th_x)Be_{13}$ system, one in the region $0 \leq x < x_{min}$ (which we refer to as the A phase) and the other at $x > x_{min}$ (B phase), where x_{min} is a function of P. Unfortunately, the $T_c(P)$ measurements yielded no information about the nature of the phase transition in the superconducting state that is responsible for the lower peak in the specific heat for $x > x_{min}$.

In order to probe the nature of the A and B superconducting phases in pure and Th-doped UBe_{13}, we have carried out $H_{c2}(T)$ measurements on $(U_{1-x}Gd_x)Be_{13}$ and $[(U_{0.971}Th_{0.029})_{1-x}Gd_x]Be_{13}$ alloys for various values of x, the results of which are shown in Figs. 1(a) and (b), respectively.[19] The purpose of this experiment was to determine whether the A and B superconducting phases respond differently to Gd^{3+} ions whose S = 7/2 spins would be expected to interact with the spins of the superconducting electrons via the exchange interaction. The data in Fig. 1(a) reveal that the Gd impurities have a strong effect on the $H_{c2}(T)$ curves for pure UBe_{13} (A phase) which becomes more pronounced with increasing concentration of Gd. Specifically, the $H_{c2}(T)$ curves develop an unusual "foot" in low fields and are generally depressed in magnitude and slope in higher fields. In contrast, the data in Fig. 1(b) show that the Gd impurities have virtually no effect on the shape of the $H_{c2}(T)$ curves for $(U_{0.971}Th_{0.029})Be_{13}$ (B phase).

The $H_{c2}(T)$ curves of Gd-doped UBe_{13} reveal that the Gd spins have a strong destructive effect on the superconductivity of pure UBe_{13}. In fact, the shapes of the $H_{c2}(T)$ curves of the Gd-doped UBe_{13} samples can be qualitatively described by the multiple pair breaking theory for a conventional type II superconductor[20,21] if the primary mechanism for "breaking" superconducting electron pairs is the Zeeman interaction of the exchange field H_J associated with the Gd spins with the superconducting electron spins. This is illustrated in Fig. 2(a) where the calculated $H_{c2}(T)$ curves for $H_J = 0$ (solid lines) and $H_J \neq 0$ (dashed lines) are compared to the data. The calculations of the $H_{c2}(T)$ curves, which will be described elsewhere in detail, were based on the following parameters: spin-orbit scattering parameter $\lambda_{so} = 2\hbar/3\pi\tau_{so}k_BT_{co} = 10$, where τ_{so} is the spin-orbit scattering lifetime; exchange interaction parameter $\mathcal{J} = 0.018$ eV [i.e., $H_J(H, T) = x\mathcal{J}(g_J - 1) <J>/2\mu_B$, where x is the concentration of paramagnetic ions, g_J and J are, respectively, the Landé g-factor and total angular momentum of the rare earth ion's Hund's rule ground state], and a temperature dependence of the orbital critical field $H_{c2}^*(T)$ that was chosen so that the pair breaking theory describes the $H_{c2}(T)$ curve of pure

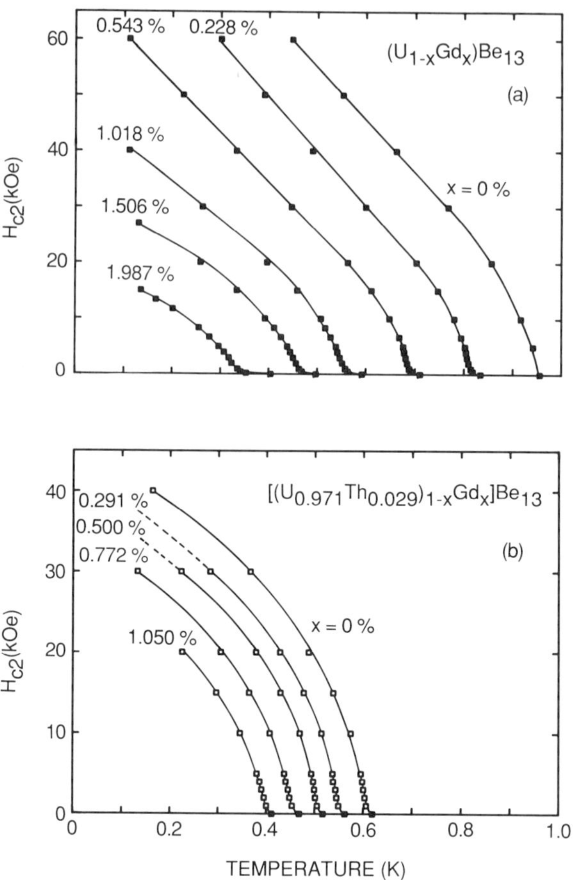

Fig. 1. Upper critical field H_{c2} vs temperature for (a) $(U_{1-x}Gd_x)Be_{13}$ and (b) $[(U_{0.971}Th_{0.029})_{1-x}Gd_x]Be_{13}$. The lines are guides to the eye.

UBe_{13}. The "foot" in the $H_{c2}(T)$ curves could then result from the saturation of the Gd spins and, in turn, the exchange field and its "pair breaking" effect, in low fields and at low temperatures. However, the calculated $H_{c2}(T)$ curves in Fig. 2(a) do not describe the low field data very well, although the discrepancy can be reduced by including a molecular field constant θ as illustrated for the $x = 1.987\%$ data. Nonetheless, within the limitations imposed by the simplifying assumptions that have been made, the multiple pair breaking theory with $H_J \neq 0$ seems to provide a satisfactory qualitative description of the $H_{c2}(T)$ data for Gd-doped UBe_{13}. Thus, the simplest interpretation of the $H_{c2}(T)$ measurements would seem to favor singlet-spin pairing, perhaps of d-wave character, or BW (Balian-Werthamer) type triplet-spin pairing of superconducting electrons in UBe_{13}

The $H_{c2}(T)$ curves of Gd-doped $(U_{0.971}Th_{0.029})Be_{13}$ indicate that the Gd spins have a negligible effect on the superconductivity of $(U_{0.971}Th_{0.029})Be_{13}$. This is depicted in Fig. 2(b) where the data are best described by the calculated $H_{c2}(T)$ curves for $H_J = 0$. The insensitivity of the superconductivity of $(U_{0.971}Th_{0.029})Be_{13}$ to the Gd exchange field suggests that this material may exhibit a qualitatively different type

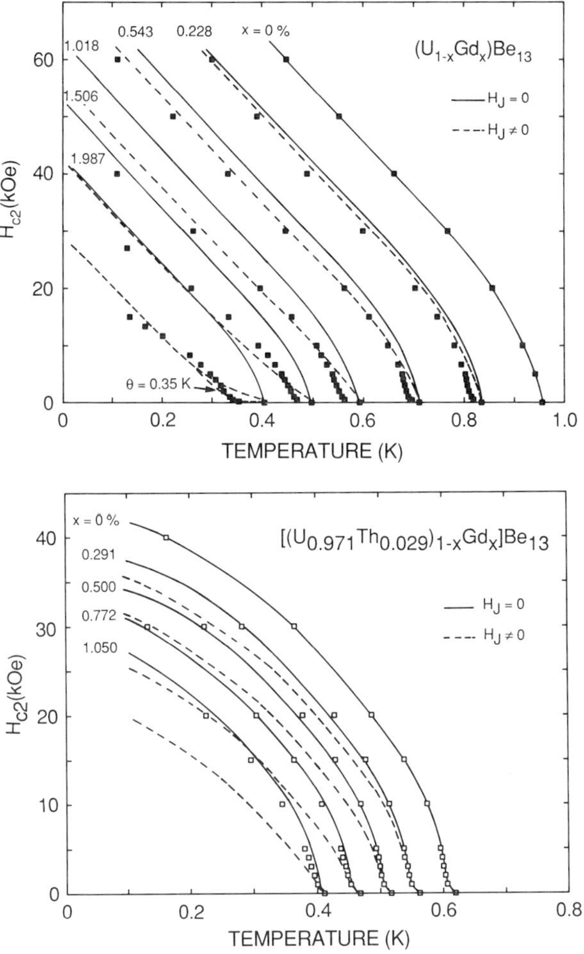

Fig. 2. Upper critical field H_{c2} vs temperature for $(U_{1-x}Gd_x)Be_{13}$ and $[(U_{0.971}Th_{0.029})_{1-x}Gd_x]Be_{13}$, compared to the multiple pair breaking theory for $H_J = 0$ (solid lines) and $H_J \neq 0$ (dashed lines).

of superconductivity, possibly involving ABM (Anderson-Brinkman-Morel) type triplet-spin pairing of superconducting electrons. On the other hand, singlet-spin and BW type triplet-spin superconductivity in $(U_{0.971}Th_{0.029})Be_{13}$ would also be insensitive to the Gd exchange field in the presence of sufficiently strong spin-orbit scattering, although unphysically large values of λ_{so} would be necessary ($\lambda_{so} \gtrsim 300$, corresponding to a spin-orbit scattering mean free path $\ell_{so} \lesssim 0.02$ Å).

HIGH T_c COPPER OXIDE SUPERCONDUCTORS

Recently, we have been involved in a detailed study of the physical properties of the $RBa_2Cu_3O_{7-\delta}$ compounds where R is rare earth element (except for Pm), in addition to $YBa_2Cu_3O_{7-\delta}$. We found that all of the compounds exhibit superconductivity, except for R = Ce, Pr and Tb.[14,22,23] Shown in Figs. 3(a) and (b) are electrical resistivity ρ,

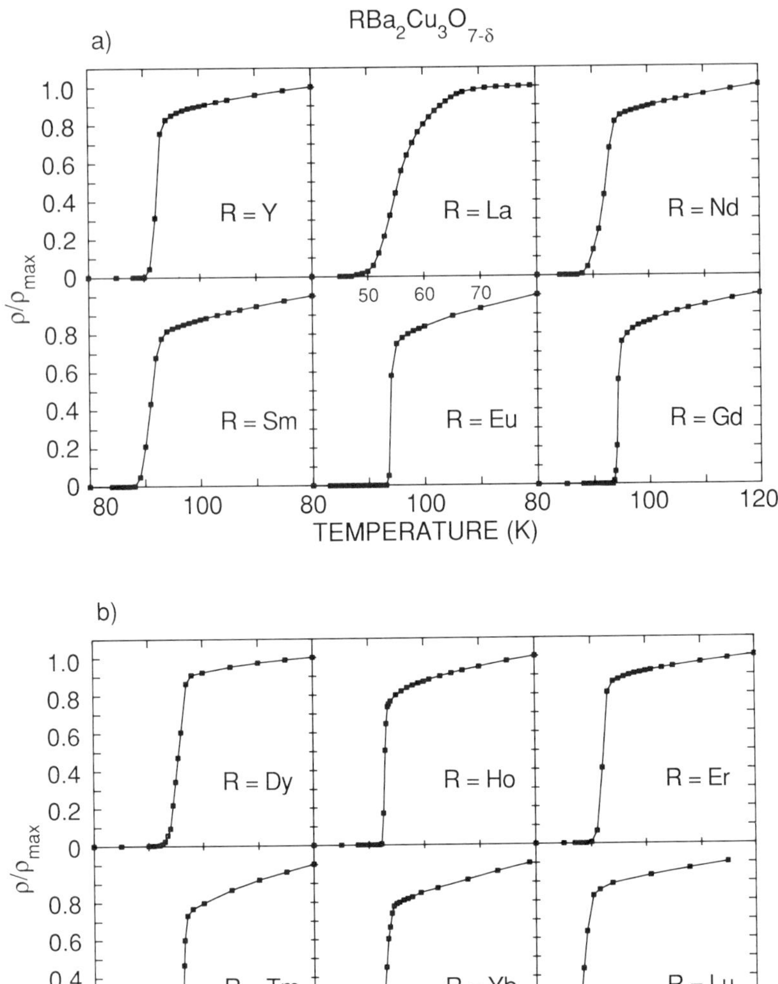

Fig. 3. Electrical resistivity ρ, normalized to its value at 120 K (80 K for La) for RBa$_2$Cu$_3$O$_{7-\delta}$ compounds with (a) R = Y, La, Nd, Sm, Eu and Gd, and (b) R = Dy, Ho, Er, Tm, Yb and Lu.

normalized to its value at 120 K (80 K for R = La), vs temperature data for the superconducting RBa$_2$Cu$_3$O$_{7-\delta}$ compounds. The ratio of the room temperature value of ρ to the value of ρ right above T_c ranges from ~1.3 for R = Dy to ~3.3 for R = Yb, and it increases when the sample quality is improved. The superconducting transition temperature T_c, defined by the temperature at which ρ drops to 50% of its extrapolated normal state value, is typically between 90 K and 94 K, while the transition width ΔT_c, defined by the temperatures at which ρ drops to 10% and 90% of its extrapolated normal state value, is between 2 K and 5 K, except for LaBa$_2$Cu$_3$O$_{7-\delta}$ for which $T_c \approx 60$ K and $\Delta T_c \approx 11$ K. In general, the more metallic samples, as indicated by a larger resistivity ratio, have narrower transition widths. The particularly low T_c for $LaBa_2Cu_3O_{7-\delta}$ is consistent with other data,[24] although there is at least one report of a T_c onset of $\sim$90 K for this compound.[25]

Plots of the inverse susceptibility χ^{-1} vs temperature are presented in Fig. 4 for $RBa_2Cu_3O_{7-\delta}$ compounds with R = Ce, Pr, Nd and Sm and in Fig. 5 for $RBa_2Cu_3O_{7-\delta}$ compounds with R = Gd, Tb, Dy, Ho, Er, Tm and Yb. The lines in Figs. 4 and 5 represent fits of the $\chi(T)$ data with the sum of a constant Pauli-like contribution and a Curie-Weiss term, i.e.,

$$\chi(T) = \chi_0 + N\mu_{eff}^2 / 3k_B(T-\theta) \tag{1}$$

where N is the Avogadro's number, μ_{eff} is the effective moment, and θ is the Curie-Weiss temperature. Values of χ_0, μ_{eff} and θ obtained from fits of Eq. (1) to the $\chi(T)$ data are listed in Table 1. The experimental values of μ_{eff} are in reasonable agreement with the theoretical R^{3+} free ion Hund's rule values of μ_{eff}, which are also given in Table 1. The weakly temperature dependent Van Vleck susceptibility of $EuBa_2Cu_3O_{7-\delta}$ (not shown in Figs. 4 or 5) reveals that Eu is trivalent in this compound with a J = 0 ground state. The magnetic susceptibility of the $RBa_2Cu_3O_{7-\delta}$ compounds of the ions R = Y, La and Lu with empty or filled 4f electron shells can also be fitted by Eq. (1), yielding the values of χ_0, μ_{eff}, and θ given in Table 1, although it is not clear whether the Curie-Weiss contribution is due to magnetic impurities or is actually an intrinsic behavior of the compounds.

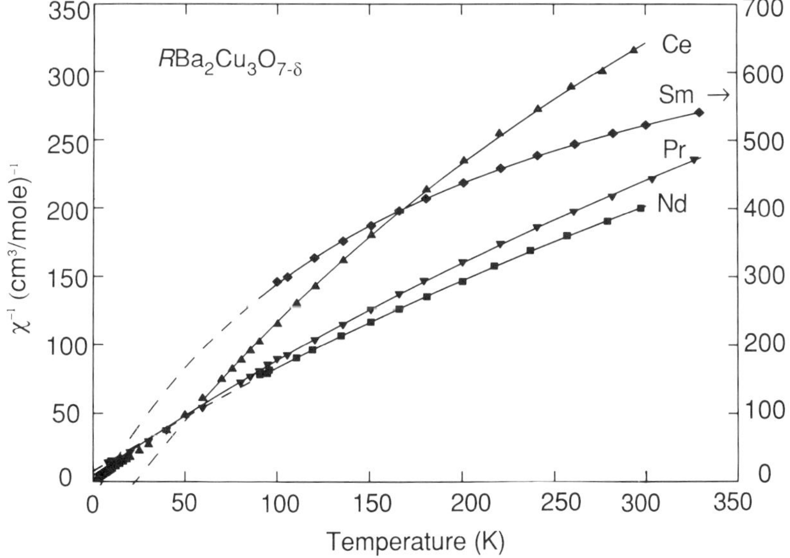

Fig. 4. Inverse magnetic susceptibility χ^{-1} vs temperature for $RBa_2Cu_3O_{7-\delta}$ compounds with R = Ce, Pr, Nd and Sm.

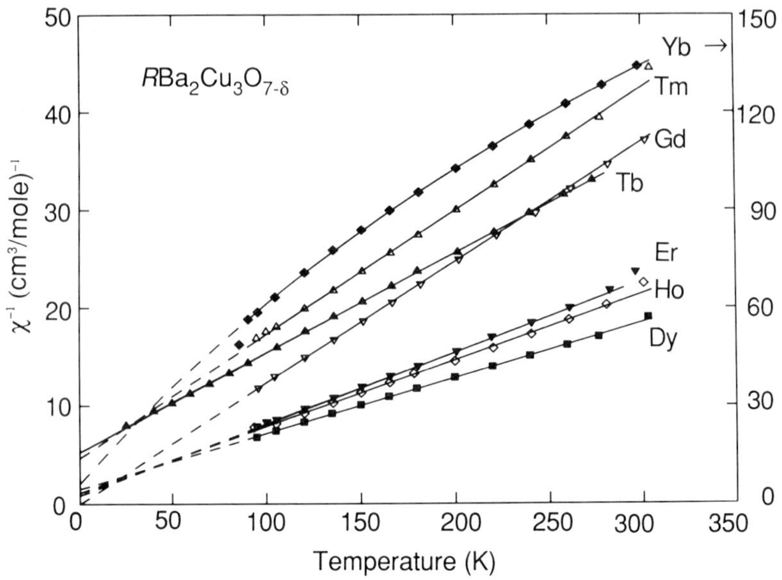

Fig. 5. Inverse magnetic susceptibility χ^{-1} vs temperature for $RBa_2Cu_3O_{7-\delta}$ compounds with R = Gd, Tb, Dy, Ho, Er, Tm and Yb.

Table 1. Magnetic Susceptibility of $RBa_2Cu_3O_{7-\delta}$ Compounds
$\chi = \chi_0 + C/(T-\theta)$; $C = N\mu_{eff}^2/3k_B$.

R	χ_0 (cm^3/mole)$\times 10^{-4}$	$\mu_{eff}^{ex}(\mu_B)$	$\mu_{eff}^{th}(\mu_B)$	θ (K)
Y	3.408	0.545	-	4.41
La	0.988	1.48	-	25.4
Ce	10.03	2.16	2.54	22.6
Pr	0.986	2.94	3.58	-5.25
Nd	1.08	3.10	3.62	-9.64
Sm	1.18	1.32	0.84	4.98
Gd[*]	-	8.05	7.94	0
Tb[*]	-	8.89	9.72	-52.4
Dy[*]	-	11.87	10.63	-27
Ho[*]	-	10.88	10.60	-17
Er[*]	-	10.48	9.59	-12
Tm[*]	-	7.96	7.57	-37
Yb	25.6	3.48	4.54	-9.49
Lu	-	2.11	-	18.5

[*]Only the Curie-Weiss term $C/(T-\theta)$ was used to fit Eq. (1) to the data.

Shown in Fig. 6 are specific heat C divided by T vs T data[22] for the related compound $La_{1.8}Sr_{0.2}CuO_{4-\delta}$ in the range $0.5\ K < T < 60\ K$. Low field magnetization measurements revealed a superconducting transition with an onset at 38 K for this sample.[22] However, since $C(T)$ is dominated by the lattice contribution near T_c, it was not possible to extract the change in the electronic specific heat due to the superconducting transition from the data of Fig. 6, although the data do show a gradual change of slope over $\sim 3\ K$ near 35 K. The C/T data below T_c (30 K < T < 35 K) and above T_c (35 K < T < 42 K) were then linearly extrapolated to 35 K, respectively, and yielded a difference $\Delta C/T_c \cong 17\ mJ/mole\text{-}K^2$. This value is comparable to the values obtained from other specific heat measurements on superconducting $La_{1-x}Sr_xCuO_{4-\delta}$ compounds.[26-28] The lattice contribution $C_\ell = \beta T^3$ can be estimated from the low temperature $C(T)$ data where the exponentially activated superconducting contribution becomes negligibly small. Shown in the inset of Fig. 6 is a plot of C/T vs T^2 at low temperatures which can be described by the equation

$$C/T = \gamma + \beta T^2 \qquad (2)$$

in the temperature range $1.5\ K \le T \le 9\ K$ with values for γ and β of 3.35 mJ/mole-K^2 and 0.256 mJ/mole-K^4, respectively. We do not know whether the non-zero γ is associated with part of the sample that remains normal down to the lowest temperatures, or whether it is an intrinsic feature of these extraordinary superconductors, e.g., due to vanishing of the superconducting energy gap over part of the Fermi surface. The Debye temperature θ_D calculated from β equals 376 K. Another interesting feature is the deviation of the specific heat from the βT^3 Debye behavior for $T > 10\ K$.

Shown in Fig. 7 are C/T vs T^2 data for the 90 K superconductor $YBa_2Cu_3O_{7-\delta}$ for $T < 60\ K$. The C/T vs T^2 data at low temperatures

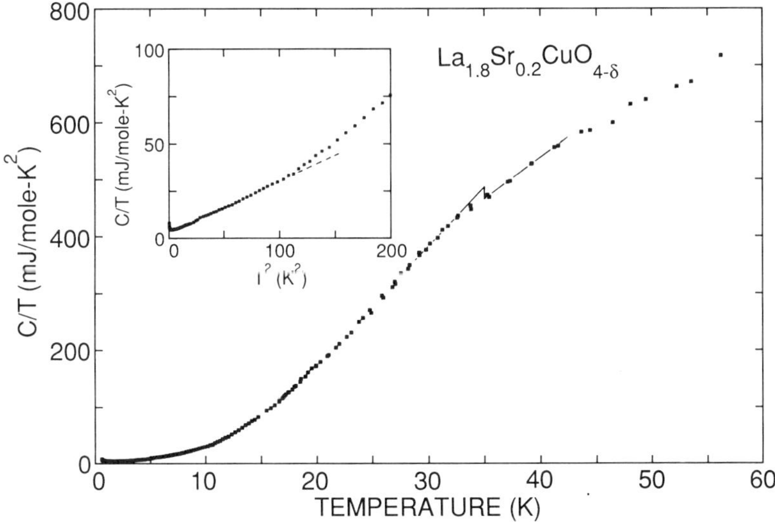

Fig. 6. Specific heat C divided by temperature T vs temperature for $La_{1.8}Sr_{0.2}CuO_{4-\delta}$. Inset: C/T vs T^2 below 14 K.

plotted in the lower inset of Fig. 7 can be described by Eq. (2) in the interval 6 K $\leq$ T $\leq$ 12 K with the parameter values $\gamma = 8.21$ mJ/mole-K^2 and $\beta = 0.449$ mJ/mole-K^4. The value of β corresponds to a Debye temperature θ_D of 383 K, which is very close to θ_D for La$_{1.8}$Sr$_{0.2}$CuO$_{4-\delta}$. An upturn in C/T for T $\leq$ 6 K can be seen in the inset which may be associated with magnetic impurity phases in the sample. A plot in the upper inset of Fig. 7 of the difference ΔC between the C(T) data and the calculated C(T) according to Eq. (2) vs T reveals a small peak in ΔC(T) at T $\cong$ 2 K which is characteristic of magnetic order. Two features of the specific heat of YBa$_2$Cu$_3$O$_{7-\delta}$ are similar to those observed in the specific heat of La$_{1.8}$Sr$_{0.2}$CuO$_{4-\delta}$, the non-zero γ in the superconducting state and the deviation of C(T) from the βT^3 Debye behavior for T $>$ 12 K.

Shown in Fig. 8 are C(T) data for HoBa$_2$Cu$_3$O$_{7-\delta}$, TmBa$_2$Cu$_3$O$_{7-\delta}$, and, for comparison, YBa$_2$Cu$_3$O$_{7-\delta}$. The peak in C(T) for the Ho compound at T $\approx$ 5 K appears to be an electronic Schottky anomaly associated with the crystalline electric field splitting of the Ho^{3+} J = 8 Hund's rule groundstate. There is no trace of a Ho nuclear Schottky anomaly in the C(T) data, which is often observed for Ho compounds, indicating that the electronic ground state of Ho is probably a nonmagnetic singlet. The C(T) data for TmBa$_2$Cu$_3$O$_{7-\delta}$ have no distinguishable peaks at low temperature, although the specific heat is several times larger than that of YBa$_2$Cu$_3$O$_{7-\delta}$ in this temperature region. The ground state of Tm in the crystalline electric field is probably a nonmagnetic singlet in this compound.

Upper critical field measurements in magnetic fields H of up to 9 tesla have been carried out on RBa$_2$Cu$_3$O$_{7-\delta}$ compounds where R = Y, Eu, Gd, Dy, Ho, Er, Tm, and Yb. These data were obtained by measuring

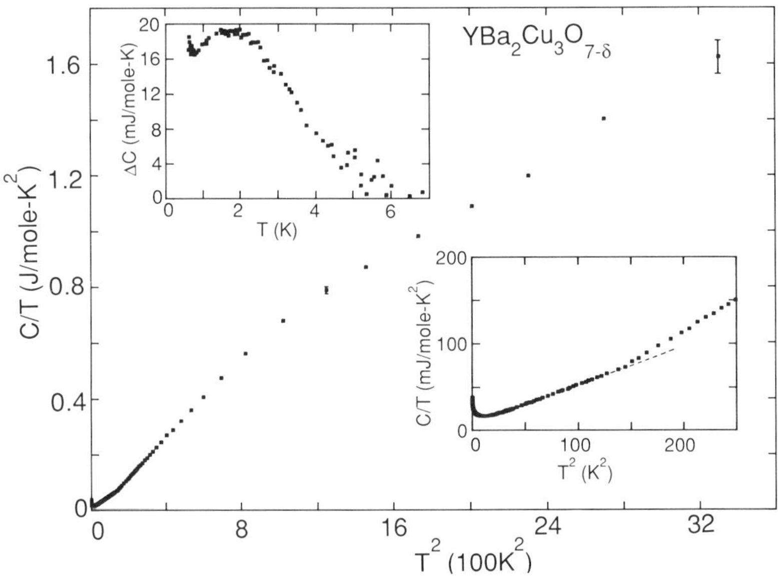

Fig. 7. Specific heat C divided by temperature T vs T^2 for YBa$_2$Cu$_3$O$_{7-\delta}$. Lower inset: C/T vs T^2 below 16 K. Upper inset: ΔC vs T below 7 K (ΔC is defined in text).

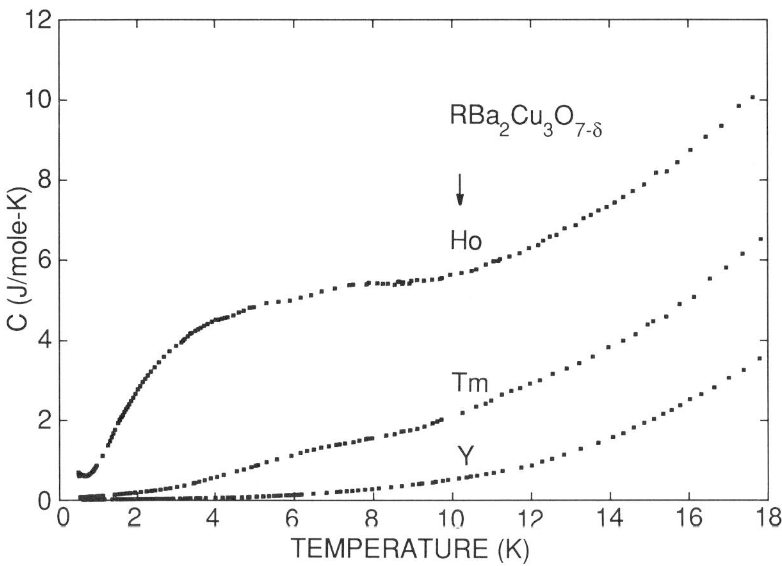

Fig. 8. Specific heat C vs temperature for RBa$_2$Cu$_3$O$_{7-\delta}$ compounds with R = Ho, Tm and Y.

the effect of H on the resistive transition between the normal and superconducting states. In Fig. 9 curves of ρ vs temperature for various applied fields are displayed for R = Tm and Y.[29] From these resistivity curves we extract the temperatures at which the resistivity drops to 0.5 of the extrapolated normal state resistivity for T $\leq$ 100 K; these temperatures vs the respective magnetic fields are plotted in Fig. 10 (square symbols). If we use the slope of a straight line drawn through these data and make extrapolations based on the standard, three dimensional, type II, dirty-limit Werthamer, Helfand, Hohenberg, and Maki (WHHM)[20] theory we obtain the remaining portion of the curves illustrated in Fig. 10.[29] Curves for minimum (λ_{so} = ∞) and maximum (λ_{so} = 0) paramagnetic limitation are shown, where λ_{so} is the spin-orbit coupling parameter. The extrapolations for R = Tm(Y) provide H$_{c2}$ (T = 0 K) = 99-175 tesla (60-71 tesla) with H$_{c2}$ (T = 77 K) $\approx$ 36 tesla (14 tesla). These extrapolations do not take into account the temperature variation of the normal-state electrical resistivity.[29] Included in this figure is the measured[30] upper critical field curve of PbGd$_{0.2}$Mo$_6$S$_8$ with its present record of H$_{c2}$ (T = 0 K) $\approx$ 60 tesla. Figure 10 represents a rather dramatic view of the "new world of superconductivity" unfolded with the discovery of superconductivity in RBa$_2$Cu$_3$O$_{7-\delta}$ compounds.

The RBa$_2$Cu$_3$O$_{7-\delta}$ compounds with R = Ce, Pr and Tb apparently fail to become superconducting because these R elements are not trivalent in this material. While PrBa$_2$Cu$_3$O$_{7-\delta}$ forms in a tetragonal phase closely related to the orthorhombic phase of YBa$_2$Cu$_3$O$_{7-\delta}$, the Ce and Tb counterparts form in a quite different crystal structure, which has not yet been identified. An attempt to substitute small amounts of Ce, Pr and Tb for Y in YBa$_2$Cu$_3$O$_{7-\delta}$ yielded multiphase samples for Ce and Tb substitutions, and single phase Y$_{1-x}$Pr$_x$Ba$_2$Cu$_3$O$_{7-\delta}$ compounds for Pr substitutions. Measurements of ρ(T) in Y$_{1-x}$Pr$_x$Ba$_2$Cu$_3$O$_{7-\delta}$ ($0 \leq x \leq 0.6$)

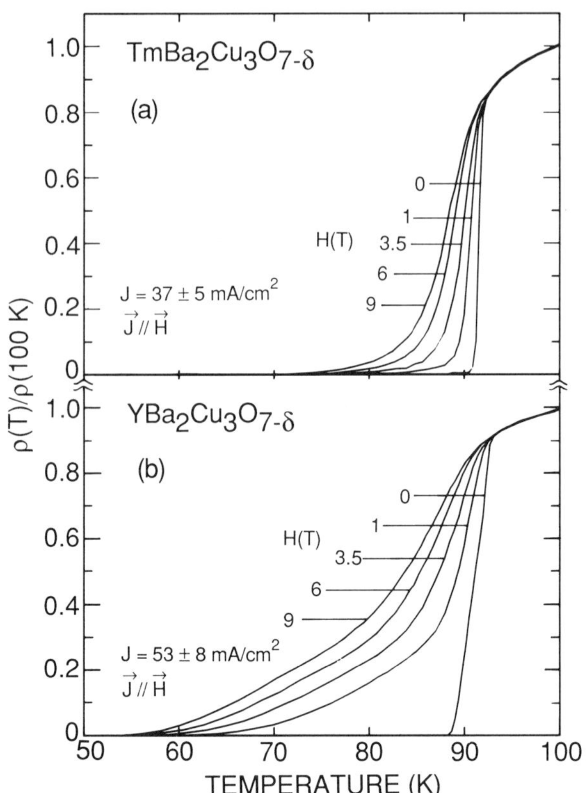

Fig. 9. Normalized resistivity ρ vs temperature for TmBa$_2$Cu$_3$O$_{7-\delta}$ and YBa$_2$Cu$_3$O$_{7-\delta}$ in various applied magnetic fields between 0 and 9 T.

compounds (not shown) reveal a gradual transition from metallic to semi-conducting behavior as Pr is substituted for Y. The resistive superconducting transition curves broadened considerably upon substitution of Pr for Y, and T_c was depressed from 93 K at x = 0 to 34 K at x = 0.5, as shown in Fig. 11, where the data points indicate transition midpoints and the vertical bars were taken between 10 and 90% of the resistive transition from the normal to the superconducting state.

Measurements of the electrical resistance under quasi-hydrostatic pressures to 150 kbar were performed on YBa$_2$Cu$_3$O$_{7-\delta}$, and a plot of the normalized resistance R/R (120 K) vs T at several pressures is shown in Fig. 12. The onset of the resistive transition from the normal to the super conducting state increases from $\sim$95 K at 8 kbar to $\sim$106 K at 149 kbar. This result suggests that yet higher values of T_c may be attainable in these oxides and related systems at ambient pressure.

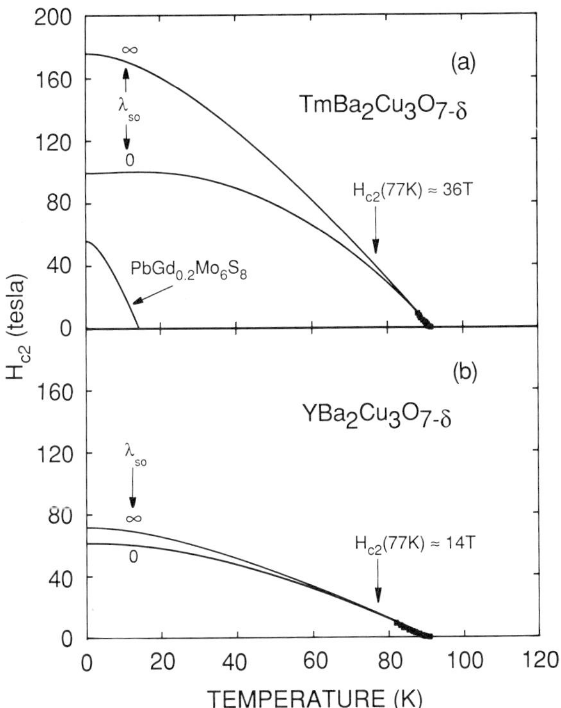

Fig. 10. Upper critical field $H_{c2}(T)$ curves for $TmBa_2Cu_3O_{7-\delta}$ and $YBa_2Cu_3O_{7-\delta}$ as extrapolated from the measured initial slope (heavy lines show extent of present measurements) in accord with standard WHHM theory.

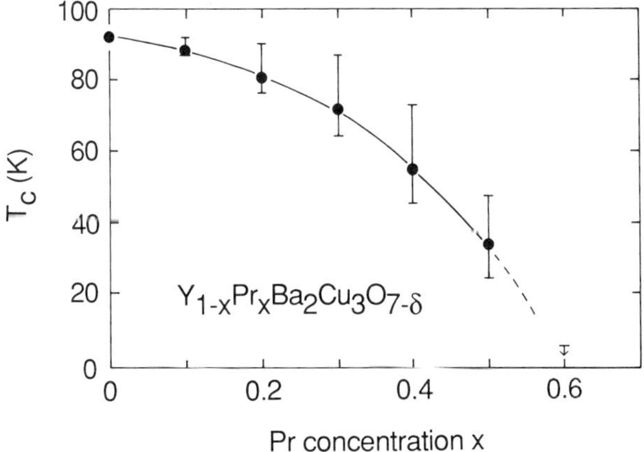

Fig. 11. Superconducting transition temperature T_c vs Pr concentration x for $Y_{1-x}Pr_xBa_2Cu_3O_{7-\delta}$ compounds. The line is a guide to the eye.

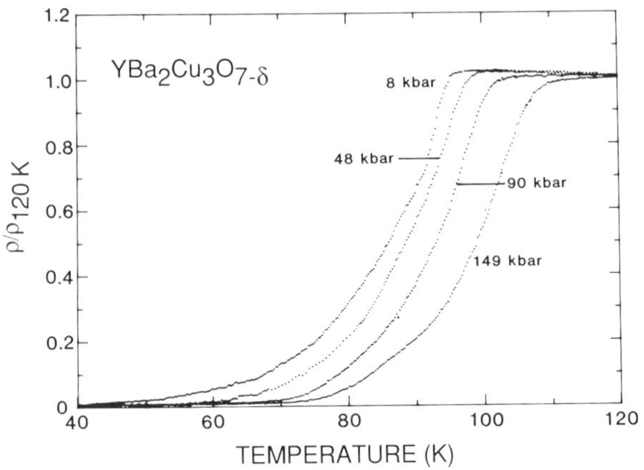

Fig. 12. Resistive superconducting transition curves for
YBa$_2$Cu$_3$O$_{7-\delta}$ at several pressures between 8 kbar and
149 kbar.

ACKNOWLEDGMENTS

This research was supported by the U.S. Department of Energy
under Grant No. DE-FG03-86ER45230. Support from U.S. National
Science Foundation-LowTemperature Physics-Grant No. DMR8411839
(MST and BWL), LLNL branch of UC IGPP (JJN), and CNPq, CAPES,
and FINEP of Brazil (JMF) are gratefully acknowledged. Work at LANL
was performed under the auspices of the U.S. Department of Energy,
Office of Basic Energy Sciences, Division of Materials Science.

REFERENCES

1. For a review, see Z. Fisk, H.R. Ott, T.M. Rice, and J.L. Smith,
   Nature 320, 124 (1984).
2. P.A. Lee, T.M. Rice, J.W. Serene, L.J. Sham, and J.W. Wilkins,
   Comments Condens. Matter Phys. 12, 99 (1986).
3. B. Batlogg, Physica 126B, 275 (1984).
4. D.C. Johnston, H. Prakash, W.H. Zachariasen, and
   R. Viswanathan, Mat. Res. Bull. 8, 777 (1973).
5. A.W. Sleight, J.L. Gillson, and P.E. Bierstedt, Solid State
   Commun. 17, 27 (1975).
6. J.G. Bednorz and K.A. Müller, Z. Physik B64, 189 (1986).
7. H. Takagi, S. Uchida, K. Kitazawa, and S. Tanaka, Japan. J.
   Appl. Phys. 26, L123 (1987).
8. R.J. Cava, R.B. van Dover, B. Batlogg, and E.A. Reitman,
   Phys. Rev. Lett. 58, 408 (1987).
9. J.M. Tarascon, L.H. Greene, W.R. McKinnon, G.W. Hull, and
   T.H. Geballe, Science 235, 1373 (1987).
10. M.K. Wu, J.R. Ashburn, C.T. Torng, P.H. Hor, R.L. Meng,
    L. Gao, Z.J. Huang, Y.Q. Wang, and C.W. Chu, Phys. Rev. Lett.
    58, 908 (1987).
11. R.J. Cava, B. Batlogg, R.B. van Dover, D.W. Murphy,
    S. Sunshine, T. Siegrist, J.P. Remeika, E.A. Rietman, S. Zahurak,
    and G.P. Espinosa, Phys. Rev. Lett. 58, 1676 (1987).

12. R. Beyers, G. Lim, E.M. Engler, R.J. Savoy, T.M. Shaw, T.R. Dinger, W.J. Gallagher, R.L. Sandstrom, preprint submitted to Appl. Phys. Lett., 1987.

13. M.A. Beno, L. Soderholm, D.W. Capone, II, D.G. Hinks, J.D. Jorgensen, I.K. Schuller, C.U. Segre, K. Zhang, and J.D. Grace, preprint submitted to Appl. Phys. Lett., 1987.

14. K.N. Yang, Y. Dalichaouch, J.M. Ferreira, B.W. Lee, J.J. Neumeier, M.S. Torikachvili, H. Zhou, M.B. Maple, and R.R. Hake, to appear in Solid State Commun.

15. P.W. Anderson, Science $\underline{235}$, 1196 (1987).

16. J.L. Smith, Z. Fisk, J.O. Willis, A.L. Giorgi, R.B. Roof, H.R. Ott, H. Rudigier, and E. Felder, Physica (Amsterdam) $\underline{135}$B, 3 (1985).

17. H.R. Ott, H. Rudigier, Z. Fisk, and J.L. Smith, Phys. Rev. B$\underline{31}$, 1651 (1985).

18. S.E. Lambert, Y. Dalichaouch, M.B. Maple, J.L. Smith, and Z. Fisk, Phys. Rev. Lett. $\underline{57}$, 1619 (1986).

19. M.B. Maple, J.W. Chen, Y. Dalichaouch, T. Kohara, S.E. Lambert, B.W. Lee, C. Rossel, M.S. Torikachvili, Z. Fisk, M.W. McElfresh, J.L. Smith, J.D. Thompson, J.O. Willis, and J.W. Allen, in Proc. 5th Int. Conf. on Valence Fluctuations, Bangalore, India, Jan. 5-9, 1987.

20. N.R. Werthamer, E. Helfand, and P.C. Hohenberg, Phys. Rev. 147, 362 (1966); K. Maki, Phys. Rev. $\underline{148}$, 362 (1966).

21. P. Fulde and K. Maki, Phys. Rev. $\underline{141}$, 275 (1966).

22. M.B. Maple, K.N. Yang, M.S. Torikachvili, J.M. Ferreira, J.J. Neumeier, H. Zhou, Y. Dalichaouch, and B.W. Lee, to appear in Solid State Commun.

23. K.N. Yang, Y. Dalichaouch, J.M. Ferreira, R.R. Hake, B.W. Lee, J.J. Neumeier, M.S. Torikachvili, H. Zhou, and M.B. Maple, in Proc. 18th Int. Conf. Low Temp. Phys., Kyoto, Japan, 1987.

24. S.-I. Lee, J.P. Golben, S.Y. Lee, X.-D. Chen, Y. Song, T.W. Noh, R.D. McMichael, Y. Cao, J. Testa, F. Zuo, J.R. Gaines A.J. Epstein, D.L. Cox, J.C. Garland, T.R. Lemberger, R. Sooryakumar, B.R. Patton, and R.T. Tettenhorst, in Proc. Symp. S, 1987 Spring Meeting Mat. Res. Soc., D.U. Gubser and M. Schlutter, eds. (Mat. Res. Soc., Pittsburgh, PA, 1987), p. 53.

25. C.W. Chu, P.H. Hor, R.L. Meng, L. Gao, Z.L. Huang, J. Bechtold, M.K. Wu, and C.Y. Huang, in Proc. Symp. S, 1987 Spring Meeting Mat. Res. Soc., D.U. Gubser and M. Schluter, eds. (Mat. Res. Soc., Pittsburgh, PA, 1987).

26. B.D. Dunlap, M.V. Nevitt, M. Slaski, T.E. Klippert, Z. Sungaila, A.G. McKale, D.W. Capone, R.B. Poeppel, and B.K. Flandermeyer, Phys. Rev. B$\underline{35}$, 7210 (1987).

27. B. Batlogg, A.P. Ramirez, R.J. Cava, R.B. van Dover, and E.A. Rietman, Phys. Rev. B$\underline{35}$, 5340 (1987).

28. M. Decroux, A. Junod, A. Bezinge, D. Cattani, J. Cors, J.L. Jorda, A. Stettler, M. Francois, K. Yvon, $\emptyset$. Fischer, and J. Muller, to appear in Europhysics Letters.

29. J.J. Neumeier, Y. Dalichaouch, J.M. Ferreira, R.R. Hake, B.W. Lee, M.B. Maple, M.S. Torikachvili, K.N. Yang, and H. Zhou, to appear in Appl. Phys. Lett.

30. $\emptyset$. Fischer, H. Jones, G. Bongi, M. Sergent, and R. Chevrel, J. Phys. C$\underline{7}$, L450 (1974).

ELECTRONIC STATES IN HIGH Tc OXIDE SUPERCONDUCTORS

S.Uchida, H.Takagi, T.Hasegawa, K.Kishio, S.Tajima,
K.Kitazawa, K.Fueki and S.Tanaka

Faculty of Engineering
University of Tokyo
Tokyo 113, Japan

INTRODUCTION

During the last several months the superconducting transition temperature (Tc) has revolutionary been raised up to 90K. The high Tc superconductivity has been realized in the Cu-oxides with perovskite-type
structures. Although quite numerous experimental data have been accumulated for these oxide superconductors, it is still far from clear why Tc is
exceptionally so high in the Cu-oxides and what is the role of the rare
earth and the alkaline earth elements which decorate the Cu-O framework and
are not involved in the states near the Fermi level. There have been
proposed almost as many theoretical models of these superconductors as
theorists, as Mott recently pointed out(1). The models are classified into
two categories: one is based on the phonon-mediated attractive force between a pair of electrons and the other invokes non-phonon mechanisms emphasizing the relevance of either spin fluctuations or charge fluctuations
arising from the interelectron Coulomb interaction.

To the above basic questions or to the question which theoretical
models best approximate the actual situation the experiments so far gave no
definite answer. Here we summarize the experimental results, mainly of our
group, going back to another oxides superconductor $Ba(Pb,Bi)O_3$. We cannot
do more than point out what is common with these oxides and which is special about the Cu oxides. The most characteristic feature of the ABO_3
-perovskite related structure is the feasibility of making the substitutional solid solutions for both A and B site cations over wide compositional range and of producing oxygen-deficient structures. So a systematic
study was possible to see the change of the properties with the composition
of the substituted elements for $BaPb_{1-x}Bi_xO_3$ and $(La_{1-x}Sr_x)_2CuO_4$ and with
the amount oxygen vacancies for $YBa_2Cu_3O_{7-y}$.

$Ba(Pb,Bi)O_3$

In most transition metal oxides the O 2p-band is well separated from
the d-band which stabilizes O^{2-} and the positively charged metal ionic
states. The band calculations (2) have indicated that the position of the
Cu 3d-band almost coincides with the O 2p-band in La_2CuO_4 as well as in
$YBa_2Cu_3O_7$. This seems quite accidental, but the same situation occurs also
in $BaPb_{1-x}Bi_xO_3$ (BPB) where the Pb(Bi) 6s-band is degenerate with the O 2p-
band. As a consequence, the states near the Fermi level (E_F) consist of O
2p orbitals hybridized with Pb(Bi) 6s orbital which leads to a strong cou-

pling between the electronic states and the O vibration (3). In BPB the coupling with the oxygen breathing mode phonon is extremely strong so that a CDW state is stablized in the end material $BaBiO_3$. The breathing-mode distortions of the oxygen octahedra surrounding Bi atoms (shown in Fig.1) are accompanied with the charge fluctuation on the Bi site, i.e. $Bi^{4+y}-Bi^{4-y}--Bi^{4+y}-Bi^{4-y}$. The semiconducting properties of $BaBiO_3$ originated from the CDW persist against Pb substitution over a wide composition range $0.35<x<1$ because the coupling with the breathing mode phonon is so strong that a local CDW is stabilized (4.5).

The neutron diffraction studies gave a support for the breathing-mode distortion (6,7) and the optical measurements, particularly the IR spectrum, performed by the authors' group provided direct evidence for CDW formation. The IR spectrum of $BaBiO_3$ shows an additional optical phonon arising from the formation of a superlattice composed of a 3D array of alternating Bi^{4+y} and Bi^{4-y} sites (8). This mode is found to persist over the whole semiconducting region, thus to give an evidence for the local CDW formation. In Raman experiments (9), an extremely strong line is observed at 570 cm^{-1} for $BaBiO_3$ and is assigned to the breathing mode. This Raman line is in strong resonance with the interband excitations across a semiconducting energy gap (2eV) created by the CDW.

The relatively high Tc of 10K range in the region $0.20<x<0.35$ is understood .as the (preserved strong) coupling of the conduction electrons with the optical breathing-mode phonon. The low carrier density which implies ineffective screening of the optical phonons, favors such strong coupling. In this compositional region, the electronic states are greatly modified from the normal metallic states. The optical conductivity exhibits a significant deviation from the Drude behavior and indicates the presence of a pseudogap in the electronic density of states (10). This is envisaged in the spectrum shown in Fig.2 as a manifestation of a superposed absorption peak at a finite energy in the 0.5–1eV range. The peak position is smoothly extrapolated to the center of the interband excitations seen in the semiconducting phase, and therefore it is originated from the strong electron–phonon interaction in this system.

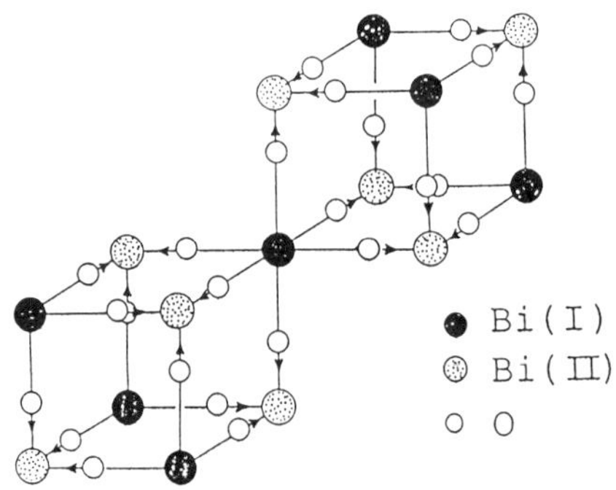

Fig.1 Breathing mode distortion of the O atoms which make alternating arrays of in equivalent Bi(I)–Bi(II) sites in $BaBiO_3$.

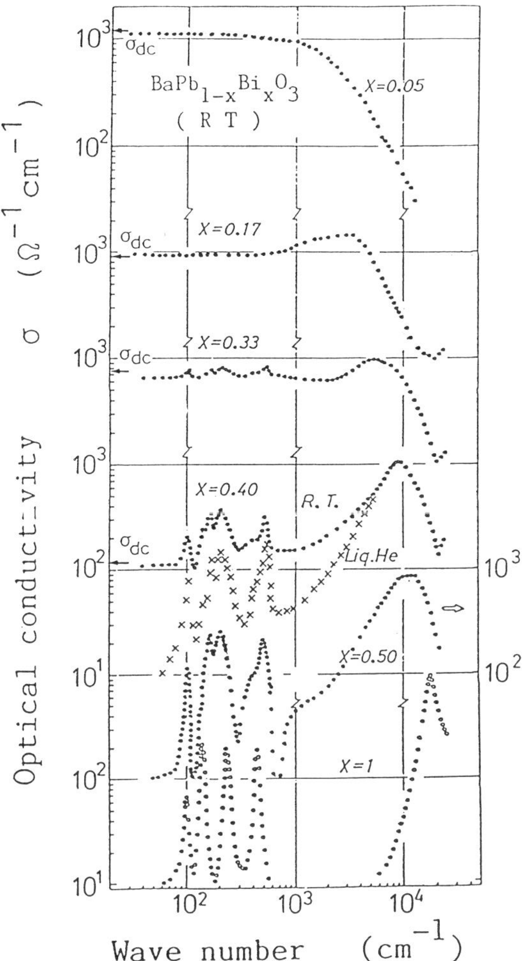

Fig.2 Real part of the optical conductivity for various compositions of $BaPb_{1-x}Bi_xO_3$. DC values of the conductivity for metallic compositions are indicated by arrows.

La_2CuO_4

The band calculation for La_2CuO_4 shows that the conduction band is composed of O_2p strongly hybridized with Cu 3d orbitals and exactly half filled (2). Although there exists some controversy on the semiconductivity of La_2CuO_4 due to the sensitiveness to the oxygen nonstoichiometry (11), it is probable that the ground state of La_2CuO_4 is nonmetallic and is in the vicinity of antiferromagnetism. Actually the antiferromagnetic order was observed by the neutron diffraction measurements (12).

In the context of the band picture, the semiconductivity can be
ascribed to the Fermi surface instability due to the perfect nesting Fermi
surface for the exactly half-filled band. In the Cu-oxides, where the
electrons originated from much less itinerant 3d orbitals, the charge
transfer between Cu sites will cost much more energy than that in BPB.
Therefore, a SDW formation will be more favorable than the CDW in the Cu-
oxides. The recent experimental finding does not support a CDW distortion
in La_2CuO_4--the structural transition at 520K is irrelevant to the
electronic properties at low temperatures. A cusp seen in the magnetic
susceptibility for La_2CuO_4 is assigned to the onset of the SDW order
(Fig.3).

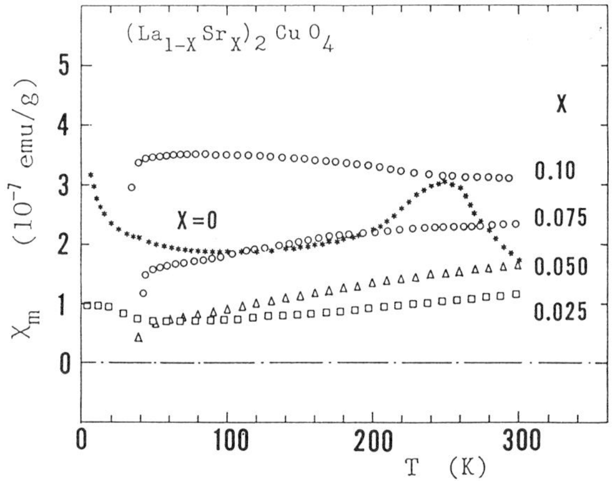

Fig.3 Normal state magnetic susceptibility of
$(La_{1-x}Sr_x)_2CuO_4$ for various compositions measured at the
field of H=5 kOe. A cusp at 250 K for La_2CuO_4 is rapidly
suppressed with increase of x, seen at 10-20 K for x=0.025

However, the experiments of the optical spectroscopies seem to suggest a breakdown of the band picture. The reflectivity spectrum shown in
Fig.4 is quite unusual in view of the band picture. The spectrum was taken
at 300K, above the presumable SDW onset ($\simeq$250K) and no significant change
of the spectrum was seen when the sample was cooled down below 250K.
Moreover, the reflectivity remains low over a wide energy range except in
the low energy region where the phonon structure dominates. This is nothing but the spectrum of nonmetallic medium with a forbidden gap extending
to the energies much higher than expected for a SDW gap ($\sim$ 0.1eV as estimated from the onset temperature).

The result of the reflectivity spectrum seems compatible with the
photo emission result shown in Fig.5. where the result for the Sr-
substituted sample is shown but no significant difference was seen in the
spectrum of La_2CuO_4(13). The occupied valence band states centered around
4eV below E_F are composed of nearly coincident Cu3d and O2p states as expected from the band calculations. The occupied states are, however,

shifted by about 2eV down relative to the band calculations and, as a
result, low state density extends to appreciable energies below E_F. This
fact was interpreted as a consequence of the localization of the Cu 3d and
O2p derived states and of a large on-site Coulomb interaction (14).

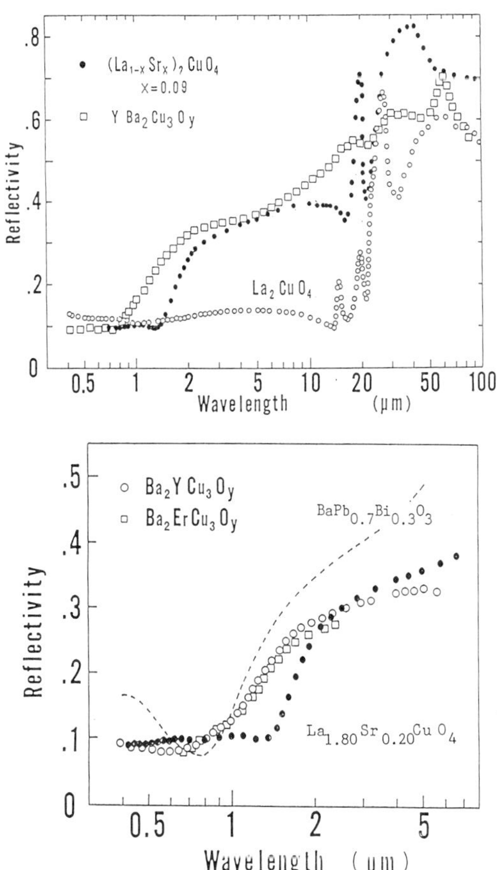

Fig. 4 (a) Reflectivity spectra of La$_2$CuO$_4$ (open circles),
(La$_{0.91}$Sr$_{0.09}$)$_2$CuO$_4$ (closed circles) and YBa$_2$Cu$_3$O$_7$
(open squares).
(b) Plasma reflectivity spectra for Y- and Er-
compounds. Also shown are those for (La,Sr)$_2$CuO$_4$
(closed circles) and Ba(Pb,Bi)O$_3$ (dashed curve).

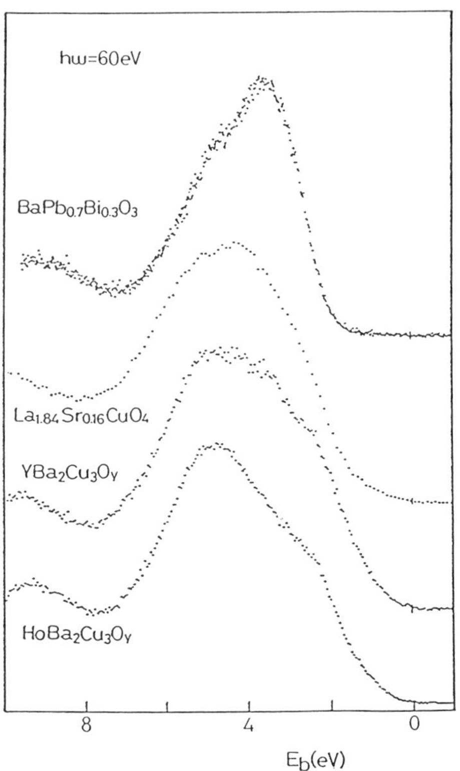

Fig. 5 Photoemission spectra at $h\nu$ =60 eV of three types
of oxide superconductors, Ba(Pb,Bi)O_3 (top),
La$_{1.84}$Sr$_{0.16}$CuO$_4$ (middle) and LnBa$_2$Cu$_3$O$_7$ (bottom).
The samples are a single crystal for BPB and polycrystalline pellets for others. They are cooled in the
vacuum chamber in order to avoid any bake out. The Cu3d
and Pb/Bi6s valence states almost coincide with O2p states
in all these oxides.

$(La,Sr)_2CuO_4$

As shown Fig.s 3 and 6, the cusp in the magnetic susceptibility
rapidly shifts to lower temperature with increase of the Sr composition x
and is not observed when x exceeds 0.025. Simultaneously, the electrical
conduction becomes metallic and a superconducting transition is observed.
The superconducting Tc rapidly rises with further increase of x, reaching
40K at around x=0.1 which coincides with the solubility limit of our series
of specimens. A phase separation is clearly seen for x=0.125 as the superconducting transition in two steps.

The critical change or the insulator-to-metal transition at x=0.025
is understandable as a suppression of a SDW giving way to a superconducting
state (15). In the reflectivity spectrum, a plasma edge becomes marked--in
the case of La$_2$CuO$_4$ only a small trace of the edge is seen at nearly the
same energy. The position of the plasma edge is rather insensitive to the
composition x(16). This is in agreement with the calculation based on the
simple tight-binding band as well as more rigorous LAPW band(17). Associated with this change, the Pauli paramagnetic contribution to the magnetic susceptibility increases, corresponding to the increase in the density of states at E$_F$(N(0)). In addition, a specific heat jump is clearly
observed at Tc for x=0.075 (Fig.7).

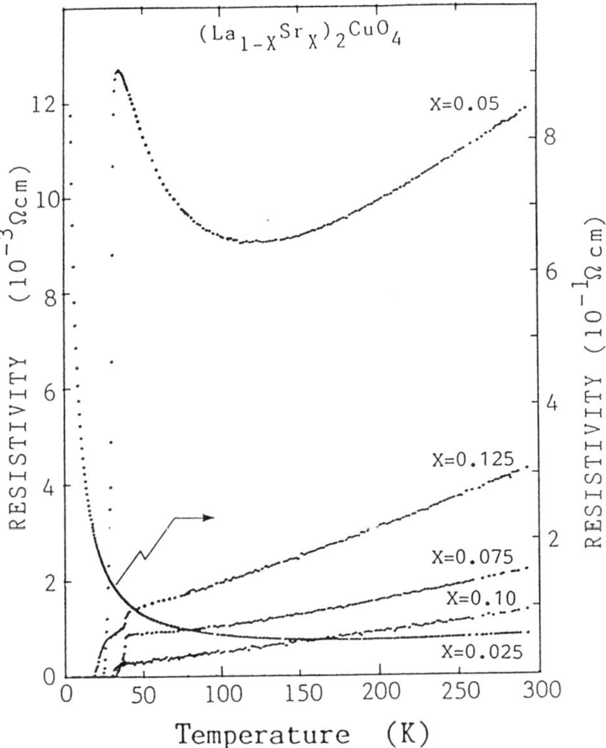

Fig.6 Resistivity vs temperature for four Sr compositions in $(La_{1-x}Sr_x)_2CuO_4$ system. The sample with x=0.0 (right hand scale) is semiconducting, showing no superconducting transition down to 4.2 K.

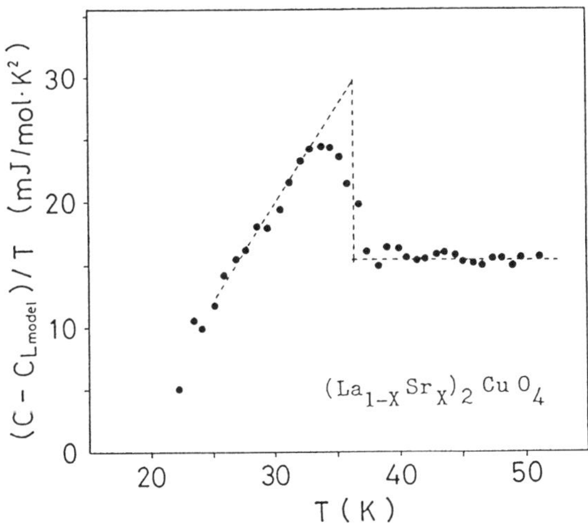

Fig.7 Estimated contribution from electronic specific heat in $La_{1.85}Sr_{0.15}CuO_4$ system. The lattice contribution, C_{Lmodel} is estimated assuming one Einstein mode with θ_E=448 K in addition to a Debye mode with θ_D=240 K
 The electronic specific heat coefficient γ=11 mJ/mol K^2 is determined from a jump at Tc=37K assuming the weak coupling limit.

These results seem to favor the normal metallic picture drawn by the
band calculations for (La,Sr)$_2$CuO$_4$. However, a serious discrepancy exists
in the value of N(0). The density of states estimated from the specific
jump is about 6 states/(eV,Cu-atom) and is coincident with that estimated
from the Pauli paramagnetism. This value is more than twice larger than
the band calculation. In addition, in order to fit the observed plasma
energy, one must assume much narrower band width, 1eV or less, for the conduction band. On the other hand, the UPS spectrum (Fig.5) for (La,Sr)$_2$CuO$_4$
is in contradiction to much high density of states at E_F, and the reflectivity spectrum (Fig.4) is quite unlike the normal metallic type. Even if
we take into account the possible anisotropy with respect to the crystalline c-axis, the plasma spectrum seriously deviates from the Drude like
behavior. As in the case of BPB where the similar situation was met, this
implies that the optical conductivity or absorption cannot be expressed
only by a Drude term but there exists some absorption bands in the energy
region lower than the plasma energy which may be more pronounced in the
spectrum of La$_2$CuO$_4$. The origin of such absorption is not clear in the Cu-
oxides. It might correspond to the Cu-O charge transfer excitations(19) or
to the SDW analogue, i.e. local SDW, to the CDW-BPB or to the Mott-Habbard
gap excitation in the strongly correlated electronic states. In either
case the density of states near Eg might be depressed appreciably, in contradiction to the "enhanced" N(0) observed by the susceptibility and
specific heat measurements.

It has been argued that Tc in the 30-40K range is posible assuming
large N(0) and strong coupling to the high frequency oxygen vibrations(20,
21). However, strong coupling to such phonons is unlikely in the case of
the Cu-oxides, since it is not in the vicinity of the CDW instability unlike BPB. As an evidence, the neutron inelastic scattering did not detect
any soft-mode behavior in the phonon spectrum of (La, Sr)$_2$CuO$_4$ system
LaCuO$_4$(22, 23). The density of states N(0) dose not correlated which Tc in
the Cu- oxides including YBa$_2$Cu$_3$O$_7$. In Fig.8 no remarkable diffrence in
the density of states in identified among three high Tc oxides,
(La,Ca)$_2$CuO$_4$(24), (La,Sr)$_2$CuO$_4$(25) and YBa$_2$Cu$_3$O$_7$(26), while Tc increases so
much from 18K for Ca, 38K for Sr to 92K for Y system. Since the phonon
cut-off frequency cannot be very different, it seems diffcult to interprete
Tc only in the frame work of the phonon mechanism.

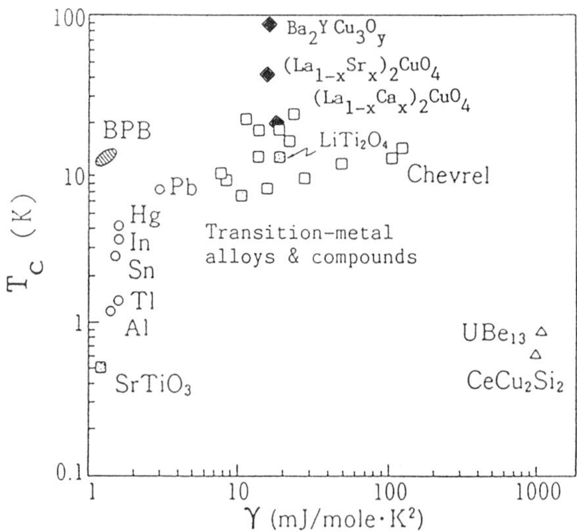

Fig.8 Superconducting critical temperature vs electronic
specific heat coefficient for various superconductors.

The outstandingly high Tc (90K) alone casts a doubt about the phonon mechanism in $YBa_2Cu_3O_7$. The non-phonon mechanisns are favored by the recent observations of the absence of the isotope effect (27) and the insensitiveness to the substitutions of the magnetic rare earth elements for Y (28). The latter is in contrast to the case of the La_2CuO_4-based materials where an appreciable depression of Tc was observed as shown in Fig.9.

The normal-state properties of $YBa_2Cu_3O_7$ are very similar to those of La_2CuO_4 based materials and are almost invariant against the substitution of the rare earth elements. The resistivity of well-qualified sample exhibits a T-linear dependence and its room temperature value is around 10^{-3}ohmcm. The magnetic susceptibility is weakly paramagnetic (10^{-7}emu/g) and shows only a slight T-dependence. The estimated N(0)-value (3states/eV,Cu atom) from the specific heat jump at Tc and from the Pauli paramagnetic susceptibility is nearly equal or rather smaller than those for $(La,Sr)_2CuO_4$. The profile of the photoemission spectrum is hardly distinguishable from other two oxide superconductors shown in Fig.5. Similarly the plasma reflectivity edge is identified in nearly the same energy region and the spectum in the lower energy region is modified from the Drude behavior by the interband excitations, possibly of the same origin as in $(La, Sr)_2CuO_4$. One may conclude that the $YBa_2Cu_3O_7$ system is electronically similar to the alkaline earth substituted La_2CuO_4. The other transport properties for polycrystalline specimens are also much the same both Hall and Seeback coefficients being positive and not so different in magnitude. A difference may be pointed out in the temperature dependence of the Hall coefficient. Whereas its room-temperature value almost coincides with that observed in the La-system, a remarkable increase is observed for the Y-system with lowering temperature (Fig.10). At present we cannot completely rule out the extrinsic origin, such as a small amount of secondary phase contained in our specimens, but the result is suggestive of a modification of the electronic states as indicated in the spectroscopic experiments.

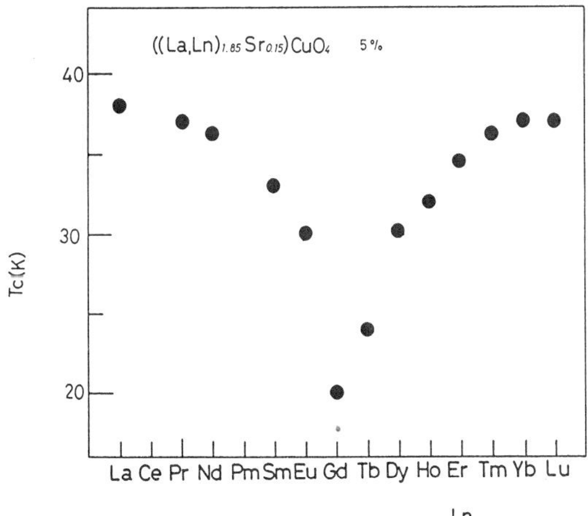

Fig.9 Variation of Tc for various lanthanide ion substitutios for $(La_{0.85}Sr_{0.1}Ln_{0.05})_2CuO_4$.

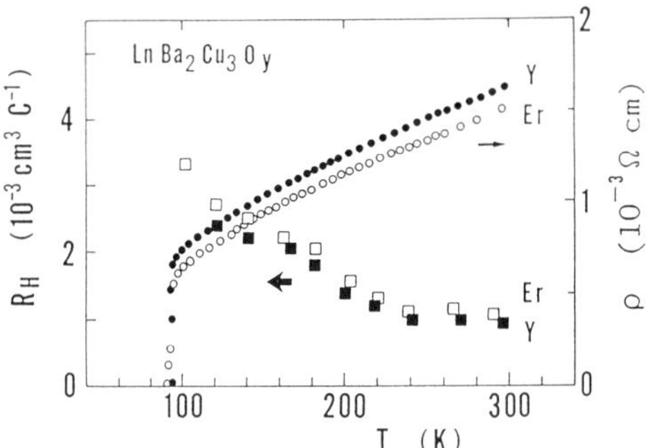

Fig.10 Temperature dependence of the Hall coefficient for
Y- and Er compounds with "1-2-3" structure.

$YBa_2Cu_3O_{7-y}$

Oxgen non-stoichiometry. It has been demonstrated that sufficient oxugen incorporation is essential for obtaining the specimens with good superconecting (29) characteristics, whereas a structural transition from non-superconducting tetragonal to superconducting or thorhombic phases is induced by the oxygen vacancies. These indicate a crucial role of oxgen nonstoichiometry in determining the properties of this oxide system. Therefore, it is dispensable to determine the exact oxygen composition under various conditions of sample preparation and heat treatment. Here we decribe the determination of oxygen nonstoichiometry in details through the thermogravimetry and chemical analysis and present the preliminary experimental results on a few samples with different y.

The specimens of $Ba_2YCu_3O_{7-y}$ were prepared either by a coprecipitation technique or by a simple powder mixing method. The measurements were made using a thermogravimetric electro-microbalance under a controled atmosphere of O_2/Ar gas mixture in the temperature range from 350 to 1000℃. The powder was put into a Pt basket and suspended with Pt wire from the arm of the balance. The absolute value of the oxygen deficiency at several reference compositions were chemically determined by the iodometric titaration technique. The value of oxygen composition was calculated using the analysed mean valence of copper ions, with a charge neutrality condition, assuming Ba^{2+}, Y^{3+} and O^{2-} take fixed valences, respectively.

Fig.11 shows the determined oxygen nonstoichiometry of the $YBa_2Cu_3O_{7-y}$ system versus partial pressure of oxygen at various temperatures. Chemical analysis were made for the specimens which had been cooled from 350 ℃ to room temperature in 1 day in air or 1 atm of oxygen. Any difference between the two different atmospheres were not recognized in the analyzed results and the finally determined oxygen composition was 6.93+0.02.

The change in stoichiometry with temperature and oxygen pressure is clearly depicted in Fig.11. The maximum oxygen deficiency, y, is about 0.9 which is 13% of the whole oxygen content, being quite a large value, compared with ordinary oxide system (30). Beyond this value, the specimen seemed to decompose or to react with the Pt basket at high temperature.

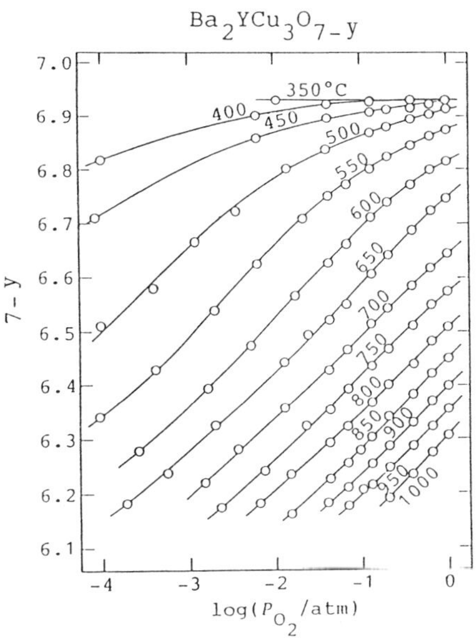

Fig.11 Nonstoichiometry of Ba$_2$YCu$_3$O$_{7-y}$ as a function of partial pressure of oxygen.

It is noted in Fig.11 that the oxygen deficiency y is minimized if specimens are annealed for a prolonged duration in air or in oxygen atmosphere below 350°C . However, to our experience, long time annealing at low temperature around 300°C to 400°C often results in the degradation of the superconductivity. The effect of the degradation is more clearly seen in the resistive superconducting transition than in the magnetic transition. We may presume that the segregation occurs due to the phase instability at lower temperatures.

Figure 12 shows the differential increment of the oxygen deficiency y with temperature, dy/dt, against y in the atmosphere of various oxygen pressures. Although the parameter dy/dT does not directly reflect physical meaning in a thermodynamical sense, it shows quite an interesting behavior. The curves in the figure experience maximum approximately at y=0.32, being independent of the oxygen pressure. Our preliminary DTA/TG measurements also suggested subtle changes in the slopes of DTA/TG traces at nearly the same composition both in air and in oxygen. These results may imply occurrence of a phase transition at this composition of oxygen deficiency. According to high temperature X-ray analyses, Schuller et al.(31) have found that the tetragonal to orthorhombic transition occurs at about 750°C in pure oxygen and Yukino et al.(32) concluded the similar transition in air to occur at 670°C and in N$_2$ at a lower temperature. If these conditions are compared with the present nonstoichiometry data, the transition under pure oxygen is found to occur at y=0.39, and that in air is at 0.42. These values are slightly larger than 0.32, the peak value in Fig.12. Therefore, the peak positions in Fig.12 should frather be interpreted as the compositions where significant changes in lattice parameters a$_0$ and b$_0$ of the orthorhombic phase are observed (31,32).

From Fig.11, one may also note a prominent feature of this transition. Since no discontinuity in y is observed against the oxygen pressure around the transition, it is concluded that this phase transition is not of a first order but of a higher one, in contrast with the previous interpretation by Nakamura et al.(33).

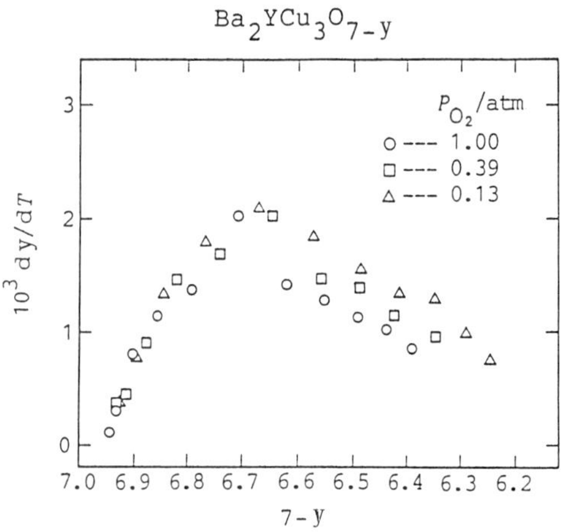

Fig.12 dy/dT vs. y of Ba$_2$YCu$_3$O$_{7-y}$ at various oxygen partial pressure.

It is also interesting to note that no clear change in the slopes of vs. log P_{O_2} is observed at y=0.5. At this composition, the mean valence of copper is calculated to be 2+ and inflection in y may be expected in this plot, if the charge state of Cu^{2+} is such that it is definitely fixed.

Physical propertis. A systematic study is now in progress in order to see the change of the physical properties as y increases. Specimens with various y's was prepared by quenching the sintered pellet into liq. N_2 from various temperatures between 300°C and 900°C.

Figure 13 shows the temperature dependence of the resistivity for the specimens with various y's. As pointed out in the earlier study (34), the transition temperature decreases with increasing y, which is accompanied with the resistivity increase and the appearance of the semiconducting temperature dependence.

The temperature dependences of the Seebeck coefficient are shown in Fig.14. The sign is positive for all the specimens. The Seebeck coefficient is almost constant above Tc and its magnitude increases as y increases. It should be noted that the similar behavior has been observed in $(La,A)_2CuO_4$(35), the seebeck coefficient of which is almost temperature independent and the magnitude decreases from semiconducting La_2CuO_4 to high–Tc metallic phase with doping alkaline earth elements

Temperature dependences of the magnetic susceptibility in the normal state are shown in Fig.15. The specimens quenched from below 400°C, i.e.y=0.1, show the almost temperature independent paramagnetic susceptibility, of the order of 10^{-7} emu/g. At low temperatures below 150K, a slight Curie-like increase with lowering temperatures is observed. As the oxygen deficiency increases, the Curie-like contribution increases and becomes dominant. The temperature dependence can be well fitted by the Curie–Weiss law. The effective moments per Cu atom estimated from the Curie constants are fairly large for the specimens with large y. For example, it reachs as large as 0.7 μ_B /Cu atom for the specimen quenched from 900 C (y is estimated to be about 0.5 from Fig.11).

We cannot detect any secondary phase in the X-ray diffiraction pattern within the instrumental resolution. Since the appearance of such a large magnetic moment cannot be explained even if we assume the presence of the secondary phase of the order of 10% which contains Cu^{2+} ions with definite spins S=3/2, The results indicate that the electrons in $YBa_2Cu_3O_{7-y}$ become lacalized and provide localized magnetic moments as y increases.

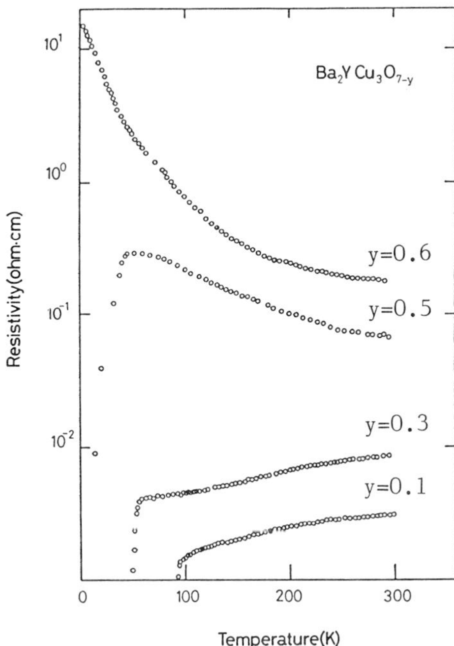

Fig.13 Temperature dependence of the resistivity for $Ba_2YCu_3O_{7-y}$ with various nonstoichiometry y.

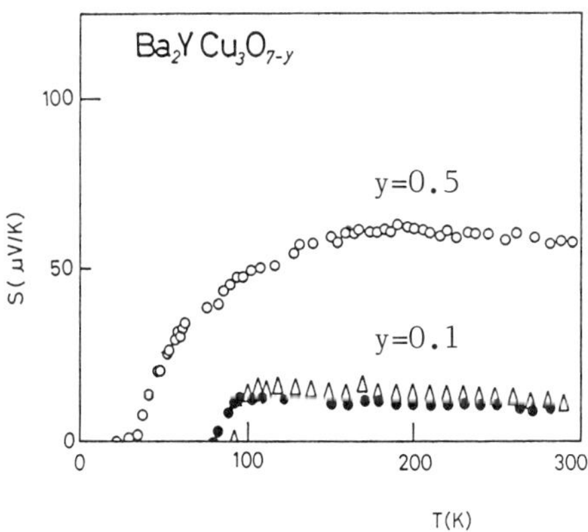

Fig.14 Temperature dependence of the Seebeck coefficient for $Ba_2YCu_3O_{7-y}$ with various nonstoichiometry y.

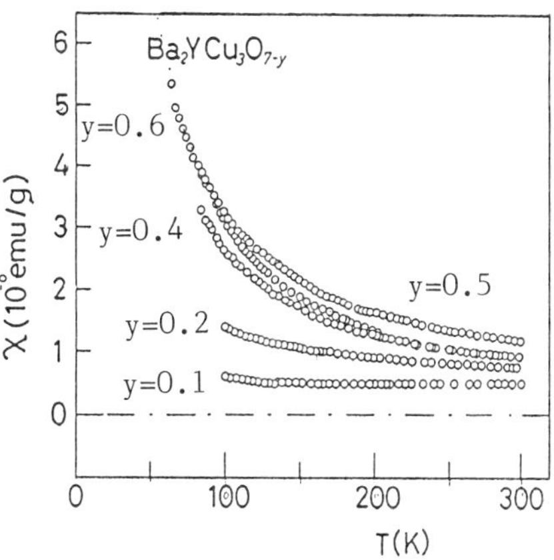

Fig.15 Temperature dependence of the magnetic susceptibility in the normal state for $Ba_2YCu_3O_{7-y}$ with various nonstoichiometry y.

Fujiwara and Hatsugai(36) performed the band calculation, using LMTO method, on the tetragonal $YBa_2Cu_3O_6$ in which oxygen atoms in the center Cu-O layer are completely missing. No significant difference in the gross profile of the conduction bands is observed as compared with $YBa_2Cu_3O_7$ except that the quasi 1-D band derived from the center Cu-O chain is missing. Thus the presence of the such large localized moment cannot be understood within the frame work of the band calculation. It may follow that the correlation between electrons is too strong to be treated in the conventional band theory, as also suggested by the photo emission experiments and that the high Tc superconductivity is located in the vicinity of magnetically ordered state. From this view point, it is suggestive that the almost temperature independent Seebeck coefficient is commonly observed in both $YBa_2Cu_3O_{7-y}$ and $(La,A)_2CuO_4$.

CONCLUSION

We have presented a review of the current activity of the University-of-Tokyo group focusing on unique electronic states in three classes of oxide superconductors. They have similar electroni structure: the valence states of Pb/Bi 6s or Cu 3d nearly coincide with the O 2p states as evidenced by the photoemission spectra and the Fermi level is located in the strongly hybridized anti-bonding band composed of such orbitals. The extensive optical studies provided evidences for the electronic states in BPB affected by the strong coupling of the electrons with the optical breathing mode phonons. Whereas the normal-state resistivity of BPB shows a semiconductor-like temperature dependence, a linear T dependence, typical of metals, is common with well qualified samples of alkaline earth substituted La_2CuO_4 and $YBa_2Cu_3O_7$. The dominant contribution to the magnetic susceptibility of the Cu-oxides is an almost T independent paramagnetic term, also typical of metals, in contrast to the dominant core diamagnetism in BPB. This implies a substantially higher density of states in the Cu-oxides The estimated values of N(0) are in the range of 3-6 states/eV Cu-atom, consistent among various experimental results, the Pauli paramagnetism, a specific heat jump at T_C and the plasma frequencies. These values are rather high as compared with the band calculations, but we cannot find any correlation between T_C and N(0) among various Cu-oxide superconductors as different from the conventional BCS theory. In view of the recent findings, the absence of isotope effect and of soft-phonon behavior, the usual phonon mechanism seems difficult to explain the high T_C superconductivity in the Cu-oxides.

An important insight into the electronic states was obtained from the spectroscopic measurements. The photoemission spectra exhibit very low intensity near the Fermi level and the optical reflectivity spectra cannot be fitted to a Drude term, indicating a modification of the states, as in the case of BPB, due either to the electron-phonon or electron-electron interaction. No evidence was so far obtained which indicates a dominance of the electron-phonon interaction, while the models emphasizing the Coulomb interaction are reinforced by the fact that an antiferromagnetic or SDW order is observed in La_2CuO_4 and that numerous paramagnetic spins appear when the oxygen content is reduced in $YBa_2Cu_3O_7$. However, it is not clear yet if these experimental observations fit to a particular theoretical model so far proposed.

References

1. N. F. Mott, Nature **327**, 185 (1987).
2. e.g. L. F. Mattheiss, Phys. Rev. Lett. **B58**, 1028 (1987);
   K. Takegahara, H. Harima, and A. Yanase, Jpn. J. Appl. Phys. **26**,
   L352 (1987).
3. L. F. Mattheiss and D. R. Hamann, Phys. Rev. **B28**, 4227 (1983).
4. D. Yoshioka and H. Fukuyama, J. Phys. Soc. Jpn. **54**, 2996 (1985).
5. E. Jurczek and T. M. Rice, Europhys, Lett. **1**, 225 (1986);
   E. Jurczek, Phys. Rev. **B35**, 6997 (1987).
6. D. E. Cox and A. W. Sleight, Proc. of the Conf. on Newtron Scattering,
   R. M. Moon (ed.), 1976.
7. C. Chaillout, Doctor Thesis (Grenoble University), 1986.
8. S. Uchida, S. Tajima, A. Masaki, S. Sugai, K. Kitazawa and S. Tanaka,
   J. Phys. Soc. Jpn. **54**, 4395 (1985).
9. S. Tajima, S. Uchida, A. Masaki, H. Takagi, K. Kitazawa, S. Tanaka,
   and S. Sugai, Phys. Rev. **B35**, 696 (1987).
10. S. Tajima, S. Uchida, A. Masaki, H. Takagi, K. Kitazawa, S. Tanaka
    and A. Katsui, Phys. Rev. **B32**, 6302 (1985).
11. e.g. D. C. Johnston, J. P. Stokes, D. P. Coshorn and J. T.
    Lewandowski, preprint
12. Y. Yamaguchi, H. Yamauchi, M. Ohashi, H. Yamamoto, N. Shimoda,
    M. Kikuchi and Y. Shono, Jpn. J. Appl. Phys. **26**, L447 (1987);
    S. Mitsuda, G. Shirane, S. K. Sinha and D. C. Johnston, preprint.
13. N. Nucker, J. Fink, B. Renker, D. Ewert, C. Politis, J. W. P. Weijs
    and J. C. Fuggle, to be published in Z. Phys. B.
14. A. Fujimori, E. Muromachi, Y. Uchida and B. Okai, preprint;
    J. A. Yarmoff, D. R. Clarke, W. Drube, U. O. Karlsson,
    A. Taleb-Ibrahimi and and F. J. Himpsel, preprint.
15. Y. Hasegawa and H. Fukuyama, Jpn. J. Appl. Phys. **26**, L322 (1987).
16. S. Tajima, S. Uchida, S. Tanaka, S. Kanbe, K. Kitazawa and K. Fueki,
    Jpn. J. Appl. Phys. **26**, L432 (1987).
17. P. B. Allen, W. E. Pichett and H. Krakauer, preprint.
18. K. Kitazawa, T. Atake, M. Sakai, S. Uchida, H. Takagi, K. Kishio,
    T. Hasegawa, K. Fueki, Y. Saito and S. Tanaka, Jpn. J. Appl. Phys.
    26, L751 (1987).
19. C. M. Varma, S. Schmitt-Rink and E. Abrahams, Solid State Commun.
    62, 681 (1987).
20. H. Fukuyama and Y. Hasegawa, J. Phys. Soc. Jpn. **56**, (1987).
21. W. Weber, Phys, Rev. Lett. **58**, 1371 (1987).
22. A. Masaki, H. Sato, S. Uchida, K. Kitazawa, S. Tanaka and K. Inoue,
    Jpn. J. Appl. Phys. **26**, L405 (1987)
23. B. Reker, F. Gompf, E. Gerin, N. Nucker, D. Ewert, W. Reichardt and
    H. Rietschel, preprint.
24. K. Kitazawa, M. Sakai, S. Uchida, H. Takagi, K. Kishio, S. Kanbe,
    S. Tanaka and K. Fueki, Jpn. J. Appl. Phys. **26**, L342 (1987).
25. K. Kitazawa, T. Atake, H. Ishii, H. Sato, H. Takagi, S. Uchida,
    Y. Saito, K. Fueki, and S. Tanaka, Jpn. J. Appl. Phys, **26**, L748 (1987)
26. A. Junod, A. Bezinge, T. Graf, J. L. Jorda, J. Muller, L. Antognazza,
    D. Cattani, J. Cors, M. Decroux, . Fischer, M. Banovski, P. Genoud,
    L. Hoffmann, AA. Manuel, M. Peter, E. Walker, M. Francois and K. Yvon,
    to be published in Europhys. Lett.
27. B.Batlogg, R. J. Cava, A. Jayaraman, R. B. van Dover, G. A. Kourouklis,
    S. Sunshine, D. W. Murphy, L. W. Rupp, H. S. Chen, A. White,
    K. T. Short, A. M. Mujsce and E. A. Rietman, Phys. Rev. Lett. **58**,
    2333 (1987); L. C. Bourne, M. F. Crommie, A. Zettl, H-C. zur Loye,
    S. W. Keller, K. L. Leary, A. Stacy, K. J. Chang, M. L. Cohen and
    D. E. Morris, Phys. Rev. Lett. **58**, 2337 (1987).

28. Several groups reported in APS meeting, New York, March 1987.
29. E. Takayama-Muromachi, Y. Uchida, K. Yukino, T. Tanaka and K. Kato, Jpn. J. Appl. Phys. **26**,L665 (1987).
30. P. Kotstad, "Nonstoichiometry, Diffusion and Electrical Conductivity in Binry Metal Oxides", John Wily and Sons, Inc. New York, (1972).
31. I. K. Shuller, D. G. Hinks, M. A. Beno, D. W. Capone II, L. Soderholn, J.-P. Locquet, Y. Bruynseraede, C. W. Segre and K. Zhang, to be published in Solid State Commun.
32. K. Yukino, T. Sato, S. Ooba, M. Ohta, F. P. Okamura and A. Ono, Jpn. J. Appl. Phys. **26**, L869 (1987).
33. K. Nakamura, T. Hatano, A. Matsushita, T. Oguchi, T. Matsumoto and K. Ogawa, Jpn. J. Appl. Phys. **26**, L791 (1987).
34. E. Takayama-Muromachi, Y. Uchida, M. Ishii, T. Tanaka and K. Kato, to be published in Jpn. J. Appl. Phys.
35. S. Uchida, H. Takagi, H. Ishii, H. Eisaki, T. Yabe, S. Tajima and S. Tanaka, Jpn. J. Appl. Phys. **26**, L440 (1987).
36. T. Fujiwara and Y. Hatsugai, Jpn. J. Appl. Phys. **26**, L716 (1987).

FLUX QUATUM AND TUNNEL CHARACTERISTICS OF $La_{2-x}Sr_xCuO_4$).05-0.2)

(x=0.05-0.2) AND $MBa_2Cu_3O_7$ (M-Lu,Y) CERAMICS

N.V. Zavaritsky*, V.N. Zavararitsky**, and S.V. Petrov

*Institute for Physical Problems, Academy of Sciences
**General Physics Institute, Academy of Sciences
Moscow, USSR

ABSTRACT

Macroscopic scale phase coherence in high T_C ceramics was
observed both in magnetic field and/or current superconductivity
destruction and during flux quantization tests. Temperature
dependence of superconducting energy gap was obtained from point
contact tunnel experiments. The maximum values of $2\Delta_0/kT_C$
obtained for $La_{1.85}Sr_{0.15}CuO4$ and $YBa_2Cu_3O_7$ are 5 and 7.5
respectively.

Recently series of high T_C ceramic superconductors have been
discovered[1,2,3]. Here we present some results obtained during
investigation of $La_{2-x}Sr_xCuO_4$ x=0.05-0.2 (LaSr) and
$Y(Lu)Ba_2Cu_3O_7$ (123) ceramics prepared by solid phase chemical
reactions.

In the case of LaSr we have obtained that initial components
of a charge can be used in different forms like oxides, hy-
droxides, oxalates, fluorides, nitrides; and at x=0.15 $T_C\cong42K$ on
the center of the transition curve with δT ~2K was obtained.

At $T>T_C$ LaSr samples resistance changes with temperature
in a different manner depending on its preparation technology:
sharp distinct drop at $T\simeq250$ K (not connected with superconductivity) in the case of excess of C and O (by XPS analysis),
pretty linear resistance drop (up to 3.5 times from 300 to T_C)
for good samples; sharp semiconductor type increase of resistance
the region $T>T_C$ for samples with inclusions.

In the presence of magnetic field transition curve shape
for LaSr drastically changes (fig.1). In small fields B < 0.2T
there is no change in the onset point position, while the temperature of vanishing resistance sharply drops. In higher
magnetic fields the curves are displaced as a whole. Values of
dHc/dT determined from different parts of such curves (for
B > 0.2 T) are presented in the left part of fig.1. As seen from
these results high T_C ceramic consists of superconducting grains
with $dHc_2/dT \simeq 2\pm0.2$ TK^{-1} (for LaSr) connected by Josefson junctions[4].

Superconducting state in LaSr is being destroyed by current
with a value (at 4.2 K) larger than $10^2A/cm^{-2}$. Current desctruc-

tion curves reversibly displaced by external magnetic field $\sim 10^{-2}$ T and displayed hysteresis in larger fields.

In $LuBa_2Cu_3O_7$ ($T_C \simeq 99$ K) at 77 K critical current exceeds 70 A·cm^{-2} and stronlgy depends on magnetic field. Solid and dashed lines in fig.2 are I_C-B characteristics for the sample without and with flux trapped.

Electrical voltage along the sample in a resistive state changes irregularly with field or current(fig.2, right). The oscillations period is approximately equal to 10^{-5}T as for LaSr (curve 1) at 30 K as for $LuBa_2Cu_3O_7$ (curve 2) at 77 K and corresponds to the flux quantization area of $(10^{-3}$cm$)^2$, of the

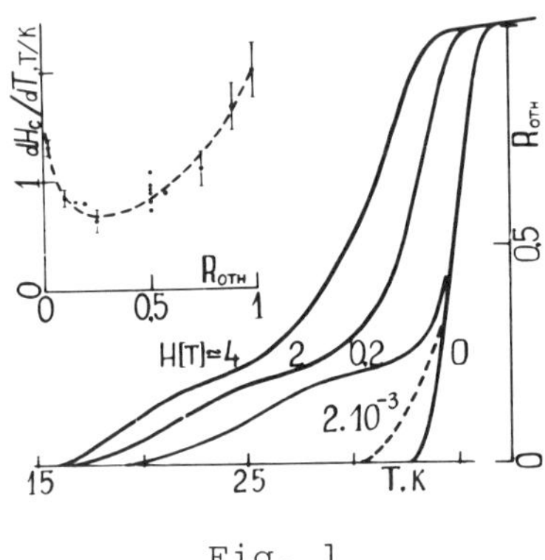

Fig. 1

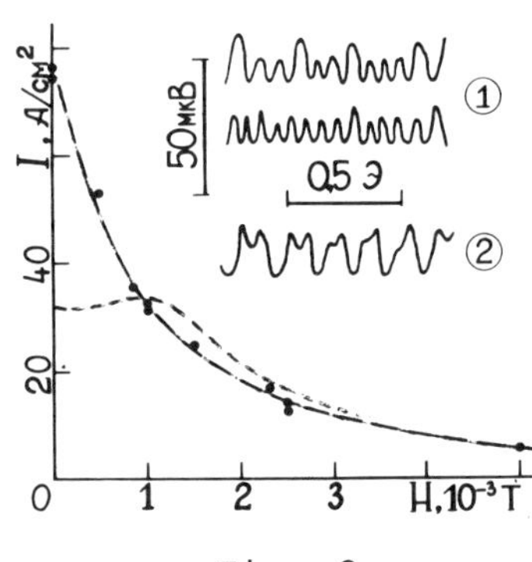

Fig. 2

order of charge particle dimension. Period of $\sim 10^{-9}$T have been observed in a loop of $\sim 10^{-2}$cm^2 with Josefson contacts. These results obviously prove that there is a macroscopic scale phase coherence in a high T_C ceramic.

The characteristics of point contact between two identical superconductors have been investigated from 4.2 K to T_C. The dV/dJ-V chatacteristics for LaSr ceramic at 4.2 K have a typical S-S tunnel junction form (fig. 3, right). The distance between main minima corresponds to 4Δ and gives the value of $2\Delta_0/kT_C \simeq 5$. During heating such characteristics often change (9K-curve, fig. 3) and additional minima appear. The temperature dependence of the positions (V) of the main (black dots) and additional (circuits) minima are presented on the left side of fig. 3.

More complex characteristics were observed in the case of $YBa_2Cu_3O_7$ point contact (fig. 4), the temperature dependence of the minima positions observed here is presented on the right side of fig. 4. In this case $2\Delta_0/kT_C \simeq 7.5$.

The obtained complicated picture seems to occur due to the large gap anysotropy in high T_C ceramic and the ambiguity of the contact, because real junction can arise between superconducting grains of different orientation.

It is of importance that all the peculiarities observed vanish at the sampe temperature T_C like in a usual superconductors [5]. Moreover, all the peculiarities positions can be explained in the assumption of two gap existence in the media - by combinations like $2\Delta_1$, $2\Delta_2$, $\Delta_1+\Delta_2$, Δ_1, Δ_2. All known values obtained by different methods [6] are placed between our Δ_{min} and Δ_{max}.

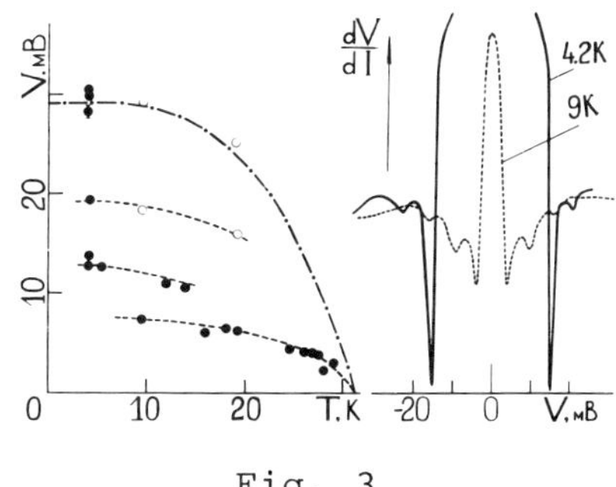

Fig. 3

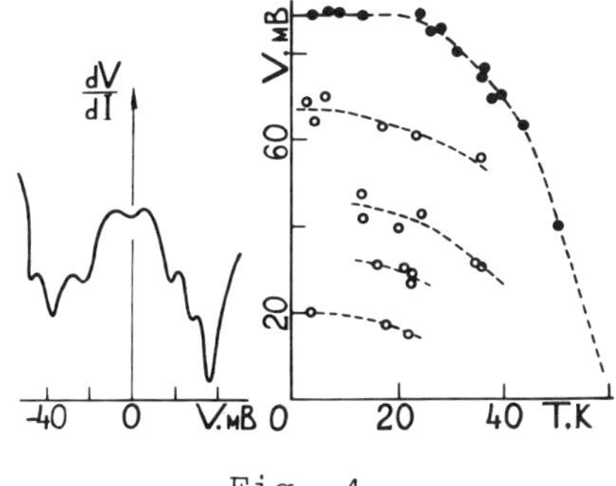

Fig. 4

In both systems LaSr and $LuBa_2Cu_3O_7$ charge transfer in a normal state produced by holes which is evident from thermo-emf α measurements carried out by A.Yurgens; at $T > T_c$ the value of αT^{-1} is equal to $3.3 \cdot 10^{-7}$ VK^{-2} and $1.2 \cdot 10^{-7} VK^{-2}$ for LaSr and $LuBa_2Cu_3O_7$ correspondingly. Large value of αT^{-1} obtained is due to the one order lower amount of carriers in ceramics than in common metals.

REFERENCES

1. J.C. Bednorz, K.A. Muller, Possible high T_c superconductivity in the Ba-La-Cu-O system, Z. Phys., B64: 189 (1986)

2. R.J. Cava, R.B. van Dover, B. Batlogg and E.A. Rietman, Bulk Superconductivity at 36 K in $La_{1.8}Sr_{0.2}CuO_4$, Phys. Rev. Lett., 58:408 (1987).

3. M.K. Wu, J.R. Ashburn, C.J. Torng, P.H.Hor, R.L. Meng, L. Gao, Z.J.Huang, Y.Q. Wang and C.W. Chu, Superconductivity at 93 K in a new mixed phase Y-Ba-Cu-O compound system at ambient pressure, Phys. Rev. Lett. 58:908 (1987).

4. V.M.Vinokur, L.F. Yoffe, A.I.Larkin, M.V.Feigelman, Set of Josefson junctions as a model of spin glass, Sov. JETF, 93:343 (1987).

5. N.V.Zavaritskii, Tunnel effect in a massive tin, Sov. Phys. JETP, 18:1260 (1964).

6. U.Walter, M.S.Sherwin, A. Stacy, P.I. Richards, A.Zettl, Energy gap in the high-T_c superconductor $La_{1.85}Sr_{0.15}CuO_4$, Phys. Rev. Lett. 58:5327 (1987).
J.R. Kirtley, C.C.Tsuei, Sung I. Park, C.C. Chi, J. Rozen, M.W. Shafer, Local tunelling measurements of the high-T_c superconductor $La_{2-x}Sr_xCuO_{4-y}$, Phys. Rev. B35:7216 (1987).
M.E.Hawler, K.E.Gray, D.W.Capone II, and D.C. Hinks, Energy gap in $La_{1.85}Sr_{0.15}CuO_{4-y}$ from point contact tunneling, Phys. Rev. B35:7224 (1987).

EFFECT OF DISORDERING ON THE PROPERTIES OF HIGH-TEMPERATURE

CERAMIC SUPERCONDUCTORS

V. I. Voronin, B. N. Goshchitskii,
S. A. Davydov, A. E. Karkin, V. L. Kozhevnikov,
A. V. Mirmelshtein, V. D. Parkhomenko,
and S. M. Cheshnitskii

Institute of Metal Physics and
Institute of Chemistry
Academy of Sciences of the USSR
Urals Branch, Sverdlovsk, USSR

The physical properties of ordered crystals in many cases are greatly
dependent upon the degree of the crystal lattice. That is why various methods of disordering crystals are widely used to study what role the crystal
structure and its features play in determining the limits of behavior of a
solid. Radiation disordering under the influence of fast-neutron irradiation
is the most "pure," in a physical sense, method for this type of investigation.
It allows one to produce different types of crystal structure defects uniformly distributed over massive crystals in a given concentration and to
follow how they affect electronic and phonon subsystems and related properties.

The fruitfulness of this approach was strikingly manifested, in particular, while studying the nature of superconductivity in intermetallic
compounds with A-15 structure, Laves and Chevrel phases [1]. These investigations, widely used in USA, USSR, FRG, Japan, France, and other countries, made it possible to obtain, along with unique scientific information, important practical data concerning the radiation resistance of prospective materials (i.e., candidates for use in superconducting magnetic
systems of fusion reactors) and ways of raising it.

Here we report the first data on radiation effects in the new ceramic
superconductors $La_{1.83}Sr_{0.17}CuO_{4-y}$ and $R_1Ba_2Cu_3O_{7-y}$ (R = Y, Ho, Er) irradiated by fast neutrons ($E_n > 1$ MeV, $T_{irr} \simeq 350$ K) with flux ranging from
5×10^{17} to 1×10^{19} cm^{-2}. For comparison we have chosen V_3Si, the most
extensively studied superconducting intermetallic compound, about the physical nature and radiation effects of which there is virtually no doubt.

Irradiation was carried out in a water cavity of an atomic reactor in
sealed aluminum tubes placed in cadmium cases. The samples were prepared
by standard ceramic technology from copper, lanthanum, or rare-earth metal
oxides and carbonates of strontium and barium [2, 3]. T_c and H_{c2} were measured by the four-contact resistivity method. The content of a superconducting phase was evaluated from inductive measurements of T_c by the value
of the change in inductance of a coil, calibrated by a lead standard with
an accuracy of ±5%.

Table 1. Characteristics of Unirradiated Samples

No.	Parameter	$La_{0.83}Sr_{0.17}CuO_{4-y}$	$YBa_2Cu_3O_{7-y}$	$ErBa_2Cu_3O_{7-y}$	$HoBa_2Cu_3O_{7-y}$	Method
1	ρ^*, mΩ/cm	1	0.4	2.0	0.7	
2	ρ_{300}/ρ^*	4.0	2.3	1.8	2.5	
3	T_c, K	37	92.7	92.6	93.3	Resistance
4	ΔT_c, K	1	1.4	1.7	0.6	(10-90)%
5	a_0, Å	3.775(1)	3.8206(4)	3.8124(3)	3.8179(6)	
6	b_0, Å	3.775(1)	3.8911(4)	3.8725(2)	3.8807(5)	
7	c_0, Å	13.246(1)	11.673(1)	11.6187(9)	11.642(1)	
8	H'_{c2}, T/K	2.2	1.6	1.4	1.7	Midpoint of T_c
9	Θ_0, K	365 ± 10	375 ± 15	—	—	Heat capcity
10	$\Delta C_p/T_c$, mJ/ g-at·K^2	2.0 ± 0.2	2.5 ± 0.5	3 ± 0.5	3 ± 0.5	Heat capacity
11	$2\Delta_0^{max}/T_c$	3.6-3.8	8.0	—	—	NMR
12	C_s, %	95.5	95.5	95.5	95.5	Inductance

*Here and elsewhere the asterisk denotes parameters corresponding to the initial transition temperature.

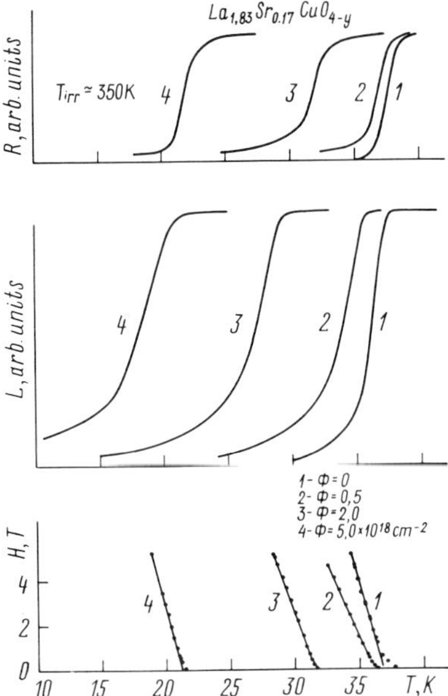

Fig. 1. Temperature dependence of electrical resistance R and inductance of a probe coil L and H_{c2} for variously irradiated samples.

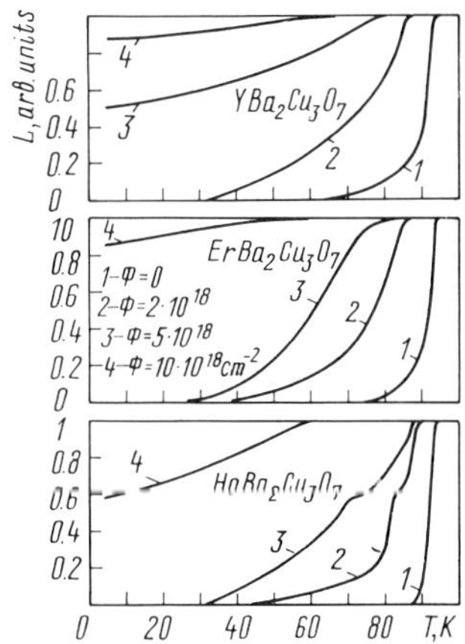

Fig. 2. Inductive transitions of $R_1Ba_2Cu_2O_{4-y}$ compounds.

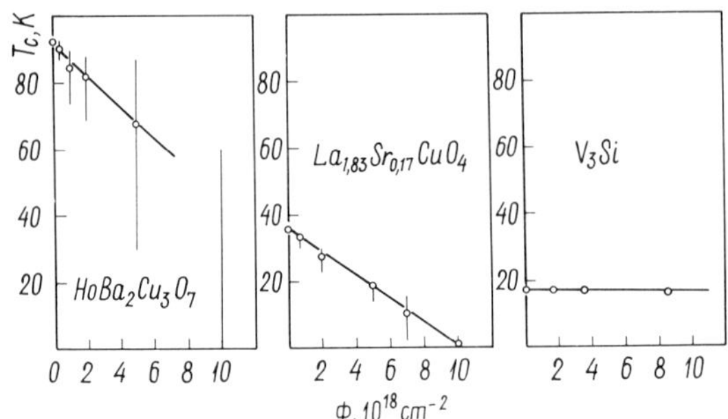

Fig. 3. Dependence of T_c and ΔT_c (vertical intercepts) upon fast neutron flux.

Table 1 shows some physical characteristics for unirradiated materials under study. Superconducting transitions were also studied in magnetic fields up to 5.1 T, the current density being of the order 0.1 A/cm^2. The temperature in magnetic fields to 6 T was measured with an accuracy of 0.1 K.

The inductive transitions were displaced toward low temperatures compared with the resistive ones. The bulk concentration of the superconducting phase (including voids) amounted to 95% at T = 30 K and T = 80 K for the La and Y systems, respectively. Irradiation leads to a rapid decrease in T_c for all of the compounds under study. The results are shown in Figs. 1, 2, and 3. It is seen that the rate of decreasing T_c with neutron flux in ceramic superconductors is sufficiently higher than in A-15 intermetallic compounds.

In La$_{1.83}$Sr$_{0.17}$CuO$_{4-y}$ T_c decreases to about one-half at flux values of 5×10^{18} cm^{-2}, while at 1×10^{19} cm^{-2} for T > 1.7 K there is no superconductivity at all. Under irradiation the width of transition markedly increases, while in V$_3$Si it practically does not change within the same flux range. The temperature derivative of the second critical field defined from the middle of the resistive transition changed slightly with flux. It should be noted that at comparable changes of T_c the A-15 intermetallic compounds always show a significant growth in H'_{c2}. Isochronous annealing of the samples irradiated by a neutron flux of 1×10^{19} cm^{-2} in air for 10 min has shown that recovery of T_c starts at 400°C and stops at 800°C.

In the R$_1$Ba$_2$Cu$_3$O$_{7-y}$ compounds (R = Y, Er, Ho), the initial superconducting transition temperature (94 K for the irradiated samples according to inductive measurements) decreases under irradiation at about the same rate (4 K/10^{18} cm^{-2}) as for La$_{1.83}$Sr$_{0.17}$CuO$_4$-y. These compounds are characterized by sharper deformation of transitions compared to La ceramics, the low-temperature part being formed more strongly than the high-temperature part. Therefore, the narrower the transition in the initial state, the less the increase in ΔT_c. It is to be noted that at fluxes of 5×10^{18} cm^{-2} and 10×10^{18} cm^{-2} the inductive transitions are not complete.

At small fluxes the temperature dependence on electrical resistance above T_c remains linear, while the ratio ρ_{300}/ρ^* decreases, for instance, in HoBa$_2$Cu$_3$O$_{7-y}$ from 2.5 to 1.8 at a flux of 2×10^{18} cm^{-2}. In R$_1$Ba$_2$Cu$_3$O$_{7-y}$ the change in H'_{c2} is much smaller than in the A-15 intermetallic compounds for comparable levels of T_c degradation.

878

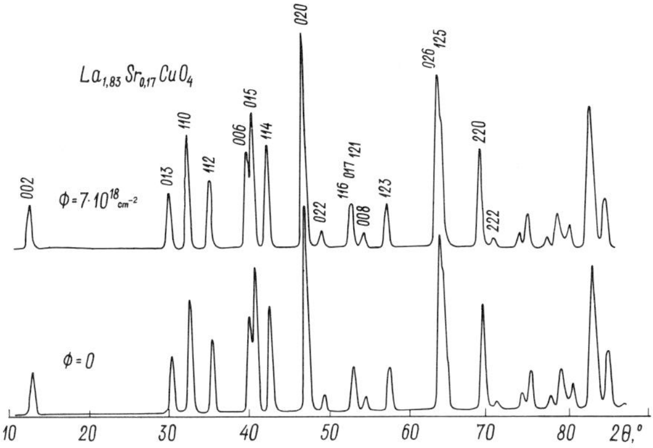

Fig. 4. Neutron diffraction patterns for $La_{1.83}Sr_{0.17}CuO_{4-y}$.

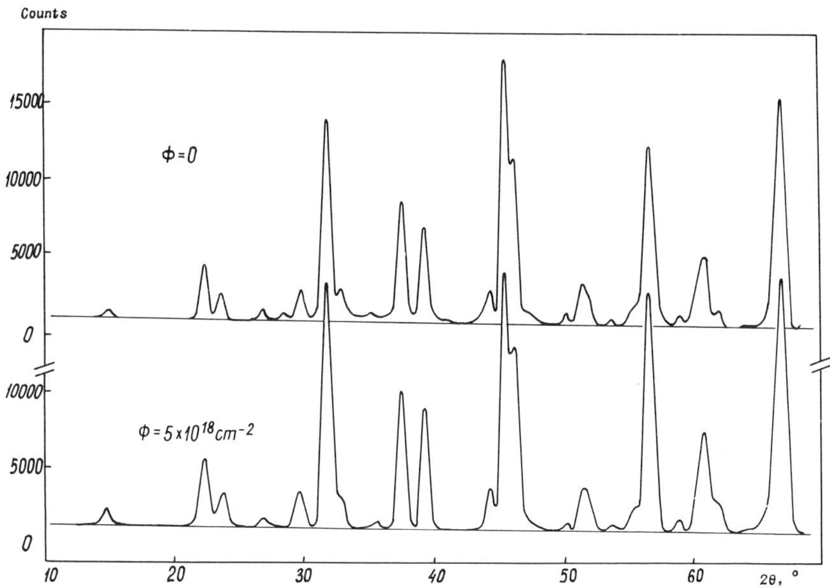

Fig. 5. Neutron diffraction patterns for $HoBa_2Cu_3O_{4-y}$.

Figures 4 and 5 show the neutron diffraction patterns (λ = 1.513 Å, $\Delta d/d \sim 0.5\%$) for the samples before and after irradiation. In $La_{1.83}Sr_{0.17}CuO_{4-y}$ the absolute intensities of structural reflexes decrease slightly, their widths and relative intensities remaining unchanged. Oxygen occupation numbers and lattice parameters did not change within the experimental errors. The Debye-Waller factor (in the isotropic approximation) changed from 0.4 ± 0.1 to 0.6 ± 0.1. X-ray analysis of a sample irradiated by a flux of 5×10^{19} cm^{-2} shows an increase in the lattice parameters of $\sim 1\%$. At this flux the initial crystal structure of type K_2NiF_4 remains (the widths of reflexes and their relative intensities practically do not change). In V_3Si an essential redistribution of the V and Si atoms over the lattice sites was observed, the crystal structure remaining unchanged ($\Phi = 5 \times 10^{19}$ cm^{-2}).

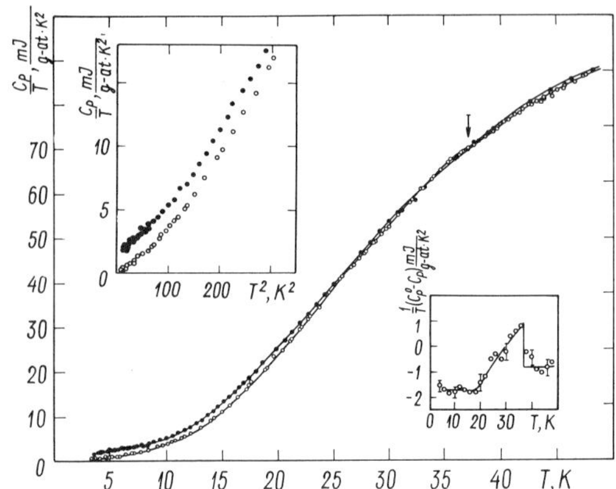

Fig. 6.   Heat capacity of $La_{1.83}Sr_{0.17}CuO_4-y$:
o) $\Phi = 0$, •) $\Phi = 7 \times 10^{18}$ cm^{-2};
$T_{irr} = 350$ K.

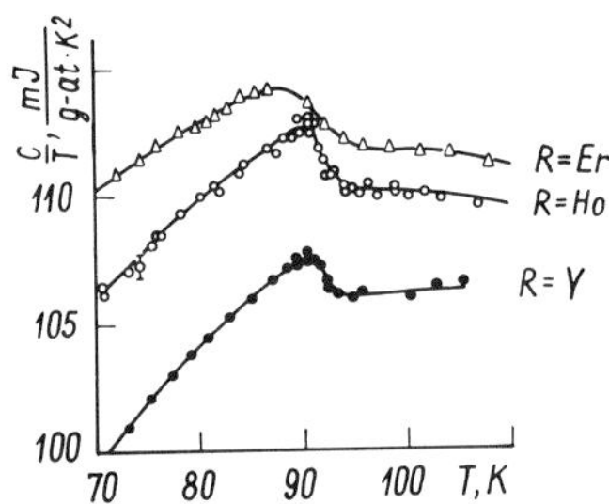

Fig. 7.   Heat capacity of $R_1Ba_2Cu_3O_7-y$.

In contrast to the La system, $HoBa_2Cu_3O_7-y$ irradiated by a flux of 5×10^{18} cm^{-2} displayed changes not only in the absolute but also in the relative intensities of different reflexes.

We also measured the heat capacity C_p for $La_{1.83}Sr_{0.17}CuO_4-y$ before and after irradiation by a flux of 7×10^{18} cm^{-2} (Fig. 6). In the irradiated sample ($T_c \simeq 10$ K) the transition is very much broadened and weakly seen in the dependence $C_p(T)$. The lower insert in Fig. 6 shows the difference in heat capacities before and after irradiation in the temperature range 3-50 K.

The above structural data and the behavior of $C_p(T)$ in 10-20 K range (see upper insert in Fig. 6) suggest that the phonon spectrum does not change greatly under irradiation. In this case the difference in heat capacities before and after irradiation yields the temperature dependence of electronic heat capacity of the unirradiated sample (see the lower insert in Fig. 6).

This procedure allows one to determine more accurately the value of the heat capacity jump while going from the normal to the superconducting state, $\Delta C_p/T_c = 2.0 \pm 0.2$ mJ/g-at·K^2. Figure 7 shows the heat capacity jumps for unirradiated $R_1Ba_2Cu_3O_7-y$. An analogous procedure may also be done for these compounds after lowering the level of induced activity. From

the data in the lower insert of Fig. 6, assuming that the entropies in the
normal and the superconducting states are equal, we obtained the value of
the electronic heat capacity factor for an unirradiated sample, $\gamma \simeq 0.8$ mJ/
g-at$\cdot$K^2.

The results obtained show that the mechanism of electron coupling in
new high-temperature superconductors is more sensitive to radiation-induced
distortions of a crystal structure than in the ordered systems studied previously.

In conclusion, we note that in real experiments only a combined effect
of radiation and temperature of irradiation (in our case, $T_{irr} = 350$ K) is
displayed and is more pronounced in compounds of the type $R_1Ba_2Cu_3O_{7-y}$.
Hence, unlike V_3Si, irradiation at lower temperatures is needed to discover
the effects of radiation proper.

We hope that experiments at $T_{irr} \simeq 80$ K, which we are now undertaking,
will give more information.

The authors thank I. F. Berger for his help in carrying out the structural investigations.

REFERENCES

1.   B. N. Goshchitskii, V. E. Arkhipov, and Yu. G. Chukalkin, Sov. Sci.
      Rev., Phys. Rev. edited by I. M. Khalatnikov, 1987, Vol. 8, Academic
      Press, New York.
2.   V. L. Kozhevnikov, S. M. Cheshnitskii, S. A. Davydov, et al., Fiz.
      Metallov i Metallovedenie, 1987, Vol. 63, No. 3, p. 625.
3.   V. R. Galakhov, B. N. Goshchitskii, V. A. Gubanov, et al., Fiz.
      Metallov i Metallovedenie, 1987, Vol. 63, No. 4, p. 829.

ANOMALOUS BEHAVIOR OF THE ELASTIC CHARACTERISTICS OF $YBa_2Cu_3O_7$ NEAR T_c

A. I. Golovashkin,[1] V. A. Danilov,[2] O. M. Ivanenko,[1]
G. M. Leitus,[3] K. V. Mitsen,[1] I. I. Perepechko,[2]
O. G. Karpinskii,[3] and V. F. Shamray[3]

[1]Lebedev Physics Institute, Moscow, Academy of Sciences
 of the USSR
[2]Moscow Automechanical Institute, Moscow, USSR
[3]Baikov, Institute of Metallurgy, Moscow, Academy of Sciences
 of the USSR

The temperature dependence of the elastic constants and lattice parameters of the $YBa_2Cu_3O_7$ samples has been studied[1,2] to find the mechanism
responsible for the superconductivity of a new class of high-T_c superconducting oxides.

The samples were obtained with the help of ceramic technology[2] by sintering finely ground powders of the oxides of the corresponding metals for
12 hours at 900°C. Special attention was paid to the uniformity of the
samples, which was checked by the width of the x-ray diffraction lines and
the diamagnetic transition sharpness. The x-ray investigation showed that
the samples were single-phase and of orthorhombic structure $YBa_2Cu_3O_7$.

Figure 1 shows the temperature dependence of resistance R and susceptibility χ of one of the samples near the critical temperature T_c. In

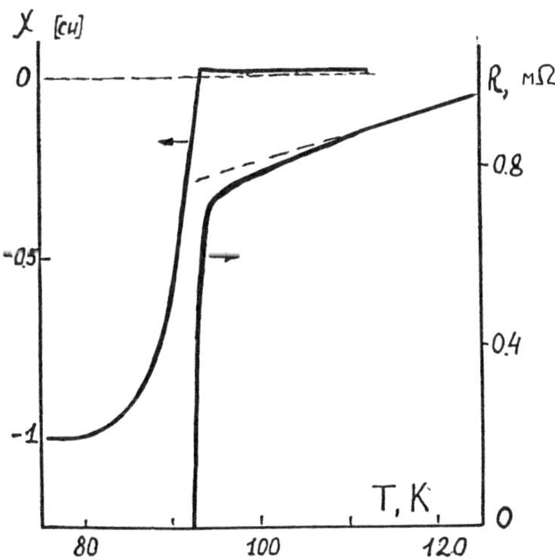

Fig. 1. Dependence of the resistance R and susceptibility χ on temperature
for $YBa_2Cu_3O_7$ near T_c.

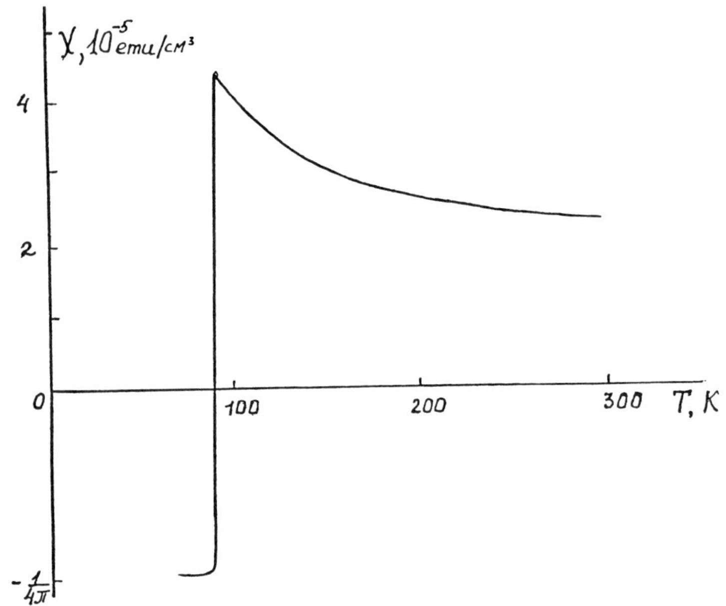

Fig. 2.   Dependence $\chi(T)$ in the complete range of temperatures (the scale
in the regions of negative and positive values of χ is different).

Fig. 2 the dependence of χ on T over the entire range of temperatures is
shown. Measurements of T_c were performed by the four-terminal method. The
value T_c (in the middle of the resistive transition) was 93 K, with transition sharpness $\Delta T_c \lesssim 1$ K. Complete superconductivity was observed at
$T_c = 92$ K. The deviation of the dependence R(T) from linearity starts at
113 K. Measurements of χ were made by a SQUID-magnetometer in a field of

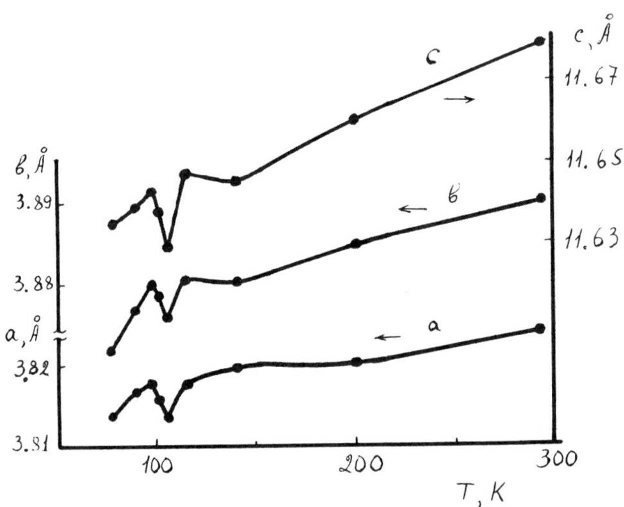

Fig. 3.   Dependence of the lattice constants a, b, and c on the temperature.

884

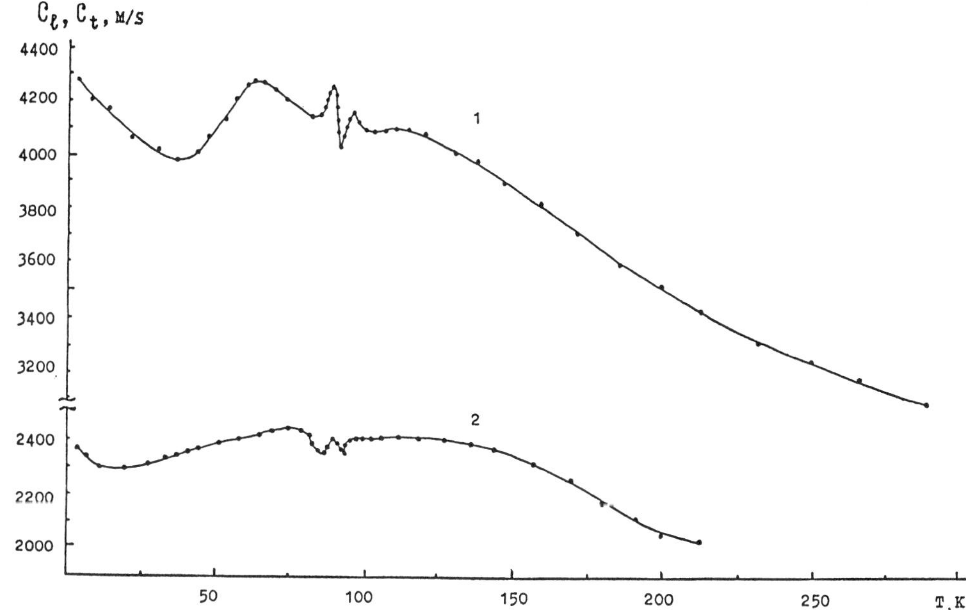

Fig. 4. Temperature dependences of the velocities of longitudinal C_1 (I)
and transverse C_t (2) ultrasonic waves.

~5 Oe. The value of χ for the best samples reached $-1/4\pi$ at 89 K, with
transition sharpness $\Delta T_c \approx 2$ K. Attention is drawn to a noticeable
increase of χ as the temperature T decreases, approaching T_c.

Low-temperature x-ray studies were carried out by the powder technique with a DRON-2 diffractometer (Fe K_α lines). The lattice parameters
were determined by a least-squares analysis of angle positions of twenty-
eight unimposed reflexes. The measurements were performed with a slow
temperature increase of the sample from 77 to 300 K. The temperature
dependences of a, b, c, the constants of the orthorhombic lattice, are
shown in Fig. 3. An anomalous behavior of the lattice parameters is
observed in the vicinity of T_c. The maximum deviation of the parameters
(0.1-0.2%) exceeds the measurement error.

Measurements of the velocity of both longitudinal C_1 and transverse
C_t ultrasonic waves were performed at the arrangement[3] with a frequency of
2.4 MHz (the accuracy is better than 1%). The thermal stabilization of
the sample was carried out to an accuracy of 0.1 K. We show the temperature dependences of C_1 and C_t in Fig. 4. An unusual increase of C_1 (by
30%) is seen when the temperature decreases to 120 K. A peculiar behavior
of C_1 and C_t is observed in the vicinity of T_c. The oscillation value of
the sound velocity in this region significantly exceeds the measurement
error and is abnormally large (5%). A nonmonotonic dependence $C_1(T)$ is
observed in the range of low temperatures (T < 65 K).

Using the experimental data (the density is assumed constant), we
estimate the temperature dependences of dynamic moduli: shear modulus
$G' = \rho_0 \cdot C_t^2$, modulus of dilatation $K_S' = G' \cdot (C_1^2/C_t^2 - 4/3)$,
Young's modulus $E' = 2(I + \sigma') \cdot G$, and Poisson's coefficient
$\sigma' \approx 0.5(C_1^2/C_t^2 - 2)/C_1^2/C_t^2 - 1)$ (here ρ_0 is the density)[4] (Fig. 5).
Some of these moduli show additional peculiarities, e.g., at 200 K.

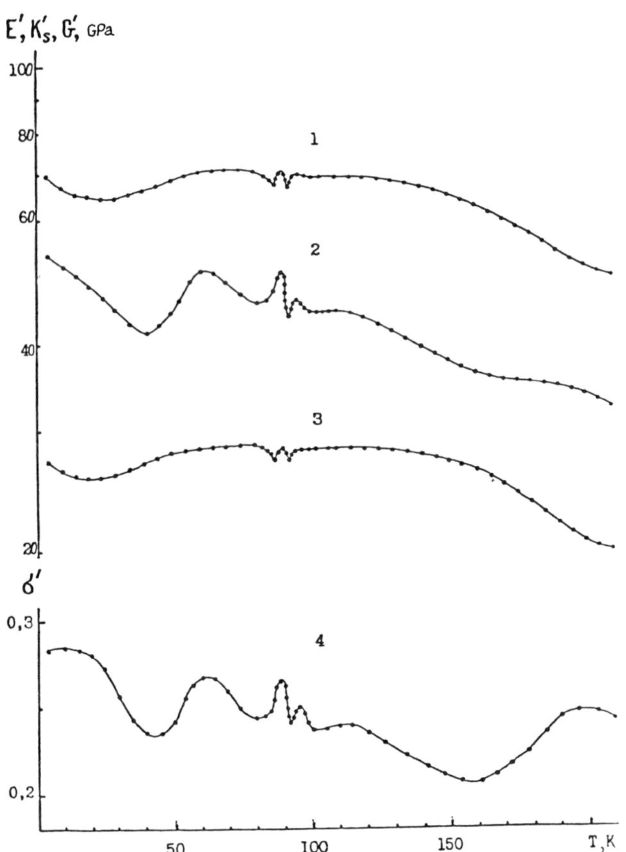

Fig. 5. Temperature dependence of the dynamic Young's modulus E' (1),
the modulus of dilatation K_s' (2), the shear modulus G' (3),
and the Poisson coefficient σ' (4).

The results obtained allow us to estimate the Debye temperature θ_D and the coefficient of linear expansion α [4] (Fig. 6). The values of α are estimated in terms of the Barker empirical formula $E' \cdot \alpha^2 \approx const$ and are in good agreement with the results of direct measurements in the vicinity of T_c.

The results indicate that the transition into the superconducting state of ceramics of composition $YBa_2Cu_3O_7$ is accompanied by anomalies in the behavior of elastic, structural, and other features. The variation of the corresponding parameters within a narrow temperature range by a number of orders exceeds similar variations in the usual superconductors. An unusual behavior of structural parameters and elastic moduli associated with a specific character of the phase transition is observed at T_c.

Shaplygin et al. [5] report for the structure K_2NiF_4 that when the ratio of the lattice parameters c/a decreases, there occurs a sharp drop in conductivity, changing the system from a metallic state to a semiconductor one (Fig. 7). This is associated with an increase in the bond length Cu-O (d_{Cu-O}) and a change in the nature of the bond. Figure 8 shows the value of log ρ plotted versus d_{Cu-O} on the basis of Shaplygin's data. [5] The transition from a metal to a dielectric takes place in the range 1.90 Å < d_{Cu-O} < 1.95 Å. For the structure $YBa_2Cu_3O_7$ the minimum bond lengths are

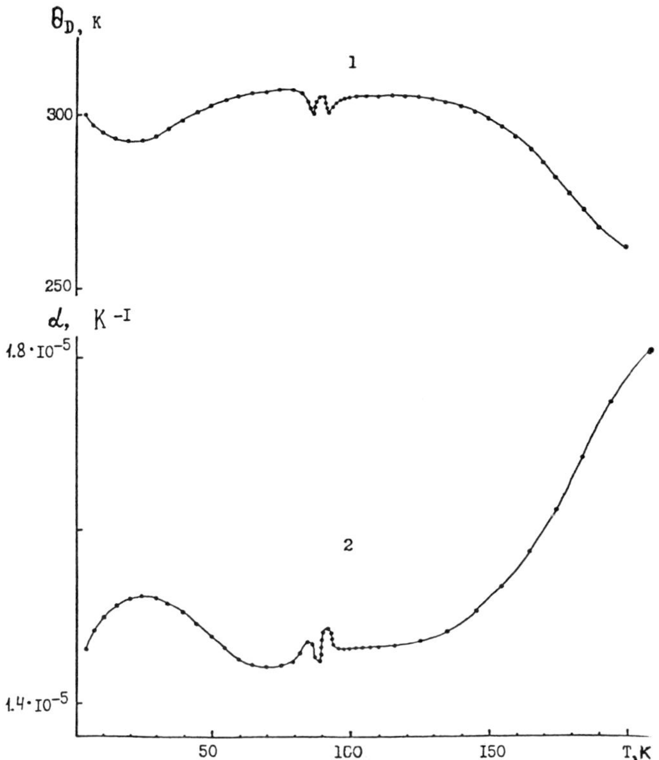

Fig. 6. Temperature dependences of θ_D (1) and the coefficient of
linear expansion α (2).

1.93 - 1.94 Å, coinciding with the interval that corresponds exactly to
the metal-dielectric transition. Small variations of the lattice parameters in the vicinity of T_c (e.g., a decrease of c/a) can drive the system

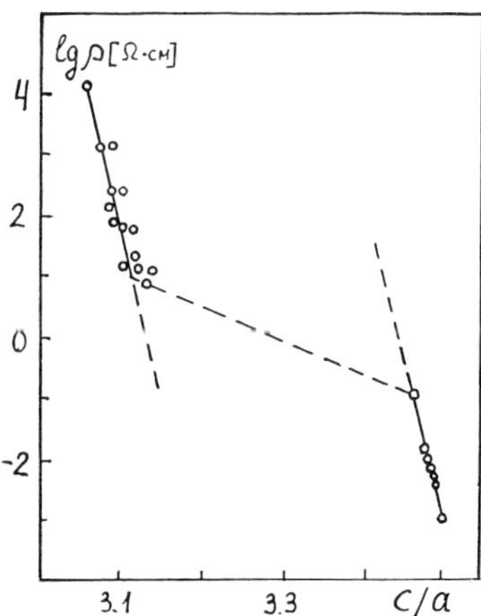

Fig. 7. Dependence of the logarithm of the resistivity ρ on the ratio c/a
for the system K_2NiF_4 (from Shaplygin et al.[5]).

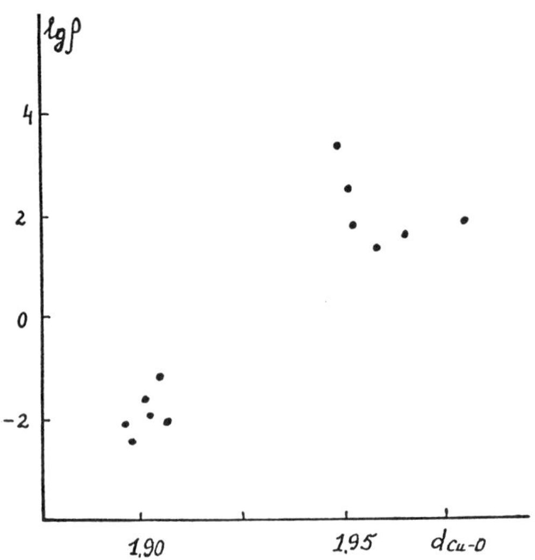

Fig. 8. Dependence of the resistivity on bond length (based on the data of Shaplygin et al.[5]).

to the onset of the metal-dielectric phase transition and the lattice change. The d-zone splitting, associated with an antiferromagnetic ordering, may be one of the reasons for the transition. The dependence $\chi(T)$ at $T > T_c$ (Fig. 2) confirms this suggestion. Structural changes bring the system to a state where electric pairing arises, causing high-temperature superconductivity. The transition to the superconducting state removes the developing instability of the lattice. A specific mechanism of the superconductivity in such a system can be of various types, e.g., of the type suggested by Rusinov et al.[6] Anomalous behavior of the structural parameters of ceramics of composition $La_{1.85}Ba_{0.15}CuO_4$ was observed by Paul et al.,[7] who attribute such a behavior to the existence of spatial structures of almost equal energies in the orthorhombic system and their mutual transformation in local regions because of stacking faults. A similar situation may be found for yttrium ceramics as well.

REFERENCES

1. J. G. Bednorz and K. A. Muller, Z. Phys. B64, 189 (1986).
2. M. K. Wu, J. R. Ashburn, C. J. Torng, P. H. Hor, R. L. Meng, L. Gao, Z. J. Huang, Y. Q. Wang, and C. W. Chu, Phys. Rev. Lett., 58, 908 (1987).
3. I. I. Perepechko and P. D. Golub, Pribory Tekhn. Eksp., No. 6, 216 (1972).
4. I. I. Perepechko, "Properties of polymers at low temperatures," Khimiya, Moscow, 1977.
5. I. S. Shaplygin, B. G. Kakhan, and V. B. Lazarev, Zh. Neorg. Khim., 24, 1478 (1979).
6. A. I. Rusinov, Do-Chan Kat, and Yu. V. Kopaev, Zh. Eksp. Teor. Fiz., 65, 1984 (1973).
7. D. McK. Paul, G. Balakrishnan, N. R. Bernhoeft, W. I. F. David, and W. T. A. Harrison, Phys. Rev. Lett. 58, 1976 (1987).

POINT CONTACT STUDY OF METAL OXIDE

SUPERCONDUCTOR Y-Ba-Cu-O

N.A. Tulina, V.A. Borodin, V.F. Kondakov,
and L.I. Chernyshova

Institute of Solid State Physics Academy of Sciences
of the USSR, 142432 Chernogolovka Moscow distr., USSR

High quality tunneling junctions made of superconductors [1] and pure point contacts made of normal metals [2] are perfect objects for investigations allowing one to obtain some quantitative information about the superconducting energy gap $\Delta(\omega)$ and to restore the function of the electron-phonon interaction $\alpha^2(\omega)\mathcal{F}(\omega)$, where $\alpha^2(\omega)$ is the square of the electron-phonon interaction matrix element, $\mathcal{F}(\omega)$ is the density of phonon states. These two experimentally observable values are of great importance from the point of view of the origin of superconductivity in now existing compounds and today they receive high priority in connection with the discovery of high-T_c superconductors La-Sr-Cu-O, Y-Ba-Cu-O [3,4]. It is still difficult to obtain perfect tunneling junctions and pure point contacts from these compounds because in reality tunneling junctions have short-circuits and point contacts have contaminations in the contact region. The BTK theory [5] (Blonder, Timkham, Klapwijk) describes a transition from the tunneling contact to a metal-like one in superconductor – constriction – normal metal structures ScN. In the present work the point contacts of ceramics with aluminium, copper, lead heterocontacts and of ceramics Y-Ba-Cu-O monocontacts have been studied. Ceramics samples were prepared from powders by a standard technique and revealed both multiphase structures $Y_{1,2}Ba_{0,8}CuO_4$ and "1,2,3" structure (uniphase as controlled by the X-ray analysis). The point contacts were produced by mechanically pressing of two electrodes in a needle-anvil configuration. Current-voltage characteristics (CVC) and dynamic resistance dv/dI were taken versus the voltage applied to the point contact. Critical temperatures of the samples investigated varied within 88-95 K. The results of the study can be treated as follows. According to the CVC behaviour one can divide all the investigated point contacts into three groups: type I – the resistance $R > 1k\Omega$, type II – $1k\Omega > R > 10\Omega$ and type III – $R < 10\Omega$. The CVC for different type of pointcontacts are given in Fig. 1.

The dynamic resistance of type I contacts didn't reveal any singularities. With an increase of the pressing force on the contact the resistance reduced, the CVC became metal-like and some distinctions appeared on dependence dv/dI (V) (Fig. 2). The CVC for type III contacts were metal-like and some of them revealed a supercurrent. According to the BTK theory the contact resistance R is determined by an effective tunneling barrier Z_{ef}, $R = (1 + Z_{ef})R_o$, where R_o is the resistance of an ideal contact in the pure limit which is in its turn determined by the Fermi parameters of the material according to the Sharvin relation $R_o = \rho l / d^2$, where ρ is the specific resistance, l is the mean free path, is the contact diameter

$$Z_{ef} = \left[Z^2 + \frac{(1-r)^2}{4r} \right]^{1/2}, \quad r = V_{F1}/V_{F2},$$

V_{Fi} – Fermi velosity of electrode, Z is the barrier determined by the contaminations. In our point contacts Z_{ef} varied within 0–2. For $Z_{ef} > 1$ the characteristics were of tunneling type without any singularities in the dynamic resistance. This is likely to be

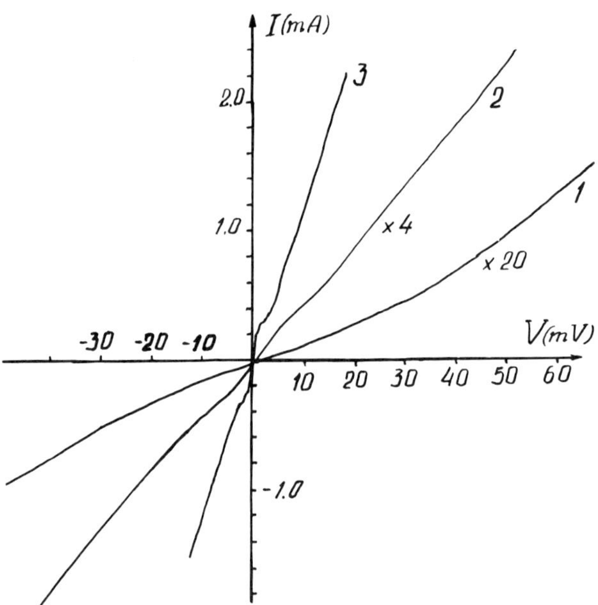

Fig. 1. Current–voltage characteristics of different type contacts T=4.2 K.

due to a break of superconducting properties in the surface layer of a few tens Å thicknes. Gap-type singularities appear as the contact diameter increases but in different contacts these singularities varied in shape and position due to the sample inhomogeneity and, probably, due to the effects associated with an electron life time. As was shown in [6] for a decreasing electron free path the tunneling characteristics are smeared out and the density of tunneling states depends on the decay factor.

The maximum gap value observed in the present work is 15,5 meV if one supposes that the tunneling takes place within

the structure SIS (within the ceramics grains) and it is two times larger if the structure SIN is carried. For such objects the
question is still open as was also pointed out in [7].

The presentation of singularities observed for the heterocontacts is much more complex (Fig. 2). Pronounced resonant
lines are the gap singularities. The origin of out-of-gap lines

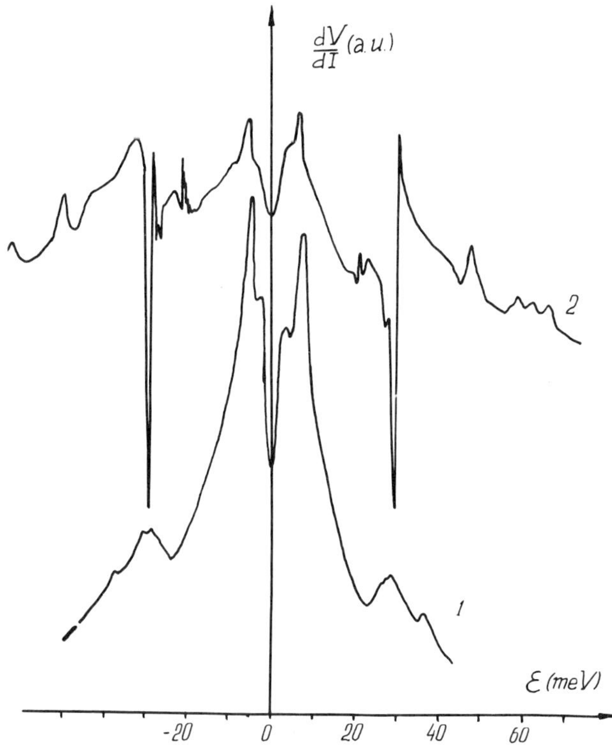

Fig. 2. Example of point contact dynamic resistance on bias voltage. T=4,2 K. 1. Pointcontact of ceramic Y-Ba-Cu-O
(monocontact) R=240 Ω ; 2. Hetero contact of ceramic
Y-Ba-Cu-O-Al. R=213 Ω.

is not clear and it is difficult to bound them up with the electron-
phonon interaction because the shape and width of these lines
most likely indicate their resonant character. We suppose this
to be a display of the Tomash oscillations [8] in a complex multilayered structure of the heterocontact. Such a structure can be
formed as a result of the proximity effects as was predicted in

[9]. To make more definitive conclusions more detailed and directed study of the above-mentioned distinctions is needed.

REFERENCES

1. Tunneling phenomena in solids. "Mir", Moscow, 1973.
2. I.K.Yanson. Point contact spectroscopy of electron-phonon interaction in pure metals. Soviet Journal "Low Temperature Physics", 9, 676 (1983).
3. J.G.Bednorz, K.A.Muller, Z.Phys. B64, 189 (1986).
4. M.K.Wu, J.R.Ashburn, C.J.Torng, P.H.Hor, R.L.Meng, L.Gao, Z.I.Huang, Y.Q.Wang, C.W.Chu. Phys.Rev.Lett. 58; 908 (1987).
5. G.E.Blonder, M.Tinkham, T.M.Klapwijk. Transition from metallic to tunneling regimes in superconducting microkonstrictions. Phys.Rev., 25, N7, 4515 (1982).
6. R.C.Dynes, J.P.Carno, G.B.Hertel, T.C.Orlando. Tunneling Study of Superconductivity near the Metal-Insulator Transitions. Phys.Rev.Lett., 53, 2437 (1984).
7. Toshikasu Ekino, Jun Akimitsu. Superconducting tunneling in Y-Ba-Cu-O system. Jap. J.A.Phys., 26; L452 (1987).
8. D.Vezzetti. Tomash oscillation in multilayered superconducting systems. Phys.Rev., 31, N5, 2707 (1985).
9. A.Hahn. Geometrical resonances in a high injection current nonexequilibrium state of superconductor-normal metal contacts. Phys.Rev.B, 31, N5, 2816 (1985).

INFRARED REFLECTIVITY, INELASTIC LIGHT SCATTERING
AND ENERGY GAP IN A Y-Ba-Cu-O SUPERCONDUCTORS

A.V. Bazhenov, A.V. Gorbunov, N.V. Klassen, S.F. Kondakov,
I.V. Kukushkin, V.D. Kulakovskii, O.V. Misochko,
V.B. Timofeev, L.I. Chernyshova, and B.N. Shepel'

Institute of Solid State Physics of the Academy of Sciences
of the USSR, 142432 Chernogolovka, Moscow district, USSR

Below 93K the spectra of FIR reflection and inelastic light
scattering of Y-Ba-Cu-O samples reveal the peculiarities related
with superconducting gap and optical phonons. The magnitude
of the gap is found to be $2\Delta = (30\pm2)$ meV.

Polycrystal samples of $Y_{1.2}Ba_{0.8}CuO_4$ (type I) and
$YBa_2Cu_3O_{9-x}$ (type II) have been produced by the high-temperature solid-phase synthesis. The type I samples were multiphase
with a resistive transition to a superconducting state at T=90.5K
and the transition width $\Delta T = (2-3)K$. The type II samples showed
a more pronounced and sharper transition (T=93K, $\Delta T=1K$). According to diamagnetic susceptibility measurements, the fraction of
the superconducting phase in type II samples amounts to about 40%,
that much exceeds the value obtained for type I samples. The
spectra were gauged with respect to the signal of reflection from
a gold mirror.

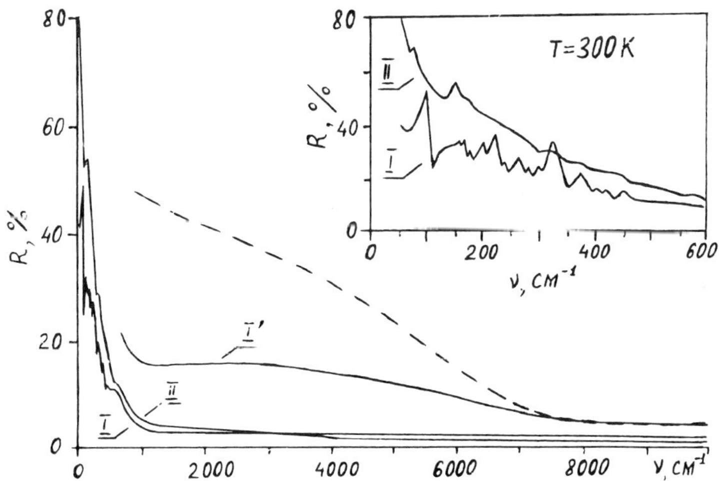

Fig. 1. Reflectivity spectra of the type I and II ceramics at T=300K.
The inset showed the FIR reflectivity spectrum.

tation spot on the sample was 0.1–1 mm (at $T < 300K$). In same LS measurements performed at $T \geqslant 300K$ 5 μm diameter laser beam spot was used to record the LS spectra from separate microcrystals.

The LS spectra of type I samples measured at T=1.5 K (λ =5145 A) involve a few narrow peaks, observed on the background of a wide band. (Fig. 3a).

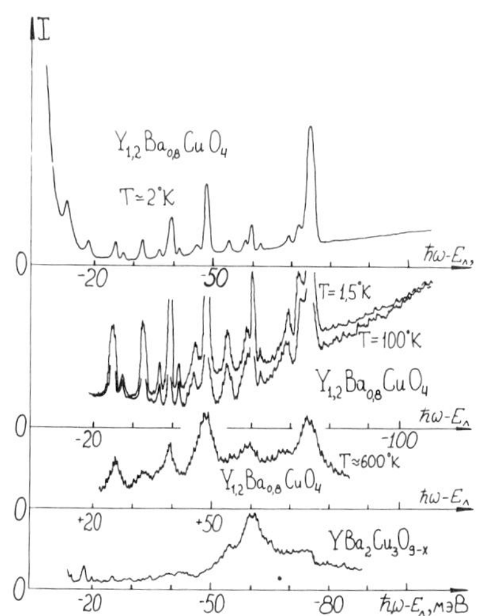

Fig. 3. The inelastic light scattering spectra registered at different temperatures in stokes (a,b,d) and antistokes (c) region.

A comparison of spectra, obtained with different laser lines has shown that this wide band is mainly due to hot luminescence with the maximum at $\hbar\omega \approx 1.2$ eV. Narrow lines are likely to be associated with phonon excitation. They remain in the stokes spectrum at all temperatures, and can be registered at $T > 200$ K in the antistokes region too (see Fig. 3c). It should be noted that the phonon associated LS in $YBa_2Cu_3O_{9-x}$ samples (Fig. 3d) turned out to be much weaker, than in the $Y_{1.2}Ba_{0.8}CuO_4$ ceramics. The samples of both types involved several different phases, each displaying a characteristics set of LS lines. The type II ceramics was more homogeneous. In $Y_{1.2}Ba_{0.8}CuO_4$ phonon associated LS is maximal at λ =5145 A whereas in $YBa_2Cu_3O_{9-x}$ it grows with decreasing λ down to 4579 A. This fact correlates with the position of minima in reflectivity spectra corresponding to interband transitions. (They are at   5000 and 3900 A for $Y_{1.2}Ba_{0.8}CuO_4$ and $YBa_2Cu_3O_{9-x}$, respectively). Therefore the observed spectra are likely to be resonant.

To separate the LS contribution caused by electron excitations through the superconducting gap, we have studied the difference of LS spectra registered under the identical conditions at temperatures below T_C (1.5K) and above T_C(100–200K). In the difference spectra ($I_{1.5K} - I_{100 K}$) presented in Fig.4 the phonon lines, weakly depending on temperature, vanish almost completely. These spectra are similar for both type samples and independent of the wavelength of an exciting light and choise of temperature above T_C. The line observed in the difference spectrum (Fig.4) has a well pronounced low–frequency edge ($h\nu_0 \approx 30$ meV) and

Fig. 1 shows reflectivity spectra at T=293 K, the measured area being 4x4mm^2. The discrete structure in the 50–500 cm^{-1} region related to optical phonons, is more distinct for type I samples. In $\nu > 500$ cm^{-1} region the reflection decreases monotonously with increasing frequency. Reflectivity measurements with an IR–microscope have showed five times higher reflectivity of the 50x50 μm^2 areas, occupied by crystallites (spectrum I'). The general character of the spectrum retained. The dotted line depicts approximation of the reflection spectrum in the framework of the Drude theory. The contribution of anisotropy was taken into account and the following parameters were used: ε_∞ =4.5, the screened plasma frequency ν_{ps}=5830 cm^{-1} and the reciprocal of the transport relaxation time γ =4000 cm^{-1}. An agreement with experiment can be improved, if one assumes that at $\nu < \nu_{ps}$ an essential contribution comes from interband electron transitions[1].

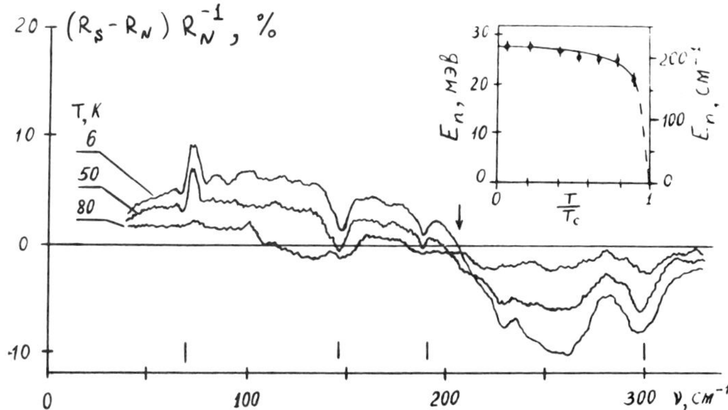

Fig. 2. Experimental superconducting–normal reflectivity difference R_S–R_N as a function of frequency for $YBa_2Cu_3O_{9-x}$. The inset showes the temperature dependence of the superconducting gap.

Fig. 2 shows superconducting–normal reflectivity difference R_S–R_N in samples II, R_N being measured at T=110K. Inside of the superconducting gap ($h\nu \lesssim 2\Delta$) R_S–R_N is positive and reaches 7% with respect to R_N, whereas outside of this region R_S–R_N <0. Such a behaviour agrees with the Bardin–Matiss theory [2]. The spectral position of the point R_S–R_N=0 is determined by the value of the superconducting gap. This point moves to the smaller energies with increasing temperature (see the inset to Fig.2), as predicted by BCS–theory [3] . The additional structure related to phonons complicates a comparison with the theory. The magnitude of the superconducting gap can be estimated from the reflectivity difference as $2\Delta \sim 28\pm2$ meV. In this estimation we take into account the small fraction of the superconducting phase ($\lesssim 40\%$) in samples studied.

Inelastic light scattering (LS) spectra of ceramic samples of both types were measured in the geometry of back scattering, using the lines of Ar$^+$ and Kr$^+$ lasers (5145 A, 4880 A, 4579 A, 6471 A) at T=1.5 K ÷ 600 K. The typical diameter of the exci-

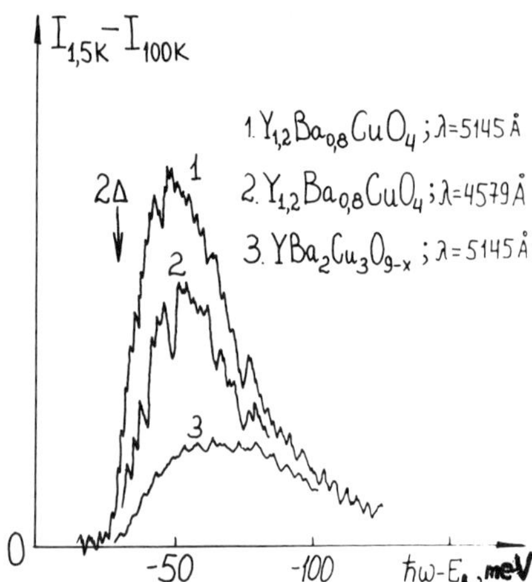

Fig. 4. The difference of LS spectra registered under the identical conditions at temperatures below T_c (1.5 K) and above T_c (100 K).

halfwidth $\Gamma \sim 2h\nu_0$. Such shape of the LS spectrum well agrees with that predicted for superconductor by the calculations [4–6]. The quantity $h\nu_0$ defines the width of the superconducting gap $h\nu_0 = 2\Delta = 30\pm2$ meV (or $\sim 3.5\ kT_c$), which is in good agreement with the above data on the FIR reflection.

REFERENCES

1. S.Tajima et al. Phys.Rev. B32, 1029 (1985).
   G.A.Thomas et al. Preprint (1987).
2. D.C.Mattis, J.Bardeen. Phys.Rev. 111, 412 (1958).
3. J.Bardeen, L.N.Cooper, J.R.Schrieffer. Phys.Rev. 108, 1175 (1957).
4. A.A.Abrikosov, I.A.Falkovskii. Zh.E.T.F. (USSR) 40, 262 (1961).
5. A.A.Abrikosov, V.M.Genkin. Zh.E.T.F. (USSR) 65, 842 (1973).
6. G.B.Guden. Phys.Rev. B13, 1993 (1976), B18, 3156 (1978).

SPECIFIC FEATURES OF MICROWAVE ABSORPTION OF

SUPERCONDUCTING CERAMICS IN A MAGNETIC FIELD

V.V. Kveder, T.R. Mchedlidze, Yu.A. Ossipyan, and
A.I. Shalynin

Institute of Solid State Physics Academy of Sciences
of the USSR, 142432 Chernogolovka Moscow distr., USSR

Presently, different properties of ceramics on the base of metal oxide superconductors $(Ba-Pb-Bi-O$ [1], $Sr-La-Cu-O$ and
$Ba-Y-Cu-O)$ are being studied intensively. This work deals with
the investigation of the microwave absorption in the $Ba-Y-Cu-O$
base ceramics in the range of 9 GHz. The $\simeq 1$ mm^3 sample was
placed in a maximum of the microwave magnetic field of the EPR
spectrometer cavity, and the microwave power, R, absorbed by the
sample and its derivative dR/dH_O were measured with respect to the
magnetic field.

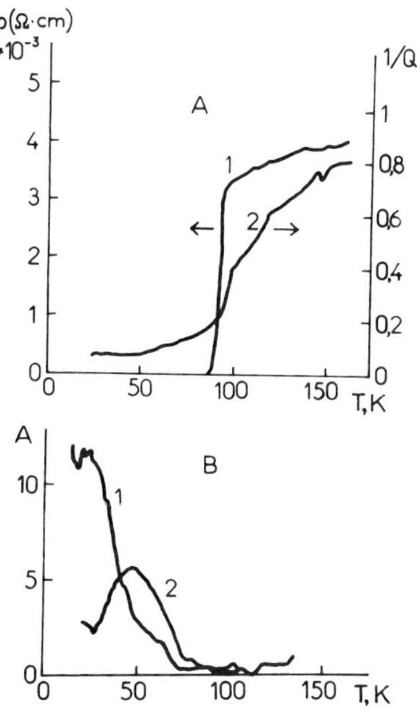

Fig. 1.   A – temperature dependence of dc resistance (curve 1)
and microwave absorption (curve 2); B – temperature
dependence of the "oscillations" (curve 1) and the
"hysteresis" (curve 2).

Fig. 1 A presents the dependence of R on T for $Y_{1.2}Ba_{0.8}CuO_x$ sample, measured in the field $H_O=0$ [2], and the temperature dependence of the d c resistance of the sample (1). The appearance of macroscopic superconductivity in the sample at $T \approx 93$ K is confirmed by the effect of stable soaring of the magnet over the ceramic tablet – "Mohammed's coffin", that, also, indicates a strong pinning of the magnetic flux. At $T < 90$ K there appears a strong nonlinear dependence $R(H_O)$, fig. 2. This dependence is enhanced with decreasing temperature.

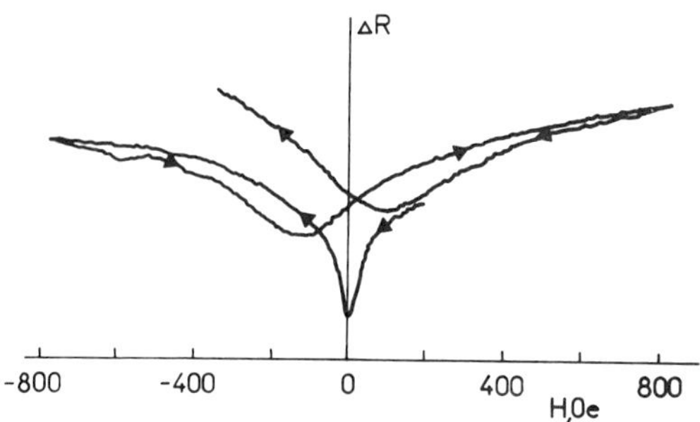

Fig. 2. Dependence of the microwave absorption on external magnetic field.

With modulation of H_O at a frequency $f_m=(80Hz \div 100$ KHz$)$ and a slow scanning of H_O, there appears modulation of the microwave power, reflected from the cavity, at a frequency f_m, whose phase is determined by the sign of dH_O/dt, whereas the amplitude is practically independent of $|dH_O/dt|$ ("hysteresis"). The temperature dependence of the "hysteresis" amplitude is shown in fig. 1B, curve 2. In the region of the superconductivity at the background of a monotonic alteration of $R(H_O)$ one can observe noise-type variations of the dR/dH_O curves (the modulation amplitude of the magnetic field is < 1 Oe). It is important that a characteristic period with respect to H_O is practically independent of dH_O/dt. The amplitude of these "oscillations" grows with decreasing T (curve 1 in fig. 1B). With a sufficiently low $T(T < 40$ K) the picture of the "oscillations" is accurately reproduced at a repeated scanning of the field H_O (see curves 1,2 in fig. 3). At recording in the field H_O, that differs dramatically (by more than 700 Oe at 15 K) from the field H_O at which cooling was performed, nonreproducible jumps are superimposed on the recurring picture of the "oscillations". At heating and repeated cooling the "oscillations" picture changes, but their Fourier spectrum changes insignificantly. Fig. 4 represents the Fourier spectrum of $R(H_O)$. Its important feature is the presence of distinct peaks.

The investigation of $YBa_2Cu_3O_{(x>7)}$ ceramics, that is characterized by a strong critical current, has shown that all the above said effects are also observed here, however, the dependence $R(H_O)$ is much weaker, the "oscillations" amplitude and the "hysteresis" are smaller.

We briefly discuss the results obtained. We might try to attribute the observed "oscillations" of the sample impedance with a varying external magnetic field to the break-away of the lattices of superconducting (Abrikosov) vortices from pinning centres or to their penetration into the pores, which could cause a drastic change of microwave absorption. However, in this case, a good recurrency of $R(H_O)$ is unlikely to occur at a repeated scanning of the field. Therefore, we assume, the other interpretation is plausible: the ceramics is a network of the superconducting phase with a great number of pores and nonsuperconducting inclusions in the cells.

898

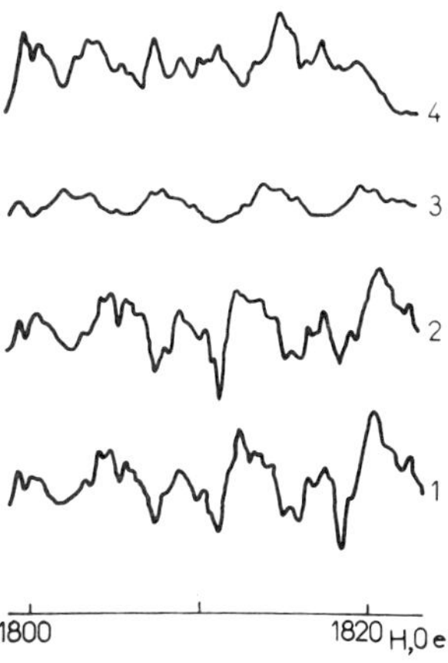

Fig. 3. Picture of the "oscillations". Curve 1 – T=15K; curve 2 – T=15 K, repeated scanning of the field H_o; curve 3 – T=50K; curve 4 – T=15K, after heating at T=50K.

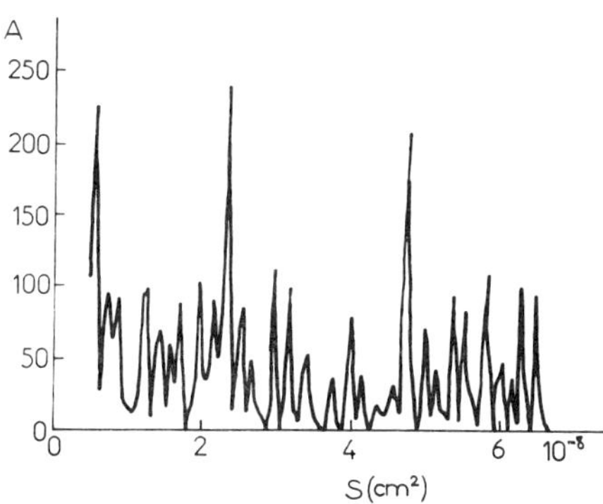

Fig. 4. The Fourier spectrum of the microwave absorption $R(H_o)$.

The junctions of superconducting grains may have the properties of "weak links", that is, fluxoids can slip without pinning through them into the cells of the network. Then, we are dealing with the network of "SQUIDs", and the observed dependences $R(H_o)$ correspond to a quantized penetration of the flux into this network. The Fourier transform of $R(H_o)$ has a number of peaks, corresponding to areas $S = \Phi_o/\Delta H_o$ of the network cells. Astonishingly, that a sufficiently large sample can exhibit such narrow singularities in the distribution (fig. 4). However, firstly, only a part of the network cells satisfy the necessary condition of having "weak" junctions. Secondly, this system may demonstrate the effect of synchronization. Its nature is not definitely clear, but experimentally the synchronization effects were observed at the investigation of the current-voltage characteristics of Ba–Bi–Pb–O ceramics [1]. Analogously, one can explain a strong reversible nonlinear magnetoresistance in small fields (fig. 2) and the "hysteresis". At an increase of H_o above 500–700 Oe the minima of $R(H_o)$ in fig. 2 are shifted and become less prominent. This may be explained by freezing of the magnetic flux in superconducting grains and strongly pinned cells. Strong pinning leads to under-estimated concentration of the superconducting phase obtained from the measurement of the Meissner effect. Alternatively, the measurement of the Meissner effect at very small H_o may yield an over-estimated concentration of the superconducting phase due to weak superconductivity of normal grains owing to the proximity effect to the superconducting phase.

In conclusion the authors express their gratitude to Borodin V.A., Kondakov S.F., Chernyshova L.I. for the preparation of the ceramic samples.

REFERENCE

1. A.M.Gabovich, D.P.Moiseev. "Metal oxide superconductor $BaPb_{1-x}Bi_xO_3$: unusual properties and new applications", Usp.Fiz.Nauk, 150: 599 (1986).

LOW TEMPERATURE X-RAY ANALYSIS AND ELECTRON

MICROSCOPY OF A NEW FAMILY OF SUPERCONDUCTING

MATERIALS

Yu.A. Ossipyan, V.A. Borodin, V.A. Goncharov, S.F. Kondakov,
S.S. Khasanov, L.M. Chernyshova, V.Sh. Shekhtman,
I.M. Shmyt'ko, and N.F. Stchegolev

Institute of Solid State Physics Academy of Sciences of the
USSR, 142432 Chernogolovka Moscow distr., USSR

Recent findings in the field of high temperature superconductivity [1] requirte that structural aspects of the behaviour of this class of materials be investigated in detail in a wide temperature interval. A series of superconducting ceramics on the base of lanthanum and yttrium oxides have been obtained in the solid state Physics Institute of the Academy of Sciences of the USSR. This paper presents the results of the analysis of powder and sintered materials, using X-ray diffractometers (DRON), scanning electron microscope and special devices, enabling the investigations to be carried out within 4.2 K–573 K.

1. SPECIFIC FEATURES OF TEMPERATURAL ALTERATIONS OF THE STRUCTURE

The samples used had the compositions $La_{2-x}Sr_4CuO_4$ (subsequently LSCO) and YBaCuO (subsequently YBCO). A preliminary analysis has shown that the LSCO samples have as the main component the tetragonal structure of K_2NiF_4-type, a=3.761 A, c=13.161 A; the superconducting transition was observed at T_c=38K. The YBCO samples are characterized by a complicated phase composition, involving the lines of $YBa_2Cu_3O_7$ compound with the orthorhombic structure , a=3.84 A, b=3.89 A, c=11.70 A. The superconducting transition was observed at T_c=90 K.

The experiments with the samples of undoped La_2CuO_4 have shown that in the initial state the material corresponds to the orthorhombic modification with the parameters a=5286 A, b=5.336 A, c=12,958 A, in accord with the published data . As the temperature is increased from room up to 563 K, the "a" and "b" parameters of the orthorhombic cell are drawing gradually together, which is completed by the transition to the tetragonal phase (fig. 1). In the low temperature interval an increase of the axial ratio a/b is observed. In this case splitting of the like (110) and its absence for (200) indicates unambigously that the transition $I_{tetr} \rightleftharpoons F_{rhomb}$ occurs by means of twinning in the system (100)/(010), (see the insert in fig. 1). At doping by strontium the structural transition temperature decreases significantly already at X=0.1, and at X=0.2 the phase remains tetragonal to 4.2 K (fig. 1).

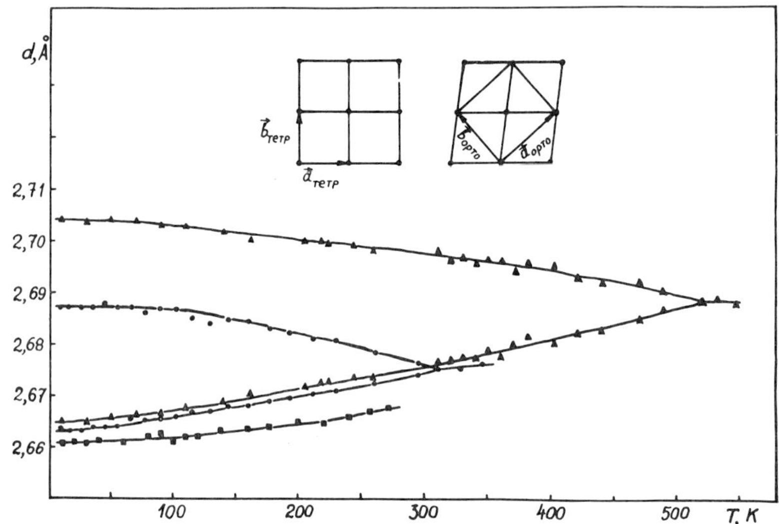

Fig. 1. Temperature dependence of atomic distances: (110) tet (■) for $X=0,2$; and (200)ort, (020)ort for $X=0$ (▲) and $X=0,1$ (●).

Detailed measurements of the diffraction angles and integral intensities yielded for LSCO a considerable temperature interval of invariability of said characteristics. Anomalous dynamic properties of the lattice are illustrated in fig. 2.

Invariability of the integral intensity in a rather large temperature interval (even for higher reflection orders) suggests that the oscillation amplitudes of heavy atoms are much smaller than static displacements from the sites of the central lattice in a disordered mixed crystal. This may imply that the atomic oscillation spectrum is determined by the properties of a local potential well alone rather than by the translation characteristics. In other words, as far as the phonon mechanism of the superconductivity initiation is concerned, it is important that the dynamical properties of the structure appear to be close to a glassy state. This fact is also supported by a considerable difference the between lattice parameters stabilization temperature $(T = 50$ K$)$ and the reflection intensity stabilization temperature $(T = 300$ K$)$ which must be in inverse relation for the ideal lattice $(T / T)^{4}$. Note, that the YBCO-base materials exhibited the analogous anomalies that overlap the transition point to the superconducting state.

We also note that in a great number of cases the LSCO and YBCO samples, that meet high T_C values, give associate-phase lines on X-ray powder photographs in addition to the above said. Whence it follows that optimal, with respect to critical characteristics, compositions should be sought for with account taken of possible equilibria between the limiting-composition phases on the state diagram. It is net excluded that the dependence "composition (strontium, barium, oxygen content) - property (T_C, H_{c2}, I_C)" can meet berthollide-type phases in this case.

2. LOW TEMPERATURE ELECTRON MICROSCOPIC CONTRAST

We investigated clean (chemically of mechanically untreated) cleavage planes of YBCO-type materials of different compositions that were preliminary tested for the presence or absence of superconductivity at T 77 K. The experiments were conducted in an electron microscope, provided with a table with cooling,

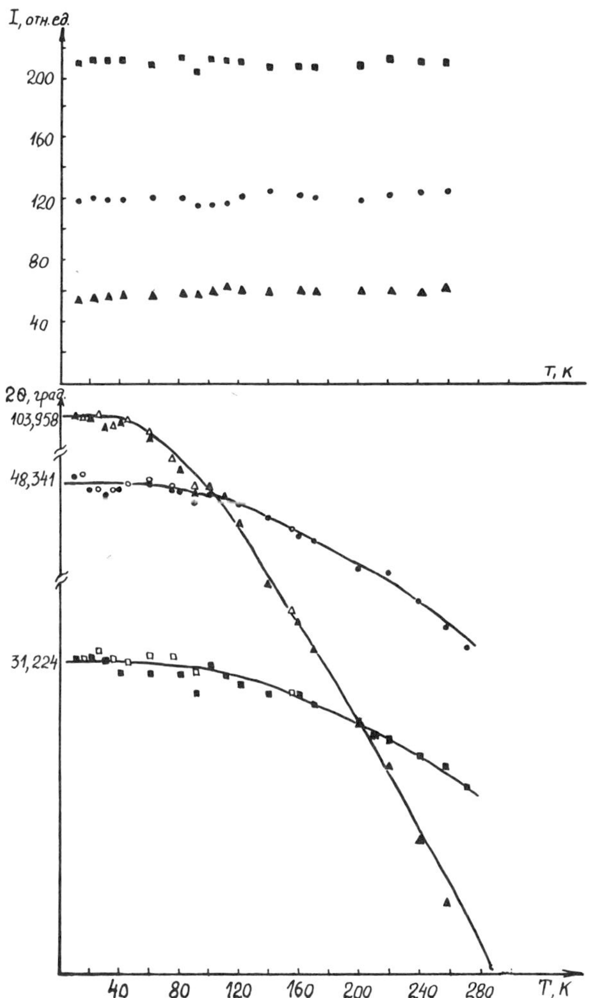

Fig. 2. Temperature dependence of integral intensities and diffraction angles for reflections (211) (▲), (200) (●) and (103) (■) for ceramics $La_{2-x}Sr_xCuO_4$ and X=0,2.

in a scanning regime of secondary electrons. We observed a new phenomenon whose gist was as follows: at a sample cooling down to some temperatural point, that is above T_c for each one composition, there arises, locally, a contrast of earlier unknown type, the usual topographic contrast being preserved in secondary electrons, revealing pores, grain structure etc. Fig. 3 presents a characteristic example, demonstrating an enhanced brightness of the region that covers tens of microns. The samples permitted repeated thermocycling, the contrast region recovering in the same area; these contrast regions disappeared with increasing temperature practically simultaneously. The samples which had no superconducting transition, did not exhibit such contrast areas. This suggests the conclusion that the singularities of the contrast are related to inhomogeneous structure of the sample, containing superconducting regions together with nonsuperconducting ones.

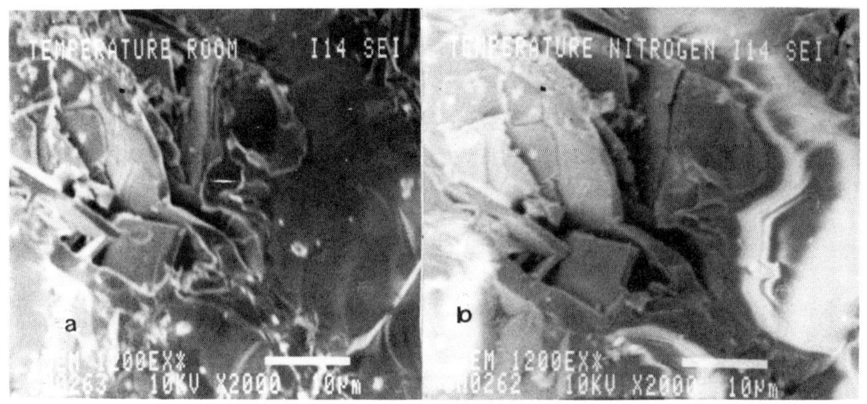

Fig. 3. Secondary electron pictures of $Y_{1.2}Ba_{0.8}CuO_4$ ceramic over (a) and under (b) superconducting transformation temperature.

It is clear that a complete interpretation of the phenomenon is restrained by insufficient understanding of the nature of superconductivity of the new materials and of the basis of interaction between moderate-energy (to 120 keV) electron beams and superconducting regions. The variation of the detector regime made it possible to establish some specific features of the process. The contrast is shown to be mediated by electrons, scattered from the sample with energies $E \lesssim 10$ eV. The contrast is lacking in the regime of backscattered electrons (the detector registers only the energies close to the primary beam electron energy), which indicates the probe electrons penetration to the superconducting regions. Then one can suppose that low energy secondary electrons may change the trajectories at the interaction with ideal dia magnetic, that the superconducting region, and provide for said enhanced "brightness". In this way the new "superconducting" contrast, that was no predicted theoretically and not observed earlier in experiments, can be formed. The fact that the contrast arises at a temperature somewhat higher than the onset of a drastic fall of the electric resistance can probably be related to local regions in which the transition occurs at higher temperatures but is not recorded on the $R(T)$ curve due to specific features of percolation. A local increase of the transition temperature is also possible because of excess of electrons introduced from the electron beam.

REFERENCES

1. M.K.Wu, J.R.Ashbern, C.J.Torng, P.H.Hor, R.L.Meng, L.Gao, Z.I.Huang, Y.Q.Wang, C.W.Chu. Phys.Rev.Lett. 58; 908 (1987).

MICROWAVE RESPONSE OF THE SUPERCONDUCTING $YBa_2Cu_3O_{9-x}$ CERAMICS

G.I. Leviev, V.G. Pogosov, and M.R. Trunin

Institute of Solid State Physics Academy of Sciences of
the USSR, 142432 Chernogolovka Moscow distr., USSR

The interaction of the $YBa_2Cu_3O_{9-x}$ ceramics with the micro-wave field of frequency $\omega/2\pi = 9.3$ GHz has been studied experimentally in a wide range of temperatures. The superconducting transition was registered according to variation of the high-frequency conductivity of the sample. Radiation of the doubled 2ω and tripled 3ω frequencies was observed on increasing the amplitude of the alternating field. Temperature dependences of harmonics of the incident-wave intensities and of the external magnetic field have been investigated. Two mechanisms of nonlinearity, describing qualitatively the experimental results are proposed.

The recent discovery of high-temperature superconductivity in ceramics have aroused great interest in the investigation of such substances.

The present paper is devoted to studying the properties of the $YBa_2Cu_3O_{9-x}$ ceramics in a microwave field. The sample was made in the following way. The required amounts Y_2O_3, $BaCO_3$ and CuO powders were mixed and ground in a planetary ball mill do the size of about 1 μm. Then the charge was placed into a hermetically sealed rubber sheath and pressed under 5 kbar in a hydrostate. The density of the pressed samples was about 4 g/cm^3. Then the samples were burned in an oxygen flow at 1000°C and held for 5 his. The density of the annealed sample was 3.47g/cm^3 (41% of the theoretical density of $YBa_2Cu_3O_{9-x}$). The specific resistance of the disk-shape sample of 4.5 mm in diameter and 1.4 mm thick made 500 $\mu\Omega\cdot$sm.

The sample was placed in a bimodal resonator which could have been tuned simultaneously to the frequencies ω and 2ω, or ω and 3ω. Rectangular and cylindrical resonators were used. Radiation of fundamental frequency $\omega/2\pi = 9.3$ GHz was applied to the input of the resonator through a 23x10 mm^2 waveguide. The harmonics were extracted through a 11x5.5 mm^2 waveguide. When the waveguides are weakly connected with the resonator, such a device allows also registering variation of the square of the quality factor of the resonator Q^2 at the frequencies $\Omega/2\pi > 15$ GHz in the experiment. For this purpose the electromagnetic wave of the frequency Ω from a klystron was applied to the resonator through a waveguide of a small cross-section and the signal of the same frequency Ω passed through the resonator and was fed to a detector through a wide waveguide. Since in normal conditions $(T > T_c)$ the losses in the resonator are mainly caused by the presence of ceramics, then $Q^2 \sim \sigma$, where σ is the conductivity of the sample. In the superconducting state the resonator Q is determined by the losses in the walls. Fig. 1 shows registering of the sample transition to a superconducting state recorded according to variation of the resonator Q at the frequency $\Omega/2\pi$ =20GHz. The amplitude of the alternating magnetic field $H_\sim$ on the sample was 0.05 Oe.

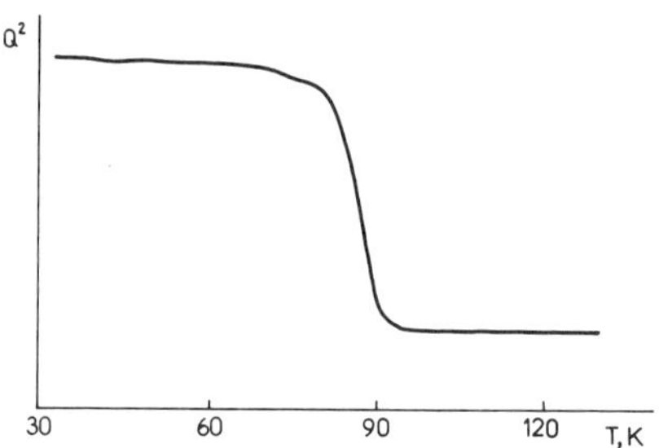

Fig. 1. Registration of the superconducting transition in the YBa$_2$Cu$_3$O$_{9-x}$ ceramics according to variation of the resonator Q at the frequency $\Omega/2\pi$ =20 GHz.

The experiments on nonlinear reflection were carried out at large amplitudes $H_\sim$ using a magnetron, operating in a pulsed regime of a long 2 μs pulse at the repetition rate 50 Hz. The radiation spectrum of magnetron involves harmonics and the latter could serve as a source of a false signal. To protect from the magnetron harmonics, a system of absorbing and reflecting filters was used. The power of the wave applied to the resonator at the frequency varied in the range 10–4000 W, and at the resonator Q equal to 2000 this corresponds to variation of the amplitude $H_\sim$ within the range 5 Oe $< H_\sim <$ 100 Oe. The harmonic signal from the sample was fed to the superheterodyne receiver, sensitive to 10^{-12} W, and then it was registered by a stroboscopic converter. The impulse power $P_{2\omega}$ (or $P_{3\omega}$) of the 2ω (or 3ω) frequency wave under examination was measured. In the experiment we used an external constant magnetic field H $<$ 300 Oe, which was created by the Helmholtz system and could rotate in the plane of the sample surface.

Fig. 2 presents the temperature dependence of the radiation power $P_{3\omega}$ in the absence of a magnetic field (H=0). The zero level of the signal $P_{3\omega}$=0 is indicated. The alternating field amplitude in this experiment is $H_\sim$=10 Oe. Near T_C there is a generation peak of the width of about 2K. Its magnitude and location on the temperature scale are independent of the static field H. At the same time the radiation $P_{3\omega}$ existing in ceramics at T$\ll$T$_C$ and almost temperature-independent vanishes in a weak field (see the inset).

We have compared the dependence $P_{3\omega}$(T,H) in the YBa$_2$Cu$_3$O$_{9-x}$ ceramics with the measurements of $P_{3\omega}$(T,H) taken on the sample of a homogeneous superconducting NbTi alloy. In a powerful alternating field ($H_\sim \sim$100 Oe) near T$\approx$T$_C$=8.2 K this sample exhibited maximum of $P_{3\omega}$(T) analogous to a narrow peak of $P_{3\omega}$ in ceramics. But far from T$_C$ in NbTi no third harmonic signal is observed at any H.

It seems to us that generation in the vicinity of T$_C$ (maximum) in YBa$_2$Cu$_3$O$_{9-x}$ and NbTi is conditioned by strong dependence of the order parameter on the alternating magnetic field H_ω [1]. According to [2], at a temperature dose to T$_C$ the current density can be given by:

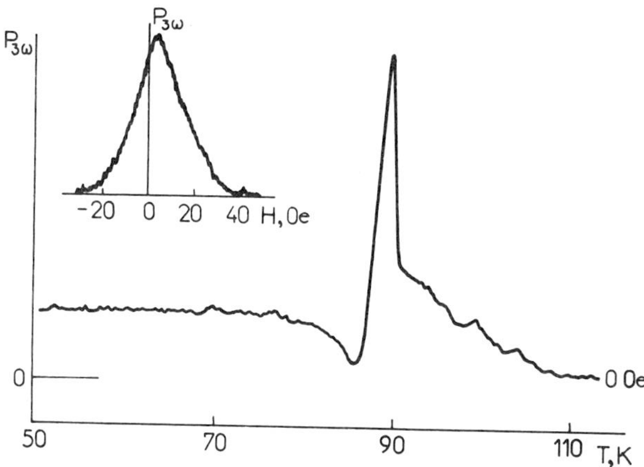

Fig. 2. Example of registration of the third harmonic power $P_{3\omega}$ as a function of temperature. The zero level $P_{3\omega}=0$ is Indicated. The alternating field amplitude on the sample is $H_{\sim}=10$ Oe. In the inset given is the dependence $P_{3\omega}(H)$ at T=60 K.

$$\vec{j} = \sigma\vec{E} - \frac{\pi\sigma|\Delta|^2}{2c\hbar kT_c}\vec{A} \tag{1}$$

Here σ is the conductivity in a normal state; $\vec{E}$, $\vec{A}$ are the electric field and the vector-potential; $|\Delta|^2$ is the square of the order parameter modulus. The current nonlinearity (1) is related to the dependence of $|\Delta|^2$ on the ratio $(H_\omega/H_c)^2$, $H_\sim < H_c$ where H_c is the critical thermodynamic field. As has been shown in [1] the account taken of the dependence $\Delta(H_\omega/H_c)$ in the absence of the static field H results in appearance of odd harmonics of current (1) in a massive superconductor. Since at $T \to T_c$ the field $H_c \to 0$, then at a given amplitude H the amplitude of harmonic $P_{3\omega}$ drastically increases while approaching T_c.

The nonlinear signal in ceramics exists at the temperatures much below T_c, is of different origin, and apparently related to generation in Josephson junctions at the boundaries of crystallites. In case the potential difference $V=V_0 + v\cos\omega t$ is applied to the Josephson element, the superconducting tunnel current $I(t)$ in terms of the junction can be written as follows [3] :

$$\bar{I}(t) = \bar{I}_c \sum_{n=-\infty}^{\infty} (-1)^n \mathfrak{J}_n\left(\frac{2ev}{\hbar\omega}\right) \sin\left[\left(\frac{2eV_0}{\hbar} - n\omega\right)t + \varphi_0\right] \tag{2}$$

where I_c is the critical current, $\mathfrak{J}_n$ is the n-th order Bessel function, φ_0 is the phase difference on the weak coupling. It can be seen from (2) that even in the absence of a constant voltage, $V_0=0$, and $\varphi_0 =0$ the current involves a constituent at the frequency 3ω. The behaviour of the complex system of Josephson junctions in a powerful high-frequency field requires special investigation, but most general properties of a single element are likely to exhibit in the system under consideration. For instance, application of a weak external field, H, results in diminishing the critical current passing through the junction and rectification of its voltampere characteristics. The amplitude of the third harmonic also drops, which has been observed in the experiment (see the inset in Fig.2).

Fig.3 shows the results of studying the non-linear signal $P_{2\omega}$ in the $YBa_2Cu_3O_{9-x}$ ceramics at the second harmonic frequency

2ω. In the registration the amplitude was always $H_\sim$ =18 Oe. To the right of the curves depicted are the values of the external field H. In a zero magnetic field no peak near T_c is observed. At $H \neq 0$ near T_C there appears a maximum of $P_{2\omega}$, its value increasing with H. Such a behaviour of the signal $P_{2\omega} \sim |j_{2\omega}|^2$ follows from (1). Indeed at $H \parallel H_\omega$ in the expansion of $|\Delta|^2$ in powers of H_ω, the term HH_ω arises, which yields the finite value of the current $j_{2\omega}$ at the doubled frequency. An analysis of the polarized dependences $P_{2\omega}$ (H, H_ω) suggests that the peak amplitude near T_C is maximal at $H \parallel H_\omega$ and becomes zero at $H \perp H_\omega$.

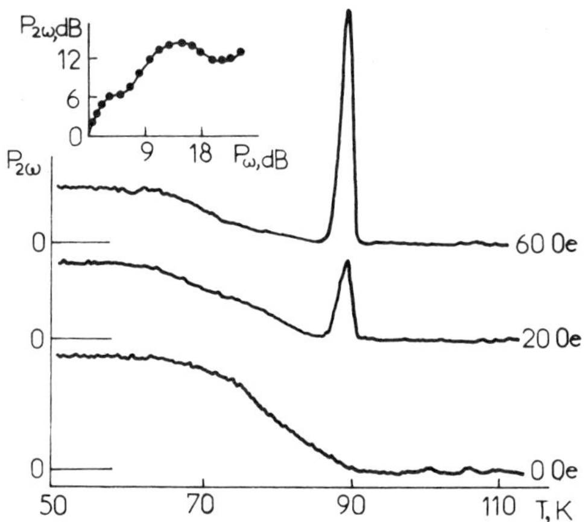

Fig. 3. Temperature-dependences of the second-harmonic signal $P_{2\omega}$ in different magnetic fields H (to the right of the curves). Zero levels $P_{2\omega}$=0 are given. The amplitude is $H_\sim$=18 Oe. In the inset depicted is the graph of $P_{2\omega}$ (P_ω) obtained at T=55 K.

The second-harmonic signal at $T \ll T_C$, as well as at the tribled frequency, is temperature-independent and suppressed by the field $H > 100$ Oe. As follows from Eq. (2), the current I(t) at V_O=0 has the component $I_{2\omega}(t) \neq 0$, provided $\varphi_o \neq 0$, that is, when a supercurrent passes through the junction. This condition can be easily satisfied, if rectification of the current on the sample inhomogeneities is taken into account. The nonmonotonous dependence of power $P_{2\omega}$ on the dropping power P_ω (see the inset) is another argument in favour of the Josephson mechanism of generation at $T \ll T_C$.

The authors thank V.F.Gantmakher, L.P.Gor'kov, B.I.Ivlev, I.F.Chshegolev and G.M.Eliashberg for a discussion of the results.

REFERENCES

1. Gor'kov L.P., Eliashberg G.M. Zh.Eksp.Teor.Fiz. 54:612 (1968) Eliashberg G.M. in the book by V.M.Fine Quantum Radiophysics, Moscow, "Sovetskoe Radio" Publishers, v.1:446 (1972 in Russ.)
2. Abrikosov A.A., Gor'kov L.P., Dzyaloshinskii I.E.Methods of quantum field theory in statistical physics, Moscow,"Fizmatgiz" Publishers, 1962 (in Russian).
3. Shapiro S. Microwave Harmonic Generation from Josephson Junctions. J.Appl.Phys. 38: 1879 (1967).

CRYSTAL PREPARATION OF $(La_{1-x}M_x)_2CuO_{4-\delta}$ (M=Sr AND Ba) AND DISCOVERY OF

MAGNETIC SUPERCONDUCTORS Ln-Ba-Cu-O SYSTEMS (Ln=LANTHANIDE ATOMS)

S. Hosoya[*], S. Shamoto, M. Onoda and M. Sato

Institute for Molecular Science, Myodaiji, Okazaki 444 JAPAN
[*]The Research Institute for Iron, Steel and Other Metals
Tohoku University, Katahira, Sendai 980 JAPAN

INTRODUCTION

The historical discovery of high-T_c superconductor La-Ba-Cu-O by Bednorz and Müller[1] has destroyed certain pessimistic idea that the T_c value does not exceed about 40K and encouraged workers in the field of searching for high-T_c superconductors. It has presented a definite problem why the T_c value of the system is much higher than those of other materials ever known. The superconducting phase was found to have the chemical formula of $(La_{1-x}Ba_x)_2CuO_{4-\delta}$[2,3] with the layers of corner-linked CuO_6 octahedra. As described in the separate paper[4], the main characteristics of the compound seemed to be common to other conductive oxides in which we had been much interested from the view point of strong electron-phonon coupling. Then, it was quite attractive for us to find what differences exist between the high-T_c oxides and other conductive oxides. To achieve this, we have been making much effort to study the physical properties of single crystal specimens.[4]

In addition, we have also been trying to synthesize similar oxides to find new superconducotrs. In this field, a large advance was first carried out by Chu et al.[5]. They discovered the Y-Ba-Cu-O system with T_c higher than 90K. Before the details of the system were known, we found the superconductivity in Ln-Ba-Cu-O systems with the magnetic lanthanide ions[6], where the transition temperalures were as high as that of Y-Ba-Cu-O system.

This paper reports the details of the preparation of single crystals of the high-T_c oxide supercondutors and also reports the method we applied in finding the Ln-Ba-Cu-O systems.

PREPARATION OF SINGLE CRYSTALS

Observation of almost complete shielding of applied magnetic field by single phase $(La_{1-x}Ba_x)_2CuO_{4-\delta}$[7] made us confident that the compound really exhibited the bulk superconductivity. Then, the preparation of the single crystals of $(La_{1-x}M_x)_2CuO_{4-\delta}$ was carried out in order to study the detailed physical properties. We adopted nonstoichiometric melting method which was expected to realize the rapid supply of crystals with a size large enough for certain kinds of experiments. Indeed, we first succeeded to prepare single crystals of $(La_{1-x}M_x)_2CuO_{4-\delta}$ which exhibited superconductivity.

In the preparation process, the largest problem we had to overcome was
to suppress the reaction $CuO \rightarrow Cu_2O + 1/2O_2$ which may occur at high
temperatures. Therefore rapid heating above the melting temperature and
rapid cooling were carried out inductively. Starting materials in Pt
crucible were the mixtures of La_2O, MO and CuO, where the condition of excess
amount of CuO was adopted to lower the melting point. In figure 1, the point

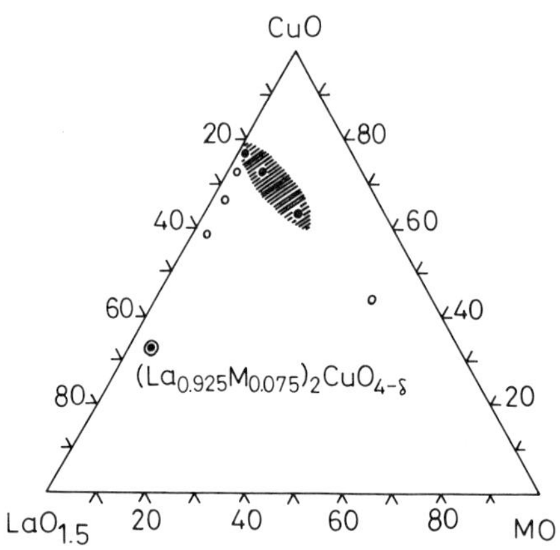

Fig. 1. Appropriate compositions of the mixtures
(shaded region) for rapid growth of
$(La_{1-x}M_x)_2CuO_{4-\delta}$ crystal.

of the metal atom fractions of $(La_{0.925}M_{0.075})_2CuO_{4-\delta}$ is indicated. On the
line which connects this point with the CuO corner, single crystals with
desired K_2NiF_4 type structure were obtained at the point indicated by the
closed circle. Then, to control the M atom concentration, the MO fraction
was changed along the line toward the MO corner. We found at the points
shown by the closed circles in the figure, K_2NiF_4 type crystals could be
obtained. At the points indicated by the open circles K_2NiF_4 type crystals
did not precipitate. The shaded area may be the best area for the crystal
preparation. The melting temperatures within the region are about 1350 °C.
The largest size of the plate-like crystals are about $2 \times 1 \times 0.05 mm^3$.[7] The
concentrations of M atoms were roughly determined by EPMA.

For as-grown crystals, only the compound with Sr atoms exhibited
superconductivity, while the compound with Ba atoms exhibited
superconductivity only after annealing at 900 C for about 15 hours in air.
Even for the crystals which exhibited superconductivity, the value of T_c was
usually much lower than that of sintered specimens as shown in the separate
paper.[4] We have also tried to prepare single crystals of $YBa_2Cu_3O_{7-\delta}$ and
$LnBa_2Cu_3O_{7-\delta}$ (Ln=Dy and Ho) by essentially same method. The obtained
crystals were found to have the δ-value of about 1.0. This large value of δ
or the large amount of oxygen deficiencies can be understood by the fact that
the rapid grenching from high temperatures (below the melting point) induces
the oxygen vacancies. These crystals do not exhibit superconductivity and
have tetragonal structures,[8] while the compounds which exhibit
superconductivity have orthorhombic structures[9] with smaller value

of δ. Then, to prepare crystals with superconducting transition, certain additional techique seems to be necessary. Such works are now underway.

HIGH-T_c SUPERCONDUCTORS WITH MAGNETIC LANTHANIDE ATOMS

In the search for new superconductors, there should be changed various parameters, for examples, the species of atomic elements, their concentrations, annealing temperature and so on, and therefore it is qutie difficult to try all possibilities even if the metal atom fractions are fixed. We tried the inhomogeneous heating method; platinum crucible was heated up inductively, where the temperature of the pellet-shaped specimens in the crucible could be made quite inhomogeneous. Then even for the specimens with the compositions far from the ideal ones for the superconducting phase, we could find the part which exhibited superconductivity. Figure 2 shows our first data, where the

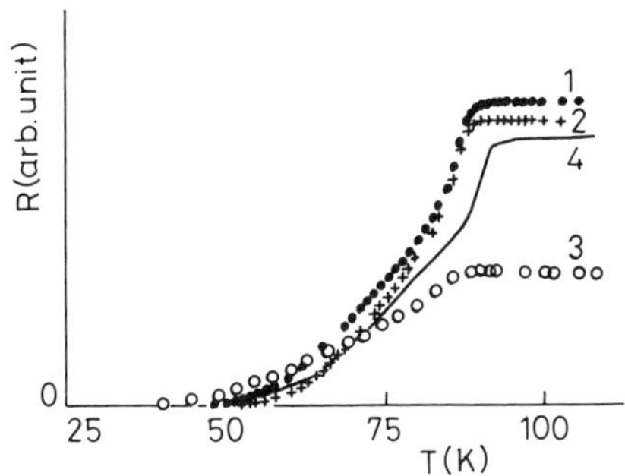

Fig. 2. Temperature dependence of the resistivities of the specimens with the nominal compositions 1. Yb_6BaCuO_x, 2. $Yb_{1.6}La_{0.2}Ba_{0.2}CuO_x$, 3. $Lu_{1.8}Ba_{0.2}CuO_x$ and 4. $Yb_{0.9}Ba_{0.1}CuO_x$.

superconductivity with T_c higher than 90K was discovered even for the compounds with magnetic lanthanide ions.[6] After this finding, the concentrations of the metal atoms were varied. Figure 3 shows the examples of the results. We have found the high-T_c superconductivity in Ln-Ba-Cu-O systems with Ln=Lu, Yb, Tm, Er, Ho, Dy, Gd, Eu, Sm and Nd. We reported[6] Ho-Sr-Cu-O also exhibited superconductivity (cuve 1 in figure 2a). However, we cannot reproduce it again. The structures of these systems were known to be almost same as that of $YBa_2Cu_3O_{7-\delta}$. Based on the resistivities shown in figure 3 and the X-ray diffraction data, we pointed out that the single phase specimens did not necessarily exhibit the sharper transition than the multiphase specimens, which suggested certain possibilities of surface superconductivity. However, this seems to be now understood by the fact that the superconductivity is very sensitive to the condition of the thermal treatments on which the δ-value or the number of oxygen vacancies depend very sensitively.

Figure 4 shows the relationship between the annealing temperatures and the compositions of the initial mixtures where the high-T_c superconductivity could be observed. From this figure it is known that in quite wide regions the superconductivity can be found by annealing at a proper temperature and therefore inhomogeneous heating is considered to be one of the powerful method to search for new superconductors.

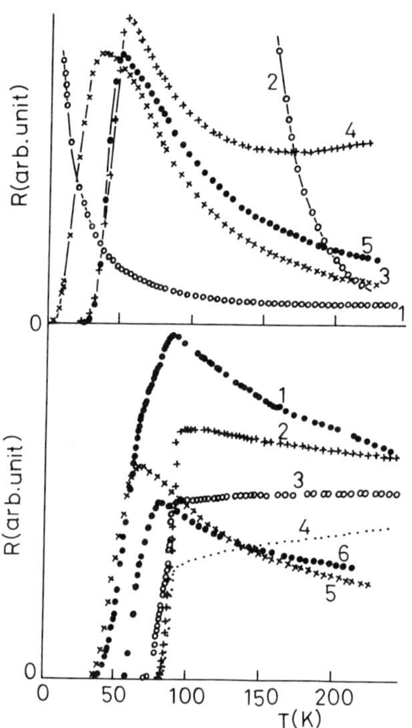

Fig. 3. Temperature dependence of the resistivities of the specimens with the nominal compositions.
(a) 1. $Ho_{1.8}Sr_{0.2}CuO_x$, 2. $Er_{0.5}Ba_{0.5}CuO_x$,
3. $Lu_{0.7}Ba_{0.3}CuO_x$ 4. $Tm_{0.5}Ba_{0.5}CuO_x$,
5. $Dy_{0.5}Ba_{0.5}CuO_x$ and 6. $Eu_{0.5}Ba_{0.5}CuO_x$.
(b) 1. $LaBa_2Cu_3O_x$, 2. $Pr_2Ba_3Cu_5O_x$,
3. $Nd_2Ba_3Cu_5O_x$, 4. $SmBa_2Cu_3O_x$ and
5. $Gd_2Ba_3Cu_5O_x$.

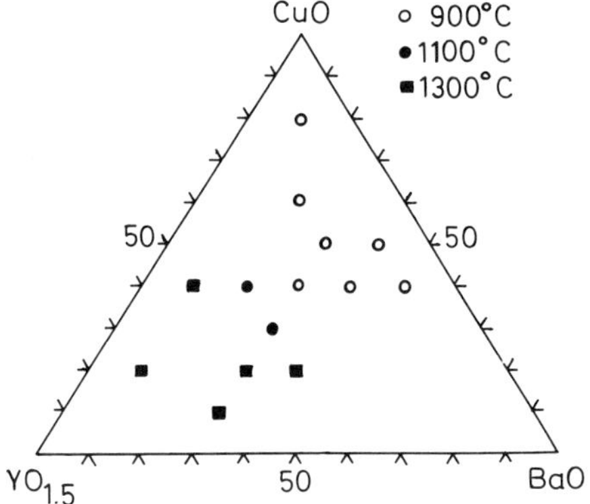

Fig. 4. Compositions where the superconducting
transition can be observed after sintering
at temperatures 900 °C (open circles),
1100 °C (closed circles) and 1200 °C
(squares).

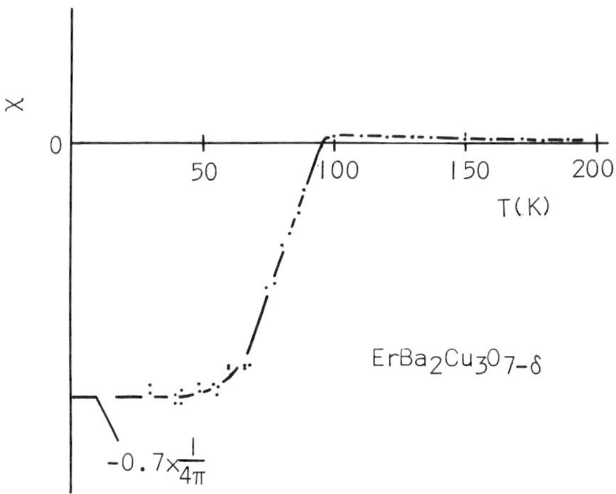

Fig. 5. Magnetic susceptibility of $ErBa_2Cu_3O_x$
taken with the field of 10 Oe. A large
volume fraction is known to be
superconducting below T_c.

A large Meissner diamagnetism observed in $ErBa_2Cu_3O_{7-\delta}$ is shown in
figure 5, which indicates the bulk nature of the high-T_c superconductivity
even for the compound with magnetic lanthanide atoms. The insensitivity of
the transition temperature to the existence of the magnetic atoms were first
pointed out by the present authors. Nowadays, the structures of the compound
has become clear[8-10], where the magnetic atoms are far from the conducting
layers of Cu and O atoms. This can qualitatively explain why the T_c value is
insensitive to the existence of the magnetic atoms.

Finally, it should be added that the temperature dependence of the
magnetic susceptibility above T_c clearly indicates the existence of Ln^{3+}
ions. In case of $ErBa_2Cu_3O_{7-\delta}$, the effective moment of Er atoms is extimated
to be 9.33 from the experiment, which should be compared with the expected
value of 9.58 for Er^{3+} ions.

REFERENCES

1. J. G. Bednorz and K. A. Müller , Z. Phys. B64: 189 (1986).
2. J. G. Bednorz, M. Takashige and K. A. Müller, Europhys. Letters 3: 379
   (1987).
3. H. Takagi, S. Uchida, K. Kitazawa and S. Tanaka, Jpn. J. Appl. Phys. 26:
   L123 (1987).
4. M. Sato, S. Shamoto, M. Onoda, M. Sera, S. Hosoya, J. Akimitsu and T.
   Ekino, this issue.
5. C. W. Chu, P. H. Hor, R. L. Meng, L. Gao, Z. J. Huang, Y. Q. Wang, J.
   Bechtold, D. Campbell, M. K. Wu, J. Ashburn and C. Y. Huang, Phys.
   Rev. Lett. to be published.
6. S. Hosoya, S. Shamoto, M. Onoda and M. Sato, Jpn. J. Appl. Phys. 26: L325
   (1987) and 26: L456 (1987) and S. Hosoya, S. Shamoto, M. Onoda,
   M. Sato and S. Hosoya, Jpn. J. Appl. Phys. 26: L642 (1987).
7. M. Sato, S. Hosoya, S. Shamoto, M. Onoda, K. Imaeda and H. Inokuchi,
   Solid State Commun. 62: 85 (1987).
8. M. Onoda, S. Shamoto, M. Sato and S. Hosoya, Jpn. J. Appl. Phys. 26: L876
   and this issue.
9. P. M. Grant, R. B. Beyers, E. M. Engler, G. Lim, S. S. P. Parkin, M. L.
   Ramirez, V. Y. Lee, A. Nazzal, J. E. Vazquez and R. J. Savoy, Phys.
   Rev. B35: 7242 (1987).
10. M. A. Beno, L. Soderholm, D. W. Capone II, D. G. Hinks, J. D. Jorgensen,
   I. K. Schuller, Appl. Phys. Lett. to be published.

OPTICAL-REFLECTANCE STUDY OF THE SINGLE-CRYSTAL SUPERCONDUCTOR

$(La_{1-x}Sr_x)_2CuO_4$

T. Koide, H. Fukutani,[*] A. Fujimori,[**] R. Suzuki,[*] T. Shidara,
T. Takahashi,[†] S. Hosoya,[††] and M. Sato[§]

Photon Factory, National Laboratory for High Energy Physics
Oho-machi, Ibaraki 305, Japan
[*] Institute of Physics, The University of Tsukuba, Sakura-mura
Ibaraki 305, Japan
[**]National Institute for Research in Inorganic Materials,
Sakura-mura, Ibaraki 305, Japan
[†] Department of Physics, Tohoku University, Sendai 980, Japan
[††]The research Institute of Iron, Steel and Other Metals
Tohoku University, Sendai 980, Japan
[§] Institute for Molecular Science, Myodaiji, Okazaki 444, Japan

INTRODUCTION

The discovery of superconductivity above 30 K in the La-Ba-Cu-O system
by Bednorz and Müller[1] has stimulated intensive studies of the superconducting members of the Cu-O perovskite family. The subsequent search for higher-
temperature superconductivity has recently led to the attainment of a critical temperature Tc of 93 K in a mixed-phase Y-Ba-Cu-O compound system.[2] The
existence of two-dimensional copper-oxide planes is an important feature of
both the layered perovskite K_2NiF_4 structure of $(La_{1-x}M_x)_2CuO_4$ (M=Ba, Sr,
and Ca) and the oxygen-deficient perovskite structure of $LnBa_2Cu_3O_{7-\delta}$ (Ln=Y,
and lanthanide elements). This high anisotropy is expected to play an important role in the physical properties of these materials. Thus, measurements of single-crystal properties[3,4] are essentially important to an understanding of their structure-property relationships and the mechanism of
high-Tc superconductivity itself. However, most of the experimental studies
reported so far have been made on polycrystalline samples. In this work
we report on room-temperature reflectance measurements from 0.55 to 6 eV
in a single crystal $(La_{1-x}Sr_x)_2CuO_4$. The measured spectrum is Kramers-
Kronig analyzed to obtain the dielectric function and related functions, and
the result is discussed with reference to the electronic properties of the
material.

EXPERIMENTAL

The single crystal of $(La_{1-x}Sr_x)_2CuO_{4-\delta}$ was prepared by the nonstoichiometric melting method, the details of which are described in ref. 5.
The sample had a composition with x = 0.075 and δ of a small positive value.
The onset temperature for superconductivity was found to be ∿ 23 K. A small
piece of the sample with the crystal face perpendicular to the c-axis was
used for the present experiment.

The near-normal-incidence reflectance was measured at room temperature from 0.55 to 6.0 eV. A tungsten-arc lamp and a hydrogen-discharge lamp were used as a light source. The detectors used were photomultipliers in the 1.0 - 6.0 eV range and a PbS cell in the 0.55 - 1.5 eV range. The time variation of the light-source intensity was cancelled out by monitoring the intensity of light reflected from a quartz-plate beam splitter which was placed behind the exit slit of the monochromator. In order to avoid a possible error arising from the position-dependent sensitivity of the detector, we used a specially-designed sliding-type sample holder; this allowed measurements of the intensities of light reflected from the sample and a quartz plate without changing the optical path. The reflectance of the sample was determined from the known values of optical constants of quartz.

RESULTS AND DISCUSSIONS

Figure 1 shows the reflectance (R) spectrum in the range between 0.55 and 6.0 eV. An expanded spectrum in the 1-6 eV range is also shown for clarity. A steep increase in reflectance is seen with decreasing energy in the infrared region below $\sim$ 1.1 eV, indicating the plasma edge of free carriers characteristic of metallic conduction. It is noted that the reflectance even exceeds 80 % at the lower energy limit of the present experiment. In the region above $\sim$ 1.1eV, in contrast, the spectrum has a rather flat profile with relatively low values of 13 $\sim$ 20 %. Nevertheless, the expanded figure reveals clear peaks located at $\sim$ 1.4, $\sim$ 2.5, $\sim$ 3.3 and $\sim$ 5.7 eV, and shoulders at $\sim$ 4.5 and $\sim$ 5.3 eV. Tajima et al.[6] have reported the plasma spectrum of a sintered, mechanically polished sample of $(La_{1-x}Sr_x)_2CuO_4$ (x = 0.09). Their spectrum shows much lower reflectances than those of ours in the infrared region; R $\leq$ 30 % over the range corresponding to the present experimental region (hν $\geq$ 0.55 eV). The high reflectance of our single crystal indicates that more metallic property manifests itself in the crystal plane perpendicular to the c-axis, as reasonably expected from its crystallographic structure.

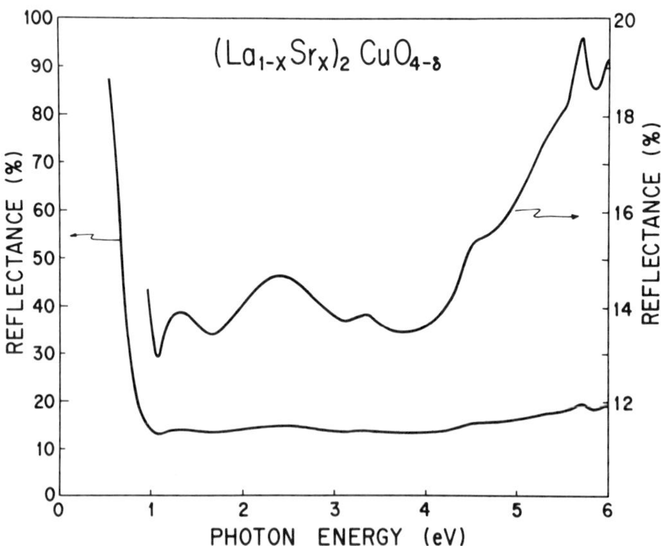

Fig. 1   Reflectance spectrum of the single-crystal $(La_{1-x}Sr_x)_2CuO_{4-\delta}$.

In order to get an insight into the electronic processes participating in these spectral features, Kramers-Kronig analysis was used to determine the optical constants from the measured spectrum with extrapolation to lower and higher energies. The reflectance below 0.6 eV was assumed to be that of free carriers, the plasma energy $\hbar\omega_p$ and the relaxation time τ of which were adjusted to give a best fit to the measured reflectance between 0.6 and 0.7 eV. Above 6 eV an extrapolation of R(E) was used with R decreasing as the $-p$ power (p>0) of the photon energy. At present, we have no more information to uniquely determine the parameter p. However, the plasma energy deduced from the Kramers-Kronig analysis was found to be independent of the choice of p $(3 \leqq p \leqq 6)$.

Figure 2 shows the optical conductivity $\sigma = \varepsilon_2 E/4\pi\hbar$ and the energy loss function $-\mathrm{Im}\varepsilon^{-1} = \varepsilon_2/(\varepsilon_1{}^2 + \varepsilon_2{}^2)$ for a parameter of p = 4, where ε_1 and ε_2 are the real and imaginary parts of the dielectric function ε. Because of uncertainty in determining the value of parameter p, the absolute values of the spectra do not bear significant meaning particularly at higher energies. Hence, we shall be mainly concerned with the spectral shapes and energies of features.

It is found that the loss function exhibits a sharp peak at the plasma energy of $\hbar\omega_p$ = 0.85 eV; this value agrees closely with that of Tajima et al.[6] The plasma frequency ω_p is related to the dc conductivity σ_0 by $\sigma_0 = (\omega_p)^2\tau/4\pi$. Using the value of $\tau = 1.27 \times 10^{-14}$ (sec) determined from the Drude fitting, we obtain $\rho_0 = 1/\sigma_0 = 0.54 \times 10^{-3}$ ($\Omega\cdot$cm), which agrees with the measured room-temperature resistivity of $\rho_0 \sim 10^{-3}$ ($\Omega\cdot$cm) within the experimental accuracy. The plasma frequency is also represented as $(\omega_p)^2 = 4\pi Ne^2/m_{op}$, where N is the carrier density and m_{op} is the optical mass. Using $N \sim 5 \times 10^{21}$ (cm^{-3}) reported for $(La_{0.9}Sr_{0.1})_2CuO_4$[7], we obtain $m_{op} \sim$ 10 m, where m is the free electron mass. This large mass is consistent with the observation by Kwok et al.[8] that the specific heat γ of the La-Sr-Cu-O system is remarkably large as compared with that of the Ba-Pb-Bi-O system.

A comparison of the reflectance and the conductivity spectra shows that there is a one-to-one correspondence of the interband structures above $\sim$ 1.1 eV. The features observed in the 4.5-6 eV region would be dominated by the transition from the O^{2-}: 2p level to the La^{3+} : 5d level. In fact, the electron-energy-loss spectrum of Strasser et al.[9] for a highly oxidized La surface exhibits the interband transition at $\sim$ 5 eV, which is attributed to

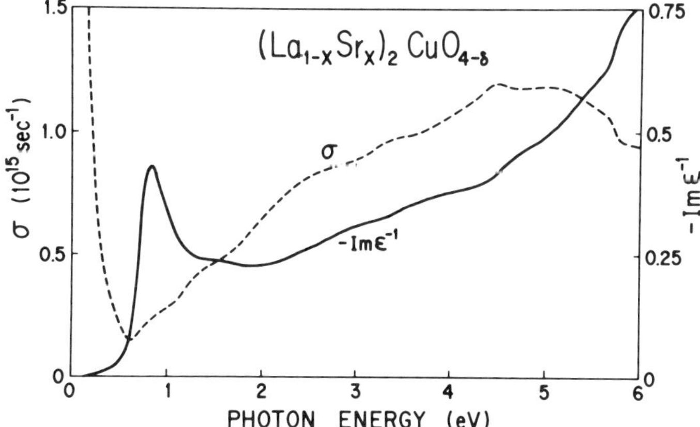

Fig. 2   The optical conductivity and the energy-loss function of $(La_{1-x}Sr_x)CuO_{4-\delta}$.

the O^{2-}: $2p \rightarrow La^{3+}$: 5d transition in La_2O_3. Band-structure calculations for La_2CuO_4 by Mattheiss[10] and Takegahara et al.[11] also indicate the existence of such a transition. Therefore, the dominant peak at $\sim$ 5.7 eV is assigned to the O^{2-} : $2p \rightarrow La^{3+}$: 5d transition. The substitution of Sr^{2+} for La^{3+} is expected to allow a small contribution of the O^{2-} : $2p \rightarrow Sr^{2+}$: 4d transition as well. Since the O^{2-} : $2p \rightarrow Sr^{2+}$: 4d interband transition has been observed at $\sim$ 5eV in the reflectance of SrO,[12] the shoulder at $\sim$ 4.5 eV or $\sim$ 5.3 eV might be ascribable to this transition. The band calculations[10,11] for La_2CuO_4 exhibit no high density of states above E_F leading to the optical features in the 1 – 3.5 eV range. On the other hand, the optical gap in CuO due to the O^{2-} : $2p \rightarrow Cu^{2+}$: 3d transition has been observed at $\sim$ 3eV.[13] It has also been reported[14] that the oxygen 2p valence band in $SrTiO_3$ shows a splitting of 0.42 eV. If a similar splitting is supposed to occur in the present $(La_{1-x}Sr_x)_2CuO_4$ compound, the peaks at $\sim$ 2.5 and $\sim$ 3.3 eV can be reasonably assigned to the O^{2-} : $2p \rightarrow Cu^{2+}$: 3d transitions. Furthermore, the O^{2-}: $2p \rightarrow Cu^{3+}$: 3d transition is expected to be observed at lower energies because of the existence of Cu^{3+} arising from the Sr^{2+} substitution. Hence, a small peak at $\sim$ 1.4 eV could be ascribed to this transition. The observation and assignment of these low-energy features would give experimental support to the validity of the localized-model picture[15] in understanding the electronic properties of this class of superconducting materials.

REFERENCES

1. J. G. Bednorz and K. A. Müller, Z. Phys. B64: 189 (1986).

2. M. K. Wu, J. R. Ashburn, C. J. Torng, P. H. Hor, R. L. Meng, L. Gao, Z. J. Huang, Y. Q. Wang, and C. W. Chu, Phys. Rev. Lett. 58: 908 (1987).

3. M. Sato, S. Hosoya, S. Shamoto, M. Onoda, K. Imaeda, and H. Inokuchi, Solid State Commun. 62: 85 (1987).

4. S. Shamoto, M. Onoda, M. Sato, and S. Hosoya, Solid State Commun. 62: 479 (1987).

5. S. Hosoya, S. Shamoto, M. Onoda, and M. Sato, this issue.

6. S. Tajima, S. Uchida, S. Tanaka, S. Kanbe, K. Kitazawa, and K. Fueki, Jpn. J. Appl. Phys. 26: L432 (1987).

7. S. Uchida, H. Takagi, K. Kishio, K. Kitazawa, K. Fueki, and S. Tanaka, Jpn. J. Appl. Phys. 26: L443 (1987).

8. W. K. Kwok, G. W. Crabtree, D. G. Hinks, D. W. Capone, J. D. Jorgensen, and K. Zhang, Phys. Rev. B35: 5343 (1987).

9. G. Strasser, G. Rosina, E. Bertel, and F. P. Netzer, Surface Science 152/153: 765 (1985).

10. L. F. Mattheiss, Phys. Rev. Lett. 58: 1028 (1987).

11. K. Takegahara, H. Harima, and A. Yanase, Jpn. J. Appl. Phys. 26: L352 (1987).

12. Y. Kaneko, K. Morimoto, and T. Koda, J. Phys. Soc. Jpn. 52: 4385 (1983).

13. S. Hüfner, Solid State Commun. 47: 943 (1983).

14. K. W. Blazey, M. Aguilar, J. G. Bednorz, and K. A. Müller, Phys. Rev. B27: 5836 (1983).

15. A. Fujimori, E. Takayama-Muromachi, Y. Uchida, and B. Okai, Phys. Rev. B (in press).

SINGLE CRYSTAL X-RAY DIFFRACTION STUDY OF $(La_{1-x}M_x)_2CuO_{4-\delta}$ (M=Sr AND Ba), $La_2CuO_{4-\delta}$ AND $LnBa_2Cu_3O_{7-\delta}$ (Ln=Y, Dy AND Ho) SYSTEMS

Masashige Onoda, Shin-ichi Shamoto, Masatoshi Sato
and Syoichi Hosoya[*]

Institute for Molecular Science, Okazaki National Research
Institutes, Myodaiji, Okazaki 444, Japan
[*]The Research Institute for Iron, Steel and Other Metals
Tohoku University, Katahira, Sendai 980, Japan

INTRODUCTION

Since the discovery of high-T_c superconductivity in a mixture of
compounds in the La-Ba-Cu-O system by Bednorz and Müller,[1] various Ln-M-Cu-O
systems have been synthesized and characterized, where Ln is one of various
lanthanide elements or yttrium and M is an element with a divalent ion such
as Ba, Sr or Ca. Up to now, two distinct superconducting phases, which are
attributed to the T_c values of 30K[1] and 90K[2-4] ranges, have been established.
The former is a K_2NiF_4-type layered structure, $(La_{1-x}M_x)CuO_{4-\delta}$, where δ is a
small positive constant.[5] The latter is an oxygen defect perovskite
structure, $LnBa_2Cu_3O_{7-\delta}$.[6] In these systems, the structural properties such
as a crystal symmetry or oxygen vacancy play a crucial role in an appearance
of the high-T_c superconductivity. Most of the studies of these systems are
for sintered specimens. But, in order to clarify their detailed electronic
and structural properties, the experimental works on single crystals are
quite important. This paper reports the results on detailed structure
analyses of the various Ln-M-Cu-O systems, that is, $(La_{1-x}M_x)_2CuO_{4-\delta}$ (M=Sr
and Ba), $La_2CuO_{4-\delta}$ and $LnBa_2Cu_3O_{7-\delta}$ (Ln=Y, Dy and Ho) by X-ray four circle
diffraction, and on a relation between the structural and superconducting
properties. A part of this report has already been published.[7]

EXPERIMENTAL METHODS

The present single crystals were prepared by nonstoichiometric melting
technique and by rapid quenching. Details on the crystal preparation are
described in a separate paper.[8] The electrical resistivity was measured with
conventional four terminal technique by an ac resistance bridge. Intensity
data for the structure analyses were collected at room temperature on Rigaku
AFC-5 and 5R four circle diffractometers and Huber off-center type goniometer
with a Rigaku rotational anode type generator, employing the θ-2θ scan
technique up to $2\theta=80°$ with a graphite-monochromatized MoKα radiation. For
$La_2CuO_{4-\delta}$, intensity data were also obtained at 528K. Lorentz-polarization
and absorption correction were applied. All the structures were solved from
three-dimensional Patterson maps calculated by UNICS III[9] and refined by the
full-matrix least-squares program RADIEL,[10] where an isotropic secondary
extinction effect was assumed. The oxygen atom positions were derived from

Table 1. Crystal Data and Final R and R_w Factors of $(La_{1-x}M_x)_2CuO_{4-\delta}$ (M=Sr, Ba and La) and $LnBa_2Cu_3O_{7-\delta}$ (Ln=Y, Dy and Ho) Systems

System	$(La_{1-x}M_x)_2CuO_{4-\delta}$ (x≈0.075)				$LnBa_2Cu_3O_{7-\delta}$		
M, Ln	Sr	Ba	La(LT)	La(HT)	Y	Ho	Dy
δ	0.0	0.0	0.0	0.0	1.0	1.0	0.75(8)
S.G.	I4/mmm	Pccn[a]	Abma	I4/mmm	P4/mmm	P4/mmm	P4/mmm
Z	2	4	4	2	1	1	1
a(Å)	3.787(1)	5.378(2)	5.406(2)	3.812(1)	3.864(1)	3.865(1)	3.868(1)
b(Å)		5.371(2)	5.357(2)				
c(Å)	13.225(4)	13.234(4)	13.145(5)	13.219(2)	11.849(2)	11.824(4)	11.811(3)
R	0.026	0.032	0.038	0.073	0.043	0.029	0.042
R_w	0.037	0.041	0.049	0.087	0.057	0.033	0.066

[a]Average structure.

difference fourier maps. Here the atomic scattering factors and the anomalous scattering corrections were taken from ref. 11. The crystal data and the final R and R_w factors of all the compounds are summarized in Table 1. The calculations were carried out on the HITAC M-680H computer at the Computer Center of the Institute for Molecular Science.

RESULTS AND DISCUSSION

$(La_{1-x}M_x)_2CuO_{4-\delta}$ System (M=Sr and Ba)

Among many prepared batches, we could find a few batches which contained crystals with superconducting transition. Figure 1 shows the several resistivity data of the Ba crystal specimens. As is seen from the figure, the temperature dependence of resistivity is strongly sample dependent. The onset transition temperatures of the Sr and Ba compounds which were used for the structure analyses were about 24K and 16K, respectively. The x values for both the compounds were fixed to be 0.075 which were roughly estimated from EPMA and the δ values were also fixed to be 0, because by their refinements we could not gain any meaningful information.

The Sr compound was found to have K_2NiF_4-type layered structure with the CuO_6 octahedra elongated along the z direction, as shown in Fig. 2(a). From the Cu-O and La(Sr)-O distances, we find that electrons of the Cu atoms are strongly hybridized with the p-orbitals of O atoms in a basal xy plane and therefore the compound exhibits two-dimensional electronic properties. For the Ba compound, we previously assigned the space group as Pccn, which has orthorhombic distortion due to the rotation of CuO_6 octahedra shown in Fig. 2(b). However, this space group is different from that (Abma) of the end-member compound La_2CuO_4 as will be discussed later and therefore we have tried to check the existence of this new type orthorhombic cell by examining an existence of twinning for the present crystal. From high resolution X-ray measurement, we have observed diffuse streaks which cross each other at high angle Bragg points. This can be explained by the twinning of orthorhombic crystals with certain distribution of the (b/a-1) values, where the lattice constants, a and b are expressed by $\sqrt{2}a_t+\eta$ and $\sqrt{2}a_t-\eta$, respectively (a_t is tetragonal lattice constant). The present Ba compound seems to be located near the tetragonal (I4/mmm)-orthorhombic (possibly Abma) boundary, which depends on the Ba and/or oxygen concentration(s). They may be subjected to have a little incomplete chemical composition which means the inhomogenity of

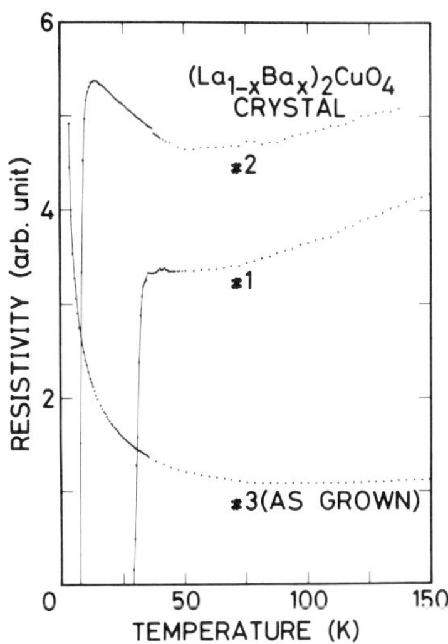

Fig. 1. Temperature dependence of the resistivity of $(La_{1-x}Ba_x)CuO_{4-\delta}$ crystals. The crystals #1 and #2 are annealed for 15 hours at 950°C in air, while the crystal #3 is as-grown one. The absolute values of resistivity at about 95K of #1, 2 and 3 are 4.38, 2.71 and 23.8mΩcm, respectively.

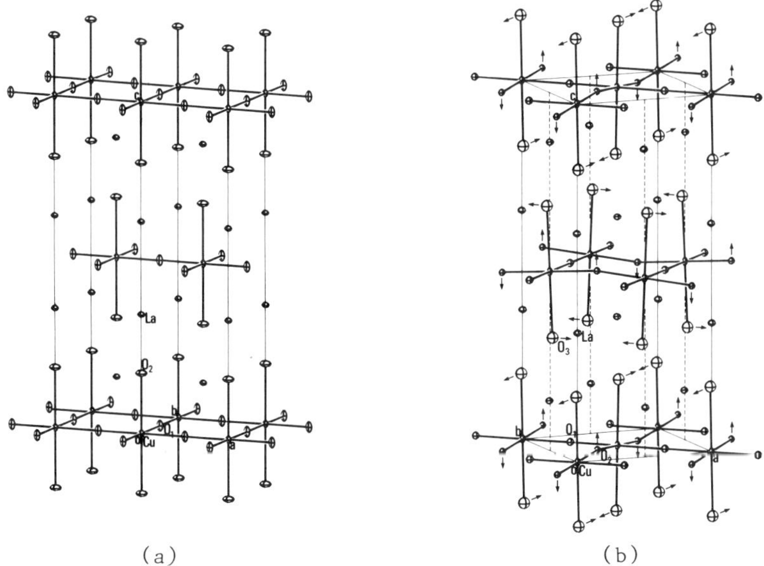

(a) (b)

Fig. 2. Crystal structures of $(La_{1-x}M_x)_2CuO_{4-\delta}$;
(a) $(La_{0.925}Sr_{0.075})_2CuO_4$ with space group I4/mmm and (b) $(La_{0.925}Ba_{0.075})_2CuO_4$ with space group Pccn (average), where the directions of oxygen displacements are indicated by arrows.

Ba and/or oxygen concentration(s). Actually the as-grown crystals of the Ba
compound did not become superconducting, while the annealed ones in air
exhibited superconducting transition as shown in Fig. 1. A drop of the T_c
values in the present crystals as compared with those in usual sintered
specimens may be understood for this reason. Therefore, the structure of the
Ba compound shown in Fig. 2(b) is average one for twinned crystals which are
formed by the orthorhombic cell (Abma). The pattern of oxygen displacements
shown in Fig. 2(b) is explained by a certain superposition of that in
La_2CuO_4 for twinned crystals. It has been suggested that the partial
substitution of La in La_2CuO_4 with Sr or Ba stabilizes the high symmetry
phase or suppresses the possible charge-density-wave (CDW) distortion.[12,13]
This idea seems to be inconsistent with the present result, because the
superconducting phase in the present Ba compound is found to be orthorhombic.

Undoped Compound $La_2CuO_{4-\delta}$

It is important to study the basic physics of undoped compound $La_2CuO_{4-\delta}$
in order to understand the origin of high-T_c in the corresponding
superconducting system. It has been reported that this compound is
orthorhombic at room temperature,[14] but changes into tetragonal at about
530K.[15] As for the physical properties, an existence of CDW[12,13] or spin-
density-wave (SDW)[16,17] or a certain kind of antiferromagnetic transition[18]
has been pointed out from both the experimental and theoretical points of
view.

To clarify whether the tetragonal-orthorhombic transition is associated
with the CDW formation, the room temperature structure and the structural

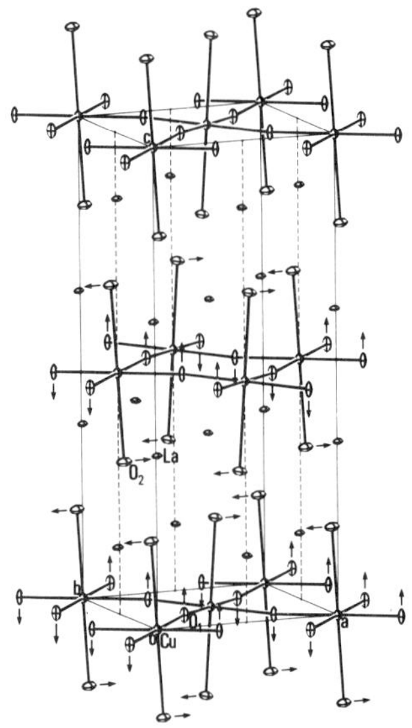

Fig. 3. Crystal structure of
La_2CuO_4 with space
group Abma at room
temperature, where
the directions of
oxygen displacements
are indicated by
arrows.

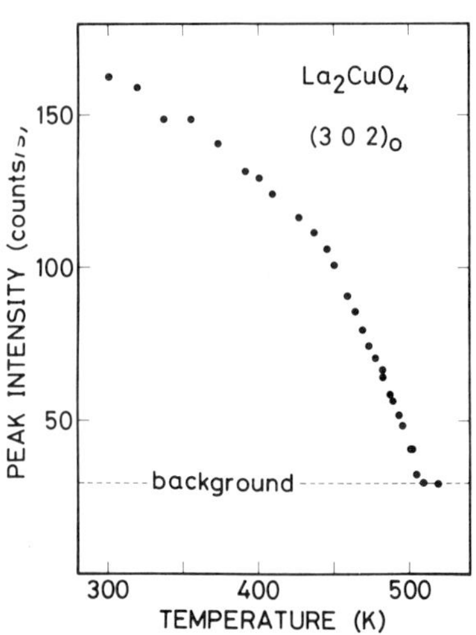

Fig. 4. Temperature dependence of
the peak intensity of the
superlattice reflection
$(3\ 0\ 2)_o$ in La_2CuO_4.

922

transition were reexamined. The reflection data at room temperature at first
sight seem to show the primitive orthorhombic cell. But high resolution X-
ray measurement indicates that the present crystal has two separated
fundamental spots. The intensity ratio is about 2:1 which agrees well with
that of superlattice reflections of (h k l) and (k h l) with h+k=2n+1. Thus
it is possible to think that the weaker superlattice reflections are
originated from the weaker domain of twinned crystals. In addition, at
several superlattice points we confirmed that the reflections do not split.
Taking these results into consideration, we may conclude that the present
crystal has really the A-face centered lattice. Then we tried to
approximately analyze the structure at room temperature by eliminating the
superlattice reflections due to the weaker domain, although the observed
intensities of fundamental reflections were affected by the two domains of
twinned crystals. The crystal structure is shown in Fig. 3. There is no
serious difference between this structure and the previously reported one.[14]
(Because of the existence of the domain structures, there exist certain error
of the order parameter.) The δ value was fixed to be 0, because its

Table 2. Atomic Coordinates ($\times 10^5$) and Equivalent Isotropic
Thermal Parameters of La_2CuO_4 at T=528K

Atom	Site	x	y	z	$B_{eq}(Å^2)$
La	4e	0	0	36137(8)	0.91(2)
Cu	2a	0	0	0	0.82(5)
O(1)	4c	0	50000	0	1.65(36)
O(2)	4e	0	0	18386(123)	1.82(26)

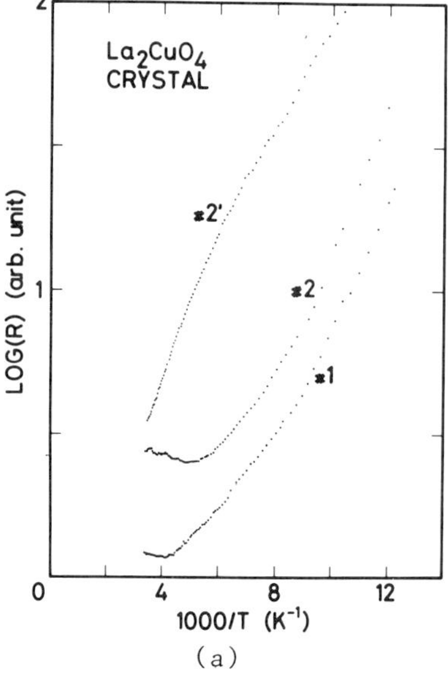

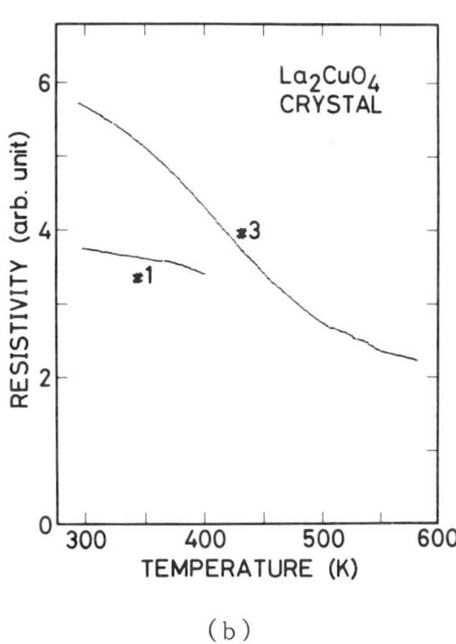

Fig. 5. Resistivity data of $La_2CuO_{4-\delta}$ in the c plane;
(a) low temperature region and (b) high temperature one.
The absolute values of resistivity at room temperature of
#1, 2, 2´ and 3 are 0.0557, 1.10, 0.970 and 2.23Ωcm,
respectively.

refinement did not give any meaningful information. The structural
transition was found to occur at about 510K and to be second-order from the
temperature dependence of the peak intensity of superlattice reflection,
shown in Fig. 4. The atomic parameters of tetragonal structure (I4/mmm)
of La_2CuO_4 at 528K are listed in Table 2. Comparing between the structures
above and below the transition, we find that the rotation of CuO_6 octahedra
about the tetragonal $\langle 1\ 1\ 0\rangle$ direction causes the orthorhombic distortion.
Following the group theoretical consideration by Takegahara et al.,[19] we can
conclude that this orthorhombic distortion is not due to the CDW formation.
Another evidence on the absence of CDW state is expected to be obtained from
the resistivity behavior. The resistivity data of several crystal specimens
from 80K to 600K are shown in Fig. 5. The temperature dependence of the
resistivity is strongly sample dependent, which may be caused by the oxygen
vacancies. It is difficult to mention if the transition is due to the CDW
formation or not., At least, the temperature dependence of resistivity seems
to be semiconducting, and here it should be noted that the real structure of
the Ba compound is possibly identical with that of La_2CuO_4 at room
temperature. We also know that the crystal #3 in Fig. 1 is metallic above
100K. These considerations suggest an absence of the CDW transition in the
$(La_{1-x}M_x)_2CuO_{4-\delta}$ system. On the other hand, it is not clear whether the
slight decrease of resistivity below room temperature is attributable to the
SDW transition. This problem is discussed in a separate paper.[20]

$LnBa_2Cu_3O_{7-\delta}$ System (Ln=Y, Dy and Ho)

The δ values determined from the structure analyses are listed in Table
1. The present structures are oxygen defect perovskite ones with the c axis
triplicated due to the ordering of Ba and Ln. Figures 6(a) and (b) show the
structures of $HoBa_2Cu_3O_6$ and $DyBa_2Cu_3O_{6.25}$, respectively. In the case of
Ln=Y and Ho, all oxygen atom sites around the Y and Ho atoms in the c plane
are vacant and those in the layers which contain the Cu(1) sites are also
vacant. In both cases, therefore, the oxygen ordering in the c plane
containing the Cu(1) site, which would induce the orthorhombic distortion can
not be expected. This is consistent with the observed tetragonal symmetry.
We have also confirmed by high resolution X-ray measurement that the twinning
of orthorhombic cell reported by Syono et al.[21] does not exist for the
present crystals. In the Dy compound, oxygen atom sites in the layers
containing the Cu(1) sites are occupied with a probability of 0.12±0.04. No

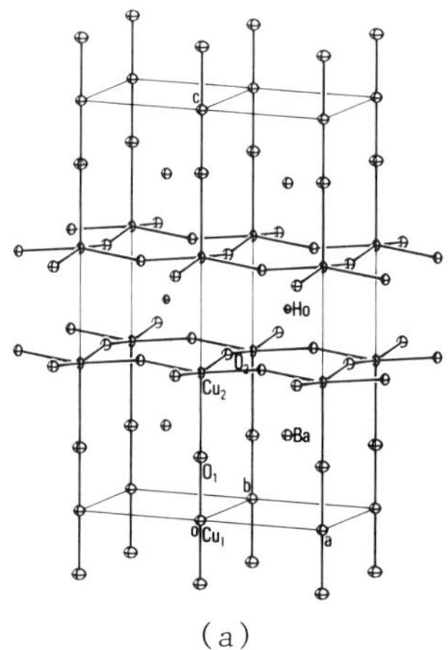

(a)

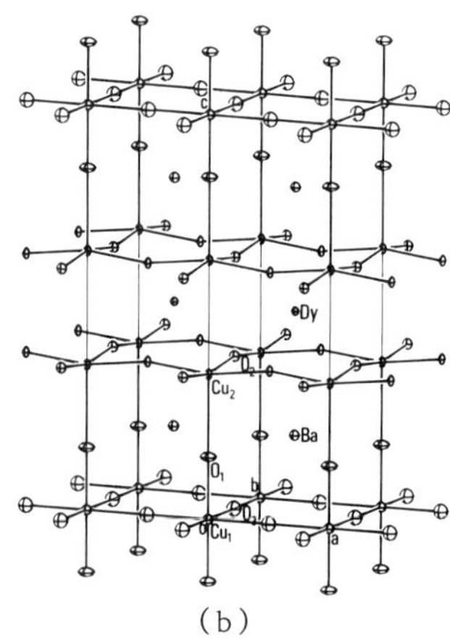

(b)

Fig. 6. Crystal structures of $LnBa_2Cu_3O_{7-\delta}$;
(a) $HoBa_2Cu_3O_6$ with space group P4/mmm and
(b) $DyBa_2Cu_3O_{6.25}$ with space group P4/mmm.

ordering of these atoms was seen in the structure analysis by the use of
orthorhombic cell (Pmmm) with close values of the lattice constants,
a and b. By the resistivity measurements, we found that the present
$YBa_2Cu_3O_6$ crystals were semiconducting. (The resistivity data of other
crystal specimens of Ho and Dy could not be taken because of very small
crystal size.) Therefore the superconductivity in the $LnBa_2Cu_3O_{7-\delta}$ system is
known to be very sensitive to the oxygen concentration. On the basis of the
present results only, we can not mention that the layers containing the Cu(1)
sites play a major role in the high-T_c mechanism, because the Cu valence or
the carrier number plays an essential role in the superconductivity in this
system. It should also be added that the lattice constant, c in the
semiconducting $YBa_2Cu_3O_6$ compound is significantly larger than that in the
high-T_c orthorhombic compound.[4]

SUMMARY

We have determined the structures of the $(La_{1-x}M_x)_2CuO_{4-\delta}$ (M=Sr and Ba),
$La_2CuO_{4-\delta}$ and $LnBa_2Cu_3O_{7-\delta}$ (Ln=Y, Dy and Ho) systems by X-ray four circle
diffraction. The present crystal specimen of $(La_{1-x}Ba_x)_2CuO_{4-\delta}$ belongs to
the orthorhombic phase (possibly Abma). It indicates that the
superconductivity in the $(La_{1-x}M_x)_2CuO_{4-\delta}$ system is not necessarily caused by
the suppression of the orthorhombic distortion by the partial substitution of
La with M. The room temperature structure of La_2CuO_4 was found to be
orthorhombic (Abma) characterized by the rotation of CuO_6 octahedra about the
b axis. The orthorhombic-tetragonal (I4/mmm) transition was seen at about
510K. It may be concluded from the structural point of view that this
transition is not associated with the CDW formation. It has become clear
that in the $LnBa_2Cu_3O_{7-\delta}$ system, considereable amount of oxygen vacancies are
introduced in the preparation of single crystals, which strongly disturbs the
occurrence of high-T_c superconductivity. From this structural study on
crystal specimens of the $(La_{1-x}M_x)_2CuO_{4-\delta}$ and $LnBa_2Cu_3O_{7-\delta}$ systems, it is
found that heat treatment in the sample preparation which does not introduce
the increase of oxygen vacancies is quite necessary to get the maximum T_c.

REFERENCES

1. J. G. Bednorz and K. A. Müller, Z. Phys. B64: 189 (1986).
2. M. K. Wu, J. R. Ashburn, C. J. Torng, P. H. Hor, R. L. Meng, L. Gao,
   Z. J. Huang, Y. Q. Wang and C. W. Chu, Phys. Rev. Lett. 58: 908
   (1987).
3. S. Hosoya, S. Shamoto, M. Onoda and M. Sato, Jpn. J. Appl. Phys. 26: L325
   and L456 (1987).
4. S. Shamoto, M. Onoda, M. Sato and S. Hosoya, Jpn. J. Appl. Phys. 26: L642
   (1987).
5. H. Takagi, S. Uchida, K. Kitazawa and S. Tanaka, Jpn. J. Appl. Phys. 26:
   L123 (1987).
6. P. M. Grant, R. B. Beyers, E. M. Engler, G. Lim, S. S. P. Parkin,
   M. L. Ramirez, V. Y. Lee, A. Nazzal, J. E. Vazquez and R. J. Savoy,
   Phys. Rev. B35: 7242 (1987).
7. M. Onoda, S. Shamoto, M. Sato and S. Hosoya, Jpn. J. Appl. Phys. 26:
   L363 and L876 (1987).
8. S. Hosoya, S. Shamoto, M. Onoda and M. Sato, this issue.
9. T. Sakurai and K. Kobayashi, Rikagaku Kenkyusho Hokoku 55: 69 (1979)
   [in Japanese].
10. P. Coppens, T. N. Guru Row, P. Leung, E. D. Stevens, P. J. Becker and
    Y. W. Yang, Acta Crystallogr. Sect. A 35: 63 (1979).
11. J. A. Ibers and W. C. Hamilton (Eds.), "International Tables for X-Ray
    Crystallography Vol. IV", Kynoch Press, Birmingham (1974).
12. J. D. Jorgensen, H.-B. Schüttler, D. G. Hinks, D. W. Capone II, K. Zhang

and M. B. Bordsky, Phys. Rev. Lett. 58: 1024 (1987).

13. L. F. Mattheiss, Phys. Rev. Lett. 58: 1028 (1987).

14. B. Grande, Hk. Müller-Buschbaum and M. Schweizer, Z. Anorg. Allg. Chem. 428: 120(1977).

15. J. M. Longo and P. M. Raccah, J. Solid State Chem. 6: 526 (1973).

16. Y. Hasegawa and H. Fukuyama, Jpn. J. Appl. Phys. 26: 322 (1987).

17. T. Fujita, Y. Aoki, Y. Maeno, J. Sakurai, H. Fukuba and H. Fujii, Jpn. J. Appl. Phys. Lett. 26: L368 (1987).

18. D. C. Johnston, J. P. Stokes, D. P. Goshorn and J. T. Lewandowski, preprint.

19. K. Takegahara, H. Harima and A. Yanase, Jpn. J. Appl. Phys. 26: L352 (1987).

20. M. Sato, S. Shamoto, M. Onoda, M. Sera, K. Fukuda, S. Hosoya, J. Akimitsu, T. Ekino and K. Imaeda, this issue.

21. Y. Syono, M. Kikuchi, K. Oh-ishi, K. Hiraga, H. Arai, Y. Matsui, N. Kobayashi, T. Sasaoka and Y. Muto, Jpn. J. Appl. Phys. 26: L498 (1987).

SINGLE CRYSTAL STUDIES AND ELECTRON TUNNELING OF $(La_{1-x}M_x)_2CuO_{4-\delta}$

(M=Ba and Sr)

M. Sato, S. Shamoto, M. Onoda, M. Sera, K. Fukuda, S. Hosoya[*],
J. Akimitsu[+], T. Ekino[+] and K. Imaeda

Institute for Molecular Science, Myodaiji, Okazaki 444 Japan
[*]The Research Insitute for Iron, Steel and Other Metals, Tohoku
University, Katahira, Sendai 980 Japan
[+]Department of Physics, Aoyama-Gakuin University, Chitosedai
Setagaya, Tokyo 157 JAPAN

INTRODUCTION

Discovery of high-T_c superconductivity in La-Ba-Cu-O system by Bednorz
and Müller[1] has stimulated an enormous number of research works not only in
the field of searching for new high-T_c superconductors but also in the
studies of basic physics. Then, the existence of another oxide system
Y-Ba-Cu-O with T_c higher than 90K was found by Chu et al[2]. The oxide systems
Ln-Ba-Cu-O with magnetic lanthanide atoms were also found by the present
authors´ group[3] to have the T_c values as high as that of Y-Ba-Cu-O system.

The chemical formulae of La-Ba-Cu-O and Y(Ln)-Ba-Cu-O systmes are now
known to be $(La_{1-x}Ba_x)_2CuO_{4-\delta}$ and $Y(Ln)Ba_2Cu_3O_{7-\delta}$, respectively. These
compounds consist of certain units such as CuO_6 octahedra. As is well known,
a number of conductive oxides A_xBO_y are characterized by a linkage of BO_n
polyhedra which forms cage-like structures with A atoms within the cage.
They usually have the small electron carrier density and often exhibit the
low dimensional nature of electron transport. We have studied their
superconductivity and charge density waves for a long time[4] from the view
point of their having a strong electron phonon interaction caused by the
large electron affinity of the oxygen atoms of the polyhedra. However, we
have not succeeded to find high-T_c superconductors among the oxides with B=W,
Mo and other transition metal atoms. We only pointed out[5] last year that in
$Li_{0.9}Mo_6O_{17}$ the electron localization effect suppressed the possibility of
high-T_c superconductivity down to 2K.

It is not easy to clarify what differences induce the large difference
of the T_c values between the present high-T_c oxides and other conductive
oxides ever known. In order to extract certain informations, it seemed
necessary to study the detailed physical properties of the high-T_c oxides.
Therefore, we have tried to prepare single crystals of these oxides. For
$(La_{1-x}M_x)_2CuO_{4-\delta}$ M=(Sr and Ba), we could get single crystals which exhibited
superconductivity and studied various physical properties using these
crystals.[6-9]

As another important technique to extract the information on the
unusually high-T_c superconductors, we adopted the electron tunneling[10,11],

where the superconducting energy gap can directly be observed.

The present paper reports some parts of the results of these experimantal studies.

EXPERIMENTS

Powder specimens were prepared by sintering the pellets as described elsewhere[11,12]. Single crystals were prepared by rapid quenching of molten nonstoichiometric mixtures. As is described in detail in the separate paper[13], single crystals of $(La_{1-x}M_x)_2CuO_{4-\delta}$ (M=Ba and Sr) and $LnBa_2Cu_3O_x$ (Ln=Ho, Dy and Y) could be obtained. However, the superconductivity could be observed only for the crystals of $(La_{1-x}M_x)_2CuO_{4-\delta}$ with M=Sr and Ba. Even for these crystals, the T_c-values were often much lower than that of the sintered specimens. We think it is due to the oxygen atom defects induced during the melting and quenching processes. The specimens with such low T_c were convenient for the experiment to study the anisotropy of H_{c2}, where the electromagnet with the maximum field of about 15kOe was used. The resistivity measurements were carried out with an ac resistance bridge using usual four terminal method. The electron tunneling experiments were carried out by point contact technique, the details of which were published elsewhere[14]. The measurements of the magnetic susceptibility were carried out by BTI SQUID magnetometer with the field up to 50kOe.

RESULTS AND DISCUSSION

Anisotropy of H_{c2}

Figure 1 shows the typical examples of the resistivities of $(La_{1-x}M_x)_2CuO_{4-}$ (M=Sr and Ba, x 0.075) as a function of applied magnetic field. It is evident that the large anisotropy of H_{c2} exists for both compounds with Sr and Ba. The ratios of the critical magnetic fields, $H_{c2\parallel}/H_{c2\perp}$ are about 13 and 6 for the compounds with Ba and Sr, respectively, where $H_{c2\parallel}$ and $H_{c2\perp}$ indicate the values of H_{c2} with the field within and perpendicular to c plane, respectively. Using the theory of dirty superconductors which is adequate for the present specimens with oxygen deficiencies, the ratio of the conductivities, $\sigma_\parallel/\sigma_\perp$ or the ratio of the electron diffusion constant D /D can be described as $(H_{c2\parallel}/H_{c2\perp})^2$, where the suffices $\parallel$ and $\perp$ indicate the values within and perpendicular to c plane. Then, $\sigma_\parallel/\sigma_\perp=D_\parallel/D_\perp$ can be estimated to be about 170 and 40 for the compounds with Ba and Sr, respectively. The anisotropy of these quantities within c plane seems to be at most minor. Then, the present high-T_c compounds have two dimensional nature of electron transport. This can be understood by the model proposed by Singh et al.[15], where $d_{x^2-y^2}$ orbitals of the Cu atoms and 2p orbitals of the O atoms in the same plane as Cu atoms form the two dimensional conduction band. Figure 2 shows the temperature dependence of $H_{c2\perp}$ determined as the field where the resistivity has the half value of the normal one. Similar curve was also obtained for the compound with Sr.

By the threory of dirty superconductors, we can also estimate the coherence length $\xi \cong (\hbar D/2\pi kT_c)^{1/2}$. For the Ba-compound, $\xi_\parallel \cong (\hbar D_\parallel/2\pi kT_c)^{1/2}$ is about 80A and for the Sr-compound, it is about 40A.

Electron Tunneling

In figure 3, the differential conductance dI/dV of $(La_{0.925}Sr_{0.075})_2CuO_{4-\delta}$-Al point contact junction obtained at 4.2K is shown as a function of the applied voltage across the junction. The transition temperature T_c of this powder specimen defined as the point of zero

resistivity is about 34K. The broken line shows the normal differential
conductance, which is determined from the curve above T_c. In the figure, one
can see the characteristic structure of the electronic density of states
expected for the superconducting phase. Only the small reduction of the
dI/dV value may be due to the existence of the leak current through the
junction.

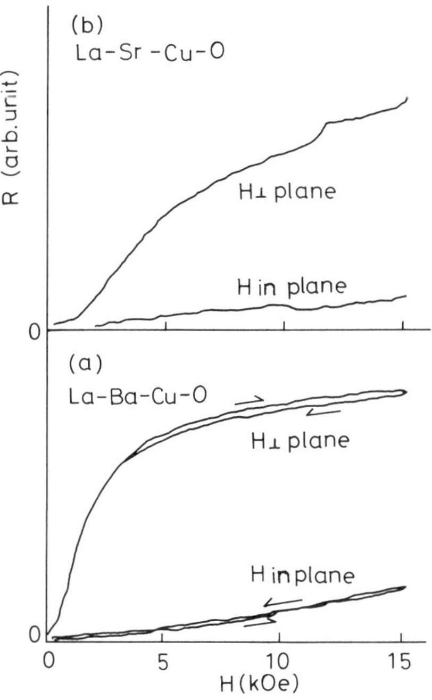

Fig. 1. top: Resistivity vs. magnetic field for
$(La_{1-x}Sr_x)_2CuO_{4-\delta}$ single crystal. T=6.3K.
bottom: same figure for $(La_{1-x}Ba_x)_2CuO_{4-\delta}$.
T=10.8K.

The energy gap 2Δ defined as the voltage distance between the points
where the curve crosses the broken line is about 14meV. Then, the value of
$2\Delta/kT_c$ is estimated to be about 4.7, which is considerably larger than the
value 3.52 expected from BCS theory. However, the present $2\Delta/kT_c$ value is
much smaller than the values observed by other groups[16].

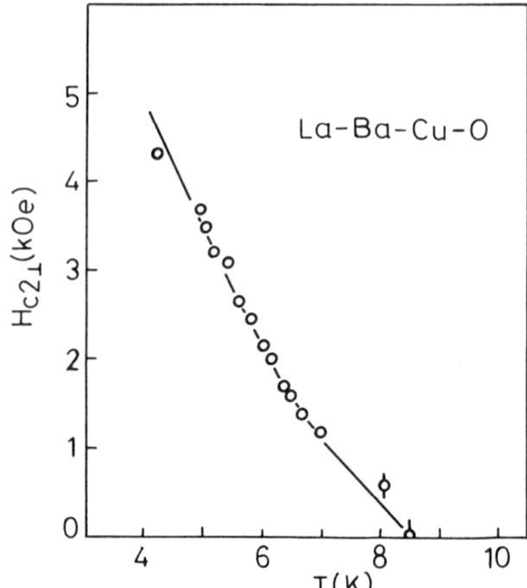

Fig. 2. Critical magnetic field $H_{c2\perp}$ with the
field perpendicular to c plane is
poltted as a function of temperature.

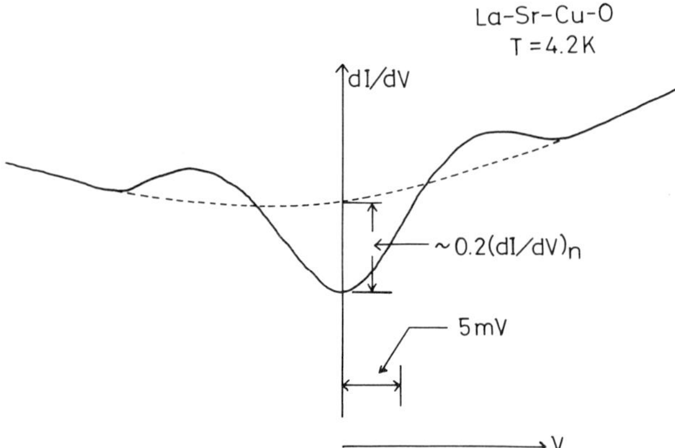

Fig. 3. The differential conductance dI/dV observed for the point
contact junction of $(La_{0.925}Sr_{0.075})_2CuO_{4-\delta}$ and Al plate. The
broken line shows the expected normal value, $(dI/dV)_n$.

There may exist the ambiguity in determining the value of 2 from the present data with rather large smearing. Figure 4 shows the example of the fitting calculation to the experimental data for $BaPb_{1-x}Bi_xO_3$ using the broadened density of states $N(E,\gamma) = Re\{(E-i\gamma)/[(E-i\gamma)^2-\Delta^2]^{1/2}\}$, where the energy gap $2\Delta = 3.45$meV and the broadening $=0.43$meV are obtained. If we simply define the 2Δ value as the voltage distance between the points where

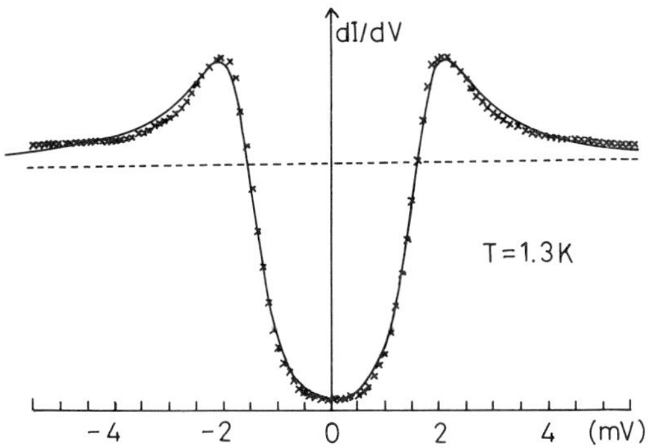

Fig. 4. Trial fitting of the observed tunneling density of
states (crosses) of $BaPb_{1-x}Bi_xO_3$ to the theoretical
line with the broadened density of states proposed
by Dynes et al.[17] The broken line shows the normal
differential conductance.

the curve crosses the normal conductance line (broken line), $2\Delta=3.25$meV can be obtained, which indicates that this simple definition gives us almost satisfactory value of 2Δ. Then, it is not easy for the present $(La_{0.925}Sr_{0.075})_2CuO_{4-\delta}$ to get the large value of 2Δ such as the one obtained in ref. 16 as long as the energy gap is isotropic. In this sense, the anisotropic energy gap proposed by Maekawa et al.[18] seems to be interesting, although Kirtley et al.[19] did not observe the significant anisotropy of energy gap in $YBa_2Cu_3O_x$.

Thermoelectric Power of $(La_{0.925}Sr_{0.075})_2CuO_{4-\delta}$ Single Crystal

The thermoelectric power has already been published by Uchida et al.[20] for sintered specimens of $(La_{1-x}Ba_x)_2CuO_{4-\delta}$. Figure 5 just shows the thermoelectric power taken for a single crystal specimens of $(La_{1-x}Sr_x)_2CuO_{4-\delta}$ (x 0.075), where the temperature gradient was applied within c plane. The qualitative features are almost same as the data for sintered specimens.

Magnetic Properties of La_2CuO_4 Single Crystal

Since Anderson[21] proposed the resonating valence bond state in La_2CuO_4, the problem of its ground state has attracted much interest. Hasegawa and Fukuyama proposed the spin density wave (SDW) state, which seems to be supported by the experiments for powder specimens by Greene et al.[23] and Fujita et al.[24] Here the preliminary experimental results of the magnetic susceptibility χ of the mosaic crystal of La_2CuO_4 with the mosaic spread of several degrees are briefly shown. Fig. 6 shows the temperature dependence of the magnetic susceptibility taken with the field of 30kOe along various

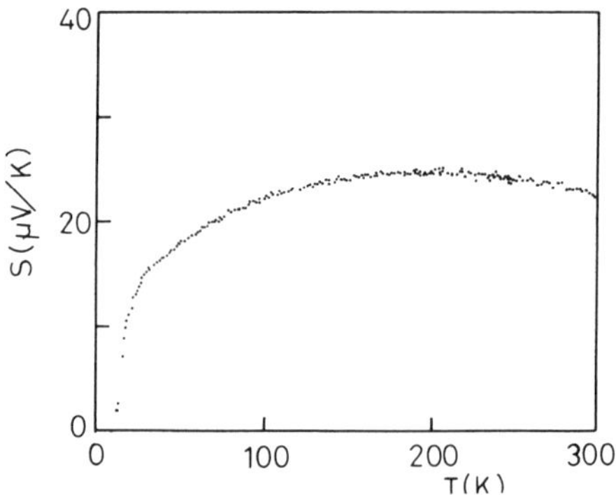

Fig. 5. Thermoelectric power of $(La_{1-x}Sr_x)_2CuO_{4-\delta}$ single crystal ($x \sim 0.075$).

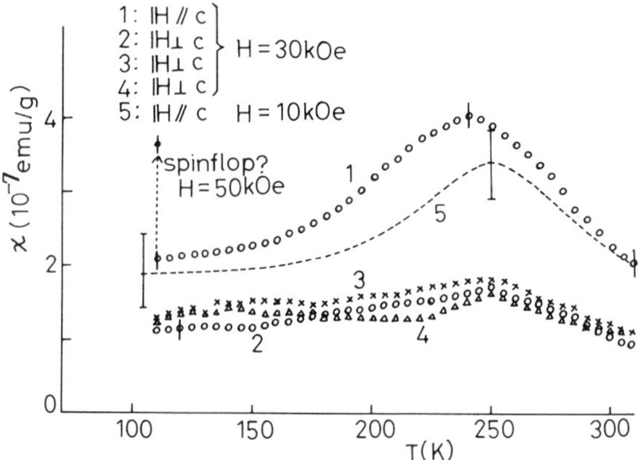

Fig. 6. Magnetic susceptibility with the field along various
directions: 1. H=30kOe, H//c, 2. H=30kOe, H⊥c and H//(Longer
edge of the crystal plane), 3. H=30kOe, H⊥c and H⊥(longer
edge of the crystal plane), 4. H=30kOe, H⊥c and H is 45 degree
apart from either directions of 3 and 4, and 5. H=10kOe
and H//c. The closed circle indicates the value taken for
H=50kOe and H//c. The vertical bars indicate the typical
estimated errors in the corresponding measurements.

directions (1-4). The broken line 5 shows the data taken with the field of
10kOe. The vertical bars indicate the typical error bars in the
measurements. The present description of the space group of La_2CuO_4 is Abma,
where c axis is defined perpendicular to the layers of the CuO_6 octahedra.
In the figure, the uniaxial anisotropy of χ can be observed even above T_0,
where χ has the peak value. The large decrease of χ in curves 1 and 5 and
little decrease of χ in curves 2-4 as T decreases below T_0 suggest the
antiferromagnetic ordering of spins or the SDW formation with the spin
direction along c axis. Indeed, under the field of 50kOe along the c
direction, χ increases as shown by the closed circle, which implies the
occurrence of the spin flop phenomenon by the external field. This increase
of χ does not take place under the field whitin c plane up to 50kOe. For
powder specimens, the similar increase was also observed at about 45kOe at
100K, while it was not observed at 300K up to 50kOe.

The spin direction suggested by the present experiment is different
from the result obtained by neutron scattering[25] and NMR[26], where the spin
direction is within c plane. To solve this disagreement, further studies may
be necessary.

ACKNOWLEDGEMENT

The authors are grateful to the members of Institute for Molecular
Science for their encouragement during the course of the present work.

REFERENCES

1. J. G. Bednorz and K. A. Müller, Z. Phys. B64: 189 (1986).
2. C. W. Chu, P. H. Hor, R. L. Meng, L. Gao, Z. J. Huang,
      Y. Q. Wang, J. Bechtold, D. Campbell, M. K. Wu, J. Ashburn and
      C. Y. Huang, Phys. Rev. Letters to be published.
3. S. Hosoya, S. Shamoto, M. Onoda and M. Sato, Jpn. J. Appl. Phys. 26: L325
      (1987), S. Hosoya, S. Shamoto, M. Onoda and M. Sato, Jpn. J. Appl.
      Phys. 26: L456 (1987) and S. Shamoto, M. Onoda, M. Sato and S. Hosoya,
      Jpn. J. Appl. Phys. 26: L642 (1987).
4. for examples, M. Sato, B. H. Grier, H. Fujishita, S. Hoshino and
      A. R. Moodenbaugh, J. Phys. C16: 5217 (1983), M. Sato, Physica 143B:
      90 (1986).
5. M. Sato, Y. Matsuda and H. Fukuyama, J. Phys. C20: L137 (1987).
6. S. Shamoto, M. Onoda, M. Sato and S. Hosoya, Solid State Commun. 62: 479
      (1987).
7. S. Sugai, M. Sato and S. Hosoya, Jpn. J. Appl. Phys. 26: L495 (1987).
8. M. Onoda, S. Shamoto, M. Sato and S. Hosoya, Jpn. J. Appl. Phys. 26: L363
      (1987).
9. K. Fukuda, M. Sato, S. Shamoto, M. Onoda and S. Hosoya, Solid State
      Commun. to be published.
10. T. Ekino, J. Akimitsu, M. Sato and S. Hosoya, Solid State Commun. to be
      published.
11. T. Ekino and J. Akimitsu, Jpn. J. Appl. Phys. 26: L452 (1987).
12. M. Sato, S. Hosoya, S. Shamoto, M. Onoda, K. Imaeda and H. Inokuchi,
      Solid State Commun. 62: 85 (1987).
13. S. Hosoya, S. Shamoto, M. Onoda and M. Sato, this issue.
14. T. Ekino, J. Akimitsu, Y. Matsuda and M. Sato: Solid State Commun. to be
      published.
15. K. K. Singh, D. Ganguly and J. B. Goodenough, J. Solid State Chem. 52:
      254 (1984).
16. For example, M. Naito, D. P. E. Smith, M. D. Kirk, B. Oh, M. R. Hahn,
      K. Char, D. B. Mitzi, J. Z. Sun, D. J. Webb, M. R. Beasley, O. Fisher,
      T. H. Geballe, R. H. Hammond, A. Kapitulnik and C. F. Quate, Phys. Rev.
      B35: 7228 (1987).
17. R. C. Dynes, J. P. Garno, G. B. Hertel and T. P. Orlando, Phys. Rev.
      Letters 53: 2437 (1984).

18. S. Maekawa, H. Ebisawa and Y. Isawa, Jpn. J. Appl. Phys. 26: L468 (1987).
19. J. R. Kirtley, C. C. Tsuei, S. I. Park, C. C. Chi, J. Rozen,
    M. W. Shafer, W. J. Gallagher, R. L. Sandstrom, T. R. Dinger and
    D. A. Chance, Proc. LT18, Kyoto (1987).
20. S. Uchida, H. Takagi, H. Ishii, H. Eisaki, T. Yabe, S. Tajima and
    S. Tanaka, Jpn. J. Appl. Phys. 26: L440 (1987).
21. P. W. Anderson, Science 235: 1196 (1987).
22. Y. Hasagawa and H. Fukuyama, Jpn. J. Appl. Phys. 26: L322 (1987).
23. R. L. Greene, H. Maletta, T. S. Plaskett, J. G. Bednorz and K. A. Müller,
    Solid State Commun. to be published.
24. T. Fujita, Y. Aoki, Y. Maeno, J. Sakurai, H. Fukuba and H. Fujii, Jpn. J.
    Appl. Phys. 26: L368 (1987).
25. D. Vaknin, S. K. Sinha, D. E. Moncton, D. C. Johnston, J. Newsam,
    C. R. Safinya, and H. E. King, Jr, preprint.
26. Y. Kitaoka, private communication.

STUDY OF SUPERCONDUCTING OXIDES AT WESTINGHOUSE

A. I. Braginski

Westinghouse R&D Center
1310 Beulah Road
Pittsburgh, PA 15235

INTRODUCTION

The purpose of this paper is to present a summary of our past and
current work on superconducting oxides having high critical temperatures,
T_c, and also of collaborative work by three other groups, at the National
Bureau of Standards in Boulder, the University of Texas in Austin and the
Boeing Corporation in Seattle. Most of our work has been of a relatively
applied character, aimed at the synthesis of new compounds and the
determination of properties related to their current carrying capability.
This activity started in January of 1987 and is continuing. The summary
represents a snapshot taken in mid-June rather than an account of completed
studies.

PREPARATION, COMPOSITION AND STRUCTURE OF OXIDE SUPERCONDUCTORS

Ceramic Bulk Samples

Preparation of bulk samples by Panson has been performed exclusively by
the now standard ceramic technique, starting from $BaCO_3$ or $SrCO_3$ and oxides
of other constituents. The characteristic features have been the avoidance
of ball milling to minimize contamination by other elements and the
three-step "react and grind" sequence with all reactions in flowing oxygen
and followed by slow cooling in O_2. Final sintering of $La_{1.8}Sr_{0.2}CuO_4$
(LSC, x = 0.2) was performed over a range of oxygen pressures up to 150 bar
(static). Reaction and sintering temperatures were held at 1100°C for this
compound. In the Y_2O_3 - BaO - CuO (YBC) ternary phase system, the
sintering temperatures depended upon the composition and were chosen just
below the softening or melting point where reaction with alumina supports
was minimal. In the case of the $Y_1Ba_2Cu_3O_x$ (1:2:3) compound, the reaction
temperature was either 1000°C or 940 ÷ 950°C, always at 1 bar of flowing
O_2. Pellets were pressed at pressures up to 4.3×10^3 Kg/cm^2. After final
sintering the density of LSC pellets was 85 to 86% of theoretical, d_x. The
density of 1:2:3 was 88% of pycnometric powder density or 73% of
$d_x = 6.35$ gm/cm^3. Data on resistivity and magnetization vs temperature
dependences were published before.[1,2] These data and derived properties
are reviewed below.

Phase Identification and Separation

In LSC with x = 0.2, the X-ray diffraction (XRD) identified phase structure was consistently that of K_2NiF_4 with occasional traces of the orthorhombic La_2CuO_4 structure. After reproducing the Wu and Chu observation of superconductivity above 90K in mixed phase YBC,[3] Panson separated the 1:2:3 superconducting phase by scanning electron microscopy (SEM). It permitted the classification of crystallites into two major groups having different growth habits and the determination of respective compositions, (1:2:3) and (2:1:1), by energy dispersive X-ray spectroscopy (EDS).[4] Preparations having these compositions were then examined by X-ray diffraction. The orthorhombic perovskite-related phase (1:2:3) was superconducting above 90K, as determined by Gavaler, Talvacchio and Pohl[2] who measured the resistivity and magnetic susceptibility vs temperature.

Oxygen Concentration

Panson determined chemically, by oxygen evolution and also iodometric titration, that YBC 1:2:3 specimens cooled slowly in O_2 contain Cu^{3+} while those quenched from above 700°C do not.[5] Oxygen evolution measurements were done for samples dissolved in HCl. The reaction involves reduction of the Cu^{3+} by water:

$$Cu^{3+} + 1/2\ H_2O \rightarrow Cu^{2+} + H^+ + 1/4\ O_2$$

This reaction is one of those likely to account for degradation of sample properties on exposure to water vapor in ambient air. Acid dissolution in the presence of KI results in reduction of Cu^{3+} by I to form I_2 which is titrated with thiosulfate. The thermogravimetry of sintered pellets at slow heating and cooling rates in O_2 and N_2 showed that the oxygen intake and loss are reversible and the maximum oxygen incorporation occurs at, approximately, 400°C, agreeing with the titration result of 7.00 ± 0.02 oxygen atoms per formula unit obtained under the assumption that at 6.5 oxygens per formula unit all copper is divalent. Thermogravimetric heating and cooling curves obtained in O_2 when starting from a quenched, oxygen-deficient state (6.5 oxygens) are shown in Figure 1. Virtually identical experiments by Manthiram et al. led to 6.96 ± 0.02 oxygen atoms per formula unit.[6]

X-Ray Fine Structure Near Absorption Edge and Structural Implications

Lytle et al. studied the X-ray Cu K-absorption near edge structure (XANES) in both LSC and YBC prepared by Panson.[7] While their energy spectra are virtually identical to those of other investigators, the interpretation is different. Based, in part, on the chemical and thermogravimetric determination of Cu^{3+} content in YBC (above), they concluded that up to 30% of La-sites in LSC and Y-sites in YBC are substituted with copper, while the displaced La or Y ions are substituting for Cu (Cu_2 site in YBC). In the La or Y sites both copper valences are present, with $Cu^{2+} > Cu^{3+}$ in YBC and $Cu^{3+} > Cu^{2+}$ in LSC. The authors' belief is that the two criteria for superconductivity are: (1) The Cu presence in the La/Y site which connects adjacent Cu-layers, and (2) the co-presence of Cu^{2+} and Cu^{3+}. Copper in La/Y sites can be thought of as a fractional population of Cu chains orthogonal to Cu-O planes which also contain unidimensional (1D) chains, leading to a vague analogy with the A15 structure where the three orthogonal Nb-chains are correlated with the relatively high T_c.

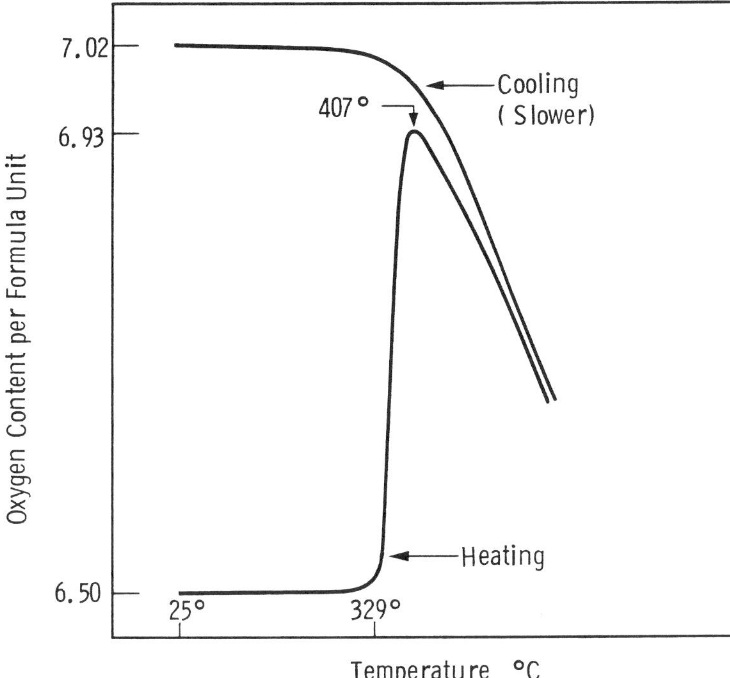

Figure 1 - Thermogravimetric heating and cooling curve for a sintered
sample of YBC. The heating rate was 3 C/min, cooling rate was
1.5 C/min.

Search for $Y_1Ba_2Cu_3O_7$ Solid Solutions

Panson et al. reported X-ray, magnetic susceptibility and Hall
constant/carrier density data which suggested the existence of a narrow
range of solid solutions with the $Y_1Ba_2Cu_3O_7$ compound.[2] The X-ray single
phase cationic compositions $Y_{1.05}Ba_{1.8}Cu_3$ and $Y_1Ba_{2.2}Cu_3$ had much lower T_c
[from susceptibility $\chi(T)$] and effective carrier density, n^*, than the
stoichiometric compound. In the Y_2O_3-BaO-CuO ternary diagram these
compositions are located off the $Y_2Cu_2O_5$ - $BaCuO_2$ join. Along the join,
however, Hinks et al. did not find any evidence of solid solutions in their
XRD, resistive T_c and J_c data. To verify this result, Wagner, Panson and
Braginski determined phase composition, $\chi(T)$, resistivity, ρ_n, and n^* in
six $Y_{1+x}Ba_{2-x}Cu_3$ compositions along the join over the range $-0.2 < x <$
$+0.2$. At the smallest x increment of ± 0.05, second phases ($BaCuO_2$ or
Y_2CuO_5) were present while the T_c onset and n^* were almost independent on
x, and characteristic of the 1:2:3 stoichiometric major phase, as shown in
Table I. Properties of previous, off-join compositions are included for
comparison. The present results confirm the conclusions of Hinks et al.
that no solid solutions exist along the Y_2CuO_5 - $BaCuO_2$ join.

Cationic and Anionic Substitutions

In the Ln_2O_3-SrO-CuO system, where the lanthanide, Ln, was Y, Dy, Yb
and Lu, no superconductivity was observed in mixed phase preparations.
Panson also tried to substitute fluorine into the 1:2:3 compound. The
purely chemical rationale, preceding the thermogravimetric and chemical
studies of oxygen vacancy content (summarized above), was to form the

hypothetical compound $Y_1Ba_2Cu_3O_4F_5$ containing only Cu^{2+} and then study the
effect of Cu^{3+} doping. Samples were prepared with various concentrations
of BaF_2 and/or CuF_2 added to the usual oxide and carbonate mixture. The
XRD indicated that up to one fluorine atom per formula unit could be
substituted into the structure. Beyond that limit BaF_2 was observed in the
X-ray patterns and no detailed phase analysis performed.

Table I. Effect of Deviation from the 1:2:3 Stoichiometry in YBC

Composition Cations Only	x	$T_c(K)$ Onset from Susceptibility	ρ_{148} μohm-cm	n^*_{-3} cm	Comment
$Y_{0.80}Ba_{2.20}Cu_3$	-0.20	91	2.8×10^3	4.2×10^{21}	
$Y_{0.90}Ba_{2.10}Cu_3$	-0.10	91	920	7.2×10^{21}	
$Y_{0.95}Ba_{2.05}Cu_3$	-0.05	91	660	5.3×10^{21}	
$Y_1Ba_2Cu_3$	0	91	390	7.6×10^{21}	on-join
$Y_{1.05}Ba_{1.95}Cu_3$	+0.05	91	530	6.0×10^{21}	
$Y_{1.10}BA_{1.90}Cu_3$	+0.10	92	620	6.1×10^{21}	
$Y_{1.20}Ba_{1.80}Cu_3$	+0.20	92	620	7.0×10^{21}	
$Y_{1.05}Ba_{1.8}Cu_3$	--	70	1.6×10^3	9.4×10^{20}	off-join
$Y_{1.0}Ba_{2.2}Cu_3$	--	70	1.1×10^4	7.1×10^{20}	

With one fluorine substituted, the critical temperature onset in the
X-ray single phase sample was 94K from resistive measurement and 92K from
dc susceptibility $\chi(T)$ data. No high-temperature anomalies were seen in
resistivity vs temperature plots. The mixed-phase sample of nominal
composition $Y_1Ba_2Cu_3O_5F_2$ exhibited upon virgin cooling a resistive anomaly
at 202K. Below this point the resistivity was steadily decreasing to 94K
where a sharper transition to zero resistivity at 83K was noted. Upon
subsequent slow heating, a hysteretic behavior qualitatively reminiscent of
that reported by Ovshinsky et al.[9] was observed above 94K. The anomaly,
due to thermoelectric electromotive forces, was rather reproducible, but we
did not see in it any evidence of superconductivity. In phase mixtures of
compositions containing more than 2.5 F per formula unit no superconducting
transitions were seen down to 4.2K. At this time we can thus acknowledge a
partial agreement with Ovshinsky et al.: an absence of resistive anomalies
at one F substitution, a resistive anomaly in $Y_1Ba_2Cu_3F_2O_x$ and absence of
superconductivity at F-substitutions higher than 2. We cannot confirm,
however, superconductivity above 100K.

Growth of Single Crystals

Hopkins, Kramer and Stewart[10] attempted to grow single crystals of the
1:2:3 compound by slowly cooling a liquid oxide solution of $BaCuO_2$ and
$Y_2Cu_2O_5$. The solution composition corresponded to $Y_{0.4}Ba_{2.4}Cu_3O_x^2$. The
large excess of $BaCuO_2$ was used to insure a relatively low melting point.
The $BaCO_3$, Y_2O_3 and CuO were reacted to synthesize the starting compounds

which were then blended, sealed in platinum envelopes and heated at 1150°C
for 48 hours under the oxygen pressure of 1 bar. The product was cooled to
room temperature at 2°C/hour. It contained crystallites in the form of
prisms, up to 0.7 X 0.7 X 3 mm in size, embedded in the the solidified flux
and difficult to separate. The prism-shaped crystals are consistent with
the tetragonal space group of the high-temperature phase and apparently
similar to those obtained by Takekawa and Iyi.[11] This growth habit,
however, differs from the plate-like habit reported by Hidaka et al.,[12]
probably due to difference in the flux composition from which the crystals
grew.

The XRD and EDS of partly separated crystal aggregates indicated the
presence of the 1:2:3 structure and an approximately correct composition.
The platinum contamination was 3 at. %. The unoriented aggregates
exhibited superconductivity at 90K after annealing in O_2 at 400°C for
100 hours. Further separation and characterization work is continuing.

Thin Film Deposition

Thin films have been deposited by both, electron beam co-evaporation
from metal sources and magnetron sputtering. To date, relatively best
success was obtained by Buhay and by Janocko and Gavaler by using the first
method in two quite different evaporators. The co-evaporation was in the
presence of very low partial pressures of oxygen $10^{-9} < p(O_2) < 10^{-5}$ torr,
on unheated substrates of Al_2O_3 (sapphire), polycrystalline and single
crystal ZrO_2 and polycrystalline $SrTiO_3$. The deposit composition was
within ±5 at. % of the 1:2:3 formula. Films were subsequently reacted at
850 to 950°C for varying lengths of time, from 1 to 30 min, and annealed at
450°C in 1 bar of O_2. The XRD patterns were dominated by the substrate
diffraction peaks but clearly indicated the presence of the 1:2:3 phase.

Figure 2 compares three resistive transitions in polycrystalline films,
600 nm thick, simultaneously co-deposited on Al_2O_3, ZrO_2 and $SrTiO_3$ and
subsequently annealed in the same conditions. The sharpest transition, 94
to 80K (95% to 5% of residual resistivity), and lowest resistivity were
obtained on $SrTiO_3$.[13] Broader transitions and correspondingly higher
resistivities were obtained on ZrO_2 and Al_2O_3. Interdiffusion with the
substrates was monitored by analyzing the top 200 nm of the film by EDS
before and after the high temperature reaction. In all cases, the
substrate components, Sr, Zr or Al, were detected only after the reaction.
Interdiffusion was thus taking place. The effect was strongest for Al and
weakest for Sr and it correlated with the transition broadening and
resistivity increase. Quantitative analyses were not performed due to EDS
peaks overlap. For the same reason a possible presence of Ti was not
verified. The results suggest that interdiffusion with substrate is one of
the causes of the broad superconducting transition or a "foot" in the $\rho(T)$
curve.[14] The effect of substrate on T_c was observed earlier by Hammond
et al.[14] who, in more recent work, have obtained sharp transitions. At
least two other groups also succeeded in eliminating the "foot" in the T_c
curve by depositing epitaxial films on $SrTiO_3$.[15,16]

PROPERTIES OF OXIDE SUPERCONDUCTORS

Critical Temperature

Critical temperatures have been determined from temperature dependences
of resistivity and magnetization, the latter measured by vibrating sample
magnetometer but verified by SQUID magnetometer.[17] We did not participate
in the search for highest T_c's but were alert to the occurrence of
resistive anomalies reported by others. Indeed, we occasionally observed

such anomalies in LSC (ref. 1) and also in Y-Ba-Cu-O phase mixtures. In no case, however, did we find convincing evidence of superconductivity above 100K. Goldfarb et al. have also determined T_c by measuring the complex susceptibility in weak magnetic fields.[18] These results are discussed below.

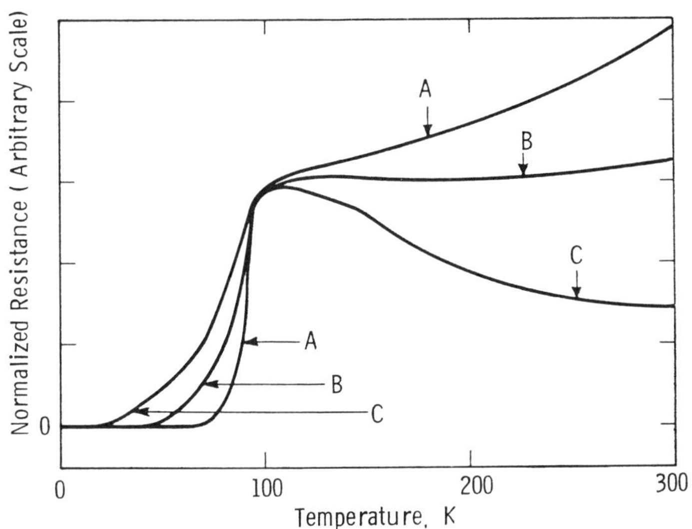

Figure 2 - Resistive transitions in polycrystalline YBC films on three different substrates: (A) on $SrTiO_3$, ρ_{100} = 680 $\mu\Omega$-cm, (B) on ZrO_2, ρ_{100} = 750 $\mu\Omega$-cm, (C) on Al_2O_3, ρ_{100} = 1200 $\mu\Omega$-cm.

Lower Critical Field

In LSC (x = 0.2), the lower critical field in bulk, polycrystalline samples was estimated by Braginski from dc magnetization curves which lead to dH_{c1}/dT = -7.9 ± 2 Oe/K near T_c.[1] Recently, Renker et al. determined $H_{c1}(T)$ in x = 0.15 by the same method and obtained dH_{c1}/dT = -8.0 Oe/K.[19]

Goldfarb et al. measured H_{c1} near T_c in YBC by the ac complex susceptibility method. Two superconducting components were identified in the X-ray single phase material. In near zero field both had the same T_c and were practically indistinguishable. With increasing field, however, one component had the T_c nearly field independent and dH_{c1}/dT = -3.2 Oe/K, while the T_c in the other decreased abruptly with increasing field, dT_c/dH = -1.5 K/Oe and an order of magnitude lower slope of dH_{c1}/dT = -0.33 Oe/K was seen. Goldfarb's plots of T_c vs H and H_{c1} vs T are shown in Figure 3. The second component is believed to represent either the "weak link" material determining the intergranular connectivity for current transfer or, alternatively, the deteriorated surface of the bulk sample. After submitting this work for publication, we learned that Kuepfer made similar measurements and observed a virtual disappearance of the degraded component upon powdering of the sample.[20] This result favors the "weak link" interpretation. The nature of such weak links is under study and, at present, we tend to assign it to the electronic anisotropy of crystallites misaligned with respect to the magnetic field and induced current flow.

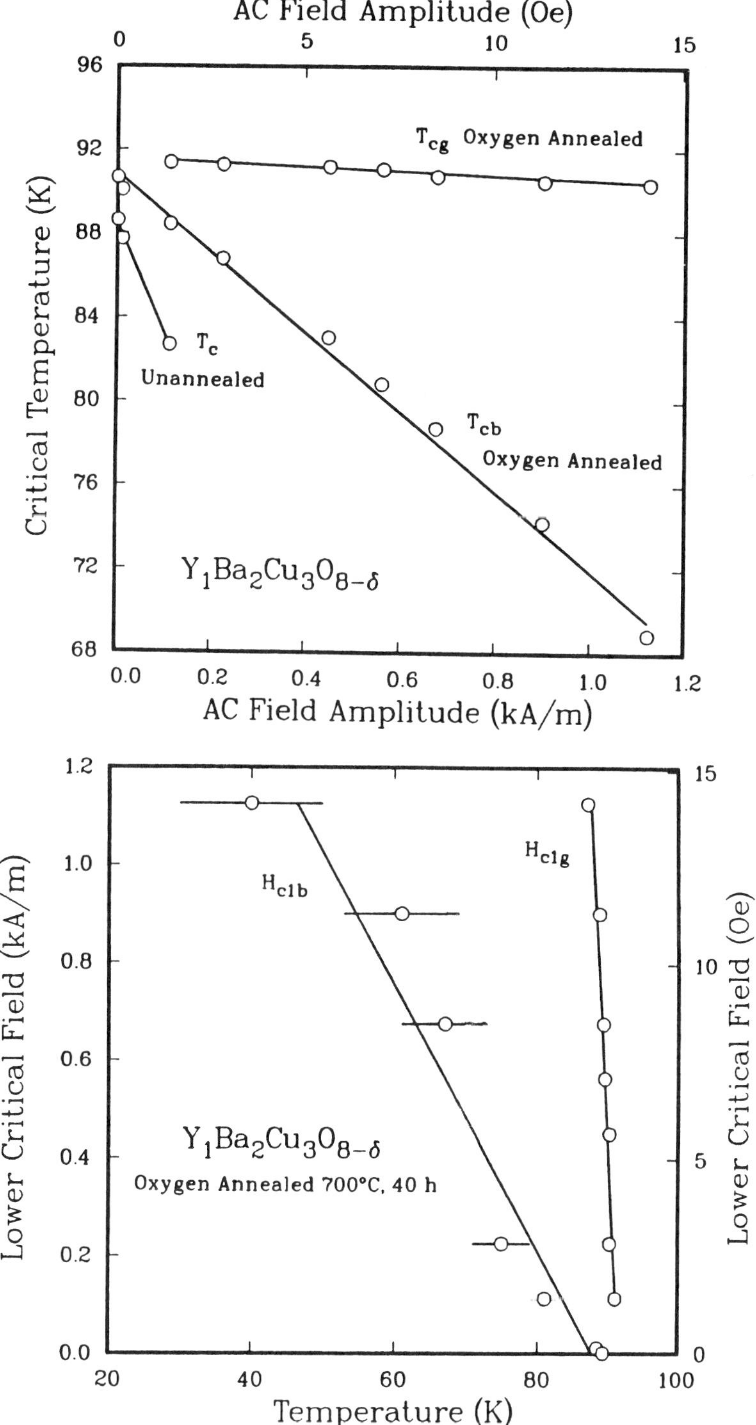

Figure 3 – Goldfarb et al. data on T_c vs H and H_{c1} vs T derived from complex susceptibity measurements in YBC. Subscripts (b) and (g) denote the two superconducting components present in the single-phase polycrystalline material. The (b) component is associated with the weak link connectivity.

The Upper Critical Field

In both LSC ($x = 0.2$) and single phase YBC the estimates of the upper
critical field, $H_{c2}(0)$, were obtained from resistive transition midpoint
values measured by Wagner in magnetic fields up to 55 kOe.[1,2] These
estimates, included in Table II, were calculated in the dirty limit
assuming zero paramagnetic limiting.[21] The very narrow ranges of
temperatures, over which $H_{c2}(T)$ could be determined, reduced the accuracy
of measurements. Furthermore, the assumption of dirty limit was not fully
justified, in view of the rather low effective carrier density (discussed
below) and very short, inferred coherence length. Nevertheless, there is
no doubt that giant $H_{c2}(0)$ values are the characteristic of at least
individual crystallites in polycrystalline, ceramic agglomerates. As shown
before, between 5 and 60 kOe the hysteresis of magnetization, ΔM, is
virtually field-independent over a wide range of reduced temperatures,
$t = T/T_c$, e.g. up to $t > 0.8$ in YBC.[2] The induced critical current
density, J_c, is proportional to ΔM and should decrease rapidly in fields
approaching H_{c2}.

Critical Current Density

The upper limit to J_c was calculated by Talvacchio as the depairing
current density following London i.e., by equating the electron pair energy
to the condensation energy.[22] In YBC it was found fully comparable to that
in conventional superconductors such as Nb-Ti or Nb_3Sn.[23] In spite of this
lack of fundamental limitation, J_c's determined from ΔM measurements
between 5 and 60 kOe were very low, in YBC between 10^4 and 10^5 A/cm^2 at
4.2K and approximately 10^2 A/cm^2 at 77K.[2] The J_c was derived assuming a
uniform, concentric current distribution in disk specimens with the
magnetic field perpendicular to the surface. We suspected this assumption
to be incorrect since loose powders of the same material exhibited ΔM lower
by only a factor of 2 to 5. A comparison of hysteresis loops of a powder,
a disk of the same preparation and a powder obtained by crushing the disk
is shown in Figure 4(a). It suggested to us that sintered disks with 25 to
27% porosity are granular materials with limited intergrain connectivity.
A comparison of ΔM in samples having markedly different grain sizes is
shown in Figure 4(b). Larger grains resulted in much larger ΔM while the
transport J_c was somewhat lower. This result led to the same
interpretation and indicated that intragrain critical currents are much
higher than those between the grains. Indeed, Ghosh et al. estimated the
intragrain current at greater than 1×10^5 A/cm^2 at 4.3K[24] while
Apfenstedt et al. gave an estimate of $> 1 \times 10^6$ A/cm^2 for a ground
sample.[25]

Ekin et al. reported transport J_c measurements at 77K over a wide range
of magnetic field intensities, up to 240 kOe.[26] Exponential decrease in
transport J_c with increasing H was observed in very weak fields (over two
orders of magnitude between 1 and 100 Oe), while a signature of
superconductivity was seen in voltage vs current characteristics even in
the 190 kOe field where the resistivity was still well below the residual
value. The weak intergrain connectivity was thus demonstrated directly.

Measurements of J_c in bulk and thin film single crystals by Dinger
et al. and Chaudhari et al. of $J_c > 10^5$ A/cm^2 at 77K and $> 10^6$ A/cm^2 proved
ultimately that the low transport J_c's and their strong dependence upon H
seen in polycrystalline samples must be due to weak connectivity.[27] The
nature of weak links, however, must be elucidated. At present, we believe
that the electronic anisotropy of misaligned crystallites is the major
cause of the observed current transfer behavior.

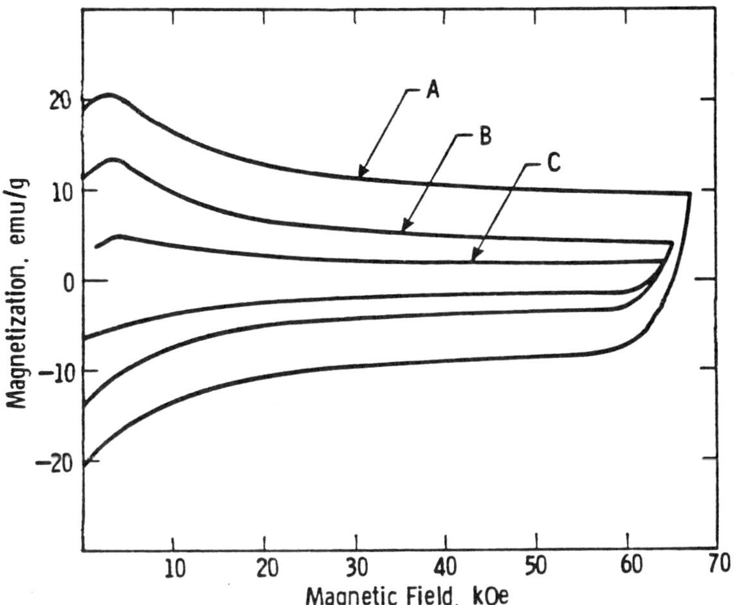

Figure 4(a) - Plots of magnetization hysteresis at 4.2K in YBC with
1:2:3 composition. Comparison of magnetization in: (A) a
sintered disk, (B) powder of which the disk was made and
(C) powder obtained by crushing the disk.

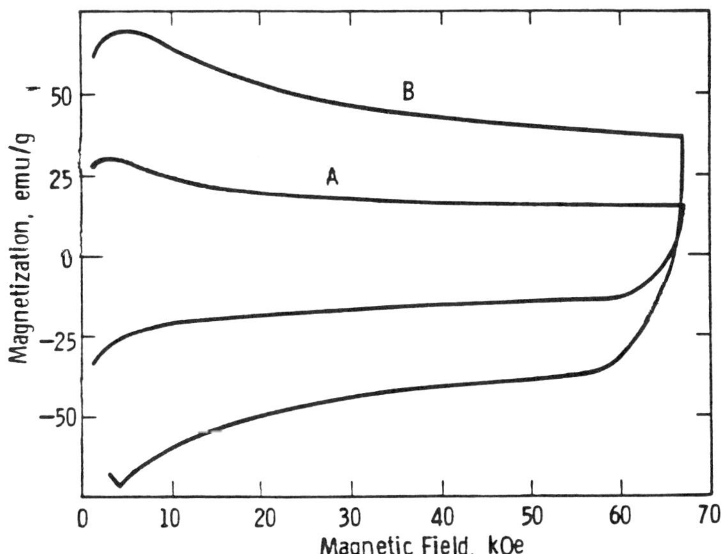

Figure 4(b) - Plots of magnetization hysteresis at 4.2K. A comparison
of data for two samples differing in grain (crystallite)
size, d: (A) d = 10 $\mu\Omega$-cm, transport J_c = 52 A/cm^2; (B)
d = 20 $\mu\Omega$-cm, transport J_c = 40 A/cm^2 (77K, H = 0).

<u>Hall Constant, Carrier Density and Mobility</u>

Wagner measured the Hall effect in magnetic fields up to 60 kOe, and at temperatures between 300K and T_c, using the dc Van der Pauw methods with square polycrystalline samples, approximately 0.1 cm thick. The Hall voltage was linear over the whole field range.

The temperature dependence of the Hall constant, R_H, and resistivity in the stoichiometric 1:2:3 YBC is shown in Figure 5. The R_H falls with increasing temperature to below the sensitivity limit indicated by a vertical bar at T = 300K. The effective carrier density, $n^* = (R_H e)^{-1}$, is thus increasing by an order of magnitude between 96 and 300K. This strong dependence would not be expected in the case of simple metallic conduction involving one band. A possible explanation is that both hole and electron conduction occurs, and the temperature dependence is due to a change in the ratio of mobilities, μ. For two carrier conduction $R_H e = (p - nr^2)/(p + nr)^2$ where $r = \mu_n/\mu_p$, and p, n are carrier concentrations, so that an increase in r with temperature would cause R_H to decrease. An alternative explanation is that with increasing T the phonon scattering reduces the mean free path to the point where electrons no longer traverse complete hole-like orbits on the Fermi surface. This effect has been seen in metals (e.g., in Al) where R_H changes sign from positive to negative with increasing T or addition of impurities. Magnetic scattering may also play an important role. A third, least likely possibility is thermal activation of holes.

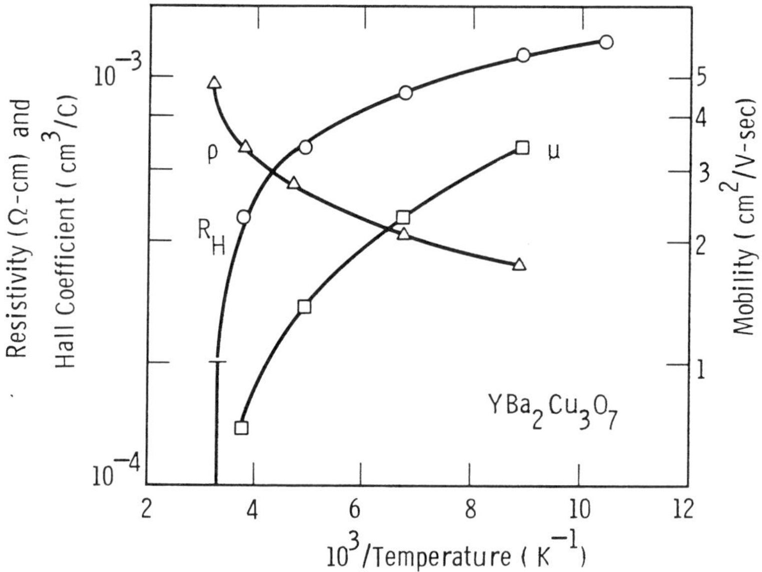

Figure 5 - Temperature dependence of Hall constant, carrier mobility and resistivity in stoichiometric YBC.

The carrier mobility, $\mu = R_H/\rho$, falls with 1/T. Figure 6 shows that the data are well fit by a $T^{-3/2}$ dependence which is expected for scattering by acoustic phonons in semiconductors but not in metals where it is T^{-1}. The strong temperature dependence is consistent with the two-carrier model but also with hole thermal activation. To further elucidate the conduction mechanism we plan R_H measurements above room temperature.

944

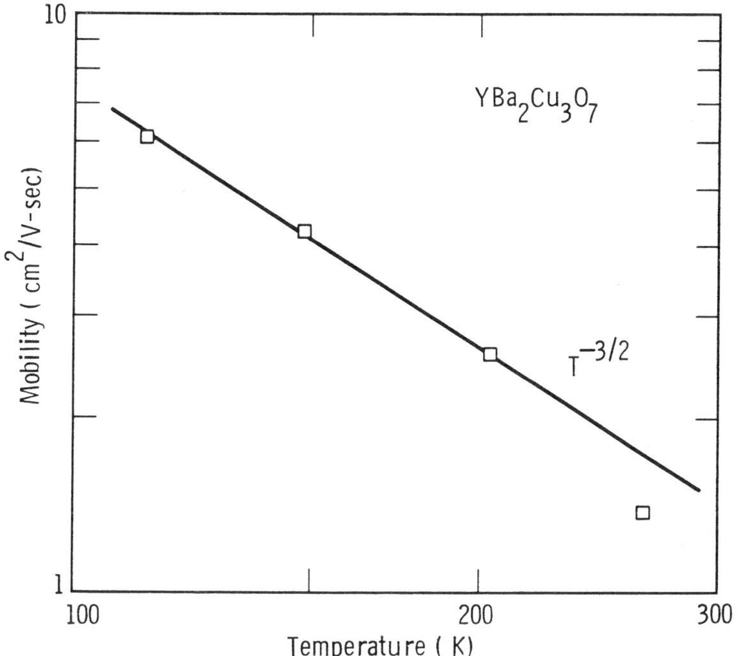

Figure 6 - A log-log plot of the carrier mobility vs temperature in stoichiometric YBC.

Tunneling into Y$_1$Ba$_2$Cu$_3$O$_7$

Tunneling spectroscopy in polycrystalline, cleaved YBC was performed by Ng et al. at 10 to 14K in a low-temperature tunneling microscope using aluminum tips to increase the effective tunneling area and thus reduce the possible depairing effects.[28] These were accounted for in fitting the data to the BCS density of states[29] by including a depairing energy term and obtaining corrected (lower) energy gap, Δ, values. The in-situ cleaving and S-I-N tunneling into fresh surfaces resulted in relatively sharp dI/dV peaks and low parabolic background, as shown in Figure 7. At various tip locations on the sample surface the energy gap was between 44.8 and 85 meV with the depairing correction of only 6 meV. The lowest value corresponds to $2\Delta/kT_c$ = 11, three times the BCS value. With such high values the possibility of S-I-S tunneling between the grains cannot be excluded. Extremely high $2\Delta/kT_c$ values in polycrystalline samples were reported by many groups but these results are not understood. Interestingly, tunneling into single crystal by Kirtley et al.[30] gave a very reasonable value for strong coupling superconductivity, $3.7 < 2\Delta/kT_c < 5.4$.

Moreland et al.[31] made the observation of Josephson effect in YBC weak link break junctions at higher temperatures, up to 85 ± 5K. Under microwave radiation the Shapiro steps gave 2.04 ± 0.05 μV/GHz, very close to the pair tunneling value of h/2e = 2.068 μV/GHz. In contrast to the tunneling data, these weak links, which were cleaved in liquid He, exhibited critical current and resistance products leading to a Δ about four to five times lower than the BCS value. This again is not understood.

<u>Summary of Measured and Derived Properties</u>

Table II gives a summary of LSC and YBC material parameters which we published earlier.[1,2] Measured material parameters used for the calculation[32] are also included. The enormous amount of analogous data which are now appearing in the literature make a detailed comparison impractical. In general, we find a fair agreement with reliable sources. The one big area of discrepancy is in γ values obtained from heat capacity measurements. These are consistently much higher than our values. For example, in YBC 1:2:3 Junod et al. obtained gamma = 13 mJ/(K^2-mole of Cu)[33] vs our 5 mJ/(K^2-mole of formula unit).

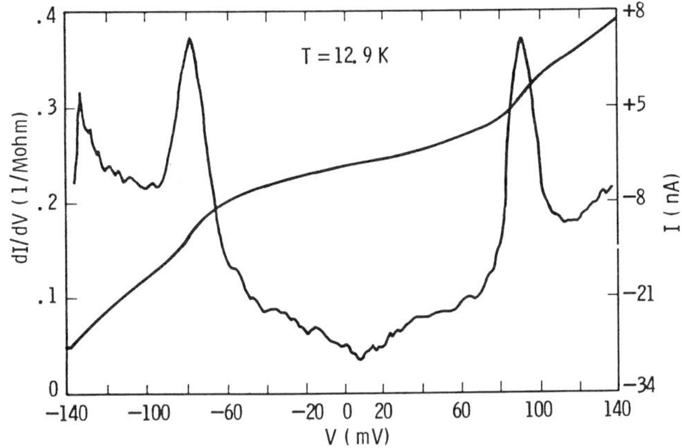

Figure 7 - The Ng et al tunneling I-V and dI/dV-V spectra in cleaved YBC at 14K. A measurement is shown which resulted in highest apparent delta.

We thought it useful to include Table II for two reasons. First, the calculations were done without assuming the dirty limit, based on the actually determined carrier density. Also, the data represent, at least, a consistent and reasonably accurate description of samples on which our studies and those of collaborating groups were based. Even if the BCS electron-phonon mechanism may not be adequate to explain the superconductivity in high-T_c oxide compounds, the phenomenological aspects of the Ginzburg-Landau theory should remain valid.

Table II. Material Parameters (Measured and Derived)

Measured	YBC	LSC
Transition temperature (midpoint) T_c (K)	94	36.5
Residual resistivity ρ_n ($\mu\Omega$cm)	400	550
Upper critical field slope dH_{c2}/dT (kOe/K)	19 ± 2	21 ± 1
Effective carrier density, n^* cm^{-3}	4.2×10^{21}	1×10^{21}
Derived		
Electronic specific heat coefficient, γ (mJ/K^2-mole fu[a])	5	1.0
Ginzburg-Landau coherence length, $\xi(0)$ (Å)	14	21
Ginzburg-Landau penetration depth, $\lambda(0)$ (Å)	2000	4100
Ginzburg-Landau κ, κ	150	200
Thermodynamic critical field, $H_c(0)$ (kOe)	8	2.7
Depairing critical current density, $J_d(0)$ (A/cm^2)	3×10^8	5×10^7

[a]fu = formula unit.

CONCLUDING REMARKS

Work at Westinghouse was aimed at determining technically important
parameters of newly discovered oxide compounds. We did not participate in
the race toward new compounds of ever higher T_c's, although our early
experiments with fluorine substitutions might be construed as an indication
to the contrary. The collaboration with Boeing Corporation resulted in a
somewhat provocative structural determination which, if it can withstand
further tests, might have important implications for understanding of
superconductivity. Transport properties and carrier density determinations
were of particular interest to us. As a consequence of joint work with
NBS, the combination of J_c data derived from magnetization and transport
measurements, also supported by complex susceptibility results, contributed
to the definition of the intergrain connectivity problem. Our intent is to
continue to investigate this problem in order to determine the feasibility
of technically useful conductors operating at higher cryogenic temperatures
up to 77K. Our recently started film synthesis work will be directed at
superconducting electronic applications.

ACKNOWLEDGEMENTS OF SUPPORT

Work at Westinghouse was supported in part, by Air Force Office of
Scientific Research, Contract No. F49620-85-C-0043. Work at NbS was
supported, in part, by the U. S. Department of Energy (DOE), Contract
No. DE-AI01-84RER52113. Work at the University of Texas was supported by
the National Science Foundation (NSF). At the Boeing Corporation, the
support was provided, in part, by NSF and, in part, by DOE.

REFERENCES

1. A. J. Panson, G. R. Wagner, A. I. Braginski, J. R. Gavaler,
    M. A. Janocko, H. C. Pohl, and J. Talvacchio, Appl. Phys. Lett. 50,
    1104 (1987).

2. A. J. Panson, A. I. Braginski, J. R. Gavaler, J. K. Hulm,
    M. A. Janocko, H. C. Pohl, A. M. Stewart, J. Talvacchio, and
    G. R. Wagner, Phys. Rev. B, June 1, 1987.

3. M. K. Wu, J. R. Ashburn, C. J. Torng, P. H. Hor, R. L. Meng, L. Gao,
    Z. J. Huang, Y. Q. Wang, and C. W. Chu, Phys. Rev. Lett., 58, 908
    (1987).

4. J. R. Gavaler, 1987 APS March Meeting, unpublished, videotaped.

5. A. J. Panson, unpublished work.

6. A. Manthiram, J. S. Swinnea, Z. T. Sui, H. Steinfink, and
    J. B. Goodenough, (University of Texas at Austin), submitted to J. Am.
    Chem. Soc., May 29, 1987.

7. F. W. Lytle, R. B. Greegor (The Boeing Corporation) and A. J. Panson,
    this Workshop.

8. D. G. Hinks, L. Soderholm, D. W. Capone II, J. D. Jorgensen, and
    I. K. Schuller, Appl. Phys. Lett. 50, 1688 (1987).

9. S. R. Ovshinsky, R. T. Young, D. D. Alfred, G. Demaggio, and
    G. A. Van der Leeden, Phys. Rev. Lett. 58, 2579 (1987).

10. R. H. Hopkins, W. E. Kramer, and A. M. Stewart, unpublished work.

11. S. Takekawa and N. Iyi, Jap. J. Appl. Phys. 26, L851 (1987).

12. Y. Hidaka, Y. Enomoto, M. Suzuki, M. Oda, A. Katsui, and T. Murakami,
    Jap. J. Appl. Phys. 26, L726 (1987).

13. M. A. Janocko and J. R. Gavaler, unpublished results.

14. R. H. Hammond, M. Naito, B. Oh, M. Hahn, P. Rosenthal, A. Marschall,
    N. Missert, M. R. Beasley, A. Kapitulnik, and T. H. Gebale, Material
    Research Society Extended Abstracts, EA-11, 169 (1987).

15. P. Chaudhari, R. H. Koch, R. B. Laibowitz, T. R. McGuire, and
    R. J. Gambino, Phys. Rev. Lett., 1987 (in press).

16. R. B. Van Dover, presented at the 1987 CEC-ICMC (unpublished data by
    J. Kwo et al., AT&T Bell Laboratories).

17. G. R. Stewart (University of Florida, Gainsville), unpublished data.

18. R. B. Goldfarb, A. F. Clark, A. I. Braginski, and A. J. Panson,
    Cryogenics 27, 475 (1987).

19. B. Renker, I. Apfenstedt, H. Kuepfer, C. Politis, H. Rietchel,
    W. Schauer, and H. Wuehl (Karlsruhe Nuclear Center), submitted to
    LT-18.

20. H. Kuepfer (Karlsruhe Nuclear Center), private communication,
    unpublished.

21. N. R. Werthamer, E. Helfand, and P. C. Hohenber, Phys Rev. $\underline{147}$, 295
    (1966).

22. H. London, Proc. Royal Soc. A $\underline{152}$, 650 (1935). In Refs. 1 and 2, the
    calculated values were too low by a factor of 100 due to numerical
    error.

23. A. I. Braginski, presented at the AGED Symposium, Washington, DC,
    April 30, 1987.

24. A. K. Ghosh, M. Suenaga, T. Asano, A. R. Moodebaugh, and
    R. L. Sabatini, 1987 CEC-ICMC, Paper FP-4, to be published in Adv.
    Cryo. Eng., Vol. 34.

25. I. Apfelstedt, R. Fluekiger, H. Kuepfer, R. Meier-Hirmer, B. Obst,
    C. Politis, W. Schauer, F. Weiss, and H. Wuehl, (Karksruhe Nuclear
    Center), submitted to LT-18.

26. J. W. Ekin, A. J. Panson, A. I. Braginski, M. A. Janocko, M. Hong,
    J. Kwo, S. H. Liou, D. W. Capone II, and B. Flandermeyer, Materials
    Research Society Extended Abstracts, Vol. EA-11, 223 (1987).

27. T. R. Dinger, T. K. Worthington, W.J. Gallagher, and R. L. Sandstrom,
    to be published in Phys. Rev. Lett., 1987.

28. K. W. Ng, S. Pan, A. L. de Lozanne, A. J. Panson, and J. Talvacchio,
    submitted to LT-18.

29. R. C. Dynes, J. P. Garno, G. B. Hertel, and T. P. Orlando, Phys. Rev.
    Lett. $\underline{53}$, 2437 (1987).

30. J. R. Kirtley, R. T. Collins, Z. Schlesinger, W. J. Gallagher,
    R. L. Sandstrom, T. R. Dinger, and D. A. Chance, Phys. Rev. B,
    June 1, 1987.

31. J. Moreland, L. F. Goodrich, J. W. Ekin, T. E. Capobianco,
    A. F. Clark, A. I. Braginski, and A. J. Panson, submitted to Appl.
    Phys. Lett.

32. Calculations based on the compilation of BCS relationships in
    T. P. Orlando, E. J. McNiff Jr., S. Foner, and M. R. Beasley, Phys.
    Rev. B $\underline{19}$, 4545 (1979), with $S/S_F = 0.5$.

33. A. Junod, A. Bezinge, T. Graf, J. L. Jorda, J. Muller, L. Antognazza,
    D. Cattani, J. Cors, M. Decroux, O. Fisher, M. Banovski, P. Genoud,
    L. Hoffmann, A. A. Manuel, M. Peter, E. Walker, M. Francois, and
    K. Yvon, submitted to Europhysics Letters, 1987.

STUDY OF THE PREPARATION AND PROPERTIES OF YBACUO-FILMS

B. Häuser and H. Rogalla

Institut für Angew. Physik der
Universität Giessen
Giessen, FRG

INTRODUCTION

After the discovery of high-T_c superconductivity in LaBaCuO by
Bednorz and Müller[1] and the subsequent discovery of superconductivity
above liquid nitrogen temperature in the YBaCuO-system by Chu et al.[2], an
immense push in the research efforts took place. Besides pure scientific
aims a vast variety of industrial applications seems to be possible. A
short while ago Nb_3Ge was supposed to be the material with the highest T_c
applicable to the superconducting electronics[3,4]. Now the new materials
seem to be applicable too. In order to fabricate integrated superconducting devices from these materials one has to develop a thin film technique which yields films with a high T_c comparable to that of the bulk
material[5]. Additionally the film properties should not be degraded during
the lithographic process.

In principle there are two ways of structuring the films : liftoff
and etching. Applying the first method limits the preparation temperature
by the maximum resist temperature, but possible film damage by etching is
prevented and the films are not exposed to aqueous solutions of resist-
developers. The second method has the advantage of allowing a wide
substrate temperature range for the preparation, but the disadvantage of
possible degradation of film properties due to ion- or wet etching.

As film preparation methods basically evaporation and sputtering are
established. Here we report on studies about the deposition of YBaCuO
thin films by sputtering from a sinter target with the aim of room temperature preparation of thin films. A post annealing treatment in oxygen
shell yield the high-T_c phase. In this case structuring of the films can
be done by liftoff prior to the anneling.

PREPARATION

The YBaCuO-films were prepared by rf-magnetron sputtering from a
sinter target in a turbo-pumped vacuum chamber. The background pressure
before sputtering was better than 10^{-3}Pa. Sapphire and MgO single
crystals were used as substrates. During the preparation the substrates
can be heated up to temperatures of 950°C in a pure argon atmosphere as

well as in an argon/oxygen-mixture. The pressure during sputtering was adjusted in the range 10^{-3}Pa to 50Pa by massflow controllers. The flow of two gases, here argon and oxygen, can be regulated independently in order to vary the oxygen content in the films during the deposition.

The targets were prepared the way commonly used for the preparation of bulk YBaCuO-superconductors. Starting from Y_2O_3, BaO, and CuO, the materials were mixed and pulverized in an agate vial. In a next step the mixture was reacted at 950°C for 14 hours and pulverized again. Then the powder was pressed at a pressure of 2kbar into the form of a disc with 50mm diameter and a thickness of 2..3mm. These discs were annealed in a pure oxygen atmosphere for 20 hours placed between two ceramic disks. After the annealing, the target was cut into parallel strips of 5 mm width with the aid of a diamond saw in order to prevent the target from breaking during the sputtering due to thermal tensions.

Each film, sputter deposited from these targets, was investigated by the following methods :
a) XES to determine the composition,
b) the x-ray diffraction, and
c) the temperature dependence of the resistance was measured in a
   continuous flow type crystat.
These methods were applied to the films prior to the annealing and after each annealing step.

DEPENDENCE OF THE FILM-COMPOSITION ON SPUTTERING PARAMETERS

In order to change the composition of the sputtered films a variety of targets with different compositions were fabricated the way described above. Aside from this way of adjusting the composition, the influence of sputter parameters like rf-power, self-bias voltage, total chamber-pressure, argon/oxygen-ratio, substrate-target distance and substrate temperature on the composition, x-ray diffration pattern, and resistance was investigated.

First the deposition conditions were varied systematically for a (non-optimum stoichiometry) model-target in order to find its influence on the resulting film composition. The Y:Ba:Cu-ratio of the used model-target is 1:5.2:4.1.

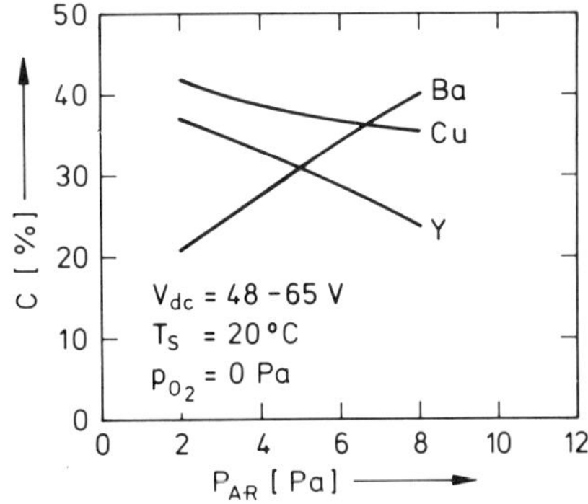

Fig. 1. Dependence of Ba-, Y-, and Cu-concentration C in a
        sputtered film on the total pressure P_{Ar} in a pure
        argon atmosphere.

Within the rf-power range of 80W to 240W there was no significant
change in the film composition. Only a slight minimum in the relative
Yttrium content was found at an rf-power level of about 180W corresponding to the onset of a saturation in the deposition rate. The cathode
self-bias voltage turned out to be practically independent of the rf-
power above a certain small rf-power level, but strongly dependent on the
composition and treatment of the target.

The dependence of the film-composition on the total pressure of a
pure argon atmosphere is plotted in Fig. 1. The films were prepared at
ambient substrate temperature T_s at an rf-power level of 130W. The target-substrate distance was 3cm, the sputter duration 45 minutes, the resulting film-thickness 70nm corresponding to a deposition rate of
1.5nm/min nearly independent of the total pressure.

The Ba-concentration increased with rising pressure from 20% at
P_{Ar}=2Pa to above 40% at P_{Ar}=8Pa. The Cu-concentration decreases less
rapidly than the Y-concentration. Thus the film composition can be varied
over a wide range by the total pressure from an initial Y:Ba:Cu-ratio of
1:0.6:1.13 at 2Pa to 1:1.7:1.5 at 8Pa. Nevertheless the necessary Cu-concentration in this film could not be achieved with the model-target.

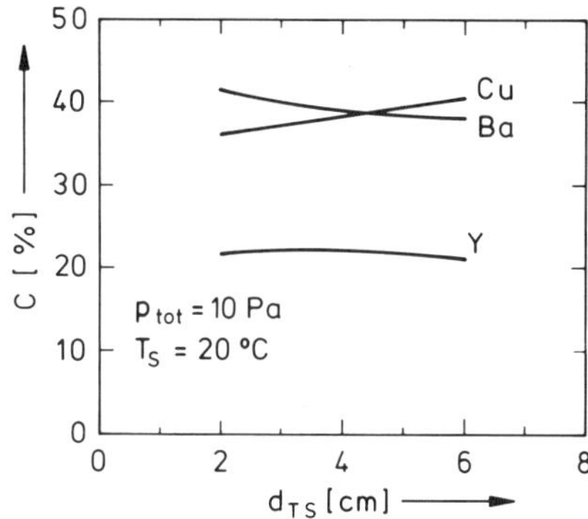

Fig. 2. Dependence of Y-, Ba-, and Cu-concentration on the
target-substrate distance d_{TS} for sputtering in a
pure argon atmosphere at a pressure p_{tot} of 10Pa.

The Y-, Ba-, and Cu-concentration C can be adjusted too by varying
the target-substrate distance d_{TS} (see Fig. 2). Compared to the pressure
dependence only a small change in the concentration is visible. The Y-
concentration remains practically constant whereas Ba and Cu reverse
their contribution. Starting from d_{TS}=2cm the Y:Ba:Cu-ratio changes from
1:1.9:1.13 to 1:1.7:1.86 at d_{TS}=6cm. Thus by increasing the target-
substrate distance to 6cm and the total pressure to 10Pa the relative Cu-
content could be increased independently of Y and Ba.

Varying the substrate temperature T_s changes the Y-, Ba-, and Cu-
concentration too (see Fig. 3). Up to about 500°C the concentrations in
the film are nearly constant. Above this temperature a pronounced change
takes place. The deposition rate for Yttrium and Barium increases,
whereas the Cu-deposition rate remains fairly constant. Starting from
room temperature the Y:Ba:Cu-ratio changes from 1:1.9:1.13 to 1:1.4:0.6
for T_S=900°C indicating the drastical relative loss of Cu.

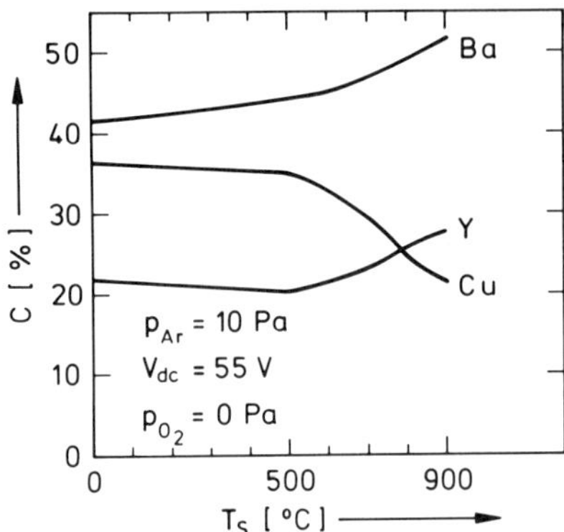

Fig. 3. Dependence of Y-, Ba-, and Cu-concentration on the
substrate temperature T_s for a target-substrate
distance of 2cm and a pure argon atmosphere with a
pressure p_{tot} of 10Pa.

Finally the oxygen partial pressure was varied for different
substrate temperatures. As an characteristic example the resulting dependence of the concentration for T_s=700°C was plotted in Fig 4. The overall
deposition rate remained constant for yttrium but decreased for barium
and copper, resulting in a decrease of the total deposition rate with increasing oxygen partial pressure. This behavior may be explained by an
increased oxide formation in the film and on the target surface.

From these measurements at a model-target the appropriate target
composition for specific sputter- and deposition conditions could be deduced. One aim was the deposition of a film with the 1:2:3-stoichiometry
at room temperature. With a target-composition of 1:6.1:8.2 this condition could be realized for sputtering in a pure argon atmosphere at
p_{Ar}=8Pa. It should be mentioned that the measured values depend on

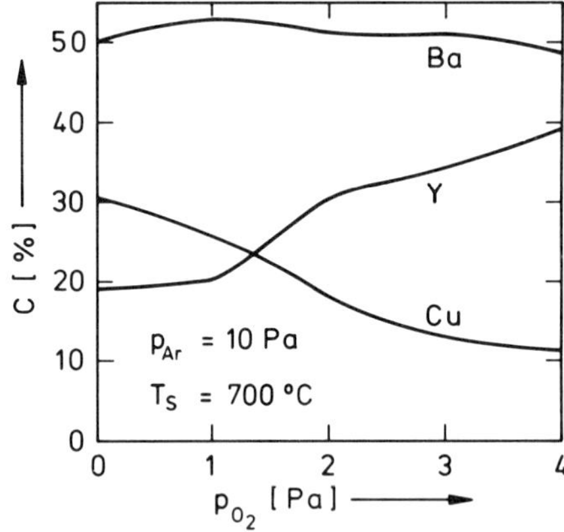

Fig. 4. Dependence of Y-, Ba-, and Cu-concentration on the
oxygen partial pressure p_{O_2} for a substrate temperature of 700°C and a target-substrate distance
of 2cm.

specific features of the sputtering equipment, e.g. magnetic field at the
surface of the magnetron cathode.

X-RAY DIFFRACTION MEASUREMENTS

As mentioned above x-ray diffraction measurements of the films were
performed prior and after the deposition. First we will describe the
diffractograms of as-deposited films.

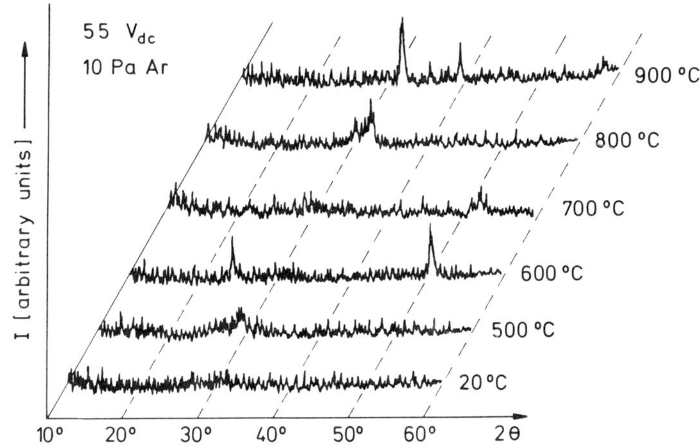

Fig. 5. X-ray diffractograms for as-deposited YBaCuO-films
of 60nm to 150nm thickness as function of the
substrate temperature during sputtering. The
preparation parameters are otherwise the same for
all films.

For an otherwise constant parameter set for the film preparation the
substrate temperature was varied between ambient temperature and 900°C.
The resulting x-ray diffractograms are shown in Fig. 5. Up to temperatures of nearly 500°C no reflexes appear – the film is practically amorphous. Starting at 500°C first reflexes can be observed indicating the
formation of a crystalline phase. For films prepared at 600°C peaks at
$2\theta=24°$ and 50° arise. If the substrate temperature is increased further
to temperatures above 700°C, these peaks nearly vanish again. Starting at
800°C the reflexes of the high-T_c phase appear[6,7]. Corresponding to the
phase transition between the orthorombic and tetragonal phase in the bulk
material at about 650°C a clear distinction into seperate regions in the
film diffractogram can be observed.

Adding oxygen to the argon atmosphere during the sputtering of the
films changes the picture significantly for films deposited at high tom-
peratures. Films prepared for instance at 900°C in an argon/oxygen-mix-
ture nearly show the same reflexes as films prepared at 600°C in a pure
argon atmosphere. This indicates, that enough oxygen during the hot
preparation favours the growth of the orthorhombic phase, whereas a small
amount of oxygen (partial pressure 2Pa O_2 and 10Pa Ar) favours the growth
of the high-T_c phase.

Independent of the preparation conditions described above no film
showed superconductivity or even a decrease in resistance with falling
temperature. Most of the films had a very high resistance which increased

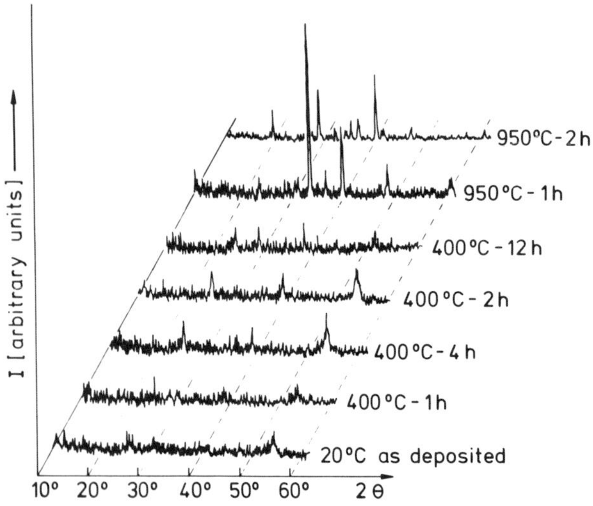

Fig. 6. X-ray diffractograms for an YBaCuO-films of 550nm
thickness after subsequently performed oxygen annealing. Parameters are the annealing temperature
and duration.

with falling temperature. Therefore these films were annealed in an oxygen atmosphere of 1bar. In Fig. 6 the x-ray diffractograms of a film deposited at room temperature were taken after subsequent annealing steps.
Immediately after the deposition the film is practically amorphous. After
annealing the film for a total of 5 hours reflexes appear, which do not
represent a high-T_c phase. After further annealing the film at 400°C for
12 hours the reflexes vanish.

Heating the sample for 1 hour at 950°C results in a spectrum with
clear high-T_c reflexes (e.g. 110+103 at 2θ=32.83°A, 014+005 at 38.54° and
200 at 47.57°). The resistance of the sample significantly drops by some
orders of magnitude at room temperature and it is nearly fully superconducting at low temperatures. After a further heat-treatment for 2 hours
at 950°C the sample became insulating. The composition did not change,
but even visually the sample changed dramatically from shiny metallic to
yellow translucent. In the diffractogram this transition is clearly visible too, especially the characteristical 110+103-reflex almost disappears.

RESISTANCE MEASUREMENTS

The resistance of the films prior to the oxygen annealing was very
high for all preparation conditions. At low temperatures the films are
insulating. Anealing these films in oxygen at temperatures above 600°C
for more than 2 hours reduced the resistance at room temperature by at
least three orders of magnitude. The reflectivity of the films decreased
slightly. Even though most of the post-treated films show good high-T_c
reflexes in the diffractogram, the resistivity shows semiconducting temperature dependence. We attribute this to a weak coupling between grains
of the right phase.

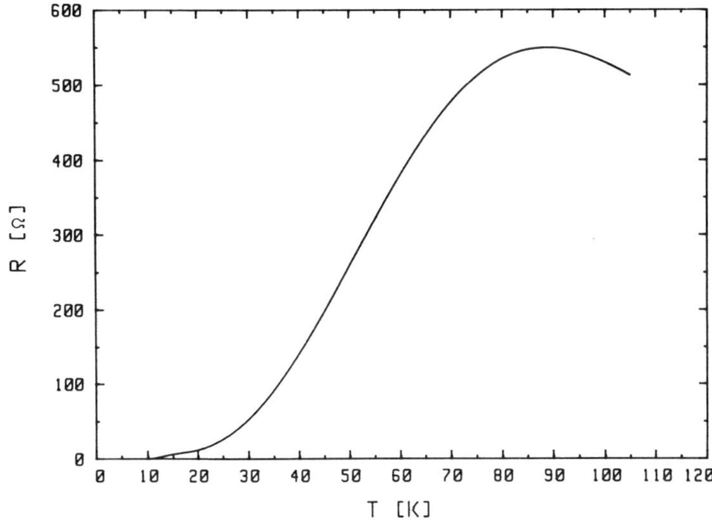

Fig. 7. Temperature dependent resistance of a 150nm film,
        which becomes fully superconducting at 12K. The
        film was deposited at ambient temperature and an-
        nealed at 650°C for 2 hours, 750°C 1 hour, and
        850°C for 1 hour.

     To understand the influence of the oxygen annealing on the re-
sistance, a series of subsequent annealing steps were performed on some
samples. At 650°C the resistance reaches a minimum after 2 hours and remains constant for longer treatments at this temperature. It is reduced
by a factor of about 50. Annealing the same film again at 750°C for one
hour further reduces the resistance by a factor of at least 50. A next
annealing step at 850°C does not change the room temperature resistance
any more. Instead a superconducting phase develops if all other preparation parameters are perfect.

Fig. 8. SEM-picture of a hot-deposited film after a full
        annealing cyclus.

In Fig. 7 temperature dependent resistivity is plotted for a film of 150nm thickness, which becomes superconducting (onset temperature) at 90K and looses its resistance fully at 12K. XES-measurements showed that the yttrium-contents of the film is slightly too large. This may be a reason for the wide transition to superconductivity.

The thermal treatment yields cristalline films with a pronounced texture in the directions of the cristallographical axes of the substrate. For our sapphire substrates the axes are parallel to its sides. Hot deposited films tend to break away from the surface in small strips parallel to an axis (see Fig. 8). The length of the strips indicates that strong tensions may be present in films deposited at high temperatures. Such a behavior was never observed in films deposited at room temperature. Further annealing at 850°C reduces the tension and the strips shorten and return to the surface.

CONCLUSION

Room temperature deposition of YBaCuO-films by rf-magnetron sputtering from a sinter target yields thin films , which become superconducting after an appropiate oxygen annealing treatment. A first film of 150nm thickness shows an onset temperature of about 90K and a broad transition to full superconductivity at 12K. An optimization of this preparation methode is currently performed in order to decrease the transition width for the very thin films.

The composition can be adjusted by changing the deposition parameters as described above. In this way a wide varity of compositions could be achieved with a single target. The post-deposition treatment turned out to be very important. Our current recepture for the post-treatment of our films is annealing in pure oxygen at 650°C for 2 hours, 750°C 1 hour, and 850°C 1 hour. There is currently no explanation available for the detailed procedure.

With this treatment cristalline films with a well defined texture parallel to the sapphire cristal axes could be prepared. Tension in the films deposited on sapphire substrates can be a problem, especially for cristalline films deposited at elevated temperatures. The choice of amorphous film-deposition as used here, and alternative substrates like $SrTiO_3$ may solve this problem.

The x-ray diffraction patterns of films without post-treatment show characteristic reflexes of the 123-phase, if the films are prepared at substrate temperatures above 800°C. A temperature of 950°C is favorable. Addition of about 20% of oxygen during the deposition further enhances the heigth of the 123-reflexes. Nevertheless, these films are of high resistivity and show no transition to superconductivity. Instead their behavior is semiconductor-like.

Films annealed at temperatures above 750°C always show the right reflexes if the film-composition is resonable near the 1:2:3-ratio. In this case the room-temparature resistance is more than 3 orders of magnitude smaller than before the treatment.

ACKNOWLEDGEMENTS

We would like to thank C. Heiden for his interest and the helpful discussions with him. Further we acknowledge the help of Mrs. K. Binzer and Mrs. G. Streit. This work was supported by a BMFT-grant.

REFERENCES

1. J.G. Bednorz and K.A. Müller, Possible High T_c Superconductivity in the Ba-La-Cu-O System, Z.Phys.B-Condensed Matter 64:189 (1986)
2. M.K. Wu, J.R. Ashburn, C.J. Torng, P.H. Hor, R.L. Meng, L.Gao, Z.J. Huang, Y.Q. Wang, and C.W. Chu, Superconductivity at 93K in a New Mixed Phase Y-Ba-Cu-O Compound System at Ambient Pressure, Phys.Rev.Lett. 58:908 (1987)
3. M.R. Beasley, Advanced Superconducting Materials for Electronic Applications, IEEE Trans.Electron Devices ED-27, 2009 (1980).
4. H. Rogalla, B. David, and M. Mück, Fabrication Techniques for Thin Film Nb_3Ge Josephson Devices, in "SQUID'85, Superconducting Quantum Interference Devices and their Applications", H.D. Hahlbohm, H. Lübbig, ed., W. de Gruyter, Berlin (1985).
5. R.B. Laibowitz, R.H. Koch, P. Chaudhari, and R.J. Gambino, Thin Superconducting Oxide Films, preprint, (1987)
6. R.J. Cava, B. Batlogg, R.B. van Dover, P.W. Murphy, S.Sunshine, T. Siegrist, J.P. Remeika, E.A. Rietman, S. Zahurak, and G.P. Espinosa, Bulk Superconductivity at 91K in Single Phase Oxygen-Defficient Perovskite $Ba_2YCu_3O_9$, Phys.Rev.Lett. 58:1972 (1987)
7. D.H.A. Blank, J. Flokstra, G.J. Gerritsma, L.J.M. van de Klundert and G.J.M. Velders, X-ray and electron spin resonance on $YBa_2Cu_3O_7$ prepared by citrate synthesis, preprint (1987)

SOME PHYSICAL PROPERTIES OF HIGH-TEMPERATURE SUPERCONDUCTORS

C. Y. Huang, L. J. Dries, F. A. Junga

Research & Development Division
Lockheed Missiles & Space Company, Inc.
Palo Alto, California 94304

P. H. Hor, R. L. Meng, C. W. Chu

Department of Physics
University of Houston
Houston, Texas 77004

ABSTRACT

We have measured the magnetization of single-phase 90-K super-
conductors, $GdBa_2Cu_3O_{6+\delta}$, $EuBa_2Cu_3O_{6+\delta}$, and $SmBa_2Cu_3O_{6+\delta}$ with
a SQUID magnetometer. We have shown that, in the superconducting state,
each magnetization-field curve exhibits a maximum at ~100 G, followed by
a linear increase of the magnetization with a slope only approximately
one-fifth of the slope for a field smaller than 50 G. We have also
investigated the effect of γ-irradiation on $YBa_2Cu_3O_{6+\delta}$, $SmBa_2Cu_3O_{6+\delta}$,
and have found that the radiation damage results in the appearance of a
tail in the superconducting transition. We have also shown that the
normal resistance decreases with increasing radiation exposure up to a
dose of 10 Mrad.

INTRODUCTION

Recently, superconductivity up to 98 K was found in the multiphased
$Y_{1.2}Ba_{0.8}CuO_{4+y}$ compound system.[1,3] It was later shown[4,5] that the
orthorhombic YBa_2CuO_{6+y} phase is responsible for this high-temperature
superconductivity. More recently, it has been demonstrated[6-9] that
all single-phased $LBa_2Cu_3O_{6+\delta}$ (L = Y, all rare earths except Ce
and Pr) are superconductors with $T_c \gtrsim 90$ K. It was surprising to find
that $GdBa_2Cu_3O_{6+\delta}$ had a T_c of approximately 94 K, because Gd^{3+}
has a large spin (S = 7/2) and is known to effectively destroy Cooper
pairs.[10] Therefore, it is important to study the magnetic properties
of this superconductor. For reasons of comparison, it is also necessary
to study other high-temperature magnetic superconductors.

For the purpose of future applications, it is important to know the
effect of radiation damage. Hence, we have measured change in the
superconducting transitions of the superconductors.

EXPERIMENTAL RESULTS

The samples GdBa$_2$Cu$_3$O$_{6+}$ (T$_c$ 92 K)[6,11], EuBa$_2$Cu$_3$O$_{6+}$ (~95 K)[6-8], SmBa$_2$Cu$_3$O$_{6+\delta}$ (~94 K)[6,7] and YBa$_2$Cu$_3$O$_{6+\delta}$ (~92 K) used in this paper were prepared by the same method as described in Ref. 1. The x-ray diffractograms have revealed that all these samples possess only the "123" single phase.[4-6]

Magnetic Measurements

The magnetic moments of the samples were measured by an SHE Model 905 SQUID magnetometer from 2 to 300 K in a field up to ~1 kG. Figure 1 displays the inverse of the magnetic susceptibility, $1/\chi$, for GdBa$_2$Cu$_3$O$_{6+\delta}$ plotted against T for T > T$_c$. As clearly demonstrated in the figure, the data can be fitted to the Curie-Weiss law, $\chi = C/(T + \Theta)$, with C = 7.39 x 10^{-2} emu-K and $\Theta \simeq 6$ K. A simple calculation leads to the conclusion that the Curie constant can solely be accounted for by the Gd^{3+} magnetic moments. The contribution to the Curie constant by the Cu^{2+} ions is negligible because of the small spin (S = 1/2). The temperature dependence of the susceptibility shows that Gd^{3+} spins interact antiferromagnetically. The Curie-Weiss temperature $\Theta \simeq 6$ K is consistent with the Neel temperature T$_N$ = 2.22 K determined calorimetrically.[12] We thus conclude that below T$_N$, superconductivity coexists with antiferromagnetism. Figure 2 shows the temperature dependence of $1/\chi$ for EuBa$_2$Cu$_3$O$_{6+\delta}$ (denoted by triangles). Based on the Mossbauer experiment,[13] the Eu ions are in the trivalent state with the nonmagnetic singlet ground state (J = 0). As expected, our data show that the susceptibility for the Eu sample is about a factor of 10 smaller than that for the Gd sample. The temperature dependence suggests that some of the Eu^{3+} excited states are occupied at T > 100 K. The Sm^{3+} ion has a small magnetic moment and, as shown in Fig. 2, the susceptibility for SmBa$_2$Cu$_3$O$_{6+\delta}$ (denoted by solid circles) is small.

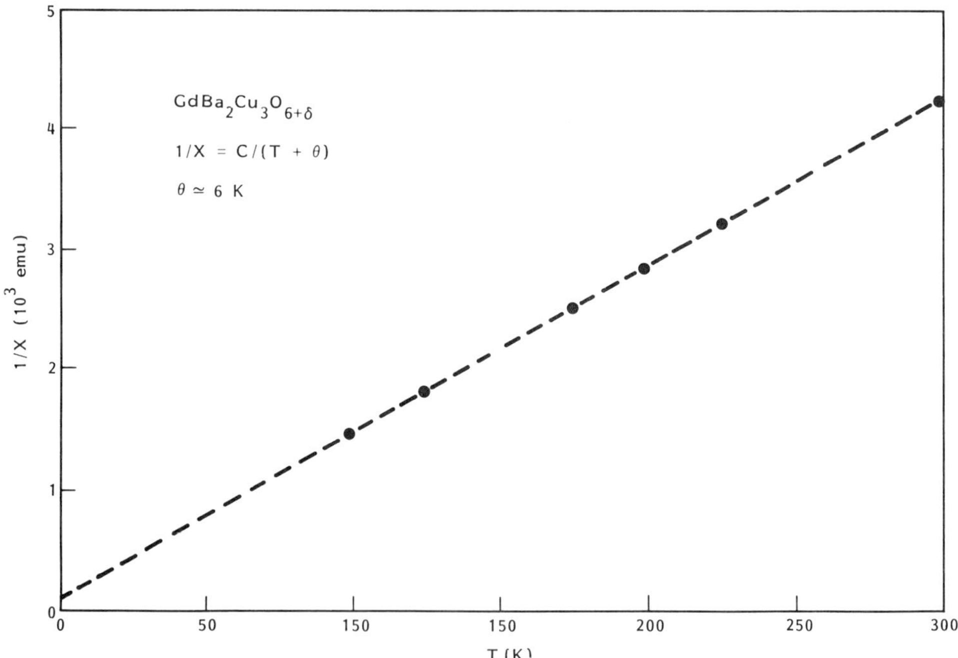

Fig. 1 Temperature dependence of the inverse of the magnetic susceptibility, $1/\chi$, for GdBa$_2$Cu$_3$O$_{6+\delta}$.

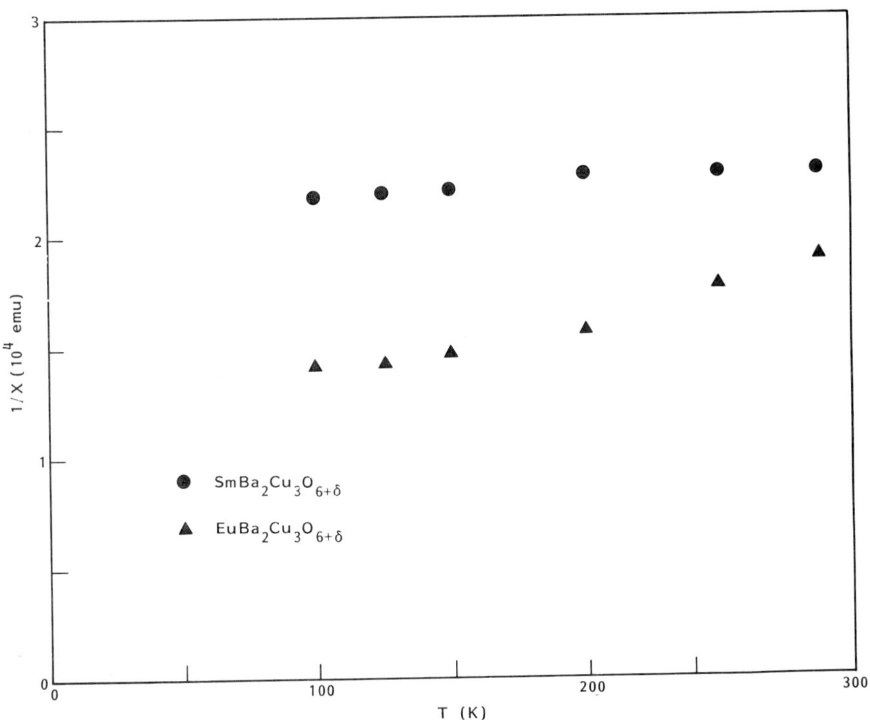

Fig. 2 Temperature dependence of the inverse of the magnetic susceptibility, $1/X$, for $EuBa_2Cu_3O_{6+\delta}$ and $SmBa_2Cu_3O_{6+\delta}$.

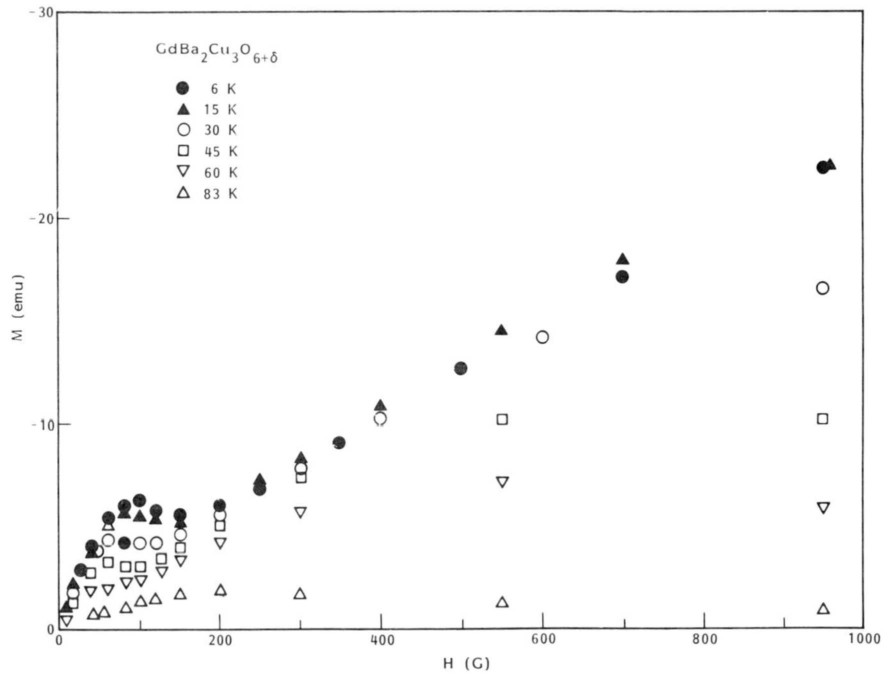

Fig. 3 Field dependence of the magnetization, M, for $GdBa_2Cu_3O_{6+\delta}$ measured at 6, 15, 30, 45, 60, and 83 K.

Figure 3 displays the field dependence of the magnetization for
$GdBa_2Cu_3O_{6+\delta}$. At 6 K, the magnetization is linear in H for
H < 50 G with a slope close to perfect diamagnetism, $\chi = -1/4\,\pi$; it
reaches a maximum at H = 100 G, above which the magnetization decreases
somewhat until H $\simeq$ 200 G, followed by the linear increase of M with
H up to ~1 kG with a slope only 1/5 of the initial slope for H < 50 G.
The data for T = 2 K are almost the same as those for 6 K. The low
field susceptibility is presumably due to the superconducting shielding,
which is destroyed by the applied field at H ~ 100 G. For H > 100 G,
the superconducting grains are responsible for the observed diamagnetic
susceptibility which, however, is greatly reduced from the low-field
diamagnetic susceptibility because the grain sizes are somewhat
comparable to the penetration depth and the sample is rather porous. As
demonstrated in Fig. 3, the maximum around 100 G becomes broader at
higher temperature. As shown in Fig. 4, a similar maximum has also been
observed in $EuBa_2Cu_3O_{6+\delta}$ at 6 K (denoted by triangles). Although
a clean maximum is absent for $SmBa_2Cu_3O_{6+\delta}$, the data (denoted by
solid circles) indicate the presence of a broadened peak as in the case
for the Gd sample at T $\gtrsim$ 75 K. The broadening of the maxima might be
related to the oxygen deficiency.

Radiation Effects

We have employed a Co^{60} gamma ray source (E = 1.17, 1.33 MeV)
to irradiate the superconductors to a dose level of 1 Mrad and 10 Mrad
(Si). The 10 Mrad dose would correspond to an estimated point defect
density of ~10^{12} cm^{-3} in these materials. After each irradiation,
resistance versus temperature data were acquired using an ac four-point
probe technique. Data for the $YBa_2Cu_3O_{6+\delta}$ sample are shown in

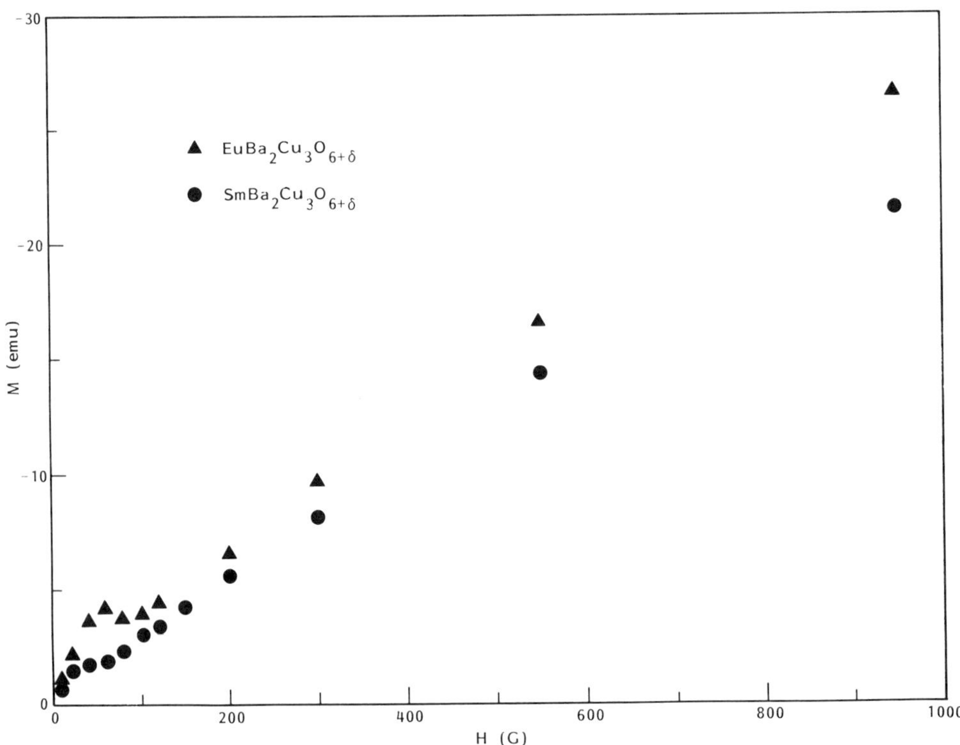

Fig. 4 Field dependence of the magnetization, M, for $EuBa_2Cu_3O_{6+\delta}$
and $SmBa_2Cu_3O_{6+\delta}$ measured at 6 K.

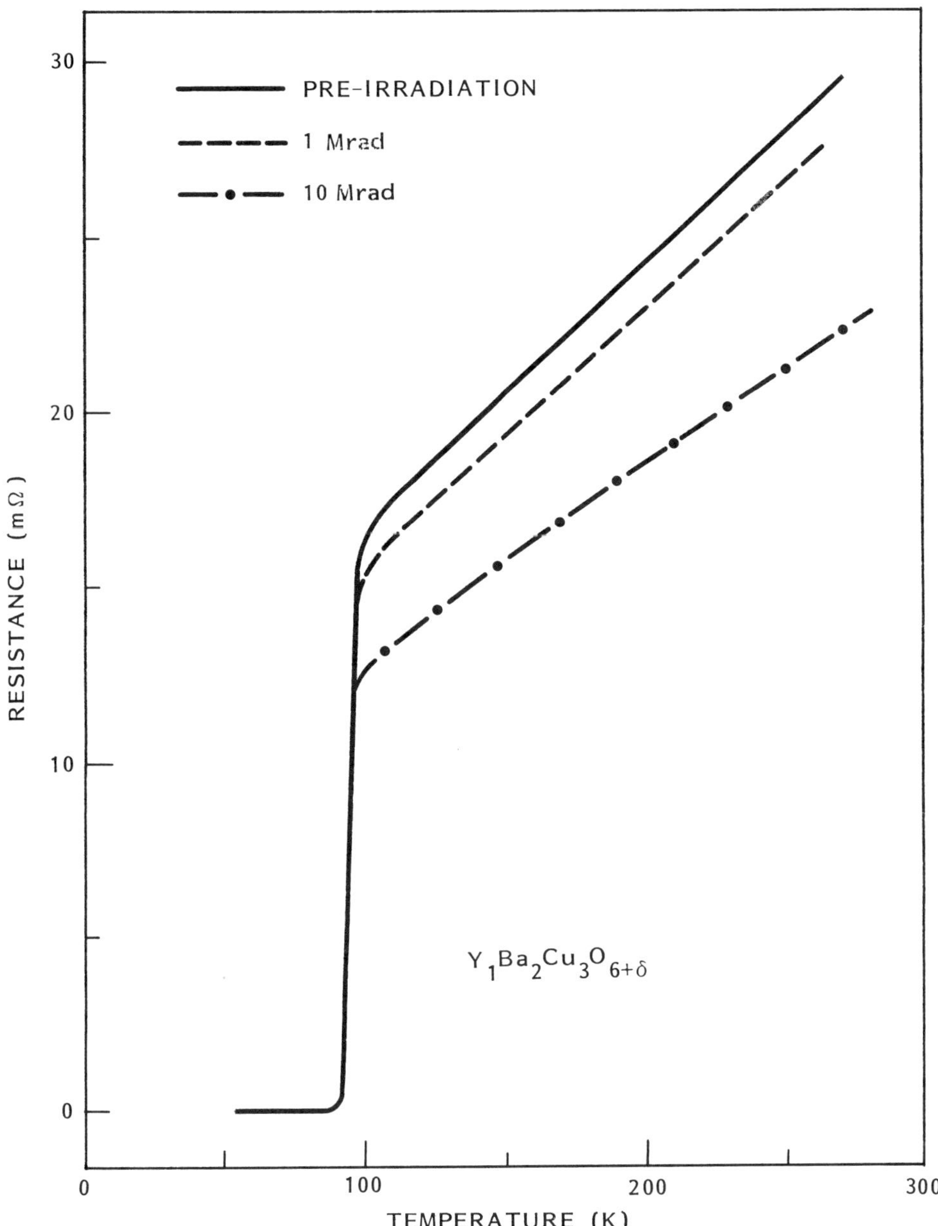

Fig. 5 Radiation dose dependence of resistance versus temperature curve
for the $YBa_2Cu_3O_{6+\delta}$ sample, showing decrease of normal
resistance with increasing gamma ray exposure.

Figs. 5 and 6. The sample resistance at high temperature shows a
decrease as the gamma dose is increased, Fig. 5. This may be due to a
vacancy-enhanced formation of short-range order which may occur in
alloys at low doses. A similar result was obtained with the
$SmBa_2Cu_3O_{6+\delta}$ material. The $GdBa_2Cu_3O_{6+\delta}$ sample also showed lower
normal resistance after 1-Mrad irradiation, but the sample became very
resistive after the 10-Mrad dose and the superconducting transition was
no longer seen. A summary of the normal resistance of these samples at
T = 273 K is shown in Table I. Additional investigation at higher dose
levels are necessary for further study. The effect of irradiation at

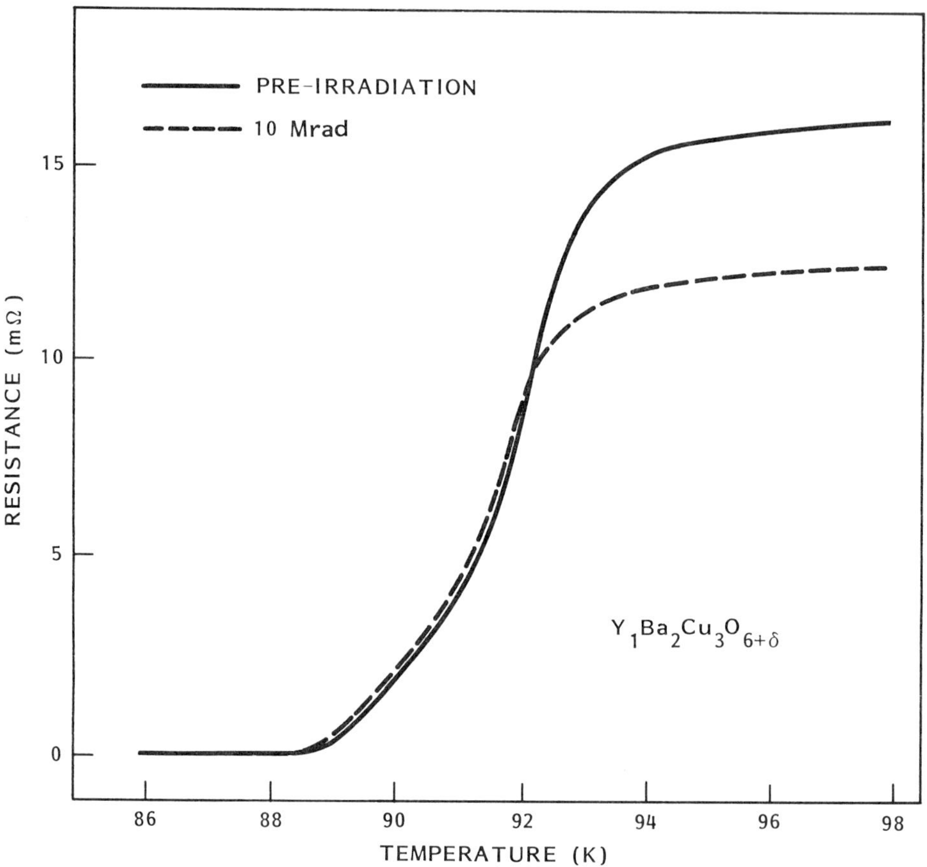

Fig. 6   Resistance versus temperature data, with expanded temperature
scale showing slight broadening of the superconducting
transition after 10-Mrad gamma-ray dose ($YBa_2Cu_3O_{6+\delta}$
sample).

these dose levels on T_c is slight, as shown in the $YBa_2Cu_3O_{6+\delta}$
sample in Fig. 6, and consists of a slight broadening of the transition
and movement of the zero resistance state to slightly lower temperature.
Again, higher dose studies are necessary to see if the superconducting
transition will broaden further or even disappear when the material
defect density is large (10^{15} cm^{-3} and higher).

CONCLUSIONS

In this paper, we have reported that the field dependencies of the
magnetization for $GdBa_2Cu_3O_{6+\delta}$, $EuBa_2Cu_3O_{6+\delta}$, and $SmBa_2Cu_3O_{6+\delta}$
in the superconducting state exhibit well-defined maxima and decreased
diamagnetic susceptibility for H > 200 G. We have interpreted the
appearance of the maxima in terms of the destruction of the shielding
current by the applied field (~100 G), and the decreased susceptibility
in terms of porous superconducting grains. However, more experimental
study has to be made to clarify this interpretation. The γ-irradiated
samples demonstrated the occurrence of a broadening of the

Table 1   RESISTANCE (mΩ) AS A FUNCTION OF γ DOSE

Dose (Mrad)	Sample		
	$YBa_2Cu_3O_{6+\delta}$	$SmBa_2Cu_3O_{6+\delta}$	$GdBa_2Cu_3O_{6+\delta}$
0	29.6	25.0	22.3
1	27.9	23.6	21.9
10	22.4	18.8	630

superconducting transitions of the sample. The most striking results are the radiation-induced reduction of the resistance in the normal state, because radiation damage generally gives rise to higher resistance. More experimental work is apparently needed in order to understand the causes of the resistance decrease.

REFERENCES

1.   M. K. Wu et al., Phys. Rev. Lett., 58:908 (1987).
2.   P. H. Hor et al., Phys. Rev. Lett., 58:911 (1987).
3.   C. W. Chu et al., to appear in Phys. Rev. B.
4.   R. H. Hazen et al., to appear in Phys. Rev. B.
5.   R. J. Cava et al., Phys. Rev. Lett., 58:1676 (1987).
6.   P. H. Hor et al., Phys. Rev. Lett., 58:1891 (1987).
7.   S. Tanaka, presented at the Am. Phys. Soc. Meeting, Mar 18, 1987.
8.   B. Batlogg, presented at the Am. Phys. Soc. Meeting, Mar 18, 1987.
9.   A. R. Moodenbaugh et al., Phys. Rev. Lett., 58:1885 (1987).
10.   A. A. Abrikosov and L. P. Gorkov, Sov. Phys. JETP, 12:1243 (1961).
11.   Z. Fisk et al., to appear in Solid State Commun.
12.   J. C. Ho et al., to appear in Solid State Commun.
13.   P. Boolchand et al., to appear in Solid State Commun.

RESISTANCE DEPENDENCE AND THERMOGRAVIMETRIC ANALYSIS OF THE

$Er_1Ba_2Cu_3O_{9-\delta}$ SUPERCONDUCTOR ABOVE ROOM TEMPERATURE

Yi Song, John P. Golben, Sung-Ik Lee, R.D. McMichael
Xiao-Dong Chen, and J.R. Gaines

Department of Physics
The Ohio State University
Columbus, OH 43210

Recently an unprecedented interest has been focused on high T_c
superconductors. Progressively higher transition temperatures have been
reported for the La-Ba-Cu-O series, [1] the Y-Ba-Cu-O series, [2] and
recently a flouridated version of this latter series. [3] A suggestion of
superconductivity at 240 K based upon the observation of an inverse AC
Josephson effect, [4] in a multiphase samples fuels the search for materials with even higher transition temperatures. Interest in the behavior
of these compounds is not concentrated just in the region below room
temperature. An understanding of what occurs in the sintering and
annealing processes is particularly imformative towards optimizing the
preparation conditions. In this paper we report resistance and thermogravimetric analysis (TGA) measurements on $Er_1Ba_2Cu_3O_{9-\delta}$ between room
temperature and 920 C. The incorporation of oxygen into the furnace
allows us to simulate the whole sintering and annealing processes.
Several resistive and oxygen content features will be discussed.

The samples were synthesized by the coprecipitation method. The
addition of Na_2CO_3 into a nitrate solution of the constituents pre-
cipitated the carbonate forms. The precipitate powder was repeatedly
heat treated to 900 C and pulverized. The powder was then pressed into
pellet form. A detailed discussion of these preliminary procedures
appears elsewhere. [5,6]

A DC technique with current reversal was used for the resistance
measurements. Four platinum leads were attached to the pressed pellet
sample with silver epoxy. The epoxy was covered with a layer of pro-
tective ceramic paste. [7] To compensate for induced thermal emf's, an
average of the forward and reversed current directions was used for
each measurement. The measurements were conducted slowly so as to obtain
thermal equilibrium at each temperature.

Figure 1 shows the resistance data upon heating to 920 C for
$Er_1Ba_2Cu_3O_{9-\delta}$, with $Y_1Ba_2Cu_3O_{9-\delta}$ shown for comparison. Initially the
resistance is large, but it decreases in general as a function of temper-
ture up to 500 C. A slight dip near 130 C is observed upon heating but
is absent in the cooling curve. This dip is present in both curves
and is likely due to the curing of the silver epoxy contacts. A very
slight bump in the Er data occurs near 360 C as well as a gradual bump

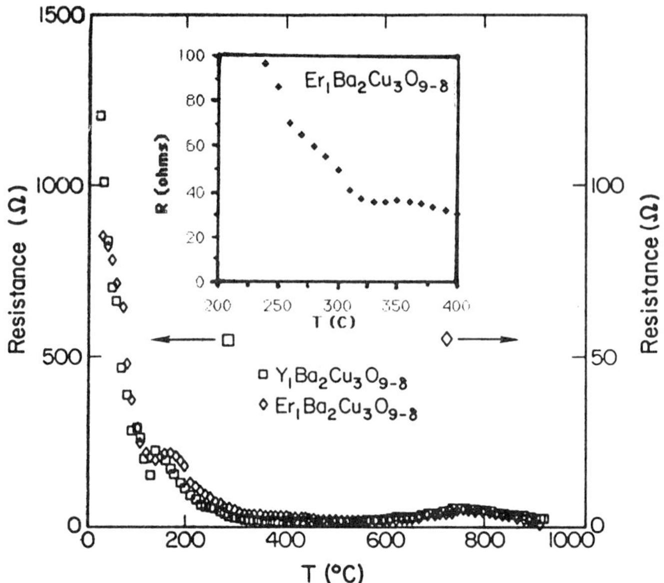

Figure 1. The Resistance above room temperature for nominal compositions of $Er_1Ba_2Cu_3O_{9-\delta}$ (triangles) and $Y_1Ba_2Cu_3O_{9-\delta}$ (squares). The extended bump between 600 C and 800 C in the $Er_1Ba_2Cu_3O_{9-\delta}$ curve is most likely a tetragonal to orthorhombic transition. The slight bump near 360 C (inset) correlates with the maximum uptake of oxygen in the TGA analysis.

from 600 to 800 C. The resistance along the cooling curve (not shown) is considerably smaller than the initial resistance upon heating, however an indication of the resistive bump at high temperatures still remains.

The data can be qualitatively understood when compared with the corresponding TGA data (Figure 2). On heating the sample in air, a major uptake of oxygen begins at about 340 C and maximizes at 410 C. The maximum uptake upon cooling is at around 360 C. This oxygen inhaling corresponds to the slight resitance bump at 360 C (Figure 1, inset). One of the possible locations for the extra oxygen incorporated would be on the Cu planes between the Ba layers of adjacent unit cells. In a recent neutron diffraction study on $Y_1Ba_2Cu_3O_{9-\delta}$ [8] it was concluded that there were ordered oxygen vacancies on these outer planes such that the remaining oxygen atoms formed "lines" with the Cu atoms. If a portion of the conductivity in the sample were associated with these lines, then the added incorporation of oxygen into these planes might have a disruptive effect. The slight resistive bump could be the result of this, or any other oxygen incorporation effect.

The extended bump between 600 C and 800 C can be interpreted similarly. In this case though, the resistive anomaly is pronounced, but the TGA curves only show a slight deviation near 680 C. The resistance curve for $Y_1Ba_2Cu_3O_{9-\delta}$ demonstrates a bump at 750 C which correlates well with the reported tetragonal to orthorhombic phase transition in $Y_1Ba_2Cu_3O_{9-\delta}$ at this temperature. [9,10] The resistance bump for $Er_1Ba_2Cu_3O_{9-\delta}$ indicates that this type of transition also occurs in this compound near this temperature. It is notable that the maximum oxygen uptake in $Er_1Ba_2Cu_3O_{9-\delta}$ at 410 C is below the reported maximum uptake for $Y_1Ba_2Cu_3O_{9-\delta}$ at 500 C. [11]

970

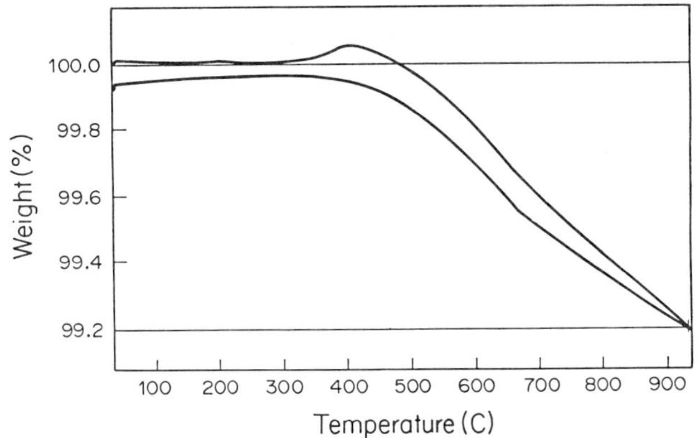

Figure 2. The TGA curves for $Er_1Ba_2Cu_3O_{9-\delta}$ upon warming in air (top) and cooling in oxygen (bottom). The slight dip near 680 C and the maximum uptake near 410 C for warming and 360 C for cooling correlate with resistive features.

The general resistive trends seen in the warming and cooling cycles can be informative about the details of sample formation. When the sample powder is prebaked to 900 C, small grains of the proper phase are formed. However the powder is pulverized before the final pressing and sintering. At first, upon initial heating the resistance of the pellet is large due to the small particles and the large number of grain boundaries. As the temperature is increased, the grains grow by merging with other grains, eliminating boundaries. This results in the formation of more conductive paths as well as fewer high resistance grain boundaries. Scanning electron microscope (SEM) pictures confirm these observations. Some of the resistive anomalies have been attributed to oxygen interaction, as noted ealier. Once the larger grains are formed, they do not break upon cooling so the total resistance is reduced.

The $Er_1Ba_2Cu_3O_{9-\delta}$ compound has been singled out in this study in order to compare and contrast its behavior with $Y_1Ba_2Cu_3O_{9-\delta}$. We emphasize that the general resistance curve upon warming is about an order of magnitude less for the $Er_1Ba_2Cu_3O_{9-\delta}$ sample. SEM photograph comparisons of $Y_1Ba_2Cu_3O_{9-\delta}$ and $Er_1Ba_2Cu_3O_{9-\delta}$ demonstrate that the grain size of the former compound is smaller than the latter. This is in support of the sample formation argument above. The temperature at which an incongruent phase is formed in $Y_1Ba_2Cu_3O_{9-\delta}$ is also much higher than the $Er_1Ba_2Cu_3O_{9-\delta}$ compound. This fact and the difference in the grains sizes of these two compounds can be expected to alter their respective oxygen access and absorption. This may explain the differences in the resistive behavior of these compounds.

In summary, the resistance dependence of $Er_1Ba_2Cu_3O_{9-\delta}$ and $Y_1Ba_2Cu_3O_{9-\delta}$ while being sintered and annealed have been compared. In light of a recent TGA study on $Y_1Ba_2Cu_3O_{9-\delta}$ and our own TGA study on $Er_1Ba_2Cu_3O_{9-\delta}$, the observed resistance features can be correlated with oxygen uptake features. A tetragonal to orthorhombic phase transition is likely in $Er_1Ba_2Cu_3O_{9-\delta}$ near 750 C, while the maximum uptake feature is located around 360 C. Oxygen content may induce the structural changes in both of these compounds.

We wish to thank Roy S. Tucker and Sang Young Lee for help in sample preparation, Margarita Rohklin for her work on the SEM, and M. S. Wong for help with the TGA. The financial support of the National Science Foundation through a grant DMR 83-16989 to the Ohio State University Materials Research Laboratory and grant DMR 84-05403 is gratefully acknowledged.

REFERENCES

1. J.G. Bednorz and K.A.Muller, Z.Phys. B 64, 189 (1986).
2. M.K. Wu, J.R.Ashburn, C.J.Torng, P.H.Hor, R.L.Meng, L.Gao, Z.J.Huang, Y.Q.Wang, and C.W.Chu, Phys. Rev. Lett. 58, 908 (1987).
3. S.R. Ovshinsky, R.T. Young, D.D. Allred, G. DeMaggio, and G.A. Van der Leeden, Phys. Rev. Lett. 58, 2579 (1987).
4. J.T. Chen, L.E. Wenger, C.J. McEwan and E.M. Logothetis, Phys. Rev. Lett. 58, 1972 (1987).
5. John P. Golben, Sung-Ik Lee, Sang Young Lee, Yi Song, Tae W. Noh, Xiao-Dong Chen, J.R. Gaines and Rodney T. Tettenhorst, Phys. Rev. B. (to be published).
6. Xiao-Dong Chen, Sang Young Lee, John P. Golben, Sung-Ik Lee, R.D. McMichael, Yi Song, Tae W. Noh, and J.R. Gaines, Rev. Sci. Instrum. (to be published).
7. We use Acme E-Solder 3022 with hardener #18 available from Acme Chemicals & Insulation Company; a Division of Allied Products Corp.; New Haven Conn. 06505, and Alumina Cement available from L.H. Marshall Co.; 270 W. Lane Ave; Columbus Ohio 43201. We emphasize that no particular endorsement of products is implied.
8. J.E. Greedan, A. O'Reilly and C.V. Stager, (submitted to Phys. Rev. Lett.).
9. Ivan K. Schuller, D.G. Hinks, M.A. Beno, D.W. Capone II, L. Soderholm, J.-P. Locquet, Y. Bruynseraede, C.U. Segre, and K. Zhang, (submitted to Sol. State Comm.).
10. P. Strobel, J.J. Capponi, C. chaillout, M. Marezio, and J.L. Tholence, Nature 327, 306 (1987).
11. J.M. Tarascon, W.R. McKinnon, L.H. Greene, G.W. Hull and E.M. Vogel (submitted to Phys. Rev. Lett.).

ANOMALOUS RESISTIVE BEHAVIOR IN $Er_1Ba_2Cu_3O_{9-\delta}$ AT 2290 K

J.R. Gaines, Sung-Ik Lee, John P. Golben, Yi Song
R.D. McMichael, Xiao-Dong Chen, S. Chittipeddi
M. Selover, and A.J. Epstein

Department of Physics
The Ohio State University
Columbus, OH 43210

Interest in high T_c superconductivity has increased on a par with the reports of higher transition temperatures. These reports range from T_c's near 30 K in the La-Ba-Cu-O system [1], to 90 K in the Y-Ba-Cu-O system [2] and rare earth substituted compounds of the form $R_1Ba_2Cu_3O_{9-\delta}$ where R = Nd, Sm, Eu, Gd, Ho, Er, and Lu. [3-5] Recently an indication of superconductivity was observed at 240 K in $Y_1Ba_2Cu_3O_{9-\delta}$ by the reverse AC Josephson effect, [6] and zero resistances as high as 155 K were reported in the system $Y_1Ba_2Cu_3O_{9-\delta}$:F. [7]

The structure of the 1-2-3 compound $Y_1Ba_2Cu_3O_{9-\delta}$ is now well established. This compound is stacked in a Ba-Y-Ba sequential cell, and contains a slight orthorhombic distortion, [8-12] possibly due to ordered oxygen vacancies. [13] The oxygen content and degree of ordering in this compound is generally believed to be crucial to the superconducting mechanism. The high T_c observed for the fluoridated system of $Y_1Ba_2Cu_3O_{9-\delta}$ noted above can possibly be understood in terms of these phenomena. In this paper we report resistive anomalies near 290 K in $Er_1Ba_2Cu_3O_{9-\delta}$ which are reproducible from sample to sample but time dependent. Oxygen deficiencies and structure may also play an important role in this compound.

The $Er_1Ba_2Cu_3O_{9-\delta}$ samples were prepared by the coprecipitation method. In this procedure, the constituents are mixed on the atomic scale and the stoichiometric ratio is preserved. The samples were pulverized and heat treated to 900 C repeatedly. Extensive details of these preliminary procedures appear elsewhere. [14,15] The final samples were pressed and heated to 915 K for 12 hours. Higher sintering temperatures resulted in the formation of an incongruent liquid phase generally thought to be unfavorable to the superconducting properties. The samples were cooled to 500 K and oxygen was incorporated into the furnace at the rate of 2 ft^3/hr. At this point one group of samples (I) was slow cooled to room temperature in the presence of this oxygen flow rate. Another group of samples (II) was quenched from the oxygen environment at 400 C.

The group (I) samples demonstrated an unusual resistive decrease from room temperature to about 240 K. Below this temperature the resistance decreased linearly until the samples became totally superconducting at about 90 K. A slight dip to a value below this line of linear resistance

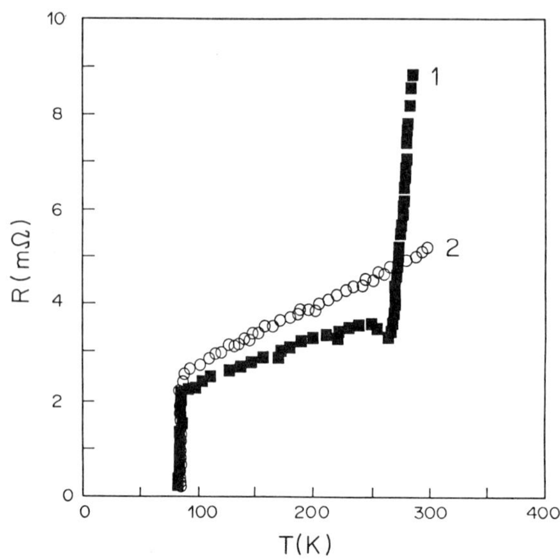

Figure 1. The resistance vs. temperature for the
group I samples that were slowly cooled to room
temperature in the presence of oxygen. (1) The
dip at 260 K and the resistive step to room temperature existed for a period less than one hour.
(2) The final resistance after a period of one
hour is shown in this plot. The room temperature
resistance is lower than what is usually observed
for our 1-2-3 superconducting samples.

at 260 K, linked the two behaviors. After a period of about 40 minutes,
the resistive dip and step disappeared and the resistance above 260 K
settled to a continuation of the linear dependence above the 90 K transition temperature (Figure 1). The settled value of the room temperature
resistance was quite small compared to the usual room temperature resistances observed by us for other 1-2-3 superconductors. However, zero
resistance was never observed at 260 K but only near 90 K.

The second group (II) of samples also demonstrated a resistive step
at 300 K (Figure 2). However, the resistivity of the second group between
the temperature of 90 K and the anomalous region is about an order of
magnitude greater than the corresponding resistivity observed for the
group I samples. The resistive anomaly for these samples remained for a
period of about 5 to 6 hours. The resistance then relaxed to a linear
dependence with room temperature value close to the peak of the original
anomaly, rather than from a value near the foot of the anomaly as in the
group I samples.

An attempt was made to coordinate the resistance measurements with a
structural analysis. Previously it had been confirmed that the
$Er_1Ba_2Cu_3O_{9-\delta}$ possessed a sequentially stacked unit cell similar to
$Y_1Ba_2Cu_3O_{9-\delta}$. [15] For samples sintered at 900 K, the x-ray diffraction
data demonstrated that these samples were single-phased and orthorhombic.
As indicated earlier this orthorhombic distortion seems to be a common
feature of 90 K superconductors, and is linked with defined oxygen vacancies in the outer Cu planes of the unit cell. [13] A recent study of
$La_1Ba_2Cu_3O_{9-\delta}$, where T_c varied in tetragonal samples that appeared single-
phased, pointed out that orthorhombic shouldering was observed in the

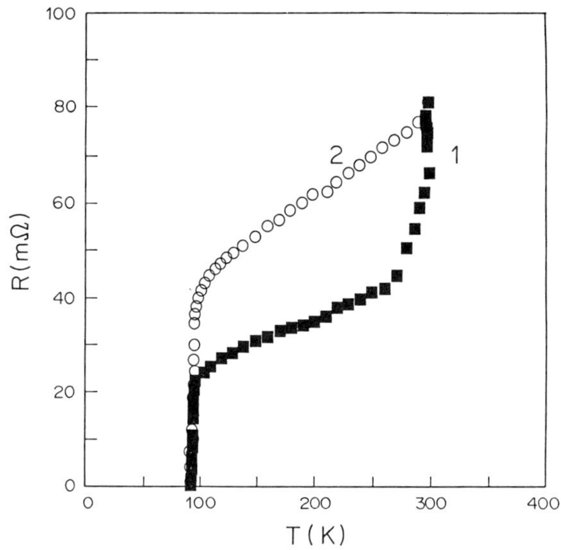

Figure 2. The resistance vs. temperature for the
group II samples that were quenched from the oxygen
atmosphere at 500 C. Visually, this plot could overlap the corresponding plot in Figure 1, however the
magnitude of the resistances are about an order of
magnitude larger. (1) Measured resistance just
after the sample is removed from the furnace (2)
After a period of about 6 hours the resistance
relaxes to a value dictated by the peak of the
original anomaly.

major peaks as T_c approached the 90 K plateau. [16,17]

X-ray diffraction spectra taken of the group I samples showed little
change over a period of several days after their removal from the oven.
However x-ray data taken within 40 minutes of the removal from the oven
for samples sintered at 910 C showed a slight structural change. These
samples were single phased with a pronounced orthorhombic distortion
similar to the regular 1-2-3 superconductors. The group II samples, however, demonstrated a structural time dependence that lasted at least 6
hours from the time of their quenching. The major movement, as observed
in the comparison of the (013) and (103)/(110) line intensities, occured
between 4 and 6 hours from the quenching time. After about 6 hours the
structure stabilized.

A thermogravimetric analysis (TGA) of these samples, taken under heat
treatments and oxygen flow rates that simulated the actual sintering and
annealing processes, showed that the oxygen content of the cooled sample
was initially less than the original room temperature content. Over a
period of 6 hours the oxygen content in the sample approached the original
value. The major increase occurred within the period of 4 to 6 hours from
the quenching time. This result seems to correlate with the observed
structural change in these samples.

We note that since these results were time dependent, the various
measurements were all conducted simultaneously on different samples from
the same batch. Several previous runs confirmed the reproducibility of
the anomalies from batch to batch and of consistent behavior between

samples within each batch. It is in fact this reproducibility and the
subsequent correlative measurements that can be conducted, that make the
observed anomaly in this compound particularly interesting.

A definite conclusion as to the source of the anomaly and its time
dependence is presently difficult. Certainly the structure and the oxygen
content seem to be intricately related and governed by the same time scale
as the observed relative anomaly. The resistive dip at 240 K which is
below the usual linear behavior, and the magnetic fluctuations maybe
indicative of a superconducting mechanism attempting to short the sample.
Erbium samples, prepared in the same way, but sintered at 900 C do not
show the resistive dips nor the temperature instability. More measurements such as the reverse AC Josephson effect, magnetic susceptibility and
high resolution x-ray diffraction are being conducted at the Ohio State
University. The financial support of the National Science Foundation
through a grant DMR 83-16989 to the Ohio State University Materials
Research Laboratory and grant DMR 84-05403 is gratefully acknowledged.

REFERENCES

1. J.G. Bednorz and K.A.Muller, Z.Phys. B 64, 189 (1986).
2. M.K. Wu, J.R.Ashburn, C.J.Torng, P.H.Hor, R.L.Meng, L.Gao, Z.J.Huang,
       Y.Q.Wang, and C.W.Chu, Phys. Rev. Lett. 58, 908 (1987).
3. D.W. Murphy, S. Sunshine, R.B. van Dover, R.J. Cava, B. Batlogg, S.M.
       Zahurak, and L.F. Schneemeyer, Phys. Rev. Lett. 58, 1888 (1987).
4. P.H. Hor, R.L. Meng, Y.Q. Wang, L. Gao, Z.J. Huang, J. Bechtold, K.
       Forster, and C.W. Chu, Phys. Rev. Lett. 58, 1891 (1987).
5. J.M. Tarascon, W.R. McKinnon, L.H. Greene, G.W. Hull and E.M. Vogel
       (submitted to Phys. Rev. Lett.).
6. J.T. Chen, L.E. Wenger, C.J. McEwan and E.M. Logothetis, Phys. Rev.
       Lett. 58, 1972 (1987).
7. S.R. Ovshinsky, R.T. Young, D.D. Allred, G. DeMaggio, and G.A. Van der
       Leeden, Phys. Rev. Lett. 58, 2579 (1987).
8. Y. Le Page, W.R. Mckinnon, J.M. Tarascon, L.H. Greene (submitted to
       Phys. Rev. Lett.).
9. R. Beyers, G. Lim, E.M. Engler and R.J. Savoy, T.M. Shaw, T.R. Dinger,
       W.J. Gallagher and R.L. Sandstrom (unpublished).
10.   P.M. Grant, R.B. Beyers, E.M. Engler, G. Lim, S.S.P. Parkin, M.L.
       Ramirez, V.Y. Lee, A. Nazzal, J.E. Vazquez and R.J. Savoy
       (unpublished).
11. T. Siegrist, S. Sunshine, D. W. Murphy, R.J. Cava and S.M. Zahurak
       (submitted to Phys. Rev. Lett.).
12. M. Hirabayashi, H. Ihara, N. Terada, K. Senzaki, K. Hayashi, S. Waki,
       K. Murata, M. Tokumoto, and Y. Kimura, Jap. J. Appl. Phys. 26,
       1454 (1987).
13. J.E. Greedan, A. O'Reilly and C.V. Stager, (submitted to Phys. Rev.
       Lett.).
14. Xiao-Dong Chen, Sang Young Lee, John P. Golben, Sung-Ik Lee, R.D.
       McMichael, Yi Song, Tae W. Noh, and J.R. Gaines, Rev. Sci.
       Instrum. (to be published.).
15. John P. Golben, Sung-Ik Lee, Sang Young Lee, Yi Song, Tae W. Noh,
       Xiao-Dong Chen, J.R. Gaines and Rodney T. Tettenhorst, Phys. Rev.
       B. (to be published).
16. Sung-Ik Lee, John P. Golben, Sang Young Lee, Xiao-Dong Chen, Yi Song,
       Tae W. Noh, R.D. McMichael, Yue Cao, Joe Testa, Fulin Zuo, J.R.
       Gaines, A.J. Epstein, D.L. Cox, J.C. Garland, T.R. Lemberger, R.
       Sooryakumar, Bruce R. Patton, and Rodney T. Tettenhorst, Extended
       Abstract Materials Research Society (April, 1987).
17. Sung-Ik Lee, John P. Golben, Sang Young Lee, Xiao-Dong Chen, Yi
       Song, Tae W. Noh, R.D. McMichael, J.R. Gaines, D.L. Cox, and Bruce
       R. Patton (submitted to Phys. Rev. Lett.).

STRUCTURE AND THE (014)/(005) X-RAY LINE INTENSITY

IN 1-2-3 SUPERCONDUCTORS

Sung-Ik Lee, John P. Golben, Yi Song, Xiao-Dong Chen
R.D. McMichael, J.R. Gaines

Department of Physics
The Ohio State University
Columbus, OH 43210

The discovery of high T_c superconductivity in the La-Ba-Cu-O system [1] initiated an accelerated interest in this and related systems. To date the highest zero resistance temperatures have been observed in compounds of the form $R_1Ba_2Cu_3O_{9-\delta}$ where R = Y, Nd, Sm, Eu, Gd, Ho, Er, and Lu. [2-6] The $Y_1Ba_2Cu_3O_{9-\delta}$ compound has been studied extensively by x-ray diffraction [7-11] neutron scattering [12,13] and thermogravimetric analysis. [2,4] The unit cell is found to have a sequential Ba-Y-Ba stacking, an ordered oxygen deficiency, [13] and a slight orthorhombic distortion. [7-11] The similarity of the x-ray diffraction data and lattice constants of the compounds in this series strongly indicates that these compounds have the same structure, with the specific rare earth element simply substituting for Y.

Ordering in these "90 K" superconductors is particularly emphasized. The Y-based compound seems to be the most stable as evidenced by the fact that the 1-2-3 superconducting phase can percolate through samples that are considerably off stoichiometry. [14] This stability can be attributed to the size difference of the Y and Ba atoms. [15] However considerable amount of ordering also seems to exist in samples substituted by the rare earth elements in the La-series. In previously studied superconductors, doping of a percent or so of magnetic ions was sufficient to quench superconductivity. But in the current class of copper oxide supercond-uctors, incorporation of magnetic ions such as Ho [5], Er [6], and Gd do not seem to have any effect on T_c. Thus it is believed that the magnetic ions are screened from the superconducting mechanism by assuming a strict position in the unit cell.

As the atomic number of the substituted atom approaches that of Ba, some increased randomness is expected. This has been postulated to occur in the superconductor $La_1Ba_2Cu_3O_{9-\delta}$ [16-18] and in the non-superconductor $Pr_1Ba_2Cu_3O_{9-\delta}$. The La^{+3} ion is isoelectronic with the Ba^{+2} ion, but their ionic radii are different. While La and Ba are neighbors in the periodic table, Pr is not far away. Because of the similarities in the atomic numbers then, the 1-2-3 x-ray spectra for these compounds appear nearly cubic, with the major lines in these spectra corresponding to the major lines in the spectra of the other 1-2-3 compounds.

However, ordering to some extent beyond a "random" cubic phase must exist in these latter two compounds as evidenced by the finite intensity

of the (014)/(005) superlattice line. This line occurs in all of the
1-2-3 compounds, and seems to vary independently from sample to sample,
while the other intensities in the spectrum change to a much lesser
degree. In this paper we present an overview of this (014)/(005) line
intensity in these compounds. In a sense, this line is the internal
structural indicator.

Most of our 1-2-3 samples have been synthesized by the coprecipitation method. This preparation method is well described elsewhere. [19]
The major advantage of this method is that the constituents are uniformly
mixed on the atomic scale, thereby enhancing the formation of the desired
phase in the solid solution. Another advantage is that this method tended
to insure the stoichiometric ratio throughout the preparation process. As
a result, the x-ray spectra of these samples were clearly single phased
with no evidence of the $BaCuO_2$ phase that seemed to plague the solid state
reacted samples.

The x-ray spectra of the 1-2-3 superconductors with the substitution
of Y, Er, Ho, and Gd, demonstrated an anomalous intensity for the
(014)/(005) line. This line was usually a factor of 2 or more greater
than the (113) line next to it. The data presented by other groups for
$Y_1Ba_2Cu_3O_{9-\delta}$ show different intensities for this line. Beyers et al. [8]
report a result similar to what has been observed by us, while at least
two other papers show a (014)/(005) intensity that is less than the
intensity of the (113) line. [11,20]

We can attempt to understand these observations by considering the
source of the intensity. In the unit cell, the reciprocal lattice planes
represented by the (014) and (005) indices pass near the Ba position.
Thus the intensity of this line is especially sensitive to the Ba (and Cu)
position in the unit cell. For "ideal" Ba positions at (1/2,1/2,1/6) and
(1/2,1/2,5/6) the atoms are situated at body-centered locations in their
respective subsections of the unit cell. From standard x-ray theory, such
placements negate the intensities of the superlattice lines which correspond to the indices h,k,l, where h+k+l is equal to an odd number. The
expected intensity of the (014)/(005) line is then zero. This is one
reason why we are not particularly concerned with the scattering from the
central rare earth atom since by symmetry it remains centralized in spite
of the atomic movements around it.

Oxygen deficiencies will cause displacements in the unit cell, most
notably in the position of the Ba and Cu atoms (Figure 1). To account for
this effect, a computer program was written to iterate the Ba and Cu
positions in the unit cell. Each situation was evaluated by the variable
$S(\beta,\Delta)$, where

$$S(\beta,\Delta) = \frac{\sum |I_{exp} - I_{theory}|}{\sum I_{exp}}$$

and β and Δ represent the respective Ba and Cu movements from the "ideal"
positions. The positional changes obtained by this method varied
according to the assumed oxygen placements. However, in general the data
were consistent with the atomic positions reported by neutron diffraction
data. [12,13] The calculated (014)/(005) line intensity increased by two
orders of magnitude to about 7% of the major line in the spectrum. It is
clear then that the various oxygen deficiencies and placements affect a
major portion of this line intensity. This may help to explain the
conflicting reports noted earlier.

For the 90 K superconductors, the observed (014)/(005) line inten-

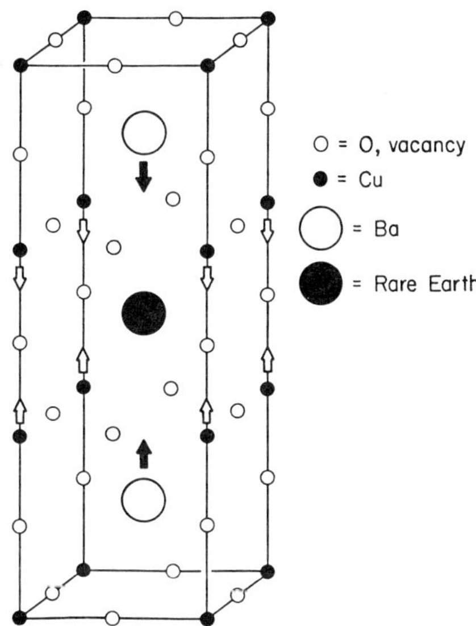

Figure 1. The 1-2-3 unit cell is shown. The intensity of the (014)/(005) line is particularly dependent upon the movement of Ba and Cu atoms in the cell. This movement is indicated by the arrows.

sity in our samples is still a factor of 2 or more greater than this new calculated value. Thus in spite of the atomic movements, an unexplained discrepancy remains. It should also be noted that the presence of other phases in these samples seemed to have a pronounced effect on the intensity of this line. For samples prepared by the solid state reaction method, a fair amount of $BaCuO_2$ phase is usually unavoidable. The x-ray diffraction spectra of samples prepared in this manner often show a reduction in the (014)/(005) line intensity to a value below the intensity of the (113) line. [11,20]

In the $La_1Ba_2Cu_3O_{9-\delta}$ compound mentioned earlier, the (014)/(005) line intensity was correlated to the value of T_c. [16,17] The samples used in this correlation all appeared to be single-phased as determined from the x-ray spectra. The first explanation considered for the change in the (014)/(005) line intensity, was that there were differences in the oxygen amount and locations in these samples, thereby causing a movement of Ba and Cu atoms in the cell, a corresponding intensity change in the (014)/(005) superlattice line, and a change in the superconducting properties. The second explanation considered was that there was a sequentially ordered superconducting phase that competed in the same sample with a random phase, and that x-ray spectra alone could not distinguish the relative presence of each. The fact that the La^{+3} and Ba^{+2} ions are isoelectronic made it very feasible that these elements could substitute for each other in this random way. Randomness could also account for the fact that the (014)/(005) line intensity in this compound was generally much reduced from the observed value in the other 1-2-3 superconductors.

The x-ray spectra for the non-superconducting compound of $Pr_1Ba_2Cu_3O_{9-\delta}$ shows that this compound easily forms the 1-2-3 phase. The intensity of the (014)/(005) line is more comparable to that observed for single phased samples of $La_1Ba_2Cu_3O_{9-\delta}$ than for the other 1-2-3 supercon- ductors. This indicates that randomness may exist in this compound.

In summary, the intensity of the (014)/(005) line observed in the
x-ray spectra of 1-2-3 compounds is dependent upon the movement of Ba and
Cu atoms in the unit cell. These atoms may move as a result of oxygen
deficiencies. In compounds where 90 K zero resistances have been
documented, the (014)/(005) line is usually anomalously large to the point
where atomic movements alone cannot account for the discrepancy between
the observed and theoretical values. The intensity of this superlattice
line may also act as an indicator of randomness in the unit cell structure
of 1-2-3 compounds.

REFERENCES

1. J.G. Bednorz and K.A.Muller, Z.Phys. B 64, 189 (1986).
2. D.W. Murphy, S. Sunshine, R.B. van Dover, R.J. Cava, B. Batlogg, S.M.
    Zahurak, and L.F. Schneemeyer, Phys. Rev. Lett. 58, 1888 (1987).
3. P.H. Hor, R.L. Meng, Y.Q. Wang, L. Gao, Z.J. Huang, J. Bechtold, K.
    Forster, and C.W. Chu, Phys. Rev. Lett. 58, 1891 (1987).
4. J.M. Tarascon, W.R. McKinnon, L.H. Greene, G.W. Hull and E.M. Vogel
    (submitted to Phys. Rev. Lett.).
5. Sung-Ik Lee, John P. Golben, Yi Song, Sang Young Lee, Tae W. Noh,
    Xiao-Dong Chen, Joe Testa, J.R. Gaines and Rodney T. Tettenhorst,
    Appl. Phys. Lett. (to be published).
6. John P. Golben, Sung-Ik Lee, Sang Young Lee, Yi Song, Tae W. Noh,
    Xiao-Dong Chen, J.R. Gaines and Rodney T. Tettenhorst, Phys. Rev. B
    (to be published).
7. Y. Le Page, W.R. Mckinnon, J.M. Tarascon, L.H. Greene (submitted to
    Phys. Rev. Lett.).
8. R. Beyers, G. Lim, E.M. Engler and R.J. Savoy, T.M. Shaw, T.R. Dinger,
    W.J. Gallagher and R.L. Sandstrom (unpublished).
9. P.M. Grant, R.B. Beyers, E.M. Engler, G. Lim, S.S.P. Parkin, M.L.
    Ramirez, V.Y. Lee, A. Nazzal, J.E. Vazquez and R.J. Savoy
    (unpublished).
10. T. Siegrist, S. Sunshine, D. W. Murphy, R.J. Cava and S.M. Zahurak
    (submitted to Phys. Rev. Lett.).
11. M. Hirabayashi, H. Ihara, N. Terada, K. Senzaki, K. Hayashi, S. Waki,
    K. Murata, M. Tokumoto, and Y. Kimura, Jap. J. Appl. Phys. 26,
    1454 (1987).
12. M.A.Beno, L.Soderholm, D.W. Capone II, D.G. Hinks, J.D.Jorgensen,
    Ivan K. Schuller (unpublished).
13. J.E. Greedan, A. O'Reilly and C.V. Stager, (submitted to Phys. Rev.
    Lett.).
14. M.K. Wu, J.R.Ashburn, C.J.Torng, P.H.Hor, R.L.Meng, L.Gao, Z.J.Huang,
    Y.Q.Wang, and C.W.Chu, Phys. Rev. Lett. 58, 908 (1987).
15. Francis S. Galasso, _Structure, Properties, and Preparation of
    Perovskite-Type Compounds_ (Pergamon Press, New York, 1969).
16. Sung-Ik Lee, John P. Golben, Sang Young Lee, Xiao-Dong Chen, Yi Song,
    Tae W. Noh, R.D. McMichael, Yue Cao, Joe Testa, Fulin Zuo, J.R.
    Gaines, A.J. Epstein, D.L. Cox, J.C. Garland, T.R. Lemberger, R.
    Sooryakumar, Bruce R. Patton, and Rodney T. Tettenhorst, Extended
    Abstract Materials Research Society (April, 1987).
17. Sung-Ik Lee, John P. Golben, Sang Young Lee, Xiao-Dong Chen, Yi Song,
    Tae W. Noh, R.D. McMichael, J.R. Gaines, D.L. Cox, and Bruce R.
    Patton (submitted to Phys. Rev. Lett.).
18. D.B. Mitzi, A.F. Marshall, J.Z. Sun, D.J. Webb, M.R. Beasley, T.H.
    Geballe, and A. Kapitulnik (submitted to Phys. Rev. B).
19. Xiao-Dong Chen, Sang Young Lee, John P. Golben, Sung-Ik Lee, R.D.
    McMichael, Yi Song, Tae W. Noh, and J.R. Gaines, Rev. Sci.
    Instrum. (to be published).
20. E. Takayama-Muromachi, Y. Uchida, Y. Matsui, and K. Kato, Jap. J.
    Appl. Phys. 26, 1476 (1987).

MEASUREMENT OF FLUCTUATION-ENHANCED CONDUCTIVITY ABOVE T_c IN Y-Ba-Cu-O.

M.A.Dubson, J.J.Calabrese, S.T.Herbert, D.C.Harris,
B.R.Patton, and J.C.Garland

Department of Physics
The Ohio State University
Columbus, OH

We have made careful measurements of the resistivity vs. temperature in
a sample of polycrystalline $YBa_2Cu_3O_{7-\delta}$ and have fit the data to a
theoretical expression for the fluctuation-enhanced conductivity in the
normal state. From this fit, we infer a value for the zero-temperature
Ginzberg-Landau coherence length $\xi(0)$ of 21±5Å.

The sample was prepared by the solid-state reaction technique[1], with the
final pressed pellet sample sintered in air at 950C for 12 hrs and then
annealed in flowing O_2 for 12 hrs. Grain size in the polycrystalline
sample, as deduced from SEM photos, is 5-20 microns. Density measurements
indicate a porosity of 30%. Contact pads for lead attachment were prepared
by ion-milling the sample and then sputtering gold pads through a mechanical
mask, resulting in a contact resistance of 0.1Ω. Sample temperature was
measured with a platinum resistance thermometer with an absolute accuracy of
0.1K.

Figure 1 shows resistivity vs. temperature. The data were obtained by
slowly cooling and then warming the sample in a liquid nitrogen cryostat
over a several hour period. Coincidence of the cooling and warming curves
indicates negligible thermal lag between sample and thermometer. The room-
temperature resistivity of the sample is about 1000 $\mu\Omega$-cm dropping linearly
to about 400$\mu\Omega$-cm at 100K. The absolute value of the resistivity is
uncertain by some 30% because of uncertain lead geometry and sample
porosity. The inset on Fig.1 shows a transition about 1.5K wide with a midpoint at 91.7K.

The sample conductivity over the temperature range 92K to 150K was fit
to an expression of the form

$$\sigma(T) = 1/\rho(T) = \sigma_n(T) + \sigma'(T) \tag{1}$$

where the normal state conductivity σ_n is assumed to have the form $1/\rho_n =$
$1/(a+bT)$, with a and b constants, and the fluctuation contribution to the
conductivity σ' has the form[2], with $\epsilon\equiv(T-T_c)/T_c$,

$$\sigma'(T)=(1/32)[e^2/(\hbar\xi(0))]\epsilon^{-1/2}[1+(36\epsilon/\pi^2)^2]^{-1/2} \tag{2}$$

The fit shown in Fig.2 yields $\xi(0)$=21±5Å, b=3.00x10^{-6} Ω-cm/K, a=147x10^{-6} Ω-
cm, and T_c=91.8K. The uncertainty in $\xi(0)$ is due mostly to the uncertainty
in converting from measured sample resistance to absolute resistivity.

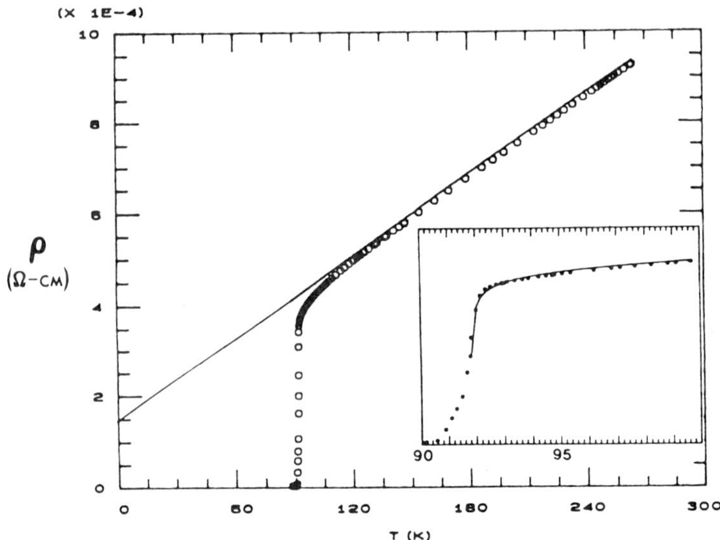

Figure 1. Resistivity vs. temperature for polycrystalline $YBa_2Cu_3O_{7-\delta}$. The straight line is the fitted normal state resistivity, $\rho_n(T)$. The line drawn through the data in the inset is the fit to Eq.1.

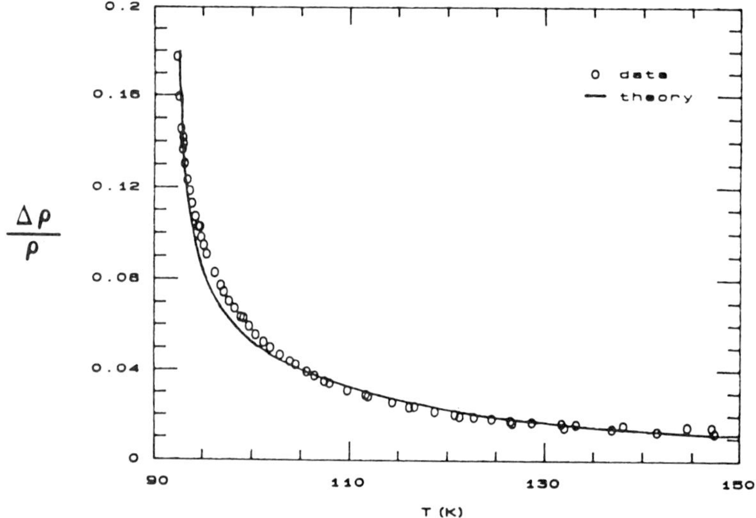

Figure 2. Plot of $\Delta\rho/\rho = (\rho_n(T) - \rho(T))/\rho(T)$ vs. T for $YBa_2Cu_3O_{7-\delta}$.

References

[1] X.D.Chen, S.Y.Lee, J.P.Golben, S.I.Lee, R.D.McMichael, Y.Song, T.W.Noh, and J.R.Gaines, Rev.Sci.Instrum., in press.

[2] B.R.Patton, to be published. Eq.(2) interpolates between the classical and quantum fluctuation regions above T_c.

Note added: It has come to our attention that investigators at IBM previously performed a similar analysis with similar findings. See P.P.Freitas, C.C.Tsuei, and T.S.Plaskett, Phys.Rev.B1, July 1987 (in press).

982

GRAIN DECOUPLING AT LOW MAGNETIC FIELDS IN CERAMIC $YBa_2Cu_3O_{7-\delta}$

J. F. Kwak, E. L. Venturini, D. S. Ginley, and W. Fu

Sandia National Laboratories
P. O. Box 5800, Albuquerque, NM 87185

INTRODUCTION

The ongoing series of discoveries of new materials with ever higher
transition temperatures has generated an unprecedented flood of research
activity in recent months, particularly on the 90 K superconductors typified
by $YBa_2Cu_3O_{7-\delta}$ (123 compound).[1-3] All these materials have Perovskite-based
nominally layered structures,[4] and thus may be expected to be anisotropic
uniaxial superconductors.[5] Unfortunately, the synthesis of crystalline
forms of useful size with good superconducting properties has proven quite
difficult, so that almost all measurements reported to date have been performed on ceramic pellets best described as randomly oriented granular
aggregates. The finding at many laboratories of very low critical current
densities (J_c's) of 10^3 A/cm^2 or less at 4 K in these materials has been a
cause of concern for their potential use in most applications.[6,7] However,
recent measurements show high J_c's (over 10^5 A/cm^2 at 77 K) in thin crystalline films and at least 10:1 anisotropy in H_{c1} in small single crystals.[8,9]
It is commonly assumed that transport in the ceramic materials is limited by
intergain impedances so that the grains are only weakly coupled in the
superconducting state, probably by Josephson currents. However, to date
there has been no direct test of this assumption.

We report here the results of J_c and magnetization measurements versus
magnetic field on ceramic $YBa_2Cu_3O_{7-\delta}$ in the superconducting state. The
directly measured J_c's in our material are several 100 A/cm^2 at 76 K in zero
field. In a longitudinal field J_c falls very rapidly at low fields, dropping by a factor of 5 at 20 Oe and reaching essentially zero below 250 Oe.
We show this field dependence to be consistent with Josephson tunneling
between grains. We show that J_c, despite its "extrinsic" nature, gives a
sensitive measure of the bulk H_{c1} because of its extreme sensitivity to
trapped flux.

Our magnetization data show almost perfect diamagnetic shielding at
very low fields (below about 5 Oe at 76 K), but then exhibit a fairly sharp
transition to significantly lower shielding (on the order of 70%) as field
is increased further (above 15 Oe at 76 K). The shielding then remains
fairly constant up to much higher fields (about 100 Oe at 76 K) before the
magnetization gradually turns over and begins to fall off.

The collapse of J_c and the change in slope of the magnetization both imply decoupling of the grains at fields well below H_{c1} in our material due to inability of intergrain junctions to carry screening currents. This conclusion is strongly supported by our finding that the magnetization of samples with large aspect ratios is quite insensitive to field orientation for fields above the decoupling field.

The finding of grain decoupling has several important consequences: (1) The critical current density derived from magnetization hysteresis data at high fields reflects the properties of the grain interiors and is on the order of 10^6 A/cm^2 at 5 K and 10^5 A/cm^2 at 77 K, consistent with the crystalline film results[8] and suggestive of moderate to high pinning forces. (2) Diamagnetic shielding data taken on ceramic 123 at low fields may have little relationship to the "intrinsic" properties of the compound. (3) The combination of randomly oriented decoupled grains and highly anisotropic uniaxial superconductivity produces a large reduction in the magnitude of the Meissner effect observed in the ceramic materials at all fields and temperatures, as proposed by other authors.[10,11]

SAMPLE PREPARATION

The ceramic 123 pellets used for the present studies were synthesized from mixed powders of CuO, BaO, and Y_2O_3. The powders were ground, pelletized at 5.6 kbar, and sintered in air for 4 hr at 950 C . Samples were then oxygen annealed (640-700 torr) at 900 C for 12 hours with a 5 hour slow cool. The average mass density was about 70% of the theoretical value of 6.4 g/cm^3. Care was taken to prevent exposure of the samples to water vapor throughout this processing sequence. X-ray powder patterns of identically prepared samples indicated single phase material, with all observed lines accounted for by an orthorhombic unit cell of a_0= 3.825, b = 3.883, and c = 11.680 Å. Energy dispersive x-ray studies showed the materials to be very uniform in composition, while SEM micrographs revealed granular structure on a 10μm scale. Further details of sample preparation and characterization are given in ref. 12. Measurements of the resistive and magnetic transitions showed the samples used in the present studies to be our best material, and comparable to the best ceramic samples reported elsewhere.

EXPERIMENTAL TECHNIQUES

Critical current measurements were obtained on samples cut from pellets in the shape of rectangular parallelepipeds with typical dimensions 10×1×1 mm^3. Copper pads were evaporated in a linear 4-probe configuration and wires attached with conductive silver paint. In order not to heat the sample and/or destroy the contacts it was usually necessary to supply the current in low duty cycle pulses. In this manner, current densities up to 1000 A/cm^2 could be generated. J_c was determined from the observation of a voltage of about 2 μV on an oscilloscope with differential inputs and a noise level of 1 μV. Resistivity (ρ) was measured from the ratio of voltage pulse amplitude to current pulse amplitude. The ratio of the resistivity at J_c to the normal state resistivity ranged from 10^{-5} at large current densities up to 1% at the lowest current densities. No difference in J_c or ρ was found between decreasing or increasing the current. Magnetic fields were applied longitudinal to the current direction using either a copper solenoid (up to 800 Oe with the sample at 76 K) or a superconducting solenoid (up to 100 kOe with the sample at 4 K).

The static magnetization was measured in a commercial SQUID magnetometer (Biomagnetic Technologies, San Diego, CA). The rectangular parallelepiped samples were suspended with their long axis parallel to the applied magnetic field except as noted. Flux exclusion (diamagnetic

shielding or zero-field-cooling) measurements were performed by first
cooling the sample from above the transition temperature down to 5 K in
nearly zero field, then applying the measurement field and taking data
during controlled warming. Field cycles were performed by cooling in nearly
zero field to the desired temperature, then taking data during a controlled
field sweep up to a maximum value and back down to nominal zero (limited by
about 5 gauss of trapped flux in the superconducting magnet). Meissner
(flux expulsion) data were obtained by cooling from above the transition
temperature down to 5K in constant field.

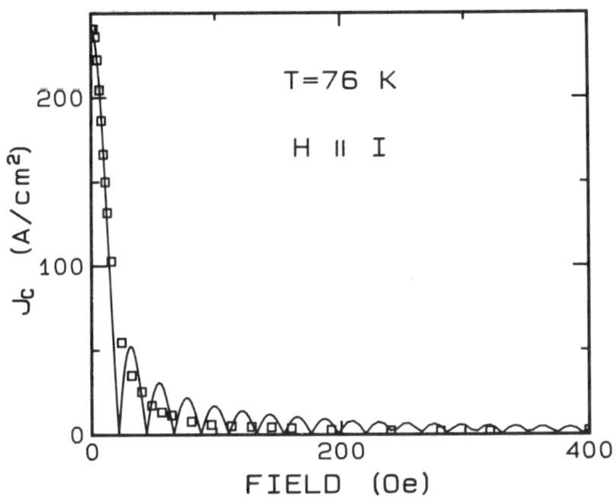

Fig. 1. Typical magnetic field dependence of the critical
current density of ceramic $YBa_2Cu_3O_{7-\delta}$ at 76 K. The
solid curve is a fit to the behavior of a Josephson
junction in a magnetic field.

CRITICAL CURRENT RESULTS AND DISCUSSION

Fig. 1 shows a typical curve of J_c versus increasing longitudinal magnetic field at 76 K for a sample cooled in zero applied field. We have not
yet done this measurement at 4 K because of problems arising from the larger
values of J_c and the necessity of using a superconducting magnet with poor
resolution at low fields. Note that J_c is only 240 A/cm^2 at H=0, goes down
by an order of magnitude at 40 Oe (Fig. 1 inset), and is basically gone
above 200 Oe. These data are qualitatively similar to J_c versus field data
reported elsewhere for transverse fields.[21,22] We actually measure a very
small J_c which persists up to very high fields, but feel this to be an artifact arising from the necessity of identifying J_c by the observation of a
finite voltage. The striking field dependence and overall low magnitude of
J_c suggest interpretation in terms of an array of Josephson junctions.

Following Clem et al,[13] we assume that the bulk J_c of our three dimensional sample can be identified with the J_c due to an individual Josephson
tunnel junction:

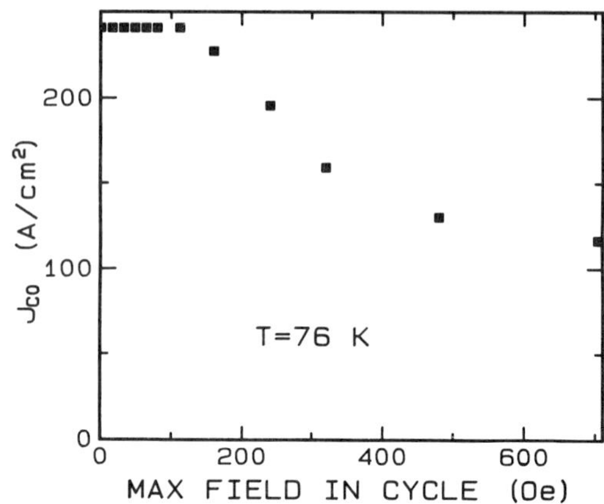

Fig. 2. Critical current density at zero field and 76 K for the sample of Fig. 1, plotted versus the maximum field seen by the sample. H_{c1} is the field where J_c begins to fall (see text).

$$J_0 \equiv J_c(H=0,T) = \{\pi\Delta(T)/2e\rho_n d\}\tanh\{\Delta(T)/2k_BT\} \tag{1}$$

where $\Delta(T)$ is the superconducting order parameter at temperature T, e is the electronic charge, ρ_n is the normal state bulk resistivity, d is an average grain dimension, and k_B is Boltzmann's constant. It is interesting to calculate J_0 from Eq. 1 for the sample of Fig. 1: We take d = 10 μm from SEM data and ρ_n = 1.2 mΩ-cm by extrapolating the measured normal state resistivity to 76 K. Assuming a BCS gap and temperature dependence for Δ with T_c = 92 K, we have $\Delta(0)$ = 1.76k_BT_c and $\Delta(76K)$ = .7$\Delta(0)$,[14] so J_0 = 7000 A/cm², compared to our measured value of 240. Eq. 1 predicts a fivefold increase in J_0 at 4 K, assuming extrapolation of the resistivity behavior is valid. This compares well with the ratio of about 4:1 we have observed on two samples. The low value of the measured J_c compared to theory indicates that the intergrain contacts do not form pure tunnel junctions, for which there should be quantitative agreement. Other forms of Josephson "weak links" (point contacts, constrictions, or metal layer junctions) are more consistent with the data.[14] We will not treat this issue further, but will use the measured value for J_0.

J_c for a Josephson junction in a transverse magnetic field is:

$$J_c(H,T) = J_0|\lceil\sin(\pi\Phi/\Phi_0)\rceil/(\pi\Phi/\Phi_0)| \tag{2}$$

where Φ_0 is the flux quantum, 2.07×10^{-7} Oe-cm², Φ = Hb(2λ+d) is the flux in the junction, b² is the junction area, d is the junction thickness, and λ is the effective penetration depth.[14] Φ = 2bλH is an excellent approximation in the present case (λ > 1000 Å   d). Eg. 2, of course, contains the well-known Josephson junction interference pattern. Because of the random orientations and dimensions of the junctions in ceramic 123 material, we expect this pattern to be washed out in the measured bulk J_c. The important point

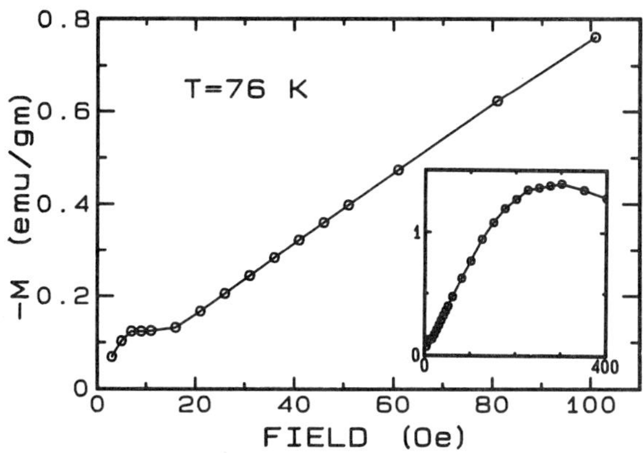

Fig. 3. Typical diamagnetic shielding magnetization versus field
for ceramic $YBa_2Cu_3O_{7-\delta}$. The regime 5-15 Oe marks a transition from coupled to decoupled grain behavior as field
increases. The inset shows a continuation of the same
data to higher fields.

here is the rapid drop of J_c at low fields. The percolative nature of the
current in these materials, which guarantees that the operative junctions
are randomly oriented, is the reason that Eq. 2 is applicable to our measurements in a nominally longitudinal field and that our results are similar
to those for a transverse field. The curve in Fig. 1 shows an excellent fit
of Eq. 2 to the data for λ = 2800 Å. $H_0 = \Phi_0/(2b\lambda)$ ($\approx$ 7 Oe in the present
case) provides a scale for the field dependence of J_c.

One interesting consequence of the strong field dependence of J_c is its
potential use as a sensitive probe of H_{c1}. To do this we ramp the field up
to some value H_{max} and back to zero before measuring what we will call J_{c0}.
For $H_{max} < H_{c1}$, there will be no flux in the sample and J_{c0} will remain at
its zero-field value. For $H_{max} > H_{c1}$, however, J_{c0} will fall at a rate reflecting the degree of flux trapping. In Fig. 2 we show the results of such
an experiment at 76 K on the sample of Fig. 1. The behavior is exactly as
described above, and clearly indicates H_{c1} = 120 $\pm$ 10 Oe.

MAGNETIZATION RESULTS AND DISCUSSION

The collapse of J_c at low fields has important implications for the
magnetic properties of ceramic 123 compound. At very low fields screening
currents flow through the junctions, causing the sample to show virtually
complete diamagnetic shielding. As field is increased, a point is reached
where the junctions are no longer capable of sustaining the screening currents, causing the grains to decouple and the magnetization to switch to
behavior reflecting screening by the individual grains (ie., flux is no
longer excluded from intergrain regions);[5] this behavior should persist up
to H_{c1}. While the decoupling cannot be sudden in these ceramic materials,
it can be characterized by an average field value which we will call H_d.

Fig. 3 shows diamagnetic shielding magnetization versus increasing field, $M(H)$, for the sample of Figs. 1 & 2 at 76 K. The behavior expected from decoupling clearly appears in the range 5-15 Oe, which compares well with the value of 7 Oe obtained above for H_0, which we therefore identify with H_d. Below 5 Oe the slope corresponds to $-4\pi\chi = 0.97$ using the actual density of the material, which is equivalent to assuming that the coupled grains completely exclude flux from the sample interior. Immediately above 15 Oe, $-4\pi\chi = 0.61$ using the theoretical density (flux everywhere outside grains). Upon grain decoupling, however, the susceptibility of ceramic material should be reduced to about 50% $[(1/3)\times(3/2)=1/2$, with 3/2 from the demagnetization for a sphere] that of a perfect diamagnet, due to the combination of uniaxial anisotropy of 123 compound and random orientation of the grains.[14] The observed fraction larger than 1/2 suggests that decoupling is not yet perfect below 100 Oe, consistent with the observed nonzero J_c at those fields. Moreover, $M(H)$ deviates slightly from linearity at fields well below H_{c1}, where flux flow into the sample should not occur. In the inset to Fig. 3, a progressively increasing deviation from linearity is apparent beginning at 50 Oe. This behavior is consistent with the observed continued decay in J_c above H_d.

In contrast to the complicated situation for shielding, the Meissner effect will always reflect nearly perfectly decoupled grains if H_d remains well below H_{c1} at all temperatures: As T is lowered in constant field, H_{c1} is encountered first and flux is expelled (or trapped) by the grains before significant junction currents can set up. On the basis of the low value of H_d at 76 K, and the theoretical temperature dependence for the Josephson J_c of a granular superconductor,[13] we are confident that the condition $H_d \ll H_{c1}$ is easily satisfied at all temperatures in our material. In fact, the typical Meissner signal at low temperatures is about 40% of the theoretical value. The reduction from the 50% value expected is consistent with the high degree of flux trapping observed in field cycles, although penetration depth effects may also have an effect.[11]

We directly demonstrated magnetic decoupling of the grains by the following simple experiment: We measured the diamagnetic shielding at 95 Oe and 5 K of a sample cut in the shape of a rectangular parralelepiped, with dimensions $4.65\times0.9\times0.9$ mm^3, for fields applied along the three axes of the sample. We found $-M = .89$ emu/gm parallel to the long axis (70% shielding) and 1.02 and 1.1 emu/gm in the transverse directions. On the basis of the sample geometry, demagnetization should have resulted in a transverse toparallel ratio of 1.5 (down from 1.85 for perfect shielding), rather than 1.15 or 1.24 as we have here. Moreover, the direction with the highest moment was that normal to the plane of the original pellet. According to Watanabe et al,[15] grains have some tendency to orient with their crystalline c axis in this direction, which would cause a higher moment to be measured for this direction than for the others. While the difference between 1.02 and 1.1 is small and could due to sample misalignment, it goes the right way and is well outside our error bars. The remaining ratio of 1.15 is consistent with slightly less than perfect decoupling at 100 Oe and 4 K, as discussed above.

Fig. 4 shows magnetization loops at (a) 5 K and (b) 77 K for a typical sample. The large hysteresis apparent at 5 K indicates a high degree of flux trapping. In fact, the remanent magnetization is almost exactly 1/2 of the peak magnetization seen in increasing field, as predicted by the critical state model of Bean.[16] This suggests essentially optimal flux pinning. At 77 K, on the other hand, the hysteresis is much smaller and the remanent magnetization is only 1/7 of the peak value, consistent with relatively weak pinning as found by other authors.[11,17]

The most interesting aspect of these data, however, is the critical current density derived from the remanent magnetization in light of grain

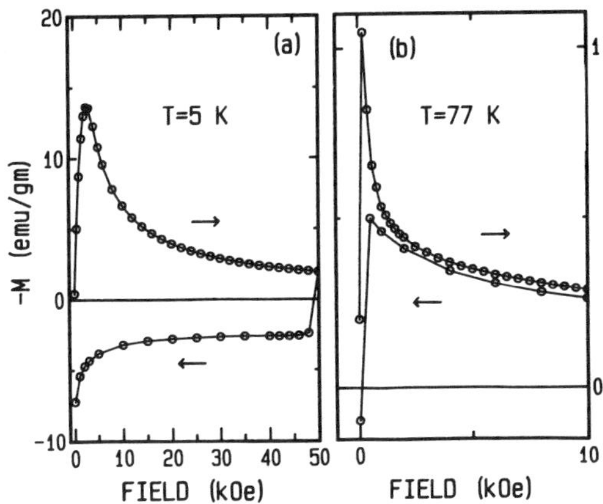

Fig. 4. Typical magnetization loops for ceramic $YBa_2Cu_3O_{7-\delta}$ at (a) 5 K, and (b) 77 K, showing considerable flux trapping effects. Arrows indicate the direction of field sweeps. The remanent magnetizations imply critical current densities within decoupled grains of 3×10^6 A/cm^2 at 5 K and 10^5 A/cm^2 at 77 K.

decoupling. In a critical state model,[16] the remanent magnetization is due to a uniform current density J_c flowing throughout the material. Since the grains are decoupled, currents flow only within grains so the bulk magnetization M (moment m per unit volume from the moment/gram multiplied by the theoretical density) is the same as the average grain magnetization. From basic magnetostatics, the moment for a circular plate of radius R and thickness ∂z is $\partial m = \int \partial r\ \pi r^2 \partial I/10$ where $\partial I = J_c \partial z \partial r$, or $\partial m = \pi J_c R^3 \partial z/30$ in practical units (A/cm^2, emu). For spherical grains, $m = \pi^2 J_c R^4/80$ per grain, so

$$M = m/(4\pi R^3/3) = 3\pi J_c R/320 \quad \text{or} \quad J_c = 320M/(3\pi R) \qquad (3)$$

From the magnetization data at 5 K, M = 7 emu/gm or 45 emu/cc. Using 5 μm for the grain radius, $J_c \approx 3\times10^6$ A/cm^2! At 76 K we have M = -.22 emu/gm, or $J_c \approx 10^5$ A/cm^2. These results are in excellent agreement with the crystalline film data.[8]

CONCLUSIONS

A survey of magnetization, J_c, and resistivity data in the literature quickly reveals tremendous variability in the properties of ceramic 123.[2–12] While we claim only that our material is representative, we have found only two cases in the literature which seem inconsistent with extensive grain decoupling. Cava et al[2] report a Meissner effect at 60 K and 18.5 Oe that is 61% of the theoretical value, and a J_c in excess of 1100 A/cm^2 at 77 K (presumably only a lower limit). However, the resistivity of their material

at 100 K is five times lower than that of our best material. This suggests
much stronger grain coupling and the possible existence of direct
superconducting shorts between grains rather than Josephson junctions, and
thus is very encouraging for the possibility of improving the properties of
ceramic material. Yamada et al[7] report a J_c for wire fabricated from
ceramic 123 powder of similar magnitude to ours at zero field, but which
persists to much higher fields (about a factor of 5 drop at 100 kOe). They
do not give the normal resistivity of their material.

The present analysis of J_c and magnetization does not take into account
the possible existence of grains composed of normal metal surrounded by superconducting shells, such as we recently proposed.[12] While grain decoupling in our material is a fact, uncertainties in the data analysis mean
that we cannot preclude the existence of such shells, for which there seems
to be indirect experimental evidence based on IR[18,19] and heat capacity[20]
studies.

In summary, we observe evidence for grain decoupling at low magnetic
fields in ceramic $YBa_2Cu_3O_{7-\delta}$. We find the bulk critical current density of
ceramic 123 compound to be unambiguously due to Josephson coupling between
grains, based on a theoretical fit to the measaured field dependence of J_c.
At very low fields the junctions can no longer maintain screening currents
and the grains' magnetic response is decoupled. This decoupling is seen in
the magnetization versus increasing field: At very low fields the
diamagnetic screening appears to be perfect, but drops considerably upon
decoupling. The Meissner effect always reflects decoupled grains and the
uniaxial anisotropy of the superconductivity. We show that the zero-field
critical current density interior to the grains is actually quite high
($>10^6$ A/cm^2 at 5 K) from an analysis of the observed remanent magnetization
in terms of decoupled grains. Finally, we note that the bulk J_c provides a
sensitive measure of H_{c1}.

ACKNOWLEDGMENTS

We take pleasure in acknowledging informative discussions with B.
Morosin, and appreciate the expert technical assistance of T. Castillo, M.
Mitchell, and R. Hellmer. Research at Sandia supported in part by the U.S.
Department of Energy under Contract No. DE-AC04-76DP00789.

REFERENCES

1.   J. G. Bednorz and K. A. Muller, Possible high T_c superconduc-
     tivity in the Ba-La-Cu-O system, <u>Z. Phys.</u>, B64:189 (1986).
2.   M. K. Wu, J. R. Ashburn, C. J. Torng, P. H. Hor, R.L.Meng, L.
     Gao, Z. J. Huang, Y. Q. Wang, and C. W. Chu, Superconducti-
     vity at 93 K in a new mixed-phase Y-Ba-Cu-O compound system
     at ambient pressure, <u>Phys. Rev. Lett.</u>, 58:908 (1987).
3.   R. J. Cava, B. Batlogg, R. B. van Dover, D. W. Murphy, S. Sun-
     shine, T. Siegrist, J. P. Remeika, E. A. Reitman, S. Zahurak
     and G. P. Espinosa, Bulk superconductivity at 91 K in
     single-phase oxygen-deficient perovskite $Ba_2YCu_3O_{9-\delta}$, <u>Phys.
     Rev. Lett.</u>, 58:1676 (1987).
4.   P. H. Hor, R.L.Meng, Y. Q. Wang, L.Gao, Z. J. Huang, J. Becht-
     old, K. Forster, and C. W. Chu, Superconductivity above 90 K
     in the square-planar compound system $ABa_2Cu_3O_{6+x}$ with A = Y,
     La, Nd, Sm, Eu, Gd, Ho, Er, and Lu, <u>Phys. Rev. Lett.</u>,
     58:1891 (1987).

5.   V. G. Kogan and J. R. Klem, Anisotropy effects upon the magnet-
     ic properties of hight-T_c superconductors, and J. R. Klem
     and V. G. Kogan, Theory of the magnetization of granular
     superconductors: Application to high-T_c superconductors, to
     be published.
6.   J. W. Ekin, A. J. Panson, A. I. Braginski, M. A. Janocko, M
     Hong, J. Kwo, S. H. Liou, D. W. Capone, and B. Flandermeyer,
     Transport critical-current characteristics of $Y_1Ba_2Cu_3O_x$, to
     be published.
7.   Y. Yamada, N. Fukushima, S. Nakayama, H. Yoshino, and S.
     Murase, Critical current density of wire type Y-Ba-Cu oxide
     superconductor, <u>Jap. J. Appl. Phys.</u>, 26:L865 (1987).
8.   P. Chaudhari, R. H. Koch, R. B. Laibowitz, T. R. McGuire, and
     R. J. Gambino, Critical current measurements in epitaxial
     films of $YBa_2Cu_3O_{7-x}$ compound, to be published.
9.   T. R. Dinger, T. K. Worthington, W. J. Gallagher, and R. L.
     Sandstrom, Direct observation of electronic anisotropy in
     single-crystal $Y_1Ba_2Cu_3O_{7-x}$, to be published.
10. H. Maletta, A. P. Malozemoff, D. C. Cronemyer, C. C. Tsuei, R.
     L. Greene, J. G. Bednorz, and K. A. Muller, Diamagnetic
     shielding and Meissner effect in the high T_c superconductor
     $Sr_{0.2}La_{1.8}CuO_4$, <u>Solid State Comm.</u>, 62:323 (1987).
11. D. K. Finnemore, R. N. Shelton, J. R. Clem, R. W. McCallum, H.
     C. Ku, R. E. McCarley, S. C. Chen, P. Klavins, and V. Kogan,
     Magnetization of superconducting lanthanum copper oxides,
     <u>Phys. Rev.</u>, B35:5319 (1987).
12. D. S. Ginley, E. L. Venturini, J. F. Kwak, R. J. Baughman, B.
     Morosin, and J. E. Schirber, Superconducting shells in ce-
     ramic $YBa_2Cu_3O_7$, <u>Phys. Rev.</u>, in press.
13. J. R. Clem, B. Bumble, S. I. Raider, W. J. Gallagher, and Y. C.
     Shih, Ambegaokar-Baratoff-Ginzburg-Landau crossover effects
     on the critical current density of granular superconductors,
     <u>Phys. Rev.</u>, B35:6637 (1987).
14. M. Tinkham, "Introduction to Superconductivity," McGraw-Hill,
     New York (1975).
15. H. Watanabe, Y. Kasai, T. Mochiku, A. Sugishita, I. Iguchi, and
     E. Yamaka, Electrical resistivity, critical current and
     crystal orientation of the sintered Y-Ba-Cu-O compounds,
     <u>Jap. J. Appl. Phys.</u>, 26:L657 (1987).
16. C. P. Bean, Magnetization of hard superconductors, <u>Phys. Rev.
     Lett.</u>, 8:250 (1962).
17. H. Kumakura, K. Togano, M. Fukutomi, M. Uehara, and K. Tachi-
     kawa, Magnetization measurements in Y-Ba-Cu-O compound sys-
     tem, <u>Jap. J. Appl. Phys.</u>, 26:L655 (1987).
18. D. A. Bonn, J. E. Greedan, C. V. Stager, T. Timusk, M. G. Doss,
     S. L. Herr, K. Kamaras, and D. B. Tanner, Far-infrared
     conductivity of the high-T_c superconductor $YBa_2Cu_3O_7$, <u>Phys.
     Rev. Lett.</u>, 58:2249 (1987).
19. S. Perkowitz, G. L. Carr, B. Lou, S. S. Yom, R. Sudharsanan,
     and D. S. Ginley, Phonon, plasmon and gap behavior in high-
     T_c superconducting $YBa_2Cu_3O_7$ and $Gd_{1.5}Ba_{1.5}Cu_3O_7$, to be
     published.
20. N. Phillips, private comm.

MAGNETIZATION OF SUPERCONDUCTING Ba(Y, Nd, Sm, Gd, Dy, Er, Yb)CuO SYSTEMS

Sang Boo Nam*, Sae Woo Nam, and Jean Ok Nam

7735 Peters Pike, Dayton, OH 45414 U.S.A.
*University Research Center, Wright State University
Dayton, OH 45435 and AFWAL/POOC, WPAFB, OH 45433

INTRODUCTION

The recently discovered superconductors of Tc in 90K range have been drawing great attention in scientific as well as social communities. The long dream of room temperature superconductors is becoming a reality.

Many researchers have reported high temperature superconductors and been making lots progress in their applications (1-9)

In this paper, we are happy to report a summary of our quest of various high Tc superconductors (10). The strategy is to test if the system is in the Meissner state. Mainly magnetization is discussed here.

The observed magnetization can be understood in the frame work of the BCS theory. The $2\Delta(0)/k_B$ Tc were deduced to be 3. - 4.4 ($\pm$.3) via studying the magnetization near Tc. However, the magnetic hysteresis was found to be present even in a low field 30G at 10K. The magnetization was found to be sensitive to the thermal fluctuations near Tc. Various ideas of mechanisms for superconductivity in these systems are suggested.

SAMPLE PREPARATION

The superconductors Ba (X= Y, Nd, Sm, Gd, Dy, Er, Yb) CuO were made independently by the solid reaction method via mixing barium carbonate, (X) oxides, and copper oxide with atomic ion ratio Ba:X:Cu=R:1:R+1 with R = 1 - 4, and sintering at 900 - 1000°C for 24 - 72hr. The sample via R=2 and sintering near 950°C was found to be the best superconductor. All samples studied were found to have layered perovskite X-ray diffraction (XRD) as given in Fig. 1. Samples via R=1,3, and 4 were also found to have structure similar to those via R=2, and found to have two phases by XRD. The lattice constants a, b, and c were determined by XRD and are given in Table I. The atomic compositions were analyzed by SEM and are given in Table I. The pictures via SEM given in Fig. 2 - 3 have sevaeral types: crystal formation, layered structure, stalagmite formatation, and irregular formation. This suggests that a homogeneous sample may appear under more stringent conditions. The magnetization of samples with large crystal sizes was found to have the temperature dependence similar to that of Pb.

Table I Measured and Estimated data for High Temperature Superconductors

BA(X)CUO DATA**

X	Y	Nd	Sm	Gd	Dy	Er	Yb
Cu(%)	62.08	56.05	56.45	56.97	55.67	55.36	56.05
Ba(%)	28.56	28.24	27.48	27.09	29.26	29.49	25.66
X (%)	9.36	15.71	16.07	15.94	15.07	15.15	18.29
T_c(K)	91	78	85	87	88	89	88
*	92	80	90	90	90	91	90
T_{c1}(K)	94	70	82	92		87	
T_{cr}(K)	91.5	92	88.3	92.2	91.2	91.5	85.6
a(A)	3.8867	3.9171	3.9002	3.9086	3.8901	3.8800	3.8867
b(A)	3.8257	3.8568	3.8667	3.8867	3.8176	3.8015	3.8128
c(A)	11.634	11.945	11.789	11.787	11.632	11.632	11.787
d	6.61	7.13	7.22	7.26	7.30	7.35	7.40
$T_c\, d^{1/2}$	234.0	208.3	228.4	238.7	237.8	241.3	239.4
#(%)	100	64	80	100	96	100	82
\$(%)	17	16	19	18	16	20	13
@	4.4	<	<	3.2	3.2	4.4	4.4

** Parts of results were reported by S. B. Nam, the Videotape of the Am.
  Phys. Soc. March 18 Meeting on high Tc Superconductors (1987);
  the Videotape of the Mat. Res. Soc. April 23-24 Symposium S on High Tc
  Superconductors (1987); Proc. of MRS ; J. Am. Cer. Soc. July (1987);
  Jpn. J. Appl. Phys. 26 (Proc. of LT18) (1987).
Atomic compositions of samples were analyzed by SEM. Samples were mixed
  via the ion ratios Ba:X:Cu=R:1:R+1 with R=1-4. The data for samples
  with R=2 were given in the above table except for the case X=Y.
Tc is determined by extrapolating the magnetization data.
* The Meissner effect onset temperature.
Tc1: the zero resistance state transition temperature by P. H. Hor et al
  Phys. Rev. Lett. 58, 1891 (1987).
Tcr: the zero resistance state transition temperature by J. M. Tarascon
  et al to be published (I thank Laura Greene for this reference).
The lattice constants a, b, and c were determined by XRD, and the
  density d's (g/cm^3) were estimated via a, b, and c.
  No attempt was made for the packing density of samples.
\$ $4\pi\chi$(magnetic susceptibility)d for the cooling case at 10G.
# The warming case at 10G after the sample was cooled at zero field.
@ $2\Delta(0)/k_B T_c$ estimated via magnetization slopes near Tc, within $\pm$.3.
  The BCS value is 3.52 and 4.4 for Pb.
< can not be estimated due perhaps to small crystal sizes of samples.
  It is highly desirable to obtain this parameter by other measurements
  such as tunnelling and infrared absorption experiments. The idea used
  here was based on the fact that the superelectron density is written
  in terms of $\Delta(T)/k_B T$ (see e.g., S. Nam, Phys. Rev. 156, 470 (1967)).

994

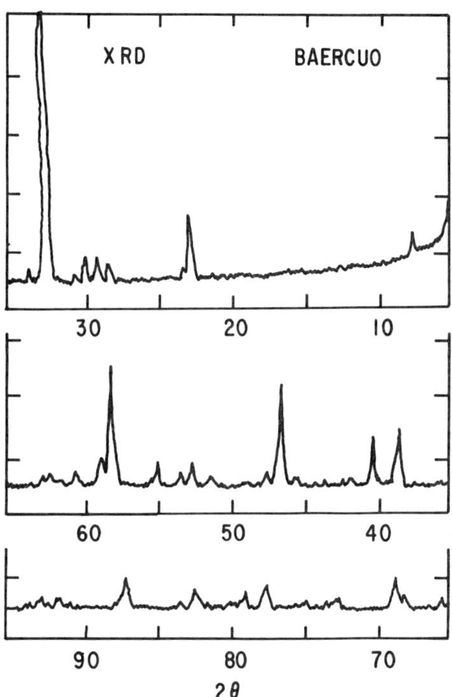

Fig. 1. For the X-ray diffration, the Cu-K$_\alpha$ line of the wave lenth
1.5418 A was used with Ni-filter, 35 kV, and 15 mA.
The above is for Ba(2)Er(1)CuO (Baercuo) sample.
The majority of samples were found to have the X-ray diffraction
patterns similar to the above.

Fig. 2a. The SEM picture for Ba(4)Er(1)Dy(1)Cu(6)O (Baerdycuo) sample. A single crystal is formed in this sample.

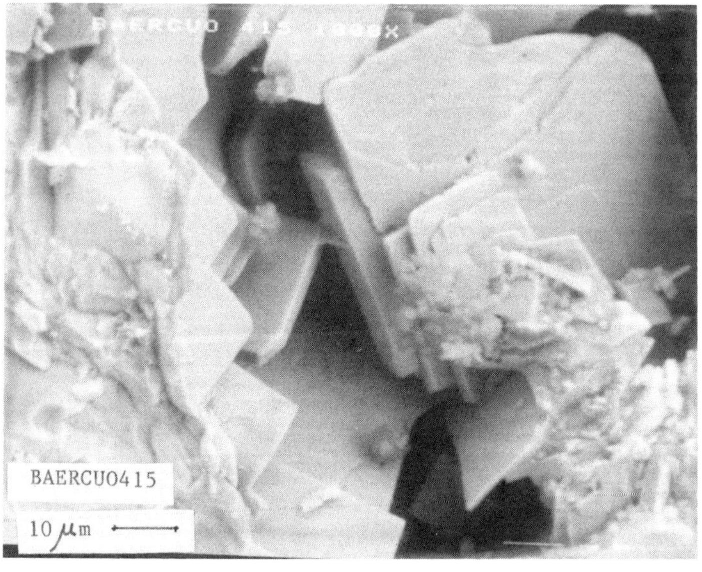

Fig. 2b. The SEM picture for Ba(4)Er(1)Cu(5)O (Baercuo) sample. Many platlets are formed in this sample.

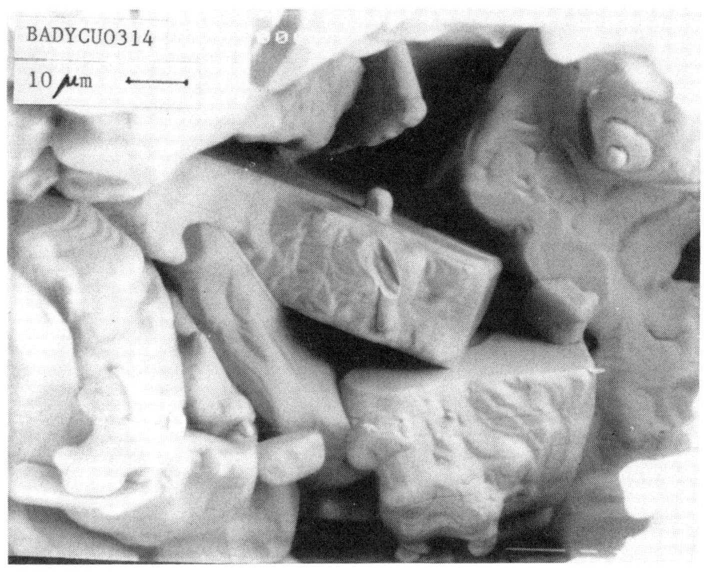

Fig. 3a. The SEM picture for Ba(3)Dy(1)Cu(4)O (Badycuo) sample.
Very interesting stalagmites are formed in this sample.

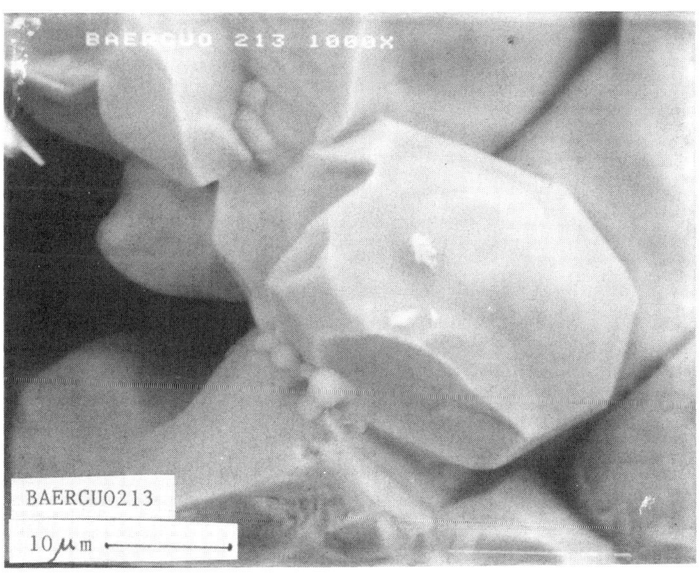

Fig. 3b. The SEM picture for Ba(2)Er(1)Cu(3)O (Baercuo) sample.
Irregular shapes are formed in this sample.

MAGNETIZATION AND MEISSNER EFFECT

The temperature dependent magnetizations M(T) were measured by the BTi SQUID magnetic susceptometer. The Meissner effect was observed up to 92K. The transition temperature Tc was determined by extrapolating M(T) such that M(Tc) = 0 and found to be sensitive to the cooling rate of the sample during its synthesis. For all the samples the fluxoids were found to be trapped during the cooling in a field at all temperatures. M(T) for Ba(x)Dy(y)Cu(z)O (Badycuo) with xyz = 213, 314, and 415 are given in the upper figure in Fig. 4, to see the atomic composition dependence. The sample with xyz=213 was found to have the strongest diamagnetic signal. However, the normalized M(T)/M(10K) was found to be same for all cases as shown in the lower figure in Fig. 4 for the warming case after the sample was cooled at zero field. It is also the same for the cooling case at 10G as well. This suggests that M(T) in these systems might be a universal function of T. To see this, in Fig. 5 we compared M(T)/M(10K) with the response function

$$S = \Delta(T)/\Delta(0) \ \tanh(\Delta(T)/2k_\beta T)$$

at zero frequency and zero momentum (11), which corresponds to the superelectron density, and is also known to be the normalized Josephson tennelling current. In Fig. 5, the upper solid line in the upper figure is for $S^{2/3}$ and the lower solid line for S. All data for the warming case were found to be in a qualitative agreement with S or $S^{2/3}$ with the BCS parameter value 3.52 for @ = $2\Delta(0)/k_\beta Tc$, where $\Delta(0)$ is the BCS order parameter (energy gap) at zero temperature.

The M(T)'s were found to be of three types: type 1 being similar to that of Pb, type 2 having a knee perhaps due to two phases in the sample, and type 3 having zero slope near Tc due to the sample crystal sizes being less than the effective penetration depth. Samples with large crystal sizes were found to be of type 1 or 2.

In Fig. 5 (the lower figure), the data for cooling case at 10G are given together with that of Pb (12) given by the solid line. The BCS parameter @'s were estimated by taking the ratios of slopes of M(T) near Tc to that of Pb and found to be 3 - 4.4 within $\pm$ .3 as given in Table I. However, Bandcuo and Basmcuo appear to have zero slope near Tc. This may be due to the sample sizes being less than the effective penetration depth.

MAGNETIC MOMENT HYSTERESIS

The magnetic moment hysyeresis with respect to magnetic field was found to be present even in low field 30G at 10K as given in Fig. 6; the upper figure for Bandcuo. It is found that once a loop of magnetization is formed then the magnetization follows the loop as indicated in the lower figure in Fig. 6 for Baercuo at 80K. Thus, it is not clear how to get the low critical field. Perhaps, one may argue that the field at which the magnetization has zero slope is to be the low critical field. The magnetic hysteresis is expected to be correlated with sample sizes, since the penetration depth plays a major role for a magnetic field being inside a superconductor. It is highly desirable to study this further.

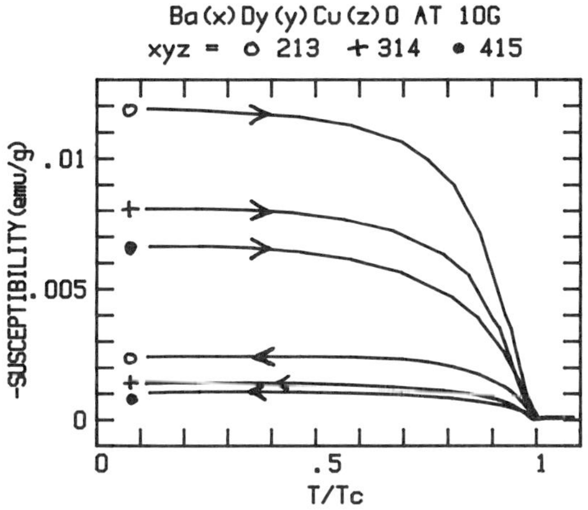

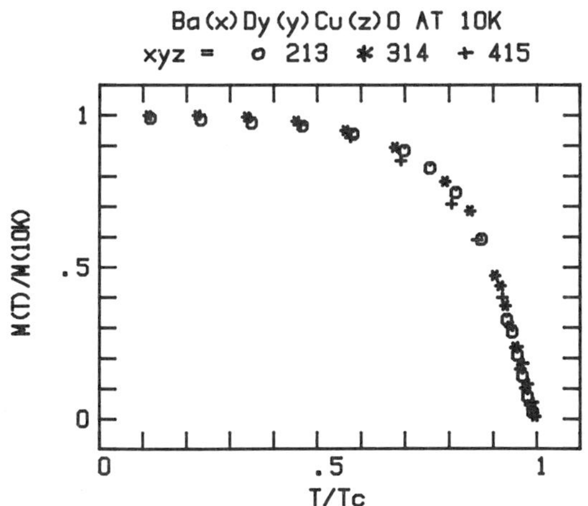

Fig. 4. The upper figure is for susceptibility of Badycuo213, 314, & 415
The lower figure is for the normalized magnetization for Badycuo
samples for the warming case after samples were cooled at zero
field. All the data is on one curve. Nearly identical
results were also found for the cooling case at 10G.

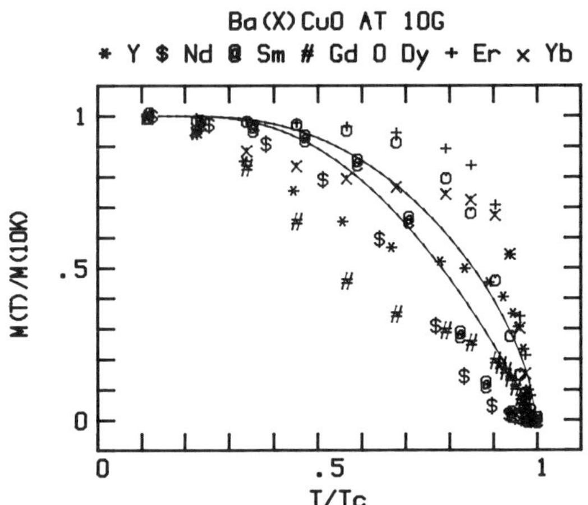

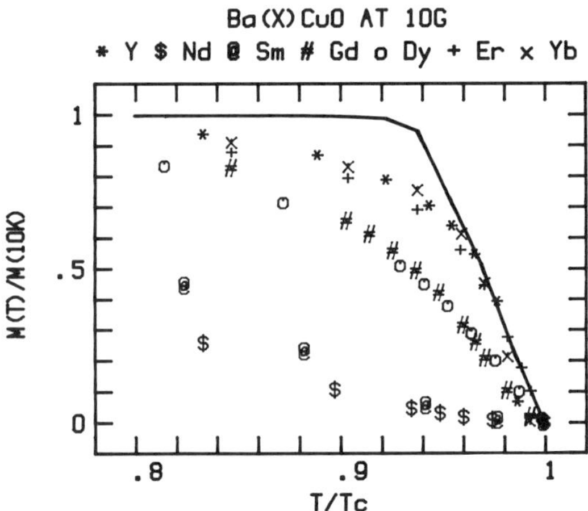

Fig. 5. The upper figure is for the normalized magnetization with its value at 10K for the warming case. The upper solid line is for $S^{2/3}$, and the lower solid line for S. The lower figure is for the cooling case. The solid line is for that of Pb.

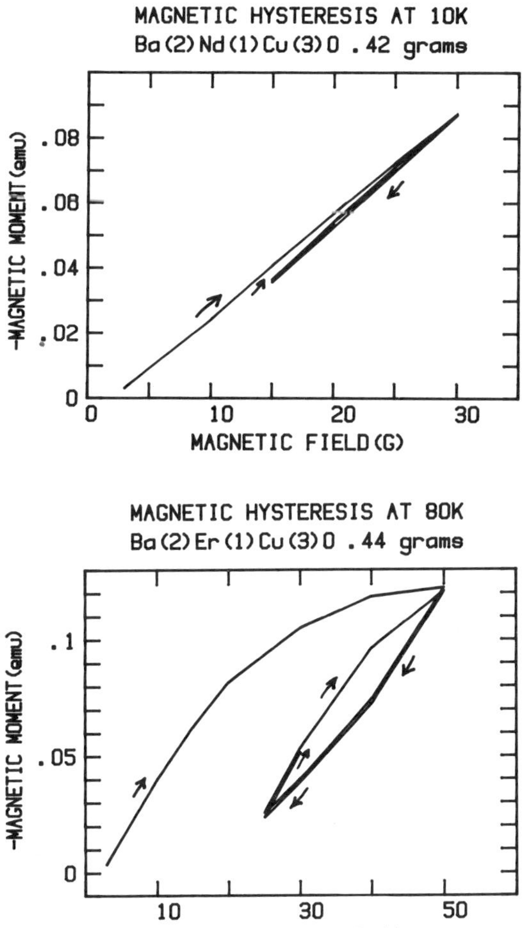

Fig. 6. The upper figure is the magnetic hysteresis in Bandcuo213 at 10K.
The lower figure is the magnetic hysteresis in Baercuo213 at 80K.

SUGGESTIVE MECHANISMS FOR SUPERCONDUCTIVITY

The fact that samples with magnetic and nonmagnetic X's have the same order of Tc suggests that in these systems the superconductivity may come from the layered structure, quassi-2D as indicated in Fig. 2. Thus one expects that the excitonic (13) and interface states would certainly play a major role for superconductivity. On the other hand, Tc were found to be sensitive to the cooling rates of samples during their systheses and oxygen pressure. Since oxygen's mass is small, one certainly expects a large electron phonon interaction. However, the recent observation of zero oxygen isotope effect (14,15) appears to rule out this view[†]. In other word, the interaction would be electron-electron interaction. Many other mechanisms such as resonance state (16), plasmonic and two band (17) are suggested. [†] For a large electron-phonon coupling case, it is ok.

SUMMARY

The superconductors of Tc in the range of 90K were independently made by the solid reaction method and their magnetizations were measured by the BTi SQUID susceptometer. The meaured magnetization can be understood in the frame work of the BCS theory, i.e., the BCS wave function. The BCS parameters @'s were deduced by taking the ratios of magnetization slopes near Tc to that of Pb, and found to be in the range of 3.2 - 4.4. It is higly desirable to obtain @ by the tunnelling and infrared absorption measurements. The magnetic hysteresis was found to be present even in low field 30G at 10K and remains to be studied. M(T) was found to be sensitive to the thermal fluctuations near Tc. This might be a reflection of the fact that the two component carrier pairing (18) may exist in these systems. The mechanism for superconductivity in these systems remains a challenging problem.

ACKNOWLEDGMENTS

Many people at WPAFB, WSU, UD, UES, and others are acknowledged for help. Without their help this work could not have been possible.

REFERENCES

1. J. Bednorz et al, Z. Phys. B26, 189 (1986).
2. S. Uchida et al, Jpn. J. Appl. Phys. 26, L1 (1987).
3. M. Wu et al, Phys. Rev. Lett. 58, 908 (1987).
4. J. Sun et al, Phys. Rev. Lett. 58, 1574 (1987).
5. R. Cava et al, Phys. Rev. Lett. 58, 1676 (1987).
6. See the Videotape on the APS High Tc Superconductors Mar 18, (1987).
7. See the Videotape on the MRS High Tc Superconductors Apr 23, (1987).
8. P. Hor et al, Phys. Rev. Lett. 58, 1891 (1987).
9. J. Tarascon et al , to be published.
10. S. Nam, NAS-NRC Report (Dec 1986); Refs 6 and 7; Announcements of
    Jan 23, Mar 6, and Mar 21 (1987); J. Cer. Soc. to be published.
11. S. Nam, Phys. Rev. 156, 470; 487 (1967).
12. S. Nam et al, Bull. Am. Phys. Soc. 31, 239 (1986).
13. D. Allender et al, Phys. Rev. B7, 1020 (1973).
14. B. Batlogg et al, Phys. Rev. Lett. 58, 2333 (1987).
15. L. Bourne et al, Phys. Rev. Lett. 58, 23337 (1987).
16. P. Anderson, Science 235, 1196 (1987).
17. V. Kresin, to be published.
18. S. Nam, Physica 107B, 715 (1981); LT17, 867 (1984);
    MRS Proc. 22, 209 (1984).

남상부

JOSEPHSON EFFECT AND ENERGY GAP MEASUREMENTS FROM Nb/YBCO POINT

CONTACT STRUCTURES

A. Barone, A. Di Chiara, G. Peluso, and U. Scotti di Uccio

Dipartimento di Fisica Nucleare, Struttura della Materia
e Fisica Applicata, Università di Napoli, Napoli, Italy
Istituto di Cibernetica, CNR, Arco Felice, Naples, Italy

A.M. Cucola and R. Vaglio

Dipartimento di Fisica, Università di Salerno, Italy

F.C. Matacotta and E. Olzi

Istituto Tecnologia Materiali Metallici non Tradizionali
del CNR, Cinisello Balsamo, Milano, Italy

INTRODUCTION

Besides its intrinsic interest, weak coupling between superconductors is widely recognized as a powerful probe to infer important informations on the underlying physics of superconductivity in
general[1]. Accordingly many authors involved in the study of the
new classes of high Tc superconductors [2,3] have focused their attention on careful investigations of various types of superconductive junction structures [4-7]. The difficulty of preparing YBCO materials in films (though at IBM in particular there were successful attempts [8]) has suggested to begin with point contact configuration rather than with sandwich tunneling barrier type junction.
However, even point contacts appear to be rather complicated. The
main problem stems from the strong normal-conducting surface layer
present on the YBCO material. This circumstance implies that in

order to establish a real link the contact pressure has to be very
high producing very likely drastic damages to the sample. In the
present work we have realized point contact structures very reliable and stable, made by a Nb point pressed on a bulk high-T superconductor ($Y Ba_2Cu_3O_7$) by using a suitable apparatus. Current-voltage characteristics and dV/dI curves have been obtained showing
both the occurrence of Josephson effect and providing an estimate
of the gap of the YBCO.

SAMPLE PREPARATION

The point contact structure was realized by a Nb needle pressed on a pellet of YBCO by means of a suitable and micrometrically
adjustable system.

The specimen of the high-Tc superconductor was prepared by
sintering at 950°C in flowing oxygen a compacted pellet of the pre-
reacted phase. For the preparation of the phase, Y_2O_3, $BaCO_3$ and
CuO powders were ground up with mortar and pestle and mixed in order to get the composition $YBa_2Cu_3O_7$. The mixture was reacted in
flowing oxygen in a furnace in SiO_2 crucible at 950°C for 16h.

Before sintering, the phase was compacted at about 10Kbar to
obtain a disc 16mm diameter and 3mm thick.

The pellet was first polished by a diamond disc to obtain a
smooth surface, then four copper pads (1mm diameter and about 4μm
thick) were sputtered to perform resistive four lead measurements.
For the sample considered it was found a Tc = 92.2K (mid-point of
resistive transition) and Δ Tc = 0.4K (90%-10%).

Before testing the "point contact" in the He, the Nb point
was chemically etched by usual procedures and the YBCO disc was
gently cleaned in vacuum using an ion gun system.

EXPERIMENTAL RESULTS

Measurements were performed in liquid helium to get information from a superconductor-superconductor contact. Current-voltage
characteristics show the expected nonlinear structure,(Fig.1) with

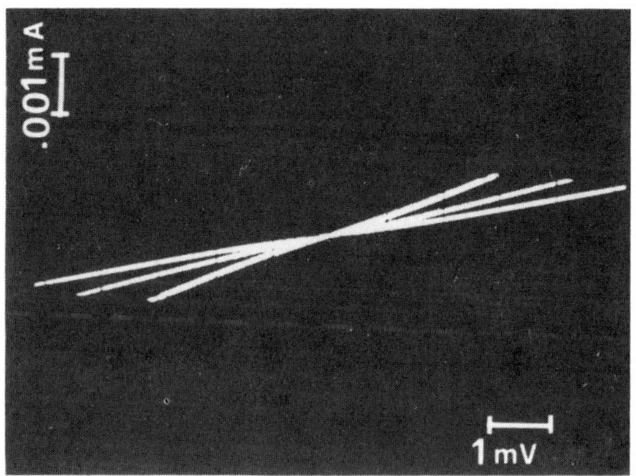

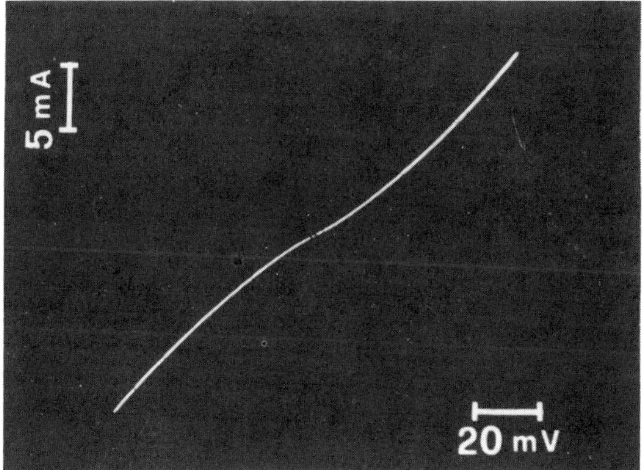

Fig. 1 - Current-voltage curves of a Nb/YBCO point contact showing the reduction of the resistance for increasing contact pressure and the appearance of the nonlinear structure.

almost zero contact pressure indicating that the surface of the
YBCO was well cleaned and therefore resonably representative of
the bulk material.   Very clear Josephson behavior was observed
by a slight pressure. Typical current-voltage characteristics
are reported in Fig. 2. The two curves show increasing zero-volt-
age Josephson current for increasing contact pressure. In Fig.4
a resonant structure is shown which represents a further feature of a typical Josephson junction. Furthermore the magnetic
field modulation of the Josephson current (a mixture of diffraction and interference behavior) was also observed as expected.

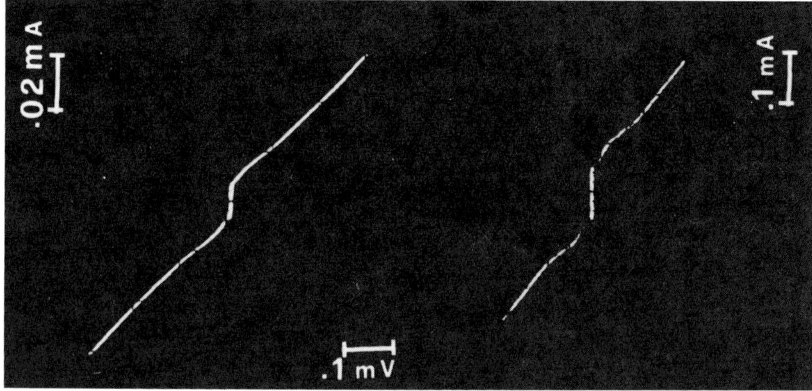

Fig.2 - Current-voltage characteristics of a Nb/YBCO point
contact displaying the Josephson effect. The two curves correspond to different contact pressures. P_1 = 427 (a.u.) and
P_2 = 435 (a.u.) on the left and on the right respectively.

In Fig. 3 dV/dI curves are reported. Although complicated by
the circumstance that point contacts include both tunneling and
superconducting link paths, it can be clearly estimated a value of
the gap for the YBCO as Δ = 19.5 mV $\pm$ 2.0 mV which leads to a

1006

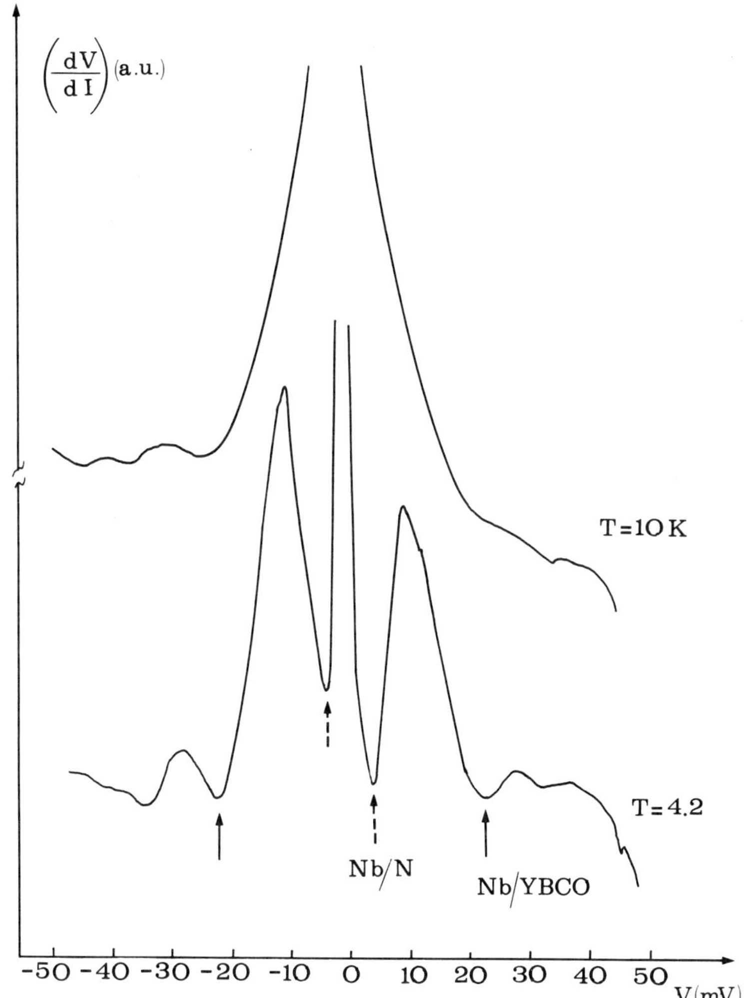

Fig. 3 - dV/dI curves for a Nb/YBCO point contact at T = 4.2K and T = 10K.

a value $\dfrac{2\Delta(0)}{K_B T_C} = 4.9 \pm 0.6$ in agreement with YBCO/YBCO contacts

investigated in Ref. 1) by the so-called break-junction technique.

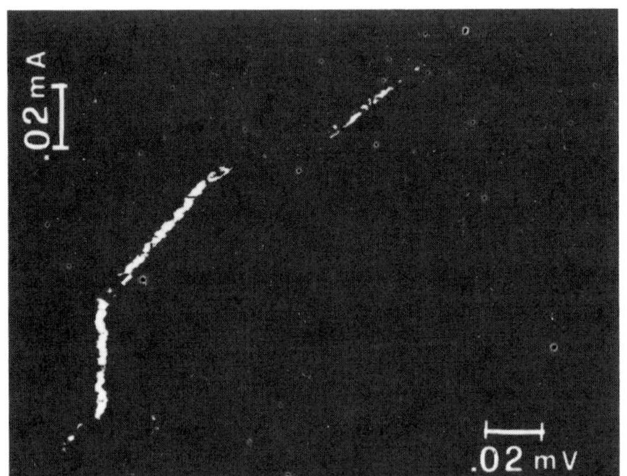

Fig. 4 – Current-voltage characteristics of a Josephson Nb /YBCO contact structure displaying a resonant current structure.

ACKNOWLEDGMENTS

Thanks are due to A. Maggio, M. Gesini and C. Salvia for the technical assistance.

This work has been partially supported by I.N.F.N.

REFERENCES

1) A. Barone and G. Paternò "Physics and Applications of the Josephson Effect", John Wiley (1982)

2) J. G. Bednorz and K.A. Müller, Z. Phys. <u>64</u>, 189 (1986)

3) M.K. Wu, J.R. Ashburn, C.J. Torng, P.H. Hor, R.L. Meng, L.Gao,
    Z.J. Huang, Y.Q. Zang and C.W. Chu, Phys. Rev. Lett. $\underline{58}$, 908
    (1986)

4) M. E. Hawley, K.E. Gray, D.W. Capone II and D. G. Hinks
    (preprint)

5) A. Th. A.M. de Waele, R.T. M. Smokers, R.W. van der Heijden,
    K. Kadowaki, Y.K. Huang, M. van Sprang and A.A. Menovsky
    (preprint)

6) J. S. Tsai, Y. Kubo, J. Tabuchi (preprint)

7) J. M.Moreland, J. W. Ekin, L. F. Goodrich, T.E. Capobianco,
    A.F. Clark, J. Kwo, M. Hong, and S.H. Liou (preprint)

8) R.B. Laibowitz, R.H. Koc-, P. Chaudari and R.J. Gambino,
    Phys. Rev. B, June (1987)

EVIDENCE OF HIGH ENERGY EXCITATIONS IN

HIGH T_c SUPERCONDUCTORS

R. Escudero, T. Akachi, R.A. Barrio
and J. Tagüeña-Martinez

Instituto de Investigaciones en Materiales
Universidad Nacional Autónoma de México
Apartado Postal 70-360, 04510 México D.F., Mécixo

ABSTRACT

Tunneling experiments were performed in Y-Ba-Cu-O
samples that show a granular composition that allow the
measurements of differential resistance using a conventional
modulation technique. Typical tunneling spectra are produced
from the samples. These curves show clear structure that
could be associated to excitations coupled to the tunneling
electrons. The energies of these excitations are of the
order of hundreds of millivolts, far to high to be due to
phonons. It is not clear yet if these excitations play a
role in the superconducting properties of the material.
However, some theoretical ideas are presented in order to
decide about the possibility for a new mechanism that
couples electrons and causes superconductivity.

EXPERIMENT AND RESULTS

Two kinds of samples were prepared with nominal
compositions: sample (a) $Y_{1.26}Ba_{.74}CuO_4$, and sample (b)
$Y_{1.20}Ba_{.80}CuO_4$, made by the conventional mixing of powders,
calcination and sintering processes.[1] Disks of 1.2 cm of
diameter and 0.2 cm thick were examined and characterized by
different probes. They showed to be different from the
usual optimum samples in the superconducting phase
$Y_1Ba_2Cu_3O_{7-x}$. Scanning electron micrographs (see Figure 1)
show a particularly messy appearance of grains and pores
that suggest a wider separation of the superconducting
grains in these samples. This characteristic is of paramount
importance for the present experiment, as it will be made
clear below. X-ray patterns show that the superconducting
phase is present in both samples, although minor changes are
detected in the intensities of certain peaks.

The central idea of this experiment[2] is to take
advantage of the granular nature of the samples in order to
make tunneling experiments with them, without the need of
fabricating a conventional tunnel junction, but benefitting

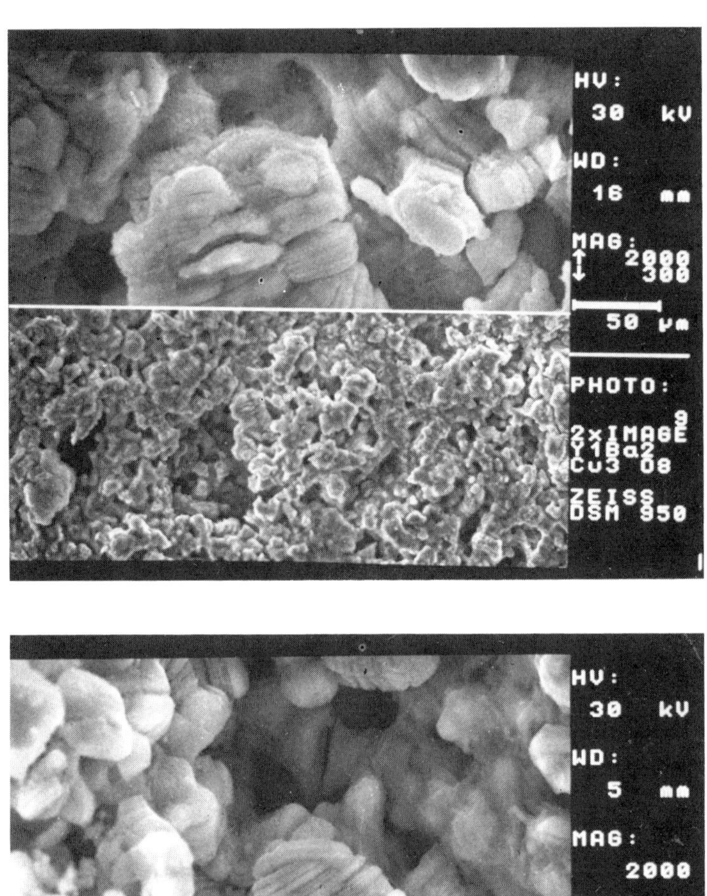

Figure 1. Electron micrographs of the granular
superconducting materials at different magnifications.

from the naturally formed barriers between the
superconducting grains. By varying the temperature one can
couple the grains when the Josephson coupling energy exceeds
k_BT. There are various facts that show that there are
tunneling junctions in the sample, although the ultimate
proof is the typical tunneling spectrum extracted from these
granular samples. Figure 2 shows the R vs. T plot from
samples (a) and (b). It is clear that the width of the
transition is larger in the more granular sample. It should
be noticed that the resistance, just before the transition,
is about 4 Ω, three orders of magnitude larger that in the
optimum 1-2-3-phase samples (see Figure 3).

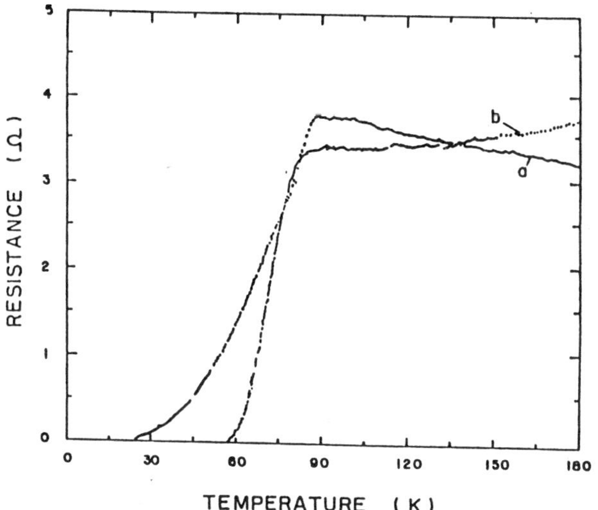

Figure 2. Resistance vs. temperature of (a) granular sample
and (b) homogeneous sample.

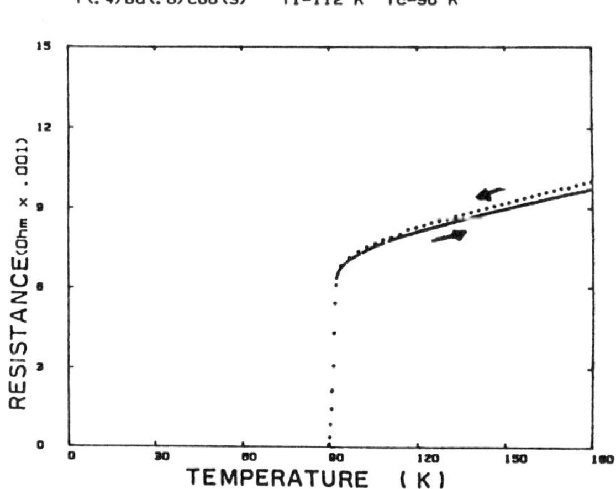

Figure 3. Resistance vs. temperature of an optimized
sample.

Differential resistance measurements were taken from sample (a) (see Figure 4), arranging the contacts in the geometry shown in the inset of Figure 4, that is, current and voltage were measured from the same flat face of the disk. The remarkable absence of noise allows the clear identification of the structure outside the gap region. There are sharp peaks, slightly asymmetrical, at 210, 360 and 520 mV. Note the presence of a supercurrent at zero voltage, and a superconducting gap that could be crudely estimated to be about 125 mV.

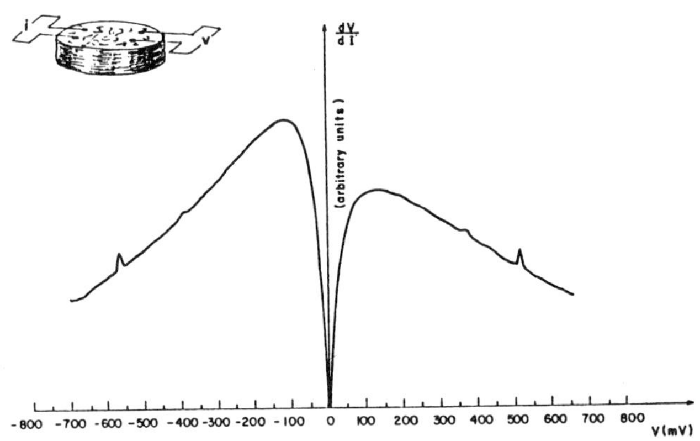

Figure 4. Differential resistance (dV/dI) vs. V of the granular sample (a) taken in the geometry shown in the inset.

If the measurement is made from opposite faces of the disk, one obtains a spectrum like the one shown in Figure 5, which resembles the former one, except that it is much noisier and the apparent gap is reduced to about 45 mV. The reason for the noise could be the anisotropy in the arrangement of the superconducting grains due to the sintering process, therefore the grains are more packed in the direction perpendicular to the flat face.

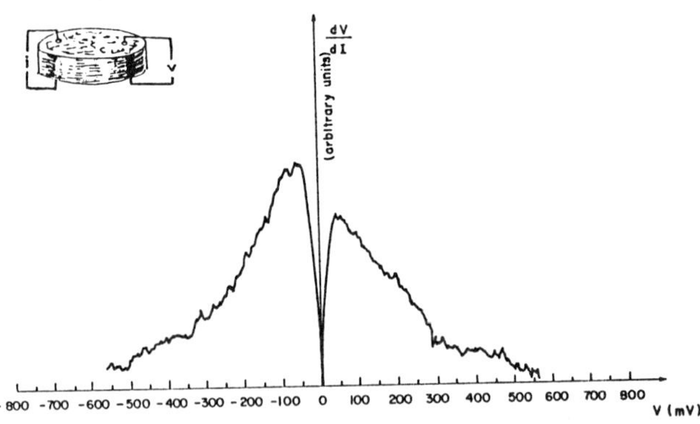

Figure 5. Differential resistance (dV/dI) vs. V of the granular sample (a) taken in a different geometry as shown in the inset.

Figure 6 shows the equivalent experiment performed with
sample (b), which is more homogeneus. Clearly the tunneling
has been suppressed, because the grains are more coupled
(the differential resistance is much smaller than before),
and a typical contact characteristic is observed.

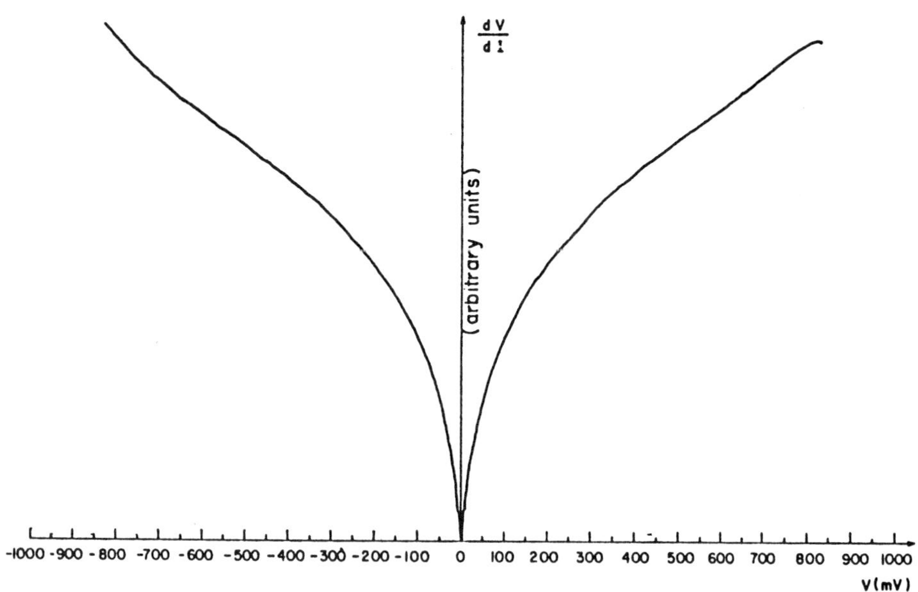

Figure 6. Differential resistance (dV/dI) vs. V of the
homogeneous sample (b).

From the point of view of normal BCS coupling there are
two puzzling results from the experiments: the magnitude of
the superconducting gap that leads to a ratio $2\Delta_o/k_BT_c$ of
about 13 (normally ~3.5 in BCS), and the presence of the
sharp peaks at energies far above the phonon cut off.
Infrared and neutron scattering results have shown that the
O-Cu breathing mode is at about 90 mV. Therefore, it seems
that we are facing a new type of excitation responsible for
electron pairing. One possible source of attractive Coulomb
interaction between electrons is the polarization induced in
the lattice by impurities,[3] or the excitonic modulation in
low dimensionality systems of the type described by Little
and Bardeen.[4] It is just possible to envisage a
polarization mechanism that induces coupling of electrons in
a low dimensional system by destroying the charge
instabilities proper of the latter situation. There has
been an enormous amount of discussion on the competing role
of charge (spin) density waves and superconductivity[5] that
in certain conditions could enhance one transition at the
expense of destroying the other. In any case, as far as we
know, there are no convincing arguments that prohibit the

presence of rapidly polarizable electrons that, in certain
conditions, could be responsible for electron pairing. The
low population at the Fermi level and the nesting of the
two-dimensional Fermi surface have to be removed in order to
give superconductivity. A very simple estimate of the
energy of the excitations that couple the electrons in these
systems can be made by assuming a perfect synchronization
between the frequency associated with the excitation and the
relaxation time of electrons at the Fermi level. The latter
can be written in terms of the effective mass of the
electrons $m^*{\approx}5m_e$, $k_f={\pi}/a$; (a=3.78Å), and the coherence
length ${\xi}_0{\approx}30$Å. The numbers are reasonable values extracted
from several experiments, and give 242 meV for the energy of
the excitation. The same argument leads to an expression
$h{\omega}={\pi}^2{\Delta}_0/2$ that involves only the superconducting gap.
Assuming ${\Delta}_0{\approx}100$ meV, which is reasonable, from the present
experiment one obtains 246 meV. It is extraordinary that
this very simple argument predicts an energy for the
excitations of the same order of magnitude of the features
observed in the spectra.

Therefore one is tempted to conclude that the
mechanisms that produce superconductivity in these systems
are similar to the BCS cooper pairs but without involving
phonons. Experiments performed on Josephson systems with RF
radiation demonstrate without a doubt that charge is carried
in pairs.[6] This is not the time to make definite
conclusions, but the present experiment will contribute to a
better understanding of the physical phenomena taking place
in high T_C superconductors.

ACKNOWLEDGEMENT

We acknowledge the help of David Rios with SEM
pictures.

REFERENCES

1. R. Escudero, L. E. Rendón-DiazMirón, T. Akachi, J.
     Heiras, C. Vázquez, L. Baños, F. Estrada, and G.
     González, to be published in Jap. J. Appl. Phys.
2. R. Escudero, L. Rendón, T. Akachi, R. A. Barrio, and J.
     Tagüeña-Martínez to be published.
3. W. B. Fowler and R. J. Elliott, Phys. Rev. B. 34:5525
     (1986).
4. D. Allender, J. Bray, and J. Bardeen, Phys. Rev. B
     7:1020 (1973). D. Davis, H. Gutfreund, and W. A.
     Little, Phys. Rev. B 13:4766 (1976).
5. K. Machida, J. Phys. Soc. of Jap., 53:712 (1984).
6.   J. S. Tsai, Y. Kubo, J. Tabuchi, Phys. Rev. Lett.
     58:1979 (1987).

STRUCTURAL AND CHARGE-TRANSFER DESCRIPTION OF HIGH-T_c SUPERCONDUCTIVITY

IN $Y_1Ba_2Cu_3O_{7-\delta}$

Gary C. Vezzoli,* Lt. Richard Benfer and William Spurgeon

US Army Materials Technology Laboratory
Ceramics Research Division
Watertown, Massachusetts 02172

Abstract: The observation of high-T_c superconductivity is described in terms of the unique structure of the defect ceramic $Y_1Ba_2Cu_3O_{7-\delta}$ and treated as a special defect case of the A_2BX_4 archtypes normally represented by K_2MnF_4 or K_2NiF_4. The recently elucidated[1] structure of the ceramic superconductor (known as the 1-2-3 material) can be shown to be a substitutive distortion defect modification of the well-known antiferromagnetic K_2NiF_4 structure[2] if it is written in the defect $A_4B_2X_8$ form $(Ba_2^{2+} Cu_2^{2+})(Y_1^{3+} Cu_1^{3+})O_{8-w}$. This material is superconducting for $1<w<1.5$. The presence of the corner Cu^{3+} in the basal planes of the 1-2-3 unit cell coupled with the linking region of an oxygen-vacancy, capable of forming an F center, and further influenced by the distorted square planar coordinated oxygens of the corner copper ions (CuO_3 units) allows for a mediation of BCS theory via exciton interactions, and influenced by a Jahn-Teller distortion. The edge Cu^{2+} ions are hybridized in CuO_2 layers, and the five-fold coordination allows for a shared oxygen ion with the corner copper ions establishing an interactive element between a-b plane and the z axis.

I. Introduction and Background

(1) Crystal Chemistry of Previous Mixed Valence Candidate High-T_c Materials:

The early experimental work of Chu et al[3] and Vezzoli and Bera[4] in CuCl at high-pressure showed conductivity and magnetic data representative of collective quantum phenomena that suggested the characteristics of high -T_c. Conditions for charge transfer and a Boson mediation of normal electron-electron repulsion seemed favorable when copper exists in more than a single valence state in mixed phase intimacy and in compound with a high electron affinity anion. In the case of CuCl, the resulting material after high pressure treatment contained Cu^0, Cu^{1+} and Cu^{2+} cations owing to the disproportionation: $2Cu^{1+}Cl^{1-} \rightarrow Cu^0 + Cu^{2+}Cl_2^{1-}$. Interaction could thus be established between $d^{10}s^1$, $d^{10}s^0$ and d^9s^0 states mediated by the distorted tetrahedral symmetry of CuCl and the cubic symmetry of free Cu. At the zincblende to tetragonal transformation in CuCl (40 kbar) the conditions were highly favorable in the solid state to effectuate this disproportionation.[4] The work of Wilson[5] showed how these interfaced orbitals in a disproportionate state can lead to the presence of excitons and how interface between a unique distorted tetrahedral and octahedral coordination can

establish an interaction which at least qualitatively can result in an
electron-electron attractive coupling leading to excitonic mediated
superconductivity. This conceptualization was in consonance with two long-
standing themes of Russian literature, namely the excitonic route to high-
T_c superconductivity given by Ginzburg[6] and the Bose condensation of
excitons at high density given by Keldysh and Kozlov[7] via an excitonic-
insulator band model. Further early ideas regarding excitonic high-T_c were
given by Russakov[8], Abrikosov[9], Little[10] and Allender et al.[11] The Bardeen
-Cooper-Schreiffer (BCS) theory[12] of course invokes phonon coupling for
superconductivity via relationship to the Debye temperature θ_D. The latter
interaction frequencies cannot fully explain high-T_c ($>20^{\circ}$K). However, BCS
allows for a coupling agent other than phonons to establish Cooper pairs.
Such agents can include bipolorons, plasmons, excitons and spin fluctuations. A leading proponent of the exciton based model was I. Lefkowitz[13]
who experimentally verified at lower pressure the earlier[3] work. It thus
seems more than coincidental to note the correlation between the earlier
pure theoretical predictions (Refs 6-12) and the recent extraordinary
finding of $T_c > 90^{\circ}$ by Chu et al in $Y_1Ba_2Cu_3O_{7-\delta}$ (Ref 14), based on the
earlier work of Bednorz and Muller (Ref 15). Interesting correlation also
exists between concepts of Wilson[5] (as a physicist) regarding adjacency
of distorted tetrahedral and octahedral coordination to produce high-T_c
via disproportionation-induced excitons and the recent work of Raveau who
as a solid state chemist searched for compounds with tetrahedral, octahedral, and pyramidal regions in order to achieve anisotropy.[16] Indications
that the charge transfer mechanism in CuCl was indeed not stable and
depended upon unique disproportionate phase conditions, thereby being not
consistently reproducible, was suggested by the <u>oscillatory</u> behavior of
the conductivity transition at liquid nitrogen conditions given by Vezzoli
and Bera.[4]

(2)   Epitaxial Studies for the Existence of High-T_c

It was recognized by Mattes and Foiles[17a] that the effect of the high-
pressure work of Vezzoli and Bera[4] (and in the work of Lefkowitz[13]) was to
create a critical degree of <u>strain</u> that caused the disproportionation central to the ideas of Wilson.[5] At that time the concepts about strained
interfaces and strained layer superlattices were just about beginning to
germinate. Mattes and Foiles[17a] deposited CuCl on silicon and reported
conductivity and diamagnetic measurements akin to those reported by Chu
et al,[3] by Vezzoli and Bera[4] and by Lefkowitz.[13] Recently the epitaxial
deposition was performed in a magnetic field, and yielded similar
results.[17b]

Very recently the possibility of epitaxial high-T_c superconductivity
was theoretically assessed by Arthur J. Freeman[18a] in a study suggested by
Vezzoli[18b] at the CuSl/Si:Si interface. The results were obtained by
employing the all-electron self-consistent full potential linearized augmented plane wave (FLAPW) method within the local density approximation.
The most distinctive feature of the two-dimensional bands for the CuCl/Si
superlattice treated by Freeman[18] is the existence of 2D <u>metallic interface states</u> which are localized within the Cu - Si and (mainly) Cl-Si
interface layers. This establishes a high density-of-states condition at
at the Fermi energy which allows for the large transfer effects (found by
Freeman[18]) from Si to CuCl which is consistent with alternate evidence
indicating space charge inversion at the interface between CuCl and Si.
Calculations of the electron-phonon interaction yield only moderate T_c
values such as 20°K for this assembly whereas experiments yield $T > 77^{\circ}$K
for the electrical and magnetic transitions, proving that an alternate
mechanism must be invoked. Freeman[18] reports analogy between the existence
of 2D conducting planes in the new 1-2-3 high-T_c Cu-O superconductors[14,15]
and in the CuCl/Si interface study.

Additional recent work by S. Williams[19] has experimentally proved that
(1) the interface between CuCl and Si is <u>unstable</u>; (2) the inherent <u>strain</u>
causes the disproportionation of free copper (Cu^O-d^OS^O) and (3) to stabilize the interface requires a 50Å buffer of CaF_2. The combination of the
Williams,[19] Freeman[18] and Mattes and Foiles[16] work lends credence to the
original works of Chu,[3] Vezzoli and Bera,[4] and Lefkowitz,[13] and shows correlation with the recent 1-2-3 Cu-O superconductors.[14,15] The FLAPW work
presented by Yu and Freeman[20] at this conference on the 1-2-3 material
shows quantitative correlation.

If an exciton coupling theory were to be embraced to explain the CuCl
conductivity and diamagnetism in terms of high-T_c superconductivity, then
crude calculations indicate a T_c maximum of about 200°K.

(3) Early Data on CdS as Candidate High-T_c

Similarly to CuCl, the work on pressure-quenched CdS by Homan et al[21]
and by Vezzoli and Otooni[22] proved that a high-T_c-like effect occurred
(at liquid N_2 temperatures) only when a mixed phase condition existed
between the sphalerite-wurtzite phases (4-coordination) and the rocksalt
structure (6-coordinated), <u>the latter being recoverable to room conditions</u>
<u>only when in the presence of the Cl^- ion</u>[23a] (for example, CdS prepared
from $CdCl_2$). Again a tetrahedral – octahedral interface seemed to be
important; however, the presence of more than a single valence state of
Cd was not established (although it is conjectured that the Cd^{4+} ion
could exist under rare circumstances). The conductance data at the high-T_c
-like transition in CdS was far less unstable than in CuCl but was neither
<u>fully</u> stable or always repeatable.[22] It was believed that almost 1%
chlorine was necessary to observe the high-T_c-like effect.[23b]

The importance of the early CuCl and CdS work was that the conductivity transition and the Meissner exclusion observed at T > 77°K could not
readily be explained via any phenomenon other than high-T_c yet could not
be purely phonon coupled.

Because of inconsistency in reproducibility and lack of understanding
of BCS-mediation, this behavior was labelled as "a collective quantum
phenomena" rather than clearcut superconductivity. Although many purists
in the field were not comfortable with even the above inferences, the
established data and the Soviet and American theoretical works, nonetheless, were embryos to pursue further study. It was clear that a condensed
exciton plasma could <u>not</u> be ruled out as a coupling mechanism which could
still satisfy the Cooper-pair requirement.

II. <u>Crystallography and Crystal   Chemistry of $Y_1Ba_2Cu_3O_{7-}$</u>

1. Structure

The recent neutron diffraction[1] study of the ceramic oxide superconductor (T_c > 95°K) $Y_1Ba_2Cu_3O_{7-\delta}$ is given in Fig 1 and is a refinement of
the original x-ray work of Hazen et al[24]. Clearly the phonon-coupled version of BCS cannot explain a T_c above about 20°K, and furthermore, the
absence of any significant isotope effect on T_c in $Y_1Ba_2Cu_3O_{7-\delta}$ argues most
strongly against purely phonon coupling which <u>must</u> be related to atomic
mass.

We compare the structure given in Fig 1 with the much earlier work of
Bergeneau et al[2] giving the A_2BX_4 structure of antiferromagnetic K_2NiF_4
(Fig 2) also from neutron diffraction data. If in the K_2NiF_4 structure we
substitute: Cu^{3+} for the corner Ni^{2+}; Cu^{2+} for the edge K^+; Ba^{2+} for the
upper and lower interior K^+; Y^{3+} for the central Ni^{2+}; remove the 100

direction basal plane anions; remove the flourine anions octahedrally
coordinated by the central Ni^{2+}; substitute O^{2-} for the remaining F-
anions;and finally introduce two puckered and two non-puckered O^{2-} ions
in a distorted dimpled (or puckered) plane that includes the <u>edge</u> Cu^{2+}
ions, we establish the structure of the defect ceramic oxide $Y_1Ba_2Cu_3O_7$
in the form of an $A_4B_2X_8$ structural type, namely $(Ba_2^{2+}Cu_2^{2+})(Y_1^{3+}Cu_1^{3+})$
O_{8-u} where $u<1$. (See Table 1 for ionic compositional ratios). If we
re-introduce the O^{2-} ions in the basal planes (between the corner copper
ions) we establish $Y_1Ba_2Cu_3O_8$ (<u>not</u> a superconductor). We have divided the
Cu_3 into 2 Cu^{2+} and 1 Cu^{3+} to achieve charge balance with O_7^{2-}, and to
introduce two valence states of copper for charge transfer. The ionic
radii criteria are:

$$Ba^{2+} = 1.34\text{Å} \qquad\qquad Y^3 = 0.893\text{Å}$$
$$Cu^{2+} = 0.72\text{Å} \qquad\qquad Cu^{3+} \quad 0.62\text{Å}$$

The radius ratio for Cu^{2+} to O^{2-} is 0.55 which is greater than for
tetrahedral or square planar coordination and somewhat less than for
<u>typical</u> octahedral coordination, thus in accord with the observed almost
five-fold coordination (not to be confused with five-fold symmetry out-
lawed in space group theory).

Although the trivalent positive ionic state of copper is uncommon, it
clearly exists in K_3CuF_6 and $KCuO_2$,and in these materials exists respectively in a four-coordinated square or a six-coordinated octahedral structure existing in a d^8 orbital configuration. The $M_2^I CuX_4$ compounds (M^I =
univalent cation: and $X = Cl^{1-}$ or Br^{1-}) contains planar $(CuX_4)^{2-}$ ions.
Squashed tetrahedra of $Cu-X_4$ can also exist as in Cs_2CuCl_4.[25]

The $(Cu^{3+}O_3^{2-})^{3-}$ ions in the bc plane thus are neither square planar
non-tetrahedral non-octahedral and must be interpreted in terms of chains
of charge centers (ions). If the two oxygen vacancies were occupied the
molecule would be $(CuO_4)^{5-}$ and could be tetrahedral. The $(Cu^{2+}O_2)^{2-}$ ions
constituting the ab slightly puckered planes are most unusual and cannot
be square planar in a conventional sense, but form planes by virtue of
being closed chains (as in S_2N_2).

We recognize that at anion stoichiometry less or equal to $O_{6.5}$, the
phenomenon of high-T_c superconductivity clearly disappears. Charge balance
requirements for this chemical reduction (from the favorable $O_{6.9}$ to the
unfavorable $O_{6.5}$ for example) suggests the composition: $(Ba_2^{2+}Cu_2^{2+})_1$
$(Y_1^{3+}Cu_1^{2+})_1O_{6.5}$. In such a compound the corner copper ions have <u>gained</u> an
electron to become Cu^{2+} and have lost distorting, polarizing, and charge-
transfer-agent ability or character. The additional missing oxygens are
most likely to be derived from one more oxygen vacant from the chain
structure, probably from each basal plane. This is favored as the missing
anion site instead of the apex anion of the inverted pyramid in the Ba^{2+}
plane because bond-length of the former is 1.94Å and the latter 1.85Å. The
shorter bond length should correspond in this case to the higher bond
strength. $O_{6.5}$ being a critical occurrence leading to the elimination of
high-T_c lends credence to the hypothesis that superconductivity is derived
from the copper chains mediated by the common edge anion between plane and
chain and by the anion vacancies at the edges. The absence of the additional oxygen (for $O_{6.5}$) can be accomplished in an ordered or disordered
manner.

We address the optimal $O_{6.9}$ condition, and make note that disorder in
the "chains" affects the collective phenomena and write $(Ba_2^{2+}Cu_2^{2+})_1$
$(Y_1^{3+}Cu_x^{2+}Cu_y^{3+})_1O_{6.9}$ where x = 0.2 and y = 0.8. The simplest manner to
envision the reduction of the oxygens is to assume a loss of one oxygen
in each of the two o,1/2,o sites in the basal lane for every fifth unit
cell.[25] Again this vacancy condition can be ordered or disordered. Lower

temperature processing and consequent ordering of vacancies has definitively been shown to alter the $O_{6.5}$ cutoff of superconductivity.[26]

2. Role of Each Ion

a) Yttrium

Yttrium itself under the above argument cannot itself be critical
because sustitutions using Sc^{3+} and rare earths cause no major change
in T_c or stability. However, a trivalent cation is required to retain
charge balance. The paramagnetic yttrium ion must coordinate eight anions
in inverted pyramidal form. However, yttrium shows a radius ration with
respect to O^{2-} of $r_+ cation/r_- anion = 0.68$. Although r_+/r_1 of 0.732 to
1.00 normally leads to eight-coordination in cubic form, the <u>puckered</u>
anionic geometry allows more available ligand-space and may compensate
for the radius ratio being less than 0.732. The magnetic isolation of the
yttrium ion is also favorable for the perfect diamagnetism developed at
T_c. Cu^{3+} in at least small molar percents should be substitutionable for
Y^{3+}. This is indicated by the work of Lytle.[27]

b) Barium

Ba^{2+} is an essential ion. In $Y_1Ba_2Cu_3O_{7-\delta}$ the Ba^{2+} must coordinate
twelve anions. Radius ratio with oxygen gives $r_+/r_- = 1.02$ just compatible with 12-coordination. The large field strength (charge of cation
divided by interatomic spacing) of Ba^{2+} (shown by its ability to coordinate twelve oxygens in perovskite structures such as BaT_1O_3) influences
the basal plane anions and any electrons trapped in the oxygen vacancies
of the basal plane. This 12-coordinating ability is precisely why Ba^{2+} is
the ideal alkaline earth metal since it allows a radius ratio > 1. Neither
Sr^{2+} nor especially Ca^{2+} is suitable for as high a T_c because they are not
sufficiently large to properly coordinate twelve oxygens. Radius ratio
with oxygen for calcium is 0.75 which is borderline for 8-coordination
especially for a puckered anion geometry, Ca^{2+} prefers 6-coordination as
in the calcite and aragonite structures. Radius ratio for strontium is
about 0.85 which normally suggests eight-coordinated cubic CsCl structure and is less than borderline for twelve coordination except in the
perovskite structure. The charge transfer interaction involving ions in
the puckered planes coordinated by Ba^{2+} is weaker than in the Cu^{3+} chains
as described below; however, 2-D metallic properties are nevertheless
predicted.

c) Divalent Copper

The edge copper ions (believed to be $Cu^{2+}-d^9$) may be important as a
5-fold Jahn-Teller ion and has been described most recently in quantum
mechanical treatments of high-T_c by Johnson et al[28] showing oxygen $p\pi$ at
ϵf suggesting a "tubular" polarized $\Psi+$ and $\Psi-$ propagation. Calculations
for T_c using a Jahn-Teller format seemingly are compatible with T_c data
and suggest a maximum T_c near 240°K at which temperature fluctuations in
conductivity have been recently reported. However, the tubes set up by
d_{xy} orbitals transfer charge in the z-direction, rather than the expected
x-direction.

Specifically the <u>Jahn-Teller theorem</u> states that if a subshell (t_{2g}
or e_g) is neither filled, half-filled, or empty, a distortion will occur
to remove any possible degeneracy. This removal of degeneracy is due to
the split of orbital energies. In the Cu^{2+} d^9 case, the three electrons
in e_g could be arranged in either $(d_{z^2})^2(d_{x^2-y^2})^1$ or $(d_{z^2})^1(d_{x^2-y^2})^2$. In
the $(d_{z^2})^2(d_{x^2-y^2})^1$ state there would exist more repulsion between the
Cu^{2+} and the ligand (oxygen) electrons along d_{z^2} than along $d_{x^2-y^2}$ simply

because of the additional electron in the former. The net result would be
a lengthening of the two bonds along d_{z2} relative to the bonds in the
d_{x2-y2} direction. The reverse situation would lead to a lengthening of the
bonds in the d_{x2-y2} directions, this being very rare in nature but observed in crystals of the CuF_4^{2-} and CuF_3^{1-} complexes. In the Cu^{2+} edge
ions, the condition is somewhat different because the coordination is
five-fold rather than 6- or 4-fold. In oxygen stoichiometries between O_7
and $O_{6.5}$ when charge balance requires Cu^{2+} at one or more corner site per
x unit cells, then the situation is still unique because the symmetry is
not truly O_h symmetry due to the missing oxygens, and is not tetrahedral
because the four oxygens in the O_7 stoichiometry are essentially square
planar. In the $O_{7-\delta}$ stoichiometry, there is another oxygen vacancy probably in the plane of the corner copper ions. Hence, the data shown in
Fig 1 support the existence of a Jahn-Teller distortion due to $(d_{z2})^2$
$(d_{x2-y2})^1$ in the above setting.

The Cu^{2+} ions are the major contributor to the 2D charge density of
the central planes. Because the edge copper ions are so much smaller than
the central barium it is unlikely that they would substitute for each
other. $Cu^{2+} \leftrightarrow Cu^{3+}$ excitations can be enhanced by optical breathing modes
which can elevate T_c.

d) Trivalent Copper

The corner copper ions (believed to be $Cu^{3+}d^8$) are suggested to be
the key ingredient along with the missing oxygens to participate in
superconductivity via coupling through <u>bound</u> (localized) excitons (such
as $Cu^{3+} \rightarrow Cu^{4+}$ excitations). These copper ions constitute the copper-
oxygen-copper chains and the copper-vacancy-copper chains. Cu^{3+} ions are
essential to the charge transfer mechanism to support an exciton mediated
BCS in a mixed valence phase which also contains Cu^{2+}. The conditions of
all Cu^{3+} ions or all Cu^{2+} ions ($Y_1Ba_2Cu_3O_8$ or $Y_1Ba_2Cu_3O_{6.5}$, respectively)
<u>cannot or does not show superconductivity</u>.

e) Oxygen

The absence of oxygens in the Cu^{3+} chains, combined with the Y^{3+}
coordination of inverted bi-pyramids may explain why the Cu^{2+} planes are
puckered. The oxygen vacancies can also act as attractive centers for
electrons otherwise normally localized on copper ions. The holes present
in the Cu-O hybridized states are candidates for forming excitons which
can cause electron Cooper pair formations. As mentioned earlier the
oxygen vacancies can be ordered[26] or disordered which means that in some
local neighborhood, even for the $O_{6.5}$ condition in an ordered state,
systematic Cu^{3+} ions can exist and participate in charge transfer. Contrasting this condition, the disordered vacancy regime gives a "smeared"
averaged state such that no continuity of Cu^{2+} (edge) - Cu^{3+} (corner)
exists.

3. Stability Considerations

The $Y_1Ba_2Cu_3O_7$ structure shows reproducible superconductivity in the
neighborhood of 88 to 98°K, compared to the inconsistent data on CuCl,
because the former does not require disproportionation, the degree and
mosaic of which is uncontrollable.

The oxygen vacancies in the 1-2-3 structure which are suggested to be
very cogent to the superconductivity mechanism may also be responsible
for rendering the material vulnerable to attack by water vapor (similar
to Cu Cl) and causing a change in T_c. However a passivating film (perhaps
nitride, not oxide) and/or proper encapsulation should overcome this

problem. Thin film deposition as from rapidly solidified melts may also
assist in stabilizing T_c as suggested by recent work by O'Handley.[29]
Stability of the 1-2-3 structure against radiation and charged particle
fields must be carefully investigated because of possible effects regarding the Cu^{3+}/Cu^{2+} concentrations, as well as induced vacancies. Radiation
testing is presently underway in our laboratory.[30]

4. Energy Band Gap

Recent scanning tunneling microscope studies of the 1-2-3 structure
surface by Meservey et al[31] show a minimum band gap of $2\Delta = 3.53$ kT_c at
$T<T_c$ (for a tungsten junction) in consonance with recent predictions of
Bardeen,[32] but indicating zero gap at $T>T_c$. This observation is at present
being studied and implies the cooling-induced metal -> semi-metal or semiconductor transition that creates a gap. STM work should in the future
study copper - oxygen clusters. Much additional work is also needed to
determine a band structure model of this system. An attempt to present a
working model is furnished in our follow-up work also presented at the
Berkeley conference.[33] This will include recent preliminary ESCA and Auger
results showing data not interpretable as due to Cu^{1+}, Cu^{2+} or impurities,
and suggesting confirmation of the herein proposed ratio of Cu^{3+} to Cu^{2+}
(corner Cu/edge Cu ions) being 1:2.

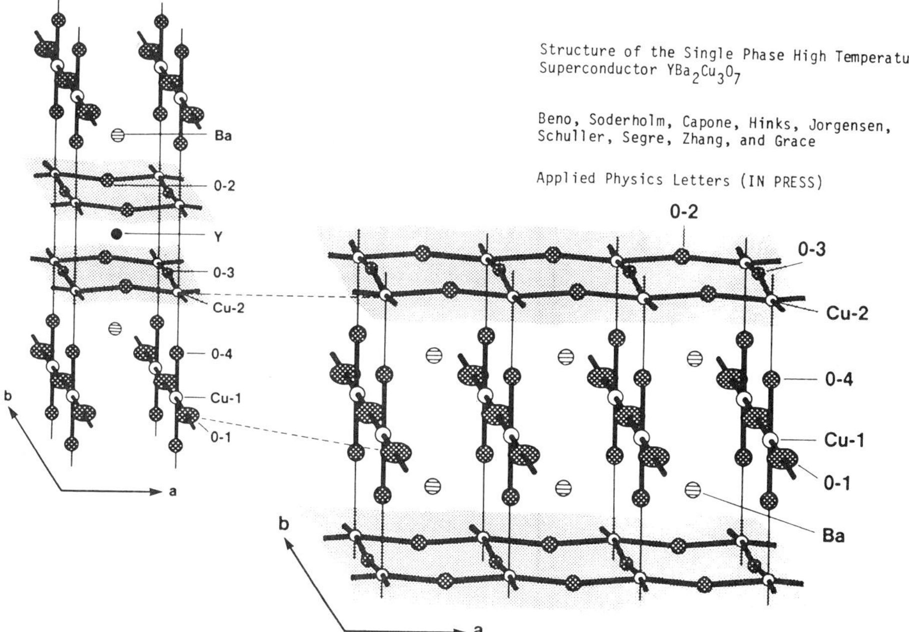

Fig. 1) The structure of $Y_1Ba_2Cu_3O_{7-\delta}$ given in Ref. 1 showing lattice
parameters: a=3.8231A, b=3.8864A,   c=11.680A, space group P_{mmm}, orthorhombic Bravais lattice. Chains in the b-c planes are CuO_3.
Layers in the a-b plane are CuO_2. Bond lengths are:

$Cu^{+3}_3 - O^{-2}(1)$ (in planc)=   1.943Å

$Cu^{+3}_2 - O94)$ (below plane)=1.850Å

$Cu^{+2}_2 - O(2)$ (in a-direction of puckered plane)=1.928Å

$Cu^{+2}_2 - O(3)$ (in b-direction of puckered plane)= 1.962Å

$Cu^{+2} - O(4)$ (in c-direction)=2.303Å

The numbers 1,2,3,4 refer to the identified oxygen ion. The
above lattice parameters suggest Jahn-Teller distortion
in d^9 (Cu^{+2}) with elongation in dz^2 hence favoring
the e_g state to be in the configuration $(dz^2)^2(d_{x^2-y^2})^1$.

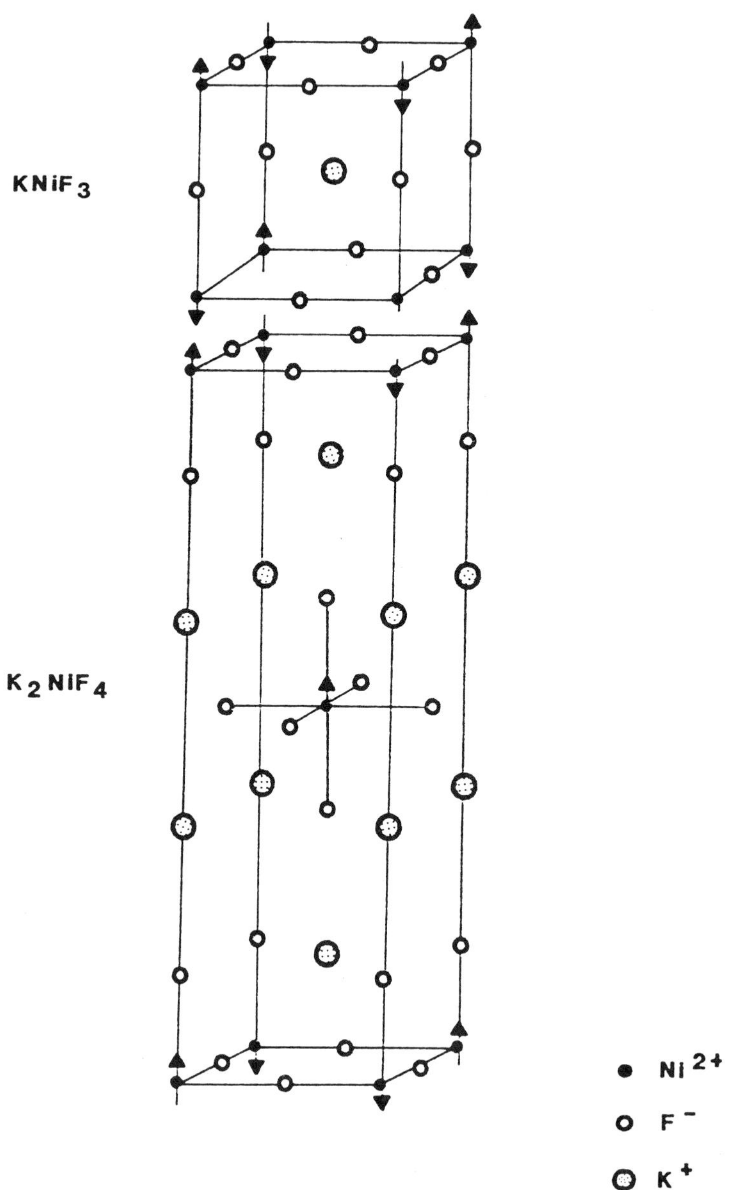

Fig 2)   The structure of the A$_2$BX$_4$ archtype K$_2$MnF$_4$ or Li$_2$NiF$_4$ as given in Ref 2.

Table 1

CALCULATION OF OXYGEN STOICHIOMETRY

a.

Plane and Oxygen Designation	Number of O^{2-} ions	Contribution of each to unit cell	Total in Plane
1. Top Cu^{3+} Plane (01)	2	1/4	1/2
2. Ba^{2+} Plane (04)	4	1/4	1
3. Cu^{2+} edge Plane (03 in b-axis) (02 in a axis)	4	1/2	2
4. Y^{3+} Plane	0	0	0
5. Cu^{2+} edge Plane	4	1/2	2
6. Ba^{2+} Plane (04)	4	1/4	1
7 Bottom Cu^{3+} Plane (01)	2	1/4	1/2

TOTAL=7

Calculation of
Cu^{2+} and Cu^{3+}
Stoichiometry

b. Cu^{2+}

C.N.5 Plane	Number of ions	Contribution of each to unit cell	Total
1	0	------	0
2	0	------	0
3	4	1/4	1
4	0	------	0
5	4	1/4	1
6	0	------	0
7	0	------	0

Total= 2

(continued)

Table 1. (Continued)

c. Cu^{3+}

c.n.4 Plane	Number	X	Contribution	=	No. Ions
1	4		1/8		1/2
2	0		--------		
3	0		--------		
4	0		--------		
5	0		--------		
6	0		--------		
7	4		1/8		1/2
					TOTAL 1

REFERENCES

*   Adjunct Professor at Rutgers University
Department of Electrical Engineering
Busch Campus
Piscataway, New Jersey 08854
        and
Visiting Scientist
Massachusetts Institute of Technology
Materials Science Department
Cambridge, Massachusetts 02139

1. M.A. Beno, L. Soderholm, D.W. Capone II, D.G. Hinks, J.D. Jorgensen, I.K. Schuller, Appl, Phys Lett. March 27, 1987 (in press).

2. A.J. Birgeneau, H.J. Guggenheim, and G. Shirane, Phys. Rev B, 1(5), 1 March 1970.

3. C.W. Chu, A.P. Russakov, S. Huang, S. Early, T.H. Geballe, and C.Y. Huang, PHys Rev 18 2116 (1978). It was for the purpose of doing this work with C.W. Chu that the Soviet scientist, A.P. Russakov, brought his ideas on self-induced excitonic materials as candidate high-T_c superconductors to this country.

4. a) G.C. Vezzoli and J. Bera, Phys. Rev B, 23(6), 3022, 15 March 1981.
    b) G.C. Vezzoli, Phys Rev B, February 1982

5. J.A. Wilson, Phil Mag B, 38(5), 427 (1978).

6. V.L. Ginzburg, Contemp. Phys. 9, 355 (1968).
    V.L. Ginzburg and D.A. Kirzhnits, Phys. Rev 4, 343 (1972).

7. L.V. Keldysh and A.N. Kozlov, JETP Lett 5, 190 (1967)
    L.V. Keldysh and Yu V. Kopaev, Sov. Phys. Sol. St. 6, 2219 (1965).

8. A.P. Russakov, Phys. Status Solidi B72, 503 (1975).

9. A.A. Abrikosov, Zh. Eksp. Teor. Fiz Pis'ma Red 27, 37, (1978); JETP Lett. 27, 219 (1978)

10. W.A. Little, Phys. Rev 157, 396 (1967).

11. D. Allender, J. Bray and J. Bardeen Phys. Rev B7, 1020 (1973); Phys
    Rev B8, 4433 (1973).
    N.B. Brandt, S.V. Kuvshinnikov, A.P. Russakov, and V.M. Semenov,
    Zh. Eksp. Teor. Fiz Pis'ma Red 27, 37 (1978); JETP Lett. 27, 33 (1978)

12. J. Bardeen, L.N. Cooper, and J.R. Schreiffer, Phys Rev 108(5),73(1957)

13. I. Lefkowitz, Univ North Carolina, Frankford Arsenal, and the Army
    Research Office private communication; I. Lefkowitz, J.S. Manning,
    and P.E. Bloomfield Phys. Rev. B $\underline{20}$, 4506 (1979).

14. C.W. Chu, P.H. Hor, R.L. Meng, L. Gao, Z.J. HUang, and Y.Q. Wang,
    Phys. Rev. Lett 58 (4), 405 (26 Jan 1987); M.K. Wu, J.R. Ashburn, and
    L. Gao, P.H. Hor, R.L. Meng,C.W.Chu,Z.J. Huang, Y.Q. Wang, and
    C.J. Torng, Phys Rev Lett 58 (9), 908 (2 March 1987); C.W. Chu, P.H.
    Hor, R.L. Meng, L. Gao, Z.J. Hueng, and Y.Q. Wang, 58 (4), 405
    (26 Jan 1987)

15. J.G. Bednorz and K.A. Muller, Z.Phys. B 64,189 (1986).

16. B. Raveau: Int. Workshop on Novel Mechanisms of Superconductivity,
    June 22-27, 1987, Berkeley, California; C. Michel, L. Er. Rakho, and
    B. Raveau, Mat. Ros. Bull. 20, 667 (1985).

17. a) B. Mattes and W. Foiles, Univ Michigan, private communications.
    b) B. Mattes, Ann Arbor, Univ Michigan, private communications.

18. a) A. Freemen, Final Report TCN 86-706, Battelle Columbus Labora-
    tories, Research Triangle Park, NC.
    b) G.C. Vezzoli, Technical Statement of Work, US Army Armament
    Research Development and Engineering Center, Picatinny Arsenal,
    Dover, NJ, August 1986.

19. S.Williams, UCLA, private communication.

20. J. Yu and A.J. Freeman, Poster Session 24, Int. Workshop on Novel
    Mechanisms of Superconductivity, June 22-26 Berkeley, Cal; also
    J. Yu and A.J. Freeman, Bonds, Bands Charge Transfer Excitations and
    Superconductivity of $Y-Na_2Cu_3O_7$ preprint.

21. C.G. Homan, D.P. Kendall and R.K. MacCrone. Sol. State Comm 32,521
    (1979): E. Brown, C.G. Homan and R.K. MacCrone, Phys. Rev. Lett 45
    (6), 478 (1980)

22. G.C. Vezzoli, M. Otooni, P. Houser and S. Krasner, Mat Res, Bull. 17
    (4) 485 (1982).

23. a) R. Miller, F. Dachille and R. Roy, Phys. 37, 4913 (1966).
    b) S. Block and G. Piermarine, National Bureau of Standards, private
    communications.

25. a) F.A. Cotton and G. Wilkinson, Advanced Inorganic Chemistry Inter-
       science, New York (1966) p. 893-901.
    b) Although the chains run along the edge of the ab basal planes, they
    are considered to be in the bc plane.

24. R.M. Hazen, L.W. Finger, R.J. Angel, C.T. Prewitt, N.L. Ross, H.K.Mao,
    C.G. Hadidiacos, P.H. Hor, R.L. Meng, C.W. Chu. This work gives space
    group P4m2 (tetragonal) $Y_1Ba_2Cu_3O_{6.5}$ with ordered oxygen vacancies
    in the BaO an YO planes. The position of the oxygen vacancies in the
    $O_{6.5}$ phase is extremely important because the phase is not normally
    superconducting at oxygen 6.5.

26. B. Batlogg "Research at High-T_c" Int. Workshop on Novel Mechanisms of
    Superconductivity June 22-26, 1987, Berkeley, Cal

27. F. Lytle "Cu Substitution in La/Y site in $La_{1.8}Sr_{0.2}CuO_4$ and
    $YBa_2Cu_3O_7$. Determination by x-ray Absorption Spectroscopy" Ibid.

28. M.E. McHenry, A. Collins, M.M. Donovan, C.Counterman, K.H. Johnson,
    R.C. O'Handley, and G. Kalonji "Molecular-Orbital Basis for High-T_c
    Superconductivity in Oxide Superconductors." Ibid.
29. R. OHandley, MIT, private communication

30. Work to be conducted by M. Shoga of Hughes Aircraft, S. Bernacki of
    Raytheon. T.P. Ma of Yale and S.T. Peng of NY Institute of Technology
    (Old Westbury)

31. R. Merservey, R. Moodera, C Hao and H Hansa, private communications.

32. J. Bardeen, "Theory of High-T_c Superconductors," Int. Workshop on
    Novel Mechanisms of Superconductivity, June 22-26, 1987, Berkeley, Cal.

34. Gratitude is extended to Cadet John Hansch of the US Military Academy
    at West Point for processing assistance and to Simon Foner and Brian
    Schwartz of MIT, Robert N Katz of MTL, Joseph Budnick of the Univ of
    Conn, Charles Gallo of the 3M Co., Peter J. Walsh of NASA Langley,
    Allen Goldman of the Univ of Minnesota, Carl Rau of Rice University,
    H. Fritzche of the Univ of Chicago, and finally to the late Isaac
    Lefkowitz for very fruitful discussions.

TUNNELING SPECTROSCOPY OF HIGH T_c OXIDE SUPERCONDUCTORS

WITH A SCANNING TUNNELING MICROSCOPE

S. Pan, K. W. Ng, and A. L. de Lozanne

Department of Physics
University of Texas
Austin, Texas 78712

We present a brief summary of our tunneling measurements on $La_{1.85}Sr_{0.15}CuO_{4-y}$ and $YBa_2Cu_3O_{6.91}$, including comments about preliminary results on the temperature dependence of the gap in the latter material. These were taken with a scanning tunneling microscope that operates at temperatures down to 10K in an ultra-high vacuum chamber. An important unique feature of this apparatus is the ability to exchange samples or tips without venting the chamber or warming up the microscope, thus giving a fast turnaround time. We also have the capability to break these samples in-situ at low temperature which we believe is important in obtaining a surface that is representative of the bulk and, in particular, minimizes the loss of oxygen.

$La_{1.85}Sr_{0.15}CuO_{4-y}$ made at Bell Communications Research showed surprisingly large values for the gap,[1] which we estimate at 12±3mV. Taking 40K as a conservative estimate for the onset T_c from the measured resistive transition we obtain $2\Delta/kT_c$= 7±2. Other authors[2-6] have obtained similar values. In ref. 1 we showed that the ubiquitous parabolic background in the conductance as a function of voltage dissapears when tunneling through a vacuum barrier into a clean, in-situ broken surface. This background is therefore likely due to low-barrier-height material on the surface of the sample. On the other hand, vacuum tunneling can easily achieve current densities in the order of 10^6 A/cm^2 so that the gap features are smeared out by depairing effects. We have succesfully fitted[7] the observed conductance with a phenomenological model of depairing[8].

In order to minimize depairing effects we are now using soft aluminum tips, so that the tip deforms when touching the surface thus increasing the tunneling area and decreasing the current density. Furthermore, we do not remove the native aluminum oxide from the tip since it serves to stabilize the tunneling barrier width and is well known for its good properties as a barrier material.

YBa$_2$Cu$_3$O$_{6.91}$ made at Westinghouse also shows a range of gap values[7]. One sample showed 40-50 meV gaps at 14K, while another sample from the same batch gave gaps clustered around 90 meV and 180 meV (these are "single gap" IV curves; in addition, we observe "double gap" IV curves as other authors,[9,10] where the second gap is twice the first).

We are currently attempting to measure the temperature dependence of the gap in the second sample mentioned above. This is difficult because a change of temperature causes a change of the position of the tip with respect to the sample due to differential thermal expansion. Since the gap we observe at any given constant temperature is not uniform over the surface, we cannot guarantee that when we change the temperature any change in the observed gap is not merely due to a change in position. Therefore we are performing many measurements at different positions at a fixed temperature, then changing the temperature, letting all thermal relaxations take place, and then repeating the process. The data is sketchy and scattered at this time, but there seems to be a group of gaps clustered around 90 meV and another around 180 meV. Both of these show a weak increase as the temperature is lowered from 90K; we have been unable to see the gap go to zero at T_C. It seems as if the gap has its full value just below T_C and is zero above, i. e. a step function. The problems associated with these measurements would be greatly reduced if we could use single crystals of a few microns in size, either bulk or thin film, since then the gap would presumably be constant over the region shifted by thermal expansion.

REFERENCES

1. S. Pan, K. W. Ng, A. L. de Lozanne, J. M. Tarascon, and L. H. Greene, Measurements of the superconducting gap of La-Sr-Cu-O with a scanning tunneling microscope, Phys. Rev. B 35:7220 (1987).
2. John Moreland, A. F. Clark, H. C. Ku, and R. N. Shelton, Electron tunneling measurement of the energy gap in a La-Sr-Cu-O superconductor, Cryogenics 27:277 (1987).
3. J. Moreland, A. F. Clark, L. F. Goodrich, H. C. Ku, and R. N. Shelton, Tunneling spectroscopy of a La-Sr-Cu-O break junction: evidence for strong-coupling superconductivity, preprint.
4. M. Naito, D. P. E. Smith, M. D. Kirk, B. Oh, M. R. Hahn, K. Char, D. B. Mitzi, J. Z. Sun, D. J. Webb, M. R. Beasley, O. Fischer, T. H. Geballe, R. H. Hammond, A. Kapitulnik, and C. F. Quate, Phys. Rev. B 35:7228 (1987).
5. M. E. Hawley, K. E. Gray, D. W. Capone II, and D. G. Hinks, Energy gap in La$_{1.85}$Sr$_{0.15}$CuO$_{4-y}$ from point contact tunneling, Phys. Rev. B 35:7224 (1987).
6. J. R. Kirtley, C. C. Tsuei, Sung I. Park, C. C. Chi, J. Rozen, and M. W. Shafer, Local tunneling measurements of the high T_C superconductor La$_{1.85}$Sr$_{0.15}$CuO$_{4-y}$, Phys. Rev. B 35:7216 (1987).
7. K. W. Ng, S. Pan, A. L. de Lozanne, A. J. Panson, and J. Talvacchio, Tunneling spectroscopy of high T_C oxide superconductors with a scanning tunneling microscope,

submitted to 18-th Low Temp. Phys. Conf., Aug. 22,
1987, Osaka, Japan.

8. R. C. Dynes et. al., Phys. Rev. Lett. 53:2437 (1984).

9. M. D. Kirk, et. al., preprint.

10. J. Moreland, J. W. Elkin, L. F. Goodrich, T. E.
    Capobianco, A. F. Clark, J. Kwo, M. Hong, and S. H.
    Liou, Break-junction tunneling measurements of the
    high T_C superconductor $YBa_2Cu_3O_{6.91}$, preprint.

OXYGEN ORDERING AND INTERFACIAL SUPERCONDUCTIVITY AT TWIN

BOUNDARIES IN A LANDAU-GINZBURG SUPERCONDUCTOR OXIDE MODEL

C. Varea and A. Robledo

Facultad de Quimica, Universidad Nacional Autónoma
de México.
México 04510, D.F.

ABSTRACT

We analyse the superconducting critical behavior of a
model $YBa_2Cu_3O_{9-y}$ twinned orthorhombic crystal. For a Landau-
Ginzburg free energy with coupled superconducting and oxygen-
vacancy order parameters, the symmetry-breaking twin boundaries induce a purely interfacial superconducting transition
at $T_c^{twin} > T_{co}$, within a range of oxygen bulk compositions.
Bulk superconductivity occurs at $T < T_{co}$ below the oxygen-
vacancy ordering temperature T_o.

The discovery of high-temperature superconductivity has
recently led to the attainment of critical temperatures as
high as 93K in mixed phase Y-Ba-Cu-O systems[1]. The superconducting single phase appears to be a heavily twinned
orthorhombic ordered phase with composition $YBa_2Cu_3O_{9-y}$[2-4]
Microtwinning[3] with small orthorhombic domains 100 Å in
lenght induce local strains as well as disorder near the
domain boundaries, a fact that is likely to be of importance
in determining the observed properties of polycrystalline
samples. The orthorhombically distorted perovskite structure
of $YBa_2Cu_3O_{9-y}$[2-4] has a unit cell (a=3.824Å, b=3.891Å and
c=11.685Å) three times that of the standard perovskite
cell due to the ordering of Ba and Y atoms along the c axis.

The suggested structure of alternating CuO_2, BaO and Y planes perpendicular to c contains copper atoms in square-planar coordination and square pyramidal coordination with the oxygen on the BaO planes. Oxygen content depends on synthetic conditions[2-4] and a value of y = 2.1 has been reported for single phase samples with zero resistance at 91 K.[2] Oxygen deficiency is almost completely accomodated on the Cu between Ba planes (Y planes have no oxygen) and an ordering of oxygen deficiency sets in along the a-b planes.[5,6] The observed twinning results from 90° rotation of the reciprocal lattice around c and the number of twin lamellae per unit grain boundary area depends on the number of new grain contacts made during growth of recrystallized structures.[7] Such twinning is often encountered when two lattice constants have similar values and surface energy is small.

Since the partial oxidation of Cu^{+2} is crucial for the superconducting properties of the perovskite related oxides, and oxidation is controlled by oxygen deficiency, which in turn is related to the ordering of oxygen vacancies, one is led to conclude that there is a coupling between the superconductivity order parameter Δ and that for spatial ordering of oxygen vacancies. This is possible given that the measured coherence length ξ at low temperature is of the order of a lattice constant,[2] an indication that superconductivity and ordering of oxygen vacancies coincide in length scale.

In the present work we model some of the properties of $YBa_2Cu_3O_{9-y}$ by means of a phenomenological Landau-Ginzburg free energy. The main feature in the theory is the coupling of the superconductivity order parameter Δ, with that for the spatial ordering of oxygen vacancies, δ. We assume that the relevant Cu atoms are those on the CuO planes between BaO layers and we have a lattice of vacancy positions formed by disconnected square plane lattices along the a,b directions. The Landau-Ginzburg free energy associated to ordering of vacancies within one such plane is:

$$F_{vac} = \sum_i \left[(T-\Lambda x(1-x))\, \delta_i^2 + \frac{B(x)}{2}\, \delta_i^2 + C(x)\, (\delta_i - \delta_{i+1})^2 \right] \qquad (1)$$

where $B(x) = \Lambda(1-x)/x$ when $x < \tfrac{1}{2}$, $B(x) = \Lambda x/(1-x)$ when $x > \tfrac{1}{2}$, $C(x) = 2\Lambda\, x(1-x)$, x is the lattice oxygen global occupation number, and δ_i is the oxygen order parameter on row i according to Fig. 1a. There, every row i is divided into two interwoven sublattices with occupation numbers $x+\delta_i$ and $x-\delta_i$, respectively. In a mean field calculation Λ is the nearest neighbor oxygen-oxygen effective interaction. The Landau-Ginzburg free energy difference between superconducting and normal states in the absence of an external magnetic field is writen in the form:

$$F_{sup} = A \sum_i \left[(T-\tau(x,\delta_i))\, \Delta_i^2 + \tfrac{1}{2}\beta(x,\delta_i)\, \Delta_i^4 + [\gamma(x,\delta_i)\, \Delta_i - \gamma(x,\delta_{i+1})\, \Delta_{i+1}]^2 \right] \qquad (2)$$

here Δ_i is the superconductivity order parameter along row i. The dependence of τ, β, and γ on x and δ_i is taken to be $\tau = b^{-1}\beta = \gamma^2 = \mu[x(1-x) - \delta_i^2]$ and A, b and μ are constants. The form chosen for F_{sup} assumes and ideal concentration of $x = \tfrac{1}{2}$ for the onset of superconductivity in the disordered state and the onset of vacancy ordering depresses the appearance of superconducting order.

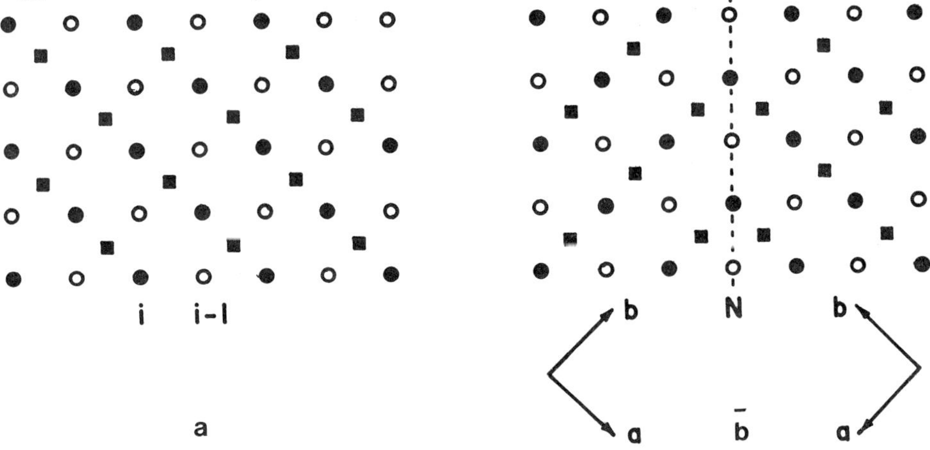

Fig.1. (a) Ordering of oxygen ($\bullet$) and oxygen vacancies ($\circ$) in the copper ($\square$) planes between Ba planes in the orthorhombic crystal structure. (b) Twin boundary along oxygens.

The Euler-Lagrange equation associated to $F_{vac} + F_{sup}$ are

$$[T-\Lambda x(1-x)]\delta_i + B(x)\delta_i^3 + C(x)[2\delta_i - \delta_{i+1} - \delta_{i-1}] +$$

$$8 A\gamma_i^{-4} [T\eta_i^2 + \tfrac{1}{2}b\eta_i^4]\delta_i = 0 \qquad (3)$$

and

$$[T-\tau(x,\delta_i)]\eta_i + b\eta_i^3 + \gamma^2(x,\delta_i)[2\eta_i - \eta_{i+1} - \eta_{i-1}] = 0 \qquad (4)$$

where $\eta_i = \gamma_i \Delta_i$, $\gamma_i \equiv \gamma(x,\delta_i)$.

The last term in Eq(3), coupling δ_i to η_i, is zero at temperatures when the system is in the normal state. The ordering of vacancies is then described by a second order transition at all compositions $0<x<1$ with an ordering temperature T_o given by $T_o = \Lambda x(1-x)$. If vacancy ordering is suppressed, according to Eq. (4), the temperature T_{cd} at which the disordered system of composition x turns superconducting is given by $T_{cd} = \mu x(1-x)$. This could be achieved only if $T_o < T_{cd}$. When $T_o > T_{cd}$ superconductivity arises in an ordered state and the coupled Eqs. (3) and (4) give $T_{co} = \mu[x(1-x) - \delta_x^2]$ with $\delta_x^2 = x(1-x)(\Lambda-\mu) / (B(x) - \mu)$. In Fig. 2 we show the critical temperatures T_o, T_{cd} and T_{co} when $\Lambda/\mu = 1.6$. At $x = 0.5$, T_o and T_{cd} take their maximum values but T_{co} vanishes. If we assign the value T_o $(x=0.5) = 240$ K, then $T_{cd}(x=0.5) = 150$ K and T_{co} $(x=0.3) = T_{co}(x=0.7) = 99$ K.

Our model 90°K model twin boundary is represented in Fig. 1b. This defect also generates a tetragonal environment on the oxygen ordering where the effective ordering interaction of the oxygen at the twin row N with its neighboring rows is $\Lambda_N = 0$. The oxygen vacancy profiles generated by the twin are obtained through the resolution of Eq. (3) with $\eta_i = 0$. The presence of twin boundaries in the crystal introduce some degree of disorder, which may be large enough to allow for the onset of superconductivity at the interfacial regions and at a temperature $T_c^{twin} > T_{co}$. This possibility is investigated by linearizing Eqs. (3) and (4) in η_i and finding the zeroes of the determinant associated to the ensuing homogeneous system of linear equations with

26 rows around the model twin boundary. The results are shown
in Fig. 2. Pure interfacial transitions only occur between
$x = 0.3$ and $x = 0.7$, the locations for the maxima of T_{co}.
There, interfacial and bulk transitions merge at the points
labelled Sp in Fig. 2. The minimum T_c^{twin} occurs at T_c^{twin}
$(x=0.5) = 81$ K and there are two equal maxima at T_c^{twin}
$(x= 0.5 \pm 0.17) = 100$ K.

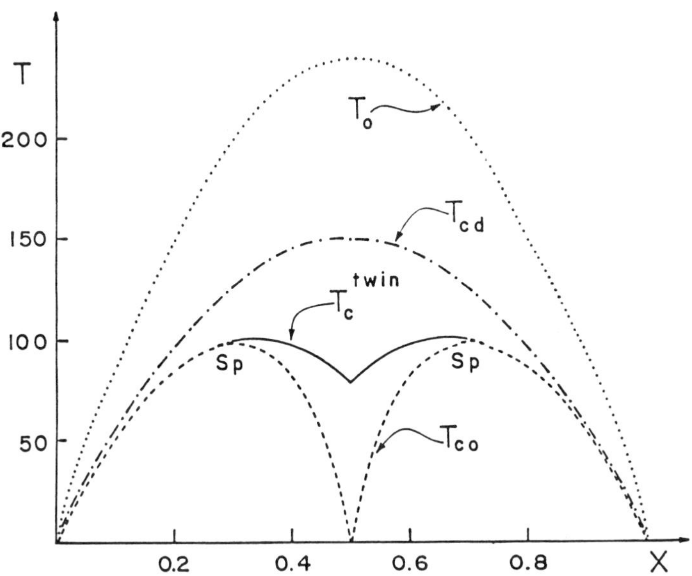

Fig. 2. Phase diagram for the model superconducting oxide.
T_o is the oxygen vacancy ordering temperature, T_{cd} is the
critical superconducting temperature when vacancy ordering
is suppressed, while T_{co} is the true critical temperature
for the perfect crystal. Disorder generated at a twin
boundary enhances the critical temperature to T_c^{twin}.

The behavior of our model superconducting oxide may be
relevant in understanding some of the observed properties
in Y-Ba-Cu-O samples. According to the phase diagram in
Fig. 2, superconductivity is depressed at very low and
very high oxygen concentration x, and the highest T_c is
known to be obtained by optimization of oxygen treatment.
Heat treatment and cooling procedure influence the abundance and spacings of twin lamellae in pure phase imperfect
crystals. Since crystallographic shear imposes upper and
lower bounds to oxygen deficiency, we believe that our

results are applicable only in a small region near $x = 0.5$. It has been found[4] that an increment of the Y/Ba ratio induces an orthorhombic to tetragonal transition where twins disappear. The onset of superconductivity is shifted to 50 K [4] which would correspond to our predicted T_{co}. Stepwise resistive transitions enhanced by the application of small magnetic field have been observed in some samples at temperatures around 90 K and 50 K [8] indicating the presence of two superconductive transitions one at T_c^{twin} and the other at T_{co}. The presence of the two transitions T_{co} and T_c^{twin} is also indicated by thin film resistivity measurements.[9] The observed sharp resistivity reductions at 150 K [10] may indicate the presence of domains where disorder of vacancies has been temporarily preserved due to the cooling history and an indication on the existence of T_{cd}. The extent of the disorder generated around the twin boundaries added to the smallness of the crystalographic domains may explain a large Meissner effect even when $T_{co} < T < T_c^{twin}$.

The possibility for a surface to form a separate phase from the bulk has been discussed by many authors[11]. In the case of a simple ferromagnet this phenomenon is induced by a surface coupling between spins different from that of the bulk. In more complex systems, with two or more coupled order parameters, inhomogeneities imposed by a wall or an antiphase boundary may produce similar interfacial critical behavior. It has been found that such is the case in ^{3}He $-$ ^{4}He mixtures in the proximity of the λ-line in contact with a wall[12] and also in the magnetic properties of the ilmenite-haematite solid solution series at symmetry-breaking antiphase boundaries.[13]

The model presented here has not taken into account the likely possibility of segregation of oxygen vacancies to the twin boundaries. Work is in progress to examine in more detail the microscopic structure of the oxygen-vacancy network near such crystal defects.

REFERENCES

1. M.K. Wu, J.R. Ashburn, C.J. Torgn, P.H. Hor, R.L. Meng,
    L.Gao, Z.J. Huang, Y.O. Wang, and C.W. Chu, Phys. Rev.
    Lett. 58: 908 (1987); P.H. Hor, L. Gao, R.L. Meng, Z.J.
    Huang, Y.Q. Wang, K. Foster, J. Vassilious, C.W. Chu,
    M.K. Wu, J.R. Ashburn, and C.J. Torgn, Phys. Rev. Lett.
    58: 911 (1987).

2. R.J. Cava, B. Batlogg, R.B. van Dover, D.W. Murphy, S.
    Sunshine, T. Siegrist, J.P. Remeika, E.A. Rietman, S.
    Zahurak, and G.P. Espinosa, Phys. Rev. Lett. 58: 1676
    (1987).

3. R.M. Hazen, L.W. Finger, R.J. Angel, C.T. Prewitt, N.L.
    Ross, H.K. Mao, C.G. Hadidiacos, P.H. Hor, R.L. Meng
    and C.W. Chu, Phys. Rev. B35: 7238 (1987).

4. Y. Syono, M. Kikuchi, K. Oh-ishi, K. Hirage, H. Arai,
    Y. Matsui, N. Kobayashi, T. Sasaoka, and Y. Muto,
    Japan J. Appl. Phys. Lett. 26: L498 (1987).

5. A. Ourmazd, J.A. Rentschler, J.C.H. Spence, M.O'Keeffe,
    R.J. Graham, D.W. Johnson Jr., W.W. Rhodes, Nature
    327: 308 (1987).

6. J.E. Greedan, A.O' Reilly and C.V. Stager, Preprint.

7. G.S. Ansell, in Physical Metalurgy, R. W. Cahn, ed.,
    (North Holland, Amsterdam, 2nd ed. 1977) p. 1184.

8. K. Murata, H. Ihara, M. Tokumoto, M. Hirabayashi, N.
    Terada, K. Senzaki, and Y. Kimmura, Japan J. Appl.
    Phys. Lett. 26: L473 (1987).

9. R.E. Somekh, M.G. Blanure, Z.H. Barber, K. Butler,
    J.H. James, G.W. Morris, E.J. Tomlinson, A.P.
    Schwarzenberger, W.M. Stobbs and J.E. Evetts, Nature
    326:    (1987).

10. K. Takita, T. Ippsohi, T. Uchino, T. Gochou and K.
    Masuda, Japan J. Appl. Phys. Lett. 26: L506 (1987).

11. K. Binder, in Phase Transitions and Critical Phenomena,
    C. Domb and J.L. Lebowitz Eds. (Academic, N.Y., 1986)
    Vol. 8, Cap. 1.

12. S. Leibler and L. Politi, Phys. Rev B29 : 1253 (1984).

13. C. Varea and A. Robledo, Phys. Rev. B36:   (1987).

SUPERCONDUCTIVITY IN $YBa_{2-x}Sr_xCu_3O_{7-y}$ AND $Y_{1-x}Sr_xCuO_{3-y}$

Yu Mei, S.M. Green, C. Jiang and H.L. Luo

Department of Electrical and Computer Engineering
University of California, San Diego
La Jolla, CA 92093

Abstract

Studies were conducted on $YBa_{2-x}Sr_xCu_3O_{7-y}$ for $0 \leqslant x \leqslant 2.0$. Single phase samples with $T_c \sim 90K$ were obtained for $x \leqslant 1.0$. $Y_{1-x}Sr_xCuO_{3-y}$ was also studied for $0.1 \leqslant x \leqslant 0.8$. Surprisingly, for $x = 0.7$, we obtained the single-phase K_2NiF_4 structure with $T_c \sim 40K$.

Introduction

The tentative announcement of possible superconductivity at $\sim 30K$ in the La-Ba-Cu-O system[1] opened a new era for superconducting materials research.[2-4] Shortly thereafter, the bulk superconducting phase was identified as $La_{2-x}M_xCuO_{4-y}$ ($x \leqslant 0.3$). The highest transition temperatures, T_c, were obtained for $x \sim 0.15$. For $M = Ba$, $T_c \sim 35K$ was obtained; however, $M = Sr$ yielded $T_c \sim 45K$, a substantial improvement.[5-9]

Thus, when the new but related class of superconductors, $YBa_2Cu_3O_{7-y}$ with $T_c \sim 95K$, was first reported,[10] the Sr substitution was attempted immediately.[11,12] The results were not encouraging because this substitution did not raise T_c. We report here the results of our experimental study of $YBa_{2-x}Sr_xCu_3O_{x-y}$ ($0 \leqslant x \leqslant 2.0$) and $Y_{1-x}Sr_xCuO_{3-y}$ ($0.1 \leqslant x \leqslant 0.8$), including the unexpected observation that $Y_{0.3}Sr_{0.7}CuO_{3-y}$ possesses the K_2NiF_4 structure and becomes superconducting at $\sim 40K$.

Experimental Procedure

Two series of compounds were prepared with the nominal compositions $YBa_{2-x}Sr_xCu_3O_{7-y}$ with $x = 0, 0.3, 1, 1.5, 2$ and $Y_{1-x}Sr_xCuO_{3-y}$ with $x = 0.1, 0.3, 0.5, 0.667, 0.7, 0.8$ as follows. Extremely fine, dried powders of Y_2O_3 (99.9%), BaO (99.999%), SrO (99.999%) and CuO (99.999%) were mixed in appropriate proportions. The mixtures were pulverized in an agate mortar and pressed into 2g blocks with dimensions $25 \times 6.5 \times 4$ mm at 30 kpsi. The blocks were heated in air at $800\,^\circ C$ for 6 hours and then were reground for at least 1 hour to assure sample homogeneity. Final sintering of the repressed blocks was performed at 920 °C for 16 hours in air or flowing O_2, then furnace-cooled to room temperature.

X-ray powder diffraction patterns were obtained using Ni-filtered CuKα radiation ($\lambda = 0.15418$ nm). For samples which were predominantly single phase, lattice parameters were calcualted by a computerized least-square fit.

The room temperature resistance of each block was measured by a quick 2-point probe across the length. Samples in the Ω, kΩ, and MΩ range will be referred to as low, high, and very high resistances, respectively. T_c measurements were performed on samples with low resistance, while samples with high or very high resistances were not analyzed further.

Resistive T_c measurements were performed by the standard 4-point technique. 40 AWG copper wires were secured with silver paint. A 0.5 mA rms current was supplied at 40 Hz with the voltage drop detected by a lock-in amplifier (PAR 5207). DC magnetic susceptibility measurements were obtained using a variable temperature SQUID magnetometer (SHE VTS-905). The Meissner effect was observed with samples cooled in a 100 Oe field.

Results and Discussion

Based on the x-ray diffraction patterns, we observe that $YBa_{2-x}Sr_xCu_3O_{7-x}$ is single-phase for $x \leqslant 1.0$ and possesses the orthorhombic-distorted perovskite structure of $YBa_2Cu_3O_{7-y}$.[11] For $x > 1.0$, the patterns display many other peaks indicating that the samples are multi-phase. All of the single-phase specimens are low resistance, while those containing multiple phases are high resistance.

In analyzing the x-ray data for the $Y_{1-x}Sr_xCuO_{3-y}$ series, we were surprised to observe the K_2NiF_4 structure emerge. However, with the exception of $x = 0.7$ and 0.8, these specimens were multi-phase and had very high resistance. (For $x = 0.667$, the resistance was high). For $x = 0.7$ and 0.8, the resistance was low. The $x = 0.7$ sample was single-phase and precisely matched the tetragonal pattern obtained for $La_{2-x}(Ba,Sr)_xCuO_{4-y}$. For $x = 0.8$, we also obtained this pattern in addition to several unidentified peaks.

The lattice parameters of the single-phase specimens from both series are listed in Table 1. Evidently, the smaller ionic radius of Sr^{2+} as compared to Ba^{2+} results in a contraction of the $YBa_{2-x}Sr_xCuO_{7-y}$ unit cell with increasing x. For $x > 1.0$, the requisite contraction is too large, and single-phase material is not obtained. The x-ray powder diffraction pattern of $Y_{0.3}Sr_{0.7}CuO_{3-y}$ is presented in Figure 1 and summarized in Table 2.

The DC susceptibility results for $YBa_{1.7}Sr_{0.3}Cu_3O_{7-y}$ and $YBaSrCu_3O_{7-y}$ obtained at H = 100 Oe are displayed in Figures 2 and 3, respectively. The T_c for both samples is ~ 90K indicating that the Sr substitution does not adversely affect the superconducting state. However, based on the strength of the diamagnetic signal, we conclude that samples annealed in O_2 contain a larger fraction of superconductor than those annealed in air.

TABLE 1. T_c Onsets and Lattice Parameters for $YBa_{2-x}Sr_xCu_3O_{7-y}$ ($x = 0$, 0.3, 1) and $Y_{0.3}Sr_{0.7}CuO_{3-y}$

Sample	T_c Onset (K)	Heating Condition	a_o(nm)	b_o(nm)	c_o(nm)
			Lattice Parameters		
$YBa_2Cu_3O_{7-y}$	92	920 °C, O_2	0.3824	0.3889	1.1675
$YBa_{1.7}Sr_{0.3}Cu_3O_{7-y}$	91	920 °C, O_2	0.3823	0.3885	1.1654
	91	920 °C, air	0.3824	0.3887	1.1657
$YBaSrCu_3O_{7-y}$	89	920 °C, O_2	0.3809	0.3875	1.1625
	88	920 °C, air	0.3817	0.3880	1.1638
$Y_{0.3}Sr_{0.7}CuO_{3-y}$	40	920 °C, air	0.3770	0.3770	1.3253

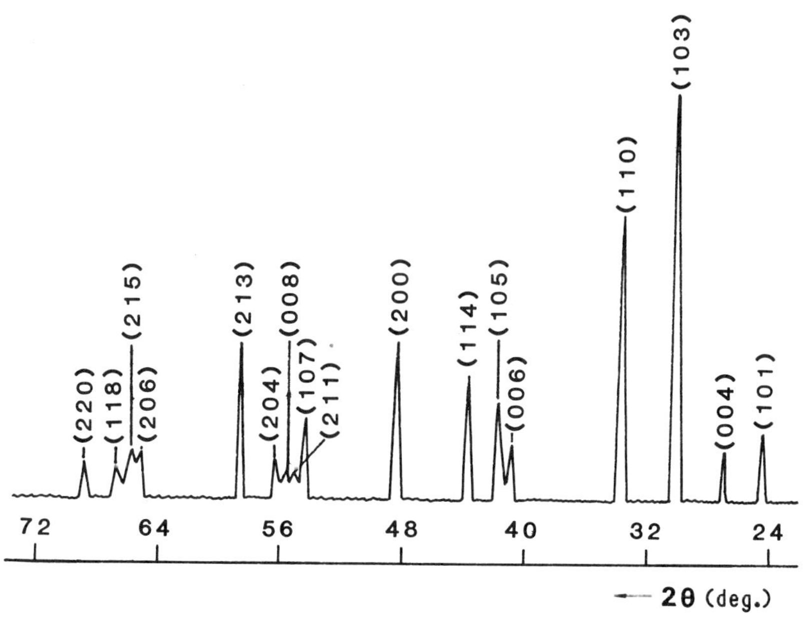

Figure 1: X-ray powder diffraction pattern of $Y_{0.3}Sr_{0.7}CuO_{3-y}$ obtained with CuKα radiation. All peaks are identified with the K_2NiF_4 structure.

TABLE 2. X-Ray Powder Diffraction Data of Single Phase $Y_{0.3}Sr_{0.7}CuO_{3-y}$

Sample: $Y_{0.3}Sr_{0.7}CuO_{3-y}$

Structure: tetragonal I4/mmm

$a_o = 0.3770$ nm, $c_o = 1.3253$ nm

Radiation: Ni-filtered $CuK\alpha$ ($\lambda = 0.15418$ nm)

2θ	h	k	l	$d_{obs.}$ (nm)	$I_{obs.}$
24.58	1	0	1	0.3622	17
26.94	0	0	4	0.3309	10
31.18	1	0	3	0.2868	100
33.63	1	1	0	0.2665	71
36.30	1	1	2	0.2475	4
40.88	0	0	6	0.2207	15
41.65	1	0	5	0.2168	24
43.59	1	1	4	0.2076	31
48.24	2	0	0	0.1886	39
53.90	1	1	6	0.1701	15
54.20	1	0	7	0.1692	17
54.84	2	1	1	0.1674	5
55.45	0	0	8	0.1657	6
56.16	2	0	4	0.1638	9
58.61	2	1	3	0.1575	38
65.05	2	0	6	0.1434	12
65.62	2	1	5	0.1423	10
66.43	1	1	8	0.1407	8
70.68	2	2	0	0.1333	8

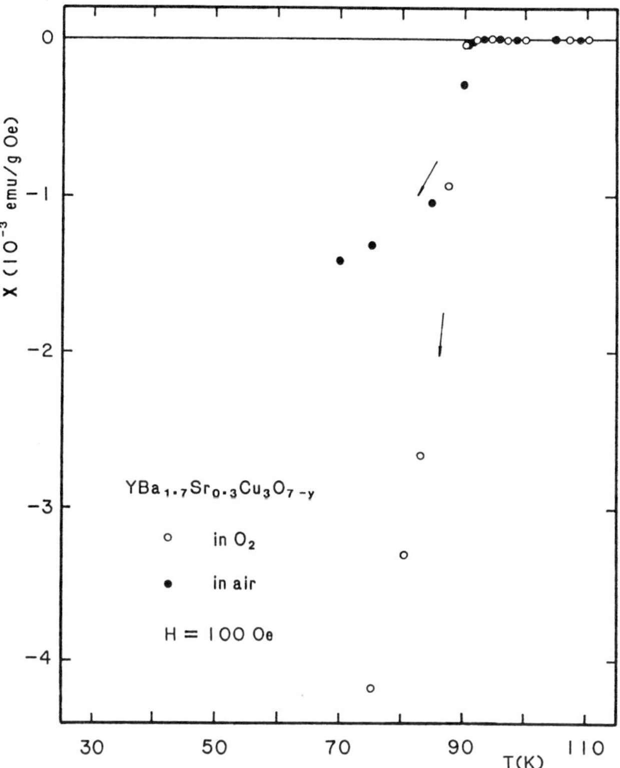

Figure 2: DC magnetic susceptibility versus temperature of $YBa_{1.7}Sr_{0.3}Cu_3O_{7-y}$.

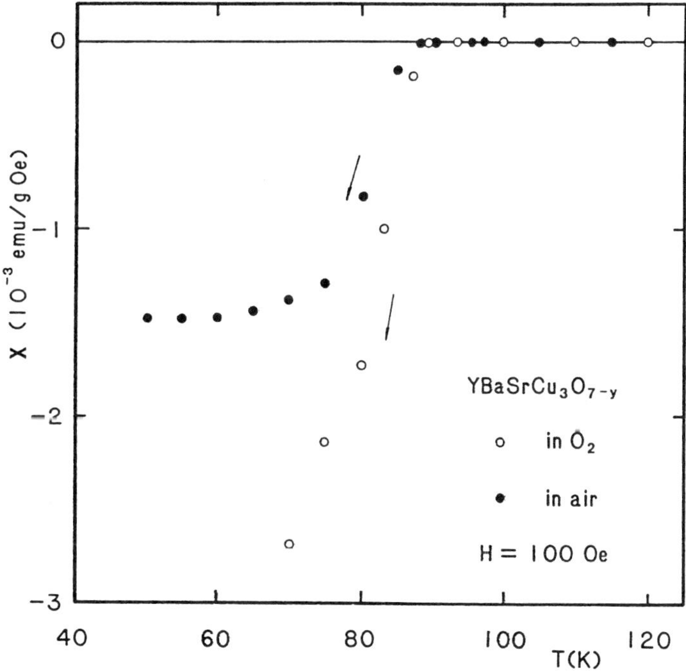

Figure 3: DC magnetic susceptibility versus temperature of $YBaSrCu_3O_{7-y}$.

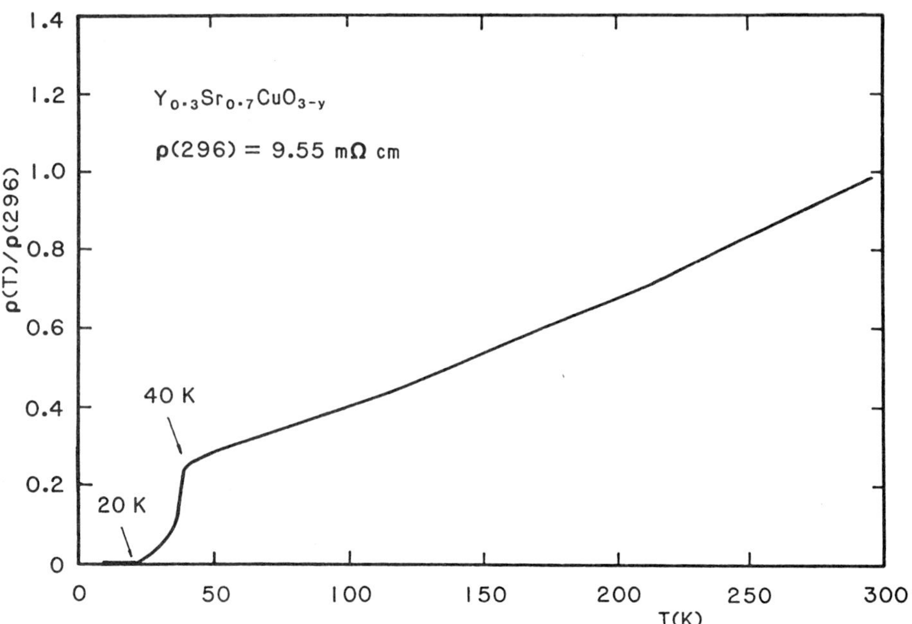

Figure 4: Resistivity versus temperature curve of $Y_{0.3}Sr_{0.7}CuO_{3-y}$.

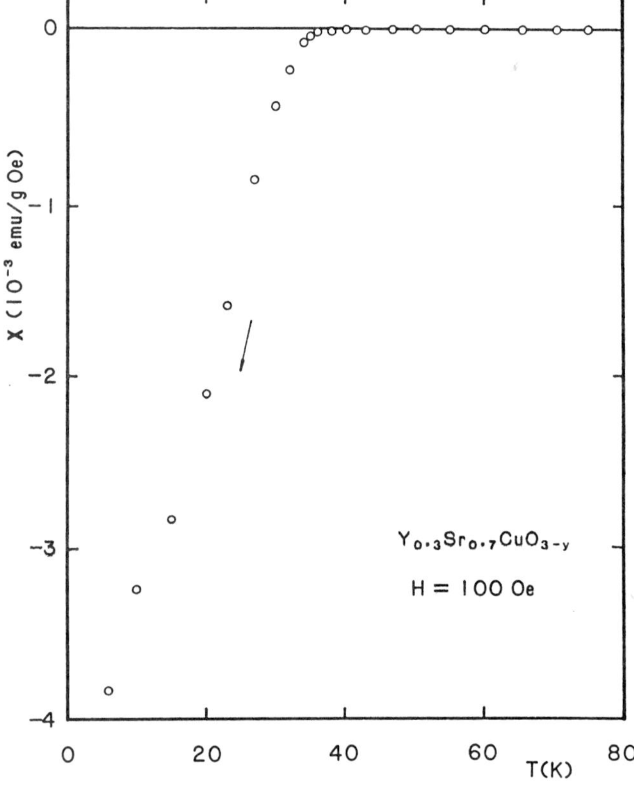

Figure 5: DC magnetic susceptibility versus temperature of $Y_{0.3}Sr_{0.7}CuO_{3-y}$.

Figure 4 displays the resistivity versus temperature curve for $Y_{0.3}Sr_{0.7}CuO_{3-y}$. The onset of superconductivity occurs at 40 K with resistance below the detection limit of 10 $\mu\Omega$ at 20K. The susceptibility curve for this sample is presented in Figure 5 and confirms the bulk superconductivity with $T_c \sim 40K$. Similar results were obtained for the multi-phase $Y_{0.2}Sr_{0.8}CuO_{3-y}$.

The striking similarity of $Y_{0.3}Sr_{0.7}CuO_{3-y}$ to $La_{2-x}(Ba,Sr)_xCuO_{4-y}$ presents some perplexing questions. When heated beyond 950°C, it becomes multi-phase and closely resembles $YSr_2Cu_3O_{7-y}$, indicating a degree of consistency in the results. It is possible that a large number of vacancies at the Y^{3+}/Sr^{2+} sites can be tolerated if the sintering temperature is not too high. Alternatively, a significant amount of CuO may not enter the lattice if the sintering temperature is low enough. Yet, we were unable to identify CuO in the x-ray pattern.

References

1. J.G. Bednorz, K.A. Müller, *Z. Phys.*, **B64**, 189 (1986).

2. C.W. Chu, P.H. Hor, R. L. Meng, L. Gao, Z.J. Huang and Y.Q. Wang, *Phys. Rev. Lett.*, **58**, 405 (1987).

3. S. Uchida, H. Takagi, K. Kitazawa and S. Tanaka, *Japan J. Appl. Phys. Lett.*, **26**, L1 (1987).

4. H. Takagi, S. Uchida, K. Kitazowa and S. Tanaka, *Japan J. Appl. Phys. Lett.*, **26**, L123 (1987).

5. J.M. Tarascon, L.H. Greene, W.R. McKinnon, G.W. Hull, T.H. Geballe, *Science*, **235**, 1373 (1987).

6. R.B. van Dover, R. Cava, B. Batlogg, E. Rietman, *Phys. Rev.*, **B35**, 5337 (1987).

7. Z.X. Zhao, L.Q. Chen, C.G. Cui, Y.Z. Huang, J.X. Liu, C.H. Chen, S.L. Li, S.G. Gao and Y.Y. He, *KEXUE TONGBAO*, No. 3 (1987).

8. H. Takagi, S. Uchida, H. Obara, K. Kishio, K. Kitazawa, K. Fueki and S. Tanaka, *Japan. J. Appl. Phys. Lett.*, **26** L434 (1987).

9. C. Politis, J. Geerk, M. Dietrich and B. Obst, *Z. Phys.*, **B66**, 141 (1987).

10. M.K. Wu, J.R. Ashburn, C.J. Torng, P.H. Hoi, R.L. Meng, L. Gao, Z.J. Huang, Y.Q. Wang and C.W. Chu, *Phys. Rev. Lett.*, **58**, 908 (1987).

11. E.M. Engler, V.Y. Lee, A.I. Nazzal, R.B. Beyers, G. Lim, P.M. Grant, S.S.P. Parkin, M.L. Ramirez, J.E. Vazquez, and R.J. Savoy, *J. Amer. Chem. Soc.* (1987).

12. J.M. Tarascon, L.H. Greene, W.R. McKinnon and G.W. Hull, *Phys. Rev. Lett.* (1987).

Cu SUBSTITUTION INTO THE La/Y SITE IN $La_{1.8}Sr_{.2}CuO_4$ AND $YBa_2Cu_3O_7$:
DETERMINATION BY X-RAY ABSORPTION SPECTROSCOPY

F. W. Lytle, R. B. Greegor and A. J. Panson*

The Boeing Company, Seattle, WA 98124
*Westinghouse R & D Center, Pittsburgh, PA 15235

INTRODUCTION

The usual techniques for investigating the structure of materials, x-ray, neutron or electron diffraction, consist of scattering
from all the material. Each diffraction peak is composed of a contribution from all the different atomic species in the sample. It is
then up to the skill and intuition of the scientist to postulate the
atomic arrangement and solve the structure. On the other hand x-ray
absorption spectroscopy has an element specific advantage in that the
x-ray beam energy can be tuned to an absorption edge of each kind of
atom in the material. The structure probe is then the quantum interference of the ejected photoelectrons as they scatter from atoms surrounding the absorbing species. This effect has been much used in
recent years by Fourier analysis of the extended x-ray absorption
fine structure (EXAFS). The region near the onset of absorption, the
x-ray absorption near edge structure (XANES) has not been as well understood but has been used as a fingerprint for structural identification.

Recently, we demonstrated that XANES can be understood in exactly the same context as EXAFS, i.e. back-scattering from nearby atoms, but with a special simplifying characteristic due to the match
of the electron wavelength with the dimensions of the crystal lattice
(1). First we develop the model and then state an axiom for XANES.

The previous work of Sette(2), Stöhr(3) and Bianconi(4) et al. identified specific features in the XANES with scattering resonances from first neighbor atoms. By selecting spectra from carefully chosen reference compounds we were able to identify and assign XANES peaks to specific bond distances, out to the 4th coordination sphere in some cases (1). An example of the data is shown in Fig. 1. The data were obtained at the Stanford Synchrotron Radiation Laboratory and details of the experimental technique have been described (1). The spectra for 1 molar MnO_4^- solution, NiO and Cu metal are shown at the bottom. The MnO_4^- anion consists of Mn^{7+} tetrahedrally coordinated by O^{-2} at 1.63 Å, any higher coordination is smeared by the water. The NaCl-type structure of NiO consists of Ni surrounded by 6 oxygen at 2.08 Å in the first coordination shell and 12 Ni at 2.95 Å in the second coordination shell. The fcc Cu metal has 12 nearest neighbors at 2.56 Å and 6 next nearest at 3.62 Å. In Fig. 1 the first one or two peaks above the edge are identified as to their origin from R_1 or R_2. Data from these and a more extensive collection of data of the same general type are collected in Table 1 and plotted in Fig. 2. This correlation can be used as a ruler (at the top of Fig. 1) for determining bond distances from XANES. To state the AXIOM: "X-ray absorption near edge structure is a scattering image, $E \propto 1/R^2$ of the lattice surrounding the absorbing atom superimposed upon the underlying, smooth transitions to the continuum". The XANES can be used as a direct measure of bond distances accurate to ± 0.1Å to atoms surrounding the absorbing atom without recourse to Fourier transforms or other data analysis. This region of the spectrum above the bound states and up to $\lambda = R_{min}$. i.e. from 10-50 eV in the spectra of Fig. 1 has no overlapping peaks from higher orders of scattering and may be viewed as a direct image of the lattice.

In Fig. 2 and table 1 the zero of energy has been designated as the 1s to 3d bound state pre-edge peak (4). The pre-edge transitions have been discussed in detail by (5, 6, 7), for example. The close correspondence of the experimental data and the plot of the free electron wavelength $\lambda = h/mv = R_n$ is noted. This identifies the scattering as $180°$ backscattering. Multiple scattering around the atoms of the first coordination shell and back to the absorbing atom can

also occur but should always be of lesser magnitude than simple
backscattering. Multiple scattering cancels out in lattices (or molecules) with inversion symmetry (8) but does occur in the MnO_4 anion
at 1.815 R_1. This has been discussed and demonstrated (1). In any
event there is a window from R_1 to approximately 1.7-2 R_1 where only
single scattering can occur. This helps in the interpretation of unknown spectra.

APPLICATION TO HIGH T_C SUPERCONDUCTORS

At the top of Fig. 1 the Cu K-edge spectra from $La_{1.8}Sr_{.2}CuO_4$
(LSC) and $YBa_2Cu_3O_7$ (YBC) are plotted. They are obviously very
similar to each other, i.e. the environment of Cu in each material
must be similar. Also note that the bond distances which cause the
XANES peaks all arise from the Cu atom and average over all the different Cu-X bond distances due to different Cu sites in the material.
With the distance scale at the top of Fig. 1 the various XANES features may be identified with specific distances which have been determined by other techniques (8, 9). Most of the expected Cu distances can be identified except for the one prominent peak at 25 eV,
which is common to both LSC and YBC. This feature corresponds to a
Cu-X distance of ~ 2.5 Å. This distance is unique to the Y or La
site and therefore a significant fraction of Cu must be in the Y/La
site. The EXAFS of Cu can also be analyzed to confirm a distance
near 2.5Å. This is done in Fig. 4. The intensity of peaks corresponding to the Cu in the Y/La site in both the XANES and the Fourier
transform of the EXAFS were used to estimate the fraction of Cu in
this site. For YBC, LSC or La_2CuO_4 (by reference to the data of
Tranquada et al.(11)) we estimate approximately 0.2 - 0.3 mole fraction of Cu in the Y/La site. Conversely, for charge balance Y/La
must substitute into the Cu site. This can be checked from a Y/La
absorption edge. Tranquada et al.(12) measured the La L_1 edges in
the $La_{2-x}Sr_xCuO_4$ series which are shown in Fig. 3. It is significant
that all the L_1 edges nearly overlap with or without Sr present, i.e.
the La environment is similar in all the compounds. Cu K-edge data
for LSC is overplotted in Fig. 3. Both the Cu 1s and La 2s transitions should observe the same selection rules. Both should show weak
transitions to unfilled d states. In Fig. 3 the Cu 3d prepeak is
aligned with the La L_1 5d prepeak. This peak is weak and difficult

to define; however, when this is done the distance scale at the top
applies and the peak attributed to Cu on the La site lines up well
with the peak assigned to the primary La-O distance of $\sim$2.6 $\overset{o}{A}$.
Equally significant some of the La has a shorter bond at $\sim$1.8 $\overset{o}{A}$, i.e.
La substituted into the Cu site. In the La L_2 and L_3 edges the transition to the empty 5d states becomes dipole allowed resulting in a
very large resonance (12). When this resonance is aligned at the position noted by "5d" in Fig. 3 the peaks at higher energy match the
energy position and peak assign ments of Fig. 3. All the Cu K and La
L spectra are consistent with these peak assignments. Data was not
available for the Y edge.

The oxygen K spectrum offers a different perspective. The energy broadening of oxygen 1s is very small and (compared to Cu) we
expect only one valence. Therefore the edges and scattering
resonances are sharp. Since metal atoms are the oxygen first neighbors many peaks are especially strong . For calibration the NiO data
of Davoli et al.(13) was used. This is shown at the top of Fig. 5
again aligning all spectra at the position of the first prepeak. The
coordination spheres in the NaCl-type lattice of NiO are indicated by
R_1 - R_5 and consist of 6 Ni, 12 O, 8 Ni, 6 O and 24 Ni, con-
secutively. The labeled verticals along the horizontal line are
scaled by $\sum N/R_2$ where N = atomic number, in a rough attempt to estimate scattering intensity. The peaks which can be clearly assigned,
R_1, R_2 and R_5 are plotted on Fig. 2 and were used as a distance scale
calibration for O K-edges i.e., the $\overset{o}{A}$ scale across the top. The
large peak labeled σ^* is a bound state transition of O 1s and appears
at approximately the same position in O_2 (14). At the bottom in Fig.
5 the O and Cu K-edges of $YBa_2Cu_3O_7$ are aligned at their prepeaks and
overplotted. The O 1s data are from Yarmoff et al.(15). Again the
verticals along the horizontal labeled line have been scaled by $\sum N/R^2$
and identified as to the origin of the scattering atom. Note the
close agreement between the O and Cu edges of the three shortest distances, i.e. Cu-O bonds viewed from different ends of the same bond.
From the perspective of the O atoms Ba is the biggest scatterer in
the lattice so these peaks are most intense. The small bumps to the
right of the σ^* peak do not fit any known distance in the lattice
and are probably due to multiple scattering. They are at approximately the distance where multiple scattering could occur. The

Table 1. Bond Distances and XANES Peak Assignments.

Sample	R_1, Å	ΔE, eV	R_2, Å	ΔE, eV
$Ti^{4+}O_4/SiO_2$	1.79	37.1		
V^5O_4/H_2O	1.74	38.7		
$Cr^{6+}O_4/H_2O$	1.68	41.5		
$Mn^{7+}O_4/H_2O$	1.63	44.3		
TiO	2.09	32.4	2.95	18.5
VO	2.05	34.4	2.89	19.0
MnO	2.22	28.2	3.14	14.3
FeO	2.16	29.7	3.05	15.1
CoO	2.13	31.9	3.02	16.4
NiO	2.08	33.0	2.95	16.6
Cu	2.56	21.6		
Ni	2.49	22.8		
Co	2.51	22.0		

Bond distances from R.W.G Wyckoff, "Crystal Structures" 2nd Ed., Interscience, New York, 1963.

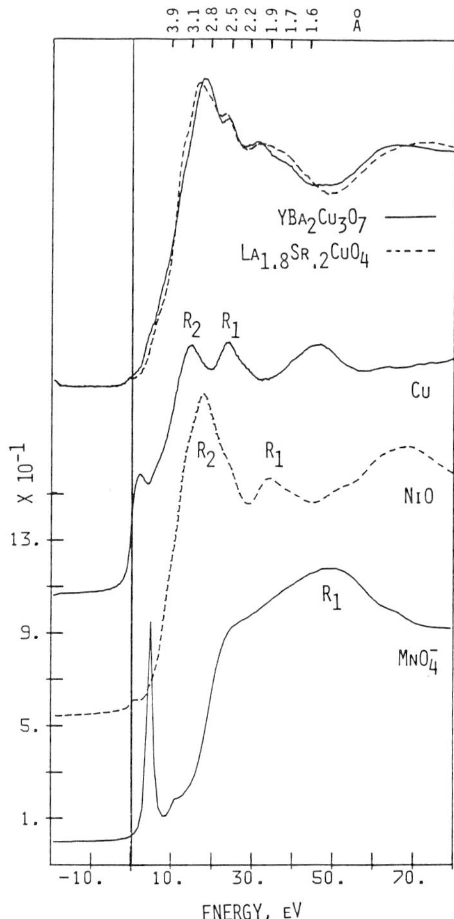

Fig. 1.   XANES spectra from KMnO$_4$ solution, NiO and Cu. Peaks labeled R$_1$ and R$_2$ are scattering resonances from the specified coordination spheres. Top, data from YBa$_2$Cu$_3$O$_7$ and La$_{1.8}$Sr$_{.2}$CuO$_4$ with a distance scale in Å derived from Fig. 2.

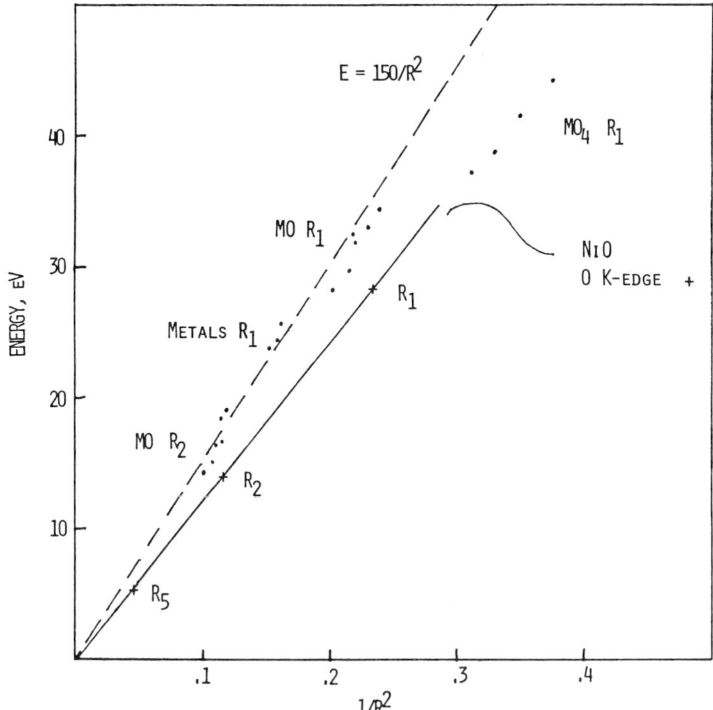

Fig. 2. Correlation of XANES peak energy and bond distance, R. The dashed line corresponds to the free electron wavelength λ=R. The solid points are from table 1, the + are O K-edge data for NiO (13) from Fig. 5.

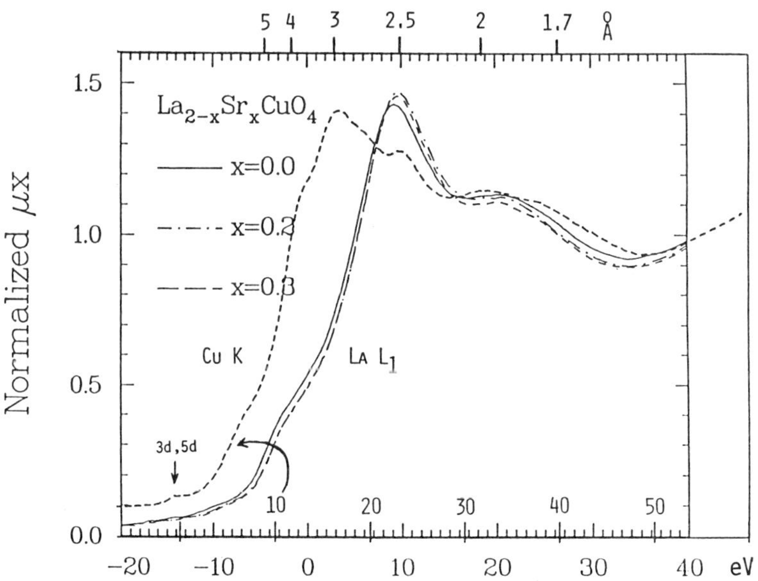

Fig. 3. Comparison of Cu K and La L_1 edges (12) for $La_{2-x}Sr_xCuO_4$. Both edges are aligned on the 3d or 5d prepeak. Note the common distances from 1.8-2.5 Å.

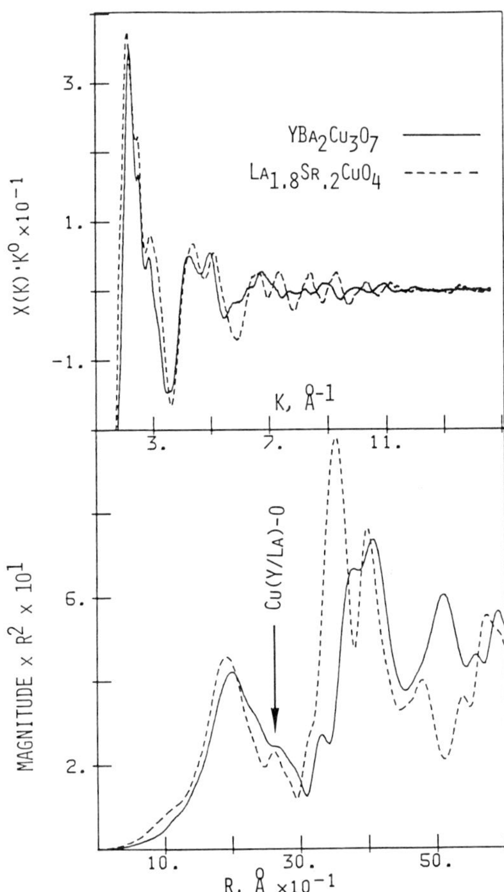

Fig. 4 The Cu K-edge EXAFS and K^0, Cu-O phase–corrected Fourier transforms
for $YBa_2Cu_3O_7$ and $La_{1.8}Sr_{.2}CuO_4$. The Cu-O peak in the La/Y site
is indicated. Multiplication of the transforms by R^2 places the
peak amplitudes in correct perspective.

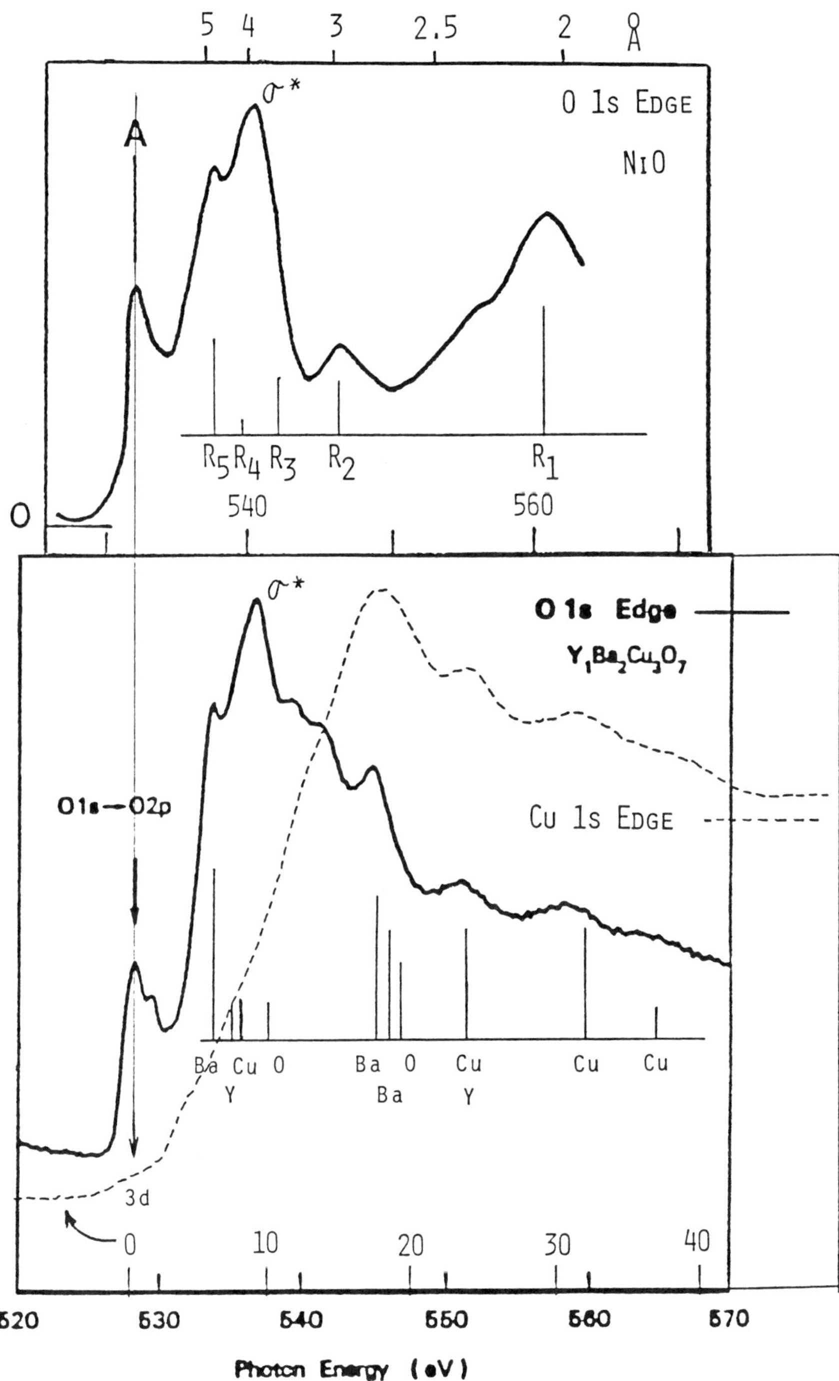

Fig. 5 Top, O K-edge spectrum from NiO (13). Near neighbor distances are indicated R_1-R_5 and used to construct the calibration of Fig. 2. Bottom, comparison of O (15) and Cu K spectra of $YBa_2Cu_3O_7$, both edges aligned on the prepeak. Where common distances occur both spectra have the same peaks. The labeled verticals estimate the position and scattered intensity.

$YBa_2Cu_3O_7$ distance information in Fig. 5 was obtained from LePage et al. (9)

DISCUSSION

Why have previous structure determinations not detected Cu in the Y or La sites? Neutron diffraction is relatively insensitive to the kind of metal atom. The neutron scattering lengths are Cu =.770, Y = .775, La = .827, Sr = .702 and O = .580. The x-ray scattering factors for the metal atoms are different from each other but not so different that the structure wouldn't refine with the metal atoms in their traditional locations. Rietveld refinements of powder diffraction data are particularly biased toward the initial lattice assignments. These materials are also complicated by a variable number of vacancies and probably a variable number of Cu to Y/La substitutions depending upon the preparation. The difficulty of the structural refinement is evident in the plethora of results all differing slightly in detail. X-ray absorption spectroscopy has an element specific advantage compared to all diffraction techniques. The environment of each kind of atom is determined separately.

The implications of Cu substitution into the Y/La site are considerable and change our concepts of these materials and how they conduct. The small Cu radius relative to Y/La makes it a loose fit in the site where it should vibrate freely, i.e. a soft mode. The disorder associated with this Cu-O distance should be very sensitive to temperature variations in contrast to the tightly-bonded shorter Cu-O distances (11, 16). Indeed the data of Boyce et al.(16) show the effect although not noted in the paper. When Cu occupies the Y/La site in either LSC or YBC a body-centered Cu sublattice is constructed which has some similarity to the A15 structure. Y/La substitution into the Cu site must be the limiting factor in how much substitution can occur and is dependent upon charge balance and lattice stability. The puckered "Cu-O" layers in the structures of LePage et al.(9) or Beno et al.(17) are the result of the larger Y ion in the planar Cu site. Also, the Cu/Y or La sublattice probably orders but that is beyond the range of our technique. The general idea of a diffusion driven substitution mechanism helps to explain some of the non-specific trends which have been observed. Since most

preparation procedures involve a solid state calcination the particle
size and degree of mixing of the constituents would be important
variables which vary from laboratory to laboratory. Also, the observation (18,19) that most of the rare earth and group IIIB elements
are equally effective in the YBC form can be explained by the fact
that the rare earth ions are about the same size and below a certain
maximum they are all equivalent for diffusing into the Cu site.

An additional requirement for superconductivity appears to be
simultaneous Cu^{2+} and Cu^{3+} valences (20). In principle this can be
determined from the X-ray absorption edge. As was noted previously
(1) the complexity of these Cu edges is such that Cu valence can't be
easily determined by comparison to other reference compounds with
different crystal structures. It would be possible to demonstrate
changing valence, in YBC for example, by comparing compunds with differing oxygen concentration (hence, different combinations of
Cu^{2+}/Cu^{3+}) to each other. Traditional chemical techniques are probably a more accurate method for valence determination for these compounds.

ACKNOWLEDGEMENT

We thank Stanford Synchrotron Radiation Laboratory for beam
time. SSRL is funded by DOE.

REFERENCES

1. F. W. Lytle, R., B. Greegor and A. J. Panson, "Discussion of
    X-ray absorption Near Edge Structure: Application to Cu in the
    High T_c Superconductors $La_{1.8}Sr_{.2}CuO_4$ and $YBa_2Cu_3O_7$", Phys.
    Rev. B (submitted).
2. F. Sette, J. Stöhr and A. P. Hitchcock, Chem. Phys. Lett. <u>110</u>,
    517 (1984).
3. J. Stöhr, F. Sette and A. L. Johnson, Phys. Rev. Lett. <u>53</u>, 1684
    (1984).
4. A. Bianconi, E. Fritsch, G. Calas and J. Petiau, Phys. Rev. <u>B32</u>,
    4292 (1985).
5. J. E. Hahn, R. A. Scott, K. O. Hodgson, S. Doniach, S. Desjarins
    and E. I. Solomon, Chem. Phys. Lett. <u>88</u>, 595 (1982).

6. F. W. Kutzler, R. A. Scott, J. M. Berg, K. O. Hodgson, S. Doniach, S. P. Cramer and C. H. Chang, J. Am. Chem. Soc. $\underline{103}$, 6083 (1981).

7. J. Wong, F. W. Lytle, R. P. Messmer and D. H. Maylotte, Phys. Rev. B$\underline{30}$, 5596 (1984).

8. J. J. Boland, S. E. Crane and J. D. Baldeschweiler, J. Chem. Phys. $\underline{77}$, 142 (1982).

9. Y. LePage, W. R. McKinnon, J. M. Tarascon, L. H. Greene, G. W. Hull and D. M. Hwang, Phys. Rev. B$\underline{35}$, 7245 (1987).

10. R. J. Cava, A. Santoro, D. W. Johnson Jr. and W. W. Rhodes, Phys. Rev. B$\underline{35}$, 6716 (1987).

11. J. M. Tranquada, S. M. Heald, A. Moodenbaugh and M. Suenaga, Phys. Rev. B$\underline{35}$, 7187 (1987).

12. J. M. Tranquada, S. M. Heald and A. R. Moodenbaugh, "X-ray Absorption Near Edge Structure Study of $La_{2-x}(BaSr)_xCuO_4$ Superconductors" Phys. Rev. B (submitted).

13. I. Davoli, M. Tomellini and M. Fanfoni, Journal de Physique, Colloque C8, Sup. No. 12, Tome 47, C8-517 (1986).

14. A. P. Hitchcock and C. E. Brion, J. Elec. Spec. and Rel. Phen. $\underline{18}$, 1 (1980).

15. J. A. Yarmoff, D. R. Clarke, W. Drube, U. V. Karlsson, A. Taleb-Ibrahimi and F. J. Himpsel, "Valence Electronic Structure of $YBa_2Cu_3O_7$" Phys. Rev. B (accepted).

16. J. B. Boyce, F. Bridges, T. Claeson, T. H. Gaballe, C. W. Chu and J. M. Tarascon, Phys. Rev. B$\underline{35}$, 7203 (1987).

17. M. A. Beno, L. Soderholm. D. W. Capone, D. G. Houks, J. D. Jorgensen, I. K. Schuller, C. U. Segre, K. Zhang and J. D. Grace, "Structure of the Single Phase High Temperature Superconductor $YBa_2Cu_3O_7$", Appl. Phys. Lett. (submitted).

18. P. H. Hor, R. L. Meng, Y. O. Wang, L. Gao, Z. J. Haung, J. Bechtold, K. Forster and C. W. Chu, Phys. Rev. Lett. $\underline{58}$ 1891 (1987).

19. D. W. Murphy, S. Sunshine, R. B. vanDover, R. J. Cava, B. Batlogg, S. M. Zahurak and L. F. Schneemeyer, Phys. Rev. Lett. $\underline{58}$, 1888 (1987).

20. J. G. Bednorz and K. A. Muller, Z. Phys. B$\underline{64}$, 189 (1986).

HALL EFFECT IN $YBa_2Cu_3O_{7-x}$ vs OXYGEN CONTENT x:
OBSERVATION OF A SHARP TRANSITION IN R_H vs. x

N.P. Ong, Z.Z. Wang and J. Clayhold

Department of Physics, Princeton University

Princcton, N.J. 08544

J.M. Tarascon, L.H. Greene, and W.R. McKinnon

Bell Communications Research, Red Bank, N.J. 07701

By changing the oxygen stoichiometry x of the superconducting perovskite $YBa_2Cu_3O_{7-x}$ we have found that a sharp transition in the Hall constant R_H occurs when the oxygen content is reduced by $\Delta x = 0.47 \pm 0.02$. At this value of Δx the oxidation of the Cu ions has been reduced by one electron per unit cell. R_H is found to be temperature dependent for all samples studied. At 100 K the carrier density corresponds to 1/6 itinerant hole per Cu ion in the $\Delta x = 0$ sample. A discussion of the electronic structure is given.

Introduction

The demonstration of superconductivity in $YBa_2Cu_3O_{7-x}$ at temperatures above 90 K by Wu et al[1] has intensified the pace of research on the family of high-T_c oxides based on the cuprates. Many mechanisms based on electron-electron interaction have been proposed[2-4] to account for the unexpectedly high temperature at which superconductivity is observed. An important step in discriminating among the mechanisms is the clarification of the normal state electronic structure of these oxides. In contrast to the earlier 40 K oxides based on La_2CuO_4 which can be "tuned"[5-9] by changing the Sr or Ba concentration, the 90 K superconductors cannot be doped in a straightforward way. However, because O is readily removed and re-introduced into $YBa_2Cu_3O_{7-x}$ the oxidation state of the metal ions (Cu) can be controlled[10]. The oxidation state of Cu strongly influences the number of carriers in the system which can be monitored by Hall measurements.

Thermogravimetric analysis

The oxygen content of the samples was changed by annealing in an atmosphere of Ar. Changes in the oxygen stoichiometry Δx (relative to the starting material) were determined by thermogravimetric analysis (TGA) which has a relative accuracy of ± 0.005. TGA was also used to determine the oxygen content of the starting material. However, the precision is poorer

Hall measurements and results

Samples with a range of Δx from 0.11 to 0.59 were used for the Hall effect studies. A sample of the starting material ($\Delta x=0$) was also studied. As in a previous work[9] we polished the samples to a thickness of between 290 and 330 μm and attached leads with Ag paint. The magnetic field B from a Bitter magnet was swept between $\pm$ 15 T in 2 minutes and the Hall voltage measured with a nanovoltmeter. Our minimum detectable Hall voltage V_H is 1 μV which corresponds to an R_H of 2 x 10^{-10} m^3/C with an excitation current I between 50 and 100 mA. Contact noise was a major factor in degrading the sensitivity. The samples were thermally anchored onto a large copper substrate, laminated to reduce transient eddy current heating. The temperature T was monitored (at zero field) by a platinum sensor while a capacitance thermometer was used for regulation at high fields. Both sensors are buried in the copper substrate. Liquid N_2 is used to cool the space around the sample chamber. The thermal stability was $\pm$ 0.2 K during a sweep. In samples with weak signals we measured V_H at three or four values of I. The Hall constant is considered to be observable only if V_H scales linearly with I and B.

Figure 1 shows the temperature dependence of R_H for samples with $\Delta x=0$, 0.11, 0.43, 0.51, 0.59 and 0.66. On general grounds the removal of O reduces the oxidation state of the metal ions, thus raising the Fermi level ε_F of the carriers. Because the carriers are holes this implies that R_H increases with increasing Δx, as indicated by the data in Fig. 1. Note that in all samples R_H increases with decreasing T. Even the samples with $\Delta x =0$ and 0.11 show a pronounced increase in R_H (by a factor of 5 and 3 respectively) as T decreases from 290 K to 90 K. This variation with T indicates a T-dependent carrier concentration if only one species of carriers is assumed to be present. However, if carriers exist in two different sheets of the Fermi Surface (FS) whose inelastic scattering rates vary with T in markedly different ways it is possible to account for the T dependence. From the conductivity σ of the $\Delta x=0$ sample at 100 K [~ 1,000 $(\Omega cm)^{-1}$] we calculate the Hall mobility ($=R_H \sigma$) to be 1 to 2 cm^2/Vs. Such low values are also found in the 40 K superconductors. The T-dependence complicates the direct interpretation of R_H in terms of carrier concentration. However, taking $1/(R_H e)$ (where e is the electronic charge) as an effective carrier density n_{eff} we find that n_{eff} equals (1.5 ± 0.1) x 10^{22} cm^{-3} at 290 K for the $\Delta x=0$ sample. As T decreases to 100K, n_{eff} steadily decreases to (3.1 ± 0.3) x 10^{21} cm^3 or 1/2 carrier/cell. This corresponds to a rather low density of carriers at the superconducting transition. Interestingly n_{eff} normalised to one Cu ion is rather similar in both $YBa_2Cu_3O_{7-x}$ and $La_{2-x}Sr_xCuO_4$ at x=0.15 (1/6 and 1/7 respectively.) For the $\Delta x = 0.11$ sample R_H increases by a factor of 3 between 290 K and 100 K. At 100 K going from the starting material to $\Delta x = 0.11$ increases R_H somewhat more than than expected from a simple stoichiometric count. The removal of 0.11 oxygen atoms/cell removes 0.22 holes. Instead n_{eff} /cell is observed to decrease by a factor of 3 at 90 K from 0.5 to 0.16.

A sharp transition in R_H vs. x

As we remove even more oxygen (increasing Δx) two interesting features appear. In Fig. 2 we report the variation of R_H vs. Δx for 6 samples at 95 K. Also shown is the variation of R_H at room temperature. As Δx increases from 0, R_H increases by a factor of 3 as previously noted. On further increase R_H remains *pinned* near the value 4.5 x 10^{-9} m^3/C (corresponding because of uncertainties in the end-products. (A full report of the TGA work appears elsewhere[10].) We estimate the stoichiometry to be 7.0 ± 0.1 per unit formula for the starting material. $YBa_2Cu_3O_{7-x}$ is orthorhombic (a<b) when the oxygen stoichiometry is 7. We have found that vacuum annealing at 420°C reduces the stoichiometry by 0.53 and converts the structure to a tetragonal symmetry (a=b.) Gallagher et al (preprint) have found in their samples that a sharp orthorhombic-to-tetragonal transition occurs when the oxygen content equals 6.5. From resistivity measurements we have found that T_c decreases with increasing Δx. The superconducting transition becomes very broad when Δx exceeds 0.4.

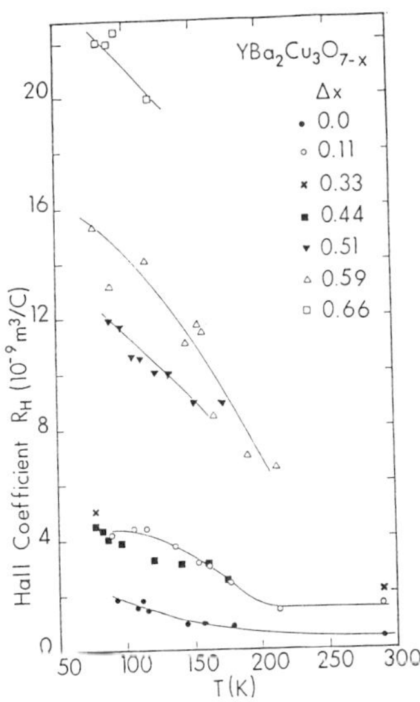

Fig.1. Variation of the Hall coefficient R_H with temperature for samples of $YBa_2Cu_3O_{7-x}$ with $\Delta x = 0.0, 0.11, 0.44, 0.51, 0.59$ and 0.66. R_H is T dependent in all samples. Note that for T below 100 K R_H increases rapidly as Δx exceeds 0.43. The data were obtained with a dc current between 50 and 100 mA in a field of 15 T. Solid lines are drawn to guide the eye.

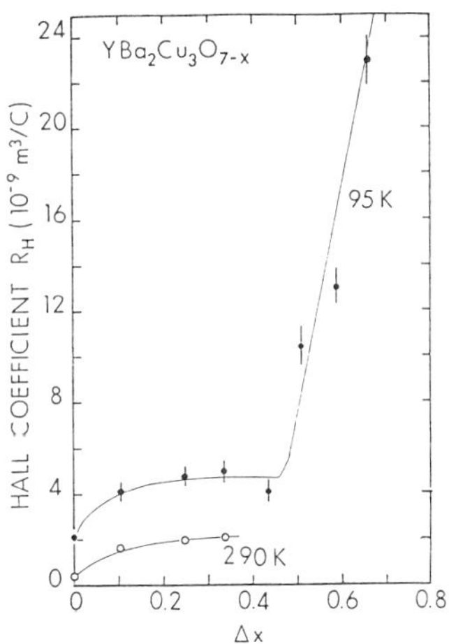

Fig. 2 Variation of the Hall coefficient R_H in $YBa_2Cu_3O_{7-x}$ versus Δx at fixed tem perature. The upper curve is taken at 95 K. R_H is pinned at the value 4.5×10^{-9} m³/C (corresponding to a density/cell of 0.24) until Δx exceeds 0.47 ± 0.02. For larger Δx, R_H increases rapidly as Δx exceeds Δx_c. The data at 290 K are for samples in which V_H can be observed. (Solid lines are drawn to guide the eye.)

$La_{2-x}Sr_xCuO_4$ with x. Two main features of our results are the linear variation of $1/R_H$ vs. x when x < 0.15. This linear variation is equivalent to 1 carrier/Sr dopant. For x in the range 0.15 to 0.2 the Hall constant R_H decreases by a factor of 30, becoming undetectable for x > 0.2. On the basis of the data we argued that the original gap in the undoped compound La_2CuO_4 is unaffected by doping. As the chemical potential decreases with increasing Sr concentration the gap persists so that the volume of hole states in k-space scales linearly with x. This scenario is in conflict with the existence or destruction of a Peierls state in the undoped compound. However, it is consistent with a Hubbard system[2,11] in which the electron-electron interaction energy U is substantial.

In the x=0 compound stoichiometry requires the valency of the Cu ions to be 2+ (configuration $(3d)^9$.) Thus there is one hole per Cu site. This "half-filled" case is an antiferromagnetic[11] (AF) state. The large intrasite repulsion keeps the holes localised so that the resistivity is thermally activated with an energy gap U-4t where t is the hopping term. The introduction of holes leads to current transport at low T because the injected holes can hop from site to site. However, because of strong correlation the hole mobility is low[11]. The Hall data for $La_{2-x}Sr_xCuO_4$ are consistent with this scenario. The number of itinerant holes is equal to the number of holes introduced by the dopants. The holes have mobilities of the order 1 cm^2/Vs and mean free paths the order of a lattice spacing (or less.) Light doping[6] also destroys the AF state. Anderson, Baskaran and Zou[12] have considered this unusual sensitivity of the AF state to light doping as strong evidence for the resonating valence bond (RVB) state which is a spin-liquid state. Thus for x > 0.02 the holes introduced by the Sr dopants are fully itinerant and it is their density that is measured by the Hall effect. For the subsequent discussion on $YBa_2Cu_3O_{7-x}$ we note that the "half-filled" case, when all Cu ions are in the $(3d)^9$ state, the system is insulating. This is the case in stoichiometric La_2CuO_4. Doping with divalent ions or (by La vacancies) introduces charge carriers which are hole-like. These may be viewed as Cu^{3+} (or $[Cu-O]^+$)[13] ions (although the holes are clearly not localised on any particular site.)

The rapid decrease in R_H below x=0.2 has not been satisfactorily accounted for. Because it coincides with the orthorhombic to tetragonal (O-T) transition at 77 K it is tempting to relate the drop in R_H to a change in electronic structure induced by the O-T transition. Recently Shafer, Penney and Olson[13] have proposed that compensation from oxygen vacancies occurs when x exceeds 0.15 in $La_{2-x}Sr_xCuO_4$.

Discussion of the electronic structure in $YBa_2Cu_3O_{7-x}$

What is the situation in the $YBa_2Cu_3O_{7-x}$ (and related) compounds ? The neutron scattering data of Greedan, O'Reilly and Stager[14] indicate that in the as-grown compound the O vacancies order in chains parallel to the b-axis. They are restricted to the middle Cu plane between the two Ba ions. This implies that the Cu atoms in the middle plane are inequivalent to the other two sites. Since the Y and Ba atoms are fully ionised an oxygen stoichiometry of 7 requires the Cu ions to have an average valency of +2.33. Therefore we expect that there exists one itinerant hole per unit cell in the x=0 case. This hole is shared among the 3 Cu ions. From the viewpoint of the Hubbard model the system is metallic since the chemical potential lies below the half-filled level. As the oxygen content is reduced (chemical potential raised), the hole density drops to zero when the average Cu valency is decreased to 2.0+. This clearly requires adding one electron per cell or the reduction of the O stoichiometry to 6.5. The system then becomes insulating.

The Hall data provide some intriguing support for this picture although there exist some to n_{eff}/cell = 0.24) until Δx exceeds a critical value Δx_c which we estimate to be 0.47 ± 0.02 whereupon R_H increases rapidly with $\Delta x - \Delta x_c$. The pinned behavior below Δx_c and the rapid increase above Δx_c indicate a sharp transition which affects the electrons near the Fermi Surface. The observed transition is tantalizingly close to the oxygen content 6.5 at which the O-T transition is observed. However, we do not have enough information to pin this down with certainty. Work on this point is in progress.

The most interesting feature of Fig. 2 is that the value of Δx_c corresponds closely to reducing the oxidation state by 1 electron/cell. This suggests that the transition in R_H occurs when the average valence state changes from $Cu^{2.33+}$ to Cu^{2+}. In other words stoichiometries in which all the Cu ions are in 2+ or lower states of oxidation are found to have a sharply reduced itinerant carrier density in the normal state. A more detailed discussion of this point is given below.

Electronic structure of $La_{2-x}Sr_xCuO_4$

In an earlier work[9] we measured the variation of the carrier density at 77 K in difficulties as well. First, Fig. 2 shows that the transition in R_H occurs when the oxygen stoichiometry is reduced by 0.47. This is in accord with the scenario sketched above, i.e. raising the chemical potential by adding one electron/cell drives the system "insulating." If the transition is indeed caused by raising the chemical potential into the Hubbard gap then the relationship between the electronic structures in $La_{2-x}Sr_xCuO_4$ and $YBa_2Cu_3O_{7-x}$ is very close. In the first case the pure compound has to be doped to move it away from half-filled. The creation of itinerant holes leads to superconductivity[2]. In the second case the as-grown compound with seven O atoms/cell already has 1 hole per 6 Cu ions (at 100K.) Removing the itinerant holes drives the system through a metal-insulator (MI) transition.

In our starting material of $YBa_2Cu_3O_{7-x}$ we measure x = 0.0 $\pm$ 0.1. Taken literally this implies that our transition in R_H occurs at 6.5. It should be noted, however, that x in the starting material is difficult to determine with precision. Other rare-earth substituted compounds in the as-grown state have x ranging from +0.2 to -0.16. It may well be that the starting x is somewhat smaller than measured. Clearly more work is required to pin down with confidence the value of x when the transition in R_H occurs. A related difficulty is that the annealed samples are never truly semiconducting (unless the content drops to 6.) In our most heavily depleted sample ($\Delta x = 0.66$) neither the resistace nor R_H shows any sign of activated behavior. Thus the transition observed at $\Delta x = 0.5$ is from a high density metal to one with much smaller densities. The question of sample inhomogeneity clearly complicates the issue here, since the current always seeks the most highly conductive path. However, this is less of a problem in Hall measurements than in resistance studies.

A second complication is the occurence of the O-T transition. In the insulating phase $YBa_2Cu_3O_{7-x}$ is tetragonal. Does the O-T transition cause the observed transition in R_H through a simple band-structure effect, or does the electronic transition seen in R_H drive the O-T transition ? These are questions that also occur in the $La_{2-x}Sr_xCuO_4$ case. If the second case is correct then the tetragonal state may be the phase which stabilizes the insulating state. (As Δx exceeds 0.5 the system remains insulating. In the Hubbard model with non-overlapping bands this is unexpected since the chemical potential is now in the upper Hubbard band.) Thirdly, the pinning of n_{eff}/cell to the value 0.24 for Δx between 0.11 and 0.5 is also a surprise. This may indicate that our arguments based on direct transfer of electrons to the Cu ions are too simple. Instead of a linear increase of R_H vs. Δx we observe an R_H which rises rapidly, and then becomes pinned until Δx exceeds Δx_c. The system appears to compensate for a decreasing oxygen content by maintaining a constant n_{eff} until Δx_c is exceeded.

Finally we comment on the T-dependence of R_H at $\Delta x=0$ and 0.11. A probable reason for the large variation with T is that the carriers are distributed between two inequivalent bands in k-space. Band-structure results[16] (with U=0) indicate that the FS is made up of sheets with 1d and 2d dispersion. If the inelastic scattering rates vs. T are very different in the two sheets then R_H (in the Drude model) is expected to be T-dependent. In a large-U model in which Cu ions coexist in both 1d and 2d lattices it is likely that a T-dependent R_H obtains as well. However, we are not aware of calculations which address this point.

Comments by P.W. Anderson and G. Baskaran were very helpful. This research was done at the National Magnet Laboratory (which is a facility supported by the National Science Foundation.) The help of Bruce Brandt at the NML was invaluable.

References

[1] M.K. Wu, J.R. Ashburn, C.J. Torng, P.H. Hor, R.L. Meng, L. Gao, Z.J. Huang, Y.Q. Wang, and C.W. Chu, Phys. Rev. Lett. **58**, 908 (1987).

[2] P.W. Anderson, Science **235**, 1196 (1987).

[3] C.M. Varma, S. Schmitt-Rink, and E. Abrahams, Solid State Commun. (submitted.)

[4] P.A. Lee and N. Read, preprint.

[5] J.M. Tarascon, L.H. Greene, W.R. McKinnon, G.W. Hull and T.H. Geballe, Science **235**, 1373 (1987).

[6] T. Fujita, Y. Aoki, Y. Maeno, J.Sakurai, H. Fukuba, and H. Fujii, Jpn. J. Appl. Phys. **26**, L368 (1987).

[7] R.M. Fleming, B.Batlogg, R.J. Cava, and E.A. Rietman, Phys. Rev. B **35**, 7191 (1987).

[8] S. Uchida, H. Takagi, H. Ishii, H. Eisaki, T. Yabe, S. Tajima, and S. Tanaka, Jpn. J. Appl. Phys. Pt. 2, **26**, L445 (1987).

[9] N.P. Ong, Z.Z. Wang, J. Clayhold, J.M. Tarascon, L.H. Greene, and W.R. McKinnon, Phys. Rev. B **35**, xxx (1987).

[10] J.M. Tarascon, W.R. McKinnon, L.H. Greene, G.W. Hull, and E.M. Vogel, Phys. Rev. B, in press; W.R. McKinnon et al, unpublished.

[11] N.F. Mott in <u>Metal-Insulator Transitions</u> (Taylor and Francis, London 1974); David Adler in <u>Solid State Physics</u> **21**, 1 (1968), edited by Seitz, Turnbull and Ehrenreich (Academic Press, N.Y.)

[12] P.W. Anderson, G. Baskaran, and Z. Zou, Phys. Rev. B, submitted.

[13] M.W. Shafer, T. Penney, and B.L. Olson, preprint.

[14] J.E. Greedan, A.O'Reilly and C.V. Stager, Phys. Rev. Lett. (submitted.)

[15] Y. Yamaguchi, H. Yamauchi, M. Ohashi, H. Yamamoto, N.Shimoda, M. Kikuchi, and Y. Syono, Jpn. J. Appl. Phys., Pt. 2, **26**, L311 (1987).

[16] S. Massidda, Jaejun Yu, A.J. Freeman, and D.D. Koelling, Phys. Lett., to appear.

OXYGEN STOICHIOMETRY OF $YBa_2Cu_3O_{7-x}$

H. Steinfink, J. S. Swinnea, A. Manthiram,
Z. T. Sui, and J. B. Goodenough
Center for Materials Science & Engineering
The University of Texas at Austin
Austin, TX 78712

It is evident that the promise of the new high
temperature ceramic superconductors will not be realized
without an intense materials science and engineering
effort focused on converting these oxide powders into
useful bulk forms. The general limitations of the
conventional press and sinter ceramic processing approach
(1) are likely to be amplified by the sensitivity of the
superconductivity transitions of $YBa_2Cu_3O_{7-x}$ to oxygen
content (2). The crystal structure of the superconducting
phase has been determined from Rietveld refinement of
powder neutron diffraction data (3-6) and we have
determined the tetragonal crystal structure of $YBa_2Cu_3O_6$
from single crystal x-ray diffraction data (7). Fig. 1
illustrates the two structures and Figs. 2 (a, b) show the
respective calculated x-ray powder diffraction patterns.
We report here the extent of the orthorhombic and
tetragonal phase fields as a function of the oxygen
composition.

$YBa_2Cu_3O_{7-x}$ was prepared by mixing Y_2O_3, $BaCO_3$ and
CuO, and firing at 950 °C in air. The product was
examined by powder x-ray diffraction techniques using a
Philips diffractometer fitted with a diffracted beam
monochromator and CuK_α radiation and was judged to be the
superconducting, single-phase, orthorhombic material on
the basis of the x-ray diffraction pattern. This material
was subjected to annealing procedures in air, oxygen,
argon and nitrogen at various temperatures. The oxygen
content of each specimen was determined by the iodometric
analysis of the oxidation state of Cu. The average
oxidation state, and hence the total oxygen content, was
determined by dissolving the sample in dilute HCl
containing KI. The liberated I_2 was titrated with

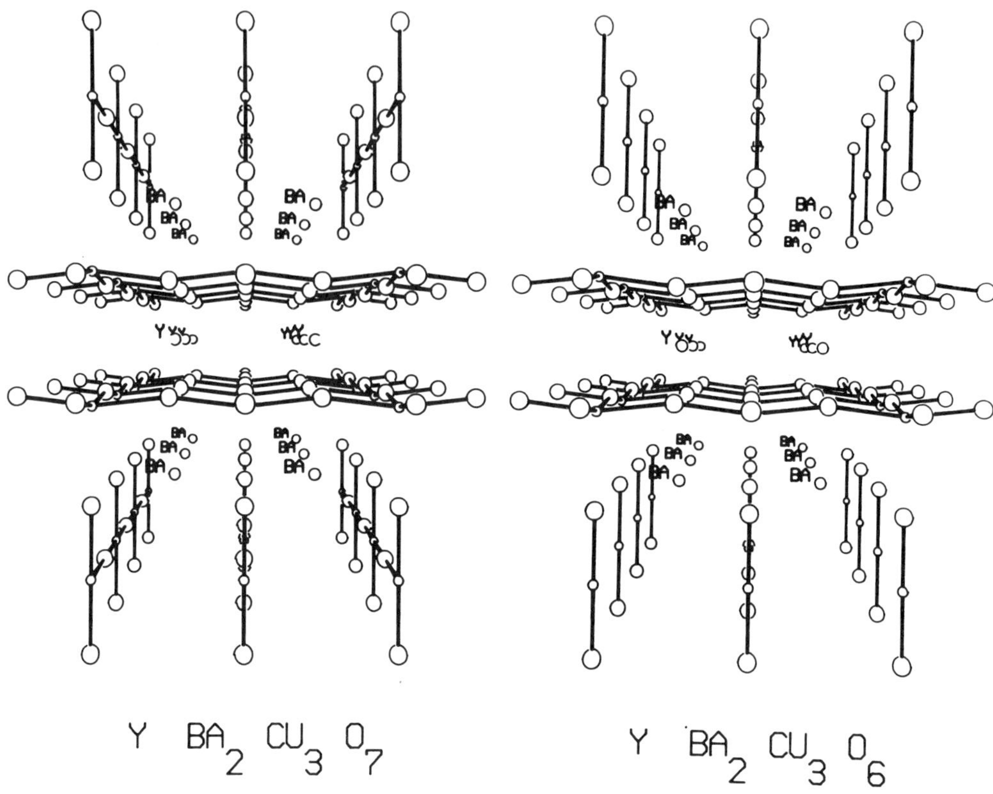

Fig. 1. The crystal structures of $YBa_2Cu_3O_{7-x}$. The a-axes are horizontal and the c-axes vertical.

standardized sodium thiosulfate using starch as the end-point indicator.

The specimens with known oxygen content were used in TGA experiments. Heating and cooling at 1 °C were carried out in O_2 and N_2, Figs. 3 (a-d). The results show that the orthorhombic phase field extends over the oxygen composition range 7.00-6.60(5) and the tetragonal phase field over the range 6.00-6.12(3). Compositions with nominal oxygen contents between 6.12-6.60 are mixtures of the orthorhombic and tetragonal phases (4). Fig. 2 (c) illustrates the x-ray powder diffraction pattern of a two-phase material with the iodometrically determined oxygen composition 6.39.

We gratefully acknowledge the research support by NSF Grant DMR 8520028 and the R. A. Welch Foundation, Houston, Texas.

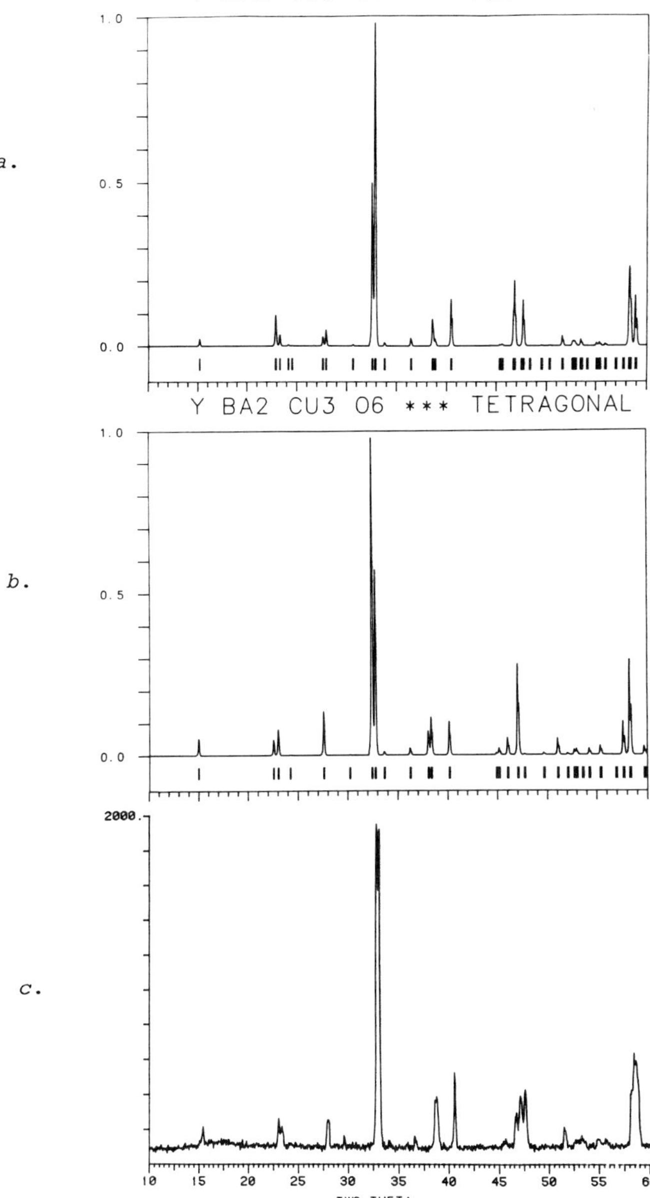

Fig. 2. Powder x-ray diffraction patterns, CuKα: (a) YBa$_2$Cu$_3$O$_7$, caculated; (b) YBa$_2$Cu$_3$O$_6$, calculated; (c) YBa$_2$Cu$_3$O$_{6.39}$, experimental, showing the presence of both phases.

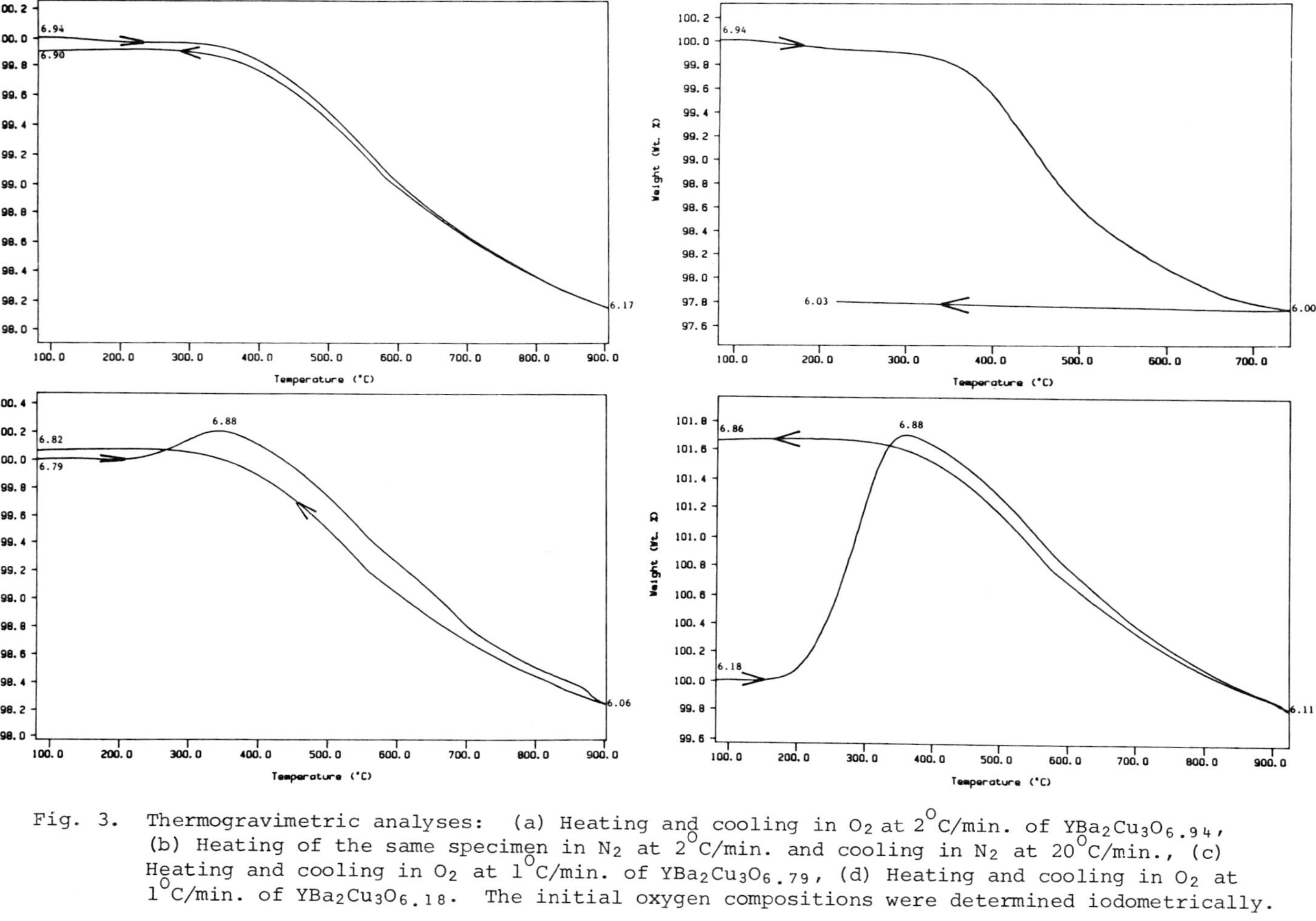

Fig. 3. Thermogravimetric analyses: (a) Heating and cooling in O_2 at 2°C/min. of $YBa_2Cu_3O_{6.94}$, (b) Heating of the same specimen in N_2 at 2°C/min. and cooling in N_2 at 20°C/min., (c) Heating and cooling in O_2 at 1°C/min. of $YBa_2Cu_3O_{6.79}$, (d) Heating and cooling in O_2 at 1°C/min. of $YBa_2Cu_3O_{6.18}$. The initial oxygen compositions were determined iodometrically.

REFERENCES

1.   M. Paulus, "The Influence of Powder Synthesis
      Techniques on Processes Occurring During Compact
      Formation and Its Sintering" in _Emergent Process
      Methods for High-Technology Ceramics_,
      eds. R. F. Davis, H. Palmour III, and
      R. L. Porter, Plenum Press, New York (1984)
      pp. 177-192.
2.   J. M. Tarascon, W. R. McKinnon, L. H. Greene,
      G. W. Hull, B. G. Bagley, E. M. Vogel, and
      Y. LePage. Proceedings of the Symposium "High
      Temperature Superconductors with T_c over
      30 K", Materials Research Society, 1987.
      In press.
3.   D. E. Cox, A. R. Moodenbaugh, J. J. Hurst and R. H.
      Jones. J. Phys. Chem. Solids, 1987. In press.
4.   J. J. Cappone, C. Chaillout, A. W. Hewat, P. Lejay, M.
      Marezio, N. Nguyen, B. Raveau, J. Soubeyroux, J. L.
      Tholence and R. Tournier. Europhs. Lett., 1987. In
      press.
5.   M. A. Beno, L. Soderholm, D. W. Capone, D. G. Hinks, J.
      D. Jorgensen, I. K. Schuller, C. U. Segre, K. Zhang
      and J. D. Grace. Appl. Phys. Lett., 1987. In
      press.
6.   F. Beech, S. Miraglia, A. Santoro and R. S. Roth.
      Phys. Rev. Lett., 1987. In press.
7.   J. S. Swinnea and H. Steinfink, J. Materials Research,
      July/August, 1987. In press.
8.   A. Manthiram, J. S. Swinnea, Z. T. Sui, H. Steinfink,
      and J. B. Goodenough, J. American Chemical
      Society, 1987. In press.

SUBSTITUTION FOR COPPER IN HIGH-T_c OXIDES

Yoshiteru Maeno and Toshizo Fujita

Department of Physics, Faculty of Science
Hiroshima University, Hiroshima 730, Japan

1. INTRODUCTION

The crystal structures of recently-discovered superconductive oxides
with high transition temperatures are all related to perovskite. Among
the superconductive oxides known previously, $BaPb_{1-x}Bi_xO_3$(BPBO) is a
perovskite and exhibits a high transition temperature, $T_c \approx 12$ K, for $x \approx$
0.3.[1] Strongly hybridized 6s(Pb/Bi) and 2p(O) electrons provide three-
dimensional (3D) conduction leading to superconductivity. The Pb site at
the center of the oxygen octahedra contains a large degree of randomness
due to Bi ions.

The 30 K superconductor, $(La_{1-x}Ba_x)_2CuO_{4-\delta}$, is in a layered perovskite
K_2NiF_4 structure.[2,3] Hybridized 3d(Cu) and 2p(O) electrons are believed
to carry superconductivity. The Cu site is at the center of the oxygen
octahedra which are connected in 2D. In contrast to the Pb site in BPBO,
the Cu site contains essentially no randomness. However, introduction of
randomness (Ba) into the La-frame, which is necessary to provide
conduction to the system, would induce random distribution of oxygen
vacancies in the 2D plane.

In the 90 K superconductor, $YBa_2Cu_3O_{7-\delta}$, ordering of Y and Ba is
responsible for the long unit cell tripled along the c-axis.[4] The
ordering of Y and Ba implies that there are two distinctive sites for Cu:
one between a pair of the Ba layers (Cu1) and the other between the Ba
and Y layers (Cu2). The strong anisotropy[5] of the occupation factors of
oxygen in the Cu1 layer is what induces the orthorhombic deformation of
the crystal structure. As a result, 1D Cu1-O chains are formed in the Cu1
plane. On the other hand, Cu2 at the base of the oxygen pyramid
constitutes 2D Cu2-O sheets.

To our knowledge, it has not been definitively made clear yet which of
the two Cu sites is essential for the occurance of the high-T_c
superconductivity in $YBa_2Cu_3O_{7-\delta}$. In other words, it is not known whether
the dimensionality (2D or 1D) or the reduced randomness intrinsic to the
structure is crucial to the high-T_c superconductivity in $YBa_2Cu_3O_{7-\delta}$.
There have been experimental findings which are used to speculate in
favor of 1D superconductivity along Cu1-O chains. Complete replacement of

Y, which is located adjacent to the Cu2-O sheet, by various magnetic rare-earth (RE) ions leaves T_c essentially unchanged.[6] Partial substitution of Sr at the Ba site, $Y(Ba_{1-x}Sr_x)_2CuO_{7-\delta}$, with the Sr concentration x up to 0.5 causes T_c to decrease by about 10 K.[7] Speculations based on these facts rely on indirect effects on the Cu sites. To discriminate which of the two Cu sites mainly carries superconducting electrons, however, we believe it necessary to investigate the effects of direct substitution for Cu sites. In this paper, we will review the results[8,9] obtained at Hiroshima University of substitution for copper in $YBa_2Cu_3O_{7-\delta}$ and $(La_{1-x}Ba_x)_2CuO_{4-\delta}$.

2. SUBSTITUTION FOR COPPER SITES IN $YBa_2Cu_3O_{7-\delta}$

It is of crucial importance to determine whether any metal is indeed introduced into the Cu sites in the oxygen-deficient perovskite structure. To make this point clear, we will first outline in this section the preparation and characterization of our samples $YBa_2(Cu_{1-x}Ni_x)_3O_{7-\delta}$,[8] and then will describe the results of resistivity mesurements.

Ceramic samples were prepared by solid-state reaction. Appropriate amounts of Y_2O_3, $BaCO_3$, CuO and NiO powders with nominal cation ratios Y:Ba:Cu:Ni = 1:2:3(1-x):3x are mixed rigorously, pressed into pellets, and heated in air at various temperatures between 880 and 1000°C for 14~27 hours. We performed electron probe microanalysis (EPMA) in various spots of a few μm^2 on a number of microcrystals and determined x, the actual content of Ni in the crystals of $YBa_2(Cu_{1-x}Ni_x)_3O_{7-\delta}$, which is plotted in Fig. 1 against the nomminal compositon x_N. For samples prepared at relatively high temperature, crystals contained in the sample show very uniform cation ratios of Y:Ba:(Cu,Ni) close to 1:2:3. For example, for x_N = 0.25, Y:Ba:Cu:Ni = 0.98±0.03 : 2.00±0.02 : 2.76 : 0.24+0.02/-0.01 (the ratio is normalized such that the sum of the Cu and Ni contents gives 3.00), indicating x = 0.08. We did not detect any crystal of $YBa_2(Cu,Ni)_3O_{7-\delta}$ without Ni content. EPMA and X-ray diffraction analysis indicate that samples prepared at relatively high temperatures are multi-phased, but with uniform cation ratios, including Ni content, in the crystals of $YBa_2(Cu,Ni)_3O_{7-\delta}$. On the other hand, samples prepared at relatively low temperature show X-ray diffraction peaks all indexed by those for $YBa_2Cu_3O_{7-\delta}$-type structure, but with premature growth of crystals which contain Ni. The information obtained is used to optimize the heat-treatment process, as described in the next section. At any rate, it is concluded that a significant amount of Ni is substituted for Cu.

Figure 2 shows the temperature dependence of the resistivity for selected samples, corresponding to those plotted in Fig. 1. In Fig. 3, the mid-points of the superconducting transition are plotted with the 10-90 % widths indicated by vertical bars. Considering the sharpness of the transitions, these results appropriately characterize the variations of T_c by substituted Ni in the Cu sites. It is interesting to note that for x(Ni) greater than about 0.05, the increase in x tends to reduce the suppression in T_c. The significance of this tendency needs to be further investigated both experimentally and theoretically.

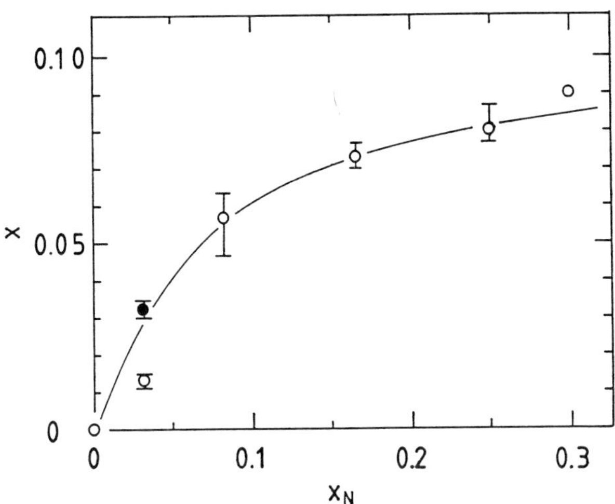

Fig.1. Concentration of Ni in $YBa_2(Cu_{1-x}Ni_x)_3CuO_{7-\delta}$ determined by EPMA, plotted vs the nominal concentration x_N.

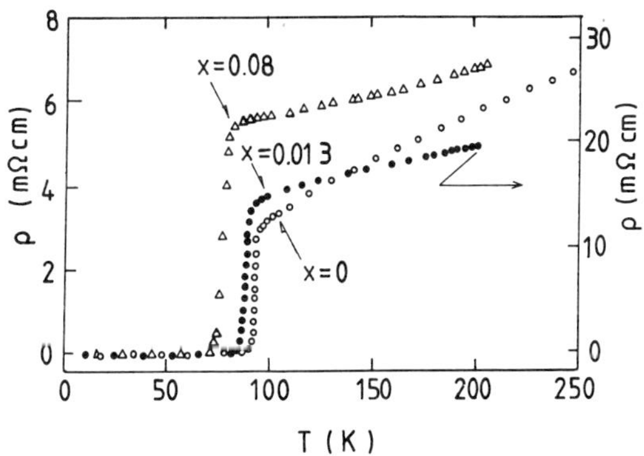

Fig.2. Temperature dependence of the resistivity of $YBa_2(Cu_{1-x}Ni_x)_3CuO_{7-\delta}$ with various Ni contents.

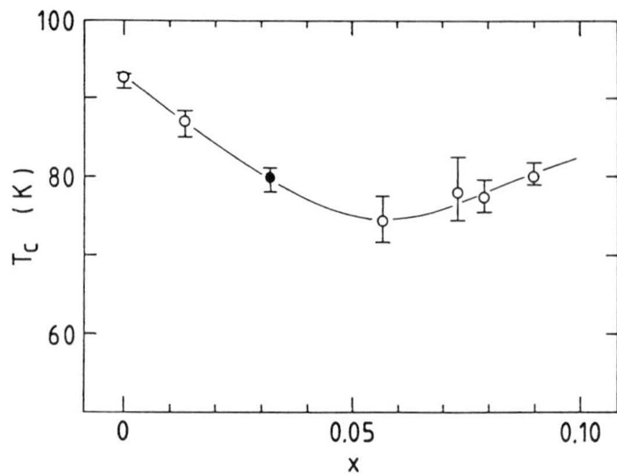

Fig.3. Variation of the mid-point temperature of superconducting
transition for YBa$_2$(Cu$_{1-x}$Ni$_x$)$_3$O$_{7-\delta}$ with the Ni-content x. The
10 - 90 % widths of the transition are shown by vertical bars.

3. MAGNETIC VS. NON-MAGNETIC SUBSTITUTIONS

Metals other than Ni are also substituted partially in the Cu sites.
Samples of YBa$_2$(Cu$_{1-x}$M$_x$)$_3$O$_{7-\delta}$ with M = Fe, Co, Ni, Zn and Ga are prepared
in a manner similar to that described above. For the starting materials
of the dopants, powders of Fe$_2$O$_3$, Co$_2$O$_3$, NiO, ZnO and Ga$_2$O$_3$ are used.
After heating at 900°C for 12 hours in air, the pellets were reground,
reformed into pellets, heated again at 900°C for 30 hours, and furnace-
cooled to room temperature in 24 hours. Composition of the crystals with
x_N = 0.033 and M = Fe, Co, Ni and Zn are determined by EPMA. The results,
along with those of X-ray analysis, show that the metals M were indeed
substituted for Cu in the crystals with the concentration x essentially
equal to the nominal concentration x_N.

The temperature dependence of the resistivity for an undoped (x = 0)
sample and selected samples with x = 0.033 are shown in Fig. 4. In Fig.
5, the transition mid-points as determined by the resistivity
measurements for x = 0.033 are summarized for various substituted
elements with vertical bars indicating the 10-90 % widths. Doping with Al
has also been tried, by using Al$_2$O$_3$ and Al metal powders as starting
materials. However, EPMA indicates that the Al content varies
significantly for different crystals and that there is precipitation of
Al$_2$O$_3$. We therefore conclude that Al is not properly substituted for Cu
in the crystals of YBa$_2$Cu$_3$O$_{7-\delta}$ by the method described here. The results
for Al are shown in Figs. 5-7 for comparison.

The decrease in T$_c$(x) is observed to be approximately linear in x for
small x. The initial slopes are $-T_c^{-1}(0)[dT_c/dx]$ = 4.4 and 6.9,
respectively for Ni and Zn, and 10.3, 11.7 and 13.0, respectively for Fe,
Co and Ga. The reduction in T$_c$ is insensitive to whether the dopant is
magnetic or non-magnetic. For example, magnetic ions Fe and Co have
essentially the same effect as a non-magnetic Ga ion.

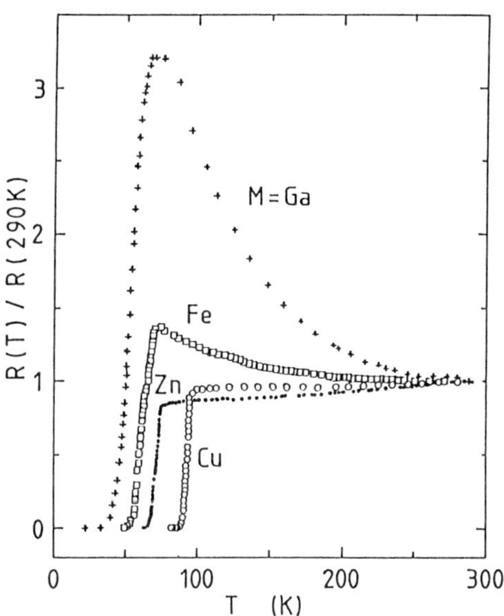

Fig.4. Temperature dependence of the resistance of $YBa_2(Cu_{1-x}M_x)_3O_{7-\delta}$ with $x = 0.033$ and M = Cu, Zn, Fe and Ga.

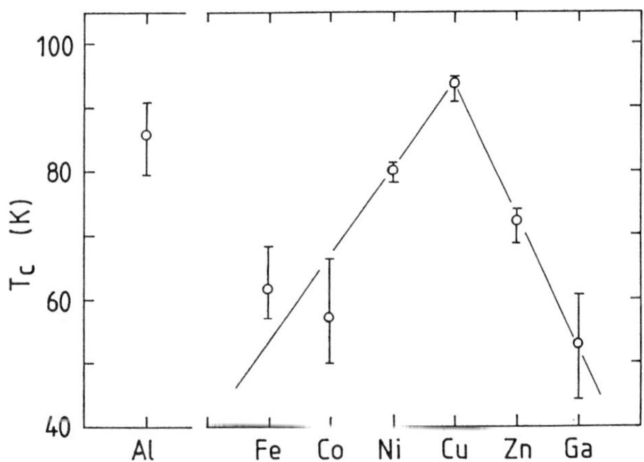

Fig.5. Superconducting transition temperatures, resistively measured, for $YBa_2(Cu_{1-x}M_x)_3O_{7-\delta}$ with $x=0.033$ and M = Fe, Co, Ni, Cu, Zn and Ga. The 10 - 90 % widths of the transition are indicated by vertical bars. See text for the sample with nominal content of Al.

4. ORTHORHOMBIC TO TETRAGONAL PHASE TRANSITION INDUCED BY IMPURITIES

The lattice parameters at room temperature for the undoped and various doped samples, $YBa_2(Cu_{1-x}M_x)_3O_{7-\delta}$ with x = 0.033, are shown in Fig. 6. For the undoped sample, they are a = 3.82 Å, b = 3.89 Å and c = 11.67 Å. Doping with Fe, Co or Ga induces an orthorhombic to tetragonal phase transition, whereas doping with Ni or Zn leaves the orthorhombic lattice parameters almost unchanged. For the undoped Y-Ba-Cu oxide, the orthorhombic to tetragonal phase transition is known to occur at 943 K in air.[10] The tetragonal phase induced by the impurities, however, is much different from that known for the undoped oxide with a large amount of oxygen deficiency. The c-axis does not change much across the phase

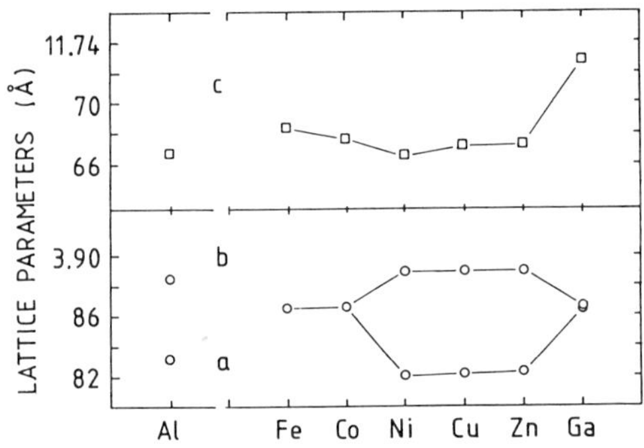

Fig.6. Lattice parameters for $YBa_2(Cu_{1-x}M_x)_3O_{7-\delta}$ with x=0.033 and M = Fe, Co, Ni, Cu, Zn and Ga. The result for the sample with the nominal Al content is also shown.

transition induced by the impurities. In contrast, it becomes longer by 0.09 Å accompanied by greater oxygen deficiency in the undoped tetragonal phase.[5] The occurance of the phase transition by a small amount of impurities suggests that Fe, Co and Ga are substituted at least into Cu1 site, since the symmetry of the environment of Cu1 is what determines the crystal structure. It is natural to assume that the other dopants, Ni and Zn, are also introduced at least into Cu1 site. Then, it must be the valencies of the doped ions, 2+ for Ni and Zn, 3+ for Ga, and 3+ (or 4+) also for Fe and Co, that characterize the degree of disturbance introduced into Cu1 site.

Figure 7 shows the critical current density J_c at 4.2 K for various
substituted elements with x = 0.033. The magnitudes of J_c confirms that
superconductivity observed in the tetragonal phases with M = Fe, Co and
Ga is of bulk nature.

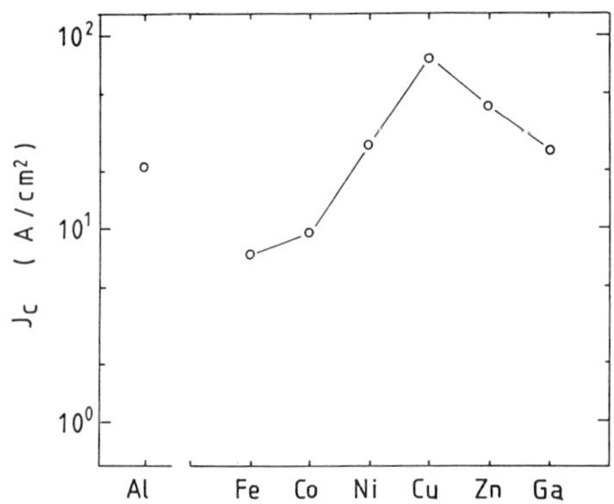

Fig.7. Critical current densities at 4.2 K for $YBa_2(Cu_{1-x}M_x)_3O_{7-\delta}$
with x=0.033 and M = Fe, Co, Ni, Cu, Zn and Ga. The result
for the sample with the nominal Al content is also shown.

The structural change induced by impurities are further investigated as
a function of x. In Fig. 8 the superconducting transition temperature,
with vertical bars indicating the 10 - 90 % drops in resistivity, and
the lattice parameters at room temperature are plotted against x for
$YBa_2(Cu_{1-x}Fe_x)_2O_{7-\delta}$. The critical concentration of Fe for the
orthorhombic to tetragonal transition is $x_c \approx 0.02$. The X-ray analysis

indicates that for x = 0.017, the structure remains close to the phase
boundary down to 77 K; a = 3.84 Å, b = 3.87 Å and c = 11.65 Å at 77 K,
whereas they are 3.85 Å, 3.88 Å and 11.67 Å at room temperature. The
reduction in T_c with x is smooth and is not sensitive to the orthorhombic
to tetragonal phase transition, at which the ordering of oxygen in the
form of 1D Cu1-O chains is lost.

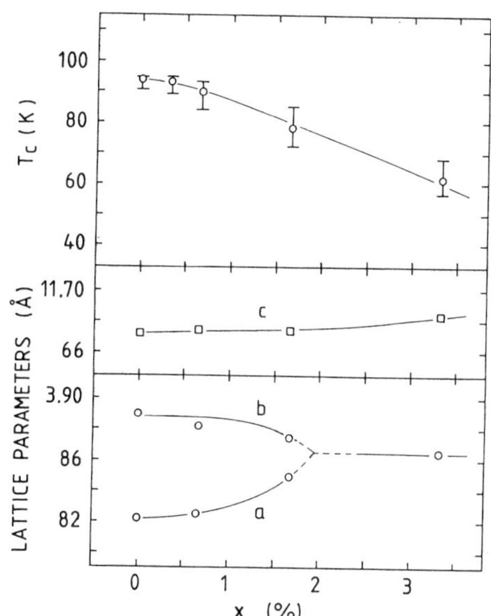

Fig.8. Concentration dependence of the superconducting transition
temperature T_c and the lattice parameters at room temperature
for $YBa_2(Cu_{1-x}Fe_x)_3O_{7-\delta}$. Doping with Fe induces an
orthorhombic to tetragonal phase transition at x ≈ 0.02. The
variation of T_c with x is smooth and insensitive to the
structural phase transition.

5. IMPURITY EFFECTS IN A La-Ba-Cu OXIDE

We have shown that Cu1 site in the Y-Ba-Cu oxide accepts impurity
metals. It has not been determined, however, whether substitution occurs
equivarently into both of the Cu sites, or preferentially only into Cu1
site. If the substitution is purely into Cu1 site, which is not principal
to superconductivity, the effect of impurities shown in Fig. 5 does not
necessarily characterize the intrinsic properties of superconductivity in
the system.

Intrinsic effects of impurities on superconductivity may be
unambiguously probed in a K_2NiF_4-type oxide, $La_{1.85}Ba_{0.15}CuO_{4-\delta}$, in which
all Cu sites are equivalent. Ceramic samples of $La_{1.85}Ba_{0.15}(Cu_{1-x}M_x)O_{4-\delta}$
with a variety of impurities M are prepared in a manner similar to that
described previously.[11] The reduction in T_c is plotted in Fig. 9 against
x. The slope $-T_c^{-1}(0)[dT_c/dx] = 22$ and 40, respectively for Ni and Zn,
is by a factor of about five larger than that for the corresponding
yttrium compounds. It should be noted that in both the yttrium and
lanthanum compounds, doping with non-magnetic Zn has a slightly stronger
effect on the suppression in T_c than that with Ni. These results lead us
to speculate that in the yttrium compound, the substitution occurs not
purely but at least predominanly into Cu1 site.

6. CONCLUSION

We have shown that elements M = Fe, Co, Ni, Zn, Ga can be substituted
partially in the Cu sites to constitute a series of alloyed compounds
$YBa_2(Cu_{1-x}M_x)_3O_{7-\delta}$. The reduction in T_c in both the yttrium and lanthanum
compounds seems insensitive to whether the dopant is magnetic or non-
magnetic. It is the valency of the dopant or properties related to it,
such as differences in ionic radii, which seems to be important to
characterize the impurity effect.

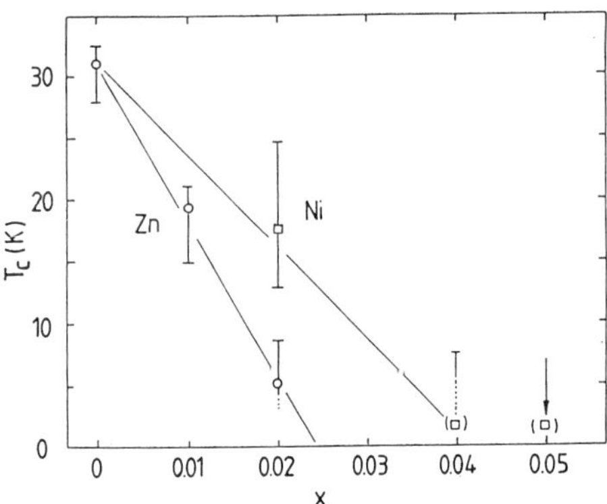

Fig.9. Variation of T_c with x for $La_{1.85}Ba_{0.15}(Cu_{1-x}M_x)O_{4-\delta}$ with M =
Ni and Zn.

The variation in T_c is smooth across the orthorhombic to tetragonal phase transition induced by impurities in the yttrium compound. This observation makes the superconductivity along 1D Cu1-O chains highly improbable. We thus speculate that the superconductivity in $YBa_2Cu_3O_{7-\delta}$ is essentially of 2D nature, similar to that in $(La_{1-x}Ba_x)_2CuO_{4-\delta}$.

Comparison of the normalized reduction of T_c with x, $-T_c^{-1}(0)[dT_c/dx]$, in the yttrium and lanthanum compounds implies that the impurities are substituted predominantly into Cu1 site of $YBa_2Cu_3O_{7-\delta}$. To explain a large enhancement of T_c in the yttrium compound, compared with the lanthanum compound, it should be noted, in addition to the ordering of Y and Ba in the frame surrounding the Cu-O complex, that the oxygen vacancies in the yttrium compound are concentrated within the Cu1 layer and there are few oxygen vacancies in the Cu2-O sheets.

ACKNOWLEDGEMENTS

The authors wish to thank their students, M.Kato, Y.Aoki, T.Nojima, S.Awaji, Y.Iguchi, M.Kyogoku and T.Tomita, for their contributions throughout the course of the experiments. They also would like to acknowledge K.Hoshino and A.Minami for their cautious analysis with an electron probe microanalyzer. They have benefitted from discussions with Prof. H.Fukuyama.

REFERENCES

1. A.W.Sleight, J.L.Gillson and P.E.Bierstedt: Solid State Commun. **17** (1975) 27.
2. J.G.Bednorz, M.Takashige and K.A.Muller: Europhys. Lett., in press.
3. H.Takagi, S.Uchida, K.Kitazawa and S.Tanaka: Jpn. J. Appl. Phys. **26** (1987) L123.
4. R.J.Cava, B.Batlogg, R.B.van Dover, D.W.Murphy, S.Sunshine, T.Siegrist, J.P.Remeika, E.A.Rietmann, S.Zahurak and G.P.Espinosa: Phys. Rev. Lett. **58** (1987) 1676.
5. F.Izumi, H.Asano, T.Ishigaki, E.Takayama-Muromachi, Y.Uchida, N.Watanabe and Tetsuji Nishikawa: Jpn. J. Appl. Phys. **26** (1987) L649.
6. S.Tsurumi, M.Hikita, T.Iwata, K.Semba and S.Kurihara: Jpn. J. Appl. Phys. **26** (1987) L856.
7. B.W.Veal, W.K.Kwok, A.Umezawa, G.W.Crabtree, J.D.Jorgensen, J.W.Downey, L.J.Nowicki, A.W.Mitchell, A.P.Paulikas and C.H.Sowers: submitted to Appl. Phys. Lett.
8. Y.Maeno, T.Nojima, Y.Aoki, M.Kato, K.Hoshino, A.Minami and T.Fujita: Jpn. J. Appl. Phys. **26** (1987) L774.
9. Y.Maeno, T.Tomita, M.Kyogoku, S.Awaji, Y.Aoki, K.Hoshino, A.Minami and T.Fujita: submitted to Nature.
10. Y.Yukino, T.Sato, S.Ooba, M.Ohta, F.P.Okamura and A.Ono: Jpn. J. Appl. Phys. **26** (1987) L869.
11. Y.Maeno, Y.Aoki, H.Kamimura, J.Sakurai and T.Fujita: Jpn. J. Appl. Phys. **26** (1987) L402.

SUPERCONDUCTIVITY IN PURE La_2CuO_4

S.A. Shaheen,[*] N. Jisrawi,[*] Y.H. Min,[*] H. Zhen,[**]
L. Rebelsky,[**] M. Croft,[*] W.L. McLean,[*] and S. Horn[**]

[*]Rutgers University
Serin Physics Laboratory
Piscataway, NJ 08855-0849 U.S.A.

[**] New York University
Department of Physics
4 Washington Place
New York, NY 10003

Detailed investigtions of the ground state properties of
La_2CuO_4 have been made. Our data indicate traces of
superconductivity in La_2CuO_4, with onset temperatues of 45K or
higher. Results on three samples prepared from variable purity
CuO (99.999%, 99.99%, 96%) indicate that superconductivity is
an intrinsic property of La_2CuO_4 and not an artifact of
impurity phase(s). Meisner signal as high as 0.3% has been
observed. Indication for antiferromagnetic ordering near 210K
is also apparent from both resitivity and magnetic
susceptibility data.

Since the reports of high temperature superconductivity in
$La_{2-x}B_xCuO_{4-y}$ type compounds,[1-4] where B is Ba, Sr or another divalent ion,
a considerable interest has been generated in the study of the ground state
properties of the parent compound La_2CuO_4.[5-10] High temperature
superconductivity in Ba and Sr doped $La_{2-x}B_xCuO_{4-y}$ type compounds has been
attributed to the layered perovskite like K_2NiF_4 structure (tetragonal,
I_4/mmm). In contrast, the ground state of La_2CuO_4 is reportedly nonmetallic, even though the crystal structure (orthorhombically distorted
form of K_2NiF_4 structure, space group Cmca) is nearly the same as of the
superconducting $La_{2-x}B_xCuO_{4-y}$ type compounds. The non metallic behavior of
undoped La_2CuO_4 should probably not be attributed to the slight
orthorhormbic distortion from the K_2NiF_4 structure because it has been
shown[11] that for various Sr compositions, $La_{2-x}Sr_x CuO_{4-y}$ samples undergo
an orthorhormbic transformation below room temperature before becoming

superconducting near ~35 K. Moreover, some band-structure calculations[12] and some early experimental data[12-13] favor the metallic ground state of La_2CuO_4. Perhaps more interesting, are recent observations of traces of superconductivity in pure La_2CuO_4 samples.[6-8] On the other hand, spin density waves and an antiferromagnetic transition occurring between 200 and 290 K in La_2CuO_4, based on magnetic susceptibility data, have been claimed.[6,7,8,15] The antiferromagnetic transition at 220 K has also been confirmed by neutron diffraction data.[10,16]

The objective of the present work has been to help understand the origin and nature of superconductivity in La_2CuO_4. The initial reports[6,7] claiming traces of superconductivity in La_2CuO_4 were based on sharp drops in resistivity and magnetic susceptibility near 30-40k. But the magnetic susceptibility for those samples remained positive down to the lowest temperatures measured. Here we report results on samples for which the magnetic susceptibility drops from a definite positive value to zero at the superconducting transition temperatures and the magnitude of the diamagnetic susceptibility increases with the decreasing temperature (see Fig. 2). The diamagnetic signal at 4.2 K is only about 0.3% of the ideal Meisner signal, indicating that superconductivity in La_2CuO_4 is not a bulk phenomena.

An obvious question is whether superconductivity is due to the impurity phase(s) because the $La_{2-x}B_xCuO_{4-y}$ type compounds also show bulk superconductivity in the same temperature regime. To get some insight in this regard, three samples of La_2CuO_4 with variable purity of CuO were prepared. The purity of CuO was 99.999% for sample I, 99.99% for sample II, and 96% for sample III. The purity of La_2O_3 was 99.99% in each case.

The samples were prepared by solid state reaction of La_2O_3 and CuO. Nominal compositions were obtained by mixing appropriate amounts of constituent powders. Thoroughly mixed powders were fired in air at 980°C for eight hours. At this stage, an intermediate grinding was applied. The temperature of the furnace was then raised to 1050°C and held there for 6 hours. The samples were cooled to 200°C in two hours. Microscope examination indicated a homogenously mixed composite and X-ray spectra at this state showed the peaks characteristic only of the orthohombically distorted K_2NiF_4 structure. Pellets were then pressed at a pressure of ~4kbars. Compressed pellets were annealed at 950°C in a flowing oxygen atmosphere for sixteen hours and then cooled to 200°C in eight hours by having the temperature of the furnace at 520°C for five hours. X-ray spectra at this stage were not different from those at the first stage with the exception of an increased sharpness of the lines.

X-ray measurements were made using a Scintag diffractometer. Four probe resistivity measurements were made on rectangular bars of typical

dimensions 2mm x 2mm x 6 mm. The contacts were made by indium soldering;
for better contacts indium enriched mercury amalgum was applied on the
surface of the samples before soldering. Temperature measurements were
made using calibrated Pt, Ge, Si diode, and carbon glass thermometers in
different cryostats. The standard Faraday method was used for magnetic
susceptibility measurements.

The resistivity of three La_2CuO_4 samples is displayed in fig. 1. The
resistivity increases slowly from its room temperature value with the
decreasing temperature, rising sharply below 100K like a semiconductor.
There is a sudden drop at ~45 K for sample I, at 41K for sample II. The
resistivity of sample III flattens between 45 and 39 K before dropping
sharply. The resistivity of samples I and II, which are the relatively
high purity samples, drops below the instrumental noise at ~18K and 16K,
respectively--at current densities below $2mA/cm^2$. For the lowest purity
sample III the resistivity remains non-zero down to 4.2K. This would not
support the hypothesis of a superconducting fraction from the impurity

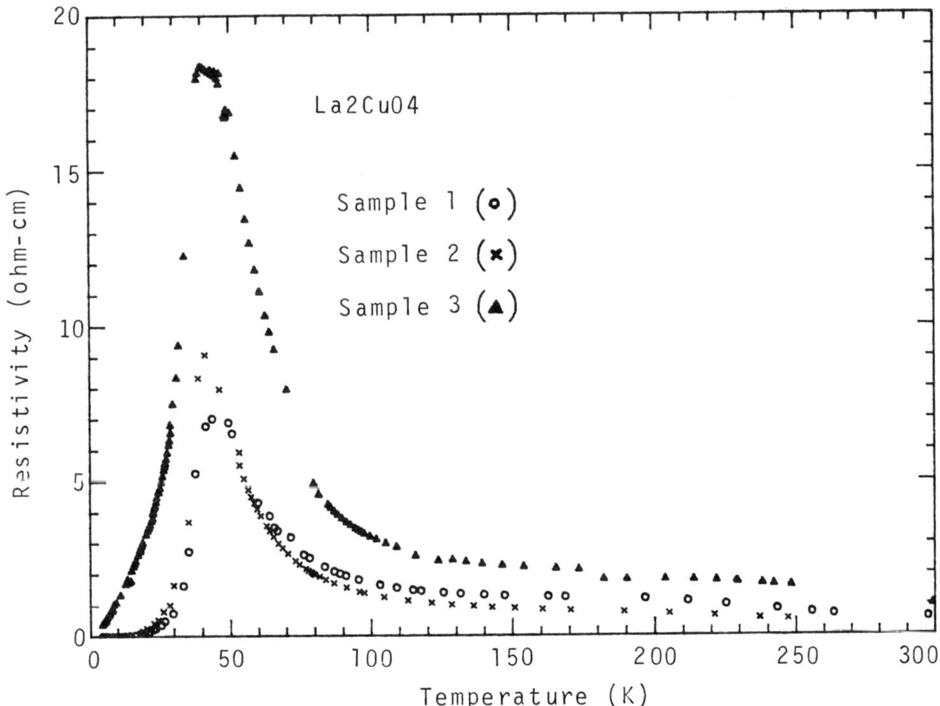

Fig. 1 Resistivity data for three La_2CuO_4 samples. Sample 1,2, and 3
are prepared from CuO of purity 99.999%, 99.99%, and 96%
respectively.

phase(s) and lends support to the idea that superconductivity is an intrinsic property of the La_2CuO_4 compound. With increasing current densities, the maxima in resistivity shift towards lower temperatures. For current densities exceeding 20 mA/cm^2, the resistivity remains very low but non zero down to the lowest temperature measured (4.2K). The resistivity below 100K is apparently current dependent, suggesting that some superconducting fraction is present at temperatures even higher than 45K. The resitivity data shows an inflection point near 210K which is hardly apparent in Fig. 1. However, inflection point becomes more pronounced in a log R versus $[T]^{-1}$ plot. This inflection point actually reflects the antiferromagnetic ordering temperature.

Fig. 2 shows the static magnetic susceptibility data for sample II in a field of 75 Oe. The flat maxima near 210K reflects the antiferromagnetic ordering. The magnetic susceptibility data below 110 K was noisy, probably due to the very small field values used for the measurements; but it may be partly due to a superconducting fraction at higher temperature. The magnetic susceptibility shows a fairly sharp drop at the superconducting transition temperature. The magnetic susceptibility starts decreasing at

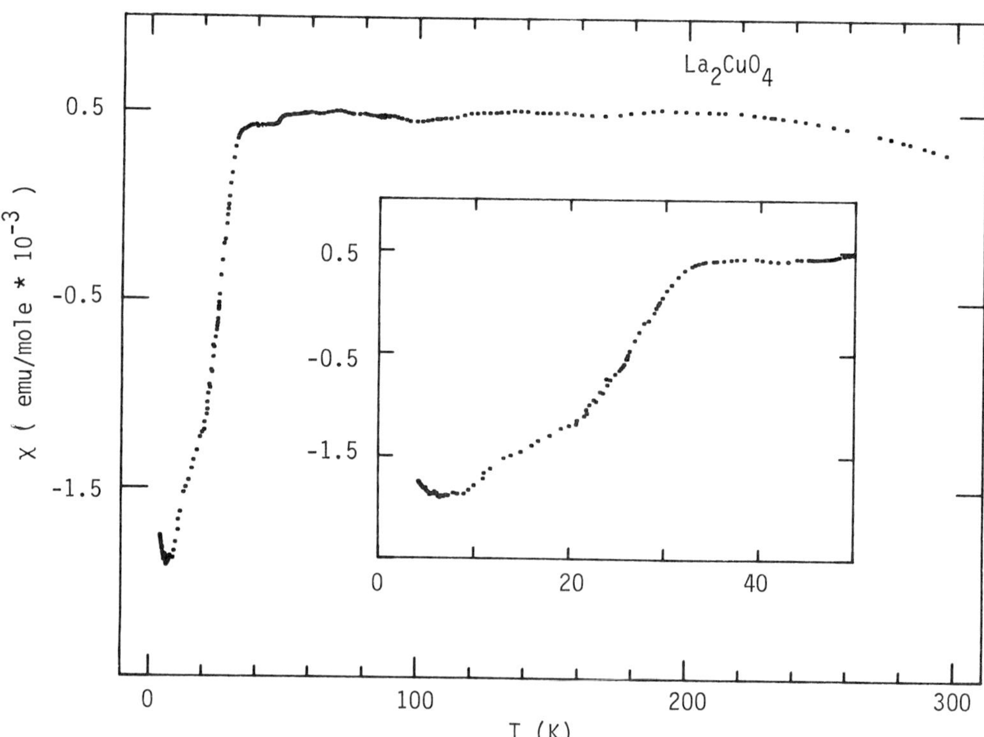

Fig. 2   Static magnetic susceptibility data in a magnetic field of 75 Oe, Insert shows an expanded view of the superconducting transition region.

~39K (which is the temperature at which the maximum in resistivity occurs) and attains its zero value at 30K. The magnetic susceptibility continously decreases down to 5.5K, below which a slight upturn due to paramagnetic inpurties appears. The Meisner signal at 4.2K corresponds to 0.32% of ideal Meisner effect for bulk superconductivity. The temperature at which the magnetic susceptibility changes sign shifts towards lower temperatures in higher fields and the Meisner signal decreases with increasing strength of the magnetic field. The Meisner signal reduces to 0.17% at 138 Oe and 0.11% at 263 Oe. We compare these values with very recently reported values of 0.016% at 250 Oe by Grant et al.[17]

Though the intrinsic nature of superconductivity in La_2CuO_4 is reflected by this and recently published work. The nature of the superconductivity remains unclear. It has been speculated [17] that superconductivity in La_2CuO_4 is filamentary and probably is caused by off stoichiometries at La and oxygen sites. The situation in La_2CuO_4 is quite complex. The authors of earlier mentioned ,theoretical calculations[12] speculate that a transformation from a metallic orthohombic to a semi-conducting monoclinic structure might be occurring at a lower temperature in La_2CuO_4. First principle energy hand calculations[18] for pure La_2CuO_4 suggest the occurrence of a strong Peierls-type Fermi-surface instability involving as breathing type displacement of oxygen atoms in the based plane of the K_2NiF_4 structure. In such a disordered structure the assumption of superconductivity due to non stoichiometrics is not unrealistic. On the other hand, resistivity data on La_2CuO_4 samples bear striking resemblance with granular superconductors. We discuss these aspects in a separate paper[19] where we report measurements on La-deficient samples, and discuss magnetic field and current depences of superconductivity in La_2CuO_4 samples.

ACKNOWLEDGEMENTS

This work was supported by grants # SNJCST 86-240040-13,
                                          87-240090-13,
                                   DMR 85-11982

REFERENCES

1. J.G. Bednorz and K.A. Muller, Z. Phys. B64, 189 (1986).
2. S. Uchida, H. Takagi, K. Kitazawa, and S. Tanaka, Jpn. J. App. Phys. Lett. 26, L1, (1987); H. Takagi, S. Uchida, K. Kitazawa and S. Tanaka, Jpn. J. Appl. Phys. 26, L123 (1987).
3. C.W. Chu, P.H. Hor, R.L. Meng, L. Gao, Z.H. Huang, and Y.O. Wang, Phys. Rev. Lett. 58, 405 (1987).
4. R.J. Cava, R.B. van Dover, B. Batlogg, and E.A. Rietman, Phys. Rev. Lett. 58, 408 (1987).

5. J.D. Jorgensen, H.B. Schutler, D.G. Hinks, D.W. Capone, II, K. Zhang, M.B. Brodsky and D.J. Scalapino, Phys. Rev. Lett. __58__, 1024 (1987).
6. R.L. Greene, H. Malelta, T.S. Plaskelt, J.G. Bednorz, and K.A. Muller, (unpublished preprint).
7. D.C. Johnston, J. P. Stokes, D.P. Goshorn and J.T. Lewandowski-preprint.
8. D. Vaknin, S.K. Sinha, D.E. Moncton, D.C. Johnston, J. Newsam, C.R. Safinya and H.E. King Jr, preprint.
9. S. Usida, H. Tagaki, H. Yangisawa, K. Kishio, K. Titagawa, K. Fueki, and S. Tanaka, Jpn. J. Appl. Phys. 26, L445 (1987).
10. T. Fujita, Y. Aoki, Y. Maeno, J. Sakurai, H. Fukuba and H. Fugii, Jpn, J. App. Phys. 26, L368 (1987).
11. R.M. Fleming, B. Batlogg, R.J. Cava and E.A. Rietman, Phys. Rev. B. __35__, 7191 (1987).
12. R.V. Kasowski, W.Y. Hsu and F. Herman, Solid State Commun. (to be published).
13. J.M. Longo and P.M. Raccah, J. Solid State Chemistry 6, 526 (1973).
14. P. Ganguly and C.N.R. Rao, Mater Res. Buld. 8, 405 (1973).
15. K.K. Singh, P. Ganguly and J.B. Gvodenough, Solid State Chemistry 52, 254 (19874).
16. Y. Yamaguchi, H. Yamauchi, M. Ohasmi, H. Yamamot, N. Shimoda, M. Kikuchi, and Y. Syono, Jpn. J. Appl. Phys. Lett. 26, L447 (1987).
17. P.M. Grant, S.S. P. Parkin, V.Y. Lee, E. M. Engler, M.L. Ramirez, J.E. Varquez, G. Liu, R.D. Jacowitz and R.L. Greene, Phys. Rev. Lett. 58, 2482 (1987).
18. L.F. Matheiss, Physics Rev. Lett. 58, 405 (1987).
19. S.A. Shaheen et.al. (unpublished).

OXYGEN-INTERCALATION EFFECT UPON TETRAGONAL-LaBa$_2$Cu$_{3-x}$O$_y$ COMPOUND SAMPLES

Ryozo Yoshizaki, Hideaki Sawada, Toshiaki Iwazumi*,
Yosuke Saito*, Yoshihito Abe*, Hiroshi Ikeda**,
and Izumi Nakai***

Institute of Applied Physics, *Institute of Physics,
**Cryogenics Center, ***Department of Chemistry, University
of Tsukuba, Sakura-mura, Ibaraki 305 Japan

INTRODUCTION

In order to clarify mechanism of high-temperature superconductivity
of the oxygen-deficient triperovskite YBa$_2$Cu$_3$O$_z$, it will be important to
investigate the relation between a crystal structure and superconducting
properties. The structural analyses of the YBa$_2$Cu$_3$O$_z$ compound have been
studied in many works and their results are summarized by Okamura et al.[1]
In this compound the superconducting transition temperature (T_c) has
found to be strongly dependent upon its crystal phase, orthorhombic or
tetragonal, and upon oxygen deficiencies.[2] In this system, however, the
structural phase transition (orthorhombic-to-tetragonal transformation)
has made it difficult to understand the superconducting properties as a
function of the oxygen content.

In the present study, semiconductor-to-superconductor transition has
been found to occur in a tetragonal-LaBa$_2$Cu$_{3-x}$O$_y$ compound by means of
annealing the sample in O$_2$ atmosphere. In this compound a tetragonal form
is stable to the oxygen intercalation by annealing. The atomic site of
intercalated oxygens has been determined by a Rietveld analysis of X-ray
powder diffraction patterns of the samples.

EXPERIMENTS

Preparation method of the tetragonal-LaBa$_2$Cu$_{3-x}$O$_y$ compound is as
follows. Pure (99.99 %) La$_2$O$_3$, BaO, and CuO were mixed in atomic ratio of
La:Ba:Cu=1:2:3. For the sample L1, well-mixed powder was pressed into a
pellet and calcined at 900 °C for 12 h. The pellet was reground and
pressed into a pellet again. Then it was sintered at 1000 °C for 12 h in
air. The sample (L1-1) was obtained by annealing the L1 sample at 820 °C
for 24 h in flowing O$_2$. These samples were used for the study of the
structural analysis. For the sample L2, BaCO$_3$ was used instead of BaO and
calcination was done at 900 °C for 24 h. A pellet was sintered at 880 °C
for 15 h in air. The annealing effect on superconducting properties was

measured for an identical specimen in resistivity and magnetization measurements. The first annealing was carried out at 880 $^{\circ}$C for 12 h (the sample L2-1). The subsequent annealing was done at 820 $^{\circ}$C for 24 h (the sample L2-2) and the third annealing was at 400 $^{\circ}$C for 24 h (the sample L2-3). Each annealing was processed in flowing O_2.

Resistivity measurement has been performed by employing a conventional four probe method and electrodes have been provided by silver paste. Magnetic susceptibility has been measured by a SQUID magnetometer (SHE model VTS-50). X-ray powder diffraction measurement was carried out by a conventional 2-circle diffractometer.

EXPERIMENTAL RESULTS

Temperature dependence of the resistivity before and after annealing is shown in Fig. 1 for the sample L1 by the broken curves and for the sample L2 by the dotted curves. The resistivity of each as-grown sample (L1 or L2) increases as the temperature is lowered like semiconductors. After annealing, both samples become superconductors; T_c=56.5 K for L1-1 and 47 K for L2-1. The T_c has increased by the subsequent annealing, i.e., 60 K for L2-2 and 65 K for L2-3. A differential thermal analysis and a thermogravimetric analysis have been performed for the as-grown samples. According to the results, no endothermic and exothermic reaction was observed below 1070 $^{\circ}$C, but a little reversible weight change was observed in heating and cooling processes. Thus it can be concluded only oxygen atoms are intercalated or deintercalated by the heat treatment of the sample below 1000 $^{\circ}$C.

Static magnetic susceptibilities have been determined for the sample L2 and the results are shown in Fig. 2(a). For the as-grown sample L2, Curie-like paramagnetic susceptibility is observed at the temperatures till 5 K. By annealing the identical sample, prominent bulk shielding effect has appeared. Both the fractional volume of the apparent superconducting and the T_c are found to increase by the subsequent annealing. The apparent superconducting fractional volume is estimated from the diamagnetic behavior as about 10 % of the total volume for L2-1, 18 % for L2-2, and 20 % for L2-2. The onset temperature of the superconducting transition is obtained as 48 K for L2-1, 60 K for L2-2, and 65 K for L2-3. Thus these results indicate that the oxygen intercalation by means of annealing make the onset temperature of the superconducting transition higher as well as the increase of the fractional volume of the apparent superconducting. In order to clarify the annealing effect on the normal state above T_c, $1/\chi$ is plotted in Fig. 2(b). Weak diamagnetic deviation from Curie-Weiss law below 20 K is evident for the L2 sample. This fact suggests that small portions of the sintered sample become superconductor below 20 K even for the as-grown sample. The paramagnetic Curie temperature of the normal state is almost unchanged by annealing while the Curie constant is weakly dependent on it.

The Rietveld analyses of the X-ray powder patterns at room temperature have been performed for the semiconductor sample L1 and for the superconducting one L1-1. The structural parameters of the two samples have been determined and the results are summarized in Table I. The details of the structural analyses are described elsewhere.[3] It can be emphasized that structural differences between the two samples are induced only by the intercalation of oxygen atoms. One of the most remarkable differences between the two structures is in the site occupancy of O2, i.e., 0.2 for the L1 sample and 0.4 for the L1-1 sample. Remind the occupancy of Cu1 is 0.88, the occupation ratio of oxygen atoms around a Cu1 atom is evaluated as 0.23 for L1 and 0.45 for L1-1. It is noted that the latter value is close to the one estimated for the orthorhombic-$YBa_2Cu_3O_y$ compound.[4] Another characteristic feature of the annealing effect appears in shrinking of the Ba–O2 distance, from 2.965 Å to 2.930 Å. Consequently, the temperature factors B of most atoms become small due to annealing. These behaviors are similar to the results obtained in the $YBa_2Cu_3O_z$ system.[1] Another noteworthy effect of annealing is the change of the valency of Cu1 evaluated from the valence sum calculation, i.e., divalence before annealing becomes much larger than 2.

DISCUSSION

The present results of the structural analysis indicate that the crystal structure of $LaBa_2Cu_{3-x}O_y$ is isostructural with that of tetragonal-$YBa_2Cu_3O_z$. High T_c has been observed in the early works for the $La_2Ba_4Cu_6O_{14+y}$ compound sample ($T_c \simeq 55$ K)[5] and for the $La_{0.8}Ba_{1.2}Cu_6O_{14-y}$ sample ($T_c=50$ K).[6] Each structure may be identical with the present result. The oxygen intercalation effect on the $LaBa_2Cu_{3-x}O_y$ compound is found to be similar to the case observed for the $YBa_2Cu_3O_z$ samples by Sawada et al.[2] In the latter case, the tetragonal crystal was semiconductor for oxygen poor samples and became superconductor ($T_c=52$–58 K) for oxygen rich samples. Referring to the facts that $T_c=80$–90 K superconductors have been synthesized for the tetragonal-La-Ba-Cu-O system,[7,8] it will be concluded that the orthorhombic phase is not the necessary condition for high-T_c superconductor. Consequently, 1-D conduction in the Cu1-O2 linear chain for the orthorhombic structure will not play an essential role on superconductivity. The L1-1 sample is superconductor whose $T_c=57$ K, although the Cu1-O2 mid-layer is disordered due to the small occupancy of Cu1(0.88) and O2(0.4). This fact suggests that the 2-D conduction in the Cu1-O2 layer will be difficult to explain high-T_c superconductor. In contrast, the Cu2-O3 plane is completely occupied even though the compound is a semiconductor (L1) or a superconductor (L1-1). Taking these discussions into account, it must be emphasized that the coupled three-layer structure is important for high-T_c superconductor[9] and the interaction among the three layers will be increased as oxygens are intercalated to the O2 site.

ACKNOWLEDGMENTS

The authors thank Professors Masahiro Inoue and Takuji Kawashima for occasional and fruitful discussions. They are also grateful to Mr. Katsuhiro Imai for his help in the structural analyses. Most of this work was performed at the Cryogenics Center of the University of Tsukuba.

Table I. Atomic Positional Parameters, Site Occupancy (p), and Isotropic Temperature Factors (B) for L1 and L1-1 Samples

Atom	L1: $LaBa_2Cu_{2.88}O_{6.4}$ space group P4/mmm a=3.9257(1), c=11.833(1) Å					L1-1: $LaBa_2Cu_{2.88}O_{6.8}$ space group P4/mmm a=3.9232(1), c=11.802(1) Å				
	x	y	z	p	B ($Å^2$)	x	y	z	p	B ($Å^2$)
Ba	0.5	0.5	0.1878(5)	1.0	0.8(2)	0.5 0.5 0.1844(6)			1.0	0.8(2)
La	0.5	0.5	0.5	1.0	0.4(3)	0.5 0.5 0.5			1.0	0.1(3)
Cu1	0.0	0.0	0.0	0.88(4)	0.5(2)	0.0 0.0 0.0			0.88*	0.3(2)
Cu2	0.0	0.0	0.348(1)	1.0	0.5	0.0 0.0 0.346(1)			1.0	0.3
O1	0.0	0.0	0.148(7)	1.0	0.5(8)	0.0 0.0 0.144(6)			1.0	0.1(9)
O2	0.0	0.5	0.0	0.2(1)	0.5	0.0 0.5 0.0			0.4(1)	0.1
O3	0.0	0.5	0.368(3)	1.0	0.5	0.0 0.5 0.364(4)			1.0	0.1

* fixed parameter (see Ref. 3).

REFERENCES

1. F. P. Okamura, S. Sueno, I. Nakai and A. Ono: Proc. Jpn. Academy. **63**, Ser. B (1987) in press.
2. H. Sawada, T. Iwazumi, Y. Saito, Y. Abe, H. Ikeda and R. Yoshizaki: to be published in Jpn. J. Appl. Phys. **26**, No. 6 (1987).
3. I. Nakai, K. Imai, T. Kawashima, and R. Yoshizaki: submitted to Jpn. J. Appl. Phys.
4. F. Izumi, H. Asano, T. Ishigaki, E. Takayama-Muromachi, Y. Uchida, N. Watanabe, and T. Nishikawa: Jpn. J. Appl.Phys. **26**, L649 (1987).
5. M. Hikita, S. Tsurumi, K. Semba, T. Iwata, and S. Kurihara: Jpn. J. Appl. Phys. **26**, L615 (1987).
6. T. Iwazumi, R. Yoshizaki, M. Inoue, H. Sawada, H. Ikeda, and E. Matsuura: Jpn. J. Appl. Phys. **26**, L621 (1987).
7. S. Tsurumi, M. Hikita, T. Iwata, K. Semba, and S. Kurihara: Jpn. J. Appl. Phys. **26**, L856 (1987).
8. D. B. Mitzi, A. F. Marshall, J. Z. Sun, D. J. Webb, M. R. Beasley, T. H. Geballe, and A. Kapitulnik: to be published in Phys. Rev.
9. M. Inoue, T. Takemori, K. Ohtaka, R. Yoshizaki, and T. Sakudo: to be published in Solid State Commun.

TRENDS AND FUTURE: THEORY

Marvin L. Cohen

Department of Physics, University of California, and
Materials and Chemical Sciences Division,
Lawrence Berkeley Laboratory, Berkeley, CA 94720

ABSTRACT

A discussion is presented of theoretical approaches which may be
appropriate for explaining the properties of high transition temperature
superconductors. The focus is on models and proposals discussed at this
conference.

INTRODUCTION

When this conference was first conceived, the problem of
superconductivity at high temperatures was an academic one. The original
plan for the meeting was to have a workshop on novel mechanisms to examine
theoretical ideas and some experimental data with the goal of determining
areas with potential promise. This perspective changed after the
discovery[1] of superconductivity above 30K in La-Ba-Cu-O and the subsequent
increase[2] in the transition temperature T_c above liquid nitrogen
temperatures in Y-Ba-Cu-O. Before these discoveries, no experimental
tests were possible for the existing theoretical methods proposed for high
T_c. Now the situation has changed dramatically. Transition temperatures
near 100K are common and large drops in resistance above 200K, which were
observed early[3], are being reproduced and studied. There are even
reports[4] of possible zero resistance metastable states around 300K, but
further experiment is necessary to clarify these data. At this point,
many researchers feel that superconductivity near room temperature is
possible and probable. We all hope that this optimistic view (which I
share) has merit. However, as most of you know, before the current
breakthrough there was only a 20K rise in T_c over a period of 75 years.
The five-fold increase over the past year is encouraging, but there is
fear that another plateau in T_c may have been reached.

The theoretical history is also sobering. After the discovery of zero
resistance by Onnes[5] in 1911, theorists had little chance of explaining
superconductivity even on a macroscopic scale. Zero resistance alone was
not a sufficient condition for characterizing the superconducting state.
The Meissner[6] effect discovered in 1933 provided crucial information, and
a macroscopic theory appeared[7] in 1935. This theory was extended and

refined[8] in 1950, but it wasn't until 1957 that a complete microscopic theory was formulated.[9] Hence, it took 46 years for a microscopic explanation of superconductivity and important experiments such as heat capacity[10] and the isotope effect[11] were critical to the development of the theory.

If the theory of the new high T_c superconductors is basically different from the BCS theory and requires sweeping new concepts, then it may be some time before it is developed. Perhaps new experiments are necessary to nucleate new ideas. I don't think this is the view of the majority of the theorists here. There is optimism that theoretical progress will occur at a rate comparable to the experimental advances of the past year and that the modern theoretical tools developed for explaining the properties of solids are sufficient for providing the fundamental theory of high T_c phenomena.

BACKGROUND

At this point, most theorists believe that the high T_c phenomena are caused by electron pairing correlations. Electron pairing allows zero resistivity, and these correlations explain a large body of data on superconductors. The nature of the pairs and the mechanisms responsible for the attractive interaction causing pairing are the central ideas being discussed and debated.

In the BCS description of conventional superconductors, the coherence length $\xi_o = \hbar v_F/\pi\Delta$ where v_F and Δ are the Fermi velocity and the superconducting energy gap. This length is typically a fraction of a micron. For a specific concentration of electrons n, a measure of the number of electrons affected by the superconductivity is $n_p \sim n\Delta/E_F$ where E_F is the Fermi energy. If we assume that the number of pairs scales like n_p, then the ratio of the pair size to the distance between pairs is $\xi_o/r_p \sim (E_F/\Delta)^{2/3}$. The distance between electrons in a metal scales inversely with k_F, the Fermi wavevector, hence the ratio of the pair size to the distance between electrons is $k_f\xi_o \sim E_F/\Delta$. In both cases, for conventional superconductors the ratios are very large, and typically $\sim 10^6$ pairs are found between the two electrons forming a pair. Large ratios are found even for very low density superconductors like $SrTiO_3$[12] because of the low T_c even though superconductivity is observed in transparent samples with $n \sim 7 \times 10^{17}$ cm^{-3}.

Above we described the case where $E_F/\Delta \gg 1$ implies BCS pairs characterized by zero momentum pairing correlations in momentum space and real space overlapping pairs. A contrasting situation exists when $E/\Delta \ll 1$ and the pairs form small Boson-like bi-polarons. The transition to the superconducting state is expected to be Bose-Einstein-like where T_c scales with E_F (i.e., $T_c \sim n^{2/3}$). It has been suggested[13] that in the high T_c oxides $E_F/\Delta \sim 1$ and that the spatial extent of the pairs is of the same order as the average distance between pairs. A model of this kind can result in zero resistance and a Meissner effect, but the pair wavefunction should differ considerably from BCS.

Hence, it is important to obtain accurate measurements of the normal state and superconducting parameters to determine the characteristics of the paired state.

In BCS theory, the mechanism for the attractive pairing interaction is phonon exchange. A prototype equation for T_c is the two square-well solution[14],

$$T_c \sim E_p e^{-\frac{1}{\lambda^* - \mu^*}} \qquad (1)$$

where E_p is a phonon energy, λ^* is the renormalized ($\lambda^* = \lambda/1+\lambda$) attractive dimensionless coupling constant and μ^* represents the renormalized $\left(\mu^* = \mu(1 + \mu\ln E_F/E_D)^{-1}\right)$ Coulomb coupling constant. A form similar to Eq. (1) is expected for most pairing interactions, that is,

$$T_c \sim E_0 f(\lambda_0) \qquad (2)$$

where E_0 and λ_0 are the energy and dimensionless coupling constant for the excitation causing the electron-electron pairing and f is a function which increases monotonically with λ_0. Weak and strong coupling conditions are defined by λ_0 being small and large respectively. For example, Eq. (1) is appropriate for weak coupling (and medium coupling $\lambda < 1.5$). For strong coupling $T_c \sim E_p\sqrt{\lambda}$.

Using the generic form for the T_0 equation $\left(\text{Eq. (2)}\right)$, we can divide most of the high T_c theories into two classes. The first is weak coupling where the Boson excitation energy E_0 is large and $f(\lambda_0)$ is small. This class is appropriate for excitations like high frequency plasmons or excitons since E_0 scales like eV's or fractions of an eV. To obtain T_c's in a reasonable range, $f(\lambda_0)$ must be small. In contrast, for low frequency excitations like phonons, spin waves (Pines[15]), some forms of plasmon excitations, etc., $f(\lambda_0)$ must be large to compensate for the small E_0. For example, if $T_c \sim 100K$, then for phonons where $E_0 \sim 300K$, $f(E_0) \sim 1/3$ while an electronic excitation of ~1 eV implies an $f(E_0) \sim 1/100$. It is difficult to argue <u>a priori</u> using theory to decide on the preferred range of E_0 and $f(E_0)$. However, it has been suggested[16] that an optimal situation occurs when $T_c \sim 1/10 \ E_0$ which favors large E_0 and weak coupling for high T_c. Experimental limits on the parameters together with the arguments above will put significant constraints on possible theories.

A general Feynman diagram describing the pairing of electrons via Boson exchange has the form shown in Fig. 1.
We have been cautioned in this conference to use exchange corrections (Gutfreund[17]), to beware of the random phase approximation (Sham[18]), and to include higher order diagrams (Ashcroft[19]). Acknowledging these caveats, we can use the diagram of Fig. 1 to examine pairing and, in principle, can consider phonons, 2d and 3d plasmons, demons (acoustic plasmons), excitons, magnons, superexchange, interband transitions, satellites, and other excitations for pairing discussed at this conference. Suggestions for

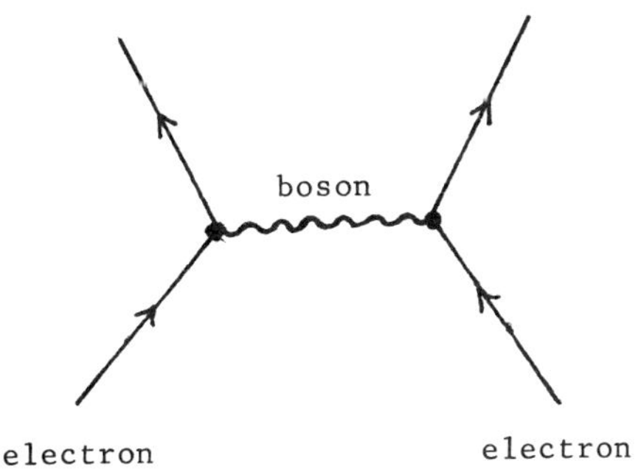

Fig. 1. Electron pairing through Boson exchange.

coupling mechanisms have appeared in the literature with only demonstrations of an attractive electron-electron interaction for a particular frequency ω and wavevector q. This is insufficient to demonstrate pair formation. Both the ω and q dependence of the interaction must be included to evaluate the coupling and ultimately T_c. A q-independent generic form of the attractive kernel $K(\omega)$ of the BCS energy gap equation

$$K(\omega) = \frac{E_o^2 \lambda_o}{\omega^2 - E_o^2}$$

(3)

is often used for computing Δ or T_c. Because many pairing kernels have this form, Eq. (3) is a useful form to explore, but ulitimately, it is necessary to fit this expression to a rigorous calculation of the kernel after including q-dependence.

A convenient way to deal with both ω and q-dependence is to express the interaction when possible in terms of a screened Coulomb interaction with a total dielectric function $\varepsilon(q,\omega)$ containing all of the electron-electron interaction. This approach is possible even for phonons when the phonon polarizability $\pi_p(q,\omega)$ is added to the electronic polarizability $\pi_e(q,\omega)$,

$$\varepsilon(q,\omega) = 1 + \pi_e(q,\omega) + \pi_p(q,\omega)$$

(4)

The kernel of the BCS equation is now

$$K(\omega) \sim \int_{LD} \frac{V(q)}{\varepsilon(q,\omega)} \, qdq$$

(5)

where $V(q) = 4\pi e^2/q^2$ is the bare Coulomb interaction and the integration is over the Landau damping (LD) region of the (q,ω) plane. A net attractive kernel results only if the integration over the positive and negative regions of $\varepsilon(q,\omega)$ add appropriately. This approach is useful for investigating exciton, plasmon, demon, and other pairing schemes.

SOME PAIRING INTERACTIONS

Phonons

With the possible exception of heavy fermion superconductors, before 1986 phonon-induced attraction was considered by most theorists to be the predominant pairing mechanism in all superconducting solids. Within the framework of the BCS theory and its extensions, this interaction explained virtually all the properties of superconductors. Tunneling measurements[20] were particularly important in verifying the details of the theory, and it was demonstrated that the Eliashberg[21] formulation of the theory was appropriate for strong or weak coupling.

Despite the enormous success of this approach, successful theoretical predictions of new superconductors are rare and probably number less than ten. The reason for the paucity of successful predictions of this kind is the sensitivity of T_c to the normal state parameters of the solid. Calculations of λ and μ need to be extremely accurate if expressions like Eq. (1) are to be used. Even when the Eliashberg equations are solved numerically for stronger coupling, where the exponential dependence on coupling constants is removed, the sensitivity to the normal state properties is a limiting factor for estimations of T_c.

It is even difficult to make accurate predictions of the maximum transition temperature for arbitrary coupling. Studies[16] do indicate what features are desirable and some of the limiting factors. Two main examples are that local fields or covalent-like bonds allow larger electron-phonon couplings than free electron charge distributions and a restriction on λ arises because of lattice instabilities which are induced for large coupling cases.

Recently there have been significant advances in formulating theories and using them for computing electronic, structural, and vibrational properties of materials starting with limited input such as the atomic number and atomic mass.[22] Total energy calculations for different structural configurations can yield a determination of the most favorable crystal structure from a set of candidates and the various parameters describing ground state properties. The approach is general, but it is presently limited to simple systems because of computational complexity. Recently, electron-phonon couplings have been included among the outputs of the scheme,[22] and together with standard estimates of μ^*, first principles calculations of λ are used to predict T_c. Silicon can be used as a prototype system, and new high pressure metallic forms were predicted to exist and to be superconducting. The successful predictions are successes of the total energy scheme and of the BCS-phonon theory. An illustrative example[23] of the behavior of a typical system for large λ is the behavior of T_c with pressure in the highly compressed ph (primitive hexagonal) and hcp (hexagonal closed packed) structures of Si.

The theoretical calculations predicted that in the ph phase T_c would be ~ 5-10K, T_c would decrease with pressure and at high pressures T_c would increase with pressure until the transformation from ph to hcp occurred. After the transition to hcp, T_c was expected to be insensitive to pressure. All of the above properties were confirmed experimentally.[23] However, even though the predicted rise in T_c near the transition pressure from ph to hcp was found, the structural transition occurred before λ and T_c became very large. This illustrates the expected behavior[16] described earlier where strong electron-phonon coupling can result in lattice instabilities.

The above example demonstrates that detailed calculations for T_c are possible for simple systems. At this point, a similar highly accurate calculation is not possible for the superconducting oxides. However, considerable progress has been made. One electron band structures[24,25] and models[26] for fitting these with a few parameters exist. A calculation[27] of the electron-phonon coupling in La-Ba-Cu-O based on the band structure results has also been done. This work estimates λ and puts limits on the T_c's available in this system.

It is instructive to explore the constraints that the current experimental results put on the phonon mechanism explanation of the observed superconductivity. If we consider T_c ~ 100K for an Y-Ba-Cu-O system, then it is possible to examine the range of parameters E_p, λ, and μ^* which are consistent with T_c. This is done by solving the Eliashberg[21] equations numerically and fitting the resulting curves. For $\mu^* = 0$ and E_p represented by an average phonon energy or Debye temperature of 400K or 600K, solutions are obtained for $1 < \lambda < 3$. Finite μ^* results in large λ values.

There is, however, another constraint on T_c. Substitution of ^{18}O for ^{16}O in Y-Ba-Cu-O has yielded no isotope effect,[27,28] and at this conference similar results were reported[29] for isotope substitutions for Ba and Cu. These results limit λ and μ^*. This can be seen analytically[30,31] using the two-square model of Eq. (1). A zero isotope effect is possible within this model when $\mu^* = \lambda^*/2$ as shown by these equations

$$T_c \sim M^{-\alpha}$$

$$\alpha = \frac{1}{2}\left[1 - \left(\frac{\mu^*}{\lambda^* - \mu^*}\right)^2\right] \ . \tag{6}$$

Although the BCS value, $\alpha = 1/2$, can be reduced significantly, in weak coupling it is not possible to fit both $T_c \sim 100K$ and $\alpha \sim 0$ within this model. Defect models[32] or strong coupling together with selective movements of parts of the phonon spectrum are necessary[33] for consistent behavior with phonon-induced pairing. The latter result depends sensitively on the choice of E_p, λ, and μ^*.

In summary, not enough is known about the normal state properties of the high T_c oxides to verify or refute the phonon mechanism. This is not a problem with BCS theory since for degenerate semiconductors[34] or simple systems,[22] it is possible to use the normal state properties to compute superconducting parameters. However, at this point without consideration of the special properties of these materials,[32] a standard application of Eliasberg theory yields very restrictive results. More experimental work is necessary; for example, a demonstration that superconductivity in Y-Ba-Cu-O is of the weak coupling class would probably rule out phonon-induced pairing.

Plasmons, Demons, and Exitons

A standard theme in the literature for non-phonon mechanisms is to replace the phonon by an electronic excitation. Several authors have pointed out that an advantage of most of these excitations is their high E_o (Eq. (2)). Even though a $q \sim 0$ dielectric function $\varepsilon(\omega) = 1 - \omega_p^2/\omega^2$ where ω_p is the plasma frequency yields negative values and the overscreening will cause electron-electron attraction, this contributes nothing to the kernel of Eq. (5) because it's outside the LD region. Takada[35] and Sham[18] have included q-dependence and achieve plasmon pairing even for free-electron metals like Na. However, as was pointed out at this conference, the RPA approximation is highly suspect for these applications, and alkali metals have not been observed to be superconducting. Kresin[36] has shown that for two-dimensional plasmons, the dispersion $\omega(q)$ enhances their effectiveness. He also emphasizes that their role may be to enhance existing phonon pairing.

Another type of plasma excitation which is effective in pairing electrons is the demon. The demon or d-mon was introduced by Pines[37] as a possible excitation in a semiconductor with a sound-like dispersion relation. In analogy with gas plasmas all that is required is two different masses $M_1 \gg M_2$ and the screening of the heavy mass by the light mass results in an acoustic plasmon or demon. Garland[38] considered the role of this mode in transition metal superconductivity and examined the pairing of light s-electrons through the excitation of demons associated with the heavier d-electrons. Frolich,[39] Ruvalds,[40] and Ihm, Cohen, and Tuan[41] did further studies of this mechanisms, and a fairly good understanding of its properties has resulted.

A convenient way to examine the role of demons is to use Eqs. (4) and (5). The electronic polarizability π_e now contains the polarizabilities arising from s, d, and s-d transitions. Interband effects[41] can also be included, and these have been explored further by Lee and Ihm[42] using other methods. Generally, the attraction is large because $\varepsilon(q,\omega)$ is negative over large regions of the (q,ω) plane. This is a more rigorous approach to computing the pairing interaction than making the analogy between the demon and the phonon and using Eq. (3) and the diagram of Fig. 1. A similar statement is true for plasmons and most other electronic excitations.

One important point of confusion found in the literature in the application of this mechanism to pairing is the fact that it is the light particles which pair with the larger superconducting gap. In discussions of the role of this mechanism in transition metals, the view that it is the d-electrons which have the large energy gap is not consistent with the theory. For high T_C oxides, application of this approach is straightforward using the dielectric function method.[41] In this case, assuming p and d electrons near E_F coming from oxygen and copper respectively, pairing occurs with the larger gap in the p-like bands or in the parts of the hybrid bands which are more p-like.

It was pointed out at this conference that demons themselves have not been observed. My view is that "this is an experimental problem." Damping may significantly broaden these modes, but they must occur. Phonon modes can be viewed as a lattice of heavy ions screened by light electrons giving rise to acoustic modes. This is demon-like. Demons should exist, but their role in superconductivity is not clear.

A convenient aspect of plasmons and demons is that they usually correspond to zeros of the electronic dielectric function and therefore, they are generally longitudinal in character. This allows direct coupling between pairing electrons, and the application of Eqs. (4) and (5) is straightforward. Excitons are usually transverse excitations which are characterized optically by poles in the dielectric function. Longitudinal excitons do exist, but most applications to superconductivity are viewed in terms of transverse excitons.

Since the pairing kernel is inversely proportional to the dielectric function, poles give zero coupling. This effect is similar to the coupling of electrons to transverse phonons where the interacton is zero without the inclusion of umklapp processes. Another caution is to be careful not to count the Coulomb interaction twice.[43] A correct coupling scheme should involve the use of a dielectric matrix which would include local field corrections. Many of these points were not appreciated in early studies but have since been discussed in the literature,[44] and most of the discussions of excitonic pairing at this conference[45,46,47] have reflected these considerations.

Bardeen[47] examined the properties of his earlier excitonic model[48] using parameters appropriate to the high T_C oxides. His numerical values for the parameters differed somewhat from those of Pines,[15] and he concluded that a weak coupling description was appropriate. Varma[45] analyzed optical spectra in terms of excitonic excitations and examined specific models of charge transfers between nearest neighbor Cu and O ions to produce excitonic resonances. Both of these studies and Little's[46] analysis of the electronic structure of superconducting oxides have added to our understanding of the roles excitons can play. Unfortunately, at this point a first principles calculation of λ_0 for excitons is not available. Even Johnson's[49] molecular-orbital approach which spans the entire range of electronic interaction between phonon-like and exciton-like excitations required fitting. A dielectric matrix calculation for a simple system could help clarify many controversial aspects of this pairing scheme. In addition, more normal states studies of the oxides could limit the range of parameters.

Hubbard U Theories

One of the most popular avenues being explored is the role of strong electron correlation and antiferromagnetic type ordering. Anderson[50] described the latest developments in his theory based on a resonating valence bond (RVB) approach. The original proposal by Anderson was introduced to examine a two-dimensional triangular lattice of $s = 1/2$ spins coupled via nearest neighbor antiferromagnetic Heisenberg coupling. The RVB state involves singlet pairs which resonate with different spatial configurations. The system is expected to behave as a fluid, and Anderson emphasized the gapless nature of his model and the expected linear heat capacity.

Appel,[51] Emery,[52] and Doniach[53] also discussed positive Hubbard U models. In some examples, superexchange[51,53] couplings and analogies with solid helium were described. In this model, superconductivity occurs when the band deviates from half filling and the low energy excitations or spinons have no gap in agreement with Anderson's assertion. However, detailed descriptions of the superconducting state and its wavefunction are not yet available. Heat capacity and tunneling experiments are likely to be central to verifying the validity of this approach.

Let me mention another positive Hubbard U mechanism based on the two-hole bound state mode.[54] A large U produces satellites in photoemission spectra of solids which are interpreted in terms of excited states involving two bound holes usually associated with d-electrons. These holes affect the electronic susceptibility and yield[55] negative regions in $\varepsilon(q,\omega)$. This results in pairings for electrons and large self-energy or polaron effects. Strong photoemission satellites have been seen[56] for Y-Ba-Cu-O suggesting that this mechanism may be important for this system.

CONCLUSIONS

I have covered only part of the oral presentations at this conference and have not discussed the interesting poster presentations. Also, bipolarons and the polaron couplings discussed by Scalapino[57] were not described here, but this information and the papers corresponding to the theory posters are provided in the proceedings.

During the conference, it was stated several times that the theorists have revived their old theories to see if they now fit the observations for the high T_c oxides. However, it is clear that new approaches have been born since the experimental discoveries were made. The new versions of the RVB approach are a good example of the latter. It's clear that we need more experimental measurements to decide on the validity of the various proposals, but we also need further calculations to make specific predictions.

In the past, we had one pairing mechanism (phonons) which explained almost all of the superconducting materials. Now we have phonons, plasmons, excitons, demons, the RVB, bipolarons, magnons, satellites, superexchange, interband pairing, etc., to explain just two superconductors. I think most of us feel that in the future only one (or maybe two) of the proposed pairing interactions will join the phonon interaction as valid mechanisms.

I'm reminded of a statement attributed to Einstein, "The most important tool of the theoretical physicist is his wastebasket."

ACKNOWLEDGEMENTS

This work was supported by National Science Foundation Grant No. DMR8319024 and by the Director, Office of Energy Research, Office of Basic Energy Sciences, Materials Sciences Division of the U.S. Department of Energy under Contract No. DE-AC03-76SF00098.

REFERENCES

1. J. G. Bednorz and K. A. Muller, Z. Phys. B 64:189 (1986).
2. W. K. Wu, J. R. Ashburn, C. J. Torng, P. H. Hor, R. L. Meng, L. Gao, Z. J. Huang, Y. Q. Wang, and C. W. Chu, Phys. Rev. Lett. 58:908 (1987).
3. L. C. Bourne, M. L. Cohen, W. N. Creager, M. F. Crommie, A. M. Stacy, and A. Zettl, Phys. Lett. 120:494 (1987).
4. L. C. Bourne, M. L. Cohen, W. N. Creager, M. F. Crommie, and A. Zettl, Phys. Lett. (in press).
5. H. Kamerlingh Onnes, Akad. van Wetenschappen (Amsterdam) 14:113:818 (1911).
6. W. Meissner and R. Ochsenfeld, Naturwiss. 21:787 (1933).
7. F. London, "Superfluids," Vols. I and II, Wiley, New York (1950).
8. V. L. Ginsburg and L. D. Landau, JETP (USSR) 20:1064 (1950).
9. J. Bardeen, L. N. Cooper, and J. R. Schrieffer, Phys. Rev. 1-6:162 (1957); 108:1175 (1957).
10. W. H. Keesom and J. A. Kok, Physica I:175 (1934).
11. E. Maxwell, Phys. Rev. 78:477 (1950); C. A. Reynolds, B. Serin, W. H. Wright, and L. B. Nesbitt, Phys. Rev. 78:487 (1950).
12. J. R. Schooley, W. R. Hosler, and M. L. Cohen, Phys. Rev. Lett. 12:474 (1964).
13. C. Noguera and P. Garoche, to be published.
14. W. L. McMillan, Phys. Rev. 167:331 (1968).
15. D. Pines, this conference.
16. M. L. Cohen and P. W. Anderson, in: "Superconductivity in d- and f-Band Metals," D. Douglass, ed., American Institute of Physics, New York (1972).
17. H. Gutfreund, this conference.
18. L. Sham, this conference.
19. N. Ashcroft, this conference.
20. W. L. McMillan and J. M. Rowell, in: "Superconductivity," R. D. Parks, ed., Marcel Dekker, Inc., New York (1969).
21. G. M. Eliashberg, JETP 11:696 (1960).
22. M. L. Cohen, Science 234:549 (1986).
23. D. Erskine, P. Y. Yu, K. J. Chang, and M. L. Cohen, Phys. Rev. Lett. 57:2741 (1986).
24. L. F. Mattheiss, Phys. Rev. Lett. 58:1028 (1987); J. Yu, A. J. Freeman, and J.-H. Xu, Phys. Rev. Lett. 58:1035 (1987).
25. J. Yu, S. Massidda, A. J. Freeman, and D. D. Koelling, to be published; L. F. Mattheiss and D. R. Hamann, to be published.
26. W. Harrison, this conference.
27. B. Batlogg, R. J. Cava, A. Jayaraman, R. B. van Dover, G. A. Kourouklis, S. Sunshine, D. W. Murphy, L. W. Rupp, H. S. Chen, A. White, K. T. Short, A. M. Mujsce, and E. A. Rietman, Phys. Rev. Lett. 58:2333 (1987).
28. L. C. Bourne, M. F. Crommie, A. Zettl, H. zur Loye, S. W. Keller, H. L. Leary, A. M. Stacy, K. J. Chang, M. L. Cohen, and D. E. Morris, Phys. Rev. Lett. 58:2337 (1987).

29. A. Zettl, this conference.
30. J. C. Swihart, IBM J. of Res. and Dev. 6:14 (1962).
31. J. W. Garland, Phys. Rev. Lett. 11:111:114 (1963).
32. J. C. Phillips, Phys. Rev. (in press).
33. L. C. Bourne, A. Zettl, T. W. Barbee III, and M. L. Cohen, to be published.
34. M. L. Cohen, in: "Superconductivity," R. D. Parks, ed., Marcel Dekker, Inc., New York (1969).
35. Y. Takada, this conference.
36. V. Kresin, this conference.
37. D. Pines, Can. J. Phys. 34:1379 (1956).
38. J. W. Garland, in: "Proceedings of the Eighth International Conference on Low Temperature Physics," R. O. Davies, ed., Butterworths, Massachusetts (1963).
39. H. Frolich, J. Phys. C 1:544 (1968).
40. J. Ruvalds, Advances in Physics 30:677 (1981).
41. J. Ihm, M. L. Cohen, and S. F. Tuan, Phys. Rev. B 23:3258 (1981).
42. D. H. Lee and J. Ihm, to be published.
43. J. C. Inkson and P. W. Anderson, Phys. Rev. B 8:4429 (1973).
44. M. L. Cohen and S. G. Louie, in: "Superconductivity in d- and f-Band Metal," Plenum, New York (1976).
45. C. Varma, this conference.
46. W. Little, this conference.
47. J. Bardeen, this conference.
48. D. Allender, J. Bray, and J. Bardeen, Phys. Rev. B 8:4433 (1973).
49. K. Johnson, this conference.
50. P. W. Anderson, this conference.
51. J. Appel, this conference.
52. V. Emery, this conference.
53. S. Doniach, this conference.
54. D. R. Penn, Phys. Rev. Lett. 42:921 (1979).
55. K. J. Chang, M. L. Cohen, and D. R. Penn, to be published.
56. P. Thiry, G. Rossi, Y. Petroff, A. Revcolevschi, and J. Jegoudez, to be published.
57. D. Scalapino, this conference.

TRENDS AND FUTURE - AS SEEN AT THE BERKELEY WORKSHOP

T. H. Geballe

Applied Physics Department
Stanford University
Stanford, CA 94305

INTRODUCTION

The Berkeley workshop exposed those attending to a wide range of experiments, analyses, model calculations, and theories. One reasonably might have expected a concensus in terms of key results and future directions to emerge. What happened, however, is that interpretations of many of the results presented lead to almost completely contradictory models. P. W. Anderson took note of this state of confusion in his opening remarks when he urged experimentalists to ignore theory and present raw data, in an unbiased form not geared to fit any particular model. There are enough data to support any of the wildly diverse models and pairing mechanisms - ranging from conventional electron-phonon to highly correlated electron-electron interactions - which are advocated in the papers given in these proceedings and elsewhere at the present time [see the accompanying article by M. Cohen].   Thus we have been presented with optical data which support weak coupling models, some heat capacity data which also support a weak coupling and others more in favor of strong coupling, tunneling data which indicate a very strong coupling, etc. A partial isotope effect in the $LaSr_2CuO_4$ compound [A. Zettl, Berkeley; B. Battlogg, AT&T] in contrast to the zero isotope effect in $YBa_2Cu_3O_7$, can be understood in different ways - as a cancellation or partial cancellation of competing attractive and repulsive pairing interactions (as is the case, respectively, in the transition metals Ru and Mo), as the result of two attractive pairing mechanisms acting in parallel (with a non-phonon mechanism being completely dominant for the 90K material), or, as suggested by R. Laughlin, because the Fermi energy lies below the Debye energy. A major reason for this present state of uncertainty is, I believe, due to extreme sensitivity of these defect layered-perovskite copper-oxide compounds to changes in composition, ordering of oxygen vacancies and other subtle defects. Part of what makes the field of high temperature superconductivity so fascinating at this time is in determining how the pairing mechanism is dependent upon specific structural features and upon the presence or absence of the various kinds of imperfections. Another important issue which must be taken into account when making quantitative comparisons is anisotropy, and so is the short coherence length which puts many experiments in the clean limit and makes fluctuations important.

The key to resolving these problems lies in being able to prepare well defined samples. The sensitivity of the superconductivity of $YBa_2Cu_3O_{7-x}$ to oxygen concentration and ordering is illustrated by the behavior (observed by many groups) of T_c as the oxygen stoichiometry is reduced from 7 (i.e. $x = 0.0$) to 6.6 (i.e. $x = 0.4$). T_c drops to 50K and then disappears discontinuously as the oxygen is reduced further and an orthorhombic-tetragonal transition takes place. In fact more quantitative data [B. Batlogg, AT&T] show T_c varying discontinuously from 90K to 50K as a function of oxygen concentration. Such behavior is most likely a manifestation of the fact that oxygen vacancies always form ordered compounds

(Magneli phases) appropriate to whatever composition of oxygen is present - unless prevented from doing so by quenching. It is therefore essential to be able to characterize the surfaces as to the ordering and concentration of oxygen vacancies, for example, before surface-sensitive properties such as XPS, UPS, and tunneling can be related in the quantitative way to the superconductivity, which is necessary to distinquish between the various models. There is a trend at present to prepare surfaces by scraping or breaking polycrystalline samples in high vacuum to assure good surfaces. But I am afraid that such procedures, while in the right direction, are insufficient in themselves. The scraping or breaking is likely to expose areas at grain boundaries where the material is the poorest. A way to restore the oxygen content and order may be to subsequent exposure to a low-temperature plasma discharge [J. M. Tarascon, Bellcore]. Other advances have been made and undoubtedly will continue in the preparation of bulk ceramic polycrystalline samples, bulk single crystal samples, and thin films with preferred orientation.

In what follows I will attempt to give some of the highlights of the meeting. It was possible to absorb only a small fraction of what went on. The time constraints imposed by the publications committee - which are quite proper for the rapidly progressing field - make it impossible to attempt to be more complete. I apologize for neglecting important work. (Also, there is no adequate way to reference the material in this camera-ready manuscript; hence I will quote the talks by the name of the speaker and the group represented. In many cases the work was presented at a poster session. The purpose of referencing it here with the name of the speaker is simply to enable the reader to trace the paper in these proceedings, and definitely is not intended to slight the real authors.) As to the future - that's easy! The key experiments will be done, and the relevant models will survive. I happen to believe that the delicate balance between magnetism and superconductivity in the prototype compound, La_2CuO_4 - which can be changed from an antiferromagnet to a superconductor by doping on the La sites as discovered by J. G. Bednorz and K. A. Müller, or by (presumably) removing a few oxygen vacancies [J. M. Tarascon, Bellcore] - is symptomatic. But it is up to future experiments to tilt the delicate balance and to observe what happens. The substantial progress in the preparation of ceramic samples, single crystals and thin films which is discussed below is encouraging.

BULK SAMPLES

A concensus seems to have been arrived at on the procedure for making ceramic samples of $YBa_2Cu_3O_{7-x}$ which consists of at least one exposure to temperatures between 850°C and 950°C, followed by a series of oxygen annealings, grindings and sinterings, below the tetragonal to orthorhombic transition (700°C). The former is to carry out the reaction, and the latter is to break up grain boundaries and to introduce oxygen. The resulting samples have sharp superconducting transitions just above 90K and have Meissner signals indicative of bulk superconductivity. Such characterization is insufficient to insure that further research on the samples will yield intrinsic properties. There is glassy-like behavior [K. A. Müller] associated with what is believed to be frustration in the ordering of superconducting clusters. Introducing a non-resonant microwave-absorption technique, Müller was able to estimate the area within a cluster or loop as being a few tenths of a square micron. The intragranular defects responsible for the clusters have not been identified - in particular, whether they are associated with structural defects which are seen in TEM [B. Raveau], with local variations in oxygen content and ordering, or with some other defects remains to be determined. Defects (in-between grains as well as intragranular ones) which produce weak links can degrade properties such as critical current. Other defects of course can be effective pinning centers and can increase the critical current. The highest critical currents I am aware of for bulk polycrystalline ceramic material were reported from the NRL group [D. Gubser, Naval Res. Lab.].

SINGLE CRYSTALS

A number of groups have produced small single crystals and thus have begun to address the importance of anisotropy. A crystal with $T_c = 88.8K$ [W. J. Gallagher, IBM, Yorktown] allowed an estimate for the coherence length ~ 7Å parallel to the c-axis, which is larger than half the unit cell and thus puts the transitions into the three-dimensional limit (barely). Large critical currents of $> 10^6 A/cm^2$ are reported [see also I. Schuller, Argonne] for current flow perpendicular to the c-axis and more than an order of magnitude less for parallel flow. There are evidently strong anistropic pinning forces either intrinsic or extrinsic in the "single crystals" which can be twinned (compare with thin film results below). Earlier reports by the IBM Yorktown group [Proceedings of the MRS Meeting, Anaheim, 1987] that tunneling into single crystals showed no gap anisotropy, is likely to be due to the fact that anisotropy was destroyed in the construction of the tunnel junction or that the surface itself becomes degraded and isotropic.

THIN FILMS

A number of different groups have succeeded in making films with reasonably sharp transitions near or above 90K. The sharpness is a more stringent criterion than in the bulk because the thin films are columnar, i.e., the grain size and the film thickness are roughly the same, so that the zero-resistance occurs by percolation through a two-dimensional path, and three times more material must be superconducting to achieve zero-resistance than in a three-dimensional system. Different vapor deposition methods have been used, namely, electron beam evaporation (Stanford, IBM, Cornell); molecular beam epitaxy (Stanford-Varian, AT&T, Bell, Westinghouse), sputtering (Stanford), and pulsed-laser evaporation (Bellcore). The latter is from a single source and has the useful characteristic that the film has the same (metal) composition as the source. In spite of adequate oxygen being present during deposition in all the different methods, a post-deposition heat treatment in oxygen was necessary to achieve satisfactory results. In only one instance [R. Buhrman, Cornell] superconductivity was obtained in situ using a deposition temperature of 700°C.

In no case were single crystal films reported, however, highly preferred orientation is found when $SrTiO_3$ substrates are used. Some films [M. Beasley, Stanford; B. Battlogg, AT&T] show a strongly predominant a-axis orientation (i.e. b- and c-axes in the plane of the film), as indeed expected from lattice-matching considerations. Other films show mixed a and c-axis orientation (Stanford), whereas others show primarily c-axis orientation [R. Greene, IBM-Yorktown; M. Tarascon, Bellcore]. The Stanford results show that even polycrystalline films with random orientation can have high critical currents, $\sim 10^4 A/cm^2$ at 77K (induced by the application of a field of $< 1T$ perpendicular to the film). They further show that with the field applied parallel to an a-axis oriented film, the added pinning of the surface barrier increases the current, averaged over the 1 micron thickness of the film, to $> 5 \times 10^7 A/cm^2$ at 4.2K and $1 \times 10^6 A/cm^2$ at 78K [B. Oh, M. Naito, et al., submitted to App. Phys. Lett.]. It seems likely that the depairing velocity is approached within a penetration length of the surface. While the high current will not scale with film thickness, of course, it is a more relevant parameter for the performance of thin film superconducting elements than the bulk critical current. High-frequency behavior characteristics remain to be evaluated.

NEW SUPERCONDUCTING PHASES, $T_c < 100K$

The structure and composition of the superconducting phases which exist have been confirmed by numerous single crystal X-ray and neutron powder diffraction experiments, by TEM and various analytical techniques. The enhanced (90K) superconductivity in the $YBa_2Cu_3O_7$ compound is attributed by several groups to the existence of the one-dimensional chains of CuO which are sandwiched between the two-dimensional CuO_2 planes. An essential feature of the argument rests upon the fact that when the samples are quenched from 900°C, they remain tetragonal; the oxygen vacancies remain disordered, one-dimensional chains are not formed, and the superconductivity disappears [I. Schuller, Argonne]. There is a correlation between the ordering of the vacancies giving rise to the one-dimensional chains and the occurrence of superconductivity. In the face of all this evidence the suggestion [K. A. Müller], based upon intuitive physico-chemical arguments that an alternate structure with only octahedrally and square-planar coordinated Cu should exist for $YBa_2Cu_3O_7$, seems unlikely.

However, the Stanford group [D. Mitzi et al., preprint] has replaced Y by La and obtained a different crystallographic structure, while T_c still can reach 90K. This latter structure is tetragonal and presumably is the one described by C. Michel and B. Raveau [Rev. Chim. Minerale, 21, 407 (1984)] as a '3-3-6' structure which has no one-dimensional features and no oxygen vacancies in the copper-containing planes. Neutron diffraction experiments are needed to confirm the structure.

NEW SUPERCONDUCTING PHASES, $T_c > 100K$

It is not surprising when such an unprecedently large number of multiphase samples are being scrutinized by so many researchers to find sudden drops in resistance. All that is needed to find "zero-resistance" in a four-point measurement is to have one of the phases topologically distributed so that it separates the voltage probes from the current probes, and to undergo a metal-insulator transition as the sample is cooled. What is really measured then is a zero voltage, because current no longer flows between the voltage leads. What is surprising is that so few researchers substantiate claims of possible superconductivity with simple tests such as permuting the current and voltage leads, and making two point measurements in which only the contact resistance should remain. Flourine-substituted YBaCuO was reported [A. Braginski, Westinghouse] to show resistance anomalies, but no superconductivity, in contrast to earlier reports of Ovshinski et al.[Phys. Rev. Lett. 58, 2579(1987)]. Zero-resistance observations in other multiphase samples were interpreted [C. W. Chu, Houston; A. Zettl, Berkeley] as indicating superconductivity at temperatures of 250K and higher. More convincing evidence for the existence of superconductivity (metastable and in trace amounts) comes from the magnetic measurements [C. Y. Huang, Lockheed]. These results, together with a few previous ones, indicate that superconductivity could exist under some highly metastable conditions at temperatures approaching room temperature. The fact that the indications are only found in multiphase systems means that concentration gradients, strain and other non-equilibrium conditions which can be trapped between different coexisting phases might be important. A way to explore this possibility would be to make controlled interfaces by utilizing the techniques which have been developed in growing multilayered and strain layered superlattices.

IMPORTANT UNRESOLVED ISSUES

There are conflicting data arising from three of the most important methods for investigating superconducting properties, namely heat capacity, tunneling, and far-infrared spectroscopy.

Heat capacity results were reported by the groups from AT&T, Berkeley, Geneva, Illinois, and Tokyo. A term linear in temperature is found in the superconducting state by all researchers. This contribution can be due to gapless superconductivity [as predicted e.g. by resonance-valence-bond theory of P. W. Anderson], or by any other excitations that have a finite density of states at the Fermi level, such as the Bloch states of a normal metal, but also, e.g. magnetic excitations, or ferroelectric-type excitations which are rather common in the perovskite materials. Some evidence was presented [O. Fischer, Geneva] that a substantial contribution to the linear term [> ten percent of the estimated Sommerfeld constant of the normal metal] is due to some non-superconducting metal present in the sample. If so, it demonstrates concretely that the usual criteria used to establish the credentials of a good sample such as diffraction patterns, sharp transitions and large shielding and Meissner signals are not sufficient.

John Bardeen interpreted the thermodynamic data as being consistent with BCS in the weak coupling limit. Agreement is found between the measured Ginzburg-Landau parameters and those calculated from simple theory and measurements. On the other hand, the stiffening of the lattice seen in elastic constant measurements [B. Battlogg, AT&T] below T_c, and in related lattice heat capacity data [O. Fischer, Geneva] indicates that the gap is roughly equal to the Fermi energy, i.e., the coupling is very strong.

There have been many tunneling experiments done using the high T_c perovskites. It is not surprising in view of the sensitivity to oxygen that no reliable prescription for making junctions has been developed. The review by K. Gray illustrates the wide variety of structure found in the I-V and tunneling conductances which have been reported. Values of the gaps which are deduced range from BCS-like to so large as to rule out weak coupling models. There is a preponderance of data on YBa_2CuO_7 which suggests a gap of 20 meV. Gaps as high as 180 meV are reported [A. L. deLozanne, Texas] using a surface which is cleaved in situ and a point contact. Results reported [Zavaritsky, Moscow] for SIS (superconducting-insulator-superconducting) point contact tunneling show multiple gaps which can be explained in different ways, including, as the authors suggest, anisotropic tunneling in the clean limit. As a function of temperature they find the gap disappears near 50K for YBaCuO, as did Iguchi, et al. [Japan J. Appl. Phys. 26, L645 (1987)]. These are the only two cases where the T_c corresponding to the tunnel junction material itself has been directly measured, and in both cases the T_c of a 50K makes it seem likely the tunneling is from a region where the oxygen content is 6.6. If degradation of T_c at the junction from 90 to 50K turns out to be general, then reported values of $2\Delta/kTc$ [see Gray's review] will have to be increased by a factor of 9/5. There are other uncertainties connected with whether the tunneling is SIS, SIN, or multiples of the above, all of which can introduce factors of two in Δ. Estimates, at present, place the values of $2\Delta/kTc$ at the high end of the range, or even beyond the range, which can be understood in terms of strong coupling theory. Additional structure seen in the conductance curves at multiples of the gap can contain a wealth of clues, but further experimental control is needed and should be forthcoming now that oriented thin films are available.

Optical studies in the far-infrared, in contrast to the tunneling results have tended to give energy gaps close to the weak coupled BCS value of $2\Delta/kTc = 3.5$. The non-Drude-like behavior makes the analysis of the absorption ambiguous. Optical reflectivity data presented [B. Battlogg, AT&T; R. Greene, IBM-Yorktown] agree in so far as the raw data are concerned; however, the respective data analyses lead to different models. The lack of weight at low frequencies results after a Kramers-Kronig analysis [AT&T] in a large oscillator strength associated with an excitation near 0.5 ev, perhaps related to charge transfer or spin excitations. On the other hand, the IBM analysis takes the strong anisotropy into account (the c-axis is taken to be non-conducting) and finds after averaging over orientation that the behavior is Drude-like, and there is no need to bring in oscillator strength at higher frequency. Further work on single crystals or oriented films is needed. Preliminary work presented by the Tokyo group on single crystals has not yet been analyzed.

FUTURE TRENDS

There has been tremendous activity since the discoveries of the existence of superconductivity in the 40K and the 90K regimes. It is far from clear what new physics is needed to describe the behavior and to gain a quantitative understanding. What has been evident at the Berkeley workshop is that there is as yet no convergence of the general ideas of what is going on and of what the most important interactions are. The fluidity of the present state is extraordinary - thus we are presented with evidence, mentioned above, for no energy gap, for a conventional energy gap, and for an unprecedentedly high energy gap. Other measurements of the gap, utilizing Raman difference spectroscopy, NMR relaxation, and acoustic absorption, will be helpful. During the next period of research there will undoubtedly be better films and bulk material available and as spurious results are eliminated we can look forward toward the emergence, at least, of what the correct model must entail.

It is not possible to predict on rational grounds that there will, or will not, be a much higher temperature superconductor (defined as such in Tanaka's terms, which includes identification of the responsible material). The hopes of discovering one would become brighter if it turns out that the unique structural features of the orthorhombic $YBa_2Cu_3O_7$ phase are not essential. The study of oxygen vacancy concentration and ordering in the other copper oxide perovskites is a fertile field, as Mitzi's results discussed above illustrate.

ACKNOWLEDGEMENTS

I would like to thank many colleagues, particularly Ivan Bozovic, Mac Beasley, Aharon Kapitulnik, and Oystein Fischer for helpful discussions. Support for the Stanford program has been made available by funds from the AFOSR, the NSF-MRL program through Stanford's Center for Materials Research, and the ONR.

CONCLUDING REMARKS

Vladimir Z.Kresin

Chairman

The final remarks do not present a difficult problem to me. After
the detailed summary talks by M.L.Cohen and T.Geballe there is not much left
for me to say.

Our conference is an unusual one. It is filled with the spirit of a
great discovery. Many great scientists, such as B.Mattias, P.Kapitza and
B.Geilikman, dreamed of high T_c; we are fortunate that this discovery
occured in our lifetime.

The conference has gathered leading specialists in the field of superconductivity. But it has been more than a celebration. The time has come
for serious discussions, and our conference was the first one devoted to a
detailed analysis of the new materials and new physics. We were fortunate
in having been able to listen to two talks by K.A.Müller, the first one
about the "Road to High T_c" and the second one describing the glass state
in the new materials.

The state of the theory deserves a special mention. There are a number
of interesting theories, but neither one can be considered generally accepted.
There is nothing strange in this. After all, it took 50 years for the BCS
theory to appear. Of course, today the situation is quite different, in that
we have an understanding of the phenomenon of superconductivity and the
intensity of research is remarkably high, so that the wait will not be as
long. But many key experimental facts essential for a theory have become
known only rather recently. Much depends, and Prof. J.Bardeen has stressed
this point, on the strength of the electron-electron interaction. For
instance, the phonon mechanism corresponds to the weak coupling limit, and
the exciton mechanism to the strong. I am a partisan of the coexistence of
the phonon and plasmon mechanisms, which is the case of intermediate coupling.
(A very convenient position for a conference chairman!)

I think that acceptance of a theory results from an experimental
confirmation of a non-trivial prediction. Of course, it is important for
theory to be able to obtain the correct value of T_c. But this is not
sufficient; this is not a prediction, because we already know the critical
temperature.

There has been a lot of excitement in the media lately (we are happy
that our conference also has been widely covered; there is hope that many

young people will become interested in physics). This excitement is connected
mostly with possible applications of high temperature superconductivity.
But I would like to stress that the recent progress illustrates the importance
of basic research. This is what our conference has been devoted to.

From the point of view of basic research, I believe that one of the
main problems is to carry out tunneling spectroscopy experiments. These are
very difficult because one must have good films, and also because of the
smallness of the coherence length. But one can hope that good data are coming.
The talk by the Stanford group has been encouraging in this respect.

The discovery of the new materials is not the end but the beginning
of a new period. The unique intensity of research is due not only to the
relative simplicity of preparing the materials and to the promise of practical
applications, but also to the new physics. Indeed, usually the inequality
$\Delta \ll \tilde{\Omega} \ll \varepsilon_f$ holds, where Δ is the gap, $\tilde{\Omega}$ is the characteristic phonon
frequency, and ε_f is the Fermi energy. In the new materials everything is
different. These quantities turn out to be comparable, a large fraction of
carriers are paired, and we are dealing with a new state.

In conclusion, I wish to thank the administration of LBL and MCSD for
hosting the workshop and ONR for its support. We thank all the participants
for coming, and we will always be happy to see you in Berkeley. The process
of studying the new materials, unique in its importance and intensity,
continues, and I hope that this workshop will help to speed it up.

D. Albrecht
U.S. Patent and
Trademark Office

P.B. Allen
Naval Research Lab

J.S. Anandan
Univ. of S. Carolina

P.W. Anderson
Princeton Univ.

J.C. Appel
UC, San Diego

G.B. Arnold
Univ. of Notre Dame

P.R. Aron
NASA, Lewis RC

J. Ashburn
Univ. of Alabama

N.W. Ashcroft
Cornell Univ.

W.T. Bakker
Electric Power
Research Institute

P.J. Barboux
Bellcore

J. Bardeen
Univ. of Ill., Urbana

R. Baron
Hughes Research Lab

A. Barone
CEIN, ITALY

R. Barrio
Instituto de
Investigaciones en Mater.

G. Baskaran
Princeton Univ.

B. Batlogg
Bell Labs

M.R. Beasley
Stanford Univ.

J.J. Bechtold
Univ. of Houston

F.D. Bedard
DIRNSA

R.N. Bhatt
Bell Labs

J.L. Birman
City College

D.J. Bishop
Bell Labs

P. Boolchand
Univ. of Cincinnati

A. Borshchevsky
Jet Propulsion Lab

C. Bourbonnais
Univ. de Sherbrooke

I. Bozovic
Stanford Univ.

A.I. Braginski
Westinghouse Labs

R.G. Brandt
Off. of Naval Research

I.G. Brown
Lawrence Berkeley Lab

S.E. Brown
Los Alamos Lab

R.A. Buhrman
Cornell Univ.

W.H. Butler
Oak Ridge Lab

J.P. Carbotte
McMaster Univ.

D.M. Ceperley
Lawrence Livermore Lab

M. Chou
Exxon Res. & Engin. Co.

P.W. Chu
Univ. of Houston

J. Clarke
UC, Berkeley

T. Clem
Dept. of Navy

W.L. Clinton
Nat'l Science Found.

D. Coffey
UC, San Diego

L. Coffey
Ohio State Univ.

M.L. Cohen
UC, Berkeley

T.C. Collins
Univ. of Tennessee

J.P. Collman
Stanford Univ.

L. Cooper
Brown Univ.

L.R. Cooper
Off. of Naval Research

D.L. Cox
Ohio State Univ.

G.B. Cvijanovich
AMP, Inc.

T. Datta
Univ. of S. Carolina

L. DeLong
Univ. of Kentucky

G. Deutscher
Tel Aviv Univ.

M. Devoret
PRMA, France

L.W.R. Dicks
Shell Development Co.

D.R. Dietderich
Lawrence Berkeley Lab

S. Doniach
Stanford Univ.

M.M. Doria
Los Alamos Lab

T. Van Duzer
UC,Berkeley

R.C. Dynes
Bell Labs

E.A. Edelsack
Off.of Navval Research

V.J. Emery
Brookhaven Lab

R. Escudero
UNAM, Mexico

E.W. Fenton
Nat. Res. Council Canada

H.J. Fink
UC, Davis

D. Finnemore
Ames Lab.

O. Fischer
Univ. de Geneve

R.A. Fisher
UC, Berkeley

T.L. Francavilla
Naval Research Lab

H. Frohlich
Univ. of Liverpool

H. Fukuyama
Univ. of Tokyo

W.J. Gallagher
IBM, T.J.Watson Ctr.

C.F. Gallo
3M Corp.

P.L. Gammel
Bell Labs

L. Gao
Univ. of Houston

E.L. Garwin
Stanford Linear
Accelerator Ctr.

T.H. Geballe
Stanford Univ.

U.W. Geiser
Argonne

A.D. Gerber
Rutgers Univ.

W.I. Glabersom
Nat'l. Science Found.

E.S. Goldburt
Philips Labs

A.M. Goldman
Univ. of Minn.

A. Golovashkin
P.N. Lebedev Instit.

R.G. Goodrich
Louisianna. State Univ.

L. Gor'kov
L. Landau Instit.

B. Goschitskiy
Academy of Sciences
of the U.S.S.R.

W.G. Gottenberg
Shell Devel. Co.

P.M. Grant
IBM, Almaden Res. Ctr.

K.E. Gray
Argonne Lab

L.H. Greene
Bellcore

R.L. Greene
IBM, T.J. Watson Res. Ctr

R. Gronsky
Lawrence Berkeley Lab

G. Gruner
UC, Los Angeles

D. Gubser
Naval Research Lab

M. Gurvitch
Bell Labs

H. Gutfreund
Hebrew Univ.
ISREAL

E.L. Haase
Institut fur Kernphysik
GERMANY

J. Halbritter
Univ. Karlsruhe,
GERMANY

E.E. Haller
Lawrence Berkeley Lab

R.H. Hammond
Stanford Univ.

J. Bindslev Hansen
The Technical Univ.
DENMARK

W.L. Hansen
Lawrence Berkeley Lab

W.A. Harrison
Stanford Univ.

Y. Hasegawa
Univ. of Tokyo

K.B. Hathaway
Off. of Naval Research

A.F. Hebard
Bell Labs

Per Hedegard
Nordisk Inst. DENMARK

R.H. Heffner
Los Alamos Lab

E.S. Hellman
Stanford Univ.

F. Herman
IBM, Almaden Res. Ctr.

S. Hikami
Univ. of Tokyo

D.L. Hildenbrand
SRI

J.E. Hirsch
UC, San Diego

J.R. Holzrichter
Lawrence Livermore Lab

P. Hor
Univ. of Houston

S. Hosoya
Tohoku Univ., JAPAN

W.Y. Hsu
E.I. Du Pont

C.Y. Huang
Lockheed Lab

O. Hudak
Inst. of Experimental
Physics, CZECHO.

I. Iguchi
Univ.of Tsukuba
JAPAN

A.E. Jacobs
Univ. of Toronto
CANADA

L. Janos
Unv.Erlangen-Nuernberg
GERMANY

C.D. Jeffries
UC, Berkeley

D. Jerome
Orsay, FRANCE

J. Ihm
Univ, of Seoul, KOREA

S. John
Princeton Univ.

K.H. Johnson
MIT

P.E. Johnson
Lwarence Berkeley Lab

G. Kalonji
MIT

Huey-Chuen I. Kao
Argonne Lab

R.V. Kasowski
E.I. Du Pont

A. Khurana
Amer. Inst. of Physics

A.M. Kini
Argonne Lab

K. Kitazawa
Univ. of Tokyo

C. Kittel
UC, Berkeley

B.M. Klein
Naval Research Lab

W.D. Knight
UC, Berkeley

R. Koch
IBM,T.J.Watson Res.Ctr.

K.E. Kihlstrom
Westmont College

Y. Kopayez
P.N.Lebedev Inst,USSR

V.Z. Kresin
Lawrence Berkeley Lab

M.M. Kuchment
Harvard Univ.

J.M. Kwak
Sandia Labs

J. Kwo
Bell Labs

R. Laughlin
Stanford Univ.

T.R. Lemberger
Ohio State Univ.

W.A. Lester
Lawrence Berkeley Lab

M. Levy
Univ. of Wisconsin

S. Liang
Princeton Univ

Li-Jen T. Lin
Bellcore

P. Lindenfeld
Rutgers Univ.

W.A. Little
Stanford Univ.

B.H. Loo
Univ. of Alabama

T.L. Loucks
Rothschild

S.G. Louie
UC, Berkeley

A.L. de Lozanne
Univ. of Texas

G.B. Lubkin
Amer. Inst. of Physics

F.W. Lytle
Boeing Aerospace Co.

S. Maekawa
Tohoku Univ.

Y. Maeno
Hiroshima Univ.

A.P. Malozemoff
IBM,T.J.Watson Res.Ctr.

M.B. Maple
UC, San Diego

R.M. Martin
Xerox Research Center

D.L. Matthies
SRI

R.W. McCallum
Ames Laboratory

M.E. McHenry
MIT

W.L. McLean
Rutgers

A.K. McMahan
Lawrence Livermore Lab

M. Melich
Naval Postgraduate School

R. Meng
Univ. of Houston

N. Milleron
Neven Corp.

A.J. Millis
Bell Labs

J.L. Moll
Hewlett-Packard Co.

H. Morawitz
IBM, Almaden Res. Ctr.

D. Morris
Lawrence Berkeley Lab

D.R. Mueller
Naval Research Lab

K.A. Muller
IBM, Zurich Res. Lab

B. Murdock
Tektronix, Inc

Y. Muto
Tohoku Univ.

M. Naito
Stanford Univ.

Sang Boo Nam
Wright State UNiv.

Tsu-Wei Nee
Naval Center

W.J. Nellis
Lawrence Livermore Lab

D.M Newns
IBM,T.J.Watson Res.Ctr.

K.L. Ngai
Naval Research Lab

P. Nigrey
Sandia Labs

M. Nisenoff
Naval Research Lab

K. Noto
Tohoku Univ.

M.M. Olmstead
UC, Davis

N. Phuan Ong
Princeton Univ.

J.W. Orenstein
Bell Labs

S. Oseroff
San Diego State Univ.

M.S. Osofsky
Naval Research Lab

H.R. Ott
ETH, SWITZERLAND

A.J. Panson
Westinghouse Res. Labs

J. Paterno
ENEA-Cre-Frascati, ITALY

M.A. St.Peters
Mankato State Univ.

N.E. Phillips
UC, Berkeley

W.E. Pickett
Naval Research Lab

D. Pines
Los Alamos Lab

A.K. Rajagopal
Naval Research Lab

A.P. Ramirez
Bell Labs

B. Raveau
Univ. de Caen, FRANCE

V. Rehn
Naval Center

A. Ricca
CISE, ITALY

P. Richards
UC, Berkeley

H. Rogalla
Univ. GieBen, GERMANY

H. Rosen
IBM, Almaden Res. Ctr.

F. Rothwarf
BDM

J. Ruvalds
Harvard Univ.

J. Sarfatti
P.O. Box 26548

M. Sato
Instit. for Molecular
Science, JAPAN

S.S. Satpathy
Xerox Res. Ctr.

M.L. Sattler
Lawrence Livermore Lab

D.J. Scalapino
UC, Santa Barbara

B. Schechter
Simon & Schuster

J. E. Schirber
Sandia Labs

Z. Schlesinger
IBM,T.J.Watson Res.Ctr

M.A. Schluter
Bell Labs

I.K. Schuller
Argonne Labs

S. Schultz
UC, San Diego

H.B. Schuttler
Argonne Lab

J.W. Serene
Nat'l Science Found.

Lu J. Sham
UC, San Diego

I. Shchegolev
Acad.of Sciences,USSR

J.L. Sheiman
Shell Development Co.

A.T. Shih
Naval Research Lab

D. Shirley
Lawrence Berkeley Lab

B.S. Shivaram
Univ of Virginia

Y. Shoham
Shell Development Co.

A.H. Silver
TRW

R.N. Silver
Los Alamos Lab

R.W. Simon
TRW

L. Sniadower
Raychem Corps.

W.A. Soffa
Lawrence Berkeley Lab

J.W. Spargo
Hughes Co.

A. Stacy
UC, Berkeley

H. Steinfink
Univ. of Texas

M. Strongin
Brookhaven Lab

Wu=Pei Su
Univ. of Houston

H. Suhl
UC, San Diego

P.E. Sulewski
Cornell Univ.

W. Sung
Pohan Institute

J.C. Swihart
Indiana Univ.

J.S. Swinnea
Univ. of Texas

J.T. Martinez
UNAM, Mexico

Y. Takada
Univ of Tokyo

H. Takagi
Univ. of Tokyo

J. Talvacchio
Westinghouse Res. Lab

S. Tanaka
Univ. of Tokyo

J.M. Tarascon
Bellcore

E. Teller
Lawrence Livermore Lab

G.A. Thomas
Bell Labs

R.S. Thompson
Univ. of S. Carolina

C.S. Ting
Univ. of Houston

M. Tinkman
Harvard Univ.

J.B. Torrance
IBM, Almaden Res. Ctr.

S.A. Trugman
Los Alamos Lab

S. Uchida
Univ of Tokyo

R. Upendra
U.S. Patent and
Trademark Office

C.M. Varma
Bell Labs

T. Venkatesan
Bellcore

G.C. Vezzoli
U.S. Tech. Lab

H.J. Vinegar
Shell Development Co.

C.E. Violet
Lawrence Livermore Lab

P.J. Visuri
Consulate General
of Finland

Y. Wang
Univ. of Houston

E.R. Weber
UC, Berkeley

M. Weger
Hebrew Univ., ISREAL

H. Weinstock
Air Force Office of
Scientific Research

S.D. Wijeyesekera
DOW

J.W. Wilkins
Cornell Univ.

M.S. Wire

J. Wong
Lawrence Livermore Lab

M.K. Wu
Univ. of Alabama

Ryozo Yoshizaki
Univ. of Tsukuba

A. Zettl
UC, Berkeley

J.M. Williams
Argonne Lab

E.L. Wolf
Polytech.Inst.,Brooklyn

C. Wood
Jet Propulsion Lab

G. Xiao
Johns Hopkins Univ.

P.Y. Yu
UC, Berkeley

G.O. Zimmerman
Boston Univ.

N. Winter
Lawrence Livermore Lab

S.A. Wolf
Naval Research Lab

T.K. Worthington
IBM,T.J.Watson Res. Ctr.

D. Xing
Univ. of Houston

N. Zavaritskiy
Acad.of Sciences, USSR

Z. Zou
Princeton Univ.

I4/mmm, 925
IEX mechanism, 161-169
"impedence match", 325
Impurity, 144, 193, 380, 1084
  band, 29
  effect, 206, 1078-1082
  scattering, 54, 277
incomplete inner shells, 473
incongruent melting, 635
independent particle Green's function, 324
inductive transition. 759
inelastic light scattering, 893-896
inelastic scattering rate, 623
Infared
  reflectivity, 893-896
  scattering, 1015
instability, 169
interband
  effects, 1100
  scattering, 451
  transitions
interchain
  kinetic coupling, 161
  pair tunneling, 164-169
  pairing, 113
"interface dominated superconductivity", 9-10
interface phonon modes, 324-331
interfacial superconductivity, 1033
interference effect, 58
inverse magnetic susceptibility, 188
ion implantation, 29-37
IR
  conductivity, 103-133
  response, 460
isotope effect, 1, 297, 341, 386-388, 567, 653, 656,
        733-738, 1002, 1099
isotropic free electron model, 334
itinerant electron, 143
itinerant holes, 1066

Jahn-Teller
  distortion, 296
  effect, 565-567
  theorem, 1021
jellium shell model, 48
Josephson
  coupling, 588
  effect, 1003
  junctions, 67
  tunneling, 613, 614

K_2MnF_4, 1024
K_2NiF_4, 885, 910, 915, 919,
  structure, 1083
kernel function, 442
KMK scheme, 435-436